LAWLER, LENSTRA, RINNOOY KAN, AND SHMOYS, EDITORS
The Traveling Salesman Program: A Guided Tour
of Combinatorial Optimization

MAHMOUD
Evolution of Random Search Trees

MARTELLO AND TOTH
Knapsack Problems: Algorithms and Computer Implementations

MINC
Nonnegative Matrices

MINOUX
Mathematical Programming: Theory and Algorithms
(*Translated by S. Vajda*)

MIRCHANDANI AND FRANCIS, EDITORS
Discrete Location Theory

NEMHAUSER AND WOLSEY
Integer and Combinatorial Optimization

NEMIROVSKY AND YUDIN
Problem Complexity and Method Efficiency in Optimization
(*Translated by E. R. Dawson*)

PLESS
Introduction to the Theory of Error-Correcting Codes
Second Edition

SCHRIJVER
Theory of Linear and Integer Programming

TOMESCU
Problems in Combinatorics and Graph Theory
(*Translated by R. A. Melter*)

TUCKER
Applied Combinatorics
Second Edition

Nonlinear Programming

Theory and Algorithms

Second Edition

Nonlinear Programming

Theory and Algorithms

Mokhtar S. Bazaraa
American Airlines Decision Technologies
Dallas, Texas

Hanif D. Sherali
Department of Industrial and Systems Engineering
Virginia Polytechnic Institute and State University
Blacksburg, Virginia

C. M. Shetty
School of Industrial and Systems Engineering
Georgia Institute of Technology
Atlanta, Georgia

John Wiley & Sons, Inc.
New York • Chichester • Brisbane • Toronto • Singapore

Acquisitions Editor Charity Robey
Marketing Manager Susan Elbe
Production Supervisor Sandra Russell
Designer Pedro A. Noa
Manufacturing Manager Andrea Price
Copy Editing Supervisor Deborah Herbert
Illustration Coordinator Jaime Perea

This book was set in 10/12 Times Roman by Monotype Composition and printed and bound by Hamilton Printing. The cover was printed by Hamilton Printing.

Library of Congress Cataloging in Publication Data:

Bazaraa, M. S. 1943
Nonlinear programming : theory and algorithms / Mokhtar S. Bazaraa, Hanif D. Sherali, C.M. Shetty.
p. cm.
Includes bibliographical references and index.
ISBN 0-471-55793-5
1. Nonlinear programming. I. Sherali, Hanif D., 1952– . II. Shetty, C. M., 1929– . III. Title.
T57.8.B39 1993
519.7′6 — dc20 92.30957
CIP

Printed in the United States of America

10 9 8 7 6 5 4 3 2 1

Dedicated to
OUR PARENTS

Preface

Nonlinear programming deals with the problem of optimizing an objective function in the presence of equality and inequality constraints. If all the functions are linear, we obviously have a *linear program*. Otherwise, the problem is called a *nonlinear program*. The development of highly efficient and robust algorithms and software for linear programming, the advent of high-speed computers, and the education of managers and practitioners in regard to the advantages and profitability of mathematical modelling and analysis, have made linear programming an important tool for solving problems in diverse fields. However, many realistic problems cannot be adequately represented or approximated as a linear program owing to the nature of the nonlinearity of the objective function and/or the nonlinearity of any of the constraints. Efforts to solve such nonlinear problems efficiently have made rapid progress during the past three decades. This book presents these developments in a logical and self-contained form.

The book is divided into three major parts dealing, respectively, with convex analysis, optimality conditions and duality, and computational methods. Convex analysis involves convex sets and convex functions and is central to the study of the field of optimization. The ultimate goal in optimization studies is to develop efficient computational schemes for solving the problem at hand. Optimality conditions and duality can be used not only to develop termination criteria but also to motivate and design the computational method itself.

In preparing this book, a special effort has been made to make certain that it is self-contained and that it is suitable both as a text and as a reference. Within each chapter, detailed numerical examples and graphical illustrations have been provided to aid the reader in understanding the concepts and methods discussed. In addition, each chapter contains many exercises. These include (1) simple numerical problems to reinforce the material discussed in the text, (2) problems introducing new material related to that developed in the text, and (3) theoretical exercises meant for advanced students. At the end of each chapter, extensions, references, and material related to that covered in the text are presented. These notes should be useful to the reader for further study. The book also contains an extensive bibliography.

Chapter 1 gives several examples of problems from different engineering disciplines that can be viewed as nonlinear programs. Problems involving optimal control, both discrete and continuous, are discussed and illustrated by examples from production and inventory control and from highway design. Examples of a two-bar truss design and a two-bearing journal design are given. Steady-state conditions of an electrical network are discussed from the point of view of obtaining an optimal solution to a quadratic program. A large-scale nonlinear model arising in the management of water resources is developed, and nonlinear models arising in stochastic programming and in location theory are discussed. Finally, we provide an important discussion on modelling and on formulating nonlinear programs from the viewpoint of favorably influencing the performance of algorithms that will ultimately be used for solving them.

The remaining chapters are divided into three parts. Part 1, consisting of Chapters 2 and 3, deals with convex sets and convex functions. Topological properties of convex sets, separation and support of convex sets, polyhedral sets, extreme points and extreme directions of polyhedral sets, and linear programming are discussed in Chapter 2. Properties of convex functions, including subdifferentiability, and minima and maxima over a convex set are discussed in Chapter 3. Generalization of convex functions and their interrelationships are also included, since nonlinear programming algorithms suitable for convex functions can be used for a more general class involving pseudoconvex and quasiconvex functions. The appendix provides additional tests for checking generalized convexity properties, and we discuss the concept of convex envelopes and their uses in global optimization methods through the exercises.

Part 2, which includes Chapter 4 through 6, covers optimality conditions and duality. In Chapter 4, the classical Fritz John (FJ) and the Karush–Kuhn–Tucker (KKT) optimality conditions are developed both for inequality- and equality-constrained problems. Both first-order and second-order optimality conditions are developed, and higher-order conditions are discussed along with some cautionary examples. The nature, interpretation, and value of FJ and KKT points are also discussed and emphasized. Some foundational material on both first-order and second-order constraint qualifications is presented in Chapter 5. We discuss interrelationships between various proposed constraint qualifications and provide insights through many illustrations. Chapter 6 deals with Lagrangian duality and saddle point optimality conditions. Duality theorems, properties of the dual function, and both differentiable and nondifferentiable methods for solving the dual problem are discussed. We also discuss necessary and sufficient conditions for the absence of a duality gap, and interpret this in terms of a suitable perturbation function. In addition, we relate Lagrangian duality to other special forms of duals for linear and quadratic programming problems. Besides Lagrangian duality, there are several other duality formulations in nonlinear programming, such as conjugate duality, min–max duality, surrogate duality, composite Lagrangian and surrogate duality, and symmetric duality. Among these, the Lagrangian duality seems to be the most promising in the areas of theoretical and algorithmic developments. Moreover, the results that can be obtained via these alternative duality formulations are closely related. In view of this, and for brevity, we have elected to discuss Lagrangian duality in the text and to only introduce other duality formulations in the exercises.

Part 3, consisting of Chapters 7 through 11, presents algorithms for solving both unconstrained and constrained nonlinear programming problems. Chapter 7 deals exclusively with convergence theorems, viewing algorithms as point-to-set maps. These theorems are used actively throughout the remainder of the book to establish the convergence of the various algorithms. Likewise, we discuss the issue of rates of convergence and provide a brief discussion on criteria that can be used to evaluate algorithms.

Chapter 8 deals with the topic of unconstrained optimization. To begin, we discuss several methods for performing both exact and inexact line searches, as well as methods for minimizing a function of several variables. Methods using both derivative and derivative-free information are presented. Newton's method and its variants based on trust region and the Levenberg–Marquardt approaches are discussed. Methods that are based on the concept of conjugacy are also covered. In particular, we present quasi-Newton (variable metric) and conjugate gradient (fixed metric) algorithms that have gained a great deal of popularity in practice. We also introduce the subject of subgradient optimization methods for nondifferentiable problems, and discuss variants fashioned in the spirit of conjugate gradient and variable metric methods. Throughout, we address the issue of convergence and rates of convergence for the various algorithms, as well as practical implementation aspects.

Chapter 9 discusses penalty and barrier function methods for solving nonlinear programs, in which the problem is essentially solved as a sequence of unconstrained problems. We describe general exterior penalty function methods, as well as the particular exact absolute value and the Augmented Lagrangian penalty function approaches, along with the method of multipliers. We also present interior barrier function penalty approaches. In all cases, implementation issues and convergence rate characteristics are addressed. We conclude this chapter by describing a polynomial-time primal–dual path following algorithm for linear programming based on a logarithmic barrier function approach. This method can also be extended to polynomially solve convex quadratic programs.

Chapter 10 deals with the method of feasible directions in which, given a feasible point, a feasible improving direction is first found and, then, a new improved feasible point is determined by minimizing the objective function along that direction. The original methods proposed by Zoutendijk and subsequently modified by Topkis and Veinott to assure convergence are presented. This is followed by the popular successive linear and quadratic programming approaches, including the use of l_1 penalty functions either directly in the direction-finding subproblems, or as merit functions to assure global convergence. Convergence rates and the Maratos effect are also discussed. This chapter also describes the gradient projection method of Rosen along with its convergent variants, the reduced gradient method of Wolfe and the generalized reduced gradient method, along with its specialization to Zangwill's convex simplex method. In addition, we also unify and extend the reduced gradient and the convex simplex methods through the concept of suboptimization and the superbasic–basic–nonbasic partitioning scheme. Effective first-order and second-order variants of this approach are discussed.

Finally, Chapter 11 deals with some special problems that arise in different applications as well as in the solution of other nonlinear programming problems. In particular, we present the linear complementary, quadratic, separable, linear fractional, and geometric programming problems. Methodologies used for solving these problems as, for example, the use of Lagrangian duality concepts in the algorithmic development for geometric programs, serve to strengthen the ideas described in the preceding chapters.

This book can be used both as a reference for topics on nonlinear programming and as a text in the fields of operations research, management science, industrial engineering, and applied mathematics, and in engineering disciplines that deal with analytical optimization techniques. The material discussed requires some mathematical maturity and a working knowledge of linear algebra and calculus. For the convenience of the reader, Appendix A summarizes some mathematical topics frequently used in the book, including matrix factorization techniques.

As a text, the book can be used (1) in a course on foundations of optimization and (2) in a course on computational methods as detailed below. It can also be used in a two-course sequence covering all the topics.

1. *Foundations of Optimization*

This course is meant for undergraduate students in applied mathematics and graduate students in other disciplines. The suggested coverage is given schematically below, and it can be covered in the equivalent of a one-semester course. Chapter 5 on constraint qualifications could be omitted without loss of continuity. A reader familiar with linear programming may also skip Section 2.7.

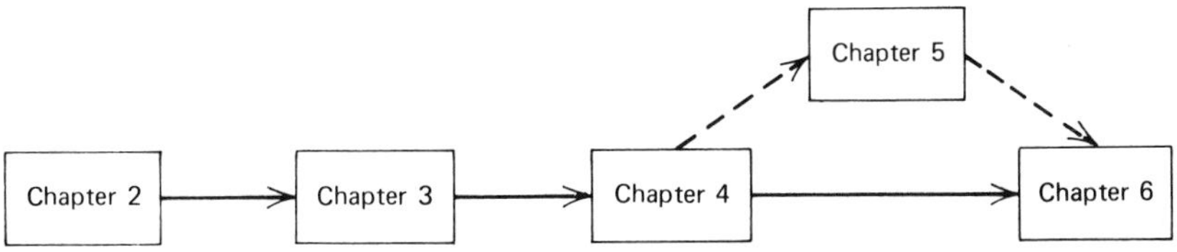

2. *Computational Methods in Nonlinear Programming*

This course is meant for graduate students who are interested in algorithms for solving nonlinear programs. The suggested coverage is given schematically below, and it can be covered in the equivalent of a one-semester course. The reader who is not interested in convergence analyses may skip Chapter 7 and the discussion related to convergence in Chapters 8 through 11. The minimal background on convex analysis and optimality conditions needed to study Chapters 8 through 11 is summarized in Appendix B for the convenience of the reader. Chapter 1, which gives many examples of nonlinear programming problems, provides a good introduction to the course, but no continuity will be lost if this chapter is skipped.

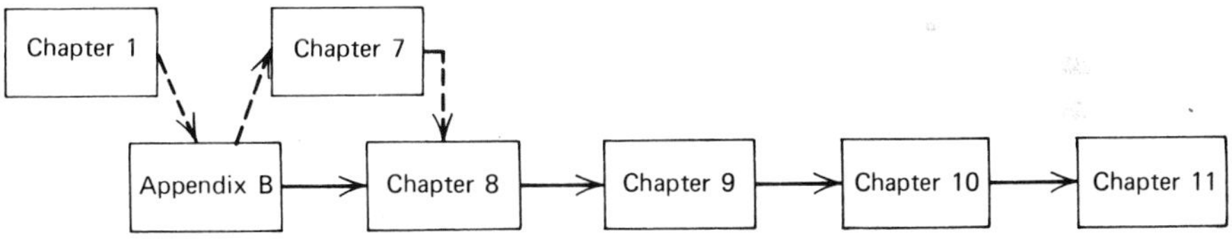

We again express our thanks to Dr. Robert N. Lehrer, former director of the School of Industrial and Systems Engineering at the Georgia Institute of Technology, for his support in the preparation of the first edition of this book; to Dr. Jamie J. Goode of the School of Mathematics, Georgia Institute of Technology, for his friendship and active cooperation; and to Mrs. Carolyn Piersma, Mrs. Joene Owen, and Ms. Kaye Watkins for their painstaking typing of the first edition of this book.

In the preparation of the second edition of this book, we thank Professor Robert D. Dryden, Chairman of the Department of Industrial and Systems Engineering at Virginia Polytechnic Institute and State University, for his support. We thank Mr. Gyunghyun Choi, Mr. Ravi Krishnamurthy, and Mrs. Semeen Sherali for their diligent typing efforts, and Ms. Joanna Leleno for her painstaking preparation of the solutions manual.

Mokhtar S. Bazaraa
Hanif D. Sherali
C. M. Shetty

Contents

Chapter 1 Introduction

Operations research analysts, engineers, managers, and planners are traditionally confronted by problems that need solving. The problems may involve arriving at an optimal design, allocating scarce resources, planning industrial operations, or finding the trajectory of a rocket. In the past, a wide range of solutions was considered acceptable. In engineering design, for example, it was common to include a large safety factor. However, because of continued competition, it is no longer adequate to develop only an acceptable design. In other instances, such as in space vehicle design, the acceptable designs themselves may be limited. Hence, there is a real need to answer such questions as: Are we making the most effective use of our scarce resources? Can we obtain a more economical design? Are we taking risks within acceptable limits? During the past four decades, in response to these pressures, there has been a very rapid growth of optimization models and techniques. Fortunately, the growth of large and fast computing facilities has aided substantially in the use of the techniques developed.

Another aspect that has stimulated the use of a systematic approach to problem solving is the rapid increase in the size and complexity of problems as a result of the technological growth since World War II. Engineers and managers are called upon to study all facets of a problem and their complicated interrelationships. Some of these interrelationships may not even be well understood. Before a system can be viewed as a whole, it is necessary to understand how the components of the system interact. Advances in the techniques of measurement, coupled with statistical methods to test out hypotheses, have significantly aided in this process of studying the interaction between components of the system.

The acceptance of the field of operations research in the study of industrial, business, military, and governmental activities can be attributed, at least in part, to the extent to which the operations research approach and methodology have aided the decision makers.

Early postwar applications of operations research in the industrial context were mainly in the area of linear programming and the use of statistical analysis. By that time, efficient procedures and computer codes were available to handle such problems. This book is concerned with nonlinear programming, including the characterization of optimal solutions and the development of computational procedures.

In this chapter, we introduce the nonlinear programming problem and discuss some simple situations that give rise to such a problem. Our purpose is only to provide some background on nonlinear problems; indeed, an exhaustive discussion of potential applications of nonlinear programming can be the subject matter of an entire book. We also provide some guidelines here for constructing models and problem formulations from the viewpoint of enhancing algorithmic efficiency and problem solvability. Although many of these remarks will be better appreciated as the reader progresses through the book, it is best to bear these general fundamental comments in mind at the very onset.

1.1 Problem Statement and Basic Definitions

Consider the following *nonlinear programming problem*:

$$\begin{array}{lll} \text{Minimize} & f(\mathbf{x}) & \\ \text{subject to} & g_i(\mathbf{x}) \leq 0 & \text{for } i = 1, \ldots, m \\ & h_i(\mathbf{x}) = 0 & \text{for } i = 1, \ldots, l \\ & \mathbf{x} \in X & \end{array}$$

where $f, g_1, \ldots, g_m, h_1, \ldots, h_l$ are functions defined on E_n, X is a subset of E_n, and $\mathbf{x}$ is a vector of n components $x_1, \ldots, x_n$. The above problem must be solved for the values of the variables $x_1, \ldots, x_n$ that satisfy the restrictions and meanwhile minimize the function f.

The function f is usually called the *objective function*, or the *criterion function*. Each of the constraints $g_i(\mathbf{x}) \leq 0$ for $i = 1, \ldots, m$ is called an *inequality constraint*, and each of the constraints $h_i(\mathbf{x}) = 0$ for $i = 1, \ldots, l$ is called an *equality constraint*. The set X might typically include lower and upper bounds on the variables, which, even if implied by the other constraints, can play a useful role in some algorithms. Alternatively, this set might represent some specially structured constraints that are highlighted to be exploited by the optimization routine, or it might represent certain regional containment or other complicating constraints that are to be handled separately via a special mechanism. A vector $\mathbf{x} \in X$ satisfying all the constraints is called a *feasible solution* to the problem. The collection of all such solutions forms the *feasible region*. The nonlinear programming problem, then, is to find a feasible point $\bar{\mathbf{x}}$ such that $f(\mathbf{x}) \geq f(\bar{\mathbf{x}})$ for each feasible point $\mathbf{x}$. Such a point $\bar{\mathbf{x}}$ is called an *optimal solution*, or simply a *solution*, to the problem. If more than one optimum exist, they are collectively referred to as *alternative optimal solutions*.

Needless to say, a nonlinear programming problem can be stated as a maximization problem, and the inequality constraints can be written in the form $g_i(\mathbf{x}) \geq 0$ for $i = 1, \ldots, m$. In the special case when the objective function is linear and when all the constraints, including the set X, can be represented by linear inequalities and/or linear equations, the above problem is called a *linear program*.

To illustrate, consider the following problem:

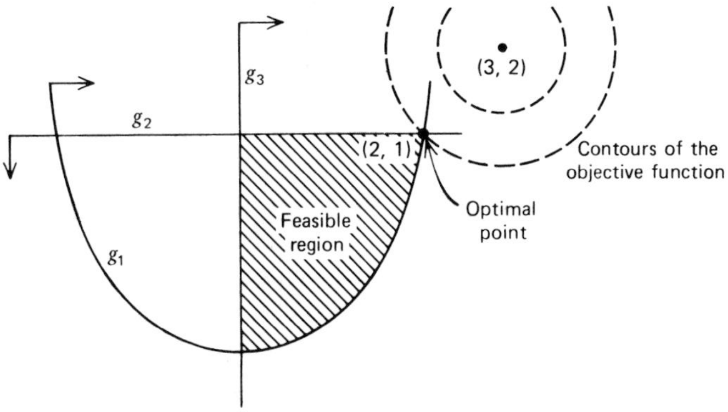

Figure 1.1 Geometric solution of a nonlinear problem.

$$\begin{array}{ll} \text{Minimize} & (x_1 - 3)^2 + (x_2 - 2)^2 \\ \text{subject to} & x_1^2 - x_2 - 3 \leq 0 \\ & x_2 - 1 \leq 0 \\ & -x_1 \leq 0 \end{array}$$

The objective function and the three inequality constraints are

$$f(x_1, x_2) = (x_1 - 3)^2 + (x_2 - 2)^2$$

$$g_1(x_1, x_2) = x_1^2 - x_2 - 3$$

$$g_2(x_1, x_2) = x_2 - 1$$

$$g_3(x_1, x_2) = -x_1$$

Figure 1.1 illustrates the feasible region. The problem, then, is to find a point in the feasible region with the smallest possible value of $(x_1 - 3)^2 + (x_2 - 2)^2$. Note that points (x_1, x_2) with $(x_1 - 3)^2 + (x_2 - 2)^2 = c$ represent a circle with radius $\sqrt{c}$ and center (3, 2). This circle is called the *contour* of the objective function having the value c. Since we wish to minimize c, we must find the circle with the smallest radius that intersects the feasible region. As shown in Figure 1.1, the smallest such circle has $c = 2$ and intersects the feasible region at the point (2, 1). Therefore, the optimal solution occurs at the point (2, 1) and has an objective value equal to 2.

The approach used above is to find an optimal solution by determining the objective contour with the smallest objective value that intersects the feasible region. Obviously, this approach of solving the problem geometrically is only suitable for small problems and is not practical for problems with more than two variables or those with complicated objective and constraint functions.

Notation

The following notation will be used throughout the book. Vectors are denoted by boldface lowercase Roman letters, such as **x**, **y**, and **z**. All vectors are column vectors unless explicitly stated otherwise. Row vectors are the transpose of column vectors; for example, $\mathbf{x}^t$ denotes the row vector $(x_1, \ldots, x_n)$. The n-dimensional *real Euclidean space*, composed of all real vectors of dimension n, is denoted by E_n. Matrices are denoted by boldface capital Roman letters, such as **A** and **B**. Scalar-valued functions are denoted by lowercase Roman or Greek letters, such as f, g, and θ. Vector-valued functions are denoted by boldface lowercase Roman or Greek letters, such as $\mathbf{g}$ and $\boldsymbol{\psi}$. Point-to-set maps are denoted by boldface capital Roman letters such as $\mathbf{A}$ and $\mathbf{B}$. Scalars are denoted by lowercase Roman and Greek letters, such as k, λ, and α.

1.2 Some Illustrative Examples

In this section, we discuss some example problems that can be formulated as nonlinear programs. In particular, we discuss optimization problems in the following areas:

A. Optimal control
B. Structural design
C. Mechanical design
D. Electrical networks
E. Water resources management
F. Stochastic resource allocation
G. Location of facilities

A. Optimal Control Problems

As we will learn shortly, a discrete control problem can be stated as a nonlinear programming problem. Furthermore, a continuous optimal control problem can be approximated by a nonlinear programming problem. Hence, the procedures discussed later in the book can be used to solve some optimal control problems.

Discrete Optimal Control

Consider a fixed-time discrete optimal control problem of duration K periods. At the beginning of period k, the system is represented by the *state vector* $\mathbf{y}_{k-1}$. A *control vector* $\mathbf{u}_k$ changes the state of the system from $\mathbf{y}_{k-1}$ to $\mathbf{y}_k$ at the end of period k according to the following relationship:

$$\mathbf{y}_k = \mathbf{y}_{k-1} + \boldsymbol{\phi}_k(\mathbf{y}_{k-1}, \mathbf{u}_k) \qquad \text{for } k = 1, \ldots, K$$

Given the initial state $\mathbf{y}_0$, applying the sequence of controls $\mathbf{u}_1, \ldots, \mathbf{u}_K$ would result in a sequence of state vectors $\mathbf{y}_1, \ldots, \mathbf{y}_K$ called the *trajectory*. This process is illustrated in Figure 1.2.

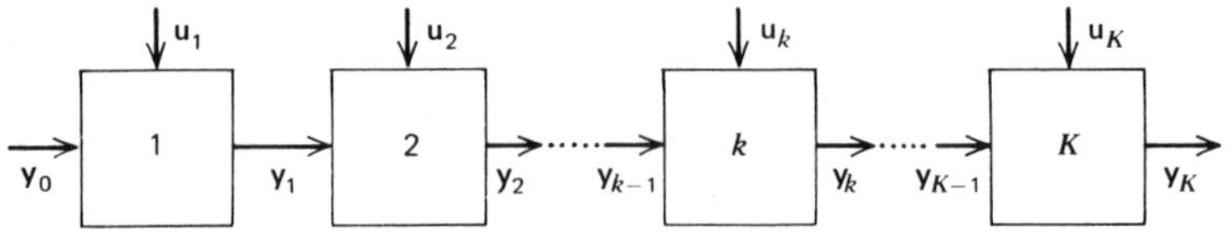

Figure 1.2 Illustration of a discrete control system.

A sequence of controls $\mathbf{u}_1, \ldots, \mathbf{u}_K$ and a sequence of state vectors $\mathbf{y}_0, \mathbf{y}_1, \ldots, \mathbf{y}_K$ are called *admissible* or *feasible* if they satisfy the following restrictions:

$$\mathbf{y}_k \in Y_k \qquad \text{for } k = 1, \ldots, K$$

$$\mathbf{u}_k \in U_k \qquad \text{for } k = 1, \ldots, K$$

$$\boldsymbol{\psi}(\mathbf{y}_0, \ldots, \mathbf{y}_K, \mathbf{u}_1, \ldots, \mathbf{u}_K) \in D$$

where $Y_1, \ldots, Y_K, U_1, \ldots, U_K$, and D are specified sets, and $\boldsymbol{\psi}$ is a known function, usually called the *trajectory constraint function*. Among all feasible controls and trajectories, one seeks a control and a corresponding trajectory that optimizes a certain objective function. The discrete control problem can thus be stated as follows:

$$\begin{array}{lll}
\text{Minimize} & \alpha(\mathbf{y}_0, \mathbf{y}_1, \ldots, \mathbf{y}_K, \mathbf{u}_1, \ldots, \mathbf{u}_K) & \\
\text{subject to} & \mathbf{y}_k = \mathbf{y}_{k-1} + \boldsymbol{\phi}_k(\mathbf{y}_{k-1}, \mathbf{u}_k) & \text{for } k = 1, \ldots, K \\
 & \mathbf{y}_k \in Y_k & \text{for } k = 1, \ldots, K \\
 & \mathbf{u}_k \in U_k & \text{for } k = 1, \ldots, K \\
 & \boldsymbol{\psi}(\mathbf{y}_0, \ldots, \mathbf{y}_K, \mathbf{u}_1, \ldots, \mathbf{u}_K) \in D &
\end{array}$$

Combining $\mathbf{y}_1, \ldots, \mathbf{y}_K, \mathbf{u}_1, \ldots, \mathbf{u}_K$ as the vector $\mathbf{x}$, and by suitable choices of $\mathbf{g}$, $\mathbf{h}$, and X, it can be easily verified that the above problem can be stated as the nonlinear programming problem introduced in Section 1.1.

A Production–Inventory Example We illustrate the formulation of a discrete control problem with the following production–inventory example. Suppose that a company produces a certain item to meet a known demand, and suppose that the production schedule must be determined over a total of K periods. The demand during any period can be met from the inventory at the beginning of the period and the production during the period. The maximum production during any period is restricted by the production capacity of the available equipment so that it cannot exceed b units. Assume that adequate temporary labor can be hired when needed and fired if superfluous. However, to discourage heavy labor fluctuations, a cost proportional to the square of the difference in the labor force during any two successive periods is assumed. Also, a cost proportional to the inventory carried forward from one period to another is incurred. Find the labor force and inventory during periods $1, \ldots, K$ such that the demand is satisfied and the total cost is minimized.

In this problem, there are two state variables, the inventory level I_k and the labor force L_k at the end of period k. The control variable u_k is the labor force acquired during period k ($u_k < 0$ means that the labor is reduced by an amount $-u_k$). The production–inventory problem can thus be stated as follows:

$$\begin{array}{lll}
\text{Minimize} & \sum_{k=1}^{K} (c_1 u_k^2 + c_2 I_k) & \\
\text{subject to} & L_k = L_{k-1} + u_k & \text{for } k = 1, \ldots, K \\
 & I_k = I_{k-1} + pL_{k-1} - d_k & \text{for } k = 1, \ldots, K \\
 & 0 \le L_k \le b/p & \text{for } k = 1, \ldots, K \\
 & I_k \ge 0 & \text{for } k = 1, \ldots, K
\end{array}$$

where the initial inventory I_0 and the initial labor force L_0 are known, d_k is the known demand during period k, and p is the number of units produced per worker during any given period.

Continuous Optimal Control

In the case of a discrete control problem, the controls are exercised at discrete points. We shall now consider a fixed-time continuous control problem in which a control function, $\mathbf{u}$, is to be exerted over the planning horizon $[0, T]$. Given the initial state $\mathbf{y}_0$, the relationship between the state vector $\mathbf{y}$ and the control vector $\mathbf{u}$ is governed by the following differential equation:

$$\dot{\mathbf{y}}(t) = \boldsymbol{\phi}[\mathbf{y}(t), \mathbf{u}(t)] \qquad \text{for } t \in [0, T]$$

The control function and the corresponding trajectory function are called admissible if the following restrictions hold true:

$$\begin{aligned} &\mathbf{y}(t) \in Y && \text{for } t \in [0, T] \\ &\mathbf{u}(t) \in U && \text{for } t \in [0, T] \\ &\boldsymbol{\psi}(\mathbf{y}, \mathbf{u}) \in D && \end{aligned}$$

A typical example of the set U is the collection of piecewise continuous functions on $[0, T]$ such that $\mathbf{a} \leq \mathbf{u}(t) \leq \mathbf{b}$ for $t \in [0, T]$. The optimal control problem can be stated as follows, where the initial state vector $\mathbf{y}(0) = \mathbf{y}_0$ is given:

$$\begin{aligned} \text{Minimize} \quad & \int_0^T \alpha[\mathbf{y}(t), \mathbf{u}(t)]dt && \\ \text{subject to} \quad & \dot{\mathbf{y}}(t) = \boldsymbol{\phi}[\mathbf{y}(t), \mathbf{u}(t)] && \text{for } t \in [0, T] \\ & \mathbf{y}(t) \in Y && \text{for } t \in [0, T] \\ & \mathbf{u}(t) \in U && \text{for } t \in [0, T] \\ & \boldsymbol{\psi}(\mathbf{y}, \mathbf{u}) \in D && \end{aligned}$$

A continuous optimal control problem can be approximated by a discrete problem. In particular, suppose that the planning region $[0, T]$ is divided into K periods, each of duration Δ, such that $K\Delta = T$. Denoting $\mathbf{y}(k\Delta)$ by $\mathbf{y}_k$ and $\mathbf{u}(k\Delta)$ by $\mathbf{u}_k$, for $k = 1, \ldots, K$, the above problem can be approximated as follows, where the initial state $\mathbf{y}_0$ is given:

$$\begin{aligned} \text{Minimize} \quad & \sum_{k=1}^{K} \alpha(\mathbf{y}_k, \mathbf{u}_k) && \\ \text{subject to} \quad & \mathbf{y}_k = \mathbf{y}_{k-1} + \Delta\boldsymbol{\phi}(\mathbf{y}_{k-1}, \mathbf{u}_k) && \text{for } k = 1, \ldots, K \\ & \mathbf{y}_k \in Y && \text{for } k = 1, \ldots, K \\ & \mathbf{u}_k \in U && \text{for } k = 1, \ldots, K \\ & \boldsymbol{\psi}(\mathbf{y}_0, \ldots, \mathbf{y}_K, \mathbf{u}_1, \ldots, \mathbf{u}_K) \in D && \end{aligned}$$

An Example of Rocket Launching Consider the problem of a rocket that is to be moved from ground level to a height $\bar{y}$ in time T. Let $y(t)$ denote the height from the ground at time t, and let $u(t)$ denote the force exerted in the vertical direction at time t. Assuming that the rocket has mass m, the equation of motion is given by

$$m\ddot{y}(t) + mg = u(t) \qquad \text{for } t \in [0, T]$$

where $\ddot{y}(t)$ is the acceleration at time t and g is the deceleration due to gravity. Further, suppose that the maximum force that could be exerted at any time cannot exceed b. If the objective is to expend the smallest possible energy so that the rocket reaches an altitude $\bar{y}$ at time T, the problem can be formulated as follows:

$$\begin{aligned} \text{Minimize} \quad & \int_0^T |u(t)|dt && \\ \text{subject to} \quad & m\ddot{y}(t) + mg = u(t) && \text{for } t \in [0, T] \\ & |u(t)| \leq b && \text{for } t \in [0, T] \\ & y(T) = \bar{y} && \end{aligned}$$

where $y(0) = 0$. The rocket problem with a second-order differential equation can be transformed into an equivalent problem with two first-order differential equations. This can be done by the following substitution: $y_1 = y$ and $y_2 = \dot{y}$. Therefore, $\ddot{y} + mg = u$ is equivalent to $\dot{y}_1 = y_2$ and $m\dot{y}_2 + mg = u$. Hence, the problem can be restated as follows:

$$\begin{array}{lll} \text{Minimize} & \displaystyle\int_0^T |u(t)|dt & \\ \text{subject to} & \dot{y}_1(t) = y_2(t) & \text{for } t \in [0, T] \\ & m\dot{y}_2(t) = u(t) - mg & \text{for } t \in [0, T] \\ & |u(t)| \leq b & \text{for } t \in [0, T] \\ & y_1(T) = \bar{y} & \end{array}$$

where $y_1(0) = y_2(0) = 0$. Suppose that we divide the interval $[0, T]$ into K periods. To simplify the notation, suppose that each period has length 1. Denoting the force, altitude, and velocity at the end of period k by u_k, $y_{1,k}$, and $y_{2,k}$ respectively, for $k = 1, \ldots, K$, the above problem can be approximated by the following nonlinear program, where $y_{1,0} = y_{2,0} = 0$:

$$\begin{array}{lll} \text{Minimize} & \displaystyle\sum_{k=1}^{K} |u_k| & \\ \text{subject to} & y_{1,k} - y_{1,k-1} = y_{2,k-1} & \text{for } k = 1, \ldots, K \\ & m(y_{2,k} - y_{2,k-1}) = u_k - mg & \text{for } k = 1, \ldots, K \\ & |u_k| \leq b & \text{for } k = 1, \ldots, K \\ & y_{1,K} = \bar{y} & \end{array}$$

The interested reader may refer to Luenberger [1969, 1983] for this problem and other continuous optimal control problems.

An Example of Highway Construction Suppose that a road is to be constructed over an uneven terrain. The construction cost is assumed to be proportional to the amount of dirt added or removed. Let T be the length of the road, and let $c(t)$ be the known height of the terrain at any given $t \in [0, T]$. The problem is to formulate an equation describing the height of the road $y(t)$ for $t \in [0, T]$.

To avoid excessive slopes on the road, the maximum slope must not exceed b_1 in magnitude, that is, $|\dot{y}(t)| \leq b_1$. In addition, to reduce the roughness of the ride, the rate of change of the slope of the road must not exceed b_2 in magnitude, that is, $|\ddot{y}(t)| \leq b_2$. Furthermore, the end conditions $y(0) = a$ and $y(T) = b$ must be observed. The problem can thus be stated as follows:

$$\begin{array}{lll} \text{Minimize} & \displaystyle\int_0^T |y(t) - c(t)|dt & \\ \text{subject to} & |\dot{y}(t)| \leq b_1 & \text{for } t \in [0, T] \\ & |\ddot{y}(t)| \leq b_2 & \text{for } t \in [0, T] \\ & y(0) = a & \\ & y(T) = b & \end{array}$$

Note that the control variable is the amount of dirt added or removed, that is, $u(t) = y(t) - c(t)$.

Now let $y_1 = y$ and $y_2 = \dot{y}$, and divide the road length into K intervals. For simplicity, suppose that each interval has length 1. Denoting $c(k)$, $y_1(k)$, and $y_2(k)$ by c_k, $y_{1,k}$, and $y_{2,k}$, respectively, the above problem can be approximated by the following nonlinear program:

$$\begin{array}{lll} \text{Minimize} & \sum_{k=1}^{K} |y_{1,k} - c_k| & \\ \text{subject to} & y_{1,k} - y_{1,k-1} = y_{2,k-1} & \text{for } k = 1, \ldots, K \\ & -b_1 \leq y_{2,k} \leq b_1 & \text{for } k = 0, \ldots, K-1 \\ & -b_2 \leq y_{2,k} - y_{2,k-1} \leq b_2 & \text{for } k = 1, \ldots, K-1 \\ & y_{1,0} = a & \\ & y_{1,K} = b & \end{array}$$

The interested reader may refer to Citron [1969] for more details of this example.

B. Structural Design

Structural designers have traditionally endeavoured to develop designs that could safely carry the projected loads. The concept of optimality was implicit only through the standard practice and experience of the designer. Recently, the design of sophisticated structures, such as aerospace structures, has called for more explicit consideration of optimality.

The main approaches used for minimum weight design of structural systems are based on the use of mathematical programming or other rigorous numerical techniques combined with structural analysis methods. Linear programming, nonlinear programming, and Monte Carlo simulation have been the principal techniques used for this purpose.

As noted by Batt and Gellatly [1974],

> The total process for the design of a sophisticated aerospace structure is a multistage procedure which ranges from consideration of overall systems performance down to the detailed design of individual components. While all levels of the design process have some greater or lesser degree of interaction with each other, the past state-of-the-art in design has demanded the assumption of a relatively loose coupling between the stages. Initial work in structural optimization has tended to maintain this stratification of design philosophy, although this state of affairs has occurred, possibly, more as a consequence of the methodology used for optimization than from any desire to perpetuate the delineations between design stages.

The following example illustrates how structural analysis methods can be used to yield a nonlinear programming problem involving a minimum weight design of a two-bar truss.

Two-Bar Truss Consider the planar truss shown in Figure 1.3. The truss consists of two steel tubes pinned together at one end and fixed at two pivot points at the other end. The span, that is, the distance between the two pivots, is fixed at $2s$. The design problem is to choose the height of the truss and the thickness and average diameter of the steel tubes so that the truss will support a load of $2W$ while minimizing the total weight of the truss.

Denote the average tube diameter, the tube thickness, and the truss height by x_1, x_2, and x_3, respectively. The weight of the steel truss is then given by $2\pi\rho x_1 x_2 (s^2 + x_3^2)^{1/2}$, where ρ is the density of the steel tube. The following constraints must be observed:

1. Because of space limitations, the height of the truss must not exceed b_1, that is, $x_3 \leq b_1$.
2. The ratio of the diameter of the tube to the thickness of the tube must not exceed b_2, that is, $x_1/x_2 \leq b_2$.

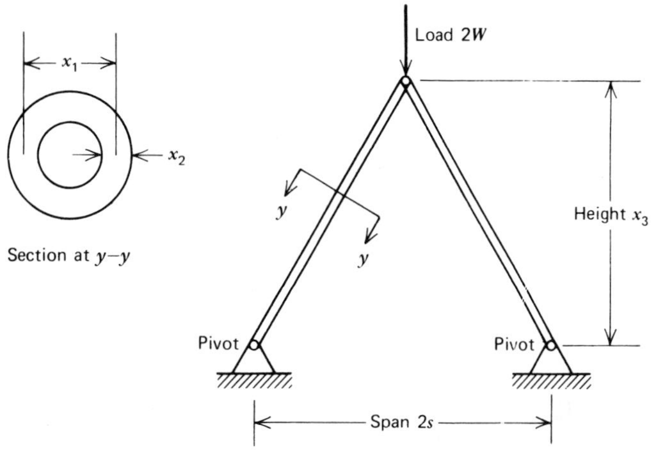

Figure 1.3 Example of a two-bar truss.

3. The compression stress in the steel tubes must not exceed the steel yield stress. This gives the following constraint, where b_3 is a constant:

$$W(s^2 + x_3^2)^{1/2} \leq b_3 x_1 x_2 x_3$$

4. The height, diameter, and thickness must be chosen such that the tubes will not buckle under the load. This constraint can be expressed mathematically as follows, where b_4 is a known parameter:

$$W(s^2 + x_3^2)^{3/2} \leq b_4 x_1 x_3 (x_1^2 + x_2^2)$$

From the above discussion, the truss design problem can be stated as the following nonlinear programming problem:

$$\begin{array}{ll}
\text{Minimize} & x_1 x_2 (s^2 + x_3^2)^{1/2} \\
\text{subject to} & x_3 - b_1 \leq 0 \\
& x_1 - b_2 x_2 \leq 0 \\
& W(s^2 + x_3^2)^{1/2} - b_3 x_1 x_2 x_3 \leq 0 \\
& W(s^2 + x_3^2)^{3/2} - b_4 x_1 x_3 (x_1^2 + x_2^2) \leq 0 \\
& x_1, x_2, x_3 \geq 0
\end{array}$$

C. Mechanical Design

In mechanical design, the concept of optimization can be used in conjunction with the traditional use of statics, dynamics, and the properties of materials. Asimov [1962], Johnson [1971], and Fox [1971] give several examples of optimum mechanical design using mathematical programming. As noted by Johnson [1971], in designing mechanisms for high-speed machines, significant dynamic stresses and vibrations are inherently unavoidable. Hence, it is necessary to design certain mechanical elements on the basis of minimizing these undesirable characteristics. The following example illustrates an optimum design for a bearing journal.

A Journal Design Problem Consider a two-bearing journal, each of length L, supporting a flywheel of weight W mounted on a shaft of diameter D, as shown in Fig-

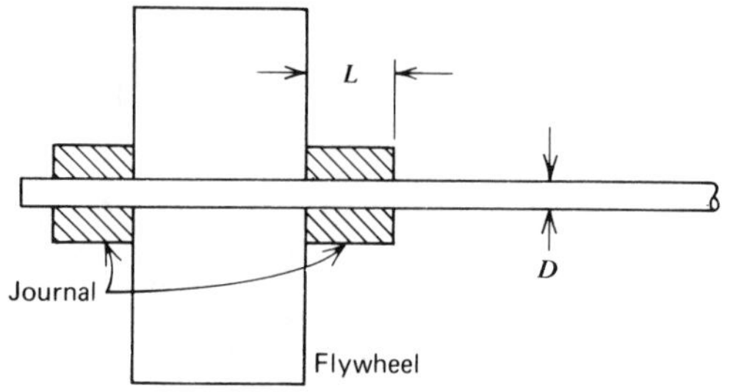

Figure 1.4 Journal bearing assembly.

ure 1.4. We wish to determine L and D that minimize frictional moment while keeping the shaft twist angle and clearances within acceptable limits.

A layer of oil film between the journal and the shaft is maintained by forced lubrication. The oil film serves to minimize the frictional moment and to limit the heat rise, thereby increasing the life of the bearing. Let h_0 be the smallest oil film thickness under steady-state operation. Then, we must have

$$\hat{h}_0 \leq h_0 \leq \delta$$

where

$\hat{h}_0$ = minimum oil film thickness to prevent metal-to-metal contact
δ = radial clearance specified as the difference between the journal radius and the shaft radius

A further limitation on h_0 is imposed by the following inequality:

$$0 \leq e \leq \hat{e}$$

where e is the *eccentricity ratio* defined by $e = 1 - (h_0/\delta)$, and $\hat{e}$ is a prespecified upper limit.

Depending on the point at which the torque is applied on the shaft, or the nature of the torque impulses, and on the ratio of the shear modulus of elasticity to the maximum shear stress, a constant k_1 can be specified such that the angle of twist of the shaft is given by

$$\theta = \frac{1}{k_1 D}$$

Furthermore, the frictional moment for the two bearings is given by

$$M = k_2 \frac{\omega}{\delta\sqrt{1 - e^2}} D^3 L$$

where k_2 is a constant that depends on the viscosity of the lubricating oil and ω is the rotational speed. Also, based on hydrodynamic considerations, the safe load-carrying capacity of a bearing is given by

$$c = k_3 \frac{\omega}{\delta^2} D L^3 \phi(e)$$

where k_3 is a constant depending upon the viscosity of the oil and

$$\phi(e) = \frac{e}{(1 - e^2)^2}[\pi^2(1 - e^2) + 16e^2]^{1/2}$$

Obviously, we need to have $2c \geq W$ to carry the weight W of the flywheel.

Thus, if δ, $\hat{h}_0$ and $\hat{e}$ are specified, one typical design problem is to find D, L, and h_0 to minimize the frictional moment while keeping the twist angle within an acceptable limit α. The model is thus given by

$$\begin{aligned}
\text{Minimize} \quad & \frac{\omega}{\delta\sqrt{1-e^2}} D^3L \\
\text{subject to} \quad & \frac{1}{k_1 D} \leq \alpha \\
& 2\frac{k_3\omega}{\delta^2} DL^3\phi\left(1 - \frac{h_0}{\delta}\right) \geq W \\
& \hat{h}_0 \leq h_0 \leq \delta \\
& 0 \leq \left(1 - \frac{h_0}{\delta}\right) \leq \hat{e} \\
& D \geq 0 \\
& L \geq 0
\end{aligned}$$

For a thorough discussion of this problem, the reader may refer to Asimov [1962]. The reader can also formulate the model to minimize the twist angle subject to the frictional moment being within a given maximum limit M'. One could also conceive of an objective function involving both the frictional moment and the angle of twist, if a proper weight for these factors is selected to reflect their relative importance.

D. Electrical Networks

It has been well recognized for over a century that the equilibrium conditions of an electrical or a hydraulic network are attained as the total energy loss is minimized. Dennis was perhaps the first to investigate the relationship between electrical circuit theory, mathematical programming, and duality. The following discussion is based on the pioneering work of Dennis [1959].

An electrical circuit can be described by, for example, n *branches* connecting m *nodes*. In the following, we consider a direct current network and assume that the nodes and each connecting branch are defined so that only one of the following electrical devices is encountered:

1. A *voltage source* that maintains a constant branch voltage v_s irrespective of the branch current c_s. Such a device absorbs power equal to $-v_s c_s$.
2. A *diode* that permits the branch current c_d to flow only in one direction and consumes zero power regardless of the branch current or voltage. Denoting the latter by v_d, this can be stated as

$$c_d \geq 0 \qquad v_d \geq 0 \qquad v_d c_d = 0 \tag{1.1}$$

3. A *resistor* that consumes power and whose branch current c_r and branch voltage v_r are related by

$$v_r = -rc_r \tag{1.2}$$

where r is the *resistance* of the resistor. The power consumed is given by

$$-v_r c_r = \frac{v_r^2}{r} = rc_r^2 \tag{1.3}$$

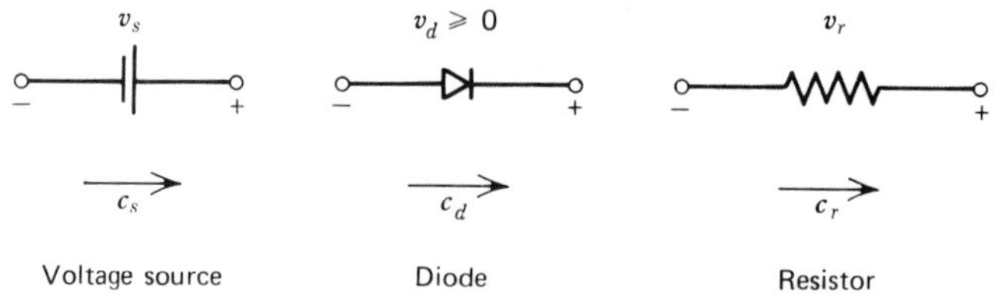

Figure 1.5 Typical electrical devices in a circuit.

The three devices are shown schematically in Figure 1.5. The current flow in the diagram is shown from the negative terminal of the branch to the positive terminal of the branch. The former is called the *origin* node, and the latter is the *ending* node of the branch. If the current flows in the opposite direction, the corresponding branch current will have a negative value, which, incidentally, is not permissible for the diode. The same sign convention will be used for branch voltages.

A network with a number of branches can be described by a *node-branch incidence* matrix $\mathbf{N}$ whose rows correspond to the nodes and whose columns correspond to the branches. A typical element n_{ij} of $\mathbf{N}$ is given by

$$n_{ij} = \begin{cases} -1 & \text{if branch } j \text{ has node } i \text{ as its origin} \\ 1 & \text{if branch } j \text{ ends in node } i \\ 0 & \text{otherwise} \end{cases}$$

For a network with several voltage sources, diodes, and resistors, let $\mathbf{N}_S$ denote the node-branch incidence matrix for all the branches with voltage sources, $\mathbf{N}_D$ denote the node-branch incidence matrix for all branches with diodes, and $\mathbf{N}_R$ denote the node-branch incidence matrix for all branches with resistors. Then, without loss of generality, we can partition $\mathbf{N}$ as

$$\mathbf{N} = [\mathbf{N}_S, \mathbf{N}_D, \mathbf{N}_R]$$

Likewise, the column vector $\mathbf{c}$, representing the branch currents, can be partitioned as

$$\mathbf{c}^t = [\mathbf{c}_S^t, \mathbf{c}_D^t, \mathbf{c}_R^t]$$

and the column vector $\mathbf{v}$, representing the branch voltages, can be written as

$$\mathbf{v}^t = [\mathbf{v}_S^t, \mathbf{v}_D^t, \mathbf{v}_R^t]$$

Associated with each node i is a *node potential* p_i. The column vector $\mathbf{p}$, representing node potentials, can be written as

$$\mathbf{p}^t = [\mathbf{p}_S^t, \mathbf{p}_D^t, \mathbf{p}_R^t]$$

The following basic laws govern the equilibrium conditions of the network:

Kirchhoff's node law. The sum of all currents entering a node is equal to the sum of all currents leaving the node. This can be written as $\mathbf{Nc} = \mathbf{0}$, or

$$\mathbf{N}_S\mathbf{c}_S + \mathbf{N}_D\mathbf{c}_D + \mathbf{N}_R\mathbf{c}_R = \mathbf{0} \tag{1.4}$$

Kirchhoff's loop law. The difference between the node potentials at the ends of each branch is equal to the branch voltage. This can be written as $\mathbf{N}^t\mathbf{p} = \mathbf{v}$, or

$$\begin{aligned} \mathbf{N}_S^t\mathbf{p} &= \mathbf{v}_S \\ \mathbf{N}_D^t\mathbf{p} &= \mathbf{v}_D \\ \mathbf{N}_R^t\mathbf{p} &= \mathbf{v}_R \end{aligned} \tag{1.5}$$

In addition, we have the equations representing the characteristics of the electrical devices. From (1.1), for the set of diodes, we have

$$\mathbf{v}_D \geq \mathbf{0} \qquad \mathbf{c}_D \geq \mathbf{0} \qquad \mathbf{v}_D^t\mathbf{c}_D = 0 \tag{1.6}$$

and from (1.2), for the resistors, we have

$$\mathbf{v}_R = -\mathbf{R}\mathbf{c}_R \tag{1.7}$$

where $\mathbf{R}$ is a diagonal matrix whose diagonal elements are the resistance values.

Thus, (1.4) through (1.7) represent the equilibrium conditions of the circuit, and we wish to find $\mathbf{v}_D$, $\mathbf{v}_R$, $\mathbf{c}$, and $\mathbf{p}$ satisfying these conditions.

Now consider the following quadratic programming problem, which is discussed in Section 11.2:

$$\begin{aligned} \text{Minimize} \quad & \tfrac{1}{2}\mathbf{c}_R^t\mathbf{R}\mathbf{c}_R - \mathbf{v}_S^t\mathbf{c}_S \\ \text{subject to} \quad & \mathbf{N}_S\mathbf{c}_S + \mathbf{N}_D\mathbf{c}_D + \mathbf{N}_R\mathbf{c}_R = \mathbf{0} \\ & -\mathbf{c}_D \leq \mathbf{0} \end{aligned}$$

Here, we wish to determine the branch currents $\mathbf{c}_S$, $\mathbf{c}_D$, and $\mathbf{c}_R$ to minimize the sum of half the energy absorbed in the resistors and the energy loss of the voltage source. From Section 4.3, the optimality conditions for this problem are

$$\begin{aligned} \mathbf{N}_S^t\mathbf{u} - \mathbf{v}_S &= \mathbf{0} \\ \mathbf{N}_D^t\mathbf{u} - \mathbf{I}\mathbf{u}_0 &= \mathbf{0} \\ \mathbf{N}_R^t\mathbf{u} + \mathbf{R}\mathbf{c}_R &= \mathbf{0} \\ \mathbf{N}_S\mathbf{c}_S + \mathbf{N}_D\mathbf{c}_D + \mathbf{N}_R\mathbf{c}_R &= \mathbf{0} \\ \mathbf{c}_D^t\mathbf{u}_0 &= 0 \\ \mathbf{c}_D, \mathbf{u}_0 &\geq \mathbf{0} \end{aligned}$$

where $\mathbf{u}$ and $\mathbf{u}_0$ are column vectors representing the Lagrangian multipliers. It can be readily verified that, letting $\mathbf{v}_D = \mathbf{u}_0$, $\mathbf{p} = \mathbf{u}$ and noting (1.7), the above conditions are precisely the equilibrium conditions (1.4) through (1.7). Note that the Lagrangian multiplier vector $\mathbf{u}$ is precisely the node potential vector $\mathbf{p}$.

Associated with the above problem is another problem, referred to as the *dual problem* (given below), where $\mathbf{G} = \mathbf{R}^{-1}$ is a diagonal matrix whose elements are the conductances and where $\mathbf{v}_S$ is fixed.

$$\begin{aligned} \text{Maximize} \quad & -\tfrac{1}{2}\mathbf{v}_R^t\mathbf{G}\mathbf{v}_R \\ \text{subject to} \quad & \mathbf{N}_S^t\mathbf{p} = \mathbf{v}_S \\ & \mathbf{N}_D^t\mathbf{p} - \mathbf{v}_D = \mathbf{0} \\ & \mathbf{N}_R^t\mathbf{p} - \mathbf{v}_R = \mathbf{0} \\ & \mathbf{v}_D \geq \mathbf{0} \end{aligned}$$

Here, $\mathbf{v}_R^t\mathbf{G}\mathbf{v}_R$ is the power absorbed by the resistors, and we wish to find the branch voltages $\mathbf{v}_D$ and $\mathbf{v}_R$ and the potential vector $\mathbf{p}$.

The optimality conditions for this problem also are precisely (1.4) through (1.7). Furthermore, the Lagrangian multipliers for this problem are the branch currents.

It is interesting to note by Theorem 6.2.4, the main Lagrangian duality theorem, that the objective function values of the above two problems are equal at optimality, that is,

$$\tfrac{1}{2}\mathbf{c}_R^t\mathbf{R}\mathbf{c}_R + \tfrac{1}{2}\mathbf{v}_R^t\mathbf{G}\mathbf{v}_R - \mathbf{v}_S^t\mathbf{c}_S = 0$$

Since $\mathbf{G} = \mathbf{R}^{-1}$ and noting (1.6) and (1.7), the above equation reduces to

$$\mathbf{v}_R^t\mathbf{c}_R + \mathbf{v}_D^t\mathbf{c}_D + \mathbf{v}_S^t\mathbf{c}_S = 0$$

which is precisely the principle of energy conservation.

The reader may be interested in other applications of mathematical programming for solving problems associated with generation and distribution of electrical power. A brief discussion, along with suitable references, is given in the Notes and References section at the end of the chapter.

E. Water Resources Management

We now develop an optimization model for the conjunctive use of water resources for both hydropower generation and agricultural use. Consider the river basin depicted schematically in Figure 1.6.

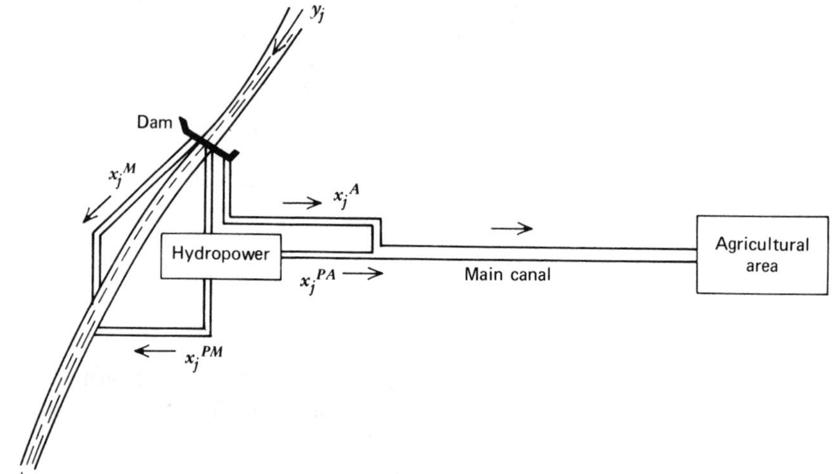

Figure 1.6 A typical river basin.

A dam across the river provides the surface water storage facility to provide water for power generation and agriculture. The power plant is assumed to be close to the dam, and water for agriculture is conveyed from the dam, directly or after power generation, through a canal.

There are two classes of variables associated with the problem:

1. ***Design variables:*** What should be the optimal capacity S of the reservoir, the capacity U of the canal supplying agricultural water, and the capacity E of the power plant?

2. ***Operational variables:*** How much water should be released for agriculture, power generation, and other purposes?

From Figure 1.6, the following operational variables can readily be identified for the jth period:

x_j^A = Water released from the dam for agriculture
x_j^{PA} = Water released for power generation and then for agricultural use
x_j^{PM} = Water released for power generation and then returned downstream
x_j^M = Water released from the dam directly downstream

For the purpose of a planning model, we shall adopt a planning horizon of N periods corresponding to the life span of major capital investments, such as that for the dam. The objective is to minimize the total discounted costs associated with the reservoir, power plant, and canal minus the revenues from power generation and agriculture. These costs and revenues are discussed below.

Power Plant Associated with the power plant, we have a cost of

$$C(E) + \sum_{j=1}^{N} \beta_j \hat{C}_e(E) \tag{1.8}$$

where $C(E)$ is the cost of the power plant, associated structures, and transmission facilities if the power plant capacity is E, and $\hat{C}_e(E)$ is the annual operation, maintenance, and replacement costs of the power facilities. Here, β_j is a discounting factor to give the present worth of the cost in period j. See Mobasheri [1968] for the nature of the functions $C(E)$ and $\hat{C}_e(E)$.

Furthermore, the discounted revenues associated with the energy sales can be expressed as

$$\delta\left\{\sum_{j=1}^{N} \beta_j[p_f F_j + p_d(f_j - F_j)]\right\} + (1 - \delta)\left\{\sum_{j=1}^{n} \beta_j[p_f f_j - p_s(F_j - f_j)]\right\} \tag{1.9}$$

where F_j is the known firm power demand that can be sold at p_f and f_j is the power production. Here, $\delta = 1$ if $f_j > F_j$, and the excess power $f_j - F_j$ can be sold at a dump price of p_d. On the other hand, $\delta = 0$ if $f_j < F_j$, and a penalty of $p_s(F_j - f_j)$ is incurred since power has to be bought from adjoining power networks.

Reservoir and Canal The discounted capital costs are given by

$$C_r(S) + \alpha C_l(U) \tag{1.10}$$

where $C_r(S)$ is the cost of the reservoir if its capacity is S, and $C_l(U)$ is the capital cost of the main canal if its capacity is U. Here, α is a scalar to account for the lower life span of the canal compared to that of the reservoir.

The discounted operational costs are given by

$$\sum_{j=1}^{N} \beta_j[\hat{C}_r(S) + \hat{C}_l(U)] \tag{1.11}$$

The interested reader may refer to Mobasheri [1968] and Maass et al. [1967] for a discussion on the nature of the functions discussed here.

Irrigation Revenues The crop yield from irrigation can be expressed as a function R of the water used for irrigation during period j as shown by Minhas, Parikh, and Srinivasan [1974]. Thus, the revenue from agriculture is given by

$$\sum_{j=1}^{N} \beta_j R(x_j^A + x_j^{PA}) \tag{1.12}$$

Here, for convenience, we have neglected the water supplied through rainfall.

Thus far, we have discussed the various terms in the objective function. The model must also consider the constraints imposed on the design and decision variables.

Power Generation Constraints Clearly, the power generated cannot exceed the energy potential of the water supplied, so that

$$f_j \leq (x_j^{PM} + x_j^{PA})\psi(s_j)\gamma e \tag{1.13}$$

where $\psi(s_j)$ is the head created by the water s_j stored in the reservoir at period j, γ is the power conversion factor, and e is the efficiency of the power system. Refer to O'Laoghaire and Himmelblau [1974] for the nature of the function ψ.

Likewise, the power generated cannot exceed the generating capacity of the plant, so that

$$f_j \leq \alpha_j E e H_j \tag{1.14}$$

where α_j is the load factor defined as the ratio of the average daily production to the daily peak production and H_j is the number of operational hours.

Finally, the capacity of the plant has to be within known acceptable limits, that is,

$$E' \leq E \leq E'' \tag{1.15}$$

Reservoir Constraints If we neglect the evaporation loss, then the amount of water y_j flowing into the dam must be equal to the change in the amount stored in the dam and the water released for different purposes. This can be expressed as

$$s_{j+1} - s_j + x_j^A + x_j^M + x_j^{PM} + x_j^{PA} = y_j \tag{1.16}$$

A second set of constraints states that the storage of the reservoir should be adequate and be within acceptable limits, that is,

$$S \geq s_j \tag{1.17}$$

$$S' \leq S \leq S'' \tag{1.18}$$

Mandatory Water Release Constraint It is usually necessary to specify that a certain amount of water M_j is released to meet the downstream water requirements. This mandatory release requirement may be specified as

$$x_j^M + x_j^{PM} \geq M_j \tag{1.19}$$

Canal Capacity Finally, we need to specify that the canal capacity U should be adequate to handle the agricultural water. Hence,

$$x_j^A + x_j^{PA} \leq U \tag{1.20}$$

The objective is, then, to minimize the net costs represented by the sum of (1.8),

(1.10), and (1.11) minus the revenues given by (1.9) and (1.12). The constraints are given by (1.13) to (1.20), together with the restriction that all variables are nonnegative.

F. Stochastic Resource Allocation

Consider the following linear programming problem:

$$\begin{array}{ll} \text{Maximize} & \mathbf{c}^t\mathbf{x} \\ \text{subject to} & \mathbf{A}\mathbf{x} \leq \mathbf{b} \\ & \mathbf{x} \geq \mathbf{0} \end{array}$$

where $\mathbf{c}$ and $\mathbf{x}$ are n vectors, $\mathbf{b}$ is an m vector, and $\mathbf{A} = [\mathbf{a}_1, \ldots, \mathbf{a}_n]$ is an $m \times n$ matrix. The above problem can be interpreted as a resource allocation model as follows. Suppose that we have m resources represented by the vector $\mathbf{b}$. Column $\mathbf{a}_j$ of $\mathbf{A}$ represents an activity j, and the variable x_j represents the level of the activity to be selected. Activity j at level x_j consumes $\mathbf{a}_j x_j$ of the available resources; hence the constraint, $\mathbf{A}\mathbf{x} = \sum_{j=1}^n \mathbf{a}_j x_j \leq \mathbf{b}$. If the unit profit of activity j is c_j, then the total profit is $\sum_{j=1}^n c_j x_j = \mathbf{c}^t\mathbf{x}$. Thus, the problem can be interpreted as finding the best way of allocating the resource vector $\mathbf{b}$ to the various available activities so that the total profit is maximized.

For some practical problems, the above deterministic model is not adequate because the profit coefficients $c_1, \ldots, c_n$ are not fixed but are random variables. We shall thus assume that $\mathbf{c}$ is a random vector with *mean* $\bar{\mathbf{c}} = (\bar{c}_1, \ldots, \bar{c}_n)^t$ and *covariance matrix* $\mathbf{V}$. The objective function, denoted by z, will thus be a random variable with mean $\bar{\mathbf{c}}^t\mathbf{x}$ and variance $\mathbf{x}^t\mathbf{V}\mathbf{x}$.

If we want to maximize the expected value of z, we must solve the following problem:

$$\begin{array}{ll} \text{Maximize} & \bar{\mathbf{c}}^t\mathbf{x} \\ & \mathbf{A}\mathbf{x} \leq \mathbf{b} \\ & \mathbf{x} \geq \mathbf{0} \end{array}$$

which is a linear programming problem discussed in Section 2.6. On the other hand, if the variance of z is to be minimized, we have to solve the problem

$$\begin{array}{ll} \text{Minimize} & \mathbf{x}^t\mathbf{V}\mathbf{x} \\ & \mathbf{A}\mathbf{x} \leq \mathbf{b} \\ & \mathbf{x} \geq \mathbf{0} \end{array}$$

which is a quadratic program discussed in Section 11.2.

Satisficing Criteria and Chance Constraints

In maximizing the expected value, we have completely neglected the variance of the gain z. On the other hand, while minimizing the variance, we did not take into account the expected value of z. In a realistic problem, one would perhaps like to maximize the expected value and, at the same time, minimize the variance. This is a multiple objective problem, and some research has been done on dealing with such problems (see Zeleny [1974], and Zeleny and Cochrane [1973], and Steur [1986]). However, there are several other ways of considering both the expected value and the variance simultaneously.

Suppose one is interested in ensuring that the expected value should be at least equal to a certain value $\bar{z}$, frequently referred to as the *aspiration level*, or the *satisficing level*. The problem can then be stated as

$$\begin{array}{ll} \text{Minimize} & \mathbf{x}'\mathbf{V}\mathbf{x} \\ \text{subject to} & \mathbf{A}\mathbf{x} \le \mathbf{b} \\ & \bar{\mathbf{c}}'\mathbf{x} \ge \bar{z} \\ & \mathbf{x} \ge \mathbf{0} \end{array} \tag{1.21}$$

which is again a quadratic programming problem.

Another approach that can be adopted is as follows. Let $\alpha = \text{Prob}\,(\mathbf{c}'\mathbf{x} \ge \bar{z})$; that is, α gives the probability that the aspiration level $\bar{z}$ will be attained. Clearly, one would like to maximize α. Now suppose the vector of random variables $\mathbf{c}$ can be expressed as the function $\mathbf{d} + y\mathbf{f}$, where $\mathbf{d}$ and $\mathbf{f}$ are fixed vectors and y is a random variable. Then,

$$\begin{aligned} \alpha &= \text{Prob}\,(\mathbf{d}'\mathbf{x} + y\mathbf{f}'\mathbf{x} \ge \bar{z}) \\ &= \text{Prob}\left(y \ge \frac{\bar{z} - \mathbf{d}'\mathbf{x}}{\mathbf{f}'\mathbf{x}}\right) \end{aligned}$$

if $\mathbf{f}'\mathbf{x} > 0$. Hence, in this case, the problem of maximizing α reduces to

$$\begin{array}{ll} \text{Minimize} & \dfrac{\bar{z} - \mathbf{d}'\mathbf{x}}{\mathbf{f}'\mathbf{x}} \\ \text{subject to} & \mathbf{A}\mathbf{x} \le \mathbf{b} \\ & \mathbf{x} \ge \mathbf{0} \end{array}$$

This is a linear *fractional programming problem*, a solution procedure for which is discussed in Section 11.4.

Alternatively, if we wished to minimize the variance, but we also wanted to include a constraint that required the probability of the profit $\mathbf{c}'\mathbf{x}$ exceeding the desired value $\bar{z}$ to be at least some specified value q, then this could be incorporated by using the following *chance constraint*.

$$\text{Prob}(\mathbf{c}'\mathbf{x} \ge \bar{z}) = \text{Prob}\left(y \ge \frac{\bar{z} - \mathbf{d}'\mathbf{x}}{\mathbf{f}'\mathbf{x}}\right) \ge q$$

Now, assuming that y is a continuously distributed random variable for which ϕ_q denotes the upper $100q$ percentile value, that is, $\text{Prob}(y \ge \phi_q) = q$, the foregoing constraint can be equivalently written as

$$\frac{\bar{z} - \mathbf{d}'\mathbf{x}}{\mathbf{f}'\mathbf{x}} \le \phi_q \qquad \text{or} \qquad \mathbf{d}'\mathbf{x} + \phi_q\mathbf{f}'\mathbf{x} \ge \bar{z}$$

This linear constraint can then be used to replace the expected value constraint in the model (1.21).

Risk Aversion Model

The above approaches for handling the variance and the expected value of the return do not take into account the risk aversion behavior of individuals. For example, an individual who wants to avoid risk may prefer a gain with an expected value of \$100 and a variance of 10 to a gain with an expected value of \$110 with variance of 30. The individual who chooses the expected value of \$100 is more aversive to risk than the individual who might choose the alternative with expected value of \$110. This difference in risk-taking behavior can be taken into account by considering the utility of money for the individual.

For most individuals, the value of an additional dollar decreases as their total net worth increases. The value associated with a net worth z is called the *utility* of z.

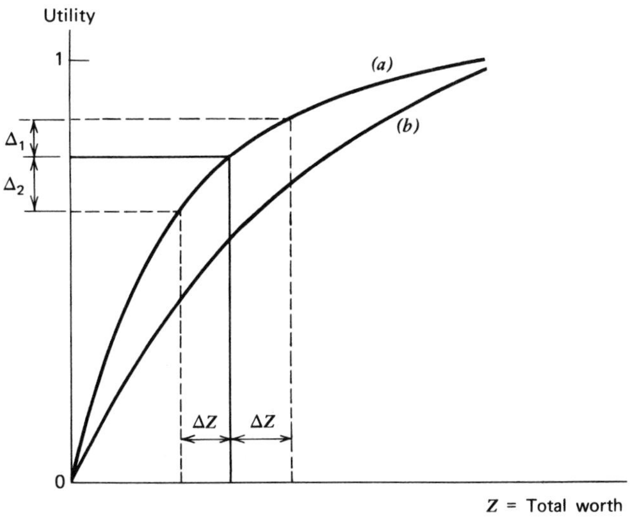

Figure 1.7 Utility functions.

Frequently, it is convenient to normalize the utility u so that $u = 0$ for $z = 0$ and $u = 1$ as z approaches the value ∞. The function u is called the individual's utility function and is usually a nondecreasing continuous function. Figure 1.7 gives two typical utility functions for two different individuals. For individual (a), a gain of Δz increases the utility by Δ_1 and a loss of Δz decreases the utility by Δ_2. Since Δ_2 is larger than Δ_1, this individual would prefer a lower variance. Such an individual is more aversive to risk than the individual whose utility function is as in (b) in Figure 1.7.

Different curves such as (a) or (b) in Figure 1.7 can be expressed mathematically as

$$u(z) = 1 - e^{-kz}$$

where $k > 0$ is called a *risk aversion constant*. Note that a larger value of k results in a more risk-aversive behavior.

Now suppose that the current worth is zero, so that the total worth is equal to the gain z. Suppose that $\mathbf{c}$ is a normal random vector with mean $\bar{\mathbf{c}}$ and covariance matrix $\mathbf{V}$. Then, z is a normal random variable with mean $\bar{z} = \bar{\mathbf{c}}'\mathbf{x}$ and variance $\sigma^2 = \mathbf{x}'\mathbf{V}\mathbf{x}$. In particular, the density function ϕ of the gain is given by

$$\phi(z) = \frac{1}{\sigma\sqrt{2\pi}} \exp -\frac{1}{2}\left(\frac{z - \bar{z}}{\sigma}\right)^2$$

We wish to maximize the expected value of the utility given by

$$\begin{aligned}
\int_{-\infty}^{\infty} (1 - e^{-kz})\phi(z)dz &= 1 - \frac{1}{\sigma\sqrt{2\pi}} \int_{-\infty}^{\infty} \exp\left[-kz - \frac{1}{2}\left(\frac{z - \bar{z}}{\sigma}\right)^2\right] dz \\
&= 1 - \frac{1}{\sigma\sqrt{2\pi}} \int_{-\infty}^{\infty} \exp\left[-\frac{1}{2}\left(\frac{z - \bar{z} + k\sigma^2}{\sigma}\right)^2\right] \exp\left(-k\bar{z} + \tfrac{1}{2}k^2\sigma^2\right) dz \\
&= 1 - \frac{\exp\left(-k\bar{z} + \tfrac{1}{2}k^2\sigma^2\right)}{\sigma\sqrt{2\pi}} \int_{-\infty}^{\infty} \exp\left[-\frac{1}{2}\left(\frac{z - \bar{z}\, k\sigma^2}{\sigma}\right)^2\right] dz \\
&= 1 - \exp\left(-k\bar{z} + \tfrac{1}{2}k^2\sigma^2\right)
\end{aligned}$$

Hence, maximizing the expected value of the utility is equivalent to maximizing $k\bar{z} - \frac{1}{2}k^2\sigma^2$. Substituting for $\bar{z}$ and σ^2, we get the following quadratic program:

$$\begin{array}{ll} \text{Maximize} & k\bar{\mathbf{c}}^t\mathbf{x} - \frac{1}{2}k^2\mathbf{x}^t\mathbf{V}\mathbf{x} \\ \text{subject to} & \mathbf{A}\mathbf{x} \le \mathbf{b} \\ & \mathbf{x} \ge \mathbf{0} \end{array}$$

Again, this can be solved by using the methods discussed in Chapter 11, depending on the nature of $\mathbf{V}$.

G. Location of Facilities

A frequently encountered problem is the optimal location of centers of activities. This may involve the location of machines or departments in a factory, the location of factories or warehouses from which goods can be shipped to retailers or consumers, or the location of emergency facilities (i.e., fire or police stations) in an urban area.

Consider the following simple case. Suppose that there are n markets with known demands and locations. These demands are to be met from m warehouses of known capacities. The problem is to determine the locations of the warehouses so that the total distance weighted by the shipment from the warehouses to the markets is minimized. More specifically, let

(x_i, y_i) = unknown location of warehouse i for $i = 1, \ldots, m$
c_i = capacity of warehouse i for $i = 1, \ldots, m$
(a_j, b_j) = known location of market j for $j = 1, \ldots, n$
r_j = known demand at market j for $j = 1, \ldots, n$
d_{ij} = distance from warehouse i to market area for j for $i = 1, \ldots, m$; $j = 1, \ldots, n$
w_{ij} = units shipped from warehouse i to market area j for $i = 1, \ldots, m$; $j = 1, \ldots, n$

The problem of locating the warehouses and determining the shipping pattern can be stated as follows:

$$\begin{array}{lll} \text{Minimize} & \sum_{i=1}^{m}\sum_{j=1}^{n} w_{ij}\, d_{ij} & \\ \text{subject to} & \sum_{j=1}^{n} w_{ij} \le c_i & \text{for } i = 1, \ldots, m \\ & \sum_{i=1}^{m} w_{ij} = r_j & \text{for } j = 1, \ldots, n \\ & w_{ij} \ge 0 & \text{for } i = 1, \ldots, m; j = 1, \ldots, n \end{array}$$

Note that both w_{ij} and d_{ij} are to be determined, and hence the above problem is a nonlinear programming problem. Different measures of distance can be chosen, using the *rectilinear, Euclidean*, or l_p *norm* metrics, where the value of p is chosen to approximate city travel distances. These are respectively given by

$$d_{ij} = |x_i - a_j| + |y_i - b_j|$$

$$d_{ij} = [(x_i - a_j)^2 + (y_i - b_j)^2]^{1/2}$$

$$d_{ij} = [(x_i - a_j)^p + (y_i - b_j)^p]^{1/p}$$

Each choice leads to a particular nonlinear problem in the variables $x_1, \ldots, x_m$, $y_1, \ldots, y_m$, $w_{11}, \ldots, w_{mn}$. If the locations of the warehouses are fixed, then the d_{ij}'s are known, and the above problem reduces to a special case of a linear programming problem known as the *transportation problem*. On the other hand, for fixed values of the transportation variables, the problem reduces to a (pure) *location problem*. Consequently, the above problem is also known as a *location–allocation problem*.

H. Miscellaneous Applications

There are a host of other applications to which nonlinear programming models and techniques have been applied. These include the problems of chemical equilibrium and process control; gasoline blending; oil extraction, blending, and distribution; forest thinning and harvest scheduling; economic equilibration of supply and demand interactions under various market behavioral phenomena; pipe network design for reliable water distribution systems; electric utility capacity expansion planning and load management; production and inventory control in manufacturing concerns; least squares estimation of statistical parameters and data fitting; and the design of aircraft, ships, bridges, and other structures. The Notes and References section cites several references that provide details on these and other applications.

1.3 Some Guidelines for Model Construction

The modeling process is concerned with the construction of a mathematical abstraction of a given problem that can be analyzed to produce meaningful answers which guide the decisions to be implemented. Central to this process is the *identification* or the *formulation* of the problem. By the nature of human activities, a problem is seldom isolated and crisply defined, but rather interacts with various other problems at the fringes and encompasses various details obfuscated by uncertainty. For example, a problem of scheduling jobs on machines interacts with the problems of acquiring raw materials, forecasting uncertain demand, and planning for inventory storage and dissipation; and it must contend with machine reliability, worker performance and absenteeism, and insertions of spurious or rush jobs. A modeler must therefore identify the particular scope and aspect of the problem to be explicitly considered in formulating the problem, and must make suitable simplifying assumptions so that the resulting model is a balanced compromise between *representability* and *mathematical tractability*. The model, being only an abstraction of the real problem, will yield answers that are only as meaningful as the degree of accuracy with which it represents the actual physical system. On the other hand, an unduly complicated model might be too complex to be analyzed mathematically for obtaining any credible solution for consideration at all! This compromise, of course, need not be achieved at a single attempt. Often, it is instructive to begin with a simpler model representation, to test it to gain insights into the problem, and then to guide the direction in which the model should be further refined to make it more representative while maintaining adequate tractability. While accomplishing this, it should be borne in mind that the answers from the model are meant to provide guidelines for making decisions rather than to replace the decision maker. The model is only an abstraction of reality and is not necessarily an equivalent representation of reality itself. At the same time, these guidelines need to be well founded and meaningful.

Moreover, one important function of a model is to provide more information on system behavior through *sensitivity analyses*, in which the response of the system is studied under various scenarios related to perturbations in different problem parameters. To obtain reliable insights through such an analysis, it is important that a careful balance be struck between problem representation and tractability.

Accompanying the foregoing process is the actual *construction of a mathematical statement* of the problem. Often, there are several ways in which an identified problem can be mathematically modeled. Although these alternative forms may be mathematically equivalent, they might differ substantially in the felicity they afford to solution algorithms. Hence, some foresight into the operation and limitations of algorithms is necessary. For example, the restriction that a variable x should take on the values 0, 1, or 2 can be modeled "correctly" via the constraint $x(x - 1)(x - 2) = 0$. However, the nonconvex structure of this constraint will impose far more difficulty for most algorithms than if this discrete restriction was handled separately and explicitly as in a branch-and-bound framework, for instance (see Nemhauser and Wolsey [1988] or Parker and Rardin [1988]). As another example, a feasible region defined by the inequalities $g_i(\mathbf{x}) \leq 0$ for $i = 1, \ldots, m$ can be equivalently stated as the set of equality constraints $g_i(\mathbf{x}) + s_i^2 = 0$ for $i = 1, \ldots, m$ by introducing new (unrestricted) variables s_i, $i = 1, \ldots, m$. Although this is done sometimes to extend a theory or technique for equality constraints to one for inequality constraints, a blind application of this strategy can be disastrous. Besides increasing the dimension with respect to nonlinearly appearing variables, it injects nonconvexities into the problem by virtue of which the optimality conditions of Chapter 4 can be satisfied at nonoptimal points, even though this was not the case with the original inequality constrained problem.

In the same spirit, the inequality and equality constraints of the nonlinear program stated in Section 1.1 can be equivalently written as the *single* equality constraint

$$\sum_{i=1}^{m} [g_i(\mathbf{x}) + s_i^2]^2 + \sum_{j=1}^{l} h_j^2(\mathbf{x}) = 0$$

or as

$$\sum_{i=1}^{m} \max\{g_i(\mathbf{x}), 0\} + \sum_{j=1}^{l} |h_j(\mathbf{x})| = 0$$

or

$$\sum_{i=1}^{m} \max{}^2\{g_i(\mathbf{x}), 0\} + \sum_{j=1}^{l} h_j^2(\mathbf{x}) = 0$$

These different statements have different structural properties; and if they are not matched properly with algorithmic capabilities, one can obtain meaningless or arbitrary solutions, if any at all. However, although such a constraint is rarely adopted in practice, the construct of these equivalent formulations is indeed very useful in devising penalty functions when accommodated with the objective function, as will be seen later in Chapter 9.

Generally speaking, there are some guidelines that one can follow to construct a suitable mathematical formulation that will be amenable to most algorithms. Some experience and forethought is necessary in applying these guidelines, and the process is more of an art than a science. We provide some suggestions below, but caution the reader that these are only general recommendations and guiding principles rather than a universal set of instructions.

Foremost among these guidelines are the requirements to construct an adequate statement of the problem, to identify any inherent special structures, and to exploit these structures in the algorithmic process. Such structures might simply be linearity of constraints, or the presence of tight lower and upper bounds on the variables, dictated either by practice or by some knowledge of the neighborhood containing an optimum. Most existing powerful algorithms require differentiability of the functions involved, and so a smooth representation with derivative information is useful wherever possible. Although higher, second-order, derivative information is usually expensive to obtain and might require excessive storage for use in relatively large-sized problems, it can enhance algorithmic efficiency substantially if available. Hence, many efficient algorithms use approximations of this information, assuming second-order differentiability. Besides linearity and differentiability, there are many other structures afforded by either the nature of the constraints themselves (such as network flow constraints) or, generally, by the manner in which the nonzero coefficients appear in the constraints (e.g., in a block diagonal fashion over a substantial set of constraints; see Lasdon [1970]). Such structures can enhance algorithmic performance and, therefore, can increase the size of problems solvable within a reasonable amount of computational effort.

In contrast with special structures that are explicitly identified and exploited, the problem function being optimized might be a complex "black box" of an implicit unknown form whose evaluation itself might be an expensive task, perhaps requiring experimentation. In such instances, a response surface fitting methodology as described in Myers [1976] or some discretized grid approximations of such functions might be useful devices.

In the modeling process, it is also useful to distinguish between *hard* constraints, which must necessarily be satisfied without any compromise, and *soft* constraints, for which mild violations can be tolerated, albeit at some incurred cost. For example, the expenditure $g(\mathbf{x})$ for some activities $\mathbf{x}$ might be required to be no more than a budgeted amount B, but violations within limits might be permissible if economically justifiable. Hence, this constraint can be modeled as $g(\mathbf{x}) - B = y^+ - y^-$, where y^+ and y^- are nonnegative variables, and where the "violation" y^+ is bounded above by a limit on the capital that can be borrowed or raised and, accordingly, also accompanied by a cost term $c(y^+)$ in the objective function.

It is insightful to note that permitting mild violations in some constraints, if tolerable, can have a significant impact on the solution obtained. For example, imposing a pair of constraints $h_1(\mathbf{x}) = 0$ and $h_2(\mathbf{x}) = 0$ as hard constraints might cause the feasible region defined by their intersection to be far removed from attractively valued solutions, while such solutions only mildly violate these constraints. Hence, by treating them as soft constraints and rewriting them as $-\Delta_i \le h_i(\mathbf{x}) \le \Delta_i$, where Δ_i is a small positive tolerance factor for $i = 1, 2$, we might be able to obtain far better solutions which, from a managerial viewpoint, compromise more judiciously between solution quality and feasibility. These concepts are related to *goal programming* (see Ignizio [1976]), where the soft constraints represent goals to be attained, along with accompanying penalties or rewards for under- or over-achievements.

We conclude this section by addressing the all-important, but often neglected, practice of problem *scaling*, which can have a profound influence on algorithmic performance. This operation involves both the scaling of constraints by multiplying through with a (positive) constant, and the scaling of variables through a simple linear transformation that replaces $\mathbf{x}$ by $\mathbf{y} = \mathbf{D}\mathbf{x}$, where $\mathbf{D}$ is a nonsingular diagonal matrix. The end result sought is to try and make the magnitudes of the variables, and the magnitudes of the constraint coefficients (as they dictate the values of the dual variables or Lagrange multipliers; see Chapter 4), vary within similar or compatible ranges. This tends to reduce numerical accuracy problems and to alleviate ill-conditioning effects associated with severely skewed or highly ridgelike function contours encountered during the optimization process. As can well be imagined, if a pipe network design problem, for example, contains variables representing pipe thicknesses, pipe lengths, and rates of flows, all in diversely varying dimensional magnitudes, this can play havoc with numerical computations. Besides, many algorithms base their termination cirteria on prespecified tolerances on constraint satisfaction and on objective value improvements obtained over a given number of most recent iterations. Evidently, for such checks to be reliable, it is necessary that the problem be reasonably well scaled. This is true even for *scale-invariant* algorithms, which are designed to produce the same sequence of iterates regardless of problem scaling, but for which similar feasibility and objective improvement termination tests are used. Overall, although a sufficiently badly scaled problem can undoubtedly benefit by problem scaling, the effect of the scaling mechanism used on reasonably well-scaled problems can be mixed. As pointed out by Lasdon and Beck [1981], the scaling of nonlinear programs is as yet a "black art" that needs further study and refinement.

Exercises

1.1 Consider the following nonlinear programming problem:

$$\begin{aligned} \text{Minimize} \quad & (x_1 - 3)^2 + (x_2 - 3)^2 \\ \text{subject to} \quad & 4x_1^2 + 9x_2^2 \leq 36 \\ & x_1^2 + 3x_2 = 3 \\ & \mathbf{x} = (x_1, x_2) \in X \equiv \{\mathbf{x} : x_1 \geq -1\} \end{aligned}$$

a. Sketch the feasible region and the contours of the objective function. Hence, identify the optimum graphically.

b. Repeat part a by replacing minimization with maximization in the problem statement.

1.2 Consider the following *portfolio selection problem*. An investor must choose a portfolio $\mathbf{x} = (x_1, x_2, \ldots, x_n)'$, where x_j is the proportion of the assets allocated to the jth security. The return on the portfolio has mean $\bar{\mathbf{c}}'\mathbf{x}$ and variance $\mathbf{x}'\mathbf{V}\mathbf{x}$, where $\bar{\mathbf{c}}$ is the vector denoting mean returns and $\mathbf{V}$ is the matrix of covariances of the returns. The investor would like to increase his or her expected return while decreasing the variance and hence the risk. A portfolio is called *efficient* if there exists no other portfolio with a larger expected return and a smaller variance. Formulate the problem of finding an efficient portfolio, and suggest some procedures for choosing among efficient portfolios.

1.3 A rectangular heat storage unit of length L, width W, and height H will be used to store heat energy temporarily. The rate of heat losses h_c due to convection and h_r due to radiation are given by

$$h_c = k_c A(T - T_a)$$

$$h_r = k_r A(T^4 - T_a^4)$$

where k_c and k_r are constants, T is the temperature of the heat storage unit, A is the surface area, and T_a is the ambient temperature. The heat energy stored in the unit is given by

$$Q = kV(T - T_a)$$

where k is a constant and V is the volume of the storage unit. The storage unit should have the ability to store at least Q'. Furthermore, suppose that space availability restricts the dimensions of the storage unit to

$$0 \leq L \leq L' \qquad 0 \leq W \leq W' \qquad 0 \leq H \leq H'$$

a. Formulate the problem of finding the dimensions L, W, and H to minimize the total heat losses.

b. Suppose that the constants k_c and k_r are linear functions of t, the insulation thickness. Formulate the problem of determining the optimal dimensions L, W, and H to minimize the insulation cost.

1.4 Formulate the model for Exercise 1.3 if the storage unit is a cylinder of diameter D and height H.

1.5 An office room of length 60 feet and width 35 feet is to be illuminated by n light bulbs of wattage W_i, $i = 1, \ldots, n$. The bulbs are to be located 7 feet above the working surface. Let (x_i, y_i) denote the x and y coordinates of the ith bulb. To ensure adequate lighting, illumination is checked at the working surface level at grid points of the form (α, β), where

$$\alpha = 10p \qquad p = 0, 1, \ldots, 6$$

$$\beta = 5q \qquad q = 0, 1, \ldots, 7$$

The illumination at (α, β) resulting from a bulb of wattage W_i located at (x_i, y_i) is given by

$$E_i(\alpha, \beta) = k\frac{W_i\,\|(\alpha, \beta) - (x_i, y_i)\|}{\|(\alpha, \beta, 7) - (x_i, y_i, 0)\|^3}$$

where k is a constant reflecting the efficiency of the bulb. The total illumination at (α, β) can be taken to be $\sum_{i=1}^{n} E_i(\alpha, \beta)$. At each of the points checked, an illumination of between 2.6 and 3.2 units is required. The wattage of the bulbs used is between 40 W and 300 W. Assume that the W_i's are continuous variables.

a. Construct a mathematical model to minimize the number of bulbs used and to determine their location and wattage, assuming that the cost of installation and of periodic bulb replacement is a function of the number of bulbs used.

b. Construct a mathematical model similar to that of part a, with the added restriction that all bulbs must be of the same wattage.

c. Select a suitable value of k from the literature. Verify whether the lighting in your classroom is reasonably close to the answer obtained in part b above.

1.6 A household with budget b purchases n commodities. The unit price of commodity j is c_j, and the minimal amount to be purchased of the commodity is l_j. After the minimal amounts of the n products are consumed, a function a_j of the remaining budget is allocated to commodity j. The behavior of the household is observed over m months for the purpose of estimating $l_1, \ldots, l_n$ and $a_1, \ldots, a_n$. Develop a regression model for estimating these parameters if

a. The sum of the squares of the error is to be minimized.

b. The maximum absolute value of the error is to be minimized.

c. The sum of the absolute values of the error is to be minimized.

d. For both parts b and c, reformulate the problems as linear programs.

1.7 A steel company manufactures crankshafts. Previous research indicates that the mean shaft diameter may assume the value μ_1 or μ_2, where $\mu_2 > \mu_1$. Furthermore, the probability that the mean is equal to μ_1 is p. To test whether the mean is μ_1 or μ_2, a sample of size n is chosen, and the diameters $x_1, \ldots, x_n$ are recorded. If $\bar{x} = \sum_{j=1}^{n} x_j/n$ is less than or equal to K, the hypothesis $\mu = \mu_1$ is accepted; otherwise the hypothesis $\mu = \mu_2$ is accepted. Let $f(\bar{x}|\mu_1)$ and $f(\bar{x}|\mu_2)$ be the probability density functions of the sample mean if the population mean is μ_1 and μ_2, respectively. Furthermore, suppose that the penalty cost of accepting $\mu = \mu_1$ when $\mu = \mu_2$ is α and that the penalty cost of accepting $\mu = \mu_2$ when $\mu = \mu_1$ is β. Formulate the problem of choosing K such that the expected total cost is minimized. Show how the problem could be reformulated as a nonlinear program.

1.8 Consider the following problem of a regional effluent control along a river. Currently, n manufacturing facilities dump their refuse into the river. The current rate of dumping by facility j is μ_j, $j = 1, \ldots, n$. The water quality is examined along the river at m control points. The minimum desired quality improvement at point i is b_i, $i = 1, \ldots, m$. Let x_j be the amount of waste to be removed from source j at a cost of $f_j(x_j)$, and let a_{ij} be the quality improvement at control point i for each unit of waste removed at source j.

a. Formulate the problem of improving the water quality at a minimum cost as a nonlinear program.

b. In the above formulation, it is possible that certain sources would have to remove substantial amounts of waste, whereas others would be only required to remove small amounts of waste or none at all. Reformulate the problem so that a measure of equity among the sources is attained.

1.9 An elevator has a vertical acceleration $u(t)$ at time t. Passengers would like to move from the ground level at zero altitude to the sixteenth floor at altitude 50 as fast as possible but dislike fast acceleration. Suppose that the passenger's time is valued at $\$\alpha$ per unit time, and furthermore suppose that the passenger is willing to pay at a rate of $\$\beta u^2(t)$ per unit time to avoid fast acceleration. Formulate the problem of determining the acceleration from the time the elevator starts ascending until it reaches the sixteenth floor as an optimal control problem. Can you formulate the problem as a nonlinear program?

1.10 Consider a linear program to minimize $\mathbf{c}^t\mathbf{x}$ subject to $\mathbf{A}\mathbf{x} \leq \mathbf{b}$, $\mathbf{x} \geq \mathbf{0}$. Suppose that the components c_j's of the vector $\mathbf{c}$ are random variables distributed independently of each other and of the variables x_j's, and that the expected value of c_j is $\bar{c}_j$, $j = 1, \ldots, n$.

a. Show that the minimum expected cost is obtained by solving the problem to minimize $\bar{\mathbf{c}}^t\mathbf{x}$ subject to $\mathbf{A}\mathbf{x} \leq \mathbf{b}$, $\mathbf{x} \geq \mathbf{0}$, where $\bar{\mathbf{c}} = (\bar{c}, \ldots, \bar{c}_n)^t$.

b. Suppose that a firm makes two products that consume a common resource, which is expressed as follows:

$$3x_1 + 4x_2 \leq 20$$

where x_j is the amount of product j produced. The unit profit for product 1 is normally distributed with mean 3 and variance 2. The unit profit for product 2 is given by a χ^2-distribution with 2 degrees of freedom. Assume that the random variables are independently distributed and that they are not dependent upon x_1 and x_2. Find the quantities of each product that must be produced to maximize expected profit. Will your answer differ if the variance for the first product were 4?

1.11 Suppose that the demand for a certain product is a normally distributed random variable with mean 100 and variance 36, and that the production function is given by $p(\mathbf{x}) = \boldsymbol{\alpha}^t\mathbf{x}$, where $\mathbf{x}$ represents a set of n activity levels. Formulate the chance constraint that the probability of production falling short of demand by more than 5 units should be no more than 1% as a linear constraint.

1.12 Suppose that the demand $d_1, \ldots, d_n$ for a certain product over n periods is known. The demand during period j can be met from the production x_j during the period or from the warehouse stock. Any excess production can be stored at the warehouse. However, the warehouse has capacity K, and it would cost $\$c$ to carry over one unit from one period to another. The cost of production during period j is given by $f(x_j)$ for $j = 1, \ldots, n$. If the initial inventory is I_0, formulate the production scheduling problem as a nonlinear program.

1.13 A manufacturing firm produces four different products. One of the necessary raw materials is in short supply, and only R pounds are available. The selling price of product i is $\$S_i$ per pound. Furthermore, each pound of product i uses a_i pounds of the critical raw material. The variable cost, excluding the raw material cost, of producing x_i pounds of product i is $k_i x_i^2$, where $k_i > 0$ is known. Develop a mathematical model for the problem.

1.14 Suppose that the daily demand for product j is d_j, for $j = 1, 2$. The demand should be met from inventory, and the latter is replenished from production whenever the inventory reaches zero. Here, the production time is assumed to be insignificant. During each product run, Q_j units can be produced at a fixed setup cost of $\$k_j$ and a variable cost of $\$c_j Q_j$. Also, a variable inventory-holding cost of $\$h_j$ per unit per day is also incurred, based on the average inventory. Thus, the total cost associated with product j during T days is $\$Td_jk_j/Q_j + Tc_jd_j + TQ_jh_j/2$. Adequate storage area for handling the maximum inventory Q_j has to be reserved for each product j. Each unit of product j needs s_j square feet of storage space, and the total space available is S.

a. We wish to find the optimal production quantities Q_1 and Q_2 to minimize the total cost. Write the model for the problem.

b. Now, suppose that shortages are permitted and production need not start when inventory reaches zero level. During the period when inventory is zero and demand is not met, sales are lost. The loss per unit thus incurred is $\$l_j$. On the other hand, if a sale is made, the profit per unit is $\$P_j$. Reformulate the mathematical model.

Notes and References

The advent of high-speed computers has considerably increased our ability to apply iterative procedures in solving large-scale problems, both linear and nonlinear. Although our ability to obtain global minimal solutions to nonconvex problems of realistic size is still rather limited, hopefully new theoretical breakthroughs will overcome this problem.

Section 1.2 gives some simplified examples of problems that could be solved by the nonlinear programming methods discussed in the book. Our purpose was not to give complete details but only a flavor of the diverse problem areas that can be attacked. See Lasdon and Waren [1980] for further applications.

Optimal control is closely linked with mathematical programming. Dantzig [1966] has shown how certain optimal control problems can be solved by applying the simplex method. For further details of the application of mathematical programming to control problems, refer to Bracken and McCormick [1968], Canon and Eaton [1966], Canon, Cullum, and Polak [1970], Cutler and Perry [1983], and Tabak and Kuo [1971].

With the recent developments and interest in aerospace and related technology, optimum design in this area has taken on added importance. In fact, since 1969 the Advisory Group for Aerospace Research and Development under NATO has sponsored several symposia on structural optimization. With improved materials being used for special purposes, optimum mechanical design has also increased in importance. The works of Cohn [1969], Fox [1969, 1971], Johnson [1971], Majid [1974], and Siddal [1972] are of interest in understanding how design concepts are integrated with optimization concepts in mechanical and structural design.

Mathematical programming has also been successfully used for over a decade to solve various problems associated with the generation and distribution of electrical power and the operation of the system. These problems include the study of load flow, substation switching, expansion planning, maintenance scheduling, and the like. In the load flow problem, one is concerned with the flow of power through a transmission network to meet a given demand. The power distribution is governed by the well-known Kirchhoff's laws, and the equilibrium power flows satisfying these conditions can be computed by nonlinear programming. In other situations, the power output from hydroelectric plants is considered fixed, and the objective is to minimize the cost of fuel at the thermal plants. This problem, referred to as the economic dispatch problem, is usually solved on-line every few minutes, with appropriate power adjustments made. The generation capacity expansion problems study a minimum cost equipment purchase and dispatchment plan that can satisfy the demand load at a specified reliability level over a given time horizon. For more details, refer to Abou-Taleb et al. [1974], Adams et al. [1972], Anderson [1972], Beglari and Laughton [1975], Bloom [1983], Bloom et al. [1984], Kirchmayer

[1958], Sasson [1969a and 1969b], Sasson et al. [1971], Sasson and Merrill [1974], Sherali [1985], Sherali and Soyster [1983], and Sherali and Staschus [1985].

The field of water resources systems analysis has shown a spectacular growth during the last two decades. As in many fields of science and technology, the rapid growth of water resources engineering and systems analysis was accompanied by an information explosion of considerable proportions. The problem discussed in Section 1.2 is concerned with rural water resources management for which an optimal balance between the use of water for hydropower generation and agriculture is sought. Some typical studies in this area can be found in Haimes [1973, 1977], Haimes and Nainis [1974], and Yu and Haimes [1974].

As a result of the rapid growth of urban areas, city managers are also concerned with integrating urban water distribution and land use. Some typical quantitative studies on urban water distribution and disposal can be found in Argaman, Shamir, and Spivak [1973], Dajani, Gemmel, and Morlok [1972], Deb and Sarkar [1971], Fujiwara et al. [1987], Jacoby [1968], Loganathan et al. [1990], Shamir [1974], Walsh and Brown [1973], and Wood and Charles [1973].

In his classic study on portfolio allocation, Markowitz [1952] showed how the variance of the returns on the portfolio can be incorporated in the optimal decision. In Exercise 1.2, the portfolio allocation problem is briefly introduced.

From 1955 to 1959, numerous studies were undertaken to incorporate uncertainty in the parameter values of a linear program. Refer to Charnes and Cooper [1959], Dantzig [1955], Freund [1956], and Madansky [1959] for some of the early work in this area. Since then, many other studies have been undertaken. The approaches, referred to in the literature as *chance constrained problems* and programming with recourse, seem particularly attractive. The interested reader may refer to Charnes and Cooper [1961, 1963], Charnes, Kirby, and Raike [1967], Dantzig [1963], Elmaghraby [1960], Evers [1967], Geoffrion [1967c], Madansky [1962], Mangasarian [1964], Parikh [1970], Sengupta [1972], Sengupta and Portillo-Campbell [1970], Sengupta, Tintner, and Millham [1963], Vajda [1970, 1972], Wets [1966a, 1966b, 1972], Williams [1965, 1966], and Ziemba [1970, 1971, 1974, 1975].

For a description of other applications, the interested reader is referred to Ali et al. [1978] for an oil resource management problem; to Lasdon [1985] and Prince et al. [1983] for Texaco's OMEGA gasoline blending problem; to Rothfarb et al. [1970] for the design of offshore natural gas pipeline distribution systems; to Berna et al. [1980], Heyman [1990], Sarma and Reklaitis [1979], and Wall et al. [1986] for chemical process optimization and equilibrium problems; to Intriligator [1971], Murphy et al. [1982], Sherali [1984] and Sherali et al. [1983] for mathematical economics problems; to Adams and Sherali [1984], Francis et al. [1991], Love et al. [1988], Shetty and Sherali [1980] and Sherali and Tuncbilek [1992] for location–allocation problems; to Bullard et al. [1985] for forest harvesting problems; to Myers [1976] for response surface methodologies; and to Dennis and Schnabel [1983], Fletcher [1987], and Sherali et al. [1988] for a discussion on least squares estimation problems with applications to data fitting and statistical parameter estimation.

For further discussion on problem scaling we refer the reader to Bauer [1963], Curtis and Reid [1972], Lasdon and Beck [1981], and Tomlin [1973]. Gill et al. [1981, 1984d, 1985] provide a good discussion on guidelines for model building and their influence on algorithms.

Finally, we mention that various modeling languages, such as GAMS (see Brooke et al., 1985), LINGO (see Cunningham and Schrage, 1989), and AMPL (see Fourer et al., 1990), are available to assist in the implementation of models and algorithms. Various nonlinear programming software packages, such as MINOS (see Murtagh and Saunders, 1982), GINO (see Liebman et al., 1986), GRG2 (see Lasdon et al., 1978), CONOPT (see Drud, 1985), SQP (see Mahidhara and Lasdon, 1990), and LSGRG (see Smith and Lasdon, 1992), among others, are also available to facilitate implementation.

PART 1
Convex Analysis

Chapter 2 Convex Sets

The concept of convexity is of great importance in the study of optimization problems. Convex sets, polyhedral sets, and separation of disjoint convex sets are used frequently in the analysis of mathematical programming problems, the characterization of their optimal solutions, and the development of computational procedures.

The following is an outline of the chapter. The reader is encouraged to first review the mathematical preliminaries given in Appendix A.

Section 2.1: Convex Hulls This section is elementary. It presents some examples of convex sets and defines convex hulls. Readers with previous knowledge of convex sets may skip this section (possibly with the exception of the Carathéodory theorem).

Section 2.2: Closure and Interior of a Set Some topological properties of sets, related to interior, boundary, and closure points, are discussed.

Section 2.3: Weierstrass' Theorem We discuss the concepts of min, max, inf, and sup, and present an important result relating to the existence of minimizing or maximizing solutions.

Section 2.4: Separation and Support of Sets This section is important, since the notions of separation and support of convex sets are frequently used in optimization. A careful study of this section is recommended.

Section 2.5: Convex Cones and Polarity This is a short section dealing mainly with polar cones. This section may be skipped without loss of continuity.

Section 2.6: Polyhedral Sets, Extreme Points, and Extreme Directions This section treats the special important case of polyhedral sets. Characterization of extreme points and extreme directions of polyhedral sets is developed. Also, the representation of a polyhedral set in terms of its extreme points and extreme directions is proved.

Section 2.7: Linear Programming and the Simplex Method The well-known simplex method is developed as a natural extension of the material in the previous section. Readers who are familiar with the simplex method may skip this section. A polynomial-time algorithm for linear programming problems is discussed later in Chapter 9.

2.1 Convex Hulls

In this section, we first introduce the notions of convex sets and convex hulls. We then demonstrate that any point in the convex hull of a set S can be represented in terms of $n + 1$ points in the set S.

2.1.1 Definition

A set S in E_n is said to be *convex* if the *line segment* joining any two points of the set also belongs to the set. In other words, if $\mathbf{x}_1$ and $\mathbf{x}_2$ are in S, then $\lambda\mathbf{x}_1 + (1 - \lambda)\mathbf{x}_2$ must also belong to S for each $\lambda \in [0, 1]$. Weighted averages of the form $\lambda\mathbf{x}_1 + (1 - \lambda)\mathbf{x}_2$, where $\lambda \in [0, 1]$, are referred to as *convex combinations* of $\mathbf{x}_1$ and $\mathbf{x}_2$. Inductively, weighted averages of the form $\sum_{j=1}^{k} \lambda_j\mathbf{x}_j$, where $\sum_{j=1}^{k} \lambda_j = 1$, $\lambda_j \geq 0$, $j = 1, \ldots, k$, are also called *convex combinations* of $\mathbf{x}_1, \ldots, \mathbf{x}_k$. In this definition, if the nonnegativity conditions on the multipliers λ_j is dropped, $j = 1, \ldots, k$, then the combination is known as an *affine combination*. Finally, a combination $\sum_{j=1}^{k} \lambda_j\mathbf{x}_j$, where the multipliers λ_j, $j = 1, \ldots, k$, are simply required to be in E_1, is known as a *linear combination*.

Figure 2.1 below illustrates the notion of a convex set. Note that in Figure 2.1*b*, the line segment joining $\mathbf{x}_1$ and $\mathbf{x}_2$ does not entirely lie in the set.

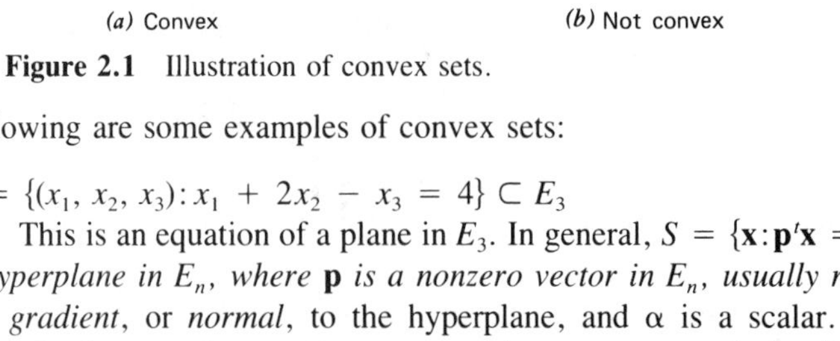

(*a*) Convex (*b*) Not convex

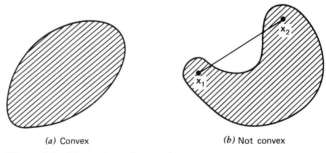

Figure 2.1 Illustration of convex sets.

The following are some examples of convex sets:

1. $S = \{(x_1, x_2, x_3): x_1 + 2x_2 - x_3 = 4\} \subset E_3$

   This is an equation of a plane in E_3. In general, $S = \{\mathbf{x}: \mathbf{p}'\mathbf{x} = \alpha\}$ is called a *hyperplane in* E_n, *where* $\mathbf{p}$ *is a nonzero vector in* E_n, *usually referred to as the gradient*, or *normal*, to the hyperplane, and α is a scalar. Note that if $\bar{\mathbf{x}} \in S$, then we have $\mathbf{p}'\bar{\mathbf{x}} = \alpha$, so that we can equivalently write $S =$

$\{\mathbf{x}:\mathbf{p}'(\mathbf{x} - \bar{\mathbf{x}}) = 0\}$. Hence, the vector $\mathbf{p}$ is orthogonal to all vectors $(\mathbf{x} - \bar{\mathbf{x}})$ for $\mathbf{x} \in S$, and so, it is perpendicular to the surface of the hyperplane S.

2. $S = \{(x_1, x_2, x_3): x_1 + 2x_2 - x_3 \leq 4\} \subset E_3$

   These are points on one side of the hyperplane defined above. These points form a *half-space*. In general, a half-space $S = \{\mathbf{x}:\mathbf{p}'\mathbf{x} \leq \alpha\}$ in E_n is a convex set.

3. $S = \{(x_1, x_2, x_3): x_1 + 2x_2 - x_3 \leq 4, \quad 2x_1 - x_2 + x_3 \leq 6\} \subset E_3$

   This set is the intersection of two half-spaces. In general, the set $S = \{\mathbf{x}:\mathbf{A}\mathbf{x} \leq \mathbf{b}\}$ is a convex set, where $\mathbf{A}$ is an $m \times n$ matrix and $\mathbf{b}$ is an m vector. This set is the intersection of m half-spaces and is usually called a *polyhedral set*.

4. $S = \{(x_1, x_2): x_2 \geq |x_1|\} \subset E_2$

   This set represents a *convex cone* in E_2, and is treated more fully in Section 2.4.

5. $S = \{(x_1, x_2): x_1^2 + x_2^2 \leq 4\} \subset E_2$

   This set represents points on and inside a circle with center (0, 0) and radius 2.

6. $S = \{\mathbf{x}:\mathbf{x} \text{ solves Problem P below}\}$

   Problem P

   Minimize $\quad \mathbf{c}'\mathbf{x}$

   subject to $\quad \mathbf{A}\mathbf{x} = \mathbf{b}$

   $\qquad\qquad \mathbf{x} \geq \mathbf{0}$

Here, $\mathbf{c}$ is an n vector, $\mathbf{b}$ is an m vector, $\mathbf{A}$ is an $m \times n$ matrix, and $\mathbf{x}$ is an n vector. The set S gives all optimal solutions to the *linear programming problem* of minimizing the linear function $\mathbf{c}'\mathbf{x}$ over the polyhedral region defined by $\mathbf{A}\mathbf{x} = \mathbf{b}$ and $\mathbf{x} \geq \mathbf{0}$. This set itself happens to be a polyhedral set, being the intersection of $\mathbf{c}'\mathbf{x} = \nu^*$ with $\mathbf{A}\mathbf{x} = \mathbf{b}$, $\mathbf{x} \geq 0$, where ν^* is the optimal value of P.

The following lemma is an immediate consequence of the definition of convexity. It states that the intersection of two convex sets is convex and that the algebraic sum of two convex sets is also convex. The proof is elementary and is left as an exercise.

2.1.2 Lemma

Let S_1 and S_2 be convex sets in E_n. Then,

1. $S_1 \cap S_2$ is convex.
2. $S_1 \oplus S_2 = \{\mathbf{x}_1 + \mathbf{x}_2 : \mathbf{x}_1 \in S_1, \mathbf{x}_2 \in S_2\}$ is convex.
3. $S_1 \ominus S_2 = \{\mathbf{x}_1 - \mathbf{x}_2 : \mathbf{x}_1 \in S_1, \mathbf{x}_2 \in S_2\}$ is convex.

Convex Hulls

Given an arbitrary set S in E_n, different convex sets can be generated from S. In particular, we discuss below the convex hull of S.

2.1.3 Definition

Let S be an arbitrary set in E_n. The *convex hull* of S, denoted by $H(S)$, or conv (S), is the collection of all convex combinations of S. In other words, $\mathbf{x} \in H(S)$ if and only if $\mathbf{x}$ can be represented as

$$\mathbf{x} = \sum_{j=1}^{k} \lambda_j \mathbf{x}_j$$

$$\sum_{j=1}^{k} \lambda_j = 1$$

$$\lambda_j \geq 0 \qquad \text{for } j = 1, \ldots, k$$

where k is a positive integer and $\mathbf{x}_1, \ldots, \mathbf{x}_k \in S$.

Figure 2.2 shows some examples of convex hulls. Actually, we see that in each case, $H(S)$ is the minimal convex set that contains S. This is indeed the case in general, as given in Lemma 2.1.4. The proof is left as an exercise.

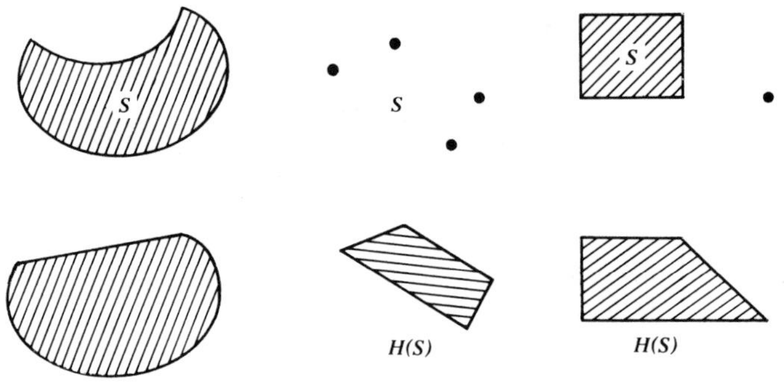

Figure 2.2 Examples of convex hulls.

2.1.4 Lemma

Let S be an arbitrary set in E_n. Then, $H(S)$ is the smallest convex set containing S. Indeed, $H(S)$ is the intersection of all convex sets containing S.

Similar to the foregoing discussion, we can define the *affine hull* of S as the collection of all affine combinations of points in S. This is the smallest dimensional affine subspace that contains S. For example, the affine hull of two distinct points is the one-dimensional line containing these two points. Similarly, the *linear hull* of S is the collection of all linear combinations of points in S.

We have discussed above the convex hull of an arbitrary set S. The convex hull of a finite number of points leads to the definitions of a polytope and a simplex.

2.1.5 Definition

The convex hull of a finite number of points $\mathbf{x}_1, \ldots, \mathbf{x}_{k+1}$ in E_n is called a *polytope*. If $\mathbf{x}_1, \mathbf{x}_2, \ldots, \mathbf{x}_k$, and $\mathbf{x}_{k+1}$ are *affinely independent*, which means that $\mathbf{x}_2 - \mathbf{x}_1, \mathbf{x}_3 - \mathbf{x}_1, \ldots, \mathbf{x}_{k+1} - \mathbf{x}_1$ are linearly independent, then $H(\mathbf{x}_1, \ldots, \mathbf{x}_{k+1})$, the convex hull of $\mathbf{x}_1, \ldots, \mathbf{x}_{k+1}$, is called a *simplex* with vertices $\mathbf{x}_1, \ldots, \mathbf{x}_{k+1}$.

Figure 2.3 shows examples of a polytope and a simplex in E_n. Note that the maximum number of linearly independent vectors in E_n is n and, hence, there could be no simplex in E_n with more than $n + 1$ vertices.

Carathéodory Theorem

By definition, a point in the convex hull of a set can be represented as a convex combination of a finite number of points in the set. The following theorem shows that

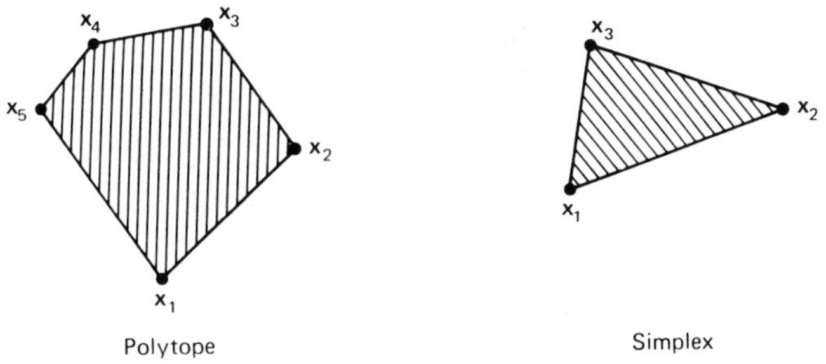

Figure 2.3 Examples of a polytope and a simplex.

any point $\mathbf{x}$ in the convex hull of a set S can be represented as a convex combination of, at most, $n + 1$ points in S. The theorem is trivially true for $\mathbf{x} \in S$.

2.1.6 Theorem

Let S be an arbitrary set in E_n. If $\mathbf{x} \in H(S)$, then $\mathbf{x} \in H(\mathbf{x}_1, \ldots, \mathbf{x}_{n+1})$, where $\mathbf{x}_j \in S$ for $j = 1, \ldots, n + 1$. In other words, $\mathbf{x}$ can be represented as

$$\mathbf{x} = \sum_{j=1}^{n+1} \lambda_j \mathbf{x}_j$$

$$\sum_{j=1}^{n+1} \lambda_j = 1$$

$$\lambda_j \geq 0 \qquad \text{for } j = 1, \ldots, n + 1$$

$$\mathbf{x}_j \in S \qquad \text{for } j = 1, \ldots, n + 1$$

Proof

Since $\mathbf{x} \in H(S)$, then $\mathbf{x} = \sum_{j=1}^{k} \lambda_j \mathbf{x}_j$, where $\lambda_j > 0$ for $j = 1, \ldots, k$, $\mathbf{x}_j \in S$ for $j = 1, \ldots, k$, and $\sum_{j=1}^{k} \lambda_j = 1$. If $k \leq n + 1$, the result is at hand. Now, suppose that $k > n + 1$. A reader familiar with basic feasible solutions and extreme points (see Theorem 2.5.4) will now immediately notice that at an extreme point of the set $\{\lambda : \sum_{j=1}^{k} \lambda_j \mathbf{x}_j = \mathbf{x}, \sum_{j=1}^{k} \lambda_j = 1, \lambda \geq 0\}$, no more than $n + 1$ components of λ are positive, hence proving the result. However, let us continue to provide an independent argument.

Toward this end, note that $\mathbf{x}_2 - \mathbf{x}_1, \mathbf{x}_3 - \mathbf{x}_1, \ldots, \mathbf{x}_k - \mathbf{x}_1$ are linearly dependent. Thus, there exist scalars $\mu_2, \mu_3, \ldots, \mu_k$ not all zero such that $\sum_{j=2}^{k} \mu_j(\mathbf{x}_j - \mathbf{x}_1) = \mathbf{0}$. Letting $\mu_1 = -\sum_{j=2}^{k} \mu_j$, it follows that $\sum_{j=1}^{k} \mu_j \mathbf{x}_j = \mathbf{0}$, $\sum_{j=1}^{k} \mu_j = 0$, and not all the μ_j's are equal to zero. Note that at least one μ_j is larger than 0. Then,

$$\mathbf{x} = \sum_{j=1}^{k} \lambda_j \mathbf{x}_j + \mathbf{0} = \sum_{j=1}^{k} \lambda_j \mathbf{x}_j - \alpha \sum_{j=1}^{k} \mu_j \mathbf{x}_j = \sum_{j=1}^{k} (\lambda_j - \alpha\mu_j)\mathbf{x}_j$$

for any real α. Now choose α as follows:

$$\alpha = \underset{1 \leq j \leq k}{\text{minimum}} \left\{ \frac{\lambda_j}{\mu_j} : \mu_j > 0 \right\} = \frac{\lambda_i}{\mu_i} \qquad \text{for some } i \in \{1, \ldots, k\}$$

Note that $\alpha > 0$. If $\mu_j \leq 0$, then $\lambda_j - \alpha\mu_j > 0$, and if $\mu_j > 0$, then $\lambda_j/\mu_j \geq \lambda_i/\mu_i = \alpha$

and, hence, $\lambda_j - \alpha\mu_j \geq 0$. In other words, $\lambda_j - \alpha\mu_j \geq 0$ for all $j = 1, \ldots, k$. In particular, $\lambda_i - \alpha\mu_i = 0$ by definition of α. Therefore, $\mathbf{x} = \sum_{j=1}^{k} (\lambda_j - \alpha\mu_j)\mathbf{x}_j$, where $\lambda_j - \alpha\mu_j \geq 0$ for $j = 1, \ldots, k$, $\sum_{j=1}^{k} (\lambda_j - \alpha\mu_j) = 1$ and, furthermore, $\lambda_i - \alpha\mu_i = 0$. In other words, $\mathbf{x}$ is represented as a convex combination of at most $k - 1$ points in S. The process is repeated until $\mathbf{x}$ is represented as a convex combination of at most $n + 1$ points in S. This completes the proof.

2.2 Closure and Interior of a Set

In this section, we develop some topological properties of sets in general and of convex sets in particular. As a preliminary, given a point $\mathbf{x}$ in E_n, an ε-neighborhood around it is the set $N_\varepsilon(\mathbf{x}) = \{\mathbf{y}: \|\mathbf{y} - \mathbf{x}\| < \varepsilon\}$. Let us first review the definitions of closure, interior, and boundary of an arbitrary set in E_n, using the concept of an ε-neighborhood.

2.2.1 Definition

Let S be an arbitrary set in E_n. A point $\mathbf{x}$ is said to be in the *closure* of S, denoted by cl S, if $S \cap N_\varepsilon(\mathbf{x}) \neq \varnothing$ for every $\varepsilon > 0$. If $S =$ cl S, then S is called *closed*. $\mathbf{x}$ is said to be in the *interior* of S, denoted by int S, if $N_\varepsilon(\mathbf{x}) \subset S$ for some $\varepsilon > 0$. A *solid* set $S \subseteq E_n$ is one with a nonempty interior. If $S =$ int S, then S is called *open*. Finally, $\mathbf{x}$ is said to be in the *boundary* of S, denoted by ∂S, if $N_\varepsilon(\mathbf{x})$ contains at least one point in S and one point not in S for every $\varepsilon > 0$. A set S is *bounded* if it can be contained in a ball of a sufficiently large radius. A *compact* set is one that is both closed and bounded.

To illustrate, consider $S = \{(\mathbf{x}_1, \mathbf{x}_2): x_1^2 + x_2^2 \leq 1\}$, which represents all points within a circle with center (0, 0) and radius 1. It can be easily verified that S is closed; that is, $S =$ cl S. Furthermore, int S consists of all points inside the circle; that is, int $S = \{(x_1, x_2): x_1^2 + x_2^2 < 1\}$. Finally, ∂S consists of points on the circle; that is, $\partial S = \{(x_1, x_2): x_1^2 + x_2^2 = 1\}$.

Hence, a set S is closed if and only if it contains all its boundary points. Moreover, cl $S \equiv S \cup \partial S$ is the smallest closed set containing S. Similarly, a set is open if and only if it does not contain any of its boundary points. Clearly, a set may be neither open nor closed, and the only sets in E_n that are both open and closed are the empty set and E_n itself. Also, note that any point $\mathbf{x} \in S$ must be either an interior or a boundary point of S. However, $S \neq$ int $S \cup \partial S$, since S need not contain its boundary points. But since int $S \subseteq S$, we have, int $S = S - \partial S$, while $\partial S \neq S -$ int S necessarily.

There is another equivalent definition of a closed set, which is often important from the viewpoint of demonstrating that a set is closed. This definition is based on sequences of points contained in S (review Appendix A for related mathematical concepts). A set S is closed if and only if, for any convergent sequence of points $\{\mathbf{x}_k\}$ contained in S with limit point $\bar{\mathbf{x}}$, we also have $\bar{\mathbf{x}} \in S$. The equivalence of this and the previous definition of closedness is easily seen by noting that the limit point $\bar{\mathbf{x}}$ of any convergent sequence of points in S must either lie in the interior or on the boundary of S, since, otherwise, there would exist an $\varepsilon > 0$ such that $\{\mathbf{x}: \|\mathbf{x} - \bar{\mathbf{x}}\| < \varepsilon\} \cap S = \phi$, contradicting that $\bar{\mathbf{x}}$ is the limit point of a sequence contained in S. Hence, if S is closed, then $\bar{\mathbf{x}} \in S$. Conversely, if S satisfies the above sequence property, then it is closed, since, otherwise, there would exist some boundary point $\bar{\mathbf{x}}$ not contained in S. But by the definition of a boundary point, the set $N_{\varepsilon^k}(\bar{\mathbf{x}}) \cap S \neq \varnothing$ for each $k = 1, 2, \ldots$, where $0 < \varepsilon < 1$ is some scalar. Hence, selecting some $\mathbf{x}_k \in N_{\varepsilon^k}(\bar{\mathbf{x}}) \cap S$ for each $k = 1, 2, \ldots$, we will have $\{\mathbf{x}_k\} \subseteq S$; and, clearly, $\{\mathbf{x}_k\} \to \bar{\mathbf{x}}$, which means that we must have $\bar{\mathbf{x}} \in S$ by our hypothesis. This is a contradiction.

To illustrate, note that the polyhedral set $S = \{\mathbf{x}: \mathbf{Ax} \leq \mathbf{b}\}$ is closed, since, given any convergent sequence $\{\mathbf{x}_k\} \subseteq S$, with $\{\mathbf{x}_k\} \to \bar{\mathbf{x}}$, we also have $\bar{\mathbf{x}} \in S$. This follows because $\mathbf{Ax}_k \leq \mathbf{b}$ for all k; and so, by the continuity of linear functions, we have in the limit that $\mathbf{A}\bar{\mathbf{x}} \leq \mathbf{b}$ as well, or that $\bar{\mathbf{x}} \in S$.

Line Segment Between Points in the Closure and the Interior of a Set

Given a convex set with a nonempty interior, the line segment (excluding the endpoints) joining a point in the interior of the set and a point in the closure of the set belongs to the interior of the set. This result is proved below. (Exercise 2.57 suggests a means for constructing a simpler proof based on the concept of supporting hyperplanes introduced in Section 2.4).

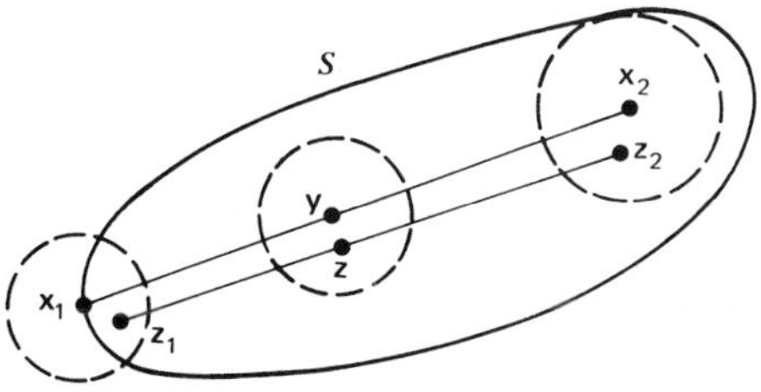

Figure 2.4 Line segment joining closure and interior points.

2.2.2 Theorem

Let S be a convex set in E_n with a nonempty interior. Let $\mathbf{x}_1 \in \text{cl } S$ and $\mathbf{x}_2$ int S. Then, $\lambda\mathbf{x}_1 + (1 - \lambda)\mathbf{x}_2 \in \text{int } S$ for each $\lambda \in (0, 1)$.

Proof

Since $\mathbf{x}_2 \in \text{int } S$, there exists an $\varepsilon > 0$ such that $\{\mathbf{z}: \|\mathbf{z} - \mathbf{x}_2\| < \varepsilon\} \subset S$. Let $\mathbf{y}$ be such that

$$\mathbf{y} = \lambda\mathbf{x}_1 + (1 - \lambda)\mathbf{x}_2 \tag{2.1}$$

where $\lambda \in (0, 1)$. To prove that $\mathbf{y}$ belongs to int S, it suffices to construct a neighborhood about $\mathbf{y}$ that also belongs to S. In particular, we show that $\{\mathbf{z}: \|\mathbf{z} - \mathbf{y}\| < (1 - \lambda)\varepsilon\} \subset S$. Let $\mathbf{z}$ be such that $\|\mathbf{z} - \mathbf{y}\| < (1 - \lambda)\varepsilon$ and refer to Figure 2.4. Since $\mathbf{x}_1 \in \text{cl } S$, then

$$\left\{\mathbf{x}: \|\mathbf{x} - \mathbf{x}_1\| < \frac{(1 - \lambda)\varepsilon - \|\mathbf{z} - \mathbf{y}\|}{\lambda}\right\} \cap S$$

is not empty. In particular, there exists a $\mathbf{z}_1 \in S$ such that

$$\|\mathbf{z}_1 - \mathbf{x}_1\| < \frac{(1 - \lambda)\varepsilon - \|\mathbf{z} - \mathbf{y}\|}{\lambda} \tag{2.2}$$

Now, let $\mathbf{z}_2 = \dfrac{\mathbf{z} - \lambda\mathbf{z}_1}{1 - \lambda}$. From (2.1), the Schwartz inequality, and (2.2), we get

$$\|\mathbf{z}_2 - \mathbf{x}_2\| = \left\| \frac{\mathbf{z} - \lambda\mathbf{z}_1}{1 - \lambda} - \mathbf{x}_2 \right\| = \left\| \frac{(\mathbf{z} - \lambda\mathbf{z}_1) - (\mathbf{y} - \lambda\mathbf{x}_1)}{1 - \lambda} \right\|$$

$$= \frac{1}{1 - \lambda} \|(\mathbf{z} - \mathbf{y}) + \lambda(\mathbf{x}_1 - \mathbf{z}_1)\|$$

$$\leq \frac{1}{1 - \lambda} (\|\mathbf{z} - \mathbf{y}\| + \lambda \|\mathbf{x}_1 - \mathbf{z}_1\|)$$

$$< \varepsilon$$

Therefore, $\mathbf{z}_2 \in S$. By definition of $\mathbf{z}_2$, note that $\mathbf{z} = \lambda\mathbf{z}_1 + (1 - \lambda)\mathbf{z}_2$; and since both $\mathbf{z}_1$ and $\mathbf{z}_2$ belong to S, then $\mathbf{z}$ also belongs to S. We have shown that any $\mathbf{z}$ with $\|\mathbf{z} - \mathbf{y}\| < (1 - \lambda)\varepsilon$ belongs to S. Therefore, $\mathbf{y} \in \text{int } S$ and the proof is complete.

Corollary 1

Let S be a convex set. Then, int S is convex.

Corollary 2

Let S be a convex set with a nonempty interior. Then, cl S is convex.

Proof

Let $\mathbf{x}_1, \mathbf{x}_2 \in \text{cl } S$. Pick $\mathbf{z} \in \text{int } S$ (by assumption int $S \neq \emptyset$). By the theorem, $\lambda\mathbf{x}_2 + (1 - \lambda)\mathbf{z} \in \text{int } S$ for each $\lambda \in (0, 1)$. Now fix $\mu \in (0, 1)$. By the theorem, $\mu\mathbf{x}_1 + (1 - \mu)[\lambda\mathbf{x}_2 + (1 - \lambda)\mathbf{z}] \in \text{int } S \subset S$ for each $\lambda \in (0, 1)$. If we take the limit as λ approaches 1, it follows that $\mu\mathbf{x}_1 + (1 - \mu)\mathbf{x}_2 \in \text{cl } S$, and the proof is complete.

Corollary 3

Let S be a convex set with a nonempty interior. Then, cl (int S) = cl S.

Proof

Clearly, cl (int S) $\subset$ cl S. Now let $\mathbf{x} \in \text{cl } S$, and pick $\mathbf{y} \in \text{int } S$ (by assumption int $S \neq \emptyset$). Then, $\lambda\mathbf{x} + (1 - \lambda)\mathbf{y} \in \text{int } S$ for each $\lambda \in (0, 1)$. Letting $\lambda \to 1^-$, it follows that $\mathbf{x} \in$ cl (int S).

Corollary 4

Let S be a convex set with a nonempty interior. Then, int (cl S) = int S.

Proof

Note that int $S \subset$ int (cl S). Let $\mathbf{x}_1 \in$ int (cl S). We need to show that $\mathbf{x}_1 \in \text{int } S$. There exists an $\varepsilon > 0$ such that $\|\mathbf{y} - \mathbf{x}_1\| < \varepsilon$ implies that $\mathbf{y} \in \text{cl } S$. Now let $\mathbf{x}_2 \neq \mathbf{x}_1$ belong to int S and let $\mathbf{y} = (1 + \Delta)\mathbf{x}_1 - \Delta\mathbf{x}_2$, where $\Delta = \varepsilon/(2\,\|\mathbf{x}_1 - \mathbf{x}_2\|)$. Since $\|\mathbf{y} - \mathbf{x}_1\| = \varepsilon/2$, then $\mathbf{y} \in \text{cl } S$. But $\mathbf{x}_1 = \lambda\mathbf{y} + (1 - \lambda)\mathbf{x}_2$, where $\lambda = 1/(1 + \Delta) \in (0, 1)$. Since $\mathbf{y} \in \text{cl } S$ and $\mathbf{x}_2 \in \text{int } S$, then, by the theorem, $\mathbf{x}_1 \in \text{int } S$, and the proof is complete.

The above theorem and its corollaries can be considerably strengthened by using the notion of relative interiors (see the Notes and References section at the end of the chapter).

2.3 Weierstrass' Theorem

A very important and widely used result is based on the foregoing concepts. This result relates to the existence of a minimizing solution for an optimization problem. Here, we say that $\bar{\mathbf{x}}$ is a minimizing solution for the problem min $\{f(\mathbf{x}):\mathbf{x} \in S\}$, provided that $\bar{\mathbf{x}} \in S$ and $f(\bar{\mathbf{x}}) \leq f(\mathbf{x})$ for all $\mathbf{x} \in S$. In such a case, we say that a minimum exists. On the other hand, we say that $\alpha = \inf \{f(\mathbf{x}):\mathbf{x} \in S\}$ if α is the greatest lower bound of f on S, that is, $\alpha \leq f(\mathbf{x})$ for all $\mathbf{x} \in S$ and there is no $\bar{\alpha} > \alpha$ such that $\bar{\alpha} \leq f(\mathbf{x})$ for all $\mathbf{x} \in S$. Similarly, $\alpha = \max \{f(\mathbf{x}):\mathbf{x} \in S\}$ if there exists a solution $\bar{\mathbf{x}} \in S$ such that $\alpha = f(\bar{\mathbf{x}}) \geq f(\mathbf{x})$ for all $\mathbf{x} \in S$. On the other hand, $\alpha = \sup \{f(\mathbf{x}):\mathrm{x} \in S\}$ if α is the least upper bound of f on S, that is, $\alpha \geq f(\mathbf{x})$ for all $\mathbf{x} \in S$, and there is no $\bar{\alpha} < \alpha$ such that $\bar{\alpha} \geq f(\mathbf{x})$ for all $\mathbf{x} \in S$.

Figure 2.5 illustrates three instances where a minimum does not exist. In Figure 2.5*a*, the infimum of f over (a, b) is given by $f(b)$ but, since S is not closed and, in particular, b L S, a minimum does not exist. In Figure 2.5*b*, we have that inf $\{f(x):x \in [a, b]\}$ is given by the limit of $f(x)$ as x approaches b from the "left," denoted $\lim_{x \to b^-} f(x)$. However, since f is discontinuous at b, a minimizing solution does not exist. Finally, Figure 2.5*c* illustrates a situation in which f is unbounded over the unbounded set $S = \{x:x \geq a\}$.

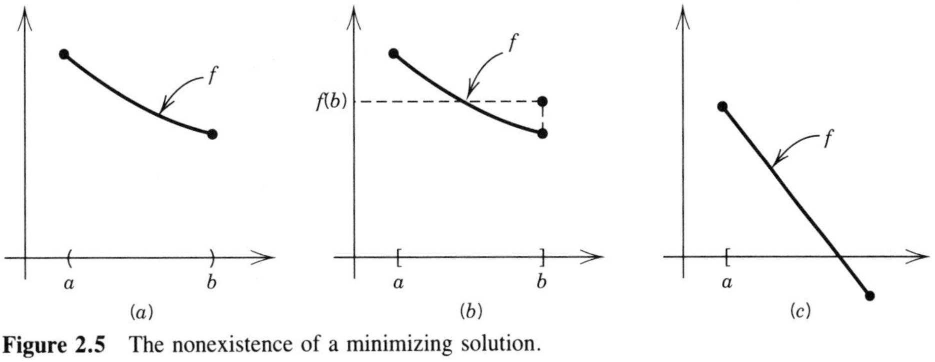

Figure 2.5 The nonexistence of a minimizing solution.

We now formally state and prove the result that if S is nonempty, closed, and bounded, and if f is continuous on S, then, unlike the various situations of Figure 2.5, a minimum exists. The reader is encouraged to study how these different assumptions guarantee the different assertions made in the following proof.

2.3.1 Theorem

Let S be a nonempty, compact set, and let $f:S \to E_1$ be continuous on S. Then, the problem min $\{f(\mathbf{x}):\mathbf{x} \in S\}$ attains its minimum, that is, there exists a minimizing solution to this problem.

Proof

Since f is continuous on S and S is both closed and bounded, f is bounded below on S. Consequently, since $S \neq \emptyset$, there exists a greatest lower bound $\alpha \equiv \inf \{f(\mathbf{x}): \mathbf{x} \in S\}$. Now, let $0 < \varepsilon < 1$, and consider the set $S_k = \{\mathbf{x} \in S : \alpha \leq f(\mathbf{x}) \leq \alpha + \varepsilon^k\}$ for each $k = 1, 2, \ldots$. By the definition of an infimum, $S_k \neq \emptyset$ for each k, and so we may construct a sequence of points $\{\mathbf{x}_k\} \subseteq S$ by selecting a point $\mathbf{x}_k \in S_k$ for each $k = 1, 2, \ldots$. Since S is bounded, there exists a convergent subsequence $\{\mathbf{x}_k\}_K \rightarrow \bar{\mathbf{x}}$, indexed by the set K. By the closedness of S, we have $\bar{\mathbf{x}} \in S$; and by the continuity of f, since $\alpha \leq f(\mathbf{x}_k) \leq \alpha + \varepsilon^k$ for all k, we have $\alpha = \lim_{k \to \infty,\, k \in K} f(\mathbf{x}_k) = f(\bar{\mathbf{x}})$. Hence, we have shown that there exists a solution $\bar{\mathbf{x}} \in S$ such that $f(\bar{\mathbf{x}}) = \alpha = \inf \{f(\mathbf{x}) : \mathbf{x} \in S\}$, and so $\bar{\mathbf{x}}$ **is** a minimizing solution. This completes the proof.

2.4 Separation and Support of Sets

The notions of supporting hyperplanes and separation of disjoint convex sets are very important in optimization. Almost all optimality conditions and duality relationships use some sort of separation or support of convex sets. The results of this section are based on the following geometric fact: given a closed convex set S and a point $\mathbf{y}$ L S, there exists a unique point $\bar{\mathbf{x}} \in S$ with minimum distance from $\mathbf{y}$ and a hyperplane that separates $\mathbf{y}$ and S.

Minimum Distance from a Point to a Convex Set

To establish the above important result, the following *parallelogram law* is needed. Let $\mathbf{a}$ and $\mathbf{b}$ be two vectors in E_n. Then,

$$\|\mathbf{a} + \mathbf{b}\|^2 = \|\mathbf{a}\|^2 + \|\mathbf{b}\|^2 + 2\mathbf{a}'\mathbf{b}$$

$$\|\mathbf{a} - \mathbf{b}\|^2 = \|\mathbf{a}\|^2 + \|\mathbf{b}\|^2 - 2\mathbf{a}'\mathbf{b}$$

By adding, we get

$$\|\mathbf{a} + \mathbf{b}\|^2 + \|\mathbf{a} - \mathbf{b}\|^2 = 2\,\|\mathbf{a}\|^2 + 2\,\|\mathbf{b}\|^2$$

This result is illustrated in Figure 2.6 and can be interpreted as follows: the sum of squared norms of the diagonals of a parallelogram is equal to the sum of squared norms of its sides.

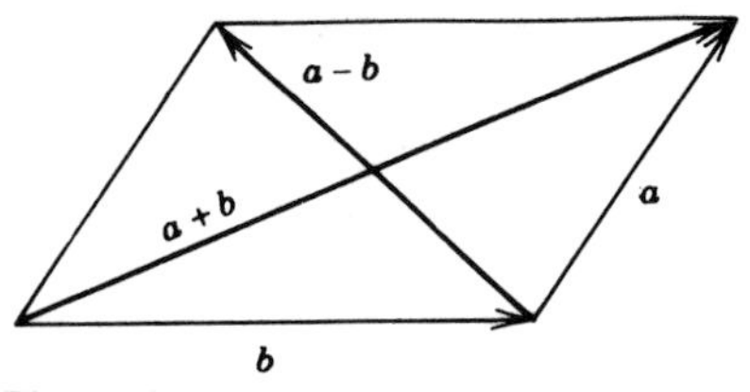

Figure 2.6 Parallelogram law.

We now state and prove the closest point theorem. Again, the reader is encouraged to investigate how the various assumptions play a role in guaranteeing various assertions.

2.4.1 Theorem

Let S be a nonempty, closed convex set in E_n and $\mathbf{y} \notin S$. Then, there exists a unique point $\bar{\mathbf{x}} \in S$ with minimum distance from $\mathbf{y}$. Furthermore, $\bar{\mathbf{x}}$ is the minimizing point if and only if $(\mathbf{y} - \bar{\mathbf{x}})^t(\mathbf{x} - \bar{\mathbf{x}}) \leq 0$ for all $\mathbf{x} \in S$.

Proof

First, let us establish the existence of a closest point. Since $S \neq \emptyset$, there exists a point $\hat{\mathbf{x}} \in S$, and we can confine our attention to the set $\bar{S} = S \cap \{\mathbf{x}: \|\mathbf{y} - \mathbf{x}\| \leq \|\mathbf{y} - \hat{\mathbf{x}}\|\}$ in seeking the closest point. In other words, the closest point problem inf $\{\|\mathbf{y} - \mathbf{x}\|: \mathbf{x} \in S\}$ is equivalent to inf $\{\|\mathbf{y} - \mathbf{x}\|: \mathbf{x} \in \bar{S}\}$. But the latter problem involves finding the minimum of a continuous function over a nonempty, compact set $\bar{S}$, and so by Weierstrass' theorem, 2.3.1, we know that there exists a minimizing point $\bar{\mathbf{x}}$ in S that is closest to the point $\mathbf{y}$.

To show uniqueness, suppose that there is an $\bar{\mathbf{x}}' \in S$ such that $\|\mathbf{y} - \bar{\mathbf{x}}\| = \|\mathbf{y} - \bar{\mathbf{x}}'\| = \gamma$. By convexity of S, $(\bar{\mathbf{x}} + \bar{\mathbf{x}}')/2 \in S$. By the Schwartz inequality, we get

$$\left\|\mathbf{y} - \frac{\bar{\mathbf{x}} + \bar{\mathbf{x}}'}{2}\right\| \leq \tfrac{1}{2}\|\mathbf{y} - \bar{\mathbf{x}}\| + \tfrac{1}{2}\|\mathbf{y} - \bar{\mathbf{x}}'\| = \gamma$$

If strict inequality holds, we have a contradiction to $\bar{\mathbf{x}}$ being the closest point to $\mathbf{y}$. Therefore, equality holds, and we must have $\mathbf{y} - \bar{\mathbf{x}} = \lambda(\mathbf{y} - \bar{\mathbf{x}}')$ for some λ. Since $\|\mathbf{y} - \bar{\mathbf{x}}\| = \|\mathbf{y} - \bar{\mathbf{x}}'\| = \gamma$, $|\lambda| = 1$. Clearly, $\lambda \neq -1$, because otherwise $\mathbf{y} = (\bar{\mathbf{x}} + \bar{\mathbf{x}}')/2 \in S$, contradicting the assumption that $\mathbf{y} \notin S$. So, $\lambda = 1$, $\bar{\mathbf{x}}' = \bar{\mathbf{x}}$, and uniqueness is established.

To complete the proof, we need to show that $(\mathbf{y} - \bar{\mathbf{x}})^t(\mathbf{x} - \bar{\mathbf{x}}) \leq 0$ for all $\mathbf{x} \in S$ is both a necessary and sufficient condition for $\bar{\mathbf{x}}$ to be the point in S closest to $\mathbf{y}$.

To prove sufficiency, let $\mathbf{x} \in S$. Then,

$$\|\mathbf{y} - \mathbf{x}\|^2 = \|\mathbf{y} - \bar{\mathbf{x}} + \bar{\mathbf{x}} - \mathbf{x}\|^2 = \|\mathbf{y} - \bar{\mathbf{x}}\|^2 + \|\bar{\mathbf{x}} - \mathbf{x}\|^2 + 2(\bar{\mathbf{x}} - \mathbf{x})^t(\mathbf{y} - \bar{\mathbf{x}})$$

Since $\|\bar{\mathbf{x}} - \mathbf{x}\|^2 \geq 0$ and $(\bar{\mathbf{x}} - \mathbf{x})^t(\mathbf{y} - \bar{\mathbf{x}}) \geq 0$ by assumption, $\|\mathbf{y} - \mathbf{x}\|^2 \geq \|\mathbf{y} - \bar{\mathbf{x}}\|^2$, and $\bar{\mathbf{x}}$ is the minimizing point. Conversely, assume that $\|\mathbf{y} - \mathbf{x}\|^2 \geq \|\mathbf{y} - \bar{\mathbf{x}}\|^2$ for all $\mathbf{x} \in S$. Let $\mathbf{x} \in S$ and note that $\bar{\mathbf{x}} + \lambda(\mathbf{x} - \bar{\mathbf{x}}) \in S$ for $0 \leq \lambda \leq 1$ by the convexity of S. Therefore,

$$\|\mathbf{y} - \bar{\mathbf{x}} - \lambda(\mathbf{x} - \bar{\mathbf{x}})\|^2 \geq \|\mathbf{y} - \bar{\mathbf{x}}\|^2 \tag{2.3}$$

Also

$$\|\mathbf{y} - \bar{\mathbf{x}} - \lambda(\mathbf{x} - \bar{\mathbf{x}})\|^2 = \|\mathbf{y} - \bar{\mathbf{x}}\|^2 + \lambda^2\|\mathbf{x} - \bar{\mathbf{x}}^2\| - 2\lambda(\mathbf{y} - \bar{\mathbf{x}})^t(\mathbf{x} - \bar{\mathbf{x}}) \tag{2.4}$$

From (2.3) and (2.4), we get

$$2\lambda(\mathbf{y} - \bar{\mathbf{x}})^t(\mathbf{x} - \bar{\mathbf{x}}) \leq \lambda^2\|\mathbf{x} - \bar{\mathbf{x}}\|^2 \tag{2.5}$$

for all $0 \leq \lambda \leq 1$. Dividing (2.5) by any such $\lambda > 0$ and letting $\lambda \to 0^+$, the result follows.

The above theorem is illustrated in Figure 2.7*a*. Note that the angle between $(\mathbf{y} - \bar{\mathbf{x}})$ and $(\mathbf{x} - \bar{\mathbf{x}})$ for any point $\mathbf{x}$ in S is greater than or equal to 90° and, hence, $(\mathbf{y} - \bar{\mathbf{x}})^t(\mathbf{x} - \bar{\mathbf{x}}) \leq 0$ for all $\mathbf{x} \in S$. This says that the set S lies in the half-space $\boldsymbol{\alpha}^t(\mathbf{x} - \bar{\mathbf{x}}) \leq 0$ relative to the hyperplane $\boldsymbol{\alpha}^t(\mathbf{x} - \bar{\mathbf{x}}) = 0$ passing through $\bar{\mathbf{x}}$ and having a normal $\boldsymbol{\alpha} = (\mathbf{y} - \bar{\mathbf{x}})$. Note also by referring to Figure 2.7*b* that this feature does not necessarily hold even over $N_\varepsilon(\bar{\mathbf{x}}) \cap S$ if S is not convex.

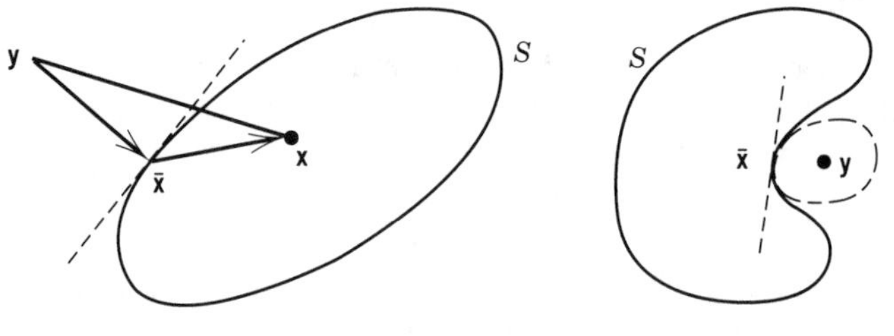

Figure 2.7 Minimum distance to a closed convex set.

Hyperplanes and Separation of Two Sets

Since we shall be dealing with separating and supporting hyperplanes, precise definitions of hyperplanes and half-spaces are reiterated below.

2.4.2 Definition

A *hyperplane* H in E_n is a collection of points of the form $\{\mathbf{x}:\mathbf{p}'\mathbf{x} = \alpha\}$, where $\mathbf{p}$ is a nonzero vector in E_n and α is a scalar. The vector $\mathbf{p}$ is called the *normal* vector of the hyperplane. A hyperplane H defines two *closed half-spaces* $H^+ = \{\mathbf{x}:\mathbf{p}'\mathbf{x} \geq \alpha\}$ and $H^- = \{\mathbf{x}:\mathbf{p}'\mathbf{x} \leq \alpha\}$ and the two *open half-spaces* $\{\mathbf{x}:\mathbf{p}'\mathbf{x} > \alpha\}$ and $\{\mathbf{x}:\mathbf{p}'\mathbf{x} < \alpha\}$.

Note that any point in E_n lies in H^+, in H^-, or in both. Also, a hyperplane H and the corresponding half-spaces can be written in reference to a fixed point, say, $\bar{\mathbf{x}} \in H$. If $\bar{\mathbf{x}} \in H$, then $\mathbf{p}'\bar{\mathbf{x}} = \alpha$ and, hence, any point $\mathbf{x} \in H$ must satisfy $\mathbf{p}'\mathbf{x} - \mathbf{p}'\bar{\mathbf{x}} = \alpha - \alpha = 0$; that is, $\mathbf{p}'(\mathbf{x} - \bar{\mathbf{x}}) = 0$. Accordingly, $H^+ = \{\mathbf{x}:\mathbf{p}'(\mathbf{x} - \bar{\mathbf{x}}) \geq 0\}$ and $H^- = \{\mathbf{x}:\mathbf{p}'(\mathbf{x} - \bar{\mathbf{x}}) \leq 0\}$. Figure 2.8 shows a hyperplane H passing through $\bar{\mathbf{x}}$ and having a normal vector $\mathbf{p}$.

As an example, consider $H = \{(x_1, x_2, x_3, x_4):x_1 + x_2 - x_3 + 2x_4 = 4\}$. The normal vector is $\mathbf{p} = (1, 1, -1, 2)'$. Alternatively, the hyperplane can be written in reference to any point in H, for example, $\bar{\mathbf{x}} = (0, 6, 0, -1)'$. In this case, we write $H = \{(x_1, x_2, x_3, x_4) : x_1 + (x_2 - 6) - x_3 + 2(x_4 + 1) = 0\}$.

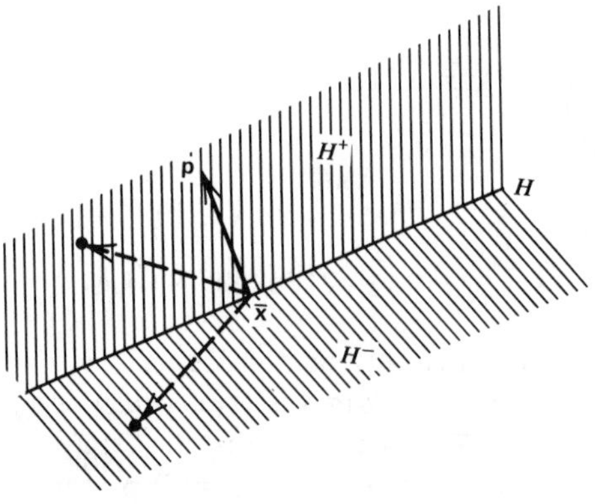

Figure 2.8 An example of a hyperplane and corresponding half-spaces.

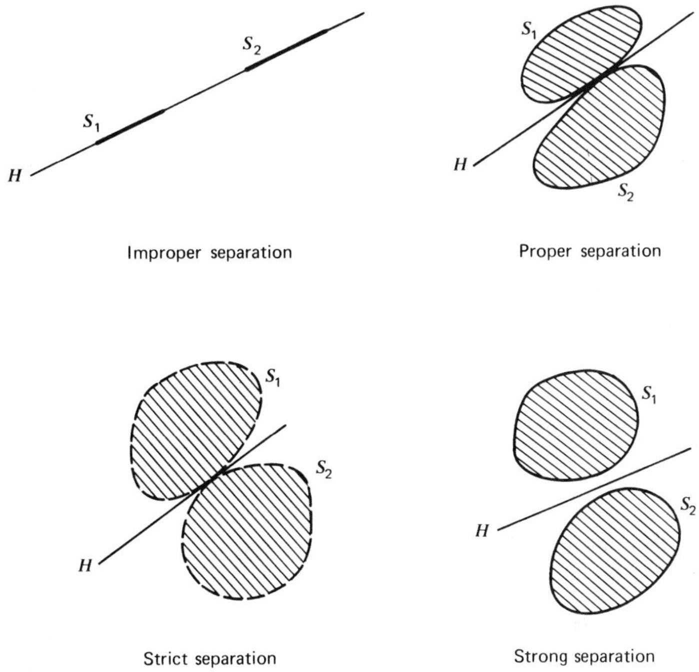

Figure 2.9 Various types of separation.

2.4.3 Definition

Let S_1 and S_2 be nonempty sets in E_n. A hyperplane $H = \{\mathbf{x}:\mathbf{p}'\mathbf{x} = \alpha\}$ is said to *separate* S_1 and S_2 if $\mathbf{p}'\mathbf{x} \geq \alpha$ for each $\mathbf{x} \in S_1$ and $\mathbf{p}'\mathbf{x} \leq \alpha$ for each $\mathbf{x} \in S_2$. If, in addition, $S_1 \cup S_2 \not\subset H$, then H is said to *properly separate* S_1 and S_2. The hyperplane H is said to *strictly separate* S_1 and S_2 if $\mathbf{p}'\mathbf{x} > \alpha$ for each $\mathbf{x} \in S_1$ and $\mathbf{p}'\mathbf{x} < \alpha$ for each $\mathbf{x} \in S_2$. The hyperplane H is said to *strongly separate* S_1 and S_2 if $\mathbf{p}'\mathbf{x} \geq \alpha + \varepsilon$ for each $\mathbf{x} \in S_1$ and $\mathbf{p}'\mathbf{x} \leq \alpha$ for each $\mathbf{x} \in S_2$, where ε is a positive scalar.

Figure 2.9 shows various types of separation. Of course, strong separation implies strict separation, which implies proper separation, which in turn implies separation. *Improper* separation is usually of little value, since it corresponds to a hyperplane containing both S_1 and S_2 as shown in Figure 2.9.

Separation of a Convex Set and a Point

We shall now present the first and most fundamental separation theorem. Other separation and support theorems will follow from this basic result.

2.4.4 Theorem

Let S be a nonempty closed convex set in E_n and $\mathbf{y} \notin S$. Then, there exists a nonzero vector $\mathbf{p}$ and a scalar α such that $\mathbf{p}'\mathbf{y} > \alpha$ and $\mathbf{p}'\mathbf{x} \leq \alpha$ for each $\mathbf{x} \in S$.

Proof

The set S is a nonempty closed convex set and $\mathbf{y} \notin S$. Hence, by Theorem 2.4.1, there exists a unique minimizing point $\bar{\mathbf{x}} \in S$ that $(\mathbf{x} - \bar{\mathbf{x}})'(\mathbf{y} - \bar{\mathbf{x}}) \leq 0$ for each $\mathbf{x} \in S$.

Letting $\mathbf{p} = (\mathbf{y} - \bar{\mathbf{x}}) \neq 0$ and $\alpha = \bar{\mathbf{x}}'(\mathbf{y} - \bar{\mathbf{x}}) \equiv \mathbf{p}'\bar{\mathbf{x}}$, we get $\mathbf{p}'\mathbf{x} \leq \alpha$ for each $\mathbf{x} \in S$, while $\mathbf{p}'\mathbf{y} - \alpha = (\mathbf{y} - \bar{\mathbf{x}})'(\mathbf{y} - \bar{\mathbf{x}}) = \|\mathbf{y} - \bar{\mathbf{x}}\|^2 > 0$. This completes the proof.

Corollary 1

Let S be a closed convex set in E_n. Then, S is the intersection of all half-spaces containing S.

Proof

Obviously, S is contained in the intersection of all half-spaces containing it. To the contrary of the desired result, suppose that there is a point $\mathbf{y}$ in the intersection of these half-spaces but not in S. By the theorem, there exists a half-space that contains S but not $\mathbf{y}$. This contradiction proves the corollary.

Corollary 2

Let S be a nonempty set, and let $\mathbf{y} \notin \text{cl conv } S$, the closure of the convex hull of S. Then, there exists a strongly separating hyperplane for S and $\mathbf{y}$.

Proof

The result follows by letting cl conv S play the role of S in Theorem 2.4.4.

The following statements are equivalent to the conclusion of the theorem. The reader is asked to verify this equivalence. Note that statements 1 and 2 are equivalent only in this special case since $\mathbf{y}$ is a point. Note also that in the above theorem, we have $\alpha \equiv \mathbf{p}'\bar{\mathbf{x}} = \max \{\mathbf{p}'\mathbf{x}:\mathbf{x} \in S\}$, since for any $\mathbf{x} \in S$, $\mathbf{p}'(\bar{\mathbf{x}} - \mathbf{x}) = (\mathbf{y} - \bar{\mathbf{x}})'(\bar{\mathbf{x}} - \mathbf{x}) \geq 0$.

1. There exists a hyperplane that *strictly* separates S and $\mathbf{y}$.
2. There exists a hyperplane that *strongly* separates S and $\mathbf{y}$.
3. There exists a vector $\mathbf{p}$ such that $\mathbf{p}'\mathbf{y} > \sup \{\mathbf{p}'\mathbf{x}:\mathbf{x} \in S\}$.
4. There exists a vector $\mathbf{p}$ such that $\mathbf{p}'\mathbf{y} < \inf \{\mathbf{p}'\mathbf{x}:\mathbf{x} \in S\}$.

Farkas' Theorem as a Consequence of Theorem 2.4.4

Farkas' theorem is used extensively in the derivation of optimality conditions of linear and nonlinear programming problems. The theorem can be stated as follows. Let $\mathbf{A}$ be an $m \times n$ matrix, and let $\mathbf{c}$ be an n vector. Then, exactly one of the following two systems has a solution:

System 1 $\mathbf{A}\mathbf{x} \leq \mathbf{0}$ and $\mathbf{c}'\mathbf{x} > 0$ for some $\mathbf{x} \in E_n$

System 2 $\mathbf{A}'\mathbf{y} = \mathbf{c}$ and $\mathbf{y} \geq \mathbf{0}$ for some $\mathbf{y} \in E_m$

If we denote the columns of $\mathbf{A}'$ by $\mathbf{a}_1, \ldots, \mathbf{a}_m$, then System 2 has a solution if $\mathbf{c}$ lies in the convex cone generated by $\mathbf{a}_1, \ldots, \mathbf{a}_m$. System 1 has a solution if the closed convex cone $\{\mathbf{x}:\mathbf{A}\mathbf{x} \leq \mathbf{0}\}$ and the open half-space $\{\mathbf{x}:\mathbf{c}'\mathbf{x} > 0\}$ have a nonempty intersection. These two cases are illustrated geometrically in Figure 2.10.

2.4.5 Theorem (Farkas' Theorem)

Let $\mathbf{A}$ be an $m \times n$ matrix and $\mathbf{c}$ be an n vector. Then, exactly one of the following two systems has a solution:

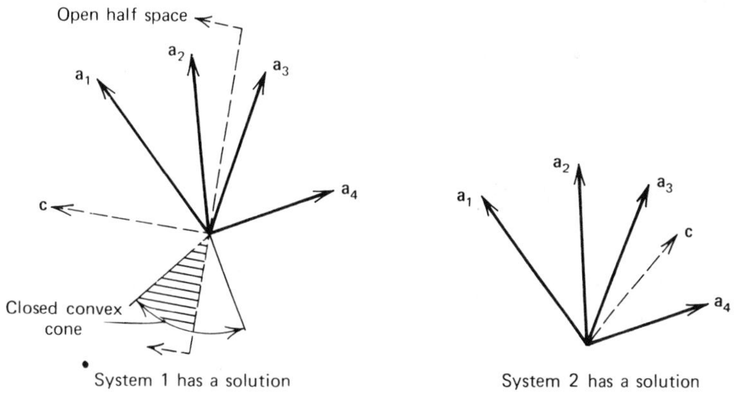

Figure 2.10 Illustration of Farkas' theorem.

System 1 $\mathbf{Ax} \leq \mathbf{0}$ and $\mathbf{c}'\mathbf{x} > 0$ for some $\mathbf{x} \in E_n$

System 2 $\mathbf{A}'\mathbf{y} = \mathbf{c}$ and $\mathbf{y} \geq \mathbf{0}$ for some $\mathbf{y} \in E_m$

Proof

Suppose that System 2 has a solution; that is, there exists a $\mathbf{y} \geq \mathbf{0}$ such that $\mathbf{A}'\mathbf{y} = \mathbf{c}$. Let $\mathbf{x}$ be such that $\mathbf{Ax} \leq \mathbf{0}$. Then $\mathbf{c}'\mathbf{x} = \mathbf{y}'\mathbf{Ax} \leq 0$. Hence, System 1 has no solution. Now suppose that System 2 has no solution. Form the set $S = \{\mathbf{x}:\mathbf{x} = \mathbf{A}'\mathbf{y},\ \mathbf{y} \geq \mathbf{0}\}$. Note that S is a closed convex set and that $\mathbf{c}$ L S. By Theorem 2.4.4, there exists a vector $\mathbf{p} \in E_n$ and a scalar α such that $\mathbf{p}'\mathbf{c} > \alpha$ and $\mathbf{p}'\mathbf{x} \leq \alpha$ for all $\mathbf{x} \in S$. Since $\mathbf{0} \in S$, then $\alpha \geq 0$ and so $\mathbf{p}'\mathbf{c} > 0$. Also, $\alpha \geq \mathbf{p}'\mathbf{A}'\mathbf{y} = \mathbf{y}'\mathbf{Ap}$ for all $\mathbf{y} \geq \mathbf{0}$. Since $\mathbf{y} \geq \mathbf{0}$ can be made arbitrarily large, the last inequality implies that $\mathbf{Ap} \leq \mathbf{0}$. We have therefore constructed a vector $\mathbf{p} \in E_n$ such that $\mathbf{Ap} \leq \mathbf{0}$ and $\mathbf{c}'\mathbf{p} > 0$. So System 1 has a solution, and the proof is complete.

Corollary 1 (Gordan's Theorem)

Let $\mathbf{A}$ be an $m \times n$ matrix. Then, exactly one of the following two systems has a solution:

System 1 $\mathbf{Ax} < 0$ for some $\mathbf{x} \in E_n$

System 2 $\mathbf{A}'\mathbf{y} = \mathbf{0},\ \mathbf{y} \geq \mathbf{0}$ for some nonzero $\mathbf{y} \in E_m$

Proof

Note that System 1 can be equivalently written as $\mathbf{Ax} + \mathbf{e}s \leq \mathbf{0}$ for some $\mathbf{x} \in E_n$ and $s > 0$, $s \in E_1$, where $\mathbf{e}$ is a vector of m ones. Rewriting this in the form of System 1 of Theorem 2.4.5, we get $[\mathbf{A}\ \ \mathbf{e}]\begin{bmatrix}\mathbf{x}\\ s\end{bmatrix} \leq \mathbf{0}$, and $(0, \ldots, 0, 1)\begin{pmatrix}\mathbf{x}\\ s\end{pmatrix} > 0$ for some $\begin{pmatrix}\mathbf{x}\\ s\end{pmatrix} \in E_{n+1}$. By Theorem 2.4.5, the associated System 2 states that $\begin{bmatrix}\mathbf{A}'\\ \mathbf{e}'\end{bmatrix}\mathbf{y} = (0, \ldots, 0, 1)'$ and $\mathbf{y} \geq \mathbf{0}$ for some $\mathbf{y} \in E_m$; that is, $\mathbf{A}'\mathbf{y} = \mathbf{0}$, $\mathbf{e}'\mathbf{y} = 1$ and $\mathbf{y} \geq \mathbf{0}$ for some $\mathbf{y} \in E_m$. This is equivalent to System 2 of the corollary and, hence, the result follows.

Corollary 2

Let **A** be an $m \times n$ matrix and **c** be an n vector. Then, exactly one of the following two systems has a solution:

System 1 $\mathbf{Ax} \leq \mathbf{0}$, $\mathbf{x} \geq \mathbf{0}$, $\mathbf{c}^t\mathbf{x} > 0$ for some $\mathbf{x} \in E_n$

System 2 $\mathbf{A}^t\mathbf{y} \geq \mathbf{c}$, $\mathbf{y} \geq \mathbf{0}$ for some $\mathbf{y} \in E_m$

Proof

The result follows by writing the first set of constraints of System 2 as equalities and, accordingly, replacing $\mathbf{A}^t$ in the theorem by $[\mathbf{A}^t, \ -\mathbf{I}]$.

Corollary 3

Let **A** be an $m \times n$ matrix, **B** be an $l \times n$ matrix, and **c** be an n vector. Then, exactly one of the following two systems has a solution:

System 1 $\mathbf{Ax} \leq \mathbf{0}$, $\mathbf{Bx} = \mathbf{0}$, $\mathbf{c}^t\mathbf{x} > 0$ for some $\mathbf{x} \in E_n$

System 2 $\mathbf{A}^t\mathbf{y} + \mathbf{B}^t\mathbf{z} = \mathbf{c}$, $\mathbf{y} \geq \mathbf{0}$ for some $\mathbf{y} \in E_m$ and $\mathbf{z} \in E_l$

Proof

The result follows by writing $\mathbf{z} = \mathbf{z}_1 - \mathbf{z}_2$, where $\mathbf{z}_1 \geq 0$ and $\mathbf{z}_2 \geq 0$ in System 2 and, accordingly, replacing $\mathbf{A}^t$ in the theorem by $[\mathbf{A}^t, \ \mathbf{B}^t, \ -\mathbf{B}^t]$.

Support of Sets at Boundary Points

2.4.6 Definition

Let S be a nonempty set in E_n, and let $\bar{\mathbf{x}} \in \partial S$. A hyperplane $H = \{\mathbf{x}: \mathbf{p}^t(\mathbf{x} - \bar{\mathbf{x}}) = 0\}$ is called a *supporting hyperplane* of S at $\bar{\mathbf{x}}$ if either $S \subseteq H^+$, that is, $\mathbf{p}^t(\mathbf{x} - \bar{\mathbf{x}}) \geq 0$ for each $\mathbf{x} \in S$, or else, $S \subseteq H^-$ such that $\mathbf{p}^t(\mathbf{x} - \bar{\mathbf{x}}) \leq 0$ for each $\mathbf{x} \in S$. If, in addition, S N H, then H is called a *proper supporting hyperplane* of S at $\bar{\mathbf{x}}$.

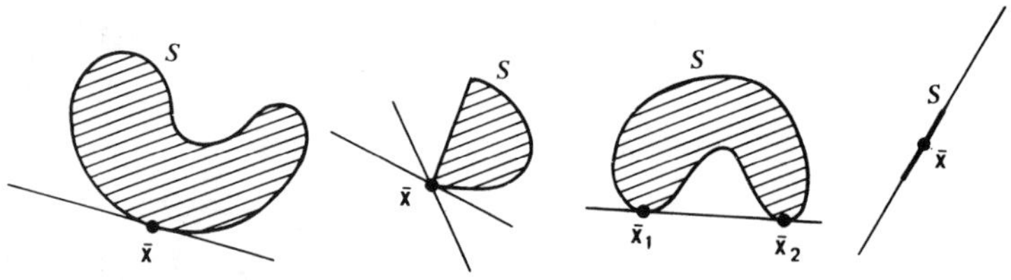

Figure 2.11 Examples of supporting hyperplanes.

Note that the above definition can be equivalently stated as follows. The hyperplane $H = \{\mathbf{x}: \mathbf{p}^t(\mathbf{x} - \bar{\mathbf{x}}) = 0\}$ is a supporting hyperplane of S at $\bar{\mathbf{x}} \in \partial S$ if $\mathbf{p}^t\bar{\mathbf{x}} = \inf \{\mathbf{p}^t\mathbf{x}: \mathbf{x} \in S\}$ or else $\mathbf{p}^t\bar{\mathbf{x}} = \sup \{\mathbf{p}^t\mathbf{x}: \mathbf{x} \in S\}$. This follows by noting that either $\bar{\mathbf{x}} \in S$; or if $\bar{\mathbf{x}}$ L S, then, since $\bar{\mathbf{x}} \in \partial S$, there exist points in S arbitrarily close to $\bar{\mathbf{x}}$ and, hence, arbitrarily close in the value of the function $\mathbf{p}^t\mathbf{x}$ to the value $\mathbf{p}^t\bar{\mathbf{x}}$.

Figure 2.11 shows some examples of supporting hyperplanes. The figure illustrates the cases of a unique supporting hyperplane at a boundary point, an infinite number of supporting hyperplanes at a boundary point, a hyperplane that supports the set at more

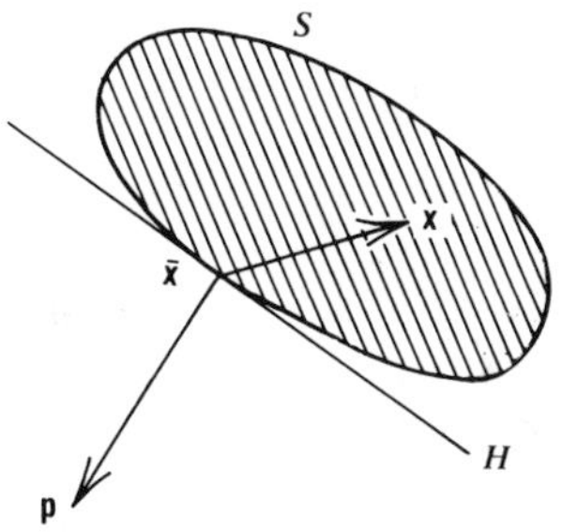

Figure 2.12 Supporting hyperplane.

than one point, and finally an improper supporting hyperplane that contains the whole set.

We now prove that a convex set has a supporting hyperplane at each boundary point (see Figure 2.12). As a corollary, a result similar to Theorem 2.4.4, where S is not required to be closed, follows.

2.4.7 Theorem

Let S be a nonempty convex set in E_n, and let $\bar{\mathbf{x}} \in \partial S$. Then, there exists a hyperplane that supports S at $\bar{\mathbf{x}}$; that is, there exists a nonzero vector $\mathbf{p}$ such that $\mathbf{p}^t(\mathbf{x} - \bar{\mathbf{x}}) \leq 0$ for each $\mathbf{x} \in \text{cl } S$.

Proof

Since $\bar{\mathbf{x}} \in \partial S$, there exists a sequence $\{\mathbf{y}_k\}$ not in cl S such that $\mathbf{y}_k \to \bar{\mathbf{x}}$. By Theorem 2.4.4, corresponding to each $\mathbf{y}_k$ there exists a $\mathbf{p}_k$ with norm 1 such that $\mathbf{p}_k^t\mathbf{y}_k > \mathbf{p}_k^t\mathbf{x}$ for each $\mathbf{x} \in \text{cl } S$. (In Theorem 2.4.4, the normal vector can be normalized by dividing it by its norm, so that $\|\mathbf{p}_k\| = 1$.) Since $\{\mathbf{p}_k\}$ is bounded, it has a convergent subsequence $\{\mathbf{p}_k\}_{\mathcal{K}}$ with limit $\mathbf{p}$ whose norm is also equal to 1. Considering this subsequence, we have $\mathbf{p}_k^t\mathbf{y}_k > \mathbf{p}_k^t\mathbf{x}$ for each $\mathbf{x} \in \text{cl } S$. Fixing $\mathbf{x} \in \text{cl } S$ and taking limits as $k \in \mathcal{K}$ approaches ∞, we get, $\mathbf{p}^t(\mathbf{x} - \bar{\mathbf{x}}) \leq 0$. Since this is true for each $\mathbf{x} \in \text{cl } S$, the result follows.

Corollary 1

Let S be a nonempty convex set in E_n and $\bar{\mathbf{x}} \notin \text{int } S$. Then, there is a nonzero vector $\mathbf{p}$ such that $\mathbf{p}^t(\mathbf{x} - \bar{\mathbf{x}}) \leq 0$ for each $\mathbf{x} \in \text{cl } S$.

Proof

If $\bar{\mathbf{x}} \notin \text{cl } S$, then the corollary follows from Theorem 2.4.4. On the other hand, if $\bar{\mathbf{x}} \in \partial S$, the corollary reduces to the above theorem.

Corollary 2

Let S be a nonempty set in E_n, and let $\mathbf{y} \notin \text{int conv } S$. Then, there exists a hyperplane that separates S and $\mathbf{y}$.

Proof

The result follows by identifying conv S and $\mathbf{y}$ with S and $\mathbf{x}$, respectively, in Corollary 1.

Corollary 3

Let S be a nonempty set in E_n, and let $\bar{\mathbf{x}} \in \partial S \cap \partial \text{conv } S$. Then, there exists a hyperplane that supports S at $\bar{\mathbf{x}}$.

Proof

The result follows by treating conv S as the set of Theorem 2.4.7.

Separation of Two Convex Sets

Thus far, we have discussed the separation of a convex set and a point not in the set and have also discussed the support of convex sets at boundary points. In addition, if we have two disjoint convex sets, they can be separated by a hyperplane H such that one of the sets belongs to H^+ and the other set belongs to H^-. In fact, this result holds true even if the two sets have some points in common, as long as their interiors are disjoint. This result is made precise by the following theorem.

2.4.8 Theorem

Let S_1 and S_2 be nonempty convex sets in E_n and suppose that $S_1 \cap S_2$ is empty. Then there exists a hyperplane that separates S_1 and S_2; that is, there exists a nonzero vector $\mathbf{p}$ in E_n such that

$$\inf \{\mathbf{p}'\mathbf{x} : \mathbf{x} \in S_1\} \geq \sup \{\mathbf{p}'\mathbf{x} : \mathbf{x} \in S_2\}$$

Proof

Let $S = S_1 \ominus S_2 = \{\mathbf{x}_1 - \mathbf{x}_2 : \mathbf{x}_1 \in S_1 \text{ and } \mathbf{x}_2 \in S_2\}$. Note that S is convex. Furthermore, $\mathbf{0}$ L S, because otherwise $S_1 \cap S_2$ will be nonempty. By Corollary 1 of Theorem 2.4.7, there exists a nonzero $\mathbf{p} \in E_n$ such that $\mathbf{p}'\mathbf{x} \geq 0$ for all $\mathbf{x} \in S$. This means that $\mathbf{p}'\mathbf{x}_1 \geq \mathbf{p}'\mathbf{x}_2$ for all $\mathbf{x}_1 \in S_1$ and $\mathbf{x}_2 \in S_2$, and the result follows.

Corollary 1

Let S_1 and S_2 be nonempty convex sets in E_n. Suppose that int S_2 is not empty and that $S_1 \cap \text{int } S_2$ is empty. Then, there exists a hyperplane that separates S_1 and S_2, that is, there exists a nonzero $\mathbf{p}$ such that

$$\inf \{\mathbf{p}'\mathbf{x} : \mathbf{x} \in S_1\} \geq \sup \{\mathbf{p}'\mathbf{x} : \mathbf{x} \in S_2\}$$

Proof

Replace S_2 by int S_2, apply the theorem, and note that

$$\sup \{\mathbf{p}'\mathbf{x} : \mathbf{x} \in S_2\} = \sup \{\mathbf{p}'\mathbf{x} : \mathbf{x} \in \text{int } S_2\}$$

Corollary 2

Let S_1 and S_2 be nonempty sets in E_n such that int conv $S_i \neq \varnothing$, for $i = 1, 2$, but int conv $S_1 \cap$ int conv $S_2 = \varnothing$. Then, there exists a hyperplane that separates S_1 and S_2.

Note the importance of assuming nonempty interiors in Corollary 2. Otherwise, for example, two crossing lines in E_2 can be taken as S_1 and S_2 (or as conv S_1 and conv S_2), and we would have int conv $S_1 \cap$ int conv $S_2 = \varnothing$. But there does not exist a hyperplane that separates S_1 and S_2.

Gordan's Theorem as a Consequence of Theorem 2.4.8

We shall now prove Gordan's theorem (see Corollary 1 to Theorem 2.4.5) using the existence of a hyperplane that separates two disjoint convex sets. This theorem is important in deriving optimality conditions for nonlinear programming.

2.4.9 Theorem (Gordan's Theorem)

Let $\mathbf{A}$ be an $m \times n$ matrix. Then, exactly one of the following systems has a solution:

System 1	$\mathbf{Ax} < \mathbf{0}$	for some $\mathbf{x} \in E_n$
System 2	$\mathbf{A}'\mathbf{p} = \mathbf{0}$ and $\mathbf{p} \geq \mathbf{0}$	for some nonzero $\mathbf{p} \in E_m$

Proof

We shall first prove that if System 1 has a solution $\hat{\mathbf{x}}$, then we cannot have a solution to $\mathbf{A}'\mathbf{p} = \mathbf{0}$, $\mathbf{p} \geq \mathbf{0}$, $\mathbf{p}$ nonzero. Suppose, on the contrary, that a solution $\hat{\mathbf{p}}$ exists. Then, since $\mathbf{A}\hat{\mathbf{x}} < \mathbf{0}$, $\hat{\mathbf{p}} \geq \mathbf{0}$ and $\hat{\mathbf{p}} \neq \mathbf{0}$, we have $\hat{\mathbf{p}}'\mathbf{A}\hat{\mathbf{x}} < 0$, that is, $\hat{\mathbf{x}}'\mathbf{A}'\hat{\mathbf{p}} < 0$. But this contradicts the hypothesis that $\mathbf{A}'\hat{\mathbf{p}} = \mathbf{0}$. Hence, System 2 cannot have a solution.

Now assume that System 1 has no solution. Consider the following two sets:

$$S_1 = \{\mathbf{z} : \mathbf{z} = \mathbf{Ax}, \mathbf{x} \in E_n\}$$

$$S_2 = \{\mathbf{z} : \mathbf{z} < \mathbf{0}\}$$

Note that S_1 and S_2 are nonempty convex sets such that $S_1 \cap S_2 = \varnothing$. Then, by Theorem 2.4.8, there exists a hyperplane that separates S_1 and S_2; that is, there exists a nonzero vector $\mathbf{p}$ such that

$$\mathbf{p}'\mathbf{Ax} \geq \mathbf{p}'\mathbf{z} \qquad \text{for each } \mathbf{x} \in E_n \quad \text{and} \quad \mathbf{z} \in \text{cl } S_2$$

Since each component of $\mathbf{z}$ can be made an arbitrarily large negative number, we must have $\mathbf{p} \geq \mathbf{0}$. Also, by letting $\mathbf{z} = \mathbf{0}$, we must have $\mathbf{p}'\mathbf{Ax} \geq 0$ for each $\mathbf{x} \in E_n$. By choosing $\mathbf{x} = -\mathbf{A}'\mathbf{p}$, it then follows that $-\|\mathbf{A}'\mathbf{p}\|^2 \geq 0$, and thus $\mathbf{A}'\mathbf{p} = \mathbf{0}$. Hence, System 2 has a solution, and the proof is complete.

The separation theorem 2.4.8 can be strengthened to avoid trivial separation where both S_1 and S_2 are contained in the separating hyperplane.

2.4.10 Theorem (Strong Separation)

Let S_1 and S_2 be closed convex sets, and suppose that S_1 is bounded. If $S_1 \cap S_2$ is empty, then there exists a hyperplane that strongly separates S_1 and S_2, that is, there exists a nonzero $\mathbf{p}$ and $\varepsilon > 0$ such that

$$\inf \{\mathbf{p}'\mathbf{x} : \mathbf{x} \in S_1\} \geq \varepsilon + \sup \{\mathbf{p}'\mathbf{x} : \mathbf{x} \in S_2\}$$

Proof

Let $S = S_1 \ominus S_2$ and note that S is convex and that $\mathbf{0}$ L S. We shall show that S is closed. Let $\{\mathbf{x}_k\}$ in S converge to $\mathbf{x}$. By definition of S, $\mathbf{x}_k = \mathbf{y}_k - \mathbf{z}_k$, where $\mathbf{y}_k \in S_1$ and $\mathbf{z}_k \in S_2$. Since S_1 is compact, there is a subsequence $\{\mathbf{y}_k\}_{\mathcal{K}}$ with limit $\mathbf{y}$ in S_1. Since

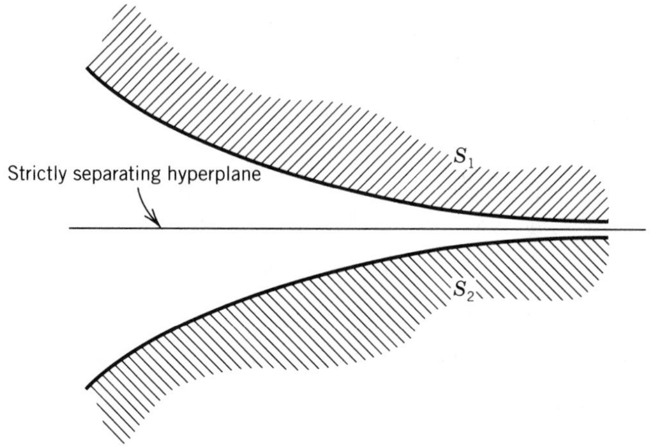

Figure 2.13 Nonexistence of a strongly separating hyperplane.

$\mathbf{y}_k - \mathbf{z}_k \to \mathbf{x}$ and $\mathbf{y}_k \to \mathbf{y}$ for $k \in \mathcal{K}$, then $\mathbf{z}_k \to \mathbf{z}$ for $k \in \mathcal{K}$. Since S_2 is closed, $\mathbf{z} \in S_2$. Therefore, $\mathbf{x} = \mathbf{y} - \mathbf{z}$ with $\mathbf{y} \in S_1$ and $\mathbf{z} \in S_2$. Therefore, $\mathbf{x} \in S$ and, hence, S is closed. By Theorem 2.4.4, there is a nonzero $\mathbf{p}$ and an ε such that $\mathbf{p}'\mathbf{x} \geq \varepsilon$ for each $\mathbf{x} \in S$ and $\mathbf{p}'\mathbf{0} < \varepsilon$. Therefore, $\varepsilon > 0$. By the definition of S, we conclude that $\mathbf{p}'\mathbf{x}_1 \geq \varepsilon + \mathbf{p}'\mathbf{x}_2$ for each $\mathbf{x}_1 \in S_1$ and $\mathbf{x}_2 \in S_2$, and the result follows.

Note the importance of assuming the boundedness of, at least, one of the sets S_1 and S_2 in Theorem 2.4.10. Figure 2.13 illustrates a situation in E_2, where the boundaries of S_1 and S_2 asymptotically approach the strictly separating hyperplane shown therein. Here, S_1 and S_2 are closed convex sets, and $S_1 \cap S_2 = \emptyset$, but there does not exist a hyperplane that strongly separates S_1 and S_2. However, if we bound one of the sets, then we can obtain a strongly separating hyperplane.

As a direct consequence of Theorem 2.4.10, the following corollary gives a strengthened restatement of the theorem.

Corollary 1

Let S_1 and S_2 be nonempty sets in E_n, and suppose that S_1 is bounded. If cl conv $S_1 \cap$ cl conv $S_2 = \emptyset$, then there exists a hyperplane that strongly separates S_1 and S_2.

2.5 Convex Cones and Polarity

In this section, we briefly discuss the notions of convex cones and polar cones. Except for the definition of a (convex) cone, this section may be skipped without loss of continuity.

2.5.1 Definition

A nonempty set C in E_n is called a *cone* with vertex zero if $\mathbf{x} \in C$ implies that $\lambda\mathbf{x} \in C$ for all $\lambda \geq 0$. If, in addition, C is convex, then C is called a *convex cone*.

Figure 2.14 shows an example of a convex cone and an example of a nonconvex cone.

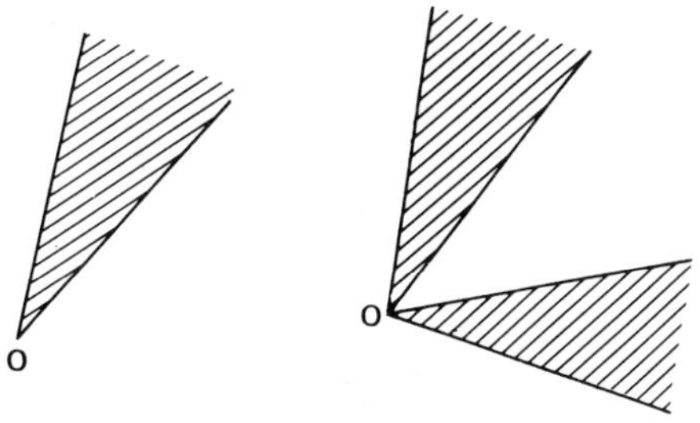

Figure 2.14 Examples of cones.

An important special class of convex cones is that of polar cones defined below and illustrated in Figure 2.15.

2.5.2 Definition

Let S be a nonempty set in E_n. Then, the *polar cone* of S, denoted by S^*, is given by $\{\mathbf{p}:\mathbf{p}'\mathbf{x} \leq 0 \text{ for all } \mathbf{x} \in S\}$. If S is empty, S^* will be interpreted as E_n.

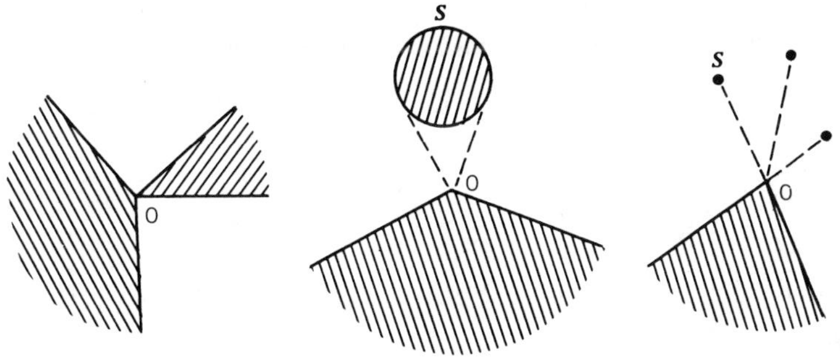

Figure 2.15 Examples of polar cones.

The following lemma, the proof of which is left as an exercise, summarizes some facts about polar cones.

2.5.3 Lemma

Let S, S_1, and S_2 be nonempty sets in E_n. Then, the following statements hold true.

1. S^* is a closed convex cone.
2. $S \subseteq S^{**}$, where S^{**} is the polar cone of S^*.
3. $S_1 \subseteq S_2$ implies that $S_2^* \subseteq S_1^*$.

We now prove an important theorem for closed convex cones. As an application of the theorem, we give another derivation of Farkas' theorem.

2.5.4 Theorem

Let C be a nonempty closed convex cone. Then, $C = C^{**}$.

Proof

Clearly, $C \subseteq C^{**}$. Now let $\mathbf{x} \in C^{**}$, and suppose, by contradiction, that $\mathbf{x}$ L C. By Theorem 2.4.4, there exists a nonzero vector $\mathbf{p}$ and a scalar α such that $\mathbf{p}'\mathbf{y} \leq \alpha$ for all $\mathbf{y} \in C$ and $\mathbf{p}'\mathbf{x} > \alpha$. But since $\mathbf{y} = \mathbf{0} \in C$, then $\alpha \geq 0$, and so $\mathbf{p}'\mathbf{x} > 0$. We now show that $\mathbf{p} \in C^*$. If not, then $\mathbf{p}'\bar{\mathbf{y}} > 0$ for some $\bar{\mathbf{y}} \in C$, and $\mathbf{p}'(\lambda\bar{\mathbf{y}})$ can be made arbitrarily large by choosing λ arbitrarily large. This contradicts the fact that $\mathbf{p}'\mathbf{y} \leq \alpha$ for all $\mathbf{y} \in C$. Therefore, $\mathbf{p} \in C^*$. Since $\mathbf{x} \in C^{**} \equiv \{\mathbf{u}:\mathbf{u}'\mathbf{v} \leq 0 \text{ for all } \mathbf{v} \in C^*\}$, then $\mathbf{p}'\mathbf{x} \leq 0$. This contradicts the fact that $\mathbf{p}'\mathbf{x} > 0$, and we conclude that $\mathbf{x} \in C$. This completes the proof.

Farkas' Theorem as a Consequence of Theorem 2.5.4

Let $\mathbf{A}$ be an $m \times n$ matrix, and let $C = \{\mathbf{A}'\mathbf{y}:\mathbf{y} \geq \mathbf{0}\}$. Note that C is a closed convex cone. It can be easily verified that $C^* = \{\mathbf{x}:\mathbf{A}\mathbf{x} \leq \mathbf{0}\}$. By the theorem, $\mathbf{c} \in C^{**}$ if and only if $\mathbf{c} \in C$. But $\mathbf{c} \in C^{**}$ means that whenever $\mathbf{x} \in C^*$, then $\mathbf{c}'\mathbf{x} \leq 0$ or, equivalently, $\mathbf{A}\mathbf{x} \leq \mathbf{0}$ implies that $\mathbf{c}'\mathbf{x} \leq 0$. By the definition of C, $\mathbf{c} \in C$ means that $\mathbf{c} = \mathbf{A}'\mathbf{y}$ and $\mathbf{y} \geq \mathbf{0}$. Thus, the result $C = C^{**}$ could be stated as follows: System 1 below is consistent if and only if System 2 has a solution $\mathbf{y}$.

System 1 $\quad \mathbf{A}\mathbf{x} \leq \mathbf{0}$ implies $\mathbf{c}'\mathbf{x} \leq 0$.

System 2 $\quad \mathbf{A}'\mathbf{y} = \mathbf{c},\ \mathbf{y} \geq \mathbf{0}$.

This statement can be put in the more usual and equivalent form of Farkas' theorem. Exactly one of the following two systems has a solution:

System 1 $\quad \mathbf{A}\mathbf{x} \leq \mathbf{0},\ \mathbf{c}'\mathbf{x} > 0 \qquad$ (that is, $\mathbf{c} \notin C^{**} = C$)

System 2 $\quad \mathbf{A}'\mathbf{y} = \mathbf{c},\ \mathbf{y} \geq \mathbf{0} \qquad$ (that is, $\mathbf{c} \in C$)

2.6 Polyhedral Sets, Extreme Points, and Extreme Directions

In this section, we introduce the notions of extreme points and extreme directions for convex sets. We then discuss in more detail their use for the special important case of polyhedral sets.

Polyhedral Sets

Polyhedral sets represent an important special case of convex sets. We have seen from the corollary to Theorem 2.4.4 that any closed convex set is the intersection of all closed half-spaces containing it. In the case of polyhedral sets, only a finite number of half-spaces is needed to represent the set.

2.6.1 Definition

A set S in E_n is called a *polyhedral set* if it is the intersection of a finite number of closed half-spaces; that is, $S = \{\mathbf{x}:\mathbf{p}_i'\mathbf{x} \leq \alpha_i \text{ for } i = 1, \ldots, m\}$, where $\mathbf{p}_i$ is a nonzero vector and α_i is a scalar for $i = 1, \ldots, m$.

Note that a polyhedral set is a closed convex set. Since an equation can be represented by two inequalities, a polyhedral set can be represented by a finite number of inequalities and/or equations. The following are some typical examples of polyhedral sets, where $\mathbf{A}$ is an $m \times n$ matrix and $\mathbf{b}$ is an m vector:

$$S = \{\mathbf{x}: \mathbf{Ax} \leq \mathbf{b}\}$$
$$S = \{\mathbf{x}: \mathbf{Ax} = \mathbf{b}, \mathbf{x} \geq \mathbf{0}\}$$
$$S = \{\mathbf{x}: \mathbf{Ax} \geq \mathbf{b}, \mathbf{x} \geq \mathbf{0}\}$$

Figure 2.16 illustrates the polyhedral set

$$S = \{(x_1, x_2): -x_1 + x_2 \leq 2, x_2 \leq 4, x_1 \geq 0, x_2 \geq 0\}$$

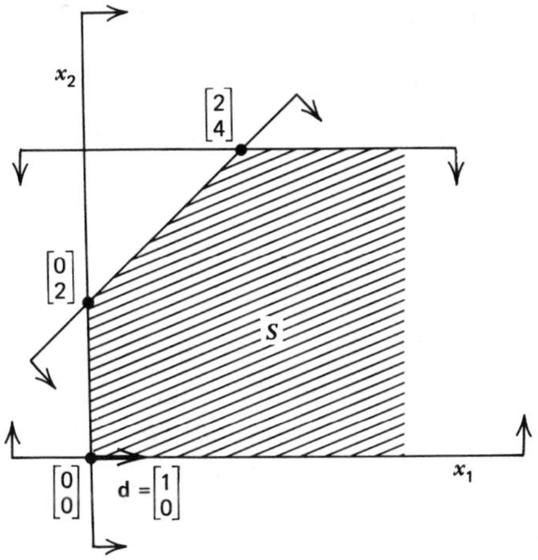

Figure 2.16 A polyhedral set.

Extreme Points and Extreme Directions

We now introduce the concepts of extreme points and extreme directions for convex sets. We then give their full characterizations in the case of polyhedral sets.

2.6.2 Definition

Let S be a nonempty convex set in E_n. A vector $\mathbf{x} \in S$ is called an *extreme point* of S if $\mathbf{x} = \lambda\mathbf{x}_1 + (1 - \lambda)\mathbf{x}_2$ with $\mathbf{x}_1, \mathbf{x}_2 \in S$, and $\lambda \in (0, 1)$ implies that $\mathbf{x} = \mathbf{x}_1 = \mathbf{x}_2$.

The following are some examples of extreme points of convex sets. We denote the set of extreme points by E and illustrate them in Figure 2.17 by dark points or dark lines as indicated.

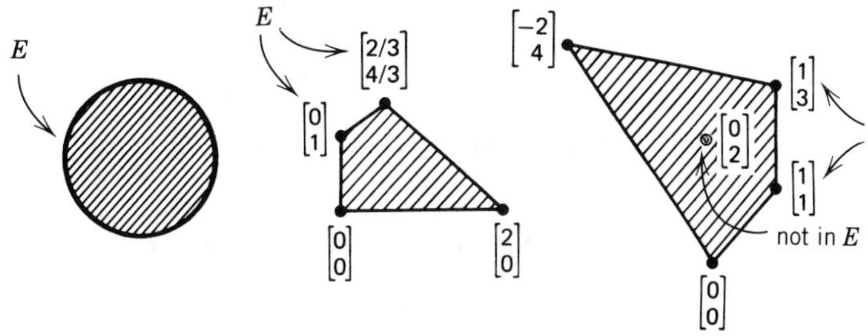

Figure 2.17 Examples of extreme points.

1. $S = \{(x_1, x_2): x_1^2 + x_2^2 \leq 1\}$
   $E = \{(x_1, x_2): x_1^2 + x_2^2 = 1\}$

2. $S = \{(x_1, x_2): x_1 + x_2 \leq 2,\ -x_1 + 2x_2 \leq 2,\ x_1, x_2 \geq 0\}$
   $E = \{(0, 0)', (0, 1)', (2/3, 4/3)', (2, 0)'\}$
3. S is the polytope generated by $(0, 0)'$, $(1, 1)'$, $(1, 3)'$, $(-2, 4)'$, and $(0, 2)'$
   $E = \{(0, 0)', (1, 1)', (1, 3)', (-2, 4)'\}$

From Figure 2.17, we see that any point of the convex set S can be represented as a convex combination of the extreme points. This turns out to be true for compact convex sets. However, for unbounded sets, we may not be able to represent every point in the set as a convex combination of its extreme points. To illustrate, let $S = \{(x_1, x_2): x_2 \geq |x_1|\}$. Note that S is convex and closed. However, S contains only one extreme point, namely, the origin, and obviously S is not equal to the collection of convex combinations of its extreme points. To deal with unbounded sets, the notion of extreme directions is needed.

2.6.3 Definition

Let S be a nonempty, closed convex set in E_n. A nonzero vector $\mathbf{d}$ in E_n is called a *direction*, or a *recession direction*, of S if for each $\mathbf{x} \in S$, $\mathbf{x} + \lambda\mathbf{d} \in S$ for all $\lambda \geq 0$. Two directions $\mathbf{d}_1$ and $\mathbf{d}_2$ of S are called *distinct* if $\mathbf{d}_1 \neq \alpha\mathbf{d}_2$ for any $\alpha > 0$. A direction $\mathbf{d}$ of S is called an *extreme direction* if it cannot be written as a positive linear combination of two distinct directions, that is, if $\mathbf{d} = \lambda_1\mathbf{d}_1 + \lambda_2\mathbf{d}_2$ for $\lambda_1, \lambda_2 > 0$, then $\mathbf{d}_1 = \alpha\mathbf{d}_2$ for some $\alpha > 0$.

To illustrate, consider $S = \{(x_1, x_2): x_2 \geq |x_1|\}$ shown in Figure 2.18. The directions of S are nonzero vectors that make an angle less than or equal to 45° with the vector $(0, 1)'$. In particular, $\mathbf{d}_1 = (1, 1)'$ and $\mathbf{d}_2 = (-1, 1)'$ are two extreme directions of S. Any other direction of S can be represented as a positive linear combination of $\mathbf{d}_1$ and $\mathbf{d}_2$.

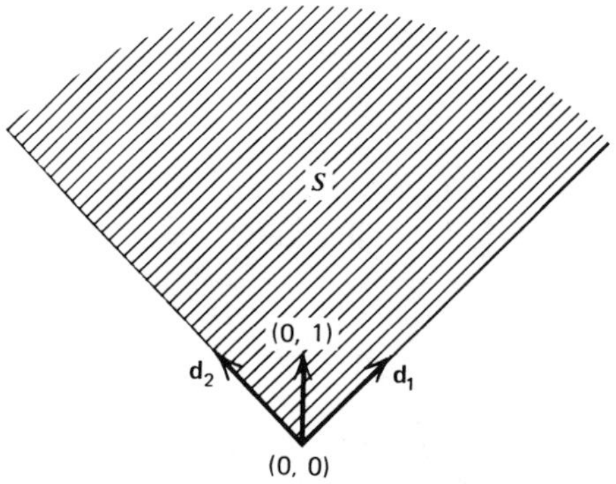

Figure 2.18 Example of extreme directions.

Characterization of Extreme Points and Extreme Directions for Polyhedral Sets

Consider the polyhedral set $S = \{\mathbf{x}: \mathbf{A}\mathbf{x} = \mathbf{b},\ \mathbf{x} \geq \mathbf{0}\}$, where $\mathbf{A}$ is an $m \times n$ matrix and $\mathbf{b}$ is an m vector. We assume that the rank of $\mathbf{A}$ is m. If not, then assuming that $\mathbf{A}\mathbf{x} = \mathbf{b}$ is consistent, we can throw away any redundant equations to obtain a full row rank matrix.

Extreme Points Rearrange the columns of $\mathbf{A}$ so that $\mathbf{A} = [\mathbf{B}, \mathbf{N}]$, where $\mathbf{B}$ is an $m \times m$ matrix of full rank, and $\mathbf{N}$ is an $m \times (n - m)$ matrix. Let $\mathbf{x}_B$ and $\mathbf{x}_N$ be the vectors corresponding to $\mathbf{B}$ and $\mathbf{N}$, respectively. Then, $\mathbf{Ax} = \mathbf{b}$ and $\mathbf{x} \geq \mathbf{0}$ can be rewritten as follows:

$$\mathbf{Bx}_B + \mathbf{Nx}_N = \mathbf{b} \quad \text{and} \quad \mathbf{x}_B \geq \mathbf{0},\ \mathbf{x}_N \geq \mathbf{0}$$

The following theorem gives a necessary and sufficient characterization of an extreme point of S.

2.6.4 Theorem (Characterization of Extreme Points)

Let $S = \{\mathbf{x}: \mathbf{Ax} = \mathbf{b},\ \mathbf{x} \geq \mathbf{0}\}$, where $\mathbf{A}$ is an $m \times n$ matrix of rank m, and $\mathbf{b}$ is an m vector. A point $\mathbf{x}$ is an extreme point of S if and only if $\mathbf{A}$ can be decomposed into $[\mathbf{B}, \mathbf{N}]$ such that

$$\mathbf{x} = \begin{bmatrix} \mathbf{x}_B \\ \mathbf{x}_N \end{bmatrix} = \begin{bmatrix} \mathbf{B}^{-1}\mathbf{b} \\ \mathbf{0} \end{bmatrix}$$

where $\mathbf{B}$ is an $m \times m$ invertible matrix satisfying $\mathbf{B}^{-1}\mathbf{b} \geq \mathbf{0}$. Any such solution is called a *basic feasible solution* (*BFS*) for S.

Proof

Suppose that $\mathbf{A}$ can be decomposed into $[\mathbf{B}, \mathbf{N}]$ with $\mathbf{x} = \begin{bmatrix} \mathbf{B}^{-1}\mathbf{b} \\ \mathbf{0} \end{bmatrix}$ and $\mathbf{B}^{-1}\mathbf{b} \geq \mathbf{0}$. It is obvious that $\mathbf{x} \in S$. Now suppose that $\mathbf{x} = \lambda\mathbf{x}_1 + (1 - \lambda)\mathbf{x}_2$ with $\mathbf{x}_1, \mathbf{x}_2 \in S$ for some $\lambda \in (0, 1)$. In particular, let $\mathbf{x}_1^t = (\mathbf{x}_{11}^t, \mathbf{x}_{12}^t)$ and $\mathbf{x}_2^t = (\mathbf{x}_{21}^t, \mathbf{x}_{22}^t)$. Then,

$$\begin{bmatrix} \mathbf{B}^{-1}\mathbf{b} \\ \mathbf{0} \end{bmatrix} = \lambda \begin{bmatrix} \mathbf{x}_{11} \\ \mathbf{x}_{12} \end{bmatrix} + (1 - \lambda) \begin{bmatrix} \mathbf{x}_{21} \\ \mathbf{x}_{22} \end{bmatrix}$$

Since $\mathbf{x}_{12}, \mathbf{x}_{22} \geq \mathbf{0}$ and $\lambda \in (0, 1)$, it follows that $\mathbf{x}_{12} = \mathbf{x}_{22} = \mathbf{0}$. But this implies that $\mathbf{x}_{11} = \mathbf{x}_{21} = \mathbf{B}^{-1}\mathbf{b}$ and, hence, $\mathbf{x} = \mathbf{x}_1 = \mathbf{x}_2$. This shows that $\mathbf{x}$ is an extreme point of S. Conversely, suppose that $\mathbf{x}$ is an extreme point of S. Without loss of generality, suppose that $\mathbf{x} = (x_1, \ldots, x_k, 0, \ldots, 0)^t$, where $x_1, \ldots, x_k$ are positive. We shall first show that $\mathbf{a}_1, \ldots, \mathbf{a}_k$ are linearly independent. By contradiction, suppose that there exist scalars $\lambda_1, \ldots, \lambda_k$ not all zero such that $\sum_{j=1}^{k} \lambda_j \mathbf{a}_j = \mathbf{0}$. Let $\boldsymbol{\lambda} = (\lambda_1, \ldots, \lambda_k, 0, \ldots, 0)^t$. Construct the following two vectors where $\alpha > 0$ is chosen such that $\mathbf{x}_1, \mathbf{x}_2 \geq \mathbf{0}$:

$$\mathbf{x}_1 = \mathbf{x} + \alpha\boldsymbol{\lambda} \quad \text{and} \quad \mathbf{x}_2 = \mathbf{x} - \alpha\boldsymbol{\lambda}$$

Note that

$$\mathbf{Ax}_1 = \mathbf{Ax} + \alpha\mathbf{A}\boldsymbol{\lambda} = \mathbf{Ax} + \alpha \sum_{j=1}^{k} \lambda_j \mathbf{a}_j = \mathbf{b}$$

and, similarly, $\mathbf{Ax}_2 = \mathbf{b}$. Therefore, $\mathbf{x}_1, \mathbf{x}_2 \in S$ and, since $\alpha > 0$ and $\lambda \neq 0$, $\mathbf{x}_1$ and $\mathbf{x}_2$ are distinct. Moreover, $\mathbf{x} = \frac{1}{2}\mathbf{x}_1 + \frac{1}{2}\mathbf{x}_2$. This contradicts the fact that $\mathbf{x}$ is an extreme point. Thus, $\mathbf{a}_1, \ldots, \mathbf{a}_k$ are linearly independent and, since $\mathbf{A}$ has rank m, $m - k$ columns out of the last $n - k$ columns may be chosen such that they, together with the first k columns, form a linearly independent set of m vectors. To simplify the notation, suppose that these columns are $\mathbf{a}_{k+1}, \ldots, \mathbf{a}_m$. Thus, $\mathbf{A}$ can be written as $\mathbf{A} = [\mathbf{B}, \mathbf{N}]$, where $\mathbf{B} = [\mathbf{a}_1, \ldots, \mathbf{a}_m]$ is of full rank. Furthermore, $\mathbf{B}^{-1}\mathbf{b} = (x_1, \ldots, x_k, 0, \ldots, 0)^t$, and, since $x_j > 0$ for $j = 1, \ldots, k$, then $\mathbf{B}^{-1}\mathbf{b} \geq \mathbf{0}$. This completes the proof.

Corollary

The number of extreme points of S is finite.

Proof

The number of extreme points is less than or equal to

$$\binom{n}{m} = \frac{n!}{m!\,(n-m)!}$$

which is the maximum number of possible ways to choose m columns of $\mathbf{A}$ to form $\mathbf{B}$.

While the above corollary proves that a polyhedral set of the form $\{\mathbf{x}:\mathbf{Ax} = \mathbf{b}, \mathbf{x} \geq \mathbf{0}\}$ has a finite number of extreme points, the following theorem shows that every nonempty polyhedral set of this form must have at least one extreme point.

2.6.5 Theorem (Existence of Extreme Points)

Let $S = \{\mathbf{x}:\mathbf{Ax} = \mathbf{b}, \mathbf{x} \geq \mathbf{0}\}$ be nonempty, where $\mathbf{A}$ is an $m \times n$ matrix of rank m and $\mathbf{b}$ is an m vector. Then, S has at least one extreme point.

Proof

Let $\mathbf{x} \in S$ and, without loss of generality, suppose that $\mathbf{x} = (x_1, \ldots, x_k, 0, \ldots, 0)^t$, where $x_j > 0$ for $j = 1, \ldots, k$. If $\mathbf{a}_1, \ldots, \mathbf{a}_k$ are linearly independent, then $k \leq m$, and $\mathbf{x}$ is an extreme point. Otherwise, there exist scalars $\lambda_1, \ldots, \lambda_k$ with at least one positive component such that $\sum_{j=1}^{k} \lambda_j \mathbf{a}_j = \mathbf{0}$. Define $\alpha > 0$ as follows:

$$\alpha = \underset{1 \leq j \leq k}{\text{minimum}} \left\{ \frac{x_j}{\lambda_j} : \lambda_j > 0 \right\} = \frac{x_i}{\lambda_i}$$

Consider the point $\mathbf{x}'$ whose jth component x_j' is given by

$$x_j' = \begin{cases} x_j - \alpha\lambda_j & \text{for } j = 1, \ldots, k \\ 0 & \text{for } j = k+1, \ldots, n \end{cases}$$

Note that $x_j' \geq 0$ for $j = 1, \ldots, k$ and $x_j' = 0$ for $j = k+1, \ldots, n$. Moreover, $x_i' = 0$, and

$$\sum_{j=1}^{n} \mathbf{a}_j x_j' = \sum_{j=1}^{k} \mathbf{a}_j (x_j - \alpha\lambda_j) = \sum_{j=1}^{k} \mathbf{a}_j x_j - \alpha \sum_{j=1}^{k} \mathbf{a}_j \lambda_j = \mathbf{b} - \mathbf{0} = \mathbf{b}$$

Thus so far, we have constructed a new point $\mathbf{x}'$ with, at most, $k - 1$ positive components. The process is continued until the positive components correspond to linearly independent columns, which results in an extreme point. Thus, we have shown that S has at least one extreme point, and the proof is complete.

Extreme Directions Let $S = \{\mathbf{x}:\mathbf{Ax} = \mathbf{b}, \mathbf{x} \geq \mathbf{0}\} \neq \varnothing$, where $\mathbf{A}$ is an $m \times n$ matrix of rank m. By definition, a nonzero vector $\mathbf{d}$ is a direction of S if $\mathbf{x} + \lambda\mathbf{d} \in S$

for each $\mathbf{x} \in S$ and each $\lambda \geq 0$. Noting the structure of S, it is clear that $\mathbf{d} \neq \mathbf{0}$ is a direction of S if and only if

$$\mathbf{A}\mathbf{d} = \mathbf{0} \qquad \mathbf{d} \geq \mathbf{0}$$

In particular, we are interested in the characterization of extreme directions of S.

2.6.6 Theorem (Characterization of Extreme Directions)

Let $S = \{\mathbf{x}:\mathbf{A}\mathbf{x} = \mathbf{b},\ \mathbf{x} \geq \mathbf{0}\} \neq \varnothing$, where $\mathbf{A}$ is an $m \times n$ matrix of rank m, and $\mathbf{b}$ is an m vector. A vector $\bar{\mathbf{d}}$ is an extreme direction of S if and only if $\mathbf{A}$ can be decomposed into $[\mathbf{B}, \mathbf{N}]$ such that $\mathbf{B}^{-1}\mathbf{a}_j \leq \mathbf{0}$ for some column $\mathbf{a}_j$ of $\mathbf{N}$, and $\bar{\mathbf{d}}$ is a positive multiple of $\mathbf{d} = \begin{pmatrix} -\mathbf{B}^{-1}\mathbf{a}_j \\ \mathbf{e}_j \end{pmatrix}$, where $\mathbf{e}_j$ is an $n - m$ vector of zeros except for a 1 in position j.

Proof

If $\mathbf{B}^{-1}\mathbf{a}_j \leq \mathbf{0}$, then $\mathbf{d} \geq \mathbf{0}$. Furthermore, $\mathbf{A}\mathbf{d} = \mathbf{0}$, so that $\mathbf{d}$ is a direction of S. We now show that $\mathbf{d}$ is indeed an extreme direction. Suppose that $\mathbf{d} = \lambda_1\mathbf{d}_1 + \lambda_2\mathbf{d}_2$, where $\lambda_1, \lambda_2 > 0$ and $\mathbf{d}_1$, $\mathbf{d}_2$ are directions of S. Noting that $n - m - 1$ components of $\mathbf{d}$ are equal to zero, then the corresponding components of $\mathbf{d}_1$ and $\mathbf{d}_2$ must also be equal to zero. Thus, $\mathbf{d}_1$ and $\mathbf{d}_2$ could be written as follows:

$$\mathbf{d}_1 = \alpha_1\begin{pmatrix} \mathbf{d}_{11} \\ \mathbf{e}_j \end{pmatrix} \qquad \mathbf{d}_2 = \alpha_2\begin{pmatrix} \mathbf{d}_{21} \\ \mathbf{e}_j \end{pmatrix}$$

where $\alpha_1, \alpha_2 > 0$. Noting that $\mathbf{A}\mathbf{d}_1 = \mathbf{A}\mathbf{d}_2 = \mathbf{0}$, it can easily be verified that $\mathbf{d}_{11} = \mathbf{d}_{21} = -\mathbf{B}^{-1}\mathbf{a}_j$. Thus, $\mathbf{d}_1$ and $\mathbf{d}_2$ are not distinct, which implies that $\mathbf{d}$ is an extreme direction. Since $\bar{\mathbf{d}}$ is a positive multiple of $\mathbf{d}$, it is also an extreme direction.

Conversely, suppose that $\bar{\mathbf{d}}$ is an extreme direction of S. Without loss of generality, suppose that

$$\bar{\mathbf{d}} = (\bar{d}_1, \ldots, \bar{d}_k, 0, \ldots, \bar{d}_j, \ldots, 0)^t$$

where $\bar{d}_i > 0$ for $i = 1, \ldots, k$ and for $i = j$. We claim that $\mathbf{a}_1, \ldots, \mathbf{a}_k$ are linearly independent. By contradiction, suppose that this were not the case. Then, there would exist scalars $\lambda_1, \ldots, \lambda_k$ not all zero such that $\sum_{i=1}^{k} \lambda_i\mathbf{a}_i = \mathbf{0}$. Let $\boldsymbol{\lambda} = (\lambda_1, \ldots, \lambda_k, \ldots, 0, \ldots, 0)^t$ and choose $\alpha > 0$ sufficiently small such that both

$$\mathbf{d}_1 = \bar{\mathbf{d}} + \alpha\boldsymbol{\lambda} \qquad \text{and} \qquad \mathbf{d}_2 = \bar{\mathbf{d}} - \alpha\boldsymbol{\lambda}$$

are nonnegative. Note that

$$\mathbf{A}\mathbf{d}_1 = \mathbf{A}\bar{\mathbf{d}} + \alpha\mathbf{A}\boldsymbol{\lambda} = \mathbf{0} + \alpha\sum_{i=1}^{k} \mathbf{a}_i\lambda_i = \mathbf{0}$$

Similarly, $\mathbf{A}\mathbf{d}_2 = \mathbf{0}$. Since $\mathbf{d}_1, \mathbf{d}_2 \geq 0$, they are both directions of S. Note also that they are distinct, since $\alpha > 0$ and $\boldsymbol{\lambda} \neq \mathbf{0}$. Furthermore, $\bar{\mathbf{d}} = \frac{1}{2}\mathbf{d}_1 + \frac{1}{2}\mathbf{d}_2$, contradicting the assumption that $\bar{\mathbf{d}}$ is an extreme direction. Thus, $\mathbf{a}_1, \ldots, \mathbf{a}_k$ are linearly independent and, since rank $\mathbf{A}$ is equal to m, it is clear that $k \leq m$. Then, there must exist $m - k$ vectors from among the set of vectors $\{\mathbf{a}_i : i = k + 1, \ldots, n;\ i \neq j\}$ which, together with $\mathbf{a}_1, \ldots, \mathbf{a}_k$, form a linearly independent set of m vectors. Without loss of generality, suppose that these are $\mathbf{a}_{k+1}, \ldots, \mathbf{a}_m$. Denote $[\mathbf{a}_1, \ldots, \mathbf{a}_m]$ by $\mathbf{B}$, and note that $\mathbf{B}$ is invertible. Thus, $\mathbf{0} = \mathbf{A}\bar{\mathbf{d}} = \mathbf{B}\hat{\mathbf{d}} + \mathbf{a}_j\bar{d}_j$, where $\hat{\mathbf{d}}$ is a vector of the first m components of

$\bar{\mathbf{d}}$. Therefore, $\hat{\mathbf{d}} = -\bar{d}_j\mathbf{B}^{-1}\mathbf{a}_j$ and, hence, the vector $\bar{\mathbf{d}}$ is of the form $\bar{\mathbf{d}} = \bar{d}_j\begin{pmatrix} -\mathbf{B}^{-1}\mathbf{a}_j \\ \mathbf{e}_j \end{pmatrix}$. Noting that $\bar{\mathbf{d}} \geq \mathbf{0}$ and that $\bar{d}_j > 0$, then $\mathbf{B}^{-1}\mathbf{a}_j \leq \mathbf{0}$, and the proof is complete.

Corollary

The number of extreme directions of S is finite.

Proof

For each choice of a matrix **B** from **A**, there are $n - m$ possible ways to extract the column $\mathbf{a}_j$ from **N**. Therefore, the maximum number of extreme directions is bounded by

$$(n - m)\binom{n}{m} = \frac{n!}{m!\,(n - m - 1)!}$$

The Representation of Polyhedral Sets in Terms of Extreme Points and Extreme Directions

By definition, a polyhedral set is the intersection of a finite number of half-spaces. This representation may be thought of as an *outer representation*. A polyhedral set can also be described fully by an *inner representation* by means of its extreme points and extreme directions. This fact is fundamental to several linear and nonlinear programming procedures.

The main result can be stated as follows. Let S be a nonempty polyhedral set of the form $\{\mathbf{x}: \mathbf{A}\mathbf{x} = \mathbf{b},\ \mathbf{x} \geq \mathbf{0}\}$. Then, any point in S can be represented as a convex combination of its extreme points plus a nonnegative linear combination of its extreme directions. Of course, if S is bounded, then it contains no directions, and so any point in S can be described as a convex combination of its extreme points.

In Theorem 2.6.7 below, it is implicitly assumed that the extreme points and the extreme directions of S are finite in number. This fact follows from the corollaries to Theorems 2.6.4 and 2.6.6. (See Exercises 2.45–2.47 for an alternative, constructive derivation of this theorem.)

2.6.7 Theorem (Representation Theorem)

Let S be a nonempty polyhedral set in E_n of the form $\{\mathbf{x}: \mathbf{A}\mathbf{x} = \mathbf{b} \text{ and } \mathbf{x} \geq \mathbf{0}\}$, where **A** is an $m \times n$ matrix with rank m. Let $\mathbf{x}_1, \ldots, \mathbf{x}_k$ be the extreme points of S and $\mathbf{d}_1, \ldots, \mathbf{d}_l$ be the extreme directions of S. Then, $\mathbf{x} \in S$ if and only if $\mathbf{x}$ can be written as

$$\mathbf{x} = \sum_{j=1}^{k} \lambda_j \mathbf{x}_j + \sum_{j=1}^{l} \mu_j \mathbf{d}_j$$

$$\sum_{j=1}^{k} \lambda_j = 1 \tag{2.6}$$

$$\lambda_j \geq 0 \qquad \text{for } j = 1, \ldots, k \tag{2.7}$$

$$\mu_j \geq 0 \qquad \text{for } j = 1, \ldots, l \tag{2.8}$$

Proof

Construct the following set:

$$\wedge = \left\{ \sum_{j=1}^{k} \lambda_j \mathbf{x}_j + \sum_{j=1}^{l} \mu_j \mathbf{d}_j : \sum_{j=1}^{k} \lambda_j = 1, \lambda_j \geq 0 \text{ for all } j, \mu_j \geq 0 \text{ for all } j \right\}$$

Note that $\wedge$ is a closed convex set. Furthermore, by Theorem 2.6.5, S has at least one extreme point and, hence, $\wedge$ is not empty. Also note that $\wedge \subseteq S$. To show that $S \subseteq \wedge$, suppose by contradiction that there is a $\mathbf{z} \in S$ such that $\mathbf{z} \notin \wedge$. By Theorem 2.4.4, there exists a scalar α and a nonzero vector $\mathbf{p}$ in E_n such that

$$\mathbf{p}^t\mathbf{z} > \alpha \tag{2.9}$$

$$\mathbf{p}^t\left(\sum_{j=1}^{k} \lambda_j \mathbf{x}_j + \sum_{j=1}^{l} \mu_j \mathbf{d}_j \right) \leq \alpha$$

for λ_j's and μ_j's, satisfying (2.6), (2.7), and (2.8). Since μ_j can be made arbitrarily large, (2.9) holds true only if $\mathbf{p}^t\mathbf{d}_j \leq 0$ for $j = 1, \ldots, l$. From (2.9), by letting $\mu_j = 0$ for all j, $\lambda_j = 1$, and $\lambda_i = 0$ for $i \neq j$, it follows that $\mathbf{p}^t\mathbf{x}_j \leq \alpha$ for each $j = 1, \ldots, k$. Since $\mathbf{p}^t\mathbf{z} > \alpha$, we have $\mathbf{p}^t\mathbf{z} > \mathbf{p}^t\mathbf{x}_j$, for all j. Summarizing, there exists a nonzero vector $\mathbf{p}$ such that

$$\mathbf{p}^t\mathbf{z} > \mathbf{p}^t\mathbf{x}_j \qquad \text{for } j = 1, \ldots, k \tag{2.10}$$

$$\mathbf{p}^t\mathbf{d}_j \leq 0 \qquad \text{for } j = 1, \ldots, l \tag{2.11}$$

Consider the extreme point $\bar{\mathbf{x}}$ defined as follows:

$$\mathbf{p}^t\bar{\mathbf{x}} = \underset{1 \leq j \leq k}{\text{maximum}}\ \mathbf{p}^t\mathbf{x}_j \tag{2.12}$$

Since $\bar{\mathbf{x}}$ is an extreme point, by Theorem 2.6.4, $\bar{\mathbf{x}} = \begin{pmatrix} \mathbf{B}^{-1}\mathbf{b} \\ \mathbf{0} \end{pmatrix}$ where $\mathbf{A} = [\mathbf{B}, \mathbf{N}]$ and $\mathbf{B}^{-1}\mathbf{b} \geq \mathbf{0}$. Without loss of generality assume that $\mathbf{B}^{-1}\mathbf{b} > \mathbf{0}$ (see Exercise 2.44). Since $\mathbf{z} \in S$, then $\mathbf{Az} = \mathbf{b}$ and $\mathbf{z} \geq \mathbf{0}$. Therefore, $\mathbf{Bz}_B + \mathbf{Nz}_N = \mathbf{b}$ and, hence, $\mathbf{z}_B = \mathbf{B}^{-1}\mathbf{b} - \mathbf{B}^{-1}\mathbf{Nz}_N$, where $\mathbf{z}^t$ is decomposed into $(\mathbf{z}_B^t, \mathbf{z}_N^t)$. From (2.10), we have $\mathbf{p}^t\mathbf{z} - \mathbf{p}^t\bar{\mathbf{x}} > 0$, and decomposing $\mathbf{p}^t$ into $(\mathbf{p}_B^t, \mathbf{p}_N^t)$, we get

$$\begin{aligned} 0 &< \mathbf{p}^t\mathbf{z} - \mathbf{p}^t\bar{\mathbf{x}} \\ &= \mathbf{p}_B^t(\mathbf{B}^{-1}\mathbf{b} - \mathbf{B}^{-1}\mathbf{Nz}_N) + \mathbf{p}_N^t\mathbf{z}_N - \mathbf{p}_B^t\mathbf{B}^{-1}\mathbf{b} \\ &= (\mathbf{p}_N^t - \mathbf{p}_B^t\mathbf{B}^{-1}\mathbf{N})\mathbf{z}_N \end{aligned} \tag{2.13}$$

Since $\mathbf{z}_N \geq \mathbf{0}$, from (2.13) it follows that there is a component $j \geq m + 1$ such that $z_j > 0$ and $p_j - \mathbf{p}_B^t\mathbf{B}^{-1}\mathbf{a}_j > 0$. We first show that $\mathbf{y}_j = \mathbf{B}^{-1}\mathbf{a}_j \not\leq \mathbf{0}$. By contradiction, suppose that $\mathbf{y}_j \leq \mathbf{0}$. Consider the vector $\mathbf{d}_j = \begin{bmatrix} -\mathbf{y}_j \\ \mathbf{e}_j \end{bmatrix}$, where $\mathbf{e}_j$ is an $n - m$ dimensional unit vector with 1 at position j. By Theorem 2.6.6, $\mathbf{d}_j$ is an extreme direction of S. From (2.11), $\mathbf{p}^t\mathbf{d}_j \leq 0$, that is, $-\mathbf{p}_B^t\mathbf{B}^{-1}\mathbf{a}_j + p_j \leq 0$, which contradicts the assertion that $p_j - \mathbf{p}_B^t\mathbf{B}^{-1}\mathbf{a}_j > 0$. Therefore, $\mathbf{y}_j \not\leq \mathbf{0}$, and we can construct the following vector:

$$\mathbf{x} = \begin{pmatrix} \bar{\mathbf{b}} \\ \mathbf{0} \end{pmatrix} + \lambda \begin{pmatrix} -\mathbf{y}_j \\ \mathbf{e}_j \end{pmatrix}$$

where $\bar{\mathbf{b}}$ is given by $\mathbf{B}^{-1}\mathbf{b}$ and λ is given by

$$\lambda = \underset{1 \le i \le m}{\text{minimum}} \left\{ \frac{\bar{b}_i}{y_{ij}} : y_{ij} > 0 \right\} = \frac{\bar{b}_r}{y_{rj}} > 0$$

Note that $\mathbf{x} \ge \mathbf{0}$ has, at most, m positive components, where the rth component drops to zero and the jth component is given by λ. The vector $\mathbf{x}$ belongs to S, since $\mathbf{Ax} = \mathbf{B}(\mathbf{B}^{-1}\mathbf{b} - \lambda \mathbf{B}^{-1}\mathbf{a}_j) + \lambda \mathbf{a}_j = \mathbf{b}$. Since $y_{rj} \neq 0$, it can be shown that the vectors $\mathbf{a}_1, \ldots, \mathbf{a}_{r-1}, \mathbf{a}_{r+1}, \ldots, \mathbf{a}_m, \mathbf{a}_j$ are linearly independent. Therefore, by Theorem 2.6.4, $\mathbf{x}$ is an extreme point; that is, $\mathbf{x} \in \{\mathbf{x}_1, \mathbf{x}_2, \ldots, \mathbf{x}_k\}$. Furthermore,

$$\begin{aligned} \mathbf{p}^t\mathbf{x} &= (\mathbf{p}_B^t, \mathbf{p}_N^t)\begin{pmatrix} \bar{\mathbf{b}} - \lambda \mathbf{y}_j \\ \lambda \mathbf{e}_j \end{pmatrix} \\ &= \mathbf{p}_B^t\bar{\mathbf{b}} - \lambda \mathbf{p}_B^t\mathbf{y}_j + \lambda p_j \\ &= \mathbf{p}^t\bar{\mathbf{x}} + \lambda(p_j - \mathbf{p}_B^t\mathbf{B}^{-1}\mathbf{a}_j) \end{aligned}$$

Since $\lambda > 0$ and $p_j - \mathbf{p}_B^t\mathbf{B}^{-1}\mathbf{a}_j > 0$, then $\mathbf{p}^t\mathbf{x} > \mathbf{p}^t\bar{\mathbf{x}}$. Thus, we have constructed an extreme point $\mathbf{x}$ such that $\mathbf{p}^t\mathbf{x} > \mathbf{p}^t\bar{\mathbf{x}}$, which contradicts (2.12). This contradiction asserts that $\mathbf{z}$ must belong to Λ, and the proof is complete.

Corollary (Existence of Extreme Directions)

Let S be a nonempty polyhedral set of the form $\{\mathbf{x} : \mathbf{Ax} = \mathbf{b}, \mathbf{x} \ge \mathbf{0}\}$ where $\mathbf{A}$ is an $m \times n$ matrix with rank m. Then, S has at least one extreme direction if and only if it is unbounded.

Proof

If S has an extreme direction, then it is obviously unbounded. Now suppose that S is unbounded and, by contradiction, suppose that S has no extreme directions. Using the theorem and the Schwartz inequality, it follows that

$$\|\mathbf{x}\| = \left\| \sum_{j=1}^{k} \lambda_j \mathbf{x}_j \right\| \le \sum_{j=1}^{k} \lambda_j \|\mathbf{x}_j\| \le \sum_{j=1}^{k} \|\mathbf{x}_j\|$$

for any $\mathbf{x} \in S$. However, this violates the unboundedness assumption. Therefore, S has at least one extreme direction, and the proof is complete.

2.7 Linear Programming and the Simplex Method

A linear programming problem is the minimization or the maximization of a linear function over a polyhedral set. Many problems can be formulated as, or approximated by, linear programs. Also, linear programming is often used in the process of solving nonlinear and discrete problems. In this section, we describe the well-known simplex method for solving linear programming problems. The method is mainly based on exploiting the extreme points and directions of the polyhedral set defining the problem. Several other algorithms developed in this book can also be specialized to solve linear programming problems. In particular, Chapter 9 describes an efficient (polynomial-time) primal-dual, interior-point, path-following algorithm that competes favorably with the simplex method.

Consider the following linear programming problem:

$$\begin{array}{ll} \text{Minimize} & \mathbf{c}'\mathbf{x} \\ \text{subject to} & \mathbf{x} \in S \end{array}$$

where S is a polyhedral set in E_n. The set S is called the *constraint set*, or the *feasible region*, and the linear function $\mathbf{c}'\mathbf{x}$ is called the *objective function*.

The optimum objective function value of a linear programming problem may be finite or unbounded. We give below a necessary and sufficient condition for a finite optimal solution. The importance of the concepts of extreme points and extreme directions in linear programming will be evident from the theorem.

2.7.1 Theorem (Optimality Conditions in Linear Programming)

Consider the following linear programming problem: Minimize $\mathbf{c}'\mathbf{x}$, subject to $\mathbf{A}\mathbf{x} = \mathbf{b}$, $\mathbf{x} \geq \mathbf{0}$. Here, $\mathbf{c}$ is an n vector, $\mathbf{A}$ is an $m \times n$ matrix of rank m, and $\mathbf{b}$ is an m vector. Suppose that the feasible region is not empty, and let $\mathbf{x}_1, \mathbf{x}_2, \ldots, \mathbf{x}_k$ be the extreme points and $\mathbf{d}_1, \ldots, \mathbf{d}_l$ be the extreme directions of the feasible region. A necessary and sufficient condition for a finite optimal solution is that $\mathbf{c}'\mathbf{d}_j \geq 0$ for $j = 1, \ldots, l$. If this condition holds true, then there exists an extreme point $\mathbf{x}_i$ that solves the problem.

Proof

By Theorem 2.6.7, $\mathbf{A}\mathbf{x} = \mathbf{b}$ and $\mathbf{x} \geq \mathbf{0}$ if and only if

$$\mathbf{x} = \sum_{j=1}^{k} \lambda_j \mathbf{x}_j + \sum_{j=1}^{l} \mu_j \mathbf{d}_j$$

$$\sum_{j=1}^{k} \lambda_j = 1$$

$$\lambda_j \geq 0 \qquad \text{for } j = 1, \ldots, k$$

$$\mu_j \geq 0 \qquad \text{for } j = 1, \ldots, l$$

Therefore, the linear programming problem can be stated as follows:

$$\begin{array}{lll} \text{Minimize} & \mathbf{c}'\left(\sum_{j=1}^{k} \lambda_j \mathbf{x}_j + \sum_{j=1}^{l} \mu_j \mathbf{d}_j\right) & \\ \text{subject to} & \sum_{j=1}^{k} \lambda_j = 1 & \\ & \lambda_j \geq 0 & \text{for } j = 1, \ldots, k \\ & \mu_j \geq 0 & \text{for } j = 1, \ldots, l \end{array}$$

Note that if $\mathbf{c}'\mathbf{d}_j < 0$ for some j, then μ_j can be chosen arbitrarily large, leading to an unbounded optimal objective value. This shows that a necessary and sufficient condition for a finite optimum is $\mathbf{c}'\mathbf{d}_j \geq 0$ for $j = 1, \ldots, l$. If this condition holds true, then, in order to minimize the objective function, we may choose $\mu_j = 0$ for $j = 1, \ldots, l$, and the problem reduces to minimizing $\mathbf{c}'(\sum_{j=1}^{k} \lambda_j \mathbf{x}_j)$ subject to $\sum_{j=1}^{k} \lambda_j = 1$ and $\lambda_j \geq 0$ for $j = 1, \ldots, k$. It is clear that the optimal solution to this latter problem is finite and found by letting $\lambda_i = 1$ and $\lambda_j = 0$ for $j \neq i$, where the index i is given by $\mathbf{c}'\mathbf{x}_i = \text{minimum}_{1 \leq j \leq k}\, \mathbf{c}'\mathbf{x}_j$. Thus, there exists an optimal extreme point, and the proof is complete.

From the above theorem, at least for the case in which the feasible region is bounded, one may be tempted to calculate $\mathbf{c}'\mathbf{x}_j$ for $j = 1, \ldots, k$ and then find minimum$_{1 \le j \le k}$ $\mathbf{c}'\mathbf{x}_j$. Even though this is theoretically possible, it is computationally not advisable because the number of extreme points is usually prohibitively large.

The Simplex Method

The simplex method is a systematic procedure for solving a linear programming problem by moving from an extreme point to an extreme point with a better (at least not worse) objective function value. This process continues until an optimal extreme point is reached and recognized, or else, until an extreme direction $\mathbf{d}$ with $\mathbf{c}'\mathbf{d} < 0$ is found. In the latter case, we conclude that the objective value is unbounded, and we declare the *problem* to be "unbounded." Note that the unboundedness of the feasible region is a necessary, but not sufficient, condition for the problem to be unbounded.

Consider the following linear programming problem in which the polyhedral set is defined in terms of equations and variables that are restricted to be nonnegative:

$$\begin{aligned} &\text{Minimize} && \mathbf{c}'\mathbf{x} \\ &\text{subject to} && \mathbf{A}\mathbf{x} = \mathbf{b} \\ & && \mathbf{x} \ge \mathbf{0} \end{aligned}$$

Note that any polyhedral set can be put in the above *standard format*. For example, an inequality of the form $\sum_{j=1}^{n} a_{ij}x_j \le b_i$ can be transformed into an equation by adding the nonnegative *slack variable* s_i, so that $\sum_{j=1}^{n} a_{ij}x_j + s_i = b_i$. Also, an unrestricted variable x_j can be replaced by the difference of two nonnegative variables; that is, $x_j = x_j^+ - x_j^-$, where $x_j^+, x_j^- \ge 0$. These and other manipulations could be used to put the problem in the above format. We shall assume for the time being that the constraint set admits at least one feasible point and that the rank of $\mathbf{A}$ is equal to m.

By Theorem 2.7.1, at least in the case of a finite optimal solution, it suffices to concentrate on extreme points. Suppose that we have an extreme point $\bar{\mathbf{x}}$. By Theorem 2.6.4, this point is characterized by a decomposition of $\mathbf{A}$ into $[\mathbf{B}, \mathbf{N}]$, where $\mathbf{B} = [\mathbf{a}_{B_1}, \ldots, \mathbf{a}_{B_m}]$ is an $m \times m$ matrix of full rank called the *basis*, and $\mathbf{N}$ is an $m \times (n - m)$ matrix. By Theorem 2.6.4, note that $\bar{\mathbf{x}}$ could be written as $\bar{\mathbf{x}}' = (\bar{\mathbf{x}}_B', \bar{\mathbf{x}}_N') = (\bar{\mathbf{b}}, \mathbf{0}')$, where $\bar{\mathbf{b}} = \mathbf{B}^{-1}\mathbf{b} \ge \mathbf{0}$. The variables corresponding to the basis $\mathbf{B}$ are called *basic variables* and are denoted by $x_{B_1}, \ldots, x_{B_m}$, whereas the variables corresponding to $\mathbf{N}$ are called *nonbasic variables*. Now let us consider a point $\mathbf{x}$ satisfying $\mathbf{A}\mathbf{x} = \mathbf{b}$ and $\mathbf{x} \ge \mathbf{0}$. Decompose $\mathbf{x}'$ into $(\mathbf{x}_B', \mathbf{x}_N')$ and note that $\mathbf{x}_B, \mathbf{x}_N \ge \mathbf{0}$. Also, $\mathbf{A}\mathbf{x} = \mathbf{b}$ can be written as $\mathbf{B}\mathbf{x}_B + \mathbf{N}\mathbf{x}_N = \mathbf{b}$. Hence,

$$\mathbf{x}_B = \mathbf{B}^{-1}\mathbf{b} - \mathbf{B}^{-1}\mathbf{N}\mathbf{x}_N \tag{2.14}$$

Then, using (2.14),

$$\begin{aligned} \mathbf{c}'\mathbf{x} &= \mathbf{c}_B'\mathbf{x}_B + \mathbf{c}_N'\mathbf{x}_N \\ &= \mathbf{c}_B'\mathbf{B}^{-1}\mathbf{b} + (\mathbf{c}_N' - \mathbf{c}_B'\mathbf{B}^{-1}\mathbf{N})\mathbf{x}_N \\ &= \mathbf{c}'\bar{\mathbf{x}} + (\mathbf{c}_N' - \mathbf{c}_B'\mathbf{B}^{-1}\mathbf{N})\mathbf{x}_N \end{aligned} \tag{2.15}$$

Hence, if $\mathbf{c}_N' - \mathbf{c}_B'\mathbf{B}^{-1}\mathbf{N} \ge \mathbf{0}$, then, since $\mathbf{x}_N \ge \mathbf{0}$, we have $\mathbf{c}'\mathbf{x} \ge \mathbf{c}'\bar{\mathbf{x}}$ and that $\bar{\mathbf{x}}$ is an optimal extreme point. On the other hand, suppose that $\mathbf{c}_N' - \mathbf{c}_B'\mathbf{B}^{-1}\mathbf{N} \not\ge \mathbf{0}$. In particular, suppose that the jth component $\mathbf{c}_j - \mathbf{c}_B'\mathbf{B}^{-1}\mathbf{a}_j$ is negative. Consider $\mathbf{x} = \bar{\mathbf{x}} + \lambda\mathbf{d}_j$, where

$$\mathbf{d}_j = \begin{pmatrix} -\mathbf{B}^{-1}\mathbf{a}_j \\ \mathbf{e}_j \end{pmatrix}$$

where $\mathbf{e}_j$ is an $n - m$ unit vector with a 1 at position j. Then, from (2.15),

$$\mathbf{c}^t\mathbf{x} = \mathbf{c}^t\bar{\mathbf{x}} + \lambda(c_j - \mathbf{c}_B^t\mathbf{B}^{-1}\mathbf{a}_j) \tag{2.16}$$

and we get $\mathbf{c}^t\mathbf{x} < \mathbf{c}^t\bar{\mathbf{x}}$ for $\lambda > 0$, since $c_j - \mathbf{c}_B^t\mathbf{B}^{-1}\mathbf{a}_j < 0$. We now consider the following two cases, where $\mathbf{y}_j = \mathbf{B}^{-1}\mathbf{a}_j$.

Case 1: $\mathbf{y}_j \leq \mathbf{0}$. Note that $\mathbf{A}\mathbf{d}_j = \mathbf{0}$, and since $\mathbf{A}\bar{\mathbf{x}} = \mathbf{b}$, then $\mathbf{A}\mathbf{x} = \mathbf{b}$ for $\mathbf{x} = \bar{\mathbf{x}} + \lambda\mathbf{d}_j$ and for all values of λ. Hence, $\mathbf{x}$ is feasible if and only if $\mathbf{x} \geq \mathbf{0}$. This obviously holds true for all $\lambda \geq 0$ if $\mathbf{y}_j \leq \mathbf{0}$. Thus, from (2.16), the objective function value is unbounded. In this case, we have found an extreme direction $\mathbf{d}_j$ with $\mathbf{c}^t\mathbf{d}_j = c_j - \mathbf{c}_B^t\mathbf{B}^{-1}\mathbf{a}_j < 0$ (see Theorems 2.7.1 and 2.6.6).

Case 2: $\mathbf{y}_j$ P $\mathbf{0}$. Let $\mathbf{B}^{-1}\mathbf{b} = \bar{\mathbf{b}}$, and let λ be defined by

$$\lambda = \underset{1 \leq i \leq m}{\text{minimum}} \left\{ \frac{\bar{b}_i}{y_{ij}} : y_{ij} > 0 \right\} = \frac{\bar{b}_r}{y_{rj}} \geq 0 \tag{2.17}$$

where y_{ij} is the ith component of $\mathbf{y}_j$. In this case, the components of $\mathbf{x} = \bar{\mathbf{x}} + \lambda\mathbf{d}_j$ are given by

$$x_{B_i} = \bar{b}_i - \frac{\bar{b}_r}{y_{rj}} y_{ij} \qquad \text{for } i = 1, \ldots, m$$

$$x_j = \frac{\bar{b}_r}{y_{rj}} \tag{2.18}$$

And all other x_i's are equal to zero.

The positive components of $\mathbf{x}$ can only be $x_{B_1}, \ldots, x_{B_{r-1}}, x_{B_{r+1}}, x_{B_m}$ and x_j. Hence, at most, m components of $\mathbf{x}$ are positive. It is easy to verify that their corresponding columns in $\mathbf{A}$ are linearly independent. Therefore, by Theorem 2.6.4, the point $\mathbf{x}$ is itself an extreme point. In this case, we say that the basic variable x_{B_r} left the basis and the nonbasic variable x_j entered the basis in exchange.

Thus far, we have shown that, given an extreme point, we can check its optimality and stop, or find an extreme direction leading to an unbounded solution, or find an extreme point with a better objective value. The process is then repeated.

Summary of the Simplex Algorithm

Outlined below is a summary of the simplex algorithm for a minimization problem of the form to minimize $\mathbf{c}^t\mathbf{x}$ subject to $\mathbf{A}\mathbf{x} = \mathbf{b}$, $\mathbf{x} \geq \mathbf{0}$. A maximization problem can be either transformed into a minimization problem or else we have to modify step 1 such that we stop if $\mathbf{c}_B^t\mathbf{B}^{-1}\mathbf{N} - \mathbf{c}_N^t \geq \mathbf{0}$, and introduce x_j into the basis if $\mathbf{c}_B^t\mathbf{B}^{-1}\mathbf{a}_j - c_j < 0$.

Initialization Step Find a starting extreme point $\mathbf{x}$ with basis $\mathbf{B}$. If such a point is not readily available, then use artificial variables as discussed later in the section.

Main Step 1. Let $\mathbf{x}$ be an extreme point with basis $\mathbf{B}$. Calculate $\mathbf{c}_B^t\mathbf{B}^{-1}\mathbf{N} - \mathbf{c}_N^t$. If this vector is nonpositive, then stop; $\mathbf{x}$ is an optimal extreme point. Otherwise pick the most positive component $\mathbf{c}_B^t\mathbf{B}^{-1}\mathbf{a}_j - c_j$. If $\mathbf{y}_j = \mathbf{B}^{-1}\mathbf{a}_j \leq \mathbf{0}$, then stop; the objective value is unbounded along the *ray*

$$\left\{\mathbf{x} + \lambda\begin{pmatrix} -\mathbf{y}_j \\ \mathbf{e}_j \end{pmatrix} : \lambda \geq 0\right\}$$

where $\mathbf{e}_j$ is a vector of zeros except for a 1 in position j. If, on the other hand, $\mathbf{y}_j \nleq \mathbf{0}$, then go to step 2.

2. Compute the index r from (2.17) and form the new extreme point $\mathbf{x}$ in (2.18). Form the new basis by deleting the column $\mathbf{a}_{B_r}$ from $\mathbf{B}$ and introducing $\mathbf{a}_j$ in its place. Repeat step 1.

Finite Convergence of the Simplex Method

If at each iteration, that is, one pass through the main step, we have $\bar{\mathbf{b}} = \mathbf{B}^{-1}\mathbf{b} > \mathbf{0}$, then λ, defined by (2.17), would be strictly positive, and the objective value at the current extreme point would be strictly less than that at any of the previous iterations. This would imply that the current point is distinct from those previously generated. Since we have a finite number of extreme points, the simplex algorithm must stop in a finite number of iterations. If, on the other hand, $\bar{b}_r = 0$, then $\lambda = 0$, and we would remain at the same extreme point but with a different basis. In theory, this could happen an infinite number of times and may cause nonconvergence. This phenomenon is called *cycling* and rarely occurs in practice. The problem of cycling can be overcome, but this topic will not be discussed here. Most textbooks on linear programming give detailed procedures for avoiding cycling (see the Notes and References section at the end of this chapter).

Tableau Format of the Simplex Method

Suppose that we have the starting basis $\mathbf{B}$ corresponding to an initial extreme point. The objective function and the constraints can be written as

$$\text{Objective row:} \qquad f - \mathbf{c}_B^t\mathbf{x}_B - \mathbf{c}_N^t\mathbf{x}_N = 0$$

$$\text{Constraint rows:} \qquad \mathbf{B}\mathbf{x}_B + \mathbf{N}\mathbf{x}_N = \mathbf{b}$$

These equations can be displayed in the following *simplex tableau*, where the entries in the RHS column are the right-hand-side constants of the equations.

f	$\mathbf{x}_B^t$	$\mathbf{x}_N^t$	RHS
1	$-\mathbf{c}_B^t$	$-\mathbf{c}_N^t$	0
$\mathbf{0}$	$\mathbf{B}$	$\mathbf{N}$	$\mathbf{b}$

The constraint rows are updated by multiplying by $\mathbf{B}^{-1}$, and the objective row is updated by adding it to $\mathbf{c}_B^t$ times the new constraint rows. We then get the following updated tableau. Note that the basic variables are indicated on the left-hand side, and that $\bar{\mathbf{b}} = \mathbf{B}^{-1}\mathbf{b}$.

	f	$\mathbf{x}_B^t$	$\mathbf{x}_N^t$	RHS
f	1	$\mathbf{0}^t$	$\mathbf{c}_B^t\mathbf{B}^{-1}\mathbf{N} - \mathbf{c}_N^t$	$\mathbf{c}_B^t\bar{\mathbf{b}}$
$\mathbf{x}_B$	$\mathbf{0}$	$\mathbf{I}$	$\mathbf{B}^{-1}\mathbf{N}$	$\bar{\mathbf{b}}$

Note that the values of the basic variables and that of f are recorded on the right-hand side of the tableau. Also, the vector $\mathbf{c}_B^t\mathbf{B}^{-1}\mathbf{N} - \mathbf{c}_N^t$ and the matrix $\mathbf{B}^{-1}\mathbf{N}$ are conveniently stored under the nonbasic variables.

The above tableau displays all the information needed to perform step 1 of the simplex method. If $\mathbf{c}_B^t\mathbf{B}^{-1}\mathbf{N} - \mathbf{c}_N^t \le \mathbf{0}$, then we stop; the current extreme point is optimal. Otherwise, upon examining the objective row, we can pick a nonbasic variable with negative $\mathbf{c}_B^t\mathbf{B}^{-1}\mathbf{a}_j - c_j$. If $\mathbf{B}^{-1}\mathbf{a}_j \le \mathbf{0}$, then we stop; the problem is unbounded. Now suppose that $\mathbf{y}_j = \mathbf{B}^{-1}\mathbf{a}_j$ P $\mathbf{0}$. Since $\bar{\mathbf{b}}$ and $\mathbf{y}_j$ are recorded under RHS and x_j, respectively, then λ in (2.17) can be easily calculated from the tableau. The basic variable x_{B_r}, corresponding to the minimum ratio of (2.17), leaves the basis and x_j enters the basis.

Now we would like to update the tableau to reflect the new basis. This can be done by *pivoting* at the x_{B_r} row and the x_j column, that is, at y_{rj}, as follows:

1. Divide the rth row corresponding to x_{B_r} by y_{rj}.
2. Multiply the new rth row by y_{ij} and subtract from the ith constraint row, for $i = 1, \ldots, m, i \neq r$.
3. Multiply the new rth row by $\mathbf{c}_B^t\mathbf{B}^{-1}\mathbf{a}_j - c_j$ and subtract from the objective row.

The reader can easily verify that the above pivoting operation will update the tableau to reflect the new basis (see Exercise 2.53).

2.7.2 Example

$$\begin{array}{ll} \text{Minimize} & x_1 - 3x_2 \\ \text{subject to} & -x_1 + 2x_2 \le 6 \\ & x_1 + x_2 \le 5 \\ & x_1, \quad x_2 \ge 0 \end{array}$$

The problem is illustrated in Figure 2.19. It is clear that the optimal point is $(\frac{4}{3}, \frac{11}{3})$ and that the corresponding value of the objective function is $-\frac{29}{3}$.

To use the simplex method, we now introduce the two *slack variables* $x_3 \ge 0$ and $x_4 \ge 0$. This leads to the following standard format:

$$\begin{array}{ll} \text{Minimize} & x_1 - 3x_2 \\ \text{subject to} & -x_1 + 2x_2 + x_3 \qquad = 6 \\ & x_1 + x_2 \qquad + x_4 = 5 \\ & x_1, \quad x_2, \quad x_3, \quad x_4 \ge 0 \end{array}$$

Note that $\mathbf{c} = (1, -3, 0, 0)^t$, $\mathbf{b} = \begin{pmatrix} 6 \\ 5 \end{pmatrix}$, and $\mathbf{A} = \begin{bmatrix} -1 & 2 & 1 & 0 \\ 1 & 1 & 0 & 1 \end{bmatrix}$. By choosing $\mathbf{B} = [\mathbf{a}_3, \mathbf{a}_4] = \begin{bmatrix} 1 & 0 \\ 0 & 1 \end{bmatrix}$, we note that $\mathbf{B}^{-1}\mathbf{b} = \mathbf{b} \ge \mathbf{0}$ and, hence, we have a starting extreme point. The corresponding tableau is displayed below:

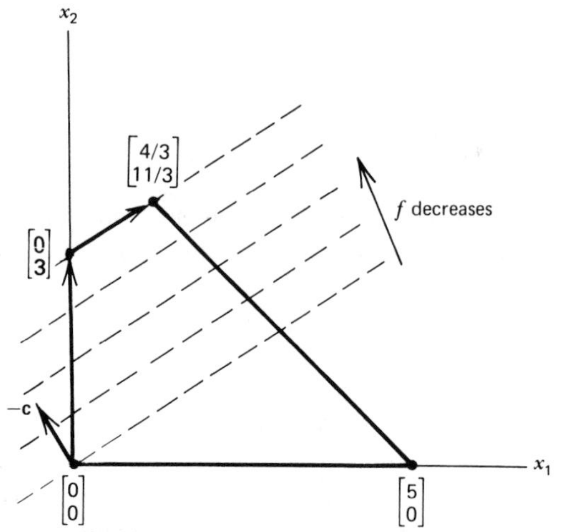

Figure 2.19 A linear programming example.

	f	x_1	x_2	x_3	x_4	RHS
f	1	-1	3	0	0	0
x_3	0	-1	⓶ (2, circled)	1	0	6
x_4	0	1	1	0	1	5

Note that x_2 enters and x_3 leaves the basis. The new basis is $\mathbf{B} = [\mathbf{a}_2, \mathbf{a}_4]$.

	f	x_1	x_2	x_3	x_4	RHS
f	1	$\frac{1}{2}$	0	$-\frac{3}{2}$	0	-9
x_2	0	$-\frac{1}{2}$	1	$\frac{1}{2}$	0	3
x_4	0	$\frac{3}{2}$ (circled)	0	$-\frac{1}{2}$	1	2

Now x_1 enters and x_4 leaves the basis. The new basis is $\mathbf{B} = [\mathbf{a}_2, \mathbf{a}_1]$.

	f	x_1	x_2	x_3	x_4	RHS
f	1	0	0	$-\frac{4}{3}$	$-\frac{1}{3}$	$-\frac{29}{3}$
x_2	0	0	1	$\frac{1}{3}$	$\frac{1}{3}$	$\frac{11}{3}$
x_1	0	1	0	$-\frac{1}{3}$	$\frac{2}{3}$	$\frac{4}{3}$

This solution is optimal since $\mathbf{c}_B^t\mathbf{B}^{-1}\mathbf{N} - \mathbf{c}_N^t \leq \mathbf{0}$. The three points corresponding to the three tableaux are shown in the (x_1, x_2) space in Figure 2.19. We see that the simplex method moved from one extreme point to another until the optimal point was reached.

The Initial Extreme Point

Recall that the simplex method starts with an initial extreme point. From Theorem 2.6.4, finding an initial extreme point of the set $S = \{\mathbf{x}: \mathbf{A}\mathbf{x} = \mathbf{b}, \mathbf{x} \geq \mathbf{0}\}$ involves decomposing

$\mathbf{A}$ into $\mathbf{B}$ and $\mathbf{N}$ with $\mathbf{B}^{-1}\mathbf{b} \geq \mathbf{0}$. In Example 2.7.2 above, an initial extreme point was immediately available. However, in many cases, an initial extreme point may not be conveniently available. This difficulty can be overcome by introducing *artificial variables*.

We discuss briefly two procedures for obtaining the initial extreme point. These are the two-phase method and the big-*M* method. For both methods, the problem is first put in the standard format $\mathbf{Ax} = \mathbf{b}$ and $\mathbf{x} \geq \mathbf{0}$, with the additional requirement that $\mathbf{b} \geq \mathbf{0}$ (if $b_i < 0$, then the ith constraint is multiplied by -1).

The Two-Phase Method In this method, the constraints of the problem are altered by the use of artificial variables so that an extreme point of the new system is at hand. In particular, the constraint system is modified to

$$\mathbf{Ax} + \mathbf{x}_a = \mathbf{b}$$

$$\mathbf{x}, \mathbf{x}_a \geq \mathbf{0}$$

where $\mathbf{x}_a$ is an artificial vector. Obviously, $\mathbf{x} = \mathbf{0}$ and $\mathbf{x}_a = \mathbf{b}$ represent an extreme point of the above system. Since a feasible solution of the original system will be obtained only if $\mathbf{x}_a = \mathbf{0}$, we can use the simplex method itself to minimize the sum of the artifical variables starting from the above extreme point. This leads to the following *Phase I* problem:

$$\begin{array}{ll} \text{Minimize} & \mathbf{e}^t\mathbf{x}_a \\ \text{subject to} & \mathbf{Ax} + \mathbf{x}_a = \mathbf{b} \\ & \mathbf{x}, \mathbf{x}_a \geq \mathbf{0} \end{array}$$

where $\mathbf{e}$ is a vector of 1's. At the end of Phase I, either $\mathbf{x}_a \neq \mathbf{0}$ or $\mathbf{x}_a = \mathbf{0}$. In the former case we conclude that the original system is inconsistent; that is, the feasible region is empty. In the latter case, the artificial variables would drop from the basis† and, hence, we would obtain an extreme point of the original system. Starting with this extreme point, *Phase II* of the simplex method minimizes the original objective $\mathbf{c}^t\mathbf{x}$.

The Big-M Method As in the two-phase method, the constraints are modified by the use of artificial variables so that an extreme point of the new system is immediately available. A large positive cost coefficient M is assigned to each artificial variable so that they will drop to zero level. This leads to the following problem:

$$\begin{array}{ll} \text{Minimize} & \mathbf{c}^t\mathbf{x} + M\mathbf{e}^t x_a \\ \text{subject to} & \mathbf{Ax} + \mathbf{x}_a = \mathbf{b} \\ & \mathbf{x}, \mathbf{x}_a \geq \mathbf{0} \end{array}$$

One can execute the simplex method without actually specifying a numerical value for M by carrying the objective coefficients of M for the nonbasic variables as a separate vector. These coefficients precisely identify with the reduced objective coefficients for the Phase I problem, hence, directly relating the two-phase and the big-*M* methods.

If, at termination, $\mathbf{x}_a = \mathbf{0}$, then we have an optimal solution to the original problem. Otherwise, if $\mathbf{x}_a \neq \mathbf{0}$ at termination of the simplex method, and provided that the variable entering the basis is the one with the most positive coefficient in the objective row, that

†It is possible that some of the artificial variables remain in the basis at zero level at the end of Phase I. This case can be easily treated (see Charnes and Cooper [1961] and Dantzig [1963]).

is, we give priority attention to the coefficients of M or to the Phase I component of the big-M objective function, we can conclude that the system $\mathbf{Ax} = \mathbf{b}$ and $\mathbf{x} \geq \mathbf{0}$ admits no feasible solutions.

Duality in LP

The simplex method affords a simple derivation of a duality theorem for linear programming. Consider the linear program in standard form to minimize $\mathbf{c}^t\mathbf{x}$ subject to $\mathbf{Ax} = \mathbf{b}$ and $\mathbf{x} \geq \mathbf{0}$. Let us refer to this as the primal problem P. The following linear program is called the *dual* of the foregoing primal problem:

$$
\begin{array}{lll}
\text{D:} & \text{Maximize} & \mathbf{b}^t\mathbf{y} \\
& \text{subject to} & \mathbf{A}^t\mathbf{y} \leq \mathbf{c} \\
& & \mathbf{y} \text{ unrestricted}
\end{array}
$$

We then have the following result that intimately relates the pair of linear programs P and D, and permits the solution of one problem to be recovered from that to the other. As evident from the proof given below, this is possible via the simplex method, for example.

2.7.3 Theorem

Let the pair of linear programs P and D be as defined above. Then we have

(a) ***Weak Duality Result.*** $\mathbf{c}^t\mathbf{x} \geq \mathbf{b}^t\mathbf{y}$ for any feasible solution $\mathbf{x}$ to P and any feasible solution $\mathbf{y}$ to D.

(b) ***Unbounded–Infeasible Relationship.*** If P is unbounded, then D is infeasible, and vice versa.

(c) ***Strong Duality Result.*** If both P and D are feasible, then they both have optimal solutions with the same objective value.

Proof

For any pair of feasible solutions $\mathbf{x}$ and $\mathbf{y}$ to P and D, respectively, we have $\mathbf{c}^t\mathbf{x} \geq \mathbf{y}^t\mathbf{Ax} = \mathbf{y}^t\mathbf{b}$. This proves part a of the theorem. Also, if P is unbounded, then D must be infeasible, or else, a feasible solution to D would provide a lower bound on the objective value for P by part a. Similarly, if D is unbounded, then P is infeasible, and this proves part b. Finally suppose that both P and D are feasible. Then, by part b, neither could be unbounded, and so they both have optimal solutions. In particular, let $\bar{\mathbf{x}}^t = (\bar{\mathbf{x}}_B^t, \bar{\mathbf{x}}_N^t)$ be an optimal basic feasible solution to P, where $\bar{\mathbf{x}}_B = \mathbf{B}^{-1}\mathbf{b}$ and $\bar{\mathbf{x}}_N = \mathbf{0}$ by Theorem 2.6.4. Now consider the solution $\bar{\mathbf{y}}^t = \mathbf{c}_B^t\mathbf{B}^{-1}$, where $\mathbf{c}^t = (\mathbf{c}_B^t, \mathbf{c}_N^t)$. We have $\bar{\mathbf{y}}^t\mathbf{A} = \mathbf{c}_B^t\mathbf{B}^{-1}[\mathbf{B}, \mathbf{N}] = [\mathbf{c}_B^t, \mathbf{c}_B^t\mathbf{B}^{-1}\mathbf{N}] \leq [\mathbf{c}_B^t, \mathbf{c}_N^t]$, since $\mathbf{c}_B^t\mathbf{B}^{-1}\mathbf{N} - \mathbf{c}_N^t \leq \mathbf{0}$ by the optimality of the given basic feasible solution. Hence, $\bar{\mathbf{y}}$ is feasible to D. Moreover, we have $\bar{\mathbf{y}}^t\mathbf{b} = \mathbf{c}_B^t\mathbf{B}^{-1}\mathbf{b} = \mathbf{c}^t\bar{\mathbf{x}}$; and so, by part a, since $\mathbf{b}^t\mathbf{y} \leq \mathbf{c}^t\bar{\mathbf{x}}$ for all $\mathbf{y}$ feasible to D, we have that $\mathbf{y}$ solves D with the same optimal objective value as that for P. This completes the proof.

Corollary 1

If D is infeasible, then P is unbounded or infeasible, and vice versa.

Proof

If D is infeasible, then P could not have an optimal solution, or else, as in the proof of the theorem, we would be able to obtain an optimal solution for D, a contradiction.

Hence, P must be either infeasible or unbounded. Similarly, we can argue that if P is infeasible, then D is either unbounded or infeasible.

Corollary 2 (Farkas' Theorem as a Consequence of Theorem 2.7.3)

Let $\mathbf{A}$ be an $m \times n$ matrix and let $\mathbf{c}$ be an n-vector. Then exactly one of the following two systems has a solution:

System 1	$\mathbf{Ax} \leq \mathbf{0}$ and $\mathbf{c}^t\mathbf{x} > 0$	for some $\mathbf{x} \in E_n$
System 2	$\mathbf{A}^t\mathbf{y} = \mathbf{c}$ and $\mathbf{y} \geq \mathbf{0}$	for some $\mathbf{y} \in E_m$

Proof

Consider the linear program P to minimize $\{\mathbf{0}^t\mathbf{y}: \mathbf{A}^t\mathbf{y} = \mathbf{c},\ \mathbf{y} \geq \mathbf{0}\}$. The dual D to this problem is given by maximize $\{\mathbf{c}^t\mathbf{x}: \mathbf{Ax} \leq \mathbf{0}\}$. Then, System 2 has no solution if and only if P is infeasible, and this happens by part a of the theorem and Corollary 1 if and only if D is unbounded, since D is feasible. (For example, $\mathbf{x} = \mathbf{0}$ is a feasible solution.) But because $\mathbf{Ax} \leq \mathbf{0}$ defines a cone, this happens in turn if and only if there exists an $\mathbf{x} \in E_n$ such that $\mathbf{Ax} \leq \mathbf{0}$ and $\mathbf{c}^t\mathbf{x} > 0$. Hence, System 2 has no solution if and only if System 1 has a solution, and this completes the proof.

Corollary 3 (Complementary Slackness Conditions and Characterization of Optimality)

Consider the pair of primal and dual linear programs P and D given above. Let $\bar{\mathbf{x}}$ be a primal feasible solution, and let $\bar{\mathbf{y}}$ be a dual feasible solution. Then, $\bar{\mathbf{x}}$ and $\bar{\mathbf{y}}$ are respectively optimal to P and D if and only if $\bar{v}_j\bar{x}_j = 0$ for $j = 1, \ldots, n$, where $\bar{\mathbf{v}} = (\bar{v}_1, \bar{v}_2, \ldots, \bar{v}_n)^t = \mathbf{c} - \mathbf{A}^t\bar{\mathbf{y}}$ is the vector of slack variables in the dual constraints for the dual solution $\bar{\mathbf{y}}$. (The latter conditions are called *Complementary Slackness Conditions*, and when they hold true, the primal and dual solutions are called complementary slack solutions.) In particular, a given feasible solution is optimal for P if and only if there exists a complementary slack dual feasible solution, and vice versa.

Proof

Since, $\bar{\mathbf{x}}$ and $\bar{\mathbf{y}}$ are, respectively, primal and dual feasible, we have $\mathbf{A}\bar{\mathbf{x}} = \mathbf{b}$, $\bar{\mathbf{x}} \geq \mathbf{0}$, and $\mathbf{A}^t\bar{\mathbf{y}} + \bar{\mathbf{v}} = \mathbf{c}$, $\bar{\mathbf{v}} \geq \mathbf{0}$, where $\bar{\mathbf{v}}$ is the vector of dual slack variables corresponding to $\bar{\mathbf{y}}$. Hence, $\mathbf{c}^t\bar{\mathbf{x}} - \mathbf{b}^t\bar{\mathbf{y}} = (\mathbf{A}^t\bar{\mathbf{y}} + \bar{\mathbf{v}})^t\bar{\mathbf{x}} - (\mathbf{A}\bar{\mathbf{x}})^t\bar{\mathbf{y}} = \bar{\mathbf{v}}^t\bar{\mathbf{x}}$. By Theorem 2.7.3, the solutions $\bar{\mathbf{x}}$ and $\bar{\mathbf{y}}$ are, respectively, optimal to P and D if and only if $\mathbf{c}^t\bar{\mathbf{x}} = \mathbf{b}^t\bar{\mathbf{y}}$. The foregoing statement asserts that this happens if and only if $\bar{\mathbf{v}}^t\bar{\mathbf{x}} = \mathbf{0}$. But $\bar{\mathbf{v}} \geq \mathbf{0}$ and $\bar{\mathbf{x}} \geq \mathbf{0}$. Hence, $\bar{\mathbf{v}}^t\bar{\mathbf{x}} = 0$ if and only if $\bar{v}_j\bar{x}_j = 0$ for all $j = 1, \ldots, n$. We have shown that $\bar{\mathbf{x}}$ and $\bar{\mathbf{y}}$ are, respectively, optimal to P and D if and only if the complementary slackness conditions hold. The final statement of the theorem can now be readily verified using this result along with Theorem 2.7.3. and Corollary 1. This completes the proof.

Exercises

2.1 Let S be a nonempty set in E_n. Show that S is convex if and only if for each integer $k \geq 2$, the following holds true: $\mathbf{x}_1, \ldots, \mathbf{x}_k \in S$ implies that $\sum_{j=1}^{k} \lambda_j\mathbf{x}_j \in S$, where $\sum_{j=1}^{k} \lambda_j = 1$ and $\lambda_j \geq 0$ for $j = 1, \ldots, k$.

2.2 Let S be a convex set in E_n, $\mathbf{A}$ be an $m \times n$ matrix, and α be a scalar. Show that the following two sets are convex.

a. $\mathbf{A}S = \{\mathbf{y}:\mathbf{y} = \mathbf{A}\mathbf{x}, \mathbf{x} \in S\}$
b. $\alpha S = \{\alpha\mathbf{x}:\mathbf{x} \in S\}$

2.3 Let $S_1 = \{\mathbf{x}:x_1 = 0, 0 \le x_2 \le 1\}$ and $S_2 = \{\mathbf{x}:0 \le x_1 \le 1, x_2 = 2\}$. Describe $S_1 \oplus S_2$ and $S_1 \ominus S_2$.

2.4 Prove Lemma 2.12.

2.5 Let S be a closed set. Give an example to show that $H(S)$ is not necessarily closed. Specify a sufficient condition so that $H(S)$ is closed, and prove your assertion.
(*Hint*: Suppose that S is compact.)

2.6 Let S_1 and S_2 be nonempty sets in E_n. Show that $H(S_1 \cap S_2) \subseteq H(S_1) \cap H(S_2)$. Is $H(S_1 \cap S_2) = H(S_1) \cap H(S_1)$ true in general? If not, give a counterexample.

2.7 Prove Lemma 2.1.4.

2.8 Let S be a polytope in E_n. Show that S is a closed, bounded convex set.

2.9 Let S_1 and S_2 be closed convex sets. Prove that $S_1 \oplus S_2$ is convex. Show by an example that $S_1 \oplus S_2$ is not necessarily closed. Prove that compactness of S_1 or S_2 is a sufficient condition for $S_1 \oplus S_2$ to be closed.

2.10 Let $S_1 = \{\lambda\mathbf{d}_1:\lambda \ge 0\}$ and $S_2 = \{\lambda\mathbf{d}_2:\lambda \ge 0\}$, where $\mathbf{d}_1$ and $\mathbf{d}_2$ are nonzero vectors in E_n. Show that $S_1 \oplus S_2$ is a closed convex set.

2.11 A *linear subspace* L of E_n is a subset of E_n such that $\mathbf{x}_1, \mathbf{x}_2 \in L$ implies that $\lambda_1\mathbf{x}_1 + \lambda_2\mathbf{x}_2 \in L$ for all scalars λ_1 and λ_2. The *orthogonal complement* $L^\perp$ is defined by $L^\perp = \{\mathbf{y}:\mathbf{y}'\mathbf{x} = 0 \text{ for all } \mathbf{x} \in L\}$. Show that any vector $\mathbf{x}$ in E_n could be represented uniquely as $\mathbf{x}_1 + \mathbf{x}_2$, where $\mathbf{x}_1 \in L$ and $\mathbf{x}_2 \in L^\perp$. Illustrate by writing the vector (1, 2, 3) as the sum of two vectors in L and $L^\perp$, respectively, where $L = \{(x_1, x_2, x_3):2x_1 + x_2 - x_3 = 0\}$.

2.12 Let S be a polytope in E_n and let $S_j = \{\mu_j\mathbf{d}_j:\mu_j \ge 0\}$, where $\mathbf{d}_j$ is a nonzero vector in E_n for $j = 1, 2, \ldots, k$. Show that $S \oplus S_1 \oplus \ldots \oplus S_k$ is a closed convex set.
(Note that Exercises 2.8 and 2.12 show that the set $\wedge$ in the proof of Theorem 2.6.7 is closed.)

2.13 Identify the closure, interior, and boundary of each of the following convex sets:
a. $S = \{\mathbf{x}:x_1^2 + x_2^2 \le x_3\}$
b. $S = \{\mathbf{x}:1 \le x_1 \le 2, x_2 = 3\}$
c. $S = \{\mathbf{x}:x_1 + x_2 \le 3, -x_1 + x_2 + x_3 \le 5, x_1, x_2, x_3 \ge 0\}$
d. $S = \{\mathbf{x}:x_1 + x_2 = 3, x_1 + x_2 + x_3 \le 6\}$
e. $S = \{\mathbf{x}:x_1^2 + x_2^2 + x_3^2 \le 4, x_1 + x_3 = 1\}$

2.14 Let $S_1 = \{\mathbf{x}:\mathbf{A}_1\mathbf{x} \le \mathbf{b}_1\}$ and $S_2 = \{\mathbf{x}:\mathbf{A}_2\mathbf{x} \le \mathbf{b}_2\}$ be nonempty. Define $S = S_1 \cup S_2$, and let $\bar{S} = \{\mathbf{x}:\mathbf{x} = \mathbf{y} + \mathbf{z}, \mathbf{A}_1\mathbf{y} \le \mathbf{b}_1\lambda_1, \mathbf{A}_2\mathbf{z} \le \mathbf{b}_2\lambda_2, \lambda_1 + \lambda_2 = 1, (\lambda_1, \lambda_2) \ge \mathbf{0}\}$.
a. Assuming that S_1 and S_2 are bounded, show that $H(S) = \bar{S}$.
b. In general, show that $\text{cl } H(S) = \bar{S}$.

2.15 Let $S = \{\mathbf{x}:x_1^2 + x_2^2 + x_3^2 \le 1, x_1^2 - x_2 \le 0\}$ and $\mathbf{y} = (1, 0, 2)'$. Find the minimum distance from $\mathbf{y}$ to S, the unique minimizing point, and a separating hyperplane.

2.16 Prove that exactly one of the following two systems has a solution:
a. $\mathbf{A}\mathbf{x} \ge \mathbf{0}$, $\mathbf{x} \ge \mathbf{0}$, and $\mathbf{c}'\mathbf{x} > 0$
b. $\mathbf{A}'\mathbf{y} \ge \mathbf{c}$ and $\mathbf{y} \le \mathbf{0}$
(*Hint*: Use Farkas' theorem.)

2.17 Show that the system $\mathbf{A}\mathbf{x} \le \mathbf{0}$ and $\mathbf{c}'\mathbf{x} > 0$ has a solution $\mathbf{x}$ in E_3, where $\mathbf{A} = \begin{bmatrix} 1 & -1 & -1 \\ 2 & 2 & 0 \end{bmatrix}$ and $\mathbf{c} = (1, 0, 5)'$.

2.18 Let $\mathbf{A}$ be an $m \times n$ matrix. Using Farkas' theorem, prove that exactly one of the following two systems has a solution:
System 1 $\mathbf{A}\mathbf{x} > \mathbf{0}$
System 2 $\mathbf{A}'\mathbf{y} = \mathbf{0}$, $\mathbf{y} \ge \mathbf{0}$, $\mathbf{y} \ne \mathbf{0}$
(This is Gordan's theorem developed in the text using Theorem 2.4.8.)

2.19 Let $\mathbf{A}$ be an $m \times n$ matrix and $\mathbf{c}$ be an n vector. Show that exactly one of the following two systems has a solution:

System 1 $\mathbf{Ax} = \mathbf{c}$
System 2 $\mathbf{A}'\mathbf{y} = \mathbf{0},\ \mathbf{c}'\mathbf{y} = 1$
(This is a theorem of the alternative credited to Gale.)

2.20 Let $\mathbf{A}$ be an $m \times n$ matrix. Show that the following two systems have solutions $\bar{\mathbf{x}}$ and $\bar{\mathbf{y}}$ such that $\mathbf{A}\bar{\mathbf{x}} + \bar{\mathbf{y}} > \mathbf{0}$:
System 1 $\mathbf{Ax} \geq \mathbf{0}$
System 2 $\mathbf{A}'\mathbf{y} = \mathbf{0},\ \mathbf{y} \geq \mathbf{0}$
(This is an existence theorem credited to Tucker.)

2.21 Let $\mathbf{A}$ be a $p \times n$ matrix and $\mathbf{B}$ be a $q \times n$ matrix. Show that if System 1 below has no solution, then System 2 has a solution:
System 1 $\mathbf{Ax} < \mathbf{0}$ $\mathbf{Bx} = \mathbf{0}$ for some $\mathbf{x} \in E_n$
System 2 $\mathbf{A}'\mathbf{u} + \mathbf{B}'\mathbf{v} = \mathbf{0}$ for some nonzero $(\mathbf{u}, \mathbf{v})$ with $\mathbf{u} \geq \mathbf{0}$
Furthermore, show that if $\mathbf{B}$ has full rank, then exactly one of the systems has a solution. Is this necessarily true if $\mathbf{B}$ is not of full rank? Prove, or give a counterexample.

2.22 Let $\mathbf{A}$ be a $p \times n$ matrix and $\mathbf{B}$ be a $q \times n$ matrix. Show that exactly one of the following systems has a solution:
System 1 $\mathbf{Ax} < \mathbf{0}$ $\mathbf{Bx} = \mathbf{0}$ for some $\mathbf{x} \in E_n$
System 2 $\mathbf{A}'\mathbf{u} + \mathbf{B}'\mathbf{v} = \mathbf{0}$ for some $(\mathbf{u}, \mathbf{v}),\ \mathbf{u} \neq \mathbf{0},\ \mathbf{u} \geq \mathbf{0}$

2.23 Prove Gordan's Theorem 2.4.9 using the linear programming duality approach of Corollary 2 to Theorem 2.7.3.

2.24 Let S_1 and S_2 be convex sets in E_n. Show that there exists a hyperplane that strongly separates S_1 and S_2 if and only if

$$\inf \{\|\mathbf{x}_1 - \mathbf{x}_2\| : \mathbf{x}_1 \in S_2,\ \mathbf{x}_2 \in S_2\} > 0$$

2.25 Let $S_1 = \{\mathbf{x} : x_2 \geq e^{-x_1}\}$ and $S_2 = \{\mathbf{x} : x_2 \leq -e^{-x_1}\}$. Show that S_1 and S_2 are disjoint convex sets, and then find a hyperplane that separates them. Does there exist a hyperplane that strongly separates S_1 and S_2?

2.26 Let S_1 and S_2 be nonempty, disjoint convex sets in E_n. Prove that there exist two nonzero vectors $\mathbf{p}_1$ and $\mathbf{p}_2$ such that

$$\mathbf{p}_1'\mathbf{x}_1 + \mathbf{p}_2'\mathbf{x}_2 \geq 0 \qquad \text{for all } \mathbf{x}_1 \in S_1 \text{ and all } \mathbf{x}_2 \in S_2$$

Can you generalize the result for three or more disjoint convex sets?

2.27 Consider $S = \{\mathbf{x} : x_1^2 + x_2^2 \leq 1\}$. Represent S as the intersection of a collection of half-spaces. Find the half-spaces explicitly.

2.28 Let C be a nonempty set in E_n. Show that C is a convex cone if and only if $\mathbf{x}_1, \mathbf{x}_2 \in C$ implies that $\lambda_1\mathbf{x}_1 + \lambda_2\mathbf{x}_2 \in C$ for all $\lambda_1, \lambda_2, \geq 0$.

2.29 Let C_1 and C_2 be convex cones in E_n. Show that $C_1 \oplus C_2$ is also a convex cone and that $C_1 \oplus C_2 = H(C_1 \cup C_2)$.

2.30 Let S be a nonempty set in E_n and let $\bar{\mathbf{x}} \in S$. Consider the set $C = \{\mathbf{y} : \mathbf{y} = \lambda(\mathbf{x} - \bar{\mathbf{x}}),\ \lambda \geq 0,\ \mathbf{x} \in S\}$.

a. Show that C is a cone and interpret it geometrically.
b. Show that C is convex if S is convex.
c. Suppose that S is closed. Is it necessarily true that C is closed? If not, under what conditions would C be closed?

2.31 Let $C_\varepsilon = \{\mathbf{y} : \mathbf{y} = \lambda(\mathbf{x} - \bar{\mathbf{x}}),\ \lambda \geq 0,\ \mathbf{x} \in S \cap N_\varepsilon(\bar{\mathbf{x}})\}$, where $N_\varepsilon(\bar{\mathbf{x}})$ is an ε-neighborhood around $\bar{\mathbf{x}}$. Let T be the intersection of all such cones, that is, $T = \cap \{C_\varepsilon : \varepsilon > 0\}$. Interpret the cone T geometrically. (T is called the *cone of tangents* of S at $\bar{\mathbf{x}}$ and is discussed in more detail in Chapter 5.)

2.32 Derive an explicit form of the polar C^* of the following cones:

a. $C = \{(x_1, x_2) : 0 \leq x_2 \leq x_1\}$
b. $C = \{(x_1, x_2) : x_2 \geq -|x_1|\}$
c. $C = \{\mathbf{x} : \mathbf{x} = \mathbf{Ap},\ \mathbf{p} \geq \mathbf{0}\}$

2.33 Let S be a nonempty set in E_n. The *polar set* of S, denoted by S_p, is given by $\{\mathbf{y}:\mathbf{y}^t\mathbf{x} \leq 1$ for all $\mathbf{x} \in S\}$.

a. Find the polar sets of the following two sets:

$$\{(x_1, x_2): x_1^2 + x_2^2 \leq 1\}, \text{ and } \{(x_1, x_2): x_1 + x_2 \leq 2, -x_1 + 2x_2 \leq 1, x_1, x_2 \geq 0\}$$

b. Show that S_p is a convex set. Is it necessarily closed?

c. If S is a polyhedral set, is it necessarily true that S_p is also a polyhedral set?

d. Show that if S is a polyhedral set containing the origin then $S = S_{pp}$.

2.34 Let C be a nonempty convex cone in E_n. Show that $C + C^* = E_n$; that is, any point in E_n can be written as a point in the cone C plus a point in its polar cone C^*. Is the representation unique? What if C is a linear subspace?

2.35 Identify the extreme points and extreme directions of the following sets.

a. $S = \{\mathbf{x}: x_2 \geq x_1^2, x_1 + x_2 + x_3 \leq 1\}$

b. $S = \{\mathbf{x}: x_1 + x_2 + x_3 \leq 2, x_1 + x_2 = 1, x_1, x_2, x_3 \geq 0\}$

c. $S = \{\mathbf{x}: x_2 \geq |x_1|, x_1^2 + x_2^2 \leq 1\}$

2.36 Consider the set $S = \{\mathbf{x}: -x_1 + 2x_2 \leq 3, x_1 + x_2 \leq 2, x_2 \leq 1, x_1, x_2 \geq 0\}$. Identify all extreme points and extreme directions. Represent the point $(1, \frac{1}{2})$ as a convex combination of the extreme points plus a nonnegative combination of the extreme directions.

2.37 Let S be a simplex in E_n with vertices $\mathbf{x}_1, \mathbf{x}_2 \ldots, \mathbf{x}_{k+1}$. Show that the extreme points of S consist of its vertices.

2.38 Establish the set of directions for each of the following convex sets:

a. $S = \{(x_1, x_2): x_2 \geq x_1^2\}$

b. $S = \{(x_1, x_2): x_1x_2 \geq 1, x_1 > 0\}$

c. $S = \{(x_1, x_2): |x_1| + |x_2| \leq 1\}$

2.39 Let S be a closed convex set in E_n and let $\bar{\mathbf{x}} \in S$. Suppose that $\mathbf{d}$ is a nonzero vector in E_n and that $\bar{\mathbf{x}} + \lambda\mathbf{d} \in S$ for all $\lambda \geq 0$. Show that $\mathbf{d}$ is a direction of S.

2.40 Find the extreme points and directions of the following polyhedral sets:

a. $S = \{\mathbf{x}: x_1 + x_2 + x_3 \leq 10, -x_1 + 2x_2 = 4, x_1, x_2, x_3 \geq 0\}$

b. $S = \{\mathbf{x}: x_1 + 2x_2 \geq 2, -x_1 + x_2 = 4, x_1, x_2 \geq 0\}$

2.41 Show that $C = \{\mathbf{x}:\mathbf{A}\mathbf{x} \leq \mathbf{0}\}$, where $\mathbf{A}$ is an $m \times n$ matrix, has at most one extreme point, namely, the origin.

2.42 Let $S = \{\mathbf{x}: x_1 + x_2 \leq 1\}$. Find the extreme points and directions of S. Can you represent any point in S as a convex combination of its extreme points plus a nonnegative linear combination of its extreme directions? If not, discuss in relation to Theorem 2.6.7.

2.43 Consider the nonempty unbounded polyhedral set $S = \{\mathbf{x}:\mathbf{A}\mathbf{x} = \mathbf{b}, \mathbf{x} \geq \mathbf{0}\}$, where $\mathbf{A}$ is an $m \times n$ matrix of rank m. Starting with a direction of S, use the characterization of Theorem 2.6.6 to show how an extreme direction of S can be constructed.

2.44 Prove Theorem 2.6.7 if the nondegeneracy assumption $\mathbf{B}^{-1}\mathbf{b} > 0$ is dropped.

2.45 Consider the polyhedral set $S = \{\mathbf{x}:\mathbf{A}\mathbf{x} = \mathbf{b}, \mathbf{x} \geq \mathbf{0}\}$, where $\mathbf{A}$ is an $m \times n$ matrix of rank m. Then, show that $\bar{\mathbf{x}}$ is an extreme point of S as defined by Theorem 2.6.4 if and only if there exist some n linearly independent hyperplanes defining S that are binding at $\bar{\mathbf{x}}$.

2.46 Consider the polyhedral set $S = \{\mathbf{x}:\mathbf{A}\mathbf{x} = \mathbf{b}, \mathbf{x} \geq \mathbf{0}\}$, where $\mathbf{A}$ is an $m \times n$ matrix of rank m and define $D = \{\mathbf{d}:\mathbf{A}\mathbf{d} = \mathbf{0}, \mathbf{d} \geq \mathbf{0}, \mathbf{e}^t\mathbf{d} = 1\}$, where $\mathbf{e}$ is a vector of n 1's. Using the characterization of Theorem 2.6.6, show that $\mathbf{d} \neq \mathbf{0}$ is an extreme direction of S if and only if, when it is normalized to satisfy $\mathbf{e}^t\mathbf{d} = 1$, it is an extreme point of D. Hence, show using Exercise 2.45 that the number of extreme directions is bounded above by $n!/(n - m - 1)!(m + 1)!$. Compare this with the corollary to Theorem 2.6.6.

2.47 Let S be a nonempty polyhedral set defined by $S = \{\mathbf{x}:\mathbf{A}\mathbf{x} = \mathbf{b}, \mathbf{x} \geq \mathbf{0}\}$, where $\mathbf{A}$ is an $m \times n$ matrix of rank m. Consider any nonextreme point feasible solution $\bar{\mathbf{x}} \in S$.

a. Show, using the definition of Exercise 2.45, that starting at $\bar{\mathbf{x}}$, one can constructively recover an extreme point $\hat{\mathbf{x}}$ of S at which the hyperplanes binding at $\bar{\mathbf{x}}$ are also binding at $\hat{\mathbf{x}}$.

b. Assume that S is bounded, and compute $\lambda_{\max} = \max\{\lambda:\hat{\mathbf{x}} + \lambda(\bar{\mathbf{x}} - \hat{\mathbf{x}}) \in S\}$. Show that $\lambda_{max} > 0$ and that at the point $\tilde{\mathbf{x}} = \hat{\mathbf{x}} + \lambda_{max}(\bar{\mathbf{x}} - \hat{\mathbf{x}})$, all the hyperplanes binding at $\bar{\mathbf{x}}$ are also binding and, that in addition, at least one more linearly independent defining hyperplane of S is binding.

c. Assuming that S is bounded and noting how $\bar{\mathbf{x}}$ is represented as a convex combination of the vertex $\hat{\mathbf{x}}$ of S and the point $\tilde{\mathbf{x}} \in S$ at which the number of linearly independent binding hyperplanes is, at least, one more than at $\bar{\mathbf{x}}$, show how $\bar{\mathbf{x}}$ can be constructively represented in terms of the extreme points of S.

d. Now suppose that S is unbounded. Define the nonempty, bounded polytope $\bar{S} = \{\mathbf{x} \in S: \mathbf{e}^t\mathbf{x} \leq M\}$, where $\mathbf{e}$ is a vector of n 1's, and M is large enough so that any extreme point $\hat{\mathbf{x}}$ of S satisfies $\mathbf{e}^t\hat{\mathbf{x}} < M$. Applying part c to $\bar{S}$ and simply using the definitions of extreme points and extreme directions as given in Exercises 2.45 and 2.46, prove the Representation Theorem 2.6.7. (This exercise is based on Sherali [1987].)

2.48 Consider the following problem:

$$\begin{array}{ll} \text{Minimize} & \mathbf{c}^t\mathbf{x} \\ \text{subject to} & \mathbf{A}\mathbf{x} = \mathbf{b} \\ & \mathbf{x} \geq \mathbf{0} \end{array}$$

where $\mathbf{A}$ is an $m \times n$ matrix of rank m. Let $\mathbf{x}$ be an extreme point with corresponding basis $\mathbf{B}$. Furthermore, suppose that $\mathbf{B}^{-1}\mathbf{b} > \mathbf{0}$. Use Farkas' theorem to show that $\mathbf{x}$ is an optimal point if and only if $\mathbf{c}_N^t - \mathbf{c}_B^t\mathbf{B}^{-1}\mathbf{N} \geq \mathbf{0}$.

2.49 Consider the following problem:

$$\begin{array}{ll} \text{Minimize} & \mathbf{c}^t\mathbf{x} \\ \text{subject to} & \mathbf{A}\mathbf{x} = \mathbf{b} \\ & \mathbf{x} \geq \mathbf{0} \end{array}$$

where $\mathbf{A}$ is an $m \times n$ matrix of rank m. Let $\mathbf{x}$ be an extreme point with basis $\mathbf{B}$, and let $\bar{\mathbf{b}} = \mathbf{B}^{-1}\mathbf{b}$. Furthermore, suppose that $\bar{b}_i = 0$ for some component i. Is it possible that $\mathbf{x}$ is an optimal solution even if $c_j - \mathbf{c}_B^t\mathbf{B}^{-1}\mathbf{a}_j < 0$ for some nonbasic x_j? Discuss and give an example if this is possible.

2.50 Solve the following problem by the simplex method:

$$\begin{array}{lrcrcrl} \text{Maximize} & x_1 & + & 3x_2 & + & x_3 & \\ \text{subject to} & x_1 & + & 4x_2 & + & 3x_3 & \leq 12 \\ & -x_1 & + & 2x_2 & - & x_3 & \leq 4 \\ & x_1, & & x_2, & & x_3 & \geq 0 \end{array}$$

What is an optimal dual solution to this problem?

2.51 Consider the set $\{\mathbf{x}: \mathbf{A}\mathbf{x} \leq \mathbf{b}, \mathbf{x} \geq \mathbf{0}\}$, where $\mathbf{A}$ is an $m \times n$ matrix and $\mathbf{b}$ is an m vector. Show that a nonzero vector $\mathbf{d}$ is a direction of the set if and only if $\mathbf{A}\mathbf{d} \leq \mathbf{0}$ and $\mathbf{d} \geq \mathbf{0}$. Show how the simplex method can be used to generate such a direction.

2.52 Consider the following problem:

$$\begin{array}{lrcrl} \text{Minimize} & x_1 & - & 6x_2 & \\ \text{subject to} & x_1 & + & x_2 & \leq 12 \\ & -x_1 & + & 2x_2 & \leq 4 \\ & & & x_2 & \leq 6 \end{array}$$

Find the optimal solution geometrically and verify its optimality by showing that $\mathbf{c}_N^t - \mathbf{c}_B^t\mathbf{B}^{-1}\mathbf{N} \geq \mathbf{0}$. What is an optimal dual solution to this problem?

2.53 Show in detail that pivoting at y_{rj} updates the simplex tableau.

2.54 Solve the following problem by the two-phase simplex method and by the big-M method:

$$\begin{array}{lrcrcrl} \text{Maximize} & -x_1 & - & 2x_2 & + & x_3 & \\ \text{subject to} & x_1 & + & 3x_2 & + & x_3 & \geq 4 \\ & x_1 & + & 2x_2 & - & x_3 & \geq 6 \\ & x_1 & & & + & x_3 & \leq 12 \\ & x_1, & & x_2, & & x_3 & \geq 0 \end{array}$$

Also, identify an optimal dual solution to this problem from the final tableau obtained.

2.55 Let P and D be a pair of primal and dual linear programs as in Theorem 2.7.3. Show that P is infeasible if and only if the homogeneous version of D (with right-hand sides replaced by zeros) is unbounded, and vice versa.

2.56 Let P: minimize $\{\mathbf{c}'\mathbf{x}:\mathbf{A}\mathbf{x} \geq \mathbf{b},\ \mathbf{x} \geq \mathbf{0}\}$ and D: maximize $\{\mathbf{b}'\mathbf{y}:\mathbf{A}'\mathbf{y} \leq \mathbf{c},\ \mathbf{y} \geq \mathbf{0}\}$. Show that P and D are a pair of primal and dual linear programs in the same sense as the pair of primal and dual programs of Theorem 2.7.3. (This pair is sometimes referred to as a symmetric pair of problems in *canonical form.*)

2.57 Use Theorem 2.4.7 to construct an alternative proof for Theorem 2.2.2, by showing how the assumption that $\lambda\mathbf{x}_1 + (1 - \lambda)\mathbf{x}_2 \in \partial S$ readily leads to a contradiction.

Notes and References

In this chapter, we treat the topic of convex sets. This subject was first studied systematically by Minkowski [1991], whose work contains the essence of the important results in this area. The topic of convexity is fully developed in a variety of good texts, and the interested reader may refer to Eggleston [1958], Rockafellar [1970], Stoer and Witzgall [1970], and Valentine [1964] for a more detailed analysis of convex sets.

Section 2.1 presents some basic definitions and develops the Carathéodory theorem, which states that each point in the convex hull of any given set can be represented as the convex combination of $n + 1$ points in the set. This result can be sharpened by using the notion of *dimension of the set*. Using this notion, several Carathéodory-type theorems can be developed. See, for example, Bazaraa and Shetty [1976], Eggleston [1958], and Rockafellar [1970].

In Section 2.2, we develop some topological properties of convex sets related to interior and closure points. Exercise 2.14 gives an important result due to Balas [1974] (also, see Sherali and Shetty [1980c]) that is used to algebraically construct the closure convex hull for disjunctive programs. Section 2.3 presents an important theorem due to Weierstrass that is widely used to establish the existence of optimal solutions.

Section 2.4 presents various types of theorems that separate disjoint convex sets. Support and separation theorems are of special importance in the area of optimization, and are also widely used in game theory, functional analysis, and optimal control theory. An interesting application is the use of these results in coloring problems in graph theory. For further reading on support and separation of convex sets, see Eggleston [1958], Klee [1969], Mangasarian [1969a], Rockafellar [1970], Stoer and Witzgall [1970], and Valentine [1964].

Many of the results in Sections 2.2 and 2.4 can be strengthened by using the notion of *relative interior*. For example, every nonempty convex set has a nonempty relative interior. Furthermore, a hyperplane that properly separates two convex sets exists provided that they have disjoint relative interiors. Also, Theorem 2.2.2 and several of its corollaries can be sharpened using this concept. For a good discussion of relative interiors, see Eggleston [1958], Rockafellar [1970], and Valentine [1964].

In Section 2.5, a brief introduction to polar cones is given. For more details, see Rockafellar [1970]. In Section 2.6 we treat the important special case of polyhedral sets and prove the representation theorem, which states that every point in the set can be represented as a convex combination of the extreme points plus a nonnegative linear combination of the extreme directions. This result was first provided by Motzkin [1936] using a different approach. The representation theorem is also true for closed convex sets that contain no lines. For a proof of this result, see Bazaraa and Shetty [1976] and Rockafellar [1970]. An exhaustive treatment of convex polytopes is given by Grünbaum [1967]. Akgul [1986] and Sherali [1987] provide geometrically motivated constructive proofs for the representation theorem based on definitions of extreme points and directions (see Exercises 2.45–2.47).

In Section 2.7, we present the simplex algorithm for solving linear programming problems. The simplex algorithm was developed by Dantzig in 1947. The efficiency of the simplex algorithm, the advances in computer technology, and the ability of linear programming to model large and complex problems led to the popularity of the simplex method and linear programming. The presentation of the simplex method in Section 2.7 is a natural extension of the material in Section 2.6 on polyhedral sets. For further study of linear programming, see Bazaraa, Jarvis, and Sherali [1990], Charnes and Cooper [1961], Chvatal [1980], Dantzig [1963], Hadley [1962], Murty [1983], and Simonnard [1966].

Chapter 3 Convex Functions and Generalizations

Convex and concave functions have many special and important properties. For example, any local minimum of a convex function over a convex set is also a global minimum. In this chapter, we introduce the important topics of convex and concave functions and develop some of their properties. As we shall learn in this and later chapters, these properties can be utilized in developing suitable optimality conditions and computational schemes for optimization problems that involve convex and concave functions.

The following is an outline of the chapter:

Section 3.1: Definitions and Basic Properties We introduce convex and concave functions and develop some of their basic properties. Continuity of convex functions is proved, and the concept of a directional derivative is introduced.

Section 3.2: Subgradients of Convex Functions A convex function has a convex epigraph and hence has a supporting hyperplane. This leads to the important notion of a subgradient of a convex function.

Section 3.3: Differentiable Convex Functions This section gives some characterizations of differentiable convex functions. These are helpful tools for checking convexity of simple differentiable functions.

Section 3.4: Minima and Maxima of Convex Functions This section is important, since it deals with the questions of minimizing and maximizing a convex function over a convex set. A necessary and sufficient condition for a minimum is developed, and we provide a characterization for the set of alternative optimal solutions. We also show that the maximum occurs at an extreme point. This fact is particularly important if the convex set is polyhedral.

Section 3.5: Generalizations of Convex Functions Various relaxations of convexity and concavity are possible. We present quasiconvex and pseudoconvex functions and develop some of their properties. We then discuss various types of convexity at a point. These types of convexity are sometimes sufficient for optimality, as will be shown in Chapter 4. (This section can be omitted by beginning readers, and later references to generalized convexity properties can largely be substituted simply by convexity.)

3.1 Definitions and Basic Properties

This section deals with some basic properties of convex and concave functions. In particular, we investigate their continuity and differentiability properties.

3.1.1 Definition

Let $f:S \rightarrow E_1$, where S is a nonempty convex set in E_n. The function f is said to be *convex* on S if

$$f(\lambda \mathbf{x}_1 + (1 - \lambda)\mathbf{x}_2) \leq \lambda f(\mathbf{x}_1) + (1 - \lambda) f(\mathbf{x}_2)$$

for each $\mathbf{x}_1, \mathbf{x}_2 \in S$ and for each $\lambda \in (0, 1)$. The function f is called *strictly convex* on S if the above inequality is true as a strict inequality for each distinct $\mathbf{x}_1$ and $\mathbf{x}_2$ in S and for each $\lambda \in (0, 1)$. The function $f:S \rightarrow E_1$ is called *concave* (*strictly concave*) on S if $-f$ is convex (strictly convex) on S.

Now let us consider the geometric interpretation of convex and concave functions. Let $\mathbf{x}_1$ and $\mathbf{x}_2$ be two distinct points in the domain of f, and consider the point $\lambda \mathbf{x}_1 + (1 - \lambda)\mathbf{x}_2$, with $\lambda \in (0, 1)$. Note that $\lambda f(\mathbf{x}_1) + (1 - \lambda) f(\mathbf{x}_2)$ gives the weighted average of $f(\mathbf{x}_1)$ and $f(\mathbf{x}_2)$, while $f[\lambda \mathbf{x}_1 + (1 - \lambda)\mathbf{x}_2]$ gives the value of f at the point $\lambda \mathbf{x}_1 + (1 - \lambda)\mathbf{x}_2$. So, for a convex function f, the value of f at points on the line segment $\lambda \mathbf{x}_1 + (1 - \lambda)\mathbf{x}_2$, is less or equal to the height of the chord joining the points $[\mathbf{x}_1, f(\mathbf{x}_1)]$ and $[\mathbf{x}_2, f(\mathbf{x}_2)]$. For a concave function, the chord is below the function itself. Hence, a function is both convex and concave if and only if it is *affine*. Figure 3.1 shows some examples of convex and concave functions.

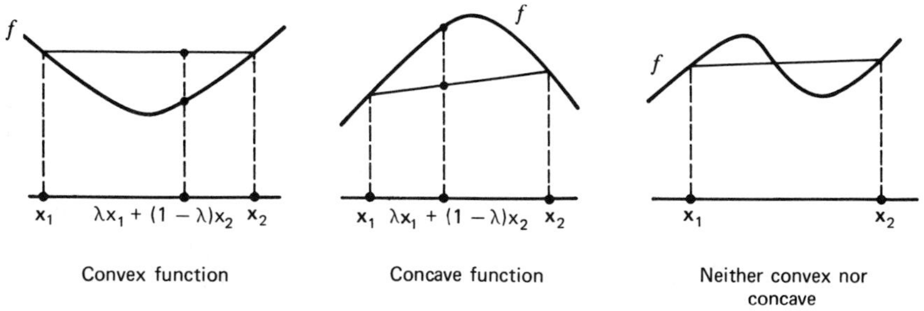

Figure 3.1 Examples of convex and concave functions.

The following are some examples of convex functions. By taking the negatives of these functions, we get some examples of concave functions.

1. $f(x) = 3x + 4$
2. $f(x) = |x|$
3. $f(x) = x^2 - 2x$
4. $f(x) = -x^{1/2}$ if $x \geq 0$
5. $f(x_1, x_2) = 2x_1^2 + x_2^2 - 2x_1x_2$
6. $f(x_1, x_2, x_3) = x_1^4 + 2x_2^2 + 3x_3^2 - 4x_1 - 4x_2x_3$

Note that in each of the above examples, except for example 4, the function f is convex over E_n. In example 4, the function is not defined for $x < 0$. One can readily construct examples of functions that are convex over a region but not over E_n. For instance, $f(x) = x^3$ is not convex over E_1 but is convex over $S = \{x:x \geq 0\}$.

The examples above cite some arbitrary illustrative instances of convex functions. In contrast, we give below some particularly important instances of convex functions that arise very often in practice and that are useful to remember.

1. Let $f_1, f_2, \ldots, f_k: E_n \to E_1$ be convex functions. Then,
   (a) $f(\mathbf{x}) = \sum_{j=1}^{k} \alpha_j f_j(\mathbf{x})$, where $\alpha_j > 0$ for $j = 1, 2, \ldots, k$, is a convex function (see Exercise 3.8). (b) $f(\mathbf{x}) = \text{maximum } \{f_1(\mathbf{x}), f_2(\mathbf{x}), \ldots, f_k(\mathbf{x})\}$ is a convex function (see Exercise 3.9).
2. Suppose that $g:E_n \to E_1$ is a concave function. Let $S = \{\mathbf{x}:g(\mathbf{x}) > 0)$, and define $f:S \to E_n$ as $f(\mathbf{x}) = 1/g(\mathbf{x})$. Then, f is convex over S (see Exercise 3.7).
3. Let $g:E_1 \to E_1$ be a nondecreasing, univariate, convex function, and let h: $E_n \to E_1$ be a convex function. Then, the composite function $f:E_n \to E_1$ defined as $f(\mathbf{x}) = g[h(\mathbf{x})]$ is a convex function (see Exercise 3.12).
4. Let $g:E_m \to E_1$ be a convex function, and let $\mathbf{h}:E_n \to E_m$ be an affine function of the form $\mathbf{h}(\mathbf{x}) = \mathbf{A}\mathbf{x} + \mathbf{b}$, where $\mathbf{A}$ is an $m \times n$ matrix and $\mathbf{b}$ is an $m \times 1$ vector. Then, the composite function $f: E_n \to E_1$ defined as $f(\mathbf{x}) = g[\mathbf{h}(\mathbf{x})]$ is a convex function (see Exercise 3.13).

From now on, we will concentrate only on convex functions. Results for concave functions can be easily obtained by noting that f is concave if and only if $-f$ is convex.

Associated with a convex function f is the set $S_\alpha = \{\mathbf{x} \in S : f(\mathbf{x}) \leq \alpha\}$, $\alpha \in E_1$, usually referred to as a *level set*. Sometimes, this set is called a *lower-level set*, to differentiate it from the *upper-level set* $\{\mathbf{x} \in S : f(\mathbf{x}) \geq \alpha\}$, which has similar properties for concave functions. Lemma 3.1.2 below shows that S_α is convex for each real number α. Hence, if $g_i:E_n \to E$ is convex for $i = 1, \ldots, m$, the set $\{\mathbf{x} : g_i(\mathbf{x}) \leq 0, i = 1, \ldots, m\}$ is a convex set.

3.1.2 Lemma

Let S be a nonempty convex set in E_n and let $f:S \to E_1$ be a convex function. Then, the level set $S_\alpha = \{\mathbf{x} \in S : f(\mathbf{x}) \leq \alpha\}$, where α is a real number, is a convex set.

Proof

Let $\mathbf{x}_1, \mathbf{x}_2 \in S_\alpha$. Thus, $\mathbf{x}_1, \mathbf{x}_2 \in S$, and $f(\mathbf{x}_1) \leq \alpha$ and $f(\mathbf{x}_2) \leq \alpha$. Now let $\lambda \in (0, 1)$ and $\mathbf{x} = \lambda\mathbf{x}_1 + (1 - \lambda)\mathbf{x}_2$. By convexity of S, $\mathbf{x} \in S$. Furthermore, by convexity of f,

$$f(\mathbf{x}) \leq \lambda f(\mathbf{x}_1) + (1 - \lambda)f(\mathbf{x}_2) \leq \lambda\alpha + (1 - \lambda)\alpha = \alpha$$

Hence, $\mathbf{x} \in S_\alpha$, and therefore, S_α is convex.

Continuity of Convex Functions

An important property of convex and concave functions is that they are continuous on the interior of their domain. This fact is proved below.

3.1.3 Theorem

Let S be a nonempty convex set in E_n, and let $f: S \rightarrow E_1$ be convex. Then f is continuous on the interior of S.

Proof

Let $\bar{\mathbf{x}} \in \text{int } S$. To prove continuity of f at $\bar{\mathbf{x}}$, we need to show that given $\varepsilon > 0$, there exists a $\delta > 0$ such that $\|\mathbf{x} - \bar{\mathbf{x}}\| \leq \delta$ implies that $|f(\mathbf{x}) - f(\bar{\mathbf{x}})| \leq \varepsilon$. Since $\bar{\mathbf{x}} \in \text{int } S$, there exists a $\delta' > 0$ such that $\|\mathbf{x} - \bar{\mathbf{x}}\| \leq \delta'$ implies that $\mathbf{x} \in S$. Construct θ as follows.

$$\theta = \underset{1 \leq i \leq n}{\text{maximum}} \{\text{maximum } [f(\bar{\mathbf{x}} + \delta'\mathbf{e}_i) - f(\bar{\mathbf{x}}), f(\bar{\mathbf{x}} - \delta'\mathbf{e}_i) - f(\bar{\mathbf{x}})]\} \tag{3.1}$$

where $\mathbf{e}_i$ is a vector of zeros except for a 1 at the ith position. Note that $0 \leq \theta < \infty$. Let

$$\delta = \text{minimum}\left(\frac{\delta'}{n}, \frac{\varepsilon\delta'}{n\theta}\right) \tag{3.2}$$

Choose an $\mathbf{x}$ with $\|\mathbf{x} - \bar{\mathbf{x}}\| \leq \delta$. If $x_i - \bar{x}_i \geq 0$, let $\mathbf{z}_i = \delta'\mathbf{e}_i$; otherwise, let $\mathbf{z}_i = -\delta'\mathbf{e}_i$. Then, $\mathbf{x} - \bar{\mathbf{x}} = \sum_{i=1}^n \alpha_i \mathbf{z}_i$, where $\alpha_i \geq 0$ for $i = 1, 2, \ldots, n$. Furthermore,

$$\|\mathbf{x} - \bar{\mathbf{x}}\| = \delta'\left(\sum_{i=1}^n \alpha_i^2\right)^{1/2} \tag{3.3}$$

From (3.2), and since $\|\mathbf{x} - \bar{\mathbf{x}}\| \leq \delta$, it follows that $\alpha_i \leq 1/n$ for $i = 1, 2, \ldots, n$. Hence, by convexity of f, and since $0 \leq n\alpha_i \leq 1$, we get

$$\begin{aligned}
f(\mathbf{x}) = f\left(\bar{\mathbf{x}} + \sum_{i=1}^n \alpha_i \mathbf{z}_i\right) &= f\left[\frac{1}{n}\sum_{i=1}^n (\bar{\mathbf{x}} + n\alpha_i \mathbf{z}_i)\right] \\
&\leq \frac{1}{n}\sum_{i=1}^n f(\bar{\mathbf{x}} + n\alpha_i \mathbf{z}_i) \\
&= \frac{1}{n}\sum_{i=1}^n f[(1 - n\alpha_i)\bar{\mathbf{x}} + n\alpha_i(\bar{\mathbf{x}} + \mathbf{z}_i)] \\
&\leq \frac{1}{n}\sum_{i=1}^n [(1 - n\alpha_i)f(\bar{\mathbf{x}}) + n\alpha_i f(\bar{\mathbf{x}} + \mathbf{z}_i)]
\end{aligned}$$

Therefore, $f(\mathbf{x}) - f(\bar{\mathbf{x}}) \leq \sum_{i=1}^n \alpha_i[f(\bar{\mathbf{x}} + \mathbf{z}_i) - f(\bar{\mathbf{x}})]$. From (3.1) it is obvious that $f(\bar{\mathbf{x}} + \mathbf{z}_i) - f(\bar{\mathbf{x}}) \leq \theta$ for each i; and since $\alpha_i \geq 0$, it follows that

$$f(\mathbf{x}) - f(\bar{\mathbf{x}}) \leq \theta \sum_{i=1}^n \alpha_i \tag{3.4}$$

Noting (3.3) and (3.2), it follows that $\alpha_i \leq \varepsilon/n\theta$, and (3.4) implies that $f(\mathbf{x}) - f(\bar{\mathbf{x}}) \leq \varepsilon$. So far, we have shown that $\|\mathbf{x} - \bar{\mathbf{x}}\| \leq \delta$ implies that $f(\mathbf{x}) - f(\bar{\mathbf{x}}) \leq \varepsilon$. By definition, this establishes the *upper semicontinuity* of f at $\bar{\mathbf{x}}$. To finish the proof, we need to establish the *lower semicontinuity* of f at $\bar{\mathbf{x}}$ as well, that is, to show that $f(\bar{\mathbf{x}}) - f(\mathbf{x}) \leq \varepsilon$. Let $\mathbf{y} = 2\bar{\mathbf{x}} - \mathbf{x}$ and note that $\|\mathbf{y} - \bar{\mathbf{x}}\| \leq \delta$. Therefore, as above,

$$f(\mathbf{y}) - f(\bar{\mathbf{x}}) \leq \varepsilon \tag{3.5}$$

But $\bar{\mathbf{x}} = \frac{1}{2}\mathbf{y} + \frac{1}{2}\mathbf{x}$, and by the convexity of f, we have

$$f(\bar{\mathbf{x}}) \leq \tfrac{1}{2}f(\mathbf{y}) + \tfrac{1}{2}f(\mathbf{x}) \tag{3.6}$$

Combining (3.5) and (3.6) above, it follows that $f(\bar{\mathbf{x}}) - f(\mathbf{x}) \leq \varepsilon$, and the proof is complete.

Note that convex and concave functions may not be continuous everywhere. However, by the above theorem, points of discontinuity are only allowed at the boundary of S, as illustrated by the following convex function defined on $S = \{x: -1 \leq x \leq 1\}$:

$$f(x) = \begin{cases} x^2 & \text{for} \quad |x| < 1 \\ 2 & \text{for} \quad |x| = 1 \end{cases}$$

Directional Derivative of Convex Functions

The concept of directional derivatives is particularly useful in the motivation and development of some optimality criteria and computational procedures in nonlinear programming where one is interested in finding a direction along which the function decreases or increases.

3.1.4 Definition

Let S be a nonempty set in E_n, and let $f: S \rightarrow E_1$. Let $\bar{\mathbf{x}} \in S$ and $\mathbf{d}$ be a nonzero vector such that $\bar{\mathbf{x}} + \lambda\mathbf{d} \in S$ for $\lambda > 0$ and sufficiently small. The *directional derivative* of f at $\bar{\mathbf{x}}$ along the vector $\mathbf{d}$, denoted by $f'(\bar{\mathbf{x}}; \mathbf{d})$, is given by the following limit if it exists:

$$f'(\bar{\mathbf{x}}; \mathbf{d}) = \lim_{\lambda \to 0^+} \frac{f(\bar{\mathbf{x}} + \lambda\mathbf{d}) - f(\bar{\mathbf{x}})}{\lambda}$$

In particular, the limit in Definition 3.1.4 exists for globally defined convex and concave functions as shown below. As evident from the proof of the following lemma, if $f: S \rightarrow E_1$ is convex on S, then the limit exists if $\bar{\mathbf{x}} \in \text{int } S$, but might be $-\infty$ if $\bar{\mathbf{x}} \in \partial S$, even if f is continuous at $\bar{\mathbf{x}}$, as seen in Figure 3.2.

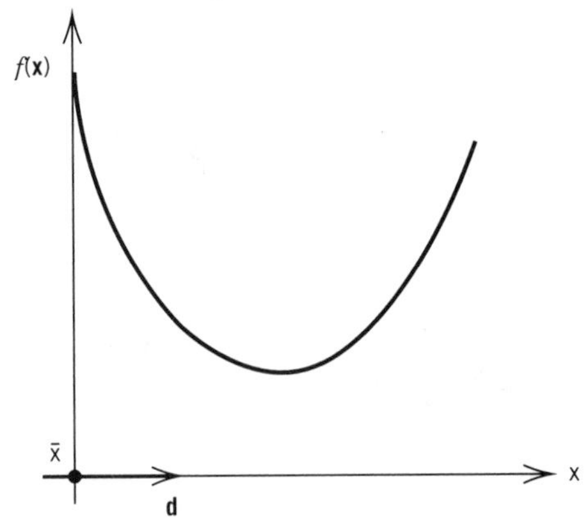

Figure 3.2 Nonexistence of the directional derivative of f at $\bar{\mathbf{x}}$ in the direction $\mathbf{d}$.

3.1.5 Lemma

Let $f: E_n \to E_1$ be a convex function. Consider any point $\bar{\mathbf{x}} \in E_n$ and a nonzero direction $\mathbf{d} \in E_n$. Then, the directional derivative $f'(\bar{\mathbf{x}}; \mathbf{d})$, of f at $\bar{\mathbf{x}}$ in the direction $\mathbf{d}$, exists.

Proof

Let $\lambda_2 > \lambda_1 > 0$. Noting the convexity of f, we have

$$f(\bar{\mathbf{x}} + \lambda_1\mathbf{d}) = f\left[\frac{\lambda_1}{\lambda_2}(\bar{\mathbf{x}} + \lambda_2\mathbf{d}) + \left(1 - \frac{\lambda_1}{\lambda_2}\right)\bar{\mathbf{x}}\right]$$
$$\leq \frac{\lambda_1}{\lambda_2} f(\bar{\mathbf{x}} + \lambda_2\mathbf{d}) + \left(1 - \frac{\lambda_1}{\lambda_2}\right) f(\bar{\mathbf{x}})$$

This inequality implies that

$$\frac{f(\bar{\mathbf{x}} + \lambda_1\mathbf{d}) - f(\bar{\mathbf{x}})}{\lambda_1} \leq \frac{f(\bar{\mathbf{x}} + \lambda_2\mathbf{d}) - f(\bar{\mathbf{x}})}{\lambda_2}$$

Thus, the difference quotient $[f(\bar{\mathbf{x}} + \lambda\mathbf{d}) - f(\bar{\mathbf{x}})]/\lambda$ is monotone decreasing (nonincreasing) as $\lambda \to 0^+$.

Now, given any $\lambda > 0$, we also have, by the convexity of f, that

$$f(\bar{\mathbf{x}}) = f\left[\frac{\lambda}{(1 + \lambda)}(\bar{\mathbf{x}} - \mathbf{d}) + \frac{1}{(1 + \lambda)}(\bar{\mathbf{x}} + \lambda\mathbf{d})\right]$$
$$\leq \frac{\lambda}{(1 + \lambda)} f(\bar{\mathbf{x}} - \mathbf{d}) + \frac{1}{(1 + \lambda)} f(\bar{\mathbf{x}} + \lambda\mathbf{d})$$

And so,

$$\frac{f(\bar{\mathbf{x}} + \lambda\mathbf{d}) - f(\bar{\mathbf{x}})}{\lambda} \geq f(\bar{\mathbf{x}}) - f(\bar{\mathbf{x}} - \mathbf{d})$$

Hence, the monotone decreasing sequence of values $[f(\bar{\mathbf{x}} + \lambda\mathbf{d}) - f(\bar{\mathbf{x}})]/\lambda$, as $\lambda \to 0^+$, is bounded from below by the constant $f(\bar{\mathbf{x}}) - f(\bar{\mathbf{x}} - \mathbf{d})$. Hence, the limit in the theorem exists and is given by

$$\lim_{\lambda \to 0^+} \frac{f(\bar{\mathbf{x}} + \lambda\mathbf{d}) - f(\bar{\mathbf{x}})}{\lambda} = \inf_{\lambda > 0} \frac{f(\bar{\mathbf{x}} + \lambda\mathbf{d}) - f(\bar{\mathbf{x}})}{\lambda}$$

3.2 Subgradients of a Convex Function

In this section, we introduce the important concept of subgradients of convex and concave functions via supporting hyperplanes to the epigraphs of convex functions and to the hypographs of concave functions.

Epigraph and Hypograph of a Function

A function f on S can be fully described by the set $\{[\mathbf{x}, f(\mathbf{x})]: \mathbf{x} \in S\} \subset E_{n+1}$, which is referred to as the *graph* of the function. One can construct two sets that are related to

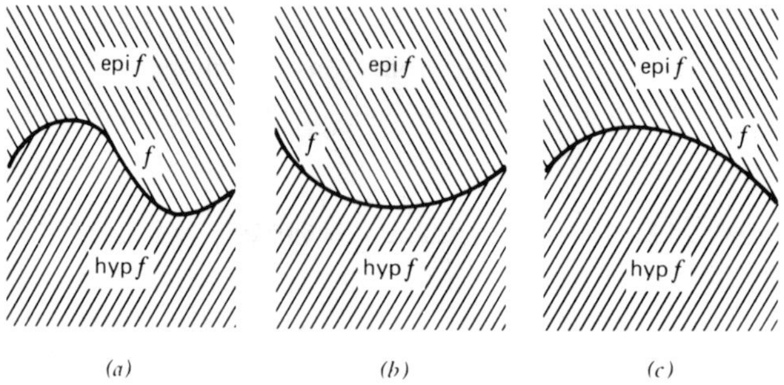

Figure 3.3 Examples of epigraphs and hypographs.

the graph of f: the epigraph, which consists of points above the graph of f, and the hypograph, which consists of points below the graph of f. These notions are clarified by Definition 3.2.1 below.

3.2.1 Definition

Let S be a nonempty set in E_n, and let $f:S \to E_1$. The *epigraph* of f, denoted by epi f, is a subset of E_{n+1} defined by

$$\{(\mathbf{x}, y):\mathbf{x} \in S,\ y \in E_1,\ y \geq f(\mathbf{x})\}$$

The *hypograph* of f, denoted by hyp f, is a subset of E_{n+1} defined by

$$\{(\mathbf{x}, y):\mathbf{x} \in S,\ y \in E_1,\ y \leq f(\mathbf{x})\}$$

Figure 3.3 illustrates the epigraphs and hypographs of several functions. In Figure 3.3*a*, neither the epigraph nor the hypograph of f is a convex set. But in Figures 3.3*b* and *c*, respectively, the epigraph and hypograph of f are convex sets. It turns out that a function is convex if and only if its epigraph is a convex set and, equivalently, that a function is concave if and only if its hypograph is a convex set.

3.2.2 Theorem

Let S be a nonempty convex set in E_n, and let $f:S \to E_1$. Then, f is convex if and only if epi f is a convex set.

Proof

Assume that f is convex, and let $(\mathbf{x}_1, y_1)$ and $(\mathbf{x}_2, y_2) \in$ epi f; that is, $\mathbf{x}_1, \mathbf{x}_2 \in S$, $y_1 \geq f(\mathbf{x}_1)$, and $y_2 \geq f(\mathbf{x}_2)$. Let $\lambda \in (0, 1)$. Then,

$$\lambda y_1 + (1 - \lambda)y_2 \geq \lambda f(\mathbf{x}_1) + (1 - \lambda)f(\mathbf{x}_2) \geq f(\lambda\mathbf{x}_1 + (1 - \lambda)\mathbf{x}_2)$$

where the last inequality follows by the convexity of f. Note that $\lambda\mathbf{x}_1 + (1 - \lambda)\mathbf{x}_2 \in S$. Thus, $[\lambda\mathbf{x}_1 + (1 - \lambda)\mathbf{x}_2, \lambda y_1 + (1 - \lambda)y_2] \in$ epi f and, hence, epi f is convex. Conversely, assume that epi f is convex, and let $\mathbf{x}_1, \mathbf{x}_2 \in S$. Then, $[\mathbf{x}_1, f(\mathbf{x}_1)]$ and $[\mathbf{x}_2, f(\mathbf{x}_2)]$ belong to epi f, and by the convexity of epi f, we must have

$$[\lambda\mathbf{x}_1 + (1 - \lambda)\mathbf{x}_2, \lambda f(\mathbf{x}_1) + (1 - \lambda)f(\mathbf{x}_2)] \in \text{epi } f \qquad \text{for } \lambda \in (0, 1)$$

In other words, $\lambda f(\mathbf{x}_1) + (1 - \lambda)f(\mathbf{x}_2) \geq f[\lambda\mathbf{x}_1 + (1 - \lambda)\mathbf{x}_2]$ for each $\lambda \in (0, 1)$; that is, f is convex. This completes the proof.

The above theorem can be used to verify convexity or concavity of a given function f. Making use of this result, it is clear that the functions illustrated in Figure 3.3 are (*a*) neither convex nor concave, (*b*) convex, and (*c*) concave.

Since the epigraph of a convex function and the hypograph of a concave function are convex sets, they have supporting hyperplanes at points of their boundary. These supporting hyperplanes lead to the notion of subgradients, which is defined below.

3.2.3 Definition

Let S be a nonempty convex set in E_n, and let $f:S \to E_1$ be convex. Then, $\boldsymbol{\xi}$ is called a *subgradient of* f at $\bar{\mathbf{x}} \in S$ if

$$f(\mathbf{x}) \geq f(\bar{\mathbf{x}}) + \boldsymbol{\xi}^t(\mathbf{x} - \bar{\mathbf{x}}) \qquad \text{for all } \mathbf{x} \in S$$

Similarly, let $f:S \to E_1$ be concave. Then, $\boldsymbol{\xi}$ is called a *subgradient* of f at $\bar{\mathbf{x}} \in S$ if

$$f(\mathbf{x}) \leq f(\bar{\mathbf{x}}) + \boldsymbol{\xi}^t(\mathbf{x} - \bar{\mathbf{x}}) \qquad \text{for all } \mathbf{x} \in S$$

From the above definition, it immediately follows that the collection of subgradients of f at $\bar{\mathbf{x}}$ (known as the *subdifferential* of f at $\bar{\mathbf{x}}$) is a convex set. Figure 3.4 shows examples of subgradients of convex and concave functions. From the figure, we see that the function $f(\bar{\mathbf{x}}) + \boldsymbol{\xi}^t(\mathbf{x} - \bar{\mathbf{x}})$ corresponds to a supporting hyperplane of the epigraph or the hypograph of the function f. The subgradient vector $\boldsymbol{\xi}$ corresponds to the slope of the supporting hyperplane.

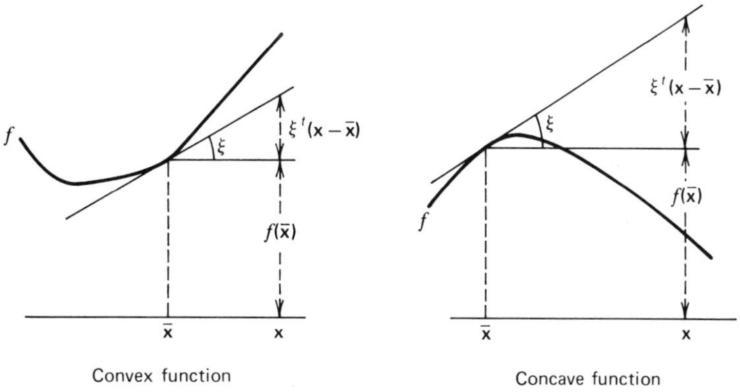

Figure 3.4 Geometric interpretation of subgradients.

3.2.4 Example

Let $f(x)$ = minimum $[f_1(x), f_2(x)]$, where f_1 and f_2 are defined below:

$$f_1(x) = 4 - |x| \qquad x \in E_1$$

$$f_2(x) = 4 - (x - 2)^2 \qquad x \in E_1$$

Since $f_2(x) \geq f_1(x)$ for $1 \leq x \leq 4$, f can be represented as follows:

$$f(x) = \begin{cases} 4 - x & 1 \leq x \leq 4 \\ 4 - (x - 2)^2 & \text{otherwise} \end{cases}$$

In Figure 3.5, the concave function f is shown in dark lines. Note that $\xi = -1$ is the slope and hence the subgradient of f at any point x in the open interval $(1, 4)$. If

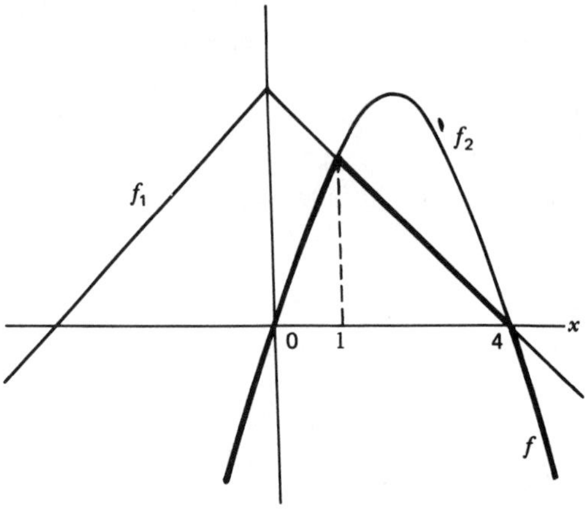

Figure 3.5 Illustration of Example 3.2.4.

$x < 1$ or $x > 4$, then $\xi = -2(x - 2)$ is a unique subgradient of f. At the points $x = 1$ and $x = 4$, the subgradients are not unique because many supporting hyperplanes exist. At $x = 1$, the family of subgradients is characterized by $\lambda\nabla f_1(1) + (1 - \lambda)\nabla f_2(1) = \lambda(-1) + (1 - \lambda)(2) = 2 - 3\lambda$ for $\lambda \in [0, 1]$. In other words, any ξ in the interval $[-1, 2]$ is a subgradient of f at $x = 1$, and this corresponds to the slopes of the family of supporting hyperplanes of f at $x = 1$. At $x = 4$, the family of subgradients is characterized by $\lambda\nabla f_1(4) + (1 - \lambda)\nabla f_2(4) = \lambda(-1) + (1 - \lambda)(-4) = -4 + 3\lambda$ for $\lambda \in [0, 1]$. In other words, any ξ in the interval $[-4, -1]$ is a subgradient of f at $x = 4$. Exercise 3.25 addresses the general characterization of subgradients of functions of the form $f(\mathbf{x}) = \text{minimum } [f_1(\mathbf{x}), f_2(\mathbf{x})]$.

The following theorem shows that every convex or concave function has at least one subgradient at points in the interior of its domain. The proof relies on the fact that a convex set has a supporting hyperplane at points of the boundary.

3.2.5 Theorem

Let S be a nonempty convex set in E_n, and let $f: S \to E_1$ be convex. Then, for $\bar{\mathbf{x}} \in \text{int } S$, there exists a vector $\boldsymbol{\xi}$ such that the hyperplane

$$H = \{(\mathbf{x}, y) : y = f(\bar{\mathbf{x}}) + \boldsymbol{\xi}^t(\mathbf{x} - \bar{\mathbf{x}})\}$$

supports epi f at $[\bar{\mathbf{x}}, f(\bar{\mathbf{x}})]$. In particular,

$$f(\mathbf{x}) \geq f(\bar{\mathbf{x}}) + \boldsymbol{\xi}^t(\mathbf{x} - \bar{\mathbf{x}}) \qquad \text{for each } \mathbf{x} \in S$$

that is, $\boldsymbol{\xi}$ is a subgradient of f at $\bar{\mathbf{x}}$.

Proof

By Theorem 3.2.2, epi f is convex. Noting that $[\bar{\mathbf{x}}, f(\bar{\mathbf{x}})]$ belongs to the boundary of epi f, by Theorem 2.3.7, there exists a nonzero vector $(\boldsymbol{\xi}_0, \mu) \in E_n \times E_1$ such that

$$\boldsymbol{\xi}_0^t(\mathbf{x} - \bar{\mathbf{x}}) + \mu[y - f(\bar{\mathbf{x}})] \leq 0 \qquad \text{for all } (\mathbf{x}, y) \in \text{epi} f \tag{3.7}$$

Note that μ is not positive, because otherwise, the above inequality will be contradicted by choosing y sufficiently large. We now show that $\mu < 0$. By contradiction, suppose that $\mu = 0$. Then, $\boldsymbol{\xi}_0^t(\mathbf{x} - \bar{\mathbf{x}}) \leq 0$ for all $\mathbf{x} \in S$. Since $\bar{\mathbf{x}} \in \text{int } S$, there exists a $\lambda > 0$ such that $\bar{\mathbf{x}} + \lambda\boldsymbol{\xi}_0 \in S$ and, hence, $\lambda\boldsymbol{\xi}_0^t\boldsymbol{\xi}_0 \leq 0$. This implies that $\boldsymbol{\xi}_0 = \mathbf{0}$ and $(\boldsymbol{\xi}_0, \mu) = (\mathbf{0}, 0)$, contradicting the fact that $(\boldsymbol{\xi}_0, \mu)$ is a nonzero vector. Therefore, $\mu < 0$. Denoting $\boldsymbol{\xi}_0/|\mu|$ by $\boldsymbol{\xi}$ and dividing the inequality in (3.7) by $|\mu|$, we get

$$\boldsymbol{\xi}^t(\mathbf{x} - \bar{\mathbf{x}}) - y + f(\bar{\mathbf{x}}) \leq 0 \qquad \text{for all } (\mathbf{x}, y) \in \text{epi} f \tag{3.8}$$

In particular, the hyperplane $H = \{(\mathbf{x}, y) : y = f(\bar{\mathbf{x}}) + \boldsymbol{\xi}^t(\mathbf{x} - \bar{\mathbf{x}})\}$ supports epi f at $[\bar{\mathbf{x}}, f(\bar{\mathbf{x}})]$. By letting $y = f(\mathbf{x})$ in (3.8), we get $f(\mathbf{x}) \geq f(\bar{\mathbf{x}}) + \boldsymbol{\xi}^t(\mathbf{x} - \bar{\mathbf{x}})$ for all $\mathbf{x} \in S$, and the proof is complete.

Corollary

Let S be a nonempty convex set in E_n, and let $f: S \to E_1$ be strictly convex. Then, for $\bar{\mathbf{x}} \in \text{int } S$, there exists a vector $\boldsymbol{\xi}$ such that

$$f(\mathbf{x}) > f(\bar{\mathbf{x}}) + \boldsymbol{\xi}^t(\mathbf{x} - \bar{\mathbf{x}}) \qquad \text{for all } \mathbf{x} \in S, \mathbf{x} \neq \bar{\mathbf{x}}$$

Proof

By the theorem, there exists a vector $\boldsymbol{\xi}$ such that

$$f(\mathbf{x}) \geq f(\bar{\mathbf{x}}) + \boldsymbol{\xi}^t(\mathbf{x} - \bar{\mathbf{x}}) \qquad \text{for all } \mathbf{x} \in S \tag{3.9}$$

By contradiction, suppose that there is an $\hat{\mathbf{x}} \neq \bar{\mathbf{x}}$ such that $f(\hat{\mathbf{x}}) = f(\bar{\mathbf{x}}) + \boldsymbol{\xi}^t(\hat{\mathbf{x}} - \bar{\mathbf{x}})$. Then, by the strict convexity of f for $\lambda \in (0, 1)$, we get

$$f[\lambda\bar{\mathbf{x}} + (1 - \lambda)\hat{\mathbf{x}}] < \lambda f(\bar{\mathbf{x}}) + (1 - \lambda)f(\hat{\mathbf{x}}) = f(\bar{\mathbf{x}}) + (1 - \lambda)\boldsymbol{\xi}^t(\hat{\mathbf{x}} - \bar{\mathbf{x}}) \tag{3.10}$$

But letting $\mathbf{x} = \lambda\bar{\mathbf{x}} + (1 - \lambda)\hat{\mathbf{x}}$ in (3.9), we must have

$$f[\lambda\bar{\mathbf{x}} + (1 - \lambda)\hat{\mathbf{x}}] \geq f(\bar{\mathbf{x}}) + (1 - \lambda)\boldsymbol{\xi}^t(\hat{\mathbf{x}} - \bar{\mathbf{x}})$$

contradicting (3.10). This proves the corollary.

The converse of the above theorem is not true in general. In other words, if corresponding to each point $\mathbf{x} \in \text{int } S$ there is a subgradient of f, then f is not necessarily a convex function. To illustrate, consider the following example, where f is defined on $S = \{(x_1, x_2): 0 \leq x_1, x_2 \leq 1\}$:

$$f(x_1, x_2) = \begin{cases} 0 & 0 \leq x_1 \leq 1, \quad 0 < x_2 \leq 1 \\ \frac{1}{4} - (x_1 - \frac{1}{2})^2 & 0 \leq x_1 \leq 1, \quad x_2 = 0 \end{cases}$$

For each point in the interior of the domain, the zero vector is a subgradient of f. However, f is not convex on S since epi f is clearly not a convex set. However, as the following theorem shows, f is indeed convex on int S.

3.2.6 Theorem

Let S be a nonempty convex set in E_n, and let $f: S \to E_1$. Suppose that for each point $\bar{\mathbf{x}} \in \text{int } S$ there exists a subgradient vector $\boldsymbol{\xi}$ such that

$$f(\mathbf{x}) \geq f(\bar{\mathbf{x}}) + \boldsymbol{\xi}^t(\mathbf{x} - \bar{\mathbf{x}}) \qquad \text{for each } \mathbf{x} \in S$$

Then, f is convex on int S.

Proof

Let $\mathbf{x}_1, \mathbf{x}_2 \in \text{int } S$, and let $\lambda \in (0, 1)$. By Corollary 1 to Theorem 2.2.2, int S is convex, and we must have $\lambda\mathbf{x}_1 + (1 - \lambda)\mathbf{x}_2 \in \text{int } S$. By assumption, there exists a subgradient $\boldsymbol{\xi}$ of f at $\lambda\mathbf{x}_1 + (1 - \lambda)\mathbf{x}_2$. In particular, the following two inequalities hold true:

$$f(\mathbf{x}_1) \geq f[\lambda\mathbf{x}_1 + (1 - \lambda)\mathbf{x}_2] + (1 - \lambda)\boldsymbol{\xi}^t(\mathbf{x}_1 - \mathbf{x}_2)$$

$$f(\mathbf{x}_2) \geq f[\lambda\mathbf{x}_1 + (1 - \lambda)\mathbf{x}_2] + \lambda\boldsymbol{\xi}^t(\mathbf{x}_2 - \mathbf{x}_1)$$

Multiplying the above two inequalities by λ and $(1 - \lambda)$, respectively, and adding, we obtain

$$\lambda f(\mathbf{x}_1) + (1 - \lambda)f(\mathbf{x}_2) \geq f[\lambda\mathbf{x}_1 + (1 - \lambda)\mathbf{x}_2]$$

and the result follows.

3.3 Differentiable Convex Functions

We now focus on differentiable convex and concave functions. First, consider the following definition of differentiability.

3.3.1 Definition

Let S be a nonempty set in E_n, and let $f: S \rightarrow E_1$. Then, f is said to be *differentiable* at $\bar{\mathbf{x}} \in \text{int } S$ if there exist a vector $\nabla f(\bar{\mathbf{x}})$, called the *gradient vector*, and a function $\alpha: E_n \rightarrow E_1$ such that

$$f(\mathbf{x}) = f(\bar{\mathbf{x}}) + \nabla f(\bar{\mathbf{x}})^t(\mathbf{x} - \bar{\mathbf{x}}) + \|\mathbf{x} - \bar{\mathbf{x}}\|\, \alpha(\bar{\mathbf{x}}; \mathbf{x} - \bar{\mathbf{x}}) \qquad \text{for each } \mathbf{x} \in S$$

where $\lim_{\mathbf{x}\rightarrow\bar{\mathbf{x}}} \alpha(\bar{\mathbf{x}}; \mathbf{x} - \bar{\mathbf{x}}) = 0$. The function f is said to be differentiable on the open set $S' \subseteq S$ if it is differentiable at each point in S'. The above representation of f is called a *first-order (Taylor series) expansion* of f at $\bar{\mathbf{x}}$; and without the implicitly defined *remainder term* involving the function α, the resulting representation is called a *first-order (Taylor series) approximation* of f.

Note that if f is differentiable at $\bar{\mathbf{x}}$, then there could only be one gradient vector, and this vector is given by

$$\nabla f(\bar{\mathbf{x}}) = \left(\frac{\partial f(\bar{\mathbf{x}})}{\partial x_1}, \ldots, \frac{\partial f(\bar{\mathbf{x}})}{\partial x_n}\right)^t \equiv (f_1(\bar{\mathbf{x}}), \ldots, f_n(\bar{\mathbf{x}}))^t$$

where $\partial f(\bar{\mathbf{x}})/\partial x_i \equiv f_i(\bar{\mathbf{x}})$ is the partial derivative of f with respect to x_i at $\bar{\mathbf{x}}$ (see Exercise 3.28, and review Appendix A.4).

The following lemma shows that a differentiable convex function has only one subgradient, namely, the gradient vector. Hence, the results of the previous section can be easily specialized to the differentiable case, in which the gradient vector replaces subgradients.

3.3.2 Lemma

Let S be a nonempty convex set in E_n, and let $f: S \rightarrow E_1$ be convex. Suppose that f is differentiable at $\bar{\mathbf{x}} \in \text{int } S$. Then, the collection of subgradients of f at $\bar{\mathbf{x}}$ is the singleton set $\{\nabla f(\bar{\mathbf{x}})\}$.

Proof

By Theorem 3.2.5, the set of subgradients of f at $\bar{\mathbf{x}}$ is not empty. Now let $\boldsymbol{\xi}$ be a subgradient of f at $\bar{\mathbf{x}}$. As a result of Theorem 3.2.5 and the differentiability of f at $\bar{\mathbf{x}}$, for any vector $\mathbf{d}$ and for λ sufficiently small, we get

$$f(\bar{\mathbf{x}} + \lambda\mathbf{d}) \geq f(\bar{\mathbf{x}}) + \lambda\boldsymbol{\xi}^t\mathbf{d}$$

$$f(\bar{\mathbf{x}} + \lambda\mathbf{d}) = f(\bar{\mathbf{x}}) + \lambda\nabla f(\bar{\mathbf{x}})^t\mathbf{d} + \lambda\,\|\mathbf{d}\|\,\alpha(\bar{\mathbf{x}}; \lambda\mathbf{d})$$

Subtracting the equation from the inequality, we obtain

$$0 \geq \lambda[\boldsymbol{\xi} - \nabla f(\bar{\mathbf{x}})]^t\mathbf{d} - \lambda\,\|\mathbf{d}\|\,\alpha(\bar{\mathbf{x}}; \lambda\mathbf{d})$$

If we divide by $\lambda > 0$ and let $\lambda \to 0^+$, it follows that $[\boldsymbol{\xi} - \nabla f(\bar{\mathbf{x}})]^t\mathbf{d} \leq 0$. Choosing $\mathbf{d} = \boldsymbol{\xi} - \nabla f(\bar{\mathbf{x}})$, the last inequality implies that $\boldsymbol{\xi} = \nabla f(\bar{\mathbf{x}})$. This completes the proof.

In the light of the above lemma, we give the following important characterization of differentiable convex functions. The proof is immediate from Theorems 3.2.5 and 3.2.6 and Lemma 3.3.2.

3.3.3 Theorem

Let S be a nonempty open convex set in E_n, and let $f: S \to E_1$ be differentiable on S. Then f is convex if and only if for any $\bar{\mathbf{x}} \in S$, we have

$$f(\mathbf{x}) \geq f(\bar{\mathbf{x}}) + \nabla f(\bar{\mathbf{x}})^t(\mathbf{x} - \bar{\mathbf{x}}) \qquad \text{for each } \mathbf{x} \in S$$

Similarly, f is strictly convex if and only if for each $\bar{\mathbf{x}} \in S$, we have

$$f(\mathbf{x}) > f(\bar{\mathbf{x}}) + \nabla f(\bar{\mathbf{x}})^t(\mathbf{x} - \bar{\mathbf{x}}) \qquad \text{for each } \mathbf{x} \neq \bar{\mathbf{x}} \text{ in } S$$

There are two evident implications of the above result that find use in various contexts. The first is that if we have an optimization problem to minimize $f(\mathbf{x})$ subject to $\mathbf{x} \in X$, where f is a convex function, then, given any point $\bar{\mathbf{x}}$, the affine function $f(\bar{\mathbf{x}}) + \nabla f(\bar{\mathbf{x}})^t(\mathbf{x} - \bar{\mathbf{x}})$ bounds f from below. Hence, the minimum of $f(\bar{\mathbf{x}}) + \nabla f(\bar{\mathbf{x}})^t(\mathbf{x} - \bar{\mathbf{x}})$ over X (or over a relaxation of X) yields a lower bound on the optimum value of the given optimization problem, which can prove to be useful in an algorithmic approach. A second point in the same spirit is that this affine bounding function can be used to derive polyhedral outer approximations. For example, consider the set $X = \{\mathbf{x} : g_i(\mathbf{x}) \leq 0, i = 1, \ldots, m\}$, where g_i is a convex function for each $i = 1, \ldots, m$. Given any point $\bar{\mathbf{x}}$, construct the polyhedral set $\bar{X} = \{\mathbf{x} : g_i(\bar{\mathbf{x}}) + \nabla g_i(\bar{\mathbf{x}})^t(\mathbf{x} - \bar{\mathbf{x}}) \leq 0, i = 1, \ldots, m\}$. Note that the polyhedral set $\bar{X}$ contains X and, hence, affords an *outer linearization* of this set, since for any $\mathbf{x} \in X$, we have $0 \geq g_i(\mathbf{x}) \geq g_i(\bar{\mathbf{x}}) + \nabla g_i(\bar{\mathbf{x}})^t(\mathbf{x} - \bar{\mathbf{x}})$ for $i = 1, \ldots, m$ by Theorem 3.3.3. Such representations play a central role in many successive approximation algorithms for various nonlinear optimization problems.

The following theorem gives another necessary and sufficient characterization of differentiable convex functions. For a function of one variable, the characterization reduces to the slope being nondecreasing.

3.3.4 Theorem

Let S be a nonempty open convex set in E_n and let $f: S \to E_1$ be differentiable on S. Then, f is convex if and only if, for each $\mathbf{x}_1, \mathbf{x}_2 \in S$, we have

$$[\nabla f(\mathbf{x}_2) - \nabla f(\mathbf{x}_1)]'(\mathbf{x}_2 - \mathbf{x}_1) \geq 0$$

Similarly, f is strictly convex if and only if, for each distinct $\mathbf{x}_1, \mathbf{x}_2 \in S$, we have

$$[\nabla f(\mathbf{x}_2) - \nabla f(\mathbf{x}_1)]'(\mathbf{x}_2 - \mathbf{x}_1) > 0$$

Proof

Assume that f is convex, and let $\mathbf{x}_1, \mathbf{x}_2 \in S$. By Theorem 3.3.3, we have

$$f(\mathbf{x}_1) \geq f(\mathbf{x}_2) + \nabla f(\mathbf{x}_2)'(\mathbf{x}_1 - \mathbf{x}_2)$$

$$f(\mathbf{x}_2) \geq f(\mathbf{x}_1) + \nabla f(\mathbf{x}_1)'(\mathbf{x}_2 - \mathbf{x}_1)$$

Adding the two inequalities, we get $[\nabla f(\mathbf{x}_2) - \nabla f(\mathbf{x}_1)]'(\mathbf{x}_2 - \mathbf{x}_1) \geq 0$. To show the converse, let $\mathbf{x}_1, \mathbf{x}_2 \in S$. By the mean value theorem,

$$f(\mathbf{x}_2) - f(\mathbf{x}_1) = \nabla f(\mathbf{x})'(\mathbf{x}_2 - \mathbf{x}_1) \tag{3.11}$$

where $\mathbf{x} = \lambda \mathbf{x}_1 + (1 - \lambda)\mathbf{x}_2$ for some $\lambda \in (0, 1)$. By assumption, $[\nabla f(\mathbf{x}) - \nabla f(\mathbf{x}_1)]'(\mathbf{x} - \mathbf{x}_1) \geq 0$; that is, $(1 - \lambda)[\nabla f(\mathbf{x}) - \nabla f(\mathbf{x}_1)]'(\mathbf{x}_2 - \mathbf{x}_1) \geq 0$. This implies that $\nabla f(\mathbf{x})'(\mathbf{x}_2 - \mathbf{x}_1) \geq \nabla f(\mathbf{x}_1)'(\mathbf{x}_2 - \mathbf{x}_1)$. By (3.11), we get $f(\mathbf{x}_2) \geq f(\mathbf{x}_1) + \nabla f(\mathbf{x}_1)'(\mathbf{x}_2 - \mathbf{x}_1)$, and so by Theorem 3.3.3, f is convex. The strict case is similar and the proof is complete.

Even though Theorems 3.3.3 and 3.3.4 provide necessary and sufficient characterizations of convex functions, checking these conditions is difficult from a computational standpoint. A simple and more manageable characterization, at least for quadratic functions, can be obtained, provided that the function is twice differentiable.

Twice Differentiable Convex and Concave Functions

A function f that is differentiable at $\bar{\mathbf{x}}$ is said to be twice differentiable at $\bar{\mathbf{x}}$ if the *second-order (Taylor series) expansion* representation of Definition 3.3.5 below is true.

3.3.5 Definition

Let S be a nonempty set in E_n and let $f: S \rightarrow E_1$. Then, f is said to be *twice differentiable* at $\bar{\mathbf{x}} \in \text{int } S$ if there exist a vector $\nabla f(\bar{\mathbf{x}})$, and an $n \times n$ symmetric matrix $\mathbf{H}(\bar{\mathbf{x}})$, called the *Hessian* matrix, and a function $\alpha: E_n \rightarrow E_1$ such that

$$f(\mathbf{x}) = f(\bar{\mathbf{x}}) + \nabla f(\bar{\mathbf{x}})'(\mathbf{x} - \bar{\mathbf{x}}) + \tfrac{1}{2}(\mathbf{x} - \bar{\mathbf{x}})'\mathbf{H}(\bar{\mathbf{x}})(\mathbf{x} - \bar{\mathbf{x}}) + \|\mathbf{x} - \bar{\mathbf{x}}\|^2 \alpha(\bar{\mathbf{x}}; \mathbf{x} - \bar{\mathbf{x}})$$

for each $\mathbf{x} \in S$, where $\lim_{\mathbf{x} \to \bar{\mathbf{x}}} \alpha(\bar{\mathbf{x}}; \mathbf{x} - \bar{\mathbf{x}}) = 0$. The function f is said to be twice differentiable on the open set $S' \subseteq S$ if it is twice differentiable at each point in S'.

It may be noted that for twice-differentiable functions the Hessian matrix $\mathbf{H}(\bar{\mathbf{x}})$ is comprised of the second-order partial derivatives $\partial^2 f(\bar{\mathbf{x}})/\partial x_i \, \partial x_j \equiv f_{ij}(\bar{\mathbf{x}})$ for $i = 1, \ldots, n$, $j = 1, \ldots, n$, and is given as follows:

$$\mathbf{H}(\bar{\mathbf{x}}) = \begin{bmatrix} \dfrac{\partial^2 f(\bar{\mathbf{x}})}{\partial x_1^2} & \dfrac{\partial^2 f(\bar{\mathbf{x}})}{\partial x_1 x_2} & \cdots & \dfrac{\partial^2 f(\bar{\mathbf{x}})}{\partial x_1 x_n} \\ \dfrac{\partial^2 f(\bar{\mathbf{x}})}{\partial x_2 x_1} & \dfrac{\partial^2 f(\bar{\mathbf{x}})}{\partial x_2^2} & \cdots & \dfrac{\partial^2 f(\bar{\mathbf{x}})}{\partial x_2 x_n} \\ \vdots & \vdots & & \vdots \\ \dfrac{\partial^2 f(\bar{\mathbf{x}})}{\partial x_n x_1} & \dfrac{\partial^2 f(\bar{\mathbf{x}})}{\partial x_n x_2} & \cdots & \dfrac{\partial^2 f(\bar{\mathbf{x}})}{\partial x_n^2} \end{bmatrix} \equiv \begin{bmatrix} f_{11}(\bar{\mathbf{x}}) & f_{12}(\bar{\mathbf{x}}) & \cdots & f_{1n}(\bar{\mathbf{x}}) \\ f_{21}(\bar{\mathbf{x}}) & f_{22}(\bar{\mathbf{x}}) & \cdots & f_{2n}(\bar{\mathbf{x}}) \\ & \cdots & & \cdots \\ f_{n1}(\bar{\mathbf{x}}) & f_{n2}(\bar{\mathbf{x}}) & \cdots & f_{nn}(\bar{\mathbf{x}}) \end{bmatrix}$$

In expanded form, the foregoing representation can be written as

$$f(\mathbf{x}) = f(\bar{\mathbf{x}}) + \sum_{j=1}^{n} f_j(\bar{\mathbf{x}})(x_j - \bar{x}_j) + \frac{1}{2}\sum_{i=1}^{n}\sum_{j=1}^{n} (x_i - \bar{x}_i)(x_j - \bar{x}_j)f_{ij}(\bar{\mathbf{x}}) + \|\mathbf{x} - \bar{\mathbf{x}}\|^2 \alpha(\bar{\mathbf{x}}; \mathbf{x} - \bar{\mathbf{x}})$$

Again, without the remainder term associated with the function α, this representation is known as a *second-order (Taylor Series) approximation.*

3.3.6 Examples

Example 1. Let $f(x_1, x_2) = 2x_1 + 6x_2 - 2x_1^2 - 3x_2^2 + 4x_1x_2$. Then, we have

$$\nabla f(\bar{\mathbf{x}}) = \begin{bmatrix} 2 - 4\bar{x}_1 + 4\bar{x}_2 \\ 6 - 6\bar{x}_2 + 4\bar{x}_1 \end{bmatrix} \text{ and } \mathbf{H}(\bar{\mathbf{x}}) = \begin{bmatrix} -4 & 4 \\ 4 & -6 \end{bmatrix}$$

For example, taking $\bar{\mathbf{x}} = (0, 0)^t$, the second-order expansion of this function is given by

$$f(x_1, x_2) = (2, 6)\begin{pmatrix} x_1 \\ x_2 \end{pmatrix} + \frac{1}{2}(x_1, x_2)\begin{bmatrix} -4 & 4 \\ 4 & -6 \end{bmatrix}\begin{pmatrix} x_1 \\ x_2 \end{pmatrix}$$

Note that there is no remainder term here since the given function is quadratic, and so the above representation is exact.

Example 2. Let $f(x_1, x_2) = e^{2x_1 + 3x_2}$. Then, we get

$$\nabla f(\bar{\mathbf{x}}) = \begin{bmatrix} 2e^{2\bar{x}_1 + 3\bar{x}_2} \\ 3e^{2\bar{x}_1 + 3\bar{x}_2} \end{bmatrix} \text{ and } \mathbf{H}(\bar{\mathbf{x}}) = \begin{bmatrix} 4e^{2\bar{x}_1 + 3\bar{x}_2} & 6e^{2\bar{x}_1 + 3\bar{x}_2} \\ 6e^{2\bar{x}_1 + 3\bar{x}_2} & 9e^{2\bar{x}_1 + 3\bar{x}_2} \end{bmatrix}$$

Hence, the second-order expansion of this function about the point $\bar{\mathbf{x}} = (2, 1)^t$ is given by

$$f(\bar{\mathbf{x}}) = e^7 + (2e^7, 3e^7)\begin{pmatrix} x_1 - 2 \\ x_2 - 1 \end{pmatrix} + \frac{1}{2}(x_1 - 2, x_2 - 1)\begin{bmatrix} 4e^7 & 6e^7 \\ 6e^7 & 9e^7 \end{bmatrix}\begin{pmatrix} x_1 - 2 \\ x_2 - 1 \end{pmatrix}$$
$$+ \|\mathbf{x} - \bar{\mathbf{x}}\|^2 \alpha(\bar{\mathbf{x}}; \mathbf{x} - \bar{\mathbf{x}})$$

The following theorem shows that f is convex on S if and only if its Hessian matrix is *positive semidefinite* (PSD) everywhere in S; that is, for any $\bar{\mathbf{x}}$ in S, we have $\mathbf{x}^t\mathbf{H}(\bar{\mathbf{x}})\mathbf{x} \geq 0$ for all $\mathbf{x} \in E_n$. Symmetrically, a function f is concave on S if and only if its Hessian matrix is *negative semidefinite (NSD)* everywhere in S, that is, for any $\bar{\mathbf{x}} \in S$, we have $\mathbf{x}^t\mathbf{H}(\bar{\mathbf{x}})\mathbf{x} \leq 0$ for all $\mathbf{x} \in E_n$. A matrix that is neither positive nor negative semidefinite is called *indefinite (ID)*.

3.3.7 Theorem

Let S be a nonempty open convex set in E_n, and let $f: S \rightarrow E_1$ be twice differentiable on S. Then, f is convex if and only if the Hessian matrix is positive semidefinite at each point in S.

Proof

Suppose that f is convex and let $\bar{\mathbf{x}} \in S$. We need to show that $\mathbf{x}^t\mathbf{H}(\bar{\mathbf{x}})\mathbf{x} \geq 0$ for each $\mathbf{x} \in E_n$. Since S is open, then, for any given $\mathbf{x} \in E_n$, $\bar{\mathbf{x}} + \lambda\mathbf{x} \in S$ for $|\lambda| \neq 0$ and sufficiently small. By Theorem 3.3.3 and by the twice-differentiability of f, we get the following two expressions:

$$f(\bar{\mathbf{x}} + \lambda\mathbf{x}) \geq f(\bar{\mathbf{x}}) + \lambda\nabla f(\bar{\mathbf{x}})^t\mathbf{x} \tag{3.12}$$

$$f(\bar{\mathbf{x}} + \lambda\mathbf{x}) = f(\bar{\mathbf{x}}) + \lambda\nabla f(\bar{\mathbf{x}})^t\mathbf{x} + \tfrac{1}{2}\lambda^2\mathbf{x}^t\mathbf{H}(\bar{\mathbf{x}})\mathbf{x} + \lambda^2\,\|\mathbf{x}\|^2\,\alpha(\bar{\mathbf{x}}; \lambda\mathbf{x}) \tag{3.13}$$

Subtracting (3.13) from (3.12), we get

$$\tfrac{1}{2}\lambda^2\mathbf{x}^t\mathbf{H}(\bar{\mathbf{x}})\mathbf{x} + \lambda^2\,\|\mathbf{x}\|^2\,\alpha(\bar{\mathbf{x}}; \lambda\mathbf{x}) \geq 0$$

Dividing by $\lambda^2 > 0$ and letting $\lambda \to 0$, it follows that $\mathbf{x}^t\mathbf{H}(\bar{\mathbf{x}})\mathbf{x} \geq 0$. Conversely, suppose that the Hessian matrix is positive semidefinite at each point in S. Consider $\mathbf{x}$ and $\bar{\mathbf{x}}$ in S. Then, by the mean value theorem, we have

$$f(\mathbf{x}) = f(\bar{\mathbf{x}}) + \nabla f(\bar{\mathbf{x}})^t(\mathbf{x} - \bar{\mathbf{x}}) + \tfrac{1}{2}(\mathbf{x} - \bar{\mathbf{x}})^t\mathbf{H}(\hat{\mathbf{x}})(\mathbf{x} - \bar{\mathbf{x}}) \tag{3.14}$$

where $\hat{\mathbf{x}} = \lambda\bar{\mathbf{x}} + (1 - \lambda)\mathbf{x}$ for some $\lambda \in (0, 1)$. Note that $\hat{\mathbf{x}} \in S$ and, hence, by assumption, $\mathbf{H}(\hat{\mathbf{x}})$ is positive semidefinite. Therefore, $(\mathbf{x} - \bar{\mathbf{x}})^t\mathbf{H}(\hat{\mathbf{x}})(\mathbf{x} - \bar{\mathbf{x}}) \geq 0$, and from (3.14), we conclude that

$$f(\mathbf{x}) \geq f(\bar{\mathbf{x}}) + \nabla f(\bar{\mathbf{x}})^t(\mathbf{x} - \bar{\mathbf{x}})$$

Since the above inequality is true for each $\mathbf{x}$, $\bar{\mathbf{x}}$ in S, f is convex by Theorem 3.3.3. This completes the proof.

The above theorem is useful in checking convexity or concavity of a twice differentiable function. In particular, if the function is quadratic, then the Hessian matrix is independent of the point under consideration. Hence, checking convexity reduces to checking positive semidefiniteness for a constant matrix.

Results analogous to Theorem 3.3.7 can be obtained for the strict convex and concave cases. It turns out that if the Hessian matrix is positive definite at each point in S, then the function is strictly convex. In other words, if for any given point $\bar{\mathbf{x}}$ in S, we have $\mathbf{x}^t\mathbf{H}(\bar{\mathbf{x}})\mathbf{x} > 0$ for all $\mathbf{x} \neq \mathbf{0}$ in E_n, then f is strictly convex. This follows readily from the proof of Theorem 3.3.7. However, if f is strictly convex, then its Hessian matrix is *positive semidefinite*. It is not necessarily true that the Hessian matrix is positive definite everywhere in S, unless if f is quadratic. The latter is seen by writing (3.12) as a strict inequality for $\lambda\mathbf{x} \neq 0$, and noting that the remainder term in (3.13) is then absent. To illustrate, consider the strictly convex function defined by $f(x) = x^4$. The Hessian matrix $\mathbf{H}(x) = 12x^2$ is positive definite for all nonzero x but is *positive semidefinite* at $x = 0$. The following theorem records this fact.

3.3.8 Theorem

Let S be a nonempty open convex set in E_n, and let $f: S \to E_1$ be twice differentiable on S. If the Hessian matrix is positive definite at each point in S, then f is strictly convex. Conversely, if f is strictly convex, then the Hessian matrix is positive semidefinite at each point in S. However, if f is strictly convex and quadratic, then its Hessian is positive definite.

The foregoing result can be somewhat strengthened while providing some additional insights into the second-order characterization of convexity. Consider, for example, the univariate function $f(x) = x^4$ addressed above, and let us show how we can argue that this function is strictly convex despite the fact that $f''(0) = 0$. Since $f''(x) \geq 0$ for all $x \in E_1$, we have by Theorem 3.3.7 that f is convex. Hence, by Theorem 3.3.3, all that we need to show is that for any point $\bar{x}$, the supporting hyperplane $y = f(\bar{x}) + \nabla f(\bar{\mathbf{x}})^t (\mathbf{x} - \bar{\mathbf{x}})$ to the epigraph of the function touches this epigraph only at the given point $(x, y) = (\bar{x}, f(\bar{x}))$. On the contrary, if this supporting hyperplane also touches the epigraph at some other point $(\hat{x}, f(\hat{x}))$, then we have $f(\hat{x}) = f(\bar{x}) + \nabla f(\bar{x})^t(\hat{x} - \bar{x})$. But this means that for any $x_\lambda = \lambda\bar{x} + (1 - \lambda)\hat{x}$, $0 \leq \lambda \leq 1$, we have, on using Theorem 3.3.3 and the convexity of f,

$$\lambda f(\bar{x}) + (1 - \lambda)f(\hat{x}) = f(\bar{x}) + \nabla f(\bar{x})^t(x_\lambda - \bar{x}) \leq f(x_\lambda) \leq \lambda f(\bar{x}) + (1 - \lambda)f(\hat{x})$$

Hence, equality holds true throughout, and the supporting hyperplane touches the graph of the function at all convex combinations $(x_\lambda, f(x_\lambda))$ as well. In fact, we obtain $f(x_\lambda) = \lambda f(\bar{x}) + (1 - \lambda)f(\hat{x})$ for all $0 \leq \lambda \leq 1$, and so $f''(x_\lambda) = 0$ at the uncountably infinite number of points x_λ for all $0 < \lambda < 1$ as well. This contradicts the fact that $f''(x) = 0$ only at $x = 0$ from the above example, and so the function is strictly convex. Therefore, we can lose positive definiteness of a univariate convex function at only a finite (or countably infinite) number of points and still claim strict convexity.

Staying with univariate functions for the time being, if the function is infinitely differentiable, then we can derive a necessary and sufficient condition for the function to be strictly convex. (By an *infinitely differentiable* function f, we mean one for which for any $\bar{\mathbf{x}}$ in E_n derivatives of all orders exist, and so are continuous; are uniformly bounded in values; and for which the infinite Taylor series expansion of $f(\mathbf{x})$ about $f(\bar{\mathbf{x}})$ gives an infinite series representation of the value of f. Of course, this infinite series can possibly have only a finite number of terms as, for example, when derivatives of order exceeding some value all vanish.)

3.3.9 Theorem

Let S be a nonempty open convex set in E_1 and let $f: S \to E_1$ be infinitely differentiable. Then, f is strictly convex on S if and only if, for each $\bar{x} \in S$, there exists an even n such that $f^{(n)}(\bar{x}) > 0$, while $f^{(j)}(\bar{x}) = 0$ for any $1 < j < n$, where $f^{(j)}$ denotes the jth order derivative of f.

Proof

Let $\bar{x}$ be any point in S, and consider the infinite Taylor series expansion of f about $\bar{x}$ for a perturbation $h \neq 0$ and small enough:

$$f(\bar{x} + h) = f(\bar{x}) + hf'(\bar{x}) + \frac{h^2}{2!}f''(\bar{x}) + \frac{h^3}{3!}f'''(\bar{x}) + \cdots$$

If f is strictly convex, then, by Theorem 3.3.3, we have that $f(\bar{x} + h) > f(\bar{x}) + hf'(\bar{x})$ for $h \neq 0$. Using this above, we get that, for all $h \neq 0$ and sufficiently small,

$$\frac{h^2}{2!}f''(\bar{x}) + \frac{h^3}{3!}f'''(\bar{x}) + \frac{h^4}{4!}f^{(4)}(\bar{x}) + \cdots > 0$$

Hence, not all derivatives of order greater than or equal to 2 at $\bar{x}$ can be zero. Moreover, since by making h sufficiently small, we can make the first nonzero term above dominate the rest of the expansion; and since h can be of either sign, it follows that this first nonzero derivative must be of an even order and positive for the inequality to hold true.

Conversely, suppose that given any $\bar{\mathbf{x}} \in S$, there exists an even n such that $f^{(n)}(\bar{\mathbf{x}}) > 0$, while $f^{(j)}(\bar{\mathbf{x}}) = 0$ for $1 < j < n$. Then, as above, we have $(\bar{x} + h) \in S$ and $f(\bar{x} + h) > f(\bar{x}) + hf'(\bar{x})$ for all $-\delta < h < \delta$, for some $\delta > 0$ and sufficiently small. Now the given hypothesis also asserts that $f''(\bar{x}) \geq 0$ for all $\bar{x} \in S$, and so, by Theorem 3.3.7, we know that f is convex. Consequently, for any $\bar{h} \neq 0$, $(\bar{x} + \bar{h}) \in S$, we get $f(\bar{x} + \bar{h}) \geq f(\bar{x}) + \bar{h}f'(\bar{x})$ by Theorem 3.3.3. To complete the proof, we must show that this inequality is indeed strict. On the contrary, if $f(\bar{x} + \bar{h}) = f(\bar{x}) + hf'(\bar{x})$, we get

$$\lambda f(\bar{x} + \bar{h}) + (1 - \lambda)f(\bar{x}) = f(\bar{x}) + \lambda \bar{h}f'(\bar{x}) \leq f(\bar{x} + \lambda \bar{h})$$

$$= f[\lambda(\bar{x} + \bar{h}) + (1 - \lambda)\bar{x}] \leq \lambda f(\bar{x} + \bar{h}) + (1 - \lambda)f(x^{-})$$

for all $0 \leq \lambda \leq 1$. But this means that equality holds throughout and that $f(\bar{x} + \lambda\bar{h}) = f(\bar{x}) + \lambda\bar{h}f'(\bar{x})$ for all $0 \leq \lambda \leq 1$. By taking λ close enough to zero, we can contradict the statement that $f(\bar{x} + h) > f(\bar{x}) + hf'(\bar{x})$ for all $-\delta < h < \delta$, and this completes the proof.

To illustrate, when $f(x) = x^4$, we have $f'(x) = 4x^3$ and $f''(x) = 12x^2$. Hence, for $\bar{x} \neq 0$, the first nonzero derivative of Theorem 3.3.9 is of order 2 and is positive. Furthermore, for $\bar{x} = 0$, we have $f'''(\bar{x}) = 0$ and $f^{(4)}(\bar{x}) = 24 > 0$; and so, by Theorem 3.3.9, we can conclude that f is strictly convex.

Now let us turn to the multivariate case. The following result provides an insightful connection between the univariate and the multivariate cases and permits us to derive results for the latter case from those for the former case. For notational simplicity, we have stated this result for $f: E_n \to E_1$, although one can readily restate it for $f: S \to E_1$, where S is some nonempty convex subset of E_n.

3.3.10 Theorem

Consider a function $f: E_n \to E_1$, and for any point $\bar{\mathbf{x}} \in E_n$ and a nonzero direction $\mathbf{d} \in E_n$, define $F_{(\bar{\mathbf{x}};\mathbf{d})}(\lambda) = f(\bar{\mathbf{x}} + \lambda\mathbf{d})$ as a function of $\lambda \in E_1$. Then, f is (strictly) convex if and only if $F_{(\bar{\mathbf{x}};\mathbf{d})}$ is (strictly) convex for all $\bar{\mathbf{x}}$ and $\mathbf{d} \neq \mathbf{0}$ in E_n.

Proof

Given any $\bar{\mathbf{x}}$ and $\mathbf{d} \neq \mathbf{0}$ in E_n, let us write $F_{(\bar{\mathbf{x}};\mathbf{d})}(\lambda)$ simply as $F(\lambda)$ for convenience. If f is convex, then, for any λ_1 and λ_2 in E_1 and for any $0 \leq \alpha \leq 1$, we have

$$F(\alpha\lambda_1 + (1 - \alpha)\lambda_2) = f(\alpha[\bar{\mathbf{x}} + \lambda_1\mathbf{d}] + (1 - \alpha)[\bar{\mathbf{x}} + \lambda_2\mathbf{d}])$$

$$\leq \alpha f(\bar{\mathbf{x}} + \lambda_1\mathbf{d}) + (1 - \alpha)f(\bar{\mathbf{x}} + \lambda_2\mathbf{d}) = \alpha F(\lambda_1) + (1 - \alpha)F(\lambda_2)$$

Hence, F is convex. Conversely, suppose that $F_{(\bar{\mathbf{x}};\mathbf{d})}(\lambda)$, $\lambda \in E_1$, is convex for all $\bar{\mathbf{x}}$ and $\mathbf{d} \neq \mathbf{0}$ in E_n. Then, for any $\mathbf{x}_1$ and $\mathbf{x}_2$ in E_n and $0 \leq \lambda \leq 1$, we have

$$\lambda f(\mathbf{x}_1) + (1 - \lambda)f(\mathbf{x}_2) = \lambda f[\mathbf{x}_1 + 0(\mathbf{x}_2 - \mathbf{x}_1)] + (1 - \lambda)f[\mathbf{x}_1 + 1(\mathbf{x}_2 - \mathbf{x}_1)]$$

$$= \lambda F_{[\mathbf{x}_1;(\mathbf{x}_2 - \mathbf{x}_1)]}(0) + (1 - \lambda)F_{[\mathbf{x}_1;(\mathbf{x}_2 - \mathbf{x}_1)]}(1) \geq F_{[\mathbf{x}_1;(\mathbf{x}_2 - \mathbf{x}_1)]}(1 - \lambda)$$

$$= f[\mathbf{x}_1 + (1 - \lambda)(\mathbf{x}_2 - \mathbf{x}_1)] = f[\lambda\mathbf{x}_1 + (1 - \lambda)\mathbf{x}_2]$$

and so f is convex. The argument for the strictly convex case is similar, and this completes the proof.

This insight of examining $f: E_n \to E_1$ via its univariate cross sections $F_{(\bar{\mathbf{x}};\mathbf{d})}$ can be very useful both as a conceptual tool for viewing f and as an analytical tool for deriving various results. For example, writing $F(\lambda) = F_{(\bar{\mathbf{x}};\mathbf{d})}(\lambda) = f(\bar{\mathbf{x}} + \lambda\mathbf{d})$, for any given $\bar{\mathbf{x}}$ and $\mathbf{d} \neq \mathbf{0}$ in E_n, we have from the univariate Taylor series expansion (assuming infinite differentiability) that

$$F(\lambda) = F(0) + \lambda F'(0) + \frac{\lambda^2}{2!}F''(0) + \frac{\lambda^3}{3!}F'''(0) + \cdots$$

By using the chain rule for differentiation, we obtain

$$F'(\lambda) = \nabla f(\bar{\mathbf{x}} + \lambda\mathbf{d})^t\mathbf{d} \equiv \sum_i f_i(\bar{\mathbf{x}} + \lambda\mathbf{d})d_i$$

$$F''(\lambda) = \mathbf{d}^t\mathbf{H}(\bar{\mathbf{x}} + \lambda\mathbf{d})\mathbf{d} \equiv \sum_i\sum_j f_{ij}(\bar{\mathbf{x}} + \lambda\mathbf{d})$$

$$F'''(\lambda) = \sum_i\sum_j\sum_k f_{ijk}(\bar{\mathbf{x}} + \lambda\mathbf{d})d_id_jd_k$$

Substituting above, this gives the corresponding multivariate Taylor series expansion as

$$f(\bar{\mathbf{x}} + \lambda\mathbf{d}) = f(\bar{\mathbf{x}}) + \lambda\nabla f(\bar{\mathbf{x}})^t\mathbf{d} + \frac{\lambda^2}{2!}\mathbf{d}^t\mathbf{H}(\bar{\mathbf{x}})\mathbf{d} + \frac{\lambda^3}{3!}\sum_i\sum_j\sum_k f_{ijk}(\bar{\mathbf{x}})d_id_jd_k + \cdots$$

As another example, using the second-order derivative result for characterizing the convexity of a univariate function along with Theorem 3.3.10, we can derive that $f: E_n \to E_1$ is convex if and only if $F''_{(\bar{\mathbf{x}};\mathbf{d})}(\lambda) \geq 0$ for all $\lambda \in E_1$ $\bar{\mathbf{x}} \in E_n$ and $\mathbf{d} \in E_n$. But since $\bar{\mathbf{x}}$ and $\mathbf{d}$ can be chosen arbitrarily, this is equivalent to requiring that $F''_{(\bar{\mathbf{x}};\mathbf{d})}(0) \geq 0$ for all $\bar{\mathbf{x}}$ and $\mathbf{d}$ in E_n. From above, this translates to the statement that $\mathbf{d}^t\mathbf{H}(\bar{\mathbf{x}})\mathbf{d} \geq 0$ for all $\mathbf{d} \in E_n$, for each $\bar{\mathbf{x}} \in E_n$, or that $\mathbf{H}(\bar{\mathbf{x}})$ is positive semidefinite for all $\bar{\mathbf{x}} \in E_n$, as in Theorem 3.3.7. In a likewise manner, or by directly using the multivariate Taylor series expansion as in the proof of Theorem 3.3.9, we can assert that an infinitely differentiable function $f: E_n \to E_1$ is strictly convex if and only if, for each $\bar{\mathbf{x}}$ and $\mathbf{d} \neq \mathbf{0}$ in E_n, the first nonzero derivative term of order greater than or equal to 2 in the above Taylor series expansion exists, is of even order, and is positive. We leave the details of exploring this result to the reader in Exercise 3.31.

We now present below an efficient (polynomial-time) algorithm for checking the definiteness of a (symmetric) Hessian matrix $\mathbf{H}(\bar{\mathbf{x}})$ using elementary Gauss–Jordan operations. The appendix cites a characterization of definiteness in terms of eigenvalues which finds use in some analytical proofs but is not an algorithmically convenient alternative. Moreover, if one needs to check for the definiteness of a matrix $\mathbf{H}(\mathbf{x})$ that is a function of $\mathbf{x}$, then this eigenvalue method is very cumbersome, if not virtually impossible, to use. Although the method presented below can also get messy in such instances, it is overall a more simple and efficient approach.

We begin by considering a 2×2 Hessian matrix $\mathbf{H}$ in Lemma 3.3.11 below, where the argument $\bar{\mathbf{x}}$ has been suppressed for convenience. This is then generalized in an inductive fashion to an $n \times n$ matrix in Theorem 3.3.12.

3.3.11 Lemma

Consider a symmetric matrix $\mathbf{H} = \begin{bmatrix} a & b \\ b & c \end{bmatrix}$. Then, $\mathbf{H}$ is positive semidefinite if and only if $a \geq 0$, $c \geq 0$, and $ac - b^2 \geq 0$, and is positive definite if and only if the foregoing inequalities are all strict.

Proof

By definition, H is positive semidefinite if and only if $\mathbf{d}^t\mathbf{H}\mathbf{d} = ad_1^2 + 2bd_1d_2 + cd_2^2 \geq 0$ for all $(d_1, d_2)^t \in E_2$. Hence, if $\mathbf{H}$ is positive semidefinite, then we must clearly have $a \geq 0$ and $c \geq 0$. Moreover, if $a = 0$, then we must have $b = 0$, and so $ac - b^2 = 0$; or else, by taking $d_2 = 1$ and $d_1 = -Mb$ for $M > 0$ and large enough, we would obtain $\mathbf{d}^t\mathbf{H}\mathbf{d} < 0$, a contradiction. On the other hand, if $a > 0$, then, completing the squares, we get

$$\mathbf{d}^t\mathbf{H}\mathbf{d} = a[d_1^2 + \frac{2bd_1d_2}{a} + \frac{b^2}{a^2}d_2^2] + d_2^2\left(c - \frac{b^2}{a}\right) = a\left(d_1 + \frac{b}{a}d_2\right)^2 + d_2^2\left(\frac{ac - b^2}{a}\right)$$

Hence, we must again have $(ac - b^2) \geq 0$ since, otherwise, by taking $d_2 = 1$ and $d_1 = -b/a$, we would get $\mathbf{d}^t\mathbf{H}\mathbf{d} = (ac - b^2)/a < 0$, a contradiction. Hence, the condition of the theorem holds. Conversely, suppose that $a \geq 0$, $c \geq 0$, and $ac - b^2 \geq 0$. If $a = 0$, this gives $b = 0$, and so $\mathbf{d}^t\mathbf{H}\mathbf{d} = cd_2^2 \geq 0$. On the other hand, if $a > 0$, then, by completing the squares as above, we get

$$\mathbf{d}^t\mathbf{H}\mathbf{d} = a\left(d_1 + \frac{b}{a}d_2^2\right) + d_2^2\left(\frac{ac - b^2}{a}\right) \geq 0$$

Hence, $\mathbf{H}$ is positive semidefinite. The proof of positive definiteness is similar, and this completes the proof.

We remark here that, since a matrix $\mathbf{H}$ is negative semidefinite (negative definite) if and only if $-\mathbf{H}$ is positive semidefinite (positive definite), we get from Lemma 3.3.11 that $\mathbf{H}$ is negative semidefinite if and only if $a \leq 0$, $c \leq 0$, and $ac - b^2 \geq 0$, and that $\mathbf{H}$ is negative definite if and only if these inequalities are strict. The theorem below is stated for checking positive semidefiniteness or positive definiteness of $\mathbf{H}$. By replacing $\mathbf{H}$ by $-\mathbf{H}$, we could symmetrically test for negative semidefiniteness or negative definiteness. If the matrix turns out to be neither positive semidefinite nor negative semidefinite, then it is indefinite. Also, we assume below that $\mathbf{H}$ is symmetric, being the Hessian of a twice-differentiable function for our purposes. In general, if $\mathbf{H}$ is not symmetric, then, since $\mathbf{d}^t\mathbf{H}\mathbf{d} = \mathbf{d}^t\mathbf{H}^t\mathbf{d} = \mathbf{d}^t[(\mathbf{H} + \mathbf{H}^t)/2]\mathbf{d}$, we can check for the definiteness of $\mathbf{H}$ by using the symmetric matrix $(\mathbf{H} + \mathbf{H}^t)/2$ below.

3.3.12 Theorem (Checking for PSD/PD)

Let $\mathbf{H}$ be a symmetric $n \times n$ matrix with elements h_{ij}.

(a) If $h_{ii} \leq 0$ for any $i \in \{1, \ldots, n\}$, then $\mathbf{H}$ is not positive definite; and if $h_{ii} < 0$ for any $i \in \{1, \ldots, n\}$, then $\mathbf{H}$ is not positive semidefinite.

(b) If $h_{ii} = 0$ for any $i \in \{1, \ldots, n\}$, then we must have $h_{ij} = h_{ji} = 0$ for all $j = 1, \ldots, n$ as well, or else $\mathbf{H}$ is not positive semidefinite.

(c) If $n = 1$, then $\mathbf{H}$ is positive semidefinite (positive definite) if and only if $h_{11} \geq 0$ (>0). Otherwise, if $n \geq 2$, let

$$\mathbf{H} = \begin{bmatrix} h_{11} & \mathbf{q}^t \\ \mathbf{q} & \mathbf{G} \end{bmatrix}$$

in partitioned form, where $\mathbf{q} = \mathbf{0}$ if $h_{11} = 0$ and, otherwise, $h_{11} > 0$. Perform elementary Gauss–Jordan operations using the first row of $\mathbf{H}$ to reduce it to the following matrix in either case:

$$\mathbf{H} = \begin{bmatrix} \mathbf{h}_{11} & \mathbf{q}^t \\ \mathbf{0} & \mathbf{G}_{\text{new}} \end{bmatrix}$$

Then, $\mathbf{G}_{\text{new}}$ is a symmetric $(n - 1) \times (n - 1)$ matrix, and $\mathbf{H}$ is positive semidefinite if and only if $\mathbf{G}_{\text{new}}$ is positive semidefinite. Moreover, if $h_{11} > 0$, then $\mathbf{H}$ is positive definite if and only if $\mathbf{G}_{\text{new}}$ is positive definite.

Proof

(a) Since $\mathbf{d}^t\mathbf{H}\mathbf{d} = d_i^2 h_{ii}$ whenever $d_j = 0$ for all $j \neq i$, part a of the theorem is obviously true.

(b) Suppose that for some $i \neq j$, we have $h_{ii} = 0$ and $h_{ij} \neq 0$. Then, by taking $d_k = 0$ for all $k \neq i$ or j, we get $\mathbf{d}^t\mathbf{H}\mathbf{d} = 2h_{ij}d_id_j + d_j^2h_{jj}$, which can be made negative as in the proof of Lemma 3.3.11 by taking $d_j = 1$ and $d_i = -h_{ij}M$ for $M > 0$ and sufficiently large. This establishes part b.

(c) Finally, suppose that $\mathbf{H}$ is given in partitioned form as in part c. If $n = 1$, then the result is trivial. Otherwise, for $n \geq 2$, let $\mathbf{d}^t = (d_1, \boldsymbol{\delta}^t)$. If $h_{11} = 0$, then, by assumption, we also have $\mathbf{q} = \mathbf{0}$, and then $\mathbf{G}_{\text{new}} = \mathbf{G}$. Moreover, in this case, $\mathbf{d}^t\mathbf{H}\mathbf{d} = \boldsymbol{\delta}^t\mathbf{G}_{\text{new}}\boldsymbol{\delta}$, and so $\mathbf{H}$ is positive semidefinite if and only if $\mathbf{G}_{\text{new}}$ is positive semidefinite. On the other hand, if $h_{11} > 0$, then we get

$$\mathbf{d}^t\mathbf{H}\mathbf{d} = (d_1, \boldsymbol{\delta}^t)\begin{bmatrix} h_{11} & \mathbf{q}^t \\ \mathbf{q} & \mathbf{G} \end{bmatrix}\begin{pmatrix} d_1 \\ \boldsymbol{\delta} \end{pmatrix} = d_1^2h_{11} + 2d_1(\mathbf{q}^t\boldsymbol{\delta}) + \boldsymbol{\delta}^t\mathbf{G}\boldsymbol{\delta}$$

But by the Gauss–Jordan reduction process, we have

$$\mathbf{G}_{\text{new}} = \mathbf{G} - \frac{1}{h_{11}}\begin{bmatrix} q_1\mathbf{q}^t \\ q_2\mathbf{q}^t \\ \vdots \\ q_n\mathbf{q}^t \end{bmatrix} = \mathbf{G} - \frac{1}{h_{11}}\mathbf{q}\mathbf{q}^t$$

which is a symmetric matrix. By substituting this above, we get

$$\mathbf{d}^t\mathbf{H}\mathbf{d} = d_1^2h_{11} + 2d_1(\mathbf{q}^t\boldsymbol{\delta}) + \boldsymbol{\delta}^t\left[\mathbf{G}_{\text{new}} + \frac{1}{h_{11}}\mathbf{q}\mathbf{q}^t\right]\boldsymbol{\delta} = \boldsymbol{\delta}^t\mathbf{G}_{\text{new}}\boldsymbol{\delta} + h_{11}\left[d_1 + \frac{\mathbf{q}^t\boldsymbol{\delta}}{h_{11}}\right]^2$$

Hence, it can be readily verified that $\mathbf{d}^t\mathbf{H}\mathbf{d} \geq 0$ for all $\mathbf{d} \in E_n$ if and only if $\boldsymbol{\delta}^t\mathbf{G}_{\text{new}}\boldsymbol{\delta} \geq 0$ for all $\boldsymbol{\delta} \in E_{n-1}$, since $h_{11}[d_1 + \mathbf{q}^t\boldsymbol{\delta}/h_{11}]^2 \geq 0$; and this last term can be made zero by selecting $d_1 = -\mathbf{q}^t\boldsymbol{\delta}/h_{11}$, if necessary. By the same argument, $\mathbf{d}^t\mathbf{H}\mathbf{d} > 0$ for all $\mathbf{d} \neq \mathbf{0}$ in E_n if and only if $\boldsymbol{\delta}^t\mathbf{G}_{\text{new}}\boldsymbol{\delta} > 0$ for all $\boldsymbol{\delta} \neq \mathbf{0}$ in E_{n-1}, and this completes the proof.

Observe that Theorem 3.3.12 prompts a polynomial-time algorithm for checking the PSD/PD of a symmetric $n \times n$ matrix $\mathbf{H}$. We first scan the diagonal elements to see if either condition (a) or (b) leads to the conclusion that the matrix is not PSD/PD. If this does not terminate the process, we perform the Gauss–Jordan reduction as in part (c), and arrive at a matrix $\mathbf{G}_{\text{new}}$ of one lesser dimension for which we may now perform the same test as on $\mathbf{H}$. When $\mathbf{G}_{\text{new}}$ is finally a 2×2 matrix, we can use Lemma 3.3.11, or we can continue to reduce it to a 1×1 matrix and, hence, determine the PSD/PD of **H**. Since each pass through the inductive step of the algorithm is of complexity $O(n^2)$ (read as "*of order* n^2" and meaning that the number of elementary arithmetic operations, comparison, etc., involved are bounded above by Kn^2 for some constant K) and the number of steps is of $O(n)$, the overall process is of polynomial complexity $O(n^3)$. Because the algorithm basically works toward reducing the matrix to an upper triangular one, it is sometimes called a *superdiagonalization algorithm*. This algorithm affords a proof for the following useful result, which can alternatively be proved using the eigenvalue characterization of definiteness (see Exercise 3.33).

Corollary

Let **H** be an $n \times n$ symmetric matrix. Then **H** is positive definite if and only if it is positive semidefinite and nonsingular.

Proof

If **H** is positive definite, then it is positive semidefinite; and since the superdiagonalization algorithm reduces the matrix **H** to an upper triangular matrix with positive diagonal elements via elementary row operations, **H** is nonsingular. Conversely, if **H** is positive semidefinite and nonsingular, the superdiagonalization algorithm must always encounter nonzero elements along the diagonal because **H** is nonsingular, and these must be positive because **H** is positive semidefinite. Hence, **H** is positive definite.

3.3.13 Examples

Example 1. Consider Example 1 of Section 3.3.6. Here, we have

$$\mathbf{H}(\bar{\mathbf{x}}) = \begin{bmatrix} -4 & 4 \\ 4 & -6 \end{bmatrix}$$

and so

$$-\mathbf{H}(\bar{\mathbf{x}}) = \begin{bmatrix} 4 & -4 \\ -4 & 6 \end{bmatrix}$$

By Lemma 3.3.11, we conclude that $-\mathbf{H}(\bar{\mathbf{x}})$ is positive definite, and so $\mathbf{H}(\bar{\mathbf{x}})$ is negative definite and the function f is strictly concave.

Example 2. Consider the function $f(x_1, x_2) = x_1^3 + 2x_2^2$. Here, we have

$$\nabla f(\bar{\mathbf{x}}) = \begin{bmatrix} 3x_1^2 \\ 4x_2 \end{bmatrix} \quad \text{and} \quad \mathbf{H}(\bar{\mathbf{x}}) = \begin{bmatrix} 6x_1 & 0 \\ 0 & 4 \end{bmatrix}$$

By Lemma 3.3.11, whenever $x_1 < 0$, $\mathbf{H}(\mathbf{x})$ is indefinite. However, $\mathbf{H}(\mathbf{x})$ is positive definite for $x_1 > 0$, and so f is strictly convex over $\{\mathbf{x} : x_1 > 0\}$.

Example 3. Consider the matrix

$$\mathbf{H} = \begin{bmatrix} 2 & 1 & 2 \\ 1 & 2 & 3 \\ 2 & 3 & 4 \end{bmatrix}$$

Note that the matrix is not negative semidefinite. To check PSD/PD, apply the superdiagonalization algorithm and reduce H to

$$\begin{bmatrix} 2 & 1 & 2 \\ 0 & \frac{3}{2} & 2 \\ 0 & 2 & 2 \end{bmatrix} \quad \text{which gives} \quad \mathbf{G}_{\text{new}} = \begin{bmatrix} \frac{3}{2} & 2 \\ 2 & 2 \end{bmatrix}$$

Now, the diagonals of $\mathbf{G}_{\text{new}}$ are positive, but $\det(\mathbf{G}_{\text{new}}) = -1$. Hence, $\mathbf{H}$ is not positive semidefinite. Alternatively, we could have verified this by continuing to reduce $\mathbf{G}_{\text{new}}$ to obtain the matrix

$$\begin{bmatrix} \frac{3}{2} & 2 \\ 0 & -\frac{2}{3} \end{bmatrix}$$

Since the diagonal element is negative, $\mathbf{H}$ is not positive semidefinite. Since $\mathbf{H}$ is not negative semidefinite either, it is indefinite.

3.4 Minima and Maxima of Convex Functions

In this section, we consider the problems of minimizing and maximizing a convex function over a convex set and develop necessary and/or sufficient conditions for optimality.

Minimizing a Convex Function

The case of maximizing a concave function is similar to that of minimizing a convex function. We develop the latter in detail and ask the reader to draw the analogous results for the concave case.

3.4.1 Definition

Let $f: E_n \to E_1$, and consider the problem to minimize $f(\mathbf{x})$ subject to $\mathbf{x} \in S$. A point $\mathbf{x} \in S$ is called a *feasible solution* to the problem. If $\bar{\mathbf{x}} \in S$ and $f(\mathbf{x}) \geq f(\bar{\mathbf{x}})$ for each $\mathbf{x} \in S$, then $\bar{\mathbf{x}}$ is called an *optimal solution*, a *global optimal* solution, or simply a *solution* to the problem. The collection of optimal solutions are called *alternative optimal solutions*. If $\bar{\mathbf{x}} \in S$ and if there exists an ε-neighborhood $N_\varepsilon(\bar{\mathbf{x}})$ around $\bar{\mathbf{x}}$ such that $f(\mathbf{x}) \geq f(\bar{\mathbf{x}})$ for each $\mathbf{x} \in S \cap N_\varepsilon(\bar{\mathbf{x}})$, then $\bar{\mathbf{x}}$ is called a *local optimal solution*. Similarly, if $\bar{\mathbf{x}} \in S$ and if $f(\mathbf{x}) > f(\bar{\mathbf{x}})$ for all $\mathbf{x} \in S \cap N_\varepsilon(\bar{\mathbf{x}})$, $\mathbf{x} \neq \bar{\mathbf{x}}$, for some $\varepsilon > 0$, then $\bar{\mathbf{x}}$ is called a *strict local optimal solution*. On the other hand, if $\bar{\mathbf{x}} \in S$ is the *only* local minimum in $S \cap N_\varepsilon(\bar{\mathbf{x}})$, for some ε-neighborhood $N_\varepsilon(\bar{\mathbf{x}})$ around $\bar{\mathbf{x}}$, then $\bar{\mathbf{x}}$ is called a *strong* or *isolated local optimal solution*. All these types of local optima or minima are sometimes also referred to as *relative minima*.

Figure 3.6 illustrates instances of local and global minima for the problem of minimizing $f(\mathbf{x})$ subject to $\mathbf{x} \in S$, where f and S are shown in the figure.

The points in S corresponding to A, B, and C are also both strict and strong local minima, whereas those corresponding to the flat segment of the graph between D and E are local minima that are neither strict nor strong. Note that if $\bar{\mathbf{x}}$ is a strong or isolated local minimum, then it is also a strict minimum. To see this, consider the ε-neighborhood $N_\varepsilon(\bar{\mathbf{x}})$ characterizing the strong local minimum nature of $\bar{\mathbf{x}}$. Then, we must also have $f(\mathbf{x}) > f(\bar{\mathbf{x}})$ for all $x \in S \cap N_\varepsilon(\bar{\mathbf{x}})$, because otherwise, suppose that there exists a $\hat{\mathbf{x}} \in S \cap N_\varepsilon(\bar{\mathbf{x}})$ such that $f(\hat{\mathbf{x}}) = f(\bar{\mathbf{x}})$. Note that $\hat{\mathbf{x}}$ is an alternative optimal solution within

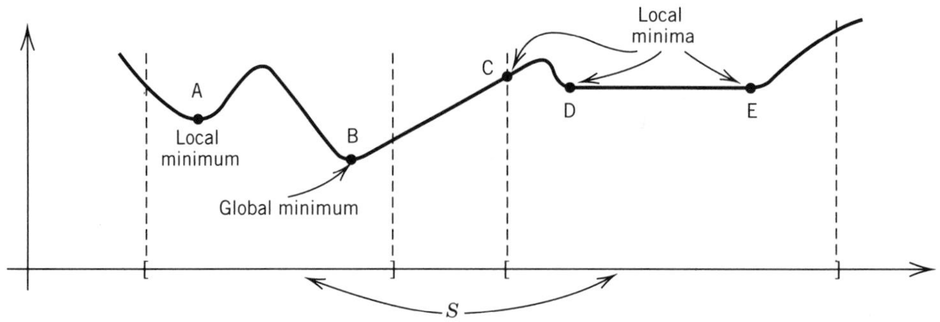

Figure 3.6 Local and global minima.

$S \cap N_\varepsilon(\bar{\mathbf{x}})$; and so there exists some $0 < \varepsilon' < \varepsilon$ such that $f(\mathbf{x}) \geq f(\hat{\mathbf{x}})$ for all $\mathbf{x} \in S \cap N_{\varepsilon'}(\hat{\mathbf{x}})$. But this contradicts the isolated local minimum status of $\bar{\mathbf{x}}$ and, hence, $\bar{\mathbf{x}}$ must also be a strict local minimum. On the other hand, a strict local minimum need not be an isolated local minimum. Figure 3.7 illustrates two such instances. In Figure 3.7*a*, $S = E_1$ and $f(x) = 1$ for $x = 1$ and is equal to 2 otherwise. Note that the point of discontinuity $\bar{x} = 1$ of f is a strict local minimum but is not isolated, since any ε-neighborhood about $\bar{x}$ contains points other than $\bar{x} = 1$, all of which are also local minima. Figure 3.7*b* illustrates another case in which $f(x) = x^2$, a strictly convex function; but $S = \{1/2^k, k = 0, 1, 2, \ldots,\} \cup \{0\}$ is a nonconvex set. Here, the point $\bar{\mathbf{x}} = 1/2^k$ for any integer $k \geq 0$ is an isolated and, therefore, a strict local minimum because it can be captured as the unique feasible solution in $S \cap N_\varepsilon(\bar{\mathbf{x}})$ for some sufficiently small $\varepsilon > 0$. However, while $\bar{x} = 0$ is clearly a strict local minimum (it is, in fact, the unique global minimum), it is not isolated because any ε-neighborhood about $\bar{x} = 0$ contains other local minima of the foregoing type.

Nonetheless, for optimization problems min $\{f(\mathbf{x}) : \mathbf{x} \in S\}$, where f is a convex function and S is a convex set, which are known as *convex programming problems*, and which are of interest to us in this section, a strict local minimum is also a strong local minimum as shown in Theorem 3.4.2 below (see Exercise 3.36 for a weaker sufficient condition). The principal result here is that each local minimum of a convex program is also a global minimum. This fact is quite useful in the optimization process, since it enables us to stop with a global optimal solution if the search in the vicinity of a feasible point does not lead to an improving feasible solution.

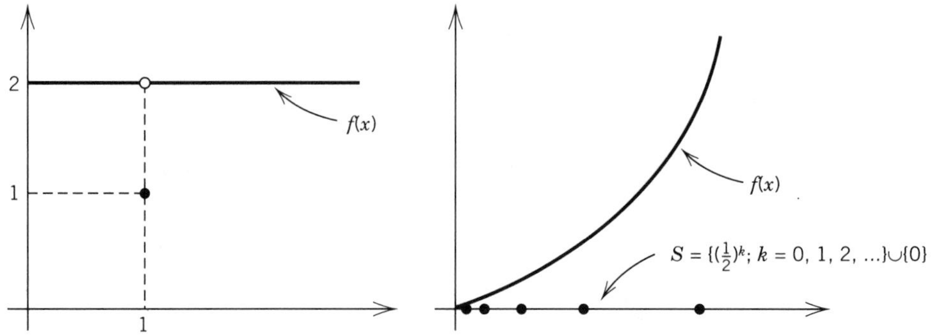

Figure 3.7 Strict local minima are not necessarily strong minima.

3.4.2 Theorem

Let S be a nonempty convex set in E_n, and let $f:S \to E_1$ be convex on S. Consider the problem to minimize $f(\mathbf{x})$ subject to $\mathbf{x} \in S$. Suppose that $\bar{\mathbf{x}} \in S$ is a local optimal solution to the problem.

1. Then, $\bar{\mathbf{x}}$ is a global optimal solution.
2. If either $\bar{\mathbf{x}}$ is a strict local minimum, or if f is strictly convex, then $\bar{\mathbf{x}}$ is the unique global optimal solution, and it is also a strong local minimum.

Proof

Since $\bar{\mathbf{x}}$ is a local optimal solution, there exists an ε-neighborhood $N_\varepsilon(\bar{\mathbf{x}})$ around $\bar{\mathbf{x}}$ such that

$$f(\mathbf{x}) \geq f(\bar{\mathbf{x}}) \text{ for each } \mathbf{x} \in S \cap N_\varepsilon(\bar{\mathbf{x}}) \tag{3.15}$$

By contradiction, suppose that $\bar{\mathbf{x}}$ is not a global optimal solution so that $f(\hat{\mathbf{x}}) < f(\bar{\mathbf{x}})$ for some $\hat{\mathbf{x}} \in S$. By the convexity of f, the following is true for each $\lambda \in (0, 1)$:

$$f(\lambda\hat{\mathbf{x}} + (1 - \lambda)\bar{\mathbf{x}}) \leq \lambda f(\hat{\mathbf{x}}) + (1 - \lambda)f(\bar{\mathbf{x}}) < \lambda f(\bar{\mathbf{x}}) + (1 - \lambda)f(\bar{\mathbf{x}}) = f(\bar{\mathbf{x}})$$

But for $\lambda > 0$ and sufficiently small, $\lambda\hat{\mathbf{x}} + (1 - \lambda)\bar{\mathbf{x}} \in S \cap N\varepsilon(\bar{\mathbf{x}})$. Hence, the above inequality contradicts (3.15), and part 1 is proved.

Next, suppose that $\bar{\mathbf{x}}$ is a strict local minimum. By part 1, it is a global minimum. Now, on the contrary, if there exists a $\hat{\mathbf{x}} \in S$ such that $f(\hat{\mathbf{x}}) = f(\bar{\mathbf{x}})$, then, defining $\mathbf{x}_\lambda = \lambda\hat{\mathbf{x}} + (1 - \lambda)\bar{\mathbf{x}}$ for $0 \leq \lambda \leq 1$, we have, by the convexity of f and S, that $f(\mathbf{x}_\lambda) \leq \lambda f(\hat{\mathbf{x}}) + (1 - \lambda)f(\bar{\mathbf{x}}) = f(\bar{\mathbf{x}})$, and $\mathbf{x}_\lambda \in S$ for all $0 \leq \lambda \leq 1$. By taking $\lambda \to 0^+$, since we can make $\mathbf{x}_\lambda \in N_\varepsilon(\bar{\mathbf{x}}) \cap S$ for any $\varepsilon > 0$, this contradicts the strict local optimality of $\bar{\mathbf{x}}$. Hence, $\bar{\mathbf{x}}$ is a unique global minimum. Therefore, it must also be an isolated local minimum, since any other local minimum in $N_\varepsilon(\bar{\mathbf{x}}) \cap S$ for any $\varepsilon > 0$ would also be a global minimum, which is a contradiction.

Finally, suppose that $\bar{\mathbf{x}}$ is a local optimal solution and that f is strictly convex. Since strict convexity implies convexity, then, by part 1, $\bar{\mathbf{x}}$ is a global optimal solution. By contradiction, suppose that $\bar{\mathbf{x}}$ is not the unique global optimal solution so that there exists a $\mathbf{x} \in S$, $\mathbf{x} \neq \bar{\mathbf{x}}$, such that $f(\mathbf{x}) = f(\bar{\mathbf{x}})$. By strict convexity,

$$f(\tfrac{1}{2}\mathbf{x} + \tfrac{1}{2}\bar{\mathbf{x}}) < \tfrac{1}{2}f(\mathbf{x}) + \tfrac{1}{2}f(\bar{\mathbf{x}}) = f(\bar{\mathbf{x}})$$

By the convexity of S, $\frac{1}{2}\mathbf{x} + \frac{1}{2}\bar{\mathbf{x}} \in S$, and the above inequality violates global optimality of $\bar{\mathbf{x}}$. Hence, $\bar{\mathbf{x}}$ is the unique global minimum and, as above, is also a strong local minimum. This completes the proof.

We now develop a necessary and sufficient condition for the existence of a global solution. If such an optimal solution does not exist, then either $\inf \{f(\mathbf{x}):\mathbf{x} \in S\}$ is finite but is not achieved at any point in S, or it is equal to $-\infty$.

3.4.3 Theorem

Let $f : E_n \to E_1$ be a convex function, and S be a nonempty convex set in E_n. Consider the problem to minimize $f(\mathbf{x})$ subject to $\mathbf{x} \in S$. The point $\bar{\mathbf{x}} \in S$ is an optimal solution to this problem if and only if f has a subgradient $\boldsymbol{\xi}$ at $\bar{\mathbf{x}}$ such that $\boldsymbol{\xi}^t(\mathbf{x} - \bar{\mathbf{x}}) \geq 0$ for all $\mathbf{x} \in S$.

Proof

Suppose that $\boldsymbol{\xi}^t(\mathbf{x} - \bar{\mathbf{x}}) \geq 0$ for all $\mathbf{x} \in S$, where $\boldsymbol{\xi}$ is a subgradient of f at $\bar{\mathbf{x}}$. By the convexity of f, we have

$$f(\mathbf{x}) \geq f(\bar{\mathbf{x}}) + \boldsymbol{\xi}^t(\mathbf{x} - \bar{\mathbf{x}}) \geq f(\bar{\mathbf{x}}) \qquad \text{for } \mathbf{x} \in S$$

and, hence, $\bar{\mathbf{x}}$ is an optimal solution of the problem.

To show the converse, suppose that $\bar{\mathbf{x}}$ is an optimal solution to the problem and construct the following two sets in E_{n+1}:

$$\Lambda_1 = \{(\mathbf{x} - \bar{\mathbf{x}}, y) : \mathbf{x} \in E_n, y > f(\mathbf{x}) - f(\bar{\mathbf{x}})\}$$

$$\Lambda_2 = \{(\mathbf{x} - \bar{\mathbf{x}}, y) : \mathbf{x} \in S, y \leq 0\}$$

The reader may easily verify that both Λ_1 and Λ_2 are convex sets. Also $\Lambda_1 \cap \Lambda_2 = \varnothing$ because otherwise there would exist a point $(\mathbf{x}, y)$ such that

$$\mathbf{x} \in S \qquad 0 \geq y > f(\mathbf{x}) - f(\bar{\mathbf{x}})$$

contradicting the assumption that $\bar{\mathbf{x}}$ is an optimal solution of the problem. By Theorem 2.4.8, there is a hyperplane that separates Λ_1 and Λ_2; that is, there exist a nonzero vector $(\boldsymbol{\xi}_0, \mu)$ and a scalar α such that

$$\boldsymbol{\xi}_0^t(\mathbf{x} - \bar{\mathbf{x}}) + \mu y \leq \alpha \qquad \mathbf{x} \in E_n, y > f(\mathbf{x}) - f(\bar{\mathbf{x}}) \tag{3.16}$$

$$\boldsymbol{\xi}_0^t(\mathbf{x} - \bar{\mathbf{x}}) + \mu y \geq \alpha \qquad \mathbf{x} \in S, y \leq 0 \tag{3.17}$$

If we let $\mathbf{x} = \bar{\mathbf{x}}$ and $y = 0$ in (3.17), it follows that $\alpha \leq 0$. Next, letting $\mathbf{x} = \bar{\mathbf{x}}$ and $y = \varepsilon > 0$ in (3.16), it follows that $\mu\varepsilon \leq \alpha$. Since this is true for every $\varepsilon > 0$, then $\mu \leq 0$ and also $\alpha \geq 0$. To summarize, we have shown that $\mu \leq 0$ and $\alpha = 0$. If $\mu = 0$, then, from (3.16), $\boldsymbol{\xi}_0^t(\mathbf{x} - \bar{\mathbf{x}}) \leq 0$ for each $\mathbf{x} \in E_n$. If we let $\mathbf{x} = \bar{\mathbf{x}} + \boldsymbol{\xi}_0$, it follows that

$$0 \geq \boldsymbol{\xi}_0^t(\mathbf{x} - \bar{\mathbf{x}}) = \|\boldsymbol{\xi}_0\|^2$$

and, hence, $\boldsymbol{\xi}_0 = \mathbf{0}$. Since $(\boldsymbol{\xi}_0, \mu) \neq (\mathbf{0}, 0)$, we must have $\mu < 0$. Dividing (3.16) and (3.17) by $-\mu$, and denoting $-\boldsymbol{\xi}_0/\mu$ by $\boldsymbol{\xi}$, we get the following inequalities:

$$y \geq \boldsymbol{\xi}^t(\mathbf{x} - \bar{\mathbf{x}}) \qquad \mathbf{x} \in E_n, y > f(\mathbf{x}) - f(\bar{\mathbf{x}}) \tag{3.18}$$

$$\boldsymbol{\xi}^t(\mathbf{x} - \bar{\mathbf{x}}) - y \geq 0 \qquad \mathbf{x} \in S, y \leq 0. \tag{3.19}$$

By letting $y = 0$ in (3.19), we get $\boldsymbol{\xi}^t(\mathbf{x} - \bar{\mathbf{x}}) \geq 0$ for all $\mathbf{x} \in S$. From (3.18), it is obvious that

$$f(\mathbf{x}) \geq f(\bar{\mathbf{x}}) + \boldsymbol{\xi}^t(\mathbf{x} - \bar{\mathbf{x}}) \qquad \text{for all } \mathbf{x} \in E_n$$

Therefore, $\boldsymbol{\xi}$ is a subgradient of f at $\bar{\mathbf{x}}$ with the property that $\boldsymbol{\xi}^t(\mathbf{x} - \bar{\mathbf{x}}) \geq 0$ for all $\mathbf{x} \in S$, and the proof is complete.

Corollary 1

Under the assumptions of Theorem 3.4.3, if S is open, then $\bar{\mathbf{x}}$ is an optimal solution to the problem if and only if there exists a zero subgradient of f at $\bar{\mathbf{x}}$. In particular, if $S = E_n$, then $\bar{\mathbf{x}}$ is a global minimum if and only if there exists a zero subgradient of f at $\bar{\mathbf{x}}$.

Proof

By the theorem, $\bar{\mathbf{x}}$ is an optimal solution if and only if $\boldsymbol{\xi}^t(\mathbf{x} - \bar{\mathbf{x}}) \geq 0$ for each $\mathbf{x} \in S$, where $\boldsymbol{\xi}$ is a subgradient of f at $\bar{\mathbf{x}}$. Since S is open, $\mathbf{x} = \bar{\mathbf{x}} - \lambda\boldsymbol{\xi} \in S$ for some positive λ. Therefore, $-\lambda \|\boldsymbol{\xi}\|^2 \geq 0$, that is, $\boldsymbol{\xi} = \mathbf{0}$.

Corollary 2

In addition to the assumptions of the theorem, suppose that f is differentiable. Then, $\bar{\mathbf{x}}$ is an optimal solution if and only if $\nabla f(\bar{\mathbf{x}})^t(\mathbf{x} - \bar{\mathbf{x}}) \geq 0$ for all $\mathbf{x} \in S$. Furthermore, if S is open, then $\bar{\mathbf{x}}$ is an optimal solution if and only if $\nabla f(\bar{\mathbf{x}}) = \mathbf{0}$.

Note the important implications of Theorem 3.4.3. First, the theorem gives a necessary and sufficient characterization of optimal points. This characterization reduces to the well-known condition of vanishing derivatives if f is differentiable and S is open. Another important implication is that if we reach a nonoptimal point $\bar{\mathbf{x}}$, where $\nabla f(\bar{\mathbf{x}})^t (\mathbf{x} - \bar{\mathbf{x}}) < 0$ for some $\mathbf{x} \in S$, then there is an obvious way to proceed to an improving solution. This can be achieved by moving from $\bar{\mathbf{x}}$ in the direction $d = \mathbf{x} - \bar{\mathbf{x}}$. The actual size of the step can be determined by a one-dimensional minimization subproblem of the following form: minimize $f[\bar{\mathbf{x}} + \lambda(\mathbf{x} - \bar{\mathbf{x}})]$ subject to $\lambda \geq 0$ and $\bar{\mathbf{x}} + \lambda(\mathbf{x} - \bar{\mathbf{x}}) \in S$. This procedure is called the *method of feasible directions* and is discussed in more detail in Chapter 10.

To provide additional insights, let us dwell for a while on Corollary 2, which addresses the differentiable case for Theorem 3.4.3. Figure 3.8 illustrates the geometry of the result. Now suppose that for the problem to minimize $f(\mathbf{x})$ subject to $\mathbf{x} \in S$, we have f differentiable and convex, but S is an arbitrary set; and it turns out that the directional derivative $f'(\bar{\mathbf{x}}; \mathbf{x} - \bar{\mathbf{x}}) \equiv \nabla f(\bar{\mathbf{x}})^t(\mathbf{x} - \bar{\mathbf{x}}) \geq 0$ for all $\mathbf{x} \in S$. The proof of the theorem actually shows that $\bar{\mathbf{x}}$ is a global minimum irregardless of S, since, for any solution $\hat{\mathbf{x}}$ that improves over $\bar{\mathbf{x}}$, we have, by the convexity of f, that $f(\bar{\mathbf{x}}) > f(\hat{\mathbf{x}}) \geq f(\bar{\mathbf{x}}) + \nabla f(\bar{\mathbf{x}})^t(\hat{\mathbf{x}} - \bar{\mathbf{x}})$, which implies that $\nabla f(\bar{\mathbf{x}})^t(\hat{\mathbf{x}} - \bar{\mathbf{x}}) < 0$, whereas $\nabla f(\bar{\mathbf{x}})^t(\mathbf{x} - \bar{\mathbf{x}}) \geq 0$ for all $\mathbf{x} \in S$. Hence, the hyperplane $\nabla f(\bar{\mathbf{x}})^t(\mathbf{x} - \bar{\mathbf{x}}) = 0$ separates S from solutions that improve over $\bar{\mathbf{x}}$. (For the nondifferentiable case, the hyperplane $\boldsymbol{\xi}^t(\mathbf{x} - \bar{\mathbf{x}}) = 0$ plays a similar role.) However, if f was not convex, then the directional derivative $\nabla f(\bar{\mathbf{x}})^t (\mathbf{x} - \bar{\mathbf{x}})$ being nonnegative for all $\mathbf{x} \in S$ does not even necessarily imply that $\bar{\mathbf{x}}$ is a local

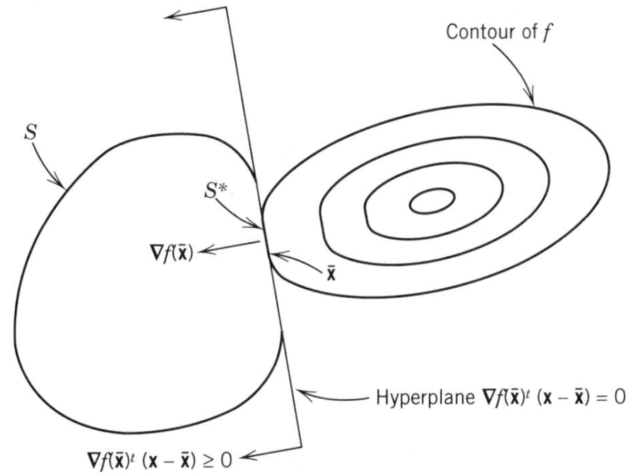

Figure 3.8 Illustrations for Theorems 3.4.3 and 3.4.4.

minimum. For example, for the problem to minimize $f(\mathbf{x}) = x^3$ subject to $-1 \le x \le 1$, we have the condition $\nabla f(\bar{x})^t(x - \bar{x}) \ge 0$ for all $x \in S$ being satisfied at $\bar{x} = 0$, since $f'(0) = 0$, but $\bar{\mathbf{x}} = 0$ is not even a local minimum for this problem.

Conversely, suppose that f is differentiable but arbitrary otherwise, and that S is a convex set. Then, if $\bar{\mathbf{x}}$ is a global minimum, we must have $f'(\bar{\mathbf{x}}; \mathbf{x} - \bar{\mathbf{x}}) = \nabla f(\bar{\mathbf{x}})^t (\mathbf{x} - \bar{\mathbf{x}}) \ge 0$. This follows because otherwise, if $\nabla f(\bar{\mathbf{x}})^t(\mathbf{x} - \bar{\mathbf{x}}) < 0$, we could move along the direction $\mathbf{d} = (\mathbf{x} - \bar{\mathbf{x}})$ and, as above, the objective value would fall for sufficiently small step lengths, whereas $\bar{\mathbf{x}} + \lambda\mathbf{d}$ would remain feasible for $0 \le \lambda \le 1$, by the convexity of S. Note that this explains a more general concept, namely, that if f is differentiable but f and S are otherwise arbitrary, and if $\bar{\mathbf{x}}$ is a local minimum of f over S, then, for any direction $\mathbf{d}$ for which $\bar{\mathbf{x}} + \lambda\mathbf{d}$ remains feasible for $0 < \lambda \le \delta$ for some $\delta > 0$, we must have a nonnegative directional derivative of f at $\bar{\mathbf{x}}$ in the direction $\mathbf{d}$, that is, we must have $f'(\bar{\mathbf{x}}; \mathbf{d}) = \nabla f(\bar{\mathbf{x}})^t\mathbf{d} \ge 0$.

Now let us turn our attention back to convex programming problems. The following result and its corollaries characterize the set of alternative optimal solutions and show, in part, that the gradient of the objective function (assuming twice-differentiability) is a constant over the optimal solution set, and that for a quadratic objective function the optimal solution set is in fact polyhedral. (See Figure 3.8 to identify the set of alternative optimal solutions S^* defined by the theorem in light of Theorem 3.4.3.)

3.4.4 Theorem

Consider the problem to minimize $f(\mathbf{x})$ subject to $\mathbf{x} \in S$, where f is a convex and twice-differentiable function, and S is a convex set, and suppose that there exists an optimal solution $\bar{\mathbf{x}}$. Then, the set of alternative optimal solutions is characterized by the set

$$S^* = \{\mathbf{x} \in S : \nabla f(\bar{\mathbf{x}})^t(\mathbf{x} - \bar{\mathbf{x}}) \le 0 \quad \text{and} \quad \nabla f(\mathbf{x}) = \nabla f(\bar{\mathbf{x}})\}$$

Proof

Denote the set of alternative optimal solutions as $\bar{S}$, say, and note that $\bar{\mathbf{x}} \in \bar{S} \neq \emptyset$. Consider any $\hat{\mathbf{x}} \in S^*$. By the convexity of f and the definition of S^*, we have $\hat{\mathbf{x}} \in S$ and

$$f(\bar{\mathbf{x}}) \ge f(\hat{\mathbf{x}}) + \nabla f(\hat{\mathbf{x}})^t(\bar{\mathbf{x}} - \hat{\mathbf{x}}) = f(\hat{\mathbf{x}}) + \nabla f(\bar{\mathbf{x}})^t(\bar{\mathbf{x}} - \hat{\mathbf{x}}) \ge f(\hat{\mathbf{x}})$$

and so we must have $\hat{\mathbf{x}} \in \bar{S}$ by the optimality of $\bar{\mathbf{x}}$. Hence, $S^* \subseteq \bar{S}$.

Conversely, suppose that $\hat{\mathbf{x}} \in \bar{S}$ so that $\hat{\mathbf{x}} \in S$ and $f(\hat{\mathbf{x}}) = f(\bar{\mathbf{x}})$. This means that $f(\bar{\mathbf{x}}) = f(\hat{\mathbf{x}}) \ge f(\bar{\mathbf{x}}) + \nabla f(\bar{\mathbf{x}})^t(\hat{\mathbf{x}} - \bar{\mathbf{x}})$ or that $\nabla f(\bar{\mathbf{x}})^t(\hat{\mathbf{x}} - \bar{\mathbf{x}}) \le 0$. But by Corollary 2 to Theorem 3.4.3, we have $\nabla f(\bar{\mathbf{x}})^t(\hat{\mathbf{x}} - \bar{\mathbf{x}}) \ge 0$. Hence, $\nabla f(\bar{\mathbf{x}})^t(\hat{\mathbf{x}} - \bar{\mathbf{x}}) = 0$. By interchanging the roles of $\bar{\mathbf{x}}$ and $\hat{\mathbf{x}}$, we symmetrically obtain $\nabla f(\hat{\mathbf{x}})^t(\bar{\mathbf{x}} - \hat{\mathbf{x}}) = 0$. Therefore,

$$[\nabla f(\bar{\mathbf{x}}) - \nabla f(\hat{\mathbf{x}})]^t(\bar{\mathbf{x}} - \hat{\mathbf{x}}) = 0 \tag{3.20}$$

Now we have

$$\begin{aligned}[\nabla f(\bar{\mathbf{x}}) - \nabla f(\hat{\mathbf{x}})] &= \nabla f[\hat{\mathbf{x}} + \lambda(\bar{\mathbf{x}} - \hat{\mathbf{x}})]_{\lambda=0}^{\lambda=1} \\ &= \int_{\lambda=0}^{\lambda=1} \mathbf{H}[\hat{\mathbf{x}} + \lambda(\bar{\mathbf{x}} - \hat{\mathbf{x}})](\bar{\mathbf{x}} - \hat{\mathbf{x}})d\lambda = \mathbf{G}(\bar{\mathbf{x}} - \hat{\mathbf{x}})\end{aligned} \tag{3.21}$$

where $\mathbf{G} = \int_0^1 \mathbf{H}[\hat{\mathbf{x}} + \lambda(\bar{\mathbf{x}} - \hat{\mathbf{x}})]d\lambda$ and where the integral of the matrix is performed componentwise. But note that $\mathbf{G}$ is positive semidefinite, because $\mathbf{d}^t\mathbf{G}\mathbf{d} = \int_0^1 \mathbf{d}^t\mathbf{H}[\hat{\mathbf{x}} + \lambda(\bar{\mathbf{x}} - \hat{\mathbf{x}})]\mathbf{d}\ d\lambda \ge 0$ for all $\mathbf{d} \in E_n$, since $\mathbf{d}^t\mathbf{H}[\hat{\mathbf{x}} + \lambda(\bar{\mathbf{x}} - \hat{\mathbf{x}})]\mathbf{d}$ is a nonnegative function of λ by the convexity of f. Hence, by (3.20) and (3.21), we get $0 = (\bar{\mathbf{x}} - \hat{\mathbf{x}})^t[\nabla f(\bar{\mathbf{x}}) - \nabla f(\hat{\mathbf{x}})] = (\bar{\mathbf{x}} - \hat{\mathbf{x}})^t\mathbf{G}(\bar{\mathbf{x}} - \hat{\mathbf{x}})$. But the positive semidefiniteness of $\mathbf{G}$ implies that $\mathbf{G}(\bar{\mathbf{x}} - \hat{\mathbf{x}}) = \mathbf{0}$ by a standard result (see Exercise 3.35). Therefore, by (3.21), we have $\nabla f(\bar{\mathbf{x}}) = \nabla f(\hat{\mathbf{x}})$. We have hence shown that $\hat{\mathbf{x}} \in S$, $\nabla f(\bar{\mathbf{x}})^t(\hat{\mathbf{x}} - \bar{\mathbf{x}}) \le 0$, and $\nabla f(\hat{\mathbf{x}}) = \nabla f(\bar{\mathbf{x}})$. This means that $\hat{\mathbf{x}} \in S^*$, and thus, $\bar{S} \subseteq S^*$. This, together with $S^* \subseteq \bar{S}$, completes the proof.

Corollary 1

The set S^* of alternative optimal solutions can equivalently be defined as

$$S^* = \{\mathbf{x} \in S : \nabla f(\bar{\mathbf{x}})^t(\mathbf{x} - \bar{\mathbf{x}}) = 0 \quad \text{and} \quad \nabla f(\mathbf{x}) = \nabla f(\bar{\mathbf{x}})\}$$

Proof

The proof follows from the definition of S^* in Theorem 3.4.4 and the fact that $\nabla f(\bar{\mathbf{x}})^t(\mathbf{x} - \bar{\mathbf{x}}) \ge 0$ for all $\mathbf{x} \in S$ by Corollary 2 to Theorem 3.4.3.

Corollary 2

Suppose that f is a quadratic function given by $f(\mathbf{x}) = \mathbf{c}^t\mathbf{x} + \frac{1}{2}\mathbf{x}^t\mathbf{H}\mathbf{x}$ and that S is polyhedral. Then, S^* is a polyhedral set given by

$$S^* = \{\mathbf{x} \in S : \mathbf{c}^t(\mathbf{x} - \bar{\mathbf{x}}) \le 0, \mathbf{H}(\mathbf{x} - \bar{\mathbf{x}}) = \mathbf{0}\} = \{\mathbf{x} \in S : \mathbf{c}^t(\mathbf{x} - \bar{\mathbf{x}}) = 0, \mathbf{H}(\mathbf{x} - \bar{\mathbf{x}}) = \mathbf{0}\}$$

Proof

The proof follows by direct substitution in Theorem 3.4.4 and Corollary 1, noting that $\nabla f(\mathbf{x}) = \mathbf{c} + \mathbf{H}\mathbf{x}$

3.4.5 Example

$$\begin{aligned} \text{Minimize} \quad & (x_1 - \tfrac{3}{2})^2 + (x_2 - 5)^2 \\ \text{subject to} \quad & -x_1 + x_2 \le 2 \\ & 2x_1 + 3x_2 \le 11 \\ & -x_1 \le 0 \\ & -x_2 \le 0 \end{aligned}$$

Clearly, $f(x_1, x_2) = (x_1 - \frac{3}{2})^2 + (x_2 - 5)^2$ is a convex function, which gives the square of the distance from the point $(\frac{3}{2}, 5)$. The convex polyhedral set S is represented by the above four inequalities. The problem is depicted in Figure 3.9. From the figure, clearly the optimal point is (1, 3). The gradient vector of f at the point (1, 3) is $\nabla f(1, 3) = (-1, -4)^t$. We see geometrically that the vector $(-1, -4)$ makes an angle of $<90°$ with each vector of the form $(x_1 - 1, x_2 - 3)$, where $(x_1, x_2) \in S$. Thus, the optimality condition of Theorem 3.4.3 is verified and, by Theorem 3.4.4, (1, 3) is the unique optimum.

To illustrate further, suppose that it is claimed that $\hat{\mathbf{x}} = (0, 0)^t$ is an optimal point. By Theorem 3.4.4, this cannot be true since we have $\nabla f(\bar{\mathbf{x}})^t(\hat{\mathbf{x}} - \bar{\mathbf{x}}) = 13 > 0$ when $\bar{\mathbf{x}} = (1, 3)^t$. Similarly, by Theorem 3.4.3, we can easily verify that $\hat{\mathbf{x}}$ is not optimal. Note that $\nabla f(0, 0) = (-3, -10)^t$ and actually, for each nonzero $\mathbf{x} \in S$, we have $-3x_1 - 10x_2 < 0$. Hence, the origin could not be an optimal point. Moreover, we can improve

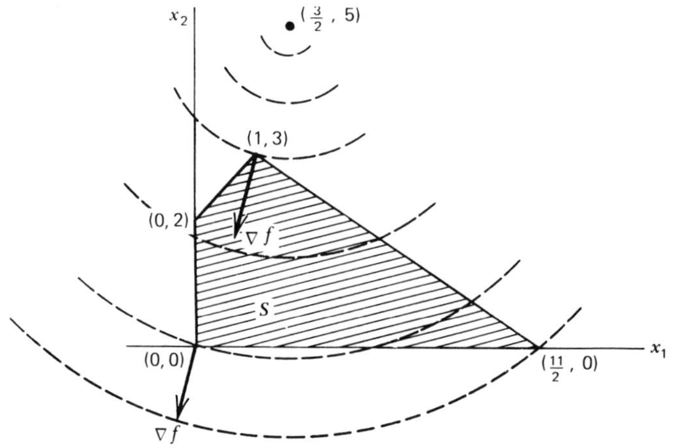

Figure 3.9 Illustration of Example 3.4.5.

f by moving from $\mathbf{0}$ in the direction $\mathbf{x} - \mathbf{0}$ for any $\mathbf{x} \in S$. In this case, the best local direction is $-\nabla f(0, 0)$, that is, the direction (3, 10). In Chapter 10, we discuss methods for finding a particular direction among many alternatives.

Maximizing a Convex Function

We now develop a necessary condition for a maximum of a convex function over a convex set. Unfortunately, this condition is not sufficient. Therefore, it is possible, and actually not unlikely, that several local maxima satisfying the condition of Theorem 3.4.6 exist. Unlike the minimization case, there exists no local information at such solutions that could lead us to better points. Hence, maximizing a convex function is usually a much harder task than minimizing a convex function. Again, minimizing a concave function is similar to maximizing a convex function and, hence, the development for the concave case is left to the reader.

3.4.6 Theorem

Let $f: E_n \rightarrow E_1$ be a convex function, and let S be a nonempty convex set in E_n. Consider the problem to maximize $f(\mathbf{x})$ subject to $\mathbf{x} \in S$. If $\bar{\mathbf{x}} \in S$ is a local optimal solution, then $\boldsymbol{\xi}^t(\mathbf{x} - \bar{\mathbf{x}}) \leq 0$ for each $\mathbf{x} \in S$, where $\boldsymbol{\xi}$ is any subgradient of f at $\bar{\mathbf{x}}$.

Proof

Suppose that $\bar{\mathbf{x}} \in S$ is a local optimal solution. Then, an ε-neighborhood $N_\varepsilon(\bar{\mathbf{x}})$ exists such that $f(\mathbf{x}) \leq f(\bar{\mathbf{x}})$ for each $\mathbf{x} \in S \cap N_\varepsilon(\bar{\mathbf{x}})$. Let $\mathbf{x} \in S$, and note that $\bar{\mathbf{x}} + \lambda(\mathbf{x} - \bar{\mathbf{x}}) \in S \cap N_\varepsilon(\bar{\mathbf{x}})$ for $\lambda > 0$ and sufficiently small. Hence,

$$f[\bar{\mathbf{x}} + \lambda(\mathbf{x} - \bar{\mathbf{x}})] \leq f(\bar{\mathbf{x}}) \tag{3.22}$$

Let $\boldsymbol{\xi}$ be a subgradient of f at $\bar{\mathbf{x}}$. By the convexity of f, we have

$$f[\bar{\mathbf{x}} + \lambda(\mathbf{x} - \bar{\mathbf{x}})] - f(\bar{\mathbf{x}}) \geq \lambda\boldsymbol{\xi}^t(\mathbf{x} - \bar{\mathbf{x}})$$

The above inequality, together with (3.20), implies that $\lambda\boldsymbol{\xi}^t(\mathbf{x} - \bar{\mathbf{x}}) \leq 0$, and, dividing by $\lambda > 0$, the result follows.

Corollary

In addition to the assumptions of the theorem, suppose that f is differentiable. If $\bar{\mathbf{x}} \in S$ is a local optimal solution, then $\nabla f(\bar{\mathbf{x}})^t(\mathbf{x} - \bar{\mathbf{x}}) \leq 0$ for all $\mathbf{x} \in S$.

Note that the above result is, in general, necessary but not sufficient for optimality. To illustrate, let $f(x) = x^2$ and $S = \{x: -1 \leq x \leq 2\}$. The maximum of f over S is equal to 4 and is achieved at $x = 2$. However, at $\bar{x} = 0$, we have $\nabla f(\bar{x}) = 0$ and, hence, $\nabla f(\bar{x})^t(x - \bar{x}) = 0$ for each $x \in S$. Clearly, the point $\bar{x} = 0$ is not even a local maximum. Referring to Example 3.4.5 discussed earlier, we have two local maxima, namely, $(0, 0)$ and $(\frac{11}{2}, 0)$. Both points satisfy the necessary condition of Theorem 3.4.6. If we are currently at the local optimal point $(0, 0)$, unfortunately no local information exists that will lead us toward the global maximum point $(\frac{11}{2}, 0)$. Also, if we are at the global maximum point $(\frac{11}{2}, 0)$, there is no convenient local criterion that tells us that we are at the optimal point.

Theorem 3.4.7 shows that a convex function achieves a maximum over a compact polyhedral set at an extreme point. This result has been utilized by several computational schemes for solving such problems. We ask the reader to think for a moment about the case when the objective function is linear and, hence, both convex and concave. Theorem 3.4.7 could be extended to the case where the convex feasible region is not polyhedral.

3.4.7 Theorem

Let $f: E_n \to E_1$ be a convex function, and let S be a nonempty compact polyhedral set in E_n. Consider the problem to maximize $f(\mathbf{x})$ subject to $\mathbf{x} \in S$. An optimal solution $\bar{\mathbf{x}}$ to the problem then exists, where $\bar{\mathbf{x}}$ is an extreme point of S.

Proof

By Theorem 3.1.3, note that f is continuous. Since S is compact, f assumes a maximum at $\mathbf{x}' \in S$. If $\mathbf{x}'$ is an extreme point of S, then the result is at hand. Otherwise, by Theorem 2.6.7, $\mathbf{x}' = \sum_{j=1}^k \lambda_j \mathbf{x}_j$, where $\sum_{j=1}^k \lambda_j = 1$, $\lambda_j > 0$, and $\mathbf{x}_j$ is an extreme point of S for $j = 1, \ldots, k$. By the convexity of f, we have

$$f(\mathbf{x}') = f\left(\sum_{j=1}^{k} \lambda_j \mathbf{x}_j\right) \leq \sum_{j=1}^{k} \lambda_j f(\mathbf{x}_j)$$

But since $f(\mathbf{x}') \geq f(\mathbf{x}_j)$ for $j = 1, \ldots, k$, the above inequality implies that $f(\mathbf{x}') = f(\mathbf{x}_j)$ for $j = 1, \ldots, k$. Thus, the extreme points $\mathbf{x}_1, \ldots, \mathbf{x}_k$ are optimal solutions to the problem, and the proof is complete.

3.5 Generalizations of a Convex Function

In this section, we present various types of functions that are similar to convex and concave functions but that share only some of their desirable properties. As we shall learn, many of the results presented later in the book do not require the restrictive assumption of convexity, but rather the less restrictive assumptions of quasiconvexity, pseudoconvexity, and convexity at a point.

Quasiconvex Functions

Definition 3.5.1 introduces quasiconvex functions. From the definition, it is apparent that every convex function is also quasiconvex.

3.5.1 Definition

Let $f: S \to E_1$, where S is a nonempty convex set in E_n. The function f is said to be *quasiconvex* if, for each $\mathbf{x}_1$ and $\mathbf{x}_2 \in S$, the following inequality is true:

$$f[\lambda \mathbf{x}_1 + (1 - \lambda)\mathbf{x}_2] \leq \text{maximum } \{f(\mathbf{x}_1), f(\mathbf{x}_2)\} \qquad \text{for each } \lambda \in (0, 1)$$

The function f is said to be *quasiconcave* if $-f$ is quasiconvex.

From the above definition, a function f is quasiconvex if, whenever $f(\mathbf{x}_2) \geq f(\mathbf{x}_1)$, $f(\mathbf{x}_2)$ is greater than or equal to f at all convex combinations of $\mathbf{x}_1$ and $\mathbf{x}_2$. Hence, if f increases from its value at a point along any direction, it must remain nondecreasing in that direction. Therefore, its univariate cross section is either monotone or unimodal (see Exercise 3.53). A function f is quasiconcave if, whenever $f(\mathbf{x}_2) \geq f(\mathbf{x}_1)$, f at all convex combinations of $\mathbf{x}_1$ and $\mathbf{x}_2$ is greater than or equal to $f(\mathbf{x}_1)$. Figure 3.10 shows some examples of quasiconvex and quasiconcave functions. We shall concentrate only on quasiconvex functions. The reader is advised to draw all the parallel results for quasiconcave functions. A function that is both quasiconvex and quasiconcave is called *quasimonotone* (see Figure 3.10*d*).

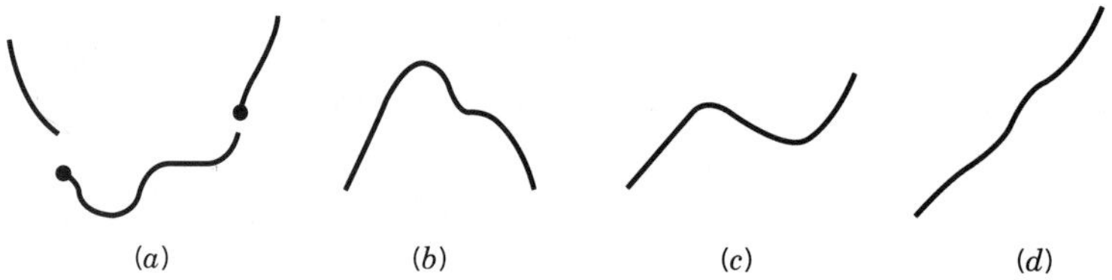

Figure 3.10 Quasiconvex and quasiconcave functions. (*a*) Quasiconvex. (*b*) Quasiconcave. (*c*) Neither quasiconvex nor quasiconcave. (*d*) Quasimonotone.

We have learned in Section 3.2 that a convex function can be characterized by the convexity of its epigraph. We now learn that a quasiconvex function can be characterized by the convexity of its level sets. This result is given in Theorem 3.5.2.

3.5.2 Theorem

Let $f: S \to E_1$, where S is a nonempty convex set in E_n. The function f is quasiconvex if and only if $S_\alpha = \{\mathbf{x} \in S : f(\mathbf{x}) \leq \alpha\}$ is convex for each real number α.

Proof

Suppose that f is quasiconvex, and let $\mathbf{x}_1, \mathbf{x}_2 \in S_\alpha$. Therefore, $\mathbf{x}_1, \mathbf{x}_2, \in S$ and maximum $\{f(\mathbf{x}_1), f(\mathbf{x}_2)\} \leq \alpha$. Let $\lambda \in (0, 1)$, and let $\mathbf{x} = \lambda \mathbf{x}_1 + (1 - \lambda)\mathbf{x}_2$. By the convexity of S, $\mathbf{x} \in S$. Furthermore, by the quasiconvexity of f, $f(\mathbf{x}) \leq$ maximum $\{f(\mathbf{x}_1), f(\mathbf{x}_2)\} \leq \alpha$. Hence, $\mathbf{x} \in S_\alpha$, and thus S_α is convex. Conversely, suppose that S_α is convex for each real number α. Let $\mathbf{x}_1, \mathbf{x}_2 \in S$. Furthermore, let $\lambda \in (0, 1)$ and $\mathbf{x} = \lambda \mathbf{x}_1 + (1 - \lambda)\mathbf{x}_2$. Note that $\mathbf{x}_1, \mathbf{x}_2 \in S_\alpha$ for $\alpha =$ maximum $\{f(\mathbf{x}_1), f(\mathbf{x}_2)\}$. By assumption, S_α is convex, so that $\mathbf{x} \in S_\alpha$. Therefore, $f(\mathbf{x}) \leq \alpha =$ maximum $\{f(\mathbf{x}_1), f(\mathbf{x}_2)\}$. Hence, f is quasiconvex, and the proof is complete.

The level set S_α defined in Theorem 3.5.2 is sometimes referred to as a *lower-level set*, to differentiate it from the *upper-level set* $\{\mathbf{x} \in S : f(\mathbf{x}) \geq \alpha\}$, which is convex for all $\alpha \in E_1$ if and only if f is quasiconcave. Also, it can be shown (see Exercise 3.54) that f is quasimonotone if and only if the *level surface* $\{\mathbf{x} \in S : f(\mathbf{x}) = \alpha\}$ is convex for all $\alpha \in R$.

We now give a result analogous to Theorem 3.4.7. Theorem 3.5.3 shows that the maximum of a continuous quasiconvex function over a compact polyhedral set occurs at an extreme point.

3.5.3 Theorem

Let S be a nonempty compact polyhedral set in E_n, and let $f: E_n \to E_1$ be quasiconvex and continuous on S. Consider the problem to maximize $f(\mathbf{x})$ subect to $\mathbf{x} \in S$. Then, an optimal solution $\bar{\mathbf{x}}$ to the problem exists, where $\bar{\mathbf{x}}$ is an extreme point of S.

Proof

Note that f is continuous on S and, hence, attains a maximum, say, at $\mathbf{x}' \in S$. If there is an extreme point whose objective is equal to $f(\mathbf{x}')$, then the result is at hand. Otherwise, let $\mathbf{x}_1, \ldots, \mathbf{x}_k$ be the extreme points of S, and assume that $f(\mathbf{x}') > f(\mathbf{x}_j)$ for $j = 1, \ldots, k$. By Theorem 2.6.7, $\mathbf{x}'$ can be represented as

$$\mathbf{x}' = \sum_{j=1}^{k} \lambda_j \mathbf{x}_j$$

$$\sum_{j=1}^{k} \lambda_j = 1$$

$$\lambda_j \geq 0 \qquad j = 1, \ldots, k$$

Since $f(\mathbf{x}') > f(\mathbf{x}_j)$ for each j, then

$$f(\mathbf{x}') > \underset{1 \leq j \leq k}{\text{maximum}}\, f(\mathbf{x}_j) = \alpha \tag{3.23}$$

Now consider the set $S_\alpha = \{\mathbf{x} : f(\mathbf{x}) \leq \alpha\}$. Note that $\mathbf{x}_j \in S_\alpha$ for $j = 1, \ldots, k$ and by the quasiconvexity of f, S_α is convex. Hence, $\mathbf{x}' = \sum_{j=1}^{k} \lambda_j \mathbf{x}_j$ belongs to S_α. This implies that $f(\mathbf{x}') \leq \alpha$, which contradicts (3.23). This contradiction shows that $f(\mathbf{x}') = f(\mathbf{x}_j)$ for some extreme point $\mathbf{x}_j$, and the proof is complete.

Differentiable Quasiconvex Functions

The following theorem gives a necessary and sufficient characterization of a differentiable quasiconvex function. (See Appendix B for a second-order characterization in terms of *bordered Hessian determinants*.)

3.5.4 Theorem

Let S be a nonempty open convex set in E_n, and let $f: S \to E_1$ be differentiable on S. Then, f is quasiconvex if and only if either one of the following equivalent statements holds:

1. If $\mathbf{x}_1, \mathbf{x}_2 \in S$ and $f(\mathbf{x}_1) \leq f(\mathbf{x}_2)$, then $\nabla f(\mathbf{x}_2)^t(\mathbf{x}_1 - \mathbf{x}_2) \leq 0$.
2. If $\mathbf{x}_1, \mathbf{x}_2 \in S$ and $\nabla f(\mathbf{x}_2)^t(\mathbf{x}_1 - \mathbf{x}_2) > 0$, then $f(\mathbf{x}_1) > f(\mathbf{x}_2)$.

Proof

Obviously, statements 1 and 2 are equivalent. We shall prove part 1. Let f be quasiconvex, and let $\mathbf{x}_1, \mathbf{x}_2 \in S$ be such that $f(\mathbf{x}_1) \le f(\mathbf{x}_2)$. By differentiability of f at $\mathbf{x}_2$, for $\lambda \in (0, 1)$, we have

$$f[\lambda\mathbf{x}_1 + (1 - \lambda)\mathbf{x}_2] - f(\mathbf{x}_2) = \lambda\nabla f(\mathbf{x}_2)^t(\mathbf{x}_1 - \mathbf{x}_2) + \lambda\,\|\mathbf{x}_1 - \mathbf{x}_2\|\,\alpha[\mathbf{x}_2; \lambda(\mathbf{x}_1 - \mathbf{x}_2)]$$

where $\alpha[\mathbf{x}_2; \lambda(\mathbf{x}_1 - \mathbf{x}_2)] \to 0$ as $\lambda \to 0$. By the quasiconvexity of f, we have $f[\lambda\mathbf{x}_1 + (1 - \lambda)\mathbf{x}_2] \le f(\mathbf{x}_2)$, and, hence, the above equation implies that

$$\lambda\nabla f(\mathbf{x}_2)^t(\mathbf{x}_1 - \mathbf{x}_2) + \lambda\|\mathbf{x}_1 - \mathbf{x}_2\|\,\alpha[\mathbf{x}_2; \lambda(\mathbf{x}_1 - \mathbf{x}_2)] \le 0$$

Dividing by λ and letting $\lambda \to 0$, we get $\nabla f(\mathbf{x}_2)^t(\mathbf{x}_1 - \mathbf{x}_2) \le 0$.

Conversely, suppose that $\mathbf{x}_1, \mathbf{x}_2 \in S$ and that $f(\mathbf{x}_1) \le f(\mathbf{x}_2)$. We need to show that given part 1, we have $f[\lambda\mathbf{x}_1 + (1 - \lambda)\mathbf{x}_2] \le f(\mathbf{x}_2)$ for each $\lambda \in (0, 1)$. We do this by showing that the set

$$L = \{\mathbf{x} : \mathbf{x} = \lambda\mathbf{x}_1 + (1 - \lambda)\mathbf{x}_2,\ \lambda \in (0, 1), f(\mathbf{x}) > f(\mathbf{x}_2)\}$$

is empty. By contradiction, suppose that there exists an $\mathbf{x}' \in L$. Therefore, $\mathbf{x}' = \lambda\mathbf{x}_1 + (1 - \lambda)\mathbf{x}_2$ for some $\lambda \in (0, 1)$ and $f(\mathbf{x}') > f(\mathbf{x}_2)$. Since f is differentiable, it is continuous, and there must exist a $\delta \in (0, 1)$ such that

$$f[\mu\mathbf{x}' + (1 - \mu)\mathbf{x}_2] > f(\mathbf{x}_2) \qquad \text{for each } \mu \in [\delta, 1] \tag{3.24}$$

and $f(\mathbf{x}') > f[\delta\mathbf{x}' + (1 - \delta)\mathbf{x}_2]$. By this inequality and the mean value theorem, we must have

$$0 < f(\mathbf{x}') - f[\delta\mathbf{x}' + (1 - \delta)\mathbf{x}_2] = (1 - \delta)\nabla f(\hat{\mathbf{x}})^t(\mathbf{x}' - \mathbf{x}_2) \tag{3.25}$$

where $\hat{\mathbf{x}} = \hat{\mu}\mathbf{x}' + (1 - \hat{\mu})\mathbf{x}_2$ for some $\hat{\mu} \in (\delta, 1)$. From (3.24), it is clear that $f(\hat{\mathbf{x}}) > f(\mathbf{x}_2)$. Dividing (3.25) by $1 - \delta > 0$, it follows that $\nabla f(\hat{\mathbf{x}})^t(\mathbf{x}' - \mathbf{x}_2) > 0$, which in turn implies that

$$\nabla f(\hat{\mathbf{x}})^t(\mathbf{x}_1 - \mathbf{x}_2) > 0 \tag{3.26}$$

But, on the other hand, $f(\hat{\mathbf{x}}) > f(\mathbf{x}_2) \ge f(\mathbf{x}_1)$, and $\hat{\mathbf{x}}$ is a convex combination of $\mathbf{x}_1$ and $\mathbf{x}_2$, say, $\hat{\mathbf{x}} = \hat{\lambda}\mathbf{x}_1 + (1 - \hat{\lambda})\mathbf{x}_2$ where $\hat{\lambda} \in (0, 1)$. By the assumption of the theorem, $\nabla f(\hat{\mathbf{x}})^t(\mathbf{x}_1 - \hat{\mathbf{x}}) \le 0$, and thus we must have

$$0 \ge \nabla f(\hat{\mathbf{x}})^t(\mathbf{x}_1 - \hat{\mathbf{x}}) = (1 - \hat{\lambda})\nabla f(\hat{\mathbf{x}})^t(\mathbf{x}_1 - \mathbf{x}_2)$$

The above inequality is not compatible with (3.26). Therefore, L is empty, and the proof is complete.

To illustrate the above theorem, let $f(x) = x^3$. To check quasiconvexity, suppose that $f(x_1) \le f(x_2)$, that is, $x_1^3 \le x_2^3$. This is true only if $x_1 \le x_2$. Now consider $\nabla f(x_2)(x_1 - x_2) = 3(x_1 - x_2)x_2^2$. Since $x_1 \le x_2$, $3(x_1 - x_2)x_2^2 \le 0$. Therefore, $f(x_1) \le f(x_2)$ implies that $\nabla f(x_2)(x_1 - x_2) \le 0$ and, by the theorem, that f is quasiconvex. As another illustration, let $f(x_1, x_2) = x_1^3 + x_2^3$. Let $\mathbf{x}_1 = (2, -2)^t$ and $\mathbf{x}_2 = (1, 0)^t$. Note that $f(\mathbf{x}_1) = 0$ and $f(\mathbf{x}_2) = 1$, so that $f(\mathbf{x}_1) < f(\mathbf{x}_2)$. But, on the other hand, $\nabla f(\mathbf{x}_2)^t(\mathbf{x}_1 - \mathbf{x}_2) = (3, 0)\,(1, -2)^t = 3$. By the necessary part of the theorem, f is not quasiconvex. This also shows that the sum of two quasiconvex functions is not necessarily quasiconvex.

Strictly Quasiconvex Functions

Strictly quasiconvex and strictly quasiconcave functions are especially important in nonlinear programming because they ensure that a local minimum and a local maximum over a convex set are, respectively, a global minimum and a global maximum.

3.5.5 Definition

Let $f: S \to E_1$, where S is a nonempty convex set in E_n. The function f is said to be *strictly quasiconvex* if, for each $\mathbf{x}_1, \mathbf{x}_2 \in S$ with $f(\mathbf{x}_1) \neq f(\mathbf{x}_2)$, we have

$$f[\lambda \mathbf{x}_1 + (1 - \lambda)\mathbf{x}_2] < \text{maximum } \{f(\mathbf{x}_1), f(\mathbf{x}_2)\} \qquad \text{for each } \lambda \in (0, 1).$$

The function f is called *strictly quasiconcave* if $-f$ is strictly quasiconvex. Strictly quasiconvex functions are also sometimes referred to as *semistrictly quasiconvex, functionally convex*, or *explicitly quasiconvex*.

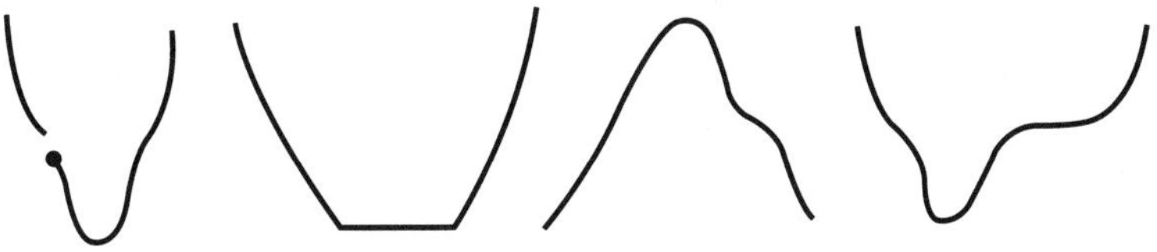

Figure 3.11 Strictly quasiconvex and strictly quasiconcave functions, (*a*) Strictly quasiconvex. (*b*) Strictly quasiconvex. (*c*) Strictly quasiconcave. (*d*) Neither strictly quasiconvex nor quasiconcave.

Note from the above definition that every convex function is strictly quasiconvex. Figure 3.11 gives examples of strictly quasiconvex and strictly quasiconcave functions. Also, the definition precludes any "flat spots" from occurring anywhere except at extremizing points. This is formalized by the following theorem, which shows that a local minimum of a strictly quasiconvex function over a convex set is also a global minimum. This property is not enjoyed by quasiconvex functions, as seen from Figure 3.10*a*.

3.5.6 Theorem

Let $f: E_n \to E_1$ be strictly quasiconvex. Consider the problem to minimize $f(\mathbf{x})$ subject to $\mathbf{x} \in S$, where S is a nonempty convex set in E_n. If $\bar{\mathbf{x}}$ is a local optimal solution, then $\bar{\mathbf{x}}$ is also a global optimal solution.

Proof

Assume, on the contrary, that there exists an $\hat{\mathbf{x}} \in S$ with $f(\hat{\mathbf{x}}) < f(\bar{\mathbf{x}})$. By the convexity of S, $\lambda\hat{\mathbf{x}} + (1 - \lambda)\bar{\mathbf{x}} \in S$ for each $\lambda \in (0, 1)$. Since $\bar{\mathbf{x}}$ is a local minimum by assumption, then $f(\bar{\mathbf{x}}) \leq f[\lambda\hat{\mathbf{x}} + (1 - \lambda)\bar{\mathbf{x}}]$ for all $\lambda \in (0, \delta)$ and for some $\delta \in (0, 1)$. But because f is strictly quasiconvex, and $f(\hat{\mathbf{x}}) < f(\bar{\mathbf{x}})$, $f[\lambda\hat{\mathbf{x}} + (1 - \lambda)\bar{\mathbf{x}}] < f(\bar{\mathbf{x}})$ for each $\lambda \in (0, 1)$. This contradicts the local optimality of $\bar{\mathbf{x}}$, and the proof is complete.

As seen from Definition 3.1.1, every strictly convex function is indeed a convex function. But every strictly quasiconvex function is not quasiconvex. To illustrate, consider the following function given by Karamardian [1967]:

$$f(x) = \begin{cases} 1 & \text{if } x = 0 \\ 0 & \text{if } x \neq 0 \end{cases}$$

By Definition 3.5.5, f is strictly quasiconvex. However, f is not quasiconvex, since, for $x_1 = 1$ and $x_2 = -1$, $f(x_1) = f(x_2) = 0$, but $f(\frac{1}{2}x_1 + \frac{1}{2}x_2) = f(0) = 1 > f(x_2)$. If f is lower semicontinuous, however, then as shown below, strict quasiconvexity implies quasiconvexity, as one would usually expect from the word "strict." (For a definition of lower semicontinuity, refer to Appendix A.)

3.5.7 Lemma

Let S be a nonempty convex set in E_n and let $f: S \rightarrow E_1$ be strictly quasiconvex and lower semicontinuous. Then, f is quasiconvex.

Proof

Let $\mathbf{x}_1$ and $\mathbf{x}_2 \in S$. If $f(\mathbf{x}_1) \neq f(\mathbf{x}_2)$, then, by the strict quasiconvexity of f, we must have $f[\lambda\mathbf{x}_1 + (1 - \lambda)\mathbf{x}_2] <$ maximum $\{f(\mathbf{x}_1), f(\mathbf{x}_2)\}$ for each $\lambda \in (0, 1)$. Now suppose that $f(\mathbf{x}_1) = f(\mathbf{x}_2)$. To show that f is quasiconvex, we need to show that $f[\lambda\mathbf{x}_1 + (1 - \lambda)\mathbf{x}_2] \leq f(\mathbf{x}_1)$ for each $\lambda \in (0, 1)$. By contradiction, suppose that $f[\mu\mathbf{x}_1 + (1 - \mu)\mathbf{x}_2] > f(\mathbf{x}_1)$ for some $\mu \in (0, 1)$. Denote $\mu\mathbf{x}_1 + (1 - \mu)\mathbf{x}_2$ by $\mathbf{x}$. Since f is lower semicontinuous, there exists a $\lambda \in (0, 1)$ such that

$$f(\mathbf{x}) > f[\lambda\mathbf{x}_1 + (1 - \lambda)\mathbf{x}] > f(\mathbf{x}_1) = f(\mathbf{x}_2) \tag{3.27}$$

Note that $\mathbf{x}$ can be represented as a convex combination of $\lambda\mathbf{x}_1 + (1 - \lambda)\mathbf{x}$ and $\mathbf{x}_2$. Hence, by the strict quasiconvexity of f and since $f[\lambda\mathbf{x}_1 + (1 - \lambda)\mathbf{x}] > f(\mathbf{x}_2)$, $f(\mathbf{x}) < f[\lambda\mathbf{x}_1 + (1 - \lambda)\mathbf{x}]$, contradicting (3.27). This completes the proof.

Strongly Quasiconvex Functions

From Theorem 3.5.6, it followed that a local minimum of a strictly quasiconvex function over a convex set is also a global optimal solution. However, strict quasiconvexity does not assert uniqueness of the global optimal solution. We shall define here another version of quasiconvexity, called strong quasiconvexity, which assures uniqueness of the global minimum.

3.5.8 Definition

Let S be a nonempty convex set in E_n, and let $f: S \rightarrow E_1$. The function f is said to be *strongly quasiconvex* if for each $\mathbf{x}_1, \mathbf{x}_2 \in S$, with $\mathbf{x}_1 \neq \mathbf{x}_2$, we have

$$f[\lambda\mathbf{x}_1 + (1 - \lambda)\mathbf{x}_2] < \text{maximum } \{f(\mathbf{x}_1), f(\mathbf{x}_2)\}$$

for each $\lambda \in (0, 1)$. The function f is said to be *strongly quasiconcave* if $-f$ is srongly quasiconvex. (We caution the reader that such a function is sometimes referred to in the literature as being "strictly quasiconvex," whereas a function satisfying Definition 3.5.5 is called "semi-strictly quasiconvex." This is done because of Karamardian's example given above and property 3 below.)

From Definition 3.5.8 above and from Definitions 3.1.1, 3.5.1, and 3.5.5, the following statements hold:

1. Every strictly convex function is strongly quasiconvex.
2. Every strongly quasiconvex function is strictly quasiconvex.
3. Every strongly quasiconvex function is quasiconvex even in the absence of any semicontinuity assumption.

Figure 3.11*a* illustrates a case where the function is both strongly quasiconvex and strictly quasiconvex, whereas the function represented in Figure 3.11*b* is strictly quasiconvex but not strongly quasiconvex. The key to strong quasiconvexity is that it enforces strict unimodality (see Exercise 3.56). This leads to the following property.

3.5.9 Theorem

Let $f: E_n \rightarrow E_1$ be strongly quasiconvex. Consider the problem to minimize $f(\mathbf{x})$ subject to $\mathbf{x} \in S$, where S is a nonempty convex set in E_n. If $\bar{\mathbf{x}}$ is a local optimal solution, then $\bar{\mathbf{x}}$ is the unique global optimal solution.

Proof

Since $\bar{\mathbf{x}}$ is a local optimal solution, then there exists an ε-neighborhood $N_\varepsilon(\bar{\mathbf{x}})$ around $\bar{\mathbf{x}}$ such that $f(\bar{\mathbf{x}}) \leq f(\mathbf{x})$ for all $\mathbf{x} \in S \cap N_\varepsilon(\bar{\mathbf{x}})$. Suppose, by contradiction to the conclusion of the theorem, that there exists a point $\hat{\mathbf{x}} \in S$ such that $\hat{\mathbf{x}} \neq \bar{\mathbf{x}}$ and $f(\hat{\mathbf{x}}) \leq f(\bar{\mathbf{x}})$. By strong quasiconvexity, it follows that

$$f[\lambda\hat{\mathbf{x}} + (1 - \lambda)\mathbf{x}] < \text{maximum } \{f(\hat{\mathbf{x}}), f(\bar{\mathbf{x}})\} = f(\bar{\mathbf{x}})$$

for all $\lambda \in (0, 1)$. But for λ small enough, $\lambda\hat{\mathbf{x}} + (1 - \lambda)\bar{\mathbf{x}} \in S \cap N_\varepsilon(\bar{\mathbf{x}})$, so that the above inequality violates local optimality of $\bar{\mathbf{x}}$. This completes the proof.

Pseudoconvex Functions

The astute reader might already have observed that differentiable strongly (or strictly) quasiconvex functions do not share the property of convex functions, which says that if $\nabla f(\bar{\mathbf{x}}) = \mathbf{0}$ at some point $\bar{\mathbf{x}}$, then $\bar{\mathbf{x}}$ is a global minimum of f. Figure 3.12*c* illustrates this fact. This motivates the definition of pseudoconvex functions which share this important property with convex functions, and leads to a generalization of various derivative based optimality conditions.

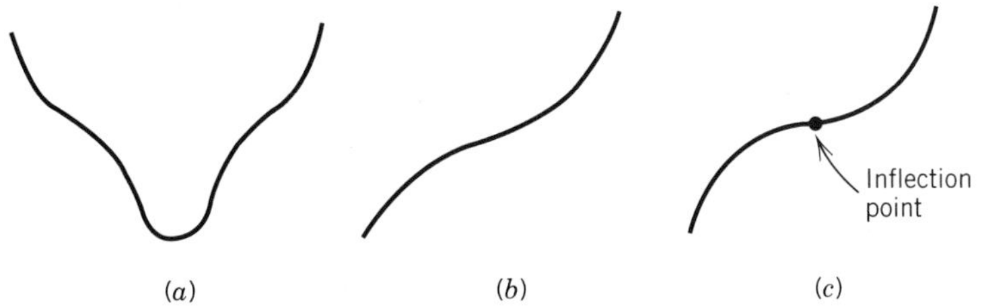

Figure 3.12 Pseudoconvex and pseudoconcave functions. (*a*) Pseudoconvex. (*b*) Both pseudoconvex and pseudoconcave. (*c*) Neither pseudoconvex nor pseudoconcave.

3.5.10 Definition

Let S be a nonempty open set in E_n, and let $f: S \rightarrow E_1$ be differentiable on S. The function f is said to be *pseudoconvex* if for each $\mathbf{x}_1, \mathbf{x}_2 \in S$ with $\nabla f(\mathbf{x}_1)^t(\mathbf{x}_2 - \mathbf{x}_1) \geq 0$ we have

$f(\mathbf{x}_2) \geq f(\mathbf{x}_1)$; or, equivalently, if $f(\mathbf{x}_2) < f(\mathbf{x}_1)$, then $\nabla f(\mathbf{x}_1)^t(\mathbf{x}_2 - \mathbf{x}_1) < 0$. The function f is said to be *pseudoconcave* if $-f$ is pseudoconvex.

The function f is said to be *strictly pseudoconvex* if, for each distinct $\mathbf{x}_1, \mathbf{x}_2 \in S$ satisfying $\nabla f(\mathbf{x}_1)^t(\mathbf{x}_2 - \mathbf{x}_1) \geq 0$, we have $f(\mathbf{x}_2) > f(\mathbf{x}_1)$; or, equivalently, if for each distinct $\mathbf{x}_1, \mathbf{x}_2 \in S$, $f(\mathbf{x}_2) \leq f(\mathbf{x}_1)$ implies that $\nabla f(\mathbf{x}_1)^t(\mathbf{x}_2 - \mathbf{x}_1) < 0$. The function f is said to be *strictly pseudoconcave* if $-f$ is strictly pseudoconvex.

Figure 3.12*a* illustrates a pseudoconvex function. From the definition of pseudoconvexity, it is clear that if $\nabla f(\bar{\mathbf{x}}) = \mathbf{0}$ at any point $\bar{\mathbf{x}}$, then $f(\mathbf{x}) \geq f(\bar{\mathbf{x}})$ for all $\mathbf{x}$; and so, $\bar{\mathbf{x}}$ is a global minimum for f. Hence, the function in Figure 3.12*c* is neither pseudoconvex nor pseudoconcave. In fact, the definition asserts that if the directional derivative of f at any point $\mathbf{x}_1$ in the direction $(\mathbf{x}_2 - \mathbf{x}_1)$ is nonnegative, then the function values are nondecreasing in that direction (see Exercise 3.63). Furthermore, observe that the pseudoconvex functions shown in Figure 3.12 are also strictly quasiconvex, which is true in general, as shown by Theorem 3.5.11 below. The reader may note that the function in Figure 3.8*c* is not pseudoconvex yet strictly quasiconvex.

3.5.11 Theorem

Let S be a nonempty open convex set in E_n, and let $f: S \to E_1$ be a differentiable pseudoconvex function on S. Then, f is both strictly quasiconvex and quasiconvex.

Proof

We first show that f is strictly quasiconvex. By contradiction, suppose that there exist $\mathbf{x}_1, \mathbf{x}_2 \in S$ such that $f(\mathbf{x}_1) \neq f(\mathbf{x}_2)$ and $f(\mathbf{x}') \geq \text{maximum } \{f(\mathbf{x}_1), f(\mathbf{x}_2)\}$, where $\mathbf{x}' = \lambda\mathbf{x}_1 + (1 - \lambda)\mathbf{x}_2$ for some $\lambda \in (0, 1)$. Without loss of generality, assume that $f(\mathbf{x}_1) < f(\mathbf{x}_2)$, so that

$$f(\mathbf{x}') \geq f(\mathbf{x}_2) > f(\mathbf{x}_1) \tag{3.28}$$

Note, by the pseudoconvexity of f, that $\nabla f(\mathbf{x}')^t(\mathbf{x}_1 - \mathbf{x}') < 0$. Now, since $\nabla f(\mathbf{x}')^t(\mathbf{x}_1 - \mathbf{x}') < 0$ and $\mathbf{x}_1 - \mathbf{x}' = -(1 - \lambda)(\mathbf{x}_2 - \mathbf{x}')/\lambda$, then $\nabla f(\mathbf{x}')^t(\mathbf{x}_2 - \mathbf{x}') > 0$; and, hence, by the pseudoconvexity of f, we must have $f(\mathbf{x}_2) \geq f(\mathbf{x}')$. Therefore, by (3.28), we get $f(\mathbf{x}_2) = f(\mathbf{x}')$. Also, since $\nabla f(\mathbf{x}')^t(\mathbf{x}_2 - \mathbf{x}') > 0$, there exists a point $\hat{\mathbf{x}} = \mu\mathbf{x}' + (1 - \mu)\mathbf{x}_2$ with $\mu \in (0, 1)$ such that

$$f(\hat{\mathbf{x}}) > f(\mathbf{x}') = f(\mathbf{x}_2)$$

Again, by the pseudoconvexity of f, we have $\nabla f(\hat{\mathbf{x}})^t(\mathbf{x}_2 - \hat{\mathbf{x}}) < 0$. Similarly, $\nabla f(\hat{\mathbf{x}})^t(\mathbf{x}' - \hat{\mathbf{x}}) < 0$. Summarizing, we must have

$$\nabla f(\hat{\mathbf{x}})^t(\mathbf{x}_2 - \hat{\mathbf{x}}) < 0$$

$$\nabla f(\hat{\mathbf{x}})^t(\mathbf{x}' - \hat{\mathbf{x}}) < 0$$

Note that $\mathbf{x}_2 - \hat{\mathbf{x}} = \mu(\hat{\mathbf{x}} - \mathbf{x}')/(1 - \mu)$ and, hence, the above two inequalities are not compatible. This contradiction shows that f is strictly quasiconvex. By Lemma 3.5.7, then, f is also quasiconvex, and the proof is complete.

In Theorem 3.5.12 below, we see that every strictly pseudoconvex function is strongly quasiconvex.

3.5.12 Theorem

Let S be a nonempty open convex set in E_n, and let $f:S \to E_1$ be a differentiable strictly pseudoconvex function. Then, f is strongly quasiconvex.

Proof

By contradiction, suppose that there exist distinct $\mathbf{x}_1$, $\mathbf{x}_2 \in S$ and $\lambda \in (0, 1)$ such that $f(\mathbf{x}) \geq \text{maximum } \{f(\mathbf{x}_1), f(\mathbf{x}_2)\}$, where $\mathbf{x} = \lambda\mathbf{x}_1 + (1 - \lambda)\mathbf{x}_2$. Since $f(\mathbf{x}_1) \leq f(\mathbf{x})$, we have, by the strict pseudoconvexity of f, that $\nabla f(\mathbf{x})^t(\mathbf{x}_1 - \mathbf{x}) < 0$ and, hence,

$$\nabla f(\mathbf{x})^t(\mathbf{x}_1 - \mathbf{x}_2) < 0 \tag{3.29}$$

Likewise, since $f(\mathbf{x}_2) \leq f(\mathbf{x})$,

$$\nabla f(\mathbf{x})^t(\mathbf{x}_2 - \mathbf{x}_1) < 0 \tag{3.30}$$

The two inequalities (3.29) and (3.30) are not compatible and, hence, f is strongly quasiconvex. This completes the proof.

We remark here in connection with Theorems 3.5.11 and 3.5.12, for the special case in which f is quadratic, that f is pseudoconvex if and only if f is strictly quasiconvex, which holds true if and only if f is quasiconvex. Moreover, we also have that f is strictly pseudoconvex if and only if f is strongly quasiconvex. Hence, all these properties become equivalent to each other for quadratic functions (see Exercise 3.51). Also, Appendix B provides a bordered Hessian determinant characterization for checking pseudoconvexity and strict pseudoconvexity of quadratic functions.

Thus, far, we have discussed various types of convexity and concavity. Figure 3.13 summarizes the implications among these types of convexity. These implications either follow from the definitions or from the various results proved in this section. A similar figure can be constructed for the concave case.

Convexity at a Point

Another useful concept in optimization is the notion of convexity or concavity at a point. In some cases, the requirement of a convex or concave function may be too strong and really not essential. Instead, convexity or concavity at a point may be all that is needed.

3.5.13 Definition

Let S be a nonempty convex set in E_n, and $f:S \to E_1$. The following are relaxations of various forms of convexity presented in this chapter:

Convexity at $\bar{\mathbf{x}}$. The function f is said to be convex at $\bar{\mathbf{x}} \in S$ if

$$f[\lambda\bar{\mathbf{x}} + (1 - \lambda)\mathbf{x}] \leq \lambda f(\bar{\mathbf{x}}) + (1 - \lambda)f(\mathbf{x})$$

for each $\lambda \in (0, 1)$ and each $\mathbf{x} \in S$.

Strict convexity at $\bar{\mathbf{x}}$. The function f is said to be strictly convex at $\bar{\mathbf{x}} \in S$ if

$$f[\lambda\bar{\mathbf{x}} + (1 - \lambda)\mathbf{x}] < \lambda f(\bar{\mathbf{x}}) + (1 - \lambda)f(\mathbf{x})$$

for each $\lambda \in (0, 1)$ and for each $\mathbf{x} \in S$, $\mathbf{x} \neq \bar{\mathbf{x}}$.

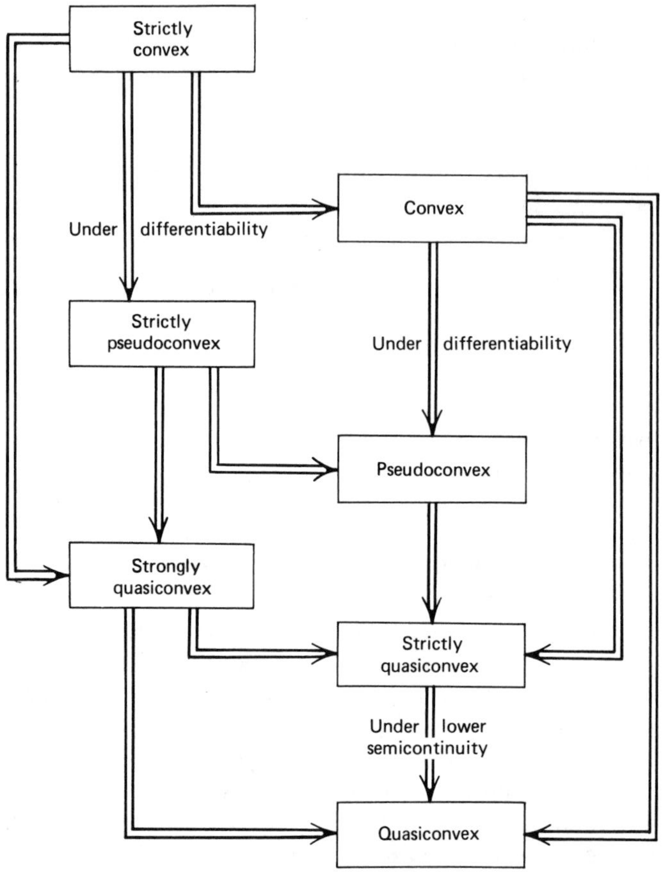

Figure 3.13 Relationship among various types of convexity.

Quasiconvexity at $\bar{x}$. The function f is said to be quasiconvex at $\bar{\mathbf{x}} \in S$ if

$$f[\lambda\bar{\mathbf{x}} + (1 - \lambda)\mathbf{x}] \leq \text{maximum } \{f(\mathbf{x}), f(\bar{\mathbf{x}})\}$$

for each $\lambda \in (0, 1)$ and each $\mathbf{x} \in S$.

Strict quasiconvexity at $\bar{x}$. The function f is said to be strictly quasiconvex at $\bar{\mathbf{x}} \in S$ if

$$f[\lambda\bar{\mathbf{x}} + (1 - \lambda)\mathbf{x}] < \text{maximum } \{f(\mathbf{x}), f(\bar{\mathbf{x}})\}$$

for each $\lambda \in (0, 1)$ and each $\mathbf{x} \in S$ such that $f(\mathbf{x}) \neq f(\bar{\mathbf{x}})$.

Strong quasiconvexity at $\bar{x}$. The function f is said to be strongly quasiconvex at $\bar{\mathbf{x}} \in S$ if

$$f[\lambda\bar{\mathbf{x}} + (1 - \lambda)\mathbf{x}] < \text{maximum } \{f(\mathbf{x}), f(\bar{\mathbf{x}})\}$$

for each $\lambda \in (0, 1)$ and each $\mathbf{x} \in S$, $\mathbf{x} \neq \bar{\mathbf{x}}$.

Pseudoconvexity at $\bar{x}$. The function f is said to be pseudoconvex at $\bar{\mathbf{x}} \in S$ if $\nabla f(\bar{\mathbf{x}})^t(\mathbf{x} - \bar{\mathbf{x}}) \geq 0$ for $\mathbf{x} \in S$ implies that $f(\mathbf{x}) \geq f(\bar{\mathbf{x}})$.

Strict pseudoconvexity at $\bar{x}$. The function f is said to be strictly pseudoconvex at $\bar{\mathbf{x}} \in S$ if $\nabla f(\bar{\mathbf{x}})^t(\mathbf{x} - \bar{\mathbf{x}}) \geq 0$ for $\mathbf{x} \in S$, $\mathbf{x} \neq \bar{\mathbf{x}}$, implies that $f(\mathbf{x}) > f(\bar{\mathbf{x}})$.

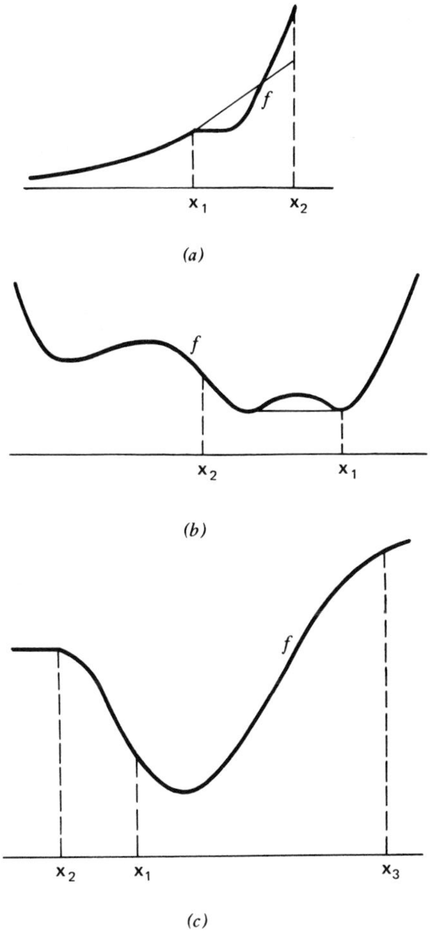

Figure 3.14 Various types of convexity at a point. (*a*) Convexity and strict convexity: f is convex but not strictly convex at $\mathbf{x}_1$, f is both convex and strictly convex at $\mathbf{x}_2$. (*b*) Pseudoconvexity and strict pseudoconvexity: f is pseudoconvex but not strictly pseudoconvex at $\mathbf{x}_1$, f is both pseudoconvex and strictly pseudoconvex at $\mathbf{x}_2$. (*c*) Quasiconvexity, strict quasiconvexity and strong quasiconvexity: f is quasiconvex but neither strictly quasiconvex nor strongly quasiconvex at $\mathbf{x}_1$, f is both quasiconvex and strictly quasiconvex at $\mathbf{x}_2$ but not strongly quasiconvex at $\mathbf{x}_2$, f is quasiconvex, strictly quasiconvex and strongly quasiconvex at $\mathbf{x}_3$.

Various types of concavity at a point can be stated in a similar fashion. Figure 3.14 shows some types of convexity at a point. As the figure suggests, these types of convexity at a point represent a significant relaxation of the concept of convexity.

We specify below some important results related to convexity of a function f at a point, where $f:S \to E_1$ and S is a nonempty convex set in E_n. Of course, not all the results developed throughout this chapter hold true. However, several of these results hold true and are summarized below. The proofs are similar to the corresponding theorems in this chapter.

1. Let f be both convex and differentiable at $\bar{\mathbf{x}}$. Then, $f(\mathbf{x}) \geq f(\bar{\mathbf{x}}) + \nabla f(\bar{\mathbf{x}})^t(\mathbf{x} - \bar{\mathbf{x}})$ for each $\mathbf{x} \in S$. If f is strictly convex, then strict inequality holds for $\mathbf{x} \neq \bar{\mathbf{x}}$.
2. Let f be both convex and twice differentiable at $\bar{\mathbf{x}}$. Then, the Hessian matrix $\mathbf{H}(\bar{\mathbf{x}})$ is positive semidefinite.
3. Let f be convex at $\bar{\mathbf{x}} \in S$, and let $\bar{\mathbf{x}}$ be an optimal solution to the problem to minimize $f(\mathbf{x})$ subject to $\mathbf{x} \in S$. Then, $\bar{\mathbf{x}}$ is a global optimal solution.
4. Let f be convex and differentiable at $\bar{\mathbf{x}} \in S$. Then, $\bar{\mathbf{x}}$ is an optimal solution to the problem to minimize $f(\mathbf{x})$ subject to $\mathbf{x} \in S$ if and only if $\nabla f(\bar{\mathbf{x}})^t(\mathbf{x} - \bar{\mathbf{x}}) \geq 0$ for each $\mathbf{x} \in S$. In particular, if $\bar{\mathbf{x}} \in \text{int } S$, then $\bar{\mathbf{x}}$ is an optimal solution if and only if $\nabla f(\bar{\mathbf{x}}) = \mathbf{0}$.
5. Let f be convex and differentiable at $\bar{\mathbf{x}} \in S$. Suppose that $\bar{\mathbf{x}}$ is an optimal solution to the problem to maximize $f(\mathbf{x})$ subject to $\mathbf{x} \in S$. Then, $\nabla f(\bar{\mathbf{x}})^t(\mathbf{x} - \bar{\mathbf{x}}) \leq 0$ for each $\mathbf{x} \in S$.
6. Let f be both quasiconvex and differentiable at $\bar{\mathbf{x}}$, and let $\mathbf{x} \in S$ be such that $f(\mathbf{x}) \leq f(\bar{\mathbf{x}})$. Then, $\nabla f(\bar{\mathbf{x}})^t(\mathbf{x} - \bar{\mathbf{x}}) \leq 0$.
7. Suppose that $\bar{\mathbf{x}}$ is a local optimal solution to the problem to minimize $f(\mathbf{x})$ subject to $\mathbf{x} \in S$. If f is strictly quasiconvex at $\bar{\mathbf{x}}$, then $\bar{\mathbf{x}}$ is a global optimal solution. If f is strongly quasiconvex at $\bar{\mathbf{x}}$, then $\bar{\mathbf{x}}$ is the unique global optimal solution.
8. Consider the problem to minimize $f(\mathbf{x})$ subject to $\mathbf{x} \in S$, and let $\bar{\mathbf{x}} \in S$ be such that $\nabla f(\bar{\mathbf{x}}) = \mathbf{0}$. If f is pseudoconvex at $\bar{\mathbf{x}}$, then $\bar{\mathbf{x}}$ is a global optimal solution; and if f is strictly pseudoconvex at $\bar{\mathbf{x}}$, then $\bar{\mathbf{x}}$ is the unique global optimal solution.

Exercises

3.1 Let S be a nonempty convex set in E_n, and let $f:S \to E_1$. Show that f is concave if and only if hyp f is convex.

3.2 Let S be a nonempty convex set in E_n, and let $f:S \to E_1$. Show that f is convex if and only if for any integer $k \geq 2$, the following holds true: $\mathbf{x}_1, \ldots, \mathbf{x}_k \in S$ implies that $f(\sum_{j=1}^k \lambda_j \mathbf{x}_j) \leq \sum_{j=1}^k \lambda_j f(\mathbf{x}_j)$, where $\sum_{j=1}^k \lambda_j = 1$, $\lambda_j \geq 0$ for $j = 1, \ldots, k$.

3.3 Which of the following functions is convex, concave, or neither? Why?

a. $f(x_1, x_2) = x_1^2 + 2x_1x_2 - 10x_1 + 5x_2$
b. $f(x_1, x_2) = x_1e^{-(x_1+x_2)}$
c. $f(x_1, x_2) = -x_1^2 - 5x_2^2 + 2x_1x_2 + 10x_1 - 10x_2$
d. $f(x_1, x_2, x_3) = x_1x_2 + 2x_1^2 + x_2^2 + 2x_3^2 - 6x_1x_3$
e. $f(x_1, x_2, x_3) = -x_1^2 - 3x_2^2 - 2x_3^2 + 4x_1x_2 + 2x_1x_3 + 4x_2x_3$

3.4 Show that a function $f:E_n \to E_1$ is affine if and only if f is both convex and concave. (A function f is affine if it is of the form $f(\mathbf{x}) = \alpha + \mathbf{c}^t\mathbf{x}$, where α is a scalar and $\mathbf{c}$ is an n vector.)

3.5 Prove or disprove concavity of the following function defined over $S = \{(x_1, x_2): -1 \le x_1 \le 1, -1 \le x_2 \le 1\}$:

$$f(x_1, x_2) = 10 - 2(x_2 - x_1^2)^2$$

Repeat for a convex set $S \subseteq \{(x_1, x_2): x_1^2 \ge x_2\}$

3.6 Let F be a cumulative distribution function for a random variable b, that is, $F(y) = \text{Prob}(b \le y)$. Show that $\phi(z) = \int_{-\infty}^{z} F(y)\, dy$ is a convex function. Is ϕ convex for any nondecreasing function F?

3.7 Let $g: E_n \to E_1$ be a concave function, and let f be defined by $f(\mathbf{x}) = 1/g(\mathbf{x})$. Show that f is convex over $S = \{\mathbf{x}: g(\mathbf{x}) > 0\}$. State a symmetric result interchanging the convex and concave functions.

3.8 Let $f_1, f_2, \ldots, f_k: E_n \to E_1$ be convex functions. Consider the function f defined by $f(\mathbf{x}) = \text{maximum } \{f_1(\mathbf{x}), f_2(\mathbf{x}) \ldots, f_k(\mathbf{x})\}$. Show that f is convex. State and prove a similar result for concave functions.

3.9 Let $f_1, f_2, \ldots, f_k: E_n \to E_1$ be convex functions. Consider the function f defined by $f(\mathbf{x}) = \sum_{j=1}^{k} \alpha_j f_j(\mathbf{x})$, where $\alpha_j > 0$ for $j = 1, 2, \ldots, k$. Show that f is convex. State and prove a similar result for concave functions.

3.10 Let S be a nonempty convex set in E_n, and let $f: E_n \to E_1$ be defined as follows:

$$f(\mathbf{y}) = \inf \{\|\mathbf{y} - \mathbf{x}\|: \mathbf{x} \in S\}$$

Note that $f(\mathbf{y})$ gives the distance from $\mathbf{y}$ to the set S and is called the *distance function*. Prove that f is convex.

3.11 Let $S = \{(x_1, x_2): x_1^2 + x_2^2 \le 1\}$. Let f be the distance function defined in Exercise 3.10. Find the function f explicitly.

3.12 Let $h: E_n \to E_1$ be a convex function, and let $g: E_1 \to E_1$ be a nondecreasing convex function. Consider the composite function $f: E_n \to E_1$ defined by $f(\mathbf{x}) = g[h(\mathbf{x})]$. Show that f is convex.

3.13 Let $g: E_m \to E_1$ be a convex function, and let $\mathbf{h}: E_n \to E_m$ be an affine function of the form $\mathbf{h}(\mathbf{x}) = \mathbf{A}\mathbf{x} + \mathbf{b}$, where $\mathbf{A}$ is an $m \times n$ matrix and $\mathbf{b}$ is an $m \times 1$ vector. Then, show that the composite function $f: E_n \to E_1$ defined as $f(\mathbf{x}) = g[\mathbf{h}(\mathbf{x})]$ is a convex function. Also, assuming twice-differentiability of g, derive an expression for the Hessian of f.

3.14 Let $f: S \to E_1$ be defined as

$$f(\mathbf{x}) = \frac{(\boldsymbol{\alpha}'\mathbf{x})^2}{(\boldsymbol{\beta}'\mathbf{x})}$$

where S is a convex subset of E_n, $\boldsymbol{\alpha}$ and $\boldsymbol{\beta}$ are vectors in E_n, and where $\boldsymbol{\beta}'\mathbf{x} > 0$ for all $\mathbf{x} \in S$. Derive an explicit expression for the Hessian of f and, hence, verify that f is convex over S.

3.15 Let $f: E_n \to E_1$ be lower semicontinuous. Show that the level set $S_\alpha = \{\mathbf{x}: f(\mathbf{x}) \le \alpha\}$ is closed for all $\alpha \in E_1$.

3.16 Over what subset of $\{x : x > 0\}$ is the univariate function $f(x) = e^{-ax^\alpha}$ convex, where $a > 0$ and $\alpha \ge 1$?

3.17 Over what domain is the function $f(x) = x^2(x^2 - 1)$ convex? Is it strictly convex over the specified region(s)? Justify your answer.

3.18 Let S be a nonempty, bounded convex set in E_n, and let $f: E_n \to E_1$ be defined as follows:

$$f(\mathbf{y}) = \sup \{\mathbf{y}'\mathbf{x}: \mathbf{x} \in S\}$$

The function f is called the *support function* of S. Prove that f is convex. Also show that if $f(\mathbf{y}) = \mathbf{y}'\bar{\mathbf{x}}$, where $\bar{\mathbf{x}} \in S$, then $\bar{\mathbf{x}}$ is a subgradient of f at $\mathbf{y}$.

3.19 Let $S = A \cup B$, where

$$A = \{(x_1, x_2): x_1 < 0, x_1^2 + x_2^2 \le 1\}$$

$$B = \{(x_1, x_2): x_1 \ge 0, -1 \le x_2 \le 1\}$$

Find the support function defined in Exercise 3.18 explicitly.

3.20 A function $f:E_n \to E_1$ is called a *gauge function* if it satisfies the following equality:

$$f(\lambda\mathbf{x}) = \lambda f(\mathbf{x}) \qquad \text{for all } \mathbf{x} \in E_n \text{ and all } \lambda \geq 0$$

Further, a gauge function is said to be *subadditive* if it satisfies the following inequality:

$$f(\mathbf{x}) + f(\mathbf{y}) \geq f(\mathbf{x} + \mathbf{y}) \qquad \text{for all } \mathbf{x}, \mathbf{y} \in E_n.$$

Prove that subadditivity is equivalent to convexity of gauge functions.

3.21 Let $f:E_n \to E_1$ be convex. Show that $\boldsymbol{\xi}$ is a subgradient of f at $\bar{\mathbf{x}}$ if and only if the hyperplane $\{(\mathbf{x}, y):y = f(\bar{\mathbf{x}}) + \boldsymbol{\xi}^t(\mathbf{x} - \bar{\mathbf{x}})\}$ supports epi f at $[\bar{\mathbf{x}}, f(\bar{\mathbf{x}})]$. State and prove a similar result for concave functions.

3.22 Let f be a convex function on E_n. Prove that the set of subgradients of f at a given point forms a closed convex set.

3.23 Consider the function θ defined by the following optimization problem:

$$\begin{aligned} \theta(u_1, u_2) = \text{Minimum} \quad & x_1(1 - u_1) + x_2(1 - u_2) \\ \text{subject to} \quad & x_1^2 + x_2^2 \leq 1 \end{aligned}$$

a. Show that θ is concave.
b. Evaluate θ at the point (1, 1).
c. Find the collection of subgradients of θ at (1, 1).

3.24 Let $f:E_n \to E_1$ be defined by $f(\mathbf{x}) = \|\mathbf{x}\|$. Prove that subgradients of f are characterized as follows:

If $\mathbf{x} = \mathbf{0}$, then $\boldsymbol{\xi}$ is a subgradient of f at $\mathbf{x}$ if and only if $\|\boldsymbol{\xi}\| \leq 1$. On the other hand, if $\mathbf{x} \neq \mathbf{0}$, then $\boldsymbol{\xi}$ is a subgradient of f at $\mathbf{x}$ if and only if $\|\boldsymbol{\xi}\| = 1$ and $\boldsymbol{\xi}^t\mathbf{x} = \|\mathbf{x}\|$.

Use this result to show that f is differentiable at each $\mathbf{x} \neq \mathbf{0}$, and characterize the gradient vector.

3.25 Let $f_1, f_2:E_n \to E_1$ be differentiable convex functions. Consider the function f defined by $f(\mathbf{x}) = \text{maximum } \{f_1(\mathbf{x}), f_2(\mathbf{x})\}$. Let $\bar{\mathbf{x}}$ be such that $f(\bar{\mathbf{x}}) = f_1(\bar{\mathbf{x}}) = f_2(\bar{\mathbf{x}})$. Show that $\boldsymbol{\xi}$ is a subgradient of f at $\bar{\mathbf{x}}$ if and only if

$$\boldsymbol{\xi} = \lambda\nabla f_1(\bar{\mathbf{x}}) + (1 - \lambda)\nabla f_2(\bar{\mathbf{x}}) \qquad \text{where } \lambda \in [0, 1]$$

Generalize the result to several convex functions and state a similar result for concave functions.

3.26 Consider the function θ defined by the following optimization problem, where X is a compact polyhedral set.

$$\begin{aligned} \theta(\mathbf{u}) = \text{Minimum} \quad & \mathbf{c}^t\mathbf{x} + \mathbf{u}^t(\mathbf{A}\mathbf{x} - \mathbf{b}) \\ \text{subject to} \quad & \mathbf{x} \in X \end{aligned}$$

a. Show that θ is concave.
b. Characterize the subgradients of θ at any given $\mathbf{u}$.

3.27 In reference to Exercise 3.26, find the function θ explicitly and describe the set of subgradients at each point $\mathbf{u} \geq \mathbf{0}$ if

$$\mathbf{A} = \begin{bmatrix} 1 & 1 \\ -1 & 2 \end{bmatrix} \qquad \mathbf{b} = \begin{bmatrix} 3 \\ 4 \end{bmatrix} \qquad \mathbf{c} = \begin{bmatrix} -1 \\ -3 \end{bmatrix}$$

$$X = \{(x_1, x_2):0 \leq x_1 \leq 2, 0 \leq x_2 \leq \tfrac{3}{2}\}$$

3.28 Let $f:E_n \to E_1$ be a differentiable function. Show that the gradient vector is given by

$$\nabla f(\mathbf{x}) = \left(\frac{\partial f(\mathbf{x})}{\partial x_1}, \frac{\partial f(\mathbf{x})}{\partial x_2}, \ldots, \frac{\partial f(\mathbf{x})}{\partial x_n}\right)^t$$

3.29 Let $f: E_n \to E_1$ be a differentiable function. The *linear approximation* of f at a given point $\bar{\mathbf{x}}$ is given by

$$f(\bar{\mathbf{x}}) + \nabla f(\bar{\mathbf{x}})^t(\mathbf{x} - \bar{\mathbf{x}})$$

If f is twice differentiable at $\bar{\mathbf{x}}$, then the *quadratic approximation* of f at $\bar{\mathbf{x}}$ is given by

$$f(\bar{\mathbf{x}}) + \nabla f(\bar{\mathbf{x}})^t(\mathbf{x} - \bar{\mathbf{x}}) + \tfrac{1}{2}(\mathbf{x} - \bar{\mathbf{x}})^t\mathbf{H}(\bar{\mathbf{x}})(\mathbf{x} - \bar{\mathbf{x}})$$

Let $f(x_1, x_2) = e^{x_1^2+x_2^2} - 5x_1 + 10x_2$. Give the linear and quadratic approximations of f at $(0, 1)$. Are these approximations convex, concave, or neither? Why?

3.30 Consider the function $f: E_3 \to E_1$ given by $f(\mathbf{x}) = \mathbf{x}^t\mathbf{A}\mathbf{x}$, where

$$\mathbf{A} = \begin{bmatrix} 2 & 1 & 3 \\ 2 & 3 & 2 \\ 1 & 1 & \theta \end{bmatrix}$$

What is the Hessian of f? For what value of θ is f strictly convex?

3.31 Consider the function $f: E_n \to E_1$ and suppose that f is infinitely differentiable. Then, show that f is strictly convex if and only if, for each $\bar{\mathbf{x}}$ and $\mathbf{d}$ in E_n, the first nonzero derivative term of order greater than or equal to 2 in the Taylor series expansion exists, is of even order, and is positive.

3.32 Consider the function $f(x) = x^3$, defined over the set $S = \{x \in E_1 : x \geq 0\}$. Show that f is strictly convex over S. Noting that $f''(0) = 0$ and $f'''(0) = 6$, comment on the application of Theorem 3.3.9.

3.33 Let $\mathbf{H}$ be an $n \times n$ symmetric matrix. Using the eigenvalue characterization of definiteness, verify that $\mathbf{H}$ is positive definite if and only if it is positive semidefinite and nonsingular.

3.34 Suppose that $\mathbf{H}$ is an $n \times n$ symmetric matrix. Show how Theorem 3.3.12 demonstrates that $\mathbf{H}$ is positive definite if and only if it can be premultiplied by a series of n lower triangular Gauss–Jordan reduction matrices $\mathbf{L}_1, \ldots, \mathbf{L}_n$ to yield an upper triangular matrix $\mathbf{U}$ with positive diagonal elements. (Letting $\mathbf{L}^{-1} = \mathbf{L}_n \ldots, \mathbf{L}_1$, we obtain $\mathbf{H} = \mathbf{LU}$, where $\mathbf{L}$ is lower triangular. This is known as the *LU-decomposition of* $\mathbf{H}$; see Appendix A.2.) Furthermore, show that $\mathbf{H}$ is positive definite if and only if there exists a lower triangular matrix $\mathbf{L}$ with positive diagonal elements such that $\mathbf{H} = \mathbf{LL}^t$. (This is known as the *Cholesky factorization* of $\mathbf{H}$; see Appendix A.2.)

3.35 Let $\mathbf{H}$ be an $n \times n$ symmetric, positive semidefinite matrix, and suppose that $\mathbf{x}^t\mathbf{H}\mathbf{x} = 0$ for some $\mathbf{x} \in E_n$. Then, show that $\mathbf{H}\mathbf{x} = \mathbf{0}$. (*Hint*: Consider the diagonal of the quadratic form $\mathbf{x}^t\mathbf{H}\mathbf{x}$ via the transformation $\mathbf{x} = \mathbf{Q}\mathbf{y}$, where the columns of $\mathbf{Q}$ are the normalized eigenvectors of $\mathbf{H}$.)

3.36 Consider the problem to minimize $\{f(\mathbf{x}) : \mathbf{x} \in S\}$ and suppose that there exists an $\varepsilon > 0$ such that $N_\varepsilon(\bar{\mathbf{x}}) \cap S$ is a convex set, and that $f(\bar{\mathbf{x}}) \leq f(\mathbf{x})$ for all $\mathbf{x} \in N_\varepsilon(\bar{\mathbf{x}}) \cap S$.

a. Show that if $\mathbf{H}(\bar{\mathbf{x}})$ is positive definite, then $\bar{\mathbf{x}}$ is both a strict and a strong local minimum.

b. Show that if $\bar{\mathbf{x}}$ is a strict local minimum and f is pseudoconvex on $N_\varepsilon(\bar{\mathbf{x}}) \cap S$, then $\bar{\mathbf{x}}$ is also a strong local minimum.

3.37 Suppose that $S \neq \varnothing$ is closed and convex. Let $f: S \to E_1$ be differentiable. State if the following are true or false, justifying your answer:

a. If f is convex on S, then $f(\mathbf{x}) \geq f(\bar{\mathbf{x}}) + \nabla f(\bar{\mathbf{x}})^t(\mathbf{x} - \bar{\mathbf{x}})$ for all $\mathbf{x} \in S$, $\bar{\mathbf{x}} \in \text{int } S$.

b. If $f(\mathbf{x}) \geq f(\bar{\mathbf{x}}) + \nabla f(\bar{\mathbf{x}})^t(\mathbf{x} - \bar{\mathbf{x}})$ for all $\mathbf{x} \in S$ and $\bar{\mathbf{x}} \in \text{int } S$, then f is convex on S.

3.38 Consider the following problem:

$$\begin{aligned} \text{Minimize} \quad & (x_1 - 4)^2 + (x_2 - 6)^2 \\ \text{subject to} \quad & x_2 \geq x_1^2 \\ & x_2 \leq 4 \end{aligned}$$

Write a necessary condition for optimality and verify that it is satisfied by the point $(2, 4)$. Is this the optimal point? Why?

3.39 Use Theorem 3.4.3 to prove that every local minimum of a convex function over a convex set is also a global minimum.

3.40 Consider the following problem:

Maximize $\mathbf{c}'\mathbf{x} + \frac{1}{2}\mathbf{x}'\mathbf{H}\mathbf{x}$

subject to $\mathbf{A}\mathbf{x} \leq \mathbf{b}$

$\mathbf{x} \geq \mathbf{0}$

where $\mathbf{H}$ is a symmetric negative definite matrix, $\mathbf{A}$ is an $m \times n$ matrix, $\mathbf{c}$ is an n vector, and $\mathbf{b}$ is an m vector. Write the necessary and sufficient condition for optimality of Theorem 3.4.3, and simplify it using the special structure of this problem.

3.41 Consider the problem to minimize $f(\mathbf{x})$ subject to $\mathbf{x} \in S$, where $f: E_n \to E_1$ is a differentiable convex function, and S is a nonempty convex set in E_n. Prove that $\bar{\mathbf{x}}$ is an optimal solution if and only if $\nabla f(\bar{\mathbf{x}})'(\mathbf{x} - \bar{\mathbf{x}}) \geq 0$ for each $\mathbf{x} \in S$. Also state and prove a similar result for the maximization of a concave function.

(This result was proved in the text as Corollary 2 to Theorem 3.4.3. In this exercise, the reader is asked to give a direct proof without resort to subgradients.)

3.42 Let $f: E_n \to E_1$ be a convex function, and suppose that $f(\mathbf{x} + \lambda\mathbf{d}) \geq f(\mathbf{x})$ for all $\lambda \in (0, \delta)$, where $\delta > 0$. Show that $f(\mathbf{x} + \lambda\mathbf{d})$ is a nondecreasing function of λ. In particular, show that $f(\mathbf{x} + \lambda\mathbf{d})$ is a strictly increasing function of λ if f is strictly convex.

3.43 A vector $\mathbf{d}$ is called a *direction of descent* of f at $\bar{\mathbf{x}}$ if there exists a $\delta > 0$ such that $f(\bar{\mathbf{x}} + \lambda\mathbf{d}) < f(\bar{\mathbf{x}})$ for each $\lambda \in (0, \delta)$. Suppose that f is convex. Show that $\mathbf{d}$ is a direction of descent if and only if $f'(\bar{\mathbf{x}}; \mathbf{d}) < 0$. Does the result hold true without the convexity of f?

3.44 Consider the problem to minimize $f(\mathbf{x})$ subject to $\mathbf{x} \in S$, where $f: E_n \to E_1$ is convex and S is a nonempty convex set in E_n. The cone of feasible directions of S at $\bar{\mathbf{x}} \in S$ is defined by

$$D = \{\mathbf{d} : \text{there exists a } \delta > 0 \text{ such that } \bar{\mathbf{x}} + \lambda\mathbf{d} \in S \text{ for } \lambda \in (0, \delta)\}$$

Show that $\bar{\mathbf{x}}$ is an optimal solution to the problem if and only if $f'(\bar{\mathbf{x}}; \mathbf{d}) \geq 0$ for each $\mathbf{d} \in D$. Compare this result with the necessary and sufficient condition of Theorem 3.4.3. Specialize the result to the case where $S = E_n$.

3.45 Consider the following problem:

Maximize $f(\mathbf{x})$

subject to $\mathbf{A}\mathbf{x} = \mathbf{b}$

$\mathbf{x} \geq \mathbf{0}$

where $\mathbf{A}$ is an $m \times n$ matrix with rank m, and f is a differentiable convex function. Consider the extreme point $(\mathbf{x}_B', \mathbf{x}_N') = (\bar{\mathbf{b}}', \mathbf{0}')$, where $\bar{\mathbf{b}} = \mathbf{B}^{-1}\mathbf{b} \geq \mathbf{0}$ and $\mathbf{A} = [\mathbf{B}, \mathbf{N}]$. Decompose $\nabla f(\mathbf{x})$ accordingly into $\nabla_B f(\mathbf{x})$ and $\nabla_N f(\mathbf{x})$. Show that the necessary condition of Theorem 3.4.6 holds true if $\nabla_N f(\mathbf{x})' - \nabla_B f(\mathbf{x})'\mathbf{B}^{-1}\mathbf{N} \leq \mathbf{0}$. If this condition holds, is it necessarily true that $\mathbf{x}$ is a local maximum? Prove or give a counterexample.

If $\nabla_N f(\mathbf{x})' - \nabla_B f(\mathbf{x})'\mathbf{B}^{-1}\mathbf{N} \nleq \mathbf{0}$, choose a positive component j and increase its corresponding nonbasic variable x_j until a new extreme point is reached. Show that this process results in a new extreme point with a larger objective value. Does this method guarantee convergence to a global optimal solution? Prove or give a counterexample.

3.46 Apply the procedure of Exercise 3.45 to the following problem starting with the extreme point $(1, 3, 0, 0)$:

$$\begin{aligned}
\text{Maximize} \quad & (x_1 - \tfrac{3}{2})^2 + (x_2 - 5)^2 \\
\text{subject to} \quad & -x_1 + x_2 + x_3 = 2 \\
& 2x_1 + 3x_2 + x_4 = 11 \\
& x_1, \quad x_2, \quad x_3, \quad x_4, \geq 0
\end{aligned}$$

3.47 Let $\mathbf{c}_1$, $\mathbf{c}_2$ be nonzero vectors in E_n, and α_1, α_2 be scalars. Let $S = \{\mathbf{x} : \mathbf{c}_2'\mathbf{x} + \alpha_2 > 0\}$. Consider the function $f: S \to E_1$ defined as follows:

$$f(\mathbf{x}) = \frac{\mathbf{c}_1'\mathbf{x} + \alpha_1}{\mathbf{c}_2'\mathbf{x} + \alpha_2}$$

Show that f is both pseudoconvex and pseudoconcave. (Functions that are both pseudoconvex and pseudoconcave are called *pseudolinear.*)

3.48 Consider a quadratic function $f: E_n \to E_1$, and suppose that f is convex on S, where S is a nonempty convex set in E_n. Show that:

a. The function f is convex on $M(S)$, where $M(S)$ is the *affine manifold* containing S defined by $M(S) = \{\mathbf{y}: \mathbf{y} = \sum_{j=1}^{k} \lambda_j \mathbf{x}_j, \sum_{j=1}^{k} \lambda_j = 1, \mathbf{x}_j \in S \text{ for all } j, \text{ for } k \geq 1\}$.

b. The function f is convex on $L(S)$, the *linear subspace* parallel to $M(S)$, defined by $L(S) = \{\mathbf{y} - \mathbf{x}: \mathbf{y} \in M(S) \text{ and } \mathbf{x} \in S\}$. (This result is credited to Cottle [1967].)

3.49 Consider the quadratic function $f: E_n \to E_1$ defined by $f(\mathbf{x}) = \mathbf{x}^t\mathbf{H}\mathbf{x}$. The function f is said to be *positive subdefinite* if $\mathbf{x}^t\mathbf{H}\mathbf{x} < 0$ implies $\mathbf{H}\mathbf{x} \geq \mathbf{0}$ or $\mathbf{H}\mathbf{x} \leq \mathbf{0}$ for each $\mathbf{x} \in E_n$. Prove that f is quasiconvex on the *nonnegative orthant*, $E_n^+ = \{\mathbf{x} \in E_n: \mathbf{x} \geq \mathbf{0}\}$, if and only if it is positive subdefinite. (This result is credited to Martos [1969].)

3.50 The function f defined in Exercise 3.49 is said to be *strictly positive subdefinite* if $\mathbf{x}^t\mathbf{H}\mathbf{x} < 0$ implies $\mathbf{H}\mathbf{x} > \mathbf{0}$ or $\mathbf{H}\mathbf{x} < \mathbf{0}$ for each $\mathbf{x} \in E_n$. Prove that f is pseudoconvex on the nonnegative orthant excluding $\mathbf{x} = \mathbf{0}$ if and only if it is strictly positive subdefinite. (This result is credited to Martos [1969].)

3.51 Let $f: E_n \to E_1$ be a quadratic function. Then, show that f is quasiconvex if and only if it is strictly quasiconvex, which holds if and only if it is pseudoconvex. Furthermore, show that f is strongly quasiconvex if and only if it is strictly pseudoconvex.

3.52 Let $h: E_n \to E_1$ be a quasiconvex function, and let $g: E_1 \to E_1$ be a nondecreasing function. Then, show that the composite function $f: E_n \to E_1$ defined as $f(\mathbf{x}) = g[h(\mathbf{x})]$ is quasiconvex.

3.53 Let $f: S \subseteq E_1 \to E_1$ be a univariate function, where S is some interval on the real line. Define f as *unimodal* on S if there exists a $x^* \in S$ at which f attains a minimum and f is nondecreasing on the interval $\{x \in S: x \geq x^*\}$, whereas it is nonincreasing on the interval $\{x \in S: x \leq x^*\}$. Assuming that f attains a minimum on S, show that f is quasiconvex if and only if it is unimodal on S.

3.54 Let $f: S \to E_1$ be a continuous function, where S is a convex subset of E_n. Show that f is quasimonotone if and only if the level surface $\{\mathbf{x} \in S: f(\mathbf{x}) = \alpha\}$ is a convex set for all $\alpha \in E_1$.

3.55 Let $f: S \to E_1$ be a differentiable function, where S is an open, convex subset of E_n. Show that f is quasimonotone if and only if, for every $\mathbf{x}_1$ and $\mathbf{x}_2$ in S, $f(\mathbf{x}_1) \geq f(\mathbf{x}_2)$ implies that $\nabla f(\mathbf{x}_2)^t(\mathbf{x}_1 - \mathbf{x}_2) \geq 0$ and $f(\mathbf{x}_1) \leq f(\mathbf{x}_2)$ implies that $\nabla f(\mathbf{x}_2)^t(\mathbf{x}_1 - \mathbf{x}_2) \leq 0$. Hence, show that f is quasimonotone if and only if $f(\mathbf{x}_1) \geq f(\mathbf{x}_2)$ implies that $\nabla f(\mathbf{x}_\lambda)^t(\mathbf{x}_1 - \mathbf{x}_2) \geq 0$ for all $\mathbf{x}_1$ and $\mathbf{x}_2$ in S, and for all $\mathbf{x}_\lambda = \lambda\mathbf{x}_1 + (1 - \lambda)\mathbf{x}_2$, where $0 \leq \lambda \leq 1$.

3.56 Let $f: S \to E_1$, where f is lower-semicontinuous and where S is a convex subset of E_n. Define f as being *strongly unimodal* on S if, for each $\mathbf{x}_1$ and $\mathbf{x}_2$ in S for which the function $F(\lambda) = f[\mathbf{x}_1 + \lambda(\mathbf{x}_2 - \mathbf{x}_1)]$, $0 \leq \lambda \leq 1$, attains a minimum at a point $\lambda^* > 0$, we have $F(0) > F(\lambda) > F(\lambda^*)$ for all $0 < \lambda < \lambda^*$. Show that f is strongly quasiconvex on S if and only if it is strongly unimodal on S (see Exercise 8.4).

3.57 Let $g: S \to E_1$ and $h: S \to E_1$, where S is a nonempty convex set in E_n. Consider the function $f: S \to E_1$ defined by $f(\mathbf{x}) = g(\mathbf{x})/h(\mathbf{x})$. Show that f is a quasiconvex if the following two conditions hold true:

a. g is convex on S, and $g(\mathbf{x}) \geq 0$ for each $\mathbf{x} \in S$.

b. h is concave on S, and $h(\mathbf{x}) > 0$ for each $\mathbf{x} \in S$.

(*Hint*: Use Theorem 3.5.2.)

3.58 Show that the function f defined in Exercise 3.57 is quasiconvex if the following two conditions hold true:

a. g is convex on S, and $g(\mathbf{x}) \leq 0$ for each $\mathbf{x} \in S$.

b. h is convex on S, and $h(\mathbf{x}) > 0$ for each $\mathbf{x} \in S$.

3.59 Let $g: S \to E_1$ and $h: S \to E_1$, where S is a nonempty convex set in E_n. Consider the function $f: S \to E_1$ defined by $f(\mathbf{x}) = g(\mathbf{x})h(\mathbf{x})$. Show that f is quasiconvex if the following two conditions hold true:

a. g is convex, and $g(\mathbf{x}) \leq 0$ for each $\mathbf{x} \in S$.

b. h is concave, and $h(\mathbf{x}) > 0$ for each $\mathbf{x} \in S$.

3.60 In each of the Exercises 3.57, 3.58, and 3.59, show that f is pseudoconvex provided that S is open and g and h are differentiable.

3.61 Let $f: S \to E_1$ be a continuously differentiable convex function, where S is some open interval in E_1. Then, show that f is (strictly) pseudoconvex if and only if, whenever

$f'(\bar{x}) = 0$ for any $\bar{x} \in S$, this implies that $\bar{x}$ is a (strict) local minimum of f on S. Generalize this result to the multivariate case.

3.62 Let $f: S \to E_1$ be a twice-differentiable, univariate function, where S is some open interval in E_1. Then, show that f is (strictly) pseudoconvex if and only if, whenever $f'(\bar{x}) = 0$ for any $\bar{x} \in S$, we have that either $f''(\bar{x}) > 0$ or that $f''(\bar{x}) = 0$ and $\bar{x}$ is a (strict) local minimum of f over S. Generalize this result to the multivariate case.

3.63 Let $f: E_n \to E_1$ be pseudoconvex, and suppose that, for some $\mathbf{x}_1$ and $\mathbf{x}_2$ in E_n, we have $\nabla f(\mathbf{x}_1)^t(\mathbf{x}_2 - \mathbf{x}_1) \geq 0$. Show that the function $F(\lambda) = f[\mathbf{x}_1 + \lambda(\mathbf{x}_2 - \mathbf{x}_1)]$ is nondecreasing for $\lambda \geq 0$.

3.64 Let $\mathbf{f}: E_n \to E_m$ and $\mathbf{g}: E_n \to E_k$ be differentiable and convex. Let $\phi: E_{m+k} \to E_1$ satisfy the following inequality. If $\mathbf{a}_2 \geq \mathbf{a}_1$ and $\mathbf{b}_2 \geq \mathbf{b}_1$, then $\phi(\mathbf{a}_2, \mathbf{b}_2) \geq \phi(\mathbf{a}_1, \mathbf{b}_1)$. Consider the function $h: E_n \to E_1$ defined by $h(\mathbf{x}) = \phi(\mathbf{f}(\mathbf{x}), \mathbf{g}(\mathbf{x}))$. Show the following:

a. If ϕ is convex, then h is convex.
b. If ϕ is pseudoconvex, then h is pseudoconvex.
c. If ϕ is quasiconvex, then h is quasiconvex.

3.65 Let S be a nonempty convex set in E_n, and let $f: E_n \to E_1$ and $\mathbf{g}: E_n \to E_m$ be convex. Consider the *perturbation function* $\phi: E_m \to E_1$ defined below:

$$\phi(\mathbf{y}) = \inf \{f(\mathbf{x}): \mathbf{g}(\mathbf{x}) \leq \mathbf{y},\ \mathbf{x} \in S\}$$

a. Prove that ϕ is convex.
b. Show that if $\mathbf{y}_1 \leq \mathbf{y}_2$, then $\phi(\mathbf{y}_1) \geq \phi(\mathbf{y}_2)$.

3.66 Let $f: E_n \to E_1$ be convex, and $\mathbf{A}$ be an $m \times n$ matrix. Consider the function $h: E_m \to E_1$ defined as follows:

$$h(\mathbf{y}) = \inf \{f(\mathbf{x}): \mathbf{A}\mathbf{x} = \mathbf{y}\}$$

Show that h is convex.

3.67 Let $g_1, g_2: E_n \to E_1$, and let $\alpha \in [0, 1]$. Consider the function $G_\alpha: E_n \to E_1$ defined as

$$G_\alpha(\mathbf{x}) = \tfrac{1}{2}[g_1(\mathbf{x}) + g_2(\mathbf{x}) - \sqrt{g_1^2(\mathbf{x}) + g_2^2(\mathbf{x}) - 2\alpha g_1(\mathbf{x})g_2(\mathbf{x})}]$$

where $\sqrt{\ }$ denotes the positive square root.

a. Show that $G_\alpha(\mathbf{x}) \geq 0$ if and only if $g_1(\mathbf{x}) \geq 0$ and $g_2(\mathbf{x}) \geq 0$, that is, minimum $[g_1(\mathbf{x}), g_2(\mathbf{x})] \geq 0$.
b. If g_1 and g_2 are differentiable, show that G_α is differentiable at $\mathbf{x}$ for each $\alpha \in [0, 1)$ provided that $g_1(\mathbf{x}), g_2(\mathbf{x}) \neq 0$.
c. Now suppose that g_1 and g_2 are concave. Show that G_α is concave for α in the interval $[0, 1]$. Does this result hold for $\alpha \in (-1, 0)$?
d. Suppose that g_1 and g_2 are quasiconcave. Show that G_α is quasiconcave for $\alpha = 1$.
e. Let $g_1(\mathbf{x}) = -x_1^2 - x_2^2 + 4$, and $g_2(\mathbf{x}) = 2x_1 + x_2 - 1$. Obtain an explicit expression for G_α, and verify parts a, b, and c above.

The above exercise describes a general method for combining two constraints of the form $g_1(\mathbf{x}), g_2(\mathbf{x}) \geq 0$ into an equivalent single constraint of the form $G_\alpha(\mathbf{x}) \geq 0$. This procedure could be successively applied to reduce a problem with several constraints into an equivalent single constrained problem. The procedure is due to Rvačev [1963].

3.68 Let $g_1, g_2: E_n \to E_1$, and let $\alpha \in [0, 1]$. Consider the function $G_\alpha: E_n \to E_1$ defined by

$$G_\alpha(\mathbf{x}) = \tfrac{1}{2}[g_1(\mathbf{x}) + g_2(\mathbf{x}) + \sqrt{g_1^2(\mathbf{x}) + g_2^2(\mathbf{x}) - 2\alpha g_1(\mathbf{x})g_2(\mathbf{x})}]$$

where $\sqrt{\ }$ denotes the positive square root.

a. Show that $G_\alpha(\mathbf{x}) \geq 0$ if and only if maximum $[g_1(\mathbf{x}), g_2(\mathbf{x})] \geq 0$.
b. If g_1 and g_2 are differentiable, show that G_α is differentiable at $\mathbf{x}$ for each $\alpha \in [0, 1)$, provided that $g_1(\mathbf{x}), g_2(\mathbf{x}) \neq 0$.
c. Now suppose that g_1 and g_2 are convex. Show that G_α is convex for $\alpha \in [0, 1]$. Does the result hold for $\alpha \in (-1, 0)$?
d. Suppose that g_1 and g_2 are quasiconvex. Show that G_α is quasiconvex for $\alpha = 1$.

e. In some optimization problems, the restriction that the variable $x = 0$ or 1 arises. Show that this restriction is equivalent to maximum $[g_1(x), g_2(x)] \geq 0$, where $g_1(x) = -x^2$ and $g_2(x) = -(x - 1)^2$. Find the function G_α explicitly, and verify statements a, b, and c above.

The above exercise describes a general method for combining the *either–or* constraints of the form $g_1(\mathbf{x}) \geq 0$ or $g_2(\mathbf{x}) \geq 0$ into a single constraint of the form $G_\alpha(\mathbf{x}) \geq 0$, and is due to Rvačev [1963].

3.69 Let $f: S \to E_1$, where $S \subseteq E_n$ is a nonempty convex set. Then, the *convex envelope* of f over S, denoted $f_S(\mathbf{x})$, $\mathbf{x} \in S$, is a convex function such that $f_S(\mathbf{x}) \leq f(\mathbf{x})$ for all $\mathbf{x} \in S$; and if g is any other convex function for which $g(\mathbf{x}) \leq f(\mathbf{x})$ for all $\mathbf{x} \in S$, then $f_S(\mathbf{x}) \geq g(\mathbf{x})$ for all $\mathbf{x} \in S$. Hence, f_S is the pointwise supremum over all convex underestimators of f over S. Show that $\min \{f(\mathbf{x}): \mathbf{x} \in S\} = \min \{f_S(\mathbf{x}): \mathbf{x} \in S\}$, assuming that the minima exist, and that

$$\{\mathbf{x}^* \in S : f(\mathbf{x}^*) \leq f(\mathbf{x}), \text{ for all } \mathbf{x} \in S\} \subseteq \{\mathbf{x}^* \in S : f_S(\mathbf{x}^*) \leq f_S(\mathbf{x}), \text{ for all } \mathbf{x} \in S\}.$$

3.70 Let $f: S \to E_1$ be a concave functon, where $S \subseteq E_n$ is a nonempty polytope with vertices $\mathbf{x}_1, \ldots, \mathbf{x}_E$. Show that the convex envelope (see Exercise 3.69) of f over S is given by

$$f_S(\mathbf{x}) = \min \left\{ \sum_{i=1}^{E} \lambda_i f(\mathbf{x}_i) : \sum_{i=1}^{E} \lambda_i \mathbf{x}_i = \mathbf{x}, \sum_{i=1}^{E} \lambda_i = 1, \lambda_i \geq 0 \text{ for } i = 1 = 1, \ldots, E \right\}$$

Hence, show that if S is a simplex in E_n, then f_S is an affine function that attains the same values as f over all the vertices of S. (This result is due to Falk [1976].)

3.71 Let $f: S \to E_1$ and $f_S: S \to E_1$ be as defined in Exercise 3.69. Show that if f is continuous, then the epigraph $\{(\mathbf{x}, y): y \geq f_S(\mathbf{x}), \mathbf{x} \in S, y \in E_1\}$ of f_S over S is the closure of the convex hull of the epigraph $\{(\mathbf{x}, y): y \geq f(\mathbf{x}), \mathbf{x} \in S, y \in E_1\}$ of f over S. Give an example to show that the epigraph of the latter set is not necessarily closed.

3.72 Let $f(x, y) = xy$ be a bivariate bilinear function, and let S be a polytope in R^2 having no edge with a finite, positive slope. Define $\Lambda = \{(\alpha, \beta, \gamma) \in R^3: \alpha x_k + \beta y_k + \gamma \leq x_k y_k$, for $k = 1, \ldots, K\}$, where (x_k, y_k), $k = 1, \ldots, K$, are the vertices of S. Referring to Exercise 3.69, show that if S is two-dimensional, then the set of extreme points $(\alpha_e, \beta_e, \gamma_e)$, $e = 1, \ldots, E$, of Λ is nonempty, and that $f_S(x, y) = \max \{\alpha_e x + \beta_e y + \gamma_e, e = 1, \ldots, E\}$. On the other hand, if S is one-dimensional and given by the convex hull of (x_1, y_1) and (x_2, y_2), then show that there exists a solution $(\alpha_1, \beta_1, \gamma_1)$ to the system $\alpha x_k + \beta y_k + \gamma = x_k y_k$ for $k = 1, 2$ and, in this case, $f_S(x, y) = \alpha_1 x + \beta_1 y + \gamma_1$. Specialize this result to verify that if $S = \{(x, y): a \leq x \leq b, c \leq y \leq d\}$, where $a < b$ and $c < d$, then $f_S(x, y) = \max \{dx + by - bd, cx + ay - ac\}$. (This result is due to Sherali and Alameddine [1990].)

3.73 Consider a triangle S with vertices (0, 1), (2, 0) and (1, 2) and let $f(x, y) = xy$ be a bivariate, bilinear function. Show that the convex envelope f_S of f over S (see Exercise 3.69) is given by

$$f_S(x, y) = \begin{cases} -y + \dfrac{3y^2}{2 - x + y} & \text{for } (x, y) \neq (2, 0) \\ 0 & \text{for } (x, y) = (2, 0) \end{cases} \qquad \text{for } (x, y) \in S$$

Can you generalize your approach to finding the convex envelope of f over a triangle having a single edge that has a finite, positive slope? (This result is due to Sherali and Alameddine [1990].)

Notes and References

In this chapter, we deal with the important topic of convex and concave functions. The recognition of these functions is generally traced to Jensen [1905, 1906]. For earlier related works on the subject, see Hadamard [1893] and Hölder [1889].

In Section 3.1, several results related to continuity and directional derivatives of a convex function are presented. In particular, we showed that a convex function is continuous on the interior of the domain. See, for example, Rockafellar [1970]. Rockafellar also discusses the *convex extension* to E_n of a convex function $f: S \subset E_n \to E_1$, which takes on finite values over a convex subset S of E_n, by letting $f(\mathbf{x}) = \infty$ for $\mathbf{x}$ L S. Accordingly, a set of arithmetic operations involving ∞ also needs to be defined. In this case, S is referred to as the *effective domain* of f. Also, a *proper convex function* is then defined as a convex function for which $f(\mathbf{x}) < \infty$ for at least one point $\mathbf{x}$, and for which $f(\mathbf{x}) > -\infty$ for all $\mathbf{x}$.

In Section 3.2, we discuss subgradients of convex functions. Many of the properties of differentiable convex functions are retained by replacing the gradient vector by a subgradient. For this reason, subgradients have been used frequently in the optimization of nondifferentiable functions. See, for example, Bertsekas [1975], Demyanov and Pallaschke [1985], Demyanov and Vasilev [1985], Held and Karp [1970], Held, Wolfe and Crowder [1974], Kiwiel [1985], and Wolfe [1976].

Section 3.3 gives some properties of differentiable convex functions. For further study of these topics as well as other properties of convex functions, refer to Eggleston [1958], Fenchel [1953], Roberts and Varberg [1973], and Rockafellar [1970]. The superdiagonalization algorithm derived from Theorem 3.3.12 provides an efficient polynomial-time algorithm for checking definiteness properties of matrices. This method is intimately related with *LU* and Cholesky factorization techniques (see Exercise 3.34, and refer to Appendix A.2, Fletcher [1985], Luenberger [1983], and Murty [1983] for further details).

Section 3.4 treats the subject of minima and maxima of convex functions over convex sets. Robinson [1987] discusses the distinction between strict and strong local minima. For general functions, the study of minima and maxima is quite complicated. As shown in Section 3.4, however, every local minimum of a convex function over a convex set is also a global minimum, and the maximum of a convex function over a convex set occurs at an extreme point. For an excellent study of optimization of convex functions, see Rockafellar [1970]. The characterization of the optimal solution set for convex programs is due to Mangasarian [1988]. This paper also extends the results given in Section 3.4 to subdifferentiable convex functions.

In Section 3.5, we examine other classes of functions that are related to convex functions, namely quasiconvex and pseudoconvex functions. The class of quasiconvex functions was first studied by De Finetti [1949]. Arrow and Enthoven [1961] derived necessary and sufficient conditions for quasiconvexity on the nonnegative orthant assuming twice differentiability. Their results were extended by Ferland [1972]. Note that a local minimum of a quasiconvex function over a convex set is not necessarily a global minimum. This result holds true, however, for a strictly quasiconvex function. Ponstein [1967] introduced the concept of strongly quasiconvex functions, which ensures that the global minimum is unique, a property that is not enjoyed by strictly quasiconvex functions. The notion of pseudoconvexity was first introduced by Mangasarian [1965]. The significance of the class of pseudoconvex functions stems from the fact that every point with a zero gradient is a global minimum. Matrix theoretic characterizations (see, e.g., Exercises 3.49 and 3.50) of quadratic pseudoconvex and quasiconvex functions have been presented by Cottle and Ferland [1972] and by Martos [1965a, 1967b, 1969, 1975]. For further reading on this topic, refer to Fenchel [1953], Karamardian [1967], Mangasarian [1969], Ponstein [1967], Greenberg and Pierskalla [1971], Schaible [1981], Schaible and Ziemba [1981], and Avriel et al. [1988]. The last four references give excellent surveys on this topics, and the results of Exercises 3.51–3.56 and 3.61–3.63 are discussed in detail by Schaible [1981] and Avriel et al. [1988]. Karamardian and Schaible [1990] also present various tests for checking generalized properties for differentiable functions. Also, see Appendix B.2.

Exercises 3.69–3.72 deal with convex envelopes of nonconvex functions. This construct plays an important role in global optimization techniques for nonconvex programming problems. For additional information on this subject, we refer the reader to Al-Khayyal and Falk [1983], Falk [1976], Grotzinger [1985], Horst [1986], Horst and Tuy [1990], Pardalos and Rosen [1987], and Sherali and Alameddine [1990].

PART 2 Optimality Conditions and Duality

Chapter 4 The Fritz John and the Karush–Kuhn–Tucker Optimality Conditions

In Chapter 3, we derived an optimality condition for a problem of the following form: minimize $f(\mathbf{x})$ subject to $\mathbf{x} \in S$, where f is a convex function and S is a convex set. The necessary and sufficient condition for $\bar{\mathbf{x}}$ to solve the problem was shown to be

$$\nabla f(\bar{\mathbf{x}})^t(\mathbf{x} - \bar{\mathbf{x}}) \geq 0 \qquad \text{for all } \mathbf{x} \in S$$

In this chapter, the nature of the set S will be more explicitly specified in terms of inequality and/or equality constraints. A set of first-order necessary conditions are derived without any convexity assumptions that are sharper than the above in the sense that they explicitly consider the constraint functions and are more easily verifiable, since they deal with a system of equations. Under suitable convexity assumptions, these necessary conditions are also sufficient for optimality. These optimality conditions lead to *classical* or *direct* optimization techniques for solving unconstrained and constrained problems that construct these conditions and then attempt to find a solution to them. In contrast, we discuss several *indirect methods* in Chapters 8 through 11, which iteratively improve the current solution, converging to a point that can be shown to satisfy these optimality conditions. A discussion of second-order necessary and/or sufficient conditions for unconstrained as well as for constrained problems is also provided.

Readers who are unfamiliar with generalized convexity concepts from Section 3.5 may substitute any references to such properties by related convexity assumptions for ease in reading.

The following is an outline of the chapter.

Section 4.1: Unconstrained Problems We briefly consider optimality conditions for unconstrained problems. First-order and second-order conditions are discussed.

Section 4.2: Problems with Inequality Constraints Both the Fritz John (FJ) and the Karush–Kuhn–Tucker (KKT) conditions for problems with inequality constraints are derived. The nature and value of solutions satisfying these conditions are emphasized.

Section 4.3: Problems with Inequality and Equality Constraints This section extends the results of the previous section to problems with both inequality and equality constraints.

Section 4.4: Second-Order Necessary and Sufficient Optimality Conditions for Constrained Problems Similar to the unconstrained case discussed in Section 4.1, we develop second-order necessary and sufficient optimality conditions as an extension to the first-order conditions developed in Sections 4.2 and 4.3 for inequality and equality constrained problems. Many results and algorithms in nonlinear programming assume the existence of a local optimal solution that satisfies the second-order sufficiency conditions.

4.1 Unconstrained Problems

An unconstrained problem is a problem of the form to minimize $f(\mathbf{x})$ without any constraints on the vector $\mathbf{x}$. Unconstrained problems seldom arise in practical applications. However, we consider such problems here because optimality conditions for constrained problems become a logical extension of the conditions for unconstrained problems. Furthermore, as shown in Chapter 9, one strategy for solving a constrained problem is to solve a sequence of unconstrained problems.

We recall below the definitions of local and global minima for unconstrained problems as a special case of Definition 3.4.1, where the set S is replaced by E_n.

4.1.1 Definition

Consider the problem of minimizing $f(\mathbf{x})$ over E_n, and let $\bar{\mathbf{x}} \in E_n$. If $f(\bar{\mathbf{x}}) \leq f(\mathbf{x})$ for all $\mathbf{x} \in E_n$, then $\bar{\mathbf{x}}$ is called a *global minimum*. If there exists an ε-neighborhood $N_\varepsilon(\bar{\mathbf{x}})$ around $\bar{\mathbf{x}}$ such that $f(\bar{\mathbf{x}}) \leq f(\mathbf{x})$ for each $\mathbf{x} \in N_\varepsilon(\bar{\mathbf{x}})$, then $\bar{\mathbf{x}}$ is called a *local minimum*, while if $f(\bar{\mathbf{x}}) < f(\mathbf{x})$ for all $\mathbf{x} \in N_\varepsilon(\bar{\mathbf{x}})$, $\mathbf{x} \neq \bar{\mathbf{x}}$, for some $\varepsilon > 0$, then $\bar{\mathbf{x}}$ is called a *strict local minimum*. Clearly, a global minimum is also a local minimum.

Necessary Optimality Conditions

Given a point $\mathbf{x}$ in E_n, we wish to determine, if possible, whether or not the point is a local or a global minimum of a function f. For this purpose, we need to characterize the minimum point. Fortunately, the differentiability assumption of f provides a means of obtaining this characterization. The corollary to Theorem 4.1.2 below gives a first-order necessary condition for $\bar{\mathbf{x}}$ to be a local optimum. Theorem 4.1.3 gives a second-order necessary condition using the Hessian matrix.

4.1.2 Theorem

Suppose that $f: E_n \rightarrow E_1$ is differentiable at $\bar{\mathbf{x}}$. If there is a vector $\mathbf{d}$ such that $\nabla f(\bar{\mathbf{x}})^t\mathbf{d} < 0$, then there exists a $\delta > 0$ such that $f(\bar{\mathbf{x}} + \lambda\mathbf{d}) < f(\bar{\mathbf{x}})$ for each $\lambda \in (0, \delta)$, so that $\mathbf{d}$ is a *descent direction* of f at $\bar{\mathbf{x}}$.

Proof

By the differentiability of f at $\bar{\mathbf{x}}$, we must have

$$f(\bar{\mathbf{x}} + \lambda\mathbf{d}) = f(\bar{\mathbf{x}}) + \lambda\nabla f(\bar{\mathbf{x}})^t\mathbf{d} + \lambda\|\mathbf{d}\|\alpha(\bar{\mathbf{x}}; \lambda\mathbf{d})$$

where $\alpha(\bar{\mathbf{x}}; \lambda\mathbf{d}) \to 0$ as $\lambda \to 0$. Rearranging the terms and dividing by λ, $\lambda \neq 0$, we get

$$\frac{f(\bar{\mathbf{x}} + \lambda\mathbf{d}) - f(\bar{\mathbf{x}})}{\lambda} = \nabla f(\bar{\mathbf{x}})^t\mathbf{d} + \|\mathbf{d}\|\alpha(\bar{\mathbf{x}}; \lambda\mathbf{d})$$

Since $\nabla f(\bar{\mathbf{x}})^t\mathbf{d} < 0$ and $\alpha(\bar{\mathbf{x}}; \lambda\mathbf{d}) \to 0$ as $\lambda \to 0$, there exists a $\delta > 0$ such that $\nabla f(\bar{\mathbf{x}})^t\mathbf{d} + \|\mathbf{d}\|\alpha(\bar{\mathbf{x}}; \lambda\mathbf{d}) < 0$ for all $\lambda \in (0, \delta)$. The result then follows.

Corollary

Suppose that $f: E_n \to E_1$ is differentiable at $\bar{\mathbf{x}}$. If $\bar{\mathbf{x}}$ is a local minimum, then $\nabla f(\bar{\mathbf{x}}) = \mathbf{0}$.

Proof

Suppose that $\nabla f(\bar{\mathbf{x}}) \neq \mathbf{0}$. Then, letting $\mathbf{d} = -\nabla f(\bar{\mathbf{x}})$, we get $\nabla f(\bar{\mathbf{x}})^t\mathbf{d} = -\|\nabla f(\bar{\mathbf{x}})\|^2 < 0$; and by Theorem 4.1.2, there is a $\delta > 0$ such that $f(\bar{\mathbf{x}} + \lambda\mathbf{d}) < f(\bar{\mathbf{x}})$ for $\lambda \in (0, \delta)$, contradicting the assumption that $\bar{\mathbf{x}}$ is a local minimum. Hence, $\nabla f(\bar{\mathbf{x}}) = \mathbf{0}$.

The above condition uses the gradient vector whose components are the first partials of f. Hence, it is called a *first-order condition*. Necessary conditions can also be stated in terms of the Hessian matrix $\mathbf{H}$ whose elements are the second partials of f, and are then called *second-order conditions*. One such condition is given below.

4.1.3 Theorem

Suppose that $f: E_n \to E_1$ is twice differentiable at $\bar{\mathbf{x}}$. If $\bar{\mathbf{x}}$ is a local minimum, then $\nabla f(\bar{\mathbf{x}}) = \mathbf{0}$ and $\mathbf{H}(\bar{\mathbf{x}})$ is positive semidefinite.

Proof

Consider an arbitrary direction $\mathbf{d}$. Then, from the differentiability of f at $\bar{\mathbf{x}}$, we have

$$f(\bar{\mathbf{x}} + \lambda\mathbf{d}) = f(\bar{\mathbf{x}}) + \lambda\nabla f(\bar{\mathbf{x}})^t\mathbf{d} + \tfrac{1}{2}\lambda^2\mathbf{d}^t\mathbf{H}(\bar{\mathbf{x}})\mathbf{d} + \lambda^2\|\mathbf{d}\|^2\alpha(\bar{\mathbf{x}}; \lambda\mathbf{d}) \tag{4.1}$$

where $\alpha(\bar{\mathbf{x}}; \lambda\mathbf{d}) \to 0$ as $\lambda \to 0$. Since $\bar{\mathbf{x}}$ is a local minimum, from the corollary to Theorem 4.1.2, we have $\nabla f(\bar{\mathbf{x}}) = \mathbf{0}$. Rearranging the terms in (4.1) and dividing by $\lambda^2 > 0$, we get

$$\frac{f(\bar{\mathbf{x}} + \lambda\mathbf{d}) - f(\bar{\mathbf{x}})}{\lambda^2} = \tfrac{1}{2}\mathbf{d}^t\mathbf{H}(\bar{\mathbf{x}})\mathbf{d} + \|\mathbf{d}\|^2\alpha(\bar{\mathbf{x}}; \lambda\mathbf{d}) \tag{4.2}$$

Since $\bar{\mathbf{x}}$ is a local minimum, $f(\bar{\mathbf{x}} + \lambda\mathbf{d}) \geq f(\bar{\mathbf{x}})$ for λ sufficiently small. From (4.2), it is thus clear that $\frac{1}{2}\mathbf{d}^t\mathbf{H}(\bar{\mathbf{x}})\mathbf{d} + \|\mathbf{d}\|^2\alpha(\bar{\mathbf{x}}, \lambda\mathbf{d}) \geq 0$ for λ sufficiently small. By taking the limit as $\lambda \to 0$, it follows that $\mathbf{d}^t\mathbf{H}(\bar{\mathbf{x}})\mathbf{d} \geq 0$; and, hence, $\mathbf{H}(\bar{\mathbf{x}})$ is positive semidefinite.

Sufficient Optimality Conditions

The conditions discussed thus far are necessary conditions; that is, they must be true for every local optimal solution. On the other hand, a point satisfying these conditions need not be a local minimum. Theorem 4.1.4 gives a sufficient condition for a local minimum.

4.1.4 Theorem

Suppose that $f: E_n \rightarrow E_1$ is twice differentiable at $\bar{\mathbf{x}}$. If $\nabla f(\bar{\mathbf{x}}) = \mathbf{0}$ and $\mathbf{H}(\bar{\mathbf{x}})$ is positive definite, then $\bar{\mathbf{x}}$ is a strict local minimum.

Proof

Since f is twice differentiable at $\bar{\mathbf{x}}$, we must have, for each $\mathbf{x} \in E_n$,

$$f(\mathbf{x}) = f(\bar{\mathbf{x}}) + \nabla f(\bar{\mathbf{x}})^t(\mathbf{x} - \bar{\mathbf{x}}) + \tfrac{1}{2}(\mathbf{x} - \bar{\mathbf{x}})^t\mathbf{H}(\bar{\mathbf{x}})(\mathbf{x} - \bar{\mathbf{x}}) + \|\mathbf{x} - \bar{\mathbf{x}}\|^2\alpha(\bar{\mathbf{x}}; \mathbf{x} - \bar{\mathbf{x}}) \tag{4.3}$$

where $\alpha(\bar{\mathbf{x}}; \mathbf{x} - \bar{\mathbf{x}}) \rightarrow 0$ as $\mathbf{x} \rightarrow \bar{\mathbf{x}}$. Suppose, by contradiction, that $\bar{\mathbf{x}}$ is not a strict local minimum; that is, suppose there exists a sequence $\{\mathbf{x}_k\}$ converging to $\bar{\mathbf{x}}$ such that $f(\mathbf{x}_k) \leq f(\bar{\mathbf{x}})$, $\mathbf{x}_k \neq \bar{\mathbf{x}}$, for each k. Considering this sequence, noting that $\nabla f(\bar{\mathbf{x}}) = \mathbf{0}$ and $f(\mathbf{x}_k) \leq f(\bar{\mathbf{x}})$, and denoting $(\mathbf{x}_k - \bar{\mathbf{x}})/\|\mathbf{x}_k - \bar{\mathbf{x}}\|$ by $\mathbf{d}_k$, (4.3) then implies that

$$\tfrac{1}{2}\mathbf{d}_k^t\mathbf{H}(\bar{\mathbf{x}})\mathbf{d}_k + \alpha(\bar{\mathbf{x}}; \mathbf{x}_k - \bar{\mathbf{x}}) \leq 0 \qquad \text{for each } k \tag{4.4}$$

But $\|\mathbf{d}_k\| = 1$ for each k; and, hence, there exists an index set $\mathcal{K}$ such that $\{\mathbf{d}_k\}_{\mathcal{K}}$ converges to $\mathbf{d}$, where $\|\mathbf{d}\| = 1$. Considering this subsequence and the fact that $\alpha(\bar{\mathbf{x}}; \mathbf{x}_k - \bar{\mathbf{x}}) \rightarrow 0$ as $k \in \mathcal{K}$ approaches ∞, then (4.4) implies that $\mathbf{d}^t\mathbf{H}(\bar{\mathbf{x}})\mathbf{d} \leq 0$. This contradicts the assumption that $\mathbf{H}(\bar{\mathbf{x}})$ is positive definite since $\|\mathbf{d}\| = 1$. Therefore, $\bar{\mathbf{x}}$ is indeed a strict local minimum.

Essentially, note that assuming f to be twice continuously differentiable, since $\mathbf{H}(\bar{\mathbf{x}})$ is positive definite, we have that $\mathbf{H}(\mathbf{x})$ is positive definite in an ε-neighborhood of $\bar{\mathbf{x}}$, and so f is strictly convex in an ε-neighborhood of $\bar{\mathbf{x}}$. Therefore, as follows from Theorem 3.4.2, $\bar{\mathbf{x}}$ is a strict local minimum, that is, it is the unique global minimum over $N_\varepsilon(\bar{\mathbf{x}})$ for some $\varepsilon > 0$. In fact, noting the second part of Theorem 3.4.2, we can conclude that $\bar{\mathbf{x}}$ is also a strong or isolated local minimum in this case.

In Theorem 4.1.5 below, we show that the necessary condition $\nabla f(\bar{\mathbf{x}}) = \mathbf{0}$ is also sufficient for $\bar{\mathbf{x}}$ to be a global minimum if f is pseudoconvex at $\bar{\mathbf{x}}$. In particular if $\nabla f(\bar{\mathbf{x}}) = \mathbf{0}$ and if $\mathbf{H}(\mathbf{x})$ is positive semidefinite for all $\mathbf{x}$, then f is convex and therefore also pseudoconvex. Consequently, $\bar{\mathbf{x}}$ is a global minimum. This is also evident from Theorem 3.3.3 or from Corollary 2 to Theorem 3.4.3.

4.1.5 Theorem

Let $f: E_n \rightarrow E_1$ be pseudoconvex at $\bar{\mathbf{x}}$. Then, $\bar{\mathbf{x}}$ is a global minimum if and only if $\nabla f(\bar{\mathbf{x}}) = \mathbf{0}$.

Proof

By the corollary to Theorem 4.1.2, if $\bar{\mathbf{x}}$ is a global minimum, then $\nabla f(\bar{\mathbf{x}}) = \mathbf{0}$. Now suppose that $\nabla f(\bar{\mathbf{x}}) = \mathbf{0}$, so that $\nabla f(\bar{\mathbf{x}})^t(\mathbf{x} - \bar{\mathbf{x}}) = 0$ for each $\mathbf{x} \in E_n$. By the pseudoconvexity of f at $\bar{\mathbf{x}}$, it then follows that $f(\mathbf{x}) \geq f(\bar{\mathbf{x}})$ for each $\mathbf{x} \in E_n$, and the proof is complete.

The above theorem provides a necessary *and* sufficient optimality condition in terms of the first-order derivative alone when f is pseudoconvex. In a likewise manner, we can derive necessary *and* sufficient conditions for local optimality in terms of higher-order derivatives when f is infinitely differentiable, as an extension to the foregoing results. Toward this end, consider the following result for the *univariate* case.

4.1.6 Theorem

Let $f: E_1 \to E_1$ be an infinitely differentiable univariate function. Then, $x \in E_1$ is a local minimum if and only if either $f^{(j)}(\bar{\mathbf{x}}) = 0$ for all $j = 1, 2, \ldots,$ or else, there exists an even $n \geq 2$ such that $f^{(n)}(\bar{\mathbf{x}}) > 0$ while $f^{(j)}(\bar{\mathbf{x}}) = 0$ for all $1 \leq j < n$, where $f^{(j)}$ denotes the jth order derivative of f.

Proof

We know that $\bar{\mathbf{x}}$ is a local minimum of f if and only if $f(\bar{\mathbf{x}} + h) - f(\bar{\mathbf{x}}) \geq 0$ for all sufficiently small values of $|h|$. Using the infinite Taylor series representation of $f(\bar{\mathbf{x}} + h)$, this holds true if and only if

$$hf^{(1)}(\bar{\mathbf{x}}) + \frac{h^2}{2!}f^{(2)}(\bar{\mathbf{x}}) + \frac{h^3}{3!}f^{(3)}(\bar{\mathbf{x}}) + \frac{h^4}{4!}f^{(4)}(\bar{\mathbf{x}}) + \cdots \geq 0$$

for all $|h|$ small enough. Similar to the proof of Theorem 3.3.9, it is readily verified that the foregoing inequality holds if and only if the condition of the theorem is satisfied, and this completes the proof.

Before proceeding, we remark here that for a local maximum, the condition of Theorem 4.1.6 remains the same, except that we require $f^{(n)}(\bar{\mathbf{x}}) < 0$ in lieu of $f^{(n)}(\bar{\mathbf{x}}) > 0$. Observe also, noting Theorem 3.3.9, that the above result essentially asserts that for the case under discussion, $\bar{\mathbf{x}}$ is a local minimum if and only if f is locally convex about $\bar{\mathbf{x}}$. This result can be partially extended, at least in theory, to the case of multivariate functions. Toward this end, suppose that $\bar{\mathbf{x}}$ is a local minimum for $f: E_n \to E_1$. Then, this holds true if and only if $f(\bar{\mathbf{x}} + \lambda\mathbf{d}) \geq f(\bar{\mathbf{x}})$ for all $\mathbf{d} \in E_n$ and for all sufficiently small values of $|\lambda|$. Assuming f to be infinitely differentiable, this asserts that for all $\mathbf{d} \in E_n$, $\|\mathbf{d}\| = 1$, we must equivalently have

$$f(\bar{\mathbf{x}} + \lambda\mathbf{d}) - f(\bar{\mathbf{x}}) = \lambda\nabla f(\bar{\mathbf{x}})^t\mathbf{d} + \frac{\lambda^2}{2!}\mathbf{d}^t\mathbf{H}(\bar{\mathbf{x}})\mathbf{d} + \frac{\lambda^3}{3!}\sum_i\sum_j\sum_k f_{ijk}(\bar{\mathbf{x}})d_id_jd_k + \cdots \geq 0$$

for all $-\delta \leq \lambda \leq \delta$, for some $\delta > 0$. Consequently, the first nonzero derivative term, if it exists, must correspond to an even power of λ and must be positive in value.

Note that the foregoing concluding statement is *not* sufficient to claim local optimality of $\bar{\mathbf{x}}$. The difficulty is that it might be the case that this statement holds, implying that for any $\mathbf{d} \in E_n$, $\|\mathbf{d}\| = 1$, we have $f(\bar{\mathbf{x}} + \lambda\mathbf{d}) \geq f(\bar{\mathbf{x}})$ for all $-\delta_{\mathbf{d}} \leq \lambda \leq \delta_{\mathbf{d}}$ for some $\delta_{\mathbf{d}} > 0$ which depends on $\mathbf{d}$, but $\delta_{\mathbf{d}}$ may get vanishingly small as $\mathbf{d}$ varies, so that we cannot assert the existence of a $\delta > 0$ such that $f(\bar{\mathbf{x}} + \lambda\mathbf{d}) \geq f(\bar{\mathbf{x}})$ for all $-\delta \leq \lambda \leq \delta$. In this case, by moving along curves instead of along straight lines, improving values of f might be accessible in the immediate neighborhood of $\bar{\mathbf{x}}$. On the other hand, a valid sufficient condition by Theorem 4.1.5 is that $\nabla f(\bar{\mathbf{x}}) = \mathbf{0}$ and f be convex (or pseudoconvex) over an ε-neighborhood about $\bar{\mathbf{x}}$, for some $\varepsilon > 0$. However, this might not be easy to check, and we might need to numerically assess the situation by examining values of the function at perturbations about the point $\bar{\mathbf{x}}$ (also, refer to Exercise 4.14).

To illustrate the above point, consider the following example due to the mathematician Peano. Let $f(x_1, x_2) = (x_2^2 - x_1)(x_2^2 - 2x_1) = 2x_1^2 - 3x_1x_2^2 + x_2^4$. Then, we have, at $\bar{\mathbf{x}} = (0, 0)^t$.

$$\nabla f(\mathbf{0}) = \begin{bmatrix} 0 \\ 0 \end{bmatrix} \qquad \mathbf{H}(\mathbf{0}) = \begin{bmatrix} 4 & 0 \\ 0 & 0 \end{bmatrix} \qquad f_{122}(\mathbf{0}) = f_{212}(\mathbf{0}) = f_{221}(\mathbf{0}) = -6 \qquad f_{2222}(\mathbf{0}) = 24$$

and all other partial derivatives of f of order 3 or higher are zeros. Hence, we obtain by the Taylor series expansion

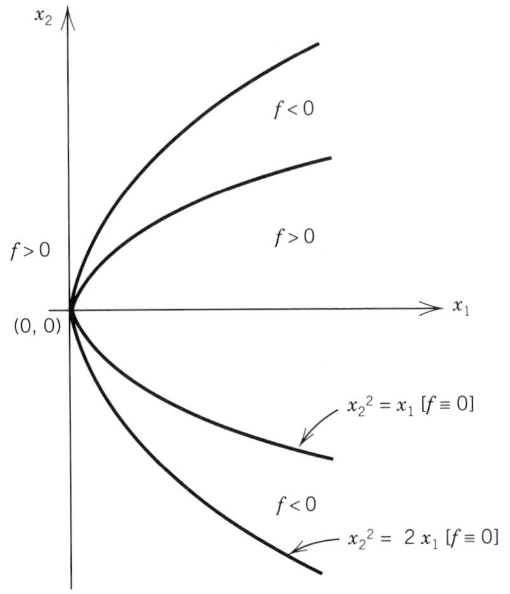

Figure 4.1 Regions of zero, positive, and negative values of $F(x_1, x_2) = (x_2^2 - x_1)(x_2^2 - 2x_1)$.

$$f(\bar{\mathbf{x}} + \lambda \mathbf{d}) - f(\bar{\mathbf{x}}) = \frac{\lambda^2}{2}(4d_1^2) + \frac{\lambda^3}{6}(-18d_1d_2^2) + \frac{\lambda^4}{24}(24d_2^4)$$

$$= 2\lambda^2\left[d_1 - \frac{3\lambda}{4}d_2^2\right]^2 - \frac{1}{8}\lambda^4 d_2^4$$

Note that for any $\mathbf{d} = (d_1, d_2)^t$, $\|\mathbf{d}\| = 1$, if $d_1 \neq 0$, the given necessary condition holds true because the second-order term is positive. On the other hand, if $d_1 = 0$, then we must have $d_2 \neq 0$, and the condition again holds true because the first nonzero term is of order 4 and is positive. However, $\bar{\mathbf{x}} = (0, 0)^t$ is not a local minimum as evident from Figure 4.1. We have $f(0, 0) = 0$, while there exist negative values of f in any ε-neighborhood about the point (0, 0). In fact, taking $\mathbf{d} = (\sin\theta, \cos\theta)^t$, we have $f(\bar{\mathbf{x}} + \lambda\mathbf{d}) - f(\bar{\mathbf{x}}) = 2\sin^2\theta\lambda^2 - 3\sin\theta\cos^2\theta\lambda^3 + \cos^4\theta\lambda^4$; and for this to be nonnegative for all $-\delta_\theta \leq \lambda \leq \delta_\theta$, $\delta_\theta > 0$, we observe that as $\theta \to 0^+$, we must have $\delta_\theta \to 0^+$ as well (see Exercise 4.38). Hence, we cannot derive a $\delta > 0$ such that $f(\bar{\mathbf{x}} + \lambda\mathbf{d}) - f(\bar{\mathbf{x}}) \geq 0$ for all $\mathbf{d} \in E_n$ and $-\delta \leq \lambda \leq \delta$, and so $\bar{\mathbf{x}}$ is not a local minimum.

To afford further insight into the multivariate case, let us examine a situation in which $f: E_n \to E_1$ is twice continuously differentiable, and at a given point $\bar{\mathbf{x}} \in E_n$, we have $\nabla f(\bar{\mathbf{x}}) = \mathbf{0}$ but $\mathbf{H}(\bar{\mathbf{x}})$ is indefinite. Hence, there exist directions $\mathbf{d}_1$ and $\mathbf{d}_2$ in E_n such that $\mathbf{d}_1^t\mathbf{H}(\bar{\mathbf{x}})\mathbf{d}_1 > 0$ and $\mathbf{d}_2^t\mathbf{H}(\bar{\mathbf{x}})\mathbf{d}_2 < 0$. Defining $F_{(\bar{\mathbf{x}};\mathbf{d}_j)}(\lambda) = f(\bar{\mathbf{x}} + \lambda\mathbf{d}_j) \equiv F_{\mathbf{d}_j}(\lambda)$, say, for $j = 1, 2$, and denoting derivatives by primes, we get

$$F'_{\mathbf{d}_j}(\lambda) = \nabla f(\bar{\mathbf{x}} + \lambda\mathbf{d}_j)^t\mathbf{d}_j \quad \text{and} \quad F''_{\mathbf{d}_j}(\lambda) = \mathbf{d}_j^t\mathbf{H}(\bar{\mathbf{x}} + \lambda\mathbf{d}_j)\mathbf{d}_j \quad \text{for } j = 1,2$$

Hence, for $j = 1$, we have $F'_{\mathbf{d}_1}(0) = 0$, $F''_{\mathbf{d}_1}(0) > 0$; and, moreover, by continuity of the second derivative, $F''_{\mathbf{d}_1}(\lambda) > 0$, for $|\lambda|$ sufficiently small. Hence, $F_{\mathbf{d}_1}(\lambda)$ is strictly convex in some ε-neighborhood of $\lambda = 0$, achieving a strict local minimum at $\lambda = 0$. Similarly, for $j = 2$, noting that $F'_{\mathbf{d}_2}(0) = 0$ and $F''_{\mathbf{d}_2}(0) < 0$, we conclude that $F_{\mathbf{d}_2}(\lambda)$ is strictly

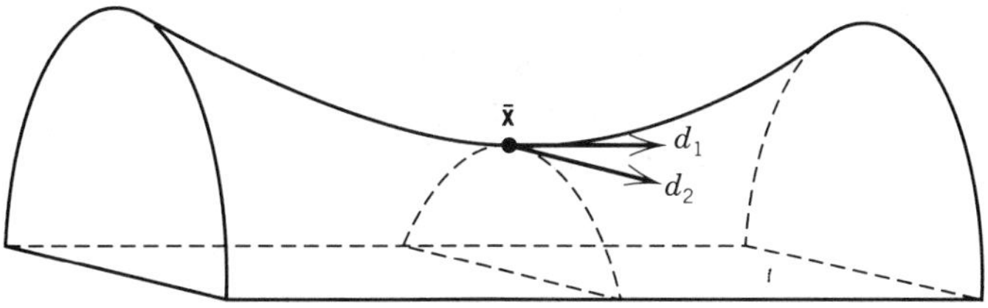

Figure 4.2 Saddle point at $\bar{\mathbf{x}}$.

concave in some ε-neighborhood of $\lambda = 0$, achieving a strict local maximum at $\lambda = 0$. Hence, as foretold by Theorem 4.1.3, $\bar{\mathbf{x}} = 0$ is neither a local minimum nor a local maximum. Such a point $\bar{\mathbf{x}}$ is called a *saddle point* (or an *inflection point*). Figure 4.2 illustrates the situation. Observe the convex and concave cross sections of the function in the respective directions $\mathbf{d}_1$ and $\mathbf{d}_2$ about the point $\bar{\mathbf{x}}$ at which $\nabla f(\bar{\mathbf{x}}) = \mathbf{0}$, which gives the function the appearance of a saddle in the vicinity of $\bar{\mathbf{x}}$.

4.1.7 Examples

Example 1: Univariate Function To illustrate the necessary and sufficient conditions of this section, consider the problem to minimize $f(x) = (x^2 - 1)^3$.

First, let us determine the candidate points for optimality satisfying the first-order necessary condition that $\nabla f(x) = 0$. Note that $\nabla f(x) \equiv f'(x) = 6x(x^2 - 1)^2$, and $\nabla f(-1) = \nabla f(0) = \nabla f(1) = 0$. Now let us examine the second-order derivatives. We have $\mathbf{H}(x) \equiv f''(x) = 24x^2(x^2 - 1) + 6(x^2 - 1)^2$ and, hence, $\mathbf{H}(1) = \mathbf{H}(-1) = 0$ and $\mathbf{H}(0) = 6$. Since $\mathbf{H}$ is positive definite at $\bar{x} = 0$, we have by Theorem 4.1.4 that $\bar{x} = 0$ is a strict local minimum. However, at $x = +1$ or -1, $\mathbf{H}$ is both positive and negative semidefinite; and although it satisfies the second-order necessary condition of Theorem 4.1.3, this is not sufficient for us to conclude anything about the behavior of f at these points. Hence, we continue and examine the third-order derivative $f'''(x) = 48x(x^2 - 1) + 48x^3 + 24x(x^2 - 1)$. Evaluating this at the two candidate points in question, we obtain $f'''(1) = 48 > 0$, $f'''(-1) = -48 < 0$. By Theorem 4.1.6, it follows that we have neither a local minimum nor a local maximum at these points, and these points are merely inflection points.

Example 2: Multivariate Function Consider the bivariate function $f(x_1, x_2) = x_1^3 + x_2^3$. Evaluating the gradient and the Hessian of f, we obtain

$$\nabla f(\mathbf{x}) = \begin{pmatrix} 3x_1^2 \\ 3x_2^2 \end{pmatrix} \quad \text{and} \quad \mathbf{H}(\mathbf{x}) = \begin{bmatrix} 6x_1 & 0 \\ 0 & 6x_2 \end{bmatrix}$$

The first-order necessary condition $\nabla f(\mathbf{x}) = \mathbf{0}$ yields $\bar{\mathbf{x}} \equiv (0, 0)^t$ as the single candidate point. However, $\mathbf{H}(\bar{\mathbf{x}})$ is the zero matrix; and although it satisfies the second-order necessary condition of Theorem 4.1.3, we need to examine higher-order derivatives to make any conclusive statement about the point $\bar{\mathbf{x}}$. Defining $F_{(\bar{\mathbf{x}};\mathbf{d})}(\lambda) = f(\bar{\mathbf{x}} + \lambda\mathbf{d}) \equiv F_{\mathbf{d}}(\lambda)$, say, we have $F'_{\mathbf{d}}(\lambda) = \nabla f(\bar{\mathbf{x}} + \lambda\mathbf{d})^t\mathbf{d}$, $F''_{\mathbf{d}}(\lambda) = \mathbf{d}^t\mathbf{H}(\bar{\mathbf{x}} + \lambda\mathbf{d})\mathbf{d}$, and $F'''_{\mathbf{d}}(\lambda) = \sum_{i=1}^{2}\sum_{j=1}^{2}\sum_{k=1}^{2} d_i d_j d_k f_{ijk}(\bar{\mathbf{x}} + \lambda\mathbf{d})$. Noting that $f_{111}(\mathbf{x}) = 6$, $f_{222}(\mathbf{x}) = 6$, and $f_{ijk}(\mathbf{x}) = 0$ otherwise, we obtain $F'''_{\mathbf{d}}(0) = 6d_1^3 + 6d_2^3$. Since there exist directions $\mathbf{d}$ for which the first nonzero derivative is $F'''_{\mathbf{d}}(0)$, which is of odd order, $\bar{\mathbf{x}} = (0, 0)^t$ is a saddle point and is therefore neither a local minimum nor a local maximum. In fact, note that $F''_{\mathbf{d}}(\lambda) = 6\lambda(d_1^3 + d_2^3)$ can be made to take on opposite signs about $\lambda = 0$ along any direction $\mathbf{d}$ for which $(d_1^3 + d_2^3) \neq 0$; and so the function switches from a convex to a concave function, or vice versa, about the point $\mathbf{0}$ along any search direction. Observe also that H is positive semidefinite over $\{\mathbf{x}: x_1 \geq 0, x_2 \geq 0\}$; and, hence, over this region, the function is convex, yielding $\bar{\mathbf{x}} = (0, 0)^t$ as a global minimum. Similarly, $\bar{\mathbf{x}} = (0, 0)^t$ is a global maximum over the region $\{\mathbf{x}: x_1 \leq 0, x_2 \leq 0\}$.

4.2 Problems with Inequality Constraints

In this section, we first develop a necessary optimality condition for the problem to minimize $f(\mathbf{x})$ subject to $\mathbf{x} \in S$ for a general set S. Later, we let S be more specifically defined as the feasible region of a nonlinear programming problem of the form to minimize $f(\mathbf{x})$ subject to $\mathbf{g}(\mathbf{x}) \leq \mathbf{0}$ and $\mathbf{x} \in X$.

Geometric Optimality Conditions

In Theorem 4.2.2 below, we develop a necessary optimality condition for the problem to minimize $f(\mathbf{x})$ subject to $\mathbf{x} \in S$, using the cone of feasible directions defined below.

4.2.1 Definition

Let S be a nonempty set in E_n, and let $\bar{\mathbf{x}} \in \text{cl } S$. The *cone of feasible directions* of S at $\bar{\mathbf{x}}$, denoted by D, is given by

$$D = \{\mathbf{d}: \mathbf{d} \neq \mathbf{0}, \text{ and } \bar{\mathbf{x}} + \lambda\mathbf{d} \in S \quad \text{for all } \lambda \in (0, \delta) \quad \text{for some } \delta > 0\}$$

Each nonzero vector $\mathbf{d} \in D$ is called a *feasible direction*. Moreover, given a function $f: E_n \to E_1$, the *cone of improving directions* at $\bar{\mathbf{x}}$, denoted by F, is given by

$$F = \{\mathbf{d}: f(\bar{\mathbf{x}} + \lambda\mathbf{d}) < f(\bar{\mathbf{x}}), \quad \text{for all } \lambda \in (0, \delta) \quad \text{for some } \delta > 0\}$$

Each direction $\mathbf{d} \in F$ is called an *improving direction*, or a *descent direction*, of f at $\bar{\mathbf{x}}$.

From the above definitions, it is clear that a small movement from $\bar{\mathbf{x}}$ along a vector $\mathbf{d} \in D$ leads to feasible points, whereas a similar movement along a $\mathbf{d} \in F$ vector leads to solutions of improving objective value. Furthermore, from Theorem 4.1.2, if $\nabla f(\bar{\mathbf{x}})^t\mathbf{d} < 0$, then $\mathbf{d}$ is an improving direction; that is, starting from $\bar{\mathbf{x}}$, a small movement along $\mathbf{d}$ will reduce the value of f. As shown in Theorem 4.2.2 below, if $\bar{\mathbf{x}}$ is a local minimum and if $\nabla f(\bar{\mathbf{x}})^t\mathbf{d} < 0$, then $\mathbf{d} \notin D$; that is, a necessary condition for local optimality is that every improving direction is not a feasible direction. This fact is illustrated in Figure 4.3, where the vertices of the cones $F_0 \equiv \{\mathbf{d} : \nabla f(\bar{\mathbf{x}})^t\mathbf{d} < 0\}$ and D are translated from the origin to $\bar{\mathbf{x}}$ for convenience.

4.2.2 Theorem

Consider the problem to minimize $f(\mathbf{x})$ subject to $\mathbf{x} \in S$, where $f: E_n \to E_1$ and S is a nonempty set in E_n. Suppose that f is differentiable at a point $\bar{\mathbf{x}} \in S$. If $\bar{\mathbf{x}}$ is a local optimal solution, then $F_0 \cap D = \emptyset$, where $F_0 = \{\mathbf{d}: \nabla f(\bar{\mathbf{x}})^t\mathbf{d} < 0\}$ and D is the cone of feasible directions of S at $\bar{\mathbf{x}}$. Conversely, suppose that $F_0 \cap D = \emptyset$, f is pseudoconvex

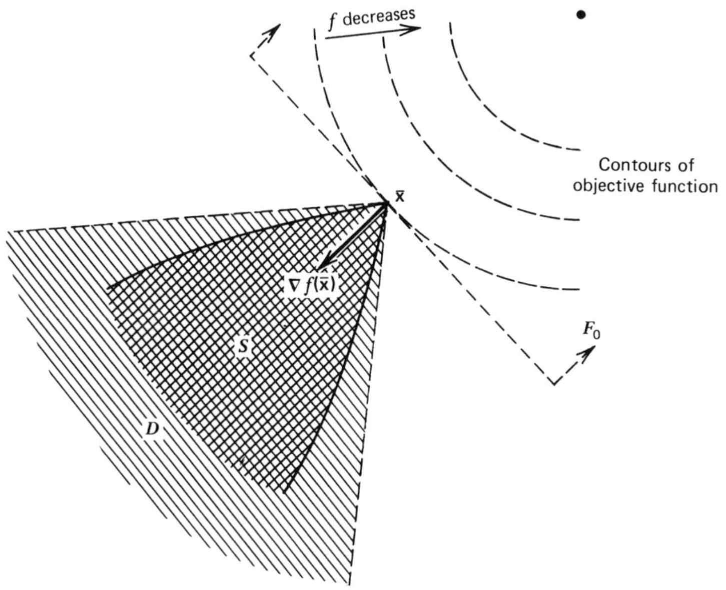

Figure 4.3 Illustration of the necessary condition $F_0 \cap D = \emptyset$.

at $\bar{\mathbf{x}}$ and that there exists an ε-neighborhood $N_\varepsilon(\bar{\mathbf{x}})$, $\varepsilon > 0$, such that $\mathbf{d} = (\mathbf{x} - \bar{\mathbf{x}}) \in D$ for any $\mathbf{x} \in S \cap N_\varepsilon(\bar{\mathbf{x}})$. Then, $\bar{\mathbf{x}}$ is a local minimum of f.

Proof

By contradiction, suppose that there exists a vector $\mathbf{d} \in F_0 \cap D$. Then, by Theorem 4.1.2, there exists a $\delta_1 > 0$ such that

$$f(\bar{\mathbf{x}} + \lambda\mathbf{d}) < f(\bar{\mathbf{x}}) \qquad \text{for each } \lambda \in (0, \delta_1) \tag{4.5a}$$

Furthermore, by Definition 4.2.1, there exists a $\delta_2 > 0$ such that

$$\bar{\mathbf{x}} + \lambda\mathbf{d} \in S \qquad \text{for each } \lambda \in (0, \delta_2) \tag{4.5b}$$

The assumption that $\bar{\mathbf{x}}$ is a local optimal solution to the problem is not compatible with (4.5). Thus, $F_0 \cap D = \emptyset$.

Conversely, suppose that $F_0 \cap D = \emptyset$ and that the given conditions in the converse statement of the theorem hold. Then, we must have $f(\mathbf{x}) \geq f(\bar{\mathbf{x}})$ for all $\mathbf{x} \in S \cap N_\varepsilon(\bar{\mathbf{x}})$. To understand this, suppose that $f(\hat{\mathbf{x}}) < f(\bar{\mathbf{x}})$ for some $\hat{\mathbf{x}} \in S \cap N_\varepsilon(\bar{\mathbf{x}})$. By the assumption on $S \cap N_\varepsilon(\bar{\mathbf{x}})$, we have $\mathbf{d} = (\hat{\mathbf{x}} - \bar{\mathbf{x}}) \in D$. Moreover, by the pseudoconvexity of f at $\bar{\mathbf{x}}$, we have $\nabla f(\bar{\mathbf{x}})^t\mathbf{d} < 0$; or else, if $\nabla f(\bar{\mathbf{x}})^t\mathbf{d} \geq 0$, then we would obtain $f(\hat{\mathbf{x}}) \equiv f(\bar{\mathbf{x}} + \mathbf{d}) \geq f(\bar{\mathbf{x}})$. We have therefore shown that, if $\bar{\mathbf{x}}$ is not a local minimum over $S \cap N_\varepsilon(\bar{\mathbf{x}})$, then there exists a direction $\mathbf{d} \in F_0 \cap D$, which is a contradiction. This completes the proof.

Observe that the set F_0 defined in Theorem 4.2.2 provides an algebraic characterization for the set of improving directions F. In fact, we have $F_0 \subseteq F$ in general by Theorem 4.1.2. Also, if $\mathbf{d} \in F$, we must have $\nabla f(\bar{\mathbf{x}})^t\mathbf{d} \leq 0$, or else, analogous to Theorem 4.1.2, $\nabla f(\bar{\mathbf{x}})^t\mathbf{d} > 0$ would imply that $\mathbf{d}$ is an ascent direction. Hence, we have

$$F_0 \subseteq F \subseteq F_0' = \{\mathbf{d} \neq \mathbf{0} : \nabla f(\bar{\mathbf{x}})^t\mathbf{d} \leq 0\} \tag{4.6}$$

Note that when $\nabla f(\bar{\mathbf{x}})^t\mathbf{d} = 0$, we are unsure about the behavior of f as we proceed from

$\bar{\mathbf{x}}$ along the direction $\mathbf{d}$, unless if we know more about the function. For example, it might very well be that $\nabla f(\bar{\mathbf{x}}) = 0$, and there might exist directions of motion that give descent or ascent, or even hold the value of f constant as we move away from $\bar{\mathbf{x}}$. Hence, it is entirely possibly to have $F_0 \subset F \subset F_0'$ (see Figure 4.1, for example). However, if f is pseudoconvex, then we know that whenever $\nabla f(\bar{\mathbf{x}})'\mathbf{d} \geq 0$, we have $f(\bar{\mathbf{x}} + \lambda\mathbf{d}) \geq f(\bar{\mathbf{x}})$ for all $\lambda \geq 0$. Hence, if f is pseudoconvex, then $\mathbf{d} \in F$ also implies $\mathbf{d} \in F_0$ as well, and so, from (4.6), we have $F_0 = F$. Similarly, if f is strictly pseudoconcave, then we know that whenever $\mathbf{d} \in F_0'$, we have $f(\bar{\mathbf{x}} + \lambda\mathbf{d}) < f(\bar{\mathbf{x}})$ for all $\lambda > 0$, and so we have $\mathbf{d} \in F$ as well. Consequently, we obtain $F = F_0'$ in this case. This establishes the following result, stated in terms of the weaker assumption of pseudoconvexity or strict pseudoconcavity at $\bar{\mathbf{x}}$ itself, rather than everywhere.

4.2.3 Lemma

Consider a differentiable function $f: E_n \to E_1$ and let F, F_0, F_0' be as defined in Definition 4.2.1, Theorem 4.2.2, and Equation (4.6), respectively. Then, we have $F_0 \subseteq F \subseteq F_0'$. Moreover, if f is pseudoconvex at $\bar{\mathbf{x}}$, then $F = F_0$, and if f is strictly pseudoconcave at $\bar{\mathbf{x}}$, then $F = F_0'$.

We now specify the feasible region S as follows:

$$S = \{\mathbf{x} \in X: g_i(\mathbf{x}) \leq 0 \text{ for } i = 1, \ldots, m\}$$

where $g_i: E_n \to E_1$ for $i = 1, \ldots, m$, and X is a nonempty open set in E_n. This gives us the following nonlinear programming problem with inequality constraints:

Problem P:

$$\begin{array}{lll} \text{Minimize} & f(\mathbf{x}) & \\ \text{subject to} & g_i(\mathbf{x}) \leq 0 & \text{for } i = 1, \ldots, m \\ & \mathbf{x} \in X & \end{array}$$

Recall that a necessary condition for local optimality at $\bar{\mathbf{x}}$ is that $F_0 \cap D = \varnothing$, where F_0 is an open half-space defined in terms of the gradient vector $\nabla f(\bar{\mathbf{x}})$, and D is the cone of feasible directions, which is not necessarily defined in terms of the gradients of the functions involved. This precludes us from converting the geometric optimality condition $F_0 \cap D = \varnothing$ into a more usable algebraic statement involving equations. As Lemma 4.2.4 below indicates, we can define an open cone G_0 in terms of the gradients of the binding constraints at $\bar{\mathbf{x}}$, such that $G_0 \subseteq D$. Since $F_0 \cap D = \varnothing$ must hold at $\bar{\mathbf{x}}$, and since $G_0 \subseteq D$, then $F_0 \cap G_0 = \varnothing$ is also a necessary optimality condition. Since F_0 and G_0 are both defined in terms of the gradient vectors, we will use the condition $F_0 \cap G_0 = \varnothing$ later in the section to develop the optimality conditions credited to Fritz John. With mild additional assumptions, the conditions reduce to the well-known Karush–Kuhn–Tucker (KKT) optimality conditions.

4.2.4 Lemma

Consider the feasible region $S = \{\mathbf{x} \in X: g_i(\mathbf{x}) \leq 0 \text{ for } i = 1, \ldots, m\}$, where X is a nonempty open set in E_n, and where $g_i: E_n \to E_1$ for $i = 1, \ldots, m$. Given a feasible point $\bar{\mathbf{x}} \in S$, let $I = \{i: g_i(\bar{\mathbf{x}}) = 0\}$ be the index set for the *binding* or *active* constraints, and assume that g_i for $i \in I$ are differentiable at $\bar{\mathbf{x}}$ and that the g_i's for $i \notin I$ are continuous at $\bar{\mathbf{x}}$. Define the sets

$$G_0 = \{\mathbf{d} : \nabla g_i(\bar{\mathbf{x}})^t\mathbf{d} < 0 \quad \text{for each } i \in I\}$$
$$G_0' = \{\mathbf{d} \neq \mathbf{0} : \nabla g_i(\bar{\mathbf{x}})^t\mathbf{d} \leq 0 \quad \text{for each } i \in I\}$$

Then, we have

$$G_0 \subseteq D \subseteq G_0' \tag{4.7}$$

Moreover, if g_i, $i \in I$, are strictly pseudoconvex at $\bar{\mathbf{x}}$, then $D = G_0$; and if g_i, $i \in I$, are pseudoconcave at $\bar{\mathbf{x}}$, then $D = G_0'$.

Proof

Let $\mathbf{d} \in G_0$. Since $\bar{\mathbf{x}} \in X$, and X is open, there exists a $\delta_1 > 0$ such that

$$\bar{\mathbf{x}} + \lambda\mathbf{d} \in X \quad \text{for } \lambda \in (0, \delta_1) \tag{4.8a}$$

Also, since $g_i(\bar{\mathbf{x}}) < 0$ and g_i is continuous at $\bar{\mathbf{x}}$ for $i \notin I$, there exists a $\delta_2 > 0$ such that

$$g_i(\bar{\mathbf{x}} + \lambda\mathbf{d}) < 0 \quad \text{for } \lambda \in (0, \delta_2) \text{ and for } i \notin I \tag{4.8b}$$

Furthermore, since $\mathbf{d} \in G_0$, then $\nabla \mathbf{g}_i(\bar{\mathbf{x}})^t\mathbf{d} < 0$ for each $i \in I$; and by Theorem 4.1.2, there exists a $\delta_3 > 0$ such that

$$g_i(\bar{\mathbf{x}} + \lambda\mathbf{d}) < g_i(\bar{\mathbf{x}}) = 0 \quad \text{for } \lambda \in (0, \delta_3) \text{ and for } i \in I \tag{4.8c}$$

From (4.8a, b, c), it is clear that points of the form $\bar{\mathbf{x}} + \lambda\mathbf{d}$ are feasible to S for each $\lambda \in (0, \delta)$, where $\delta = \text{minimum}\ (\delta_1, \delta_2, \delta_3)$. Thus, $\mathbf{d} \in D$, where D is the cone of feasible directions of the feasible region at $\bar{\mathbf{x}}$. We have shown thus far that $\mathbf{d} \in G_0$ implies that $\mathbf{d} \in D$, and, hence, $G_0 \subseteq D$.

Similarly, if $\mathbf{d} \in D$, then we must have $\mathbf{d} \in G_0'$, since otherwise, if $\nabla g_i(\bar{\mathbf{x}})^t\mathbf{d} > 0$ for any $i \in I$, we would obtain via Theorem 4.1.2 that $g_i(\bar{\mathbf{x}} + \lambda\mathbf{d}) > g_i(\bar{\mathbf{x}}) = 0$ for all $|\lambda|$ sufficiently small, contradicting the hypotheses that $\mathbf{d} \in D$. Hence, $D \subseteq G_0'$. This establishes (4.7).

Now suppose that g_i, $i \in I$, are strictly pseudoconvex at $\bar{\mathbf{x}}$, and let $\mathbf{d} \in D$. Then, we must have $\mathbf{d} \in G_0$ as well, because otherwise, if $\nabla g_i(\bar{\mathbf{x}})^t\mathbf{d} \geq 0$ for any $i \in I$, we would obtain that $g_i(\bar{\mathbf{x}} + \lambda\mathbf{d}) > g_i(\bar{\mathbf{x}}) = 0$ for all $\lambda > 0$, contradicting that $\mathbf{d} \in D$. Hence, from (4.7), we get $D = G_0$ in this case.

Finally, suppose that g_i, $i \in I$, are pseudoconcave at $\bar{\mathbf{x}}$, and consider any $\mathbf{d} \in G_0'$. We therefore have $g_i(\bar{\mathbf{x}} + \lambda\mathbf{d}) \leq g_i(\bar{\mathbf{x}}) = 0$ for all $\lambda \geq 0$ for each $i \in I$. Moreover, by the continuity of g_i, $i \notin I$, and since X is an open set, we obtain as above that $(\bar{\mathbf{x}} + \lambda\mathbf{d}) \in S$ for all $|\lambda|$ sufficiently small, and so $\mathbf{d} \in D$. This establishes that $G_0' \subseteq D$, and so, from (4.7), we obtain $D = G_0'$ in this case. This completes the proof.

As an illustration, note that in Figure 4.3, $G_0 = D \subset G_0'$, whereas in Figure 2.18 with $\bar{\mathbf{x}} = (0, 0)^t$, since the constraint functions are affine, we have $G_0 \subset D = G_0'$.

Lemma 4.2.4 directly leads to the following result.

4.2.5 Theorem

Consider Problem P to minimize $f(\mathbf{x})$ subject to $\mathbf{x} \in X$ and $g_i(\mathbf{x}) \leq 0$ for $i = 1, \ldots, m$, where X is a nonempty open set in E_n, $f{:}E_n \to E_1$, and $g_i{:}E_n \to E_1$, for $i = 1, \ldots, m$. Let $\bar{\mathbf{x}}$ be a feasible point, and denote $I = \{i{:}g_i(\bar{\mathbf{x}}) = 0\}$. Furthermore, suppose that f and g_i for $i \in I$ are differentiable at $\bar{\mathbf{x}}$ and that g_i for $i \notin I$ are continuous at $\bar{\mathbf{x}}$. If $\bar{\mathbf{x}}$ is a local optimal solution, then $F_0 \cap G_0 = \emptyset$, where $F_0 = \{\mathbf{d}: \nabla f(\bar{\mathbf{x}})^t\mathbf{d} < 0\}$ and $G_0 = \{\mathbf{d}: \nabla g_i(\bar{\mathbf{x}})^t\mathbf{d} < 0 \text{ for each } i \in I\}$. Conversely, if $F_0 \cap G_0 = \emptyset$, and if f is pseudoconvex at $\bar{\mathbf{x}}$ and g_i, $i \in I$ are strictly pseudoconvex over some ε-neighborhood of $\bar{\mathbf{x}}$, then $\bar{\mathbf{x}}$ is a local minimum.

Proof

Let $\bar{\mathbf{x}}$ be a local minimum. Then, we have the following string of implications via Theorem 4.2.2 and (4.7) of Lemma 4.2.4, which proves the first part of the theorem:

$$\bar{\mathbf{x}} \text{ is a local minimum} \Rightarrow F_0 \cap D = \emptyset \Rightarrow F_0 \cap G_0 = \emptyset \tag{4.9a}$$

Conversely, suppose that $F_0 \cap G_0 = \emptyset$ and that f and g_i, $i \in I$ are as specified in the theorem. Then, redefining the feasible region S only in terms of the binding constraints by dropping the nonbinding constraints, we have $G_0 = D$ by Lemma 4.2.4, and so we conclude that $F_0 \cap D = \emptyset$. Furthermore, since the level sets $g_i(\mathbf{x}) \leq 0$, for $i \in I$, are convex over some ε-neighborhood $N_\varepsilon(\bar{\mathbf{x}})$ of $\bar{\mathbf{x}}$, $\varepsilon > 0$, it follows that $S \cap N_\varepsilon(\bar{\mathbf{x}})$ is a convex set. Since we also have $F_0 \cap D = \emptyset$ from above, and since f is pseudoconvex at $\bar{\mathbf{x}}$, we conclude from the converse statement of Theorem 4.2.2 that $\bar{\mathbf{x}}$ is a local minimum. This statement continues to hold true by including the nonbinding constraints within S, and this completes the proof.

Observe that under the converse hypothesis of the above theorem, and assuming that g_i, i L I are continuous at $\bar{\mathbf{x}}$, we have, noting (4.9a),

$$\bar{\mathbf{x}} \text{ is a local minimum} \Leftrightarrow F_0 \cap D = \emptyset \Leftrightarrow F_0 \cap G_0 = \emptyset \tag{4.9b}$$

There is a useful insight worth deriving at this point. Note from Definition 4.2.1 that if $\bar{\mathbf{x}}$ is a local minimum then, clearly, we must have $F \cap D = \emptyset$. However, the converse is not necessarily true. That is, if $F \cap D = \emptyset$, then this does not necessarily imply that $\bar{\mathbf{x}}$ is a local minimum. For example, if $S = \{\mathbf{x} = (x_1, x_2): x_2 = x_1^2\}$ and if $f(\mathbf{x}) = x_2$, then the point $\bar{\mathbf{x}} = (1, 1)^t$ is clearly not a local minimum, since f can be decreased by reducing x_1. However, for the given point $\bar{\mathbf{x}}$, the set $D = \emptyset$, since no *straight-line directions* can lead to feasible solutions, whereas improving feasible solutions are accessible via curvilinear directions. Hence, $F \cap D = \emptyset$, but $\bar{\mathbf{x}}$ is not a local minimum. But now, if f is pseudoconvex at $\bar{\mathbf{x}}$, and if there exists an $\varepsilon > 0$ such that for any $\mathbf{x} \in S \cap N_\varepsilon(\bar{\mathbf{x}})$, we have $\mathbf{d} = (\mathbf{x} - \bar{\mathbf{x}}) \in D$ [as, for example, if $S \cap N_\varepsilon(\bar{\mathbf{x}})$ is a convex set], then $F_0 = F$ by Lemma 4.2.3; and noting (4.9a) and the converse to Theorem 4.2.2, we obtain, in this case,

$$F \cap D = \emptyset \Leftrightarrow F_0 \cap D = \emptyset \Leftrightarrow \bar{\mathbf{x}} \text{ is a local minimum}$$

4.2.6 Example

$$\begin{aligned} \text{Minimize} \quad & (x_1 - 3)^2 + (x_2 - 2)^2 \\ \text{subject to} \quad & x_1^2 + x_2^2 \leq 5 \\ & x_1 + x_2 \leq 3 \\ & x_1 \geq 0 \\ & x_2 \geq 0 \end{aligned}$$

In this case, we let $g_1(\mathbf{x}) = x_1^2 + x_2^2 - 5$, $g_2(\mathbf{x}) = x_1 + x_2 - 3$, $g_3(\mathbf{x}) = -x_1$, $g_4(\mathbf{x}) = -x_2$, and $X = E_2$. Consider the point $\bar{\mathbf{x}} = (\frac{9}{5}, \frac{6}{5})^t$, and note that the only binding constraint is $g_2(\mathbf{x}) = x_1 + x_2 - 3$. Also note that

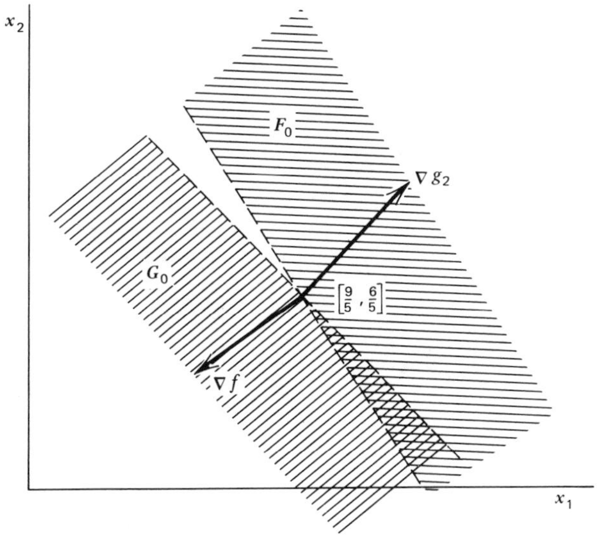

Figure 4.4 Illustration of $F_0 \cap G_0 \neq \varnothing$ at a non-optimal point.

$$\nabla f(\bar{\mathbf{x}}) = \left(\frac{-12}{5}, \frac{-8}{5}\right)^t \qquad \text{and} \qquad \nabla g_2(\bar{\mathbf{x}}) = (1, 1)^t$$

The sets F_0 and G_0, with the origin translated to $(\frac{9}{5}, \frac{6}{5})^t$ for convenience, are shown in Figure 4.4. Since $F_0 \cap G_0 \neq \varnothing$, $\bar{\mathbf{x}} = (\frac{9}{5}, \frac{6}{5})^t$ is not a local optimal solution to the above problem.

Now consider the point $\bar{\mathbf{x}} = (2, 1)^t$, and note that the first two constraints are binding. The corresponding gradients at this point are

$$\nabla f(\bar{\mathbf{x}}) = (-2, -2)^t \qquad \nabla g_1(\bar{\mathbf{x}}) = (4, 2)^t \qquad \nabla g_2(\bar{\mathbf{x}}) = (1, 1)^t$$

The sets F_0 and G_0 are shown in Figure 4.5, and indeed $F_0 \cap G_0 = \varnothing$. Note also that the sufficiency condition of Theorem 4.2.5 is not satisfied because g_2 is not strictly pseudoconvex over any neighborhood about $\bar{\mathbf{x}}$. However, from Figure 4.5, we observe that we also have $F_0 \cap G_0' = \varnothing$ in this case; and so, by (4.7), we have $F_0 \cap D = \varnothing$. By the converse to Theorem 4.2.2, we can now conclude that $\bar{\mathbf{x}}$ is a local minimum. In fact, since the given problem is a convex program with a strictly convex objective function, this in turn implies that $\bar{\mathbf{x}}$ is the unique global minimum.

It might be interesting to note that the utility of Theorem 4.2.5 also depends on how the constraint set is expressed. This is illustrated by Example 4.2.7 below.

4.2.7 Example

$$\begin{array}{ll} \text{Minimize} & (x_1 - 1)^2 + (x_2 - 1)^2 \\ \text{subject to} & (x_1 + x_2 - 1)^3 \leq 0 \\ & x_1 \geq 0 \\ & x_2 \geq 0 \end{array}$$

Note that the necessary condition of Theorem 4.2.5 holds true at each feasible point with $x_1 + x_2 = 1$. However, the constraint set can be represented equivalently by

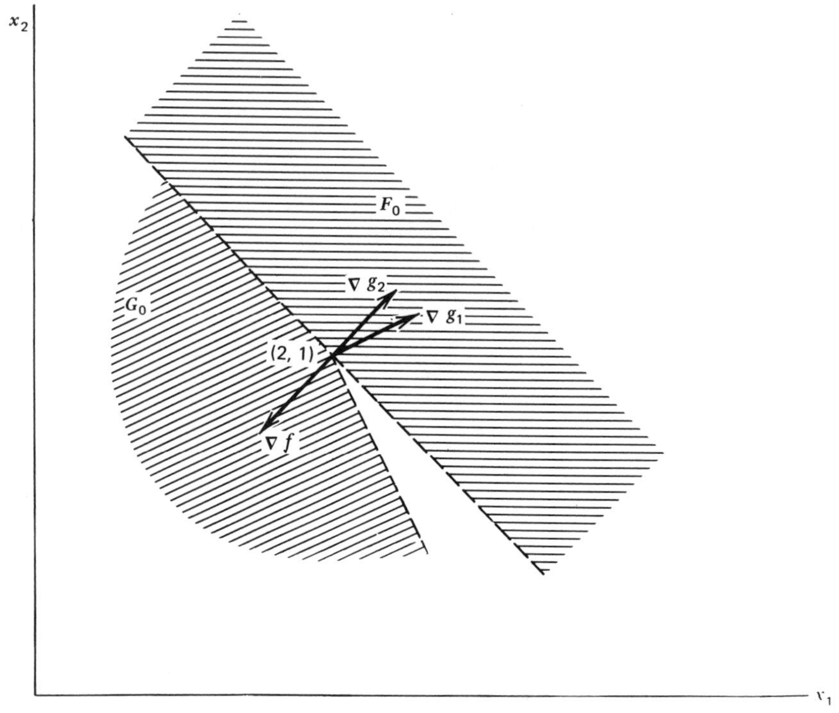

Figure 4.5 Illustration of $F_0 \cap G_0 = \emptyset$ at an optimal point.

$$x_1 + x_2 \leq 1$$
$$x_1 \geq 0$$
$$x_2 \geq 0$$

It can be easily verified that $F_0 \cap G_0 = \emptyset$ is now satisfied only at the point $(\frac{1}{2}, \frac{1}{2})$. Moreover, it can also be easily verified that $F_0 \cap G_0' = \emptyset$ in this case, and so, by (4.7), $F_0 \cap D = \emptyset$. Following the converse to Theorem 4.2.2 and noting the convexity of the feasible region and the strict convexity of the objective function, we can conclude that $\bar{\mathbf{x}} = (\frac{1}{2}, \frac{1}{2})^t$ is indeed, the unique global minimum to the given problem.

There are several cases where the necessary conditions of Theorem 4.2.5 are satisfied trivially by possibly nonoptimal points also. Some of these cases are discussed below.

Suppose that $\bar{\mathbf{x}}$ is a feasible point such that $\nabla f(\bar{\mathbf{x}}) = \mathbf{0}$. Clearly, $F_0 = \{\mathbf{d}: \nabla f(\bar{\mathbf{x}})^t\mathbf{d} < 0\} = \emptyset$ and, hence, $F_0 \cap G_0 = \emptyset$. Thus, any point $\bar{\mathbf{x}}$ with $\nabla f(\bar{\mathbf{x}}) = \mathbf{0}$ satisfies the necessary optimality conditions. Likewise, any point $\bar{\mathbf{x}}$ with $\nabla g_i(\bar{\mathbf{x}}) = \mathbf{0}$ for some $i \in I$ will also satisfy the necessary conditions. Now consider the following example with an equality constraint:

$$\begin{array}{ll} \text{Minimize} & f(\mathbf{x}) \\ \text{subject to} & g(\mathbf{x}) = 0 \end{array}$$

The equality constraint $g(\mathbf{x}) = 0$ could be replaced by the inequality constraints $g_1(\mathbf{x}) = g(\mathbf{x}) \leq 0$ and $g_2(\mathbf{x}) = -g(\mathbf{x}) \leq 0$. Let $\bar{\mathbf{x}}$ be any feasible point. Then, $g_1(\bar{\mathbf{x}}) = g_2(\bar{\mathbf{x}}) = 0$. Note that $\nabla g_1(\bar{\mathbf{x}}) = -\nabla g_2(\bar{\mathbf{x}})$, and therefore there could exist no vector $\mathbf{d}$ such that $\nabla g_1(\bar{\mathbf{x}})^t\mathbf{d} < 0$ and $\nabla g_2(\bar{\mathbf{x}})^t\mathbf{d} < 0$. Therefore, $G_0 = \emptyset$ and, hence, $F_0 \cap G_0 = \emptyset$. In other words, the necessary condition of Theorem 4.2.5 is satisfied by all feasible solutions and is hence not usable.

The Fritz John (FJ) Optimality Conditions

We now reduce the geometric necessary optimality condition $F_0 \cap G_0 = \varnothing$ to a statement in terms of the gradients of the objective function and of the binding constraints. The resulting optimality conditions, credited to Fritz John [1948], are given below.

4.2.8 Theorem (The Fritz John Necessary Conditions)

Let X be a nonempty open set in E_n and let $f: E_n \to E_1$, and $g_i: E_n \to E_1$ for $i = 1, \ldots, m$. Consider the Problem P to minimize $f(\mathbf{x})$ subject to $\mathbf{x} \in X$ and $g_i(\mathbf{x}) \le 0$ for $i = 1, \ldots, m$. Let $\bar{\mathbf{x}}$ be a feasible solution, and denote $I = \{i: g_i(\bar{\mathbf{x}}) = 0\}$. Furthermore, suppose that f and g_i for $i \in I$ are differentiable at $\bar{\mathbf{x}}$ and that g_i for $i \notin I$ are continuous at $\bar{\mathbf{x}}$. If $\bar{\mathbf{x}}$ locally solves Problem P, then there exist scalars μ_0 and μ_i for $i \in I$ such that

$$u_0 \nabla f(\bar{\mathbf{x}}) + \sum_{i \in I} u_i \nabla g_i(\bar{\mathbf{x}}) = \mathbf{0}$$

$$u_0, u_i \ge 0 \qquad \text{for } i \in I$$

$$(u_0, \mathbf{u}_I) \ne (0, \mathbf{0})$$

where $\mathbf{u}_I$ is the vector whose components are u_i for $i \in I$. Furthermore, if g_i for $i \notin I$ are also differentiable at $\bar{\mathbf{x}}$, then the foregoing conditions can be written in the following equivalent form:

$$u_0 \nabla f(\bar{\mathbf{x}}) + \sum_{i=1}^{m} u_i \nabla g_i(\bar{\mathbf{x}}) = \mathbf{0}$$

$$u_i g_i(\bar{\mathbf{x}}) = 0 \qquad \text{for } i = 1, \ldots, m$$

$$u_0, u_i \ge 0 \qquad \text{for } i = 1, \ldots, m$$

$$(u_0, \mathbf{u}) \ne (0, \mathbf{0})$$

where $\mathbf{u}$ is the vector whose components are u_i for $i = 1, \ldots, m$.

Proof

Since $\bar{\mathbf{x}}$ locally solves *Problem P*, then by Theorem 4.2.5, there exists no vector $\mathbf{d}$ such that $\nabla f(\bar{\mathbf{x}})^t \mathbf{d} < 0$ and $\nabla g_i(\bar{\mathbf{x}})^t \mathbf{d} < 0$ for each $i \in I$. Now, let $\mathbf{A}$ be the matrix whose rows are $\nabla f(\bar{\mathbf{x}})^t$ and $\nabla g_i(\bar{\mathbf{x}})^t$ for $i \in I$. The necessary optimality condition of Theorem 4.2.5 is then equivalent to the statement that the system $\mathbf{Ad} < \mathbf{0}$ is inconsistent. By Gordon's Theorem 2.4.9 there exists a nonzero vector $\mathbf{p} \ge \mathbf{0}$ such that $\mathbf{A}^t\mathbf{p} = \mathbf{0}$. Denoting the components of $\mathbf{p}$ by u_0 and u_i for $i \in I$, the first part of the result follows. The equivalent form of the necessary conditions is readily obtained by letting $u_i = 0$ for $i \notin I$, and the proof is complete.

In the conditions of Theorem 4.2.8, the scalars u_0 and u_i for $i = 1, \ldots, m$ are usually called *Lagrangian*, or *Lagrange, multipliers*. The condition that $\bar{\mathbf{x}}$ be feasible to Problem P is called the *primal feasibility (PF)* condition, whereas the requirements

$u_0\nabla f(\bar{\mathbf{x}}) + \sum_{i=1}^{m} u_i \nabla g_i(\mathbf{x}) = 0$, $(u_0, \mathbf{u}) \geq (0, \mathbf{0})$, and $(u_0, \mathbf{u}) \neq (0, \mathbf{0})$ are sometimes referred to as *dual feasibility (DF)* conditions. The condition $u_i g_i(\bar{\mathbf{x}}) = 0$ for $i = 1, \ldots, m$ is called the *complementary slackness (CS)* condition. It requires that $u_i = 0$ if the corresponding inequality is nonbinding; that is, if $g_i(\bar{\mathbf{x}}) < 0$. Likewise, it permits $u_i > 0$ only for those constraints that are binding. Together, the PF, DF, and the CS conditions are called the Fritz John (FJ) optimality conditions. Any point $\bar{\mathbf{x}}$ for which there exist Lagrangian multipliers $(\bar{u}_0, \bar{\mathbf{u}})$ such that $(\bar{\mathbf{x}}, \bar{u}_0, \bar{\mathbf{u}})$ satisfy the FJ conditions is called a *Fritz John (FJ)* point. The Fritz John conditions can also be written in vector notation as follows, in addition to the PF requirement:

$$u_0\nabla f(\bar{\mathbf{x}}) + \nabla \mathbf{g}(\bar{\mathbf{x}})^t\mathbf{u} = \mathbf{0}$$

$$\mathbf{u}^t\mathbf{g}(\bar{\mathbf{x}}) = 0$$

$$(u_0, \mathbf{u}) \geq (0, \mathbf{0})$$

$$(u_0, \mathbf{u}) \neq (0, \mathbf{0})$$

Here, $\nabla \mathbf{g}(\bar{\mathbf{x}})$ is an $m \times n$ Jacobian matrix whose ith row is $\nabla g_i(\bar{\mathbf{x}})^t$, and $\mathbf{u}$ is an m vector denoting the Lagrangian multipliers.

4.2.9 Example

$$\begin{aligned} \text{Minimize} \quad & (x_1 - 3)^2 + (x_2 - 2)^2 \\ \text{subject to} \quad & x_1^2 + x_2^2 \leq 5 \\ & x_1 + 2x_2 \leq 4 \\ & -x_1 \leq 0 \\ & -x_2 \leq 0 \end{aligned}$$

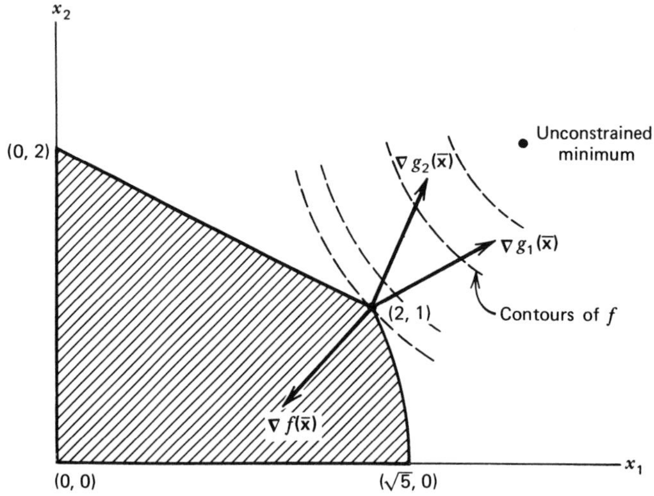

Figure 4.6 Illustration of Example 4.2.9.

The feasible region for the above problem is illustrated in Figure 4.6. We now verify that the Fritz John conditions are true at the optimal point (2, 1). First, note that the set of binding constraints I at $\bar{\mathbf{x}} = (2, 1)^t$ is given by $I = \{1, 2\}$. Thus, the Lagrangian multipliers u_3 and u_4 associated with $-x_1 \leq 0$ and $-x_2 \leq 0$, respectively, are equal to zero. Note that

$$\nabla f(\bar{\mathbf{x}}) = (-2, -2)^t \qquad \nabla g_1(\bar{\mathbf{x}}) = (4, 2)^t \qquad \nabla g_2(\bar{\mathbf{x}}) = (1, 2)^t$$

Hence, to satisfy the Fritz John conditions, we now need a nonzero vector $(u_0, u_1, u_2) \geq \mathbf{0}$ satisfying

$$u_0\begin{pmatrix}-2\\-2\end{pmatrix} + u_1\begin{pmatrix}4\\2\end{pmatrix} + u_2\begin{pmatrix}1\\2\end{pmatrix} = \begin{pmatrix}0\\0\end{pmatrix}$$

This implies that $u_1 = u_0/3$and $u_2 = 2u_0/3$. Taking u_1 and u_2 as such for any $u_0 > 0$, we satisfy the FJ conditions. As another illustration, let us check whether the point $\hat{\mathbf{x}} = (0, 0)^t$ is a FJ point. Here, the set of binding constraints is $I = \{3, 4\}$, and thus $u_1 = u_2 = 0$. Note that

$$\nabla f(\hat{\mathbf{x}}) = (-6, -4)^t \qquad \nabla g_3(\hat{\mathbf{x}}) = (-1, 0)^t \qquad \nabla g_4(\hat{\mathbf{x}}) = (0, -1)^t$$

Also note that the DF condition

$$u_0\begin{pmatrix}-6\\-4\end{pmatrix} + u_3\begin{pmatrix}-1\\0\end{pmatrix} + u_4\begin{pmatrix}0\\-1\end{pmatrix} = \begin{pmatrix}0\\0\end{pmatrix}$$

holds true if and only if $u_3 = -6u_0$ and $u_4 = -4u_0$. If $u_0 > 0$, then $u_3, u_4 < 0$, contradicting the nonnegativity restrictions. If, on the other hand, $u_0 = 0$, then $u_3 = u_4 = 0$, which contradicts the stipulation that the vector (u_0, u_3, u_4) is nonzero. Thus, the Fritz John conditions do not hold true at $\hat{\mathbf{x}} = (0, 0)^t$, which also shows that the origin is not a local optimal point.

4.2.10 Example

Consider the following problem from Kuhn and Tucker [1951]:

$$\begin{array}{ll} \text{Minimize} & -x_1 \\ \text{subject to} & x_2 - (1 - x_1)^3 \leq 0 \\ & -x_2 \leq 0 \end{array}$$

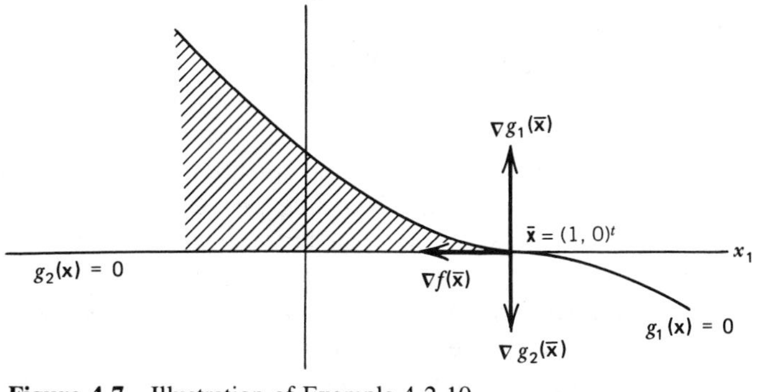

Figure 4.7 Illustration of Example 4.2.10.

The feasible region is illustrated in Figure 4.7. We now verify that the Fritz John conditions indeed hold true at the optimal point $\bar{\mathbf{x}} = (1, 0)^t$. Note that the set of binding constraints at $\bar{\mathbf{x}}$ is given by $I = \{1, 2\}$. Also,

$$\nabla f(\bar{\mathbf{x}}) = (-1, 0)^t \qquad \nabla g_1(\bar{\mathbf{x}}) = (0, 1)^t \qquad \nabla g_2(\bar{\mathbf{x}}) = (0, -1)^t$$

The DF condition

$$u_0\begin{pmatrix}-1\\0\end{pmatrix} + u_1\begin{pmatrix}0\\1\end{pmatrix} + u_2\begin{pmatrix}0\\-1\end{pmatrix} = \begin{pmatrix}0\\0\end{pmatrix}$$

is true only if $u_0 = 0$. Thus, the Fritz John conditions are true at $\bar{\mathbf{x}}$ by letting $u_0 = 0$ and $u_1 = u_2 = \alpha$, where α is a positive scalar.

4.2.11 Example

Minimize $\quad -x_1$

subject to $\quad x_1 + x_2 - 1 \le 0$

$\qquad\qquad -x_2 \le 0$

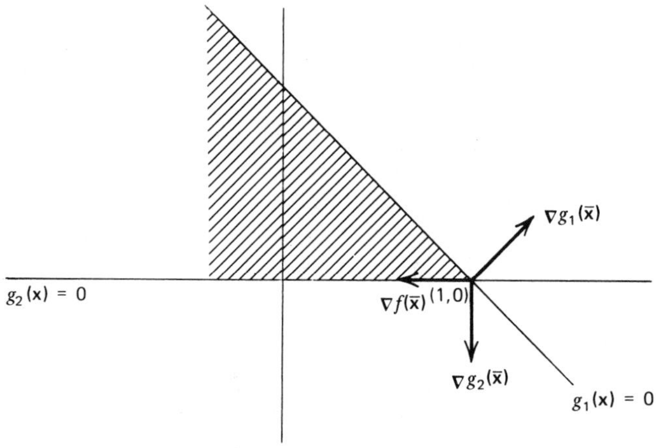

Figure 4.8 Illustration of Example 4.2.11.

The feasible region is sketched in Figure 4.8, and the optimal point is $\bar{\mathbf{x}} = (1, 0)^t$. Note that

$$\nabla f(\bar{\mathbf{x}}) = (-1, 0)^t \qquad \nabla g_1(\bar{\mathbf{x}}) = (1, 1)^t \qquad \nabla g_2(\bar{\mathbf{x}}) = (0, -1)^t$$

and the Fritz John conditions are true with $u_0 = u_1 = u_2 = \alpha$ for any positive scalar α.

As in the case of Theorem 4.2.5, there are points that satisfy the Fritz John conditions trivially. If a point $\bar{\mathbf{x}}$ satisfies $\nabla f(\bar{\mathbf{x}}) = \mathbf{0}$ or $\nabla g_i(\bar{\mathbf{x}}) = \mathbf{0}$ for some $i \in I$, then clearly we can let the corresponding Lagrangian multiplier be any positive number, set all the other multipliers equal to zero, and satisfy the conditions of Theorem 4.2.8. The Fritz John conditions of Theorem 4.2.8 also hold true trivially at each feasible point for problems with equality constraints if each equality constraint is replaced by two equivalent inequalities. Specifically, if $g(\mathbf{x}) = 0$ is replaced by $g_1(\mathbf{x}) = g(\mathbf{x}) \le 0$ and $g_2(\mathbf{x}) = -g(\mathbf{x}) \le 0$, then the Fritz John conditions are satisfied by taking $u_1 = u_2 = \alpha$, and setting all the other multipliers equal to zero, where α is a positive scalar.

In fact, given *any feasible solution* $\bar{\mathbf{x}}$ to the problem of minimizing $f(\mathbf{x})$ subject to $\mathbf{x} \in S$, we can add a redundant constraint to the problem to make $\bar{\mathbf{x}}$ a FJ point! Specifically, we can add the constraint $\|\mathbf{x} - \bar{\mathbf{x}}\|^2 \ge 0$, which holds true for all $\mathbf{x} \in E_n$. In particular, this constraint is binding at $\bar{\mathbf{x}}$ and its gradient is also zero at $\bar{\mathbf{x}}$. Consequently, we obtain $F_0 \cap G_0 = \varnothing$ at $\bar{\mathbf{x}}$ since $G_0 = \varnothing$; and so, $\bar{\mathbf{x}}$ is a FJ point!

This leads us to consider two issues. The first pertains to a set of conditions under which we can claim local optimality for a FJ point, and it is addressed in Theorem 4.2.12 below. The second consideration leads to the Karush–Kuhn–Tucker necessary optimality conditions, and this is addressed subsequently.

4.2.12 Theorem (Fritz John Sufficient Conditions)

Let X be a nonempty open set in E_n, and let $f:E_n \to E_1$ and $g_i:E_n \to E_1$, for $i = 1, \ldots, m$. Consider Problem P to minimize $f(\mathbf{x})$ subject to $\mathbf{x} \in X$, and $g_i(\mathbf{x}) \leq 0$ for $i = 1, \ldots, m$. Let $\bar{\mathbf{x}}$ be a FJ solution and denote $I = \{i:g_i(\bar{\mathbf{x}}) = 0\}$. Define S as the relaxed feasible region for Problem P in which the nonbinding constraints are dropped.

a. If there exists an ε-neighborhood $N_\varepsilon(\bar{\mathbf{x}})$, $\varepsilon > 0$ such that f is pseudoconvex over $N_\varepsilon(\bar{\mathbf{x}}) \cap S$ and g_i, $i \in I$ are strictly pseudoconvex over $N_\varepsilon(\bar{\mathbf{x}}) \cap S$, then $\bar{\mathbf{x}}$ is a local minimum for Problem P.
b. If f is pseudoconvex at $\bar{\mathbf{x}}$ and if g_i, $i \in I$ are both strictly pseudoconvex and quasiconvex at $\bar{\mathbf{x}}$, then $\bar{\mathbf{x}}$ is a global optimal solution for Problem P. In particular, if these generalized convexity assumptions hold true only by restricting the domain of f to $N_\varepsilon(\bar{\mathbf{x}})$ for some $\varepsilon > 0$, then $\bar{\mathbf{x}}$ is a local minimum for Problem P.

Proof

Suppose that the condition of part a holds. Since $\bar{\mathbf{x}}$ is a FJ point, we have equivalently by Gordon's theorem that $F_0 \cap G_0 = \varnothing$. By restricting attention to $S \cap N_\varepsilon(\bar{\mathbf{x}})$, we have, by closely following the proof to the converse statement of Theorem 4.2.5, that $\bar{\mathbf{x}}$ is a local minimum. This proves part a.

Next, consider part b. Again, we have $F_0 \cap G_0 = \varnothing$. By restricting attention to S, we have $G_0 = D$ by Lemma 4.2.4; and so we conclude that $F_0 \cap D = \varnothing$. Now, let $\mathbf{x}$ be any feasible solution in the relaxed constraint set S. (In the case when the generalized convexity assumptions hold true over $N_\varepsilon(\bar{\mathbf{x}})$ alone, let $\mathbf{x} \in S \cap N_\varepsilon(\bar{\mathbf{x}})$.) Since $g_i(\mathbf{x}) \leq g_i(\bar{\mathbf{x}}) = 0$ for all $i \in I$, we have, by the quasiconvexity of g_i at $\bar{\mathbf{x}}$ for all $i \in I$, that

$$g_i[\mathbf{x} + \lambda(\mathbf{x} - \bar{\mathbf{x}})] \equiv g_i[\lambda\mathbf{x} + (1 - \lambda)\bar{\mathbf{x}}] \leq \max\{g_i(\mathbf{x}), g_i(\bar{\mathbf{x}})\} = g_i(\bar{\mathbf{x}}) = 0$$

for all $0 \leq \lambda \leq 1$, and for each $i \in I$. This means that the direction $\mathbf{d} = (\mathbf{x} - \bar{\mathbf{x}}) \in D$. Because $F_0 \cap D = \varnothing$, we must therefore have $\nabla f(\bar{\mathbf{x}})^t\mathbf{d} \geq 0$, that is, $\nabla f(\bar{\mathbf{x}})^t(\mathbf{x} - \bar{\mathbf{x}}) \geq 0$. By the pseudoconvexity of f at $\bar{\mathbf{x}}$, this in turn implies that $f(\mathbf{x}) \geq f(\bar{\mathbf{x}})$. Hence, $\bar{\mathbf{x}}$ is a global optimum over the relaxed set S [or over $S \cap N_\varepsilon(\bar{\mathbf{x}})$ in the second case]. And since it belongs to the original feasible region, or to its intersection with $N_\varepsilon(\bar{\mathbf{x}})$, it is a global (or a local in the second case) minimum for Problem P. This completes the proof.

We remark here that as is evident from the analysis thus far, several variations of the assumptions in Theorem 4.2.12 are possible. We encourage the reader to explore this in Exercise 4.39.

The Karush–Kuhn–Tucker (KKT) Conditions

We have observed above that a point $\bar{\mathbf{x}}$ is a FJ point if and only if $F_0 \cap G_0 = \varnothing$. In particular, this condition holds true at any feasible solution $\bar{\mathbf{x}}$ at which $G_0 = \varnothing$, regardless of the objective function. Hence, for example, if the feasible region has no interior in

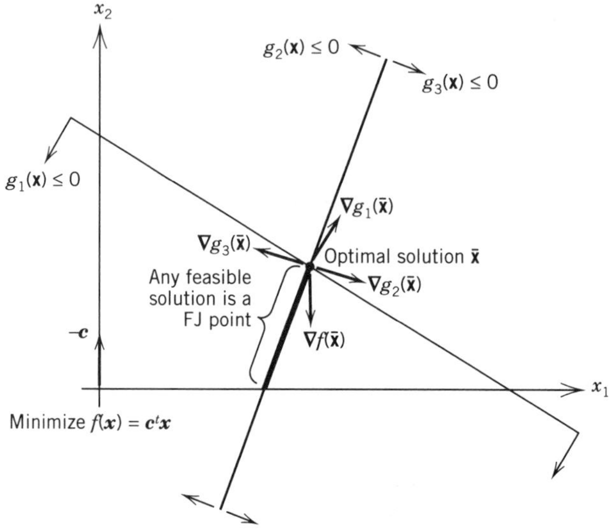

Figure 4.9 FJ conditions are not sufficient for optimality for LP problems.

the immediate vicinity of $\bar{\mathbf{x}}$, or if the gradient of some binding constraint (which might even be redundant) vanishes, then $G_0 = \varnothing$. Generally speaking, by Gordan's theorem, $G_0 = \varnothing$ if and only if the gradients of the binding constraints can be made to cancel out using nonnegative, nonzero linear combinations, and whenever this case occurs, $\bar{\mathbf{x}}$ will be a FJ point. More disturbingly, it follows that FJ points can be nonoptimal even for the well-behaved and important class of linear programming (LP) problems. Figure 4.9 illustrates this situation.

Motivated by this observation, we are led to the KKT conditions described next that encompass FJ points for which there exist Lagrange multipliers such that $u_0 > 0$ and, hence, force the objective function gradient to play a role in the optimality conditions. These conditions were independently derived by Karush [1939] and by Kuhn and Tucker [1951], and are precisely the FJ conditions with the added requirement that $u_0 > 0$. Note that when $u_0 > 0$, by scaling the dual feasibility conditions, if necessary, we can assume without loss of generality that $u_0 \equiv 1$. Hence, in Example 4.2.9, taking $u_0 = 1$ in the FJ conditions, we obtain $(u_0, u_1, u_2) = (1, \frac{1}{3}, \frac{2}{3})$ as the Lagrange multipliers corresponding to the optimal solution. Moreover, in Figure 4.9, the only FJ point that is also a KKT point is the optimal solution $\bar{\mathbf{x}}$. In fact, as we shall see later, the KKT conditions are both necessary and sufficient for optimality for linear programming problems. Example 4.2.11 gives another illustration of a linear programming problem.

Also, note from the above discussion that if $G_0 \neq \varnothing$ at a local minimum $\bar{\mathbf{x}}$, then $\bar{\mathbf{x}}$ must indeed be a KKT point, that is, it must be a FJ point with $u_0 > 0$. This follows because by Gordan's theorem, if $G_0 \neq \varnothing$, then no solution exists to FJ's dual feasibility conditions with $u_0 = 0$. Hence, $G_0 \neq \varnothing$ is a *sufficient* condition placed on the behavior of the constraints to ensure that a local minimum $\bar{\mathbf{x}}$ is a KKT point. Of course, it need not necessarily hold true whenever a local minimum $\bar{\mathbf{x}}$ turns out to be a KKT point, as in Figure 4.9, for example. Such a condition is known as a *constraint qualification (CQ)*. Several conditions of this kind are discussed in more detail subsequently and later in Chapter 5. Note that the importance of constraint qualifications is to guarantee that, by examining only KKT points, we do not lose out on local minima and, hence possibly, global optimal solutions. This can certainly occur, as is evident from Figure 4.7 of Example 4.2.10, where u_0 is necessarily zero in the FJ conditions for the optimal solution.

In Theorem 4.2.13 below, by imposing the constraint qualification that the gradient vectors of the binding constaints are linearly independent, we obtain the KKT conditions. Note that if the gradients of the binding constraints are linearly independent, then certainly they cannot be canceled by using nonzero, nonnegative, linear combinations; and hence this implies, by Gordan's theorem, that $G_0 \neq \emptyset$. Therefore, the linear independence constraint qualification implies the constraint qualification that $G_0 \neq \emptyset$; and hence, as above, it implies that a local minimum $\bar{\mathbf{x}}$ satisfies the KKT conditions. This is formalized below.

4.2.13 Theorem (Karush–Kuhn–Tucker (KKT) Necessary Conditions)

Let X be a nonempty open set in E_n, and let $f: E_n \to E_1$ and $g_i: E_n \to E_1$ for $i = 1, \ldots, m$. Consider the Problem P to minimize $f(\mathbf{x})$ subject to $\mathbf{x} \in X$ and $g_i(\mathbf{x}) \leq 0$ for $i = 1, \ldots, m$. Let $\bar{\mathbf{x}}$ be a feasible solution, and denote $I = \{i : g_i(\bar{\mathbf{x}}) = 0\}$. Suppose that f and g_i for $i \in I$ are differentiable at $\bar{\mathbf{x}}$ and that g_i for i L I are continuous at $\bar{\mathbf{x}}$. Furthermore, suppose that $\nabla g_i(\bar{\mathbf{x}})$ for $i \in I$ are linearly independent. If $\bar{\mathbf{x}}$ locally solves Problem P, then there exist scalars u_i for $i \in I$ such that

$$\nabla f(\bar{\mathbf{x}}) + \sum_{i \in I} u_i \nabla g_i(\bar{\mathbf{x}}) = \mathbf{0}$$

$$u_i \geq 0 \qquad \text{for } i \in I$$

In addition to the above assumptions, if g_i for each i L I is also differentiable at $\bar{\mathbf{x}}$, then the foregoing conditions can be written in the following equivalent form:

$$\nabla f(\bar{\mathbf{x}}) + \sum_{i=1}^{m} u_i \nabla g_i(\bar{\mathbf{x}}) = \mathbf{0}$$

$$u_i g_i(\bar{\mathbf{x}}) = 0 \qquad \text{for } i = 1, \ldots, m$$

$$u_i \geq 0 \qquad \text{for } i = 1, \ldots, m$$

Proof

By Theorem 4.2.8, there exist scalars u_0 and $\hat{u}_i$ for $i \in I$, not all equal to zero, such that

$$u_0 \nabla f(\bar{\mathbf{x}}) + \sum_{i \in I} \hat{u}_i \nabla g_i(\bar{\mathbf{x}}) = \mathbf{0} \tag{4.10}$$

$$u_0, \hat{u}_i \geq 0 \qquad \text{for } i \in I$$

Note that $u_0 > 0$, because (4.10) would contradict the assumption of linear independence of $\nabla g_i(\bar{\mathbf{x}})$ for $i \in I$ if $u_0 = 0$. The first part of the theorem then follows by letting $u_i = \hat{u}_i / u_0$ for each $i \in I$. The equivalent form of the necessary conditions follows by letting $u_i = 0$ for i L I. This completes the proof.

As in the Fritz John conditions, the scalars u_i are called the *Lagrangian*, or *Lagrange, multipliers*. The requirement that $\bar{\mathbf{x}}$ be feasible to Problem P is called the *primal feasibility (PF)* condition, whereas the condition that $\nabla f(\bar{\mathbf{x}}) + \sum_{i=1}^{m} u_i \nabla g_i(\bar{\mathbf{x}}) = 0$, $u_i \geq 0$ for $i = 1, \ldots, m$ is referred to as the *dual feasibility (DF)* condition. The restriction $u_i g_i(\bar{\mathbf{x}}) = 0$ for $i = 1, \ldots, m$ is called the *complementary slackness (CS)* condition.

Together, these PF, DF, and CS conditions are called the *Karush–Kuhn–Tucker (KKT)* conditions. Any point $\bar{\mathbf{x}}$ for which there exist Lagrangian or Lagrange multipliers $\bar{\mathbf{u}}$ such that $(\bar{\mathbf{x}}, \bar{\mathbf{u}})$ satisfies the KKT conditions is called a *KKT point*. Note that if the gradients $\nabla g_i(\bar{\mathbf{x}})$, $i \in I$, are linearly independent, then, by the DF and CS conditions, the associated Lagrange multipliers are uniquely determined at the KKT point $\bar{\mathbf{x}}$.

The KKT conditions can alternatively be written in vector form as follows, in addition to the PF requirement:

$$\nabla f(\bar{\mathbf{x}}) + \nabla \mathbf{g}(\bar{\mathbf{x}})^t \mathbf{u} = \mathbf{0}$$

$$\mathbf{u}^t \mathbf{g}(\bar{\mathbf{x}}) = 0$$

$$\mathbf{u} \geq \mathbf{0}$$

Here, $\nabla \mathbf{g}(\bar{\mathbf{x}})^t$ is an $n \times m$ matrix whose ith column is $\nabla g_i(\bar{\mathbf{x}})$, (it is the transpose of the *Jacobian* of $\mathbf{g}$ at $\bar{\mathbf{x}}$) and $\mathbf{u}$ is an m vector denoting the Lagrangian multipliers.

Now consider Examples 4.2.9, 4.2.10, and 4.2.11 discussed earleir. In Example 4.2.9, at $\bar{\mathbf{x}} = (2, 1)^t$, the reader may verify that $u_1 = \frac{1}{3}$, $u_2 = \frac{2}{3}$, and $u_3 = u_4 = 0$ will satisfy the KKT conditions. Example 4.2.10 does not satisfy the assumptions of Theorem 4.2.13 at $\bar{\mathbf{x}} = (1, 0)^t$, since $\nabla g_1(\bar{\mathbf{x}})$ and $\nabla g_2(\bar{\mathbf{x}})$ are linearly dependent. In fact, in this example, we saw that u_0 is necessarily zero in the FJ conditions. In Example 4.2.11 $\bar{\mathbf{x}} = (1, 0)^t$ and $u_1 = u_2 = 1$ satisfy the KKT conditions.

4.2.14 Example (Linear Programming Problems)

Consider the linear programming Problem P: minimize $\{\mathbf{c}^t\mathbf{x} : \mathbf{A}\mathbf{x} = \mathbf{b}, \mathbf{x} \geq \mathbf{0}\}$, where $\mathbf{A}$ is $m \times n$ and the other vectors are conformable. Writing the constraints as $-\mathbf{A}\mathbf{x} \leq -\mathbf{b}$, $\mathbf{A}\mathbf{x} \leq \mathbf{b}$, and $-\mathbf{x} \leq \mathbf{0}$, and denoting the Lagrange multiplier vectors with respect to these three sets as $\mathbf{y}^+$, $\mathbf{y}^-$, and $\mathbf{v}$, respectively, the KKT conditions are as follows:

$$\begin{array}{llll} \textit{PF:} & \mathbf{A}\mathbf{x} = \mathbf{b} & & \mathbf{x} \geq \mathbf{0} \\ \textit{DF:} & -\mathbf{A}^t\mathbf{y}^+ + \mathbf{A}^t\mathbf{y}^- - \mathbf{v} = -\mathbf{c} & & (\mathbf{y}^+, \mathbf{y}^-, \mathbf{v}) \geq \mathbf{0} \\ \textit{CS:} & (\mathbf{b} - \mathbf{A}\mathbf{x})^t\mathbf{y}^+ = 0 & (\mathbf{A}\mathbf{x} - \mathbf{b})^t\mathbf{y}^- = 0 & -\mathbf{x}^t\mathbf{v} = 0 \end{array}$$

Denoting $\mathbf{y} = \mathbf{y}^+ - \mathbf{y}^-$ as the difference of the two nonnegative variable vectors $\mathbf{y}^+$ and $\mathbf{y}^-$, we can equivalently write the KKT conditions as follows, noting the use of the PF and DF conditions in simplifying the CS conditions:

$$\begin{array}{lll} \textit{PF:} & \mathbf{A}\mathbf{x} = \mathbf{b} & \mathbf{x} \geq \mathbf{0} \\ \textit{DF:} & \mathbf{A}^t\mathbf{y} + \mathbf{v} = \mathbf{c} & \mathbf{v} \geq 0 (\mathbf{y} \text{ unrestricted}) \\ \textit{CS:} & x_j v_j = 0 & \text{for } j = 1, \ldots, n \end{array}$$

Hence, from Theorem 2.7.3 and its Corollary 3, observe that $\bar{\mathbf{x}}$ is KKT solution with associated Lagrange multipliers $(\bar{\mathbf{y}}, \bar{\mathbf{v}})$ if and only if $\bar{\mathbf{x}}$ and $\bar{\mathbf{y}}$ are, respectively, optimal to the primal and dual linear programs P and D, where D: maximize $\{\mathbf{b}^t\mathbf{y}: \mathbf{A}^t\mathbf{y} \leq \mathbf{c}\}$. In particular, observe that the DF restriction in the KKT conditions is precisely the feasibility condition to the dual D; hence, the name. This example therefore establishes that, for linear programming problems, the KKT conditions are both necessary and sufficient for optimality to the primal and dual problems.

Geometric Interpretation of the Karash–Kuhn–Tucker Conditions: Connections with Linear Programming Approximations Note that any vector of the form

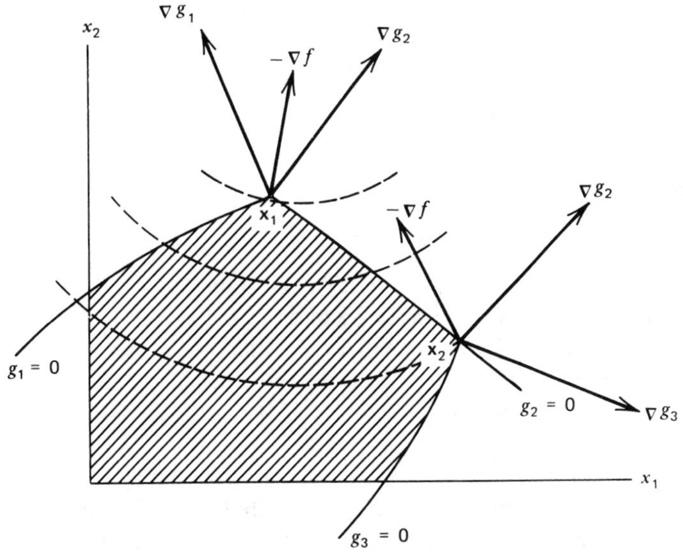

Figure 4.10 Geometric illustration of the KKT conditions.

$\sum_{i\in I} u_i \nabla g_i(\bar{\mathbf{x}})$, where $u_i \geq 0$ for $i \in I$, belongs to the cone spanned by the gradients of the binding constraints. The KKT dual feasibility conditions $-\nabla f(\bar{\mathbf{x}}) = \sum_{i\in I} u_i \nabla g_i(\bar{\mathbf{x}})$ and $u_i \geq 0$ for $i \in I$ can then be interpreted as $-\nabla f(\bar{\mathbf{x}})$ belonging to the cone spanned by the gradients to the binding constraints at a given feasible solution $\bar{\mathbf{x}}$.

Figure 4.10 illustrates two points $\mathbf{x}_1$ and $\mathbf{x}_2$. Note that $-\nabla f(\mathbf{x}_1)$ belongs to the cone spanned by the gradients of the binding constraints at $\mathbf{x}_1$ and, hence, $\mathbf{x}_1$ is a KKT point; that is, $\mathbf{x}_1$ satisfies the KKT conditions. On the other hand, $-\nabla f(\mathbf{x}_2)$ lies outside the cone spanned by the gradients of the binding constraints at $\mathbf{x}_2$ and thus contradicts the KKT conditions.

Likewise, in Figures 4.6 and 4.8, for $\bar{\mathbf{x}} = (2, 1)^t$ and $\bar{\mathbf{x}} = (1, 0)^t$, respectively, $-\nabla f(\bar{\mathbf{x}})$ is in the cone spanned by the gradients of the binding constraints at $\bar{\mathbf{x}}$. On the other hand, in Figure 4.7, for $\bar{\mathbf{x}} = (1, 0)^t$, $-\nabla f(\bar{\mathbf{x}})$ lies outside the cone spanned by the gradients of the binding constraints at $\bar{\mathbf{x}}$.

We now provide a key insight into the KKT conditions via linear programming duality and Farkas' lemma as expounded in Theorem 2.7.3 and its Corollary 2. The following result asserts that a feasible solution $\bar{\mathbf{x}}$ is a KKT point if and only if it happens to be an optimal to the linear program obtained by replacing the objective and the constraints by their first-order approximations at $\bar{\mathbf{x}}$. Not only does this provide a useful conceptual characterization of KKT points and an insight into its value and interpretation, but it also affords a useful construct in deriving algorithms that are designed to converge to a KKT solution.

4.2.15 Theorem (KKT Conditions and First-Order LP Approximations)

Let X be a nonempty open set in E_n, and let $f: E_n \to E_1$ and $g_i: E_n \to E_1$, $i = 1, \ldots, m$ be differentiable functions. Consider Problem P to minimize $f(\mathbf{x})$ subject to $\mathbf{x} \in S = \{\mathbf{x} \in X: g_i(\mathbf{x}) \leq 0,\ i = 1, \ldots, m\}$. Let $\bar{\mathbf{x}}$ be a feasible solution, and denote $I = \{i: g_i(\bar{\mathbf{x}}) = 0\}$. Define $F_0 = \{\mathbf{d}: \nabla f(\bar{\mathbf{x}})^t\mathbf{d} < 0\}$ and $G_0' = \{\mathbf{d} \neq \mathbf{0}: \nabla g_i(\bar{\mathbf{x}})^t\mathbf{d} \leq 0$, for each $i \in I\}$ as before, and let $G' = \{\mathbf{d}: \nabla g_i(\bar{\mathbf{x}})^t\mathbf{d} \leq 0$ for each $i \in I\} \equiv G_0' \cup \{\mathbf{0}\}$. Then, $\bar{\mathbf{x}}$ is a KKT solution if and only if $F_0 \cap G' = \emptyset$, which is equivalent to $F_0 \cap G'_0 = \emptyset$. Furthermore, consider the first-order linear programming approximation to Problem P given below:

$$LP(\bar{\mathbf{x}}) : \text{minimize } \{f(\bar{\mathbf{x}}) + \nabla f(\bar{\mathbf{x}})^t(\mathbf{x} - \bar{\mathbf{x}}) : g_i(\bar{\mathbf{x}}) + \nabla g_i(\bar{\mathbf{x}})^t(\mathbf{x} - \bar{\mathbf{x}}) \le 0 \quad \text{for } i = 1, \ldots, m\}$$

Then, $\bar{\mathbf{x}}$ is a KKT solution if and only if $\bar{\mathbf{x}}$ solves LP($\bar{\mathbf{x}}$).

Proof

The feasible solution $\bar{\mathbf{x}}$ is a KKT point if and only if there exists a solution **u** to the system $\sum_{i \in I} u_i \nabla g_i(\bar{\mathbf{x}}) = -\nabla f(\bar{\mathbf{x}})$, and $u_i \ge 0$ for $i \in I$. By Farkas' lemma (see, for example, Corollary 2 to Theorem 2.7.3), this holds true if and only if there does not exist a solution to the system $\nabla g_i(\bar{\mathbf{x}})^t\mathbf{d} \le 0$ for $i \in I$ and $\nabla f(\bar{\mathbf{x}})^t\mathbf{d} < 0$. Hence, $\bar{\mathbf{x}}$ is a KKT point if and only if $F_0 \cap G' = \emptyset$. Clearly, we also have that this holds true if and only if $F_0 \cap G'_0 = \emptyset$.

Now, consider the first-order linear programming approximation LP($\bar{\mathbf{x}}$) given in the theorem. The solution $\bar{\mathbf{x}}$ is obviously feasible to LP. Ignoring the constant terms in the objective function and writing LP($\bar{\mathbf{x}}$) in the form of Problem D of Theorem 2.7.3, we get that, equivalently, LP($\bar{\mathbf{x}}$):maximize $\{-\nabla f(\bar{\mathbf{x}})^t\mathbf{x} : \nabla g_i(\bar{\mathbf{x}})^t\mathbf{x} \le \nabla g_i(\bar{\mathbf{x}})^t\bar{\mathbf{x}} - g_i(\bar{\mathbf{x}}) \text{ for } i = 1, \ldots, m\}$. The dual to this problem, denoted DLP($\bar{\mathbf{x}}$), is to

$$\text{Minimize } \sum_{i=1}^{m} u_i[\nabla g_i(\bar{\mathbf{x}})^t\bar{\mathbf{x}} - g_i(\bar{\mathbf{x}})]$$

$$\text{subject to } \sum_{i=1}^{m} u_i \nabla g_i(\bar{\mathbf{x}}) = -\nabla f(\bar{\mathbf{x}}), \ u_i \ge 0 \qquad \text{for } i = 1, \ldots, m$$

Hence, by Corollary 3 to Theorem 2.7.3, we deduce that $\bar{\mathbf{x}}$ is an optimal solution to LP($\bar{\mathbf{x}}$) if and only if there exists a solution $\bar{\mathbf{u}}$ feasible to DLP($\bar{\mathbf{x}}$) that also satisfies the complementary slackness condition $\bar{u}_i[\nabla g_i(\bar{\mathbf{x}})^t\bar{\mathbf{x}} - \nabla g_i(\bar{\mathbf{x}})^t\bar{\mathbf{x}} + g_i(\bar{\mathbf{x}})] \equiv \bar{u}_i g_i(\bar{\mathbf{x}}) = 0$ for $i = 1, \ldots, m$. But this is precisely the KKT conditions. Hence, $\bar{\mathbf{x}}$ is optimal to LP($\bar{\mathbf{x}}$) if and only if $\bar{\mathbf{x}}$ is a KKT solution for P, and this completes the proof.

To illustrate, observe that in Figure 4.6 of Example 4.2.9, if we replace $g_i(\mathbf{x}) \le 0$ by its tangential first-order approximation at the point (2, 1) and the objective function by the linear objective of minimizing $\nabla f(\bar{\mathbf{x}})^t\mathbf{x}$, the given point (2, 1) is optimal to the resulting linear programming problem and, hence, is a KKT solution. On the other hand, in Figure 4.7 of Example 4.2.10, the linear programming feasible region approximation at $\bar{\mathbf{x}} = (1, 0)^t$ is the entire x_1 axis. Clearly, then, the point (1, 0) is not optimal to the underlying linear program LP($\bar{\mathbf{x}}$) of minimizing $\nabla f(\bar{\mathbf{x}})^t\mathbf{x}$ over this region, and thus, the point (1, 0) is not a KKT point. Hence, the KKT conditions, being oblivious to the nonlinear behavior of the constraints $g_1(\mathbf{x}) \le 0$ about the point $\bar{\mathbf{x}}$ other than its first-order approximation, fail to recognize the optimality of this solution for the original nonlinear problem.

Theorem 4.2.16 below shows that, under moderate convexity assumptions, the KKT conditions are also sufficient for (local) optimality.

4.2.16 Theorem (Karush–Kuhn–Tucker Sufficient Conditions)

Let X be a nonempty open set in E_n, and let $f:E_n \to E_1$ and $g_i:E_n \to E_1$ for $i = 1, \ldots, m$. Consider Problem P to minimize $f(\mathbf{x})$ subject to $\mathbf{x} \in X$ and $g_i(\mathbf{x}) \le 0$ for $i = 1, \ldots, m$. Let $\bar{\mathbf{x}}$ be a KKT solution, and denote $I = \{i : g_i(\bar{\mathbf{x}}) = 0\}$. Define S as the relaxed feasible region for Problem P in which the constraints that are not binding at $\bar{\mathbf{x}}$ are dropped. Then,

a. If there exists an ε-neighborhood $N_\varepsilon(\bar{\mathbf{x}})$ about $\bar{\mathbf{x}}$, $\varepsilon > 0$, such that f is pseudoconvex over $N_\varepsilon(\bar{\mathbf{x}}) \cap S$ and g_i, $i \in I$ are differentiable at $\bar{\mathbf{x}}$ and are quasiconvex over $N_\varepsilon(\bar{\mathbf{x}}) \cap S$, then $\bar{\mathbf{x}}$ is a local minimum for Problem P.
b. If f is pseudoconvex at $\bar{\mathbf{x}}$, and if g_i, $i \in I$ are differentiable and quasiconvex at $\bar{\mathbf{x}}$, then $\bar{\mathbf{x}}$ is a global optimal solution to Problem P. In particular, if this assumption holds true with the domain of the feasible restriction to $N_\varepsilon(\bar{\mathbf{x}})$, for some $\varepsilon > 0$, then $\bar{\mathbf{x}}$ is a local minimum for P.

Proof

First, consider part a. Since $\bar{\mathbf{x}}$ is a KKT point, we have, equivalently, by Theorem 4.2.15 that $F_0 \cap G_0' = \varnothing$. From (4.7), this means that $F_0 \cap D = \varnothing$. Since g_i, $i \in I$ are quasiconvex over $N_\varepsilon(\bar{\mathbf{x}}) \cap S$, we have that $N_\varepsilon(\bar{\mathbf{x}}) \cap S$ is a convex set. By restricting attention to $N_\varepsilon(\bar{\mathbf{x}}) \cap S$, we therefore have the condition of the converse statement of Theorem 4.2.2 holding true; and so, $\bar{\mathbf{x}}$ is a minimum over $N_\varepsilon(\bar{\mathbf{x}}) \cap S$. Hence, $\bar{\mathbf{x}}$ is a local minimum for the more restricted original Problem P. This proves part a.

Next, consider part b. Let $\mathbf{x}$ be any feasible solution to Problem P. (In case the generalized convexity definitions are restricted to $N_\varepsilon(\bar{\mathbf{x}})$, let $\mathbf{x}$ be any feasible solution to P that lies within $N_\varepsilon(\bar{\mathbf{x}})$.) Then, for $i \in I$, $g_i(\mathbf{x}) \le g_i(\bar{\mathbf{x}})$, since $g_i(\mathbf{x}) \le 0$ and $g_i(\bar{\mathbf{x}}) = 0$. By the quasiconvexity of g_i at $\bar{\mathbf{x}}$, it follows that:

$$g_i[\bar{\mathbf{x}} + \lambda(\mathbf{x} - \bar{\mathbf{x}})] = g_i[\lambda\mathbf{x} + (1 - \lambda)\bar{\mathbf{x}}] \le \text{maximum } [g_i(\mathbf{x}), g_i(\bar{\mathbf{x}})] = g_i(\bar{\mathbf{x}})$$

for all $\lambda \in (0, 1)$. This implies that g_i does not increase by moving from $\bar{\mathbf{x}}$ along the direction $\mathbf{x} - \bar{\mathbf{x}}$. Thus, by Theorem 4.1.2, we must have $\nabla g_i(\bar{\mathbf{x}})^t(\mathbf{x} - \bar{\mathbf{x}}) \le 0$. Multiplying by the Lagrange multipliers u_i corresonding to the KKT point $\bar{\mathbf{x}}$, and summing over I, we get $[\sum_{i\in I} u_i \nabla g_i(\bar{\mathbf{x}})^t](\mathbf{x} - \bar{\mathbf{x}}) \le 0$. But since $\nabla f(\bar{\mathbf{x}}) + \sum_{i\in I} u_i \nabla g_i(\bar{\mathbf{x}}) = \mathbf{0}$, it follows that $\nabla f(\bar{\mathbf{x}})^t(\mathbf{x} - \bar{\mathbf{x}}) \ge 0$. Then, by the pseudoconvexity of f at $\bar{\mathbf{x}}$, we must have $f(\mathbf{x}) \ge f(\bar{\mathbf{x}})$, and the proof is complete.

Needless to say, if f and g_i are convex at $\bar{\mathbf{x}}$ and hence both pseudoconvex and quasiconvex at $\bar{\mathbf{x}}$, then the KKT conditions are sufficient. Also, if convexity at a point is replaced by the stronger requirement of global convexity, the KKT conditions are also sufficient. (We ask the reader to explore other variations of this result in Exercises 4.39 and 4.49.)

There is one important point to note in regard to KKT conditions that is often a source of error. Namely, despite the usually well-behaved nature of convex programming problems and the sufficiency of KKT conditions under convexity assumptions, the KKT conditions are *not* necessary for optimality for convex programming problems. Figure 4.11 illustrates the situation for the convex programming problem given below:

$$\begin{aligned} &\text{Minimize} && x_1 \\ &\text{subject to} && (x_1 - 1)^2 + (x_2 - 1)^2 \le 1 \\ & && (x_1 - 1)^2 + (x_2 + 1)^2 \le 1 \end{aligned}$$

The only feasible solution $\bar{\mathbf{x}} = (1, 0)^t$ is naturally optimal. However, this is not a KKT point. Note in connection with Theorem 4.2.15 that the first-order linear programming approximation at $\bar{\mathbf{x}}$ is unbounded. However, as we shall see in Chapter 5, if there exists an interior point feasible solution to the set of constraints that are binding at an optimum $\bar{\mathbf{x}}$ to a convex programming problem, then $\bar{\mathbf{x}}$ is a KKT point and is therefore captured by the KKT conditions.

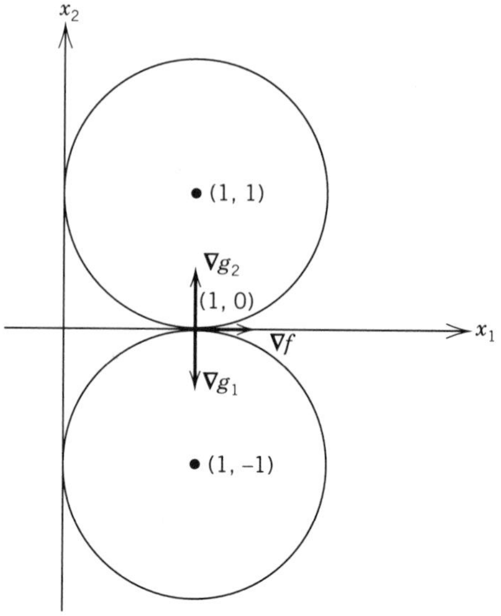

Figure 4.11 KKT conditions are not necessary for convex programming problems.

4.3 Problems with Inequality and Equality Constraints

In this section, we generalize the optimality conditions of the previous section to handle equality constraints as well as inequality constraints. Consider the following nonlinear programming Problem P.

$$\begin{aligned} \text{Minimize} \quad & f(\mathbf{x}) \\ \text{subject to} \quad & g_i(\mathbf{x}) \le 0 \quad \text{for } i = 1, \ldots, m \\ & h_i(\mathbf{x}) = 0 \quad \text{for } i = 1, \ldots, l \\ & \mathbf{x} \in X \end{aligned}$$

As a natural extension of Theorem 4.2.5, in Theorem 4.3.1 below we show that if $\bar{\mathbf{x}}$ is a local optimal solution to Problem P, then either the gradients of the equality constraints are linearly dependent at $\bar{\mathbf{x}}$, or else, $F_0 \cap G_0 \cap H_0 = \emptyset$, where $H_0 = \{\mathbf{d}: \nabla h_i(\bar{\mathbf{x}})^t\mathbf{d} = 0 \text{ for } i = 1, \ldots, l\}$. A reader with only a casual interest in the derivation of optimality conditions may skip the proof of Theorem 4.3.1, since it involves the more advanced concepts of solving a system of differential equations.

4.3.1 Theorem

Let X be a nonempty open set in E_n. Let $f: E_n \to E_1$, $g_i: E_n \to E_1$ for $i = 1, \ldots, m$ and $h_i: E_n \to E_1$ for $i = 1, \ldots, l$. Consider Problem P to

$$\begin{aligned} \text{Minimize} \quad & f(\mathbf{x}) \\ \text{subject to} \quad & g_i(\mathbf{x}) \le 0 \quad \text{for } i = 1, \ldots, m \\ & h_i(\mathbf{x}) = 0 \quad \text{for } i = 1, \ldots, l \\ & \mathbf{x} \in X \end{aligned}$$

Suppose that $\bar{\mathbf{x}}$ is a local optimal solution, and let $I = \{i: g_i(\bar{\mathbf{x}}) = 0\}$. Furthermore, suppose that each g_i for $i \notin I$ is continuous at $\bar{\mathbf{x}}$, that f and g_i for $i \in I$ are differentiable at $\bar{\mathbf{x}}$, and that each h_i for $i = 1, \ldots, l$ is continuously differentiable at $\bar{\mathbf{x}}$. If $\nabla h_i(\bar{\mathbf{x}})$ for $i = 1, \ldots, l$ are linearly independent, then $F_0 \cap G_0 \cap H_0 = \varnothing$, where

$$F_0 = \{\mathbf{d}: \nabla f(\bar{\mathbf{x}})^t\mathbf{d} < 0\}$$

$$G_0 = \{\mathbf{d}: \nabla g_i(\bar{\mathbf{x}})^t\mathbf{d} < 0 \text{ for } i \in I\}$$

$$H_0 = \{\mathbf{d}: \nabla h_i(\bar{\mathbf{x}})^t\mathbf{d} = 0 \text{ for } i = 1, \ldots, l\}$$

Conversely, suppose that $F_0 \cap G_0 \cap H_0 = \varnothing$. If f is pseudoconvex at $\bar{\mathbf{x}}$, g_i for $i \in I$ are strictly pseudoconvex over some ε neighborhood of $\bar{\mathbf{x}}$; and if h_i for $i = 1, \ldots, l$ are affine, then $\bar{\mathbf{x}}$ is a local optimal solution.

Proof

Consider the first part of the theorem. By contradiction, suppose there exists a vector $\mathbf{y} \in F_0 \cap G_0 \cap H_0$; that is, $\nabla f(\bar{\mathbf{x}})^t\mathbf{y} < 0$, $\nabla g_i(\bar{\mathbf{x}})^t\mathbf{y} < 0$ for each $i \in I$, and $\nabla \mathbf{h}(\bar{\mathbf{x}})\,\mathbf{y} = \mathbf{0}$, where $\nabla \mathbf{h}(\bar{\mathbf{x}})$ is the $l \times n$ Jacobian matrix whose ith row is $\nabla h_i(\bar{\mathbf{x}})^t$. Let us now construct a feasible arc from $\bar{\mathbf{x}}$ obtained by projecting points along $\mathbf{y}$ from $\bar{\mathbf{x}}$ onto the equality constraint surface. For $\lambda \geq 0$, define $\boldsymbol{\alpha}: E_1 \to E_n$ by the following differential equation and boundary condition:

$$\frac{d\boldsymbol{\alpha}(\lambda)}{d\lambda} = \mathbf{P}(\lambda)\mathbf{y} \qquad \boldsymbol{\alpha}(0) = \bar{\mathbf{x}} \tag{4.11}$$

where $\mathbf{P}(\lambda)$ is the matrix that projects any vector in the null space of $\nabla \mathbf{h}[\boldsymbol{\alpha}(\lambda)]$. For λ sufficiently small, the above equation is well defined and solvable because $\nabla \mathbf{h}(\bar{\mathbf{x}})$ has full rank and $\mathbf{h}$ is continuously differentiable at $\bar{\mathbf{x}}$, so that $\mathbf{P}$ is continuous in λ. Obviously, $\boldsymbol{\alpha}(\lambda) \to \bar{\mathbf{x}}$ as $\lambda \to 0^+$.

We now show that for $\lambda > 0$ and sufficiently small, $\boldsymbol{\alpha}(\lambda)$ is feasible and $f[\boldsymbol{\alpha}(\lambda)] < f(\bar{\mathbf{x}})$, thus contradicting local optimality of $\bar{\mathbf{x}}$. By the chain rule of differentiation and from (4.11), we get

$$\frac{d}{d\lambda} g_i[\boldsymbol{\alpha}(\lambda)] = \nabla g_i[\boldsymbol{\alpha}(\lambda)]^t\mathbf{P}(\lambda)\mathbf{y} \tag{4.12}$$

for each $i \in I$. In particular, $\mathbf{y}$ is in the null space of $\nabla \mathbf{h}(\bar{\mathbf{x}})$, and so for $\lambda = 0$, we have $\mathbf{P}(0)\mathbf{y} = \mathbf{y}$. Hence, from (4.12) and the fact that $\nabla g_i(\bar{\mathbf{x}})^t\mathbf{y} < 0$, we get

$$\frac{d}{d\lambda} g_i[\boldsymbol{\alpha}(0)] = \nabla g_i(\bar{\mathbf{x}})^t\mathbf{y} < 0 \tag{4.13}$$

for $i \in I$. This further implies that $g_i[\boldsymbol{\alpha}(\lambda)] < 0$ for $\lambda > 0$ and sufficiently small. For $i \notin I$, $g_i(\bar{\mathbf{x}}) < 0$, and g_i is continuous at $\bar{\mathbf{x}}$, and thus $g_i[\boldsymbol{\alpha}(\lambda)] < 0$ for λ sufficiently small. Also, since X is open, $\boldsymbol{\alpha}(\lambda) \in X$ for λ sufficiently small. To show feasibility of $\boldsymbol{\alpha}(\lambda)$, we only need to show that $h_i[\boldsymbol{\alpha}(\lambda)] = 0$ for λ sufficiently small. By the mean value theorem, we have

$$h_i[\boldsymbol{\alpha}(\lambda)] = h_i[\boldsymbol{\alpha}(0)] + \lambda \frac{d}{d\lambda} h_i[\boldsymbol{\alpha}(\mu)]$$

$$= \lambda \frac{d}{d\lambda} h_i[\boldsymbol{\alpha}(\mu)] \tag{4.14}$$

for some $\mu \in (0, \lambda)$. But by the chain rule of differentiation and similar to (4.12), we get

$$\frac{d}{d\lambda} h_i[\boldsymbol{\alpha}(\mu)] = \nabla h_i[\boldsymbol{\alpha}(\mu)]^t \mathbf{P}(\mu)\mathbf{y}$$

By construction, $\mathbf{P}(\mu)\mathbf{y}$ is in the null space of $\nabla h_i[\boldsymbol{\alpha}(\mu)]$ and, hence, from the above equation, we get $(d/d\lambda)h_i[\boldsymbol{\alpha}(\mu)] = 0$. Substituting in (4.14), it follows that $h_i[\boldsymbol{\alpha}(\lambda)] = 0$. Since this is true for each i, it then follows that $\boldsymbol{\alpha}(\lambda)$ is a feasible solution to Problem P for each $\lambda > 0$ and sufficiently small. By an argument similar to that leading to (4.13), we get

$$\frac{d}{d\lambda} f[\boldsymbol{\alpha}(0)] = \nabla f(\bar{\mathbf{x}})^t \mathbf{y} < 0$$

and, hence, $f[\boldsymbol{\alpha}(\lambda)] < f(\bar{\mathbf{x}})$ for $\lambda > 0$ and sufficiently small. This contradicts local optimality of $\bar{\mathbf{x}}$. Hence, $F_0 \cap G_0 \cap H_0 = \varnothing$.

Conversely, suppose that $F_0 \cap G_0 \cap H_0 = \varnothing$ and that the assumptions of the converse statement to the theorem hold true. Since h_i, $i = 1, \ldots, l$ are affine, we have that $\mathbf{d}$ is a feasible direction for the equality constraints if and only if $\mathbf{d} \in H_0$. Using Lemma 4.2.4, it is readily verified that, since g_i, $i \in I$ are strictly pseudoconvex over $N_\varepsilon(\bar{\mathbf{x}})$ for some $\varepsilon > 0$, we have $D = G_0 \cap H_0$, where D is the set of feasible directions at $\bar{\mathbf{x}}$ defined for the set $S = \{\mathbf{x} : g_i(\mathbf{x}) \le 0 \text{ for } i \in I,\ h_i(\mathbf{x}) = 0 \text{ for } i = 1, \ldots, l\}$. Hence, we have $F_0 \cap D = \varnothing$. Moreover, by our assumptions, we know that $S \cap N_\varepsilon(\bar{\mathbf{x}})$ is a convex set and that f is pseudoconvex at $\bar{\mathbf{x}}$. Hence, by the converse to Theorem 4.2.2, $\bar{\mathbf{x}}$ is a minimum over $S \cap N_\varepsilon(\bar{\mathbf{x}})$. Therefore, $\bar{\mathbf{x}}$ is a local minimum for the more restricted original problem as well, and this completes the proof.

The Fritz John Conditions

We now express the geometric optimality condition $F_0 \cap G_0 \cap H_0 = \varnothing$ in a more usable algebraic form. This is done in Theorem 4.3.2 below, which is a generalization of the Fritz John conditions of Theorem 4.2.6.

4.3.2 Theorem (The Fritz John Necessary Conditions)

Let X be a nonempty open set in E_n, and let $f: E_n \to E_1$, $g_i: E_n \to E_1$ for $i = 1, \ldots, m$, and $h_i: E_n \to E_1$ for $i = 1, \ldots, l$. Consider the Problem P to

$$\begin{aligned} \text{Minimize} \quad & f(\mathbf{x}) \\ \text{subject to} \quad & g_i(\mathbf{x}) \le 0 \quad \text{for } i = 1, \ldots, m \\ & h_i(\mathbf{x}) = 0 \quad \text{for } i = 1, \ldots, l \\ & \mathbf{x} \in X \end{aligned}$$

Let $\bar{\mathbf{x}}$ be a feasible solution, and let $I = \{i : g_i(\bar{\mathbf{x}}) = 0\}$. Furthermore, suppose that g_i for $i \in I$ is continuous at $\bar{\mathbf{x}}$, that f_i and g_i for $i \in I$ are differentiable at $\bar{\mathbf{x}}$, and that h_i for $i = 1, \ldots, l$ is continuously differentiable at $\bar{\mathbf{x}}$. If $\bar{\mathbf{x}}$ locally solves Problem P, then there exist scalars u_0, u_i for $i \in I$ and v_i for $i = 1, \ldots, l$ such that

$$u_0 \nabla f(\bar{\mathbf{x}}) + \sum_{i \in I} u_i \nabla g_i(\bar{\mathbf{x}}) + \sum_{i=1}^{l} v_i \nabla h_i(\bar{\mathbf{x}}) = \mathbf{0}$$

$$u_0, u_i \ge 0 \qquad \text{for } i \in I$$

$$(u_0, \mathbf{u}_I, \mathbf{v}) \ne (0, \mathbf{0}, \mathbf{0})$$

where $\mathbf{u}_I$ is the vector whose components are u_i for $i \in I$ and $\mathbf{v} = (v_1, \ldots, v_l)^t$. Furthermore, if each g_i for $i \notin I$ is also differentiable at $\bar{\mathbf{x}}$, then the Fritz John conditions can be written in the following equivalent form, where $\mathbf{u} = (u_1, \ldots, u_m)^t$ and $\mathbf{v} = (v_1, \ldots, v_l)^t$:

$$u_0 \nabla f(\bar{\mathbf{x}}) + \sum_{i=1}^{m} u_i \nabla g_i(\bar{\mathbf{x}}) + \sum_{i=1}^{l} v_i \nabla h_i(\bar{\mathbf{x}}) = \mathbf{0}$$

$$u_i g_i(\bar{\mathbf{x}}) = 0 \qquad \text{for } i = 1, \ldots, m$$

$$u_0, u_i \geq 0 \qquad \text{for } i = 1, \ldots, m$$

$$(u_0, \mathbf{u}, \mathbf{v}) \neq (0, \mathbf{0}, \mathbf{0})$$

Proof

If $\nabla h_i(\bar{\mathbf{x}})$ for $i = 1, \ldots, l$ are linearly dependent, then one can find scalars $v_1, \ldots, v_l$, not all zero, such that $\sum_{i=1}^{l} v_i \nabla h_i(\bar{\mathbf{x}}) = \mathbf{0}$. Letting u_0, u_i for $i \in I$ be equal to zero, the conditions of the first part of the theorem hold trivial.

Now suppose that $\nabla h_i(\bar{\mathbf{x}})$ for $i = 1, \ldots, l$ are linearly independent. Let $\mathbf{A}_1$ be the matrix whose rows are $\nabla f(\bar{\mathbf{x}})^t$ and $\nabla g_i(\bar{\mathbf{x}})^t$ for $i \in I$, and let $\mathbf{A}_2$ be the matrix whose rows are $\nabla h_i(\bar{\mathbf{x}})^t$ for $i = 1, \ldots, l$. Then, from Theorem 4.3.1, local optimality of $\bar{\mathbf{x}}$ implies that the system

$$\mathbf{A}_1\mathbf{d} < \mathbf{0} \qquad \mathbf{A}_2\mathbf{d} = \mathbf{0}$$

is inconsistent. Now consider the following two sets:

$$S_1 = \{(\mathbf{z}_1, \mathbf{z}_2) : \mathbf{z}_1 = \mathbf{A}_1\mathbf{d}, \mathbf{z}_2 = \mathbf{A}_2\mathbf{d}\}$$

$$S_2 = \{(\mathbf{z}_1, \mathbf{z}_2) : \mathbf{z}_1 < \mathbf{0}, \mathbf{z}_2 = \mathbf{0}\}$$

Note that S_1 and S_2 are nonempty convex sets such that $S_1 \cap S_2 = \varnothing$. Then, by Theorem 2.4.8, there exists a nonzero vector $\mathbf{p}^t = (\mathbf{p}_1^t, \mathbf{p}_2^t)$ such that

$$\mathbf{p}_1^t\mathbf{A}_1\mathbf{d} + \mathbf{p}_2^t\mathbf{A}_2\mathbf{d} \geq \mathbf{p}_1^t\mathbf{z}_1 + \mathbf{p}_2^t\mathbf{z}_2 \qquad \text{for each } \mathbf{d} \in E_n \text{ and } (\mathbf{z}_1, \mathbf{z}_2) \in \text{cl } S_2$$

Letting $\mathbf{z}_2 = \mathbf{0}$ and since each component of $\mathbf{z}_1$ can be made an arbitrarily large negative number, it follows that $\mathbf{p}_1 \geq \mathbf{0}$. Also, letting $(\mathbf{z}_1, \mathbf{z}_2) = (\mathbf{0}, \mathbf{0})$, we must have $(\mathbf{p}_1^t\mathbf{A}_1 + \mathbf{p}_2^t\mathbf{A}_2)\mathbf{d} \geq \mathbf{0}$ for each $\mathbf{d} \in E_n$. Letting $\mathbf{d} = -(\mathbf{A}_1^t\mathbf{p}_1 + \mathbf{A}_2^t\mathbf{p}_2)$, it follows that $-\|\mathbf{A}_1^t\mathbf{p}_1 + \mathbf{A}_2^t\mathbf{p}_2\|^2 \geq 0$, and thus $\mathbf{A}_1^t\mathbf{p}_1 + \mathbf{A}_2^t\mathbf{p}_2 = \mathbf{0}$.

To summarize, we have shown that there exists a nonzero vector $\mathbf{p}^t = (\mathbf{p}_1^t, \mathbf{p}_2^t)$ with $\mathbf{p}_1 \geq \mathbf{0}$ such that $\mathbf{A}_1^t\mathbf{p}_1 + \mathbf{A}_2^t\mathbf{p}_2 = \mathbf{0}$. Denoting the components of $\mathbf{p}_1$ by u_0 and u_i for $i \in I$, and letting $\mathbf{p}_2 = \mathbf{v}$, the first result follows. The equivalent form of the necessary conditions is readily obtained by letting $u_i = 0$ for $i \notin I$, and the proof is complete.

The reader may note that the Lagrangian multiplier v_i associated with the ith equality constraints is unrestricted in sign. Note also that these conditions are *not* equivalently obtained by writing each equality as two associated inequalities and then applying the FJ conditions for the inequality constrained case. The FJ conditions can also be written in vector notation as follows:

$$u_0 \nabla f(\bar{\mathbf{x}}) + \nabla \mathbf{g}(\bar{\mathbf{x}})^t\mathbf{u} + \nabla \mathbf{h}(\bar{\mathbf{x}})^t\mathbf{v} = \mathbf{0}$$

$$\mathbf{u}^t\mathbf{g}(\bar{\mathbf{x}}) = 0$$

$$(u_0, \mathbf{u}) \geq (0, \mathbf{0})$$

$$(u_0, \mathbf{u}, \mathbf{v}) \neq (0, \mathbf{0}, \mathbf{0})$$

Here, $\nabla \mathbf{g}(\bar{\mathbf{x}})$ is an $m \times n$ Jacobian matrix whose ith row is $\nabla g_i(\bar{\mathbf{x}})^t$, and $\nabla \mathbf{h}(\bar{\mathbf{x}})$ is an $l \times n$ Jacobian matrix whose ith row is $\nabla h_i(\bar{\mathbf{x}})^t$. Also, $\mathbf{u}$ and $\mathbf{v}$ are, respectively, an m vector and an l vector, denoting the Lagrangian multipliers associated with the inequality and equality constraints.

4.3.3 Example

$$\begin{array}{ll} \text{Minimize} & x_1^2 + x_2^2 \\ \text{subject to} & x_1^2 + x_2^2 \le 5 \\ & -x_1 \le 0 \\ & -x_2 \le 0 \\ & x_1 + 2x_2 = 4 \end{array}$$

Here, we have only one equality constraint. We verify below that the Fritz John conditions are true at the optimal point $\bar{\mathbf{x}} = (\frac{4}{5}, \frac{8}{5})^t$. First, note that there are no binding inequality constraints at $\bar{\mathbf{x}}$; that is, $I = \varnothing$. Hence, the multipliers associated with the inequality constraints are equal to zero. Note that

$$\nabla f(\bar{\mathbf{x}}) = \left(\tfrac{8}{5}, \tfrac{16}{5}\right)^t \quad \text{and} \quad \nabla h_1(\bar{\mathbf{x}}) = (1, 2)^t$$

Thus,

$$u_0 \begin{pmatrix} \frac{8}{5} \\ \frac{16}{5} \end{pmatrix} + v_1 \begin{pmatrix} 1 \\ 2 \end{pmatrix} = \begin{pmatrix} 0 \\ 0 \end{pmatrix}$$

is satisfied, for example, by $u_0 = 5$ and $v_1 = -8$.

4.3.4 Example

$$\begin{array}{ll} \text{Minimize} & (x_1 - 3)^2 + (x_2 - 2)^2 \\ \text{subject to} & x_1^2 + x_2^2 \le 5 \\ & -x_1 \le 0 \\ & -x_2 \le 0 \\ & x_1 + 2x_2 = 4 \end{array}$$

This example is the same as Example 4.2.9, with the inequality constraint $x_1 + 2x_2 \le 4$ replaced by $x_1 + 2x_2 = 4$. At the optimal point $\bar{\mathbf{x}} = (2, 1)^t$, we have only one inequality constraint $x_1^2 + x_2^2 \le 5$ binding. The Fritz John condition

$$u_0 \begin{pmatrix} -2 \\ -2 \end{pmatrix} + u_1 \begin{pmatrix} 4 \\ 2 \end{pmatrix} + v_1 \begin{pmatrix} 1 \\ 2 \end{pmatrix} = \begin{pmatrix} 0 \\ 0 \end{pmatrix}$$

is satisfied, for example, by $u_0 = 3$, $u_1 = 1$, and $v_1 = 2$.

4.3.5 Example

$$\begin{array}{ll} \text{Minimize} & -x_1 \\ \text{subject to} & x_2 - (1 - x_1)^3 = 0 \\ & -x_2 - (1 - x_1)^3 = 0 \end{array}$$

As shown in Figure 4.12, this problem has only one feasible point, namely, $\bar{\mathbf{x}} = (1, 0)^t$. At this point, we have

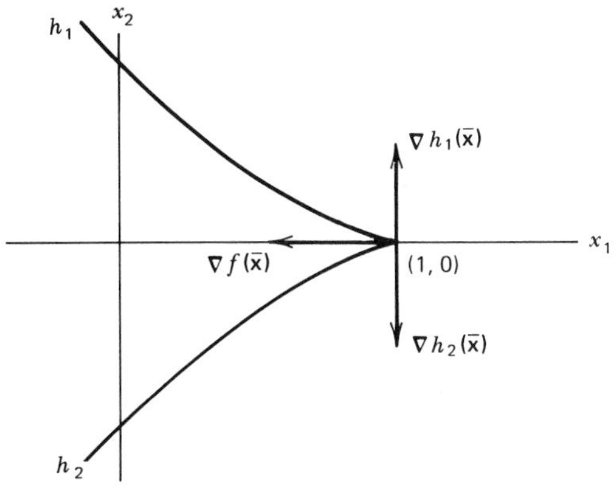

Figure 4.12 Illustration of Example 4.3.5.

$$\nabla f(\bar{\mathbf{x}}) = (-1, 0)^t \qquad \nabla h_1(\bar{\mathbf{x}}) = (0, 1)^t \qquad \nabla h_2(\bar{\mathbf{x}}) = (0, -1)^t$$

The condition

$$u_0\begin{pmatrix}-1\\0\end{pmatrix} + v_1\begin{pmatrix}0\\1\end{pmatrix} + v_2\begin{pmatrix}0\\-1\end{pmatrix} = \begin{pmatrix}0\\0\end{pmatrix}$$

is true only if $u_0 = 0$ and $v_1 = v_2 = \alpha$, where α is any scalar. Thus, the Fritz John necessary conditions are met at the point $\bar{\mathbf{x}}$.

Similar to Theorem 4.2.12, we now provide a set of sufficient conditions that enable us to guarantee that a FJ point is a local minimum. Again, as with Theorem 4.2.12, several variations of such sufficient conditions are possible. We state the result below using one such condition motivated by the converse to Theorem 4.3.1, and ask the reader to explore other conditions in Exercise 4.39.

4.3.6 Theorem (Fritz John Sufficient Conditions)

Let X be a nonempty open set in E_n, and let $f: E_n \to E_1$, $g_i: E_n \to E_1$, $i = 1, \ldots, m$, and $h_i: E_n \to E_1$, $i = 1, \ldots, l$. Consider Problem P to

$$\begin{array}{lll} \text{Minimize} & f(\mathbf{x}) & \\ \text{subject to} & g_i(\mathbf{x}) \le 0 & \text{for } i = 1, \ldots, m \\ & h_i(\mathbf{x}) = 0 & \text{for } i = 1, \ldots, l \\ & \mathbf{x} \in X & \end{array}$$

Let $\bar{\mathbf{x}}$ be a FJ solution and denote $I = \{i: g_i(\bar{\mathbf{x}}) = 0\}$. Define $S = \{\mathbf{x}: g_i(\mathbf{x}) \le 0 \text{ for } i \in I,\ h_i(\mathbf{x}) = 0, \text{ for } i = 1, \ldots, l\}$. If h_i for $i = 1, \ldots, l$ are affine and $\nabla h_i(\bar{\mathbf{x}})$, $i = 1, \ldots, l$ are linearly independent, and if there exists an ε-neighborhood $N_\varepsilon(\bar{\mathbf{x}})$ of $\bar{\mathbf{x}}$, $\varepsilon > 0$, such that f is pseudoconvex on $S \cap N_\varepsilon(\bar{\mathbf{x}})$, and g_i for $i \in I$ are strictly pseudoconvex over $S \cap N_\varepsilon(\bar{\mathbf{x}})$, then $\bar{\mathbf{x}}$ is a local minimum for Problem P.

Proof

Let us first show that $F_0 \cap G_0 \cap H_0 = \varnothing$, where these sets are as defined in Theorem 4.3.1. On the contrary, suppose that there exists a solution $\mathbf{d} \in F_0 \cap G_0 \cap H_0$. Then, by taking the inner product of the dual feasibility condition $u_0\nabla f(\bar{\mathbf{x}}) + \sum_{i\in I} u_i \nabla g_i(\bar{\mathbf{x}}) + \sum_{i=1}^{l} v_i \nabla h_i(\bar{\mathbf{x}}) = \mathbf{0}$ with $\mathbf{d}$, we obtain $u_0\nabla f(\bar{\mathbf{x}})^t\mathbf{d} + \sum_{i\in I} u_i \nabla g_i(\bar{\mathbf{x}})^t\mathbf{d} = 0$, since $\mathbf{d} \in H_0$. But $\mathbf{d} \in F_0 \cap G_0$ and $(u_0, u_i \text{ for } i \in I) \geq \mathbf{0}$ implies that $(u_0, u_i$ for $i \in I) = (0, \mathbf{0})$. Since $\bar{\mathbf{x}}$ is a FJ point, we therefore must have a solution to the system $\sum_{i=1}^{l} v_i \nabla h_i(\bar{\mathbf{x}}) = 0$, $\mathbf{v} \neq \mathbf{0}$, which contradicts the linear independence of $\nabla h_i(\bar{\mathbf{x}})$ for $i = 1, \ldots, l$. Hence, $F_0 \cap G_0 \cap H_0 = \emptyset$.

Now, closely following the proof to the converse statement of Theorem 4.3.1, and restricting attention to $S \cap N_\varepsilon(\bar{\mathbf{x}})$, we can conclude that $\bar{\mathbf{x}}$ is a local minimum for P. This completes the proof.

The Karush–Kuhn–Tucker Conditions

In the Fritz John conditions, the Lagrangian multiplier associated with the objective function is not necessarily positive. Under further assumptions on the constraint set, one can claim that at any local minimum, there exists a set of Lagrange multipliers for which u_0 is positive. In Theorem 4.3.7 below, we obtain a generalization of the KKT necessary optimality conditions of Theorem 4.2.13. This is done by imposing a qualification on the gradients of the equality and binding inequality constraints that ensure that $u_0 > 0$ necessarily holds in the Fritz John conditions. Other qualifications on the constraints to ensure the existence of $u_0 > 0$ in the FJ conditions at a local minimum are discussed in Chapter 5.

4.3.7 Theorem (Karush–Kuhn–Tucker Necessary Conditions)

Let X be a nonempty open set in E_n, and let $f: E_n \to E_1$, $g_i: E_n \to E_1$ for $i = 1, \ldots, m$, and $h_i: E_n \to E_1$ for $i = 1, \ldots, l$. Consider Problem P to

$$
\begin{array}{lll}
\text{Minimize} & f(\mathbf{x}) & \\
\text{subject to} & g_i(\mathbf{x}) \leq 0 & \text{for } i = 1, \ldots, m \\
& h_i(\mathbf{x}) = 0 & \text{for } i = 1, \ldots, l \\
& \mathbf{x} \in X &
\end{array}
$$

Let $\bar{\mathbf{x}}$ be a feasible solution, and let $I = \{i : g_i(\bar{\mathbf{x}}) = 0\}$. Suppose that f and g_i for $i \in I$ are differentiable at $\bar{\mathbf{x}}$, that each g_i for $i \,\mathrm{L}\, I$ is continuous at $\bar{\mathbf{x}}$, and that each h_i for $i = 1, \ldots, l$ is continuously differentiable at $\bar{\mathbf{x}}$. Further, suppose that $\nabla g_i(\bar{\mathbf{x}})$ for $i \in I$ and $\nabla h_i(\bar{\mathbf{x}})$ for $i = 1, \ldots, l$ are linearly independent. (Such an $\bar{\mathbf{x}}$ is sometimes called *regular*). If $\bar{\mathbf{x}}$ solves Problem P locally, then unique scalars u_i for $i \in I$ and v_i for $i = 1, \ldots, l$ exist such that

$$
\nabla f(\bar{\mathbf{x}}) + \sum_{i \in I} u_i \nabla g_i(\bar{\mathbf{x}}) + \sum_{i=1}^{l} v_i \nabla h_i(\bar{\mathbf{x}}) = \mathbf{0}
$$

$$
u_i \geq 0 \qquad \text{for } i \in I
$$

In addition to the above assumptions, if each g_i for $i \,\mathrm{L}\, I$ is also differentiable at $\bar{\mathbf{x}}$, then the KKT conditions can be written in the following equivalent form:

$$
\nabla f(\bar{\mathbf{x}}) + \sum_{i=1}^{m} u_i \nabla g_i(\bar{\mathbf{x}}) + \sum_{i=1}^{l} v_i \nabla h_i(\bar{\mathbf{x}}) = \mathbf{0}
$$

$$
u_i g_i(\bar{\mathbf{x}}) = 0 \qquad \text{for } i = 1, \ldots, m
$$

$$
u_i \geq 0 \qquad \text{for } i = 1, \ldots, m
$$

Proof

By Theorem 4.3.2, there exist scalars u_0 and $\hat{u}_i$ for $i \in I$, and $\hat{v}_i$ for $i = 1, \ldots, l$, not all zero, such that

$$u_0\nabla f(\bar{\mathbf{x}}) + \sum_{i\in I} \hat{u}_i\nabla g_i(\bar{\mathbf{x}}) + \sum_{i=1}^{l} \hat{v}_i\nabla h_i(\bar{\mathbf{x}}) = \mathbf{0} \tag{4.15}$$

$$u_0, \hat{u}_i \geq 0 \qquad \text{for } i \in I$$

Note that $u_0 > 0$, because if $u_0 = 0$, then (4.15) would contradict the assumption of linear independence of $\nabla g_i(\bar{\mathbf{x}})$ for $i \in I$ and $\nabla h_i(\bar{\mathbf{x}})$ for $i = 1, \ldots, l$. The first result then follows by letting $u_i = \hat{u}_i/u_0$ for $i \in I$, and $v_i = \hat{v}_i/u_0$ for $i = 1, \ldots, l$, and noting that the linear independence assumption implies the uniqueness of these Lagrangian multipliers. The equivalent form of the necessary conditions follows by letting $u_i = 0$ for $i \notin I$. This completes the proof.

Note that the KKT conditions of Theorem 4.3.7 can be written in vector form as follows:

$$\nabla f(\bar{\mathbf{x}}) + \nabla \mathbf{g}(\bar{\mathbf{x}})^t\mathbf{u} + \nabla \mathbf{h}(\bar{\mathbf{x}})^t\mathbf{v} = \mathbf{0}$$

$$\mathbf{u}^t\mathbf{g}(\bar{\mathbf{x}}) = 0$$

$$\mathbf{u} \geq \mathbf{0}$$

Here, $\nabla \mathbf{g}(\bar{\mathbf{x}})$ is an $m \times n$ Jacobian matrix and $\nabla \mathbf{h}(\bar{\mathbf{x}})$ is an $l \times n$ matrix whose ith rows, respectively, are $\nabla g_i(\bar{\mathbf{x}})^t$ and $\nabla h_i(\bar{\mathbf{x}})^t$. The vectors $\mathbf{u}$ and $\mathbf{v}$ are the Lagrangian multiplier vectors.

The reader might have observed that the KKT conditions of Theorem 4.3.6 are precisely the KKT conditions of the inequality case given in Theorem 4.2.13 when each equality constraint $h_i(\mathbf{x}) = 0$ is replaced by the two equivalent inequalities $h_i(\mathbf{x}) \leq 0$ and $-h_i(\mathbf{x}) \leq 0$, for $i = 1, \ldots, l$. Denoting v_i^+ and v_i^- as the nonnegative Lagrangian multipliers associated with the latter two inequalities and using the KKT conditions for the inequality case produces the KKT conditions of Theorem 4.3.7 on replacing the difference $v_i^+ - v_i^-$ of two nonnegative variables by the unrestricted variables v_i for each $i = 1, \ldots, l$. In fact, writing the equalities as equivalent inequalities, the sets G_0' and G' defined in Theorem 4.2.15 become, respectively, $G_0' \cap H_0$ and $G' \cap H_0$. Theorem 4.2.15 then asserts that for Problem P of the present section,

$$\bar{\mathbf{x}} \text{ is a KKT solution} \Leftrightarrow F_0 \cap G_0' \cap H_0 = \emptyset \Leftrightarrow F_0 \cap G' \cap H_0 = \emptyset \tag{4.16}$$

Moreover, this happens if and only if $\bar{\mathbf{x}}$ solves the first-order linear programming approximation LP($\bar{\mathbf{x}}$) at the point $\bar{\mathbf{x}}$ given by

$$LP(\bar{\mathbf{x}})\text{: minimize } \{f(\bar{\mathbf{x}}) + \nabla f(\bar{\mathbf{x}})^t(\mathbf{x} - \bar{\mathbf{x}}) : g_i(\bar{\mathbf{x}}) + \nabla g_i(\bar{\mathbf{x}})^t(\mathbf{x} - \bar{\mathbf{x}}) \leq 0$$

$$\text{for } i = 1, \ldots, m, \nabla h_i(\bar{\mathbf{x}})^t(\mathbf{x} - \bar{\mathbf{x}}) = 0 \qquad \text{for } i = 1, \ldots, l\} \tag{4.17}$$

Now consider Examples 4.3.3, 4.3.4, and 4.3.5. In Example 4.3.3, the reader can verify that $u_1 = u_2 = u_3 = 0$ and $v_1 = -\frac{8}{5}$ satisfy the KKT conditions at $\bar{\mathbf{x}} = (\frac{4}{5}, \frac{8}{5})^t$. In Example 4.3.4, the values of the multipliers satisfying the KKT conditions at $\bar{\mathbf{x}} = (2, 1)^t$ are

$$u_1 = \tfrac{1}{3} \qquad u_2 = u_3 = 0 \qquad v_1 = \tfrac{2}{3}$$

Finally, Example 4.3.5 does not satisfy the constraint qualification of Theorem 4.3.7 at $\bar{\mathbf{x}} = (1, 0)^t$, since $\nabla h_1(\bar{\mathbf{x}})$ and $\nabla h_2(\bar{\mathbf{x}})$ are linearly dependent. In fact, no constraint qualification (known or unknown!) can hold true at this point $\bar{\mathbf{x}}$ because it is not a KKT point. The first-order linear programming feasible region LP($\bar{\mathbf{x}}$) is given by the entire x_1 axis; and unless $\nabla f(\bar{\mathbf{x}})$ is orthogonal to this axis, $\bar{\mathbf{x}}$ is not an optimal solution for LP($\bar{\mathbf{x}}$).

Theorem 4.3.8 below shows that, under rather mild convexity assumptions on f, g_i and h_i, the KKT conditions are also sufficient for local optimality. Again, we fashion this result following Theorem 4.2.16 and ask the reader to investigate other variations in Exercises 4.39 and 4.49.

4.3.8 Theorem (Karush–Kuhn–Tucker Sufficient Conditions)

Let X be a nonempty open set in E_n, and let $f:E_n \to E_1$, $g_i:E_n \to E_1$ for $i = 1, \ldots, m$, and $h_i:E_n \to E_1$ for $i = 1, \ldots, l$. Consider Problem P to

$$
\begin{array}{lll}
\text{Minimize} & f(\mathbf{x}) & \\
\text{subject to} & g_i(\mathbf{x}) \le 0 & \text{for } i = 1, \ldots, m \\
 & h_i(\mathbf{x}) = 0 & \text{for } i = 1, \ldots, l \\
 & \mathbf{x} \in X &
\end{array}
$$

Let $\bar{\mathbf{x}}$ be a feasible solution, and let $I = \{i : g_i(\bar{\mathbf{x}}) = 0\}$. Suppose that the KKT conditions hold at $\bar{\mathbf{x}}$, that is, there exist scalars $\bar{u}_i \ge 0$ for $i \in I$ and $\bar{v}_i$ for $i = 1, \ldots, l$ such that

$$\nabla f(\bar{\mathbf{x}}) + \sum_{i \in I} \bar{u}_i \nabla g_i(\bar{\mathbf{x}}) + \sum_{i=1}^{l} \bar{v}_i \nabla h_i(\bar{\mathbf{x}}) = \mathbf{0} \tag{4.18}$$

Let $J = \{i : \bar{v}_i > 0\}$ and $K = \{i : \bar{v}_i < 0\}$. Further, suppose that f is pseudoconvex at $\bar{\mathbf{x}}$, g_i is quasiconvex at $\bar{\mathbf{x}}$ for $i \in I$, h_i is quasiconvex at $\bar{\mathbf{x}}$ for $i \in J$ and h_i is quasiconcave at $\bar{\mathbf{x}}$ for $i \in K$. Then, $\bar{\mathbf{x}}$ is a global optimal solution to Problem P. In particular, if the generalized convexity assumptions on the objective and constraint functions are restricted to the domain $N_\varepsilon(\bar{\mathbf{x}})$ for some $\varepsilon > 0$, then $\bar{\mathbf{x}}$ is a local minimum for P.

Proof

Let $\mathbf{x}$ be any feasible solution to Problem P. (In case the generalized convexity assumptions hold true only by restricting the domain of the objective and constraint functions to $N_\varepsilon(\bar{\mathbf{x}})$, then let $\mathbf{x}$ be any feasible solution to Problem P that also lies within $N_\varepsilon(\bar{\mathbf{x}})$.)

Then, for $i \in I$, $g_i(\mathbf{x}) \le g_i(\bar{\mathbf{x}})$, since $g_i(\mathbf{x}) \le 0$ and $g_i(\bar{\mathbf{x}}) = 0$. By the quasiconvexity of g_i at $\bar{\mathbf{x}}$ it follows that

$$g_i(\bar{\mathbf{x}} + \lambda(\mathbf{x} - \bar{\mathbf{x}})) = g_i(\lambda\mathbf{x} + (1 - \lambda)\bar{\mathbf{x}}) \le \text{maximum } (g_i(\mathbf{x}), g_i(\bar{\mathbf{x}})) = g_i(\bar{\mathbf{x}})$$

for all $\lambda \in (0, 1)$. This implies that g_i does not increase by moving from $\bar{\mathbf{x}}$ along the direction $\mathbf{x} - \bar{\mathbf{x}}$. Thus, by Theorem 4.1.2, we must have

$$\nabla g_i(\bar{\mathbf{x}})^t(\mathbf{x} - \bar{\mathbf{x}}) \le 0 \qquad \text{for } i \in I \tag{4.19}$$

Similarly, since h_i is quasiconvex at $\bar{\mathbf{x}}$ for $i \in J$, and h_i is quasiconcave at $\bar{\mathbf{x}}$ for $i \in K$, we have

$$\nabla h_i(\bar{\mathbf{x}})^t(\mathbf{x} - \bar{\mathbf{x}}) \le 0 \qquad \text{for } i \in J \tag{4.20}$$

$$\nabla h_i(\bar{\mathbf{x}})^t(\mathbf{x} - \bar{\mathbf{x}}) \ge 0 \qquad \text{for } i \in K \tag{4.21}$$

Multiplying (4.19), (4.20), and (4.21) by $\bar{u}_i \ge 0$, $\bar{v}_i > 0$, and $\bar{v}_i < 0$, respectively, and adding, we get

$$\left[\sum_{i\in I} \bar{u}_i \nabla g_i(\bar{\mathbf{x}}) + \sum_{i\in J\cup K} \bar{v}_i \nabla h_i(\bar{\mathbf{x}})\right]^t (\mathbf{x} - \bar{\mathbf{x}}) \le 0 \tag{4.22}$$

Multiplying (4.18) by $\mathbf{x} - \bar{\mathbf{x}}$, and noting that $\bar{v}_i = 0$ for $i \notin J \cup K$, then (4.22) implies that

$$\nabla f(\bar{\mathbf{x}})^t(\mathbf{x} - \bar{\mathbf{x}}) \ge 0$$

By the pseudoconvexity of f at $\bar{\mathbf{x}}$, we must have $f(\mathbf{x}) \ge f(\bar{\mathbf{x}})$, and the proof is complete.

It is instructive to note that, as is evident from Theorem 4.3.8 and its proof, the equality constraints with positive Lagrangian multipliers at $\bar{\mathbf{x}}$ can be replaced by "less than or equal to" type constraints, and those with negative Lagrangian multipliers at $\bar{\mathbf{x}}$ can be replaced by "greater than or equal to" type constraints, whereas those with zero Lagrangian multipliers can be deleted, and $\bar{\mathbf{x}}$ will still remain a KKT solution for this relaxed problem P′, say. Hence, noting Theorem 4.2.16, the generalized convexity assumptions of Theorem 4.3.8 imply that $\bar{\mathbf{x}}$ is optimal to the relaxed problem P′, and being feasible to P, it is optimal for P (globally or locally as the case might be). This argument provides an alternate simpler proof for Theorem 4.3.8 based on Theorem 4.2.16. Moreover, it asserts that under generalized convexity assumptions, the sign of the Lagrangian multipliers can be used to assess whether an equality constraint is effectively behaving as a "less than or equal to" or "greater than or equal to" constraint.

Two points of caution are worth noting here in connection with the foregoing relaxation P′ of P. First, under the (generalized) convexity assumptions, deleting an equality constraint with a zero Lagrangian multiplier can create alternative optimal solutions that are not feasible to the original problem. For example, in the problem to minimize $\{x_1 : x_1 \ge 0 \text{ and } x_2 = 1\}$, the Lagrangian multiplier associated with the equality at the unique optimum $\bar{\mathbf{x}} = (0, 1)^t$ is zero. However, deleting this constraint produces an infinite number of alternative optimal solutions.

Second, for the nonconvex case, note that even if $\bar{\mathbf{x}}$ is optimal for P, it may not even be a local optimum for P′, although it remains a KKT point for P′. For example, consider the problem to minimize $(-x_1^2 - x_2^2)$ subject to $x_1 = 0$ and $x_2 = 0$. The unique optimum is obviously $\bar{\mathbf{x}} = (0, 0)^t$, and the Lagrangian multipliers associated with the constraints at $\bar{\mathbf{x}}$ are both zeros. However, deleting either of the two constraints, or even replacing either with a "less than or equal to" or "greater than or equal to" inequality, will make the problem unbounded. In general, the reader should bear in mind that deleting even nonbinding constraints for nonconvex problems can change the optimality status of a solution. Figure 4.13 illustrates one such situation. Here, $g_2(\mathbf{x}) \le 0$ is nonbinding at the optimum $\bar{\mathbf{x}}$; but deleting it changes the global optimum to the point $\hat{\mathbf{x}}$, leaving $\bar{\mathbf{x}}$ as only a local minimum. (See Exercise 4.40 for an instance in which the optimum does not even remain locally optimum after deleting a nonbinding constraint.)

Alternative Forms of the Karush–Kuhn–Tucker Conditions for General Problems

Consider the problem to minimize $f(\mathbf{x})$ subject to $g_i(\mathbf{x}) \le 0$ for $i = 1, \ldots, m$, $h_i(\mathbf{x}) = 0$ for $i = 1, \ldots, l$, and $\mathbf{x} \in X$, where X is an open set in E_n. We have derived above the following necessary conditions of optimality at a feasible point $\bar{\mathbf{x}}$:

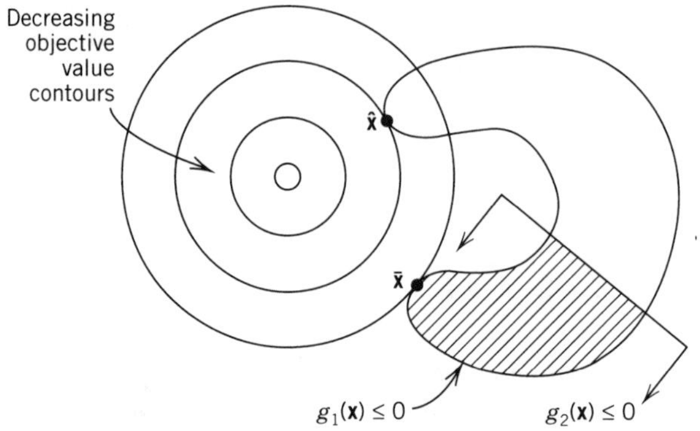

Figure 4.13 Caution on deleting nonbinding constraints for nonconvex problems.

$$\nabla f(\bar{\mathbf{x}}) + \sum_{i=1}^{m} u_i \nabla g_i(\bar{\mathbf{x}}) + \sum_{i=1}^{l} v_i \nabla h_i(\bar{\mathbf{x}}) = \mathbf{0}$$

$$u_i g_i(\bar{\mathbf{x}}) = 0 \qquad \text{for } i = 1, \ldots, m$$

$$u_i \ge 0 \qquad \text{for } i = 1, \ldots, m$$

Some authors prefer to use the multipliers $\lambda_i = -u_i \le 0$ and $\mu_i = -v_i$. In this case, the KKT conditions can be written as follows:

$$\nabla f(\bar{\mathbf{x}}) - \sum_{i=1}^{m} \lambda_i \nabla g_i(\bar{\mathbf{x}}) - \sum_{i=1}^{l} \mu_i \nabla h_i(\bar{\mathbf{x}}) = \mathbf{0}$$

$$\lambda_i g_i(\bar{\mathbf{x}}) = 0 \qquad \text{for } i = 1, \ldots, m$$

$$\lambda_i \le 0 \qquad \text{for } i = 1, \ldots, m$$

Now consider the problem to minimize $f(\mathbf{x})$ subject to $g_i(\mathbf{x}) \le 0$ for $i = 1, \ldots, m_1$, $g_i(\mathbf{x}) \ge 0$ for $i = m_1 + 1, \ldots, m$, $h_i(\mathbf{x}) = 0$ for $i = 1, \ldots, l$, and $\mathbf{x} \in X$, where X is an open set in E_n. Clearly, one can write $g_i(\mathbf{x}) \ge 0$ for $i = m_1 + 1, \ldots, m$ as $-g_i(\mathbf{x}) \le 0$ for $i = m_1 + 1, \ldots, m$, and use the results of Theorem 4.3.7. Hence, the necessary conditions for this problem can be expressed as follows:

$$\nabla f(\bar{\mathbf{x}}) + \sum_{i=1}^{m} u_i \nabla g_i(\bar{\mathbf{x}}) + \sum_{i=1}^{l} v_i \nabla h_i(\bar{\mathbf{x}}) = \mathbf{0}$$

$$u_i g_i(\bar{\mathbf{x}}) = 0 \qquad \text{for } i = 1, \ldots, m$$

$$u_i \ge 0 \qquad \text{for } i = 1, \ldots, m_1$$

$$u_i \le 0 \qquad \text{for } i = m_1 + 1, \ldots, m$$

We now consider problems of the type to minimize $f(\mathbf{x})$ subject to $g_i(\mathbf{x}) \le 0$ for $i = 1, \ldots, m$, $h_i(\mathbf{x}) = 0$ for $i = 1, \ldots, l$, and $\mathbf{x} \ge \mathbf{0}$. Such problems with nonnegativity restrictions on the variables frequently arise in practice. Clearly, the KKT conditions discussed earlier would apply as usual. However, it is sometimes convenient to eliminate the Lagrangian multipliers associated with $\mathbf{x} \ge \mathbf{0}$. The conditions then reduce to

$$\nabla f(\bar{\mathbf{x}}) + \sum_{i=1}^{m} u_i \nabla g_i(\bar{\mathbf{x}}) + \sum_{i=1}^{l} v_i \nabla h_i(\bar{\mathbf{x}}) \geq \mathbf{0}$$

$$\left[\nabla f(\bar{\mathbf{x}}) + \sum_{i=1}^{m} u_i \nabla g_i(\bar{\mathbf{x}}) + \sum_{i=1}^{l} v_i \nabla h_i(\bar{\mathbf{x}})\right]^t \mathbf{x} = 0$$

$$u_i g_i(\bar{\mathbf{x}}) = 0 \qquad \text{for } i = 1, \ldots, m$$

$$u_i \geq 0 \qquad \text{for } i = 1, \ldots, m$$

Finally, consider the problem to *maximize* $f(\mathbf{x})$ subject to $g_i(\mathbf{x}) \leq 0$ for $i = 1, \ldots, m_1$, $g_i(\mathbf{x}) \geq 0$ for $i = m_1 + 1, \ldots, m$, $h_i(\mathbf{x}) = 0$ for $i = 1, \ldots, l$, and $\mathbf{x} \in X$, where X is an open set in E_n. The necessary conditions for optimality can be written as follows:

$$\nabla f(\bar{\mathbf{x}}) + \sum_{i=1}^{m} u_i \nabla g_i(\bar{\mathbf{x}}) + \sum_{i=1}^{l} v_i \nabla h_i(\bar{\mathbf{x}}) = \mathbf{0}$$

$$u_i g_i(\bar{\mathbf{x}}) = 0 \qquad \text{for } i = 1, \ldots, m$$

$$u_i \leq 0 \qquad \text{for } i = 1, \ldots, m_1$$

$$u_i \geq 0 \qquad \text{for } i = m_1 + 1, \ldots, m$$

4.4 Second-Order Necessary and Sufficient Conditions for Constrained Problems

In Section 4.1, we considered the unconstrained problem of minimizing $f(\mathbf{x})$ subject to $\mathbf{x} \in E_n$ and, assuming differentiability, we derived the first-order necessary optimality condition that $\nabla f(\bar{\mathbf{x}}) = \mathbf{0}$ at all local optimal solutions $\bar{\mathbf{x}}$. However, when $\nabla f(\bar{\mathbf{x}}) = \mathbf{0}$, $\bar{\mathbf{x}}$ can be a local minimum, a local maximum, or a saddle point. To further reduce the candidate set of solutions produced by this first-order necessary optimality condition, and to assess the local optimality status of a given candidate solution, we developed second-order (and higher) necessary and/or sufficient optimality conditions.

Over Sections 4.2 and 4.3, we have developed first-order necessary optimality conditions for constrained problems. In particular, assuming a suitable constraint qualification, we have derived the first-order necessary KKT optimality conditions. Based on various (generalized) convexity assumptions, we have provided sufficient conditions to guarantee that a given solution that satisfies the first-order optimality conditions is globally or locally optimum. Analogous to the unconstrained case, we now derive second-order necessary and sufficient optimality conditions for constrained problems.

Toward this end, let us introduce the concept of a *Lagrangian function*. Consider the problem

$$P\text{: minimize } \{f(\mathbf{x}) : \mathbf{x} \in S\} \tag{4.23a}$$

where $S = \{\mathbf{x} : g_i(\mathbf{x}) \leq 0 \text{ for } i = 1, \ldots, m,\ h_i(\mathbf{x}) = 0 \text{ for } i = 1, \ldots, l, \text{ and } \mathbf{x} \in X\}$ (4.23b)

Assume that f, g_i for $i = 1, \ldots, m$ and h_i for $i = 1, \ldots, l$ are all defined on $E_n \to E_1$ and are twice differentiable, and that X is a nonempty open set in E_n. The *Lagrangian function* for this problem is defined as

$$\phi(\mathbf{x}, \mathbf{u}, \mathbf{v}) = f(\mathbf{x}) + \sum_{i=l}^{m} u_i g_i(\mathbf{x}) + \sum_{i=1}^{l} v_i h_i(\mathbf{x}) \tag{4.24}$$

As we shall learn in Chapter 6, this function enables us to formulate a duality theory for nonlinear programming problems, akin to that for linear programming problems as expounded in Theorem 2.7.3 and its corollaries. Now, let $\bar{\mathbf{x}}$ be a KKT point for problem P, with associated Lagrangian multipliers $\bar{\mathbf{u}}$ and $\bar{\mathbf{v}}$ corresponding to the inequality and equality constraints, respectively. Conditioned on $\bar{\mathbf{u}}$ and $\bar{\mathbf{v}}$, define the *restricted Lagrangian function*

$$L(\mathbf{x}) \equiv \phi(\mathbf{x}, \bar{\mathbf{u}}, \bar{\mathbf{v}}) \equiv f(\mathbf{x}) + \sum_{i \in I} \bar{u}_i g_i(\mathbf{x}) + \sum_{i=1}^{l} \bar{v}_i h_i(\mathbf{x}) \tag{4.25}$$

where $I = \{i : g_i(\bar{\mathbf{x}}) = 0\}$ is the index set of the binding inequality constraints at $\bar{\mathbf{x}}$.

Observe that the dual feasibility condition

$$\nabla f(\bar{\mathbf{x}}) + \sum_{i \in I} \bar{u}_i \nabla g(\bar{\mathbf{x}}) + \sum_{i=1}^{l} \bar{v}_i \nabla h_i(\bar{\mathbf{x}}) = \mathbf{0} \tag{4.26}$$

in the KKT system is equivalent to the statement that the gradient $\nabla L(\bar{\mathbf{x}})$ of L at $\mathbf{x} = \bar{\mathbf{x}}$ vanishes. Moreover, we have

$$L(\mathbf{x}) \le f(\mathbf{x}) \qquad \text{for all } \mathbf{x} \in S, \qquad \text{while } L(\bar{\mathbf{x}}) = f(\bar{\mathbf{x}}) \tag{4.27}$$

because $h_i(\mathbf{x}) = 0$ for $i = 1, \ldots, l$ and $g_i(\mathbf{x}) \le 0$ for $i \in I$ for all $\mathbf{x} \in S$, while $\bar{u}_i g_i(\bar{\mathbf{x}}) = 0$ for $i \in I$, and $h_i(\bar{\mathbf{x}}) = 0$ for $i = 1, \ldots, l$. Hence, if $\bar{\mathbf{x}}$ turns out to be a (local) minimizer for L, then it will also be a (local) minimizer for Problem P. This is formalized below.

4.4.1 Lemma

Consider problem P as defined in (4.23), where the objective and constraint defining functions are all twice differentiable, and where X is a nonempty, open set in E_n. Suppose that $\bar{\mathbf{x}}$ is a KKT point for Problem P with Lagrangian multipliers $\bar{\mathbf{u}}$ and $\bar{\mathbf{v}}$ associated with the inequality and equality constraints, respectively. Define the restricted Lagrangian function L as in (4.25), and denote its Hessian as $\nabla^2 L$.

a. If $\nabla^2 L$ is positive semidefinite for all $\mathbf{x} \in S$, then $\bar{\mathbf{x}}$ is a global minimum for Problem P. On the other hand, if $\nabla^2 L$ is positive semidefinite for all $\mathbf{x} \in S \cap N_\varepsilon(\bar{\mathbf{x}})$ for some ε-neighborhood $N_\varepsilon(\bar{\mathbf{x}})$ about $\bar{\mathbf{x}}$, $\varepsilon > 0$, then $\bar{\mathbf{x}}$ is a local minimum for Problem P.
b. If $\nabla^2 L(\bar{\mathbf{x}})$ is positive definite, then $\bar{\mathbf{x}}$ is a strict local minimum for Problem P.

Proof

From (4.25) and (4.26), we have $\nabla L(\bar{\mathbf{x}}) = \mathbf{0}$. Hence, under the first condition of part a, we obtain, by the convexity of $L(\mathbf{x})$ over S, that $L(\bar{\mathbf{x}}) \le L(\mathbf{x})$ for all $\mathbf{x} \in S$; and thus, from (4.27), we get $f(\bar{\mathbf{x}}) = L(\bar{\mathbf{x}}) \le L(\mathbf{x}) \le f(\mathbf{x})$ for all $\mathbf{x} \in S$. Therefore, $\bar{\mathbf{x}}$ solves Problem P. By restricting attention to $S \cap N_\varepsilon(\bar{\mathbf{x}})$ in the second case of part a, we conclude similarly that $f(\bar{\mathbf{x}}) \le f(\mathbf{x})$ for all $\mathbf{x} \in S \cap N_\varepsilon(\bar{\mathbf{x}})$. This proves part a.

Similarly, if $\nabla^2 L(\bar{\mathbf{x}})$ is positive definite, then by Theorem 4.1.4, since $\nabla L(\bar{\mathbf{x}}) = \mathbf{0}$, we have that $\bar{\mathbf{x}}$ is a strict local minimum for L. Hence, from (4.27), we deduce that $f(\bar{\mathbf{x}}) = L(\bar{\mathbf{x}}) < L(\mathbf{x}) \le f(\mathbf{x})$ for all $\mathbf{x} \ne \bar{\mathbf{x}}$ in $S \cap N_\varepsilon(\bar{\mathbf{x}})$ for some ε-neighborhood $N_\varepsilon(\bar{\mathbf{x}})$ of $\bar{\mathbf{x}}$, and this completes the proof.

The above result is related to the *saddle-point optimality conditons* explored more fully in Chapter 6, which establish that a KKT solution $(\bar{\mathbf{x}}, \bar{\mathbf{u}}, \bar{\mathbf{v}})$ for which $\bar{\mathbf{x}}$ minimizes $L(\mathbf{x})$ subject to $\mathbf{x} \in S$ corresponds to a certain pair of primal and dual problems having no duality gap. Indeed, observe that the global (or local) optimality claims in Lemma 4.4.1 continue to hold true under the less restrictive assumption that $\bar{\mathbf{x}}$ globally (or locally) minimizes $L(\mathbf{x})$ over S. However, our choice of stating Lemma 4.4.1 as above is motivated by the following result, which asserts that $\mathbf{d}'\nabla^2 L(\bar{\mathbf{x}})\mathbf{d}$ needs to be positive only for $\mathbf{d}$ restricted to lie in a specified cone, rather than for all $\mathbf{d} \in E_n$ as in Lemma 4.4.1b, for us to be able to claim that $\bar{\mathbf{x}}$ is a strict local minimum for P. In other words, this result is shown to hold whenever the Lagrangian function $L(\mathbf{x})$ displays a positive curvature at $\bar{\mathbf{x}}$ along directions restricted to the set given below.

4.4.2 Theorem (KKT Second-Order Sufficient Conditions)

Consider Problem P as defined in (4.23), where the objective and constraint defining functions are all twice differentiable, and where X is a nonempty, open set in E_n. Let $\bar{\mathbf{x}}$ be a KKT point for Problem P, with Lagrangian multipliers $\bar{\mathbf{u}}$ and $\bar{\mathbf{v}}$ associated with the inequality and equality constraints, respectively. Let $I = \{i : g_i(\bar{\mathbf{x}}) = 0\}$, and denote $I^+ = \{i \in I : \bar{u}_i > 0\}$, and $I^0 = \{i \in I : \bar{u}_i = 0\}$. ($I^+$ and I^0 are sometimes referred to as the set of *strongly active* and *weakly active* constraints, respectively.) Define the restricted Lagrangian function $L(\mathbf{x})$ as in (4.25), and denote its Hessian at $\bar{\mathbf{x}}$ by

$$\nabla^2 L(\bar{\mathbf{x}}) \equiv \nabla^2 f(\bar{\mathbf{x}}) + \sum_{i\in I} \bar{u}_i \nabla^2 g_i(\bar{\mathbf{x}}) + \sum_{i=1}^{l} \bar{v}_i \nabla^2 h_i(\bar{\mathbf{x}})$$

where $\nabla^2 f(\bar{\mathbf{x}})$, $\nabla^2 g_i(\bar{\mathbf{x}})$ for $i \in I$ and $\nabla^2 h_i(\bar{\mathbf{x}})$ for $i = 1, \ldots, l$ are the Hessians of f, g_i for $i \in I$, and h_i for $i = 1, \ldots, l$, respectively, all evaluated at $\bar{\mathbf{x}}$. Define the cone

$$\begin{aligned} C = \{\mathbf{d} \neq \mathbf{0} : \nabla g_i(\bar{\mathbf{x}})'\mathbf{d} &= 0 \quad \text{for } i \in I^+,\ \nabla g_i(\bar{\mathbf{x}})'\mathbf{d} \leq 0 \text{ for } i \in I^0, \\ \nabla h_i(\bar{\mathbf{x}})'\mathbf{d} &= 0 \quad \text{for } i = 1, \ldots, l\} \end{aligned}$$

Then, if $\mathbf{d}'\nabla^2 L(\bar{\mathbf{x}})\mathbf{d} > 0$ for all $\mathbf{d} \in C$, we have that $\bar{\mathbf{x}}$ is a strict local minimum for P.

Proof

Suppose that $\bar{\mathbf{x}}$ is not a strict local minimum. Then, as in Theorem 4.1.4, there exists a sequence $\{\mathbf{x}_k\}$ in S converging to $\bar{\mathbf{x}}$ such that $\mathbf{x}_k \neq \bar{\mathbf{x}}$ and $f(\mathbf{x}_k) \leq f(\bar{\mathbf{x}})$ for all k. Defining $\mathbf{d}_k = (\mathbf{x}_k - \bar{\mathbf{x}})/\|\mathbf{x}_k - \bar{\mathbf{x}}\|$ and $\lambda_k = \|\mathbf{x}_k - \bar{\mathbf{x}}\|$ for all k, we have $\mathbf{x}_k = \bar{\mathbf{x}} + \lambda_k \mathbf{d}_k$, where $\|\mathbf{d}_k\| = 1$ for all k, and we have $\{\lambda_k\} \rightarrow 0^+$ as $k \rightarrow \infty$. Since $\|\mathbf{d}_k\| = 1$ for all k, a convergent subsequence exists. Assume, without loss of generality, that the given sequence itself represents this convergent subsequence. Hence, $\{\mathbf{d}_k\} \rightarrow \mathbf{d}$, where $\|\mathbf{d}\| = 1$. Moreover, we have

$$0 \geq f(\bar{\mathbf{x}} + \lambda_k \mathbf{d}_k) - f(\bar{\mathbf{x}}) = \lambda_k \nabla f(\bar{\mathbf{x}})'\mathbf{d}_k + \tfrac{1}{2}\lambda_k^2 \mathbf{d}_k' \nabla^2 f(\bar{\mathbf{x}})\mathbf{d}_k + \lambda_k^2 \alpha_f(\bar{\mathbf{x}}; \lambda_k \mathbf{d}_k) \tag{4.28a}$$

$$\begin{aligned} 0 \geq g_i(\bar{\mathbf{x}} + \lambda_k \mathbf{d}_k) - g_i(\bar{\mathbf{x}}) &= \lambda_k \nabla g_i(\bar{\mathbf{x}})'\mathbf{d}_k \\ &+ \tfrac{1}{2}\lambda_k^2 \mathbf{d}_k' \nabla^2 g_i(\bar{\mathbf{x}})\mathbf{d}_k + \lambda_k^2 \alpha_{g_i}(\bar{\mathbf{x}}; \lambda_k \mathbf{d}_k) \qquad \text{for } i \in I \end{aligned} \tag{4.28b}$$

$$\begin{aligned} 0 \geq h_i(\bar{\mathbf{x}} + \lambda_k \mathbf{d}_k) - h_i(\bar{\mathbf{x}}) &= \lambda_k \nabla h_i(\bar{\mathbf{x}})'\mathbf{d}_k \\ &+ \tfrac{1}{2}\lambda_k^2 \mathbf{d}_k' \nabla^2 h_i(\bar{\mathbf{x}})\mathbf{d}_k + \lambda_k^2 \alpha_{h_i}(\bar{\mathbf{x}}; \lambda_k \mathbf{d}_k) \qquad \text{for } i = 1, \ldots, l \end{aligned} \tag{4.28c}$$

where α_f, α_{g_i} for $i \in I$ and α_{h_i} for $i = 1, \ldots, l$ approach zeros as $k \to \infty$. Dividing each expression in (4.28) by $\lambda_k > 0$ and taking limits as $k \to \infty$, we obtain

$$\nabla f(\bar{\mathbf{x}})^t\mathbf{d} \leq 0, \ \nabla g_i(\bar{\mathbf{x}})^t\mathbf{d} \leq 0 \quad \text{for } i \in I \quad \text{and} \quad \nabla h_i(\bar{\mathbf{x}})^t\mathbf{d} = 0 \quad \text{for } i = 1, \ldots, l \tag{4.29}$$

Now, since $\bar{\mathbf{x}}$ is a KKT point, we have $\nabla f(\bar{\mathbf{x}}) + \sum_{i \in I} \bar{u}_i \nabla g_i(\bar{\mathbf{x}}) + \sum_{i=1}^{l} \bar{v}_i \nabla h_i(\bar{\mathbf{x}}) = 0$. Taking the inner product of this with $\mathbf{d}$ and using (4.29), we conclude that

$$\begin{aligned} \nabla f(\bar{\mathbf{x}})^t\mathbf{d} = 0, \ \nabla g_i(\bar{\mathbf{x}})^t\mathbf{d} = 0 \quad \text{for } i \in I^+, \quad \nabla g_i(\bar{\mathbf{x}})^t\mathbf{d} \leq 0 \text{ for } i \in I^0 \\ \text{and} \quad \nabla h_i(\bar{\mathbf{x}})^t\mathbf{d} = 0 \quad \text{for } i = 1, \ldots, l \end{aligned} \tag{4.30}$$

Hence, in particular, $\mathbf{d} \in C$. Furthermore, multiplying each of (4.28b) by $\bar{u}_i$ for $i \in I$, and each of (4.28c) by $\bar{v}_i$ for $i = 1, \ldots, l$, and adding, we get, using $\nabla f(\bar{\mathbf{x}})^t\mathbf{d}_k + \sum_{i \in I} \bar{u}_i \nabla g_i(\bar{\mathbf{x}})^t\mathbf{d}_k + \sum_{i=1}^{l} \bar{v}_i \nabla h_i(\bar{\mathbf{x}})^t\mathbf{d}_k = 0$,

$$0 \geq \frac{\lambda_k^2}{2}\mathbf{d}_k^t\nabla^2 L(\bar{\mathbf{x}})\mathbf{d}_k + \lambda_k^2\left[\alpha_f(\bar{\mathbf{x}}; \lambda_k\mathbf{d}_k) + \sum_{i \in I} \bar{u}_i\alpha_{g_i}(\bar{\mathbf{x}}; \lambda_k\mathbf{d}_k) + \sum_{i=1}^{l} \bar{v}_i\alpha_{h_i}(\bar{\mathbf{x}}; \lambda_k\mathbf{d}_k)\right]$$

Dividing the above inequality by $\lambda_k^2 > 0$ and taking limits as $k \to \infty$, we obtain $\mathbf{d}^t\nabla^2 L(\bar{\mathbf{x}})\mathbf{d} \leq 0$, where $\|\mathbf{d}\| = 1$ and $\mathbf{d} \in C$. This is a contradiction. Therefore, $\bar{\mathbf{x}}$ must be a strict local minimum for Problem P, and the proof is complete.

Corollary

Consider Problem P as defined in the theorem, and let $\bar{\mathbf{x}}$ be a KKT point with associated Lagrangian multipliers $\bar{\mathbf{u}}$ and $\bar{\mathbf{v}}$ corresponding to the inequality and equality constraints, respectively. Furthermore, suppose that the collection $\nabla g_i(\bar{\mathbf{x}})$ for $i \in I^+ = \{i : \bar{u}_i > 0\}$ and $\nabla h_i(\bar{\mathbf{x}})$ for $i = 1, \ldots, l$ contains a set of n linearly independent vectors. Then, $\bar{\mathbf{x}}$ is a strict local minimum for P.

Proof

Under the linear independence condition of the corollary, we have $C = \varnothing$, and so Theorem 4.4.2 holds true vacuously by default. This completes the proof.

Several remarks concerning Theorem 4.4.2 above are in order at this point. First, observe that it might appear from the proof of the theorem that the result can be strengthened by further restricting the cone C to include the constraint $\nabla f(\bar{\mathbf{x}})^t\mathbf{d} = 0$. Although this is valid, it does not further restrict C, since when $\bar{\mathbf{x}}$ is a KKT point and $\mathbf{d} \in C$, we automatically have $\nabla f(\bar{\mathbf{x}})^t\mathbf{d} = 0$. Second, observe that if the problem is unconstrained, then Theorem 4.2.2 reduces to asserting that if $\nabla f(\bar{\mathbf{x}}) = \mathbf{0}$ and if $\nabla^2 f(\bar{\mathbf{x}}) \equiv H(\bar{\mathbf{x}})$ is positive definite, then $\bar{\mathbf{x}}$ is a strict local minimum. Hence, Theorem 4.1.4 is a special case of this result. Similarly, Lemma 4.4.1b is a special case of this result. Finally, observe that for linear programming problems, this sufficient condition does not necessarily hold, unless under the condition of the corollary, whence $\bar{\mathbf{x}}$ is a unique extreme point optimal solution.

We now turn our attention to the counterpart of Theorem 4.4.2 that deals with second-order necessary optimality conditions. Theorem 4.4.3 below shows that if $\bar{\mathbf{x}}$ is a local minimum, then, under a suitable second-order constraint qualification, it is a KKT point; and, moreover, $\mathbf{d}^t\nabla^2 L(\bar{\mathbf{x}})\mathbf{d} \geq 0$ for all $\mathbf{d}$ belonging to some cone $C' \supseteq C$ as defined in the theorem. The last statement indicates that the Lagrangian function L has a nonnegative curvature at $\bar{\mathbf{x}}$ along any direction in C'.

4.4.3 Theorem (KKT Second-Order Necessary Conditions)

Consider Problem P as defined in (4.23), where the objective and constraint defining functions are all twice-differentiable, and where X is a nonempty, open set in E_n. Let $\bar{\mathbf{x}}$ be a local minimum for Problem P, and denote $I = \{i: g_i(\bar{\mathbf{x}}) = 0\}$. Define the restricted Lagrangian function $L(\mathbf{x})$ as in (4.25), and denote its Hessian at $\bar{\mathbf{x}}$ by

$$\nabla^2 L(\bar{\mathbf{x}}) \equiv \nabla^2 f(\bar{\mathbf{x}}) + \sum_{i\in I} \bar{u}_i \nabla^2 g_i(\bar{\mathbf{x}}) + \sum_{i=1}^{l} \bar{v}_i \nabla^2 h_i(\bar{\mathbf{x}})$$

where $\nabla^2 f(\bar{\mathbf{x}})$, $\nabla^2 g_i(\bar{\mathbf{x}})$ for $i \in I$ and $\nabla^2 h_i(\bar{\mathbf{x}})$ for $i = 1, \ldots, l$ are the Hessians of f, g_i for $i \in I$, and h_i for $i = 1, \ldots, l$, respectively, all evaluated at $\bar{\mathbf{x}}$. Assume that $\nabla g_i(\bar{\mathbf{x}})$ for $i \in I$, and $\nabla h_i(\bar{\mathbf{x}})$ for $i = 1, \ldots, l$ are linearly independent. Then, $\bar{\mathbf{x}}$ is a KKT point and, moreover, $\mathbf{d}^t\nabla^2 L(\bar{\mathbf{x}})\mathbf{d} \geq 0$ for all $\mathbf{d} \in C' = \{\mathbf{d} \neq \mathbf{0}: \nabla g_i(\bar{\mathbf{x}})^t\mathbf{d} \leq 0$ for all $i \in I$, $\nabla h_i(\bar{\mathbf{x}})^t\mathbf{d} = 0$ for all $i = 1, \ldots, l\}$.

Proof

By Theorem 4.3.7, we directly have that $\bar{\mathbf{x}}$ is a KKT point. Now, if $C' = \emptyset$, the result is trivially true. Otherwise, consider any $\mathbf{d} \in C'$, and denote $I(\mathbf{d}) = \{i \in I : \nabla g_i(\bar{\mathbf{x}})^t\mathbf{d} = 0\}$. For $\lambda \geq 0$, define $\boldsymbol{\alpha}: E_1 \to E_n$ by the following differential equation and boundary condition:

$$\frac{d\boldsymbol{\alpha}(\lambda)}{d\lambda} = \mathbf{P}(\lambda)\mathbf{d} \qquad \boldsymbol{\alpha}(0) = \bar{\mathbf{x}}$$

where $\mathbf{P}(\lambda)$ is the matrix that projects any vector in the null space of the matrix with rows $\nabla g_i(\boldsymbol{\alpha}(\lambda))$, $i \in I(\mathbf{d})$, and $\nabla h_i(\boldsymbol{\alpha}(\lambda))$, $i = 1, \ldots, l$. Following the proof of Theorem 4.3.1 [by treating g_i for $i \in I(\mathbf{d})$, and h_i, $i = 1, \ldots, l$, as the "equations" therein, and treating g_i for $i \in I - I(\mathbf{d})$, for which $\nabla g_i(\bar{\mathbf{x}})^t\mathbf{d} < 0$, as the "inequalities" therein], we obtain that $\boldsymbol{\alpha}(\lambda)$ is feasible for $0 \leq \lambda \leq \delta$, for some $\delta > 0$.

Now, consider a sequence $\{\lambda_k\} \to 0^+$ and denote $\mathbf{x}_k = \boldsymbol{\alpha}(\lambda_k)$ for all k. By the Taylor series expansion, we have

$$L(\mathbf{x}_k) = L(\bar{\mathbf{x}}) + \nabla L(\bar{\mathbf{x}})^t(\mathbf{x}_k - \bar{\mathbf{x}}) + \tfrac{1}{2}(\mathbf{x}_k - \bar{\mathbf{x}})^t\nabla^2 L(\bar{\mathbf{x}})(\mathbf{x}_k - \bar{\mathbf{x}}) + \|\mathbf{x}_k - \bar{\mathbf{x}}\|^2 \beta[\bar{\mathbf{x}};(\mathbf{x}_k - \bar{\mathbf{x}})] \tag{4.31}$$

where $\beta(\bar{\mathbf{x}};(\mathbf{x}_k - \bar{\mathbf{x}})) \to 0$ as $\mathbf{x}_k \to \bar{\mathbf{x}}$. Since $g_i(\mathbf{x}_k) = 0$ for all $i \in I$ and $h_i(\mathbf{x}_k) = 0$ for all $i = 1, \ldots, l$, we have $L(\mathbf{x}_k) = f(\mathbf{x}_k)$ from (4.25). Similarly, $L(\bar{\mathbf{x}}) = f(\bar{\mathbf{x}})$. Also, since $\bar{\mathbf{x}}$ is a KKT point, we have $\nabla L(\bar{\mathbf{x}}) = \mathbf{0}$. Moreover, since $\mathbf{x}_k = \boldsymbol{\alpha}(\lambda_k)$ is feasible, $\mathbf{x}_k \to \bar{\mathbf{x}}$ as $\lambda_k \to 0^+$ or as $k \to \infty$, and since $\bar{\mathbf{x}}$ is a local minimum, we must have $f(\mathbf{x}_k) \geq f(\bar{\mathbf{x}})$ for k sufficiently large. Consequently, from (4.31), we get

$$\frac{f(\mathbf{x}_k) - f(\bar{\mathbf{x}})}{\lambda_k^2} = \frac{1}{2}\frac{(\mathbf{x}_k - \bar{\mathbf{x}})^t}{\lambda_k}\nabla^2 L(\bar{\mathbf{x}})\frac{(\mathbf{x}_k - \bar{\mathbf{x}})}{\lambda_k} + \left\|\frac{\mathbf{x}_k - \bar{\mathbf{x}}}{\lambda_k}\right\|^2 \beta[\bar{\mathbf{x}};(\mathbf{x}_k - \bar{\mathbf{x}})] \geq 0 \tag{4.32a}$$

for k large enough. But note that

$$\lim_{k\to\infty} \frac{(\mathbf{x}_k - \bar{\mathbf{x}})}{\lambda_k} = \lim_{k\to\infty} \frac{\boldsymbol{\alpha}(\lambda_k) - \boldsymbol{\alpha}(0)}{\lambda_k} = \boldsymbol{\alpha}'(0) = \mathbf{P}(0)\mathbf{d} = \mathbf{d} \tag{4.32b}$$

since $\mathbf{d} \in C'$ implies that $\mathbf{d}$ is already in the null space of the matrix with rows $\nabla g_i(\bar{\mathbf{x}})$ for $i \in I(\mathbf{d})$, and $\nabla h_i(\bar{\mathbf{x}})$ for $i = 1, \ldots, l$. Taking limits in (4.32a) as $k \to \infty$ and using (4.32b), we get $\mathbf{d}^t\nabla^2 L(\bar{\mathbf{x}})\mathbf{d} \geq 0$, and this completes the proof.

Observe that the set C' defined in the theorem is precisely $G_0' \cap H_0$ and that it contains the set C of Theorem 4.2.2. Hence, not only is $\mathbf{d}^t\nabla^2 L(\bar{\mathbf{x}})\mathbf{d} \geq 0$ required to hold at a local minimum $\bar{\mathbf{x}}$ for all $\mathbf{d} \in C$, but because of the more restrictive constraint qualification of linear independence of all binding constraints, the nonnegative curvature of L at $\bar{\mathbf{x}}$ is required for all $\mathbf{d} \in G_0' \cap H_0 \equiv C'$. Let us now illustrate the use of the foregoing results.

4.4.4 Example (McCormick [1967])

Consider the (nonconvex programming) problem

$$P\text{: minimize } \{[(x_1 - 1)^2 + x_2^2] : g_1(\mathbf{x}) = 2kx_1 - x_2^2 \leq 0\}$$

where k is a positive constant. Figure 4.14 illustrates two possible ways in which the optimum is determined, depending on the value of k.

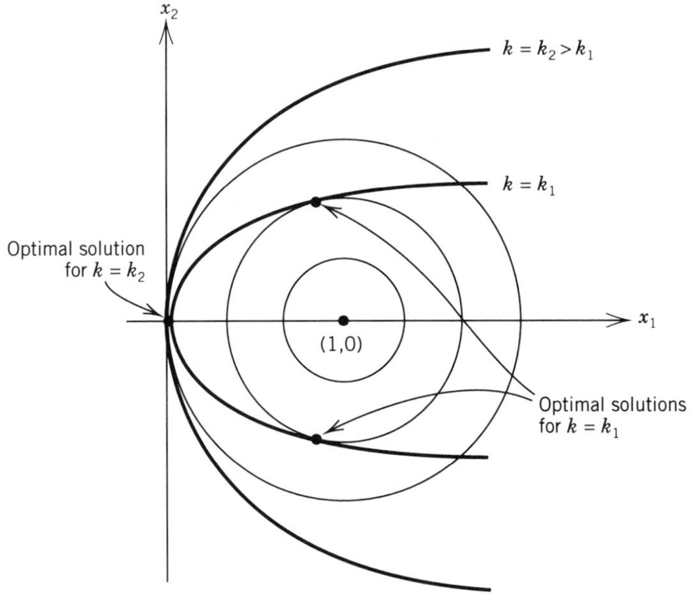

Figure 4.14 Illustration of Example 4.44.

Note that $\nabla g_1(\mathbf{x}) = (2k, -2x_2)^t \neq (0, 0)^t$, and hence the linear independence constraint qualification holds at any feasible solution $\mathbf{x}$. The KKT conditions require primal feasibility and that

$$\begin{bmatrix} 2(x_1 - 1) \\ 2x_2 \end{bmatrix} + u_1 \begin{bmatrix} 2k \\ -2x_2 \end{bmatrix} = \begin{bmatrix} 0 \\ 0 \end{bmatrix}$$

where $u_1 \geq 0$ and $u_1[2kx_1 - x_2^2] = 0$. If $u_1 = 0$, then we must have $(x_1, x_2) = (1, 0)$, which is the unconstrained minimum, and which is infeasible for any $k > 0$. Hence, u_1 must be positive for any KKT point; and so, by complementary slackness, $2kx_1 = x_2^2$ must hold true. Furthermore, by the second dual feasibility constraint, we must either have $x_2 = 0$ or $u_1 = 1$. If $x_2 = 0$, then $2kx_1 = x_2^2$ yields $x_1 = 0$, and the first dual feasibility constraint yields $u_1 = 1/k$. This gives one KKT solution. Similarly, from the KKT conditions, we obtain, when $u_1 = 1$, that $x_1 = (1 - k)$ and $x_2 = \pm\sqrt{2k(1-k)}$, which yields a different set of KKT solutions when $0 < k < 1$. Hence, the KKT solutions are $\{\bar{\mathbf{x}}^1 = (0, 0)', \bar{u}_1^1 = 1/k\}$ for any $k > 0$, and $\{\bar{\mathbf{x}}^2 = (1 - k, \sqrt{2k(1-k)}\,), \bar{u}_1^2 = 1\}$ along with $\{\bar{\mathbf{x}}^3 = (1 - k, -\sqrt{2k(1-k)}\,), \bar{u}_1^3 = 1\}$ whenever $0 < k < 1$.

By examining convex combinations of objective values at the above KKT points and at any other point $\mathbf{x}$ on the constraint surface, for example, it is readily verified that g_1 is not quasiconvex at these points; and thus, while the first-order necessary condition of Theorem 4.2.13 holds, the sufficient condition of Theorem 4.2.16 does not hold. Hence, we are uncertain about the character of the above KKT solutions using these results.

Now, let us examine the second-order necessary condition of Theorem 4.4.3. Note that $L(\bar{\mathbf{x}}) = f(\mathbf{x}) + \bar{u}g(\mathbf{x}) = (x_1 - 1)^2 + x_2^2 + \bar{u}[2kx_1 - x_2^2]$, and so

$$\nabla^2 L(\bar{\mathbf{x}}) = \begin{bmatrix} 2 & 0 \\ 0 & 2(1-\bar{u}) \end{bmatrix}$$

Furthermore, the cone C' defined in Theorem 4.4.3 is given by

$$C' = \{\mathbf{d} \neq \mathbf{0} : kd_1 \leq \bar{x}_2 d_2\}$$

For the KKT solution $(\bar{\mathbf{x}}^1, \bar{u}_1^1)$, Theorem 4.2.3 requires that $2d_1^2 + 2(1 - 1/k)d_2^2 \geq 0$ for all (d_1, d_2) such that $d_1 \leq 0$. Whenever $k \geq 1$, this obviously holds true. However, when $0 < k < 1$, this condition can be violated by making $|d_2|$ large enough. Hence, using Theorem 4.4.3, we can conclude that $\bar{\mathbf{x}}^1$ is not a local minimum for $0 < k < 1$. On the other hand, since $\bar{u}_1^2 = \bar{u}_1^3 = 1$, $\nabla^2 L(\bar{\mathbf{x}}^2)$ and $\nabla^2 L(\bar{\mathbf{x}}^3)$ are positive semidefinite, and hence the other sets of KKT solutions satisfy the second-order necessary optimality conditions.

Next, let us examine the second-order sufficient conditions of Theorem 4.4.2. The cone C defined therein is given by $C = \{\mathbf{d} \neq \mathbf{0} : kd_1 = \bar{x}_2 d_2\}$. For the KKT solution $(\bar{\mathbf{x}}^1, \bar{u}_1^1)$, whenever $k > 1$, $\nabla^2 L(\bar{\mathbf{x}}^1)$ is itself positive definite; and so, even by Lemma 4.4.1b, we have $\bar{\mathbf{x}}^1$ is a strict local minimum. However, for $k = 1$, although $\bar{\mathbf{x}}^1$ solves Problem P, we are unable to recognize this via Theorem 4.4.2, since $\mathbf{d}'\nabla^2 L(\bar{\mathbf{x}}^1)\mathbf{d} = 2d_1^2 = 0$ for $\mathbf{d} \in C = \{\mathbf{d} \neq \mathbf{0} : d_1 = 0\}$.

Next, consider the KKT solution $(\bar{\mathbf{x}}^2, \bar{u}_1^2)$ for $0 < k < 1$. Here, $C = \{\mathbf{d} \neq \mathbf{0}: kd_1 = \sqrt{2k(1-k)}\, d_2\}$; and for any $\mathbf{d}$ in C, we have $\mathbf{d}'\nabla^2 L(\bar{\mathbf{x}}^2)\mathbf{d} = 2d_1^2 > 0$. Hence, by Theorem 4.4.2, $\bar{\mathbf{x}}^2$ is a strict local minimum for $0 < k < 1$. Note that $\nabla^2 L(\bar{\mathbf{x}}^2)$ itself is not positive definite, and therefore Theorem 4.4.2 plays a critical role in concluding the local minimum status of $\bar{\mathbf{x}}^2$. Similarly, $\bar{\mathbf{x}}^3$ is a strict local minimum for $0 < k < 1$. The global minimum status of the above strict local minima must be established by other means because of the nonconvexity of the problem (see Exercise 4.34).

Exercises

4.1 Consider the following unconstrained problem:

$$\text{minimize } x_1^2 - x_1x_2 + 2x_2^2 - 2x_1 + e^{x_1+x_2}$$

a. Write the first-order necessary optimality conditions. Is this condition also sufficient for optimality? Why?

b. Is $\bar{\mathbf{x}} = (0, 0)^t$ an optimal solution? If not, identify a direction $\mathbf{d}$ along which the function would decrease.

c. Minimize the function starting from (0, 0) along the direction $\mathbf{d}$ obtained in part b above.

d. Dropping the last term in the objective function, use a classical direct optimization technique to solve this problem.

4.2 Consider the univariate function $f(x) = xe^{-x}$. Find all local minima/maxima and inflection points. Also, what can you claim about a global minimum and a global maximum for f? Give analytical justifications for your claims.

4.3 Answer the following, justifying your answer:

a. For a minimization nonlinear program, can a KKT point be a local maximum?

b. Let f be differentiable, let X be convex, and let $\bar{\mathbf{x}} \in X$ satisfy $\nabla f(\bar{\mathbf{x}})^t(\mathbf{x} - \bar{\mathbf{x}}) > 0$ for all $\mathbf{x} \in X$, $\mathbf{x} \neq \bar{\mathbf{x}}$. Then, is $\bar{\mathbf{x}}$ necessarily a local minimum?

c. What is the effect on the application of the FJ and the KKT optimality conditions by duplicating an equality constraint or an inequality constraint in the problem?

4.4 Consider the problem to minimize $\|\mathbf{Ax} - \mathbf{b}\|^2$, where $\mathbf{A}$ is an $m \times n$ matrix and $\mathbf{b}$ is an m vector.

a. Give a geometric interpretation of the problem.

b. Write a necessary condition for optimality. Is this also a sufficient condition?

c. Is the optimal solution unique? Why or why not?

d. Can you give a closed-form solution of the optimal solution? Specify any assumptions that you may need.

e. Solve the problem for $\mathbf{A}$ and $\mathbf{b}$ given below

$$\mathbf{A} = \begin{bmatrix} 1 & -1 & 0 \\ 0 & 2 & 1 \\ 0 & 1 & 0 \\ 1 & 0 & 1 \end{bmatrix} \qquad \mathbf{b} = \begin{bmatrix} 2 \\ 1 \\ 1 \\ 0 \end{bmatrix}$$

4.5 Consider the following problem:

$$\begin{aligned} \text{Maximize} \quad & 3x_1 - x_2 + x_3^2 \\ \text{subject to} \quad & x_1 + x_2 + x_3 \leq 0 \\ & -x_1 + 2x_2 + x_3^2 = 0 \end{aligned}$$

a. Write the KKT optimality conditions.

b. Test for the second-order optimality conditions.

c. Argue why this problem is unbounded.

4.6 Consider the following problem:

$$\begin{aligned} \text{Maximize} \quad & x_1^2 + 4x_1x_2 + x_2^2 \\ \text{subject to} \quad & x_1^2 + x_2^2 = 1 \end{aligned}$$

a. Using the KKT conditions, find an optimal solution to the problem.

b. Test for the second-order optimality conditions.

c. Does the problem have a unique optimal solution?

4.7 Consider the following linear program:

$$\begin{aligned} \text{Maximize} \quad & 2x_1 + 3x_2 \\ \text{subject to} \quad & x_1 + x_2 \leq 8 \\ & -x_1 + 2x_2 \leq 4 \\ & x_1, x_2 \geq 0 \end{aligned}$$

a. Write the KKT optimality conditions.

b. For each extreme point, verify whether or not the KKT conditions are true, both algebraically and geometrically. From this, find the optimal solution.

4.8 Consider the following problem:

$$\text{Minimize} \quad (x_1 - \tfrac{9}{4})^2 + (x_2 - 2)^2$$

subject to $x_2 - x_1^2 \geq 0$

$x_1 + x_2 \leq 6$

$x_1, x_2 \geq 0$

a. Write the KKT optimality conditions and verify that these conditions are true at the point $\bar{\mathbf{x}} = (\frac{3}{2}, \frac{9}{4})^t$.

b. Interpret the KKT conditions at $\bar{\mathbf{x}}$ graphically.

c. Show that $\bar{\mathbf{x}}$ is indeed the unique global optimal solution.

4.9 Consider the following problem:

Minimize $x_1^2 + 2x_2^2$

subject to $x_1 + x_2 - 1 = 0$

Find a point satisfying the KKT conditions and verify that it is indeed an optimal solution. Re-solve the problem if the objective function is replaced by $x_1^3 + x_2^3$.

4.10 Consider the following problem:

Minimize $x_1^4 + x_2^4 + 12x_1^2 + 6x_2^2 - x_1x_2 - x_1 - x_2$

subject to $x_1 + x_2 \geq 6$

$2x_1 - x_2 \geq 3$

$x_1 \geq 0, x_2 \geq 0$

Write out the KKT conditions and show that $(x_1, x_2) = (3, 3)$ is the unique optimal solution.

4.11 Consider the following problem:

Maximize $(x_1 - 2)^2 + (x_2 - 3)^2$

subject to $3x_1 + 2x_2 \geq 6$

$-x_1 + x_2 \leq 3$

$x_1 \leq 2$

a. Graphically, find all locally maximizing solutions. What is the global maximum for this problem?

b. Repeat part a analytically, using first- and second-order KKT optimality conditions along with any other formal optimality characterizations.

4.12 Write the KKT necessary optimality conditions for Exercises 1.12 and 1.13. Using these conditions, find the optimal solutions.

4.13 Consider the following one-dimensional minimization problem:

Minimize $f(\mathbf{x} + \lambda\mathbf{d})$

subject to $\lambda \geq 0$

where $\mathbf{x}$ is a given vector and $\mathbf{d}$ is a given nonzero direction.

a. Write a necessary condition for a minimum if f is differentiable. Is this condition also sufficient? If not, what assumptions on f would make the necessary condition also sufficient?

b. Suppose that f is convex but not differentiable. Can you develop a necessary optimality condition for the above problem using subgradients of f defined in Section 3.2?

4.14 Let $f: E_n \to E_1$ be infinitely differentiable, and let $\bar{\mathbf{x}} \in E_n$. For any $\mathbf{d} \in E_n$, define $F_{\mathbf{d}}(\lambda) = f(\bar{\mathbf{x}} + \lambda\mathbf{d})$ for $\lambda \in E_1$. Write out the infinite Taylor series expansion for $F_{\mathbf{d}}(\lambda)$ and compute $F''_{\mathbf{d}}(\lambda)$. Compare the nonnegativity or positivity of this expression with the necessary and sufficient Taylor series-based inequality for $\bar{\mathbf{x}}$ to be a local minimum for f. What conclusions can you draw?

4.15 Consider the following problem:

Minimize $\dfrac{x_1 + 3x_2 + 3}{2x_1 + x_2 + 6}$

subject to $2x_1 + x_2 \leq 12$

$-x_1 + 2x_2 \leq 4$

$x_1, x_2 \geq 0$

a. Show that the KKT conditions are sufficient for this problem.

b. Show that any point on the line segment joining the points (0, 0) and (6,0) is an optimal solution.

4.16 Use the KKT conditions to prove Farkas' theorem discussed in Section 2.3. (*Hint:* Consider the problem to maximize $\mathbf{c}^t\mathbf{x}$ subject to $\mathbf{A}\mathbf{x} \leq \mathbf{0}$.)

4.17 Consider the problem to minimize $f(\mathbf{x})$ subject to $g_i(\mathbf{x}) \le 0$ for $i = 1, \ldots, m$.

a. Show that verifying whether a point $\bar{\mathbf{x}}$ is a KKT point is equivalent to finding a vector $\mathbf{u}$ satisfying a system of the form $\mathbf{A}^t\mathbf{u} = \mathbf{c}$, $\mathbf{u} \ge \mathbf{0}$. This can be done using Phase I of linear programming.

b. Indicate the modifications needed in part a if the problem had equality constraints.

c. Illustrate part a by the following problem, where $\bar{\mathbf{x}} = (1, 2, 5)^t$:

$$\begin{aligned} \text{Minimize} \quad & 2x_1^2 + x_2^2 + 2x_3^2 + x_1x_3 - x_1x_2 + x_1 + 2x_3 \\ & x_1^2 + x_2^2 - x_3 \le 0 \\ & x_1 + x_2 + 2x_3 \le 16 \\ & x_1 + x_2 \ge 3 \\ & x_1, x_2, x_3 \ge 0 \end{aligned}$$

4.18 Consider the problem to minimize $f(\mathbf{x})$ subject to $g_i(\mathbf{x}) \le 0$ for $i = 1, \ldots, m$. Let $\bar{\mathbf{x}}$ be a feasible point, and let $I = \{i : g_i(\bar{\mathbf{x}}) = 0\}$. Suppose that f is differentiable at $\bar{\mathbf{x}}$ and that g_i for $i \in I$ is differentiable and concave at $\bar{\mathbf{x}}$. Furthermore, suppose that g_i for i L I is continuous at $\bar{\mathbf{x}}$. Consider the following linear problem:

$$\begin{aligned} \text{Minimize} \quad & \nabla f(\bar{\mathbf{x}})^t\mathbf{d} & \\ \text{subject to} \quad & \nabla g_i(\bar{\mathbf{x}})^t\mathbf{d} \le 0 & \text{for } i \in I \\ & -1 \le d_j \le 1 & \text{for } j = 1, \ldots, n \end{aligned}$$

Let $\bar{\mathbf{d}}$ be an optimal solution with objective function value $\bar{z}$.

a. Show that $\bar{z} \le 0$.

b. Show that if $\bar{z} < 0$, then there exists a $\delta > 0$ such that $\bar{\mathbf{x}} + \lambda\bar{\mathbf{d}}$ is feasible and $f(\bar{\mathbf{x}} + \lambda\bar{\mathbf{d}}) < f(\bar{\mathbf{x}})$ for each $\lambda \in (0, \delta)$.

c. Show that if $\bar{z} = 0$, then $\bar{\mathbf{x}}$ satisfies the KKT conditions.

4.19 Consider the problem P to minimize $f(\mathbf{x})$ subject to some "less than or equal to" type of linear inequality constraints. Let $\bar{\mathbf{x}}$ be a feasible solution, and let the binding constraints be represented as $\mathbf{Ax} = \mathbf{b}$, where $\mathbf{A}$ is an $m \times n$ matrix of rank m. Let $\mathbf{d} = -\nabla f(\bar{\mathbf{x}})$ and consider the following problem:

$$\bar{P}: \text{minimize } \{\tfrac{1}{2}\|\mathbf{x} - (\bar{\mathbf{x}} + \mathbf{d})\|^2 : \mathbf{Ax} = \mathbf{b}\}$$

Let $\hat{\mathbf{x}}$ solve $\bar{\text{P}}$.

a. Provide a geometric interpretation for $\bar{\text{P}}$ and its solution $\hat{\mathbf{x}}$.

b. Write the KKT conditions for $\bar{\text{P}}$. Discuss if these conditions are necessary and sufficient for optimality.

c. Suppose that the given point $\bar{\mathbf{x}}$ happens to be a KKT point for $\bar{\text{P}}$. Then, is $\bar{\mathbf{x}}$ also a KKT point for P? If so, why? if not, then under what additional conditions can you make this claim?

d. Determine a closed-form expression for the solution $\hat{\mathbf{x}}$ to Problem $\bar{\text{P}}$.

4.20 Let $f: E_n \to E_1$, $g_i: E_n \to E_1$ for $i = 1, \ldots, m$ be convex functions. Consider the problem to minimize $f(\mathbf{x})$ subject to $g_i(\mathbf{x}) \le 0$ for $i = 1, \ldots, m$. Let M be a proper subset of $\{1, \ldots, m\}$, and suppose that $\hat{\mathbf{x}}$ solves the problem to minimize $f(\mathbf{x})$ subject to $g_i(\mathbf{x}) \le 0$ for $i \in M$. Let $V = \{i : g_i(\hat{\mathbf{x}}) > 0\}$. If $\bar{\mathbf{x}}$ solves the original problem, and if $f(\bar{\mathbf{x}}) > f(\hat{\mathbf{x}})$, show that $g_i(\bar{\mathbf{x}}) = 0$ for some $i \in V$. Show that this is not necessarily true if $f(\bar{\mathbf{x}}) = f(\hat{\mathbf{x}})$. (This exercise also shows that if an unconstrained minimum of f is infeasible and has an objective value lesser than the optimum value, then any constrained minimum lies on the boundary of the feasible region.)

4.21 Consider the problem to minimize $f(\mathbf{x})$ subject to $\mathbf{x} \ge \mathbf{0}$, where f is a differentiable convex function. Let $\bar{\mathbf{x}}$ be a given point and denote $\nabla f(\bar{\mathbf{x}})$ by $(\nabla_1, \ldots, \nabla_n)^t$. Show that $\bar{\mathbf{x}}$ is an optimal solution if and only if $\mathbf{d} = \mathbf{0}$, where $\mathbf{d}$ is defined by

$$d_i = \begin{cases} -\nabla_i & \text{if } x_i > 0 \text{ or } \nabla_i < 0 \\ 0 & \text{if } x_i = 0 \text{ and } \nabla_i \ge 0 \end{cases}$$

4.22 Consider the following problem:

$$\text{Minimize} \quad \sum_{j=1}^{n} f_j(x_j)$$

$$\text{subject to} \quad \sum_{j=1}^{n} x_j = 1$$

$$x_j \geq 0 \qquad \text{for } j = 1, \ldots, n$$

Suppose that $\bar{\mathbf{x}} = (\bar{x}_1, \ldots, \bar{x}_n)^t \geq \mathbf{0}$ solves the above problem. Letting $\delta_j = \dfrac{\partial f_j(\bar{\mathbf{x}})}{\partial x_j}$, show that there exists a scalar k such that

$$\delta_j \geq k \quad \text{and} \quad (\delta_j - k)\bar{x}_j = 0 \quad \text{for } j = 1, \ldots, n$$

4.23 Consider the following problem, where $\mathbf{c}$ is a nonzero vector in E_n:

Maximize $\mathbf{c}^t\mathbf{d}$

subject to $\mathbf{d}^t\mathbf{d} \leq 1$

a. Show that $\bar{\mathbf{d}} = \mathbf{c}/\|\mathbf{c}\|$ is a KKT point. Furthermore, show that $\bar{\mathbf{d}}$ is indeed the unique global optimal solution.

b. Using the result of part a, show that the direction of steepest ascent of f at a point $\mathbf{x}$ is given by $\nabla f(\mathbf{x})/\|\nabla f(\mathbf{x})\|$ provided that $\nabla f(\mathbf{x}) \neq \mathbf{0}$.

4.24 Consider the following problem, where a_j, b, and c_j are positive constants:

$$\text{Minimize} \quad \sum_{j=1}^{n} \frac{c_j}{x_j}$$

$$\text{subject to} \quad \sum_{j=1}^{n} a_j x_j = b$$

$$x_j \geq 0 \qquad \text{for } j = 1, \ldots, n$$

Write the KKT conditions, and solve for the point $\bar{\mathbf{x}}$ satisfying these conditions.

4.25 In geometric programming, the following result is used. If $x_1, \ldots, x_n \geq 0$, then

$$\frac{1}{n}\sum_{j=1}^{n} x_j \geq \left(\prod_{j=1}^{n} x_j\right)^{1/n}$$

Prove the result using the KKT conditions. *Hint:* Consider one of the following problems and justify your use of it:

- $$\text{Minimize} \quad \sum_{j=1}^{n} x_j$$

  $$\text{subject to} \quad \prod_{j=1}^{n} x_j = 1, x_j \geq 0 \qquad \text{for } j = 1, \ldots, n$$

- $$\text{Maximize} \quad \prod_{j=1}^{n} x_j$$

  $$\text{subject to} \quad \sum_{j=1}^{n} x_j = 1, x_j \geq 0 \qquad \text{for } j = 1, \ldots, n$$

4.26 Consider the following problem, where $\mathbf{y}$, $\mathbf{e}$, and $\mathbf{y}_0$ belong to E_n, and where $\mathbf{y} = (y_1, \ldots, y_n)^t$, $\mathbf{y}_0 = (1/n, \ldots, 1/n)^t$, and $\mathbf{e} = (1, \ldots, 1)^t$:

$$\text{minimize } \{y_1 : \|\mathbf{y} - \mathbf{y}_0\|^2 \leq 1/n(n-1), \mathbf{e}^t\mathbf{y} = 1\}$$

Interpret this problem with respect to an inscribed sphere in the simplex defined by $\{\mathbf{y} : \mathbf{e}^t\mathbf{y} = 1, \mathbf{y} \geq \mathbf{0}\}$. Write the KKT conditions for this problem and verify that $(0, 1/(n-1), \ldots, 1/(n-1))$ is an optimal solution.

4.27 Let $\mathbf{c}$ be an n vector, $\mathbf{b}$ an m vector, $\mathbf{A}$ an $m \times n$ matrix, and $\mathbf{H}$ a symmetric $n \times n$ positive definite matrix. Consider the following two problems:

- Minimize $\mathbf{c}^t\mathbf{x} + \frac{1}{2}\mathbf{x}^t\mathbf{H}\mathbf{x}$

  subject to $\mathbf{A}\mathbf{x} \leq \mathbf{b}$
- Minimize $\mathbf{h}^t\mathbf{v} + \frac{1}{2}\mathbf{v}^t\mathbf{G}\mathbf{v}$

  subject to $\mathbf{v} \geq \mathbf{0}$

where $\mathbf{G} = \mathbf{A}\mathbf{H}^{-1}\mathbf{A}^t$ and $\mathbf{h} = \mathbf{A}\mathbf{H}^{-1}\mathbf{c} + \mathbf{b}$. Investigate the relationship between the KKT conditions of these two problems.

4.28 Consider the following problem:

Minimize $f(\mathbf{x})$

subject to $\mathbf{Ax} = \mathbf{b}$

$\mathbf{x} \geq \mathbf{0}$

Let $\bar{\mathbf{x}}^t = (\bar{\mathbf{x}}_B^t, \bar{\mathbf{x}}_N^t)$ be an extreme point, where $\bar{\mathbf{x}}_B = \mathbf{B}^{-1}\mathbf{b} > \mathbf{0}$, $\bar{\mathbf{x}}_N = \mathbf{0}$, and $\mathbf{A} = [\mathbf{B}, \mathbf{N}]$ with $\mathbf{B}$ invertible. Now consider the following direction-finding problem:

Minimize $[\nabla_N f(\bar{\mathbf{x}}) - \nabla_B f(\bar{\mathbf{x}})\mathbf{B}^{-1}\mathbf{N}]^t \mathbf{d}_N$

subject to $0 \leq d_j \leq 1$ for each nonbasic component j

where $\nabla_B f(\bar{\mathbf{x}})$ and $\nabla_N f(\bar{\mathbf{x}})$ denote the gradient of f with respect to the basic and nonbasic variables, respectively. Let $\bar{\mathbf{d}}_N$ be an optimal solution, and let $\bar{\mathbf{d}}_B = -\mathbf{B}^{-1}\mathbf{N}\bar{\mathbf{d}}_N$. Show that if $\bar{\mathbf{d}}^t = (\bar{\mathbf{d}}_B^t, \bar{\mathbf{d}}_N^t) \neq (\mathbf{0}, \mathbf{0})$, then it is an improving feasible direction. What are the implications of $\bar{\mathbf{d}} = \mathbf{0}$?

4.29 Consider the problem to minimize $f(\mathbf{x})$ subject to $\mathbf{Ax} \leq \mathbf{b}$. Suppose that $\bar{\mathbf{x}}$ is a feasible solution such that $\mathbf{A}_1\bar{\mathbf{x}} = \mathbf{b}_1$ and $\mathbf{A}_2\bar{\mathbf{x}} < \mathbf{b}_2$, where $\mathbf{A}^t = (\mathbf{A}_1^t, \mathbf{A}_2^t)$ and $\mathbf{b}^t = (\mathbf{b}_1^t, \mathbf{b}_2^t)$. Assuming that $\mathbf{A}_1$ has full rank, the matrix $\mathbf{P}$ that projects any vector unto the nullspace of $\mathbf{A}_1$ is given by

$$\mathbf{P} = \mathbf{I} - \mathbf{A}_1^t(\mathbf{A}_1, \mathbf{A}_1^t)^{-1}\mathbf{A}_1$$

a. Let $\bar{\mathbf{d}} = -\mathbf{P}\nabla f(\bar{\mathbf{x}})$. Show that if $\bar{\mathbf{d}} \neq \mathbf{0}$, then it is an improving feasible direction; that is, $\bar{\mathbf{x}} + \lambda\bar{\mathbf{d}}$ is feasible and $f(\bar{\mathbf{x}} + \lambda\bar{\mathbf{d}}) < f(\bar{\mathbf{x}})$ for $\lambda > 0$ and sufficiently small.

b. Suppose that $\bar{\mathbf{d}} = \mathbf{0}$ and that $\mathbf{u} = -(\mathbf{A}_1\mathbf{A}_1^t)^{-1}\mathbf{A}_1 \nabla f(\bar{\mathbf{x}}) \geq \mathbf{0}$. Show that $\bar{\mathbf{x}}$ is a KKT point.

c. Show that $\bar{\mathbf{d}}$ generated above is of the form $\lambda\hat{\mathbf{d}}$ for some $\lambda > 0$, where $\hat{\mathbf{d}}$ is an optimal solution to the following problem:

Minimize $\nabla f(\bar{\mathbf{x}})^t\mathbf{d}$

subject to $\mathbf{A}_1\mathbf{d} = \mathbf{0}$

$\|\mathbf{d}\|^2 \leq 1$

d. Make all possible simplifications if $\mathbf{A} = -\mathbf{I}$ and $\mathbf{b} = \mathbf{0}$, that is, if the constraints are of the form $\mathbf{x} \geq \mathbf{0}$.

4.30 Consider the following problem:

Minimize $x_1^2 - x_1x_2 + 2x_2^2 - 4x_1 - 5x_2$

subject to $x_1 + 2x_2 \leq 6$

$x_1 \leq 2$

$x_1, x_2 \geq 0$

a. Solve the problem geometrically and verify the optimality of the solution obtained by the KKT conditions.

b. Find the direction $\bar{\mathbf{d}}$ of Exercise 4.29 at the optimal solution. Verify that $\bar{\mathbf{d}} = \mathbf{0}$ and that $\mathbf{u} \geq \mathbf{0}$.

c. Find the direction $\bar{\mathbf{d}}$ of Exercise 4.29 at $\bar{\mathbf{x}} = (1, \frac{5}{2})^t$. Verify that $\bar{\mathbf{d}}$ is an improving feasible direction. Also verify that the optimal solution $\hat{\mathbf{d}}$ of part c of Exercise 4.29 indeed points along $\bar{\mathbf{d}}$.

4.31 Let $\mathbf{A}$ be an $m \times n$ matrix of rank m, and let $\mathbf{P} = \mathbf{I} - \mathbf{A}^t(\mathbf{AA}^t)^{-1}\mathbf{A}$ be the matrix that projects any vector into the null space of $\mathbf{A}$. Define $c = \{\mathbf{d} : \mathbf{Ad} = \mathbf{0}\}$, and let $\mathbf{H}$ be an $n \times n$ symmetric matrix. Show that $\mathbf{d} \in C$ if and only if $\mathbf{d} = \mathbf{Pw}$ for some $\mathbf{w} \in E_n$. Then, show that $\mathbf{d}^t\mathbf{Hd} \geq 0$ for all $\mathbf{d} \in C$ if and only if $\mathbf{P}^t\mathbf{HP}$ is positive semidefinite.

4.32 Consider Problem P to minimize $f(\mathbf{x})$ subject to $h_i(\mathbf{x}) = 0$ for $i = 1, \ldots, l$. Let $\bar{\mathbf{x}}$ be a feasible solution and define $\mathbf{A}$ as an $l \times n$ matrix whose rows represent $\nabla h_i(\bar{\mathbf{x}})^t$ for $i = 1, \ldots, l$ and assume that $\mathbf{A}$ is of rank l. Define $\mathbf{P} = \mathbf{I} - \mathbf{A}^t(\mathbf{AA}^t)^{-1}\mathbf{A}$ as in Exercise 4.29 to be the matrix that projects any vector into the nullspace of $\mathbf{A}$. Explain how Exercise 4.31 relates to checking the second-order necessary conditions for Problem P. How can you extend this to checking second-order sufficient conditions? Illustrate using Example 4.4.4.

4.33 Consider the problem to maximize $3x_1x_2 + 2x_2x_3 + 12x_1x_3$ subject to $6x_1 + x_2 + 4x_3 = 6$. Using first- and second-order KKT optimality conditions, show that $\bar{\mathbf{x}} = (\frac{1}{3}, 2, \frac{1}{2})^t$ is a local maximum. Use Exercise 4.32 to check the second-order sufficient conditions.

4.34 Consider the Problem of Example 4.4.4 for the case $k = 1$. Provide an analytical argument to show that $\bar{\mathbf{x}} = (0, 0)^t$ is an optimal solution. By examining a sequence of values of $k \to 1^-$ with respect to the point (0, 0), explain why the second-order optimality conditions are unable to resolve this case.

4.35 Investigate the relationship between the optimal solutions and the KKT conditions of the following two problems, where $\boldsymbol{\lambda} \geq \mathbf{0}$ is a given fixed vector:

Problem P: Minimize $f(\mathbf{x})$ subject to $\mathbf{x} \in X$, $\mathbf{g}(\mathbf{x}) \leq \mathbf{0}$

Problem P′: Minimize $f(\mathbf{x})$ subject to $\mathbf{x} \in X$, $\boldsymbol{\lambda}^t\mathbf{g}(\mathbf{x}) \leq 0$

(Problem P′ has only one constraint and is referred to as the *surrogate problem*.)

4.36 Consider Problem P to minimize $f(\mathbf{x})$ subject to $g_i(\mathbf{x}) \leq 0$ for $i = 1, \ldots, m$ and $h_i(\mathbf{x}) = 0$ for $i = 1, \ldots, l$. Show that P is mathematically equivalent to the following single constraint problem $\bar{\text{P}}$, where $s_1 \ldots, s_m$ are additional variables:

$$\bar{P}\text{: minimize}\left\{f(\mathbf{x}) : \sum_{i=1}^{m} [g_i(\mathbf{x}) + s_i^2]^2 + \sum_{i=1}^{l} h_i^2(\mathbf{x}) = 0\right\}.$$

Write out the FJ and the KKT conditions for $\bar{\text{P}}$. What statements can you make about the relationship between local optima, FJ, and KKT points? What is your opinion about the utility of $\bar{\text{P}}$ in solving P?

4.37 Consider Problem P to minimize $f(\mathbf{x})$ subject to $g_i(\mathbf{x}) \leq 0$ for $i = 1, \ldots, m$ and $h_i(\mathbf{x}) = 0$ for $i = 1, \ldots, l$. Suppose that this problem is reformulated as

$$\bar{P}\text{: minimize}\,\{f(\mathbf{x}) : g_i(\mathbf{x}) + s_i^2 = 0 \text{ for } i = 1, \ldots, m \text{ and } h_i(\mathbf{x}) = 0 \text{ for } i = 1, \ldots, l\}.$$

Write the KKT conditions for P and for $\bar{\text{P}}$ and compare them. Explain any difference between the two and what arguments you can use to resolve them. Express your opinion on using the formulation $\bar{\text{P}}$ to solve the given problem.

4.38 Consider the problem to minimize $f(x_1, x_2) = (x_2^2 - x_1)(x_2^2 - 2x_1)$, and let $\bar{\mathbf{x}} = (0, 0)^t$. Show that for each $\mathbf{d} \in E_n$, $\|\mathbf{d}\| = 1$, there exists a $\delta_\mathbf{d} > 0$ such that for $-\delta_\mathbf{d} \leq \lambda \leq \delta_\mathbf{d}$, we have $f(\bar{\mathbf{x}} + \lambda\mathbf{d}) \geq f(\bar{\mathbf{x}})$. However, show that $\inf\{\delta_\mathbf{d} : \mathbf{d} \in E_n, \|\mathbf{d}\| = 1\} = 0$. In reference to Figure 4.1, discuss what this entails regarding the local optimality of $\bar{\mathbf{x}}$.

4.39 Suppose that $f{:}S \to E_1$, where $S \subseteq E_n$.

a. If f is pseudoconvex over $N_\varepsilon(\bar{\mathbf{x}}) \cap S$, then does this imply that f is pseudoconvex at $\bar{\mathbf{x}}$?

b. If f is strictly pseudoconvex at $\bar{\mathbf{x}}$, then does this imply that f is quasiconvex at $\bar{\mathbf{x}}$?

c. For each of the FJ and KKT sufficiency theorems for both the equality and for the equality–inequality constraint cases, provide alternative sets of sufficient conditions to guarantee local optimality of a point satisfying these conditions. Prove your claims. Examine your proof for possibly strengthening the theorem by weakening your assumptions.

4.40 Let $\bar{\mathbf{x}}$ be an optimal solution to the problem of minimizing $f(\mathbf{x})$ subject to $g_i(\mathbf{x}) \leq 0$, $i = 1, \ldots, m$ and $h_i(\mathbf{x}) = 0$, $i = 1, \ldots, l$. Suppose that $g_k(\bar{\mathbf{x}}) < 0$ for some $k \in \{1, \ldots, m\}$. Show that if this nonbinding constraint is deleted, it is possible that $\bar{\mathbf{x}}$ is not even a local minimum for the resulting problem. (*Hint:* Consider $g_k(\bar{\mathbf{x}}) = -1$ and $g_k(\mathbf{x}) = 1$ for $\mathbf{x} \neq \bar{\mathbf{x}}$.) Show that if all problem-defining functions are continuous, then, by deleting nonbinding constraints, $\bar{\mathbf{x}}$ remains at least a local optimal solution.

4.41 Consider the following problem:

Minimize $\mathbf{c}^t\mathbf{x} + \frac{1}{2}\mathbf{x}^t\mathbf{H}\mathbf{x}$

subject to $\mathbf{A}\mathbf{x} \leq \mathbf{b}$

where $\mathbf{c}$ is an n vector, $\mathbf{b}$ is an m vector, $\mathbf{A}$ is an $m \times n$ matrix, and $\mathbf{H}$ is an $n \times n$ symmetric matrix.

a. Write the second-order necessary optimality conditions of Theorem 4.4.3. Make all possible simplifications.

b. Is it necessarily true that every local minimum to the above problem is also a global minimum? Prove, or give a counterexample.

c. Provide the first- and second-order necessary optimality conditions for the special case where $\mathbf{c} = \mathbf{0}$ and $\mathbf{H} = \mathbf{I}$. In this case, the problem reduces to finding the point in a polyhedral set closest to the origin.

(The above problem is referred to in the literature as a *least distance programming problem*.)

4.42 Consider the following problem:

Minimize $-x_1 + x_2$

subject to $x_1^2 + x_2^2 - 2x_1 = 0$

$(x_1, x_2) \in X$

where X is the convex combinations of the points $(-1, 0)$, $(0, 1)$, $(1, 0)$, and $(0, -1)$.

a. Find the optimal solution graphically.

b. Do the Fritz John or the KKT conditions hold at the optimal solution in part a? If not, explain in terms of Theorems 4.3.2 and 4.3.7.

c. Replace the set X by a suitable system of inequalities and answer part b. What are your conclusions.?

4.43 Consider the following problem to minimize $f(\mathbf{x})$ subject to $g_i(\mathbf{x}) \le 0$ for $i = 1, \ldots, m$ and $h_i(\mathbf{x}) = 0$ for $i = 1, \ldots, l$. Suppose that $\bar{\mathbf{x}}$ solves the problem locally, and let $I = \{i : g_i(\bar{\mathbf{x}}) = 0\}$. Furthermore, suppose that g_i for $i \in I$ is differentiable at $\bar{\mathbf{x}}$, g_i for i L I is continuous at $\bar{\mathbf{x}}$, $h_1, \ldots, h_l$ are affine, that is, h_i is of the form $h_i(\mathbf{x}) = \mathbf{a}_i^t\mathbf{x} - b_i$.

a. Show that $F_0 \cap G \cap H_0 = \emptyset$, where

$F_0 = \{\mathbf{d} : \nabla f(\bar{\mathbf{x}})^t\mathbf{d} < 0\}$

$H_0 = \{\mathbf{d} : \nabla h_i(\bar{\mathbf{x}})^t\mathbf{d} = 0 \text{ for } i = 1, \ldots, l\}$

$G = \{\mathbf{d} : \nabla g_i(\bar{\mathbf{x}})^t\mathbf{d} \le 0 \text{ for } i \in J \text{ and } \nabla g_i(\bar{\mathbf{x}})^t\mathbf{d} < 0 \text{ for } i \in I - J\}$

$J = \{i \in I : g_i \text{ is pseudoconcave at } \bar{\mathbf{x}}\}$

b. Show how this condition can be verified by using linear programming.

c. Dropping the equality constraints $h_i(\mathbf{x}) = 0$, $i = 1, \ldots, l$ from the problem, and letting D denote the resulting set of feasible directions, show that $G \subseteq D$ and, hence, that $F_0 \cap G = \emptyset$.

4.44 Consider the problem to minimize $f(\mathbf{x})$ subject to $g_i(\mathbf{x}) \le 0$ for $i = 1, \ldots, m$, where $f: E_n \to E_1$ and $g_i : E_n \to E_1$ for $i = 1, \ldots, m$, are all differentiable functions. We know that if $\bar{\mathbf{x}}$ is a local minimum, then $F \cap D = \emptyset$, where F and D are, respectively, the set of improving and feasible directions. Show, giving examples, that the converse is false even if f is convex or if the feasible region is convex (though not both). However, suppose that there exists an ε-neighborhood $N_\varepsilon(\bar{\mathbf{x}})$ about $\bar{\mathbf{x}}$ such that f is pseudoconvex and g_i for $i \in I = \{i : g_i(\bar{\mathbf{x}}) = 0\}$ are quasiconvex over $N_\varepsilon(\bar{\mathbf{x}})$. Then, show that $\bar{\mathbf{x}}$ is a local minimum if and only if $F \cap D = \emptyset$. (*Hint:* Examine Lemma 4.2.3 and Theorem 4.2.5.) Extend this result to include equality constraints.

4.45 Let X be a nonempty open set in E_n, and consider $f: E_n \to E_1$, $g_i : E_n \to E_1$ for $i = 1, \ldots, m$, and $h_i : E_n \to E_1$ for $i = 1, \ldots, l$. Consider Problem P to

Minimize $f(\mathbf{x})$

subject to $g_i(\mathbf{x}) \le 0$ for $i = 1, \ldots, m$

$h_i(\mathbf{x}) = 0$ for $i = 1, \ldots, l$

$\mathbf{x} \in X$

Let $\bar{\mathbf{x}}$ be a feasible solution, and let $I = \{i : g_i(\bar{\mathbf{x}}) = 0\}$. Suppose that the KKT conditions hold at $\bar{\mathbf{x}}$, that is, there exist scalars $\bar{u}_i \ge 0$ for $i \in I$ and $\bar{v}_i$ for $i = 1, \ldots, l$ such that

$$\nabla f(\bar{\mathbf{x}}) + \sum_{i \in I} \bar{u}_i \nabla g_i(\bar{\mathbf{x}}) + \sum_{i=1}^{l} \bar{v}_i \nabla h_i(\bar{\mathbf{x}}) = \mathbf{0}$$

a. Suppose that f is pseudoconvex at $\bar{\mathbf{x}}$ and that ϕ is quasiconvex at $\bar{\mathbf{x}}$, where

$$\phi(\mathbf{x}) = \sum_{i \in I} \bar{u}_i g_i(\mathbf{x}) + \sum_{i=1}^{l} \bar{v}_i h_i(\mathbf{x})$$

Show that $\bar{\mathbf{x}}$ is a global optimal solution to Problem P.

b. Show that if $f + \sum_{i \in I} \bar{u}_i g_i + \sum_{i=1}^{l} \bar{v}_i h_i$ is pseudoconvex, then $\bar{\mathbf{x}}$ is a global optimal solution to Problem P.

c. Show by means of examples that the convexity assumptions in parts a and b above and those of Theorem 4.3.8 are not equivalent to each other.

d. Relate this result to Lemma 4.4.1 and to the discussion immediately following it.

4.46 Consider Problem P to minimize $f(\mathbf{x})$ subject to $h_i(\mathbf{x}) = 0$ for $i = 1, \ldots, l$ where $f: E_n \to E_1$ and $h_i: E_n \to E_1$ for $i = 1, \ldots, l$ are all continuously differentiable functions. Let $\bar{\mathbf{x}}$ be a feasible solution and define an $l \times l$ Jacobian submatrix $\mathbf{J}$ as

$$\mathbf{J} = \begin{bmatrix} \dfrac{\partial h_1(\bar{\mathbf{x}})}{\partial x_1} & \dfrac{\partial h_1(\bar{\mathbf{x}})}{\partial x_l} \\ \vdots & \vdots \\ \dfrac{\partial h_l(\bar{\mathbf{x}})}{\partial x_1} & \dfrac{\partial h_l(\bar{\mathbf{x}})}{\partial x_l} \end{bmatrix}$$

Assume that $\mathbf{J}$ is nonsingular, and so, in particular, $\nabla h_i(\bar{\mathbf{x}})$, $i = 1, \ldots, l$ are linearly independent. Under these conditions, the *Implicit Function Theorem* (see Exercise 4.50) asserts that if we define $\mathbf{y} = (x_{l+1}, \ldots, x_n)^t \in E_{n-l}$, with $\bar{\mathbf{y}} = (\bar{x}_{l+1}, \ldots, \bar{x}_n)^t$, then there exists a neighborhood of $\bar{\mathbf{y}}$ over which the (first) l variables $x_1, \ldots, x_l$ can be (implicitly) solved for in terms of the $\mathbf{y}$ variables using the l equality constraints. More precisely, there exists a neighborhood of $\bar{\mathbf{y}}$ and a set of functions $\psi_1(\mathbf{y}), \ldots, \psi_l(\mathbf{y})$ such that over this neighborhood, we have that $\psi_1(\mathbf{y}) \ldots, \psi_l(\mathbf{y})$ are continuously differentiable, $\psi_i(\bar{\mathbf{y}}) \equiv \bar{x}_i$ for $i = 1, \ldots, l$, and $h_i[\psi_1(\mathbf{y}), \ldots, \psi_l(\mathbf{y}), \mathbf{y}] = 0$ for $i = 1, \ldots, l$.

Now, suppose that $\bar{\mathbf{x}}$ is a local minimum and that the above assumptions hold. Then, argue that $\bar{\mathbf{y}}$ must be a local minimum for the unconstrained function $F(\mathbf{y}) \equiv f[\psi_1(\mathbf{y}), \ldots, \psi_l(\mathbf{y}), \mathbf{y}]: E_{n-l} \to E_1$. Using the first-order necessary optimality conditions $\nabla F(\bar{\mathbf{y}}) = \mathbf{0}$ for unconstrained problems, derive the KKT necessary optimality conditions for Problem P. In particular, show that the Lagrangian multiplier vector $\bar{\mathbf{v}}$ in the dual feasibility condition $\nabla f(\bar{\mathbf{x}}) + [\nabla h(\bar{\mathbf{x}})]^t \bar{\mathbf{v}}$, where $\nabla h(\bar{\mathbf{x}})$ is the matrix whose rows are $\nabla h_1(\bar{\mathbf{x}})^t, \ldots, \nabla h_l(\bar{\mathbf{x}})^t$, is given uniquely by

$$\bar{\mathbf{v}} = -\mathbf{J}^{-1} \left[\frac{\partial f(\bar{\mathbf{x}})}{\partial x_1}, \ldots, \frac{\partial f(\bar{\mathbf{x}})}{\partial x_l} \right]^t$$

4.47 Consider the quadratic assignment programming to minimize $\mathbf{c}^t\mathbf{x} + \frac{1}{2}\mathbf{x}^t\mathbf{Q}\mathbf{x}$ subject to the assignment constraints $\sum_{j=1}^{m} x_{ij} = 1$ for all $i = 1, \ldots, m$, $\sum_{i=1}^{m} x_{ij} = 1$ for all $j = 1, \ldots, m$, and $\mathbf{x}$ is binary-valued. Here, the component x_{ij} of the vector $\mathbf{x}$ takes on a value of 1 if i is assigned to j and is 0 otherwise, for $i, j = 1, \ldots, m$. Show that if M exceeds the sum of absolute values of elements in any row of $\mathbf{Q}$, then the matrix $\bar{\mathbf{Q}}$ obtained from $\mathbf{Q}$ by subtracting M from each diagonal element is negative definite. Now, consider the following problem:

$$\overline{QAP}\text{: Minimize} \left\{ \mathbf{c}^t\mathbf{x} + \frac{1}{2}\mathbf{x}^t\bar{\mathbf{Q}}\mathbf{x} : \sum_{i=1}^{m} x_{ij} = 1 \text{ for all } i, \sum_{i=1}^{m} x_{ij} = 1 \text{ for all } j, \mathbf{x} \geq \mathbf{0} \right\}$$

Using the fact that the extreme points of $\overline{\text{QAP}}$ are all binary valued, show that $\overline{\text{QAP}}$ is equivalent to QAP. Moreover, show that every extreme point of $\overline{\text{QAP}}$ is a KKT point. (This exercise is due to Bazaraa and Sherali [1982].)

4.48 Consider the *bilinear program* to minimize $\mathbf{c}^t\mathbf{x} + \mathbf{d}^t\mathbf{y} + \mathbf{x}^t\mathbf{H}\mathbf{y}$ subject to $\mathbf{x} \in X$ and $\mathbf{y} \in Y$, where X and Y are bounded polyhedral sets in E_n and E_m, respectively. Let $\hat{\mathbf{x}}$ and $\hat{\mathbf{y}}$ be extreme points of the sets X and Y, respectively.

a. Verify that the objective function is neither quasiconvex nor quasiconcave.

b. Prove that there exists an extreme point $(\bar{\mathbf{x}}, \bar{\mathbf{y}})$ that solves the bilinear program.

c. Prove that the point $(\hat{\mathbf{x}}, \hat{\mathbf{y}})$ is a local minimum of the bilinear program if and only if the following are true: (i) $\mathbf{c}^t(\mathbf{x} - \hat{\mathbf{x}}) \geq 0$ and $\mathbf{d}^t(\mathbf{y} - \hat{\mathbf{y}}) \geq 0$ for each $\mathbf{x} \in X$ and $\mathbf{y} \in Y$; (ii) $\mathbf{c}^t(\mathbf{x} - \hat{\mathbf{x}}) + \mathbf{d}^t(\mathbf{y} - \hat{\mathbf{y}}) > 0$ whenever $(\mathbf{x} - \hat{\mathbf{x}})^t\mathbf{H}(\mathbf{y} - \hat{\mathbf{y}}) < 0$.

d. Show that the point $(\hat{\mathbf{x}}, \hat{\mathbf{y}})$ is a KKT point if and only if $(\mathbf{c}^t + \hat{\mathbf{y}}^t\mathbf{H})(\mathbf{x} - \hat{\mathbf{x}}) \geq 0$ for each $\mathbf{x} \in X$ and $(\mathbf{d}^t + \hat{\mathbf{x}}^t\mathbf{H})(\mathbf{y} - \hat{\mathbf{y}}) \geq 0$ for each $\mathbf{y} \in Y$.

e. Consider the problem to minimize $x_2 + y_1 + x_2y_1 - x_1y_2 + x_2y_2$ subject to $(x_1, x_2) \in X$ and $(y_1, y_2) \in Y$, where X is the polyhedral set defined by its extreme points (0, 0), (0, 1), (1, 4), (2, 4), and (3, 0), and Y is the polyhedral set defined by its extreme points (0, 0), (0, 1), (1, 5), (3, 5), (4, 4), and (3, 0). Verify that the point $(x_1, x_2, y_1, y_2) = (0, 0, 0, 0)$ is a KKT point but not a local minimum. Verify that the point $(x_1, x_2, y_1, y_2) = (3, 0, 1, 5)$ is both a KKT point and a local minimum. What is the global minimum to the problem?

4.49 A differentiable function $\psi: E_n \to E_1$ is said to be an η-*invex function* if there exists some (arbitrary) function $\eta: E_{2n} \to E_n$ such that for each $\mathbf{x}_1, \mathbf{x}_2 \in E_n$, $\psi(\mathbf{x}_2) \geq \psi(\mathbf{x}_1) + \nabla\psi(\mathbf{x}_1)^t\eta(\mathbf{x}_1, \mathbf{x}_2)$. Furthermore, ψ is said to be a η-*pseudoinvex function* if $\nabla\psi(\mathbf{x}_1)^t \eta(\mathbf{x}_1, \mathbf{x}_2) \geq 0$ implies $\psi(\mathbf{x}_2) \geq \psi(\mathbf{x}_1)$. Similarly, ψ is said to be a η-*quasiinvex function* if $\psi(\mathbf{x}_2) \leq \psi(\mathbf{x}_1)$ implies that $\nabla\psi(\mathbf{x}_1)^t\eta(\mathbf{x}_1, \mathbf{x}_2) \leq 0$.

a. When invex is replaced by convex in the usual sense, what is $\eta(\mathbf{x}_1, \mathbf{x}_2)$ defined to be?

b. Consider the problem to minimize $f(\mathbf{x})$ subject to $g_i(\mathbf{x}) \leq 0$ for $i = 1, \ldots, m$ where $f: E_n \to E_1$ and $g_i: E_n \to E_1$ for $i = 1, \ldots, m$ are all differentiable functions. Let $\bar{\mathbf{x}}$ be a KKT point. Show that $\bar{\mathbf{x}}$ is optimal if f and g_i for $i \in I = \{i: g_i(\bar{\mathbf{x}}) = 0\}$ are all η-invex.

c. Repeat part b if f is η-pseudoinvex and g_i $i \in I$ are η-quasiinvex. (The reader is referred to Hanson [1981] and to Hanson and Mond [1982, 1987] for discussions on invex functions and their uses.)

4.50 Consider Problem P to minimize $f(\mathbf{x})$ subject to $h_i(\mathbf{x}) = 0$ for $i = 1, \ldots, l$ where $\mathbf{x} \in E_n$, and where all objective and constraint functions are continuously differentiable. Suppose that $\bar{\mathbf{x}}$ is a local minimum for P, and that the gradients $\nabla h_i(\bar{\mathbf{x}}) = 0$, $i = 1, \ldots, l$ are linearly independent. Use the implicit function theorem stated below to derive the KKT optimality conditions for P. Extend this to include inequality constraints $g_i(\mathbf{x}) \leq 0$, $i = 1, \ldots, m$ as well. (*Hint:* See Exercise 4.46.)

Implicit Function Theorem (see Taylor and Mann [1983], for example). Suppose that $\phi_i(\mathbf{x})$, $i = 1, \ldots, p$ (representing the binding constraints at $\bar{\mathbf{x}}$) are continuously differentiable functions, and suppose that the gradients $\nabla\phi_i(\bar{\mathbf{x}})$, $i = 1, \ldots, p$ are linearly independent, where $p < n$. Denote $\phi \equiv (\phi_1, \ldots, \phi_p)^t: E_n \to E_p$. Hence, $\phi(\bar{\mathbf{x}}) = \mathbf{0}$, and we can partition $\mathbf{x}^t = (\mathbf{x}_\mathbf{B}^t, \mathbf{x}_\mathbf{N}^t)$, where $\mathbf{x}_\mathbf{B} \in E_p$ and $\mathbf{x}_\mathbf{N} \in E_{n-p}$, such that, for the corresponding partition $[\nabla_\mathbf{B}\phi(\mathbf{x}), \nabla_\mathbf{N}\phi(\mathbf{x})]$ of the Jacobian $\nabla\phi(\mathbf{x})$, the $p \times p$ submatrix $\nabla_\mathbf{B}\phi(\bar{\mathbf{x}})$ is nonsingular. Then, the following holds. There exists an open neighborhood $N_\varepsilon(\bar{\mathbf{x}}) \subseteq E_n$, $\varepsilon > 0$, an open neighborhood $N_{\varepsilon'}(\bar{\mathbf{x}}_\mathbf{N}) \subseteq E_{n-p}$, $\varepsilon' > 0$, and a function $\psi: E_{n-p} \to E_p$ that is continuously differentiable on $N_{\varepsilon'}(\bar{\mathbf{x}}_\mathbf{N})$ such that

(i) $\bar{\mathbf{x}}_\mathbf{B} = \psi(\bar{\mathbf{x}}_\mathbf{N})$

(ii) For every $\mathbf{x}_\mathbf{N} \in N_{\varepsilon'}(\bar{\mathbf{x}}_\mathbf{N})$, we have $\phi[\psi(\mathbf{x}_\mathbf{N}), \mathbf{x}_\mathbf{N}] = \mathbf{0}$

(iii) The Jacobian $\nabla\phi(\mathbf{x})$ has full row rank p for each $\mathbf{x} \in N_\varepsilon(\bar{\mathbf{x}})$.

(iv) For any $\mathbf{x}_\mathbf{N} \in N_{\varepsilon'}(\bar{\mathbf{x}}_\mathbf{N})$, the Jacobian $\nabla\psi(\mathbf{x}_\mathbf{N})$ is given by the (unique) solution to the linear equation system

$$\{\nabla_\mathbf{B}\phi[\psi(\mathbf{x}_\mathbf{N}), \mathbf{x}_\mathbf{N}]\}\,\nabla\psi(\mathbf{x}_\mathbf{N}) = -\nabla_\mathbf{N}\phi[\psi(\mathbf{x}_\mathbf{N}), \mathbf{x}_\mathbf{N}]$$

Notes and References

In this chapter, we begin by developing first- and second-order optimality conditions for unconstrained optimization problems in Section 4.1. These classical results can be found in most textbooks dealing with real analysis. For more details on this subject relating to higher-order necessary and sufficiency conditions, refer to Gue and Thomas [1968] and Hancock [1950]; and for information regarding the handling of equality constraints via the Lagrangian multiplier rule, refer to Bartle [1976] and Rudin [1964].

In Section 4.2, we treat the problem of minimizing a function in the presence of inequality constraints and develop the Fritz John [1948] necessary optimality conditions. A weaker form of these conditions, in which the nonnegativity of the multipliers was not asserted, was derived by Karush [1939]. Under a suitable constraint qualification, the Lagrangian multiplier associated with the objective function is positive, and the Fritz John conditions reduce to those of Kuhn and Tucker [1951], which were independently derived. Even though the latter conditions were originally derived by Karush [1939] using calculus of variations, this work has not received much attention, since it has not been published. However, we refer to these conditions as KKT conditions, recognizing Karush, Kuhn, and Tucker. An excellent historical review of optimality conditions for nonlinear programming can be found in Kuhn [1976] and Lenstra et al. [1991]. Kyparisis [1985] presents a necessary and sufficient condition for the KKT Lagrangian multipliers to be unique. Gehner [1974] extends the FJ optimality conditions to the case of *semi-infinite nonlinear programming problems*, where there are an infinite number of parametrically described equality and inequality constraints. The reader may refer to the following references for a further study of the Fritz John and KKT conditions: Abadie [1967b], Avriel [1967], Canon, Cullum, and Polak [1966], Gould and Tolle [1972], Luenberger [1973], Mangasarian [1969a] and Zangwill [1969].

Mangasarian and Fromovitz [1967] generalized the Fritz John conditions for handling both equality and inequality constraints. Their approach used the implicit function theorem. In Section 4.3, we develop the Fritz John conditions for equality and inequality constraints by constructing a feasible arc, as in the work of Fiacco and McCormick [1968].

In Sections 4.2 and 4.3, we show that the KKT conditions are indeed sufficient for optimality under suitable convexity assumptions. This result was proved by Kuhn and Tucker [1951] if the functions f, g_i for $i \in I$ are convex, the functions h_i for all i are affine, and the set X is convex. This result was generalized later, so that weaker convexity assumptions are needed to guarantee optimality, as shown in Sections 4.2 and 4.3 (see Mangasarian [1969a]). The reader may also refer to Bhatt and Misra [1975], who relaxed the condition that h_i be affine, provided that the associated Lagrangian multiplier has the correct sign. Further generalizations using invex functions can be found in Hanson [1981] and Hanson and Mond [1982].

Other generalizations and extensions of the Fritz John and KKT conditions were developed by many authors. One such extension is to relax the condition that the set X is open. In this case, we obtain necessary optimality conditions of the minimum principle type. For details on this type of optimality conditions, see Bazaraa and Goode [1972], Canon, Cullum, and Polak [1970], and Mangasarian [1969a]. Another extension is to treat the problem in an infinite-dimensional setting. The interested reader may refer to Canon, Cullum, and Polak [1970], Dubovitskii and Milyutin [1965], Gehner [1974], Guignard [1969], Halkin and Neustadt [1966], Hestenes [1966], Neustadt [1968], and Varaiya [1967].

In Section 4.4, we treat second-order necessary and sufficient optimality conditions for constrained problems, developed initially by McCormick [1967]. Since a less restrictive constraint qualification is used by us, our second-order necessary optimality condition is stronger than that presented by McCormick [1967] or Fletcher [1987] (see also Ben-Tal [1980]). For a discussion on checking these conditions based on eigenvalues computed over a projected tangential subspace, or based on bordered Hessian matrices, we refer the reader to Luenberger [1984]. Also, see Exercise 4.32 for a related approach. For an additional study of this topic, we refer the reader to Avriel [1976], Ben-Tal [1980], Ben-Tal and Zowe [1982], Fiacco [1968], Fletcher [1983], Luenberger [1984]. McCormick [1967], and Messerli and Polak [1969].

Chapter

5 Constraint Qualifications

In Chapter 4, we considered Problem P to minimize $f(\mathbf{x})$ subject to $\mathbf{x} \in X$ and $g_i(\mathbf{x}) \leq 0$, $i = 1, \ldots, m$. We obtained the Karush–Kuhn–Tucker (KKT) necessary conditions for optimality by deriving the Fritz John conditions and then asserting that the multiplier associated with the objective function is positive at a local optimum when a constraint qualification is satisfied. In this chapter, we develop the KKT conditions directly without first deriving the Fritz John conditions. This is done under various constraint qualifications for problems with inequality constraints and for problems with both inequality and equality constraints.

The following is an outline of the chapter:

Section 5.1: The Cone of Tangents We introduce the cone of tangents T and show that $F_0 \cap T = \varnothing$ is a necessary condition for local optimality. Using a constraint qualification, we derive the KKT conditions directly for problems with inequality constraints.

Section 5.2: Other Constraint Qualifications We introduce other cones contained in the cone of tangents. Making use of these cones, we present various constraint qualifications that validate the KKT conditions. Relationships among these constraint qualifications are explored.

Section 5.3: Problems with Inequality and Equality Constraints The results of Section 5.2 are extended to problems with equality and inequality constraints.

5.1 The Cone of Tangents

In Section 4.2, we discussed the KKT necessary optimality conditions for problems with inequality constraints. In particular, we showed that local optimality implies that $F_0 \cap$

$G_0 = \emptyset$, which in turn implies the Fritz John conditions. Under the linear independence constraint qualification, or, more generally, under the constraint qualification $G_0 \neq \emptyset$, we deduced that the Fritz John conditions can only be satisfied if the Lagrangian multiplier associated with the objective function is positive. This led to the KKT conditions. This process is summarized in the following flowchart.

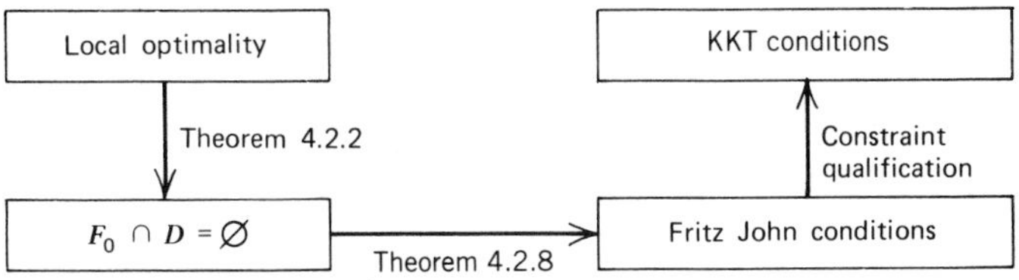

In this section, we derive the KKT conditions directly without first obtaining the Fritz John conditions. As shown in Theorem 5.1.2 below, a necessary condition for local optimality is that $F_0 \cap T = \emptyset$, where T is the cone of tangents given in Definition 5.1.1 below. Using the constraint qualification $T = G'$, where G' is as defined in Theorem 5.1.3 (see also Theorem 4.2.15), we get $F_0 \cap G' = \emptyset$. Using Farkas' theorem, this statement gives the KKT conditions. This process is summarized in the following flowchart.

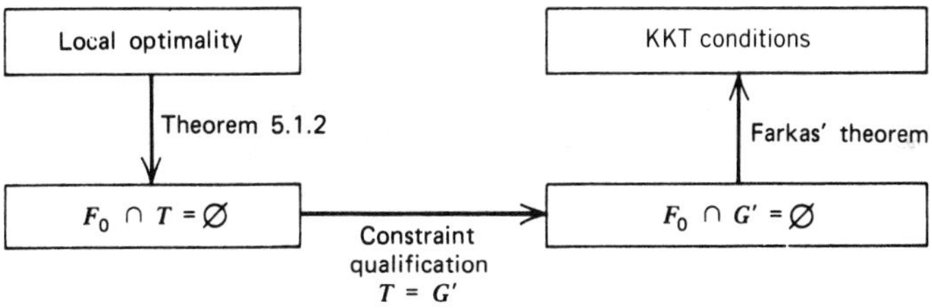

5.1.1 Definition

Let S be a nonempty set in E_n, and let $\bar{\mathbf{x}} \in \text{cl } S$. The *cone of tangents* of S at $\bar{\mathbf{x}}$, denoted by T, is the set of all directions $\mathbf{d}$ such that $\mathbf{d} = \lim_{k\to\infty} \lambda_k(\mathbf{x}_k - \bar{\mathbf{x}})$, where $\lambda_k > 0$, $\mathbf{x}_k \in S$ for each k, and $\mathbf{x}_k \to \bar{\mathbf{x}}$.

From the above definition, it is clear that $\mathbf{d}$ belongs to the cone of tangents if there is a feasible sequence $\{\mathbf{x}_k\}$ converging to $\bar{\mathbf{x}}$ such that the directions of the cords $\mathbf{x}_k - \bar{\mathbf{x}}$ converge to $\mathbf{d}$. Exercise 5.3 provides alternative equivalent descriptions for the cone of tangents T; and in Exercise 5.4, we ask the reader to show that the cone of tangents is indeed a closed cone. Figure 5.1 illustrates some examples of the cone of tangents, where the origin is translated to $\bar{\mathbf{x}}$ for convenience.

Theorem 5.1.2 below shows that for a problem of the form to minimize $f(\mathbf{x})$ subject to $\mathbf{x} \in S$, $F_0 \cap T = \emptyset$ is, indeed, a necessary condition for optimality. Later we specify S to be the set $\{\mathbf{x} \in X : g_i(\mathbf{x}) \leq 0 \text{ for } i = 1, \ldots, m\}$.

5.1.2 Theorem

Let S be a nonempty set in E_n, and let $\bar{\mathbf{x}} \in S$. Furthermore, suppose that $f: E_n \to E_1$ is differentiable at $\bar{\mathbf{x}}$. If $\bar{\mathbf{x}}$ locally solves the problem to minimize $f(\mathbf{x})$ subject to $\mathbf{x} \in S$,

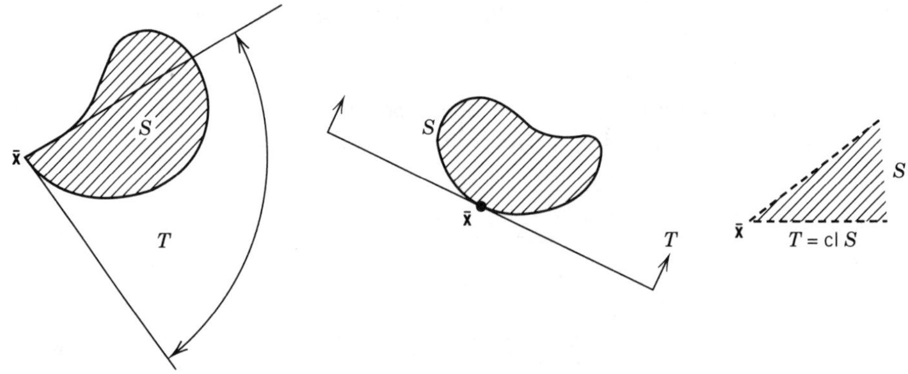

Figure 5.1 Examples of the cone of tangents.

then $F_0 \cap T = \emptyset$, where $F_0 = \{\mathbf{d}: \nabla f(\bar{\mathbf{x}})^t\mathbf{d} < 0\}$ and T is the cone of tangents of S at $\bar{\mathbf{x}}$.

Proof

Let $\mathbf{d} \in T$, that is, $\mathbf{d} = \lim_{k\to\infty} \lambda_k(\mathbf{x}_k - \bar{\mathbf{x}})$, where $\lambda_k > 0$, $\mathbf{x}_k \in S$ for each k, and $\mathbf{x}_k \to \bar{\mathbf{x}}$. By the differentiability of f at $\bar{\mathbf{x}}$, we get

$$f(\mathbf{x}_k) - f(\bar{\mathbf{x}}) = \nabla f(\bar{\mathbf{x}})^t(\mathbf{x}_k - \bar{\mathbf{x}}) + \|\mathbf{x}_k - \bar{\mathbf{x}}\|\, \alpha(\bar{\mathbf{x}}; \mathbf{x}_k - \bar{\mathbf{x}}) \tag{5.1}$$

where $\alpha(\bar{\mathbf{x}}; \mathbf{x}_k - \bar{\mathbf{x}}) \to 0$ as $\mathbf{x}_k \to \bar{\mathbf{x}}$. Noting the local optimality of $\bar{\mathbf{x}}$, we have for k large enough that $f(\mathbf{x}_k) \geq f(\bar{\mathbf{x}})$; and so, from (5.1), we get

$$\nabla f(\bar{\mathbf{x}})^t(\mathbf{x}_k - \bar{\mathbf{x}}) + \|\mathbf{x}_k - \bar{\mathbf{x}}\|\, \alpha(\bar{\mathbf{x}}; \mathbf{x}_k - \bar{\mathbf{x}}) \geq 0$$

Multiplying by $\lambda_k > 0$ and taking the limit as $k \to \infty$, the above inequality implies that $\nabla f(\bar{\mathbf{x}})^t\mathbf{d} \geq 0$. Hence, we have shown that $\mathbf{d} \in T$ implies that $\nabla f(\bar{\mathbf{x}})^t\mathbf{d} \geq 0$ and, therefore, that $F_0 \cap T = \emptyset$. This completes the proof.

It is worth noting that the condition $F_0 \cap T = \emptyset$ does not necessarily imply that $\bar{\mathbf{x}}$ is a local minimum. Indeed, this condition will hold true whenever $F_0 = \emptyset$, for example, which we know is not sufficient for local optimality. However, if there exists an ε-neighborhood $N_\varepsilon(\bar{\mathbf{x}})$ about $\bar{\mathbf{x}}$ such that $N_\varepsilon(\bar{\mathbf{x}}) \cap S$ is convex and f is pseudoconvex over $N_\varepsilon(\bar{\mathbf{x}}) \cap S$, then $F_0 \cap T = \emptyset$ is sufficient to claim that $\bar{\mathbf{x}}$ is a local minimum (see Exercise 5.16).

Abadie Constraint Qualification

In Theorem 5.1.3 below, we derive the KKT conditions under the constraint qualification $T = G'$ credited to Abadie.

5.1.3 Theorem (Karush–Kuhn–Tucker Necessary Conditions)

Let X be a nonempty set in E_n, and let $f: E_n \to E_1$ and $g_i: E_n \to E_1$ for $i = 1, \ldots, m$. Consider the problem to minimize $f(\mathbf{x})$ subject to $\mathbf{x} \in X$, and $g_i(\mathbf{x}) \leq 0$ for $i = 1, \ldots, m$. Let $\bar{\mathbf{x}}$ be a feasible solution, and let $I = \{i: g_i(\bar{\mathbf{x}}) = 0\}$. Suppose that f and g_i for $i \in I$ are differentiable at $\bar{\mathbf{x}}$. Furthermore, suppose that the constraint qualification

$T = G'$ is true, where T is the cone of tangents of the feasible region at $\bar{\mathbf{x}}$, and $G' = \{\mathbf{d}: \nabla g_i(\bar{\mathbf{x}})^t\mathbf{d} \leq 0 \text{ for } i \in I\}$. If $\bar{\mathbf{x}}$ is a local optimal solution, then there exist nonnegative scalars u_i for $i \in I$ such that

$$\nabla f(\bar{\mathbf{x}}) + \sum_{i \in I} u_i \nabla g_i(\bar{\mathbf{x}}) = \mathbf{0}$$

Proof

By Theorem 5.1.2, we have $F_0 \cap T = \emptyset$, where $F_0 = \{\mathbf{d}: \nabla f(\bar{\mathbf{x}})^t\mathbf{d} < 0\}$. By assumption, $T = G'$, so that $F_0 \cap G' = \emptyset$. In other words the following system has no solution:

$$\nabla f(\bar{\mathbf{x}})^t\mathbf{d} < 0 \qquad \nabla g_i(\bar{\mathbf{x}})^t\mathbf{d} \leq 0 \qquad \text{for } i \in I$$

Hence, by Theorem 2.4.5 (Farkas' theorem), the result follows (see also Theorem 4.2.15).

The reader may verify that in Example 4.2.10, the constraint qualification $T = G'$ does not hold true at $\bar{\mathbf{x}} = (1, 0)^t$. Note that the Abadie constraint qualification $T = G'$ could be equivalently stated as $T \supseteq G'$, since $T \subseteq G'$ is always true (see Exercise 5.13). Note that openness of the set X and continuity of g_i at $\bar{\mathbf{x}}$ for $i \notin I$ were not explicitly assumed in Theorem 5.1.3. However, without these assumptions, it is unlikely that the constraint qualification $T \supseteq G'$ would hold true (see Exercise 5.11).

Linearly Constrained Problems

Lemma 5.1.4 below shows that if the constraints are linear, then the Abadie constraint qualification is automatically true. This also implies that the KKT conditions are always necessary for problems with linear constraints whether the objective function is linear or nonlinear. As an alternate proof that does not employ the cone of tangents, note that if $\bar{\mathbf{x}}$ is a local minimum, then $F_0 \cap D = \emptyset$. Now, by Lemma 4.2.4, if the constraints are linear, then $D = G_0' \equiv \{\mathbf{d} \neq \mathbf{0}: \nabla g_i(\bar{\mathbf{x}})^t\mathbf{d} \leq 0, \text{ for each } i \in I\}$. Hence, $F_0 \cap D = \emptyset \Leftrightarrow F_0 \cap G_0' = \emptyset$, which holds if and only if $\bar{\mathbf{x}}$ is a KKT point by Theorem 4.2.15.

5.1.4 Lemma

Let $\mathbf{A}$ be an $m \times n$ matrix, let $\mathbf{b}$ be an m vector, and let $S = \{\mathbf{x}: \mathbf{A}\mathbf{x} \leq \mathbf{b}\}$. Suppose $\bar{\mathbf{x}} \in S$ is such that $\mathbf{A}_1\mathbf{x} = \mathbf{b}_1$ and $\mathbf{A}_2\mathbf{x} < \mathbf{b}_2$, where $\mathbf{A}^t = (\mathbf{A}_1^t, \mathbf{A}_2^t)$ and $\mathbf{b}^t = (\mathbf{b}_1^t, \mathbf{b}_2^t)$. Then, $T = G'$, where T is the cone of tangents of S at $\bar{\mathbf{x}}$ and $G' = \{\mathbf{d}: \mathbf{A}_1\mathbf{d} \leq \mathbf{0}\}$.

Proof

If $\mathbf{A}_1$ is vacuous, $\mathbf{G}' = E_n$. Furthermore, $\bar{\mathbf{x}} \in \text{int } S$ and, hence, $T = E_n$. Thus, $G' = T$. Now suppose that $\mathbf{A}_1$ is not vacuous. Let $\mathbf{d} \in T$, that is, $\mathbf{d} = \lim_{k\to\infty} \lambda_k(\mathbf{x}_k - \bar{\mathbf{x}})$, where $\mathbf{x}_k \in S$ and $\lambda_k > 0$ for each k. Then,

$$\mathbf{A}_1(\mathbf{x}_k - \bar{\mathbf{x}}) \leq \mathbf{b}_1 - \mathbf{b}_1 = \mathbf{0} \tag{5.2}$$

Multiplying (5.2) by $\lambda_k > 0$ and taking the limit as $k \to \infty$, it follows that $\mathbf{A}_1\mathbf{d} \leq \mathbf{0}$. Thus, $\mathbf{d} \in G'$, and $\mathbf{T} \subseteq G'$. Now let $\mathbf{d} \in G'$, that is, $\mathbf{A}_1\mathbf{d} \leq \mathbf{0}$. We need to show that $\mathbf{d} \in T$. Since $\mathbf{A}_2\bar{\mathbf{x}} < \mathbf{b}_2$, there is a $\delta > 0$ such that $\mathbf{A}_2(\bar{\mathbf{x}} + \lambda\mathbf{d}) < \mathbf{b}_2$ for all $\lambda \in (0, \delta)$. Furthermore, since $\mathbf{A}_1\bar{\mathbf{x}} = \mathbf{b}_1$ and $\mathbf{A}_1\mathbf{d} \leq \mathbf{0}$, then $\mathbf{A}_1(\bar{\mathbf{x}} + \lambda\mathbf{d}) \leq \mathbf{b}_1$ for all $\lambda > 0$.

Therefore, $\bar{\mathbf{x}} + \lambda\mathbf{d} \in S$ for each $\lambda \in (0, \delta)$. This automatically shows that $\mathbf{d} \in T$. Therefore, $T = G'$, and the proof is complete.

5.2 Other Constraint Qualifications

The KKT conditions have been developed by many authors under various constraint qualifications. In this section, we present some of the more important constraint qualifications. In Section 5.1, we learned that local optimality implies that $F_0 \cap T = \emptyset$, and that the KKT conditions follow under the constraint qualification $T = G'$. If we define a cone $C \subseteq T$, then $F_0 \cap T = \emptyset$ also implies that $F_0 \cap C = \emptyset$. Therefore, any constraint qualification of the form $C = G'$ will lead to the KKT conditions. In fact, since $C \subseteq T \subseteq G'$, the constraint qualification $C = G'$ implies that $T = G'$ and is, therefore, more restrictive than Abadie's constraint qualification. This process is illustrated in the following flowchart:

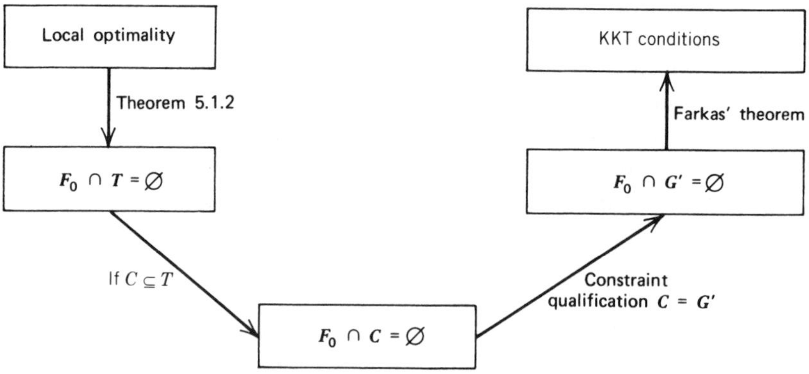

We present below several such cones whose closures are contained in T. Here the feasible region S is given by $\{\mathbf{x} \in X : g_i(\mathbf{x}) \leq 0, i = 1, \ldots, m\}$. The vector $\bar{\mathbf{x}}$ is a feasible point, and $I = \{i : g_i(\bar{\mathbf{x}}) = 0\}$.

The Cone of Feasible Directions of S *at* $\bar{x}$

This cone was introduced earlier in Definition 4.2.1. The cone of feasible directions, denoted by D, is the set of all nonzero vectors $\mathbf{d}$ such that $\bar{\mathbf{x}} + \lambda\mathbf{d} \in S$ for $\lambda \in (0, \delta)$ for some $\delta > 0$.

The Cone of Attainable Directions of S *at* $\bar{\mathbf{x}}$

A nonzero vector $\mathbf{d}$ belongs to the cone of attainable directions, denoted by A, if there exist a $\delta > 0$ and an $\boldsymbol{\alpha}: E_1 \rightarrow E_n$ such that $\boldsymbol{\alpha}(\lambda) \in S$ for $\lambda \in (0, \delta)$, $\boldsymbol{\alpha}(0) = \bar{\mathbf{x}}$, and

$$\lim_{\lambda \to 0^+} \frac{\boldsymbol{\alpha}(\lambda) - \boldsymbol{\alpha}(0)}{\lambda} = \mathbf{d}$$

In other words, $\mathbf{d}$ belongs to the cone of attainable directions if there is a feasible arc starting from $\bar{\mathbf{x}}$ that is tangential to $\mathbf{d}$.

The Cone of Interior Directions of S *at* $\bar{x}$

This cone, denoted by G_0, was introduced in Section 4.2 and is defined as $G_0 = \{\mathbf{d} : \nabla g_i(\bar{\mathbf{x}})^t\mathbf{d} < 0 \text{ for } i \in I\}$. Note that if X is open and each g_i for $i \notin I$ is continuous at

$\bar{\mathbf{x}}$, then $\mathbf{d} \in G_0$ implies that $\bar{\mathbf{x}} + \lambda\mathbf{d}$ belongs to the interior of the feasible region for $\lambda > 0$ and sufficiently small.

Lemma 5.2.1 below shows that all the above cones and their closures are contained in T.

5.2.1 Lemma

Let X be a nonempty set in E_n, and let $f: E_n \to E_1$ and $g_i: E_n \to E_1$ for $i = 1, \ldots, m$. Consider the problem to minimize $f(\mathbf{x})$ subject to $g_i(\mathbf{x}) \leq 0$ for $i = 1, \ldots, m$ and $\mathbf{x} \in X$. Let $\bar{\mathbf{x}}$ be a feasible point, and let $I = \{i: g_i(\bar{\mathbf{x}}) = 0\}$. Suppose that each g_i for $i \in I$ is differentiable at $\bar{\mathbf{x}}$, and let $G' = \{\mathbf{d}: \nabla g_i(\bar{\mathbf{x}})^t\mathbf{d} \leq 0 \text{ for } i \in I\}$. Then,

$$\text{cl } D \subseteq \text{cl } A \subseteq T \subseteq G'$$

where D, A, and T are, respectively, the cone of feasible directions, the cone of attainable directions, and the cone of tangents of the feasible region at $\bar{\mathbf{x}}$. Furthermore, if X is open and each g_i for $i \notin I$ is continuous at $\bar{\mathbf{x}}$, then $G_0 \subseteq D$, so that

$$\text{cl } G_0 \subseteq \text{cl } D \subseteq \text{cl } A \subseteq T \subseteq G'$$

where G_0 is the cone of interior directions of the feasible region at $\bar{\mathbf{x}}$.

Proof

It can be easily verified that $D \subseteq A \subseteq T \subseteq G'$ and that, since T is closed (see Exercise 5.4), $\text{cl } D \subseteq \text{cl } A \subseteq T \subseteq G'$. Now note that $G_0 \subseteq D$ by Lemma 4.2.4. Hence, the second part of the lemma follows.

To illustrate how each of the containments considered above can be strict, consider the following example. In Figure 4.9, note that $G_0 = \emptyset = \text{cl } G_0$, since there are no interior directions, whereas $D = \text{cl } D = G'$ is defined by the feasible direction along the edge incident at $\bar{\mathbf{x}}$. In regard to the cone of interior directions G_0, note that while any $\mathbf{d} \in G_0$ is a direction leading to interior feasible solutions, it is not true that any feasible direction that leads to interior points belongs to G_0. For example, consider Example 4.3.5 illustrated in Figure 4.12 with the equalities replaced by "less than or equal to" type of inequalities. The set $G_0 = \emptyset$ at $\bar{\mathbf{x}} = (1, 0)^t$, whereas $\mathbf{d} = (-1, 0)^t$ leads to interior feasible points.

To show that cl D can be a strict subset of cl A, consider the region defined by $x_1 - x_2^2 \leq 0$ and $-x_1 + x_2^2 \leq 0$. The set of feasible points lies on the parabola $x_1 = x_2^2$. At $\bar{\mathbf{x}} = (0, 0)^t$, for example, $D = \emptyset = \text{cl } D$, while $\text{cl } A = \{\mathbf{d}: \mathbf{d} = \lambda(0, 1)^t$ or $\mathbf{d} = \lambda(0, -1)^t, \lambda \geq 0\} \equiv G'$.

The possibility that $\text{cl } A \neq T$ is a little more subtle. Suppose that the feasible region S is itself the sequence $\{(1/k, 0)^t, k = 1, 2, \ldots\}$ formed by the intersection of suitable constraints (written as suitable inequalities). For example, we might have $S = \{(x_1, x_2): x_2 = h(x_1), x_2 = 0, 0 \leq x_1 \leq 1$, where $h(x_1) = x_1^3 \sin(\pi/x_1)$ if $x_1 \neq 0$, and $h(x_1) = 0$ if $x_1 = 0\}$. Then, $A = \emptyset = \text{cl } A$, since there are no feasible arcs. However, by definition, $T = \{\mathbf{d}: \mathbf{d} = \lambda(1, 0)^t, \lambda \geq 0\}$, and it is readily verified that $T = G'$.

Finally, Figure 4.7 illustrates an instance where T is a strict subset of G'. Here, $T = \{\mathbf{d}: \mathbf{d} = \lambda(-1, 0)^t, \lambda \geq 0\}$, while $G' = \{\mathbf{d}: \mathbf{d} = \lambda(-1, 0)^t$, or $\mathbf{d} = \lambda(1, 0)^t, \lambda \geq 0\}$.

We now present some constraint qualifications that validate the KKT conditions, and discuss their interrelationships.

Slater's Constraint Qualification

The set X is open, each g_i for $i \in I$ is pseudoconvex at $\bar{\mathbf{x}}$, each g_i for $i \notin I$ is continuous at $\bar{\mathbf{x}}$, and there is an $\mathbf{x} \in X$ such that $g_i(\mathbf{x}) < 0$ for all $i \in I$.

Linear Independence Constraint Qualification

The set X is open, each g_i for $i \notin I$ is continuous at $\bar{\mathbf{x}}$, and $\nabla g_i(\bar{\mathbf{x}})$ for $i \in I$ are linearly independent.

Cottle's Constraint Qualification

The set X is open and each g_i for $i \notin I$ is continuous at $\bar{\mathbf{x}}$, and $\text{cl } G_0 = G'$.

Zangwill's Constraint Qualification

$$\text{cl } D = G'$$

Kuhn–Tucker's Constraint Qualification

$$\text{cl } A = G'$$

Validity of the Constraint Qualifications and Their Interrelationships

In Theorem 5.1.3, we showed that the KKT necessary optimality conditions are necessarily true under Abadie's constraint qualification $T = G'$. We demonstrate below that all the constraint qualifications discussed above imply that of Abadie and, hence, each validate the KKT necessary conditions. From Lemma 5.2.1, it is clear that Cottle's constraint qualification implies that of Zangwill, which implies that of Kuhn and Tucker, which in turn implies Abadie's qualification. We now show that the first two qualifications imply that of Cottle.

First, suppose that Slater's constraint qualification holds true. Then, there is an $\mathbf{x} \in X$ such that $g_i(\mathbf{x}) < 0$ for $i \in I$. Since $g_i(\mathbf{x}) < 0$ and $g_i(\bar{\mathbf{x}}) = 0$, then, by the pseudoconvexity of g_i at $\bar{\mathbf{x}}$, it follows that $\nabla g_i(\bar{\mathbf{x}})^t(\mathbf{x} - \bar{\mathbf{x}}) < 0$. Thus, $\mathbf{d} = \mathbf{x} - \bar{\mathbf{x}}$ belongs to G_0. Therefore, $G_0 \neq \varnothing$ and the reader can verify that $\text{cl } G_0 = G'$ and, hence, that Cottle's constraint qualification holds true. Now suppose that the linear independence constraint qualification is satisfied. Then, $\sum_{i \in I} u_i \nabla g_i(\bar{\mathbf{x}}) = 0$ has no nonzero solution. By Theorem 2.4.9, it follows that there exists a vector $\mathbf{d}$ such that $\nabla g_i(\bar{\mathbf{x}})^t\mathbf{d} < 0$ for $i \in I$. Thus, $G_0 \neq \varnothing$, and Cottle's qualification holds true.

The relationships among the foregoing constraint qualifications are illustrated in Figure 5.2.

In discussing Lemma 5.2.1, we gave various examples in which, for each consecutive pair in the string of containments $\text{cl } G_0 \subseteq \text{cl } D \subseteq \text{cl } A \subseteq T \subseteq G'$, the containment was strict and the larger set was equal to G'. Hence, these examples also illustrate that the implications of Figure 5.2 in regard to these sets are one-way implications. Figure 5.2 accordingly illustrates for each constraint qualification an instance where it holds, whereas the preceding constraint qualification that makes a more restrictive assumption does not hold. Needless to say, if a local minimum $\bar{\mathbf{x}}$ is not a KKT point, as in Example 4.2.10 and illustrated in Figure 4.7, for instance, then no constraint qualification can possibly hold true.

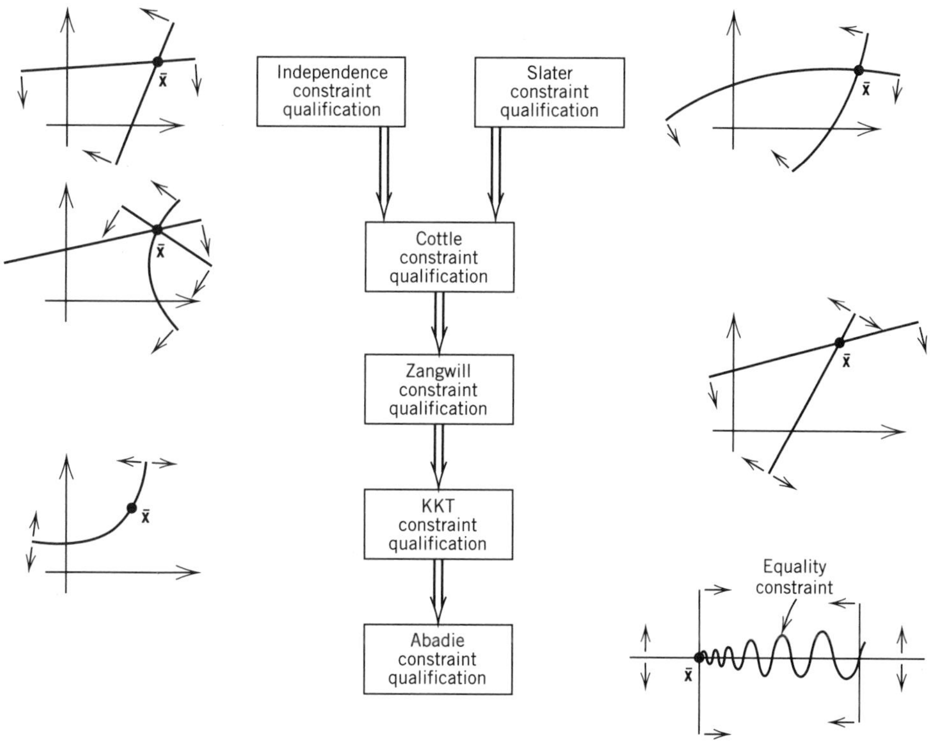

Figure 5.2 Relationships among various constraint qualifications for inequality constrained problems.

Finally, we remark that Cottle's constraint qualification is equivalent to requiring that $G_0 \neq \emptyset$ (see Exercise 5.5). Moreover, we have seen that Slater's constraint qualification and the linear independence constraint qualification both imply Cottle's constraint qualification. Hence, whenever these constraint qualifications hold at a local minimum $\bar{\mathbf{x}}$, then $\bar{\mathbf{x}}$ is a Fritz John point, with the Lagrangian multiplier u_0 associated with the objective function necessarily positive. In contrast, we might have Zangwill's, or Kuhn-Tucker's, or Abadie's constraint qualifications holding at a local minimum $\bar{\mathbf{x}}$ while u_0 might possibly be zero in some solution to the Fritz John conditions. However, since these are valid constraint qualifications, we must also have $u_0 > 0$ in some solution to the Fritz John conditions in such a case.

5.3 Problems with Inequality and Equality Constraints

In this section, we consider problems with both inequality and equality constraints. In particular, consider the following problem:

$$\begin{array}{lll} \text{Minimize} & f(\mathbf{x}) & \\ \text{subject to} & g_i(\mathbf{x}) \leq 0 & \text{for } i = 1, \ldots, m \\ & h_i(\mathbf{x}) = 0 & \text{for } i = 1, \ldots, l \\ & \mathbf{x} \in X & \end{array}$$

By Theorem 5.1.2, a necessary optimality condition is $F_0 \cap T = \emptyset$ at a local minimum $\bar{\mathbf{x}}$. By imposing the constraint qualification $T = G' \cap H_0$, where $H_0 =$

$\{\mathbf{d}: \nabla h_i(\bar{\mathbf{x}})^t\mathbf{d} = 0 \text{ for } i = 1, \ldots, l\}$, this implies that $F_0 \cap G' \cap H_0 = \varnothing$. By using Farkas' theorem [or see equation (4.16)], the KKT conditions follow. Theorem 5.3.1 below reiterates this. The process is summarized in the flowchart below.

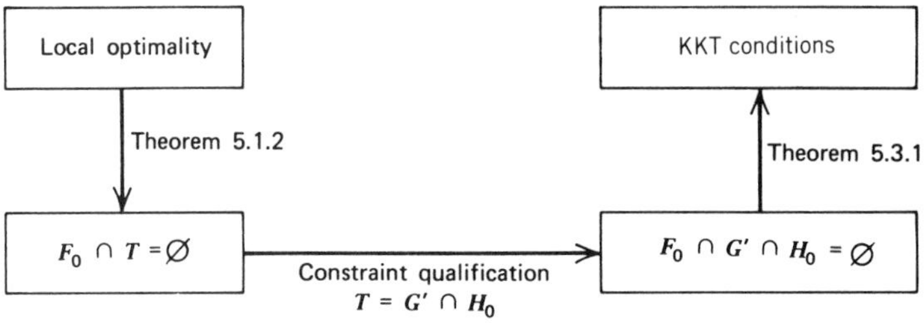

5.3.1 Theorem (Karush–Kuhn–Tucker Conditions)

Let $f: E_n \to E_1$, $g_i: E_n \to E_1$ for $i = 1, \ldots, m$, and $h_i: E_n \to E_1$ for $i = 1, \ldots, l$, and let X be a nonempty set in E_n. Consider the following problem:

$$\begin{array}{lll} \text{Minimize} & f(\mathbf{x}) & \\ \text{subject to} & g_i(\mathbf{x}) \leq 0 & \text{for } i = 1, \ldots, m \\ & h_i(\mathbf{x}) = 0 & \text{for } i = 1, \ldots, l \\ & \mathbf{x} \in X & \end{array}$$

Let $\bar{\mathbf{x}}$ locally solve the problem, and let $I = \{i: g_i(\bar{\mathbf{x}}) = 0\}$. Suppose that f, g_i for $i \in I$, and h_i for $i = 1, \ldots, l$ are differentiable at $\bar{\mathbf{x}}$. Suppose that the constraint qualification $T = G' \cap H_0$ holds true, where T is the cone of tangents of the feasible region at $\bar{\mathbf{x}}$, and

$$G' = \{\mathbf{d}: \nabla g_i(\bar{\mathbf{x}})^t \leq 0 \quad \text{for } i \in I\}.$$

$$H_0 = \{\mathbf{d}: \nabla h_i(\bar{\mathbf{x}})^t\mathbf{d} = 0 \quad \text{for } i = 1, \ldots, l\}$$

Then, $\bar{\mathbf{x}}$ is a KKT point, that is, there exist scalars $u_i \geq 0$ for $i \in I$ and v_i for $i = 1, \ldots, l$ such that

$$\nabla f(\bar{\mathbf{x}}) + \sum_{i \in I} u_i \nabla g_i(\bar{\mathbf{x}}) + \sum_{i=1}^{l} v_i \nabla h_i(\bar{\mathbf{x}}) = \mathbf{0}$$

Proof

Since $\bar{\mathbf{x}}$ solves the problem locally, by Theorem 5.1.2, we get $F_0 \cap T = \varnothing$. By the constraint qualification, we have $F_0 \cap G' \cap H_0 = \varnothing$, that is, the system $\mathbf{Ad} \leq \mathbf{0}$ and $\mathbf{c}^t\mathbf{d} > 0$ has no solution, where the rows of $\mathbf{A}$ are given by $\nabla g_i(\bar{\mathbf{x}})^t$ for $i \in I$, $\nabla h_i(\bar{\mathbf{x}})^t$ and $-\nabla h_i(\bar{\mathbf{x}})^t$ for $i = 1, \ldots, l$, and $\mathbf{c} = -\nabla f(\bar{\mathbf{x}})$. By Theorem 2.4.5, the system $\mathbf{A}^t\mathbf{y} = \mathbf{c}$ and $\mathbf{y} \geq \mathbf{0}$ has a solution. This implies that there exist nonnegative scalars u_i for $i \in I$ and α_i, β_i for $i = 1, \ldots, l$ such that

$$\nabla f(\bar{\mathbf{x}}) + \sum_{i \in I} u_i \nabla g_i(\bar{\mathbf{x}}) + \sum_{i=1}^{l} \alpha_i \nabla h_i(\bar{\mathbf{x}}) - \sum_{i=1}^{l} \beta_i \nabla h_i(\bar{\mathbf{x}}) = \mathbf{0}$$

Letting $v_i = \alpha_i - \beta_i$ for each i, the result is apparent.

We now present several constraint qualifications that validate the KKT conditions. These qualifications use several cones that were defined earlier in the chapter. By replacing each equality constraint by two equivalent inequalities, the role played by G' in the previous section is now played by the cone $G' \cap H_0$. The reader may note that Zangwill's constraint qualification is omitted here, since the cone of feasible directions is usually equal to the zero vector in the presence of nonlinear equality constraints.

Slater's Constraint Qualification

The set X is open, each g_i for $i \in I$ is pseudoconvex at $\bar{\mathbf{x}}$, each g_i for i L I is continuous at $\bar{\mathbf{x}}$, each h_i for $i = 1, \ldots, l$ is quasiconvex, quasiconcave, and continuously differentiable at $\bar{\mathbf{x}}$, and $\nabla h_i(\bar{\mathbf{x}})$ for $i = 1, \ldots, l$ are linearly independent. Furthermore, there exists an $\mathbf{x} \in X$ such that $g_i(\mathbf{x}) < 0$ for all $i \in I$ and $h_i(\mathbf{x}) = 0$ for all $i = 1, \ldots, l$.

Linear Independence Constraint Qualification The set X is open, each g_i for i L I is continuous at $\bar{\mathbf{x}}$, $\nabla g_i(\bar{\mathbf{x}})$ for $i \in I$ and $\nabla h_i(\bar{\mathbf{x}})$ for $i = 1, \ldots, l$, are linearly independent, and each h_i for $i = 1, \ldots, l$ is continuously differentiable at $\bar{\mathbf{x}}$.

Cottle's Constraint Qualification

The set X is open, each g_i for i L I is continuous at $\bar{\mathbf{x}}$, each h_i for $i = 1, \ldots, l$ is continuously differentiable at $\bar{\mathbf{x}}$, and $\nabla h_i(\bar{\mathbf{x}})$ for $i = 1, \ldots, l$ are linearly independent. Furthermore, $\text{cl}\,(G_0 \cap H_0) = G' \cap H_0$. (This is equivalent to the Mangasarian–Fromovitz constraint qualification that requires $\nabla h_i(\bar{\mathbf{x}})$, $i = 1, \ldots, l$ to be linearly independent and that $G_0 \cap H_0 \neq \varnothing$; see Exercise 5.20.)

Kuhn–Tucker's Constraint Qualification

$$\text{cl}\, A = G' \cap H_0$$

Abadie's Constraint Qualification

$$T = G' \cap H_0$$

Validity of the Constraint Qualifications and Their Interrelationships

In Theorem 5.3.1, we showed that the KKT conditions are true if Abadie's constraint qualification $T = G' \cap H_0$ is satisfied. We demonstrate below that all the constraint qualifications given above imply that of Abadie and, hence, each validates the KKT necessary conditions.

As in Lemma 5.2.1, the reader can easily verify that $\text{cl}\, A \subseteq T \subseteq G' \cap H_0$. Now suppose that X is open, g_i for each i L I is continuous at $\bar{\mathbf{x}}$, h_i for each $i = 1, \ldots, l$ is continuously differentiable, and $\nabla h_i(\bar{\mathbf{x}})$ for $i = 1, \ldots, l$ are linearly independent. From the proof of Theorem 4.3.1, it follows that $G_0 \cap H_0 \subseteq A$. Thus, $\text{cl}\,(G_0 \cap H_0) \subseteq \text{cl}\, A \subseteq T \subseteq G' \cap H_0$. In particular, Cottle's constraint qualification implies that of Kuhn and Tucker, which in turn implies Abadie's constraint qualification.

We now demonstrate that Slater's constraint qualification and the linear independence constraint qualification imply that of Cottle. Suppose that Slater's qualification is satisfied, so that $g_i(\mathbf{x}) < 0$ for $i \in I$ and $h_i(\mathbf{x}) = 0$ for $i = 1, \ldots, l$ for some $\mathbf{x} \in X$. By the pseudoconvexity of g_i at $\bar{\mathbf{x}}$, $\nabla g_i(\bar{\mathbf{x}})^t(\mathbf{x} - \bar{\mathbf{x}}) < 0$ for $i \in I$.

Also since $h_i(\mathbf{x}) = h_i(\bar{\mathbf{x}}) = 0$, quasiconvexity and quasiconcavity of h_i at $\bar{\mathbf{x}}$ imply that $\nabla h_i(\bar{\mathbf{x}})^t(\mathbf{x} - \bar{\mathbf{x}}) = 0$. Letting $\mathbf{d} = \mathbf{x} - \bar{\mathbf{x}}$, it then follows that $\mathbf{d} \in G_0 \cap H_0$. Thus, $G_0 \cap H_0 \neq \varnothing$, and the reader can verify that cl $(G_0 \cap H_0) = G' \cap H_0$. Therefore, Cottle's constraint qualification holds true.

Finally, we show that the linear independence qualification implies that of Cottle. By contradiction, suppose that $G_0 \cap H_0 = \varnothing$. Then, using a separation theorem as in the proof of Theorem 4.3.2, it follows that there exists a nonzero vector $(\mathbf{u}_I, \mathbf{v})$ such that $\sum_{i\in I} u_i \nabla g_i(\bar{\mathbf{x}}) + \sum_{i=1}^{l} v_i \nabla h_i(\bar{\mathbf{x}}) = \mathbf{0}$, where $\mathbf{u}_I \geq \mathbf{0}$ is the vector whose ith component is u_i. This contradicts the linear independence assumption. Thus, Cottle's constraint qualification holds true.

In Figure 5.3, we summarize the implications of the constraint qualifications discussed above (see also the illustrations of Figure 5.2). As mentioned earlier, these implications, together with Theorem 5.3.1, validate the KKT conditions.

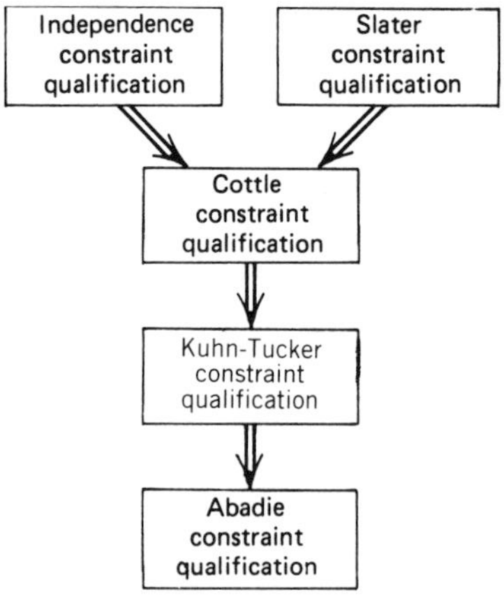

Figure 5.3 Relationships among constraint qualifications for problems with inequality and equality constraints.

Second-Order Constraint Qualifications for Inequality and Equality Constrained Problems

In Chapter 4, we developed second-order necessary KKT optimality conditions. In particular, we observed in Theorem 4.4.3 that if $\bar{\mathbf{x}}$ is a local minimum and if all problem-defining functions are twice-differentiable, with the gradients $\nabla g_i(\bar{\mathbf{x}})$ for $i \in I$ and $\nabla h_i(\bar{\mathbf{x}})$ for $i = 1, \ldots, l$ of the binding constraints being linearly independent, then $\bar{\mathbf{x}}$ is a KKT point and, additionally, $\mathbf{d}^t\nabla^2 L(\bar{\mathbf{x}})\mathbf{d} \geq 0$ must hold for all $\mathbf{d} \in G' \cap H_0$. Hence, the linear independence condition affords a *second-order constraint qualification*, which implies that, in addition to $\bar{\mathbf{x}}$ being a KKT point, a second-order type of condition must also hold true.

Alternatively, we can stipulate the following second-order constraint qualification, which is in the spirit of Abadie's constraint qualification. Suppose that all problem-defining functions are twice-differentiable and that $\bar{\mathbf{x}}$ is a local minimum at which Abadie's constraint qualification $T = G' \cap H_0$ holds true. Hence, we know from Theorem 5.3.1 that $\bar{\mathbf{x}}$ is a KKT point. Denote by $\bar{\mathbf{u}}$ and $\bar{\mathbf{v}}$ the associated set of Lagrangian multipliers corresponding to the inequality and equality constraints, respectively, and let

$I = \{i: g_i(\bar{\mathbf{x}}) = 0\}$ represent the binding inequality constraints. Now, as in Theorem 4.2.2, define $C = \{\mathbf{d} \neq \mathbf{0} : \nabla g_i(\bar{\mathbf{x}})^t\mathbf{d} = 0 \text{ for } i \in I^+, \nabla g_i(\bar{\mathbf{x}})^t\mathbf{d} \leq 0 \text{ for } i \in I^0, \text{ and } \nabla h_i(\bar{\mathbf{x}})^t\mathbf{d} = 0 \text{ for } i = 1, \ldots, l\}$, where $I^+ = \{i \in I : \bar{u}_i > 0\}$ and $I^0 = I - I^+$.

Accordingly, let T' denote the cone of tangents at $\bar{\mathbf{x}}$ when the inequality constraints with indices $i \in I^+$ are also treated as equalities. Then, the stated second-order constraint qualification asserts that if $T' = C \cup \{\mathbf{0}\}$, we must have $\mathbf{d}^t\nabla^2 L(\bar{\mathbf{x}})\mathbf{d} \geq 0$ for each $\mathbf{d} \in C$. We ask the reader to show that this assertion is valid in Exercise 5.18, using a proof similar to that of Theorem 4.4.3. Note that, in general, $T' \subseteq C \cup \{\mathbf{0}\}$, in the same manner as $T \subseteq G' \cap H_0$. However, as evident from the proof of Theorem 4.4.3, under the linear independence constraint qualification, any $\mathbf{d} \in C$ also corresponds to a limiting direction based on a feasible arc and, hence, based on a sequence of points. Therefore, $C \cup \{\mathbf{0}\} \subseteq T'$. This shows that the linear independence constraint qualification implies that $T' = C \cup \{\mathbf{0}\}$. We ask the reader to construct the precise details of this argument in Exercise 5.18. In a similar manner, we can state another second-order constraint qualification in the spirit of Kuhn-Tucker's cone of attainable directions constraint qualification. This is addressed in Exercise 5.19, where we ask the reader to justify it and to show that this is also implied by the linear independence constraint qualification.

Exercises

5.1 Find the cone of tangents for each of the following sets at the point $\bar{\mathbf{x}} = (0, 0)^t$:

a. $S = \{(x_1, x_2): x_2 \geq -x_1^3\}$

b. $S = \{(x_1, x_2): x_1 \text{ is integer}, x_2 = 0\}$

c. $S = \{(x_1, x_2): x_1 \text{ is rational}, x_2 = 0\}$

5.2 Let S be a subset of E_n, and let $\bar{\mathbf{x}} \in \text{int } S$. Show that the cone of tangents of S at $\bar{\mathbf{x}}$ is E_n.

5.3 Prove that the cone of tangents defined in Definition 5.1.1 can be equivalently characterized in either of the following ways:

a. $T = \{\mathbf{d}$: there exists a sequence $\{\lambda_k\} \to 0^+$ and a function $\boldsymbol{\alpha}: E_1 \to E_n$, where $\boldsymbol{\alpha}(\lambda) \to 0$ as $\lambda \to 0$, such that $\mathbf{x}_k = \bar{\mathbf{x}} + \lambda_k\mathbf{d} + \lambda_k\boldsymbol{\alpha}(\lambda_k) \in S$ for each $k\}$

b. $T = \left\{\mathbf{d} : \mathbf{d} = \lambda \lim_{k\to\infty} \dfrac{\mathbf{x}_k - \bar{\mathbf{x}}}{\|\mathbf{x}_k - \bar{\mathbf{x}}\|}\right.$, where $\lambda \geq 0$, $\{\mathbf{x}_k\} \to \bar{\mathbf{x}}$, and where $\mathbf{x}_k \in S$ and $\mathbf{x}_k \neq \bar{\mathbf{x}}$, for each $k\}$

5.4 Prove that the cone of tangents is a closed cone.

Hint: First show that $T = \bigcap_{N\in\mathcal{N}} \text{cl } K(S \cap N, \bar{\mathbf{x}})$, where $K(S \cap N, \bar{\mathbf{x}}) = \{\lambda(\mathbf{x} - \bar{\mathbf{x}}): \mathbf{x} \in S \cap N, \lambda > 0\}$, and $\mathcal{N}$ is the class of all open neighborhoods about $\bar{\mathbf{x}}$.

5.5 Let $\mathbf{A}$ be an $m \times n$ matrix, and consider the cones $G_0 = \{\mathbf{d}: \mathbf{A}\mathbf{d} < \mathbf{0}\}$ and $G' = \{\mathbf{d}: \mathbf{A}\mathbf{d} \leqslant \mathbf{0}\}$. Prove that

a. G_0 is an open convex cone.

b. G' is a closed convex cone.

c. $G_0 = \text{int } G'$.

d. $\text{cl } G_0 = G'$ if and only if $G_0 \neq \varnothing$.

5.6 Consider the problem to minimize $f(\mathbf{x})$ subject to $\mathbf{x} \in X$ and $g_i(\mathbf{x}) \leq 0$ for $i = 1, \ldots, m$. Let $\bar{\mathbf{x}}$ be a feasible point, and let $I = \{i: g_i(\bar{\mathbf{x}}) = 0\}$. Suppose that X is open and each g_i for i L I is continuous at $\bar{\mathbf{x}}$. Further, suppose that the set

$$\{\mathbf{d}: \nabla g_i(\bar{\mathbf{x}})^t\mathbf{d} \leq 0 \quad \text{for } i \in J, \quad \nabla g_i(\bar{\mathbf{x}})^t\mathbf{d} < 0 \quad \text{for } i \in I - J\}$$

is not empty, where $J = \{i \in I: g_i \text{ is pseudoconcave at } \bar{\mathbf{x}}\}$. Show that this condition is sufficient to validate the KKT conditions at $\bar{\mathbf{x}}$. (This is the Arrow-Hurwicz-Uzawa constraint qualification.)

5.7 Consider the problem to minimize $f(\mathbf{x})$ subject to $g_i(\mathbf{x}) \leq 0$ for $i = 1, \ldots, m$. Let $\bar{\mathbf{x}}$ be feasible, and let $I = \{i : g_i(\bar{\mathbf{x}}) = 0\}$. Let $(\bar{z}, \bar{\mathbf{d}})$ be an optimal solution to the following linear program:

$$\begin{aligned} &\text{Minimize} && z \\ &\text{subject to} && \nabla f(\bar{\mathbf{x}})^t\mathbf{d} - z \leq 0 \\ &&& \nabla g_i(\bar{\mathbf{x}})^t\mathbf{d} - z \leq 0 \quad \text{for } i \in I \\ &&& -1 \leq d_j \leq 1 \quad \text{for } j = 1, \ldots, n \end{aligned}$$

a. Show that the Fritz John conditions hold true if $\bar{z} = 0$.

b. Show that if $\bar{z} = 0$, then the KKT conditions hold true under Cottle's constraint qualification.

5.8 For each of the following sets, find the cone of feasible directions and the cone of attainable directions at $\bar{\mathbf{x}} = (0, 0)^t$:

a. $S = \{(x_1, x_2) : -1 \leq x_1 \leq 1, x_2 \geq x_1^{1/3}, x_2 \geq x_1\}$

b. $S = \{(x_1, x_2) : x_2 > x_1^2\}$

c. $S = \{(x_1, x_2) : x_2 = -x_1^3\}$

d. $S = S_1 \cup S_2$, where

$S_1 = \{(x_1, x_2) : x_1 \geq 0, x_2 \geq x_1^2\}$ and $S_2 = \{(x_1, x_2) : x_1 \leq 0, -2x_1 \leq 3x_2 \leq -x_1\}$

5.9 Let $f: E_n \to E_1$ be differentiable at $\bar{\mathbf{x}}$ with a nonzero gradient $\nabla f(\bar{\mathbf{x}})$. Let $S = \{\mathbf{x} : f(\mathbf{x}) \geq f(\bar{\mathbf{x}})\}$. Show that the cone of tangents and the cone of attainable directions of S at $\bar{\mathbf{x}}$ are both given by $\{\mathbf{d} : \nabla f(\bar{\mathbf{x}})^t\mathbf{d} \geq 0\}$. Does the result hold true if $\nabla f(\bar{\mathbf{x}}) = \mathbf{0}$? Prove or give a counterexample.

5.10 Consider the following problem:

$$\begin{aligned} &\text{Minimize} && -x_1 \\ &\text{subject to} && x_1^2 + x_2^2 \leq 1 \\ &&& (x_1 - 1)^3 - x_2 \leq 0 \end{aligned}$$

a. Show that the Kuhn-Tucker constraint qualification holds true at $\bar{\mathbf{x}} = (1, 0)^t$.

b. Show that $\bar{\mathbf{x}} = (1, 0)^t$ is a KKT point and also that it is the global optimal solution.

5.11 Consider the problem to maximize $5x - x^2$ subject to $g_1(x) \leq 0$, where $g_1(x) = x$.

a. Verify graphically that $\bar{x} = 0$ is the optimal solution.

b. Verify that each of the constraint qualifications discussed in Section 5.2 holds true at $\bar{x} = 0$.

c. Verify that the KKT necessary conditions hold true at $\bar{x} = 0$.

Now, suppose that the constraint $g_2(x) \leq 0$ is added to the above problem, where

$$g_2(x) = \begin{cases} -1 - x & \text{if } x \geq 0 \\ 1 - x & \text{if } x < 0 \end{cases}$$

Note that $\bar{x} = 0$ is still the optimal solution and that g_2 is discontinuous and nonbinding at $\bar{x}$. Check whether the constraint qualifications discussed in Section 5.2 and the KKT conditions hold true at $\bar{x}$.

(This exercise illustrates the need of the continuity assumption of the nonbinding constraints.)

5.12 Consider the feasible region $S = \{\mathbf{x} \in X : g_1(\mathbf{x}) \leq 0\}$, where $g_1(\mathbf{x}) = x_1^2 + x_2^2 - 1$, and X is the collection of all convex combinations of the four points $(-1, 0)^t$, $(0, 1)^t$, $(1, 0)^t$, and $(0, -1)^t$.

a. Find the cone of tangents T of S at $\bar{\mathbf{x}} = (1, 0)^t$.

b. Check whether $T \supseteq G'$, where $G' = \{\mathbf{d} : \nabla g_1(\bar{\mathbf{x}})^t\mathbf{d} \leq 0\}$.

c. Replace the set X by four inequality constraints. Repeat parts a and b, where $G' = \{\mathbf{d} : \nabla g_i(\bar{\mathbf{x}})^t\mathbf{d} \leq 0 \text{ for } i \in I\}$ and I is the new set of binding constraints at $\bar{\mathbf{x}} = (1, 0)^t$.

5.13 Let $S = \{\mathbf{x} \in X : g_i(\mathbf{x}) \leq 0 \text{ for } i = 1, \ldots, m\}$. Let $\bar{\mathbf{x}} \in S$, and let $I = \{i : g_i(\bar{\mathbf{x}}) = 0\}$. Show that $T \subseteq G'$, where T is the cone of tangents of S at $\bar{\mathbf{x}}$, and $G' = \{\mathbf{d} : \nabla g_i(\bar{\mathbf{x}})^t\mathbf{d} \leq 0 \text{ for } i \in I\}$.

5.14 Let $S = \{\mathbf{x} \in S : g_i(\mathbf{x}) \leq 0 \text{ for } i = 1, \ldots, m \text{ and } h_i(\mathbf{x}) = 0 \text{ for } i = 1, \ldots, l\}$. Let $\bar{\mathbf{x}} \in S$, and let $I = \{i : g_i(\bar{\mathbf{x}}) = 0\}$. Show that $T \subseteq G' \cap H_0$, where T is the cone of tangents of S at $\bar{\mathbf{x}}$, $G' = \{\mathbf{d} : \nabla g_i(\bar{\mathbf{x}})^t\mathbf{d} \leq 0 \text{ for } i \in I\}$, and $H_0 = \{\mathbf{d} : \nabla h_i(\bar{\mathbf{x}})^t\mathbf{d} = 0 \text{ for } i = 1, \ldots, l\}$.

5.15 Consider the constraints $\mathbf{Cd} \leq \mathbf{0}$ and $\mathbf{d}'\mathbf{d} \leq 1$. Let $\bar{\mathbf{d}}$ be a feasible solution such that $\bar{\mathbf{d}}'\bar{\mathbf{d}} = 1$, $\mathbf{C}_1\bar{\mathbf{d}} = \mathbf{0}$, and $\mathbf{C}_2\bar{\mathbf{d}} < \mathbf{0}$, where $\mathbf{C}' = (\mathbf{C}_1', \mathbf{C}_2')$. Show that the constraint qualification $T = G_1 = \{\mathbf{d}: \mathbf{C}_1\mathbf{d} \leq \mathbf{0},\ \mathbf{d}'\bar{\mathbf{d}} \leq 0\}$ holds true, where T is the cone of tangents of the constraint set at $\bar{\mathbf{d}}$.

5.16 For a nonlinear optimization problem, let $\bar{\mathbf{x}}$ be a feasible solution and let F be the set of improving directions, let $F_0 = \{\mathbf{d}: \nabla f(\bar{\mathbf{x}})'\mathbf{d} < 0\}$, and let T be the cone of tangents at $\bar{\mathbf{x}}$. If $\bar{\mathbf{x}}$ is a local minimum, is $F \cap T = \varnothing$? Is $F \cap T = \varnothing$ sufficient to claim that $\bar{\mathbf{x}}$ is a local minimum? Give examples to justify your answers. Show that if there exists an ε-neighborhood about $\bar{\mathbf{x}}$ over which f is pseudoconvex and the feasible region is a convex set, then $F_0 \cap T = \varnothing$ implies that $\bar{\mathbf{x}}$ is a local minimum, and so these conditions also guarantee that $\bar{\mathbf{x}}$ is a local minimum whenever $F \cap T = \varnothing$.

5.17 Consider Abadie's constraint qualification $T = G' \cap H_0$ for the case of inequality and equality constraints. Using the Kuhn-Tucker example of Figure 4.7, for instance, demonstrate by considering differentiable objective functions that it is valid and more general to instead require $F_0 \cap T = F_0 \cap G' \cap H_0$ to guarantee that if $\bar{\mathbf{x}}$ is a local minimum, then it is a KKT point. (Typically, "constraint qualifications" address only the behavior of the constraints and neglect the objective function.) Investigate the KKT conditions and Abadie's constraint qualification for the problem to minimize $\{f(\mathbf{x}): g_i(\mathbf{x}) \leq 0$ for $i = 1, \ldots, m$ and $h_i(\mathbf{x}) = 0$ for $i = 1, \ldots, l\}$ versus those for the equivalent problem to minimize $\{z: f(\mathbf{x}) \leq z,\ g_i(\mathbf{x}) \leq 0$ for $i = 1, \ldots, m$ and $h_i(\mathbf{x}) = 0$ for $i = 1, \ldots, l\}$.

5.18 Consider the problem to minimize $f(\mathbf{x})$ subject to $g_i(\mathbf{x}) \leq 0$ for $i = 1, \ldots, m$ and $h_i(\mathbf{x}) = 0$ for $i = 1, \ldots, l$, where all problem-defining functions are twice-differentiable. Let $\bar{\mathbf{x}}$ be a local minimum, and suppose that the cone of tangents $T = G' \cap H_0$, where $G' = \{\mathbf{d}: \nabla g_i(\bar{\mathbf{x}})'\mathbf{d} \leq 0$ for $i \in I\}$, $I = \{i: g_i(\bar{\mathbf{x}}) = 0\}$, and $H_0 = \{\mathbf{d}: \nabla h_i(\bar{\mathbf{x}})'\mathbf{d} = 0$ for $i = 1, \ldots, l\}$. Hence, by Theorem 5.3.1, $\bar{\mathbf{x}}$ is a KKT point. Let $\bar{u}_i$, $i = 1, \ldots, m$, and $\bar{v}_i$, $i = 1, \ldots, l$, be the associated Lagrangian multipliers in the KKT solution with respect to the inequality and the equality constraints, and define $I^+ = \{i: \bar{u}_i > 0\}$, and let $I^0 = I - I^+$. Now define T' as the cone of tangents at $\bar{\mathbf{x}}$ with respect to the region $\{\mathbf{x}: g_i(\mathbf{x}) = 0$ for $i \in I^+$, $g_i(\mathbf{x}) \leq 0$ for $i \in I^0$, $h_i(\mathbf{x}) = 0$ for $i = 1, \ldots, l\}$, and denote $C = \{\mathbf{d} \neq \mathbf{0}: \nabla g_i(\bar{\mathbf{x}})'\mathbf{d} = 0$ for $i \in I^+$, $\nabla g_i(\bar{\mathbf{x}})'\mathbf{d} \leq 0$ for $i \in I^0$, and $\nabla h_i(\bar{\mathbf{x}})'\mathbf{d} = 0$ for $i = 1, \ldots, l\}$. Show that if $T' = C \cup \{\mathbf{0}\}$, then the second-order necessary condition $\mathbf{d}'\nabla^2 L(\bar{\mathbf{x}})\mathbf{d} \geq 0$ holds for all $\mathbf{d} \in C$, where $L(\mathbf{x}) = f(\mathbf{x}) + \sum_{i \in I} \bar{u}_i g_i(\mathbf{x}) + \sum_{i=1}^{l} \bar{v}_i h_i(\mathbf{x})$. Also, show that the linear independence constraint qualification implies that $T' = C \cup \{\mathbf{0}\}$. (*Hint*: Examine the proof of Theorem 4.4.3.)

5.19 Consider the problem to minimize $f(\mathbf{x})$ subject to $g_i(\mathbf{x}) \leq 0$ for $i = 1, \ldots, m$ and $h_i(\mathbf{x}) = 0$ for $i = 1, \ldots, l$, where all problem-defining functions are twice-differentiable. Let $\bar{\mathbf{x}}$ be a local minimum that is also a KKT point. Define $\bar{C} = \{\mathbf{d} \neq \mathbf{0} : \nabla g_i(\bar{\mathbf{x}})'\mathbf{d} = 0$ for $i \in I$, and $\nabla h_i(\bar{\mathbf{x}})'\mathbf{d} = 0$ for $i = 1, \ldots, l\}$, where $I = \{i: g_i(\bar{\mathbf{x}}) = \mathbf{0}\}$. The second-order cone of attainable directions constraint qualification is said to hold at $\bar{\mathbf{x}}$ if every $\mathbf{d} \in \bar{C}$ is tangential to a twice-differentiable arc incident at $\bar{\mathbf{x}}$; that is, for every $\mathbf{d} \in \bar{C}$, there exists a twice-differentiable function $\boldsymbol{\alpha}: [0, \varepsilon] \to E_n$ for some $\varepsilon > 0$, such that for each $0 \leq \lambda \leq \varepsilon$, $\boldsymbol{\alpha}(0) = \bar{\mathbf{x}}$, $g_i[\boldsymbol{\alpha}(\lambda)] = 0$ for $i \in I$, and $h_i[\boldsymbol{\alpha}(\lambda)] = 0$ for $i = 1, \ldots, l$ and $\lim_{\lambda \to 0^+} [\boldsymbol{\alpha}(\lambda) - \boldsymbol{\alpha}(0)]/\lambda = \theta\mathbf{d}$ for some $\theta > 0$. Assuming that this condition holds, show that $\mathbf{d}'\nabla^2 L(\bar{\mathbf{x}})\mathbf{d} \geq 0$ for all $\mathbf{d} \in \bar{C}$, where $L(\bar{\mathbf{x}})$ is defined by (4.25). Also, show that this second-order constraint qualification is implied by the linear independence constraint qualification.

5.20 Consider the problem to minimize $f(\mathbf{x})$ subject to $g_i(\mathbf{x}) \leq 0$ for $i = 1, \ldots, m$, $h_i(\mathbf{x}) = 0$ for $i = 1, \ldots, l$, and $\mathbf{x} \in X$, where X is open and where all problem-defining functions are differentiable. Let $\bar{\mathbf{x}}$ be a feasible solution. The Mangasarian–Fromovitz constraint qualification requires that $\nabla h_i(\bar{\mathbf{x}})$ for $i = 1, \ldots, l$ are linearly independent, and that $G_0 \cap H_0 \neq \varnothing$, where $G_0 = \{\mathbf{d}: \nabla g_i(\bar{\mathbf{x}})'\mathbf{d} < 0$ for $i \in I\}$, $I = \{i: g_i(\bar{\mathbf{x}}) = 0\}$, and $H_0 = \{\mathbf{d}: \nabla h_i(\bar{\mathbf{x}})'\mathbf{d} = 0$ for $i = 1, \ldots, l\}$. Show that $G_0 \cap H_0 \neq \varnothing$ if and only if $\mathrm{cl}\,(G_0 \cap H_0) = G' \cap H_0$, where $G' = \{\mathbf{d}: \nabla g_i(\bar{\mathbf{x}})'\mathbf{d} \leq 0$ for $i \in I\}$ and, hence, that this constraint qualification is equivalent to that of Cottle's.

5.21 Consider the problem to minimize $f(\mathbf{x})$ subject to $g_i(\mathbf{x}) \le 0$ for $i = 1, \ldots, m$, and $h_i(\mathbf{x}) = 0$ for $i = 1, \ldots, l$, where all problem-defining functions are twice-differentiable. Let $\bar{\mathbf{x}}$ be a local minimum that also happens to be a KKT point with associated Lagrangian multipliers vectors $\bar{\mathbf{u}}$ and $\bar{\mathbf{v}}$ corresponding to the inequality and equality constraints, respectively. Define $I = \{i : g_i(\bar{\mathbf{x}}) = 0\}$, $I^+ = \{i : \bar{u}_i > 0\}$, $I^0 = I - I^+$, $\bar{G}_0 = \{\mathbf{d} : \nabla g_i(\bar{\mathbf{x}})^t\mathbf{d} < 0 \text{ for } i \in I^0, \nabla g_i(\bar{\mathbf{x}})^t\mathbf{d} = 0 \text{ for } i \in I^+\}$, and $H_0 = \{\mathbf{d} : \nabla h_i(\bar{\mathbf{x}})^t\mathbf{d} = 0 \text{ for } i = 1, \ldots, l\}$. The Strict Mangasarian–Fromovitz Constraint Qualification (SMFCQ) is said to hold at $\bar{\mathbf{x}}$ if $\nabla g_i(\bar{\mathbf{x}})$ for $i \in I^+$ and $\nabla h_i(\bar{\mathbf{x}})$ for $i = 1, \ldots, l$ are linearly independent and $\bar{G}_0 \cap H_0 \neq \varnothing$. Show that the SMFCQ condition holds at $\bar{\mathbf{x}}$ if and only if the KKT Lagrangian multiplier vector $(\bar{\mathbf{u}}, \bar{\mathbf{v}})$ is unique. Moreover, show that if the SMFCQ condition holds at $\bar{\mathbf{x}}$, then $\mathbf{d}^t\nabla^2 L(\bar{\mathbf{x}})\mathbf{d} \ge 0$ for all $\mathbf{d} \in C = \{\mathbf{d} \neq \mathbf{0} : \nabla g_i(\bar{\mathbf{x}})^t\mathbf{d} = 0 \text{ for } i \in I^+, \nabla g_i(\bar{\mathbf{x}})^t\mathbf{d} \le 0 \text{ for } i \in I^0,$ and $\nabla h_i(\bar{\mathbf{x}})^t\mathbf{d} = 0 \text{ for } i = 1, \ldots, l\}$, where $L(\mathbf{x}) = f(\mathbf{x}) + \sum_{i \in I} \bar{u}_i g_i(\mathbf{x}) + \sum_{i=1}^{l} \bar{v}_i h_i(\mathbf{x})$. (This result is due to Kyparisis [1985] and Ben-Tal [1980].)

5.22 Consider the feasible region defined by $g_i(\mathbf{x}) \le 0$ for $i = 1, \ldots, m$, where $g_i : E_n \to E_1$, $i = 1, \ldots, m$ are differentiable functions. Let $\bar{\mathbf{x}}$ be a feasible solution, and denote by Ξ the set of differentiable objective functions $f : E_n \to E_1$ for which $\bar{\mathbf{x}}$ is a local minimum, and let $D\Xi = \{\mathbf{y} : \mathbf{y} = \nabla f(\bar{\mathbf{x}}) \text{ for some } f \in \Xi\}$. Define the set $G' = \{\mathbf{d} : \nabla g_i(\bar{\mathbf{x}})^t\mathbf{d} \le 0 \text{ for } i \in I\}$, where $I = \{i : g_i(\bar{\mathbf{x}}) = 0\}$, and let T be the cone of tangents at $\bar{\mathbf{x}}$. Furthermore, for any set S, let S_* denote its reverse polar cone defined as $\{\mathbf{y} : \mathbf{y}^t\mathbf{x} \ge 0 \text{ for all } \mathbf{x} \in S\}$.

a. Show that $D\Xi = T_*$.

b. Show that the KKT conditions hold for all $f \in \Xi$ if and only if $T_* = G'_*$. (*Hint*: The statement in part b occurs if and only if $D\Xi \subseteq G'_*$. Now use part a along with the fact that $T_* \supseteq G'_*$ since $T \subseteq G'$. This result is due to Gould and Tolle [1971, 1972].)

Notes and References

In this chapter, we provide an alternative derivation of the KKT conditions for problems with inequality constraints and problems with both equality and inequality constraints. This is done directly by imposing a suitable constraint qualification, as opposed to first developing the Fritz John conditions and then the KKT conditions.

The KKT optimality conditions were originally developed by imposing the constraint qualification that, for every direction vector $\mathbf{d}$ in the cone G', there is a feasible arc whose tangent at $\bar{\mathbf{x}}$ points along $\mathbf{d}$. Since then, many authors have developed the KKT conditions under different constraint qualifications. For a thorough study of this subject, refer to the works of Abadie [1967b], Arrow, Hurwicz, and Uzawa [1961], Canon, Cullum, and Polak [1966], Cottle [1963a], Evans [1970], Evans and Gould [1970], Guignard [1969], Mangasarian [1969a], Mangasarian and Fromovitz [1967], and Zangwill [1969]. For a comparison and further study of these constraint qualifications, see the survey articles of Bazaraa, Goode, and Shetty [1972], Gould and Tolle [1972], and Peterson [1973].

Gould and Tolle [1971] showed that the constraint qualification of Guignard [1969] is the weakest possible in the sense that it is both necessary and sufficient for the validation of the KKT conditions (see Exercise 5.22 for a precise statement). For further discussion on constraint qualifications that validate second-order necessary optimality conditions, refer to Ben-Tal [1980], Ben-Tal and Zowe [1982], Fletcher [1987], Kyparisis [1985], and McCormick [1967]. Also, for an application of KKT conditions under various constraint qualifications in conducting *sensitivity analyses* in nonlinear programming, see Fiacco [1983].

Chapter 6 Lagrangian Duality and Saddle Point Optimality Conditions

Given a nonlinear programming problem, there is another nonlinear programming problem closely associated with it. The former is called the *primal problem*, and the latter is called the *Lagrangian dual problem*. Under certain convexity assumptions and suitable constraint qualifications, the primal and dual problems have equal optimal objective values and, hence, it is possible to solve the primal problem indirectly by solving the dual problem.

Several properties of the dual problem are developed in this chapter. They are used to provide general solution strategies for solving the primal and dual problems. As a by-product of one of the duality theorems, we obtain saddle point necessary optimality conditions without any differentiability assumptions.

The following is an outline of the chapter.

Section 6.1: The Lagrangian Dual Problem We introduce the Lagrangian dual problem, give its geometric interpretation, and illustrate it by several numerical examples.

Section 6.2: Duality Theorems and Saddle Point Optimality We prove the weak and strong duality theorems. The latter shows that the primal and dual objectives are equal under suitable convexity assumptions. We also develop the saddle point optimality conditions along with necessary and sufficient conditions for the absence of a duality gap, and interpret this in terms of a suitable perturbation function.

Section 6.3: Properties of the Dual Function We study several important properties of the dual function, such as concavity, differentiability, and subdifferentiability. We then give necessary and sufficient characterizations of ascent and steepest ascent directions.

Section 6.4: Formulating and Solving the Dual Problem Several procedures for solving the dual problem are discussed. In particular, we briefly describe gradient and subgradient-based methods, and present a tangential approximation cutting plane algorithm.

Section 6.5: Getting the Primal Solution We show that the points generated during the course of solving the dual problem yield optimal solutions to perturbations of the primal problem. For convex programs, we show how to obtain primal feasible solutions that are near-optimal.

Section 6.6: Linear and Quadratic Programs We give Lagrangian dual formulations for linear and quadratic programming, relating them to other standard duality formulations.

6.1 The Lagrangian Dual Problem

Consider the following nonlinear programming Problem P, which we call the *primal problem*.

Primal Problem P

$$\begin{array}{lll} \text{Minimize} & f(\mathbf{x}) & \\ \text{subject to} & g_i(\mathbf{x}) \leq 0 & \text{for } i = 1, \ldots, m \\ & h_i(\mathbf{x}) = 0 & \text{for } i = 1, \ldots, l \\ & \mathbf{x} \in X & \end{array}$$

Several problems, closely related to the above primal problem, have been proposed in the literature and are called *dual problems*. Among the various duality formulations, the Lagrangian duality formulation has perhaps attracted the most attention. It has led to several algorithms for solving large-scale linear problems, as well as convex and nonconvex nonlinear problems. It has also proved useful in discrete optimization where all or some of the variables are further restricted to be integers. The *Lagrangian dual problem D* is presented below.

Lagrangian Dual Problem D

$$\begin{array}{ll} \text{Maximize} & \theta(\mathbf{u}, \mathbf{v}) \\ \text{subject to} & \mathbf{u} \geq \mathbf{0} \end{array}$$

where $\theta(\mathbf{u}, \mathbf{v}) = \inf \{f(\mathbf{x}) + \sum_{i=1}^{m} u_i g_i(\mathbf{x}) + \sum_{i=1}^{l} v_i h_i(\mathbf{x}) : \mathbf{x} \in X\}$.

Note that the *Lagrangian dual function* θ may assume the value of $-\infty$ for some vector $(\mathbf{u}, \mathbf{v})$. The optimization problem that evaluates $\theta(\mathbf{u}, \mathbf{v})$ is sometimes referred to as the *Lagrangian dual subproblem*. In this problem, the constraints $g_i(\mathbf{x}) \leq 0$ and $h_i(\mathbf{x}) = 0$ have been incorporated in the objective function using the *Lagrangian multipliers* u_i and v_i. Also note that the multiplier u_i associated with the inequality constraint $g_i(\mathbf{x}) \leq 0$ is nonnegative, whereas the multiplier v_i associated with the equality constraint $h_i(\mathbf{x}) = 0$ is unrestricted in sign.

Since the dual problem consists of maximizing the infimum (greatest lower bound) of the function $f(\mathbf{x}) + \sum_{i=1}^{m} u_i g_i(\mathbf{x}) + \sum_{i=1}^{l} v_i h_i(\mathbf{x})$, it is sometimes referred to as the *max–min dual problem*. We remark here that, strictly speaking, we should write D as $\sup \{\theta(\mathbf{u}, \mathbf{v}) : \mathbf{u} \geq \mathbf{0}\}$, rather than $\max \{\theta(\mathbf{u}, \mathbf{v}) : \mathbf{u} \geq \mathbf{0}\}$, since the maximum may not exist (see Example 6.2.8). However, we shall specifically identify such cases wherever necessary.

The primal and Lagrangian dual problems can be written in the following form using vector notation, where $f: E_n \to E_1$, $\mathbf{g}: E_n \to E_m$ is a vector function whose ith component is g_i and $\mathbf{h}: E_n \to E_l$ is a vector function whose ith component is h_i. For the sake of convenience, we shall use this form throughout the remainder of this chapter.

Primal Problem P

$$\begin{aligned} \text{Minimize} \quad & f(\mathbf{x}) \\ \text{subject to} \quad & \mathbf{g}(\mathbf{x}) \leq \mathbf{0} \\ & \mathbf{h}(\mathbf{x}) = \mathbf{0} \\ & \mathbf{x} \in X \end{aligned}$$

Lagrangian Dual Problem D

$$\begin{aligned} \text{Maximize} \quad & \theta(\mathbf{u}, \mathbf{v}) \\ \text{subject to} \quad & \mathbf{u} \geq \mathbf{0} \end{aligned}$$

where $\theta(\mathbf{u}, \mathbf{v}) = \inf \{f(\mathbf{x}) + \mathbf{u}^t\mathbf{g}(\mathbf{x}) + \mathbf{v}^t\mathbf{h}(\mathbf{x}) : \mathbf{x} \in X\}$.

Given a nonlinear programming problem, several Lagrangian dual problems can be devised, depending on which constraints are handled as $\mathbf{g}(\mathbf{x}) \leq \mathbf{0}$ and $\mathbf{h}(\mathbf{x}) = \mathbf{0}$ and which constraints are treated by the set X. This choice can affect both the optimal value of D (as in nonconvex situations) and the effort expended in evaluating and updating the dual function θ during the course of solving the dual problem. Hence, an appropriate selection of the set X must be made, depending on the structure of the problem and the purpose for solving D (see the Notes and References section).

Geometric Interpretation of the Dual Problem

We now briefly discuss the geometric interpretation of the dual problem. For the sake of simplicity, we shall consider only one inequality constraint and assume that no equality constraints exist. Then, the primal problem is to minimize $f(\mathbf{x})$ subject to $\mathbf{x} \in X$ and $g(\mathbf{x}) \leq 0$.

In the (y, z) plane, the set $\{(y, z) : y = g(\mathbf{x}), z = f(\mathbf{x}) \text{ for some } \mathbf{x} \in X\}$ is denoted by G in Figure 6.1. Then, G is the image of X under the (g, f) map. The primal problem asks us to find a point in G with $y \leq 0$ that has a minimum ordinate. Obviously, this point is $(\bar{y}, \bar{z})$ in Figure 6.1.

Now suppose that $u \geq 0$ is given. To determine $\theta(u)$, we need to minimize $f(\mathbf{x}) + ug(\mathbf{x})$ over all $\mathbf{x} \in X$. Letting $y = g(\mathbf{x})$ and $z = f(\mathbf{x})$ for $\mathbf{x} \in X$, we want to minimize $z + uy$ over points in G. Note that $z + uy = \alpha$ is an equation of a straight line with slope $-u$ and intercept α on the z axis. To minimize $z + uy$ over G, we need to move the line $z + uy = \alpha$ parallel to itself as far down (along its negative gradient) as possible while it remains in contact with G. In other words, we move this line parallel to itself until it supports G from below, that is, the set G is above the line and touches it. Then, the intercept on the z axis gives $\theta(u)$, as seen in Figure 6.1. The dual problem is therefore equivalent to finding the slope of the supporting hyperplane such that its intercept on the z axis is maximal. In Figure 6.1, such a hyperplane has slope $-\bar{u}$ and supports the set

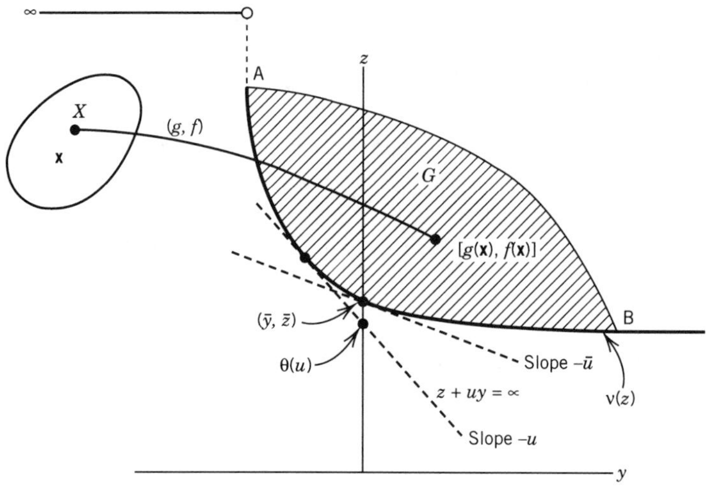

Figure 6.1 Geometric interpretation of Lagrangian duality.

G at the point $(\bar{y}, \bar{z})$. Thus, the optimal dual solution is $\bar{u}$, and the optimal dual objective value is $\bar{z}$. Furthermore, the optimal primal and dual objectives are equal in this case.

There is another related interesting interpretation that provides an important conceptual tool in this context. For the problem under consideration, define the function

$$\nu(y) = \text{minimum } \{f(\mathbf{x}) : g(\mathbf{x}) \leq y, \mathbf{x} \in X\}$$

The function ν is called a *perturbation function* since it is the optimal value function of a problem obtained from the original problem by perturbing the right-hand side of the inequality constraint $g(\mathbf{x}) \leq 0$ to y from the value of zero. Note that $\nu(y)$ is a nonincreasing function of y since, as y increases, the feasible region of the perturbed problem enlarges (or stays the same). For the present case, this function is illustrated in Figure 6.1. Observe that ν corresponds here to the lower envelope of G between points A and B because this envelope is itself monotone-decreasing. Moreover, ν remains constant at the value at point B for values of y higher than that at B, and becomes ∞ for points to the left of A because of infeasibility. In particular, if ν is differentiable at the origin, we observe that $\nu'(0) = -\bar{u}$. Hence, the marginal rate of change in objective function value with an increase in the right-hand side of the constraint from its present value of zero is given by $-\bar{u}$, the negative of the Lagrangian multiplier value at optimality. If ν is convex but is not differentiable at the origin, then $-\bar{u}$ is evidently a subgradient of ν at $y = 0$. In either case, we know that $\nu(y) \geq \nu(0) - \bar{u}y$ for all $y \in E_1$. As we shall see later, ν can be nondifferentiable and/or nonconvex, but the condition $\nu(y) \geq \nu(0) - \bar{u}y$ holds true for all $y \in E_1$, if and only if $\bar{u}$ is a KKT Lagrangian multiplier corresponding to an optimal solution $\bar{\mathbf{x}}$ such that it solves the dual problem with equal primal and dual objective values. As seen above, this happens to be the case in Figure 6.1.

6.1.1 Example

Consider the following primal problem:

$$\begin{array}{ll} \text{Minimize} & x_1^2 + x_2^2 \\ \text{subject to} & -x_1 - x_2 + 4 \leq 0 \\ & x_1, x_2 \geq 0 \end{array}$$

Note that the optimal solution occurs at the point $(x_1, x_2) = (2, 2)$, whose objective value is equal to 8.

Letting $g(x) = -x_1 - x_2 + 4$ and $X = \{(x_1, x_2) : x_1, x_2 \geq 0\}$, the dual function is given by

$$\begin{aligned}\theta(u) &= \inf\{x_1^2 + x_2^2 + u(-x_1 - x_2 + 4) : x_1, x_2 \geq 0\} \\ &= \inf\{x_1^2 - ux_1 : x_1 \geq 0\} + \inf\{x_2^2 - ux_2 : x_2 \geq 0\} + 4u\end{aligned}$$

Note that the above infima are achieved at $x_1 = x_2 = u/2$ if $u \geq 0$ and at $x_1 = x_2 = 0$ if $u < 0$. Hence,

$$\theta(u) = \begin{cases} -\frac{1}{2}u^2 + 4u & \text{for } u \geq 0 \\ 4u & \text{for } u < 0 \end{cases}$$

Note that θ is a concave function, and its maximum over $u \geq 0$ occurs at $\bar{u} = 4$. Figure 6.2 illustrates the situation. Note also that the optimal primal and dual objective values are both equal to 8.

Now let us consider the problem in the (y, z) plane, where $y = g(x)$ and $z = f(x)$. We are interested in finding G, the image of $X = \{(x_1, x_2): x_1 \geq 0, x_2 \geq 0\}$, under the (g, f) map. We do this by deriving explicit expressions for the lower and upper envelopes of G, denoted, respectively, by α and β.

Given y, note that $\alpha(y)$ and $\beta(y)$ are the optimal objective values of the following problems P_1 and P_2, respectively.

	Problem P_1		*Problem* P_2
Minimize	$x_1^2 + x_2^2$	Maximize	$x_1^2 + x_2^2$
subject to	$-x_1 - x_2 + 4 = y$	subject to	$-x_1 - x_2 + 4 = y$
	$x_1, x_2 \geq 0$		$x_1, x_2 \geq 0$

The reader can verify that $\alpha(y) = (4 - y)^2/2$ and $\beta(y) = (4 - y)^2$ for $y \leq 4$. The set G is illustrated in Figure 6.2. Note that $\mathbf{x} \in X$ implies that $x_1, x_2 \geq 0$, so that $-x_1, -x_2 + 4 \leq 4$. Thus, every point $\mathbf{x} \in X$ corresponds to $y \leq 4$.

Note that the optimal dual solution is $\bar{u} = 4$, which is the negative of the slope of the supporting hyperplane shown in Figure 6.2. The optimal dual objective value is $\alpha(0) = 8$ and is equal to the optimal primal objective value.

Again, in Figure 6.2, the perturbation function $v(y)$ for $y \in E_1$ corresponds to the lower envelope $\alpha(y)$ for $y \leq 4$, and $v(y)$ remains constant at the value 0 for $y \geq 4$. The slope $v'(0)$ equals -4, the negative of the optimal Lagrange multiplier value. Moreover, we have $v(y) \geq v(0) - 4y$ for all $y \in E_1$. As we shall see in the next section, this is a necessary and sufficient condition for the primal and dual objective values to match at optimality.

6.2 Duality Theorems and Saddle Point Optimality Conditions

In this section, we investigate the relationships between the primal and dual problems and develop saddle point optimality conditions for the primal problem.

Theorem 6.2.1 below, referred to as the *weak duality theorem*, shows that the objective value of any feasible solution to the dual problem yields a lower bound on the objective value of any feasible solution to the primal problem. Several important results follow as corollaries.

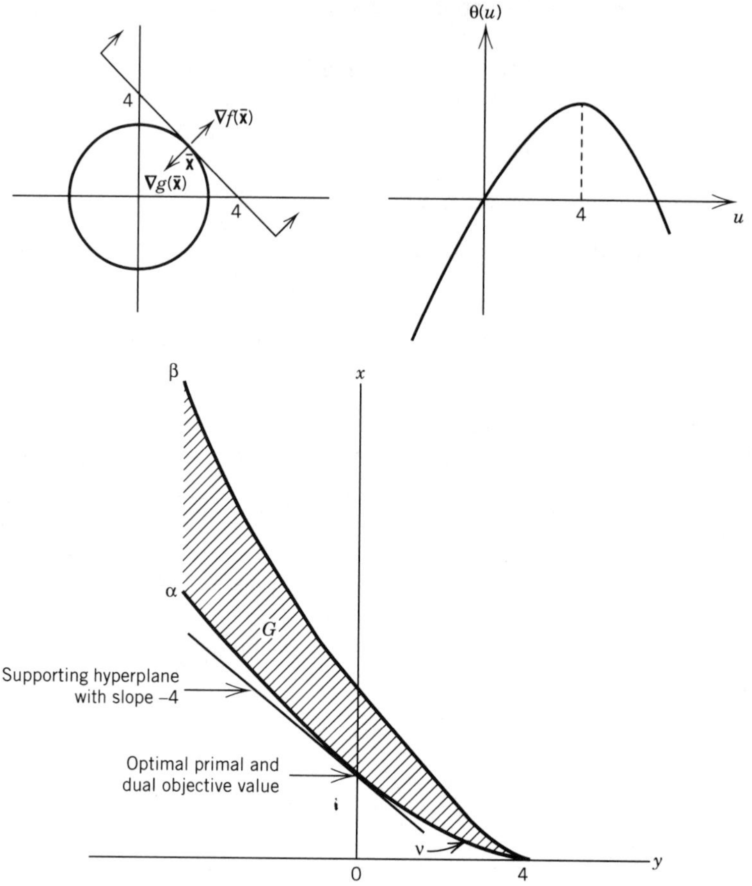

Figure 6.2 Geometric illustration of Example 6.1.1.

6.2.1 Theorem (Weak Duality Theorem)

Let $\mathbf{x}$ be a feasible solution to *Problem P*, that is $\mathbf{x} \in X$, $\mathbf{g}(\mathbf{x}) \leq \mathbf{0}$, and $\mathbf{h}(\mathbf{x}) = \mathbf{0}$. Also let $(\mathbf{u}, \mathbf{v})$ be a feasible solution to *Problem D*, that is, $\mathbf{u} \geq \mathbf{0}$. Then $f(\mathbf{x}) \geq \theta(\mathbf{u}, \mathbf{v})$.

Proof

By the definition of θ, and since $\mathbf{x} \in X$, we have

$$\begin{aligned}\theta(\mathbf{u}, \mathbf{v}) &= \inf\{f(\mathbf{y}) + \mathbf{u}^t\mathbf{g}(\mathbf{y}) + \mathbf{v}^t\mathbf{h}(\mathbf{y}) : \mathbf{y} \in X\} \\ &\leq f(\mathbf{x}) + \mathbf{u}^t\mathbf{g}(\mathbf{x}) + \mathbf{v}^t\mathbf{h}(\mathbf{x}) \leq f(\mathbf{x})\end{aligned}$$

since $\mathbf{u} \geq \mathbf{0}$, $\mathbf{g}(\mathbf{x}) \leq \mathbf{0}$, and $\mathbf{h}(\mathbf{x}) = \mathbf{0}$. This completes the proof.

Corollary 1

$$\inf\{f(\mathbf{x}) : \mathbf{x} \in X, \mathbf{g}(\mathbf{x}) \leq \mathbf{0}, \mathbf{h}(\mathbf{x}) = \mathbf{0}\} \geq \sup\{\theta(\mathbf{u}, \mathbf{v}) : \mathbf{u} \geq \mathbf{0}\}$$

Corollary 2

If $f(\bar{\mathbf{x}}) = \theta(\bar{\mathbf{u}}, \bar{\mathbf{v}})$, where $\bar{\mathbf{u}} \geq \mathbf{0}$ and $\bar{\mathbf{x}} \in \{\mathbf{x} \in X : \mathbf{g}(\mathbf{x}) \leq \mathbf{0}, \mathbf{h}(\mathbf{x}) = \mathbf{0}\}$, then $\bar{\mathbf{x}}$ and $(\bar{\mathbf{u}}, \bar{\mathbf{v}})$ solve the primal and dual problems, respectively.

Corollary 3

If $\inf \{f(\mathbf{x}) : \mathbf{x} \in X, \mathbf{g}(\mathbf{x}) \leq \mathbf{0}, \mathbf{h}(\mathbf{x}) = \mathbf{0}\} = -\infty$, then $\theta(\mathbf{u}, \mathbf{v}) = -\infty$ for each $\mathbf{u} \geq \mathbf{0}$.

Corollary 4

If $\sup \{\theta(\mathbf{u}, \mathbf{v}) : \mathbf{u} \geq \mathbf{0}\} = \infty$, then the primal problem has no feasible solution.

Duality Gap

From Corollary 1 to Theorem 6.2.1 above, the optimal objective value of the primal problem is greater than or equal to the optimal objective value of the dual problem. If strict inequality holds true, then a *duality gap* is said to exist. Figure 6.3 illustrates the case of a duality gap for a problem with a single inequality constraint and no equality constraints. The perturbation function $\nu(y)$ for $y \in E_1$ is as shown in the figure. Note that, by definition, *this is the greatest monotone nonincreasing function that envelopes* G *from below* (see Exercise 6.5). The optimal primal value is $\nu(0)$. The greatest intercept on the ordinate z axis achieved by a hyperplane that supports G from below gives the optimal dual objective value as shown. In particular, observe that there does not exist a $\bar{u}$ such that $\nu(y) \geq \nu(0) - \bar{u}y$ for all $y \in E_1$, as we had in Figures 6.1 and 6.2. Exercise 6.6 asks the reader to construct G and ν for the instance illustrated in Figure 4.13 that results in a situation similar to that of Figure 6.3.

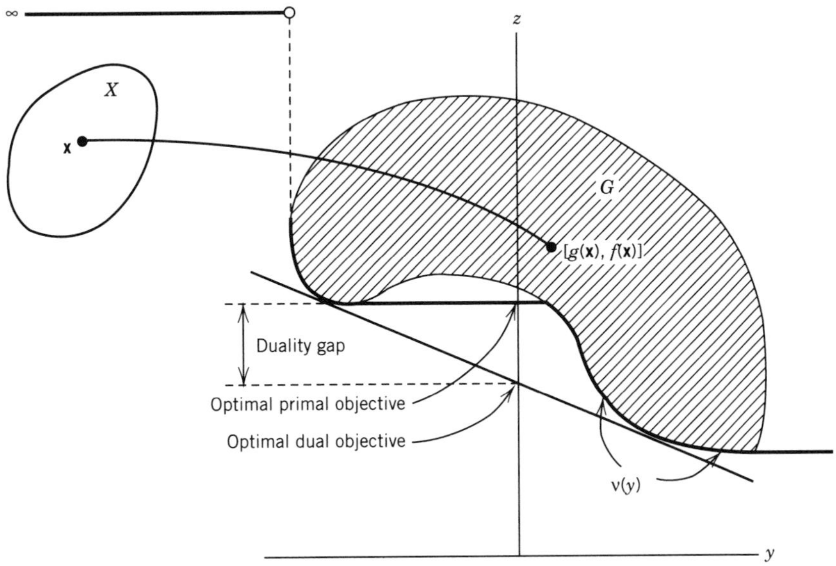

Figure 6.3 Illustration of a duality gap.

6.2.2 Example

Consider the following problem:

$$\begin{array}{ll} \text{Minimize} & f(x) = -2x_1 + x_2 \\ \text{subject to} & h(x) = x_1 + x_2 - 3 = 0 \\ & (x_1, x_2) \in X \end{array}$$

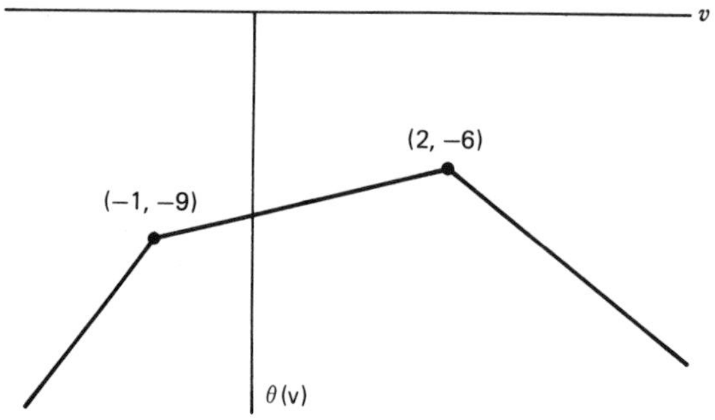

Figure 6.4 Dual function for Example 6.2.2.

where $X = \{(0, 0), (0, 4), (4, 4), (4, 0), (1, 2), (2, 1)\}$.

It is easy to verify that (2, 1) is the optimal solution to the primal problem with objective value equal to -3. The dual objective function θ is given by

$$\theta(v) = \min\{(-2x_1 + x_2) + v(x_1 + x_2 - 3) : (x_1, x_2) \in X\}$$

The reader may verify that the explicit expression for θ is given by

$$\theta(v) = \begin{cases} -4 + 5v & \text{for } v \leq -1 \\ -8 + v & \text{for } -1 \leq v \leq 2 \\ -3v & \text{for } v \geq 2 \end{cases}$$

The dual function is shown in Figure 6.4, and the optimal solution is $\bar{v} = 2$ with objective value -6. Note, in this example, that there exists a duality gap.

In this case, the set G consists of a finite number of points, each corresponding to a point in X. This is shown in Figure 6.5. The supporting hyperplane, whose intercept on the vertical axis is maximal, is shown in the figure. Note that the intercept is equal to -6 and that the slope is equal to -2. Thus, the optimal dual solution is $\bar{v} = 2$, with objective value -6. Furthermore, note that the points in the set G on the vertical axis

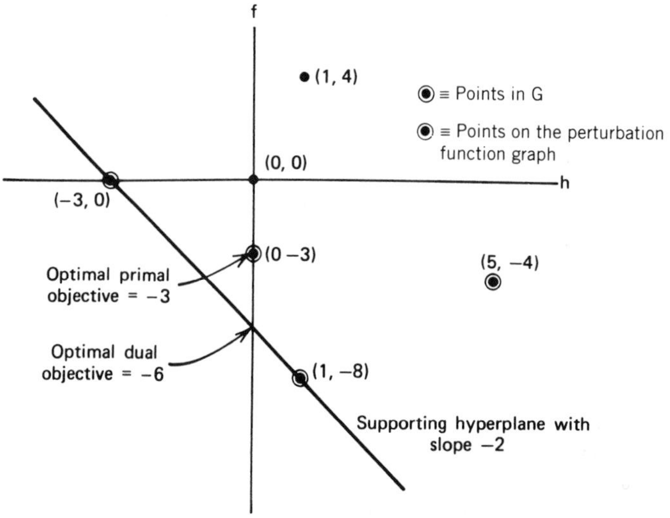

Figure 6.5 Geometric interpretation of Example 6.2.2.

correspond to the primal feasible points and, hence, the minimal primal objective value is equal to -3.

Similar to the inequality constrained case, the perturbation function here is defined as $\nu(y) = \text{minimum } \{f(\mathbf{x}) : h(\mathbf{x}) = y, \mathbf{x} \in X\}$. Because of the discrete nature of X, $h(\mathbf{x})$ can take on only a finite possible number of values. Hence, noting G in Figure 6.5, we obtain $\nu(-3) = 0$, $\nu(0) = -3$, $\nu(1) = -8$, and $\nu(5) = -4$, with $\nu(y) = \infty$ for all $y \in E_1$ otherwise. Again, the optimal primal value is $\nu(0) = -3$, and there does not exist a $\bar{v}$ such that $\nu(y) \geq \nu(0) - \bar{v}y$. Hence, a duality gap exists.

Conditions that guarantee the absence of a duality gap are given in Theorem 6.2.4. Then, Theorem 6.2.7 relates this to the perturbation function. First, however, the following lemma is needed.

6.2.3 Lemma

Let X be a nonempty convex set in E_n. Let $\alpha: E_n \to E_1$ and $\mathbf{g}: E_n \to E_m$ be convex, and let $\mathbf{h}: E_n \to E_l$ be affine; that is, $\mathbf{h}$ is of the form $\mathbf{h}(\mathbf{x}) = \mathbf{A}\mathbf{x} - \mathbf{b}$. If *System 1* below has no solution $\mathbf{x}$, then *System 2* has a solution $(u_0, \mathbf{u}, \mathbf{v})$. The converse holds if $u_0 > 0$.

System 1: $\alpha(\mathbf{x}) < 0 \quad \mathbf{g}(\mathbf{x}) \leq \mathbf{0} \quad \mathbf{h}(\mathbf{x}) = \mathbf{0}$ for some $\mathbf{x} \in X$

System 2: $u_0\alpha(\mathbf{x}) + \mathbf{u}^t\mathbf{g}(\mathbf{x}) + \mathbf{v}^t\mathbf{h}(\mathbf{x}) \geq 0$ for all $\mathbf{x} \in X$

$(u_0, \mathbf{u}) \geq \mathbf{0} \quad (u_0, \mathbf{u}, \mathbf{v}) \neq \mathbf{0}$

Proof

Suppose that System 1 has no solution, and consider the following set:

$$\wedge = \{(p, \mathbf{q}, \mathbf{r}) : p > \alpha(\mathbf{x}), \mathbf{q} \geq \mathbf{g}(\mathbf{x}), \mathbf{r} = \mathbf{h}(\mathbf{x}) \text{ for some } \mathbf{x} \in X\}$$

Noting that X, α, and $\mathbf{g}$ are convex and that $\mathbf{h}$ is affine, it can easily be shown that $\wedge$ is convex. Since System 1 has no solution, $((0, \mathbf{0}, \mathbf{0}) \notin \wedge$. By Corollary 1 to Theorem 2.4.7, there exists a nonzero $(u_0, \mathbf{u}, \mathbf{v})$ such that

$$u_0 p + \mathbf{u}^t\mathbf{q} + \mathbf{v}^t\mathbf{r} \geq 0 \qquad \text{for each } (p, \mathbf{q}, \mathbf{r}) \in \text{cl} \wedge \tag{6.1}$$

Now fix an $\mathbf{x} \in X$. Since p and $\mathbf{q}$ can be made arbitrarily large, (6.1) holds true only if $u_0 \geq 0$ and $\mathbf{u} \geq \mathbf{0}$. Furthermore, $(p, \mathbf{q}, \mathbf{r}) = [\alpha(\mathbf{x}), \mathbf{g}(\mathbf{x}), \mathbf{h}(\mathbf{x})]$ belongs to cl $\wedge$. Therefore, from (6.1), we get

$$u_0\alpha(\mathbf{x}) + \mathbf{u}^t\mathbf{g}(\mathbf{x}) + \mathbf{v}^t\mathbf{h}(\mathbf{x}) \geq 0$$

Since the above inequality is true for each $\mathbf{x} \in X$, System 2 has a solution.

To prove the converse, assume that System 2 has a solution $(u_0, \mathbf{u}, \mathbf{v})$ such that $u_0 > 0$ and $\mathbf{u} \geq \mathbf{0}$, satisfying

$$u_0\alpha(\mathbf{x}) + \mathbf{u}^t\mathbf{g}(\mathbf{x}) + \mathbf{v}^t\mathbf{h}(\mathbf{x}) \geq 0 \qquad \text{for each } \mathbf{x} \in X$$

Now let $\mathbf{x} \in X$ be such that $\mathbf{g}(\mathbf{x}) \leq \mathbf{0}$ and $\mathbf{h}(\mathbf{x}) = \mathbf{0}$. From the above inequality, since $\mathbf{u} \geq \mathbf{0}$, we conclude that $u_0\alpha(\mathbf{x}) \geq 0$. Since $u_0 > 0$, $\alpha(\mathbf{x}) \geq 0$; and, hence, System 1 has no solution. This completes the proof.

Theorem 6.2.4 below, referred to as the *strong duality theorem*, shows that under suitable convexity assumptions and under a constraint qualification, the optimal objective function values of the primal and dual problems are equal.

6.2.4 Theorem (Strong Duality Theorem)

Let X be a nonempty convex set in E_n, let $f:E_n \to E_1$ and $\mathbf{g}:E_n \to E_m$ be convex, and let $\mathbf{h}:E_n \to E_l$ be affine; that is, $\mathbf{h}$ is of the form $\mathbf{h}(\mathbf{x}) = \mathbf{Ax} - \mathbf{b}$. Suppose that the following constraint qualification holds true. There exists an $\hat{\mathbf{x}} \in X$ such that $\mathbf{g}(\hat{\mathbf{x}}) < \mathbf{0}$ and $\mathbf{h}(\hat{\mathbf{x}}) = \mathbf{0}$, and $\mathbf{0} \in \text{int } \mathbf{h}(X)$, where $\mathbf{h}(X) = \{\mathbf{h}(\mathbf{x}):\mathbf{x} \in X\}$. Then,

$$\inf\{f(\mathbf{x}):\mathbf{x} \in X, \mathbf{g}(\mathbf{x}) \le \mathbf{0}, \mathbf{h}(\mathbf{x}) = \mathbf{0}\} = \sup\{\theta(\mathbf{u}, \mathbf{v}):\mathbf{u} \ge \mathbf{0}\} \tag{6.2}$$

Furthermore, if the inf is finite, then $\sup\{\theta(\mathbf{u}, \mathbf{v}):\mathbf{u} \ge \mathbf{0}\}$ is achieved at $(\bar{\mathbf{u}}, \bar{\mathbf{v}})$ with $\bar{\mathbf{u}} \ge \mathbf{0}$. If the inf is achieved at $\bar{\mathbf{x}}$, then $\bar{\mathbf{u}}^t\mathbf{g}(\bar{\mathbf{x}}) = 0$.

Proof

Let $\gamma = \inf\{f(\mathbf{x}):\mathbf{x} \in X, \mathbf{g}(\mathbf{x}) \le \mathbf{0}, \mathbf{h}(\mathbf{x}) = \mathbf{0}\}$. By assumption, $\gamma < \infty$. If $\gamma = -\infty$, then, by Corollary 3 to Theorem 6.2.1, $\sup\{\theta(\mathbf{u}, \mathbf{v}):\mathbf{u} \ge \mathbf{0}\} = -\infty$, and, therefore (6.2) holds true. Hence, suppose that γ is finite, and consider the following system:

$$f(\mathbf{x}) - \gamma < 0 \qquad \mathbf{g}(\mathbf{x}) \le \mathbf{0} \qquad \mathbf{h}(\mathbf{x}) = \mathbf{0} \qquad \mathbf{x} \in X$$

By definition of γ, this system has no solution. Hence, from Lemma 6.2.3, there exists a nonzero vector $(u_0, \mathbf{u}, \mathbf{v})$ with $(u_0, \mathbf{u}) \ge \mathbf{0}$ such that

$$u_0[f(\mathbf{x}) - \gamma] + \mathbf{u}^t\mathbf{g}(\mathbf{x}) + \mathbf{v}^t\mathbf{h}(\mathbf{x}) \ge 0 \qquad \text{for all } \mathbf{x} \in X \tag{6.3}$$

We first show that $u_0 > 0$. By contradiction, suppose that $u_0 = 0$. By assumption, there exists an $\hat{\mathbf{x}} \in X$ such that $\mathbf{g}(\hat{\mathbf{x}}) < \mathbf{0}$ and $\mathbf{h}(\hat{\mathbf{x}}) = \mathbf{0}$. Substituting in (6.3), it follows that $\mathbf{u}^t\mathbf{g}(\hat{\mathbf{x}}) \ge 0$. Since $\mathbf{g}(\hat{\mathbf{x}}) < \mathbf{0}$ and $\mathbf{u} \ge \mathbf{0}$, $\mathbf{u}^t\mathbf{g}(\hat{\mathbf{x}}) \ge 0$ is only possible if $\mathbf{u} = \mathbf{0}$. But from (6.3), $u_0 = 0$ and $\mathbf{u} = \mathbf{0}$, which implies that $\mathbf{v}^t\mathbf{h}(\mathbf{x}) \ge 0$ for all $\mathbf{x} \in X$. But since $\mathbf{0} \in \text{int } \mathbf{h}(X)$, we can pick an $\mathbf{x} \in X$ such that $\mathbf{h}(\mathbf{x}) = -\lambda\mathbf{v}$, where $\lambda > 0$. Therefore, $0 \le \mathbf{v}^t\mathbf{h}(\mathbf{x}) = -\lambda \|\mathbf{v}\|^2$, which implies that $\mathbf{v} = \mathbf{0}$. Thus, we have shown that $u_0 = 0$ implies that $(u_0, \mathbf{u}, \mathbf{v}) = \mathbf{0}$, which is impossible. Hence, $u_0 > 0$. Dividing (6.3) by u_0 and denoting $\mathbf{u}/u_0$ and $\mathbf{v}/u_0$ by $\bar{\mathbf{u}}$ and $\bar{\mathbf{v}}$, respectively, we get

$$f(\mathbf{x}) + \bar{\mathbf{u}}^t\mathbf{g}(\mathbf{x}) + \bar{\mathbf{v}}^t\mathbf{h}(\mathbf{x}) \ge \gamma \qquad \text{for all } \mathbf{x} \in X \tag{6.4}$$

This shows that $\theta(\bar{\mathbf{u}}, \bar{\mathbf{v}}) = \inf\{f(\mathbf{x}) + \bar{\mathbf{u}}^t\mathbf{g}(\mathbf{x}) + \bar{\mathbf{v}}^t\mathbf{h}(\mathbf{x}):\mathbf{x} \in X\} \ge \gamma$. In view of Theorem 6.2.1, it is then clear that $\theta(\bar{\mathbf{u}}, \bar{\mathbf{v}}) = \gamma$, and $(\bar{\mathbf{u}}, \bar{\mathbf{v}})$ solves the dual problem.

To complete the proof, suppose that $\bar{\mathbf{x}}$ is an optimal solution to the primal problem, that is, $\bar{\mathbf{x}} \in X$, $\mathbf{g}(\bar{\mathbf{x}}) \le \mathbf{0}$, $\mathbf{h}(\bar{\mathbf{x}}) = \mathbf{0}$, and $f(\bar{\mathbf{x}}) = \gamma$. From (6.4), letting $\mathbf{x} = \bar{\mathbf{x}}$, we get $\bar{\mathbf{u}}^t\mathbf{g}(\bar{\mathbf{x}}) \ge 0$. Since $\bar{\mathbf{u}} \ge \mathbf{0}$ and $\mathbf{g}(\bar{\mathbf{x}}) \le \mathbf{0}$, we get $\bar{\mathbf{u}}^t\mathbf{g}(\bar{\mathbf{x}}) = 0$, and the proof is complete.

In the above theorem, the assumption $\mathbf{0} \in \text{int } \mathbf{h}(X)$ and that there exists an $\hat{\mathbf{x}} \in X$ such that $\mathbf{g}(\hat{\mathbf{x}}) < \mathbf{0}$ and $\mathbf{h}(\hat{\mathbf{x}}) = \mathbf{0}$ can be viewed as a generalization of Slater's constraint qualification of Chapter 5. In particular, if $X = E_n$, then $\mathbf{0} \in \text{int } \mathbf{h}(X)$ automatically holds true (if redundant equations are deleted), so that the constraint qualification asserts the existence of a point $\hat{\mathbf{x}}$ such that $\mathbf{g}(\hat{\mathbf{x}}) < \mathbf{0}$ and $\mathbf{h}(\hat{\mathbf{x}}) = \mathbf{0}$. To see this, suppose that $\mathbf{h}(\mathbf{x}) = \mathbf{Ax} - \mathbf{b}$. Without loss of generality, assume that rank $\mathbf{A} = m$, because otherwise any redundant constraints could be deleted. Now any $\mathbf{y} \in E_m$ could be represented as $\mathbf{y} = \mathbf{Ax} - \mathbf{b}$, where $\mathbf{x} = \mathbf{A}^t(\mathbf{AA}^t)^{-1}(\mathbf{y} + \mathbf{b})$. Thus, $\mathbf{h}(X) = E_m$ and, in particular, $\mathbf{0} \in \text{int } \mathbf{h}(X)$.

Saddle Point Criteria

The foregoing theorem shows that under convexity assumptions and under a suitable constraint qualification, the primal and dual objective function values match at optimality.

Actually, a necessary and sufficient condition for the latter property to hold is the existence of a saddle point, as we learn next. Given the primal Problem P, define the *Lagrangian function*

$$\phi(\mathbf{x}, \mathbf{u}, \mathbf{v}) = f(\mathbf{x}) + \mathbf{u}'\mathbf{g}(\mathbf{x}) + \mathbf{v}'\mathbf{h}(\mathbf{x})$$

A solution $(\bar{\mathbf{x}}, \bar{\mathbf{u}}, \bar{\mathbf{v}})$ is called a *saddle point* of the Lagrangian function if $\bar{\mathbf{x}} \in X$, $\bar{\mathbf{u}} \geq \mathbf{0}$, and

$$\phi(\bar{\mathbf{x}}, \mathbf{u}, \mathbf{v}) \leq \phi(\bar{\mathbf{x}}, \bar{\mathbf{u}}, \bar{\mathbf{v}}) \leq \phi(\mathbf{x}, \bar{\mathbf{u}}, \bar{\mathbf{v}}) \quad \text{for all } \mathbf{x} \in X, \text{ and all } (\mathbf{u}, \mathbf{v}) \text{ with } \mathbf{u} \geq \mathbf{0} \tag{6.5}$$

Hence, we have that $\bar{\mathbf{x}}$ minimizes ϕ over X when $(\mathbf{u}, \mathbf{v})$ is fixed at $(\bar{\mathbf{u}}, \bar{\mathbf{v}})$, and that $(\bar{\mathbf{u}}, \bar{\mathbf{v}})$ maximizes ϕ over all $(\mathbf{u}, \mathbf{v})$ with $\mathbf{u} \geq \mathbf{0}$ when $\mathbf{x}$ is fixed at $\bar{\mathbf{x}}$. Relating this to the illustration of Figure 4.2, we see why $(\bar{\mathbf{x}}, \bar{\mathbf{u}}, \bar{\mathbf{v}})$ is therefore called a "saddle point" for the Lagrangian function ϕ.

The following result characterizes a saddle point solution and shows that its existence is a necessary and sufficient condition for the absence of a duality gap.

6.2.5 Theorem (Saddle Point Optimality and Absence of a Duality Gap)

A solution $(\bar{\mathbf{x}}, \bar{\mathbf{u}}, \bar{\mathbf{v}})$ with $\bar{\mathbf{x}} \in X$ and $\bar{\mathbf{u}} \geq \mathbf{0}$ is a saddle point for the Lagrangian function $\phi(\mathbf{x}, \mathbf{u}, \mathbf{v}) = f(\mathbf{x}) + \mathbf{u}'\mathbf{g}(\mathbf{x}) + \mathbf{v}'\mathbf{h}(\mathbf{x})$ if and only if

a. $\phi(\bar{\mathbf{x}}, \bar{\mathbf{u}}, \bar{\mathbf{v}}) = \text{minimum } \{\phi(\mathbf{x}, \bar{\mathbf{u}}, \bar{\mathbf{v}}) : \mathbf{x} \in X\}$,
b. $\mathbf{g}(\bar{\mathbf{x}}) \leq \mathbf{0}$, $\mathbf{h}(\bar{\mathbf{x}}) = \mathbf{0}$, and
c. $\bar{\mathbf{u}}'\mathbf{g}(\bar{\mathbf{x}}) = 0$.

Moreover, $(\bar{\mathbf{x}}, \bar{\mathbf{u}}, \bar{\mathbf{v}})$ is a saddle point if and only if $\bar{\mathbf{x}}$ and $(\bar{\mathbf{u}}, \bar{\mathbf{v}})$ are, respectively, optimal solutions to the primal and dual problems P and D with no duality gap, that is, with $f(\bar{\mathbf{x}}) = \theta(\bar{\mathbf{u}}, \bar{\mathbf{v}})$.

Proof

Suppose that $(\bar{\mathbf{x}}, \bar{\mathbf{u}}, \bar{\mathbf{v}})$ is a saddle point for the Lagrangian function ϕ. By definition, condition (a) must be true. Furthermore, from (6.5), we have

$$f(\bar{\mathbf{x}}) + \bar{\mathbf{u}}'\mathbf{g}(\bar{\mathbf{x}}) + \bar{\mathbf{v}}'\mathbf{h}(\bar{\mathbf{x}}) \geq f(\bar{\mathbf{x}}) + \mathbf{u}'\mathbf{g}(\bar{\mathbf{x}}) + \mathbf{v}'\mathbf{h}(\bar{\mathbf{x}}) \quad \text{for all } (\mathbf{u}, \mathbf{v}) \text{ with } \mathbf{u} \geq \mathbf{0} \tag{6.6}$$

Clearly, this implies that we must have $\mathbf{g}(\bar{\mathbf{x}}) \leq \mathbf{0}$ and $\mathbf{h}(\bar{\mathbf{x}}) = \mathbf{0}$, or else (6.6) can be violated by appropriately making a component of $\mathbf{u}$ or $\mathbf{v}$ infinitely large in magnitude. Now taking $\mathbf{u} = \mathbf{0}$ in (6.6), we obtain that $\bar{\mathbf{u}}'\mathbf{g}(\bar{\mathbf{x}}) \geq \mathbf{0}$. Noting that $\bar{\mathbf{u}} \geq \mathbf{0}$ and $\mathbf{g}(\bar{\mathbf{x}}) \leq \mathbf{0}$ imply that $\bar{\mathbf{u}}'\mathbf{g}(\bar{\mathbf{x}}) \leq 0$, we must have $\bar{\mathbf{u}}'\mathbf{g}(\bar{\mathbf{x}}) = 0$. Hence, conditions (a), (b), and (c) hold.

Conversely, suppose that we are given $(\bar{\mathbf{x}}, \bar{\mathbf{u}}, \bar{\mathbf{v}})$ with $\bar{\mathbf{x}} \in X$ and $\bar{\mathbf{u}} \geq \mathbf{0}$ such that conditions (a), (b), and (c) hold. Then, $\phi(\bar{\mathbf{x}}, \bar{\mathbf{u}}, \bar{\mathbf{v}}) \leq \phi(\mathbf{x}, \bar{\mathbf{u}}, \bar{\mathbf{v}})$ for all $\mathbf{x} \in X$ by property (a). Furthermore, $\phi(\bar{\mathbf{x}}, \bar{\mathbf{u}}, \bar{\mathbf{v}}) = f(\bar{\mathbf{x}}) + \bar{\mathbf{u}}'\mathbf{g}(\bar{\mathbf{x}}) + \bar{\mathbf{v}}'\mathbf{h}(\bar{\mathbf{x}}) = f(\bar{\mathbf{x}}) \geq f(\bar{\mathbf{x}}) + \mathbf{u}'\mathbf{g}(\bar{\mathbf{x}}) + \mathbf{v}'\mathbf{h}(\bar{\mathbf{x}}) = \phi(\bar{\mathbf{x}}, \mathbf{u}, \mathbf{v})$ for all $(\mathbf{u}, \mathbf{v})$ with $\mathbf{u} \geq \mathbf{0}$, since $\mathbf{g}(\bar{\mathbf{x}}) \leq \mathbf{0}$ and $\mathbf{h}(\bar{\mathbf{x}}) = \mathbf{0}$. Hence, $(\bar{\mathbf{x}}, \bar{\mathbf{u}}, \bar{\mathbf{v}})$ is a saddle point. This proves the first part of the theorem.

Next, suppose again that $(\bar{\mathbf{x}}, \bar{\mathbf{u}}, \bar{\mathbf{v}})$ is a saddle point. By property (b), $\bar{\mathbf{x}}$ is feasible to Problem P. Since $\bar{\mathbf{u}} \geq \mathbf{0}$, we also have that $(\bar{\mathbf{u}}, \bar{\mathbf{v}})$ is feasible to D. Moreover, by properties (a), (b), and (c), $\theta(\bar{\mathbf{u}}, \bar{\mathbf{v}}) = \phi(\bar{\mathbf{x}}, \bar{\mathbf{u}}, \bar{\mathbf{v}}) = f(\bar{\mathbf{x}}) + \bar{\mathbf{u}}'\mathbf{g}(\bar{\mathbf{x}}) + \bar{\mathbf{v}}'\mathbf{h}(\bar{\mathbf{x}}) = f(\bar{\mathbf{x}})$. By Corollary 2 to Theorem 6.2.1, $\bar{\mathbf{x}}$ and $(\bar{\mathbf{u}}, \bar{\mathbf{v}})$ solve P and D, respectively, with no duality gap.

Finally, suppose that $\bar{\mathbf{x}}$ and $(\bar{\mathbf{u}}, \bar{\mathbf{v}})$ are optimal solutions to Problems P and D, respectively, with $f(\bar{\mathbf{x}}) = \theta(\bar{\mathbf{u}}, \bar{\mathbf{v}})$. Hence, we have $\bar{\mathbf{x}} \in X$, $\mathbf{g}(\bar{\mathbf{x}}) \leq \mathbf{0}$, $\mathbf{h}(\bar{\mathbf{x}}) = \mathbf{0}$, and $\bar{\mathbf{u}} \geq \mathbf{0}$. Moreover, we have by primal–dual feasibility that

$$\begin{aligned}\theta(\bar{\mathbf{u}}, \bar{\mathbf{v}}) &= \min\{f(\mathbf{x}) + \bar{\mathbf{u}}^t\mathbf{g}(\mathbf{x}) + \bar{\mathbf{v}}^t\mathbf{h}(\mathbf{x}) : \mathbf{x} \in X\} \\ &\leq f(\bar{\mathbf{x}}) + \bar{\mathbf{u}}^t\mathbf{g}(\bar{\mathbf{x}}) + \bar{\mathbf{v}}^t\mathbf{h}(\bar{\mathbf{x}}) = f(\bar{\mathbf{x}}) + \bar{\mathbf{u}}^t\mathbf{g}(\bar{\mathbf{x}}) \leq f(\bar{\mathbf{x}})\end{aligned}$$

But $\theta(\bar{\mathbf{u}}, \bar{\mathbf{v}}) = f(\bar{\mathbf{x}})$ by hypothesis. Hence, equality holds true throughout above. In particular, $\bar{\mathbf{u}}^t\mathbf{g}(\bar{\mathbf{x}}) = 0$, and so, $\phi(\bar{\mathbf{x}}, \bar{\mathbf{u}}, \bar{\mathbf{v}}) = f(\bar{\mathbf{x}}) = \theta(\bar{\mathbf{u}}, \bar{\mathbf{v}}) = \min\{\phi(\mathbf{x}, \bar{\mathbf{u}}, \bar{\mathbf{v}}): \mathbf{x} \in X\}$. Hence, properties (a), (b), and (c) hold in addition to $\bar{\mathbf{x}} \in X$ and $\bar{\mathbf{u}} \geq \mathbf{0}$; and so, $(\bar{\mathbf{x}}, \bar{\mathbf{u}}, \bar{\mathbf{v}})$ is a saddle point. This completes the proof.

Corollary

Suppose that X, f, and $\mathbf{g}$ are convex and that $\mathbf{h}$ is affine; that is, $\mathbf{h}$ is of the form $\mathbf{h}(\mathbf{x}) = \mathbf{A}\mathbf{x} - \mathbf{b}$. Further, suppose that $\mathbf{0} \in \operatorname{int} \mathbf{h}(X)$ and that there exists an $\hat{\mathbf{x}} \in X$ with $\mathbf{g}(\hat{\mathbf{x}}) < \mathbf{0}$ and $\mathbf{h}(\hat{\mathbf{x}}) = \mathbf{0}$. If $\bar{\mathbf{x}}$ is an optimal solution to the primal Problem P, then there exists a vector $(\bar{\mathbf{u}}, \bar{\mathbf{v}})$ with $\bar{\mathbf{u}} \geq \mathbf{0}$, such that $(\bar{\mathbf{x}}, \bar{\mathbf{u}}, \bar{\mathbf{v}})$ is a saddle point.

Proof

By Theorem 6.2.4, there exists an optimal solution $(\bar{\mathbf{u}}, \bar{\mathbf{v}})$, $\bar{\mathbf{u}} \geq \mathbf{0}$ to Problem D such that $f(\bar{\mathbf{x}}) = \theta(\bar{\mathbf{u}}, \bar{\mathbf{v}})$. Hence, by Theorem 6.2.5, $(\bar{\mathbf{x}}, \bar{\mathbf{u}}, \bar{\mathbf{v}})$ is a saddle point solution. This completes the proof.

There is an additional insight that can be derived in regard to the duality gap between the primal and dual problems. Note that the dual problem optimal value is given by

$$\theta^* \equiv \underset{(\mathbf{u}, \mathbf{v}):\mathbf{u} \geq \mathbf{0}}{\text{supremum}} \ \underset{\mathbf{x} \in X}{\text{infimum}} \, [\phi(\mathbf{x}, \mathbf{u}, \mathbf{v})]$$

If we interchange the order of optimization (see Exercise 6.43), we get

$$\theta^* \leq \underset{\mathbf{x} \in X}{\text{infimum}} \ \underset{(\mathbf{u}, \mathbf{v}):\mathbf{u} \geq \mathbf{0}}{\text{supremum}} \, [\phi(\mathbf{x}, \mathbf{u}, \mathbf{v})]$$

But the supremum of $\phi(\mathbf{x}, \mathbf{u}, \mathbf{v}) = f(\mathbf{x}) + \mathbf{u}^t\mathbf{g}(\mathbf{x}) + \mathbf{v}^t\mathbf{h}(\mathbf{x})$ over $(\mathbf{u}, \mathbf{v})$ with $\mathbf{u} \geq \mathbf{0}$ is infinity, unless $\mathbf{g}(\mathbf{x}) \leq \mathbf{0}$ and $\mathbf{h}(\mathbf{x}) = \mathbf{0}$, whence it is $f(\mathbf{x})$. Hence,

$$\begin{aligned}\theta^* &\leq \underset{\mathbf{x} \in X}{\text{infimum}} \ \underset{(\mathbf{u}, \mathbf{v}):\mathbf{u} \geq \mathbf{0}}{\text{supremum}} \, [\phi(\mathbf{x}, \mathbf{u}, \mathbf{v})] \\ &\equiv \inf\{f(\mathbf{x}) : \mathbf{g}(\mathbf{x}) \leq \mathbf{0}, \mathbf{h}(\mathbf{x}) = \mathbf{0}, \mathbf{x} \in \mathbf{X}\}\end{aligned}$$

which is the primal optimal value. Hence, we see that the primal and dual objective values match at optimality if and only if the interchange of the foregoing infimum and supremum operations leaves the optimal values unchanged. By Theorem 6.2.5, assuming that an optimum exists, this occurs if and only if there exists a saddle point $(\bar{\mathbf{x}}, \bar{\mathbf{u}}, \bar{\mathbf{v}})$ for the Lagrangian function ϕ.

Relationship Between the Saddlepoint Criteria and the Karush–Kuhn–Tucker (KKT) Conditions

In Chapters 4 and 5, we discussed the KKT optimality conditions for Problem P defined below:

$$\begin{aligned} &\text{Minimize} && f(\mathbf{x}) \\ &\text{subject to} && \mathbf{g}(\mathbf{x}) \leq \mathbf{0} \\ &&& \mathbf{h}(\mathbf{x}) = \mathbf{0} \\ &&& \mathbf{x} \in X \end{aligned}$$

Furthermore, in Theorem 6.2.5 above, we developed the saddle point optimality conditions for the same problem. Theorem 6.2.6 below gives the relationship between these two types of optimality conditions.

6.2.6 Theorem

Let $S = \{\mathbf{x} \in X : \mathbf{g}(\mathbf{x}) \leq \mathbf{0}, \mathbf{h}(\mathbf{x}) = \mathbf{0}\}$, and consider Problem P to minimize $f(\mathbf{x})$ subject to $\mathbf{x} \in S$. Suppose that $\bar{\mathbf{x}} \in S$ satisfies the KKT conditions, that is, there exist $\bar{\mathbf{u}} \geq \mathbf{0}$ and $\bar{\mathbf{v}}$ such that

$$\nabla f(\bar{\mathbf{x}}) + \nabla \mathbf{g}(\bar{\mathbf{x}})^t \bar{\mathbf{u}} + \nabla \mathbf{h}(\bar{\mathbf{x}})^t \bar{\mathbf{v}} = \mathbf{0} \tag{6.7}$$

$$\bar{\mathbf{u}}^t \mathbf{g}(\bar{\mathbf{x}}) = 0$$

Suppose that f, g_i for $i \in I$ are convex at $\bar{\mathbf{x}}$, where $I = \{i : g_i(\bar{\mathbf{x}}) = 0\}$. Further suppose that if $\bar{v}_i \neq 0$, then h_i is affine. Then, $(\bar{\mathbf{x}}, \bar{\mathbf{u}}, \bar{\mathbf{v}})$ is a saddle point for the Lagrangian function $\phi(\mathbf{x}, \mathbf{u}, \mathbf{v}) = f(\mathbf{x}) + \mathbf{u}^t\mathbf{g}(\mathbf{x}) + \mathbf{v}^t\mathbf{h}(\mathbf{x})$.

Conversely, suppose that $(\bar{\mathbf{x}}, \bar{\mathbf{u}}, \bar{\mathbf{v}})$ with $\bar{\mathbf{x}} \in \text{int } X$ and $\bar{\mathbf{u}} \geq \mathbf{0}$ is a saddle point solution. Then, $\bar{\mathbf{x}}$ is feasible to Problem P and, furthermore, $(\bar{\mathbf{x}}, \bar{\mathbf{u}}, \bar{\mathbf{v}})$ satisfies the KKT conditions specified by (6.7).

Proof

Suppose that $(\bar{\mathbf{x}}, \bar{\mathbf{u}}, \bar{\mathbf{v}})$, with $\bar{\mathbf{x}} \in S$ and $\bar{\mathbf{u}} \geq \mathbf{0}$, satisfies the KKT conditions specified by (6.7). By convexity at $\bar{\mathbf{x}}$ of f and g_i for $i \in I$, and since h_i is affine for $\bar{v}_i \neq 0$, we get

$$f(\mathbf{x}) \geq f(\bar{\mathbf{x}}) + \nabla f(\bar{\mathbf{x}})^t(\mathbf{x} - \bar{\mathbf{x}}) \tag{6.8a}$$

$$g_i(\mathbf{x}) \geq g_i(\bar{\mathbf{x}}) + \nabla g_i(\bar{\mathbf{x}})^t(\mathbf{x} - \bar{\mathbf{x}}) \qquad \text{for } i \in I \tag{6.8b}$$

$$h_i(\mathbf{x}) = h_i(\bar{\mathbf{x}}) + \nabla h_i(\bar{\mathbf{x}})^t(\mathbf{x} - \bar{\mathbf{x}}) \qquad \text{for } i = 1, \ldots, l, \bar{v}_i \neq 0 \tag{6.8c}$$

for all $\mathbf{x} \in X$. Multiplying (6.8b) by $\bar{u}_i \geq 0$, (6.8c) by $\bar{v}_i$, adding (6.8a) and noting (6.7), it follows from the definition of ϕ that $\phi(\mathbf{x}, \bar{\mathbf{u}}, \bar{\mathbf{v}}) \geq \phi(\bar{\mathbf{x}}, \bar{\mathbf{u}}, \bar{\mathbf{v}})$ for all $\mathbf{x} \in X$. Also, since $\mathbf{g}(\bar{\mathbf{x}}) \leq \mathbf{0}$, $\mathbf{h}(\bar{\mathbf{x}}) = \mathbf{0}$ and $\bar{\mathbf{u}}^t\mathbf{g}(\bar{\mathbf{x}}) = 0$, it follows that $\phi(\bar{\mathbf{x}}, \mathbf{u}, \mathbf{v}) \leq \phi(\bar{\mathbf{x}}, \bar{\mathbf{u}}, \bar{\mathbf{v}})$ for all $(\mathbf{u}, \mathbf{v})$ with $\mathbf{u} \geq \mathbf{0}$. Hence $(\bar{\mathbf{x}}, \bar{\mathbf{u}}, \bar{\mathbf{v}})$ satisfies the saddle point conditions given by (6.5).

To prove the converse, suppose that $(\bar{\mathbf{x}}, \bar{\mathbf{u}}, \bar{\mathbf{v}})$ with $\bar{\mathbf{x}} \in \text{int } X$ and $\bar{\mathbf{u}} \geq \mathbf{0}$ is a saddle point solution. Since $\phi(\bar{\mathbf{x}}, \mathbf{u}, \mathbf{v}) \leq \phi(\bar{\mathbf{x}}, \bar{\mathbf{u}}, \bar{\mathbf{v}})$ for all $\mathbf{u} \geq \mathbf{0}$ and all $\mathbf{v}$, we have, using (6.6) as in Theorem 6.2.5, that $\mathbf{g}(\bar{\mathbf{x}}) \leq \mathbf{0}$, $\mathbf{h}(\bar{\mathbf{x}}) = \mathbf{0}$, and $\bar{\mathbf{u}}^t\mathbf{g}(\bar{\mathbf{x}}) = 0$. This shows that $\bar{\mathbf{x}}$ is feasible to Problem P. Since $\phi(\bar{\mathbf{x}}, \bar{\mathbf{u}}, \bar{\mathbf{v}}) \leq \phi(\mathbf{x}, \bar{\mathbf{u}}, \bar{\mathbf{v}})$ for all $\mathbf{x} \in X$, then $\bar{\mathbf{x}}$ solves the problem to minimize $\phi(\mathbf{x}, \bar{\mathbf{u}}, \bar{\mathbf{v}})$ subject to $\mathbf{x} \in X$. Since $\bar{\mathbf{x}} \in \text{int } X$, then $\nabla_{\mathbf{x}}\phi(\bar{\mathbf{x}}, \bar{\mathbf{u}}, \bar{\mathbf{v}}) = \mathbf{0}$, that is, $\nabla f(\bar{\mathbf{x}}) + \nabla \mathbf{g}(\bar{\mathbf{x}})^t\bar{\mathbf{u}} + \nabla \mathbf{h}(\bar{\mathbf{x}})^t\bar{\mathbf{v}} = \mathbf{0}$; and, hence, (6.7) holds. This completes the proof.

Theorem 6.2.6 above shows that if $\bar{\mathbf{x}}$ is a KKT point, then, under certain convexity assumptions, the Lagrangian multipliers in the KKT conditions also serve as the multipliers in the saddle point criteria. Conversely, the multipliers in the saddle point conditions are the Lagrangian multipliers of the KKT conditions. Moreover, in view of Theorems 6.2.4, 6.2.5, and 6.2.6, the optimal dual variables for the Lagrangian dual problem are precisely the Lagrangian multipliers for the KKT conditions and also the multipliers for the saddle point conditions in this case.

Saddle Point Optimality Interpretation Via a Perturbation Function

While discussing the geometric interpretation of the dual problem and the associated duality gap, we introduced the concept of a perturbation function ν and illustrated this in Examples 6.1.1 and 6.1.2 (see Figures 6.1 through 6.5). As alluded to previously, the existence of a supporting hyperplane to the epigraph of this function at the point $(0, \nu(0))$ related to the absence of a duality gap in these examples. This is formalized in the discussion that follows.

Consider the primal Problem P, and define the *perturbation function* $\nu: E_{m+l} \to E_1$ as the optimal value function of the following problem, where $\mathbf{y} = (y_1, \ldots, y_m, y_{m+1}, \ldots, y_{m+l})$:

$$\begin{aligned} \nu(\mathbf{y}) = \min \{f(\mathbf{x}) : g_i(\mathbf{x}) \le y_i \text{ for } i = 1, \ldots, m, \\ h_i(\mathbf{x}) = y_{m+i} \text{ for } i = 1, \ldots, l, \mathbf{x} \in X\} \end{aligned} \tag{6.9}$$

Theorem 6.2.7 below asserts that if Problem P has an optimum, then the existence of a saddle point solution, that is, the absence of a duality gap, is equivalent to the existence of a supporting hyperplane for the epigraph of ν at the point $(\mathbf{0}, \nu(\mathbf{0}))$.

6.2.7 Theorem

Consider the primal Problem P, and assume that an optimal solution $\bar{\mathbf{x}}$ to this problem exists. Then, $(\bar{\mathbf{x}}, \bar{\mathbf{u}}, \bar{\mathbf{v}})$ is a saddle point for the Lagrangian function $\phi(\mathbf{x}, \mathbf{u}, \mathbf{v}) = f(\mathbf{x}) + \mathbf{u}^t\mathbf{g}(\mathbf{x}) + \mathbf{v}^t\mathbf{h}(\mathbf{x})$ if and only if

$$\nu(\mathbf{y}) \ge \nu(\mathbf{0}) - (\bar{\mathbf{u}}^t, \bar{\mathbf{v}}^t)\mathbf{y} \qquad \text{for all } \mathbf{y} \in E_{m+l} \tag{6.10}$$

that is, if and only if the hyperplane $z = \nu(\mathbf{0}) - (\bar{\mathbf{u}}^t, \bar{\mathbf{v}}^t)\mathbf{y}$ supports the epigraph $\{(\mathbf{y}, z): z \ge \nu(\mathbf{y}), \mathbf{y} \in E_{m+l}\}$ of ν at the point $(\mathbf{y}, z) = (\mathbf{0}, \nu(\mathbf{0}))$.

Proof

Suppose that $(\bar{\mathbf{x}}, \bar{\mathbf{u}}, \bar{\mathbf{v}})$ is a saddle point solution. Then, by Theorem 6.2.5, the absence of a duality gap asserts that

$$\begin{aligned} \nu(\mathbf{0}) &= \theta(\bar{\mathbf{u}}, \bar{\mathbf{v}}) = \min \{f(\mathbf{x}) + \bar{\mathbf{u}}^t\mathbf{g}(\mathbf{x}) + \bar{\mathbf{v}}^t\mathbf{h}(\mathbf{x}) : \mathbf{x} \in X\} \\ &= (\bar{\mathbf{u}}^t, \bar{\mathbf{v}}^t)\mathbf{y} + \min \{f(\mathbf{x}) + \sum_{i=1}^{m} \bar{u}_i[g_i(\mathbf{x}) - y_i] \\ &\qquad + \sum_{i=1}^{l} \bar{v}_i[h_i(\mathbf{x}) - y_{m+i}] : \mathbf{x} \in X\} \qquad \text{for any } \mathbf{y} \in E_{m+l} \end{aligned}$$

Applying the weak duality Theorem 6.2.1 to the perturbed problem (6.9), we obtain from the foregoing identity that $\nu(\mathbf{0}) \leq (\bar{\mathbf{u}}^t, \bar{\mathbf{v}}^t)\mathbf{y} + \nu(\mathbf{y})$ for any $\mathbf{y} \in E_{m+l}$, and so (6.10) holds true.

Conversely, suppose that (6.10) holds for some $(\bar{\mathbf{u}}, \bar{\mathbf{v}})$ and let $\bar{\mathbf{x}}$ solve Problem P. We must show that $(\bar{\mathbf{x}}, \bar{\mathbf{u}}, \bar{\mathbf{v}})$ is a saddle point solution. First, note that $\bar{\mathbf{x}} \in X$, $\mathbf{g}(\bar{\mathbf{x}}) \leq \mathbf{0}$, and $\mathbf{h}(\bar{\mathbf{x}}) = \mathbf{0}$. Furthermore, $\bar{\mathbf{u}} \geq \mathbf{0}$ must hold, because if $\bar{u}_p < 0$, say, then by selecting $\mathbf{y}$ such that $y_i = 0$ for $i \neq p$, and $y_p > 0$, we obtain $\nu(\mathbf{0}) \geq \nu(\mathbf{y}) \geq \nu(\mathbf{0}) - \bar{u}_p y_p$, which implies that $\bar{u}_p y_p \geq 0$, a contradiction.

Second observe that by fixing $\mathbf{y} = \bar{\mathbf{y}} \equiv [\mathbf{g}(\bar{\mathbf{x}})^t, \mathbf{h}(\bar{\mathbf{x}})^t]$ in (6.9), we obtain a restriction of Problem P, since $\mathbf{g}(\bar{\mathbf{x}}) \leq \mathbf{0}$ and $\mathbf{h}(\bar{\mathbf{x}}) = \mathbf{0}$. But for the same reason, since $\bar{\mathbf{x}}$ is feasible to (6.9) with $\mathbf{y}$ fixed as such, and since $\bar{\mathbf{x}}$ solves Problem P, we obtain $\nu(\bar{\mathbf{y}}) = \nu(\mathbf{0})$. By (6.10), this in turn means that $\bar{\mathbf{u}}^t\mathbf{g}(\bar{\mathbf{x}}) \geq 0$. Since $\mathbf{g}(\bar{\mathbf{x}}) \leq \mathbf{0}$ and $\bar{\mathbf{u}} \geq \mathbf{0}$, we therefore must have $\bar{\mathbf{u}}^t\mathbf{g}(\bar{\mathbf{x}}) = 0$.

Finally, we have

$$\phi(\bar{\mathbf{x}}, \bar{\mathbf{u}}, \bar{\mathbf{v}}) = f(\bar{\mathbf{x}}) + \bar{\mathbf{u}}^t\mathbf{g}(\bar{\mathbf{x}}) + \bar{\mathbf{v}}^t\mathbf{h}(\bar{\mathbf{x}}) = f(\bar{\mathbf{x}}) = \nu(\mathbf{0}) \leq \nu(\mathbf{y}) + (\bar{\mathbf{u}}^t, \bar{\mathbf{v}}^t)\mathbf{y} \tag{6.11}$$

for all $\mathbf{y} \in E_{m+l}$. Now, for any $\hat{\mathbf{x}} \in X$, denoting $\hat{\mathbf{y}} = [\mathbf{g}(\hat{\mathbf{x}})^t, \mathbf{h}(\hat{\mathbf{x}})^t]$, we obtain from (6.9) that $\nu(\hat{\mathbf{y}}) \leq f(\hat{\mathbf{x}})$, since $\hat{\mathbf{x}}$ is feasible to (6.9) with $\mathbf{y} = \hat{\mathbf{y}}$. Hence, using this in (6.11), we obtain $\phi(\bar{\mathbf{x}}, \bar{\mathbf{u}}, \bar{\mathbf{v}}) = f(\hat{\mathbf{x}}) + \bar{\mathbf{u}}^t\mathbf{g}(\hat{\mathbf{x}}) + \bar{\mathbf{v}}^t\mathbf{h}(\hat{\mathbf{x}})$ for all $\hat{\mathbf{x}} \in X$; and so, $\phi(\bar{\mathbf{x}}, \bar{\mathbf{u}}, \bar{\mathbf{v}}) = \min \{\phi(\mathbf{x}, \bar{\mathbf{u}}, \bar{\mathbf{v}}) : \mathbf{x} \in X\}$.

We have therefore shown that $\bar{\mathbf{x}} \in X$, $\bar{\mathbf{u}} \geq \mathbf{0}$ and that conditions (a), (b), and (c) of Theorem 6.2.5 hold true. Consequently, $(\bar{\mathbf{x}}, \bar{\mathbf{u}}, \bar{\mathbf{v}})$ is a saddle point for ϕ, and this completes the proof.

To illustrate, observe in Figures 6.1 and 6.2 that there does exist a supporting hyperplane for the epigraph of ν at $(0, \nu(0))$. Hence, both the primal and dual problems have optimal solutions with the same optimal objective values for these cases. However, for the situations illustrated in Figures 6.3 and 6.5, no such supporting hyperplane exists. Hence, these instances possess a positive duality gap.

In conclusion, there are two noteworthy points regarding the perturbation function ν. First, if f and $\mathbf{g}$ are convex, $\mathbf{h}$ is affine, and X is a convex set, then it can be easily shown that ν is a convex function (see Exercise 6.7). Hence, in this case, the condition (6.10) reduces to the statement that $-(\bar{\mathbf{u}}^t, \bar{\mathbf{v}}^t)$ is a subgradient of ν at $\mathbf{y} = \mathbf{0}$.

Second, suppose that, corresponding to the primal and dual Problems P and D, there exists a saddle point solution $(\bar{\mathbf{x}}, \bar{\mathbf{u}}, \bar{\mathbf{v}})$, and assume that ν is continuously differentiable at $\mathbf{y} = \mathbf{0}$. Then, we have $\nu(\mathbf{y}) = \nu(\mathbf{0}) + \nabla\nu(\mathbf{0})^t\mathbf{y} + \|\mathbf{y}\|\alpha(0; \mathbf{y})$, where $\alpha(0; \mathbf{y}) \to 0$ as $\mathbf{y} \to \mathbf{0}$. Using (6.10) of Theorem 6.2.7, this means that $-[\nabla\nu(\mathbf{0})^t + (\bar{\mathbf{u}}^t, \bar{\mathbf{v}}^t)]\mathbf{y} \leq \|\mathbf{y}\|\alpha(\mathbf{0}:\mathbf{y})$ for all $\mathbf{y} \in E_{m+l}$. Letting $\mathbf{y} = -\lambda[\nabla\nu(\mathbf{0})^t + (\bar{\mathbf{u}}^t, \bar{\mathbf{v}}^t)]$ for $\lambda \geq 0$, and letting $\lambda \to 0^+$, we readily conclude that $\nabla\nu(\mathbf{0})^t = -(\bar{\mathbf{u}}^t, \bar{\mathbf{v}}^t)$. Hence, the negative of the optimal Lagrange multiplier values give the marginal rates of change in the optimal objective value of Problem P with respect to perturbations in the right-hand sides. Assuming that the problem represents one of minimizing cost subject to various material, labor, budgetary resource limitations, and demand constraints, this yields useful *economic interpretations* in terms of the marginal change in the optimal cost with respect to perturbations in such resources or demand entities.

6.2.8 Example

Consider the following (primal) problem:

$$P\text{: Minimize } \{x_2 : x_1 \geq 1,\ x_1^2 + x_2^2 \leq 1,\ (x_1, x_2) \in E_2\}$$

As illustrated in Figure 6.6, the unique optimal solution to this problem is $(\bar{x}_1, \bar{x}_2) =$

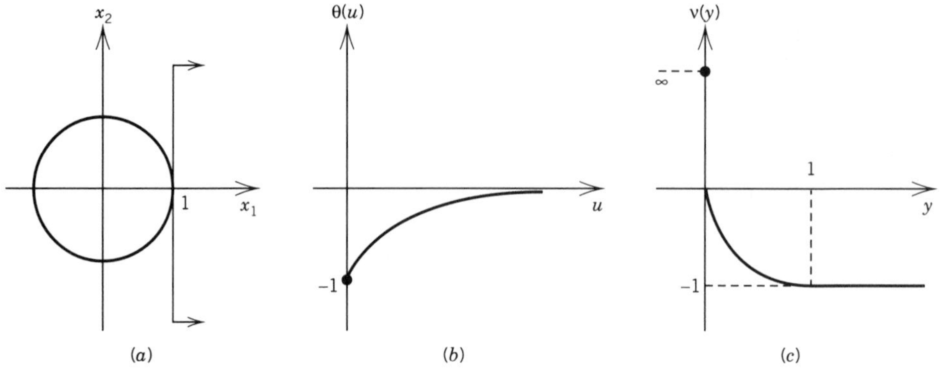

Figure 6.6 Illustration of Example 6.2.8.

(1, 0), with optimal objective function value equal to 0. However, although this is a convex programming problem, the optimum is not a KKT point, since $F_0 \cap G' \neq \emptyset$, and a saddle point solution does not exist (refer to Theorems 4.2.15 and 6.2.6).

Now let us formulate the Lagrangian dual Problem D by treating $1 - x_1 \leq 0$ as $g(\mathbf{x}) \leq 0$ and letting X denote the set $\{(x_1, x_2): x_1^2 + x_2^2 \leq 1\}$. Hence, Problem D requires us to find $\sup \{\theta(u): u \geq 0\}$, where $\theta(u) = \inf \{x_2 + u(1 - x_1): x_1^2 + x_2^2 \leq 1\}$. For any $u \geq 0$, it is readily verified that the optimum is attained at $x_1 = u/\sqrt{1 + u^2}$ and $x_2 = -1/\sqrt{1 + u^2}$. Hence, $\theta(u) = u - \sqrt{1 + u^2}$. We see that as $u \rightarrow \infty$, $\theta(u) \rightarrow 0$, the optimal primal objective value. Hence, $\sup \{\theta(u): u \geq 0\} = 0$, but this is not attained for any $\bar{u} \geq 0$, that is, a maximizing solution $\bar{u}$ does not exist.

Next, let us determine the perturbation function $\nu(y)$ for $y \in E_1$. Note that $\nu(y) = \min \{x_2: 1 - x_1 \leq y, x_1^2 + x_2^2 \leq 1\}$. Hence, we obtain $\nu(y) = \infty$ for $y < 0$, $\nu(y) = -\sqrt{y(2 - y)}$ for $0 \leq y \leq 1$, and $\nu(y) = -1$ for $y \geq 1$. This is illustrated in Figure 6.6*c*. Observe that there does not exist any supporting hyperplane at (0, 0) for the epigraph of $\nu(y)$, $y \in E_1$, since the right-hand derivative of ν with respect to y at $y = 0$ is $-\infty$.

6.3 Properties of the Dual Function

In Section 6.2, we studied the relationships between the primal and dual problems. Under certain conditions, Theorems 6.2.4 and 6.2.5 showed that the optimal objective values of the primal and dual problems are equal, and, hence, it would be possible to solve the primal problem indirectly by solving the dual problem. In order to facilitate the solution of the dual problem, we need to examine the properties of the dual function. In particular, we show that θ is concave, discuss its differentiability and subdifferentiability properties, and characterize its ascent and steepest ascent directions.

Throughout the rest of this chapter, we shall assume that the set X is compact. This will simplify the proofs of several of the theorems. Note that this assumption is not unduly restrictive, since if X were not bounded, one could add suitable lower and upper bounds on the variables such that the feasible region is not affected. For convenience, we shall also combine the vectors $\mathbf{u}$ and $\mathbf{v}$ as $\mathbf{w}$ and the functions $\mathbf{g}$ and $\mathbf{h}$ as $\boldsymbol{\beta}$. Theorem 6.3.1 below shows that θ is concave.

6.3.1 Theorem

Let X be a nonempty compact set in E_n, and let $f:E_n \to E_1$, and $\boldsymbol{\beta}:E_n \to E_{m+l}$ be continuous. Then, θ, defined by

$$\theta(\mathbf{w}) = \inf\{f(\mathbf{x}) + \mathbf{w}'\boldsymbol{\beta}(\mathbf{x}):\mathbf{x} \in X\}$$

is concave over E_{m+l}.

Proof

Since f and $\boldsymbol{\beta}$ are continuous and X is compact, θ is finite everywhere on E_{m+l}. Let $\mathbf{w}_1, \mathbf{w}_2 \in E_{m+l}$, and let $\lambda \in (0, 1)$. We then have

$$\begin{aligned}\theta[\lambda\mathbf{w}_1 + (1-\lambda)\mathbf{w}_2] &= \inf\{f(\mathbf{x}) + [\lambda\mathbf{w}_1 + (1-\lambda)\mathbf{w}_2]'\boldsymbol{\beta}(\mathbf{x}):\mathbf{x} \in X\}\\ &= \inf\{\lambda[f(\mathbf{x}) + \mathbf{w}_1'\boldsymbol{\beta}(\mathbf{x})] + (1-\lambda)[f(\mathbf{x}) + \mathbf{w}_2'\boldsymbol{\beta}(\mathbf{x})]:\mathbf{x} \in X\}\\ &\geq \lambda \inf\{f(\mathbf{x}) + \mathbf{w}_1'\boldsymbol{\beta}(\mathbf{x}):\mathbf{x} \in X\}\\ &\quad + (1-\lambda)\inf\{f(\mathbf{x}) + \mathbf{w}_2'\boldsymbol{\beta}(\mathbf{x}) : \mathbf{x} \in X\}\\ &= \lambda\theta(\mathbf{w}_1) + (1-\lambda)\theta(\mathbf{w}_2)\end{aligned}$$

Thus, θ is concave, and the proof is complete.

Since θ is concave, by Theorem 3.4.2, a local optimum of θ is also a global optimum. This makes the maximization of θ an attractive proposition. However, the main difficulty in solving the dual problem is that the dual function is not explicitly available, since θ can be evaluated at a point only after a minimization subproblem is solved. In the remainder of this section, we study differentiability and subdifferentiability properties of the dual function. These properties will aid us in maximizing the dual function.

Differentiability of θ

We now address the question of differentiability of θ defined by $\theta(\mathbf{w}) = \inf\{f(\mathbf{x}) + \mathbf{w}'\boldsymbol{\beta}(\mathbf{x}):\mathbf{x} \in X\}$. It will be convenient to introduce the following set of optimal solutions to the Lagrangian dual subproblem:

$$X(\mathbf{w}) = \{\mathbf{y}:\mathbf{y} \text{ minimizes } f(\mathbf{x}) + \mathbf{w}'\boldsymbol{\beta}(\mathbf{x}) \text{ over } \mathbf{x} \in X\}$$

The differentiability of θ at any given point $\bar{\mathbf{w}}$ depends on the elements of $X(\bar{\mathbf{w}})$. In particular, if the set $X(\bar{\mathbf{w}})$ is a singleton, then Theorem 6.3.3 below shows that θ is differentiable at $\bar{\mathbf{w}}$. First, however, the following lemma is needed.

6.3.2 Lemma

Let X be a nonempty compact set in E_n and let $f:E_n \to E_1$ and $\boldsymbol{\beta}:E_n \to E_{m+l}$ be continuous. Let $\bar{\mathbf{w}} \in E_{m+l}$, and suppose that $X(\bar{\mathbf{w}})$ is the singleton $\{\bar{\mathbf{x}}\}$. Suppose that $\mathbf{w}_k \to \bar{\mathbf{w}}$, and let $\mathbf{x}_k \in X(\mathbf{w}_k)$ for each k. Then $\mathbf{x}_k \to \bar{\mathbf{x}}$.

Proof

By contradiction, suppose that $\mathbf{w}_k \to \bar{\mathbf{w}}$, $\mathbf{x}_k \in X(\mathbf{w}_k)$, and $\|\mathbf{x}_k - \bar{\mathbf{x}}\| > \varepsilon > 0$ for $k \in \mathcal{K}$, where $\mathcal{K}$ is some index set. Since X is compact, then the sequence $\{\mathbf{x}_k\}_{\mathcal{K}}$ has a convergent subsequence $\{\mathbf{x}_k\}_{\mathcal{K}'}$, with limit $\mathbf{y}$ in X. Note that $\|\mathbf{y} - \bar{\mathbf{x}}\| \geq \varepsilon > 0$, and hence $\mathbf{y}$ and $\bar{\mathbf{x}}$ are distinct. Furthermore, for each $\mathbf{w}_k$ with $k \in \mathcal{K}'$ we have

$$f(\mathbf{x}_k) + \mathbf{w}_k^t\boldsymbol{\beta}(\mathbf{x}_k) \leq f(\bar{\mathbf{x}}) + \mathbf{w}_k^t\boldsymbol{\beta}(\bar{\mathbf{x}})$$

Taking the limit as k in $\mathcal{K}'$ approaches ∞, and noting that $\mathbf{x}_k \to \mathbf{y}$, $\mathbf{w}_k \to \bar{\mathbf{w}}$, and that f and $\boldsymbol{\beta}$ are continuous, it follows that

$$f(\mathbf{y}) + \bar{\mathbf{w}}^t\boldsymbol{\beta}(\mathbf{y}) \leq f(\bar{\mathbf{x}}) + \bar{\mathbf{w}}^t\boldsymbol{\beta}(\bar{\mathbf{x}})$$

Therefore, $\mathbf{y} \in X(\bar{\mathbf{w}})$, contradicting the assumption that $X(\bar{\mathbf{w}})$ is a singleton. This completes the proof.

6.3.3 Theorem

Let X be a nonempty compact set in E_n, and let $f:E_n \to E_1$, and $\boldsymbol{\beta}:E_n \to E_{m+l}$ be continuous. Let $\bar{\mathbf{w}} \in E_{m+l}$, and suppose that $X(\bar{\mathbf{w}})$ is the singleton $\{\bar{\mathbf{x}}\}$. Then, θ is differentiable at $\bar{\mathbf{w}}$ with gradient $\nabla\theta(\bar{\mathbf{w}}) = \boldsymbol{\beta}(\bar{\mathbf{x}})$.

Proof

Since f and $\boldsymbol{\beta}$ are continuous and X is compact, then, for any given $\mathbf{w}$, there exists an $\mathbf{x}_w \in X(\mathbf{w})$. From the definition of θ, the following two inequalities hold true:

$$\theta(\mathbf{w}) - \theta(\bar{\mathbf{w}}) \leq f(\bar{\mathbf{x}}) + \mathbf{w}^t\boldsymbol{\beta}(\bar{\mathbf{x}}) - f(\bar{\mathbf{x}}) - \bar{\mathbf{w}}^t\boldsymbol{\beta}(\bar{\mathbf{x}}) = (\mathbf{w} - \bar{\mathbf{w}})^t\boldsymbol{\beta}(\bar{\mathbf{x}}) \tag{6.12}$$

$$\theta(\bar{\mathbf{w}}) - \theta(\mathbf{w}) \leq f(\mathbf{x}_w) + \bar{\mathbf{w}}^t\boldsymbol{\beta}(\mathbf{x}_w) - f(\mathbf{x}_w) - \mathbf{w}^t\boldsymbol{\beta}(\mathbf{x}_w) = (\bar{\mathbf{w}} - \mathbf{w})^t\boldsymbol{\beta}(\mathbf{x}_w) \tag{6.13}$$

From (6.12) and (6.13) and the Schwartz inequality, it follows that

$$\begin{aligned} 0 \geq \theta(\mathbf{w}) - \theta(\bar{\mathbf{w}}) - (\mathbf{w} - \bar{\mathbf{w}})^t\boldsymbol{\beta}(\bar{\mathbf{x}}) &\geq (\mathbf{w} - \bar{\mathbf{w}})^t[\boldsymbol{\beta}(\mathbf{x}_w) - \boldsymbol{\beta}(\bar{\mathbf{x}})] \\ &\geq -\|\mathbf{w} - \bar{\mathbf{w}}\|\,\|\boldsymbol{\beta}(\mathbf{x}_w) - \boldsymbol{\beta}(\bar{\mathbf{x}})\| \end{aligned}$$

This further implies that

$$0 \geq \frac{\theta(\mathbf{w}) - \theta(\bar{\mathbf{w}}) - (\mathbf{w} - \bar{\mathbf{w}})^t\boldsymbol{\beta}(\bar{\mathbf{x}})}{\|\mathbf{w} - \bar{\mathbf{w}}\|} \geq -\|\boldsymbol{\beta}(\mathbf{x}_w) - \boldsymbol{\beta}(\bar{\mathbf{x}})\| \tag{6.14}$$

As $\mathbf{w} \to \bar{\mathbf{w}}$, then, by Lemma 6.3.2, $\mathbf{x}_w \to \bar{\mathbf{x}}$ and by the continuity of $\boldsymbol{\beta}$, $\boldsymbol{\beta}(\mathbf{x}_w) \to \boldsymbol{\beta}(\bar{\mathbf{x}})$. Therefore, from (6.14), we get

$$\lim_{\mathbf{w} \to \bar{\mathbf{w}}} \frac{\theta(\mathbf{w}) - \theta(\bar{\mathbf{w}}) - (\mathbf{w} - \bar{\mathbf{w}})^t\boldsymbol{\beta}(\bar{\mathbf{x}})}{\|\mathbf{w} - \bar{\mathbf{w}}\|} = 0$$

Hence, θ is differentiable at $\bar{\mathbf{w}}$ with gradient $\boldsymbol{\beta}(\bar{\mathbf{x}})$. This completes the proof.

Subgradients of θ

We have shown in Theorem 6.3.1 that θ is concave, and, hence, by Theorem 3.2.5, θ is subdifferentiable; that is, it has subgradients. As will be seen later, subgradients play an important role in the maximization of the dual function, since they lead naturally to the characterization of the directions of ascent. Theorem 6.3.4 below shows that each $\bar{\mathbf{x}} \in X(\bar{\mathbf{w}})$ yields a subgradient of θ at $\bar{\mathbf{w}}$.

6.3.4 Theorem

Let X be a nonempty compact set in E_n, and let $f:E_n \to E_1$ and $\boldsymbol{\beta}:E_n \to E_{m+l}$ be continuous so that for any $\bar{\mathbf{w}} \in E_{m+l}$, $X(\bar{\mathbf{w}})$ is not empty. If $\bar{\mathbf{x}} \in X(\bar{\mathbf{w}})$, then, $\boldsymbol{\beta}(\bar{\mathbf{x}})$ is a subgradient of θ at $\bar{\mathbf{w}}$.

Proof

Since f and $\boldsymbol{\beta}$ are continuous and X is compact, $X(\bar{\mathbf{w}}) \neq \emptyset$ for any $\bar{\mathbf{w}} \in E_{m+l}$. Now, let $\bar{\mathbf{w}} \in E_{m+l}$, and let $\bar{\mathbf{x}} \in X(\bar{\mathbf{w}})$. Then,

$$\begin{aligned}
\theta(\mathbf{w}) &= \inf\{f(\mathbf{x}) + \mathbf{w}'\boldsymbol{\beta}(\mathbf{x}):\mathbf{x} \in X\} \\
&\le f(\bar{\mathbf{x}}) + \mathbf{w}'\boldsymbol{\beta}(\bar{\mathbf{x}}) \\
&= f(\bar{\mathbf{x}}) + (\mathbf{w} - \bar{\mathbf{w}})'\boldsymbol{\beta}(\bar{\mathbf{x}}) + \bar{\mathbf{w}}'\boldsymbol{\beta}(\bar{\mathbf{x}}) \\
&= \theta(\bar{\mathbf{w}}) + (\mathbf{w} - \bar{\mathbf{w}})'\boldsymbol{\beta}(\bar{\mathbf{x}})
\end{aligned}$$

Therefore, $\boldsymbol{\beta}(\bar{\mathbf{x}})$ is a subgradient of θ at $\bar{\mathbf{w}}$, and the proof is complete.

6.3.5 Example

Consider the following primal problem:

$$\begin{array}{ll}
\text{Minimize} & -x_1 - x_2 \\
\text{subject to} & x_1 + 2x_2 - 3 \le 0 \\
& x_1, x_2 = 0, 1, 2, \text{ or } 3
\end{array}$$

Letting $g(x_1, x_2) = x_1 + 2x_2 - 3$ and $X = \{(x_1, x_2): x_1, x_2 = 0, 1, 2, \text{ or } 3\}$, the dual function is given by

$$\begin{aligned}
\theta(u) &= \inf\{-x_1 - x_2 + u(x_1 + 2x_2 - 3): x_1, x_2 = 0, 1, 2, \text{ or } 3\} \\
&= \begin{cases} -6 + 6u & \text{if } 0 \le u \le \frac{1}{2} \\ -3 & \text{if } \frac{1}{2} \le u \le 1 \\ -3u & \text{if } u \ge 1 \end{cases}
\end{aligned}$$

We ask the reader to plot the perturbation function for this example in Exercise 6.13, and to investigate the saddle point optimality conditions. Now, let $\bar{u} = \frac{1}{2}$. In order to find a subgradient of θ at $\bar{u}$, consider the following subproblem:

$$\begin{array}{ll}
\text{Minimize} & -x_1 - x_2 + \frac{1}{2}(x_1 + 2x_2 - 3) \\
\text{subject to} & x_1, x_2 = 0, 1, 2, \text{ or } 3
\end{array}$$

Note that the set $X(\bar{u})$ of optimal solutions to the above problem is $\{(3, 0), (3, 1), (3, 2), \text{ and } (3, 3)\}$. Thus, from Theorem 6.3.4, $g(3, 0) = 0$, $g(3, 1) = 2$, $g(3, 2) = 4$, and $g(3, 3) = 6$ are subgradients of θ at $\bar{u}$. Note, however, that $\frac{3}{2}$ is also a subgradient of θ at $\bar{u}$, but $\frac{3}{2}$ cannot be represented as $g(\bar{\mathbf{x}})$ for any $\bar{\mathbf{x}} \in X(\bar{u})$.

From the above example, it is clear that Theorem 6.3.4 gives only a sufficient characterization of subgradients. A necessary and sufficient characterization of sugradients is given in Theorem 6.3.7 below. First, however, the following important result is needed. The principal conclusion of this result is stated in the corollary and holds true for any arbitrary concave function θ (see Exercise 6.14). However, our proof of Theorem 6.3.6 below is specialized to exploit the structure of the Lagrangian dual function θ.

6.3.6 Theorem

Let X be a nonempty compact set in E_n, and let $f:E_n \to E_1$ and $\boldsymbol{\beta}:E_n \to E_{m+l}$ be continuous. Let $\bar{\mathbf{w}}, \mathbf{d} \in E_{m+l}$. Then, the directional derivative of θ at $\bar{\mathbf{w}}$ in the direction $\mathbf{d}$ satisfies

$$\theta'(\bar{\mathbf{w}}; \mathbf{d}) \geq \mathbf{d}'\boldsymbol{\beta}(\bar{\mathbf{x}}) \qquad \text{for some } \bar{\mathbf{x}} \in X(\bar{\mathbf{w}})$$

Proof

Consider $\bar{\mathbf{w}} + \lambda_k\mathbf{d}$, where $\lambda_k \to 0^+$. For each k, there exists an $\mathbf{x}_k \in X(\bar{\mathbf{w}} + \lambda_k\mathbf{d})$; and since X is compact, there is a convergent subsequence $\{\mathbf{x}_k\}_{\mathcal{K}}$ with limit $\bar{\mathbf{x}}$ in X. Given an $\mathbf{x} \in X$, note that

$$f(\mathbf{x}) + (\bar{\mathbf{w}} + \lambda_k\mathbf{d})'\boldsymbol{\beta}(\mathbf{x}) \geq f(\mathbf{x}_k) + (\bar{\mathbf{w}} + \lambda_k\mathbf{d})'\boldsymbol{\beta}(\mathbf{x}_k)$$

for each $k \in \mathcal{K}$. Taking the limit as $k \to \infty$, it follows that

$$f(\mathbf{x}) + \bar{\mathbf{w}}'\boldsymbol{\beta}(\mathbf{x}) \geq f(\bar{\mathbf{x}}) + \bar{\mathbf{w}}'\boldsymbol{\beta}(\bar{\mathbf{x}})$$

that is, $\bar{\mathbf{x}} \in X(\bar{\mathbf{w}})$. Furthermore, by the definition of $\theta(\bar{\mathbf{w}} + \lambda_k\mathbf{d})$ and $\theta(\bar{\mathbf{w}})$, we get

$$\begin{aligned}\theta(\bar{\mathbf{w}} + \lambda_k\mathbf{d}) - \theta(\bar{\mathbf{w}}) &= f(\mathbf{x}_k) + (\bar{\mathbf{w}} + \lambda_k\mathbf{d})'\boldsymbol{\beta}(\mathbf{x}_k) - \theta(\bar{\mathbf{w}}) \\ &\geq \lambda_k\mathbf{d}'\boldsymbol{\beta}(\mathbf{x}_k)\end{aligned}$$

The above inequality holds true for each $k \in \mathcal{K}$. Noting that $\mathbf{x}_k \to \bar{\mathbf{x}}$ as $k \in \mathcal{K}$ approaches ∞, we get

$$\lim_{\substack{k \in \mathcal{K} \\ k \to \infty}} \frac{\theta(\bar{\mathbf{w}} + \lambda_k\mathbf{d}) - \theta(\bar{\mathbf{w}})}{\lambda_k} \geq \mathbf{d}'\boldsymbol{\beta}(\bar{\mathbf{x}})$$

By Lemma 3.1.5,

$$\theta'(\bar{\mathbf{w}}; \mathbf{d}) = \lim_{\lambda \to 0^+} \frac{\theta(\bar{\mathbf{w}} + \lambda\mathbf{d}) - \theta(\bar{\mathbf{w}})}{\lambda}$$

exists. In view of the above inequality, the proof is complete.

Corollary

Let $\partial\theta(\bar{\mathbf{w}})$ be the collection of subgradients of θ at $\bar{\mathbf{w}}$, and suppose that the assumptions of the theorem hold true. Then,

$$\theta'(\bar{\mathbf{w}}; \mathbf{d}) = \inf\,\{\mathbf{d}'\boldsymbol{\xi} : \boldsymbol{\xi} \in \partial\theta(\bar{\mathbf{w}})\}$$

Proof

Let $\bar{\mathbf{x}}$ be as specified in the theorem. By Theorem 6.3.4, $\boldsymbol{\beta}(\bar{\mathbf{x}}) \in \partial\theta(\bar{\mathbf{w}})$; and, hence, Theorem 6.3.6 implies that $\theta'(\bar{\mathbf{w}}; \mathbf{d}) \geq \inf\,\{\mathbf{d}'\boldsymbol{\xi}:\boldsymbol{\xi} \in \partial\theta(\bar{\mathbf{w}})\}$. Now let $\boldsymbol{\xi} \in \partial\theta(\bar{\mathbf{w}})$. Since θ is concave, $\theta(\bar{\mathbf{w}} + \lambda\mathbf{d}) - \theta(\bar{\mathbf{w}}) \leq \lambda\mathbf{d}'\boldsymbol{\xi}$. Dividing by $\lambda > 0$ and taking the limit as $\lambda \to 0^+$, it follows that $\theta'(\bar{\mathbf{w}}; \mathbf{d}) \leq \mathbf{d}'\boldsymbol{\xi}$. Since this is true for each $\boldsymbol{\xi} \in \partial\theta(\bar{\mathbf{w}})$, $\theta'(\bar{\mathbf{w}}; \mathbf{d}) \leq \inf\,\{\mathbf{d}'\boldsymbol{\xi}:\boldsymbol{\xi} \in \partial\theta(\bar{\mathbf{w}})\}$, and the proof is complete.

6.3.7 Theorem

Let X be a nonempty compact set in E_n, and let $f:E_n \to E_1$, and $\boldsymbol{\beta}:E_n \to E_{m+l}$ be continuous. Then, $\boldsymbol{\xi}$ is a subgradient of θ at $\bar{\mathbf{w}} \in E_{m+l}$ if and only if $\boldsymbol{\xi}$ belongs to the convex hull of $\{\boldsymbol{\beta}(\mathbf{y}):\mathbf{y} \in X(\bar{\mathbf{w}})\}$.

Proof

Denote the set $\{\boldsymbol{\beta}(\mathbf{y}):\mathbf{y} \in X(\bar{\mathbf{w}})\}$ by $\wedge$ and its convex hull by $H(\wedge)$. By Theorem 6.3.4, $\wedge \subseteq \partial\theta(\bar{\mathbf{w}})$; and since $\partial\theta(\bar{\mathbf{w}})$ is convex, $H(\wedge) \subseteq \partial\theta(\bar{\mathbf{w}})$. Using the facts that X is compact and $\boldsymbol{\beta}$ is continuous, it can be verified that $\wedge$ is compact. Furthermore, the convex hull of a compact set is closed. Therefore, $H(\wedge)$ is a closed convex set. We shall now show that $H(\wedge) \supseteq \partial\theta(\bar{\mathbf{w}})$.

By contradiction, suppose that there is a $\boldsymbol{\xi}' \in \partial\theta(\bar{\mathbf{w}})$ but not in $H(\wedge)$. By Theorem 2.3.4, there exist a scalar α and a nonzero vector $\mathbf{d}$ such that

$$\mathbf{d}'\boldsymbol{\beta}(\mathbf{y}) \geq \alpha \qquad \text{for each } \mathbf{y} \in X(\bar{\mathbf{w}}) \tag{6.15}$$

$$\mathbf{d}'\boldsymbol{\xi}' < \alpha \tag{6.16}$$

By Theorem 6.3.6, there exists a $\mathbf{y} \in X(\bar{\mathbf{w}})$ such that $\theta'(\bar{\mathbf{w}}; \mathbf{d}) \geq \mathbf{d}'\boldsymbol{\beta}(\mathbf{y})$; and by (6.15) above, we must have $\theta'(\bar{\mathbf{w}}; \mathbf{d}) \geq \alpha$. But by the corollary to Theorem 6.3.6 and by (6.16), we get

$$\theta'(\bar{\mathbf{w}}; \mathbf{d}) = \inf\{\mathbf{d}'\boldsymbol{\xi}:\boldsymbol{\xi} \in \partial\theta(\mathbf{w})\} \leq \mathbf{d}'\boldsymbol{\xi}' < \alpha$$

which is a contradiction. Therefore, $\boldsymbol{\xi} \in H(\wedge)$, and $\partial\theta(\bar{\mathbf{w}}) = H(\wedge)$. This completes the proof.

To illustrate, consider the problem of Example 6.2.2 for which the dual function $\theta(v)$, $v \in E_1$, is sketched in Figure 6.4. Note that θ is differentiable (has a unique subgradient) for all v except for $v = -1$ and $v = 2$. Consider $v = 2$, for example. The set $X(2)$ is given by the set of alternative optimal solutions to the problem

$$\theta(2) = \min\{3x_2 - 6:(x_1, x_2) \in X\}$$

Hence, $X(2) = \{(0, 0), (4, 0)\}$, with $\theta(2) = -6$. By Theorem 6.3.4, the subgradients of the form $\beta(\bar{\mathbf{x}})$ for $\bar{\mathbf{x}} \in X(2)$ are $h(0, 0) = -3$ and $h(4, 0) = 1$. Observe that in Figure 6.4 these values are the slopes of the two affine segments defining the graph of θ that are incident at the point $(v, \theta(v)) = (2, -6)$. Therefore, as in Theorem 6.3.7, the set of subgradients of θ at $v = 2$, which is given by the slopes of the set of affine supports for the hypograph of θ, is precisely $[-3, 1]$, the set of convex combinations of -3 and 1.

For another illustration using a bivariate function θ, consider the following example.

6.3.8 Example

Consider the following primal problem:

$$\begin{aligned} \text{Minimize} \quad & -(x_1 - 4)^2 - (x_2 - 4)^2 \\ \text{subject to} \quad & x_1 - 3 \leq 0 \\ & -x_1 + x_2 - 2 \leq 0 \\ & x_1 + x_2 - 4 \leq 0 \\ & x_1, \quad x_2 \geq 0 \end{aligned}$$

In this example, we let $g_1(x_1, x_2) = x_1 - 3$, $g_2(x_1, x_2) = -x_1 + x_2 - 2$, and $X = \{(x_1, x_2): x_1 + x_2 - 4 \leq 0;\ x_1, x_2 \geq 0\}$. Thus, the dual function is given by

$$\theta(u_1, u_2) = \inf\{-(x_1 - 4)^2 - (x_2 - 4)^2 + u_1(x_1 - 3) + u_2(-x_1 + x_2 - 2):\mathbf{x} \in X\}$$

We utilize Theorem 6.3.7 above to determine the set of subgradients of θ at $\bar{\mathbf{u}} = (1, 5)'$. To find the set $X(\bar{\mathbf{u}})$, we need to solve the following problem:

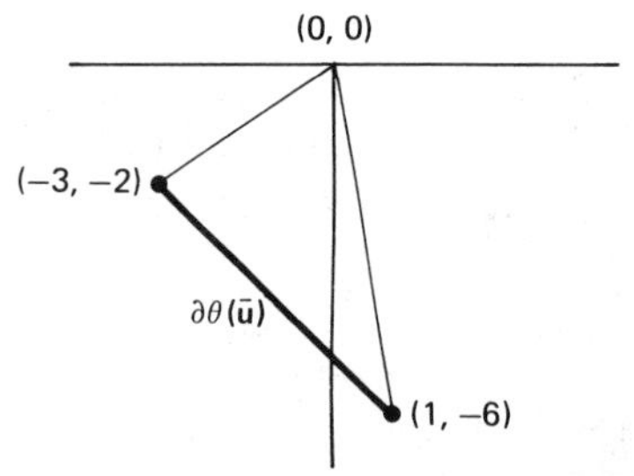

Figure 6.7 Illustration of subgradients.

$$\begin{array}{ll}\text{Minimize} & -(x_1 - 4)^2 - (x_2 - 4)^2 - 4x_1 + 5x_2 - 13 \\ \text{subject to} & x_1 + x_2 - 4 \leq 0 \\ & x_1, x_2 \geq 0\end{array}$$

The objective function of this subproblem is concave, and by Theorem 3.4.7, it assumes its minimum over a compact polyhedral set at one of the extreme points. The polyhedral set X has three extreme points, namely, (0, 0), (4, 0), and (0, 4). Noting that $f(0, 0) = f(4, 0) = -45$ and $f(0, 4) = -9$, it is evident that the optimal solutions of the above subproblem are (0, 0) and (4, 0), that is, $X(\bar{\mathbf{u}}) = \{(0, 0), (4, 0)\}$. By Theorem 6.3.7, the subgradients of θ at $\bar{\mathbf{u}}$ are thus given by the convex combinations of $\mathbf{g}(0, 0)$ and $\mathbf{g}(4, 0)$, that is, by convex combinations of the two vectors $(-3, -2)^t$ and $(1, -6)^t$. Figure 6.7 illustrates the set of subgradients.

Ascent and Steepest Ascent Directions

The dual problem is concerned with the maximization of θ subject to the constraint $\mathbf{u} \geq \mathbf{0}$. Given a point $\mathbf{w}^t = (\mathbf{u}^t, \mathbf{v}^t)$, we would like to investigate the directions along which θ increases. For the sake of clarity, first consider the following definition of an ascent direction, reiterated here for convenience.

6.3.9 Definition

A vector $\mathbf{d}$ is called an *ascent direction* of θ at $\mathbf{w}$ if there exists a $\delta > 0$ such that

$$\theta(\mathbf{w} + \lambda\mathbf{d}) > \theta(\mathbf{w}) \qquad \text{for each } \lambda \in (0, \delta)$$

Note that if θ is concave, a vector $\mathbf{d}$ is an ascent direction of θ at $\mathbf{w}$ if and only if $\theta'(\mathbf{w}; \mathbf{d}) > 0$. Furthermore, θ assumes its maximum at $\mathbf{w}$ if and only if it has no ascent directions at $\mathbf{w}$, that is, if and only if $\theta'(\mathbf{w}; \mathbf{d}) \leq 0$ for each $\mathbf{d}$.

Using the corollary to Theorem 6.3.6, it follows that a vector $\mathbf{d}$ is an ascent direction of θ at $\mathbf{w}$ if and only if $\inf \{\mathbf{d}^t\boldsymbol{\xi} : \boldsymbol{\xi} \in \partial\theta(\mathbf{w})\} > 0$, that is, if and only if the following inequality holds for some $\varepsilon > 0$.

$$\mathbf{d}^t\boldsymbol{\xi} \geq \varepsilon > 0 \qquad \text{for each } \boldsymbol{\xi} \in \partial\theta(\mathbf{w})$$

To illustrate, consider Example 6.3.8. The collection of subgradients of θ at the point (1, 5) is illustrated in Figure 6.7. A vector $\mathbf{d}$ is an ascent direction of θ if and only if $\mathbf{d}^t\boldsymbol{\xi} \geq \varepsilon$ for each subgradient $\boldsymbol{\xi}$, where $\varepsilon > 0$. In other words, $\mathbf{d}$ is an ascent direction if it makes an angle strictly less than 90° with each subgradient. The cone of ascent directions for this example is given in Figure 6.8. In this case, note that each subgradient is an ascent direction. However, this is not necessarily the case in general.

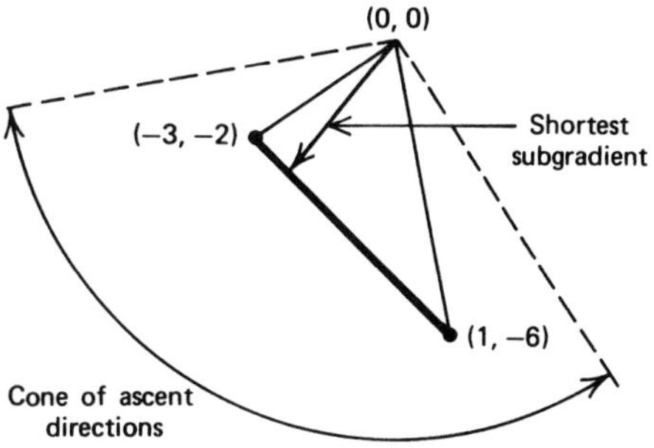

Figure 6.8 The cone of ascent directions in Example 6.3.8.

Since θ is to be maximized, we are interested not only in an ascent direction but also in the direction along which θ increases at the fastest local rate.

6.3.10 Definition

A vector $\bar{\mathbf{d}}$ is called a *direction of steepest ascent* of θ at $\mathbf{w}$ if

$$\theta'(\mathbf{w}; \bar{\mathbf{d}}) = \underset{\|\mathbf{d}\| \le 1}{\text{maximum}}\ \theta'(\mathbf{w}; \mathbf{d})$$

Theorem 6.3.11 below shows that the direction of steepest ascent of the Lagrangian dual function is given by the subgradient with the smallest Euclidean norm. As evident from the proof, this result is true for any arbitrary concave function θ.

6.3.11 Theorem

Let X be a nonempty compact set in E_n, and let $f: E_n \to E_1$, and $\boldsymbol{\beta}: E_n \to E_{m+l}$ be continuous. The direction of steepest ascent $\bar{\mathbf{d}}$ of θ at $\mathbf{w}$ is given below, where $\bar{\boldsymbol{\xi}}$ is the subgradient in $\partial\theta(\mathbf{w})$ with the smallest Euclidean norm:

$$\bar{\mathbf{d}} = \begin{cases} \mathbf{0} & \text{if } \bar{\boldsymbol{\xi}} = \mathbf{0} \\ \dfrac{\bar{\boldsymbol{\xi}}}{\|\bar{\boldsymbol{\xi}}\|} & \text{if } \bar{\boldsymbol{\xi}} \ne \mathbf{0} \end{cases}$$

Proof

By Definition 6.3.10 and by the corollary to Theorem 6.3.6, the steepest ascent direction can be obtained from the following expression:

$$\underset{\|\mathbf{d}\| \le 1}{\text{maximum}}\ \theta'(\mathbf{w}; \mathbf{d}) = \underset{\|\mathbf{d}\| \le 1}{\text{maximum}}\ \underset{\boldsymbol{\xi} \in \partial\theta(\mathbf{w})}{\text{infimum}}\ \mathbf{d}^t\boldsymbol{\xi}$$

The reader can easily verify that

$$\begin{aligned} \underset{\|\mathbf{d}\| \le 1}{\text{maximum}}\ \theta'(\mathbf{w}; \mathbf{d}) &= \underset{\|\mathbf{d}\| \le 1}{\text{maximum}}\ \underset{\boldsymbol{\xi} \in \partial\theta(\mathbf{w})}{\text{infimum}}\ \mathbf{d}^t\boldsymbol{\xi} \\ &\le \underset{\boldsymbol{\xi} \in \partial\theta(\mathbf{w})}{\text{infimum}}\ \underset{\|\mathbf{d}\| \le 1}{\text{maximum}}\ \mathbf{d}^t\boldsymbol{\xi} \\ &= \underset{\boldsymbol{\xi} \in \partial\theta(\mathbf{w})}{\text{infimum}}\ \|\boldsymbol{\xi}\| \\ &= \|\bar{\boldsymbol{\xi}}\| \end{aligned} \tag{6.17}$$

If we construct a direction $\bar{\mathbf{d}}$ such that $\theta'(\mathbf{w}; \bar{\mathbf{d}}) = \|\bar{\boldsymbol{\xi}}\|$, then, by (6.17), $\bar{\mathbf{d}}$ is the steepest ascent direction. If $\bar{\boldsymbol{\xi}} = \mathbf{0}$, then, for $\bar{\mathbf{d}} = \mathbf{0}$, we obviously have $\theta'(\mathbf{w}; \bar{\mathbf{d}}) = \|\bar{\boldsymbol{\xi}}\|$. Now suppose that $\bar{\boldsymbol{\xi}} \neq \mathbf{0}$, and let $\bar{\mathbf{d}} = \bar{\boldsymbol{\xi}}/\|\bar{\boldsymbol{\xi}}\|$. Note that

$$\begin{aligned}
\theta'(\mathbf{w}; \bar{\mathbf{d}}) &= \inf\{\bar{\mathbf{d}}'\boldsymbol{\xi} : \boldsymbol{\xi} \in \partial\theta(\mathbf{w})\} \\
&= \inf\left\{\frac{\bar{\boldsymbol{\xi}}'\boldsymbol{\xi}}{\|\bar{\boldsymbol{\xi}}\|} : \boldsymbol{\xi} \in \partial\theta(\mathbf{w})\right\} \\
&= \frac{1}{\|\bar{\boldsymbol{\xi}}\|}\inf\{\|\bar{\boldsymbol{\xi}}\|^2 + \bar{\boldsymbol{\xi}}'(\boldsymbol{\xi} - \bar{\boldsymbol{\xi}}) : \boldsymbol{\xi} \in \partial\theta(\mathbf{w})\} \\
&= \|\bar{\boldsymbol{\xi}}\| + \frac{1}{\|\bar{\boldsymbol{\xi}}\|}\inf\{\bar{\boldsymbol{\xi}}'(\boldsymbol{\xi} - \bar{\boldsymbol{\xi}}) : \boldsymbol{\xi} \in \partial\theta(\mathbf{w})\}
\end{aligned} \tag{6.18}$$

Since $\bar{\boldsymbol{\xi}}$ is the shortest vector in $\partial\theta(\mathbf{w})$, then, by Theorem 2.4.1, $\bar{\boldsymbol{\xi}}'(\boldsymbol{\xi} - \bar{\boldsymbol{\xi}}) \geq 0$ for each $\boldsymbol{\xi} \in \partial\theta(\mathbf{w})$. Hence, $\inf\{\bar{\boldsymbol{\xi}}'(\boldsymbol{\xi} - \bar{\boldsymbol{\xi}}) : \boldsymbol{\xi} \in \partial\theta(\mathbf{w})\} = 0$ is achieved at $\bar{\boldsymbol{\xi}}$. From (6.18), it then follows that $\theta'(\mathbf{w}; \bar{\mathbf{d}}) = \|\bar{\boldsymbol{\xi}}\|$. Thus, we have shown that the vector $\bar{\mathbf{d}}$ specified in the theorem is the direction of steepest ascent both when $\bar{\boldsymbol{\xi}} = \mathbf{0}$ and when $\bar{\boldsymbol{\xi}} \neq \mathbf{0}$. This completes the proof.

6.4 Formulating and Solving the Dual Problem

Given a primal Problem P to minimize $f(\mathbf{x})$ subject to $\mathbf{g}(\mathbf{x}) \leq \mathbf{0}$, $\mathbf{h}(\mathbf{x}) = \mathbf{0}$, and $\mathbf{x} \in X$, we have defined a Lagrangian dual Problem D to maximize $\theta(\mathbf{u}, \mathbf{v})$ subject to $\mathbf{u} \geq \mathbf{0}$, where $\theta(\mathbf{u}, \mathbf{v})$ is evaluated via the (Lagrangian) subproblem $\theta(\mathbf{u}, \mathbf{v}) = \min\{f(\mathbf{x}) + \mathbf{u}'\mathbf{g}(\mathbf{x}) + \mathbf{v}'\mathbf{h}(\mathbf{x}) : \mathbf{x} \in X\}$. In formulating this dual problem, we have *dualized*, that is, accommodated within the Lagrangian dual objective function, the constraints $\mathbf{g}(\mathbf{x}) \leq \mathbf{0}$ and $\mathbf{h}(\mathbf{x}) = \mathbf{0}$, maintaining any other constraints within the set X. Different *formulations* of the Lagrangian dual problem might dualize different sets of constraints in constructing the Lagrangian dual function. This choice must usually be a trade-off between the ease of evaluating $\theta(\mathbf{u}, \mathbf{v})$ for a given $(\mathbf{u}, \mathbf{v})$ versus the duality gap that might exist between P and D. For example, consider the linear discrete problem

$$\begin{aligned}
DP:\quad \text{Minimize} \quad & \mathbf{c}'\mathbf{x} \\
\text{subject to} \quad & \mathbf{A}\mathbf{x} = \mathbf{b} \\
& \mathbf{D}\mathbf{x} = \mathbf{d} \\
& \mathbf{x} \in X
\end{aligned} \tag{6.19a}$$

where $\mathbf{X}$ is some compact, discrete set. Let us define the Lagrangian dual problem

$$LDP:\quad \text{Maximize} \quad \{\theta(\boldsymbol{\pi}) : \boldsymbol{\pi} \text{ unrestricted}\} \tag{6.16b}$$

where $\theta(\boldsymbol{\pi}) = \text{minimum}\{\mathbf{c}'\mathbf{x} + \boldsymbol{\pi}'(\mathbf{A}\mathbf{x} - \mathbf{b}) : \mathbf{D}\mathbf{x} = \mathbf{d}, \mathbf{x} \in X\}$. Because of the linearity of the objective function in the latter subproblem, we equivalently have $\theta(\boldsymbol{\pi}) = \text{minimum}\{\mathbf{c}'\mathbf{x} + \boldsymbol{\pi}'(\mathbf{A}\mathbf{x} - \mathbf{b}) : \mathbf{x} \in H[\mathbf{x} \in X : \mathbf{D}\mathbf{x} = \mathbf{d}]\}$, where $H[\cdot]$ denotes the convex hull. It readily follows, then (see Exercise 6.15), that the Lagrangian dual objective value will match that of the modified Problem DP′ to minimize $\mathbf{c}'\mathbf{x}$ subject to $\mathbf{A}\mathbf{x} = \mathbf{b}$ and $\mathbf{x} \in H[\mathbf{x} \in X : \mathbf{D}\mathbf{x} = \mathbf{d}]$. Noting that DP is itself equivalent to minimizing $\mathbf{c}'\mathbf{x}$ subject to $\mathbf{x} \in H[\mathbf{x} \in X : \mathbf{A}\mathbf{x} = \mathbf{b}, \mathbf{D}\mathbf{x} = \mathbf{d}]$, we surmise how the partial convex hull operation manifested in DP′ can influence the duality gap.

In this spirit, one may sometimes wish to manipulate the primal problem itself into a special form before constructing a Lagrangian dual formulation to create exploitable structures for the subproblem. For example, the discrete Problem DP stated above can be equivalently written as minimize $\{\mathbf{c}^t\mathbf{x}: \mathbf{A}\mathbf{x} = \mathbf{b}, \mathbf{D}\mathbf{y} = \mathbf{d}, \mathbf{x} = \mathbf{y}, \mathbf{x} \in X, \mathbf{y} \in Y\}$, where Y is a copy of X in which the $\mathbf{x}$ variables have been replaced by a set of matching $\mathbf{y}$ variables. Now, we can formulate a Lagrangian dual problem

$$\overline{LDP}: \quad \text{Maximize} \quad \{\bar{\theta}(\boldsymbol{\mu}): \boldsymbol{\mu} \text{ unrestricted}\} \tag{6.20}$$

where $\bar{\theta}(\boldsymbol{\mu}) = \min \{\mathbf{c}^t\mathbf{x} + \boldsymbol{\mu}^t(\mathbf{x} - \mathbf{y}) : \mathbf{A}\mathbf{x} = \mathbf{b}, \mathbf{D}\mathbf{y} = \mathbf{d}, \mathbf{x} \in X, \mathbf{y} \in Y\}$. Observe that this subproblem decomposes into two separable problems over the $\mathbf{x}$ and $\mathbf{y}$ variables, each with a possible specially exploitable structure. Moreover, it can be shown (see Exercise 6.16) that $\max_{\boldsymbol{\mu}} \{\bar{\theta}(\boldsymbol{\mu})\} \geq \max_{\boldsymbol{\pi}} \theta(\boldsymbol{\pi})$, where θ is defined in (6.19b). Hence, the Lagrangian dual formulation $\overline{\text{LDP}}$ affords a tighter representation for the primal Problem DP in the sense that it yields a smaller duality gap than does LDP. Note that, as previously observed, the value of $\overline{\text{LDP}}$ matches that of the following partial convex hull representation of problem:

$$DP: \text{Minimize} \quad \{\mathbf{c}^t\mathbf{x} : \mathbf{x} \in H[\mathbf{x} \in X: \mathbf{A}\mathbf{x} = \mathbf{b}], \mathbf{y} \in H[\mathbf{y} \in Y: \mathbf{D}\mathbf{y} = \mathbf{d}], \mathbf{x} = \mathbf{y}\}$$

The conceptual approach leading to the formulation of $\overline{\text{LDP}}$ is called a *layering strategy* (because of the layer of constraints constructed), or a *Lagrangian decomposition strategy* (because of the separable decomposable structures generated). Refer to the Notes and References section for further reading on this subject matter.

Returning to the dual Problem D corresponding to the primal Problem P stated in Section 6.1, the reader will recall that we have described in the previous section several properties of this dual function. In particular, the dual problem requires the maximization of a concave function $\theta(\mathbf{u}, \mathbf{v})$ over the simple constraint set $\{(\mathbf{u}, \mathbf{v}): \mathbf{u} \geq \mathbf{0}\}$. If θ is differentiable due to the property stated in Theorem 6.3.3, then $\nabla\theta(\bar{\mathbf{u}}, \bar{\mathbf{v}})^t = [\mathbf{g}(\bar{\mathbf{x}})^t, \mathbf{h}(\bar{\mathbf{x}})^t]$. Various algorithms described in subsequent chapters that are applied to maximizing differentiable concave functions can be used to solve this dual problem. These algorithms involve the generation of a suitable ascent direction $\mathbf{d}$, followed by a one-dimensional line search along this direction to find a new improved solution.

To illustrate one simple scheme to find an ascent direction at a point $(\bar{\mathbf{u}}, \bar{\mathbf{v}})$, consider the following strategy. If $\nabla\theta(\bar{\mathbf{u}}, \bar{\mathbf{v}}) \neq \mathbf{0}$, then, by Theorem 4.1.2, this is an ascent direction and θ will increase by moving from $(\bar{\mathbf{u}}, \bar{\mathbf{v}})$ along $\nabla\theta(\bar{\mathbf{u}}, \bar{\mathbf{v}})$. However, if some components of $\bar{\mathbf{u}}$ are equal to zero, and any of the corresponding components of $\mathbf{g}(\bar{\mathbf{x}})$ is negative, then $\bar{\mathbf{u}} + \lambda\mathbf{g}(\bar{\mathbf{x}}) \ngeq \mathbf{0}$ for $\lambda \geq 0$, thus violating the nonnegativity restriction. In order to handle this difficulty, we can use a modified or projected direction $[\hat{\mathbf{g}}(\bar{\mathbf{x}}), \mathbf{h}(\bar{\mathbf{x}})]$, where $\hat{\mathbf{g}}(\bar{\mathbf{x}})$ is defined as

$$\hat{g}_i(\bar{\mathbf{x}}) = \begin{cases} g_i(\bar{\mathbf{x}}) & \text{if } \bar{u}_i > 0 \\ \max[0, g_i(\bar{\mathbf{x}})] & \text{if } \bar{u}_i = 0 \end{cases}$$

It can then be shown (see Exercise 6.18) that $[\hat{\mathbf{g}}(\bar{\mathbf{x}}), \mathbf{h}(\bar{\mathbf{x}})]$ is a feasible ascent direction of θ at $(\bar{\mathbf{u}}, \bar{\mathbf{v}})$. Furthermore, $[\hat{\mathbf{g}}(\bar{\mathbf{x}}), \mathbf{h}(\bar{\mathbf{x}})]$ is zero only when the dual maximum is reached. On the other hand, suppose that θ is nondifferentiable. In this case, the set of subgradients of θ are characterized by Theorem 6.3.7. For $\mathbf{d}$ to be an ascent direction of θ at $(\mathbf{u}, \mathbf{v})$, noting the corollary to Theorem 6.3.6 and the concavity of θ, we must have $\mathbf{d}^t\boldsymbol{\xi} \geq \varepsilon > 0$ for each $\boldsymbol{\xi} \in \partial\theta(\mathbf{u}, \mathbf{v})$. As a preliminary idea, the following problem can then be used for finding such a direction:

$$
\begin{array}{lll}
\text{Maximize} & \varepsilon & \\
\text{subject to} & \mathbf{d}^t\boldsymbol{\xi} \geq \varepsilon & \text{for } \boldsymbol{\xi} \in \partial\theta(\mathbf{u}, \mathbf{v}) \\
& d_i \geq 0 & \text{if } u_i = 0 \\
& -1 \leq d_i \leq 1 & \text{for } i = 1, \ldots, m + l
\end{array}
$$

Note that the constraints $d_i \geq 0$ if $u_i = 0$ ensure that the vector **d** is a feasible direction, and that the normalization constraints $-1 \leq d_i \leq 1$ guarantee a finite solution to the problem.

The reader may note the following difficulties associated with the above direction finding problem:

1. The set $\partial\theta(\mathbf{u}, \mathbf{v})$ and, hence, the constraints of the problem are not explicitly known in advance. However, Theorem 6.3.7, which fully characterizes the subgradient set, could be of use.
2. The set $\partial\theta(\mathbf{u}, \mathbf{v})$ usually admits an infinite number of subgradients, so that we have a linear program with an infinite number of constraints. However, if $\partial\theta(\mathbf{u}, \mathbf{v})$ is a compact polyhedral set, then the constraints $\mathbf{d}^t\boldsymbol{\xi} \geq \varepsilon$ for $\boldsymbol{\xi} \in \partial\theta(\mathbf{u}, \mathbf{v})$ could be replaced by the constraints

$$\mathbf{d}^t\boldsymbol{\xi}_j \geq \varepsilon \qquad \text{for } j = 1, \ldots, \gamma$$

where $\boldsymbol{\xi}_1, \ldots, \boldsymbol{\xi}_\gamma$ are the extreme points of $\partial\theta(\mathbf{u}, \mathbf{v})$. Thus, in this case the problem reduces to a finite linear program.

To alleviate some of the above problems, one might use a row generation strategy in which only a finite number of representatives of the constraint set $\mathbf{d}^t\boldsymbol{\xi} \geq \varepsilon$ for $\boldsymbol{\xi} \in \partial\theta(\mathbf{u}, \mathbf{v})$ are used, and the resulting direction $\mathbf{d}_\gamma$ is tested to ascertain if it is an ascent direction. This can be done by verifying if $\min \{\mathbf{d}_\gamma^t\boldsymbol{\xi} : \boldsymbol{\xi} \in \partial\theta(\mathbf{u}, \mathbf{v})\} > 0$. If so, then $\mathbf{d}_\gamma$ is used in the line search process. If not, then the foregoing subproblem yields a subgradient $\boldsymbol{\xi}_{\gamma+1}$ for which $\mathbf{d}_\gamma^t\boldsymbol{\xi}_{\gamma+1} \leq 0$, and thus this constraint is added to the direction-finding problem, and the operation is repeated.

We ask the reader to provide the details of this scheme in Exercise 6.25. However, this type of a procedure is fraught with computational difficulties, except for small-sized problems with simple structures. Later, in Chapter 8, we address more sophisticated and efficient subgradient-based optimization schemes that can be used to optimize θ whenever it is nondifferentiable. These procedures employ various strategies for constructing directions based on single subgradients, possibly deflected by suitable means, or based on a bundle of subgradients collected over some local neighborhood. The directions need not always be ascent directions, but ultimate convergence to an optimum is nevertheless assured. We refer the reader to Chapter 8 and its Notes and References section for further information on this topic.

We now proceed to describe in detail one particular cutting plane or outer-linearization scheme for solving the dual Problem D. The concept of this approach is important in its own right, as it constitutes a useful ingredient for many decomposition/partitioning methods.

A Cutting Plane or Outer-Linearization Method

The methods discussed in principle above for solving the dual problem generate at each iteration a direction of motion and adopt a step size along this direction, with a view to ultimately finding the maximum for the Lagrangian dual function. We now discuss a strategy for solving the dual problem in which, at each iteration, a function that approximates the dual function is optimized.

Recall that the dual function θ is defined by

$$\theta(\mathbf{u}, \mathbf{v}) = \inf \{f(\mathbf{x}) + \mathbf{u}^t\mathbf{g}(\mathbf{x}) + \mathbf{v}^t\mathbf{h}(\mathbf{x}) : \mathbf{x} \in X\}$$

Letting $z = \theta(\mathbf{u}, \mathbf{v})$, the inequality $z \leq f(\mathbf{x}) + \mathbf{u}^t\mathbf{g}(\mathbf{x}) + \mathbf{v}^t\mathbf{h}(\mathbf{x})$ must hold true for each $\mathbf{x} \in X$. Hence, the dual problem of maximizing $\theta(\mathbf{u}, \mathbf{v})$ over $\mathbf{u} \geq \mathbf{0}$ is equivalent to the following problem:

$$\begin{array}{lll} \text{Maximize} & z & \\ \text{subject to} & z \leq f(\mathbf{x}) + \mathbf{u}^t\mathbf{g}(\mathbf{x}) + \mathbf{v}^t\mathbf{h}(\mathbf{x}) & \text{for } \mathbf{x} \in X \\ & \mathbf{u} \geq \mathbf{0} & \end{array} \tag{6.21}$$

Note that the above problem is a linear program in the variables z, $\mathbf{u}$, and $\mathbf{v}$. Unfortunately, however, the constraints are infinite and are not known explicitly. Suppose that we have the points $\mathbf{x}_1, \ldots, \mathbf{x}_{k-1}$ in X, and consider the following approximating problem:

$$\begin{array}{lll} \text{Maximize} & z & \\ \text{subject to} & z \leq f(\mathbf{x}_j) + \mathbf{u}^t\mathbf{g}(\mathbf{x}_j) + \mathbf{v}^t\mathbf{h}(\mathbf{x}_j) & \text{for } j = 1, \ldots, k-1 \\ & \mathbf{u} \geq \mathbf{0} & \end{array} \tag{6.22}$$

The above problem is a linear program with a finite number of constraints and can be solved by the simplex method, for example. Let $(z_k, \mathbf{u}_k, \mathbf{v}_k)$ be an optimal solution to this approximating problem, sometimes referred to as the *master program*. If this solution satisfies (6.21), then it is an optimal solution to the Lagrangian dual problem. To check whether (6.21) is satisfied, consider the following *subproblem*:

$$\begin{array}{ll} \text{Minimize} & f(\mathbf{x}) + \mathbf{u}_k^t\mathbf{g}(\mathbf{x}) + \mathbf{v}_k^t\mathbf{h}(\mathbf{x}) \\ \text{subject to} & \mathbf{x} \in X \end{array}$$

Let $\mathbf{x}_k$ be an optimal solution to the above problem, so that $\theta(\mathbf{u}_k, \mathbf{v}_k) = f(\mathbf{x}_k) + \mathbf{u}_k^t\mathbf{g}(\mathbf{x}_k) + \mathbf{v}_k^t\mathbf{h}(\mathbf{x}_k)$. If $z_k \leq \theta(\mathbf{u}_k, \mathbf{v}_k)$, then $(\mathbf{u}_k, \mathbf{v}_k)$ is an optimal solution to the Lagrangian dual problem. Otherwise, for $(\mathbf{u}, \mathbf{v}) = (\mathbf{u}_k, \mathbf{v}_k)$ the inequality (6.21) is not satisfied for $\mathbf{x} = \mathbf{x}_k$. Thus, we add the constraint

$$z \leq f(\mathbf{x}_k) + \mathbf{u}^t\mathbf{g}(\mathbf{x}_k) + \mathbf{v}^t\mathbf{h}(\mathbf{x}_k)$$

to the constraints in (6.22), and re-solve the master linear program. Obviously, the current optimal point $(z_k, \mathbf{u}_k, \mathbf{v}_k)$ contradicts this added constraint. Thus, this point is cut away, hence the name cutting plane algorithm.

Summary of the Cutting Plane or Outer-Linearization Method

Assume that f, $\mathbf{g}$, and $\mathbf{h}$ are continuous, and that X is compact, so that the set $X(\mathbf{u}, \mathbf{v})$ is nonempty for each $(\mathbf{u}, \mathbf{v})$.

Initialization Step Find a point $\mathbf{x}_0 \in X$ such that $\mathbf{g}(\mathbf{x}_0) \leq \mathbf{0}$ and $\mathbf{h}(\mathbf{x}_0) = \mathbf{0}$. Let $k = 1$, and go to the main step.

Main Step Solve the following *master program*:

$$\begin{array}{lll} \text{Maximize} & z & \\ \text{subject to} & z \leq f(\mathbf{x}_j) + \mathbf{u}^t\mathbf{g}(\mathbf{x}_j) + \mathbf{v}^t\mathbf{h}(\mathbf{x}_j) & \text{for } j = 0, \ldots, k-1 \\ & \mathbf{u} \geq \mathbf{0} & \end{array}$$

Let $(z_k, \mathbf{u}_k, \mathbf{v}_k)$ be an optimal solution and solve the following *subproblem*:

$$\begin{array}{ll}\text{Minimize} & f(\mathbf{x}) + \mathbf{u}_k^t\mathbf{g}(\mathbf{x}) + \mathbf{v}_k^t\mathbf{h}(\mathbf{x}) \\ \text{subject to} & \mathbf{x} \in X\end{array}$$

Let $\mathbf{x}_k$ be an optimal point, and let $\theta(\mathbf{u}_k, \mathbf{v}_k) = f(\mathbf{x}_k) + \mathbf{u}_k^t\mathbf{g}(\mathbf{x}_k) + \mathbf{v}_k^t\mathbf{h}(\mathbf{x}_k)$. If $z_k = \theta(\mathbf{u}_k, \mathbf{v}_k)$, then stop; $(\mathbf{u}_k, \mathbf{v}_k)$ is an optimal dual solution. Otherwise, if $z_k > \theta(\mathbf{u}_k, \mathbf{v}_k)$, then add the constraint $z \le f(\mathbf{x}_k) + \mathbf{u}^t\mathbf{g}(\mathbf{x}_k) + \mathbf{v}^t\mathbf{h}(\mathbf{x}_k)$ to the master program, replace k by $k + 1$, and repeat the main step.

At each iteration, a cut (constraint) is added to the master problem, and, hence, the size of the master problem increases monotonically. In practice, if the size of the master problem becomes excessively large, all constraints that are not binding may be thrown away. Theoretically, this might not guarantee convergence, unless, for example, the dual value has strictly increased since the last time such a deletion was executed, and the set X has a finite number of elements. (See Exercise 6.24; and for a general convergence theorem, see Exercises 7.20 and 7.21.) Also note that the optimal solution values of the master problem form a nonincreasing sequence $\{z_k\}$. Since each z_k is an upper bound on the optimal value of the dual problem, we may stop if $z_k - \max_{1\le j\le k} \theta(\mathbf{u}_j, \mathbf{v}_j) < \varepsilon$, where ε is a small positive number.

Interpretation as a Tangential Approximation or Outer-Linearization Technique

The foregoing algorithm for maximizing the dual function can be interpreted as a tangential approximation technique. By definition of θ, we must have

$$\theta(\mathbf{u}, \mathbf{v}) \le f(\mathbf{x}) + \mathbf{u}^t\mathbf{g}(\mathbf{x}) + \mathbf{v}^t\mathbf{h}(\mathbf{x}) \qquad \text{for } \mathbf{x} \in X$$

Thus, for any fixed $\mathbf{x} \in X$, the hyperplane

$$\{(\mathbf{u}, \mathbf{v}, z) : \mathbf{u} \in E_m, \mathbf{v} \in E_l, z = f(\mathbf{x}) + \mathbf{u}^t\mathbf{g}(\mathbf{x}) + \mathbf{v}^t\mathbf{h}(\mathbf{x})\}$$

bounds the function θ from above.

The master problem at iteration k is equivalent to solving the following problem:

$$\begin{array}{ll}\text{Maximize} & \hat{\theta}(\mathbf{u}, \mathbf{v}) \\ \text{subject to} & \mathbf{u} \ge \mathbf{0}\end{array}$$

where $\hat{\theta}(\mathbf{u}, \mathbf{v}) = \min \{f(\mathbf{x}_j) + \mathbf{u}^t\mathbf{g}(\mathbf{x}_j) + \mathbf{v}^t\mathbf{h}(\mathbf{x}_j) : j = 1, \ldots, k - 1\}$. Note that θ is a piecewise linear function that approximates θ by considering only $k - 1$ of the bounding hyperplanes.

Let the optimal solution to the master problem be $(z_k, \mathbf{u}_k, \mathbf{v}_k)$. Now, the subproblem is solved yielding $\theta(\mathbf{u}_k, \mathbf{v}_k)$ and $\mathbf{x}_k$. If $z_k > \theta(\mathbf{u}_k, \mathbf{v}_k)$, then the new constraint $z \le f(\mathbf{x}_k) + \mathbf{u}^t\mathbf{g}(\mathbf{x}_k) + \mathbf{v}^t\mathbf{h}(\mathbf{x}_k)$ is added to the master problem, giving a new and tighter piecewise linear approximation to θ. Since $\theta(\mathbf{u}_k, \mathbf{v}_k) = f(\mathbf{x}_k) + \mathbf{u}_k^t\mathbf{g}(\mathbf{x}_k) + \mathbf{v}_k^t\mathbf{h}(\mathbf{x}_k)$, the hyperplane $\{(z, \mathbf{u}, \mathbf{v}) : z = f(\mathbf{x}_k) + \mathbf{u}^t\mathbf{g}(\mathbf{x}_k) + \mathbf{v}^t\mathbf{h}(\mathbf{x}_k)\}$ is tangential to the graph of θ at $(z_k, \mathbf{u}_k, \mathbf{v}_k)$.

6.4.1 Example

$$\begin{array}{lll}\text{Minimize} & (x_1 - 2)^2 + \tfrac{1}{4}x_2^2 & \\ \text{subject to} & x_1 - \tfrac{7}{2}x_2 - 1 \le 0 & \\ & 2x_1 + 3x_2 = 4 & \end{array}$$

We let $X = \{(x_1, x_2): 2x_1 + 3x_2 = 4\}$, so that the Lagrangian dual function is given by

$$\theta(u) = \min\{(x_1 - 2)^2 + \tfrac{1}{4}x_2^2 + u(x_1 - \tfrac{7}{2}x_2 - 1) : 2x_1 + 3x_2 = 4\} \tag{6.23}$$

The cutting plane method is initialized with a feasible solution $\mathbf{x}_0 = (\frac{5}{4}, \frac{1}{2})^t$. At step 1 of the first iteration, we solve the following problem:

Maximize z
subject to $z \le \frac{5}{8} - \frac{3}{2}u$
$u \ge 0$

The optimal solution is $(z_1, u_1) = (\frac{5}{8}, 0)$. At step 2, we solve (6.23) for $u = u_1 = 0$, yielding an optimal solution $\mathbf{x}_1 = (2, 0)^t$ with $\theta(u_1) = 0 < z_1$. Hence, more iterations are needed. The summary of the first four iterations are given in Table 6.1.

Table 6.1 Summary of Computations for Example 6.4.1

		Step 1 Solution	Step 2 Solution	
Iteration k	Constraint Added	(z_k, u_k)	$\mathbf{x}_k^t$	$\theta(u_k)$
1	$z \le \frac{5}{8} - \frac{3}{2}u$	$(\frac{5}{8}, 0)$	$(2, 0)$	0
2	$z \le 0 + u$	$(\frac{1}{4}, \frac{1}{4})$	$(\frac{13}{8}, \frac{1}{4})$	$\frac{3}{32}$
3	$z \le \frac{5}{32} - \frac{1}{4}u$	$(\frac{1}{8}, \frac{1}{8})$	$(\frac{29}{16}, \frac{1}{8})$	$\frac{11}{128}$
4	$z \le \frac{5}{128} + \frac{3}{8}u$	$(\frac{7}{64}, \frac{3}{16})$	$(\frac{55}{32}, \frac{3}{16})$	$\frac{51}{512}$

The approximating function $\hat{\theta}$ at the end of the fourth iteration is shown in darkened lines in Figure 6.9. The reader can easily verify that the Langrangian dual function for this problem is given by $\theta(u) = -\frac{5}{2}u^2 + u$ and that the hyperplanes added at iteration 2 onward are indeed tangential to the graph of θ at the respective points (z_k, u_k). Incidentally, the dual objective function is maximized at $\bar{u} = \frac{1}{5}$ with $\theta(\bar{u}) = \frac{1}{10}$. Note that the sequence $\{u_k\}$ converges to the optimal point $\bar{u} = \frac{1}{5}$.

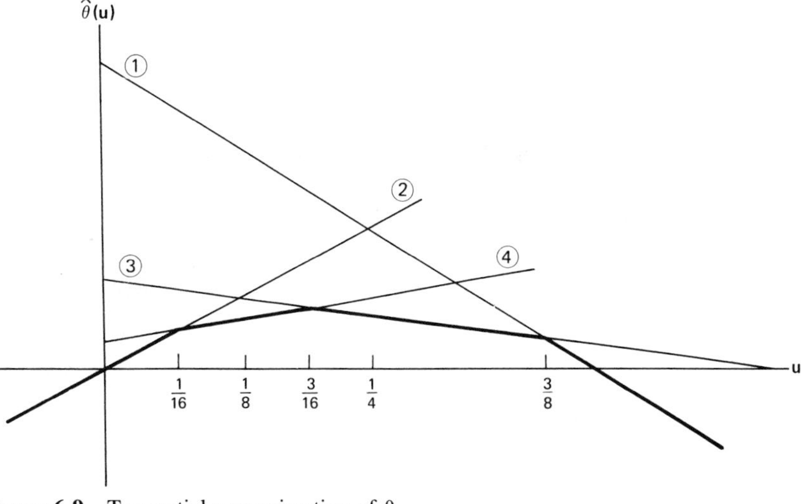

Figure 6.9 Tangential approximation of θ.

6.5 Getting the Primal Solution

Thus far, we have studied several properties of the dual function and described some procedures for solving the dual problem. However, our main concern is finding an optimal solution to the primal problem.

In this section, we develop some theorems that will aid us in finding a solution to the primal problem, as well as solutions to perturbations of the primal problem. However, for nonconvex programs, as a result of the possible presence of a duality gap, additional work is usually needed to find an optimal primal solution.

Solutions to Perturbed Primal Problems

During the course of solving the dual problem, the following problem, which is used to evaluate the function θ at $(\mathbf{u}, \mathbf{v})$, is solved frequently:

$$\begin{array}{ll} \text{Minimize} & f(\mathbf{x}) + \mathbf{u}^t\mathbf{g}(\mathbf{x}) + \mathbf{v}^t\mathbf{h}(\mathbf{x}) \\ \text{subject to} & \mathbf{x} \in X \end{array}$$

Theorem 6.5.1 shows that an optimal solution $\bar{\mathbf{x}}$ to the above problem is also an optimal solution to a problem that is similar to the primal problem, in which some of the constraints are perturbed. Specifically, $\bar{\mathbf{x}}$ evaluates $\nu[\mathbf{g}(\bar{\mathbf{x}}), \mathbf{h}(\bar{\mathbf{x}})]$, where ν is the perturbation function defined in (6.9).

6.5.1 Theorem

Let $(\mathbf{u}, \mathbf{v})$ be a given vector with $\mathbf{u} \geq \mathbf{0}$. Consider the problem to minimize $f(\mathbf{x}) + \mathbf{u}^t\mathbf{g}(\mathbf{x}) + \mathbf{v}^t\mathbf{h}(\mathbf{x})$ subject to $\mathbf{x} \in X$. Let $\bar{\mathbf{x}}$ be an optimal solution. Then, $\bar{\mathbf{x}}$ is an optimal solution to the following problem, where $I = \{i : u_i > 0\}$:

$$\begin{array}{lll} \text{Minimize} & f(\mathbf{x}) & \\ \text{subject to} & g_i(\mathbf{x}) \leq g_i(\bar{\mathbf{x}}) & \text{for } i \in I \\ & h_i(\mathbf{x}) = h_i(\bar{\mathbf{x}}) & \text{for } i = 1, \ldots, l \\ & \mathbf{x} \in X & \end{array}$$

In particular, $\bar{\mathbf{x}}$ solves the problem to evaluate $\nu[\mathbf{g}(\bar{\mathbf{x}}), \mathbf{h}(\bar{\mathbf{x}})]$, where ν is the perturbation function defined in (6.9).

Proof

Let $\mathbf{x} \in X$ be such that $h_i(\mathbf{x}) = h_i(\bar{\mathbf{x}})$ for $i = 1, \ldots, l$ and $g_i(\mathbf{x}) \leq g_i(\bar{\mathbf{x}})$ for $i \in I$. Note that

$$f(\mathbf{x}) + \mathbf{u}^t\mathbf{g}(\mathbf{x}) + \mathbf{v}^t\mathbf{h}(\mathbf{x}) \geq f(\bar{\mathbf{x}}) + \mathbf{u}^t\mathbf{g}(\bar{\mathbf{x}}) + \mathbf{v}^t\mathbf{h}(\bar{\mathbf{x}}) \tag{6.24}$$

But since $\mathbf{h}(\mathbf{x}) = \mathbf{h}(\bar{\mathbf{x}})$, and $\mathbf{u}^t\mathbf{g}(\mathbf{x}) = \sum_{i \in I} u_i g_i(\mathbf{x}) \leq \sum_{i \in I} u_i g_i(\bar{\mathbf{x}}) = \mathbf{u}^t\mathbf{g}(\bar{\mathbf{x}})$, from (6.24), we get

$$f(\mathbf{x}) + \mathbf{u}^t\mathbf{g}(\bar{\mathbf{x}}) \geq f(\mathbf{x}) + \mathbf{u}^t\mathbf{g}(\mathbf{x}) \geq f(\bar{\mathbf{x}}) + \mathbf{u}^t\mathbf{g}(\bar{\mathbf{x}})$$

which shows that $f(\mathbf{x}) \geq f(\bar{\mathbf{x}})$. Hence, $\bar{\mathbf{x}}$ solves the problem stated in the theorem. Moreover, since this problem is a relaxation of (6.9) for $\mathbf{y} = \bar{\mathbf{y}}$, where $\bar{\mathbf{y}}^t = [\mathbf{g}(\bar{\mathbf{x}})^t, \mathbf{h}(\bar{\mathbf{x}})^t]$, and since $\bar{\mathbf{x}}$ is feasible to (6.9) with $\mathbf{y} = \bar{\mathbf{y}}$, it follows that $\bar{\mathbf{x}}$ evaluates $\nu(\bar{\mathbf{y}})$. This completes the proof.

Corollary

Under the assumptions of the theorem, suppose that $\mathbf{g}(\bar{\mathbf{x}}) \leq \mathbf{0}$, $\mathbf{h}(\bar{\mathbf{x}}) = \mathbf{0}$, and $\mathbf{u}^t\mathbf{g}(\bar{\mathbf{x}}) = 0$. Then, $\bar{\mathbf{x}}$ is an optimal solution to the following problem:

$$\begin{array}{lll} \text{Minimize} & f(\mathbf{x}) & \\ \text{subject to} & g_i(\mathbf{x}) \leq 0 & \text{for } i \in I \\ & h_i(\mathbf{x}) = 0 & \text{for } i = 1, \ldots, l \\ & \mathbf{x} \in X & \end{array}$$

In particular, $\bar{\mathbf{x}}$ is an optimal solution to the original primal problem, and $(\mathbf{u}, \mathbf{v})$ is an optimal solution to the dual problem.

Proof

Note that $\mathbf{u}^t\mathbf{g}(\bar{\mathbf{x}}) = 0$ implies that $g_i(\bar{\mathbf{x}}) = 0$ for $i \in I$; and from the theorem, it follows that $\bar{\mathbf{x}}$ solves the stated problem. Also, since the feasible region of the primal problem is contained in that of the above problem, and since $\bar{\mathbf{x}}$ is a feasible solution to the primal problem, $\bar{\mathbf{x}}$ is an optimal solution to the primal problem. Furthermore, $f(\bar{\mathbf{x}}) = f(\bar{\mathbf{x}}) + \mathbf{u}^t\mathbf{g}(\bar{\mathbf{x}}) + \mathbf{v}^t\mathbf{h}(\bar{\mathbf{x}}) = \theta(\mathbf{u}, \mathbf{v})$, so that $(\mathbf{u}, \mathbf{v})$ solves the dual problem.

Of course, the conditions of the above corollary coincide precisely with the saddle point optimality conditions (a), (b), and (c) of Theorem 6.2.5, implying that $(\bar{\mathbf{x}}, \mathbf{u}, \mathbf{v})$ is a saddle point and, hence, that $\bar{\mathbf{x}}$ and $(\mathbf{u}, \mathbf{v})$ solve Problems P and D, respectively. Also, elements of the proof of Theorem 6.5.1 are evident in the proof of Theorem 6.2.7. However, our purpose in highlighting Theorem 6.5.1 and its corollary is to emphasize the role played by this result in deriving heuristic primal solutions based on solving the dual problem. As seen from the above theorem, as the dual function θ is evaluated at a given point $(\mathbf{u}, \mathbf{v})$, we obtain a point $\bar{\mathbf{x}}$ that is an optimal solution to a problem closely related to the original problem, in which the constraints are perturbed from $\mathbf{h}(\mathbf{x}) = \mathbf{0}$ and $g_i(\mathbf{x}) \leq 0$ for $i = 1, \ldots, m$ to $\mathbf{h}(\mathbf{x}) = \mathbf{h}(\bar{\mathbf{x}})$ and $g_i(\mathbf{x}) \leq g_i(\bar{\mathbf{x}})$ for $i = 1, \ldots, m$.

In particular, during the course of solving the dual problem, suppose that, for a given $(\mathbf{u}, \mathbf{v})$ with $\mathbf{u} \geq \mathbf{0}$, we have $\hat{\mathbf{x}} \in X(\mathbf{u}, \mathbf{v})$. Furthermore, for some $\varepsilon > 0$, suppose that $|g_i(\hat{\mathbf{x}})| \leq \varepsilon$ for $i \in I$, $g_i(\hat{\mathbf{x}}) \leq \varepsilon$ for $i \notin I$, and $|h_i(\hat{x})| \leq \varepsilon$ for $i = 1, \ldots, l$. Note that if ε is sufficiently small, then $\hat{\mathbf{x}}$ is *near-feasible*. Now suppose that $\bar{\mathbf{x}}$ is an optimal solution to the primal Problem P. Then, by the definition of $\theta(\mathbf{u}, \mathbf{v})$,

$$f(\hat{\mathbf{x}}) + \sum_{i\in I} u_i g_i(\hat{\mathbf{x}}) + \sum_{i=1}^{l} v_i h_i(\hat{\mathbf{x}}) \leq f(\bar{\mathbf{x}}) + \sum_{i\in I} u_i g_i(\bar{\mathbf{x}}) + \sum_{i=1}^{l} v_i h_i(\bar{\mathbf{x}}) \leq f(\bar{\mathbf{x}})$$

since $h_i(\bar{\mathbf{x}}) = 0$, $g_i(\bar{\mathbf{x}}) \leq 0$, and $u_i \geq 0$. The above inequality thus implies that

$$f(\hat{\mathbf{x}}) \leq f(\bar{\mathbf{x}}) + \varepsilon\left[\sum_{i\in I} u_i + \sum_{i=1}^{l} |v_i|\right]$$

Therefore, if ε is sufficiently small so that $\varepsilon[\sum_{i\in I} u_i + \sum_{i=1}^{l} |v_i|]$ is small enough, then $\hat{\mathbf{x}}$ is a *near-optimal* solution. In many practical problems, such a solution is often acceptable.

Note also that in the absence of a duality gap, if $\bar{\mathbf{x}}$ and $(\bar{\mathbf{u}}, \bar{\mathbf{v}})$ are, respectively, optimal primal and dual solutions, then, by Theorem 6.2.5, $(\bar{\mathbf{x}}, \bar{\mathbf{u}}, \bar{\mathbf{v}})$ is a saddle point. Hence, by property (a) of Theorem 6.2.5, $\bar{\mathbf{x}}$ minimizes $\phi(\mathbf{x}, \bar{\mathbf{u}}, \bar{\mathbf{v}})$ over $\mathbf{x} \in X$. This means that there exists an optimal solution to the primal problem among points in the set $X(\bar{\mathbf{u}}, \bar{\mathbf{v}})$, where $(\bar{\mathbf{u}}, \bar{\mathbf{v}})$ is an optimal solution to the dual problem. Of course, not any solution $\bar{\mathbf{x}} \in X(\bar{\mathbf{u}}, \bar{\mathbf{v}})$ solves the primal problem unless $\bar{\mathbf{x}}$ is feasible to P and it satisfies the complementary slackness condition $\bar{\mathbf{u}}^t\mathbf{g}(\bar{\mathbf{x}}) = 0$.

Generating Primal Feasible Solutions in the Convex Case

The foregoing discussion was concerned with general, perhaps nonconvex, problems. Under suitable convexity assumptions, we can easily obtain primal feasible solutions at each iteration of the dual problem by solving a linear program. In particular, suppose that we are given a point $\mathbf{x}_0$ which is feasible to the original problem, and let the points $\mathbf{x}_j \in X(\mathbf{u}_j, \mathbf{v}_j)$ for $j = 1, \ldots, k$ be generated by an arbitrary algorithm used to maximize the dual function. Theorem 6.5.2 below shows that a feasible solution to the primal problem can be obtained by solving the following linear programming Problem P′:

$$\begin{aligned}
\textit{Problem P}': \quad & \text{Minimize} \quad \sum_{j=0}^{k} \lambda_j f(\mathbf{x}_j) \\
& \text{subject to} \quad \sum_{j=0}^{k} \lambda_j \mathbf{g}(\mathbf{x}_j) \leq \mathbf{0} \\
& \qquad\qquad \sum_{j=0}^{k} \lambda_j \mathbf{h}(\mathbf{x}_j) = \mathbf{0} \\
& \qquad\qquad \sum_{j=0}^{k} \lambda_j = 1 \qquad\qquad (6.25) \\
& \qquad\qquad \lambda_j \geq 0 \qquad \text{for } j = 0, \ldots, k
\end{aligned}$$

6.5.2 Theorem

Let X be a nonempty convex set in E_n, let $f: E_n \to E_1$ and $\mathbf{g}: E_n \to E_m$ be convex, and let $\mathbf{h}: E_n \to E_l$ be affine; that is, $\mathbf{h}$ is of the form $\mathbf{h}(\mathbf{x}) = \mathbf{A}\mathbf{x} - \mathbf{b}$. Let $\mathbf{x}_0$ be an initial feasible solution to Problem P, and suppose that $\mathbf{x}_j \in X(\mathbf{u}_j, \mathbf{v}_j)$ for $j = 1, \ldots, k$ are generated by any algorithm for solving the dual problem. Furthermore, let $\bar{\lambda}_j$ for $j = 0, \ldots, k$ be an optimal solution to Problem P′ defined in (6.25), and let $\bar{\mathbf{x}}_k = \sum_{j=0}^{k} \bar{\lambda}_j \mathbf{x}_j$. Then, $\bar{\mathbf{x}}_k$ is a feasible solution to the primal Problem P. Furthermore, letting $z_k = \sum_{j=0}^{k} \bar{\lambda}_j f(\mathbf{x}_j)$ and $z^* = \inf \{f(\mathbf{x}): \mathbf{x} \in X, \mathbf{g}(\mathbf{x}) \leq \mathbf{0}, \mathbf{h}(\mathbf{x}) = \mathbf{0}\}$, if $z_k - \theta(\mathbf{u}, \mathbf{v}) \leq \varepsilon$ for some $(\mathbf{u}, \mathbf{v})$ with $\mathbf{u} \geq \mathbf{0}$, then $f(\bar{\mathbf{x}}_k) \leq z^* + \varepsilon$.

Proof

Since X is convex and $\mathbf{x}_j \in X$ for each j, then $\bar{\mathbf{x}}_k \in X$. Since $\mathbf{g}$ is convex and $\mathbf{h}$ is affine, and noting the constraints of Problem P′, $\mathbf{g}(\bar{\mathbf{x}}_k) \leq \mathbf{0}$ and $\mathbf{h}(\bar{\mathbf{x}}_k) = \mathbf{0}$. Thus, $\bar{\mathbf{x}}_k$ is a feasible solution to the primal problem. Now suppose that $z_k - \theta(\mathbf{u}, \mathbf{v}) \leq \varepsilon$ for some $(\mathbf{u}, \mathbf{v})$ with $\mathbf{u} \geq \mathbf{0}$. Noting the convexity of f and Theorem 6.2.1, we get

$$f(\bar{\mathbf{x}}_k) \leq \sum_{j=0}^{k} \bar{\lambda}_j f(\mathbf{x}_j) = z_k \leq \theta(\mathbf{u}, \mathbf{v}) + \varepsilon \leq z^* + \varepsilon$$

and the proof is complete.

At each iteration of the dual maximization problem, we can thus obtain a primal feasible solution by solving the linear programming Problem P′. Even though the primal objective values $\{f(\bar{\mathbf{x}}_k)\}$ of the generated primal feasible points are not necessarily decreasing, they form a sequence that is bounded from above by the nonincreasing sequence $\{z_k\}$.

Note that if z_k is close enough to the dual objective value evaluated at any dual feasible point $(\mathbf{u}, \mathbf{v})$, where $\mathbf{u} \geq \mathbf{0}$, then $\bar{\mathbf{x}}_k$ is a near-optimal primal feasible solution. Also note that we need not solve Problem P′ in the case of the cutting plane algorithm, since it is precisely the linear programming dual of the master problem stated in step 1 of this algorithm. Thus, the optimal variables $\bar{\lambda}_0, \ldots, \bar{\lambda}_k$ can be retrieved easily from the solution to the master problem and $\bar{\mathbf{x}}_k$ can be computed as $\sum_{j=0}^{k} \bar{\lambda}_j \mathbf{x}_j$. It is also worth mentioning that the termination criterion $z_k = \theta(\mathbf{u}_k, \mathbf{v}_k)$ in the cutting plane algorithm can be interpreted as letting $(\mathbf{u}, \mathbf{v}) = (\mathbf{u}_k, \mathbf{v}_k)$ and $\varepsilon = 0$ in the above theorem.

To illustrate the above procedure, consider Example 6.4.1. At the end of iteration $k = 1$, we have the points $\mathbf{x}_0 = (\frac{5}{4}, \frac{1}{2})^t$ and $\mathbf{x}_1 = (2, 0)^t$. The associated primal point $\bar{\mathbf{x}}_1$ can be obtained by solving the following linear programming problem:

$$\begin{aligned} \text{Minimize} \quad & \tfrac{5}{8}\lambda_0 \\ \text{subject to} \quad & -\tfrac{3}{2}\lambda_0 + \lambda_1 \leq 0 \\ & \lambda_0 + \lambda_1 = 1 \\ & \lambda_0, \quad \lambda_1 \geq 0 \end{aligned}$$

The optimal solution to this problem is given by $\bar{\lambda}_0 = \frac{2}{5}$ and $\bar{\lambda}_1 = \frac{3}{5}$. This yields a primal feasible solution

$$\bar{\mathbf{x}}_1 = \tfrac{2}{5}(\tfrac{5}{4}, \tfrac{1}{2})^t + \tfrac{3}{5}(2, 0)^t = (\tfrac{17}{10}, \tfrac{2}{10})^t$$

As pointed out earlier, the above linear program need not be solved separately to find the values of $\bar{\lambda}_0$ and $\bar{\lambda}_1$ since its dual has already been solved during the course of the cutting plane algorithm.

6.6 Linear and Quadratic Programs

In this section, we discuss some special cases of Lagrangian duality. In particular, we briefly discuss duality in linear and quadratic programming. For linear programming problems, we relate the Lagrangian dual to that derived in Chapter 2 (see Theorem 2.7.3 and its corollaries). In the case of quadratic programming problems, we derive the well-known Dorn's dual program via Lagrangian duality.

Linear Programming

Consider the following primal linear program:

$$\begin{aligned} \text{Minimize} \quad & \mathbf{c}^t\mathbf{x} \\ \text{subject to} \quad & \mathbf{A}\mathbf{x} = \mathbf{b} \\ & \mathbf{x} \geq \mathbf{0} \end{aligned}$$

Letting $X = \{\mathbf{x} : \mathbf{x} \geq \mathbf{0}\}$, the Lagrangian dual of this problem is to maximize $\theta(\mathbf{v})$, where

$$\theta(\mathbf{v}) = \inf\{\mathbf{c}^t\mathbf{x} + \mathbf{v}^t(\mathbf{b} - \mathbf{A}\mathbf{x}) : \mathbf{x} \geq \mathbf{0}\} = \mathbf{v}^t\mathbf{b} = \inf\{(\mathbf{c}^t - \mathbf{v}^t\mathbf{A})\mathbf{x} : \mathbf{x} \geq \mathbf{0}\}$$

Clearly,

$$\theta(\mathbf{v}) = \begin{cases} \mathbf{v}'\mathbf{b} & \text{if } (\mathbf{c}' - \mathbf{v}'\mathbf{A}) \geq \mathbf{0} \\ -\infty & \text{otherwise} \end{cases}$$

Hence, the dual problem can be stated as follows:

$$\begin{array}{ll} \text{Maximize} & \mathbf{v}'\mathbf{b} \\ \text{subject to} & \mathbf{A}'\mathbf{v} \leq \mathbf{c} \end{array}$$

Recall that this is precisely the *dual problem* discussed in Section 2.7. Thus, in the case of linear programs, the dual problem does not involve the primal variables. Furthermore, the dual problem itself is a linear program, and the reader can verify that the dual of the dual problem is the original primal program. Theorem 6.6.1 below summarizes the relationships between the primal and dual problems as established by Theorem 2.7.3 and its three corollaries.

6.6.1 Theorem

Consider the primal and dual linear problems stated above. One of the following mutually exclusive cases will occur:

1. The primal problem admits a feasible solution and has an unbounded objective value, in which case the dual problem is infeasible.
2. The dual problem admits a feasible solution and has an unbounded objective value, in which case the primal problem is infeasible.
3. Both problems admit feasible solutions, in which case both problems have optimal solutions $\bar{\mathbf{x}}$ and $\bar{\mathbf{v}}$ such that $\mathbf{c}'\bar{\mathbf{x}} = \bar{\mathbf{v}}'\mathbf{b}$ and $(\mathbf{c}' - \bar{\mathbf{v}}'\mathbf{A})\bar{\mathbf{x}} = 0$.
4. Both problems are infeasible.

Proof

See Theorem 2.7.3 and its corollaries 1 and 3.

Quadratic Programming

Consider the following quadratic programming problem:

$$\begin{array}{ll} \text{Minimize} & \tfrac{1}{2}\mathbf{x}'\mathbf{H}\mathbf{x} + \mathbf{d}'\mathbf{x} \\ \text{subject to} & \mathbf{A}\mathbf{x} \leq \mathbf{b} \end{array}$$

where $\mathbf{H}$ is symmetric and positive semidefinite, so that the objective function is convex. The Lagrangian dual problem is to maximize $\theta(\mathbf{u})$ over $\mathbf{u} \geq \mathbf{0}$, where

$$\theta(\mathbf{u}) = \inf\{\tfrac{1}{2}\mathbf{x}'\mathbf{H}\mathbf{x} + \mathbf{d}'\mathbf{x} + \mathbf{u}'(\mathbf{A}\mathbf{x} - \mathbf{b}) : \mathbf{x} \in E_n\} \tag{6.26}$$

Note that for a given $\mathbf{u}$, the function $\tfrac{1}{2}\mathbf{x}'\mathbf{H}\mathbf{x} + \mathbf{d}'\mathbf{x} + \mathbf{u}'(\mathbf{A}\mathbf{x} - \mathbf{b})$ is convex, and so a necessary and sufficient condition for a minimum is that the gradient must vanish, that is,

$$\mathbf{H}\mathbf{x} + \mathbf{A}'\mathbf{u} + \mathbf{d} = \mathbf{0} \tag{6.27}$$

Thus, the dual problem can be written as follows:

$$\begin{array}{ll} \text{Maximize} & \frac{1}{2}\mathbf{x}^t\mathbf{H}\mathbf{x} + \mathbf{d}^t\mathbf{x} + \mathbf{u}^t(\mathbf{A}\mathbf{x} - \mathbf{b}) \\ \text{subject to} & \mathbf{H}\mathbf{x} + \mathbf{A}^t\mathbf{u} = -\mathbf{d} \\ & \mathbf{u} \geq \mathbf{0} \end{array}$$

Now, from (6.27), we have $\mathbf{d}^t\mathbf{x} + \mathbf{u}^t\mathbf{A}\mathbf{x} = -\mathbf{x}^t\mathbf{H}\mathbf{x}$. Substituting this into (6.28), we derive the familiar form of Dorn's dual quadratic program given below:

$$\begin{array}{ll} \text{Maximize} & -\frac{1}{2}\mathbf{x}^t\mathbf{H}\mathbf{x} - \mathbf{b}^t\mathbf{u} \\ \text{subject to} & \mathbf{H}\mathbf{x} + \mathbf{A}^t\mathbf{u} = -\mathbf{d} \\ & \mathbf{u} \geq \mathbf{0} \end{array} \tag{6.29}$$

Again, by Lagrangian duality, if one problem is unbounded, then the other is infeasible. Moreover, following Theorem 6.2.6, if both problems are feasible, then they both have optimal solutions having the same objective value.

We now develop an alternative form of the Lagrangian dual problem under the assumption that $\mathbf{H}$ is positive definite, so that $\mathbf{H}^{-1}$ exists. In this case, the unique solution to (6.27) is given by

$$\mathbf{x} = -\mathbf{H}^{-1}(\mathbf{d} + \mathbf{A}^t\mathbf{u})$$

Substituting in (6.26), it follows that

$$\theta(\mathbf{u}) = \frac{1}{2}\mathbf{u}^t\mathbf{D}\mathbf{u} + \mathbf{u}^t\mathbf{c} - \frac{1}{2}\mathbf{d}^t\mathbf{H}^{-1}\mathbf{d}$$

where $\mathbf{D} = -\mathbf{A}\mathbf{H}^{-1}\mathbf{A}^t$ and $\mathbf{c} = -\mathbf{b} - \mathbf{A}\mathbf{H}^{-1}\mathbf{d}$. The dual problem is thus given by

$$\begin{array}{ll} \text{Maximize} & \frac{1}{2}\mathbf{u}^t\mathbf{D}\mathbf{U} + \mathbf{u}^t\mathbf{c} - \frac{1}{2}\mathbf{d}^t\mathbf{H}^{-1}\mathbf{d} \\ \text{subject to} & \mathbf{u} \geq \mathbf{0} \end{array} \tag{6.30}$$

The dual problem (6.30) can be solved relatively easily using the algorithms described in Chapters 8 through 11 of this book, noting that this problem simply seeks to maximize a concave quadratic function over the nonnegative orthant. (See Exercise 6.41 for a simplified scheme.)

Exercises

6.1 Consider the problem to minimize x_1 subject to $x_1^2 + x_2^2 = 1$. Derive explicitly the dual function, and verify its concavity. Find the optimal solutions to both the primal and dual problems, and compare their objective values.

6.2 Consider the following problem:

$$\begin{array}{lrcl} \text{Maximize} & 2x_1 + 3x_2 + x_3 & & \\ \text{subject to} & x_1 + x_2 - x_3 & \leq & 1 \\ & x_1 + x_2 & \leq & 4 \\ & x_3 & \leq & 2 \\ & x_1,\ x_2,\ x_3 & \geq & 0 \end{array}$$

a. Find explicitly the dual function, where $X = \{(x_1, x_2, x_3): x_1 + x_2 - x_3 \leq 1;\ x_1, x_2, x_3 \geq 0\}$.

b. Repeat part a for $X = \{(x_1, x_2, x_3): x_1 + x_2 \leq 4;\ x_1, x_2, x_3 \geq 0\}$.

c. In parts a and b, note that the difficulty in evaluating the dual function at a given point depends on which constraints are handled via the set X. Propose some general guidelines that could be used in selecting the set X to make the solution easier.

6.3 Consider the problem to minimize e^{-x} subject to $-x \leq 0$.

a. Solve the above primal problem.

b. Letting $X = E_1$, find the explicit form of the Lagrangian dual function, and solve the dual problem.

6.4 Consider the primal Problem P discussed in Section 6.1. Introducing the slack vector $\mathbf{s}$, the problem can be formulated as follows:

Minimize $f(\mathbf{x})$

subject to $\mathbf{g}(\mathbf{x}) + \mathbf{s} = \mathbf{0}$

$\mathbf{h}(\mathbf{x}) = \mathbf{0}$

$(\mathbf{x}, \mathbf{s}) \in X'$

where $X' = \{(\mathbf{x}, \mathbf{s}) : \mathbf{x} \in X, \mathbf{s} \geq \mathbf{0}\}$. Formulate the dual of the above problem and show that it is equivalent to the dual problem discussed in Section 6.1.

6.5 Consider the problem to minimize $f(\mathbf{x})$ subject to $g_1(\mathbf{x}) \leq \mathbf{0}$ and $g_2(\mathbf{x}) \leq \mathbf{0}$ as illustrated in Figure 4.13. Denote $X = \{\mathbf{x} : g_1(\mathbf{x}) \leq 0\}$. Sketch the perturbation function $\nu(y) = \min \{f(\mathbf{x}) : g_2(\mathbf{x}) \leq y, \mathbf{x} \in X\}$ and indicate the duality gap. Provide a possible sketch for the set $G = \{(y, z) : y = g_2(\mathbf{x}), z = f(\mathbf{x}) \text{ for some } \mathbf{x} \in X\}$ for this problem.

6.6 Consider the (singly) constrained problem to minimize $f(\mathbf{x})$ subject to $g(\mathbf{x}) \leq 0$ and $\mathbf{x} \in X$. Define $G = \{(y, z) : y = g(\mathbf{x}), z = f(\mathbf{x}) \text{ for some } \mathbf{x} \in X\}$, and let $\nu(y) = \min \{f(\mathbf{x}) : g(\mathbf{x}) \leq y, \mathbf{x} \in X\}$, $y \in E_1$, be the associated perturbation function. Show that ν is the pointwise supremum over all possible nonincreasing functions whose epigraph contains G.

6.7 Consider the problem to minimize $f(\mathbf{x})$ subject to $g_i(\mathbf{x}) \leq 0$ for $i = 1, \ldots, m$, $h_i(\mathbf{x}) = 0$ for $i = 1, \ldots, l$, and $\mathbf{x} \in X$, and let $\nu : E_{m+l} \to E_1$ be the perturbation function defined by (6.9). Assuming that f and $\mathbf{g}$ are convex, $\mathbf{h}$ is affine, and that X is a convex set, show that ν is a convex function.

6.8 In the proof of Lemma 6.2.3, show that the set $\wedge$ is convex.

6.9 Under the assumptions of Theorem 6.2.5, suppose that $\bar{\mathbf{x}}$ is an optimal solution to the primal problem, and that f and $\mathbf{g}$ are differentiable at $\bar{\mathbf{x}}$. Show that there exists a vector $(\bar{\mathbf{u}}, \bar{\mathbf{v}})$ such that

$$\left[\nabla f(\bar{\mathbf{x}}) + \sum_{i=1}^{m} \bar{u}_i \nabla g_i(\bar{\mathbf{x}}) + \sum_{i=1}^{l} \bar{v}_i \nabla h_i(\bar{\mathbf{x}})\right]^t (\mathbf{x} - \bar{\mathbf{x}}) \geq 0 \qquad \text{for each } \mathbf{x} \in X$$

$$u_i g_i(\bar{\mathbf{x}}) = 0 \qquad \text{for } i = 1, \ldots, m$$

$$\bar{\mathbf{u}} \geq \mathbf{0}$$

Show that these conditions reduce to the KKT conditions if X is open.

6.10 Prove the following saddle point optimality condition. Let X be a nonempty convex set in E_n, and let $f : E_n \to E_1$, $\mathbf{g} : E_n \to E_m$ be convex and $\mathbf{h} : E_n \to E_l$ be affine. If $\bar{\mathbf{x}}$ is an optimal solution to the problem to minimize $f(\mathbf{x})$ subject to $\mathbf{g}(\mathbf{x}) \leq \mathbf{0}$, $\mathbf{h}(\mathbf{x}) = \mathbf{0}$, $\mathbf{x} \in X$, then there exist $(\bar{u}_0, \bar{\mathbf{u}}, \bar{\mathbf{v}}) \neq \mathbf{0}$, $(\bar{u}_0, \bar{\mathbf{u}}) \geq \mathbf{0}$ such that

$$\phi(\bar{u}_0, \mathbf{u}, \mathbf{v}, \bar{\mathbf{x}}) \leq \phi(\bar{u}_0, \bar{\mathbf{u}}, \bar{\mathbf{v}}, \bar{\mathbf{x}}) \leq \phi(\bar{u}_0, \bar{\mathbf{u}}, \bar{\mathbf{v}}, \mathbf{x})$$

for all $\mathbf{u} \geq \mathbf{0}$, $\mathbf{v} \in E_l$, and $\mathbf{x} \in X$ where $\phi(u_0, \mathbf{u}, \mathbf{v}, \mathbf{x}) = u_0 f(\mathbf{x}) + \mathbf{u}^t\mathbf{g}(\mathbf{x}) + \mathbf{v}^t\mathbf{h}(\mathbf{x})$.

6.11 Consider the problem to minimize $f(\mathbf{x})$ subject to $\mathbf{g}(\mathbf{x}) \leq \mathbf{0}$, $\mathbf{x} \in X$. Theorem 6.2.4 shows that the primal and dual objective values are equal at optimality under the assumptions that f, $\mathbf{g}$ and X are convex and the constraint qualification $\mathbf{g}(\hat{\mathbf{x}}) < \mathbf{0}$ for some $\hat{\mathbf{x}} \in X$. Suppose that the convexity assumptions on f and $\mathbf{g}$ are replaced by continuity of f and $\mathbf{g}$ and that X is assumed to be convex and compact. Does the result of the theorem hold? Prove or give a counterexample.

6.12 Let P and D be the primal and dual nonlinear programs stated in Section 6.1, and denote $\mathbf{w} = (\mathbf{u}, \mathbf{v})$. Suppose that $\bar{\mathbf{w}}$ solves D. If there exists a saddle point solution to P and if $\bar{\mathbf{x}}$ uniquely solves for $\theta(\bar{\mathbf{w}})$, then show that $(\bar{\mathbf{x}}, \bar{\mathbf{w}})$ is such a saddle point solution. Correspondingly, if θ is differentiable at $\bar{\mathbf{w}}$, and if $\bar{\mathbf{x}}$ (uniquely) solves for θ at $\bar{\mathbf{w}}$, then show that $(\bar{\mathbf{x}}, \bar{\mathbf{w}})$ is a saddle point solution. (In particular, this shows that if Problem P has no saddle point solution, then θ cannot be differentiable at optimality.)

6.13 For the problem of Example 6.3.5, sketch the perturbation function ν defined by (6.9), and comment on the existence of a saddle point solution.

6.14 Let $f:E_n \to E_1$ be a concave function, and let $\partial f(\bar{\mathbf{x}})$ be the subdifferential of f at any $\bar{\mathbf{x}} \in E_n$. Show that the directional derivative of f at $\bar{\mathbf{x}}$ in the direction $\mathbf{d}$ is given by $f'(\bar{\mathbf{x}}; \mathbf{d}) = \inf \{\boldsymbol{\xi}'\mathbf{d} : \boldsymbol{\xi} \in \partial f(\bar{\mathbf{x}})\}$. what is the corresponding result if f is a convex function?

6.15 Consider the discrete optimization Problem DP: minimize $\{\mathbf{c}'\mathbf{x} : \mathbf{Ax} = \mathbf{b}, \mathbf{Dx} = \mathbf{d}, \mathbf{x} \in X\}$, where X is some compact discrete set, and assume that the problem is feasible. Define $\theta(\boldsymbol{\pi}) = \text{minimum } \{\mathbf{c}'\mathbf{x} + \boldsymbol{\pi}'(\mathbf{Ax} - \mathbf{b}) : \mathbf{Dx} = \mathbf{d}, \mathbf{x} \in X\}$ for any $\pi \in E_m$, where $\mathbf{A}$ is $m \times n$. Show that $\max \{\theta(\boldsymbol{\pi}) : \boldsymbol{\pi} \in E_m\} = \min \{\mathbf{c}'\mathbf{x} : \mathbf{Ax} = \mathbf{b}, \mathbf{x} \in H[\mathbf{x} \in X : \mathbf{Dx} = \mathbf{d}]\}$, where $H[\cdot]$ denotes the convex hull operation. Use this result to interpret the duality gap that might exist between DP and the stated Lagrangian dual problem.

6.16 Consider the problem DP given in Exercise 6.15, and rewrite this problem as minimize $\{\mathbf{c}'\mathbf{x} : \mathbf{Ax} = \mathbf{b}, \mathbf{Dy} = \mathbf{d}, \mathbf{x} = \mathbf{y}, \mathbf{x} \in X, \mathbf{y} \in Y\}$, where Y is a copy of X in which the $\mathbf{x}$ variables have been replaced by a set of matching $\mathbf{y}$ variables. Formulate a Lagrangian dual function $\bar{\theta}(\boldsymbol{\mu}) = \min \{\mathbf{c}'\mathbf{x} + \boldsymbol{\mu}'(\mathbf{x} - \mathbf{y}) : \mathbf{Ax} = \mathbf{b}, \mathbf{Dy} = \mathbf{d}, \mathbf{x} \in X, \mathbf{y} \in Y\}$. Show that $\max \{\bar{\theta}(\boldsymbol{\mu}) : \boldsymbol{\mu} \in E_n\} \geq \max \{\theta(\boldsymbol{\pi}) : \boldsymbol{\pi} \in E_m\}$, where θ is defined in Exercise 6.15. Discuss this result in relation to the respective partial convex hulls corresponding to θ and $\bar{\theta}$ as presented in Section 6.4 and Exercise 6.15.

6.17 Consider the following problem:

$$
\begin{array}{lrl}
\text{Minimize} & -2x_1 + 2x_2 + x_3 - 3x_4 & \\
\text{subject to} & x_1 + x_2 + x_3 + x_4 & \leq 8 \\
& x_1 \qquad - 2x_3 + 4x_4 & \leq 2 \\
& x_1 + x_2 \qquad\qquad & \leq 8 \\
& x_3 + 2x_4 & \leq 6 \\
& x_1, \quad x_2, \quad x_3, \quad x_4 & \geq 0
\end{array}
$$

Let $X = \{(x_1, x_2, x_3, x_4) : x_1 + x_2 \leq 8, x_3 + 2x_4 \leq 6; x_1, x_2, x_3, x_4 \geq 0\}$.

a. Find the function θ explicitly.
b. Verify that θ is differentiable at (4, 0), and find $\nabla\theta(4, 0)$.
c. Verify that $\nabla\theta(4, 0)$ is an infeasible direction, and find an improving feasible direction.
d. Starting from (4, 0), maximize θ in the direction obtained in part c.

6.18 Consider the pair of primal and dual Problems P and D stated in Section 6.1, and assume that the Lagrangian dual function θ is differentiable. Given $(\bar{\mathbf{u}}, \bar{\mathbf{v}}) \in E_{m+l}$, $\bar{\mathbf{u}} \geq \mathbf{0}$, let $\nabla\theta(\bar{\mathbf{u}}, \bar{\mathbf{v}})' = [\mathbf{g}(\bar{\mathbf{x}})', \mathbf{h}(\bar{\mathbf{x}})']$, and define $\hat{g}_i(\bar{\mathbf{x}}) = g_i(\bar{\mathbf{x}})$ if $\bar{u}_i > 0$ and $\hat{g}_i(\bar{\mathbf{x}}) = \max \{0, g_i(\bar{\mathbf{x}})\}$ if $\bar{u}_i = 0$, for $i = 1, \ldots, m$. If $(\mathbf{d_u}, \mathbf{d_v}) \equiv [\hat{\mathbf{g}}(\bar{\mathbf{x}}), \mathbf{h}(\bar{\mathbf{x}})] \neq (\mathbf{0}, \mathbf{0})$, then show that $(\mathbf{d_u}, \mathbf{d_v})$ is a feasible ascent direction of θ at $(\bar{\mathbf{u}}, \bar{\mathbf{v}})$. Hence, discuss how θ can be maximized in the direction $(\mathbf{d_u}, \mathbf{d_v})$ via the one-dimensional problem to $\text{maximize}_\lambda \{\theta(\bar{\mathbf{u}} + \lambda\mathbf{d_u}, \bar{\mathbf{v}} + \lambda\mathbf{d_v}) : \bar{\mathbf{u}} + \lambda\mathbf{d_u} \geq \mathbf{0}, \lambda \geq 0\}$. On the other hand, if $(\mathbf{d_u}, \mathbf{d_v}) = (\mathbf{0}, \mathbf{0})$, then show that $(\bar{\mathbf{u}}, \bar{\mathbf{v}})$ solves D. Consider the problem to minimize $x_1^2 + x_2^2$ subject to $g_1(\mathbf{x}) = -x_1 - x_2 + 4 \leq 0$, and $g_2(\mathbf{x}) = x_1 + 2x_2 - 8 \leq 0$. Illustrate the *gradient method* presented above by starting at the dual solution $(u_1, u_2) = (0, 0)$ and verifying that after one iteration of this method, an optimal solution is obtained in this case.

6.19 Consider the problem to minimize $x_1^2 + x_2^2$ subject to $x_1 + x_2 - 4 \geq 0$, and $x_1, x_2 \geq 0$.

a. Verify that the optimal solution is $\bar{\mathbf{x}} = (2, 2)'$ with $f(\bar{\mathbf{x}}) = 8$.
b. Letting $X = \{(x_1, x_2) : x_1 \geq 0, x_2 \geq 0\}$, write the Lagrangian dual problem. Show that the dual function is $\theta(u) = -u^2/2 - 4u$. Verify that there is no duality gap for this problem.
c. Solve the dual problem by the cutting plane algorithm of Section 6.4. Start with $\mathbf{x} = (3, 3)'$.
d. Show that θ is differentiable everywhere, and solve the problem using the gradient method of Exercise 6.18.

6.20 Consider the following problem:

$$
\begin{array}{lrl}
\text{Minimize} & x_1 + x_2 & \\
\text{subject to} & 2x_1 + x_2 & \leq 8 \\
& 3x_1 + 2x_2 & \leq 10 \\
& x_1, + x_2 & \geq 0 \\
& x_1, \quad x_2 & \text{integers}
\end{array}
$$

Let $X = \{(x_1, x_2): 3x_1 + 2x_2 \leq 10, x_1, x_2 \geq 0 \text{ and integer}\}$. At $u = 2$, is θ differentiable? If not, characterize its ascent directions.

6.21 Consider the following problem:

$$\begin{array}{llr} \text{Minimize} & (x_1 - 3)^2 + (x_2 - 5)^2 & \\ \text{subject to} & x_1^2 - x_2 \leq 0 & \\ & -x_1 \leq 1 & \\ & x_1 + 2x_2 \leq 10 & \\ & x_1, \quad x_2 \geq 0 & \end{array}$$

a. Find the optimal solution geometrically, and verify it by using the KKT conditions.
b. Formulate the dual problem in which $X = \{(x_1, x_2): x_1 + 2x_2 \leq 10; x_1, x_2 \geq 0\}$.
c. Perform three iterations of the cutting plane algorithm described in Section 6.4, starting with $(u_1, u_2) = (0, 0)$. Describe the perturbed optimization problems corresponding to the generated primal infeasible points. Also identify the primal feasible solutions generated by the algorithm.

6.22 In reference to Exercise 6.21 above, perform three iterations of the gradient method of Exercise 6.18 and compare the results with those obtained by the cutting plane algorithm.

6.23 Consider the following problem:

$$\begin{array}{ll} \text{Maximize} & 3x_1 + 6x_2 + 2x_3 + 4x_4 \\ \text{subject to} & x_1 + x_2 + x_3 + x_4 \leq 12 \\ & -x_1 + x_2 + 2x_4 \leq 4 \\ & x_1 + x_2 \leq 12 \\ & x_2 \leq 4 \\ & x_3 + x_4 \leq 6 \\ & x_1, \quad x_2, \quad x_3, \quad x_4 \geq 0 \end{array}$$

a. Formulate the dual problem in which $X = \{(x_1, x_2, x_3, x_4): x_1 + x_2 \leq 12, x_2 \leq 4, x_3 + x_4 \leq 6; x_1, x_2, x_3, x_4 \geq 0\}$.
b. Starting from the point (0, 0), solve the Lagrangian dual problem by optimizing along the direction of steepest ascent as discussed in Exercise 6.18.
c. At optimality of the dual, find the optimal primal solution.

6.24 Consider the cutting plane method described in Section 6.4, and suppose that each time the master program objective value strictly increases, we delete all the constraints of the type $z \leq f(\mathbf{x}_j) + \mathbf{u}'\mathbf{g}(\mathbf{x}_j) + \mathbf{v}'\mathbf{h}(\mathbf{x}_j)$ that are nonbinding at optimality. If X has a finite number of elements, show that this modified algorithm will converge finitely. Give some alternative conditions under which such a constraint deletion will assure convergence of the algorithm.

6.25 Consider the pair of primal and dual Problems P and D stated in Section 6.1, and suppose that the Lagrangian dual function θ is not necessarily diferentiable. Given $\bar{\mathbf{w}} = (\bar{\mathbf{u}}, \bar{\mathbf{v}}) \in E_{m+l}$, $\bar{\mathbf{u}} \geq \mathbf{0}$, let $\boldsymbol{\xi}_1, \ldots, \boldsymbol{\xi}_p$, $p \geq 1$, be some known collection of subgradients of θ at $\bar{\mathbf{w}}$. Consider the problem to maximize $\{\varepsilon : \mathbf{d}'\boldsymbol{\xi}_j \geq \varepsilon \text{ for } j = 1, \ldots, p, -1 \leq d_i \leq 1 \text{ for } i = 1, \ldots, m + l, \text{ with } d_i \geq 0 \text{ if } \bar{u}_i = 0\}$. Let $(\bar{\varepsilon}, \bar{\mathbf{d}})$ solve this problem. If $\bar{\varepsilon} = 0$, then show that $\bar{\mathbf{w}}$ solves D. Otherwise, solve the problem to maximize $\{\bar{\mathbf{d}}'\boldsymbol{\xi}: \boldsymbol{\xi} \in \partial\theta(\bar{\mathbf{w}})\}$, and let $\boldsymbol{\xi}_{p+1}$ be an optimum. If $\bar{\mathbf{d}}'\boldsymbol{\xi}_{p+1} > 0$, then show that $\bar{\mathbf{d}}$ is an ascent direction along which θ can be maximized by solving $\max \{\theta(\bar{\mathbf{w}} + \lambda\bar{\mathbf{d}}): \bar{u}_i + \lambda \bar{d}_i \geq 0 \text{ for } i = 1, \ldots, m, \lambda \geq 0\}$, and the process is repeated. Otherwise, if $\bar{\mathbf{d}}'\boldsymbol{\xi}_{p+1} \leq 0$, then increment p by 1 and re-solve the direction-finding problem given above. Discuss the possible computational difficulties associated with this scheme. How would you implement the various steps if all functions were affine and X was a nonempty polytope? Illustrate this using the example to minimize $x_1 - 4x_2$ subject to $-x_1 - x_2 + 2 \leq 0$, $x_2 - 1 \leq 0$, and $\mathbf{x} \in X = \{\mathbf{x}: 0 \leq x_1 \leq 3, 0 \leq x_2 \leq 3\}$, starting at the point $(u_1, u_2) = (0, 4)$.

6.26 Consider the problem to minimize x subject to $g(x) \leq 0$ and $x \in X = \{x: x \geq 0\}$. Derive the explicit forms of the Lagrangian dual function, and determine the collection of subgradients at $u = 0$, for each of the following cases:

a. $$g(x) = \begin{cases} -1/x & \text{for } x \neq 0 \\ 0 & \text{for } x = 0 \end{cases}$$

b. $g(x) = \begin{cases} -1/x & \text{for } x \neq 0 \\ -1 & \text{for } x = 0 \end{cases}$

c. $g(x) = \begin{cases} 1/x & \text{for } x \neq 0 \\ 1 & \text{for } x = 0 \end{cases}$

6.27 Suppose that $\theta: E_m \to E_1$ is concave.

a. Show that θ achieves its maximum at $\bar{\mathbf{u}}$ if and only if

$$\max \{\theta'(\bar{\mathbf{u}}; \mathbf{d}) : \|\mathbf{d}\| \leq 1\} = 0$$

b. Show that θ achieves its maximum over the region $U = \{\mathbf{u} : \mathbf{u} \geq \mathbf{0}\}$ at $\bar{\mathbf{u}}$ if and only if

$$\max \{\theta'(\bar{\mathbf{u}}; \mathbf{d}) : \mathbf{d} \in D, \|\mathbf{d}\| \leq 1\} = 0$$

where D is the cone of feasible directions of U at $\bar{\mathbf{u}}$. (Note that the above results can be used as stopping criteria for maximizing the Lagrangian dual function.)

6.28 Consider the following problem, in which X is a compact polyhedral set and f is a concave function:

$$\begin{array}{ll} \text{Minimize} & f(\mathbf{x}) \\ \text{subject to} & \mathbf{Ax} = \mathbf{b} \\ & \mathbf{x} \in X \end{array}$$

a. Formulate the Lagrangian dual problem.

b. Show that the dual function is concave and piecewise linear.

c. Characterize the subgradients, the ascent directions, and the steepest ascent direction for the dual function.

d. Generalize the result in part b to the case where X is not compact.

6.29 Construct a numerical problem in which a subgradient of the dual function is not an ascent direction. Is it possible that the collection of subgradients and the cone of ascent directions are disjoint at a nonoptimal solution?
(*Hint:* Consider the shortest subgradient.)

6.30 In Section 6.3, we showed that the shortest subgradient $\boldsymbol{\xi}$ of θ at $\bar{\mathbf{u}}$ is the steepest ascent direction. The following modification of $\boldsymbol{\xi}$ is proposed to maintain feasibility:

$$\bar{\xi}_i = \begin{cases} \max(0, \xi_i) & \text{if } \bar{u}_i = 0 \\ \xi_i & \text{if } \bar{u}_i \geq 0 \end{cases}$$

Is $\bar{\boldsymbol{\xi}}$ an ascent direction? Is it the direction of steepest ascent with the added nonnegativity restriction? Prove or give a counterexample.

6.31 Consider the following problem, in which X is a compact polyhedral set:

$$\begin{array}{ll} \text{Minimize} & \mathbf{c}^t\mathbf{x} \\ \text{subject to} & \mathbf{Ax} = \mathbf{b} \\ & \mathbf{x} \in X \end{array}$$

For a given vector $\mathbf{v}$, suppose that $\mathbf{x}_1, \ldots, \mathbf{x}_k$ are the extreme points in X that belong to $X(\mathbf{v})$. Show that the extreme points of $\partial\theta(\mathbf{v})$ are contained in the set $\wedge = \{\mathbf{Ax}_j - \mathbf{b} : j = 1, \ldots, k\}$. Give an example where the extreme points of $\partial\theta(\mathbf{v})$ form a proper subset of $\wedge$.

6.32 Suppose that the shortest subgradient $\bar{\boldsymbol{\xi}}$ at θ at $(\bar{\mathbf{u}}, \bar{\mathbf{v}})$ is not equal to zero. Show that there exists an $\varepsilon > 0$ such that $\|\boldsymbol{\xi} - \bar{\boldsymbol{\xi}}\| < \varepsilon$ implies that $\boldsymbol{\xi}$ is an ascent direction of θ at $(\bar{\mathbf{u}}, \bar{\mathbf{v}})$. (From the above exercise, if an iterative procedure is used to find $\bar{\boldsymbol{\xi}}$, then it would find an ascent direction after a sufficient number of iterations.)

6.33 Consider the primal and Lagrangian dual problems discussed in Section 6.1. Let $(\bar{\mathbf{u}}, \bar{\mathbf{v}})$ be an optimal solution to the dual problem. Given $(\mathbf{u}, \mathbf{v})$, suppose that $\bar{\mathbf{x}} \in X(\mathbf{u}, \mathbf{v})$. Show that there exists a $\delta > 0$ such that $\|(\bar{\mathbf{u}}, \bar{\mathbf{v}}) - (\mathbf{u}, \mathbf{v}) - \lambda[\mathbf{g}(\bar{\mathbf{x}}), \mathbf{h}(\bar{\mathbf{x}})]\|$ is a nonincreasing function of λ over the interval $[0, \delta]$. Interpret the result geometrically, and illustrate by the following problem, in which $(u_1, u_2) = (3, 1)$ are the dual variables corresponding to the first two constraints:

$$\begin{array}{ll} \text{Minimize} & -2x_1 - 2x_2 - 5x_3 \\ \text{subject to} & x_1 + x_2 + x_3 \leq 10 \\ & x_1 \qquad + 2x_3 \geq 6 \\ & x_1 \quad x_2, \quad x_3 \leq 3 \\ & x_1 \quad x_2, \quad x_3 \geq 0 \end{array}$$

6.34 From Exercise 6.33 above, it is clear that moving a small step in the direction of any subgradient leads us closer to an optimal dual solution. Consider the following algorithm for maximizing the dual of the problem to minimize $f(\mathbf{x})$ subject to $\mathbf{h}(\mathbf{x}) = \mathbf{0}$, $\mathbf{x} \in X$:

Main Step

Given $\mathbf{v}_k$, let $\mathbf{x}_k \in X(\mathbf{v}_k)$. Let $\mathbf{v}_{k+1} = \mathbf{v}_k + \lambda \mathbf{h}(\mathbf{x}_k)$, where $\lambda > 0$ is a small scalar. Replace k by $k + 1$ and repeat the main step.

a. Discuss some possible ways of choosing a suitable step size λ. Do you see any advantages in reducing the step size during later iterations? If so, propose a scheme for doing that.
b. Does the dual function necessarily increase from one iteration to another? Discuss.
c. Devise a suitable termination criterion.
d. Apply the above algorithm, starting from $\mathbf{v} = (1, 2)^t$, to solve the following problem:

$$\begin{array}{ll} \text{Minimize} & x_1^2 + x_2^2 + 2x_3 \\ \text{subject to} & x_1 + x_2 + x_3 = 6 \\ & -x_1 + x_2 + x_3 = 4 \end{array}$$

(This procedure, with a suitable step-size selection rule, is referred to as a *subgradient optimization* technique. See Chapter 8 for further details.)

6.35 Consider the problem to minimize $f(\mathbf{x})$ subject to $\mathbf{g}(\mathbf{x}) \leq \mathbf{0}$, $\mathbf{x} \in X$.

a. In Exercise 6.34 above, a subgradient optimization technique was discussed for the equality case. Modify the procedure for the above inequality constrained problem. (*Hint:* Given $\mathbf{u}$, let $\mathbf{x} \in X(\mathbf{u})$. Replace $g_i(\mathbf{x})$ by max $[0, g_i(\mathbf{x})]$ for each i with $u_i = 0$.)
b. Illustrate the procedure given in part a by solving the problem in Exercise 6.23 starting from $\mathbf{u} = (0, 0)^t$.
c. Extend the subgradient optimization technique to handle both equality and inequality constraints.

6.36 Consider the following warehouse location problem. We are given destinations $1, \ldots, k$, where the known demand for a certain product at destination j is d_j. We are also given m possible sites for building warehouses. If we decide to build a warehouse at site i, its capacity has to be s_i and it incurs a fixed cost f_i. The unit shipping cost from warehouse i to destination j is c_{ij}. The problem is to determine how many warehouses to build, where to locate them, and what shipping patterns to use so that the demand is satisfied and the total cost is minimized. The problem can be stated mathematically as follows:

$$\begin{array}{lll} \text{Minimize} & \sum_{i=1}^{m} \sum_{j=1}^{k} c_{ij}x_{ij} + \sum_{i=1}^{m} f_i\, y_i & \\ \text{subject to} & \sum_{j=1}^{k} x_{ij} \leq s_i\, y_i & \text{for } i = 1, \ldots, m \\ & \sum_{i=1}^{m} x_{ij} \geq d_j & \text{for } j = 1, \ldots, k \\ & 0 \leq x_{ij} \leq y_i \min\{s_i, d_j\} & \text{for } i = 1, \ldots, m;\ j = 1, \ldots, k \\ & y_i = 0 \text{ or } 1 & \text{for } i = 1, \ldots, m \end{array}$$

a. Formulate a suitable Lagrangian dual problem. Explain the utility of the upper bound imposed on x_{ij}.
b. Make use of the results of this chapter to devise a special scheme for maximizing the dual of the warehouse location problem.
c. Illustrate by a small numerical example.

6.37 A company wants to plan its production rate of a certain item over the planning period $[0, T]$ such that the sum of its production and inventory costs is minimized. In addition, the known demand must be met, the production rate must fall in the acceptable interval $[l, u]$, the inventory must not exceed d, and it must be at least equal to b at the end of the planning period. The problem can be formulated as follows:

$$\begin{array}{lll} \text{Minimize} & \displaystyle\int_0^T [c_1x(t) + c_2y^2(t)]\,dt & \\ \text{subject to} & x(t) = x_0 + \displaystyle\int_0^t [y(\tau) - z(\tau)]\,d\tau & \text{for } t \in [0, T] \\ & x(T) \geq b & \\ & 0 \leq x(t) \leq d & \text{for } t \in (0, T) \\ & l \leq y(t) \leq u & \text{for } t \in (0, T) \end{array}$$

where

$x(t)$ = inventory at time t

$y(t)$ = production rate at time t

$z(t)$ = known demand rate at time t

x_0 = known initial inventory

c_1, c_2 = known coefficients

a. Make the above control problem discrete, as was done in Section 1.2, and formulate a suitable Lagrangian dual problem.

b. Make use of the results of this chapter to develop a scheme for solving the primal and dual problems.

c. Apply your algorithm to the following data: $T = 6$, $x_0 = 0$, $b = 4$, $c_1 = 1$, $c_2 = 2$, $l = 2$, $u = 5$, $d = 6$, and $z(t) = 4$ over $[0, 4]$ and $z(t) = 3$ over $(4, 6]$.

6.38 Consider the primal and dual linear programming problems discussed in Section 6.6. Show directly using Farkas' lemma that if the primal is inconsistent and the dual admits a feasible solution, then the dual has an unbounded objective value.

6.39 Consider the linear program to minimize $\mathbf{c}^t\mathbf{x}$ subject to $\mathbf{Ax} = \mathbf{b}$, $\mathbf{x} \geq \mathbf{0}$. Write the dual problem. Show that the dual of the dual problem is equivalent to the primal problem.

6.40 Consider the following problem:

$$\begin{array}{lrcrcrl} \text{Minimize} & -x_1 & - & 2x_2 & - & x_3 & \\ \text{subject to} & x_1 & + & x_2 & + & x_3 & \leq 16 \\ & x_1 & - & x_2 & + & 3x_3 & \leq 12 \\ & x_1 & + & x_2 & & & \leq 4 \\ & x_1, & & x_2, & & x_3 & \geq 0 \end{array}$$

Solve the primal problem by the simplex method. At each iteration identify the dual variables from the simplex tableau. Show that the dual variables satisfy the complementary slackness conditions but violate the dual constraints. Verify that dual feasibility is attained at termination.

6.41 Consider the dual quadratic program given by (6.30). Describe a gradient based maximization scheme for this problem, following Exercise 6.18. Can you anticipate any computational difficulties? (*Hint:* See Chapter 8.) Illustrate by using the following quadratic programming problem:

$$\begin{array}{lrcrl} \text{Minimize} & 2x_1^2 + x_2^2 - 2x_1x_2 - 4x_1 - 6x_2 & & & \\ \text{subject to} & x_1 & + & x_2 & \leq 8 \\ & -x_1 & + & 2x_2 & \leq 10 \\ & x_1, & & x_2 & \geq 0 \end{array}$$

At each iteration, identify the corresponding primal infeasible point as well as the primal feasible point. Develop a suitable measure of infeasibility and check its progress. Can you draw any general conclusions?

6.42 Consider the (primal) quadratic Program PQP: Minimize $\{\mathbf{c}^t\mathbf{x} + \frac{1}{2}\mathbf{x}^t\mathbf{D}\mathbf{x} : \mathbf{Ax} \geq \mathbf{b}\}$, where $\mathbf{D}$ is an $n \times n$ symmetric matrix and $\mathbf{A}$ is $m \times n$. Let W be an arbitrary set such that $\{\mathbf{w} : \mathbf{Aw} \geq \mathbf{b}\} \subseteq W$, and consider Problem EDQP: Minimize $\{\mathbf{c}^t\mathbf{x} + \frac{1}{2}\mathbf{w}^t\mathbf{D}\mathbf{w} : \mathbf{Ax} \geq \mathbf{b}, \mathbf{Dw} = \mathbf{Dx}, \mathbf{w} \in W\}$.

a. Show that PQP and EPQP are equivalent in the sense that if $\mathbf{x}$ is feasible to PQP, then $(\mathbf{x}, \mathbf{w})$ with $\mathbf{w} = \mathbf{x}$ is feasible to EDQP with the same objective value; and, conversely, if $(\mathbf{x}, \mathbf{w})$ is feasible to EPQP, then $\mathbf{x}$ is feasible to PQP with the same objective value.

b. Construct a Lagrangian dual LD: Maximize $\{\theta(\mathbf{y})\}$, where $\theta(\mathbf{y})$ = minimum $\{(\mathbf{c} + \mathbf{D}\mathbf{y})^t\mathbf{x} + \frac{1}{2}\mathbf{w}^t\mathbf{D}\mathbf{w} - \mathbf{y}^t\mathbf{D}\mathbf{w} : \mathbf{A}\mathbf{x} \geq \mathbf{b}, \mathbf{w} \in W\}$. Show that equivalently, we have

$$LD\text{: supremum } \{\mathbf{b}^t\mathbf{u} - \tfrac{1}{2}\mathbf{y}^t\mathbf{D}\mathbf{y} + \phi(\mathbf{y}) : \mathbf{A}^t\mathbf{u} - \mathbf{D}\mathbf{y} = \mathbf{c}, \mathbf{u} \geq \mathbf{0}\}$$

where $\phi(\mathbf{y}) = \inf \{\frac{1}{2}(\mathbf{y} - \mathbf{w})^t\mathbf{D}(\mathbf{y} - \mathbf{w}) : \mathbf{w} \in W\}$.

c. Show that if $\mathbf{D}$ is positive semidefinite and $W \equiv E_n$, then $\phi(\mathbf{y}) = 0$ for all $\mathbf{y}$, and LD reduces to Dorn's dual program given in (6.29). On the other hand, if $\mathbf{D}$ is not positive semidefinite and $W \equiv E_n$, then $\phi(\mathbf{y}) = -\infty$ for all $\mathbf{y}$. Furthermore, if PQP has an optimum of objective value v_p, and if $W = \{\mathbf{w} : \mathbf{A}\mathbf{w} \geq \mathbf{b}\}$, then show that the optimum value of LD is also v_p. What does this suggest regarding the formulation of LD for nonconvex situations?

d. Illustrate part c using the problem to minimize $\{x_1x_2 : x_1 \geq 0 \text{ and } x_2 \geq 0\}$.

(This exercise is based on Sherali, 1991.)

6.43 Let $\phi(\mathbf{x}, \mathbf{y})$ be a continuous function defined for $\mathbf{x} \in X \subseteq E_n$ and $\mathbf{y} \in Y \subseteq E_m$. Show that

$$\sup_{\mathbf{y}\in Y} \inf_{\mathbf{x}\in X} \phi(\mathbf{x}, \mathbf{y}) \leq \inf_{\mathbf{x}\in X} \sup_{\mathbf{y}\in Y} \phi(\mathbf{x}, \mathbf{y})$$

6.44 Consider the problem to find

$$\underset{\mathbf{x}\in X}{\text{minimum}}\ \underset{\mathbf{y}\in Y}{\text{maximum}}\ \phi(\mathbf{x}, \mathbf{y}) \quad \text{and} \quad \underset{\mathbf{y}\in Y}{\text{maximum}}\ \underset{\mathbf{x}\in X}{\text{minimum}}\ \phi(\mathbf{x}, \mathbf{y})$$

where X and Y are nonempty compact convex sets in E_n and E_m, respectively, and ϕ is convex in $\mathbf{x}$ for any given $\mathbf{y}$, and concave in $\mathbf{y}$ for any given $\mathbf{x}$.

a. Show that $\underset{\mathbf{x}\in X}{\text{minimum}}\ \underset{\mathbf{y}\in Y}{\text{maximum}}\ \phi(\mathbf{x}, \mathbf{y}) \geqslant \underset{\mathbf{y}\in Y}{\text{maximum}}\ \underset{\mathbf{x}\in X}{\text{minimum}}\ \phi(\mathbf{x}, \mathbf{y})$ without any convexity assumptions.

b. Show that $\underset{\mathbf{y}\in Y}{\text{maximum}}\ \phi(\cdot, \mathbf{y})$ is a convex function in $\mathbf{x}$ and that $\underset{\mathbf{x}\in X}{\text{minimum}}\ \phi(\mathbf{x}, \cdot)$ is a concave function in $\mathbf{y}$.

c. Show that

$$\underset{\mathbf{x}\in X}{\text{minimum}}\ \underset{\mathbf{y}\in Y}{\text{maximum}}\ \phi(\mathbf{x}, \mathbf{y}) = \underset{\mathbf{y}\in Y}{\text{maximum}}\ \underset{\mathbf{x}\in X}{\text{minimum}}\ \phi(\mathbf{x}, \mathbf{y})$$

(*Hint:* Use part b and the necessary optimality conditions of Section 3.4.)

6.45 Let X and Y be nonempty sets in E_n, and let $f, g : E_n \to E_1$. Consider the *conjugate functions* f^* and g^* defined as follows:

$$f^*(\mathbf{u}) = \inf \{f(\mathbf{x}) - \mathbf{u}^t\mathbf{x} : \mathbf{x} \in X]$$

$$g^*(\mathbf{u}) = \sup \{g(\mathbf{x}) - \mathbf{u}^t\mathbf{x} : \mathbf{x} \in Y\}$$

a. Interpret f^* and g^* geometrically.

b. Show that f^* is concave over X^* and g^* is convex over Y^*, where $X^* = \{\mathbf{u} : f^*(\mathbf{u}) > -\infty\}$ and $Y^* = \{\mathbf{u} : g^*(\mathbf{u}) < \infty\}$.

c. Prove the following *conjugate weak duality theorem:*

$$\inf \{f(\mathbf{x}) - g(\mathbf{x}) : \mathbf{x} \in X \cap Y\} \geq \sup \{f^*(\mathbf{u}) - g^*(\mathbf{u}) : \mathbf{u} \in X^* \cap Y^*\}$$

d. Now suppose that f is convex, g is concave, int $X \cap$ int $Y \neq \emptyset$, and $\inf \{f(\mathbf{x}) - g(\mathbf{x}) : \mathbf{x} \in X \cap Y\}$ is finite. Show that equality in part c above holds true and that $\sup \{f^*(\mathbf{u}) - g^*(\mathbf{u}) : \mathbf{u} \in X^* \cap Y^*\}$ is achieved.

e. By suitable choices of f, g, X, and Y, formulate a nonlinear programming problem as follows:

Minimize $f(\mathbf{x}) - g(\mathbf{x})$
subject to $\mathbf{x} \in X \cap Y$

What is the form of the conjugate dual problem? Devise some strategies for solving the dual problem.

6.46 Consider a single constrained problem to minimize $f(\mathbf{x})$ subject to $g(\mathbf{x}) \leq 0$ and $\mathbf{x} \in X$, where X is a compact set. The Lagrangian dual problem is to maximize $\theta(u)$ subject to $u \geq 0$, where $\theta(u) = \inf \{f(\mathbf{x}) + ug(\mathbf{x}) : \mathbf{x} \in X\}$.

a. Let $\hat{u} \geq 0$, and let $\hat{\mathbf{x}} \in X(\hat{u})$. Show that if $g(\hat{\mathbf{x}}) > 0$, then $\bar{u} > \hat{u}$, and if $g(\hat{\mathbf{x}}) < 0$, then $\bar{u} < \hat{u}$, where $\bar{u}$ is an optimal solution to the Lagrangian dual.

b. Use the result of part a to find an interval $[a, b]$ that contains all the optimal solutions to the dual problem or else to conclude that the dual problem is unbounded.

c. Now consider the problem to maximize $\theta(u)$ subject to $a \leq u \leq b$. The following scheme is used to solve the problem:

   Let $\bar{u} = (a + b)/2$, and let $\bar{\mathbf{x}} \in X(\bar{u})$. If $g(\bar{\mathbf{x}}) > 0$, replace a by $\bar{u}$ and repeat the process. If $g(\bar{\mathbf{x}}) < 0$, replace b by $\bar{u}$ and repeat the process. If $g(\bar{\mathbf{x}}) = 0$, stop; $\bar{u}$ is an optimal dual solution.

   Show that the procedure converges to an optimal solution, and illustrate by solving the dual of the following problem:

   Minimize $\quad x_1^2 + x_2^2$

   subject to $\quad -x_1 - x_2 + 1 \leq 0$

d. An alternative approach to solving the problem to maximize $\theta(u)$ subject to $a \leq u \leq b$ is to specialize the tangential approximation method discussed in Section 6.4. Show that at each iteration only two supporting hyperplanes need be considered, and that the method could be stated as follows:

   Let $\mathbf{x}_a \in X(a)$ and $\mathbf{x}_b \in X(b)$. Le $\bar{u} = (f(\mathbf{x}_a) - f(\mathbf{x}_b))/(g(\mathbf{x}_b) - g(\mathbf{x}_a))$. If $\bar{u} = a$ or $\bar{u} = b$, stop; $\bar{u}$ is an optimal solution to the dual problem. Otherwise, let $\bar{\mathbf{x}} \in X(\bar{u})$. If $g(\bar{\mathbf{x}}) > 0$, replace a by $\bar{u}$ and repeat the process. If $g(\bar{\mathbf{x}}) < 0$, replace b by $\bar{u}$ and repeat the process. If $g(\bar{\mathbf{x}}) = 0$, stop; $\bar{u}$ is an optimal dual solution.

   Show that the procedure converges to an optimal solution, and illustrate by solving the problem in part c.

Notes and References

The powerful results of duality in linear programming and the saddle point optimality criteria for convex programming sparked a great deal of interest in duality in nonlinear programming. Early results in this area include the work of Cottle [1963b], Dorn [1960], Hanson [1961], Mangasarian [1962], Stoër [1963], and Wolfe [1961].

More recently, several duality formulations that enjoy many of the properties of linear dual programs have evolved. These include the Lagrangian dual problem, the conjugate dual problem, the surrogate dual problem, and the mixed Lagrangian and surrogate, or *composite dual*, problem. In this chapter, we concentrated on the Lagrangian dual formulation because, in our judgment, it is the most promising formulation from a computational standpoint and also because the results of this chapter give the general flavor of the results that one would obtain using other duality formulations. Those interested in studying the subject of conjugate duality may refer to Fenchel [1949], Rockafellar [1964, 1966, 1968, 1969, 1970], Scott and Jefferson [1984, 1989], and Whinston [1967]. For the subject of surrogate duality, where the constraints are grouped into a single constraint by the use of Lagrangian multipliers, refer to Greenberg and Pierskalla [1970b]. Several authors have developed duality formulations that retain the symmetry between the primal and dual problems. The works of Cottle [1963b], Dantzig, Eisenberg, and Cottle [1965], Mangasarian and Ponstein [1965], and Stoër [1963] are in this class. For composite duality, see Karwan and Rardin [1979, 1980].

The reader will find the work of Geoffrion [1971b] and Karamardian [1967] excellent references on various duality formulations and their interrelationships. See Everett [1963], Falk [1967, 1969], and Lasdon [1968] for further study on duality. The relationship between the Lagrangian duality formulation and other duality formulations is examined in Bazaraa, Goode, and Shetty [1971], Magnanti [1974], and Whinston [1967]. The economic interpretation of duality is covered by Balinski and Baumol [1968], Beckman and Kapur [1972], Peterson [1970], and Williams [1970].

In Sections 6.1 and 6.2, the dual problem is presented, and some of its properties are developed. As a by-product of the main duality theorem, we develop the saddle point optimality criteria for convex programs. These criteria were first developed by Kuhn and Tucker [1951]. For the related concept of min–max duality, see Mangasarian and Ponstein [1965], Ponstein [1965], Rockafellar [1968], and Stoër [1963]. For further discussions and illustrations on perturbation functions, see Geoffrion [1971b] and Minoux [1981].

In Section 6.3, we examine several properties of the dual function. We characterize the collection of subgradients at any given point, and use that to determine both ascent directions and the steepest ascent direction. We show that the steepest ascent direction is the shortest subgradient. This result is essentially given by Demyanov [1968]. In Section 6.4, we use these properties to suggest several gradient-based or outer-linearization methods for maximizing the dual function. An accelerated version of the cutting plane method that ensures the generation of ascent directions is discussed by Hearn and Lawphongpanich [1989, 1990]. For further study of this subject, see Bazaraa and Goode [1979], Demyanov [1968, 1971], Fisher et al. [1975], and Lasdon [1970]. For constraint deletion concepts in outer-linearization methods, see Eaves and Zangwill [1971] and Lasdon [1970]. There are other procedures for solving the dual problem. The cutting plane method discussed in Section 6.4 is a row generation procedure. In its dual form, it is precisely the column generation generalized programming method of Wolfe (see Dantzig [1963]). Another procedure is the subgradient optimization method, which is briefly introduced in Exercises 6.33, 6.34, and 6.35, and is discussed in more detail in Chapter 8. See Held, Wolfe, and Crowder [1974] and Polyak [1967] for validation of subgradient optimization. For other related work, see Bazaraa and Goode [1977, 1979], Bazaraa and Sherali [1981], Fisher, et al. [1975], and Held and Karp [1970].

One of the pioneering works for using the Lagrangian formulation to develop computational schemes is credited to Everett [1963]. Under certain conditions, he showed how the primal solution could be retrieved. The result and its extensions are given in Section 6.5. For duality in quadratic programming, see Dorn [1960, 1961], Cottle [1963], and Sherali [1991].

PART 3 Algorithms and Their Convergence

Chapter 7 The Concept of an Algorithm

In the remainder of the text, we describe many algorithms for solving different classes of nonlinear programming problems. This chapter introduces the concept of an algorithm. Algorithms are viewed as point-to-set maps, and the main convergence theorem is proved utilizing the concept of a closed mapping. This theorem will be utilized in the remaining chapters to analyze the convergence of several computational schemes.

The following is an outline of the chapter:

Section 7.1: Algorithms and Algorithmic Maps This section presents algorithms as point-to-set maps and introduces the concept of a solution set.

Section 7.2: Closed Maps and Convergence We first introduce the concept of a closed map and then prove the main convergence theorem.

Section 7.3: Composition of Mappings We establish closedness of composite maps by examining closedness of the individual maps. We then discuss mixed algorithms and give a condition for their convergence.

Section 7.4: Comparison Among Algorithms Some practical factors for assessing the efficiency of different algorithms are discussed.

7.1 Algorithms and Algorithmic Maps

Consider the problem to minimize $f(\mathbf{x})$ subject to $\mathbf{x} \in S$, where f is the objective function and S is the feasible region. A *solution procedure*, or an *algorithm*, for solving this problem can be viewed as an iterative process that generates a sequence of points according to a prescribed set of instructions, together with a termination criterion.

The Algorithmic Map

Given a vector $\mathbf{x}_k$ and applying the instructions of the algorithm, we obtain a new point $\mathbf{x}_{k+1}$. This process can be described by an *algorithmic map* **A**. This map is generally a point-to-set map and assigns to each point in the domain X a subset of X. Thus, given the initial point $\mathbf{x}_1$, the algorithmic map generates the sequence $\mathbf{x}_1, \mathbf{x}_2, \ldots$, where $\mathbf{x}_{k+1} \in \mathbf{A}(\mathbf{x}_k)$ for each k. The transformation of $\mathbf{x}_k$ into $\mathbf{x}_{k+1}$ through the map constitutes an *iteration* of the algorithm.

7.1.1. Example

Consider the following problem:

$$\begin{array}{ll} \text{Minimize} & x^2 \\ \text{subject to} & x \geq 1 \end{array}$$

whose optimal solution is $\bar{x} = 1$. Let the point-to-point algorithmic map be given by $\mathbf{A}(x) = \frac{1}{2}(x + 1)$. It can be easily verified that the sequence obtained by applying the map **A**, with any starting point, converges to the optimal solution $\bar{x} = 1$. With $x_1 = 4$, the algorithm generates the sequence $\{4, 2.5, 1.75, 1.375, 1.1875, \ldots\}$, as illustrated in Figure 7.1*a*.

As another example, consider the point-to-set mapping **A**, defined by

$$\mathbf{A}(x) = \begin{cases} [1, \frac{1}{2}(x + 1)] & \text{if } x \geq 1 \\ [\frac{1}{2}(x + 1), 1] & \text{if } x < 1 \end{cases}$$

As shown in Figure 7.1*b*, the image of any point x is a closed interval, and any point in that interval could be chosen as the successor of x. Starting with any point x_1, the algorithm converges to $\bar{x} = 1$. With $x_1 = 4$, the sequence $\{4, 2, 1.2, 1.1., 1.02, \ldots\}$ is a possible result of the algorithm. Unlike the previous example, other sequences could result from this algorithmic map.

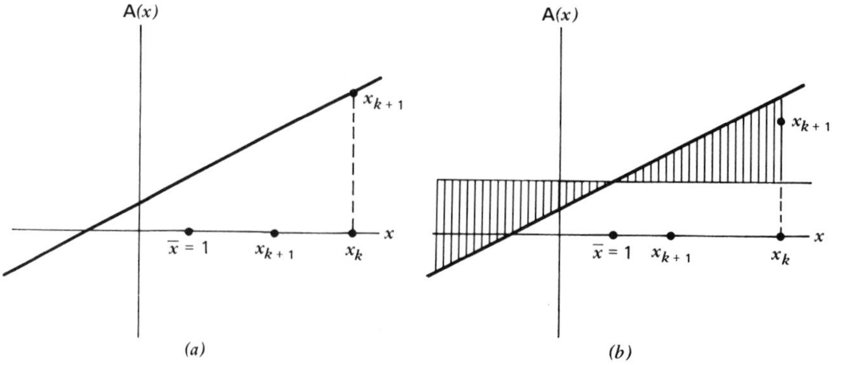

Figure 7.1 Examples of algorithmic maps.

The Solution Set and Convergence of Algorithms

Consider the following nonlinear programming problem

$$\begin{array}{ll} \text{Minimize} & f(\mathbf{x}) \\ \text{subject to} & \mathbf{x} \in S \end{array}$$

A desirable property of an algorithm for solving the above problem is that it generates a sequence of points converging to a global optimal solution. In many cases, however, we may have to be satisfied with less favorable outcomes. In fact, as a result of nonconvexity, problem size, and other difficulties, we may stop the iterative procedure if a point belonging to a prescribed set, which we call the *solution set* Ω, is reached. The following are some typical solution sets for the above mentioned problem:

1. $\Omega = \{\bar{\mathbf{x}}:\bar{\mathbf{x}}$ is a local optimal solution of the problem$\}$.
2. $\Omega = \{\bar{\mathbf{x}}:\bar{\mathbf{x}} \in S, f(\bar{\mathbf{x}}) \leq b\}$, where b is an acceptable objective value.
3. $\Omega = \{\bar{\mathbf{x}}:\bar{\mathbf{x}} \in S, f(\bar{\mathbf{x}}) < LB + \varepsilon\}$, where $\varepsilon > 0$ is a specified tolerance and LB is a lower bound on the optimal objective value. A typical lower bound is the objective value of the Lagrangian dual problem.
4. $\Omega = \{\bar{\mathbf{x}}:\bar{\mathbf{x}} \in S, f(\bar{\mathbf{x}}) - v^* < \varepsilon\}$, where v^* is the known global minimum value and $\varepsilon > 0$ is specified.
5. $\Omega = \{\bar{\mathbf{x}}:\bar{\mathbf{x}}$ satisfies the KKT optimality conditions$\}$.
6. $\Omega = \{\bar{\mathbf{x}}:\bar{\mathbf{x}}$ satisfies the Fritz John optimality conditions$\}$.

Thus, in general, convergence of algorithms is made in reference to the solution set rather than to the collection of global optimal solutions. In particular, the algorithmic map $\mathbf{A}:X \to X$ is said to *converge* over $Y \subseteq X$ if, starting with any initial point $\mathbf{x}_1 \in Y$, the limit of any convergent subsequence of the sequence $\mathbf{x}_1, \mathbf{x}_2, \ldots$, generated by the algorithm belongs to the solution set Ω. Letting Ω be the set of global optimal solutions in Example 7.1.1, it is obvious that the two stated algorithms are convergent over the real line.

7.2 Closed Maps and Convergence

In this section, we introduce the notion of closed maps and then prove a convergence theorem. The significance of the concept of closedness will be clear from the following example and the subsequent discussion.

7.2.1 Example

Consider the following problem:

$$\begin{array}{ll} \text{Minimize} & x^2 \\ \text{subject to} & x \geq 1 \end{array}$$

Let Ω be the set of global optimal solutions, that is, $\Omega = \{1\}$. Consider the algorithmic map defined by

$$\mathbf{A}(x) = \begin{cases} [\frac{3}{2} + \frac{1}{4}x, 1 + \frac{1}{2}x] & \text{if } x \geq 2 \\ \frac{1}{2}(x + 1) & \text{if } x < 2 \end{cases}$$

The map $\mathbf{A}$ is illustrated in Figure 7.2. Obviously, for any initial point $x_1 \geq 2$, any

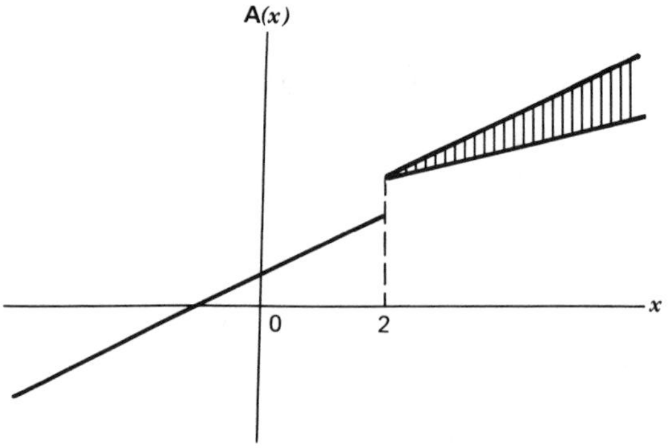

Figure 7.2 Example of a nonconvergent algorithmic map.

sequence generated by the map **A** converges to the point $\hat{x} = 2$. Note that $\hat{x} \notin \Omega$. On the other hand, for $x_1 < 2$, any sequence generated by the algorithm converges to $\bar{x} = 1$. In this example, the algorithm converges over the interval $(-\infty, 2)$ but does not converge to a point in the set Ω over the interval $[2, \infty)$.

The above example shows the significance of the initial point x_1, where convergence to a point in Ω is achieved if $x_1 < 2$ but not realized otherwise. Note that each of the algorithms in Examples 7.1.1 and 7.2.1 satisfy the following conditions:

1. Given a feasible point $x_k \geq 1$, any successor point x_{k+1} is also feasible, that is, $x_{k+1} \geq 1$.
2. Given a feasible point x_k not in the solution set Ω, any successor point x_{k+1} satisfies $f(x_{k+1}) < f(x_k)$, where $f(x) = x^2$. In other words, the objective function strictly decreases.
3. Given a feasible point x_k in the solution set Ω, that is, $x_k = 1$, the successor point is also in Ω, that is, $x_{k+1} = 1$.

Despite the above-mentioned similarities among the algorithms, the two algorithms of Example 7.1.1 converge to $\bar{x} = 1$, whereas that of Example 7.2.1 does not converge to $\bar{x} = 1$ for any initial point $x_1 \geq 2$. The reason for this is that the algorithmic map of Example 7.2.1 is not closed at $x = 2$. The notion of a closed mapping, which generalizes the notion of a continuous function, is defined below.

Closed Maps

7.2.2 Definition

Let X and Y be nonempty closed sets in E_p and E_q, respectively. Let $\mathbf{A}: X \rightarrow Y$ be a point-to-set map. The map **A** is said to be *closed* at $\mathbf{x} \in X$ if

$$\mathbf{x}_k \in X \qquad \mathbf{x}_k \rightarrow \mathbf{x}$$

$$\mathbf{y}_k \in \mathbf{A}(\mathbf{x}_k) \qquad \mathbf{y}_k \rightarrow \mathbf{y}$$

imply that $\mathbf{y} \in \mathbf{A}(\mathbf{x})$. The map **A** is said to be closed on $Z \subseteq X$ if it is closed at each point in Z.

Figure 7.2 shows an example of a point-to-set map that is not closed at $x = 2$. In particular, the sequence $\{x_k\}$ with $x_k = 2 - \frac{1}{k}$ converges to $x = 2$, and the sequence

$\{y_k\}$ with $y_k = \mathbf{A}(x_k) = \frac{3}{2} - \frac{1}{2k}$ converges to $y = \frac{3}{2}$, but $y \notin \mathbf{A}(x) = \{2\}$. Figure 7.1 shows two examples of algorithmic maps that are closed everywhere.

The Convergence Theorem

Conditions that ensure convergence of algorithmic maps are stated in Theorem 7.2.3 below. The theorem will be used in the remainder of the text to show convergence of many algorithms.

7.2.3 Theorem

Let X be a nonempty closed set in E_n, and let the nonempty set $\Omega \subseteq X$ be the solution set. Let $\mathbf{A}: X \to X$ be a point-to-set map. Given $\mathbf{x}_1 \in X$, the sequence $\{\mathbf{x}_k\}$ is generated iteratively as follows:

If $\mathbf{x}_k \in \Omega$ then stop; otherwise, let $\mathbf{x}_{k+1} \in \mathbf{A}(\mathbf{x}_k)$, replace k by $k + 1$, and repeat

Suppose that the sequence $\mathbf{x}_1, \mathbf{x}_2, \ldots$ produced by the algorithm is contained in a compact subset of X, and suppose that there exists a continuous function α, called the *descent function*, such that $\alpha(\mathbf{y}) < \alpha(\mathbf{x})$ if $\mathbf{x} \notin \Omega$ and $\mathbf{y} \in \mathbf{A}(\mathbf{x})$. If the map $\mathbf{A}$ is closed over the complement of Ω, then either the algorithm stops in a finite number of steps with a point in Ω or it generates the infinite sequence $\{\mathbf{x}_k\}$ such that

1. Every convergent subsequence of $\{\mathbf{x}_k\}$ has a limit in Ω, that is, all accumulation points of $\{\mathbf{x}_k\}$ belong to Ω.
2. $\alpha(\mathbf{x}_k) \to \alpha(\mathbf{x})$ for some $\mathbf{x} \in \Omega$.

Proof

If at any iteration a point $\mathbf{x}_k$ in Ω is generated, then the algorithm stops. Now suppose that an infinite sequence $\{\mathbf{x}_k\}$ is generated. Let $\{\mathbf{x}_k\}_{\mathcal{K}}$ be any convergent subsequence with limit $\mathbf{x} \in X$. Since α is continuous, then, for $k \in \mathcal{K}$, $\alpha(\mathbf{x}_k) \to \alpha(\mathbf{x})$. Thus, for a given $\varepsilon > 0$, there is a $k \in \mathcal{K}$ such that

$$\alpha(\mathbf{x}_k) - \alpha(\mathbf{x}) < \varepsilon \qquad \text{for } k \geq K \text{ with } k \in \mathcal{K}$$

In particular for $k = K$, we get

$$\alpha(\mathbf{x}_K) - \alpha(\mathbf{x}) < \varepsilon \tag{7.1}$$

Now let $k > K$. Since α is a descent function, $\alpha(\mathbf{x}_k) < \alpha(\mathbf{x}_K)$, and, from (7.1), we get

$$\alpha(\mathbf{x}_k) - \alpha(\mathbf{x}) = \alpha(\mathbf{x}_k) - \alpha(\mathbf{x}_K) + \alpha(\mathbf{x}_K) - \alpha(\mathbf{x}) < 0 + \varepsilon = \varepsilon$$

Since this is true for every $k > K$, and since $\varepsilon > 0$ was arbitrary, then

$$\lim_{k \to \infty} \alpha(\mathbf{x}_k) = \alpha(\mathbf{x}) \tag{7.2}$$

We now show that $\mathbf{x} \in \Omega$. By contradiction, suppose that $\mathbf{x} \notin \Omega$, and consider the sequence $\{\mathbf{x}_{k+1}\}_{\mathcal{K}}$. This sequence is contained in a compact subset of X and, hence, has a convergent subsequence $\{\mathbf{x}_{k+1}\}_{\bar{\mathcal{K}}}$ with limit $\bar{\mathbf{x}}$ in X. Noting (7.2), it is clear that $\alpha(\bar{\mathbf{x}}) = \alpha(\mathbf{x})$. Since $\mathbf{A}$ is closed at $\mathbf{x}$, and for $k \in \bar{\mathcal{K}}$, $\mathbf{x}_k \to \mathbf{x}$, $\mathbf{x}_{k+1} \in \mathbf{A}(\mathbf{x}_k)$, and $\mathbf{x}_{k+1} \to \bar{\mathbf{x}}$, then $\bar{\mathbf{x}} \in \mathbf{A}(\mathbf{x})$. Therefore, $\alpha(\bar{\mathbf{x}}) < \alpha(\mathbf{x})$, contradicting the fact that $\alpha(\bar{\mathbf{x}}) =$

$\alpha(\mathbf{x})$. Thus, $\mathbf{x} \in \Omega$ and part 1 of the theorem holds true. This, coupled with (7.2), shows that part 2 of the theorem holds true, and the proof is complete.

Corollary

Under the assumptions of the theorem, if Ω is the singleton $\{\bar{\mathbf{x}}\}$, then the whole sequence $\{\mathbf{x}_k\}$ converges to $\bar{\mathbf{x}}$.

Proof

Suppose, by contradiction, that there exists an $\varepsilon > 0$ and a sequence $\{\mathbf{x}_k\}_{\mathcal{K}}$ such that

$$\|\mathbf{x}_k - \bar{\mathbf{x}}\| > \varepsilon \qquad \text{for } k \in \mathcal{K} \tag{7.3}$$

Note that there exists $\mathcal{K}' \subset \mathcal{K}$ such that $\{\mathbf{x}_k\}_{\mathcal{K}'}$ has a limit $\mathbf{x}'$. By part 1 of the theorem, $\mathbf{x}' \in \Omega$. But $\Omega = \{\bar{\mathbf{x}}\}$, and thus $\mathbf{x}' = \bar{\mathbf{x}}$. Therefore, $\mathbf{x}_k \rightarrow \bar{\mathbf{x}}$ for $k \in \mathcal{K}'$, violating (7.3). This completes the proof.

Note that if the point at hand $\mathbf{x}_k$ does not belong to the solution set Ω, then the algorithm generates a new point $\mathbf{x}_{k+1}$ such that $\alpha(\mathbf{x}_{k+1}) < \alpha(\mathbf{x}_k)$. As mentioned before, the function α is called a *descent function*. In many cases, α is chosen as the objective function f itself, and thus the algorithm generates a sequence of points with improving objective function values. Other alternative choices of the function α are possible. For instance, if f is differentiable, α could be chosen as $\alpha(\mathbf{x}) = \|\nabla f(\mathbf{x})\|$ for an unconstrained optimization problem.

Terminating the Algorithm

As indicated in Theorem 7.2.3, the algorithm is terminated if we reach a point in the solution set Ω. In most cases, however, convergence to a point in Ω occurs only in a limiting sense, and we must resort to some practical rules for terminating the iterative procedure. The following rules are frequently used to stop a given algorithm. Here, $\varepsilon > 0$ and the positive integer N are prespecified.

1. $\|\mathbf{x}_{k+N} - \mathbf{x}_k\| < \varepsilon$
   Here, the algorithm is stopped if the distance moved after N applications of the map $\mathbf{A}$ is less than ε.

2. $\dfrac{\|\mathbf{x}_{k+1} - \mathbf{x}_k\|}{\|\mathbf{x}_k\|} < \varepsilon$
   Under this criterion, the algorithm is terminated if the relative distance moved during a given iteration is less than ε.

3. $\alpha(\mathbf{x}_k) - \alpha(\mathbf{x}_{k+N}) < \varepsilon$
   Here, the algorithm is stopped if the total improvement in the descent function value after N applications of the map $\mathbf{A}$ is less than ε.

4. $\dfrac{\alpha(\mathbf{x}_k) - \alpha(\mathbf{x}_{k+1})}{|\alpha(\mathbf{x}_k)|} < \varepsilon$
   If the relative improvement in the descent function value during any given iteration is less than ε, then the termination criterion is realized.

5. $\alpha(\mathbf{x}_k) - \alpha(\bar{\mathbf{x}}) < \varepsilon$, where $\bar{\mathbf{x}}$ belongs to Ω
This criterion for termination is suitable if $\alpha(\bar{\mathbf{x}})$ is known beforehand; for example, in unconstrained optimization, if $\alpha(\mathbf{x}) = \|\nabla f(\mathbf{x})\|$ and $\Omega = \{\bar{\mathbf{x}}: \nabla f(\bar{\mathbf{x}}) = \mathbf{0}\}$, then $\alpha(\bar{\mathbf{x}}) = 0$.

7.3 Composition of Mappings

In most nonlinear programming solution procedures, the algorithmic maps are often composed of several maps. For example, some algorithms first find a direction $\mathbf{d}_k$ to move along and then determine the step size λ_k by solving the one-dimensional problem of minimizing $\alpha(\mathbf{x}_k + \lambda \mathbf{d}_k)$. In this case, the map **A** is composed of **MD**, where **D** finds the direction $\mathbf{d}_k$, and then **M** finds an optimal step size λ_k. It is often easier to prove that the overall map is closed by examining its individual components. In this section, the notion of composite maps is stated precisely, and then a result relating closedness of the overall map to that of its individual components is given. Finally, we discuss mixed algorithms and state conditions under which they converge.

7.3.1 Definition

Let X, Y, and Z be nonempty closed sets in E_n, E_p, and E_q, respectively. Let $\mathbf{B}: X \to Y$ and $\mathbf{C}: Y \to Z$ be point-to-set maps. The *composite map* $\mathbf{A} = \mathbf{CB}$ is defined as the point-to-set map $\mathbf{A}: X \to Z$ with

$$\mathbf{A}(\mathbf{x}) = \cup \{\mathbf{C}(\mathbf{y}): \mathbf{y} \in \mathbf{B}(\mathbf{x})\}$$

Figure 7.3 illustrates the notion of a composite map, and Theorem 7.3.2 and its corollaries give several sufficient conditions for a composite map to be closed.

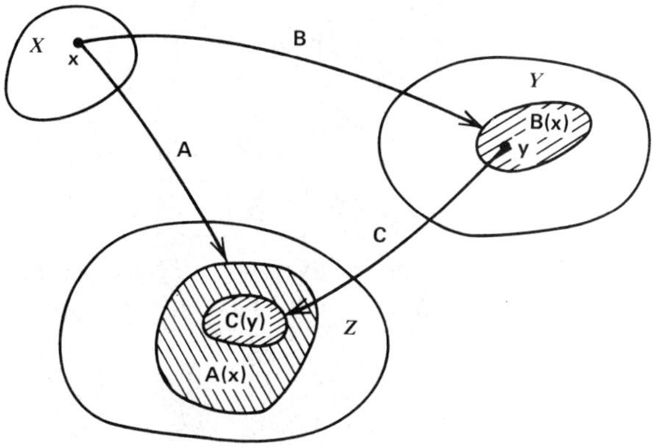

Figure 7.3 Composite maps.

7.3.2 Theorem

Let X, Y, and Z be nonempty closed sets in E_n, E_p, and E_q, respectively. Let $\mathbf{B}: X \to Y$ and $\mathbf{C}: Y \to Z$ be point-to-set maps, and consider the composite map $\mathbf{A} = \mathbf{CB}$. Suppose that **B** is closed at **x** and that **C** is closed on **B(x)**. Furthermore, suppose that if $\mathbf{x}_k \to \mathbf{x}$ and $\mathbf{y}_k \in \mathbf{B}(\mathbf{x}_k)$, then there is a convergent subsequence of $\{\mathbf{y}_k\}$. Then, **A** is closed at **x**.

Proof

Let $\mathbf{x}_k \to \mathbf{x}$, $\mathbf{z}_k \in \mathbf{A}(\mathbf{x}_k)$, and $\mathbf{z}_k \to \mathbf{z}$. We need to show that $\mathbf{z} \in \mathbf{A}(\mathbf{x})$. By definition of **A**, for each k, there is a $\mathbf{y}_k \in \mathbf{B}(\mathbf{x}_k)$ such that $\mathbf{z}_k \in \mathbf{C}(\mathbf{y}_k)$. By assumption, there is a convergent subsequent $\{\mathbf{y}_k\}_{\mathcal{K}}$ with limit **y**. Since **B** is closed at **x**, then $\mathbf{y} \in \mathbf{B}(\mathbf{x})$. Furthermore, since **C** is closed on **B(x)** it is closed at **y**, and hence $\mathbf{z} \in \mathbf{C}(\mathbf{y})$. Thus, $\mathbf{z} \in \mathbf{C}(\mathbf{y}) \in \mathbf{CB}(\mathbf{x}) = \mathbf{A}(\mathbf{x})$, and hence, **A** is closed at **x**.

Corollary 1

Let X, Y, and Z be nonempty closed sets in E_n, E_p, and E_q, respectively. Let $\mathbf{B}: X \to Y$ and $\mathbf{C}: Y \to Z$ be point-to-set maps. Suppose that **B** is closed at **x**, **C** is closed on **B(x)**, and Y is compact. Then, $\mathbf{A} = \mathbf{CB}$ is closed at **x**.

Corollary 2

Let X, Y, and Z be nonempty closed sets in E_n, E_p, and E_q, respectively. Let $\mathbf{B}: X \to Y$ be a function, and let $\mathbf{C}: Y \to Z$ be a point-to-set map. If **B** is continuous at **x**, and **C** is closed on **B(x)**, then $\mathbf{A} = \mathbf{CB}$ is closed at **x**.

Note the importance of the assumption that a convergent subsequence $\{\mathbf{y}_k\}_{\mathcal{K}}$ exists in Theorem 7.3.2. Without this assumption, even if the maps **B** and **C** are closed, the composite map $\mathbf{A} = \mathbf{CB}$ is not necessarily closed, as shown by Example 7.3.3 below. (This example is due to Professor Jamie J. Goode.)

7.3.3 Example

Consider the maps $\mathbf{B}, \mathbf{C}: E_1 \to E_1$ defined below:

$$\mathbf{B}(x) = \begin{cases} \frac{1}{x} & \text{if } x \neq 0 \\ 0 & \text{if } x = 0 \end{cases}$$

$$\mathbf{C}(y) = \{z : |z| \leq |y|\}.$$

Note that both **B** and **C** are closed everywhere. Now consider the composite map $\mathbf{A} = \mathbf{CB}$. Then, **A** is given by $\mathbf{A}(x) = \mathbf{CB}(x) = \{z : |z| \leq |\mathbf{B}(x)|\}$. From the definition of **B**, it follows that

$$\mathbf{A}(x) = \begin{cases} \{z : |z| \leq |\frac{1}{x}|\} & \text{if } x \neq 0 \\ \{0\} & \text{if } x = 0 \end{cases}$$

Note that **A** is not closed at $x = 0$. In particular, consider the sequence $\{x_k\}$, where $x_k = \frac{1}{k}$. Note that $\mathbf{A}(x_k) = \{z : |z| \leq k\}$, and hence $z_k = 1$ belongs to $\mathbf{A}(x_k)$ for each k. On the other hand, the limit point $z = 1$ does not belong to $\mathbf{A}(x) = \{0\}$. Thus, the map **A** is not closed, even though both **B** and **C** are closed. Here, Theorem 7.3.2 does not apply, since the sequence $y_k \in \mathbf{B}(x_k)$ for $x_k = \frac{1}{k}$ does not have a convergent subsequence.

Convergence of Algorithms with Composite Maps

At each iteration, many nonlinear programming algorithms use two maps **B** and **C**, say. One of the maps, **B**, is usually closed and satisfies the convergence requirements of Theorem 7.2.3. The second map, **C**, may involve any process as long as the value of the descent function does not increase. As illustrated in Exercise 7.17, the overall map may not be closed, so that Theorem 7.2.3 cannot be applied. However, as shown below, such maps do converge. Hence, such a result can be used to establish the convergence of a complex algorithm in which a step of a known convergent algorithm is interspersed at finite iteration intervals, but infinitely often over the entire algorithmic sequence. Then, by viewing the algorithm as an application of the composite map **CB**, where **B** corresponds to the step of the known convergent algorithm that satisfies the assumptions of Theorem 7.2.3 and **C** corresponds to the set of intermediate steps of the complex algorithm, the overall convergence of such a scheme would follow by Theorem 7.3.4 below. In such a context, the step of applying **B** as above is called a *spacer step*.

7.3.4 Theorem

Let X be a nonempty closed set in E_n, and let $\Omega \subseteq X$ be a nonempty solution set. Let $\alpha: E_n \to E_1$ be a continuous function, and consider the point-to-set map $\mathbf{C}: X \to X$ satisfying the following property: Given $\mathbf{x} \in X$, then $\alpha(\mathbf{y}) \le \alpha(\mathbf{x})$ for $\mathbf{y} \in \mathbf{C}(\mathbf{x})$. Let $\mathbf{B}: X \to X$ be a point-to-set map that is closed over the complement of Ω and that satisfies $\alpha(\mathbf{y}) < \alpha(\mathbf{x})$ for each $\mathbf{y} \in \mathbf{B}(\mathbf{x})$, if $\mathbf{x} \notin \Omega$. Now consider the algorithm defined by the composite map $\mathbf{A} = \mathbf{CB}$. Given $\mathbf{x}_1 \in X$, suppose that the sequence $\{\mathbf{x}_k\}$ is generated as follows:

If $\mathbf{x}_k \in \Omega$, then stop; otherwise, let $\mathbf{x}_{k+1} \in \mathbf{A}(\mathbf{x}_k)$ replace k by $k + 1$, and repeat

Suppose that the set $\Lambda = \{\mathbf{x}: \alpha(\mathbf{x}) \le \alpha(\mathbf{x}_1)\}$ is compact. Then, either the algorithm stops in a finite number of steps with a point in Ω or all accumulation points of $\{\mathbf{x}_k\}$ belong to Ω.

Proof

If at any iteration $\mathbf{x}_k \in \Omega$, then the algorithm stops finitely. Now suppose that the sequence $\{\mathbf{x}_k\}$ is generated by the algorithm, and let $\{\mathbf{x}_k\}_{\mathcal{K}}$ be a convergent subsequence, with limit $\mathbf{x}$. Thus, $\alpha(\mathbf{x}_k) \to \alpha(\mathbf{x})$ for $k \in \mathcal{K}$. Using monotonicity of α as in Theorem 7.2.3, it follows that

$$\lim_{k\to\infty} \alpha(\mathbf{x}_k) = \alpha(\mathbf{x}) \tag{7.4}$$

We want to show that $\mathbf{x} \in \Omega$. By contradiction, suppose that $\mathbf{x} \notin \Omega$, and consider the sequence $\{\mathbf{x}_{k+1}\}_{\mathcal{K}}$. By definition of the composite map $\mathbf{A}$, note that $\mathbf{x}_{k+1} \in \mathbf{C}(\mathbf{y}_k)$, where $\mathbf{y}_k \in \mathbf{B}(\mathbf{x}_k)$. Note that $\mathbf{y}_k, \mathbf{x}_{k+1} \in \Lambda$. Since Λ is compact, there exists an index set $\mathcal{K}' \subseteq \mathcal{K}$ such that $\mathbf{y}_k \to \mathbf{y}$ and $\mathbf{x}_{k+1} \to \mathbf{x}'$ for $k \in \mathcal{K}'$. Since $\mathbf{B}$ is closed at $\mathbf{x} \notin \Omega$, then $\mathbf{y} \in \mathbf{B}(\mathbf{x})$, and $\alpha(\mathbf{y}) < \alpha(\mathbf{x})$. Since $\mathbf{x}_{k+1} \in \mathbf{C}(\mathbf{y}_k)$, then, by assumption, $\alpha(\mathbf{x}_{k+1}) \le \alpha(\mathbf{y}_k)$ for $k \in \mathcal{K}'$; and, hence, by taking the limit, $\alpha(\mathbf{x}') \le \alpha(\mathbf{y})$. Since $\alpha(\mathbf{y}) < \alpha(\mathbf{x})$, then $\alpha(\mathbf{x}') < \alpha(\mathbf{x})$. Since $\alpha(\mathbf{x}_{k+1}) \to \alpha(\mathbf{x}')$ for $k \in \mathcal{K}'$, then $\alpha(\mathbf{x}') < \alpha(\mathbf{x})$ contradicts (7.4). Therefore, $\mathbf{x} \in \Omega$, and the proof is complete.

Minimizing Along Independent Directions

We now present a theorem that establishes convergence of a class of algorithms for solving a problem of the form: minimize $f(\mathbf{x})$ subject to $\mathbf{x} \in E_n$. Under mild assumptions, we show that an algorithm that generates n linearly independent search directions, and obtains a new point by sequentially minimizing f along these directions, converges to a stationary point. The theorem also establishes convergence of algorithms using linearly independent and orthogonal search directions.

7.3.5 Theorem

Let $f: E_n \to E_1$ be differentiable, and consider the problem to minimize $f(\mathbf{x})$ subject to $\mathbf{x} \in E_n$. Consider an algorithm whose map $\mathbf{A}$ is defined as follows. The vector $\mathbf{y} \in \mathbf{A}(\mathbf{x})$ means that $\mathbf{y}$ is obtained by minimizing f sequentially along the directions $\mathbf{d}_1, \ldots, \mathbf{d}_n$ starting from $\mathbf{x}$. Here, the search directions $\mathbf{d}_1, \ldots, \mathbf{d}_n$ may depend upon $\mathbf{x}$, and each has norm 1. Suppose that the following properties are true:

1. There exists an $\varepsilon > 0$ such that $\det [\mathbf{D}(\mathbf{x})] \geq \varepsilon$ for each $\mathbf{x} \in E_n$. Here, $\mathbf{D}(\mathbf{x})$ is the $n \times n$ matrix whose columns are the search directions generated by the algorithm, and $\det [\mathbf{D}(\mathbf{x})]$ denotes the determinant of $\mathbf{D}(\mathbf{x})$.
2. The minimum of f along any line in E_n is unique.

Given a starting point $\mathbf{x}_1$, suppose that the algorithm generates the sequence $\{\mathbf{x}_k\}$ as follows. If $\nabla f(\mathbf{x}_k) = \mathbf{0}$, then the algorithm stops with $\mathbf{x}_k$; otherwise, $\mathbf{x}_{k+1} \in \mathbf{A}(\mathbf{x}_k)$, k is replaced by $k + 1$, and the process is repeated. If the sequence $\{\mathbf{x}_k\}$ is contained in a compact subset of E_n, then each accumulation point $\mathbf{x}$ of the sequence $[\mathbf{x}_k\}$ must satisfy $\nabla f(\mathbf{x}) = \mathbf{0}$.

Proof

If the sequence $\{\mathbf{x}_k\}$ is finite, then the result is immediate. Now suppose that the algorithm generates the infinite sequence $\{\mathbf{x}_k\}$.

Let $\mathcal{K}$ be an infinite sequence of positive integers, and suppose that the sequence $\{\mathbf{x}_k\}_{\mathcal{K}}$ converges to a point $\mathbf{x}$. We need to show that $\nabla f(\mathbf{x}) = \mathbf{0}$. Suppose by contradiction that $\nabla f(\mathbf{x}) \neq \mathbf{0}$, and consider the sequence $\{\mathbf{x}_{k+1}\}_{\mathcal{K}}$. By assumption, this sequence is contained in a compact subset of E_n; and, hence, there exists $\mathcal{K}' \subseteq \mathcal{K}$ such that $\{\mathbf{x}_{k+1}\}_{\mathcal{K}'}$ converges to $\mathbf{x}'$. We shall first show that $\mathbf{x}'$ can be obtained from $\mathbf{x}$ by minimizing f along a set of n linearly independent directions.

Let $\mathbf{D}_k$ be the $n \times n$ matrix whose columns $\mathbf{d}_{1k}, \ldots, \mathbf{d}_{nk}$ are the search directions generated at iteration k. Thus, $\mathbf{x}_{k+1} = \mathbf{x}_k + \mathbf{D}_k\boldsymbol{\lambda}_k = \mathbf{x}_k + \sum_{j=1}^{n} \mathbf{d}_{jk}\lambda_{jk}$, where λ_{jk} is the distance moved along $\mathbf{d}_{jk}$. In particular, letting $\mathbf{y}_{1k} = \mathbf{x}_k$, $\mathbf{y}_{j+1,k} = \mathbf{y}_{jk} + \lambda_{jk}\mathbf{d}_{jk}$ for $j = 1, \ldots, n$ it follows that $\mathbf{x}_{k+1} = \mathbf{y}_{n+1,k}$, and

$$f(\mathbf{y}_{j+1,k}) \leq f(\mathbf{y}_{jk} + \lambda\mathbf{d}_{jk}) \qquad \text{for all } \lambda \in E_1 \text{ and } j = 1, \ldots, n \tag{7.5}$$

Since $\det [\mathbf{D}_k] \geq \varepsilon > 0$, $\mathbf{D}_k$ is invertible, so that $\boldsymbol{\lambda}_k = \mathbf{D}_k^{-1}(\mathbf{x}_{k+1} - \mathbf{x}_k)$. Since each column of $\mathbf{D}_k$ has norm 1, there exists $\mathcal{K}'' \subseteq \mathcal{K}'$ such that $\mathbf{D}_k \to \mathbf{D}$. Since $\det [\mathbf{D}_k] \geq \varepsilon$ for each k, $\det [\mathbf{D}] \geq \varepsilon$, so that $\mathbf{D}$ is invertible. Now, for $k \in \mathcal{K}''$, $\mathbf{x}_{k+1} \to \mathbf{x}'$, $\mathbf{x}_k \to \mathbf{x}$, $\mathbf{D}_k \to \mathbf{D}$, so that $\boldsymbol{\lambda}_k \to \boldsymbol{\lambda}$, where $\boldsymbol{\lambda} = \mathbf{D}^{-1}(\mathbf{x}' - \mathbf{x})$. Therefore, $\mathbf{x}' = \mathbf{x} + \mathbf{D}\boldsymbol{\lambda} = \mathbf{x} + \sum_{j=1}^{n} \mathbf{d}_j\lambda_j$. Let $\mathbf{y}_1 = \mathbf{x}$, and for $j = 1, \ldots, n$, let $\mathbf{y}_{j+1} = \mathbf{y}_j + \lambda_j\mathbf{d}_j$, so that $\mathbf{x}' = \mathbf{y}_{n+1}$. To show that $\mathbf{x}'$ is obtained from $\mathbf{x}$ by minimizing f sequentially along $\mathbf{d}_1, \ldots, \mathbf{d}_n$, it suffices to show that

$$f(\mathbf{y}_{j+1}) \leq f(\mathbf{y}_j + \lambda \mathbf{d}_j) \qquad \text{for all } \lambda \in E_1 \text{ and } j = 1, \ldots, n \tag{7.6}$$

Note that $\lambda_{jk} \to \lambda_j$, $\mathbf{d}_{jk} \to \mathbf{d}_j$, $\mathbf{x}_k \to \mathbf{x}$, and $\mathbf{x}_{k+1} \to \mathbf{x}'$ as $k \in \mathcal{K}''$ approaches ∞, so that $\mathbf{y}_{jk} \to \mathbf{y}_j$ for $j = 1, \ldots, n + 1$ as $k \in \mathcal{K}''$ approaches ∞. By continuity of f, then, (7.6) follows from (7.5). We have thus shown that $\mathbf{x}'$ is obtained from $\mathbf{x}$ by minimizing f sequentially along the directions $\mathbf{d}_1, \ldots, \mathbf{d}_n$.

Obviously, $f(\mathbf{x}') \leq f(\mathbf{x})$. First, consider the case $f(\mathbf{x}') < f(\mathbf{x})$. Since $\{f(\mathbf{x}_k)\}$ is a nonincreasing sequence, and since $f(\mathbf{x}_k) \to f(\mathbf{x})$ as $k \in \mathcal{K}$ approaches ∞, $\lim_{k\to\infty} f(\mathbf{x}_k) = f(\mathbf{x})$. This is impossible, however, in view of the fact that $\mathbf{x}_{k+1} \to \mathbf{x}'$ as $k \in \mathcal{K}'$ approaches ∞ and the assumption that $f(\mathbf{x}') < f(\mathbf{x})$. Now consider the case $f(\mathbf{x}') = f(\mathbf{x})$. By property 2 of the theorem, and since $\mathbf{x}'$ is obtained from $\mathbf{x}$ by minimizing f along $\mathbf{d}_1, \ldots, \mathbf{d}_n$, $\mathbf{x}' = \mathbf{x}$. This further implies that $\nabla f(\mathbf{x})^t \mathbf{d}_j = 0$ for $j = 1, \ldots, n$. Since $\mathbf{d}_1, \ldots, \mathbf{d}_n$ are linearly independent, $\nabla f(\mathbf{x}) = \mathbf{0}$, contradicting our assumption. This completes the proof.

Note that no closedness or continuity assumptions are made on the map providing the search directions. It is only required that the search directions used at each iteration be linearly independent and that as these directions converge, the limiting directions must also be linearly independent. Obviously this holds true if a fixed set of linearly independent search directions are used at every iteration. Alternatively, if the search directions used at each iteration are mutually orthogonal, and each has norm 1, then the search matrix $\mathbf{D}$ satisfies $\mathbf{D}^t\mathbf{D} = \mathbf{I}$. Therefore, det $[\mathbf{D}] = 1$, so that condition 1 of the theorem holds true.

Also note that condition 2 in the statement of the theorem is used to ensure the following property. If a differentiable function f is minmized along n independent directions starting from a point $\mathbf{x}$ and resulting in $\mathbf{x}'$, then $f(\mathbf{x}') < f(\mathbf{x})$, provided that $\nabla f(\mathbf{x}) \neq \mathbf{0}$. Without assumption 2, this is not true, as evidenced by $f(x_1, x_2) = x_2(1 - x_1)$. If $\mathbf{x} = (0, 0)^t$, then minimizing f starting from $\mathbf{x}$ along $\mathbf{d}_1 = (1, 0)^t$ and then along $\mathbf{d}_2 = (0, 1)^t$ could produce the point $\mathbf{x}' = (1, 1)^t$, where $f(\mathbf{x}') = f(\mathbf{x}) = 0$, even though $\nabla f(\mathbf{x}) = (0, 1)^t \neq (0, 0)^t$.

7.4 Comparison Among Algorithms

In the remainder of the text, we discuss several algorithms for solving different classes of nonlinear programming problems. This section discusses some important factors that must be considered when assessing the effectiveness of these algorithms and when comparing them. These factors are (1) generality, reliability, and precision; (2) sensitivity to parameters and data; (3) preparational and computational effort; and (4) convergence.

Generality, Reliability, and Precision

Different algorithms are designed for solving various classes of nonlinear programming problems, such as unconstrained optimization problems, problems with inequality constraints, problems with equality constraints, and problems with both types of constraints. Within each of these classes, different algorithms make specific assumptions about the problem structure. For example, for unconstrained optimization problems, some procedures assume that the objective function is differentiable, whereas other algorithms do not make this assumption and rely primarily on functional evaluations only. For problems with equality constraints, some algorithms can only handle linear constraints, while others can handle nonlinear constraints as well. Thus, generality of an algorithm refers to the variety of problems that the algorithm can handle and also to the restrictiveness of the assumptions required by the algorithm.

Another important factor is the reliability, or robustness, of the algorithm. Given any algorithm, it is not difficult to construct a test problem that it cannot solve effectively. *Reliability*, or *robustness*, means the ability of the procedure to solve most of the problems in the class for which it is designed with reasonable accuracy. Usually, this characteristic should hold regardless of the starting (feasible) solution used. The relationship between reliability of a certain procedure and the problem size and structure cannot be overlooked. Some algorithms are reliable if the number of variables is small or if the constraints are not highly nonlinear, and not reliable otherwise.

As implied by Theorem 7.2.3, convergence of nonlinear programming algorithms usually occurs in a limiting sense, if at all. Thus, we are interested in measuring the quality of the points produced by the algorithm after a reasonable number of iterations. Algorithms that quickly produce feasible solutions with good objective values are preferred. As discussed in Chapter 6 on duality and as will be seen in Chapter 9 on penalty functions, several procedures generate a sequence of infeasible solutions, where feasibility is achieved only at termination. Hence, at later iterations, it is imperative that the degree of infeasibility be small so that a near-feasible solution will be at hand if the algorithmic process is prematurely terminated.

Sensitivity to Parameters and Data

For most algorithms, the user must set initial values for certain parameters, such as the starting vector, the step size, the acceleration factor, and parameters for terminating the algorithm. Some procedures are quite sensitive to these parameters and to the problem data and may produce different results or stop prematurely, depending on their values. In particular, for a fixed set of selected parameters, the algorithm should solve the problem for a wide range of problem data and should be *scale invariant*, that is, insensitive to any constraint or variable scaling that might be used. Likewise, for a given set of problem data, one would prefer that the algorithm not be very sensitive to selected values of the parameters. (See Section 1.3 for a related discussion.)

Preparational and Computational Effort

Another basis for comparing algorithms is the total effort, both preparational and computational, expended for solving problems. The effort of preparing the input data should be taken into consideration when evaluating an algorithm. An algorithm that uses first- or second-order derivatives, especially if the original functions are complicated, requires a considerably larger amount of preparation time than one that only uses functional evaluations. The computational effort of an algorithm is usually assessed by the computer time, the number of iterations, or the number of functional evaluations. However, any of these measures, by itself, is not entirely satisfactory. The computer time needed to execute an algorithm depends not only on its efficiency but also on the type of machine used, the character of the measured time, the existing load on the machine, and the efficiency of coding. Also, the number of iterations cannot be used as the only measure of effectiveness of an algorithm because the effort per iteration may vary considerably from one procedure to another. Finally, the number of functional evaluations can be misleading, since it does not measure other operations, such as matrix multiplication, matrix inversion (or *factorization*; see Appendix A.2), and finding suitable directions of movement. In addition, for derivative dependent methods, we have to weigh the evaluation of first- and second-order derivatives against the evaluation of the functions themselves, and their net consequence on algorithmic performance.

Convergence

Theoretical convergence of algorithms to points in the solution set is a highly desirable property. Given two competing algorithms that converge, they could be compared theoretically on the basis of the order or speed of convergence. This notion is defined below.

7.4.1 Definition

Let the sequence $\{r_k\}$ of real numbers converge to $\bar{r}$, and assume that $r_k \neq \bar{r}$ for all k. The *order of convergence* of the sequence is the supremum of the nonnegative numbers p satisfying

$$\overline{\lim_{k\to\infty}} \frac{|r_{k+1} - \bar{r}|}{|r_k - \bar{r}|^p} = \beta < \infty$$

If $p = 1$ and the *convergence ratio* β is less than 1, the sequence is said to have a *linear convergence rate*. Since *asymptotically* we have $|r_{k+1} - \bar{r}| = \beta|r_k - \bar{r}|$, linear convergence is sometimes also referred to as *geometric convergence*, although often this terminology is reserved for situations in which the sequence is truly a geometric sequence. If $p > 1$, or if $p = 1$ and $\beta = 0$, the sequence is said to have *superlinear convergence*. In particular, if $p = 2$ and $\beta < \infty$, the sequence is said to have a second-order, or *quadratic*, rate of convergence.

For example, the sequence r_k of iterates generated by the algorithmic map of Figure 7.1*a* satisfies $r_{k+1} = (r_k + 1)/2$, where $\{r_k\} \to 1$. Hence, $(r_{k+1} - 1) = (r_k - 1)/2$; and so, with $p = 1$, the limit in Definition 7.4.1 is $\beta = \frac{1}{2}$. However, for $p > 1$, this limit is infinity. Consequently, $\{r_k\} \to 1$ only linearly.

On the other hand, suppose that we have $r_{k+1} = 1 + (r_k - 1)/2^k$, for $k = 1, 2 \ldots$, where $r_1 = 4$, say. In lieu of the sequence $\{4, 2.5, 1.75, 1.375, 1.1875, \ldots\}$ obtained above, we now produce the sequence $\{4, 2.5, 1.375, 1.046875, \ldots\}$. The sequence can readily be verified to be converging to unity. However, we now have $|r_{k+1} - 1|/|r_k - 1| = 1/2^k$, which approaches 0 as $k \to \infty$. Hence, $\{r_k\} \to 1$ superlinearly in this case.

If r_k in the above definition represents $\alpha(\mathbf{x}_k)$, the value of the descent function at the kth iteration, then the larger the value of p, the faster is the convergence of the algorithm. If the limit in Definiton 7.4.1 exists, then, for large values of k, we asymptotically have $|r_{k+1} - \bar{r}| = \beta\ |r_k - \bar{r}|^p$, which indicates faster convergence for larger values of p. For the same value of p, the smaller the convergence ratio β, the faster is the convergence. It should be noted, however, that the order of convergence and the ratio of convergence must not be solely used for evaluating algorithms that converge, since they represent the progress of the algorithm only as the number of iterations approach infinity. (See the Notes and References section for reading on *average rates* of convergence that deal with the average progress per step achieved over a large number of iterations, in contrast with the stepwise progress discussed above.

In a likewise manner, we can define convergence rates for a vector sequence $\{\mathbf{x}_k\} \to \bar{\mathbf{x}}$. Again, let us suppose that $\mathbf{x}_k \neq \bar{\mathbf{x}}$ for all k (or, alternatively, for k large enough). We can now define rates of convergence with respect to an error function that measures the separation between $\mathbf{x}_k$ and $\bar{\mathbf{x}}$, typically, the Euclidean distance function $\|\mathbf{x}_k - \bar{\mathbf{x}}\|$. Consequently, in Definition 7.4.1, we simply replace $|r_k - \bar{r}|$ by $\|\mathbf{x}_k - \bar{\mathbf{x}}\|$ for all k. In particular, if there exists a $0 < \rho < 1$ such that $\|\mathbf{x}_{k+1} - \bar{\mathbf{x}}\| \leq \rho\|\mathbf{x}_k - \bar{\mathbf{x}}\|$ for all k, then $\{\mathbf{x}_k\}$ converges to $\bar{\mathbf{x}}$ at a linear rate. On the other hand, if $\|\mathbf{x}_{k+1} - \bar{\mathbf{x}}\| \leq \rho_k\|\mathbf{x}_k - \bar{\mathbf{x}}\|$ for all k, where $\{\rho_k\} \to 0$, then the rate of convergence is superlinear. Note that these are only frequently used interpretations that coincide with Definition 7.4.1, using $\|\mathbf{x}_k - \bar{\mathbf{x}}\|$ in place of $|r_k - \bar{r}|$.

We also point out here that the foregoing rates of convergence, namely, linear, superlinear, quadratic, and so on, are sometimes referred to, respectively, as q-linear, q-superlinear, q-quadratic, and so on. The prefix q stands for the *quotient* taken in Definition 7.4.1 and is used to differ from another weaker type of *r-(root)-order* convergence rate, in which the errors $\|\mathbf{x}_k - \bar{\mathbf{x}}\|$ are only bounded above by elements of some q-order sequence converging to zero (see the Notes and References section).

Another convergence criterion frequently used in comparing algorithms is their ability to effectively minimize quadratic functions. This is used because, near the minimum, a linear approximation to a function is poor, whereas it can be adequately approximated by a quadratic form. Thus, an algorithm that does not perform well for minimizing a quadratic function is unlikely to perform well for a general nonlinear function as we move closer to the optimum.

Exercises

7.1 Which of the following maps are closed and which are not?
   a. $\mathbf{A}(x) = \{y: x^2 + y^2 \leq 1\}$
   b. $\mathbf{A}(\mathbf{x}) = \{\mathbf{y}: \mathbf{x}^t\mathbf{y} \leq 1\}$
   c. $\mathbf{A}(\mathbf{x}) = \{\mathbf{y}: \|\mathbf{y} - \mathbf{x}\| \leq 1\}$
   d. $\mathbf{A}(x) = \begin{cases} \{y : x^2 + y^2 \leq 1\} & \text{if } x \neq 0 \\ [-1, 0] & \text{if } x = 0 \end{cases}$

7.2 Consider the map $\mathbf{A}$, where $\mathbf{A}(\mathbf{x})$ is the nonnegative square root of $\mathbf{x}$. Starting from any positive x, show that the algorithm defined by the map $\mathbf{A}$ converges to $\bar{x} = 1$. (*Hint:* Let $\alpha(x) = |x - 1|$.)

7.3 Which of the following maps are closed and which are not?
   a. $(y_1, y_2) \in \mathbf{A}(x_1, x_2)$ means that $y_1 = x_1 - 1$ and $y_2 \in [x_2 - 1, x_2 + 1]$.
   b. $(y_1, y_2) \in \mathbf{A}(x_1, x_2)$ means that $y_1 = x_1 - 1$ and $y_2 \in [-x_2 + 1, x_2 + 1]$ if $x_2 \geq 0$ and $y_2 \in [x_2 + 1, -x_2 + 1]$ if $x_2 < 0$.
   c. $(y_1, y_2) \in \mathbf{A}(x_1, x_2)$ means that $y_1 \in [x_1 - \|\mathbf{x}\|, x_1 + \|\mathbf{x}\|]$ and $y_2 = x_2$.

7.4 Let X and Y be nonempty closed sets in E_p and E_q, respectively. Show that the point-to-set map $\mathbf{A}: X \to Y$ is closed if and only if the set $Z = \{(\mathbf{x}, \mathbf{y}): \mathbf{x} \in X, \mathbf{y} \in \mathbf{A}(\mathbf{x})\}$ is closed.

7.5 Let X and Y be nonempty closed sets in E_p and E_q, respectively. Let $\mathbf{A}: X \to Y$ and $\mathbf{B}: X \to Y$ be point-to-set maps. The *sum map* $\mathbf{C} = \mathbf{A} + \mathbf{B}$ is defined by $\mathbf{C}(\mathbf{x}) = \{\mathbf{a} + \mathbf{b}: \mathbf{a} \in \mathbf{A}(\mathbf{x}), \mathbf{b} \in \mathbf{B}(\mathbf{x})\}$. Show that if $\mathbf{A}$ and $\mathbf{B}$ are closed and if Y is compact, then $\mathbf{C}$ is closed.

7.6 Let $A: E_n \times E_n \to E_n$ be the point-to-set map defined as follows. Given $\mathbf{x}, \mathbf{z} \in E_n$, then $\mathbf{y} \in \mathbf{A}(\mathbf{x}, \mathbf{z})$ means that $\mathbf{y} = \bar{\lambda}\mathbf{x} + (1 - \bar{\lambda})\mathbf{z}$ for some $\bar{\lambda} \in [0, 1]$ and

$$\|\mathbf{y}\| \leq \|\lambda\mathbf{x} + (1 - \lambda)\mathbf{z}\| \qquad \text{for all } \lambda \in [0, 1]$$

Show that the map $\mathbf{A}$ is closed for each of the following cases:

a. $\| \ \|$ denotes the Euclidean norm, that is, $\|\mathbf{g}\| = (\sum_{i=1}^{n} g_i^2)^{1/2}$
b. $\| \ \|$ denotes the l_1-norm, that is $\|\mathbf{g}\| = \sum_{i=1}^{n} |g_i|$.
c. $\| \ \|$ denotes the sup-norm, that is, $\|\mathbf{g}\| = \text{maximum}_{1 \le i \le n} |g_i|$.

7.7 Let $\mathbf{A}: E_n \times E_1 \to E_n$ be the point-to-set map defined as follows. Given $\mathbf{x} \in E_n$ and $z \in E_1$, then $\mathbf{y} \in \mathbf{A}(\mathbf{x}, z)$ means that $\|\mathbf{y} - \mathbf{x}\| \le z$ and

$$\|\mathbf{y}\| \le \|\mathbf{w}\| \text{ for each } \mathbf{w} \text{ satisfying } \|\mathbf{w} - \mathbf{x}\| \le z$$

Show that the map $\mathbf{A}$ is closed for each of the norms specified in Exercise 7.6.

7.8 Let $\mathbf{A}: E_n \to E_n$ be the point-to-set map defined as follows. Given an $m \times n$ matrix $\mathbf{B}$, an m vector $\mathbf{b}$, and an n vector $\mathbf{x}$, then $\mathbf{y} \in \mathbf{A}(\mathbf{x})$ means that $\mathbf{y}$ is an optimal solution to the problem to minimize $\mathbf{x}^t\mathbf{z}$ subject to $\mathbf{Bz} = \mathbf{b}$, $\mathbf{z} \ge \mathbf{0}$. Show that the map $\mathbf{A}$ is closed.

7.9 Let $\mathbf{A}: E_m \to E_n$ be the point-to-set map defined as follows. Given an $m \times n$ matrix $\mathbf{B}$, an n vector $\mathbf{c}$, and an m vector $\mathbf{x}$, then $\mathbf{y} \in \mathbf{A}(\mathbf{x})$ means that $\mathbf{y}$ is an optimal solution to the problem to minimize $\mathbf{c}^t\mathbf{z}$ subject to $\mathbf{Bz} = \mathbf{x}$, $\mathbf{z} \ge \mathbf{0}$.
a. Show that the map $\mathbf{A}$ is closed at $\mathbf{x}$ if the set $Z = \{\mathbf{z}: \mathbf{Bz} = \mathbf{x}, \mathbf{z} \ge \mathbf{0}\}$ is compact.
b. What are your conclusions if the set Z is not compact?

7.10 Let $\mathbf{A}: E_n \times E_n \to E_1$ be defined as follows. Given $\mathbf{c}, \mathbf{d} \in E_n$, $k \in E_1$, and a compact polyhedral set $X \subseteq E_n$, then $\bar{\lambda} \in \mathbf{A}(\mathbf{c}, \mathbf{d})$ if $\bar{\lambda} = \sup\{\lambda: z(\lambda) \ge k\}$, where $z(\lambda) = \text{minimum}\ \{(\mathbf{c} + \lambda\mathbf{d})^t\mathbf{x}: \mathbf{x} \in X\}$. Show that the point-to-point map $\mathbf{A}$ is closed at $(\mathbf{c}, \mathbf{d})$.

7.11 Let $f: E_n \to E_1$ be a continuous function, and I be a closed bounded interval in E_1. Let $\mathbf{A}: E_n \times E_n \to E_n$ be the point-to-set map defined as follows. Given $\mathbf{x}, \mathbf{d} \in E_n$, where $\mathbf{d} \ne \mathbf{0}$, $\mathbf{y} \in \mathbf{A}(\mathbf{x}, \mathbf{d})$ means that $\mathbf{y} = \mathbf{x} + \bar{\lambda}\mathbf{d}$ for some $\bar{\lambda} \in I$, and furthermore $f(\mathbf{y}) \le f(\mathbf{x} + \lambda\mathbf{d})$ for each $\lambda \in I$.
a. Show that $\mathbf{A}$ is closed at $(\mathbf{x}, \mathbf{d})$.
b. Does the result hold true if $\mathbf{d} = \mathbf{0}$?
c. Does the result hold true if I is not bounded?

7.12 Let λ be a given scalar, and let $f: E_1 \to E_1$ be continuously differentiable. Let $\mathbf{A}: E_1 \to E_1$ be the point-to-point map defined as follows:

$$\mathbf{A}(x) = \begin{cases} x + \lambda & \text{if } f(x + \lambda) < f(x) \\ x - \lambda & \text{if } f(x + \lambda) \ge f(x) \text{ and } f(x - \lambda) < f(x) \\ x & \text{if } f(x + \lambda) \ge f(x) \text{ and } f(x - \lambda) \ge f(x) \end{cases}$$

a. Show that the map $\mathbf{A}$ is closed on

$$\wedge = \{x : f(x + \lambda) \ne f(x) \text{ and } f(x - \lambda) \ne f(x)\}$$

b. Starting from $x_1 = 5.3$ and letting $\lambda = 1$, apply the algorithm defined by the map $\mathbf{A}$ to minimize $f(x) = x^2 - 2x$.
c. Let $\Omega = \{x: |x - \bar{x}| \le \lambda\}$, where $df(\bar{x})/dx = 0$. Verify that if the sequence of points generated by the algorithm is contained in a compact set, then it converges to a point in Ω.
d. Is it possible that the point $\bar{x}$ in part c is a local maximum or a saddle point?

7.13 Let $f: E_1 \to E_1$ be continuously differentiable. Consider the point-to-point map $\mathbf{A}: E_1 \to E_1$ defined as follows, where $f'(x) = df(x)/dx$:

$$\mathbf{A}(x) = x - \frac{f(x)}{f'(x)} \qquad \text{if } f'(x) \ne 0$$

a. Show that $\mathbf{A}$ is closed on the set $\wedge = \{x: f'(x) \ne 0\}$.
b. Let $f(x) = x^2 - 2x - 3$, and apply the above algorithm starting from the point $x_1 = -5$. Note that the algorithm converges to $x = -1$, where $f(-1) = 0$.
c. Starting from $x_1 = \frac{3}{5}$, verify that the algorithm does not converge to a point x, where $f(x) = 0$ for the function defined by $f(x) = x^2 - |x^3|$.
d. The algorithm defined by the closed map $\mathbf{A}$ is sometimes used to find a point where f

is equal to zero. In part b, the algorithm converged, whereas in part c, it did not. Discuss in reference to Theorem 7.2.3.

7.14 The line search map $\mathbf{M}: E_n \times E_n \rightarrow E_n$ defined below is frequently encountered in nonlinear programming algorithms. The vector $\mathbf{y} \in \mathbf{M}(\mathbf{x}, \mathbf{d})$ if it solves the following problem, where $f: E_n \rightarrow E_1$:

$$\begin{array}{ll} \text{Minimize} & f(\mathbf{x} + \lambda\mathbf{d}) \\ \text{subject to} & \mathbf{x} + \lambda\mathbf{d} \geq \mathbf{0} \\ & \lambda \geq 0 \end{array}$$

To show that $\mathbf{M}$ is not closed, a sequence $(\mathbf{x}_k, \mathbf{d}_k)$ converging to $(\mathbf{x}, \mathbf{d})$ and a sequence $\mathbf{y}_k \in \mathbf{M}(\mathbf{x}_k, \mathbf{d}_k)$ converging to $\mathbf{y}$ must be exhibited such that $\mathbf{y} \notin \mathbf{M}(\mathbf{x}, \mathbf{d})$. Given that $\mathbf{x}_1 = (1, 0)^t$, $\mathbf{x}_{k+1}$ is the point on the circle $(x_1 - 1)^2 + (x_2 - 1)^2 = 1$ midway between $\mathbf{x}_k$ and $(0, 1)^t$. The vector $\mathbf{d}_k$ is defined by $(\mathbf{x}_{k+1} - \mathbf{x}_k)/\|\mathbf{x}_{k+1} - \mathbf{x}_k\|$. Letting $f(x_1, x_2) = (x_1 + 2)^2 + (x_2 - 2)^2$, show that

a. The sequence $\{\mathbf{x}_k\}$ converges to $\mathbf{x} = (0, 1)^t$.
b. The vectors $\{\mathbf{d}_k\}$ converge to $\mathbf{d} = (0, 1)^t$.
c. The sequence $\{\mathbf{y}_k\}$ converges to $\mathbf{y} = (0, 1)^t$.
d. The map $\mathbf{M}$ is not closed at $(\mathbf{x}, \mathbf{d})$.

7.15 Let $f: E_n \rightarrow E_1$ be a differentiable function. Consider the following direction-finding map $\mathbf{D}: E_n \rightarrow E_n \times E_n$ that gives the *deflected negative gradient*. Given $\mathbf{x} \geq \mathbf{0}$, then $(\mathbf{x}, \mathbf{d}) \in \mathbf{D}(\mathbf{x})$ means that

$$d_j = \begin{cases} \dfrac{-\partial f(\mathbf{x})}{\partial x_j} & \text{if } x_j > 0, \text{ or } x_j = 0 \text{ and } \dfrac{\partial f(\mathbf{x})}{\partial x_j} \leq 0 \\ 0 & \text{otherwise} \end{cases}$$

Show that $\mathbf{D}$ is not closed. (*Hint:* Let $f(x_1, x_2) = x_1 - x_2$ and consider the sequence $\{x_k\}$ converging to $(0, 1)^t$, where $\mathbf{x}_k = (\frac{1}{k}, 1)^t$.)

7.16 Let $f: E_n \rightarrow E_1$ be a differentiable function. Consider the composite map $\mathbf{A} = \mathbf{MD}$, where $\mathbf{D}: E_n \rightarrow E_n \times E_n$ and $\mathbf{M}: E_n \times E_n \rightarrow E_n$ are defined as follows. Given $\mathbf{x} \geq \mathbf{0}$, then $(\mathbf{x}, \mathbf{d}) \in \mathbf{D}(\mathbf{x})$ means that

$$d_j = \begin{cases} \dfrac{-\partial f(\mathbf{x})}{\partial x_j} & \text{if } x_j > 0, \text{ or if } x_j = 0 \text{ and } \dfrac{\partial f(\mathbf{x})}{\partial x_j} \leq 0 \\ 0 & \text{otherwise} \end{cases}$$

The vector $\mathbf{y} \in \mathbf{M}(\mathbf{x}, \mathbf{d})$ means that $\mathbf{y} = \mathbf{x} + \bar{\lambda}\mathbf{d}$ for some $\bar{\lambda} \geq 0$, and furthermore, $\bar{\lambda}$ solves the problem to minimize $f(\mathbf{x} + \lambda\mathbf{d})$ subject to $\mathbf{x} + \lambda\mathbf{d} \geq \mathbf{0}$, $\lambda \geq 0$.

a. Find an optimal solution to the following problem using the KKT conditions:

$$\begin{array}{ll} \text{Minimize} & x_1^2 + x_2^2 - x_1x_2 + 2x_1 + x_2 \\ \text{subject to} & x_1, x_2 \geq 0 \end{array}$$

b. Starting from point (2, 1), solve the problem in part a using the algorithm defined by the algorithmic map $\mathbf{A}$. Note that the algorithm converges to the optimal solution obtained in part a.

c. Starting from point (0, 0.09, 0), solve the following problem credited to Wolfe [1972] using the algorithm defined by $\mathbf{A}$:

$$\begin{array}{ll} \text{Minimize} & \frac{4}{3}(x_1^2 - x_1x_2 + x_2^2)^{3/4} - x_3 \\ \text{subject to} & x_1, x_2, x_3 \geq 0 \end{array}$$

Note that the sequence generated converges to the point $(0, 0, \bar{x}_3)$, where $\bar{x}_3 = 0.3(1 + 0.5\sqrt{2})$. Using the KKT conditions, show that this point is not an optimal solution.

Note that the algorithm converges to an optimal solution in part b but not in part c. This is because map $\mathbf{A}$ is not closed, as seen in Exercises 7.14 and 7.15.

7.17 This exercise illustrates that a map for a convergent algorithm need not be closed. Consider the following problem:

Minimize x^2
subject to $x \in E_1$
Consider the maps $\mathbf{B}, \mathbf{C}: E_1 \to E_1$ defined as

$$\mathbf{B}(x) = \frac{x}{2} \qquad \text{for all } x$$

$$\mathbf{C}(x) = \begin{cases} x & \text{if } -1 \le x \le 1 \\ x + 1 & \text{if } x < -1 \\ x - 1 & \text{if } x > 1 \end{cases}$$

Let the solution set $\Omega = \{0\}$, and let the descent function $\alpha(x) = x^2$.

a. Show that $\mathbf{B}$ and $\mathbf{C}$ satisfy all the assumptions of Theorem 7.3.4.
b. Verify that the composite map $\mathbf{A} = \mathbf{CB}$ is as given below, and verify that it is not closed:

$$\mathbf{A}(x) = \begin{cases} \frac{x}{2} & \text{if } -2 \le x \le 2 \\ \frac{x}{2} + 1 & \text{if } x < -2 \\ \frac{x}{2} - 1 & \text{if } x > 2 \end{cases}$$

c. Despite the fact that $\mathbf{A}$ is not closed, show that the algorithm defined by $\mathbf{A}$ converges to the point $\bar{x} = 0$, regardless of the starting point.

7.18 In Theorem 7.3.5, we assumed that det $[\mathbf{D}(\mathbf{x})] > \varepsilon > 0$. Could this assumption be replaced by the following?

> At each point $\mathbf{x}_k$ generated by the algorithm, the search directions $\mathbf{d}_1, \ldots,$ $\mathbf{d}_n$ generated by the algorithm are linearly independent.

7.19 Let X be a closed set in E_n, and let $f: E_n \to E_1$ and $\boldsymbol{\beta}: E_n \to E_{m+l}$ be continuous. Show that the point-to-set map $\mathbf{C}: E_{m+l} \to E_n$ defined below is closed:

$$\mathbf{y} \in \mathbf{C}(\mathbf{w}) \text{ if } \mathbf{y} \text{ solves the problem to minimize } f(\mathbf{x}) + \mathbf{w}^t\boldsymbol{\beta}(\mathbf{x}) \\ \text{subject to } \mathbf{x} \in X$$

7.20 This exercise introduces a unified approach to the class of *cutting plane methods* that are frequently used in nonlinear programming. We state the algorithm and then give the assumptions under which the algorithm converges. The symbol $\mathcal{S}$ represents the collection of polyhedral sets in E_p, and Ω is the nonempty solution set in E_q.

A General Cutting Plane Algorithm

Initialization Step
Choose a nonempty polyhedral set $Z_1 \subseteq E_p$, let $k = 1$, and go to the main step.

Main Step

1. Given Z_k, let $\mathbf{w}_k \in \mathbf{B}(Z_k)$, where $\mathbf{B}: \mathcal{S} \to E_q$. If $\mathbf{w}_k \in \Omega$, stop; otherwise, go to step 2.
2. Let $\mathbf{v}_k \in \mathbf{C}(\mathbf{w}_k)$, where $\mathbf{C}: E_q \to E_r$. Let $a: E_r \to E_1$ and $\mathbf{b}: E_r \to E_p$ be continuous functions, and let

$$Z_{k+1} = Z_k \cap \{\mathbf{x}: a(\mathbf{v}_k) + \mathbf{b}^t(\mathbf{v}_k)\mathbf{x} \ge 0\}$$

Replace k by $k + 1$, and repeat step 1.

Convergence of the Cutting Plane Algorithm

Under the following assumptions, either the algorithm stops in a finite number of steps at a point in Ω or it generates the infinite sequence $\{\mathbf{w}_k\}$ such that all of its accumulation points belong to Ω.

1. $\{\mathbf{w}_k\}$ and $\{\mathbf{v}_k\}$ are contained in compact sets in E_q and E_r, respectively.
2. For each Z, if $\mathbf{w} \in \mathbf{B}(Z)$, then $\mathbf{w} \in Z$.
3. $\mathbf{C}$ is a closed map.
4. Given $\mathbf{w} \notin \Omega$ and Z, where $\mathbf{w} \in B(Z)$, then $\mathbf{v} \in \mathbf{C}(\mathbf{w})$ implies that $\mathbf{w} \notin \{\mathbf{x}: a(\mathbf{v}) + \mathbf{b}^t(\mathbf{v})\mathbf{x} \geq 0\}$, and $Z \cap \{\mathbf{x}: a(\mathbf{v}) + \mathbf{b}^t(\mathbf{v})\mathbf{x} \geq 0\} \neq \emptyset$.

Prove the above convergence theorem.

Hint: Let $\{\mathbf{w}_k\}_{\mathcal{K}}$ and $\{\mathbf{v}_k\}_{\mathcal{K}}$ be convergent subsequences with limits $\mathbf{w}$ and $\mathbf{v}$. First, show that for any k, we must have

$$a(\mathbf{v}_k) + \mathbf{b}^t(\mathbf{v}_k)\mathbf{w}_l \geq 0 \qquad \text{for all } l \geq k + 1$$

Taking limits, show that $a(\mathbf{v}) + \mathbf{b}^t(\mathbf{v})\mathbf{w} \geq 0$. This inequality, together with assumptions 3 and 4, imply that $\mathbf{w} \in \Omega$, because otherwise a contradiction can be obtained.

7.21 Consider the dual cutting plane algorithm described in Section 6.4 for maximizing the dual function.

a. Show that the dual cutting plane algorithm is a special form of the general cutting plane algorithm discussed in Exercise 7.20.

b. Verify that the assumptions 1 through 4 of the convergence theorem stated in Exercise 7.20 hold true, so that the dual cutting plane algorithm converges to an optimal solution to the dual problem.

Hint: Referring to Exercise 7.19, note that the map $\mathbf{C}$ is closed.

7.22 This exercise describes the cutting plane algorithm of Kelley [1960] for solving a problem of the following form, where g_i for $i = 1, \ldots, m$ are convex:

$$\begin{array}{lll} \text{Minimize} & \mathbf{c}^t\mathbf{x} & \\ \text{subject to} & g_i(\mathbf{x}) \leq 0 & \text{for } i = 1, \ldots, m \\ & \mathbf{A}\mathbf{x} \leq \mathbf{b} & \end{array}$$

Kelley's Cutting Plane Algorithm

Initialization Step

Let X_1 be a polyhedral set such that $X_1 \supseteq \{\mathbf{x}: g_i(\mathbf{x}) \leq 0 \text{ for } i = 1, \ldots, m\}$. Let $Z_1 = X_1 \cap \{\mathbf{x}: \mathbf{A}\mathbf{x} \leq \mathbf{b}\}$, let $k = 1$, and go to the main step.

Main Step

1. Solve the linear program to minimize $\mathbf{c}^t\mathbf{x}$ subject to $\mathbf{x} \in Z_k$. Let $\mathbf{x}_k$ be an optimal solution. If $\mathbf{g}_i(\mathbf{x}_k) \leq 0$ for all i, stop; $\mathbf{x}_k$ is an optimal solution. Otherwise, go to step 2.
2. Let $g_j(\mathbf{x}_k) = \text{maximum}_{1 \leq i \leq m}\, g_i(\mathbf{x}_k)$ and let

$$Z_{k+1} = Z_k \cap \{\mathbf{x}: g_j(\mathbf{x}_k) + \nabla g_j(\mathbf{x}_k)^t(\mathbf{x} - \mathbf{x}_k) \leq 0\}$$

Replace k by $k + 1$, and repeat step 1.

(Obviously $\nabla g_j(\mathbf{x}_k) \neq \mathbf{0}$, because otherwise, $g_j(\mathbf{x}) \geq g_j(\mathbf{x}_k) + \nabla g_j(\mathbf{x}_k)^t(\mathbf{x} - \mathbf{x}_k) > 0$ for all $\mathbf{x}$, implying that the problem is infeasible.)

a. Apply the above algorithm to solve the following problem:

$$\begin{array}{lll} \text{Minimize} & -3x_1 - x_2 & \\ \text{subject to} & x_1^2 + x_2 + 1 \leq 0 & \\ & x_1 + x_2 \leq 3 & \\ & x_1, \quad x_2 \geq 0 & \end{array}$$

b. Show that Kelley's algorithm is a special case of the general cutting plane algorithm of Exercise 7.20.

c. Show that the above algorithm converges to an optimal solution using the convergence theorem of Exercise 7.20.

d. Consider the problem to minimize $f(\mathbf{x})$ subject to $g_i(\mathbf{x}) \leq 0$ for $i = 1, \ldots, m$ and

$\mathbf{Ax} \le \mathbf{b}$. Show how the problem can be reformulated so that the above algorithm is applicable.

(*Hint:* Consider adding the constraint $f(\mathbf{x}) - z \le 0$.)

7.23 This exercise describes the *supporting hyperplane method* of Veinott [1967] for solving a problem of the following form, where g_i for all i are pseudoconvex and where $g_i(\hat{\mathbf{x}}) < 0$ for $i = 1, \ldots, m$ for some point $\hat{\mathbf{x}} \in E_n$:

Minimize $\mathbf{c}'\mathbf{x}$

subject to $g_i(\mathbf{x}) \le 0$ for $i = 1, \ldots, m$

$\mathbf{Ax} \le \mathbf{b}$

Veinott's Supporting Hyperplane Algorithm

Initialization Step

Let X_1 be a polyhedral set such that $X_1 \supseteq \{\mathbf{x}: g_i(\mathbf{x}) \le 0 \text{ for } i = 1, \ldots, m\}$. Let $Z_1 = X_1 \cap \{\mathbf{x}: \mathbf{Ax} \le \mathbf{b}\}$, let $k = 1$, and go to the main step.

Main Step

1. Solve the linear program to minimize $\mathbf{c}'\mathbf{x}$ subject to $\mathbf{x} \in Z_k$. Let $\mathbf{x}_k$ be an optimal solution. If $g_i(\mathbf{x}_k) \le 0$ for all i, stop; $\mathbf{x}_k$ is an optimal solution to the original problem. Otherwise, go to step 2.
2. Let $\bar{\mathbf{x}}_k$ be the point on the line segment joining $\mathbf{x}_k$ and $\hat{\mathbf{x}}$, and on the boundary of the region $\{\mathbf{x}: g_i(\mathbf{x}) \le 0 \text{ for } i = 1, \ldots, m\}$. Let $g_i(\bar{\mathbf{x}}_k) = 0$ and let

$$Z_{k+1} = Z_k \cap \{\mathbf{x}: \nabla g_i(\bar{\mathbf{x}}_k)'(\mathbf{x} - \bar{\mathbf{x}}_k) \le 0\}$$

Replace k by $k + 1$, and repeat step 1.

(Note that $\nabla g_i(\bar{\mathbf{x}}_k) \ne \mathbf{0}$, because otherwise, by pseudoconvexity of g_j and since $g_j(\bar{\mathbf{x}}_k) = 0$, it follows that $g_j(\mathbf{x}) \ge 0$ for all $\mathbf{x}$, contradicting the fact that $g_j(\hat{\mathbf{x}}) < 0$.)

a. Apply the above algorithm to the problem given in part a of Exercise 7.22.

b. Show that Veinott's algorithm is a special case of the general cutting plane algorithm of Exercise 7.20.

c. Show that the above algorithm converges to an optimal solution using the convergence theorem of Exercise 7.20.

(Note that the above algorithm can handle convex objective functions by reformulating the problem as in part d of Exercise 7.22 above.)

7.24 Let $\mathbf{A}: X \to X$, where $X = \{x: x \ge \frac{1}{2}\}$ and $\mathbf{A}(x) = |\sqrt{x}|$. Use Zangwill's convergence theorem to verify that starting with any point in X, the (entire) sequence generated by $\mathbf{A}$ converges. Explicitly define Ω and α for this case. What is the rate of convergence?

Notes and References

The concept of closed maps is related to that of upper and lower semicontinuity. For a study of this subject, see Berge [1963], Hausdorff [1962], and Meyer [1970, 1976]. Hogan [1973d] studies properties of point-to-set maps from the standpoint of mathematical programming, where a number of different definitions and results are compared and integrated.

Using the notion of a closed map, Zangwill [1969] presents a unified treatment of the subject of convergence of nonlinear programming algorithms. Theorem 7.2.3, which is used to prove convergence of many algorithms, is credited to Zangwill. Polak [1970, 1971] presents several convergence theorems that are related to Theorem 7.2.3. Polak's main theorem applies to a larger number of algorithms than that of Zangwill because of its weaker assumptions. Huard [1975] has proved convergence of some general nonlinear programming algorithms using the notion of closed maps. The results of Polak and Zangwill ensure that all accumulation points of the sequence of points generated by an algorithm belong to a solution set. However, convergence of the complete sequence is not generally guaranteed.

Under the stronger assumption of closedness of the algorithmic map everywhere, and using the concept of a fixed point, Meyer [1976] proved convergence of the complete sequence of iterates to a fixed point. The utility of the result is somewhat limited, however, because many algorithmic maps are not closed at solution points.

In order to apply Theorem 7.2.3 to prove convergence of a given algorithm, we must show closedness of the overall map. Theorem 7.3.2, where the algorithmic map is viewed as the composition of several maps, may be of use here. Another approach would be to prove convergence of the algorithm directly, even though the overall map may be not closed. Theorems 7.3.4 and 7.3.5 prove convergence for two classes of such algorithms. The first relates to an algorithm that can be viewed as the composition of two maps, one of which satisfies the assumptions of Theorem 7.2.3. The second relates to an algorithm that searches along linearly independent directions.

In Section 7.4, the subject of the speed of convergence is briefly introduced. Ortega and Rheinboldt [1970] give a detailed treatment of q-order and r-order convergence rates. The parameters p and β in Definition 7.4.1 determine the order and rate of stepwise convergence to an optimal solution as the solution point is approached. For a discussion on *average convergence* rates, see Luenberger [1984]. Of particular importance is the notion of *superlinear convergence*. A great deal of research has been directed to establish rates of convergence of nonlinear programming algorithms. See Luenberger [1984] and the Notes and References section at the end of Chapter 8.

There is a class of methods for solving nonlinear programming problems that use cutting planes. An example of such a procedure is given in Section 6.4. Zangwill [1969] presents a unified treatment of cutting plane algorithms. A general theorem showing convergence of such algorithms is presented in Exercise 7.20. Exercises 7.21, 7.22, and 7.23, respectively, deal with convergence of the dual cutting plane method, Kelley's [1960] cutting plane algorithm, and Veinott's [1967] supporting hyperplane algorithm.

Chapter 8 Unconstrained Optimization

Unconstrained optimization deals with the problem of minimizing or maximizing a function in the absence of any restrictions. In this chapter, we discuss both the minimization of a function of one variable and a function of several variables. Even though most practical optimization problems have side restrictions that must be satisfied, the study of techniques for unconstrained optimization is important for several reasons. Many algorithms solve a constrained problem by converting it into a sequence of unconstrained problems via Lagrangian multipliers, as illustrated in Chapter 6, or via penalty and barrier functions, as will be discussed in Chapter 9. Furthermore, most methods proceed by finding a direction and then minimizing along this direction. This line search is equivalent to minimizing a function of one variable without constraints or with simple constraints, such as lower and upper bounds on the variables. Finally, several unconstrained optimization techniques can be extended in a natural way to provide and motivate solution procedures for constrained problems.

The following is an outline of the chapter:

Section 8.1: Line Search Without Using Derivatives We discuss several procedures for minimizing strictly quasiconvex functions of one variable without using derivatives. Uniform search, dichotomous search, the golden section method, and the Fibonacci method are covered.

Section 8.2: Line Search Using Derivatives Differentiability is assumed, and the bisecting search method and Newton's method are discussed.

Section 8.3: Some Practical Line Search Methods We describe the popular quadratic fit line search method and present the Armijo Rule for performing acceptable, inexact line searches.

Section 8.4: Closedness of the Line Search Algorithmic Map We show that the line search algorithmic map is closed, a property that is essential in convergence analyses. Readers who are not interested in convergence may skip this section.

Section 8.5: Multidimensional Search Without Using Derivatives The cyclic coordinate method, the method of Hooke and Jeeves, and Rosenbrock's method are discussed. Convergence of these methc ds is also established.

Section 8.6: Multidimensional Search Using Derivatives We develop the steepest descent method and the method of Newton and analyze their convergence properties.

Section 8.7: Modification of Newton's Method: Levenberg–Marquardt and Trust-Region Methods We describe different variants of Newton's method based on the Levenberg–Marquardt and trust region methods that ensure the global convergence of Newton's method. We also discuss some insightful connections between these methods.

Section 8.8: Methods Using Conjugate Directions: Quasi-Newton and Conjugate Gradient Methods The important concept of conjugacy is introduced. If the objective function is quadratic, then methods using conjugate directions are shown to converge in a finite number of steps. Various quasi-Newton/variable metric and conjugate gradient methods are covered based on the concept of conjugate directions, and their computational performance and convergence properties are discussed.

Section 8.9: Subgradient Optimization Methods We introduce the extension of the steepest descent algorithm to that of minimizing convex, nondifferentiable functions via subgradient-based directions. Variants of this technique that are related to conjugate gradient and variable metric methods are mentioned, and the crucial step of selecting appropriate step sizes in practice is discussed.

8.1 Line Search Without Using Derivatives

One-dimensional search is the backbone of many algorithms for solving a nonlinear programming problem. Many nonlinear programming algorithms proceed as follows. Given a point $\mathbf{x}_k$, find a direction vector $\mathbf{d}_k$ and then a suitable step size λ_k, yielding a new point $\mathbf{x}_{k+1} = \mathbf{x}_k + \lambda_k\mathbf{d}_k$; the process is then repeated. Finding the step size λ_k involves solving the subproblem to minimize $f(\mathbf{x}_k + \lambda\mathbf{d}_k)$, which is a one-dimensional search problem in the variable λ. The minimization may be over all real λ, nonnegative λ, or λ such that $\mathbf{x}_k + \lambda\mathbf{d}_k$ is feasible.

Consider a function θ of one variable λ to be minimized. One approach to minimizing θ is to set the derivative θ' equal to 0 and then solve for λ. Note, however, that θ is usually defined implicitly in terms of a function f of several variables. In particular, given the vectors $\mathbf{x}$ and $\mathbf{d}$, $\theta(\lambda) = f(\mathbf{x} + \lambda\mathbf{d})$. If f is not differentiable, then θ will not be differentiable. If f is differentiable, then $\theta'(\lambda) = \mathbf{d}^t\nabla f(\mathbf{x} + \lambda\mathbf{d})$. Therefore, to find a point λ with $\theta'(\lambda) = 0$, we have to solve the equation $\mathbf{d}^t\nabla f(\mathbf{x} + \lambda\mathbf{d}) = 0$, which is usually nonlinear in λ. Furthermore, λ satisfying $\theta'(\lambda) = 0$ is not necessarily a minimum; it may be a local minimum, a local maximum, or even a saddle point. For these reasons, and except for some special cases, we avoid minimizing θ by letting its derivative be equal to zero. Instead, we resort to some numerical techniques for minimizing the function θ.

In this section, we discuss several methods that do not use derivatives for minimizing a function θ of one variable over a closed bounded interval. These methods fall under the categories of simultaneous line search and sequential line search problems. In the former case, the candidate points are determined a priori, whereas in the sequential search, the values of the function at the previous iterations are used to determine the succeeding points.

The Interval of Uncertainty

Consider the line search problem to minimize $\theta(\lambda)$ subject to $a \le \lambda \le b$. Since the exact location of the minimum of θ over $[a, b]$ is not known, this interval is called the *interval of uncertainty*. During the search procedure if we can exclude portions of this interval that do not contain the minimum, then the interval of uncertainty is reduced. In general, $[a, b]$ is called the interval of uncertainty if a minimum point $\bar{\lambda}$ lies in $[a, b]$, though its exact value is not known.

Theorem 8.1.1 below shows that if the function θ is strictly quasiconvex, then the interval of uncertainty can be reduced by evaluating θ at two points within the interval.

8.1.1 Theorem

Let θ: $E_1 \to E_1$ be strictly quasiconvex over the interval $[a, b]$. Let $\lambda, \mu \in [a, b]$ be such that $\lambda < \mu$. If $\theta(\lambda) > \theta(\mu)$, then $\theta(z) \ge \theta(\mu)$ for all $z \in [a, \lambda)$. If $\theta(\lambda) \le \theta(\mu)$, then $\theta(z) \ge \theta(\lambda)$ for all $z \in (\mu, b]$.

Proof

Suppose that $\theta(\lambda) > \theta(\mu)$, and let $z \in [a, \lambda)$. By contradiction, suppose that $\theta(z) < \theta(\mu)$. Since λ can be written as a convex combination of z and μ, and by the strict quasiconvexity of θ, we have

$$\theta(\lambda) < \max\{\theta(z), \theta(\mu)\} = \theta(\mu)$$

contradicting $\theta(\lambda) > \theta(\mu)$. Hence, $\theta(z) \ge \theta(\mu)$. The second part of the theorem can be proved similarly.

From the above theorem, under strict quasiconvexity if $\theta(\lambda) > \theta(\mu)$, the new interval of uncertainty is $[\lambda, b]$. On the other hand, if $\theta(\lambda) \le \theta(\mu)$, the new interval of uncertainty is $[a, \mu]$. These two cases are illustrated in Figure 8.1.

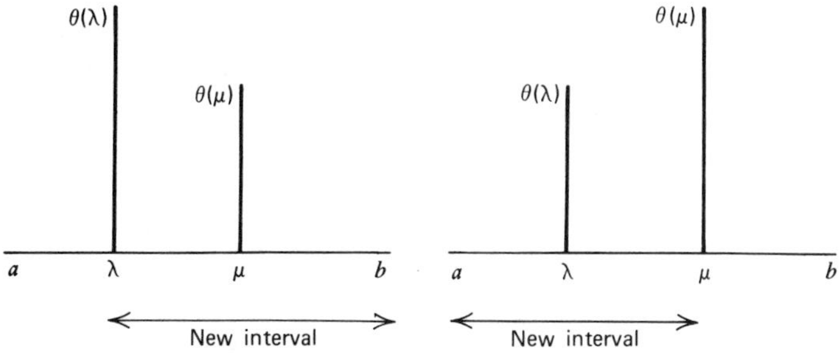

Figure 8.1 Reducing the interval of uncertainty.

Literature on nonlinear programming frequently uses the concept of *strict unimodality* of θ to reduce the interval of uncertainty (see Exercise 3.56). In this book, we are using the equivalent concept of strict quasiconvexity. (See Exercises 3.53, 3.56, and 8.4 for definitions of various forms of unimodality and their relationships with different forms of quasiconvexity.)

We now present several procedures for minimizing a strictly quasiconvex function over a closed bounded interval by iteratively reducing the interval of uncertainty.

An Example of a Simultaneous Search: Uniform Search

Uniform search is an example of simultaneous search, where we decide beforehand the points at which the functional evaluations are to be made. The interval of uncertainty $[a_1, b_1]$ is divided into smaller subintervals via the *grid points* $a_1 + k\delta$ for $k = 1, \ldots, n$, where $b_1 = a_1 + (n + 1)\delta$, as illustrated in Figure 8.2. The function θ is evaluated at each of the n grid points. Let $\hat{\lambda}$ be a grid point with the smallest value of θ. If θ is strictly quasiconvex, it follows that a minimum of θ lies in the interval $[\hat{\lambda} - \delta, \hat{\lambda} + \delta]$.

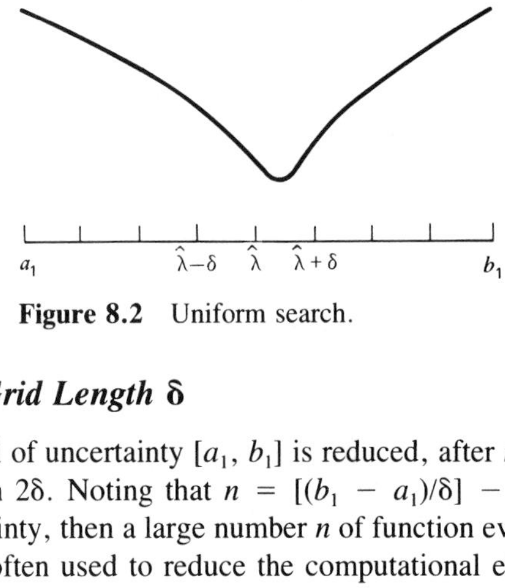

Figure 8.2 Uniform search.

The Choice of the Grid Length δ

We see that the interval of uncertainty $[a_1, b_1]$ is reduced, after n functional evaluations, to an interval of length 2δ. Noting that $n = [(b_1 - a_1)/\delta] - 1$, if we desire a small final interval of uncertainty, then a large number n of function evaluations must be made. One technique that is often used to reduce the computational effort is to utilize a large grid size first and then switch to a finer grid size.

Sequential Search

As may be expected, more efficient procedures that utilize the information generated at the previous iterations in placing the subsequent iterate can be devised. Here, we discuss the following sequential search procedures: dichotomous search, the golden section method, and the Fibonacci method.

Dichotomous Search

Consider $\theta: E_1 \to E_1$ to be minimized over the interval $[a_1, b_1]$. Suppose that θ is strictly quasiconvex. Obviously, the smallest number of functional evaluations that is needed to reduce the interval of uncertainty is two. In Figure 8.3, we consider the location of the two points λ_1 and μ_1. In Figure 8.3*a*, for $\theta = \theta_1$, note that $\theta(\lambda_1) < \theta(\mu_1)$; and, hence,

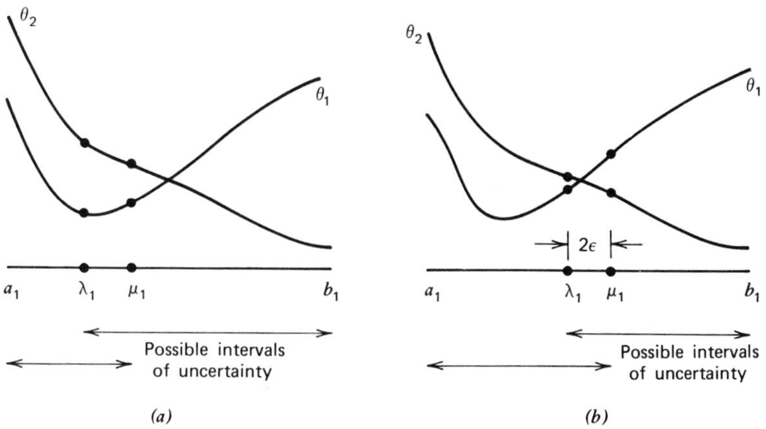

Figure 8.3 Possible intervals of uncertainty.

by Theorem 8.1.1, the new interval of uncertainty is $[a_1, \mu_1]$. However, for $\theta = \theta_2$, note that $\theta(\lambda_1) > \theta(\mu_1)$; and, hence, by Theorem 8.1.1, the new interval of uncertainty is $[\lambda_1, b_1]$. Thus, depending on the function θ, the length of the new interval of uncertainty is equal to $\mu_1 - a_1$ or $b_1 - \lambda_1$.

Note, however, that we do not know, a priori, whether $\theta(\lambda_1) < \theta(\mu_1)$ or $\theta(\lambda_1) > \theta(\mu_1)$.* Thus, the *optimal strategy* is to place λ_1 and μ_1 in such a way as to guard against the worst possible outcome, that is, to minimize the maximum of $\mu_1 - a_1$ and $b_1 - \lambda_1$. This can be accomplished by placing λ_1 and μ_1 at the midpoint of the interval $[a_1, b_1]$. If we do this, however, we would only have one trial point and would not be able to reduce the interval of uncertainty. Therefore, as shown in Figure 8.3*b*, λ_1 and μ_1 are placed symmetrically, each at a distance $\varepsilon > 0$ from the midpoint. Here, $\varepsilon > 0$ is a scalar that is sufficiently small so that the new length of uncertainty $\varepsilon + (b_1 - a_1)/2$ is close enough to the theoretical optimal value of $(b_1 - a_1)/2$ and, in the meantime, would make the functional evaluations $\theta(\lambda_1)$ and $\theta(\mu_1)$ distinguishable.

In dichotomous search, we place each of the first two observations, λ_1 and μ_1, symmetrically at a distance ε from the midpoint $(a_1 + b_1)/2$. Depending on the values of θ at λ_1 and μ_1, a new interval of uncertainty is obtained. The process is then repeated by placing two new observations.

Summary of the Dichotomous Search Method

The following is a summary of the dichotomous method for minimizing a strictly quasiconvex function θ over the interval $[a_1, b_1]$.

Initialization Step Choose the distinguishability constant, $2\varepsilon > 0$, and the allowable final length of uncertainty, $l > 0$. Let $[a_1, b_1]$ be the initial interval of uncertainty, let $k = 1$, and go to the main step.

Main Step

1. If $b_k - a_k < l$, stop; the minimum point lies in the interval $[a_k, b_k]$. Otherwise, consider λ_k and μ_k defined below, and go to step 2.

*If the equality $\theta(\lambda_1) = \theta(\mu_1)$ is true, then the interval of uncertainty can be further reduced to $[\lambda_1, \mu_1]$. It may be noted, however, that exact equality is quite unlikely to occur in practice.

$$\lambda_k = \frac{a_k + b_k}{2} - \varepsilon \qquad \mu_k = \frac{a_k + b_k}{2} + \varepsilon$$

2. If $\theta(\lambda_k) < \theta(\mu_k)$, let $a_{k+1} = a_k$ and $b_{k+1} = \mu_k$. Otherwise, let $a_{k+1} = \lambda_k$ and $b_{k+1} = b_k$. Replace k by $k + 1$, and go to step 1.

Note that the length of uncertainty at the beginning of iteration $k + 1$ is given by

$$(b_{k+1} - a_{k+1}) = \frac{1}{2^k}(b_1 - a_1) + 2\varepsilon\left(1 - \frac{1}{2^k}\right)$$

This formula can be used to determine the number of iterations needed to achieve the desired accuracy. Since each iteration requires two observations, the formula can also be used to determine the number of observations.

The Golden Section Method

To compare the various line search procedures, the following reduction ratio will be of use:

$$\frac{\text{Length of interval of uncertainty after } \nu \text{ observations are taken}}{\text{Length of interval of uncertainty before taking the observations}}$$

Obviously, more efficient schemes correspond to small ratios. In dichotomous search, the above reduction ratio is approximately $(0.5)^{\nu/2}$. We now describe the more efficient golden section method for minimizing a strictly quasiconvex function, whose reduction ratio is given by $(0.618)^{\nu-1}$.

At a general iteration k of the golden section method, let the interval of uncertainty be $[a_k, b_k]$. By Theorem 8.1.1, the new interval of uncertainty $[a_{k+1}, b_{k+1}]$ is given by $[\lambda_k, b_k]$ if $\theta(\lambda_k) > \theta(\mu_k)$ and by $[a_k, \mu_k]$ if $\theta(\lambda_k) \le \theta(\mu_k)$. The points λ_k and μ_k are selected such that

1. The length of the new interval of uncertainty $b_{k+1} - a_{k+1}$ does not depend upon the outcome of the kth iteration, that is, on whether $\theta(\lambda_k) > \theta(\mu_k)$ or $\theta(\lambda_k) \le \theta(\mu_k)$. Therefore, we must have $b_k - \lambda_k = \mu_k - a_k$. Thus, if λ_k is of the form

$$\lambda_k = a_k + (1 - \alpha)(b_k - a_k) \tag{8.1}$$

where $\alpha \in (0, 1)$, then μ_k must be of the form

$$\mu_k = a_k + \alpha(b_k - a_k) \tag{8.2}$$

so that

$$b_{k+1} - a_{k+1} = \alpha(b_k - a_k).$$

2. As λ_{k+1} and μ_{k+1} are selected for the purpose of a new iteration, either λ_{k+1} coincides with μ_k or μ_{k+1} coincides with λ_k. If this can be realized, then during iteration $k + 1$, only one extra observation is needed. To illustrate, consider Figure 8.4 and the following two cases.

Case 1: $\theta(\lambda_k) > \theta(\mu_k)$

In this case, $a_{k+1} = \lambda_k$ and $b_{k+1} = b_k$. To satisfy $\lambda_{k+1} = \mu_k$, and applying (8.1) with k replaced by $k + 1$, we get

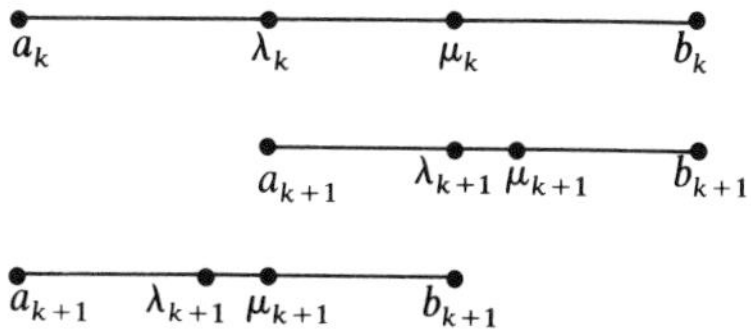

Figure 8.4 Illustration of the golden section rule.

$$\mu_k = \lambda_{k+1} = a_{k+1} + (1 - \alpha)(b_{k+1} - a_{k+1}) = \lambda_k + (1 - \alpha)(b_k - \lambda_k)$$

Substituting the expressions of λ_k and μ_k from (8.1) and (8.2) into the above equation, we get $\alpha^2 + \alpha - 1 = 0$.

Case 2: $\theta(\lambda_k) \leq \theta(\mu_k)$

In this case, $a_{k+1} = a_k$ and $b_{k+1} = \mu_k$. To satisfy $\mu_{k+1} = \lambda_k$, and applying (8.2) with k replaced by $k + 1$, we get

$$\lambda_k = \mu_{k+1} = a_{k+1} + \alpha(b_{k+1} - a_{k+1}) = a_k + \alpha(\mu_k - a_k)$$

Noting (8.1) and (8.2), the above equation gives $\alpha^2 + \alpha - 1 = 0$. The roots of the equation $\alpha^2 + \alpha - 1 = 0$ are $\alpha \cong 0.618$ and $\alpha \cong -1.618$. Since α must be in the interval (0, 1), then $\alpha \cong 0.618$. To summarize, if at iteration k, μ_k and λ_k are chosen according to (8.1) and (8.2), where $\alpha = 0.618$, then the interval of uncertainty is reduced by a factor of 0.618. At the first iteration, two observations are needed at λ_1 and μ_1, but at each subsequent iteration, only one evaluation is needed, since either $\lambda_{k+1} = \mu_k$ or $\mu_{k+1} = \lambda_k$.

Summary of the Golden Section Method

The following is a summary of the golden section method for minimizing a strictly quasiconvex function over the interval $[a_1, b_1]$.

Initialization Step Choose an allowable final length of uncertainty $l > 0$. Let $[a_1, b_1]$ be the initial interval of uncertainty, and let $\lambda_1 = a_1 + (1 - \alpha)(b_1 - a_1)$ and $\mu_1 = a_1 + \alpha(b_1 - a_1)$, where $\alpha = 0.618$. Evaluate $\theta(\lambda_1)$ and $\theta(\mu_1)$, let $k = 1$, and go to the main step.

Main Step

1. If $b_k - a_k < l$, stop; the optimal solution lies in the interval $[a_k, b_k]$. Otherwise, if $\theta(\lambda_k) > \theta(\mu_k)$, go to step 2; and if $\theta(\lambda_k) \leq \theta(\mu_k)$, go to step 3.
2. Let $a_{k+1} = \lambda_k$ and $b_{k+1} = b_k$. Furthermore, let $\lambda_{k+1} = \mu_k$, and let $\mu_{k+1} = a_{k+1} + \alpha(b_{k+1} - a_{k+1})$. Evaluate $\theta(\mu_{k+1})$ and go to step 4.
3. Let $a_{k+1} = a_k$ and $b_{k+1} = \mu_k$. Furthermore, let $\mu_{k+1} = \lambda_k$, and let $\lambda_{k+1} = a_{k+1} + (1 - \alpha)(b_{k+1} - a_{k+1})$. Evaluate $\theta(\lambda_{k+1})$ and go to step 4.
4. Replace k by $k + 1$ and go to step 1.

8.1.2 Example

Consider the following problem:

Minimize $\quad \lambda^2 + 2\lambda$
subject to $\quad -3 \leq \lambda \leq 5$

Table 8.1 Summary of Computations for the Golden Section Method

Iteration k	a_k	b_k	λ_k	μ_k	$\theta(\lambda_k)$	$\theta(\mu_k)$
1	−3.000	5.000	0.056	1.944	0.115*	7.667*
2	−3.000	1.944	−1.112	0.056	−0.987*	0.115
3	−3.000	0.056	−1.832	−1.112	−0.308*	−0.987
4	−1.832	0.056	−1.112	−0.664	−0.987	−0.887*
5	−1.832	−0.664	−1.384	−1.112	−0.853*	−0.987
6	−1.384	−0.664	−1.112	−0.936	−0.987	−0.996*
7	−1.112	−0.664	−0.936	−0.840	−0.996	−0.974*
8	−1.112	−0.840	−1.016	−0.936	−1.000*	−0.996
9	−1.112	−0.936				

Clearly, the function θ to be minimized is strictly quasiconvex, and the initial interval of uncertainty is of length 8. We reduce this interval of uncertainty to one whose length is at most 0.2. The first two observations are located at

$$\lambda_1 = -3 + 0.382(8) = 0.056 \qquad \mu_1 = -3 + 0.618(8) = 1.944$$

Note that $\theta(\lambda_1) < \theta(\mu_1)$. Hence, the new interval of uncertainty is $[-3, 1.944]$. The process is repeated, and the computations are summarized in Table 8.1. The values of θ that are computed at each iteration are indicated by an asterisk. After eight iterations involving nine observations, the interval of uncertainty is $[-1.112, -0.936]$, so that the minimum can be estimated to be the midpoint -1.024. Note that the true minimum is in fact -1.0.

The Fibonacci Search

The Fibonacci method is a line search procedure for minimizing a strictly quasiconvex function θ over a closed bounded interval. Similar to the golden section method, the Fibonacci search procedure makes two functional evaluations at the first iteration and then only one evaluation at each of the subsequent iterations. However, the procedure differs from the golden section method in that the reduction of the interval of uncertainty varies from one iteration to another.

The procedure is based on the Fibonacci sequence $\{F_\nu\}$ defined as follows:

$$\begin{aligned} F_{\nu+1} &= F_\nu + F_{\nu-1} \qquad \nu = 1, 2, \ldots \\ F_0 &= F_1 = 1 \end{aligned} \tag{8.3}$$

The sequence is therefore 1, 1, 2, 3, 5, 8, 13, 21, 34, 55, 89, 144, 233, ... At iteration k, suppose that the interval of uncertainty is $[a_k, b_k]$. Consider the two points λ_k and μ_k given below, where n is the total number of functional evaluations planned.

$$\lambda_k = a_k + \frac{F_{n-k-1}}{F_{n-k+1}}(b_k - a_k) \qquad k = 1, \ldots, n-1 \tag{8.4}$$

$$\mu_k = a_k + \frac{F_{n-k}}{F_{n-k+1}}(b_k - a_k) \qquad k = 1, \ldots, n-1 \tag{8.5}$$

By Theorem 8.1.1, the new interval of uncertainty $[a_{k+1}, b_{k+1}]$ is given by $[\lambda_k, b_k]$

if $\theta(\lambda_k) > \theta(\mu_k)$ and is given by $[a_k, \mu_k]$ if $\theta(\lambda_k) \le \theta(\mu_k)$. In the former case, noting (8.4) and letting $\nu = n - k$ in (8.3), we get

$$\begin{aligned} b_{k+1} - a_{k+1} &= b_k - \lambda_k \\ &= b_k - a_k - \frac{F_{n-k-1}}{F_{n-k+1}}(b_k - a_k) \\ &= \frac{F_{n-k}}{F_{n-k+1}}(b_k - a_k) \end{aligned} \tag{8.6}$$

In the latter case, noting (8.5), we get

$$b_{k+1} - a_{k+1} = \mu_k - a_k = \frac{F_{n-k}}{F_{n-k+1}}(b_k - a_k) \tag{8.7}$$

Thus, in either case, the interval of uncertainty is reduced by the factor F_{n-k}/F_{n-k+1}.

We now show that at iteration $k + 1$, either $\lambda_{k+1} = \mu_k$ or $\mu_{k+1} = \lambda_k$, so that only one functional evaluation is needed. Suppose that $\theta(\lambda_k) > \theta(\mu_k)$. Then, by Theorem 8.1.1, $a_{k+1} = \lambda_k$ and $b_{k+1} = b_k$. Thus, applying (8.4) with k replaced by $k + 1$, we get

$$\begin{aligned} \lambda_{k+1} &= a_{k+1} + \frac{F_{n-k-2}}{F_{n-k}}(b_{k+1} - a_{k+1}) \\ &= \lambda_k + \frac{F_{n-k-2}}{F_{n-k}}(b_k - \lambda_k) \end{aligned}$$

Substituting for λ_k from (8.4), we get

$$\lambda_{k+1} = a_k + \frac{F_{n-k-1}}{F_{n-k+1}}(b_k - a_k) + \frac{F_{n-k-2}}{F_{n-k}}\left(1 - \frac{F_{n-k-1}}{F_{n-k+1}}\right)(b_k - a_k)$$

Letting $\nu = n - k$ in (8.3), it follows that $1 - (F_{n-k-1}/F_{n-k+1}) = F_{n-k}/F_{n-k+1}$. Substituting in the above equation, we get

$$\lambda_{k+1} = a_k + \left(\frac{F_{n-k-1} + F_{n-k-2}}{F_{n-k+1}}\right)(b_k - a_k)$$

Now let $\nu = n - k - 1$ in (8.3), and, noting (8.5), it follows that

$$\lambda_{k+1} = a_k + \frac{F_{n-k}}{F_{n-k+1}}(b_k - a_k) = \mu_k$$

Similarly, if $\theta(\lambda_k) \le \theta(\mu_k)$, the reader can easily verify that $\mu_{k+1} = \lambda_k$. Thus, in either case, only one observation is needed at iteration $k + 1$.

To summarize, at the first iteration, two observations are made, and at each subsequent iteration, only one observation is necessary. Thus, at the end of iteration $n - 2$, we have completed $n - 1$ functional evaluations. Further, for $k = n - 1$, it follows from (8.4) and (8.5) that $\lambda_{n-1} = \mu_{n-1} = \frac{1}{2}(a_{n-1} + b_{n-1})$. Since either $\lambda_{n-1} = \mu_{n-2}$ or $\mu_{n-1} = \lambda_{n-2}$, theoretically no new observations are to be made at this stage. However, in order to further reduce the interval of uncertainty, the last observation is placed slightly to the right or to the left of the midpoint $\lambda_{n-1} = \mu_{n-1}$, so that $\frac{1}{2}(b_{n-1} - a_{n-1})$ is the length of the final interval of uncertainty $[a_n, b_n]$.

Choosing the Number of Observations

Unlike the dichotomous search method and the golden section procedure, the Fibonacci method requires that the total number of observations n be chosen beforehand. This is because the placement of the observations is given by (8.4) and (8.5) and, hence, is dependent on n. From (8.6) and (8.7), the length of the interval of uncertainty is reduced at iteration k by the factor F_{n-k}/F_{n-k+1}. Hence, at the end of $n - 1$ iterations, where n total observations have been made, the length of the interval of uncertainty is reduced from $b_1 - a_1$ to $b_n - a_n = (b_1 - a_1)/F_n$. Therefore, n must be chosen such that $(b_1 - a_1)/F_n$ reflects the accuracy required.

Summary of the Fibonacci Search Method

The following is a summary of the Fibonacci search method for minimizing a strictly quasiconvex function over the interval $[a_1, b_1]$.

Initialization Step Choose an allowable final length of uncertainty $l > 0$ and a distinguishability constant $\varepsilon > 0$. Let $[a_1, b_1]$ be the initial interval of uncertainty, and choose the number of observations n to be taken such that $F_n > (b_1 - a_1)/l$. Let $\lambda_1 = a_1 + (F_{n-2}/F_n)(b_1 - a_1)$ and $\mu_1 = a_1 + (F_{n-1}/F_n)(b_1 - a_1)$. Evaluate $\theta(\lambda_1)$ and $\theta(\mu_1)$, let $k = 1$, and go to the main step.

Main Step

1. If $\theta(\lambda_k) > \theta(\mu_k)$, go to step 2; and if $\theta(\lambda_k) \leq \theta(\mu_k)$, go to step 3.
2. Let $a_{k+1} = \lambda_k$ and $b_{k+1} = b_k$. Furthermore, let $\lambda_{k+1} = \mu_k$, and let $\mu_{k+1} = a_{k+1} + (F_{n-k-1}/F_{n-k})(b_{k+1} - a_{k+1})$. If $k = n - 2$, go to step 5; otherwise, evaluate $\theta(\mu_{k+1})$ and go to step 4.
3. Let $a_{k+1} = a_k$ and $b_{k+1} = \mu_k$. Furthermore, let $\mu_{k+1} = \lambda_k$, and let $\lambda_{k+1} = a_{k+1} + (F_{n-k-2}/F_{n-k})(b_{k+1} - a_{k+1})$. If $k = n - 2$, go to step 5; otherwise, evaluate $\theta(\lambda_{k+1})$ and go to step 4.
4. Replace k by $k + 1$ and go to step 1.
5. Let $\lambda_n = \lambda_{n-1}$, and $\mu_n = \lambda_{n-1} + \varepsilon$. If $\theta(\lambda_n) > \theta(\mu_n)$, let $a_n = \lambda_n$ and $b_n = b_{n-1}$. Otherwise, if $\theta(\lambda_n) \leq \theta(\mu_n)$, let $a_n = a_{n-1}$ and $b_n = \lambda_n$. Stop; the optimal solution lies in the interval $[a_n, b_n]$.

8.1.3 Example

Consider the following problem:

$$\begin{array}{ll} \text{Minimize} & \lambda^2 + 2\lambda \\ \text{subject to} & -3 \leq \lambda \leq 5 \end{array}$$

Note that the function is strictly quasiconvex on the interval and that the true minimum occurs at $\lambda = -1$. We reduce the interval of uncertainty to one whose length is, at most, 0.2. Hence, we must have $F_n > 8/0.2 = 40$, so that $n = 9$. We adopt the distinguishability constant $\varepsilon = 0.01$.

The first two observations are located at

$$\lambda_1 = -3 + \frac{F_7}{F_9}(8) = 0.054545 \qquad \mu_1 = -3 + \frac{F_8}{F_9}(8) = 1.945454$$

Note that $\theta(\lambda_1) < \theta(\mu_1)$. Hence, the new interval of uncertainty is $[-3.000000, 1.945454]$.

Table 8.2 Summary of Computations for the Fibonacci Search Method

Iteration k	a_k	b_k	λ_k	μ_k	$\theta(\lambda_k)$	$\theta(\mu_k)$
1	−3.000000	5.000000	0.054545	1.945454	0.112065*	7.675699*
2	−3.000000	1.945454	−1.109091	0.054545	−0.988099*	0.112065
3	−3.000000	0.054545	−1.836363	−1.109091	−0.300497*	−0.988099
4	−1.836363	0.054545	−1.109091	−0.672727	−0.988099	−0.892892*
5	−1.836363	−0.672727	−1.399999	−1.109091	−0.840001*	−0.988099
6	−1.399999	−0.672727	−1.109091	−0.963636	−0.988099	−0.998677*
7	−1.109091	−0.672727	−0.963636	−0.818182	−0.998677	−0.966942*
8	−1.109091	−0.818182	−0.963636	−0.963636	−0.998677	−0.998677
9	−1.109091	−0.963636	−0.963636	−0.953636	−0.998677	−0.997850*

The process is repeated, and the computations are summarized in Table 8.2. The values of θ that are computed at each iteration are indicated by an asterisk. Note that at $k = 8$, $\lambda_k = \mu_k = \lambda_{k-1}$, so that no functional evaluations are needed at this stage. For $k = 9$, $\lambda_k = \lambda_{k-1} = -0.963636$ and $\mu_k = \lambda_k + \varepsilon = -0.953636$. Since $\theta(\mu_k) > \theta(\lambda_k)$, the final interval of uncertainty $[a_9, b_9]$ is $[-1.109091, -0.963636]$, whose length l is 0.145455. We approximate the minimum to be the midpoint −1.036364. Note from Example 8.1.2 that with the same number of observations $n = 9$, the golden section method gave a final interval of uncertainty whose length is 0.176.

Comparison of Derivative-Free Line Search Methods

Given a function θ that is strictly quasiconvex on the interval $[a_1, b_1]$, obviously each of the methods discussed in this section will yield a point λ in a finite number of steps such that $|\lambda - \bar{\lambda}| \le l$, where l is the length of the final interval of uncertainty and $\bar{\lambda}$ is the minimum point over the interval. In particular, given the length l of the final interval of uncertainty, which reflects the desired degree of accuracy, the required number of observations n can be computed as the smallest positive integer satisfying the following relationships.

Uniform search method: $$n \ge \frac{b_1 - a_1}{l/2} - 1$$

Dichotomous search method: $$(1/2)^{n/2} \ge \frac{l}{b_1 - a_1}$$

Golden section method: $$(0.618)^{n-1} \ge \frac{l}{b_1 - a_1}$$

Fibonacci search method: $$F_n \ge \frac{b_1 - a_1}{l}$$

From the above expressions, we see that the number of observations needed is a function of the ratio $(b_1 - a_1)/l$. Hence, for a fixed ratio $(b_1 - a_1)/l$, the smaller the number of observations required, the more efficient is the algorithm. It should be evident that the most efficient algorithm is the Fibonacci method, followed by the golden section procedure, the dichotomous search method, and finally the uniform search method.

Also note that for n large enough, $1/F_n$ is asymptotic to $(0.618)^{n-1}$, so that the Fibonacci search method and the golden section are almost identical.

It is worth mentioning that among the derivative-free methods that minimize strict quasiconvex functions over a closed bounded interval, the Fibonacci search method is the most efficient in that it requires the smallest number of observations for a given reduction in the length of the interval of uncertainty.

General Functions

The procedures discussed above all rely on the strict quasiconvexity assumption. In many problems, this assumption does not hold true, and, in any case, it cannot be easily verified. One way to handle this difficulty, especially if the initial interval of uncertainty is large, is to divide it into smaller intervals, find the minimum over each subinterval, and then choose the smallest of the minima over the subintervals. Alternatively, one can simply apply the method assuming strict quasiconvexity, and allow the procedure to converge to some local minimum solution.

8.2 Line Search Using Derivatives

In the previous section, we discussed several line search procedures that use functional evaluations. In this section, we discuss the bisection search method and Newton's method, both of which need derivative information.

The Bisection Search Method

Suppose that we wish to minimize a function θ over a closed and bounded interval. Furthermore, suppose that θ is pseudoconvex and hence differentiable. At iteration k, let the interval of uncertainty be $[a_k, b_k]$. Suppose that the derivative $\theta'(\lambda_k)$ is known, and consider the following three possible cases:

1. If $\theta'(\lambda_k) = 0$, then, by pseudoconvexity of θ, λ_k is a minimum point.
2. If $\theta'(\lambda_k) > 0$, then, for $\lambda > \lambda_k$, we have $\theta'(\lambda_k)(\lambda - \lambda_k) > 0$; and, by pseudoconvexity of θ, it follows that $\theta(\lambda) \geq \theta(\lambda_k)$. In other words, the minimum occurs to the left of λ_k, so that the new interval of uncertainty $[a_{k+1}, b_{k+1}]$ is given by $[a_k, \lambda_k]$.
3. If $\theta'(\lambda_k) < 0$, then, for $\lambda < \lambda_k$, $\theta'(\lambda_k)(\lambda - \lambda_k) > 0$, so that $\theta(\lambda) \geq \theta(\lambda_k)$. Thus, the minimum occurs to the right of λ_k, so that the new interval of uncertainty $[a_{k+1}, b_{k+1}]$ is given by $[\lambda_k, b_k]$.

The position of λ_k in the interval $[a_k, b_k]$ must be chosen so that the maximum possible length of the new interval of uncertainty is minimized. That is, λ_k must be chosen so as to minimize the maximum of $\lambda_k - a_k$ and $b_k - \lambda_k$. Obviously, the optimal location of λ_k is the midpoint $\frac{1}{2}(a_k + b_k)$.

To summarize, at any iteration k, θ' is evaluated at the midpoint of the interval of uncertainty. Based on the value of θ', we either stop or construct a new interval of uncertainty whose length is half that at the previous iteration. Note that this procedure is very similar to the dichotomous search method except that at each iteration, only one derivative evaluation is required, as opposed to two functional evaluations for the dichotomous search method. However, the latter is akin to a finite difference derivative evaluation.

Convergence of the Bisection Search Method

Note that the length of the interval of uncertainty after n observations is equal to $(\frac{1}{2})^n(b_1 - a_1)$, so that the method converges to a minimum point within any desired degree of accuracy. In particular, if the length of the final interval of uncertainty is fixed at l, then n must be chosen to be the smallest integer such that $(\frac{1}{2})^n \leq l/(b_1 - a_1)$.

Summary of the Bisection Search Method

We now summarize the bisection search procedure for minimizing a pseudoconvex function θ over a closed and bounded interval.

Initialization Step Let $[a_1, b_1]$ be the initial interval of uncertainty, and let l be the allowable final interval of uncertainty. Let n be the smallest positive integer such that $(\frac{1}{2})^n \leq l/(b_1 - a_1)$. Let $k = 1$ and go to the main step.

Main Step

1. Let $\lambda_k = \frac{1}{2}(a_k + b_k)$ and evaluate $\theta'(\lambda_k)$. If $\theta'(\lambda_k) = 0$, stop; λ_k is an optimal solution. Otherwise, go to step 2 if $\theta'(\lambda_k) > 0$, and go to step 3 if $\theta'(\lambda_k) < 0$.
2. Let $a_{k+1} = a_k$ and $b_{k+1} = \lambda_k$. Go to step 4.
3. Let $a_{k+1} = \lambda_k$ and $b_{k+1} = b_k$. Go to step 4.
4. If $k = n$, stop; the minimum lies in the interval $[a_{n+1}, b_{n+1}]$. Otherwise, replace k by $k + 1$ and repeat step 1.

8.2.1 Example

Consider the following problem:

$$\begin{array}{ll} \text{Minimize} & \lambda^2 + 2\lambda \\ \text{subject to} & -3 \leq \lambda \leq 6 \end{array}$$

Suppose we want to reduce the interval of uncertainty to an interval whose length l is less than or equal to 0.2. Hence, the number of observations n satisfying $(\frac{1}{2})^n \leq l/(b_1 - a_1) = 0.2/9 = 0.0222$ is given by $n = 6$. A summary of the computations using the bisection search method is given in Table 8.3. Note that the final interval of uncertainty is $[-1.0313, -0.8907]$, so that the minimum could be taken as the midpoint -0.961.

Table 8.3 Summary of Computations for the Bisection Search Method

Iteration k	a_k	b_k	λ_k	$\theta'(\lambda_k)$
1	−3.0000	6.0000	1.5000	5.0000
2	−3.0000	1.5000	−0.7500	0.5000
3	−3.0000	−0.7500	−1.8750	−1.7500
4	−1.8750	−0.7500	−1.3125	−0.6250
5	−1.3125	−0.7500	−1.0313	−0.0625
6	−1.0313	−0.7500	−0.8907	0.2186
7	−1.0313	−0.8907		

Newton's Method

Newton's method is based on exploiting the quadratic approximation of the function θ at a given point λ_k. This quadratic approximation q is given by

$$q(\lambda) = \theta(\lambda_k) + \theta'(\lambda_k)(\lambda - \lambda_k) + \tfrac{1}{2}\theta''(\lambda_k)(\lambda - \lambda_k)^2$$

The point λ_{k+1} is taken to be the point where the derivative of q is equal to zero. This yields $\theta'(\lambda_k) + \theta''(\lambda_k)(\lambda_{k+1} - \lambda_k) = 0$, so that

$$\lambda_{k+1} = \lambda_k - \frac{\theta'(\lambda_k)}{\theta''(\lambda_k)} \tag{8.8}$$

The procedure is terminated when $|\lambda_{k+1} - \lambda_k| < \varepsilon$ or when $|\theta'(\lambda_k)| < \varepsilon$, where ε is a prespecified termination scalar.

Note that the above procedure can only be applied for twice-differentiable functions. Furthermore, the procedure is well defined only if $\theta''(\lambda_k) \neq 0$ for each k.

8.2.2 Example

Consider the function θ defined below:

$$\theta(\lambda) = \begin{cases} 4\lambda^3 - 3\lambda^4 & \text{if } \lambda \geq 0 \\ 4\lambda^3 + 3\lambda^4 & \text{if } \lambda < 0 \end{cases}$$

Note that θ is twice-differentiable everywhere. We apply Newton's method, starting from two different points. In the first case, $\lambda_1 = 0.40$; and, as shown in Table 8.4, the procedure produces the point 0.002807 after six iterations. The reader can verify that the procedure indeed converges to the stationary point $\lambda = 0$. In the second case, $\lambda_1 = 0.60$, and the procedure oscillates between the points 0.60 and -0.60, as shown in Table 8.5.

Table 8.4 Summary of Computations for Newton's Method Starting from $\lambda_1 = 0.4$

Iteration k	a_k	b_k	λ_k	$\theta'(\lambda_k)$
1	0.400000	1.152000	3.840000	0.100000
2	0.100000	0.108000	2.040000	0.047059
3	0.047059	0.025324	1.049692	0.022934
4	0.022934	0.006167	0.531481	0.011331
5	0.011331	0.001523	0.267322	0.005634
6	0.005634	0.000379	0.134073	0.002807

Convergence of Newton's Method

The method of Newton, in general, does not converge to a stationary point starting with an arbitrary initial point. Observe that, in general, Theorem 7.2.3 cannot be applied as a result of the unavailability of a descent function. However, as shown in Theorem 8.2.3, if the starting point is sufficiently close to a stationary point, then a suitable descent function can be devised so that the method converges.

Table 8.5 Summary of Computations for Newton's Method Starting from $\lambda_1 = 0.6$

Iteration k	λ_k	$\theta'(\lambda_k)$	$\theta''(\lambda_k)$	λ_{k+1}
1	0.600	1.728	1.440	−0.600
2	−0.600	1.728	−1.440	0.600
3	0.600	1.728	1.440	−0.600
4	−0.600	1.728	−1.440	0.600

8.2.3 Theorem

Let $\theta: E_1 \to E_1$ be continuously twice-differentiable. Consider Newton's algorithm defined by the map $\mathbf{A}(\lambda) = \lambda - \theta'(\lambda)/\theta''(\lambda)$. Let $\bar{\lambda}$ be such that $\theta'(\bar{\lambda}) = 0$ and $\theta''(\bar{\lambda}) \neq 0$. Let the starting point λ_1 be sufficiently close to $\bar{\lambda}$ so that there exists $k_1, k_2 > 0$ with $k_1 k_2 < 1$ such that

1. $$\frac{1}{|\theta''(\lambda)|} \leq k_1$$

2. $$\frac{|\theta(\bar{\lambda}) - \theta'(\lambda) - \theta''(\lambda)(\bar{\lambda} - \lambda)|}{|\bar{\lambda} - \lambda|} \leq k_2$$

for each λ satisfying $|\lambda - \bar{\lambda}| \leq |\lambda_1 - \bar{\lambda}|$. Then, the algorithm converges to $\bar{\lambda}$.

Proof

Let the solution set $\Omega = \{\bar{\lambda}\}$, and let $X = \{\lambda : |\lambda - \bar{\lambda}| \leq |\lambda_1 - \bar{\lambda}|\}$. We prove convergence by using Theorem 7.2.3. Note that X is compact and that the map $\mathbf{A}$ is closed on X. We now show that $\alpha(\lambda) = |\lambda - \bar{\lambda}|$ is indeed a descent function. Now let $\lambda \in X$ and suppose that $\lambda \neq \bar{\lambda}$. Let $\hat{\lambda} \in \mathbf{A}(\lambda)$. Then, by definition of $\mathbf{A}$ and since $\theta'(\bar{\lambda}) = 0$, we get

$$\hat{\lambda} - \bar{\lambda} = (\lambda - \bar{\lambda}) - \frac{1}{\theta''(\lambda)}[\theta'(\lambda) - \theta'(\bar{\lambda})]$$

$$= \frac{1}{\theta''(\lambda)}[\theta'(\bar{\lambda}) - \theta'(\lambda) - \theta''(\lambda)(\bar{\lambda} - \lambda)]$$

Noting the hypothesis of the theorem, it then follows that

$$|\hat{\lambda} - \bar{\lambda}| = \frac{1}{|\theta''(\lambda)|} \frac{|\theta'(\bar{\lambda}) - \theta'(\lambda) - \theta''(\lambda)(\bar{\lambda} - \lambda)|}{|\bar{\lambda} - \lambda|} |\lambda - \bar{\lambda}| \leq k_1 k_2 |\lambda - \bar{\lambda}| < |\lambda - \bar{\lambda}|$$

Therefore, α is indeed a descent function, and the result follows immediately by the corollary to Theorem 7.2.3.

8.3 Some Practical Line Search Methods

In the previous two sections, we have presented various line search methods that either use or do not use derivative-based information. Of these, the Golden Section method (which is a limiting form of Fibonacci's search method) and the bisection method are often applied in practice, sometimes in combination with other methods. However, these methods follow a restrictive pattern of placing subsequent observations and do not accelerate the process by adaptively exploiting information regarding the shape of the function. Although Newton's method tends to do this, it requires second-order derivative information and is not globally convergent. The quadratic fit technique described in the discussion that follows adopts this philosophy, enjoys global convergence under appropriate assumptions such as pseudoconvexity, and is a very popular method.

We remark here that quite often in practice, whenever ill-conditioning effects are experienced with this method, or if it fails to make sufficient progress during an iteration, a switchover to the bisection search procedure is typically made. Such a check for a possible switchover is referred to as a *safeguard technique*.

Quadratic Fit Line Search

Suppose that we are trying to minimize a continuous, strictly quasiconvex function $\theta(\lambda)$ over $\lambda \geq 0$, and assume that we have three points $0 \leq \lambda_1 < \lambda_2 < \lambda_3$ such that $\theta_1 \geq \theta_2$ and $\theta_2 \leq \theta_3$, where $\theta_j \equiv \theta(\lambda_j)$ for $j = 1, 2, 3$. Note that if $\theta_1 = \theta_2 = \theta_3$, then, by the nature of θ, it is easily verified that these must all be minimizing solutions (see Exercise 8.15). Hence, suppose that at least one of the inequalities $\theta_1 > \theta_2$ and $\theta_2 < \theta_3$ holds true. Let us refer to the conditions satisfied by these three points as the *three point pattern (TPP)*. To begin with, we can take $\lambda_1 = 0$ and examine a trial point $\hat{\lambda}$, which might be the step length of a line search at the previous iteration of an algorithm. Let $\hat{\theta} = \theta(\hat{\lambda})$. If $\hat{\theta} \geq \theta_1$, we can set $\lambda_3 = \hat{\lambda}$ and find λ_2 by repeatedly halving the interval $[\lambda_1, \lambda_3]$ until the TPP obtains. On the other hand, if $\hat{\theta} < \theta_1$, we can set $\lambda_2 = \hat{\lambda}$ and find λ_3 by doubling the interval $[\lambda_1, \lambda_2]$ until the TPP obtains.

Now, given the three points (λ_j, θ_j), $j = 1, 2, 3$, we can fit a quadratic curve passing through these points and find its minimizer $\bar{\lambda}$, which must lie in (λ_1, λ_3) by the TPP (see Exercise 8.14). There are three cases to consider. Denote $\bar{\theta} = \theta(\bar{\lambda})$ and let λ_{new} denote the revised set of three points $(\lambda_1, \lambda_2, \lambda_3)$ found as follows:

Case 1: $\bar{\lambda} > \lambda_2$ (see Figure 8.5). If $\bar{\theta} \geq \theta_2$, then we let $\lambda_{new} = (\lambda_1, \lambda_2, \bar{\lambda})$. On the other hand, if $\bar{\theta} \leq \theta_2$, we let $\lambda_{new} = (\lambda_2, \bar{\lambda}, \lambda_3)$. (Note that in case $\bar{\theta} = \theta_2$, either choice is permissible.)

Case 2: $\bar{\lambda} < \lambda_2$. Similar to case 1, if $\bar{\theta} \geq \theta_2$, we let $\lambda_{new} = (\bar{\lambda}, \lambda_2, \lambda_3)$; and if $\bar{\theta} \leq \theta_2$, we let $\lambda_{new} = (\lambda_1, \bar{\lambda}, \lambda_2)$.

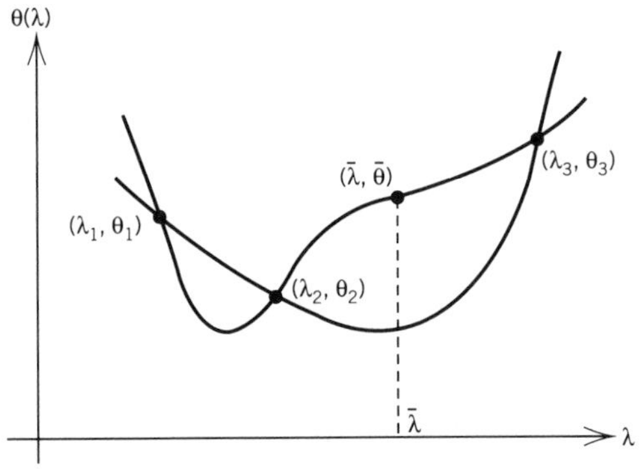

Figure 8.5 Quadratic fit line search.

Case 3: $\bar{\lambda} = \lambda_2$. In this case, we do not have a distinct point to obtain a new TPP. If $(\lambda_3 - \lambda_1) \le \varepsilon$ for some convergence tolerance $\varepsilon > 0$, we stop with λ_2 as the prescribed step length. Otherwise, we place a new observation point $\bar{\lambda}$ at a distance $\varepsilon/2$ away from λ_2 toward λ_1 or λ_3, whichever is further. This yields the situation described by case 1 or case 2 above, and, hence, a new set of points defining λ_{new} may be obtained accordingly.

Again, with respect to λ_{new} if $\theta_1 = \theta_2 = \theta_3$ or if $(\lambda_3 - \lambda_1) \le \varepsilon$ (or if $\theta'(\lambda_2) = 0$ in the differentiable case, or if some other termination criterion such as an acceptable step length in an inexact line search as described next below holds), then we can terminate this process. Otherwise, λ_{new} satisfies the TPP, and the above procedure may be repeated using $\lambda = \lambda_{new}$.

Note that in case 3 of the above procedure, when $\bar{\lambda} = \lambda_2$, the step of placing an observation in the vicinity of λ_2 is akin to evaluating $\theta'(\lambda_2)$ when θ is differentiable. In fact, if we assume that θ is pseudoconvex and continuously twice-differentiable, and we apply a modified version of the foregoing procedure that uses derivatives to represent limiting cases of coincident observation values as described in Exercise 8.16, then we can use Theorem 7.2.3 to demonstrate convergence to an optimal solution, given a starting solution $\lambda = (\lambda_1, \lambda_2, \lambda_3)$ that satisfies the TPP.

Inexact Line Searches: Armijo's Rule

Very often, in practice, we cannot afford the luxury of performing an exact line search because of the expense of excessive function evaluations, even if we terminate with some small accuracy tolerance $\varepsilon > 0$. On the other hand, if we sacrifice accuracy, then we might impair the convergence of the overall algorithm that iteratively employs such a line search. However, if we adopt a line search that guarantees a sufficient degree of accuracy or descent in the function value in a well-defined sense, then this might induce the overall algorithm to converge. Below, we describe one popular definition of an acceptable step length known as Armijo's Rule, and refer the reader to the Notes and References section and Exercise 8.17 for other such inaccurate line search criteria.

Armijo's Rule is driven by two parameters, $0 < \varepsilon < 1$ and $\alpha > 1$, which respectively manage the acceptable step length from being too large or too small. (Typical values are $\varepsilon = 0.2$ and $\alpha = 2$.) Suppose that we are minimizing some differentiable function $f: E_n \to E_1$ at the point $\bar{\mathbf{x}} \in E_n$ in the direction $\mathbf{d} \in E_n$, where $\nabla f(\bar{\mathbf{x}})^t\mathbf{d} < 0$. Hence, $\mathbf{d}$ is a descent direction. Define the line search function $\theta: E_1 \to E_1$ as $\theta(\lambda) = f(\bar{\mathbf{x}} + \lambda\mathbf{d})$ for $\lambda \ge 0$. Then, the first-order approximation of θ at $\lambda = 0$ is given by $\theta(0) + \lambda\theta'(0)$ and is depicted in Figure 8.6. Now, define

$$\hat{\theta}(\lambda) = \theta(0) + \lambda\varepsilon\theta'(0) \qquad \text{for } \lambda \ge 0$$

A step length $\bar{\lambda}$ is considered to be acceptable, provided that $\theta(\bar{\lambda}) \le \hat{\theta}(\bar{\lambda})$. However, to prevent $\bar{\lambda}$ from being too small, Armijo's Rule also requires that $\theta(\alpha\bar{\lambda}) > \hat{\theta}(\alpha\bar{\lambda})$. This gives an acceptable range for $\bar{\lambda}$, as shown in Figure 8.6.

Frequently, Armijo's Rule is adopted in the following manner. A fixed step length parameter $\bar{\lambda}$ is chosen. If $\theta(\bar{\lambda}) \le \hat{\theta}(\bar{\lambda})$, then either $\bar{\lambda}$ is itself selected as the step size, or $\bar{\lambda}$ is sequentially-doubled (assuming $\alpha = 2$) to find the largest integer $t \ge 0$ for which $\theta(2^t\bar{\lambda}) \le \hat{\theta}(2^t\bar{\lambda})$. On the other hand, if $\theta(\bar{\lambda}) > \hat{\theta}(\bar{\lambda})$, then $\bar{\lambda}$ is sequentially halved to find the smallest integer $t \ge 1$ for which $\theta(\bar{\lambda}/2^t) \le \hat{\theta}(\bar{\lambda}/2^t)$. Later, in Section 8.6, we shall analyze the convergence of a steepest descent algorithm that employs such a line search criterion.

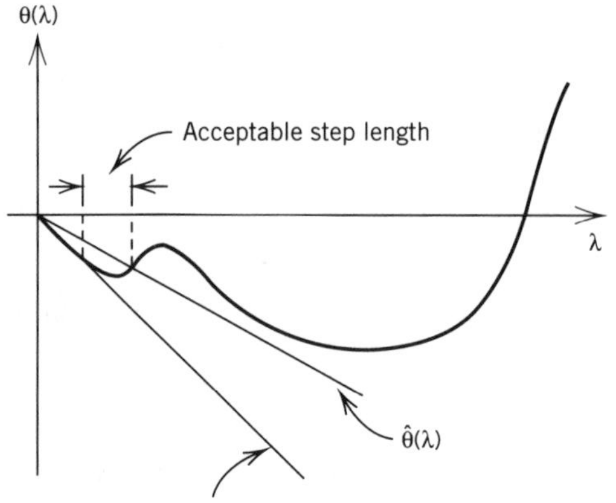

Figure 8.6 Armijo's Rule.

8.4 Closedness of the Line Search Algorithmic Map

In the previous three sections, we discussed several procedures for minimizing a function of one variable. Since the one-dimensional search is a component of most nonlinear programming algorithms, we show in this section that line search procedures define a closed map.

Consider the line search problem to minimize $\theta(\lambda)$ subject to $\lambda \in L$, where $\theta(\lambda) = f(\mathbf{x} + \lambda\mathbf{d})$ and L is a closed interval in E_1. This line search problem can be defined by the algorithmic map $\mathbf{M}: E_n \times E_n \rightarrow E_n$ defined by

$$\mathbf{M}(\mathbf{x}, \mathbf{d}) = \{\mathbf{y}: \mathbf{y} = \mathbf{x} + \bar{\lambda}\mathbf{d} \text{ for some } \bar{\lambda} \in L, \text{ and } f(\mathbf{y}) \le f(\mathbf{x} + \lambda\mathbf{d}) \text{ for each } \lambda \in L\}$$

Note that $\mathbf{M}$ is generally a point-to-set map because there can be more than one minimizing point $\mathbf{y}$. Theorem 8.4.1 below shows that the map $\mathbf{M}$ is closed. Thus, if the map $\mathbf{D}$ that determines the direction $\mathbf{d}$ is also closed, then, by Theorem 7.3.2 or its corollaries, if the stated additional conditions hold true, the overall algorithmic map $\mathbf{A} = \mathbf{MD}$ is closed.

8.4.1 Theorem

Let $f: E_n \rightarrow E_1$, and let L be a closed interval in E_1. Consider the line search map $\mathbf{M}: E_n \times E_n \rightarrow E_n$ defined by

$$\mathbf{M}(\mathbf{x}, \mathbf{d}) = \{\mathbf{y}: \mathbf{y} = \mathbf{x} + \bar{\lambda}\mathbf{d} \text{ for some } \bar{\lambda} \in L, \text{ and } f(\mathbf{y}) \le f(\mathbf{x} + \lambda\mathbf{d}) \text{ for each } \lambda \in L\}$$

If f is continuous at $\mathbf{x}$, and $\mathbf{d} \neq \mathbf{0}$, then M is closed at $(\mathbf{x}, \mathbf{d})$.

Proof

Suppose that $(\mathbf{x}_k, \mathbf{d}_k) \rightarrow (\mathbf{x}, \mathbf{d})$ and that $\mathbf{y}_k \rightarrow \mathbf{y}$, where $\mathbf{y}_k \in \mathbf{M}(\mathbf{x}_k, \mathbf{d}_k)$. We want to show that $\mathbf{y} \in \mathbf{M}(\mathbf{x}, \mathbf{d})$. First, note that $\mathbf{y}_k = \mathbf{x}_k + \lambda_k\mathbf{d}_k$, where $\lambda_k \in L$. Since $\mathbf{d} \neq \mathbf{0}$, then $\mathbf{d}_k \neq \mathbf{0}$ for k large enough, and, hence, $\lambda_k = \|\mathbf{y}_k - \mathbf{x}_k\|/\|\mathbf{d}_k\|$. Taking the limit as $k \rightarrow \infty$, then $\lambda_k \rightarrow \bar{\lambda}$, where $\bar{\lambda} = \|\mathbf{y} - \mathbf{x}\|/\|\mathbf{d}\|$, and, hence, $\mathbf{y} = \mathbf{x} + \bar{\lambda}\mathbf{d}$. Furthermore, since $\lambda_k \in L$ for each k, and since L is closed, then $\bar{\lambda} \in L$. Now let $\lambda \in L$ and note that $f(\mathbf{y}_k) \le f(\mathbf{x}_k + \lambda\mathbf{d}_k)$ for all k. Taking the limit as $k \rightarrow \infty$ and noting the continuity of f, we conclude that $f(\mathbf{y}) \le f(\mathbf{x} + \lambda\mathbf{d})$. Thus, $\mathbf{y} \in \mathbf{M}(\mathbf{x}, \mathbf{d})$, and the proof is complete.

In nonlinear programming, line search is typically performed over one of the following intervals:

$$L = \{\lambda : \lambda \in E_1\}$$

$$L = \{\lambda : \lambda \geq 0\}$$

$$L = \{\lambda : a \leq \lambda \leq b\}$$

In each of the above cases, L is closed, and the theorem applies.

In the above theorem, we required that the vector **d** be nonzero. Example 8.4.2 below presents a case in which **M** is not closed if $\mathbf{d} = \mathbf{0}$. In most cases, the direction vector **d** is nonzero over points outside the solution set Ω. Thus, **M** is closed at these points, and Theorem 7.2.3 can be applied to prove convergence.

8.4.2 Example

Consider the following problem:

$$\text{Minimize} \quad (x - 2)^4$$

Here, $f(x) = (x - 2)^4$. Now, consider the sequence $(x_k, d_k) = \left(\frac{1}{k}, \frac{1}{k}\right)$. Clearly, x_k converges to $x = 0$ and d_k converges to $d = 0$. Consider the line search map **M** defined in Theorem 8.4.1, where $L = \{\lambda : \lambda \geq 0\}$. The point y_k is obtained by solving the problem to minimize $f(x_k + \lambda d_k)$ subject to $\lambda \geq 0$. The reader can verify that $y_k = 2$ for all k, and so, its limit y equals 2. Note, however, that $\mathbf{M}(0, 0) = \{0\}$, so that $y \notin \mathbf{M}(0, 0)$. This shows that **M** is not closed.

8.5 Multidimensional Search Without Using Derivatives

In this section, we consider the problem of minimizing a function f of several variables without using derivatives. The methods described here proceed in the following manner. Given a vector **x**, a suitable direction **d** is first determined, and then f is minimized from **x** in the direction **d** by one of the techniques discussed earlier in this chapter.

Throughout the book, we are required to solve a line search problem of the form to minimize $f(\mathbf{x} + \lambda\mathbf{d})$ subject to $\lambda \in L$, where L is typically of the form $L = E_1$, $L = \{\lambda : \lambda \geq 0\}$, or $L = \{\lambda : a \leq \lambda \leq b\}$. In the statements of the algorithms, for the purpose of simplicity, we have assumed that a minimum point $\bar{\lambda}$ exists. However, this may not be the case. Here, the optimal objective value of the line search problem may be unbounded, or else the optimal objective value may be finite but not achieved at any particular λ. In the first case, the original problem is unbounded and we may stop. In the latter case, λ could be chosen as $\bar{\lambda}$ such that $f(\mathbf{x} + \bar{\lambda}\mathbf{d})$ is sufficiently close to the value $\inf\{f(\mathbf{x} + \lambda\mathbf{d}) : \lambda \in L\}$.

The Cyclic Coordinate Method

This method uses the coordinate axes as the search directions. More specifically, the method searches along the directions $\mathbf{d}_1, \ldots, \mathbf{d}_n$, where $\mathbf{d}_j$ is a vector of zeros except for a 1 at the jth position. Thus, along the search direction $\mathbf{d}_j$, the variable x_j is changed, while all other variables are kept fixed. The method is illustrated schematically in Figure 8.7 for the problem of Example 8.5.1.

Note that we are assuming here that the minimization is done in order over the dimensions $1, \ldots, n$ at each iteration. In a variant known as the *Aitken Double Sweep Method,* the search is conducted by minimizing over the dimensions $1, \ldots, n$ and then back over the dimensions $n - 1, n - 2, \ldots, 1$. This requires $n - 1$ line searches per iteration. Accordingly, if the function to be minimized is differentiable and its gradient is available, then the *Gauss–Southwell* variant recommends that one select that coordinate direction for minimizing at each step that has the largest magnitude of the partial derivative component. These types of sequential one-dimensional minimizations are sometimes referred to as *Gauss–Seidel* types of iterations, based on the Gauss–Seidel method for solving systems of equations.

Summary of the Cyclic Coordinate Method

We summarize below the cyclic coordinate method for minimizing a function of several variables without using any derivative information. As we show shortly, if the function is differentiable, then the method converges to a stationary point.

As discussed in Section 7.2, several criteria could be used for terminating the algorithm. In the statement of the algorithm below, the termination criterion $\|\mathbf{x}_{k+1} - \mathbf{x}_k\| < \varepsilon$ is used. Obviously, any of the other criteria could be used to stop the procedure.

Initialization Step Choose a scalar $\varepsilon > 0$ to be used for terminating the algorithm, and let $\mathbf{d}_1, \ldots, \mathbf{d}_n$ be the coordinate directions. Choose an initial point $\mathbf{x}_1$, let $\mathbf{y}_1 = \mathbf{x}_1$, let $k = j = 1$, and go to the main step.

Main Step

1. Let λ_j be an optimal solution to the problem to minimize $f(\mathbf{y}_j + \lambda \mathbf{d}_j)$ subject to $\lambda \in E_1$, and let $\mathbf{y}_{j+1} = \mathbf{y}_j + \lambda_j \mathbf{d}_j$. If $j < n$, replace j by $j + 1$, and repeat step 1. Otherwise, if $j = n$, go to step 2.
2. Let $\mathbf{x}_{k+1} = \mathbf{y}_{n+1}$. If $\|\mathbf{x}_{k+1} - \mathbf{x}_k\| < \varepsilon$, then stop. Otherwise, let $\mathbf{y}_1 = \mathbf{x}_{k+1}$, let $j = 1$, replace k by $k + 1$, and repeat step 1.

8.5.1 Example

Consider the following problem:

$$\text{Minimize} \quad (x_1 - 2)^4 + (x_1 - 2x_2)^2$$

Note that the optimal solution to the above problem is (2, 1) with objective value equal to zero. Table 8.6 gives a summary of computations for the cyclic coordinate method starting from the initial point (0, 3). Note that at each iteration, the vectors $\mathbf{y}_2$ and $\mathbf{y}_3$ are obtained by performing a line search in the directions (1, 0) and (0, 1), respectively. Also note that significant progress is made during the first few iterations, whereas much slower progress is made during later iterations. After seven iterations, the point (2.22, 1.11), whose objective value is 0.0023, is reached.

In Figure 8.7, the contours of the objective function are given, and the points generated above by the cyclic coordinate method are shown. Note that at later iterations, slow progress is made because of the short orthogonal movements along the valley indicated by the dotted lines. Later, we shall analyze the convergence rate of steepest

Table 8.6 Summary of Computations for the Cyclic Coordinate Method

Iteration k	$\mathbf{x}_k$ $f(\mathbf{x}_k)$	j	$\mathbf{d}_j$	$\mathbf{y}_j$	λ_j	$\mathbf{y}_{j+1}$
1	(0.00, 3.00)	1	(1.0, 0.0)	(0.00, 3.00)	3.13	(3.13, 3.00)
	52.00	2	(0.0, 1.0)	(3.13, 3.00)	−1.44	(3.13, 1.56)
2	(3.13, 1.56)	1	(1.0, 0.0)	(3.13, 1.56)	−0.50	(2.63, 1.56)
	1.63	2	(0.0, 1.0)	(2.63, 1.56)	−0.25	(2.63, 1.31)
3	(2.63, 1.31)	1	(1.0, 0.0)	(2.63, 1.31)	−0.19	(2.44, 1.31)
	0.16	2	(0.0, 1.0)	(2.44, 1.31)	−0.09	(2.44, 1.22)
4	(2.44, 1.22)	1	(1.0, 0.0)	(2.44, 1.22)	−0.09	(2.35, 1.22)
	0.04	2	(0.0, 1.0)	(2.35, 1.22)	−0.05	(2.35, 1.17)
5	(2.35, 1.17)	1	(1.0, 0.0)	(2.35, 1.17)	−0.06	(2.29, 1.17)
	0.015	2	(0.0, 1.0)	(2.29, 1.17)	−0.03	(2.29, 1.14)
6	(2.29, 1.14)	1	(1.0, 0.0)	(2.29, 1.14)	−0.04	(2.25, 1.14)
	0.007	2	(0.0, 1.0)	(2.25, 1.14)	−0.02	(2.25, 1.12)
7	(2.25, 1.12)	1	(1.0, 0.0)	(2.25, 1.12)	−0.03	(2.22, 1.12)
	0.004	2	(0.0, 1.0)	(2.22, 1.12)	−0.01	(2.22, 1.11)

descent methods. The cyclic coordinate method tends to exhibit a performance characteristic over the n coordinate line searches similar to that of an iteration of the steepest descent method.

Convergence of the Cyclic Coordinate Method

Convergence of the cyclic coordinate method to a stationary point follows immediately from Theorem 7.3.5 under the following assumptions.

1. The minimum of f along any line in E_n is unique.
2. The sequence of points generated by the algorithm is contained in a compact subset of E_n.

Note that the search directions used at each iteration are the coordinate vectors, so that the matrix of search directions $\mathbf{D} = \mathbf{I}$. Obviously, assumption 1 of Theorem 7.3.5 holds true.

As an alternative approach, Theorem 7.2.3 could have been used to prove convergence after showing that the overall algorithmic map is closed at each $\mathbf{x}$ satisfying $\nabla f(\mathbf{x}) \neq \mathbf{0}$. In this case, the descent function α is taken as f itself, and the solution set is $\Omega = \{\mathbf{x}: \nabla f(\mathbf{x}) = \mathbf{0}\}$.

Acceleration Step

We learned from the foregoing analysis that the cyclic coordinate method, when applied to a differentiable function, will converge to a point with zero gradient. In the absence of differentiability, however, the method can stall at a nonoptimal point. As shown in Figure 8.8*a*, searching along any of the coordinate axes at the point $\mathbf{x}_2$ leads to no improvement of the objective function and results in premature termination. The reason

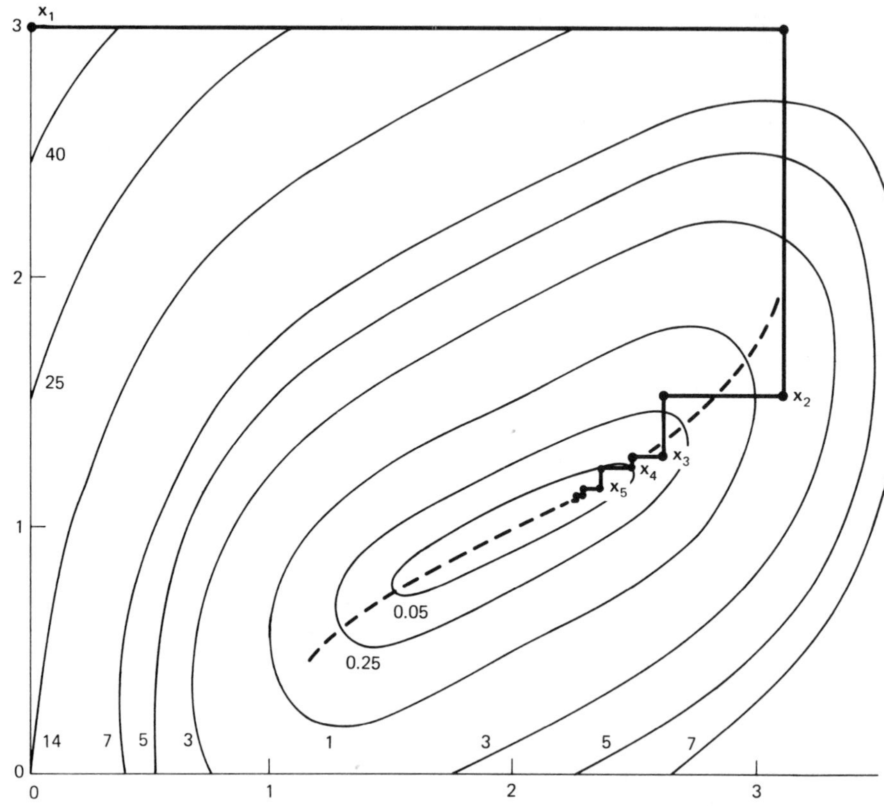

Figure 8.7 Illustration of the cyclic coordinate method.

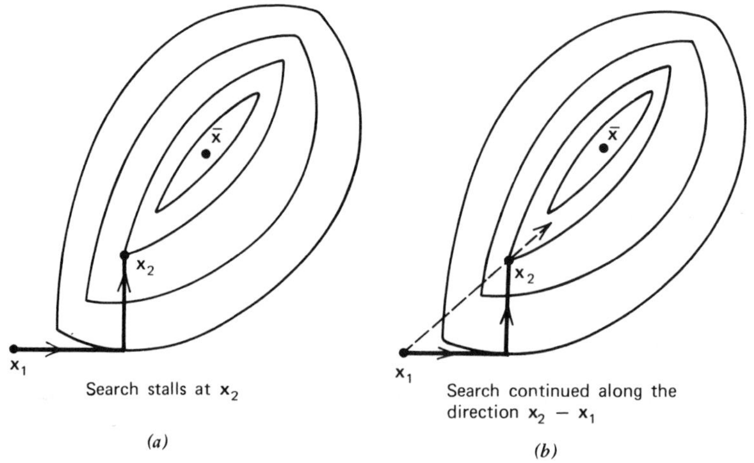

Figure 8.8 Illustration of the effect of a ridge.

for this premature termination is the presence of a valley caused by the nondifferentiability of f. As illustrated in Figure 8.8b, this difficulty could possibly be overcome by searching along the direction $\mathbf{x}_2 - \mathbf{x}_1$.

The search along a direction $\mathbf{x}_{k+1} - \mathbf{x}_k$ is frequently used in applying the cyclic coordinate method, even in the case where f is differentiable. The usual rule of thumb is to apply it at every pth iteration. This modification to the cyclic coordinate method frequently accelerates convergence, particularly when the sequence of points generated zigzags along a valley. Such a step is usually referred to as an *acceleration step*, or a *pattern search step*.

The Method of Hooke and Jeeves

The method of Hooke and Jeeves performs two types of search—exploratory search and pattern search. The first two iterations of the procedure are illustrated in Figure 8.9. Given $\mathbf{x}_1$, an exploratory search along the coordinate directions produces the point $\mathbf{x}_2$. Now a pattern search along the direction $\mathbf{x}_2 - \mathbf{x}_1$ leads to the point $\mathbf{y}$. Another exploratory search starting from $\mathbf{y}$ gives the point $\mathbf{x}_3$. The next pattern search is along the direction $\mathbf{x}_3 - \mathbf{x}_2$, yielding $\mathbf{y}'$. The process is then repeated.

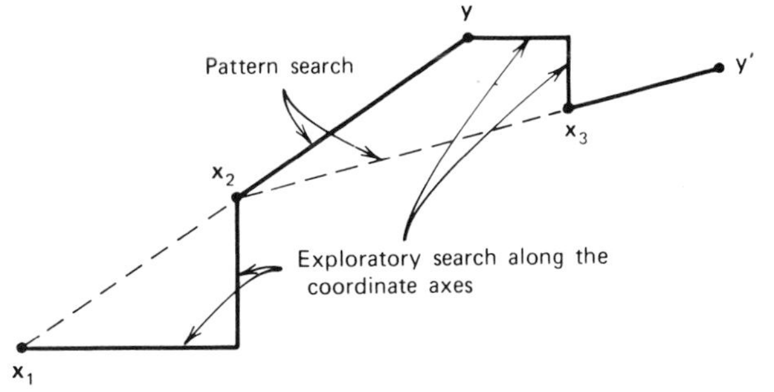

Figure 8.9 Illustration of the method of Hooke and Jeeves.

Summary of the Method of Hooke and Jeeves Using Line Searches

As originally proposed by Hooke and Jeeves, the method does not perform any line search but rather takes discrete steps along the search directions, as will be discussed later. Here, we present a continuous version of the method using line searches along the coordinate directions $\mathbf{d}_1, \ldots, \mathbf{d}_n$ and the pattern direction.

Initialization Step Choose a scalar $\varepsilon > 0$ to be used in terminating the algorithm. Choose a starting point $\mathbf{x}_1$, let $\mathbf{y}_1 = \mathbf{x}_1$, let $k = j = 1$, and go to the main step.

Main Step

1. Let λ_j be an optimal solution to the problem to minimize $f(\mathbf{y}_j + \lambda \mathbf{d}_j)$ subject to $\lambda \in E_1$, and let $\mathbf{y}_{j+1} = \mathbf{y}_j + \lambda_j \mathbf{d}_j$. If $j < n$, replace j by $j + 1$, and repeat step 1. Otherwise, if $j = n$, let $\mathbf{x}_{k+1} = \mathbf{y}_{n+1}$. If $\|\mathbf{x}_{k+1} - \mathbf{x}_k\| < \varepsilon$, stop; otherwise, go to step 2.
2. Let $\mathbf{d} = \mathbf{x}_{k+1} - \mathbf{x}_k$, and let $\hat{\lambda}$ be an optimal solution to the problem to minimize $f(\mathbf{x}_{k+1} + \lambda \mathbf{d})$ subject to $\lambda \in E_1$. Let $\mathbf{y}_1 = \mathbf{x}_{k+1} + \hat{\lambda}\mathbf{d}$, let $j = 1$, replace k by $k + 1$, and repeat step 1.

8.5.2 Example

Consider the following problem:

$$\text{Minimize} \quad (x_1 - 2)^4 + (x_1 - 2x_2)^2$$

Note that the optimal solution is (2.00, 1.00) with objective value equal to zero. Table 8.7 summarizes the computations for the method of Hooke and Jeeves, starting from the initial point (0.00, 3.00). At each iteration, an exploratory search along the coordinate directions gives the points $\mathbf{y}_2$ and $\mathbf{y}_3$, and a pattern search along the direction $\mathbf{d} = \mathbf{x}_{k+1} - \mathbf{x}_k$ gives the point $\mathbf{y}_1$, except at iteration $k = 1$, where $\mathbf{y}_1 = \mathbf{x}_1$. Note that four iterations were required to move from the initial point to the optimal point (2.00, 1.00) whose objective value is zero. At this point, $\|\mathbf{x}_5 - \mathbf{x}_4\| = 0.045$, and the procedure is terminated.

Figure 8.10 illustrates the points generated by the method of Hooke and Jeeves using line searches. Note that the pattern search has substantially improved convergence by moving along a direction that is almost parallel to the valley shown by dotted lines.

Convergence of the Method of Hooke and Jeeves

Suppose that f is differentiable, and let the solution set $\Omega = \{\bar{\mathbf{x}}: \nabla f(\bar{\mathbf{x}}) = \mathbf{0}\}$. Note that each iteration of the method of Hooke and Jeeves consists of an application of the cyclic coordinate method, in addition to a pattern search. Let the cyclic coordinate search be denoted by the map $\mathbf{B}$ and the pattern search be denoted by the map $\mathbf{C}$. Using an argument similar to that of Theorem 7.3.5, it follows that $\mathbf{B}$ is closed. If the minimum of f along any line is unique and letting $\alpha = f$, then $\alpha(\mathbf{y}) < \alpha(\mathbf{x})$ for $\mathbf{x} \notin \Omega$. By definition of $\mathbf{C}$, $\alpha(\mathbf{z}) \le \alpha(\mathbf{y})$ for $\mathbf{z} \in \mathbf{C}(\mathbf{y})$. Assuming that $\Lambda = \{\mathbf{x}: f(\mathbf{x}) \le f(\mathbf{x}_1)\}$, where $\mathbf{x}_1$ is the starting point, is compact, convergence of the procedure is established by Theorem 7.3.4.

The Method of Hooke and Jeeves with Discrete Steps

As mentioned earlier, the method of Hooke and Jeeves, as originally proposed, does not perform line searches but replaces them with a simple scheme involving functional evaluations. A summary of the method is given below.

Initialization Step Let $\mathbf{d}_1, \ldots, \mathbf{d}_n$ be the coordinate directions. Choose a scalar $\varepsilon > 0$ to be used for terminating the algorithm. Furthermore, choose an initial step size, $\Delta \ge \varepsilon$, and an acceleration factor, $\alpha > 0$. Choose a starting point $\mathbf{x}_1$, let $\mathbf{y}_1 = \mathbf{x}_1$, let $k = j = 1$, and go to the main step.

Main Step

1. If $f(\mathbf{y}_j + \Delta\mathbf{d}_j) < f(\mathbf{y}_j)$, the trial is termed a *success*; let $\mathbf{y}_{j+1} = \mathbf{y}_j + \Delta\mathbf{d}_j$, and go to step 2. If, however, $f(\mathbf{y}_j + \Delta\mathbf{d}_j) \ge f(\mathbf{y}_j)$, the trial is deemed a *failure*. In this case, if $f(\mathbf{y}_j - \Delta\mathbf{d}_j) < f(\mathbf{y}_j)$, let $\mathbf{y}_{j+1} = \mathbf{y}_j - \Delta\mathbf{d}_j$, and go to step 2; if $f(\mathbf{y}_j - \Delta\mathbf{d}_j) \ge f(\mathbf{y}_j)$, let $\mathbf{y}_{j+1} = \mathbf{y}_j$, and go to step 2.
2. If $j < n$, replace j by $j + 1$, and repeat step 1. Otherwise, go to step 3 if $f(\mathbf{y}_{n+1}) < f(\mathbf{x}_k)$, and go to step 4 if $f(\mathbf{y}_{n+1}) \ge f(\mathbf{x}_k)$.
3. Let $\mathbf{x}_{k+1} = \mathbf{y}_{n+1}$, and let $\mathbf{y}_1 = \mathbf{x}_{k+1} + \alpha(\mathbf{x}_{k+1} - \mathbf{x}_k)$. Replace k by $k + 1$, let $j = 1$, and go to step 1.
4. If $\Delta \le \varepsilon$, stop; $\mathbf{x}_k$ is the solution. Otherwise, replace Δ by $\Delta/2$. Let $\mathbf{y}_1 = \mathbf{x}_k$, $\mathbf{x}_{k+1} = \mathbf{x}_k$, replace k by $k + 1$, let $j = 1$, and repeat step 1.

The reader may note that steps 1 and 2 above describe an exploratory search. Furthermore, step 3 is an acceleration step along the direction $\mathbf{x}_{k+1} - \mathbf{x}_k$. Note that a

Table 8.7 Summary of Computations for the Method of Hooke and Jeeves Using Line Searches

Iteration k	$\mathbf{x}_k$ / $f(\mathbf{x}_k)$	j	$\mathbf{y}_j$	$\mathbf{d}_j$	λ_j	$\mathbf{y}_{j+1}$	$\mathbf{d}$	$\hat{\lambda}$	$\mathbf{y}_3 + \hat{\lambda}\mathbf{d}$
1	(0.00, 3.00)	1	(0.00, 3.00)	(1.0, 0.0)	3.13	(3.13, 3.00)	—	—	—
	52.00	2	(3.13, 3.00)	(0.0, 1.0)	−1.44	(3.13, 1.56)	(3.13, 1.44)	−0.10	(2.82, 1.70)
2	(3.13, 1.56)	1	(2.82, 1.70)	(1.0, 0.0)	−0.12	(2.70, 1.70)	—	—	—
	1.63	2	(2.70, 1.70)	(0.0, 1.0)	−0.35	(2.70, 1.35)	(−0.43, −0.21)	1.50	(2.06, 1.04)
3	(2.70, 1.35)	1	(2.06, 1.04)	(1.0, 0.0)	−0.02	(2.04, 1.04)	—	—	—
	0.24	2	(2.04, 1.04)	(0.0, 1.0)	−0.02	(2.04, 1.02)	(−0.66, −0.33)	0.06	(2.00, 1.00)
4	(2.04, 1.02)	1	(2.00, 1.00)	(1.0, 0.0)	0.00	(2.00, 1.00)	—	—	—
	0.000003	2	(2.00, 1.00)	(0.0, 1.0)	0.00	(2.00, 1.00)			
5	(2.00, 1.00)								
	0.00								

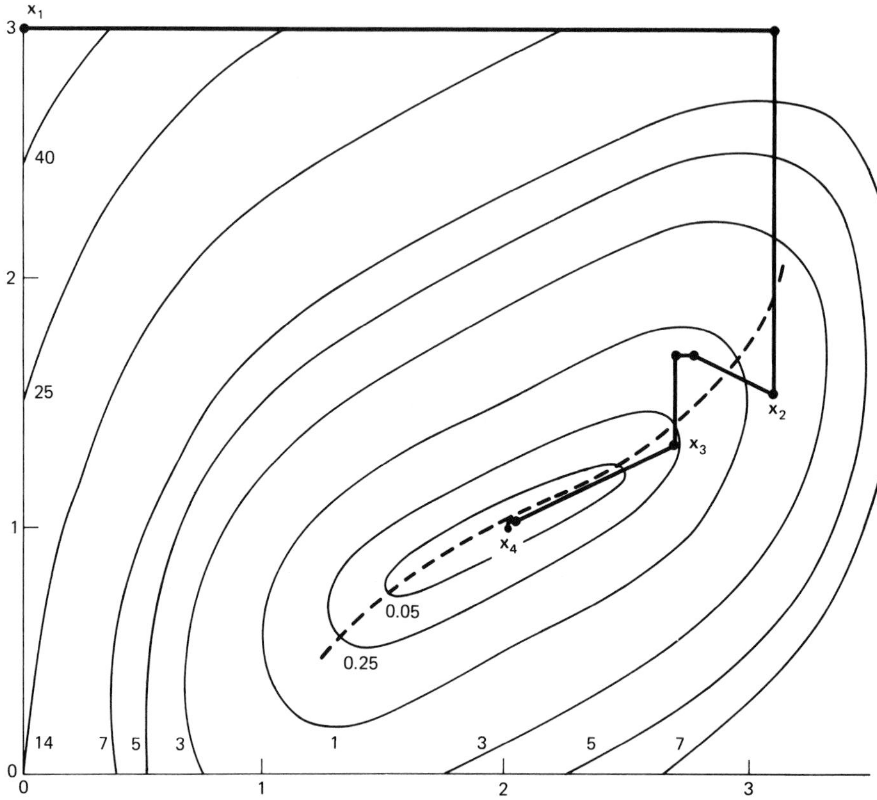

Figure 8.10 Illustration of the method of Hooke and Jeeves using line searches.

decision whether to accept or reject the acceleration step is not made until after an exploratory search is performed. In step 4, the step size Δ is reduced. The procedure could easily be modified so that different step sizes are used along the different directions. This is sometimes adopted for the purpose of scaling.

8.5.3 Example

Consider the following problem:

$$\text{Minimize} \quad (x_1 - 2)^4 + (x_1 - 2x_2)^2$$

We solve the problem using the method of Hooke and Jeeves with discrete steps. The parameters α and Δ are chosen as 1.0 and 0.2, respectively. Figure 8.11 shows the path taken by the algorithm starting from (0.0, 3.0). The points generated are numbered sequentially, and the acceleration step that is rejected is shown by the dotted lines. From this particular starting point, the optimal solution is easily reached.

To give a more comprehensive illustration, Table 8.8 summarizes the computations starting from the new initial point (2.0, 3.0). Here, (S) denotes that the trial is a success and (F) denotes that the trial is a failure. At the first iteration, and at subsequent iterations whenever $f(\mathbf{y}_3) \geq f(\mathbf{x}_k)$, the vector $\mathbf{y}_1$ is taken as $\mathbf{x}_k$. Otherwise, $\mathbf{y}_1 = 2\mathbf{x}_{k+1} - \mathbf{x}_k$. Note that at the end of iteration $k = 10$, the point (1.70, 0.80) is reached with objective value 0.02. The procedure is stopped here with the termination parameter $\varepsilon = 0.1$. If a greater degree of accuracy is required, Δ should be reduced to 0.05.

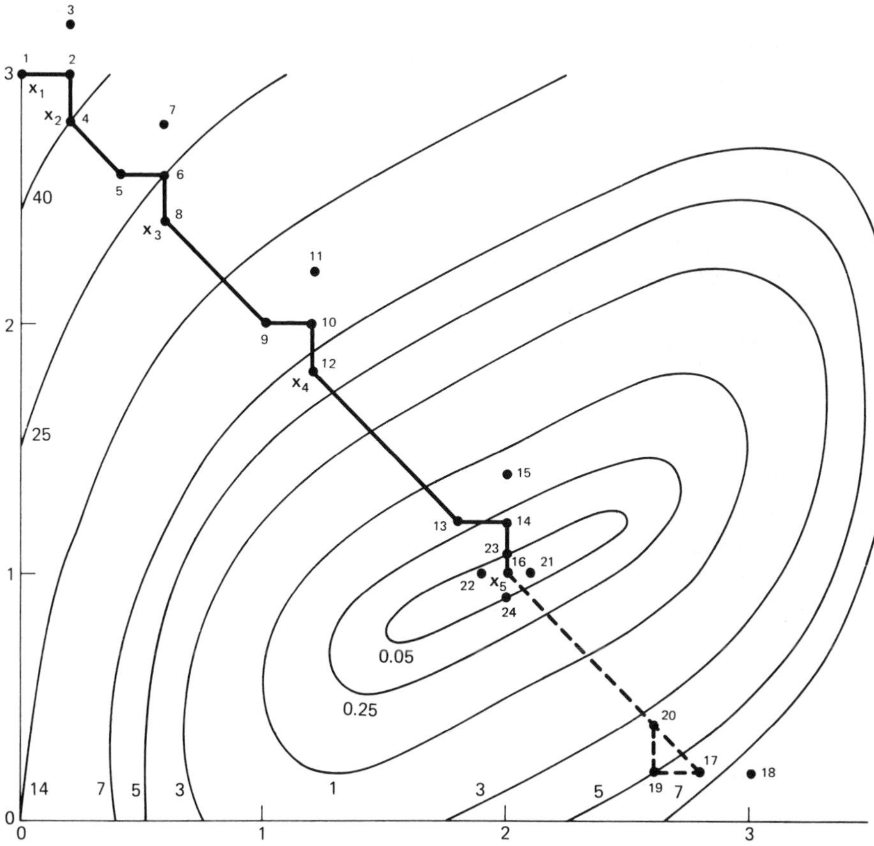

Figure 8.11 Illustration of the method of Hooke and Jeeves using discrete steps starting from (0.0, 3.0). (The numbers denote the order in which points are generated.)

Figure 8.12 illustrates the path taken by the method. The points generated are again numbered sequentially, and dotted lines represent rejected acceleration steps.

The Method of Rosenbrock

As originally proposed, the method of Rosenbrock does not employ line searches but rather takes discrete steps along the search directions. We present here a continuous version of the method that utilizes line searches. At each iteration, the procedure searches iteratively along n linearly independent and orthogonal directions. When a new point is reached at the end of an iteration, a new set of orthogonal vectors is constructed. In Figure 8.13, the new directions are denoted by $\bar{\mathbf{d}}_1$ and $\bar{\mathbf{d}}_2$.

Construction of the Search Directions

Let $\mathbf{d}_1, \ldots, \mathbf{d}_n$ be linearly independent vectors, each with a norm equal to 1. Furthermore, suppose that these vectors are mutually orthogonal, that is, $\mathbf{d}_i^t\mathbf{d}_j = 0$ for $i \neq j$. Starting from the current vector $\mathbf{x}_k$, the objective function f is minimized along each of the directions iteratively, resulting in the point $\mathbf{x}_{k+1}$. In particular, $\mathbf{x}_{k+1} - \mathbf{x}_k = \sum_{j=1}^{n} \lambda_j \mathbf{d}_j$, where λ_j is the distance moved along $\mathbf{d}_j$. The new collection of directions

Table 8.8 Summary of the Computations for the Method of Hooke and Jeeves with Discrete Steps

Iteration k	Δ	$\mathbf{x}_k$ $f(\mathbf{x}_k)$	j	$\mathbf{y}_j$ $f(\mathbf{y}_j)$	$\mathbf{d}_j$	$\mathbf{y}_j + \Delta\mathbf{d}_j$ $f(\mathbf{y}_j + \Delta\mathbf{d}_j)$	$\mathbf{y}_j - \Delta\mathbf{d}_j$ $f(\mathbf{y}_j - \Delta\mathbf{d}_j)$
1	0.2	(2.00, 3.00) 16.00	1	(2.00, 3.00) 16.00	(1.0, 0.0)	(2.20, 3.00) 14.44(S)	—
			2	(2.20, 3.00) 14.44	(0.0, 1.0)	(2.20, 3.20) 17.64(F)	(2.20, 2.80) 11.56(S)
2	0.2	(2.20, 2.80) 11.56	1	(2.40, 2.60) 7.87	(1.0, 0.0)	(2.60, 2.60) 6.89(S)	—
			2	(2.60, 2.60) 6.89	(0.0, 1.0)	(2.60, 2.80) 9.13(F)	(2.60, 2.40) 4.97(S)
3	0.2	(2.60, 2.40) 4.97	1	(3.00, 2.00) 2.00	(1.0, 0.0)	(3.20, 2.00) 2.71(F)	(2.80, 2.00) 1.85(S)
			2	(2.80, 2.00) 1.85	(0.0, 1.0)	(2.80, 2.20) 2.97(F)	(2.80, 1.80) 1.05(S)
4	0.2	(2.80, 1.80) 1.05	1	(3.00, 1.20) 1.36	(1.0, 0.0)	(3.20, 1.20) 2.71(F)	(2.80, 1.20) 0.57(S)
			2	(2.80, 1.20) 0.57	(0.0, 1.0)	(2.80, 1.40) 0.41(S)	—
5	0.2	(2.80, 1.40) 0.41	1	(2.80, 1.00) 1.05	(1.0, 0.0)	(3.00, 1.00) 2.00(F)	(2.60, 1.00) 0.49(S)
			2	(2.60, 1.00) 0.49	(0.0, 1.0)	(2.60, 1.20) 0.17(S)	—

Continued

Table 8.8 *Continued*

6	0.2	(2.60, 1.20) 0.17	1	(2.40, 1.00) 0.19	(1.0, 0.0)	(2.60, 1.00) 0.49(F)	(2.20, 1.00) 0.04(S)
			2	(2.20, 1.00) 0.04	(0.0, 1.0)	(2.20, 1.20) 0.04(F)	(2.20, 0.80) 0.36(F)
7	0.2	(2.20, 1.00) 0.04	1	(1.80, 0.80) 0.04	(1.0, 0.0)	(2.00, 0.80) 0.16(F)	(1.60, 0.80) 0.03(S)
			2	(1.60, 0.80) 0.03	(0.0, 1.0)	(1.60, 1.00) 0.19(F)	(1.60, 0.60) 0.19(F)
8	0.2	(1.60, 0.80) 0.03	1	(1.00, 0.60) 0.67	(1.0, 0.0)	(1.20, 0.60) 0.41(S)	—
			2	(1.20, 0.60) 0.41	(0.0, 1.0)	(1.20, 0.80) 0.57(F)	(1.20, 0.40) 0.57(F)
9	0.1	(1.60, 0.80) 0.03	1	(1.60, 0.80) 0.03	(1.0, 0.0)	(1.70, 0.80) 0.02(S)	—
			2	(1.70, 0.80) 0.02	(0.0, 1.0)	(1.70, 0.90) 0.02(F)	(1.70, 0.70) 0.10(F)
10	0.1	(1.70, 0.80) 0.02	1	(1.80, 0.80) 0.04	(1.0, 0.0)	(1.90, 0.80) 0.09(F)	(1.70, 0.80) 0.02(S)
			2	(1.70, 0.80) 0.02	(0.0, 1.0)	(1.70, 0.90) 0.02(F)	(1.70, 0.70) 0.10(F)

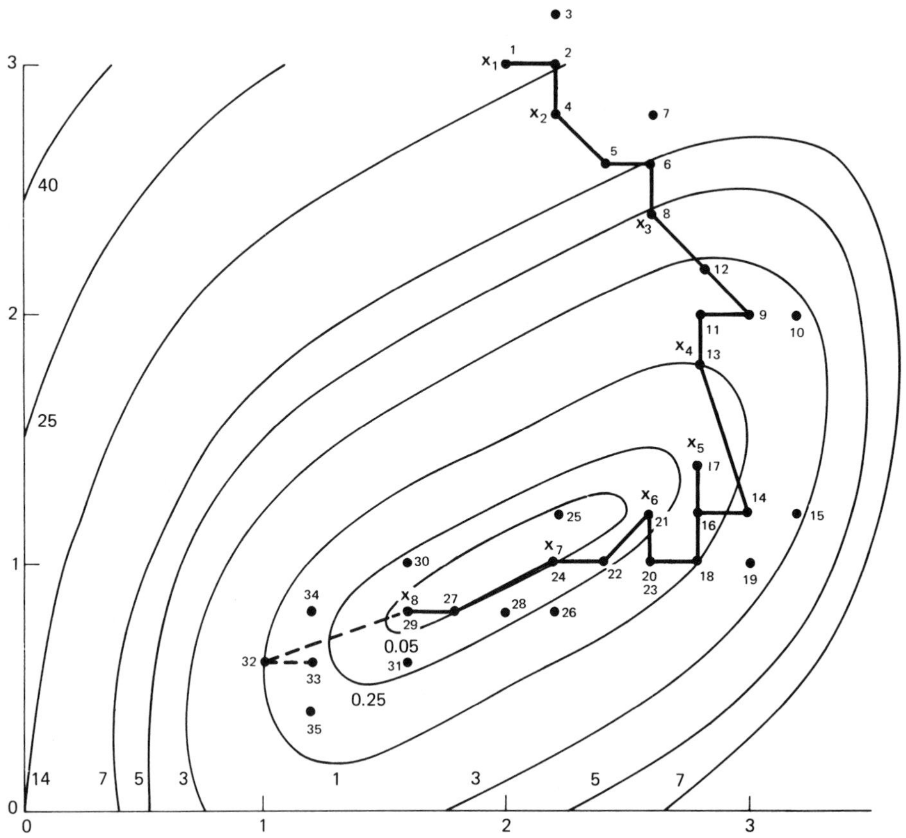

Figure 8.12 Illustration of the method of Rosenbrock using line search.

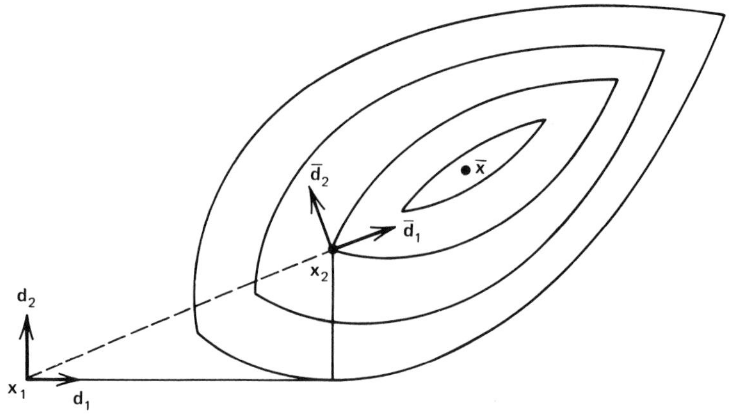

Figure 8.13 Illustration of Rosenbrock's procedure using discrete steps. (The numbers denote the order in which points are generated.)

$\bar{\mathbf{d}}_1, \ldots, \bar{\mathbf{d}}_n$ is formed by the *Gram–Schmidt procedure*, or *orthogonalization procedure*, as follows:

$$\mathbf{a}_j = \begin{cases} \mathbf{d}_j & \text{if } \lambda_j = 0 \\ \sum_{i=j}^{n} \lambda_i \mathbf{d}_i & \text{if } \lambda_j \neq 0 \end{cases}$$

$$\mathbf{b}_j = \begin{cases} \mathbf{a}_j & j = 1 \\ \mathbf{a}_j - \sum_{i=1}^{j-1} (\mathbf{a}_j^t \bar{\mathbf{d}}_i)\bar{\mathbf{d}}_i & j \geq 2 \end{cases} \tag{8.9}$$

$$\bar{\mathbf{d}}_j = \frac{\mathbf{b}_j}{\|\mathbf{b}_j\|}$$

Lemma 8.5.4 below shows that the new directions established by the Rosenbrock procedure are indeed linearly independent and orthogonal.

8.5.4 Lemma

Suppose that the vectors $\mathbf{d}_1, \ldots, \mathbf{d}_n$ are linearly independent and mutually orthogonal. Then the directions $\bar{\mathbf{d}}_1, \ldots, \bar{\mathbf{d}}_n$ defined by (8.9) are also linearly independent, and mutually orthogonal for any set of $\lambda_1, \ldots, \lambda_n$. Furthermore, if $\lambda_j = 0$ and $\bar{\mathbf{d}}_j = \mathbf{d}_j$.

Proof

We first show that $\mathbf{a}_1, \ldots, \mathbf{a}_n$ are linearly independent. Suppose that $\sum_{j=1}^{n} \mu_j \mathbf{a}_j = \mathbf{0}$. Let $I = \{j : \lambda_j = 0\}$, and let $J(j) = \{i : i \notin I, i \leq j\}$. Noting (8.9), we get

$$\begin{aligned} \mathbf{0} = \sum_{j=1}^{n} \mu_j \mathbf{a}_j &= \sum_{j \in I} \mu_j \mathbf{d}_j + \sum_{j \notin I} \mu_j \left(\sum_{i=j}^{n} \lambda_i \mathbf{d}_i \right) \\ &= \sum_{j \in I} \mu_j \mathbf{d}_j + \sum_{j \notin I} \left(\lambda_j \sum_{i \in J(j)} \mu_i \right) \mathbf{d}_j \end{aligned}$$

Since $\mathbf{d}_1, \ldots, \mathbf{d}_n$ are linearly independent, $\mu_j = 0$ for $j \in I$ and $\lambda_j \sum_{i \in J(j)} \mu_i = 0$ for $j \notin I$. But $\lambda_j \neq 0$ for $j \notin I$, and, hence, $\sum_{i \in J(j)} \mu_i = 0$ for each $j \notin I$. By the definition of $J(j)$, we therefore have $\mu_1 = \cdots = \mu_n = 0$, and, hence, $\mathbf{a}_1, \ldots, \mathbf{a}_n$ are linearly independent.

To show that $\mathbf{b}_1, \ldots, \mathbf{b}_n$ are linearly independent, we use the following induction argument. Since $\mathbf{b}_1 = \mathbf{a}_1 \neq \mathbf{0}$, it suffices to show that if $\mathbf{b}_1, \ldots, \mathbf{b}_k$ are linearly independent, then $\mathbf{b}_1, \ldots, \mathbf{b}_k, \mathbf{b}_{k+1}$ are also linearly independent. Suppose that $\sum_{j=1}^{k+1} \alpha_j \mathbf{b}_j = \mathbf{0}$. Using the definition of $\mathbf{b}_{k+1}$ in (8.9), we get

$$\begin{aligned} \mathbf{0} &= \sum_{j=1}^{k} \alpha_j \mathbf{b}_j + \alpha_{k+1} \mathbf{b}_{k+1} \\ &= \sum_{j=1}^{k} \left[\alpha_j - \frac{\alpha_{k+1}(\mathbf{a}_{k+1}^t \bar{\mathbf{d}}_j)}{\|\mathbf{b}_j\|} \right] \mathbf{b}_j + \alpha_{k+1} \mathbf{a}_{k+1} \end{aligned} \tag{8.10}$$

From (8.9), it follows that each vector $\mathbf{b}_j$ is a linear combination of $\mathbf{a}_1, \ldots, \mathbf{a}_j$. Since

$\mathbf{a}_1, \ldots, \mathbf{a}_{k+1}$ are linearly independent, it follows from (8.10) that $\alpha_{k+1} = 0$. Since $\mathbf{b}_1, \ldots, \mathbf{b}_k$ are assumed linearly independent by the induction hypotheses, from (8.10) we get $\alpha_j - \alpha_{k+1}(\mathbf{a}_{k+1}^t\bar{\mathbf{d}}_j)/\|\mathbf{b}_j\| = 0$ for $j = 1, \ldots, k$. Since $\alpha_{k+1} = 0$, $\alpha_j = 0$ for each j. This shows that $\mathbf{b}_1, \ldots, \mathbf{b}_{k+1}$ are linearly independent. By the definition of $\bar{\mathbf{d}}_j$, linear independence of $\bar{\mathbf{d}}_1, \ldots, \bar{\mathbf{d}}_n$ is immediate.

Now we show orthogonality of $\mathbf{b}_1, \ldots, \mathbf{b}_n$ and, hence, orthogonality of $\bar{\mathbf{d}}_1, \ldots, \bar{\mathbf{d}}_n$. From (8.9), $\mathbf{b}_1^t\mathbf{b}_2 = 0$; and, thus, it suffices to show that if $\mathbf{b}_1, \ldots, \mathbf{b}_k$ are mutually orthogonal, then $\mathbf{b}_1, \ldots, \mathbf{b}_k, \mathbf{b}_{k+1}$ are also mutually orthogonal. From (8.10) and noting that $\mathbf{b}_j^t\bar{\mathbf{d}}_i = 0$ for $i \neq j$, it follows that

$$\mathbf{b}_j^t\mathbf{b}_{k+1} = \mathbf{b}_j^t\left[\mathbf{a}_{k+1} - \sum_{i=1}^{k}(\mathbf{a}_{k+1}^t\bar{\mathbf{d}}_i)\bar{\mathbf{d}}_i\right]$$

$$= \mathbf{b}_j^t\mathbf{a}_{k+1} - (\mathbf{a}_{k+1}^t\bar{\mathbf{d}}_j)\mathbf{b}_j^t\bar{\mathbf{d}}_j = 0$$

Thus, $\mathbf{b}_1, \ldots, \mathbf{b}_{k+1}$ are mutually orthogonal.

To complete the proof, we show that $\bar{\mathbf{d}}_j = \mathbf{d}_j$ if $\lambda_j = 0$. From (8.9), if $\lambda_j = 0$, we get

$$\mathbf{b}_j = \mathbf{d}_j - \sum_{i=1}^{j-1}\frac{1}{\|\mathbf{b}_i\|}(\mathbf{d}_j^t\mathbf{b}_i)\bar{\mathbf{d}}_i \tag{8.11}$$

Note that $\mathbf{b}_i$ is a linear combination of $\mathbf{a}_1, \ldots, \mathbf{a}_i$, so that $\mathbf{b}_i = \sum_{r=1}^{i}\beta_{ir}\mathbf{a}_r$. From (8.9), it thus follows that

$$\mathbf{b}_i = \sum_{r\in R}\beta_{ir}\mathbf{d}_r + \sum_{r\in\bar{R}}\beta_{ir}\left(\sum_{s=r}^{n}\lambda_s\mathbf{d}_s\right) \tag{8.12}$$

where $R = \{r: r \leq i, \lambda_r = 0\}$ and $\bar{R} = \{r: r \leq i, \lambda_r \neq 0\}$. Consider $i < j$ and note that $\mathbf{d}_j^t\mathbf{d}_\nu = 0$ for $\nu \neq j$. For $r \in R$, $r \leq i < j$ and hence $\mathbf{d}_j^t\mathbf{d}_r = 0$. For $r \notin R$, $\mathbf{d}_j^t(\sum_{s=r}^{n}\lambda_s\mathbf{d}_s) = \lambda_j\mathbf{d}_j^t\mathbf{d}_j = \lambda_j$. By assumption, $\lambda_j = 0$, and thus multiplying (8.12) by $\mathbf{d}_j^t$ we get $\mathbf{d}_j^t\mathbf{b}_i = 0$ for $i < j$. From (8.11), it follows that $\mathbf{b}_j = \mathbf{d}_j$, and, hence, $\bar{\mathbf{d}}_j = \mathbf{d}_j$. This completes the proof.

From the above theorem, if $\lambda_j = 0$, then the new direction $\bar{\mathbf{d}}_j$ is equal to the old direction $\mathbf{d}_j$. Hence, we only need to compute new directions for those indices with $\lambda_j \neq 0$.

Summary of the Method of Rosenbrock Using Line Searches

We now summarize Rosenbrock's method using line searches for minimizing a function f of several variables. As we shall shortly show, if f is differentiable, then the method converges to a point with zero gradient.

Initialization Step Let $\varepsilon > 0$ be the termination scalar. Choose $\mathbf{d}_1, \ldots, \mathbf{d}_n$ as the coordinate directions. Choose a starting point $\mathbf{x}_1$, let $\mathbf{y}_1 = \mathbf{x}_1$, $k = j = 1$, and go to the main step.

Main Step

1. Let λ_j be an optimal solution to the problem to minimize $f(\mathbf{y}_j + \lambda\mathbf{d}_j)$ subject to $\lambda \in E_1$, and let $\mathbf{y}_{j+1} = \mathbf{y}_j + \lambda_j\mathbf{d}_j$. If $j < n$, replace j by $j + 1$, and repeat step 1. Otherwise, go to step 2.

Table 8.9 Summary of Computations for the Method of Rosenbrock Using Line Searches

Iteration k	$\mathbf{x}_k$ $f(\mathbf{x}_k)$	j	$\mathbf{y}_j$ $f(\mathbf{y}_j)$	$\mathbf{d}_j$	λ_j	$\mathbf{y}_{j+1}$ $f(\mathbf{y}_{j+1})$
1	(0.00, 3.00) 52.00	1	(0.00, 3.00) 52.00	(1.00, 0.00)	3.13	(3.13, 3.00) 9.87
		2	(3.13, 3.00) 9.87	(0.00, 1.00)	−1.44	(3.13, 1.56) 1.63
2	(3.13, 1.56) 1.63	1	(3.13, 1.56) 1.63	(0.91, −0.42)	−0.34	(2.82, 1.70) 0.79
		2	(2.82, 1.70) 0.79	(−0.42, −0.91)	0.51	(2.16, 1.24) 0.16
3	(2.61, 1.24) 0.16	1	(2.61, 1.24) 0.16	(−0.85, −0.52)	0.38	(2.29, 1.04) 0.05
		2	(2.29, 1.04) 0.05	(0.52, −0.85)	−0.10	(2.24, 1.13) 0.004
4	(2.24, 1.13) 0.004	1	(2.24, 1.13) 0.004	(−0.96, −0.28)	0.04	(2.20, 1.12) 0.003
		2	(2.20, 1.12) 0.003	(0.28, −0.96)	0.02	(2.21, 1.10) 0.002

2. Let $\mathbf{x}_{k+1} = \mathbf{y}_{n+1}$. If $\|\mathbf{x}_{k+1} - \mathbf{x}_k\| < \varepsilon$, then stop; otherwise, let $\mathbf{y}_1 = \mathbf{x}_{k+1}$, replace k by $k + 1$, let $j = 1$, and go to step 3.
3. Form a new set of linearly independent orthogonal search directions by (8.9). Denote these new directions by $\mathbf{d}_1, \ldots, \mathbf{d}_n$ and repeat step 1.

8.5.5 Example

Consider the following problem:

$$\text{Minimize} \quad (x_1 - 2)^4 + (x_1 - 2x_2)^2$$

We solve this problem by the method of Rosenbrock using line searches. Table 8.9 summarizes the computations starting from the point (0.00, 3.00). The point $\mathbf{y}_2$ is obtained by optimizing the function along the direction $\mathbf{d}_1$ starting from $\mathbf{y}_1$, and $\mathbf{y}_3$ is obtained by optimizing the function along the direction $\mathbf{d}_2$ starting from $\mathbf{y}_2$. After the first iteration, we have $\lambda_1 = 3.13$ and $\lambda_2 = -1.44$. Using (8.9), the new search directions are (0.91, −0.42) and (−0.42, −0.91). After four iterations, the point (2.21, 1.10) is reached, and the corresponding objective function value is 0.002. We now have $\|\mathbf{x}_4 - \mathbf{x}_3\| = 0.15$, and the procedure is stopped.

In Figure 8.14, the progress of the method is shown. It may be interesting to compare this figure with Figure 8.15, which is given later for the method of Rosenbrock using discrete steps.

Convergence of the Method of Rosenbrock

Note that, according to Lemma 8.5.4, the search directions employed by the method are linearly independent and mutually orthogonal, and each has norm 1. Thus, at any given

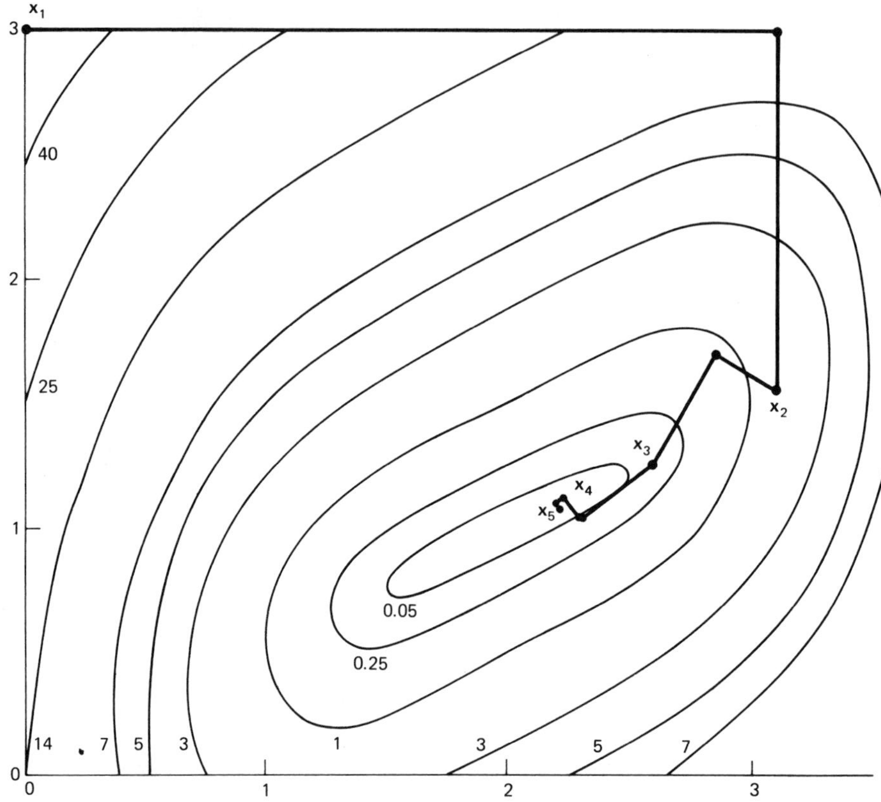

Figure 8.14 Illustration of the method of Rosenbrock using line search.

iteration, the matrix $\mathbf{D}$ denoting the search directions satisfies $\mathbf{D}^t\mathbf{D} = \mathbf{I}$. Thus, det $[\mathbf{D}] = 1$ and, hence, assumption 1 of Theorem 7.3.5 holds. By this theorem, it follows that the method of Rosenbrock using line searches converges to a stationary point if the following assumptions are true:

1. The minimum of f along any line in E_n is unique.
2. The sequence of points generated by the algorithm is contained in a compact subset of E_n.

Rosenbrock's Method with Discrete Steps

As mentioned earlier, the method proposed by Rosenbrock avoids line searches. Instead, functional values are made at specific points. Furthermore, an acceleration feature is incorporated by suitably increasing or decreasing the step lengths as the method proceeds. A summary of the method is given below.

Initialization Step Let $\varepsilon > 0$ be the termination scalar, let $\alpha > 1$ be a chosen expansion factor, and let $\beta \in (-1, 0)$ be a selected contraction factor. Choose $\mathbf{d}_1, \ldots, \mathbf{d}_n$ as the coordinate directions, and let a $\bar{\Delta}_1, \ldots, \bar{\Delta}_n > 0$ be the initial step sizes along these directions. Choose a starting point $\mathbf{x}_1$, let $\mathbf{y}_1 = \mathbf{x}_1$, $k = j = 1$, let $\Delta_j = \bar{\Delta}_j$ for each j, and go to the main step.

Main Step

1. If $f(\mathbf{y}_j + \Delta_j\mathbf{d}_j) < f(\mathbf{y}_j)$, the jth trial is deemed a *success;* set $\mathbf{y}_{j+1} = \mathbf{y}_j + \Delta_j\mathbf{d}_j$, and replace Δ_j by $\alpha\Delta_j$. If, on the other hand, $f(\mathbf{y}_j + \Delta_j\mathbf{d}_j) \geq f(\mathbf{y}_j)$, the trial is

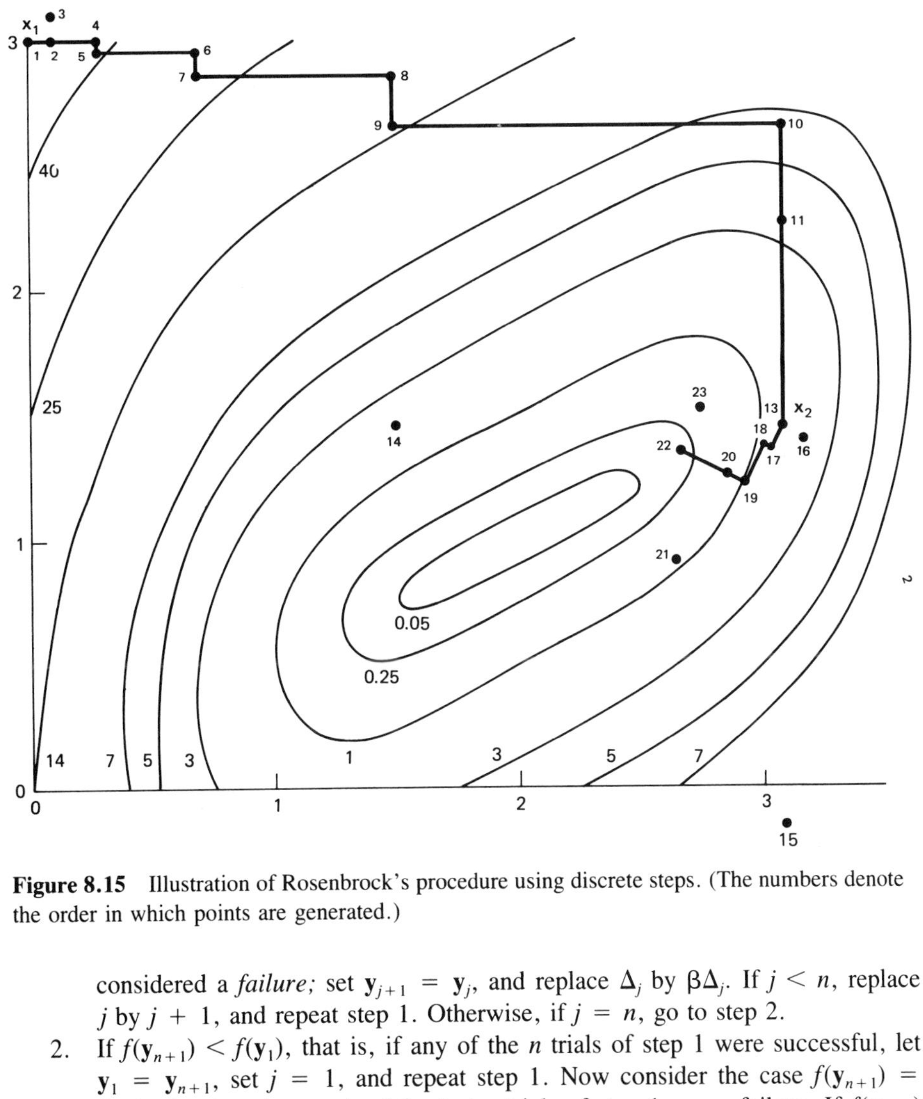

Figure 8.15 Illustration of Rosenbrock's procedure using discrete steps. (The numbers denote the order in which points are generated.)

considered a *failure;* set $\mathbf{y}_{j+1} = \mathbf{y}_j$, and replace Δ_j by $\beta\Delta_j$. If $j < n$, replace j by $j + 1$, and repeat step 1. Otherwise, if $j = n$, go to step 2.

2. If $f(\mathbf{y}_{n+1}) < f(\mathbf{y}_1)$, that is, if any of the n trials of step 1 were successful, let $\mathbf{y}_1 = \mathbf{y}_{n+1}$, set $j = 1$, and repeat step 1. Now consider the case $f(\mathbf{y}_{n+1}) = f(\mathbf{y}_1)$, that is, when each of the last n trials of step 1 was a failure. If $f(\mathbf{y}_{n+1}) < f(\mathbf{x}_k)$, that is, if at least one successful trial was encountered in step 1 during iteration k, go to step 3. If $f(\mathbf{y}_{n+1}) = f(\mathbf{x}_k)$, that is, if no successful trial is encountered, stop with $\mathbf{x}_k$ as an estimate of the optimal solution if $|\Delta_j| \leq \varepsilon$ for each j; otherwise, let $\mathbf{y}_1 = \mathbf{y}_{n+1}$, let $j = 1$, and go to step 1.
3. Let $\mathbf{x}_{k+1} = \mathbf{y}_{n+1}$. If $\|\mathbf{x}_{k+1} - \mathbf{x}_k\| < \varepsilon$, stop with $\mathbf{x}_{k+1}$ as an estimate of the optimal solution. Otherwise, compute $\lambda_1, \ldots, \lambda_n$ from the relationship $\mathbf{x}_{k+1} - \mathbf{x}_k = \sum_{j=1}^{n} \lambda_j \mathbf{d}_j$, form a new set of search directions by (8.9) and denote these directions by $\mathbf{d}_1, \ldots, \mathbf{d}_n$, let $\Delta_j = \bar{\Delta}_j$ for each j, let $\mathbf{y}_1 = \mathbf{x}_{k+1}$, replace k by $k + 1$, let $j = 1$, and repeat step 1.

Note that discrete steps are taken along the n search directions in step 1. If a success occurs along $\mathbf{d}_j$, then Δ_j is replaced by $\alpha\Delta_j$; and if a failure occurs along $\mathbf{d}_j$, then Δ_j is replaced by $\beta\Delta_j$. Since $\beta < 0$, a failure results in reversing the jth search direction during the next pass through step 1. Note that step 1 is repeated until a failure occurs along each of the search directions, in which case, if at least one success was obtained during a previous loop at this iteration, a new set of search directions is formed by the Gram–Schmidt procedure. If the loops through the search directions continue to result in failures, the step length shrinks to zero.

8.5.6 Example

Consider the following problem:

$$\text{Minimize} \quad (x_1 - 2)^4 + (x_1 - 2x_2)^2$$

We solve this problem by the method of Rosenbrock using discrete steps with $\bar{\Delta}_1 = \bar{\Delta}_2 = 0.1$, $\alpha = 2.0$, and $\beta = -0.5$. Table 8.10 summarizes the computations starting from (0.00, 3.00), where (S) denotes a succeess and (F) denotes a failure. Note that within each iteration the directions $\mathbf{d}_1$ and $\mathbf{d}_2$ are fixed. After seven passes through step 1 of Rosenbrock's method, we move from $\mathbf{x}_1^t = (0.00, 3.00)$ to $\mathbf{x}_2^t = (3.10, 1.45)$. At this point, a change of directions is required. In particular, $(\mathbf{x}_2 - \mathbf{x}_1) = \lambda_1\mathbf{d}_1 + \lambda_2\mathbf{d}_2$, where $\lambda_1 = 3.10$ and $\lambda_2 = -1.55$. Using (8.9), the reader can easily verify that the new search directions are given by (0.89, −0.45) and (−0.45, −0.89), which are used in the second iteration. The procedure is terminated during the second iteration.

In Figure 8.15, the progress of Rosenbrock's method is shown, and the points generated are numbered sequentially.

8.6 Multidimensional Search Using Derivatives

In the previous section, we described several minimization procedures that use only functional evaluations during the course of optimization. We now discuss some methods that use derivatives in determining the search directions. In particular, we discuss the steepest descent method and the method of Newton.

The Method of Steepest Descent

The method of steepest descent is one of the most fundamental procedures for minimizing a differentiable function of several variables. Recall that a vector $\mathbf{d}$ is called a direction of descent of a function f at $\mathbf{x}$ if there exists a $\delta > 0$ such that $f(\mathbf{x} + \lambda\mathbf{d}) < f(\mathbf{x})$ for all $\lambda \in (0, \delta)$. In particular, if $\lim_{\lambda \to 0^+} [f(\mathbf{x} + \lambda\mathbf{d}) - f(\mathbf{x})]/\lambda < 0$, then $\mathbf{d}$ is a direction of descent. The method of steepest descent moves along the direction $\mathbf{d}$ with $\|\mathbf{d}\| = 1$, which minimizes the above limit. Lemma 8.6.1 below shows that if f is differentiable at $\mathbf{x}$ with a nonzero gradient, then $-\nabla f(\mathbf{x})/\|\nabla f(\mathbf{x})\|$ is indeed the direction of steepest descent. For this reason, in the presence of differentiability, the method of steepest descent is sometimes called the gradient method.

8.6.1 Lemma

Suppose that $f: E_n \to E_1$ is differentiable at $\mathbf{x}$, and suppose that $\nabla f(\mathbf{x}) \neq \mathbf{0}$. Then, the optimal solution to the problem to minimize $f'(\mathbf{x}; \mathbf{d})$ subject to $\|\mathbf{d}\| \leq 1$ is given by $\bar{\mathbf{d}} = -\nabla f(\mathbf{x})/\|\nabla f(\mathbf{x})\|$; that is, $-\nabla f(\mathbf{x})/\|\nabla f(\mathbf{x})\|$ is the direction of steepest descent of f at $\mathbf{x}$.

Table 8.10 Summary of Computations for Rosenbrock's Method Using Discrete Steps

Iteration k	$\mathbf{x}_k$ $f(\mathbf{x}_k)$	j	$\mathbf{y}_j$ $f(\mathbf{y}_j)$	Δ_j	$\mathbf{d}_j$	$\mathbf{y}_j + \Delta_j\mathbf{d}_j$ $f(\mathbf{y}_j + \Delta_j\mathbf{d}_j)$
1	(0.00, 3.00) 52.00	1	(0.00, 3.00) 52.00	0.10	(1.00, 0.00)	(0.10, 3.00) 47.84(S)
		2	(0.10, 3.00) 47.84	0.10	(0.00, 1.00)	(0.10, 3.10) 50.24(F)
		1	(0.10, 3.00) 47.84	0.20	(1.00, 0.00)	(0.30, 3.00) 40.84(S)
		2	(0.30, 3.00) 40.84	−0.05	(0.00, 1.00)	(0.30, 2.95) 39.71(S)
		1	(0.30, 2.95) 39.71	0.40	(1.00, 0.00)	(0.70, 2.95) 29.90(S)
		2	(0.70, 2.95) 29.90	−0.10	(0.00, 1.00)	(0.70, 2.85) 27.86(S)
		1	(0.70, 2.85) 27.86	0.80	(1.00, 0.00)	(1.50, 2.85) 17.70(S)
		2	(1.50, 2.85) 17.70	−0.20	(0.00, 1.00)	(1.50, 2.65) 14.50(S)
		1	(1.50, 2.65) 14.50	1.60	(1.00, 0.00)	(3.10, 2.65) 6.30(S)
		2	(3.10, 2.65) 6.30	−0.40	(0.00, 1.00)	(3.10, 2.25) 3.42(S)
		1	(3.10, 2.25) 3.42	3.20	(1.00, 0.00)	(6.30, 2.25) 345.12(F)
		2	(3.10, 2.25) 3.42	−0.80	(0.00, 1.00)	(3.10, 1.45) 1.50(S)
		1	(3.10, 1.45) 1.50	−1.60	(1.00, 0.00)	(1.50, 1.45) 2.02(F)
		2	(3.10, 1.45) 1.50	−1.60	(0.00, 1.00)	(3.10, −0.15) 13.02(F)
2	(3.10, 1.45) 1.50	1	(3.10, 1.45) 1.50	0.10	(0.89, −0.45)	(3.19, 1.41) 2.14(F)
		2	(3.10, 1.45) 1.50	0.10	(−0.45, −0.89)	(3.06, 1.36) 1.38(S)
		1	(3.06, 1.36) 1.38	−0.05	(0.89, −0.45)	(3.02, 1.38) 1.15(S)
		2	(3.02, 1.38) 1.15	0.20	(−0.45, −0.89)	(2.93, 1.20) 1.03(S)

Continued

Table 8.10 *(Continued)*

Iteration k	$\mathbf{x}_k$ $f(\mathbf{x}_k)$	j	$\mathbf{y}_j$ $f(\mathbf{y}_j)$	Δ_j	$\mathbf{d}_j$	$\mathbf{y}_j + \Delta_j\mathbf{d}_j$ $f(\mathbf{y}_j + \Delta_j\mathbf{d}_j)$
		1	(2.93, 1.20) 1.03	−0.10	(0.89, −0.45)	(2.84, 1.25) 0.61(S)
		2	(2.84, 1.25) 0.61	0.40	(−0.45, −0.89)	(2.66, 0.89) 0.96(F)
		1	(2.84, 1.25) 0.61	−0.20	(0.89, −0.45)	(2.66, 1.34) 0.19(S)
		2	(2.66, 1.34) 0.19	−0.20	(−0.45, −0.89)	(2.75, 1.52) 0.40(F)

Proof

From differentiability of f at $\mathbf{x}$, it follows that

$$f'(\mathbf{x}; \mathbf{d}) = \lim_{\lambda \to 0^+} \frac{f(\mathbf{x} + \lambda \mathbf{d}) - f(\mathbf{x})}{\lambda} = \nabla f(\mathbf{x})'\mathbf{d}$$

Thus, the stated problem reduces to minimize $\nabla f(\mathbf{x})'\mathbf{d}$ subject to $\|\mathbf{d}\| \le 1$. By the Schwartz inequality, for $\|\mathbf{d}\| \le 1$, we have

$$\nabla f(\mathbf{x})'\mathbf{d} \ge -\|\nabla f(\mathbf{x})\| \, \|\mathbf{d}\| \ge -\|\nabla f(\mathbf{x})\|$$

with equality holding throughout if and only if $\mathbf{d} = \bar{\mathbf{d}} \equiv -\nabla f(\mathbf{x})/\|\nabla f(\mathbf{x})\|$. Thus, $\bar{\mathbf{d}}$ is the optimal solution, and the proof is complete.

Summary of the Steepest Descent Algorithm

Given a point $\mathbf{x}$, the steepest descent algorithm proceeds by performing a line search along the direction $-\nabla f(\mathbf{x})/\|\nabla f(\mathbf{x})\|$ or, equivalently, along the direction $-\nabla f(\mathbf{x})$. A summary of the method is given below.

Initialization Step Let $\varepsilon > 0$ be the termination scalar. Choose a starting point $\mathbf{x}_1$, let $k = 1$, and go to the main step.

Main Step If $\|\nabla f(\mathbf{x}_k)\| < \varepsilon$, stop; otherwise, let $\mathbf{d}_k = -\nabla f(\mathbf{x}_k)$, and let λ_k be an optimal solution to the problem to minimize $f(\mathbf{x}_k + \lambda \mathbf{d}_k)$ subject to $\lambda \ge 0$. Let $\mathbf{x}_{k+1} = \mathbf{x}_k + \lambda_k \mathbf{d}_k$, replace k by $k + 1$, and repeat the main step.

8.6.2 Example

Consider the following problem

Minimize $\quad (x_1 - 2)^4 + (x_1 - 2x_2)^2$

We solve this problem using the method of steepest descent, starting with the point (0.00, 3.00). The summary of the computations are given in Table 8.11. After seven

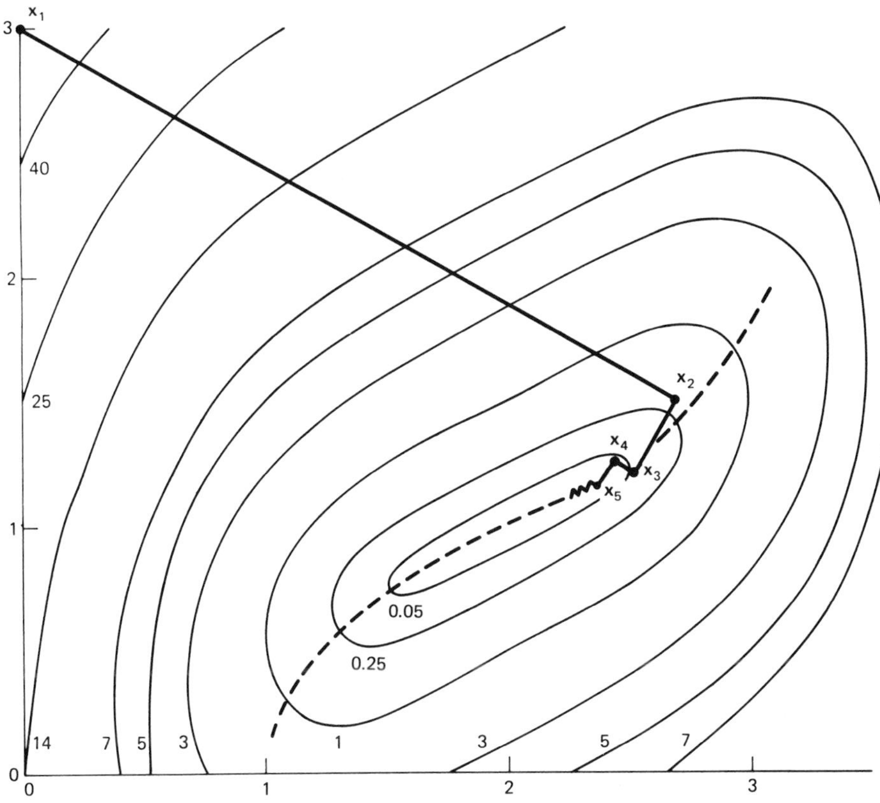

Figure 8.16 Illustration of the method of steepest descent.

iterations, the point $\mathbf{x}_8 = (2.28, 1.15)^t$ is reached. The algorithm is terminated since $\|\nabla f(\mathbf{x}_8)\| = 0.09$ is small. The progress of the method is shown in Figure 8.16. Note that the minimizing point for this problem is (2.00, 1.00).

Convergence of the Steepest Descent Method

Let $\Omega = \{\bar{\mathbf{x}}: \nabla f(\bar{\mathbf{x}}) = \mathbf{0}\}$, and let f be the descent function. The algorithmic map is $\mathbf{A} = \mathbf{MD}$, where $\mathbf{D}(\mathbf{x}) = [\mathbf{x}, \nabla f(\mathbf{x})]$, and $\mathbf{M}$ is the line search map over the closed interval $[0, \infty)$. Assuming that f is continuously differentiable, then $\mathbf{D}$ is continuous. Furthermore, $\mathbf{M}$ is closed by Theorem 8.4.1. Therefore, the algorithmic map $\mathbf{A}$ is closed by Corollary 2 to Theorem 7.3.2. Finally, if $\mathbf{x} \notin \Omega$, then $\nabla f(\mathbf{x})^t\mathbf{d} < 0$, where $\mathbf{d} = -\nabla f(\mathbf{x})$. By Theorem 4.1.2, $\mathbf{d}$ is a descent direction, and, hence, $f(\mathbf{y}) < f(\mathbf{x})$ for $\mathbf{y} \in \mathbf{A}(\mathbf{x})$. Assuming that the sequence generated by the algorithm is contained in a compact set, then, by Theorem 7.2.3, the steepest descent algorithm converges to a point with zero gradient.

Zigzagging of the Steepest Descent Method

The method of steepest descent usually works quite well during early stages of the optimization process, depending on the point of initialization. However, as a stationary

Table 8.11 Summary of Computations for the Method of Steepest Descent

Iteration k	$\mathbf{x}_k$ $f(\mathbf{x}_k)$	$\nabla f(\mathbf{x}_k)$	$\|\nabla f(\mathbf{x}_k)\|$	$\mathbf{d}_k = -\nabla f(\mathbf{x}_k)$	λ_k	$\mathbf{x}_{k+1}$
1	(0.00, 3.00) 52.00	(−44.00, 24.00)	50.12	(44.00, −24.00)	0.062	(2.70, 1.51)
2	(2.70, 1.51) 0.34	(0.73, 1.28)	1.47	(−0.73, −1.28)	0.24	(2.52, 1.20)
3	(2.52, 1.20) 0.09	(0.80, −0.48)	0.93	(−0.80, 0.48)	0.11	(2.43, 1.25)
4	(2.43, 1.25) 0.04	(0.18, 0.28)	0.33	(−0.18, −0.28)	0.31	(2.37, 1.16)
5	(2.37, 1.16) 0.02	(0.30, −0.20)	0.36	(−0.30, 0.20)	0.12	(2.33, 1.18)
6	(2.33, 1.18) 0.01	(0.08, 0.12)	0.14	(−0.08, −0.12)	0.36	(2.30, 1.14)
7	(2.30, 1.14) 0.009	(0.15, −0.08)	0.17	(−0.15, 0.08)	0.13	(2.28, 1.15)
8	(2.28, 1.15) 0.007	(0.05, 0.08)	0.09			

point is approached, the method usually behaves poorly, taking small, nearly orthogonal steps. This *zigzagging* phenomenon was encountered in Example 8.6.2 and is illustrated in Figure 8.16, in which zigzagging occurs along the valley shown by the dotted lines.

Zigzagging and poor convergence of the steepest descent algorithm at later stages can be explained intuitively by considering the following expression of the function f:

$$f(\mathbf{x}_k + \lambda\mathbf{d}) = f(\mathbf{x}_k) + \lambda\nabla f(\mathbf{x}_k)'\mathbf{d} + \lambda\,\|\mathbf{d}\|\,\alpha(\mathbf{x}_k; \lambda\mathbf{d})$$

where $\alpha(\mathbf{x}_k; \lambda\mathbf{d}) \to 0$ as $\lambda\mathbf{d} \to \mathbf{0}$, and $\mathbf{d}$ is a search direction with $\|\mathbf{d}\| = 1$. If $\mathbf{x}_k$ is close to a stationary point with zero gradient, and f is continuously differentiable, then $\|\nabla f(\mathbf{x}_k)\|$ will be small, making the coefficient of λ in the term $\lambda\nabla f(\mathbf{x}_k)'\mathbf{d}$ of a small order of magnitude. Since the steepest descent method employs the linear approximation of f to find a direction of movement, where the term $\lambda\|\mathbf{d}\|\alpha(\mathbf{x}_k; \lambda\mathbf{d})$ is essentially ignored, we should expect that the directions generated at late stages will not be very effective if the latter term contributes significantly to the description of f, even for relatively small values of λ.

As we shall learn in the remainder of the chapter, there are some ways to overcome the difficulties of zigzagging by *deflecting* the gradient. Rather than moving along $\mathbf{d} = -\nabla f(\mathbf{x})$, we can move along $\mathbf{d} = -\mathbf{D}\nabla f(\mathbf{x})$ or along $\mathbf{d} = -\nabla f(\mathbf{x}) + \mathbf{g}$, where $\mathbf{D}$ is an appropriate matrix, and $\mathbf{g}$ is an appropriate vector. These correction procedures will be discussed in more detail shortly.

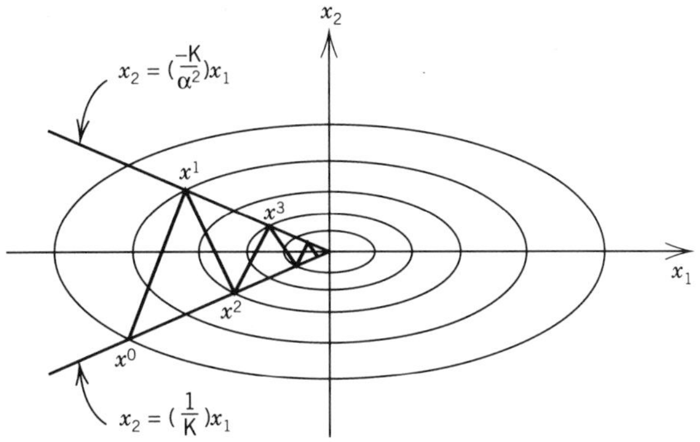

Figure 8.17 Convergence rate analysis of the steepest descent algorithm.

Convergence Rate Analysis for the Steepest Descent Algorithm

In this section, we shall try to give a more formalized analysis of the zigzagging phenomenon and the empirically observed slow convergence rate of the steepest descent algorithm. This analysis will also afford insights into possible ways of alleviating such poor algorithmic performance.

Toward this end, let us begin by considering a bivariate quadratic function $f(x_1, x_2) = \frac{1}{2}(x_1^2 + \alpha x_2^2)$, where $\alpha > 1$. Note that the Hessian matrix to this function is $H = \text{diag}\,\{1, \alpha\}$, with eigenvalues 1 and α. Let us define the *condition number* of a positive definite matrix to be the ratio of its largest to smallest eigenvalues. Hence, the condition number of H for our example is α. The contours of f are plotted in Figure 8.17. Observe that as α increases, a phenomenon that is known as *ill-conditioning*, or a *worsening of the condition number*, the contours become increasingly skewed, and the graph of the function becomes increasingly steep in the x_2 direction relative to the x_1 direction.

Now, given a starting point $\mathbf{x} = (x_1, x_2)^t$, let us apply an iteration of the steepest descent algorithm to obtain a point $\mathbf{x}_{\text{new}} = (x_{1\text{new}}, x_{2\text{new}})^t$. Note that if $x_1 = 0$ or $x_2 = 0$, the procedure converges to the optimal minimizing solution $\mathbf{x}^* = (0, 0)^t$ in one step. Hence, suppose that $x_1 \neq 0$ and $x_2 \neq 0$. The steepest descent direction is given by $\mathbf{d} = -\nabla f(\mathbf{x}) = -(x_1, \alpha x_2)^t$ resulting in $\mathbf{x}_{\text{new}} = (\mathbf{x} + \lambda\mathbf{d})$, where λ solves the line search problem to minimize $\theta(\lambda) \equiv f(\mathbf{x} + \lambda\mathbf{d}) = \frac{1}{2}[x_1^2(1 - \lambda)^2 + \alpha x_2^2(1 - \alpha\lambda)^2]$ subject to $\lambda \geq 0$. Using simple calculus, we obtain

$$\lambda = \frac{x_1^2 + \alpha^2 x_2^2}{x_1^2 + \alpha^3 x_2^2}$$

and so (8.13)

$$\mathbf{x}_{\text{new}} = \left[\frac{\alpha^2 x_1 x_2^2(\alpha - 1)}{x_1^2 + \alpha^3 x_2^2}, \frac{x_1^2 x_2(1 - \alpha)}{x_1^2 + \alpha^3 x_2^2}\right]$$

Observe that $x_{1\text{new}}/x_{2\text{new}} = -\alpha^2(x_2/x_1)$. Hence, if we begin with a solution $\mathbf{x}^0$ having $x_1^0/x_2^0 = K \neq 0$ and generate a sequence of iterates $\{\mathbf{x}^k\}$, $k = 1, 2, \ldots$, using the steepest descent algorithm, then the sequence of values $\{x_1^k/x_2^k\}$ alternate between the values K and $-\alpha^2/K$ as the sequence $\{\mathbf{x}^k\}$ converges to $\mathbf{x}^* = (0, 0)^t$. For our example, this means that the sequence zigzags between the pair of straight lines $x_2 = (1/K)x_1$ and $x_2 = (-K/\alpha^2)x_1$, as shown in Figure 8.17. Note that as the condition number α increases, this zigzagging phenomenon becomes more pronounced. On the other hand, if $\alpha = 1$, then the contours of f are circular, and we obtain $\mathbf{x}^1 = \mathbf{x}^*$ in a single iteration.

To study the rate of convergence, let us examine the rate at which $\{f(\mathbf{x}^k)\}$ converges to the value zero. From (8.13), it is easily verified that

$$\frac{f(\mathbf{x}^{k+1})}{f(\mathbf{x}^k)} = \frac{K_k^2\alpha(\alpha - 1)^2}{(K_k^2 + \alpha^3)(K_k^2 + \alpha)} \qquad \text{where } K_k = \frac{x_1^k}{x_2^k} \tag{8.14}$$

Indeed, the expression in (8.14) can be seen to be maximized when $K_k^2 = \alpha^2$ (see Exercise 8.18), so that we obtain

$$\frac{f(\mathbf{x}^{k+1})}{f(\mathbf{x}^k)} \leq \frac{(\alpha - 1)^2}{(\alpha + 1)^2} \tag{8.15}$$

Note from (8.15) that $\{f(\mathbf{x}^k)\} \to 0$ at a geometric or linear rate bounded by the ratio $(\alpha - 1)^2/(\alpha + 1)^2 < 1$. In fact, if we initialize the process with $x_1^0/x_2^0 \equiv K = \alpha$, then, since $K_k^2 = (x_1^k/x_2^k)^2 = \alpha^2$ from above (see Figure 8.17), we get from (8.14) that the convergence ratio $f(\mathbf{x}^{k+1})/f(\mathbf{x}^k)$ is precisely $(\alpha - 1)^2/(\alpha + 1)^2$. Hence, as α approaches infinity, this ratio approaches 1 from below, and the rate of convergence becomes increasingly slower.

The foregoing analysis can be extended to a general quadratic function $f(\mathbf{x}) = \mathbf{c}^t\mathbf{x} + \frac{1}{2}\mathbf{x}^t\mathbf{H}\mathbf{x}$, where $\mathbf{H}$ is an $n \times n$, symmetric, positive definite matrix. The unique minimizer $\mathbf{x}^*$ for this function is given by the solution to the system $\mathbf{H}\mathbf{x}^* = -\mathbf{c}$ obtained by setting $\nabla f(\mathbf{x}^*) = \mathbf{0}$. Also, given an iterate $\mathbf{x}_k$, the optimal step length λ and the revised iterate $\mathbf{x}_{k+1}$ are given by the following generalization of (8.13), where $\mathbf{g}_k \equiv \nabla f(\mathbf{x}_k) = \mathbf{c} + \mathbf{H}\mathbf{x}_k$:

$$\lambda = \frac{\mathbf{g}_k^t\mathbf{g}_k}{\mathbf{g}_k^t\mathbf{H}\mathbf{g}_k} \qquad \text{and} \qquad \mathbf{x}_{k+1} = \mathbf{x}_k - \lambda\mathbf{g}_k \tag{8.16}$$

Now, to evaluate the rate of convergence, let us employ a convenient measure for convergence given by the following *error function*:

$$e(\mathbf{x}) = \tfrac{1}{2}(\mathbf{x} - \mathbf{x}^*)^t\mathbf{H}(\mathbf{x} - \mathbf{x}^*) \equiv f(\mathbf{x}) + \tfrac{1}{2}\mathbf{x}^{*t}\mathbf{H}\mathbf{x}^* \tag{8.17}$$

where we have used the fact that $\mathbf{H}\mathbf{x}^* = -\mathbf{c}$. Note that $e(\mathbf{x})$ differs from $f(\mathbf{x})$ by only a constant and equals zero if and only if $\mathbf{x} = \mathbf{x}^*$. In fact, it can be shown, analogous to (8.15), that (see Exercise 8.19)

$$e(\mathbf{x}_{k+1}) = \left[1 - \frac{(\mathbf{g}_k^t\mathbf{g}_k)^2}{(\mathbf{g}_k^t\mathbf{H}\mathbf{g}_k)(\mathbf{g}_k^t\mathbf{H}^{-1}\mathbf{g}_k)}\right] e(\mathbf{x}_k) \leq \frac{(\alpha - 1)^2}{(\alpha + 1)^2} e(\mathbf{x}_k) \tag{8.18}$$

where α is the condition number of $\mathbf{H}$. Hence, $\{e(\mathbf{x}_k)\} \to 0$ at a linear or geometric convergence rate bounded above by $(\alpha - 1)^2/(\alpha + 1)^2$; and so, as before, we can expect the convergence to become increasingly slower as α increases, depending on the initial solution $\mathbf{x}_0$.

For continuously twice-differentiable nonquadratic functions $f: E_n \to E_1$, a similar result is known to hold. In such a case, if $\mathbf{x}^*$ is a local minimum to which a sequence $\{\mathbf{x}_k\}$ generated by the steepest descent algorithm converges, and if $\mathbf{H}(\mathbf{x}^*)$ is positive definite with a condition number α, then the corresponding sequence of objective values

$\{f(\mathbf{x}_k)\}$ can be shown to converge linearly to the value $f(\mathbf{x}^*)$ at a rate bounded above by $(\alpha - 1)^2/(\alpha + 1)^2$.

Convergence Analysis of the Steepest Descent Algorithm Using Armijo's Inexact Line Search

In Section 8.3, we introduced Armijo's Rule for selecting an acceptable, inexact step length during a line search process. It is instructive to observe how such a criterion still guarantees algorithmic convergence. Below, we present a convergence analysis for an inexact steepest descent algorithm applied to a function $f: E_n \to E_1$ whose gradient function $\nabla f(\mathbf{x})$ is *Lipschitz continuous with constant* $G > 0$ on $S(\mathbf{x}_0) = \{\mathbf{x}: f(\mathbf{x}) \le f(\mathbf{x}_0)\}$ for some given $\mathbf{x}_0 \in E_n$. That is, we have $\|\nabla f(\mathbf{x}) - \nabla f(\mathbf{y})\| \le G\|\mathbf{x} - \mathbf{y}\|$ for all $\mathbf{x}$, $\mathbf{y} \in S(\mathbf{x}_0)$. For example, if the Hessian of f at any point has a norm bounded above by a constant G on conv $S(\mathbf{x}_0)$ (see Appendix A for the norm of a matrix), then such a function has Lipschitz continuous gradients. This follows from the mean value theorem, noting that for any $\mathbf{x} \ne \mathbf{y} \in S(\mathbf{x}_0)$, $\|\nabla f(\mathbf{x}) - \nabla f(\mathbf{y})\| = \|H(\hat{\mathbf{x}})(\mathbf{x} - \mathbf{y})\| \le G\|\mathbf{x} - \mathbf{y}\|$.

The procedure we analyze is the often-used variant of Armijo's Rule described in Section 8.3 with parameters $0 < \varepsilon < 1$, $\alpha = 2$, and a fixed step length parameter $\bar{\lambda}$, wherein either $\bar{\lambda}$ itself is chosen, if acceptable, or is sequentially halved until an acceptable step length results. This procedure is embodied in the following result.

8.6.3 Theorem

Let $f: E_n \to E_1$ be such that its gradient $\nabla f(\mathbf{x})$ is Lipschitz continuous with constant $G > 0$ on $S(\mathbf{x}_0) = \{\mathbf{x}: f(\mathbf{x}) \le f(\mathbf{x}_0)\}$ for some given $\mathbf{x}_0 \in E_n$. Pick some fixed step length parameter $\bar{\lambda} > 0$, and let $0 < \varepsilon < 1$. Given any iterate $\mathbf{x}_k$, define the search direction $\mathbf{d}_k = -\nabla f(\mathbf{x}_k)$, and consider Armijo's function $\hat{\theta}(\lambda) = \theta(0) + \lambda\varepsilon\theta'(0)$, $\lambda \ge 0$, where $\theta(\lambda) = f(\mathbf{x}_k + \lambda\mathbf{d}_k)$, $\lambda \ge 0$, is the line search function. If $\mathbf{d}_k = 0$, then stop. Otherwise, find the smallest integer $t \ge 0$ for which $\theta(\bar{\lambda}/2^t) \le \hat{\theta}(\bar{\lambda}/2^t)$ and define the next iterate as $\mathbf{x}_{k+1} = \mathbf{x}_k + \lambda_k\mathbf{d}_k$, where $\lambda_k \equiv \bar{\lambda}/2^t$. Now, suppose that, starting with some iterate $\mathbf{x}_0$, this procedure produces a sequence of iterates $\mathbf{x}_0, \mathbf{x}_1, \mathbf{x}_2, \ldots$. Then, either the procedure terminates finitely with $\nabla f(\mathbf{x}_K) = \mathbf{0}$ for some K, or else an infinite sequence $\{\mathbf{x}_k\}$ is generated such that the corresponding sequence $\{\nabla f(\mathbf{x}_k)\} \to \mathbf{0}$.

Proof

The case of finite termination is clear. Hence, suppose that an infinite sequence $\{\mathbf{x}_k\}$ is generated. Note that the Armijo criterion $\theta(\bar{\lambda}/2^t) \le \hat{\theta}(\bar{\lambda}/2^t)$ is equivalent to $\theta(\bar{\lambda}/2^t) \equiv f(\mathbf{x}_{k+1}) \le \hat{\theta}(\bar{\lambda}/2^t) \equiv \theta(0) + (\bar{\lambda}\varepsilon/2^t)\nabla f(\mathbf{x}_k)^t\mathbf{d}_k = f(\mathbf{x}_k) - (\bar{\lambda}\varepsilon/2^t)\|\nabla f(\mathbf{x}_k)\|^2$. Hence, $t \ge 0$ is the smallest integer for which

$$f(\mathbf{x}_{k+1}) - f(\mathbf{x}_k) \le \frac{-\bar{\lambda}\varepsilon}{2^t}\|\nabla f(\mathbf{x}_k)\|^2 \tag{8.19}$$

Now, using the mean value theorem, we have, for some strict convex combination $\tilde{\mathbf{x}}$ of $\mathbf{x}_k$ and $\mathbf{x}_{k+1}$, that

$$\begin{aligned} f(\mathbf{x}_{k+1}) - f(\mathbf{x}_k) &= \lambda_k\mathbf{d}_k^t\nabla f(\tilde{\mathbf{x}}) \\ &= -\lambda_k\nabla f(\mathbf{x}_k)^t[\nabla f(\mathbf{x}_k) - \nabla f(\mathbf{x}_k) + \nabla f(\tilde{\mathbf{x}})] \\ &= -\lambda_k\|\nabla f(\mathbf{x}_k)\|^2 + \lambda_k\nabla f(\mathbf{x}_k)^t[\nabla f(\mathbf{x}_k) - \nabla f(\tilde{\mathbf{x}})] \\ &\le -\lambda_k\|\nabla f(\mathbf{x}_k)\|^2 + \lambda_k\|\nabla f(\mathbf{x}_k)\|\,\|\nabla f(\mathbf{x}_k) - \nabla f(\tilde{\mathbf{x}})\| \end{aligned}$$

But by the Lipschitz continuity of ∇f, noting from (8.19) that the descent nature of the algorithm guarantees that $\mathbf{x}_k \in S(\mathbf{x}_0)$ for all k, we have $\|\nabla f(\mathbf{x}_k) - \nabla f(\bar{\mathbf{x}})\| \leq G\|\mathbf{x}_k - \bar{\mathbf{x}}\| \leq G\|\mathbf{x}_k - \mathbf{x}_{k+1}\| = G\lambda_k\|\nabla f(\mathbf{x}_k)\|$. Substituting this above, we obtain

$$f(\mathbf{x}_{k+1}) - f(\mathbf{x}_k) \leq -\lambda_k\|\nabla f(\mathbf{x}_k)\|^2[1 - \lambda_k G] = \frac{-\bar{\lambda}}{2^t}\|\nabla f(\mathbf{x}_k)\|^2\left[1 - \frac{\bar{\lambda}G}{2^t}\right] \tag{8.20}$$

Consequently, from (8.20), we know that (8.19) will hold true when t is increased to no larger an integer value than is necessary to make $[1 - (\bar{\lambda}G/2^t)] \geq \varepsilon$, for then, (8.20) will imply (8.19). But this means that $[1 - (\bar{\lambda}G/2^{t-1})] < \varepsilon$, that is, $\bar{\lambda}\varepsilon/2^t > \varepsilon(1 - \varepsilon)/2G$. Substituting this in (8.19), we get

$$f(\mathbf{x}_{k+1}) - f(\mathbf{x}_k) < \frac{-\varepsilon(1 - \varepsilon)}{2G}\|\nabla f(\mathbf{x}_k)\|^2$$

Hence, noting that $\{f(\mathbf{x}_k)\}$ is a monotone decreasing sequence and so has a limit, taking limits as $t \to \infty$, we get

$$0 \leq \frac{-\varepsilon(1 - \varepsilon)}{2G}\lim_{k\to\infty}\|\nabla f(\mathbf{x}_k)\|^2$$

which implies that $\{\nabla f(\mathbf{x}_k)\} \to 0$. This completes the proof.

The Method of Newton

In Section 8.2, we discussed Newton's method for minimizing a function of a single variable.

The method of Newton is a procedure that deflects the steepest descent direction by premultiplying it by the inverse of the Hessian matrix. This operation is motivated by finding a suitable direction for the quadratic approximation to the function rather than by finding a linear approximation to the function, as in the gradient search. To motivate the procedure, consider the following approximation q at a given point $\mathbf{x}_k$:

$$q(\mathbf{x}) = f(\mathbf{x}_k) + \nabla f(\mathbf{x}_k)^t(\mathbf{x} - \mathbf{x}_k) + \tfrac{1}{2}(\mathbf{x} - \mathbf{x}_k)^t\mathbf{H}(\mathbf{x}_k)(\mathbf{x} - \mathbf{x}_k)$$

where $\mathbf{H}(\mathbf{x}_k)$ is the Hessian matrix of f at $\mathbf{x}_k$. A necessary condition for a minimum of the quadratic approximation q is that $\nabla q(\mathbf{x}) = \mathbf{0}$, or $\nabla f(\mathbf{x}_k) + \mathbf{H}(\mathbf{x}_k)(\mathbf{x} - \mathbf{x}_k) = \mathbf{0}$. Assuming that the inverse of $\mathbf{H}(\mathbf{x}_k)$ exists, the successor point $\mathbf{x}_{k+1}$ is given by

$$\mathbf{x}_{k+1} = \mathbf{x}_k - \mathbf{H}(\mathbf{x}_k)^{-1}\nabla f(\mathbf{x}_k) \tag{8.21}$$

The above equation gives the recursive form of the points generated by Newton's method for the multidimensional case. Assuming that $\nabla f(\bar{\mathbf{x}}) = \mathbf{0}$, that $\mathbf{H}(\bar{\mathbf{x}})$ is positive definite at a local minimum $\bar{\mathbf{x}}$, and that f is continuously twice-differentiable, it follows that $\mathbf{H}(\mathbf{x}_k)$ is positive definite at points close to $\bar{\mathbf{x}}$, and, hence, the successor point $\mathbf{x}_{k+1}$ is well defined.

It is interesting to note that Newton's method can be interpreted as a *steepest descent algorithm with affine scaling*. Specifically, given a point $\mathbf{x}_k$ at iteration k, suppose that $\mathbf{H}(\mathbf{x}_k)$ is positive definite, and that we have a Cholesky factorization (see Appendix A.2) of its inverse given by $\mathbf{H}(\mathbf{x}_k)^{-1} = \mathbf{L}\mathbf{L}^t$, where $\mathbf{L}$ is a lower triangular matrix with positive diagonal elements. Now consider the affine scaling transformation $\mathbf{x} = \mathbf{L}\mathbf{y}$. This transforms the function $f(\mathbf{x})$ to the function $F(\mathbf{y}) \equiv f[\mathbf{L}\mathbf{y}]$, and the current point in the $\mathbf{y}$ space is $\mathbf{y}_k = \mathbf{L}^{-1}\mathbf{x}_k$. Hence, we have $\nabla F(\mathbf{y}_k) = \mathbf{L}^t\nabla f[\mathbf{L}\mathbf{y}_k] = \mathbf{L}^t\nabla f(\mathbf{x}_k)$. A unit step size along the negative gradient direction in the $\mathbf{y}$ space will then take us to the point $\mathbf{y}_{k+1} = \mathbf{y}_k -$

$\mathbf{L}^t\nabla f(\mathbf{x}_k)$. Translating this to the corresponding movement in the $\mathbf{x}$ space by premultiplying throughout by $\mathbf{L}$ produces precisely equation (8.21) and, hence, yields a steepest descent interpretation of Newton's method. Observe that this comment alludes to the benefits of using an appropriate scaling transformation. Indeed, if the function f was quadratic in the above analysis, then a unit step along the steepest descent direction in the transformed space would be an optimal step along that direction, which would moreover take us directly to the optimal solution in one iteration starting from any given solution.

We also comment here that (8.21) can be viewed as an application of the *Newton–Raphson* method to the solution of the system of equations $\nabla f(\mathbf{x}) = \mathbf{0}$. Given a well-determined system of nonlinear equations, each iteration of this method adopts a first-order Taylor series approximation to this system at the current iterate and solves the resulting linear system to determine the next iterate. Applying this to the system $\nabla f(\mathbf{x}) = \mathbf{0}$ at an iterate $\mathbf{x}_k$, the first-order approximation to $\nabla f(\mathbf{x})$ is given by $\nabla f(\mathbf{x}_k) + \mathbf{H}(\mathbf{x}_k)(\mathbf{x} - \mathbf{x}_k)$. Setting this equal to zero and solving produces the solution $\mathbf{x} = \mathbf{x}_{k+1}$ as given by (8.21).

8.6.4 Example

Consider the following problem:

$$\text{Minimize} \quad (x_1 - 2)^4 + (x_1 - 2x_2)^2$$

The summary of the computations using Newton's method is given in Table 8.12. At each iteration, $\mathbf{x}_{k+1}$ is given by $\mathbf{x}_{k+1} = \mathbf{x}_k - \mathbf{H}(\mathbf{x}_k)^{-1}\nabla f(\mathbf{x}_k)$. After six iterations, the point $\mathbf{x}_7 = (1.83, 0.91)^t$ is reached. At this point, $\|\nabla f(\mathbf{x}_7)\| = 0.04$, and the procedure is terminated. The points generated by the method are shown in Figure 8.18.

In the above example, the value of the objective function decreased at each iteration. However, this will not generally be the case, so f cannot be used as a descent function. Theorem 8.6.5 below indicates that Newton's method indeed converges, provided we start from a point close enough to the optimal point.

Order-Two Convergence of the Method of Newton

In general, the points generated by the method of Newton may not converge. The reason for this is that $\mathbf{H}(\mathbf{x}_k)$ may be singular, so that $\mathbf{x}_{k+1}$ is not well defined. Even if $\mathbf{H}(\mathbf{x}_k)^{-1}$ exists, $f(\mathbf{x}_{k+1})$ is not necessarily less than $f(\mathbf{x}_k)$. However, if the starting point is close enough to a point $\bar{\mathbf{x}}$ such that $\nabla f(\bar{\mathbf{x}}) = \mathbf{0}$ and $\mathbf{H}(\bar{\mathbf{x}})$ is of full rank, then the method of Newton is well defined and converges to $\bar{\mathbf{x}}$. This is proved in Theorem 8.6.5 below by showing that all the assumptions of Theorem 7.2.3 hold, where the descent function α is given by $\alpha(\mathbf{x}) = \|\mathbf{x} - \bar{\mathbf{x}}\|$.

8.6.5 Theorem

Let $f: E_n \to E_1$ be continuously thrice-differentiable. Consider Newton's algorithm defined by the map $\mathbf{A}(\mathbf{x}) = \mathbf{x} - \mathbf{H}(\mathbf{x})^{-1}\nabla f(\mathbf{x})$. Let $\bar{\mathbf{x}}$ be such that $\nabla f(\bar{\mathbf{x}}) = \mathbf{0}$ and $\mathbf{H}(\bar{\mathbf{x}})^{-1}$ exists. Let the starting point $\mathbf{x}_1$ be sufficiently close to $\bar{\mathbf{x}}$ so that this proximity implies that there exist $k_1, k_2 > 0$ with $k_1 k_2 \|\mathbf{x}_1 - \bar{\mathbf{x}}\| < 1$ such that

1. $\|\mathbf{H}(\mathbf{x})^{-1}\|^\dagger \le k_1$

   and by the Taylor series expansion of ∇f,

---

†See Appendix A.1 for the norm of a matrix.

Table 8.12 Summary of Computations for the Method of Newton

Iteration k	$\mathbf{x}_k$ $f(\mathbf{x}_k)$	$\nabla f(\mathbf{x}_k)$	$\mathbf{H}(\mathbf{x}_k)$	$\mathbf{H}(\mathbf{x}_k)^{-1}$	$-\mathbf{H}(\mathbf{x}_k)^{-1}\nabla f(\mathbf{x}_k)$	$\mathbf{x}_{k+1}$
1	(0.00, 3.00) 52.00	(−44.0, 24.0)	$\begin{bmatrix} 50.0 & -4.0 \\ -4.0 & 8.0 \end{bmatrix}$	$\frac{1}{384}\begin{bmatrix} 8.0 & 4.0 \\ 4.0 & 50.0 \end{bmatrix}$	(0.67, −2.67)	(0.67, 0.33)
2	(0.67, 0.33) 3.13	(−9.39, −0.04)	$\begin{bmatrix} 23.23 & -4.0 \\ -4.0 & 8.0 \end{bmatrix}$	$\frac{1}{169.84}\begin{bmatrix} 8.0 & 4.0 \\ 4.0 & 23.23 \end{bmatrix}$	(0.44, 0.23)	(1.11, 0.56)
3	(1.11, 0.56) 0.63	(−2.84, −0.04)	$\begin{bmatrix} 11.50 & -4.0 \\ -4.0 & 8.0 \end{bmatrix}$	$\frac{1}{76}\begin{bmatrix} 8.0 & 4.0 \\ 4.0 & 11.50 \end{bmatrix}$	(0.30, 0.14)	(1.41, 0.70)
4	(1.41, 0.70) 0.12	(−0.80, −0.04)	$\begin{bmatrix} 6.18 & 4.0 \\ -4.0 & 8.0 \end{bmatrix}$	$\frac{1}{33.44}\begin{bmatrix} 8.0 & 4.0 \\ 4.0 & 6.18 \end{bmatrix}$	(0.20, 0.10)	(1.61, 0.80)
5	(1.61, 0.80) 0.02	(−0.22, −0.04)	$\begin{bmatrix} 3.83 & -4.0 \\ -4.0 & 8.0 \end{bmatrix}$	$\frac{1}{14.64}\begin{bmatrix} 8.0 & 4.0 \\ 4.0 & 3.83 \end{bmatrix}$	(0.13, 0.07)	(1.74, 0.87)
6	(1.74, 0.87) 0.005	(−0.07, 0.00)	$\begin{bmatrix} 2.81 & -4.0 \\ -4.0 & 8.0 \end{bmatrix}$	$\frac{1}{6.48}\begin{bmatrix} 8.0 & 4.0 \\ 4.0 & 2.81 \end{bmatrix}$	(0.09, 0.04)	(1.83, 0.91)
7	(1.83, 0.91) 0.0009	(0.0003, −0.04)				

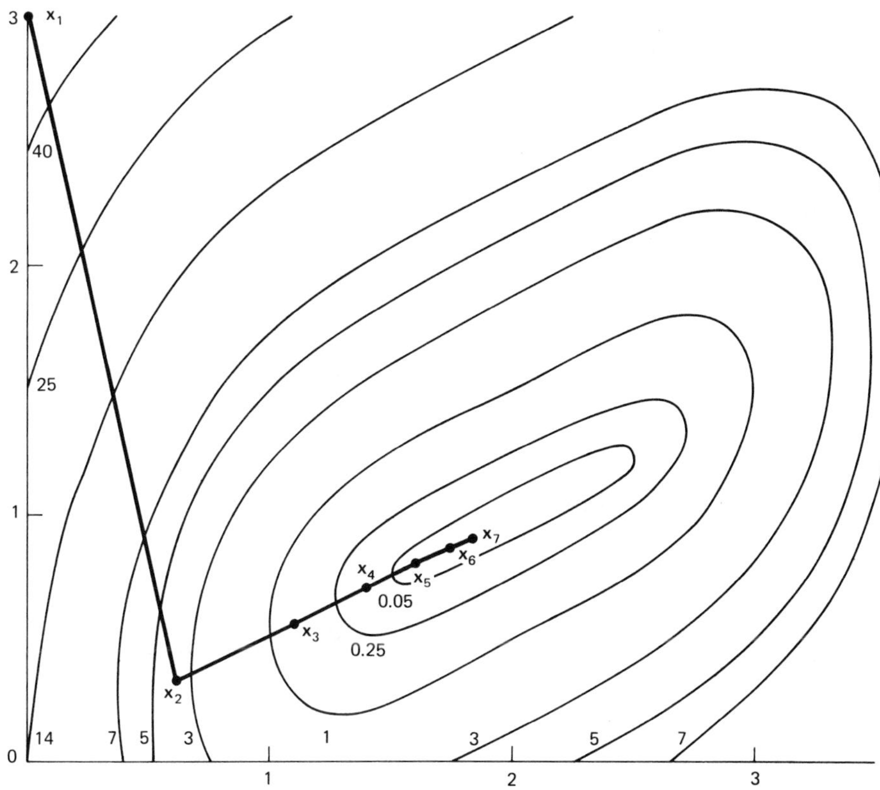

Figure 8.18 Illustration of the method of Newton.

2. $\|\nabla f(\bar{\mathbf{x}}) - \nabla f(\mathbf{x}) - \mathbf{H}(\mathbf{x})(\bar{\mathbf{x}} - \mathbf{x})\| \le k_2 \|\bar{\mathbf{x}} - \mathbf{x}\|^2$

for each $\mathbf{x}$ satisfying $\|\mathbf{x} - \bar{\mathbf{x}}\| \le \|\mathbf{x}_1 - \bar{\mathbf{x}}\|$. Then, the algorithm converges superlinearly to $\bar{\mathbf{x}}$ with, at least, an order-two or quadratic rate of convergence.

Proof

Let the solution set $\Omega = \{\bar{\mathbf{x}}\}$ and let $X = \{\mathbf{x}: \|\mathbf{x} - \bar{\mathbf{x}}\| \le \|\mathbf{x}_1 - \bar{\mathbf{x}}\|\}$. We prove convergence by using Theorem 7.2.3. Note that X is compact and that the map $\mathbf{A}$ given via (8.21) is closed on X. We now show that $\alpha(\mathbf{x}) = \|\mathbf{x} - \bar{\mathbf{x}}\|$ is indeed a descent function. Let $\mathbf{x} \in X$, and suppose that $\mathbf{x} \ne \bar{\mathbf{x}}$. Let $\mathbf{y} \in \mathbf{A}(\mathbf{x})$. Then, by definition of $\mathbf{A}$ and since $\nabla f(\bar{\mathbf{x}}) = \mathbf{0}$, we get

$$\begin{aligned} \mathbf{y} - \bar{\mathbf{x}} &= (\mathbf{x} - \bar{\mathbf{x}}) - \mathbf{H}(\mathbf{x})^{-1}[\nabla f(\mathbf{x}) - \nabla f(\bar{\mathbf{x}})] \\ &= \mathbf{H}(\mathbf{x})^{-1}[\nabla f(\bar{\mathbf{x}}) - \nabla f(\mathbf{x}) - \mathbf{H}(\mathbf{x})(\bar{\mathbf{x}} - \mathbf{x})] \end{aligned}$$

Noting 1 and 2, it then follows that

$$\begin{aligned} \|\mathbf{y} - \bar{\mathbf{x}}\| &= \|\mathbf{H}(\mathbf{x})^{-1}[\nabla f(\bar{\mathbf{x}}) - \nabla f(\mathbf{x}) - \mathbf{H}(\mathbf{x})(\bar{\mathbf{x}} - \mathbf{x})]\| \\ &\le \|\mathbf{H}(\mathbf{x})^{-1}\| \, \|\nabla f(\bar{\mathbf{x}}) - \nabla f(\mathbf{x}) - \mathbf{H}(\mathbf{x})(\bar{\mathbf{x}} - \mathbf{x})\| \\ &\le k_1 k_2 \|\mathbf{x} - \bar{\mathbf{x}}\|^2 \le k_1 k_2 \|\mathbf{x}_1 - \bar{\mathbf{x}}\| \, \|\mathbf{x} - \bar{\mathbf{x}}\| \\ &< \|\mathbf{x} - \bar{\mathbf{x}}\| \end{aligned}$$

This shows that α is indeed a descent function. By the corollary to Theorem 7.2.3, we have convergence to $\bar{\mathbf{x}}$. Moreover, for any iterate $\mathbf{x}_k \in X$, the new iterate $\mathbf{y} = \mathbf{x}_{k+1}$ produced by the algorithm satisfies $\|\mathbf{x}_{k+1} - \bar{\mathbf{x}}\| \leq k_1 k_2 \|\mathbf{x}_k - \bar{\mathbf{x}}\|^2$ from above. Since $\{\mathbf{x}_k\} \rightarrow \bar{\mathbf{x}}$, we have, at least, an order-two rate of convergence.

8.7 Modification of Newton's Method: Levenberg–Marquardt and Trust Region Methods

In Theorem 8.6.5, we have seen that if Newton's method is initialized close enough to a local minimum $\bar{\mathbf{x}}$ with a positive definite Hessian $\mathbf{H}(\bar{\mathbf{x}})$, then it converges quadratically to this solution. In general, we have observed that the method may not be defined because of the singularity of $\mathbf{H}(\mathbf{x}_k)$ at a given point $\mathbf{x}_k$, or the search direction $\mathbf{d}_k = -\mathbf{H}(\mathbf{x}_k)^{-1}\nabla f(\mathbf{x}_k)$ may not be a descent direction; or, even if $\nabla f(\mathbf{x}_k)'\mathbf{d}_k < 0$, a unit step size might not give a descent in f. To safeguard against the latter, one could perform a line search given that $\mathbf{d}_k$ is a descent direction. However, for the more critical issue of having a well-defined algorithm that converges to a point of zero gradient irrespective of the starting solution (i.e., enjoys *global convergence*), the following modification can be adopted.

We now discuss a modification of Newton's method that guarantees convergence regardless of the starting point. Given $\mathbf{x}$, consider the direction $\mathbf{d} = -\mathbf{B}\nabla f(\mathbf{x})$, where $\mathbf{B}$ is a symmetric positive definite matrix to be determined later. The successor point is $\mathbf{y} = \mathbf{x} + \hat{\lambda}\mathbf{d}$, where $\hat{\lambda}$ is an optimal solution to the problem to minimize $f(\mathbf{x} + \lambda\mathbf{d})$ subject to $\lambda \geq 0$.

We now specify the matrix $\mathbf{B}$ as $(\varepsilon\mathbf{I} + \mathbf{H})^{-1}$, where $\mathbf{H} = \mathbf{H}(\mathbf{x})$. The scalar $\varepsilon \geq 0$ is determined as follows. Fix $\delta > 0$, and let $\varepsilon \geq 0$ be the smallest scalar that would make all the eigenvalues of the matrix $(\varepsilon\mathbf{I} + \mathbf{H})$ greater or equal to δ. Since the eigenvalues of $\varepsilon\mathbf{I} + \mathbf{H}$ are all positive, $\varepsilon\mathbf{I} + \mathbf{H}$ is positive definite and invertible. In particular, $\mathbf{B} = (\varepsilon\mathbf{I} + \mathbf{H})^{-1}$ is also positive definite. Since the eigenvalues of a matrix depend continuously on its elements, ε is a continuous function of $\mathbf{x}$, and, hence, the point-to-point map $\mathbf{D}: E_n \rightarrow E_n \times E_n$ defined by $\mathbf{D}(\mathbf{x}) = (\mathbf{x}, \mathbf{d})$ is continuous. Thus, the algorithmic map is $\mathbf{A} = \mathbf{MD}$, where $\mathbf{M}$ is the usual line search map over $\{\lambda : \lambda \geq 0\}$.

Let $\Omega = \{\bar{\mathbf{x}} : \nabla f(\bar{\mathbf{x}}) = \mathbf{0}\}$, and let $\mathbf{x} \notin \Omega$. Since $\mathbf{B}$ is positive definite, $\mathbf{d} = -\mathbf{B}\nabla f(\mathbf{x}) \neq \mathbf{0}$; and, by Theorem 8.4.1, it follows that $\mathbf{M}$ is closed at $(\mathbf{x}, \mathbf{d})$. Furthermore, since $\mathbf{D}$ is a continuous function, by Corollary 2 to Theorem 7.3.2, $\mathbf{A} = \mathbf{MD}$ is closed over the complement of Ω.

In order to invoke Theorem 7.2.3, we need to specify a continuous descent function. Suppose that $\mathbf{x} \notin \Omega$, and let $\mathbf{y} \in \mathbf{A}(\mathbf{x})$. Note that $\nabla f(\mathbf{x})'\mathbf{d} = -\nabla f(\mathbf{x})'\mathbf{B}\nabla f(\mathbf{x}) < 0$ since $\mathbf{B}$ is positive definite and $\nabla f(\mathbf{x}) \neq \mathbf{0}$. Thus, $\mathbf{d}$ is a descent direction of f at $\mathbf{x}$, and by Theorem 4.1.2, $f(\mathbf{y}) < f(\mathbf{x})$. Therefore, f is indeed a descent function. Assuming that the sequence generated by the algorithm is contained in a compact set, by Theorem 7.2.3, it follows that the algorithm converges.

It should be noted that if the smallest eigenvalue of $\mathbf{H}(\bar{\mathbf{x}})$ is greater than or equal to δ, then, as the points $\{\mathbf{x}_k\}$ generated by the algorithm approach $\bar{\mathbf{x}}$, ε_k will be equal to zero. Thus, $\mathbf{d}_k = -\mathbf{H}(\mathbf{x}_k)^{-1}\nabla f(\mathbf{x}_k)$, and the algorithm reduces to that of Newton and, hence, this method also enjoys an order-two rate of convergence.

This underscores the importance of properly selecting δ. If δ is chosen to be too small to ensure the asymptotic quadratic convergence rate because of the reduction of the method to Newton's algorithm, then ill-conditioning might occur at points where the

Hessian is (near) singular. On the other hand, if δ is chosen to be very large, which would necessitate using a large value of ε and would make $\mathbf{B}$ diagonally dominant, then the method would behave similarly to the steepest descent algorithm, and only a linear convergence rate would be realized.

The foregoing algorithmic scheme of determining the new iterate $\mathbf{x}_{k+1}$ from an iterate $\mathbf{x}_k$ according to the solution of the system

$$[\varepsilon_k \mathbf{I} + \mathbf{H}(\mathbf{x}_k)](\mathbf{x}_{k+1} - \mathbf{x}_k) = -\nabla f(\mathbf{x}_k) \tag{8.22}$$

in lieu of (8.21) is generally known as a *Levenberg–Marquardt* type of method, following a similar scheme proposed for solving nonlinear least squares problems. A typical operational prescription for such a method is as follows. (The parameters 0.25, 0.75, 2, 4, etc., used below have been found to empirically work well, and the method is relatively insensitive to these parameter values.)

Given an iterate $\mathbf{x}_k$ and a parameter $\varepsilon_k > 0$, first ascertain the positive definiteness of $\varepsilon_k \mathbf{I} + \mathbf{H}(\mathbf{x}_k)$ by attempting to construct its Cholesky factorization $\mathbf{LL}^t$ (see Appendix A.2). If this is unsuccessful, then multiply ε_k by a factor of 4 and repeat until such a factorization is available. Then, solve system (8.22) via $\mathbf{LL}^t(\mathbf{x}_{k+1} - \mathbf{x}_k) = -\nabla f(\mathbf{x}_k)$, exploiting the triangularity of $\mathbf{L}$ to obtain $\mathbf{x}_{k+1}$. Compute $f(\mathbf{x}_{k+1})$ and determine R_k as the ratio of the actual decrease $f(\mathbf{x}_k) - f(\mathbf{x}_{k+1})$ in f to its predicted decrease $q(\mathbf{x}_k) - q(\mathbf{x}_{k+1})$ as foretold by the quadratic approximation q to f at $\mathbf{x} = \mathbf{x}_k$. Note that the closer R_k is to unity, the more reliable is the quadratic approximation, and the smaller we can afford ε to be. With this motivation, if $R_k < 0.25$, put $\varepsilon_{k+1} = 4\varepsilon_k$; if $R_k > 0.75$, put $\varepsilon_{k+1} = \varepsilon_k/2$; otherwise, put $\varepsilon_{k+1} = \varepsilon_k$. Furthermore, in case $R_k \leq 0$ so that no improvement in f is realized, reset $\mathbf{x}_{k+1} = \mathbf{x}_k$; or else, retain the computed $\mathbf{x}_{k+1}$. Incrementing k by 1, reiterate until convergence to a point of zero gradient is obtained.

A scheme of this type bears a close resemblance and relationship to *trust region*, or *restricted step*, types of methods for minimizing f. Note that the main difficulty with Newton's method is that the "region of trust" within which the quadratic approximation at a given point $\mathbf{x}_k$ can be considered to be sufficiently reliable might not include a point in the solution set. To circumvent this problem, we can consider the trust region subproblem

$$\text{Minimize } \{q(\mathbf{x}) : \mathbf{x} \in \Omega_k\} \tag{8.23}$$

where q is the quadratic approximation to f at $\mathbf{x} = \mathbf{x}_k$, and Ω_k is a trust region defined by $\Omega_k = \{\mathbf{x} : \|\mathbf{x} - \mathbf{x}_k\| \leq \Delta_k\}$ for some *trust region parameter* $\Delta_k > 0$. (Here, $\|\cdot\|$ is the l_2 norm; when the l_∞ norm is used instead, the method is also known as the *box step*, or *hypercube*, method.) Now let $\mathbf{x}_{k+1}$ solve (8.23) and, as before, defined R_k as the ratio of actual to predicted descent. If R_k is too small relative to unity, then the trust region needs to be reduced; but if it is sufficiently respectable in value, the trust region can actually be expanded. The following is a typical prescription for defining Δ_{k+1} for the next iteration where, again, the method is known to be relatively insensitive to the specified parameter choices. If $R_k < 0.25$, put $\Delta_{k+1} = \|\mathbf{x}_{k+1} - \mathbf{x}_k\|/4$. If $R_k > 0.75$ and $\|\mathbf{x}_{k+1} - \mathbf{x}_k\| = \Delta_k$, that is, the trust region constraint is binding in (8.23), then put $\Delta_{k+1} = 2\Delta_k$. Otherwise, retain $\Delta_{k+1} = \Delta_k$. Furthermore, in case $R_k \leq 0$ so that f did not improve at this iteration, reset $\mathbf{x}_{k+1}$ to $\mathbf{x}_k$ itself. Then, increment k by 1 and repeat until a point with a zero gradient obtains. If this does not occur finitely, it can be shown that if the sequence $\{\mathbf{x}_k\}$ generated is contained in a compact set, and if f is continuously twice-differentiable, then there exists an accumulation point $\bar{\mathbf{x}}$ of this sequence for which $\nabla f(\bar{\mathbf{x}}) = \mathbf{0}$ and $\mathbf{H}(\bar{\mathbf{x}})$ is positive semidefinite. Moreover, if $\mathbf{H}(\bar{\mathbf{x}})$ is positive definite, then,

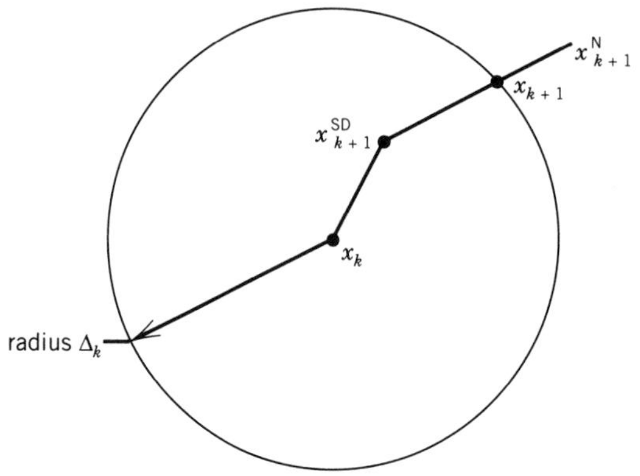

Figure 8.19 Dog-leg trajectory.

for k sufficiently large, the trust-region bound is inactive and, hence, the method reduces to Newton's method with a second-order rate of convergence (see the Notes and References section for further details).

There are two noteworthy points in relation to the foregoing discussion. First, wherever the actual Hessian has been employed above in the quadratic representation of f, an approximation to this Hessian can be used in practice, following quasi-Newton methods as discussed in the next section. Second, observe that by writing $\boldsymbol{\delta} = (\mathbf{x} - \mathbf{x}_k)$ and equivalently squaring both sides of the constraint defining Ω_k, we can write (8.23) explicitly as follows:

$$\text{Minimize}\,\{\nabla f(\mathbf{x}_k)^t\boldsymbol{\delta} + \tfrac{1}{2}\boldsymbol{\delta}^t\mathbf{H}(\mathbf{x}_k)\boldsymbol{\delta} : \tfrac{1}{2}\|\boldsymbol{\delta}\|^2 \leq \tfrac{1}{2}\Delta_k^2\} \tag{8.24}$$

The KKT conditions for (8.24) require a nonnegative Lagrange multiplier λ and a primal feasible solution $\boldsymbol{\delta}$ such that the following holds in addition to the complementary slackness condition:

$$[\mathbf{H}(\mathbf{x}_k) + \lambda\mathbf{I}]\boldsymbol{\delta} = -\nabla f(\mathbf{x}_k)$$

Note the resemblance of this to the Levenberg–Marquardt method given by (8.22). In particular, if $\Delta_k = -[\mathbf{H}(\mathbf{x}_k) + \varepsilon_k\mathbf{I}]^{-1}\nabla f(\mathbf{x}_k)$ in (8.24), where $[\mathbf{H}(\mathbf{x}_k) + \varepsilon_k\mathbf{I}]$ is positive definite, then, indeed, it is readily verified that $\boldsymbol{\delta} = (\mathbf{x}_{k+1} - \mathbf{x}_k)$ given by (8.22) and $\lambda = \varepsilon_k$ satisfy the saddle point optimality conditions for (8.24) (see Exercise 8.29). Hence, the Levenberg–Marquardt scheme described above can be viewed as a trust region type of method as well.

Finally, let us comment on a *dog-leg trajectory* proposed by Powell, which more directly follows the above philosophy of compromising between a steepest descent step and Newton's step, depending on the trust region size Δ_k. Referring to Figure 8.19, let $\mathbf{x}^{SD}_{k+1}$ and $\mathbf{x}^{N}_{k+1}$ respectively denote the new iterate obtained via a steepest descent step, (8.16), and a Newton step, (8.21) ($\mathbf{x}^{SD}_{k+1}$ is sometimes also called the *Cauchy point*). The piecewise linear curve defined by the line segments joining $\mathbf{x}_k$ to $\mathbf{x}^{SD}_{k+1}$ and $\mathbf{x}^{SD}_{k+1}$ to $\mathbf{x}^{N}_{k+1}$ is called the *dog-leg trajectory*. The proposed new iterate $\mathbf{x}_{k+1}$ is taken as the (unique) point at which the circle with radius Δ_k and centered at $\mathbf{x}_k$ intercepts this trajectory, if at all, as shown in Figure 8.19, and is taken as the Newton iterate $\mathbf{x}^{N}_{k+1}$ otherwise. Hence, when Δ_k is small relative to the dog-leg trajectory, the method behaves as a steepest descent algorithm; and with a relatively larger Δ_k, it reduces to Newton's method. Again, under suitable assumptions, as above, second-order convergence to a stationary point can be established. Moreover, the algorithmic step is simple and obviates (8.22) or (8.23). We refer the reader to the Notes and References section for further reading on this subject.

8.8 Methods Using Conjugate Directions: Quasi-Newton and Conjugate Gradient Methods

In this section, we discuss several procedures that are based on the important concept of conjugacy. Some of these procedures use derivatives, while others only use functional evaluations. The notion of conjugacy defined below is very useful in unconstrained optimization. In particular, if the objective function is quadratic, then, by searching along conjugate directions, in any order, the minimum point can be obtained in, at most, n steps.

8.8.1 Definition

Let $\mathbf{H}$ be an $n \times n$ symmetric matrix. The vectors $\mathbf{d}_1, \ldots, \mathbf{d}_k$ are called **H**-*conjugate* or simply *conjugate* if they are linearly independent, and if $\mathbf{d}_i^t\mathbf{H}\mathbf{d}_j = 0$ for $i \neq j$.

It is instructive to observe the significance of conjugacy to the minimization of quadratic functions. Consider the quadratic function $f(\mathbf{x}) = \mathbf{c}^t\mathbf{x} + \frac{1}{2}\mathbf{x}^t\mathbf{H}(\mathbf{x})$, where $\mathbf{H}$ is an $n \times n$ symmetric matrix, and suppose that $\mathbf{d}_1, \ldots, \mathbf{d}_n$ are **H**-conjugate directions. By the linear independence of these direction vectors, given a starting point $\mathbf{x}_1$, any point $\mathbf{x}$ can be uniquely represented as $\mathbf{x} = \mathbf{x}_1 + \sum_{j=1}^n \lambda_j\mathbf{d}_j$. Using this substitution to write $f(\mathbf{x})$ as a function F of λ, we get

$$F(\lambda) = \mathbf{c}^t\mathbf{x}_1 + \sum_{j=1}^n \lambda_j\mathbf{c}^t\mathbf{d}_j + \frac{1}{2}\left(\mathbf{x}_1 + \sum_{j=1}^n \lambda_j\mathbf{d}_j\right)^t\mathbf{H}\left(\mathbf{x}_1 + \sum_{j=1}^n \lambda_j\mathbf{d}_j\right)$$

Using the **H**-conjugacy of $\mathbf{d}_1, \ldots, \mathbf{d}_n$, this simplifies to

$$F(\lambda) = \sum_{j=1}^n [\mathbf{c}^t(\mathbf{x}_1 + \lambda_j\mathbf{d}_j) + \tfrac{1}{2}(\mathbf{x}_1 + \lambda_j\mathbf{d}_j)^t\mathbf{H}(\mathbf{x}_1 + \lambda_j\mathbf{d}_j)$$

Observe that F is separable in $\lambda_1, \ldots, \lambda_n$ and can be minimized by independently minimizing each term in $[\cdot]$ and then composing the net result. Note that the minimization of each such term corresponds to minimizing f from $\mathbf{x}_1$ along the direction $\mathbf{d}_j$. (In particular, if $\mathbf{H}$ is positive definite, the minimizing value of λ_j is given by $\lambda_j^* = -[\mathbf{c}^t\mathbf{d}_j + \mathbf{x}_1^t\mathbf{H}\mathbf{d}_j]/\mathbf{d}_j^t\mathbf{H}\mathbf{d}_j$ for $j = 1, \ldots, n$.) Alternatively, the foregoing derivation readily reveals that the same minimizing step lengths λ_j^*, $j = 1, \ldots, n$ result if we sequentially minimize f from $\mathbf{x}_1$ along the directions $\mathbf{d}_1, \ldots, \mathbf{d}_n$ in any order, leading to an optimal solution.

The following example illustrates the notion of conjugacy and highlights the foregoing significance of optimizing along conjugate directions for quadratic functions.

8.8.2 Example

Consider the following problem:

Minimize $\quad -12x_2 + 4x_1^2 + 4x_2^2 + 4x_1x_2$

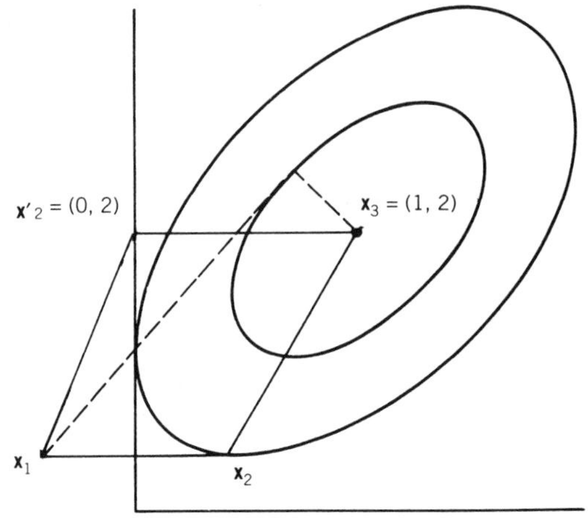

Figure 8.20 Illustration of conjugate directions.

Note that the Hessian matrix $\mathbf{H}$ is given by

$$\mathbf{H} = \begin{bmatrix} 8 & -4 \\ -4 & 8 \end{bmatrix}$$

We now generate two conjugate directions, $\mathbf{d}_1$ and $\mathbf{d}_2$. Suppose we choose $\mathbf{d}_1^t = (1, 0)$. Then, $\mathbf{d}_2^t = (a, b)$ must satisfy $0 = \mathbf{d}_1^t\mathbf{H}\mathbf{d}_2 = 8a - 4b$. In particular, we may choose $a = 1$ and $b = 2$ so that $\mathbf{d}_2^t = (1, 2)$. It may be noted that the conjugate directions are not unique.

If we minimize the objective function f starting from $\mathbf{x}_1^t = (-\frac{1}{2}, 1)$ along the direction $\mathbf{d}_1$, we get the point $\mathbf{x}_2^t = (\frac{1}{2}, 1)$. Now, starting from $\mathbf{x}_2$ and minimizing along $\mathbf{d}_2$, we get $\mathbf{x}_3^t = (1, 2)$. Note that $\mathbf{x}_3$ is the minimizing point.

The contours of the objective function and the path taken to reach the optimal point are shown in Figure 8.20. The reader can easily verify that, starting from any point and minimizing along $\mathbf{d}_1$ and $\mathbf{d}_2$, the optimal point is reached in, at most, two steps as shown, for example, by dashed lines in Figure 8.20. Furthermore, if we had started at $\mathbf{x}_1$ and then minimized along $\mathbf{d}_2$ first and next along $\mathbf{d}_1$, the optimizing step lengths along these respective directions would have remained the same as for the first case, taking the iterates from $\mathbf{x}_1$ to $\mathbf{x}_2' = (0, 2)^t$ to $\mathbf{x}_3$.

Optimization of Quadratic Functions: Finite Convergence

The above example demonstrates that a quadratic function can be minimized in, at most, n steps, provided that we search along conjugate directions of the Hessian matrix. This result is generally true for quadratic functions, as shown by Theorem 8.8.3 below. This, coupled with the fact that a general function can be closely represented by its quadratic approximation in the vicinity of the optimal point, makes the notion of conjugacy very useful for optimizing both quadratic and nonquadratic functions. Note also that this result shows that if we start at $\mathbf{x}_1$, then, at each step $k = 1, \ldots, n$, the point $\mathbf{x}_{k+1}$ obtained minimizes f over the linear subspace containing $\mathbf{x}_1$ that is spanned by the vectors $\mathbf{d}_1, \ldots, \mathbf{d}_k$. Moreover, the gradient $\nabla f(\mathbf{x}_{k+1})$, if nonzero, is orthogonal to this subspace. This is sometimes called the *expanding subspace property*, and is illustrated in Figure 8.21 for $k = 1, 2$.

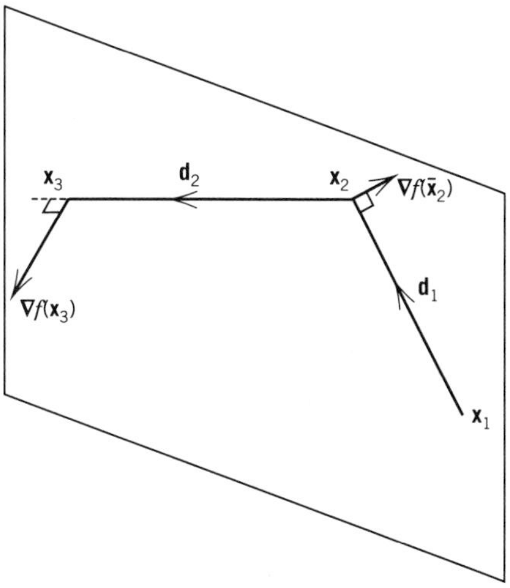

Figure 8.21 Expanding subspace property.

8.8.3 Theorem

Let $f(\mathbf{x}) = \mathbf{c}^t\mathbf{x} + \frac{1}{2}\mathbf{x}^t\mathbf{H}(\mathbf{x})$, where $\mathbf{H}$ is an $n \times n$ symmetric matrix. Let $\mathbf{d}_1, \ldots, \mathbf{d}_n$ be $\mathbf{H}$-conjugate, and let $\mathbf{x}_1$ be an arbitrary starting point. For $k = 1, \ldots, n$, let λ_k be an optimal solution to the problem to minimize $f(\mathbf{x}_k + \lambda\mathbf{d}_k)$ subject to $\lambda \in E_1$, and let $\mathbf{x}_{k+1} = \mathbf{x}_k + \lambda_k\mathbf{d}_k$. Then, for $k = 1, \ldots, n$, we must have

1. $\nabla f(\mathbf{x}_{k+1})^t\mathbf{d}_j = 0 \qquad \text{for } j = 1, \ldots, k$
2. $\nabla f(\mathbf{x}_1)^t\mathbf{d}_k = \nabla f(\mathbf{x}_k)^t\mathbf{d}_k$
3. $\mathbf{x}_{k+1}$ is an optimal solution to the problem to minimize $f(\mathbf{x})$ subject to $\mathbf{x} - \mathbf{x}_1 \in L(\mathbf{d}_1, \ldots, \mathbf{d}_k)$, where $L(\mathbf{d}_1, \ldots, \mathbf{d}_k)$ is the linear subspace formed by $\mathbf{d}_1, \ldots, \mathbf{d}_k$, that is, $L(\mathbf{d}_1, \ldots, \mathbf{d}_k) = \{\sum_{j=1}^k \mu_j\mathbf{d}_j : \mu_j \in E_1 \text{ for each } j\}$. In particular, $\mathbf{x}_{n+1}$ is a minimizing point of f over E_n.

Proof

To prove part 1, first note that $f(\mathbf{x}_j + \lambda\mathbf{d}_j)$ achieves a minimum at λ_j only if $\nabla f(\mathbf{x}_j + \lambda_j\mathbf{d}_j)^t\mathbf{d}_j = 0$, that is, $\nabla f(\mathbf{x}_{j+1})^t\mathbf{d}_j = 0$. Thus, part 1 holds for $j = k$. For $j < k$, note that

$$\begin{aligned}\nabla f(\mathbf{x}_{k+1}) = \mathbf{c} + \mathbf{H}\mathbf{x}_{j+1} &= \mathbf{c} + \mathbf{H}\mathbf{x}_{j+1} + \mathbf{H}\left(\sum_{i=j+1}^{k} \lambda_i\mathbf{d}_i\right) \\ &= \nabla f(\mathbf{x}_{j+1}) + \mathbf{H}\left(\sum_{i=j+1}^{k} \lambda_i\mathbf{d}_i\right)\end{aligned} \tag{8.25}$$

By conjugacy, $\mathbf{d}_i^t\mathbf{H}\mathbf{d}_j = 0$ for $i = j + 1, \ldots, k$. Thus, from (8.25) it follows that $\nabla f(\mathbf{x}_{k+1})^t\mathbf{d}_j = 0$, and part (1) holds.

Replacing k by $k - 1$ and letting $j = 0$ in (8.25), we get

$$\nabla f(\mathbf{x}_k) = \nabla f(\mathbf{x}_1) + \mathbf{H}\left(\sum_{i=1}^{k-1} \lambda_i\mathbf{d}_i\right) \qquad \text{for } k \geq 2$$

Multiplying by $\mathbf{d}_k^t$ and noting that $\mathbf{d}_k^t\mathbf{H}\mathbf{d}_i = 0$ for $i = 1, \ldots, k - 1$ shows that part (2) holds for $k \geq 2$. Part (2) holds true trivially for $k = 1$.

To show part (3), since $\mathbf{d}_i^t\mathbf{H}\mathbf{d}_j = 0$ for $i \neq j$, we get

$$\begin{aligned} f(\mathbf{x}_{k+1}) &= f[\mathbf{x}_1 + (\mathbf{x}_{k+1} - \mathbf{x}_1)] = f\left(\mathbf{x}_1 + \sum_{j=1}^{k} \lambda_j \mathbf{d}_j\right) \\ &= f(\mathbf{x}_1) + \nabla f(\mathbf{x}_1)^t\left(\sum_{j=1}^{k} \lambda_j \mathbf{d}_j\right) + \frac{1}{2}\sum_{j=1}^{k} \lambda_j^2 \mathbf{d}_j^t\mathbf{H}\mathbf{d}_j \end{aligned} \tag{8.26}$$

Now suppose that $\mathbf{x} - \mathbf{x}_1 \in L(\mathbf{d}_1, \ldots, \mathbf{d}_k)$, so that $\mathbf{x}$ can be written as $\mathbf{x}_1 + \sum_{j=1}^{k} \mu_j \mathbf{d}_j$. As in (8.26), we get

$$f(\mathbf{x}) = f(\mathbf{x}_1) + \nabla f(\mathbf{x}_1)^t\left(\sum_{j=1}^{k} \mu_j \mathbf{d}_j\right) + \frac{1}{2}\sum_{j=1}^{k} \mu_j^2 \mathbf{d}_j^t\mathbf{H}\mathbf{d}_j \tag{8.27}$$

To complete the proof, we need to show that $f(\mathbf{x}) \geq f(\mathbf{x}_{k+1})$. By contradiction, suppose that $f(\mathbf{x}) < f(\mathbf{x}_{k+1})$. Then by (8.26) and (8.27), we must have

$$\begin{aligned} &\nabla f(\mathbf{x}_1)^t\left(\sum_{j=1}^{k} \mu_j \mathbf{d}_j\right) + \frac{1}{2}\sum_{j=1}^{k} \mu_j^2 \mathbf{d}_j^t\mathbf{H}\mathbf{d}_j \\ &< \nabla f(\mathbf{x}_1)^t\left(\sum_{j=1}^{k} \lambda_j \mathbf{d}_j\right) + \frac{1}{2}\sum_{j=1}^{k} \lambda_j^2 \mathbf{d}_j^t\mathbf{H}\mathbf{d}_j \end{aligned} \tag{8.28}$$

By definition of λ_j, note that $f(\mathbf{x}_j + \lambda_j\mathbf{d}_j) \leq f(\mathbf{x}_j + \mu_j\mathbf{d}_j)$ for each j. Therefore

$$f(\mathbf{x}_j) + \lambda_j\nabla f(\mathbf{x}_j)^t\mathbf{d}_j + \tfrac{1}{2}\lambda_j^2 \mathbf{d}_j^t\mathbf{H}\mathbf{d}_j \leq f(\mathbf{x}_j) + \mu_j\nabla f(\mathbf{x}_j)^t\mathbf{d}_j + \tfrac{1}{2}\mu_j^2 \mathbf{d}_j^t\mathbf{H}\mathbf{d}_j$$

By part 2, $\nabla f(\mathbf{x}_j)^t\mathbf{d}_j = \nabla f(\mathbf{x}_1)^t\mathbf{d}_j$ and substituting in the above inequality, we get

$$\lambda_j\nabla f(\mathbf{x}_1)^t\mathbf{d}_j + \tfrac{1}{2}\lambda_j^2 \mathbf{d}_j^t\mathbf{H}\mathbf{d}_j \leq \mu_j\nabla f(\mathbf{x}_1)^t\mathbf{d}_j + \tfrac{1}{2}\mu_j^2 \mathbf{d}_j^t\mathbf{H}\mathbf{d}_j \tag{8.29}$$

Summing (8.29) for $j = 1, \ldots, k$ contradicts (8.28). Thus $\mathbf{x}_{k+1}$ is a minimizing point over the manifold $\mathbf{x}_1 + L(\mathbf{d}_1, \ldots, \mathbf{d}_k)$. In particular, since $\mathbf{d}_1, \ldots, \mathbf{d}_n$ are linearly independent, then $L(\mathbf{d}_1, \ldots, \mathbf{d}_n) = E_n$, and, hence, $\mathbf{x}_{n+1}$ is a minimizing point of f over E_n. This completes the proof.

Generating Conjugate Directions

In the remainder of this section, we describe several methods for generating conjugate directions of quadratic forms. These methods lead naturally to powerful algorithms for minimizing both quadratic and nonquadratic functions. In particular, we discuss the classes of quasi-Newton and conjugate gradient methods.

Quasi-Newton Methods: The Method of Davidon–Fletcher–Powell

This method was originally proposed by Davidon [1959] and later developed by Fletcher and Powell [1963]. The Davidon–Fletcher–Powell (DFP) method falls under the general class of *quasi-Newton procedures,* where the search directions are of the form $d_j = -\mathbf{D}_j\nabla f(\mathbf{y})$, in lieu of $-\mathbf{H}^{-1}(\mathbf{y})\nabla f(\mathbf{y})$, as in Newton's method. The gradient direction is thus deflected by premultiplying it by $-\mathbf{D}_j$, where $\mathbf{D}_j$ is an $n \times n$ positive definite symmetric matrix that approximates the inverse of the Hessian matrix. The positive definiteness property ensures that $\mathbf{d}_j$ is a descent direction whenever $\nabla f(\mathbf{y}) \neq 0$, since then, $\mathbf{d}_j^t\,\nabla f(\mathbf{y}) < 0$. For the purpose of the next step, $\mathbf{D}_{j+1}$ is formed by adding to $\mathbf{D}_j$ two symmetric matrices, each of rank one. Thus, this scheme is sometimes referred to as a *rank two correction procedure*.. For quadratic functions, this update scheme is shown later to produce the exact representation of the actual inverse Hessian within n steps. The DFP process is also called a *variable metric* method because it can be interpreted as adopting the steepest descent step in the transformed space based on the Cholesky factorization of the positive definite matrix $\mathbf{D}_j$, as discussed in Section 8.7, where this transformation varies with $\mathbf{D}_j$ from iteration to iteration. The quasi-Newton methods in which the quadratic approximation is permitted to be possibly indefinite are more generally called *secant methods*.

Summary of the Davidon–Fletcher–Powell (DFP) Method

We now summarize the Davidon–Fletcher–Powell (DFP) method for minimizing a differentiable function of several variables. In particular, if the function is quadratic, then, as shown later, the method yields conjugate directions and terminates in one complete iteration, that is, after searching along each of the conjugate directions as described below.

Initialization Step Let $\varepsilon > 0$ be the termination scalar. Choose an initial point $\mathbf{x}_1$ and an initial symmetric positive definite matrix $\mathbf{D}_1$. Let $\mathbf{y}_1 = \mathbf{x}_1$, let $k = j = 1$, and go to the main step.

Main Step

1. If $\|\nabla f(\mathbf{y}_j)\| < \varepsilon$, stop; otherwise, let $\mathbf{d}_j = -\mathbf{D}_j\nabla f(\mathbf{y}_j)$ and let λ_j be an optimal solution to the problem to minimize $f(\mathbf{y}_j + \lambda\mathbf{d}_j)$ subject to $\lambda \geq 0$. Let $\mathbf{y}_{j+1} = \mathbf{y}_j + \lambda_j\mathbf{d}_j$. If $j < n$, go to step 2. If $j = n$, let $\mathbf{y}_1 = \mathbf{x}_{k+1} = \mathbf{y}_{n+1}$, replace k by $k + 1$, let $j = 1$, and repeat step 1.
2. Construct $\mathbf{D}_{j+1}$ as follows:

$$\mathbf{D}_{j+1} = \mathbf{D}_j + \frac{\mathbf{p}_j\mathbf{p}_j^t}{\mathbf{p}_j^t\mathbf{q}_j} - \frac{\mathbf{D}_j\mathbf{q}_j\mathbf{q}_j^t\mathbf{D}_j}{\mathbf{q}_j^t\mathbf{D}_j\mathbf{q}_j} \tag{8.30}$$

where

$$\mathbf{p}_j = \lambda_j\mathbf{d}_j \equiv \mathbf{y}_{j+1} - \mathbf{y}_j \tag{8.31}$$

$$\mathbf{q}_j = \nabla f(\mathbf{y}_{j+1}) - \nabla f(\mathbf{y}_j) \tag{8.32}$$

Replace j by $j + 1$, and repeat step 1.

We remark here that the inner loop of the foregoing algorithm resets the procedure every n steps (whenever $j = n$ at step 1). Any variant that resets every $n' < n$ inner iteration steps is called a *partial quasi-Newton method*. This strategy can be useful from the viewpoint of conserving storage when $n' << n$, since then the inverse Hessian approximation can be stored implicitly by instead storing only the generating vectors $\mathbf{p}_j$ and $\mathbf{q}_j$ themselves within the inner loop iterations.

8.8.4 Example

Consider the following problem

$$\text{Minimize} \quad (x_1 - 2)^4 + (x_1 - 2x_2)^2$$

The summary of the computations using the DFP method is given in Table 8.13. At each iteration, for $j = 1, 2$, $\mathbf{d}_j$ is given by $-\mathbf{D}_j\nabla f(\mathbf{y}_j)$, where $\mathbf{D}_1$ is the identity matrix and

Table 8.13 Summary of Computations for the Davidon–Fletcher–Powell Method

Iteration k	$\mathbf{x}_k$ $f(\mathbf{x}_k)$	j	$\mathbf{y}_j$ $f(\mathbf{y}_j)$	$\nabla f(\mathbf{y}_j)$	$\lVert\nabla f(\mathbf{y}_j)\rVert$	$\mathbf{D}_j$	$\mathbf{d}_j$	λ_j	$\mathbf{y}_{j+1}$
1	(0.00, 3.00) (52.00)	1	(0.00, 3.00) (52.00)	(−44.00, 24.00)	50.12	$\begin{bmatrix} 1 & 0 \\ 0 & 1 \end{bmatrix}$	(44.00, −24.00)	0.062	(2.70, 1.51)
		2	(2.70, 1.51) (0.34)	(0.73, 1.28)	1.47	$\begin{bmatrix} 0.25 & 0.38 \\ 0.38 & 0.81 \end{bmatrix}$	(−0.67, −1.31)	0.22	(2.55, 1.22)
2	(2.55, 1.22) (0.1036)	1	(2.55, 1.22) (0.1036)	(0.89, −0.44)	0.99	$\begin{bmatrix} 1 & 0 \\ 0 & 1 \end{bmatrix}$	(−0.89, 0.44)	0.11	(2.45, 1.27)
		2	(2.45, 1.27) (0.0490)	(0.18, 0.36)	0.40	$\begin{bmatrix} 0.65 & 0.45 \\ 0.45 & 0.46 \end{bmatrix}$	(−0.28, −0.25)	0.64	(2.27, 1.11)
3	(2.27, 1.11) (0.008)	1	(2.27, 1.11) (0.008)	(0.18, −0.20)	0.27	$\begin{bmatrix} 1 & 0 \\ 0 & 1 \end{bmatrix}$	(−0.18, 0.20)	0.10	(2.25, 1.13)
		2	(2.25, 1.13) (0.004)	(0.04, 0.04)	0.06	$\begin{bmatrix} 0.80 & 0.38 \\ 0.38 & 0.31 \end{bmatrix}$	(−0.05, −0.03)	2.64	(2.12, 1.05)
4	(2.12, 1.05) (0.0005)	1	(2.12, 1.05) (0.0005)	(0.05, −0.08)	0.09	$\begin{bmatrix} 1 & 0 \\ 0 & 1 \end{bmatrix}$	(−0.05, 0.08)	0.10	(2.115, 1.058)
		2	(2.115, 1.058) (0.0002)	(0.004, 0.004)	0.006				

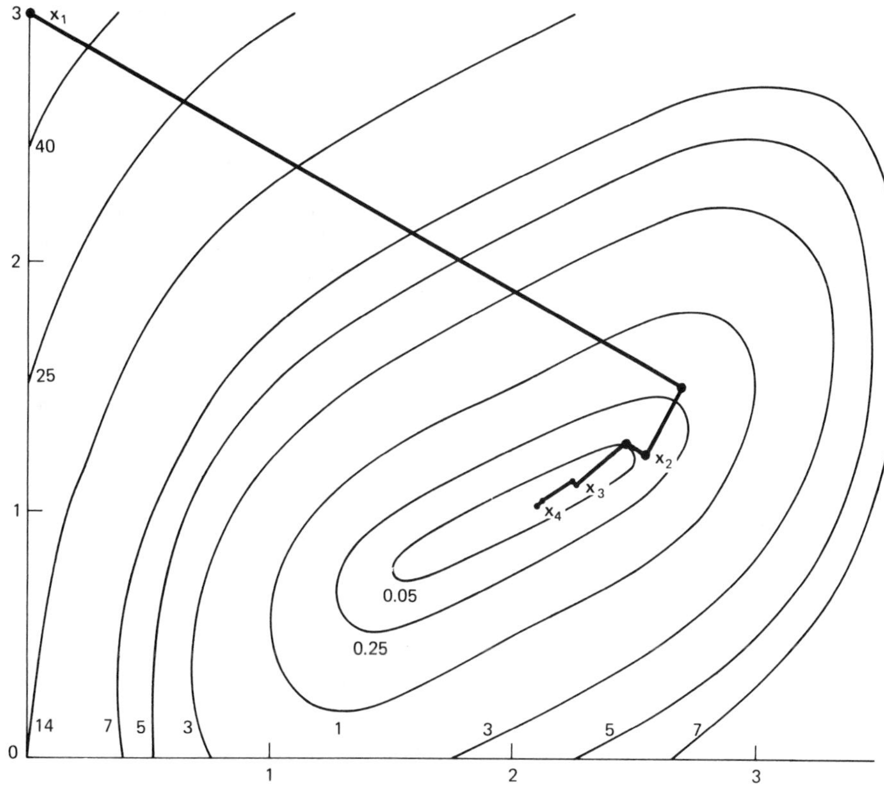

Figure 8.22 Illustration of the Davidon–Fletcher–Powell method.

$\mathbf{D}_2$ is computed from (8.30), (8.31), and (8.32). At iteration $k = 1$, we have $\mathbf{p}_1 = (2.7, -1.49)^t$ and $\mathbf{q}_1 = (44.73, -22.72)^t$ in (8.30). At iteration 2, we have $\mathbf{p}_1 = (-0.1, 0.05)^t$ and $\mathbf{q}_1 = (-0.7, 0.8)^t$, and finally at iteration 3, we have $\mathbf{p}_1 = (-0.02, 0.02)^t$ and $\mathbf{q}_1 = (-0.14, 0.24)^t$. The point $\mathbf{y}_{j+1}$ is computed by optimizing along the direction $\mathbf{d}_j$ starting from $\mathbf{y}_j$ for $j = 1, 2$. The procedure is terminated at the point $\mathbf{y}_2 = (2.115, 1.058)^t$ in the fourth iteration, since $\|\nabla f(\mathbf{y}_2)\| = 0.006$ is quite small. The path taken by the method is depicted in Figure 8.22.

Lemma 8.8.5 shows that each matrix $\mathbf{D}_j$ is positive definite and $\mathbf{d}_j$ is a direction of descent.

8.8.5 Lemma

Let $\mathbf{y}_1 \in E_n$, and let $\mathbf{D}_1$ be an initial positive definite symmetric matrix. For $j = 1, \ldots, n$, let $\mathbf{y}_{j+1} = \mathbf{y}_j + \lambda_j \mathbf{d}_j$, where $\mathbf{d}_j = -\mathbf{D}_j \nabla f(\mathbf{y}_j)$, and λ_j solves the problem to minimize $f(\mathbf{y}_j + \lambda \mathbf{d}_j)$ subject to $\lambda \geq 0$. Furthermore, for $j = 1, \ldots, n - 1$, let $\mathbf{D}_{j+1}$ be given by (8.30), (8.31), and (8.32). If $\nabla f(\mathbf{y}_j) \neq \mathbf{0}$ for $j = 1, \ldots, n$, then $\mathbf{D}_1, \ldots, \mathbf{D}_n$ are symmetric and positive definite so that $\mathbf{d}_1, \ldots, \mathbf{d}_n$ are descent directions.

Proof

We prove the result by induction. For $j = 1$, $\mathbf{D}_1$ is symmetric and positive definite by assumption. Furthermore, $\nabla f(\mathbf{y}_1)^t \mathbf{d}_1 = -\nabla f(\mathbf{y}_1)^t \mathbf{D}_1 \nabla f(\mathbf{y}_1) < 0$, since $\mathbf{D}_1$ is positive definite. By Theorem 4.1.2, $\mathbf{d}_1$ is a descent direction. We shall assume that the result holds true for $j \le n - 1$ and then show that it holds for $j + 1$. Let $\mathbf{x}$ be a nonzero vector in E_n; then, by (8.30), we have

$$\mathbf{x}^t\mathbf{D}_{j+1}\mathbf{x} = \mathbf{x}^t\mathbf{D}_j\mathbf{x} + \frac{(\mathbf{x}^t\mathbf{p}_j)^2}{\mathbf{p}_j^t\mathbf{q}_j} - \frac{(\mathbf{x}^t\mathbf{D}_j\mathbf{q}_j)^2}{\mathbf{q}_j^t\mathbf{D}_j\mathbf{q}_j} \tag{8.33}$$

Since $\mathbf{D}_j$ is a symmetric positive definite matrix, there exists a positive definite symmetric matrix $\mathbf{D}_j^{1/2}$ such that $\mathbf{D}_j = \mathbf{D}_j^{1/2}\mathbf{D}_j^{1/2}$. Let $\mathbf{a} = \mathbf{D}_j^{1/2}\mathbf{x}$ and $\mathbf{b} = \mathbf{D}_j^{1/2}\mathbf{q}_j$. Then, $\mathbf{x}^t\mathbf{D}_j\mathbf{x} = \mathbf{a}^t\mathbf{a}$, $\mathbf{q}_j^t\mathbf{D}_j\mathbf{q}_j = \mathbf{b}^t\mathbf{b}$, and $\mathbf{x}^t\mathbf{D}_j\mathbf{q}_j = \mathbf{a}^t\mathbf{b}$. Substituting in (8.33), we get

$$\mathbf{x}^t\mathbf{D}_{j+1}\mathbf{x} = \frac{(\mathbf{a}^t\mathbf{a})(\mathbf{b}^t\mathbf{b}) - (\mathbf{a}^t\mathbf{b})^2}{\mathbf{b}^t\mathbf{b}} + \frac{(\mathbf{x}^t\mathbf{p}_j)^2}{\mathbf{p}_j^t\mathbf{q}_j} \tag{8.34}$$

By the Schwartz inequality, $(\mathbf{a}^t\mathbf{a})(\mathbf{b}^t\mathbf{b}) \ge (\mathbf{a}^t\mathbf{b})^2$. Thus, to show that $\mathbf{x}^t\mathbf{D}_{j+1}\mathbf{x} \ge 0$, it suffices to show that $\mathbf{p}_j^t\mathbf{q}_j > 0$ and that $\mathbf{b}^t\mathbf{b} > 0$. From (8.31) and (8.32), it follows that

$$\mathbf{p}_j^t\mathbf{q}_j = \lambda_j\mathbf{d}_j^t[\nabla f(\mathbf{y}_{j+1}) - \nabla f(\mathbf{y}_j)]$$

The reader may note that $\mathbf{d}_j^t\nabla f(\mathbf{y}_{j+1}) = 0$, and by definition, $\mathbf{d}_j = -\mathbf{D}_j\nabla f(\mathbf{y}_j)$. Substituting in the above equation, it follows that

$$\mathbf{p}_j^t\mathbf{q}_j = \lambda_j\nabla f(\mathbf{y}_j)^t\mathbf{D}_j\nabla f(\mathbf{y}_j) \tag{8.35}$$

Note that $\nabla f(\mathbf{y}_j) \ne \mathbf{0}$ by assumption, and that $\mathbf{D}_j$ is positive definite, so that $\nabla f(\mathbf{y}_j)^t\,\mathbf{D}_j\nabla f(\mathbf{y}_j) > 0$. Furthermore, $\mathbf{d}_j$ is a descent direction and hence $\lambda_j > 0$. Therefore, from (8.35), $\mathbf{p}_j^t\mathbf{q}_j > 0$. Furthermore, $\mathbf{q}_j \ne \mathbf{0}$ and, hence, $\mathbf{b}^t\mathbf{b} = \mathbf{q}_j^t\mathbf{D}_j\mathbf{q}_j > 0$.

We now show that $\mathbf{x}^t\mathbf{D}_{j+1}\mathbf{x} > 0$. By contradiction, suppose that $\mathbf{x}^t\mathbf{D}_{j+1}\mathbf{x} = 0$. This is only possible if $(\mathbf{a}^t\mathbf{a})(\mathbf{b}^t\mathbf{b}) = (\mathbf{a}^t\mathbf{b})^2$ and $\mathbf{p}_j^t\mathbf{x} = 0$. First, note that $(\mathbf{a}^t\mathbf{a})(\mathbf{b}^t\mathbf{b}) = (\mathbf{a}^t\mathbf{b})^2$ only if $\mathbf{a} = \lambda\mathbf{b}$; that is, $\mathbf{D}_j^{1/2}\mathbf{x} = \lambda\mathbf{D}_j^{1/2}\mathbf{q}_j$. Thus, $\mathbf{x} = \lambda\mathbf{q}_j$. Since $\mathbf{x} \ne \mathbf{0}$, we have $\lambda \ne 0$. Now, $0 = \mathbf{p}_j^t\mathbf{x} = \lambda\mathbf{p}_j^t\mathbf{q}_j$ contradicts the fact that $\mathbf{p}_j^t\mathbf{q}_j > 0$ and $\lambda \ne 0$. Therefore, $\mathbf{x}^t\mathbf{D}_{j+1}\mathbf{x} > 0$, so that $\mathbf{D}_{j+1}$ is positive definite.

Since $\nabla f(\mathbf{y}_{j+1}) \ne \mathbf{0}$ and since $\mathbf{D}_{j+1}$ is positive definite, $\nabla f(\mathbf{y}_{j+1})^t\mathbf{d}_{j+1} = -\nabla f(\mathbf{y}_{j+1})^t\,\mathbf{D}_{j+1}\nabla f(\mathbf{y}_{j+1}) < 0$. By Theorem 4.1.2, then, $\mathbf{d}_{j+1}$ is a descent direction. This completes the proof.

The Quadratic Case

If the objective function f is quadratic, then by Theorem 8.8.6 below, the directions $\mathbf{d}_1, \ldots, \mathbf{d}_n$ generated by the DFP method are conjugate. Therefore, by part 3 of Theorem 8.8.3, the method stops after one complete iteration with an optimal point. Furthermore, the matrix $\mathbf{D}_{n+1}$ obtained at the end of the iteration is precisely the inverse of the Hessian matrix $\mathbf{H}$.

8.8.6 Theorem

Let $\mathbf{H}$ be an $n \times n$ symmetric positive definite matrix, and consider the problem to minimize $f(\mathbf{x}) = \mathbf{c}^t\mathbf{x} + \frac{1}{2}\mathbf{x}^t\mathbf{H}\mathbf{x}$ subject to $\mathbf{x} \in E_n$. Suppose that the problem is solved by the DFP method, starting with an initial point $\mathbf{y}_1$ and a symmetric positive definite matrix $\mathbf{D}_1$. In particular, for $j = 1, \ldots, n$, let λ_j be an optimal solution to the problem to minimize $f(\mathbf{y}_j + \lambda\mathbf{d}_j)$ subject to $\lambda \ge 0$, and let $\mathbf{y}_{j+1} = \mathbf{y}_j + \lambda_j\mathbf{d}_j$, where $\mathbf{d}_j = -\mathbf{D}_j\nabla f(\mathbf{y}_j)$ and $\mathbf{D}_j$ is determined by (8.30), (8.31), and (8.32). If $\nabla f(\mathbf{y}_j) \ne \mathbf{0}$ for each j, then the directions $\mathbf{d}_1, \ldots, \mathbf{d}_n$ are $\mathbf{H}$-conjugate and $\mathbf{D}_{n+1} = \mathbf{H}^{-1}$. Furthermore, $\mathbf{y}_{n+1}$ is an optimal solution to the problem.

Proof

We first show that, for any j with $1 \le j \le n$, we must have the following conditions:

1. $\mathbf{d}_1, \ldots, \mathbf{d}_j$ are linearly independent.
2. $\mathbf{d}_i^t\mathbf{H}\mathbf{d}_k = 0$ for $i \neq k$; $i, k \le j$. (8.36)
3. $\mathbf{D}_{j+1}\mathbf{H}\mathbf{p}_k = \mathbf{p}_k$ or, equivalently, $\mathbf{D}_{j+1}\mathbf{H}\mathbf{d}_k = \mathbf{d}_k$ for $1 \le k \le j$, where $\mathbf{p}_k = \lambda_k\mathbf{d}_k$.

We prove this result by induction. For $j = 1$, parts 1 and 2 are obvious. To prove part 3, first note that, for any k, we have

$$\mathbf{H}\mathbf{p}_k = \mathbf{H}(\lambda_k\mathbf{d}_k) = \mathbf{H}(\mathbf{y}_{k+1} - \mathbf{y}_k) = \nabla f(\mathbf{y}_{k+1}) - \nabla f(\mathbf{y}_k) = \mathbf{q}_k \tag{8.37}$$

In particular, $\mathbf{H}\mathbf{p}_1 = \mathbf{q}_1$. Thus, letting $j = 1$ in (8.30), we get

$$\mathbf{D}_2\mathbf{H}\mathbf{p}_1 = \left[\mathbf{D}_1 + \frac{\mathbf{p}_1\mathbf{p}_1^t}{\mathbf{p}_1^t\mathbf{q}_1} - \frac{\mathbf{D}_1\mathbf{q}_1\mathbf{q}_1^t\mathbf{D}_1}{\mathbf{q}_1^t\mathbf{D}_1\mathbf{q}_1}\right]\mathbf{q}_1 = \mathbf{p}_1$$

so that part 3 holds true for $j = 1$.

Now suppose that parts 1, 2, and 3 hold for $j \le n - 1$. To show that they also hold true for $j + 1$, first recall by part 1 of Theorem 8.8.3 that $\mathbf{d}_i^t\nabla f(\mathbf{y}_{j+1}) = 0$ for $i \le j$. By the induction hypothesis of part 3, $\mathbf{d}_i = \mathbf{D}_{j+1}\mathbf{H}\mathbf{d}_i$ for $i \le j$. Thus, for $i \le j$, we have

$$0 = \mathbf{d}_i^t\nabla f(\mathbf{y}_{j+1}) + \mathbf{d}_i^t\mathbf{H}\mathbf{D}_{j+1}\nabla f(\mathbf{y}_{j+1}) = -\mathbf{d}_i^t\mathbf{H}\mathbf{d}_{j+1}$$

In view of the induction hypothesis on part 2, the above equation shows that part 2 also holds for $j + 1$.

Now we show that part 3 holds for $j + 1$. Letting $k \le j + 1$,

$$\mathbf{D}_{j+2}\mathbf{H}\mathbf{p}_k = \left[\mathbf{D}_{j+1} + \frac{\mathbf{p}_{j+1}\mathbf{p}_{j+1}^t}{\mathbf{p}_{j+1}^t\mathbf{q}_{j+1}} - \frac{\mathbf{D}_{j+1}\mathbf{q}_{j+1}\mathbf{q}_{j+1}^t\mathbf{D}_{j+1}}{\mathbf{q}_{j+1}^t\mathbf{D}_{j+1}\mathbf{q}_{j+1}}\right]\mathbf{H}\mathbf{p}_k \tag{8.38}$$

Noting (8.37) and letting $k = j + 1$ in (8.38), it follows that $\mathbf{D}_{j+2}\mathbf{H}\mathbf{p}_{j+1} = \mathbf{p}_{j+1}$. Now let $k \le j$. Since part 2 holds for $j + 1$,

$$\mathbf{p}_{j+1}^t\mathbf{H}\mathbf{p}_k = \lambda_k\lambda_{j+1}\mathbf{d}_{j+1}^t\mathbf{H}\mathbf{d}_k = 0 \tag{8.39}$$

Noting the induction hypothesis on part 3, (8.37), and the fact that part 2 holds true for $j + 1$, we get

$$\mathbf{q}_{j+1}^t\mathbf{D}_{j+1}\mathbf{H}\mathbf{p}_k = \mathbf{q}_{j+1}^t\mathbf{p}_k = \mathbf{p}_{j+1}^t\mathbf{H}\mathbf{p}_k = \lambda_{j+1}\lambda_k\mathbf{d}_{j+1}^t\mathbf{H}\mathbf{d}_k = 0 \tag{8.40}$$

Substituting (8.39) and (8.40) in (8.38), and noting the induction hypothesis on part 3, we get

$$\mathbf{D}_{j+2}\mathbf{H}\mathbf{p}_k = \mathbf{D}_{j+1}\mathbf{H}\mathbf{p}_k = \mathbf{p}_k$$

Thus, part 3 holds for $j + 1$.

To finish the induction argument, we only need to show that part 1 holds true for $j + 1$. Suppose that $\sum_{i=1}^{j+1} \alpha_i\mathbf{d}_i = \mathbf{0}$. Multiplying by $\mathbf{d}_{j+1}^t\mathbf{H}$ and noting that part 2 holds for $j + 1$, it follows that $\alpha_{j+1}\mathbf{d}_{j+1}^t\mathbf{H}\mathbf{d}_{j+1} = 0$. By assumption $\nabla f(\mathbf{y}_{j+1}) \neq \mathbf{0}$ and by Lemma 8.8.5 $\mathbf{D}_{j+1}$ is positive definite, so that $\mathbf{d}_{j+1} = -\mathbf{D}_{j+1}\nabla f(\mathbf{y}_{j+1}) \neq \mathbf{0}$. Since $\mathbf{H}$ is positive definite, then $\mathbf{d}_{j+1}^t\mathbf{H}\mathbf{d}_{j+1} \neq 0$, and, hence, $\alpha_{j+1} = 0$. This in turn implies that $\sum_{i=1}^{j} \alpha_i \mathbf{d}_i = \mathbf{0}$; and since $\mathbf{d}_1, \ldots, \mathbf{d}_j$ are linearly independent by the induction hypothesis, $\alpha_i = 0$ for $i = 1, \ldots, j$. Thus, $\mathbf{d}_1, \ldots, \mathbf{d}_{j+1}$ are linearly independent and part 1 holds for $j + 1$. Thus, parts 1, 2, and 3 hold. In particular, conjugacy of $\mathbf{d}_1, \ldots, \mathbf{d}_n$ follows from parts 1 and 2 by letting $j = n$.

Now let $j = n$ in part 3. Then, $\mathbf{D}_{n+1}\mathbf{H}\mathbf{d}_k = \mathbf{d}_k$ for $k = 1, \ldots, n$. If we let $\mathbf{D}$ be the matrix whose columns are $\mathbf{d}_1, \ldots, \mathbf{d}_n$, then $\mathbf{D}_{n+1}\mathbf{H}\mathbf{D} = \mathbf{D}$. Since $\mathbf{D}$ is invertible, then $\mathbf{D}_{n+1}\mathbf{H} = \mathbf{I}$, which is only possible if $\mathbf{D}_{n+1} = \mathbf{H}^{-1}$. Finally, $\mathbf{y}_{n+1}$ is an optimal solution by Theorem 8.8.3.

An Insightful Derivation of the DFP Method

At each step of the DFP method, we have seen that, given some approximation $\mathbf{D}_j$ to the inverse Hessian matrix, we computed the search direction $\mathbf{d}_j \equiv -\mathbf{D}_j\nabla f(\mathbf{y}_j)$ by deflecting the negative gradient of f at the current solution $\mathbf{y}_j$ using this approximation $\mathbf{D}_j$ in the spirit of Newton's method. We then performed a line search along this direction and, based on the resulting solution $\mathbf{y}_{j+1}$ and the gradient $\nabla f(\mathbf{y}_{j+1})$ at this point, we obtained an updated approximation $\mathbf{D}_{j+1}$ according to (8.30), (8.31), and (8.32). As seen in Theorem 8.8.6, if f is a quadratic function given by $f(\mathbf{x}) = \mathbf{c}^t\mathbf{x} + \frac{1}{2}\mathbf{x}^t\mathbf{H}\mathbf{x}$, $\mathbf{x} \in E_n$, where $\mathbf{H}$ is symmetric and positive definite; and if $\nabla f(\mathbf{y}_j) \neq \mathbf{0}$, $j = 1, \ldots, n$, then we indeed obtain $\mathbf{D}_{n+1} = \mathbf{H}^{-1}$. In fact, observe from parts 1 and 3 of Theorem 8.8.6 that, for each $j \in \{1, \ldots, n\}$, the vectors $\mathbf{p}_1, \ldots, \mathbf{p}_j$ are linearly independent eigenvectors of $\mathbf{D}_{j+1}\mathbf{H}$ with eigenvalues equal to 1. Hence, at each step of the method, the revised approximation accumulates one additional linearly independent eigenvector, with a unit eigenvalue for the product $\mathbf{D}_{j+1}\mathbf{H}$, until $\mathbf{D}_{n+1}\mathbf{H}$ finally has all its n eigenvalues equal to 1, giving $\mathbf{D}_{n+1}\mathbf{H}\mathbf{P} = \mathbf{P}$, where $\mathbf{P}$ is the nonsingular matrix of the eigenvectors of $\mathbf{D}_{n+1}\mathbf{H}$. Hence, $\mathbf{D}_{n+1}\mathbf{H} = \mathbf{I}$, or $\mathbf{D}_{n+1} = \mathbf{H}^{-1}$.

Based on the foregoing observation, let us derive the update scheme (8.30) for the DFP method and use this derivation to motivate other more prominent updates. Toward this end, suppose that we have some symmetric, positive definite approximation $\mathbf{D}_j$ of the inverse Hessian matrix for which $\mathbf{p}_1, \ldots, \mathbf{p}_{j-1}$ are the eigenvectors of $\mathbf{D}_j\mathbf{H}$ with unit eigenvalues. (For $j = 1$, no such vectors exist.) Adopting the inductive scheme of Theorem 8.8.6, assume that these eigenvectors are linearly independent and are $\mathbf{H}$-conjugate. Now, given the current point $\mathbf{y}_j$, we conduct a line search along the direction $\mathbf{d}_j = -\mathbf{D}_j\nabla f(\mathbf{y}_j)$ to obtain the new point $\mathbf{y}_{j+1}$ and, accordingly, we define

$$\begin{aligned} \mathbf{p}_j &= (\mathbf{y}_{j+1} - \mathbf{y}_j) \qquad \text{and} \\ \mathbf{q}_j &= \nabla f(\mathbf{y}_{j+1}) - \nabla f(\mathbf{y}_j) \equiv \mathbf{H}(\mathbf{y}_{j+1} - \mathbf{y}_j) = \mathbf{H}\mathbf{p}_j \end{aligned} \tag{8.41}$$

Following the argument in the proof of Theorem 8.8.6, the vectors $\mathbf{p}_k \equiv \lambda_k\mathbf{d}_k$, $k = 1, \ldots, j$, are easily shown to be linearly independent and $\mathbf{H}$-conjugate.

We now want to construct a matrix

$$\mathbf{D}_{j+1} = \mathbf{D}_j + \mathbf{C}_j$$

where $\mathbf{C}_j$ is some symmetric correction matrix that ensures that $\mathbf{p}_1, \ldots, \mathbf{p}_j$ are eigenvectors of $\mathbf{D}_{j+1}\mathbf{H}$ with unit eigenvalues. Hence, we want $\mathbf{D}_{j+1}\mathbf{H}\mathbf{p}_k = \mathbf{p}_k$ or, from (8.41), that $\mathbf{D}_{j+1}\mathbf{q}_k = \mathbf{p}_k$ for $k = 1, \ldots, j$. For $1 \le k < j$, this translates to requiring $\mathbf{p}_k = \mathbf{D}_j\mathbf{q}_k + \mathbf{C}_j\mathbf{q}_k = \mathbf{D}_j\mathbf{H}\mathbf{p}_k + \mathbf{C}_j\mathbf{q}_k = \mathbf{p}_k + \mathbf{C}_j\mathbf{q}_k$, or that

$$\mathbf{C}_j\mathbf{q}_k = \mathbf{0} \qquad \text{for } k = 1, \ldots, j-1 \tag{8.42}$$

For $k \equiv j$, the aforementioned condition

$$\mathbf{D}_{j+1}\mathbf{q}_j = \mathbf{p}_j \tag{8.43}$$

is called the *quasi-Newton condition,* or the *secant equation,* the latter term leading to the alternate name of *secant updates* for this type of a scheme. This condition translates to the requirement that

$$\mathbf{C}_j\mathbf{q}_j = \mathbf{p}_j - \mathbf{D}_j\mathbf{q}_j \tag{8.44}$$

Now, if $\mathbf{C}_j$ had a symmetric rank-one term $\mathbf{p}_j\mathbf{p}_j^t/\mathbf{p}_j^t\mathbf{q}_j$, then $\mathbf{C}_j\mathbf{q}_j$ operating on this term would yield $\mathbf{p}_j$, as required in (8.44). Similarly, if $\mathbf{C}_j$ had a symmetric rank-one term $-(\mathbf{D}_j\mathbf{q}_j)(\mathbf{D}_j\mathbf{q}_j)^t/(\mathbf{D}_j\mathbf{q}_j)^t\mathbf{q}_j$, then $\mathbf{C}_j\mathbf{q}_j$ operating on this term would yield $-\mathbf{D}_j\mathbf{q}_j$, as required in (8.44). This therefore leads to the *rank-two DFP update* (8.30) via the correction term,

$$\mathbf{C}_j = \frac{\mathbf{p}_j\mathbf{p}_j^t}{\mathbf{p}_j^t\mathbf{q}_j} - \frac{\mathbf{D}_j\mathbf{q}_j\mathbf{q}_j^t\mathbf{D}_j}{\mathbf{q}_j^t\mathbf{D}_j\mathbf{q}_j} \equiv \mathbf{C}_j^{\text{DFP}} \tag{8.45}$$

which satisfies the quasi-Newton condition (8.43) via (8.44). (Note that, as in Lemma 8.8.5, $\mathbf{D}_{j+1} = \mathbf{D}_j + \mathbf{C}_j$ is symmetric and positive definite.) Moreover, (8.42) also holds since, for any $k \in \{1, \ldots, j-1\}$, we have from (8.45) and (8.41) that

$$\mathbf{C}_j\mathbf{q}_k = \mathbf{C}_j\mathbf{H}\mathbf{p}_k = \frac{\mathbf{p}_j\mathbf{p}_j^t\mathbf{H}\mathbf{p}_k}{\mathbf{p}_j^t\mathbf{q}_j} - \frac{\mathbf{D}_j\mathbf{q}_j\mathbf{p}_j^t\mathbf{H}\mathbf{D}_j\mathbf{H}\mathbf{p}_k}{\mathbf{q}_j^t\mathbf{D}_j\mathbf{q}_j} = \mathbf{0}$$

since $\mathbf{p}_j^t\mathbf{H}\mathbf{p}_k = 0$ in the first term and $\mathbf{p}_j^t\mathbf{H}\mathbf{D}_j\mathbf{H}\mathbf{p}_k = \mathbf{p}_j^t\mathbf{H}\mathbf{p}_k = 0$ in the second term as well. Hence, following this sequence of corrections, we shall ultimately obtain $\mathbf{D}_{n+1}\mathbf{H} = \mathbf{I}$ or $\mathbf{D}_{n+1} = \mathbf{H}^{-1}$.

The Broyden Family and the Broyden–Fletcher–Goldfarb–Shanno (BFGS) Updates

The reader might have observed in the foregoing derivation of $\mathbf{C}_j^{\text{DFP}}$ that there was a degree of flexibility in prescribing the correction matrix $\mathbf{C}_j$, the restriction being to satisfy the quasi-Newton condition (8.44) along with (8.42), and to maintain symmetry and positive definiteness of $\mathbf{D}_{j+1} = \mathbf{D}_j + \mathbf{C}_j$. In light of this, the Broyden updates suggest the use of the correction matrix $\mathbf{C}_j = \mathbf{C}_j^{\text{B}}$ given by the following family parameterized by ϕ:

$$\mathbf{C}_j^{\text{B}} = \mathbf{C}_j^{\text{DFP}} + \frac{\phi\tau_j\mathbf{v}_j\mathbf{v}_j^t}{\mathbf{p}_j^t\mathbf{q}_j} \tag{8.46}$$

where $\mathbf{v}_j = \mathbf{p}_j - (1/\tau_j)\mathbf{D}_j\mathbf{q}_j$ and where τ_j is chosen so that the quasi-Newton condition (8.44) holds by virtue of $\mathbf{v}_j^t\mathbf{q}_j$ being zero. This implies that $[\mathbf{p}_j - \mathbf{D}_j\mathbf{q}_j/\tau_j]^t\mathbf{q}_j = 0$, or that

$$\tau_j = \frac{\mathbf{q}_j^t\mathbf{D}_j\mathbf{q}_j}{\mathbf{p}_j^t\mathbf{q}_j} > 0 \tag{8.47}$$

Note that, for $1 \le k < j$, we have

$$\mathbf{v}_j^t\mathbf{q}_k = \mathbf{p}_j^t\mathbf{q}_k - \frac{1}{\tau_j}\mathbf{q}_j^t\mathbf{D}_j\mathbf{q}_k = \mathbf{p}_j^t\mathbf{H}\mathbf{p}_k - \frac{1}{\tau_j}\mathbf{p}_j^t\mathbf{H}\mathbf{D}_j\mathbf{H}\mathbf{p}_k = 0$$

because $\mathbf{p}_j^t\mathbf{H}\mathbf{p}_k = 0$ by conjugacy and $\mathbf{p}_j^t[\mathbf{D}_j\mathbf{H}\mathbf{p}_k] = \mathbf{p}_j^t\mathbf{H}\mathbf{p}_k = 0$, since $\mathbf{p}_k$ is an eigenvector of $\mathbf{D}_j\mathbf{H}$ with a unit eigenvalue. Hence, (8.42) also continues to hold. Moreover, it is clear that $\mathbf{D}_{j+1} = \mathbf{D}_j + \mathbf{C}_j^B$ continues to be symmetric and, at least for $\phi \geq 0$, also positive definite. Hence, the correction matrix (8.46)–(8.47) in this case yields a valid sequence of updates satisfying the assertion of Theorem 8.8.6.

For the value $\phi = 1$, the Broyden family gives a very useful special case which coincides with that derived independently by Broyden, Fletcher, Goldfarb, and Shanno. This update, known as the *BFGS update*, or the *positive definite secant update*, has been consistently shown in many computational studies to dominate other updating schemes in its overall performance. In contrast, the DFP update has been observed to have numerical difficulties, having the tendency to sometimes produce near-singular Hessian approximations. The additional correction term in (8.46) seems to alleviate this propensity.

To derive this update correction $\mathbf{C}_j^{BFGS}$, say, we simply substitute (8.47) into (8.46), and simplify (8.46) using $\phi = 1$ to get

$$\mathbf{C}_j^{BFGS} \equiv \mathbf{C}_j^B(\phi = 1) = \frac{\mathbf{p}_j\mathbf{p}_j^t}{\mathbf{p}_j^t\mathbf{q}_j}\left[1 + \frac{\mathbf{q}_j^t\mathbf{D}_j\mathbf{q}_j}{\mathbf{p}_j^t\mathbf{q}_j}\right] - \frac{[\mathbf{D}_j\mathbf{q}_j\mathbf{p}_j^t + \mathbf{p}_j\mathbf{q}_j^t\mathbf{D}_j]}{\mathbf{p}_j^t\mathbf{q}_j} \tag{8.48}$$

Since with $\phi = 0$, we have $\mathbf{C}_j^B \equiv \mathbf{C}_j^{DFP}$, we can write (8.46) as

$$\mathbf{C}_j^B = (1 - \phi)\mathbf{C}_j^{DFP} + \phi\mathbf{C}_j^{BFGS} \tag{8.49}$$

The above discussion assumes the use of a constant value ϕ in (8.46). This is known as a *pure* Broyden update. However, for the analytical results to hold, it is not necessary to work with a constant value of ϕ. A variable value ϕ_j can be chosen from one iteration to the next if so desired. However, there is a value of ϕ in (8.46) that will make $\mathbf{d}_{j+1} = -\mathbf{D}_{j+1}\nabla f(\mathbf{y}_{j+1})$ identically zero (see Exercise 8.33), namely,

$$\phi = \frac{-[\nabla f(\mathbf{y}_j)^t\mathbf{D}_j\nabla f(\mathbf{y}_j)][\mathbf{q}_j^t\mathbf{D}_j\mathbf{q}_j]}{\nabla f(\mathbf{y}_{j+1})^t\mathbf{D}_j\nabla f(\mathbf{y}_{j+1})} \tag{8.50}$$

Hence, the algorithm stalls and, in particular, $\mathbf{D}_{j+1}$ becomes singular and loses positive definiteness. Such a value of ϕ is said to be *degenerate*, and should be avoided. For this reason, as a safeguard, ϕ is usually taken to be nonnegative, although, sometimes, admitting negative values seems to be computationally attractive. In this connection, note that for a general differentiable function, if *perfect* line searches are performed (that is, either an exact minimum, or in the nonconvex case, the first local minimum along a search direction is found), then it can be shown (see the Notes and References section) that the sequence of iterates generated by the Broyden family is invariant with respect to the choice of the parameter ϕ, so long as nondegenerate ϕ values are chosen. Hence, the choice of ϕ becomes critical only with inexact line searches. Also, if inaccurate line searches are used, then maintaining the positive definiteness of the Hessian approximations becomes a matter of concern. In particular, this motivates the following strategy.

Updating Hessian Approximations

Note that in a spirit similar to the foregoing derivations, we could have alternatively started with a symmetric, positive definite approximation $\mathbf{B}_1$ to the Hessian $\mathbf{H}$ itself, and then updated this to produce a sequence of symmetric, positive definite approximations according to $\mathbf{B}_{j+1} = \mathbf{B}_j + \bar{\mathbf{C}}_j$ for $j = 1, \ldots, n$. Again, for each $j = 1, \ldots, n$, we would like $\mathbf{p}_1, \ldots, \mathbf{p}_j$ to be eigenvectors of $\mathbf{H}^{-1}\mathbf{B}_{j+1}$ with eigenvalues of 1, so that for $j = n$ we would obtain $\mathbf{H}^{-1}\mathbf{B}_{n+1} = \mathbf{I}$ or $\mathbf{B}_{n+1} = \mathbf{H}$ itself. Proceeding inductively as before, given that $\mathbf{p}_1, \ldots, \mathbf{p}_{j-1}$ are eigenvectors of $\mathbf{H}^{-1}\mathbf{B}_j$ with unit eigenvalues, we need to construct a correction matrix $\bar{\mathbf{C}}_j$ such that $\mathbf{H}^{-1}(\mathbf{B}_j + \bar{\mathbf{C}}_j)\mathbf{p}_k = \mathbf{p}_k$ for $k = 1, \ldots, j$. In other words, multiplying throughout by $\mathbf{H}$ and noting that $\mathbf{q}_k = \mathbf{H}\mathbf{p}_k$ for $k = 1, \ldots, j$ by (8.41), if we are given that

$$\mathbf{B}_j\mathbf{p}_k = \mathbf{q}_k \qquad \text{for } k = 1, \ldots, j-1 \tag{8.51}$$

we are required to ensure that $(\mathbf{B}_j + \bar{\mathbf{C}}_j)\mathbf{p}_k = \mathbf{q}_k$ for $k = 1, \ldots, j$ or, using (8.51), that

$$\bar{\mathbf{C}}_j\mathbf{p}_k = \mathbf{0} \text{ for } 1 \le k \le j-1 \qquad \text{and} \qquad \bar{\mathbf{C}}_j\mathbf{p}_j = \mathbf{q}_j - \mathbf{B}_j\mathbf{p}_j \tag{8.52}$$

Comparing (8.51) with the condition $\mathbf{D}_j\mathbf{q}_k = \mathbf{p}_k$ for $k = 1, \ldots, j-1$ and, similarly, comparing (8.52) with (8.42) and (8.44), we observe that the present analysis differs from the foregoing analysis involving an update of inverse Hessians in that the role of $\mathbf{D}_j$ and $\mathbf{B}_j$, and that of $\mathbf{p}_j$ and $\mathbf{q}_j$, are interchanged. By symmetry, we can derive a formula for $\bar{\mathbf{C}}_j$ by simply replacing $\mathbf{D}_j$ by $\mathbf{B}_j$ and by interchanging $\mathbf{p}_j$ and $\mathbf{q}_j$ in (8.45). An update obtained in this fashion is called a *complementary update,* or *dual update,* to the previous one. Of course, the dual of the dual formula will naturally yield the original formula. The $\bar{\mathbf{C}}_j$ derived as the dual to $\mathbf{C}_j^{\text{DFP}}$ was actually obtained independently by Broyden, Fletcher, Goldfarb, and Shanno in 1970, and the update is therefore known as the BFGS update. Hence, we have

$$\bar{\mathbf{C}}_j^{\text{BFGS}} = \frac{\mathbf{q}_j\mathbf{q}_j^t}{\mathbf{q}_j^t\mathbf{p}_j} - \frac{\mathbf{B}_j\mathbf{p}_j\mathbf{p}_j^t\mathbf{B}_j}{\mathbf{p}_j^t\mathbf{B}_j\mathbf{p}_j} \tag{8.53}$$

In Exercise 8.34, we ask the reader to directly derive (8.53) following the derivation of (8.41) through (8.45).

Note that the relationship between $\bar{\mathbf{C}}_j^{\text{BFGS}}$ and $\mathbf{C}_j^{\text{BFGS}}$ is as follows:

$$\mathbf{D}_{j+1} \equiv \mathbf{D}_j + \mathbf{C}_j^{\text{BFGS}} = \mathbf{B}_{j+1}^{-1} = (\mathbf{B}_j + \bar{\mathbf{C}}_j^{\text{BFGS}})^{-1} \tag{8.54}$$

That is to say, $\mathbf{D}_{j+1}\mathbf{q}_k = \mathbf{p}_k$ for $k = 1, \ldots, j$ implies that $\mathbf{D}_{j+1}^{-1}\mathbf{p}_k = \mathbf{q}_k$ or that $\mathbf{B}_{j+1} = \mathbf{D}_{j+1}^{-1}$ indeed satisfies (8.51) (written for $j + 1$). In fact, the inverse relationship (8.54) between (8.48) and (8.53) can be readily verified (see Exercise 8.35) by using two sequential applications of the *Sherman–Morrison–Woodbury formula* given below, which is valid for any general $n \times n$ matrix $\mathbf{A}$ and $n \times 1$ vectors $\mathbf{a}$ and $\mathbf{b}$, given that the inverse exists (or equivalently given that $(1 + \mathbf{b}^t\mathbf{A}^{-1}\mathbf{a}) \neq 0$):

$$(\mathbf{A} + \mathbf{a}\mathbf{b}^t)^{-1} = \mathbf{A}^{-1} - \frac{\mathbf{A}^{-1}\mathbf{a}\mathbf{b}^t\mathbf{A}^{-1}}{1 + \mathbf{b}^t\mathbf{A}^{-1}\mathbf{a}} \tag{8.55}$$

Note that if the Hessian approximations $\mathbf{B}_j$ are generated as above, then the search direction $\mathbf{d}_j$ at any step needs to be obtained by solving the system of equations $\mathbf{B}_j\mathbf{d}_j = -\nabla f(\mathbf{y}_j)$. This can be done more conveniently by maintaining a Cholesky factorization $\mathscr{L}_j\mathscr{D}_j\mathscr{L}_j^t$ of $\mathbf{B}_j$, where $\mathscr{L}_j$ is a lower triangular matrix and $\mathscr{D}_j$ is a diagonal matrix. Besides the numerical benefits of adopting this procedure, it can also be helpful in that the condition number of $\mathscr{D}_j$ can be useful in assessing the ill-conditioning status of $\mathbf{B}_j$, and the positive definiteness of $\mathbf{B}_j$ can be verified by checking the positiveness of the diagonal elements of $\mathscr{D}_j$. Hence, when an update of $\mathscr{D}_j$ reveals a loss of positive definiteness, alternative steps can be taken by restoring the diagonal elements of $\mathscr{D}_{j+1}$ to be positive.

Scaling of Quasi-Newton Algorithms

Let us conclude our discussion on quasi-Newton methods by making a brief, but important, comment on adopting a proper scaling of the updates generated by these methods.

In our discussion leading to the derivation of (8.41) through (8.45), we learned that at each step j, the revised update $\mathbf{D}_{j+1}$ had an additional eigenvector with a unit eigenvalue for the matrix $\mathbf{D}_{j+1}\mathbf{H}$. Hence, if, for example, $\mathbf{D}_1$ is chosen such that the eigenvalues of $\mathbf{D}_1\mathbf{H}$ are all significantly larger than unity, then, since these eigenvalues are transformed to unity one at a time as the algorithm proceeds, one can expect an unfavorable ratio of the largest to smallest eigenvalues of $\mathbf{D}_j\mathbf{H}$ at the intermediate steps. When minimizing nonquadratic functions and/or employing inexact line searches, in particular, such a phenomenon can result in ill-conditioning effects and exhibit poor convergence performance. To alleviate this, it is useful to multiply each $\mathbf{D}_j$ by some scale factor $s_j > 0$ before using the update formula. With exact line searches, this can be shown to preserve the conjugacy property in the quadratic case, although we may no longer have $\mathbf{D}_{n+1} = \mathbf{H}^{-1}$. However, the focus here is to improve the single-step rather than the n-step convergence behavior of the algorithm. Methods that automatically prescribe scale factors in a manner such that, if the function is quadratic, then the eigenvalues of $s_j\mathbf{D}_j\mathbf{H}$ tend to be spread above and below unity are called *self-scaling* methods. We refer the reader to the Notes and References section for further reading on this subject.

Conjugate Gradient Methods

Conjugate gradient methods were proposed by Hestenes and Stiefel in 1952 for solving systems of linear equations. The use of this method for unconstrained optimization was prompted by the fact that the minimization of a positive definite quadratic function is equivalent to solving the linear equation system that results when its gradient is set at zero. Actually, the extension of conjugate gradient methods to solving nonlinear equation systems and its use in solving general unconstrained minimization problems was first done in 1964 by Fletcher and Reeves. Although these methods are typically less efficient and less robust than quasi-Newton methods, they have very modest storage requirements (only three n-vectors are required for the method of Fletcher and Reeves described below) and are quite indispensable for large-sized problems (n exceeding about 100) when quasi-Newton methods become impractical because of the size of the Hessian matrix. Some very successful applications are reported by Fletcher [1987] in the context of atomic structures where problems with 3000 variables were solved using only about 50 gradient evaluations, and by Reid [1971] who solved some linear partial differential equations having some 4000 variables in about 40 iterations. Moreover, conjugate gradient methods have the advantage of simplicity, being *gradient deflection* methods that deflect the negative gradient direction using the previous direction. This deflection can alternatively be viewed as an update of a *fixed*, symmetric, positive definite matrix, usually the identity matrix, in the spirit of quasi-Newton methods. For this reason, they are sometimes referred to as *fixed-metric* methods in contrast with the term *variable-metric* methods, which applies to quasi-Newton procedures. Again, these are conjugate direction methods that converge in, at most, n iterations for unconstrained quadratic optimization problems in E_n when using exact line searches. In fact, for the latter case, they generate directions identical to the BFGS method, as will be shown later.

The basic scheme of conjugate gradient methods for minimizing a differentiable function $f: E_n \rightarrow E_1$ is to generate a sequence of iterates $\mathbf{y}_j$ according to

$$\mathbf{y}_{j+1} = \mathbf{y}_j + \lambda_j \mathbf{d}_j \tag{8.56a}$$

where $\mathbf{d}_j$ is the search direction and λ_j is the step length which minimizes f along $\mathbf{d}_j$ from the point $\mathbf{y}_j$. For $j = 1$, the search direction $\mathbf{d}_1 = -\nabla f(\mathbf{y}_1)$ can be used, and for subsequent iterations, given $\mathbf{y}_{j+1}$ with $\nabla f(\mathbf{y}_{j+1}) \neq \mathbf{0}$ for $j \geq 1$, we use

$$\mathbf{d}_{j+1} = -\nabla f(\mathbf{y}_{j+1}) + \alpha_j \mathbf{d}_j \tag{8.56b}$$

where α_j is a suitable deflection parameter that characterizes a particular conjugate gradient method. Note that we can write $\mathbf{d}_{j+1}$ in (8.56b) whenever $\alpha_j \geq 0$ as

$$\mathbf{d}_{j+1} = \frac{1}{\mu}[\mu[-\nabla f(\mathbf{y}_{j+1})] + (1 - \mu)\mathbf{d}_j]$$

where $\mu = 1/(1 + \alpha_j)$, and so $\mathbf{d}_{j+1}$ can then be essentially viewed as a convex combination of the current steepest descent direction and the direction used at the last iteration.

Now suppose that we assume f to be a quadratic function with a positive definite Hessian $\mathbf{H}$, and that we require $\mathbf{d}_{j+1}$ and $\mathbf{d}_j$ to be $\mathbf{H}$-conjugate. From (8.56a) and (8.41), $\mathbf{d}_{j+1}^t\mathbf{H}\mathbf{d}_j = 0$ amounts to requiring that $0 = \mathbf{d}_{j+1}^t\mathbf{H}\mathbf{p}_j = \mathbf{d}_{j+1}^t\mathbf{q}_j$. Using this in (8.56b) gives Hestenes and Stiefel's [1952] choice for α_j, used even in nonquadratic situations by assuming a local quadratic behavior, as

$$\alpha_j^{HS} = \frac{\nabla f(\mathbf{y}_{j+1})^t\mathbf{q}_j}{\mathbf{d}_j^t\mathbf{q}_j} \equiv \frac{\lambda_j \nabla f(\mathbf{y}_{j+1})^t\mathbf{q}_j}{\mathbf{p}_j^t\mathbf{q}_j} \tag{8.57}$$

When exact line searches are performed, we have $\mathbf{d}_j^t\nabla f(\mathbf{y}_{j+1}) = 0 = \mathbf{d}_{j-1}^t\nabla f(\mathbf{y}_j)$, leading to $\mathbf{d}_j^t\mathbf{q}_j = -\mathbf{d}_j^t\nabla f(\mathbf{y}_j) = [\nabla f(\mathbf{y}_j) - \alpha_{j-1}\mathbf{d}_{j-1}]^t\nabla f(\mathbf{y}_j) = \|\nabla f(\mathbf{y}_j)\|^2$. Substituting this into (8.57) yields Polak and Ribiere's [1969] choice for α_j as

$$\alpha_j^{PR} = \frac{\nabla f(\mathbf{y}_{j+1})^t\mathbf{q}_j}{\|\nabla f(\mathbf{y}_j)\|^2} \tag{8.58}$$

Furthermore, if f is quadratic and if exact line searches are performed, we have, using (8.56) along with $\nabla f(\mathbf{y}_{j+1})^t\mathbf{d}_j = 0 = \nabla f(\mathbf{y}_j)^t\mathbf{d}_{j-1}$ as above, that

$$\begin{aligned}\nabla f(\mathbf{y}_{j+1})^t\nabla f(\mathbf{y}_j) &= \nabla f(\mathbf{y}_{j+1})^t[\alpha_{j-1}\mathbf{d}_{j-1} - \mathbf{d}_j] \\ &= \alpha_{j-1}\nabla f(\mathbf{y}_{j+1})^t\mathbf{d}_{j-1} = \alpha_{j-1}[\nabla f(\mathbf{y}_j) + \lambda_j\mathbf{H}\mathbf{d}_j]^t\mathbf{d}_{j-1} \\ &= \alpha_{j-1}\lambda_j\mathbf{d}_j^t\mathbf{H}\mathbf{d}_{j-1} = 0\end{aligned}$$

by the $\mathbf{H}$-conjugacy of $\mathbf{d}_j$ and $\mathbf{d}_{j-1}$ (where $\mathbf{d}_0 \equiv \mathbf{0}$). Hence,

$$\nabla f(\mathbf{y}_{j+1})^t\nabla f(\mathbf{y}_j) = 0 \tag{8.59}$$

Substituting this into (8.58) and using (8.41) gives Fletcher and Reeves' [1964] choice of α_j as

$$\alpha_j^{FR} = \frac{\|\nabla f(\mathbf{y}_{j+1})\|^2}{\|\nabla f(\mathbf{y}_j)\|^2} \tag{8.60}$$

We now proceed to present and formally analyze the conjugate gradient method using Fletcher and Reeves' choice (8.60) for α_j. A similar discussion holds for other choices as well.

Summary of the Conjugate Gradient Method of Fletcher and Reeves

A summary of this conjugate gradient method for minimizing a general differentiable function is given below.

Initialization Step Choose a termination scalar $\varepsilon > 0$ and an initial point $\mathbf{x}_1$. Let $\mathbf{y}_1 = \mathbf{x}_1$, $\mathbf{d}_1 = -\nabla f(\mathbf{y}_1)$, $k = j = 1$, and go to the main step.

Main Step

1. If $\|\nabla f(\mathbf{y}_j)\| < \varepsilon$, stop. Otherwise, let λ_j be an optimal solution to the problem to minimize $f(\mathbf{y}_j + \lambda \mathbf{d}_j)$ subject to $\lambda \geq 0$, and let $\mathbf{y}_{j+1} = \mathbf{y}_j + \lambda_j \mathbf{d}_j$. If $j < n$, go to step 2; otherwise, go to step 3.
2. Let $\mathbf{d}_{j+1} = -\nabla f(\mathbf{y}_{j+1}) + \alpha_j \mathbf{d}_j$, where

$$\alpha_j = \frac{\|\nabla f(\mathbf{y}_{j+1})\|^2}{\|\nabla f(\mathbf{y}_j)\|^2}$$

   Replace j by $j + 1$, and go to step 1.
3. Let $\mathbf{y}_1 = \mathbf{x}_{k+1} = \mathbf{y}_{n+1}$, and let $\mathbf{d}_1 = -\nabla f(\mathbf{y}_1)$. Let $j = 1$, replace k by $k + 1$, and go to step 1.

8.8.7 Example

Consider the following problem:

$$\text{Minimize} \quad (x_1 - 2)^4 + (x_1 - 2x_2)^2$$

The summary of the computations using the method of Fletcher and Reeves is given in Table 8.14. At each iteration, $\mathbf{d}_1$ is given by $-\nabla f(\mathbf{y}_1)$, and $\mathbf{d}_2$ is given by $\mathbf{d}_2 = -\nabla f(\mathbf{y}_2) + \alpha_1 \mathbf{d}_1$, where $\alpha_1 = \|\nabla f(\mathbf{y}_2)\|^2 / \|\nabla f(\mathbf{y}_1)\|^2$. Furthermore, $\mathbf{y}_{j+1}$ is obtained by optimizing along $\mathbf{d}_j$, starting from $\mathbf{y}_j$. At iteration 4, the point $\mathbf{y}_2 = (2.185, 1.094)^t$, which is very close to the optimal point (2.00, 1.00), is reached. Since the norm of the gradient is equal to 0.02, which is small, we stop here. The progress of the algorithm is shown in Figure 8.23.

The Quadratic Case

If the function f is quadratic, Theorem 8.8.8 below shows that the directions $\mathbf{d}_1, \ldots, \mathbf{d}_n$ generated are indeed conjugate, and, hence, by Theorem 8.8.3, the conjugate gradient algorithm produces an optimal solution after one complete application of the main step, that is, after, at most, n line searches have been performed.

8.8.8 Theorem

Consider the problem to minimize $f(\mathbf{x}) = \mathbf{c}^t\mathbf{x} + \frac{1}{2}\mathbf{x}^t\mathbf{H}\mathbf{x}$ subject to $\mathbf{x} \in E_n$. Suppose that the problem is solved by the conjugate gradient method, starting with $\mathbf{y}_1$ and $\mathbf{d}_1 = -\nabla f(\mathbf{y}_1)$. In particular, for $j = 1, \ldots, n$, let λ_j be an optimal solution to the problem to minimize $f(\mathbf{y}_j + \lambda \mathbf{d}_j)$ subject to $\lambda \geq 0$. Let $\mathbf{y}_{j+1} = \mathbf{y}_j + \lambda_j \mathbf{d}_j$, and let $\mathbf{d}_{j+1} = -\nabla f(\mathbf{y}_{j+1}) + \alpha_j \mathbf{d}_j$, where $\alpha_j = \|\nabla f(\mathbf{y}_{j+1})\|^2 / \|\nabla f(\mathbf{y}_j)\|^2$. If $\nabla f(\mathbf{y}_j) \neq \mathbf{0}$ for $j = 1, \ldots, n$, then the following statements are true:

1. $\mathbf{d}_1, \ldots, \mathbf{d}_n$ are $\mathbf{H}$-conjugate.
2. $\mathbf{d}_1, \ldots, \mathbf{d}_n$ are descent directions.
3. $\alpha_j = \dfrac{\|\nabla f(\mathbf{y}_{j+1})\|^2}{\|\nabla f(\mathbf{y}_j)\|^2} = \dfrac{\mathbf{d}_j^t \mathbf{H} \nabla f(\mathbf{y}_{j+1})}{\mathbf{d}_j^t \mathbf{H} \mathbf{d}_j} \quad \text{for } j = 1, \ldots, n$

Table 8.14 Summary of Computations for the Method of Fletcher and Reeves

Iteration k	$\mathbf{x}_k$ $f(\mathbf{x}_k)$	j	$\mathbf{y}_j$ $f(\mathbf{y}_j)$	$\nabla f(\mathbf{y}_j)$	$\lVert\nabla f(\mathbf{y}_j)\rVert$	α_1	$\mathbf{d}_j$	λ_j	$\mathbf{y}_{j+1}$
1	(0.00, 3.00) 52.00	1	(0.00, 3.00) 52.00	(−44.00, 24.00)	50.12	—	(44.00, −24.00)	0.062	(2.70, 1.51)
		2	(2.70, 1.51) 0.34	(0.73, 1.28)	1.47	0.0009	(−0.69, −1.30)	0.23	(2.54, 1.21)
2	(2.54, 1.21) 0.10	1	(2.54, 1.21) 0.10	(0.87, −0.48)	0.99	—	(−0.87, 0.48)	0.11	(2.44, 1.26)
		2	(2.44, 1.26) 0.04	(0.18, 0.32)	0.37	0.14	(−0.30, −0.25)	0.63	(2.25, 1.10)
3	(2.25, 1.10) 0.008	1	(2.25, 1.10) 0.008	(0.16, −0.20)	0.32	—	(−0.16, 0.20)	0.10	(2.23, 1.12)
		2	(2.23, 1.12) 0.003	(0.03, 0.04)	0.05	0.04	(−0.036, −0.032)	1.02	(2.19, 1.09)
4	(2.19, 1.09) 0.0017	1	(2.19, 1.09) 0.0017	(0.05, −0.04)	0.06	—	(−0.05, 0.04)	0.11	(2.185, 1.094)
		2	(2.185, 1.094) 0.0012	(0.02, 0.01)	0.02				

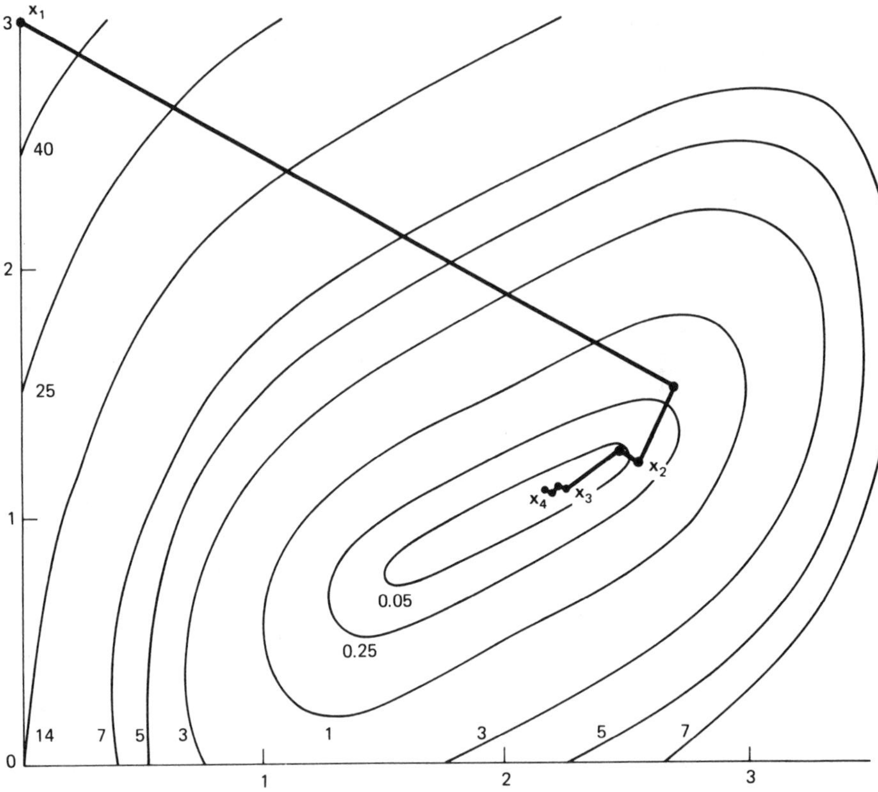

Figure 8.23 Illustration of the method of Fletcher and Reeves.

Proof

First, suppose that parts 1, 2, and 3 hold for j. We show that they also hold for $j + 1$. To show that part 1 holds for $j + 1$, we first demonstrate that $\mathbf{d}_k^t\mathbf{H}\mathbf{d}_{j+1} = 0$ for $k \le j$. Since $\mathbf{d}_{j+1} = -\nabla f(\mathbf{y}_{j+1}) + \alpha_j\mathbf{d}_j$, noting the induction hypothesis in part 3, and letting $k = j$, we get

$$\mathbf{d}_j^t\mathbf{H}\mathbf{d}_{j+1} = \mathbf{d}_j^t\mathbf{H}\left[-\nabla f(\mathbf{y}_{j+1}) + \frac{\mathbf{d}_j^t\mathbf{H}\nabla f(\mathbf{y}_{j+1})}{\mathbf{d}_j^t\mathbf{H}\mathbf{d}_j}\mathbf{d}_j\right] = 0 \tag{8.61}$$

Now let $k < j$. Since $\mathbf{d}_{j+1} = -\nabla f(\mathbf{y}_{j+1}) + \alpha_j\mathbf{d}_j$, and since $\mathbf{d}_k^t\mathbf{H}\mathbf{d}_j = 0$ by the induction hypothesis of part 1,

$$\mathbf{d}_k^t\mathbf{H}\mathbf{d}_{j+1} = -\mathbf{d}_k^t\mathbf{H}\nabla f(\mathbf{y}_{j+1}) \tag{8.62}$$

Since $\nabla f(\mathbf{y}_{k+1}) = \mathbf{c} + \mathbf{H}\mathbf{y}_{k+1}$ and $\mathbf{y}_{k+1} = \mathbf{y}_k + \lambda_k\mathbf{d}_k$, note that

$$\begin{aligned}\mathbf{d}_{k+1} &= -\nabla f(\mathbf{y}_{k+1}) + \alpha_k\mathbf{d}_k \\ &= -[\nabla f(\mathbf{y}_k) + \lambda_k\mathbf{H}\mathbf{d}_k] + \alpha_k\mathbf{d}_k \\ &= -[-\mathbf{d}_k + \alpha_{k-1}\mathbf{d}_{k-1} + \lambda_k\mathbf{H}\mathbf{d}_k] + \alpha_k\mathbf{d}_k\end{aligned}$$

By the induction hypothesis of part 2, $\mathbf{d}_k$ is a descent direction, and, hence, $\lambda_k > 0$. Therefore,

$$\mathbf{d}_k^t\mathbf{H} = \frac{1}{\lambda_k}[-\mathbf{d}_{k+1}^t + (1 + \alpha_k)\mathbf{d}_k^t - \alpha_{k-1}\mathbf{d}_{k-1}^t] \tag{8.63}$$

From (8.62) and (8.63), it follows that

$$\begin{aligned}\mathbf{d}_k^t\mathbf{H}\mathbf{d}_{j+1} &= -\mathbf{d}_k^t\mathbf{H}\nabla f(\mathbf{y}_{j+1})\\ &= -\frac{1}{\lambda_k}[-\mathbf{d}_{k+1}^t\nabla f(\mathbf{y}_{j+1}) + (1 + \alpha_k)\mathbf{d}_k^t\nabla f(\mathbf{y}_{j+1}) - \alpha_{k-1}\mathbf{d}_{k-1}^t\nabla f(\mathbf{y}_{j+1})]\end{aligned}$$

By part 1 of Theorem 8.8.3, and since $\mathbf{d}_1, \ldots, \mathbf{d}_j$ are assumed conjugate, $\mathbf{d}_{k+1}^t\nabla f(\mathbf{y}_{j+1}) = \mathbf{d}_k^t\nabla f(\mathbf{y}_{j+1}) = \mathbf{d}_{k-1}^t\nabla f(\mathbf{y}_{j+1}) = 0$. Thus, the above equation implies that $\mathbf{d}_k^t\mathbf{H}\mathbf{d}_{j+1} = 0$ for $k < j$. This, together with (8.61), shows that $\mathbf{d}_k^t\mathbf{H}\mathbf{d}_{j+1} = 0$ for all $k \leq j$.

To show that $\mathbf{d}_1, \ldots, \mathbf{d}_{j+1}$ are $\mathbf{H}$-conjugate, it thus suffices to show that they are linearly independent. Suppose that $\sum_{i=1}^{j+1} \gamma_i\mathbf{d}_i = \mathbf{0}$. Then, $\sum_{i=1}^{j} \gamma_i\mathbf{d}_i + \gamma_{j+1}[-\nabla f(\mathbf{y}_{j+1}) + \alpha_j\mathbf{d}_j] = \mathbf{0}$. Multiplying by $\nabla f(\mathbf{y}_{j+1})^t$, and noting part 1 of Theorem 8.8.3, it follows that $\gamma_{j+1}\ \|\nabla f(\mathbf{y}_{j+1})\|^2 = 0$. Since $\nabla f(\mathbf{y}_{j+1}) \neq \mathbf{0}$, $\gamma_{j+1} = 0$. This implies that $\sum_{i=1}^{j} \gamma_i\mathbf{d}_i = \mathbf{0}$, and in view of conjugacy of $\mathbf{d}_1, \ldots, \mathbf{d}_j$, it follows that $\gamma_1 = \cdots = \gamma_j = 0$. Thus, $\mathbf{d}_1, \ldots, \mathbf{d}_{j+1}$ are linearly independent and $\mathbf{H}$-conjugate, so that part 1 holds true for $j + 1$.

Now we show that part 2 holds for $j + 1$; that is, $\mathbf{d}_{j+1}$ is a descent direction. Note that $\nabla f(\mathbf{y}_{j+1}) \neq \mathbf{0}$ by assumption, and that $\nabla f(\mathbf{y}_{j+1})^t\mathbf{d}_j = 0$ by part 1 of Theorem 8.8.3. Then, $\nabla f(\mathbf{y}_{j+1})^t\mathbf{d}_{j+1} = -\|\nabla f(\mathbf{y}_{j+1})\|^2 + \alpha_j\nabla f(\mathbf{y}_{j+1})^t\mathbf{d}_j = -\|\nabla f(\mathbf{y}_{j+1})\|^2 < 0$. By Theorem 4.1.2, $\mathbf{d}_{j+1}$ is a descent direction.

Now we show that part 3 holds for $j + 1$. By letting $k = j + 1$ in (8.63) and multiplying by $\nabla f(\mathbf{y}_{j+2})$, it follows that

$$\begin{aligned}\lambda_{j+1}\mathbf{d}_{j+1}^t\mathbf{H}\nabla f(\mathbf{y}_{j+2}) &= [-\mathbf{d}_{j+2}^t + (1 + \alpha_{j+1})\mathbf{d}_{j+1}^t - \alpha_j\mathbf{d}_j^t]\nabla f(\mathbf{y}_{j+2})\\ &= [\nabla f(\mathbf{y}_{j+2})^t + \mathbf{d}_{j+1}^t - \alpha_j\mathbf{d}_j^t]\nabla f(\mathbf{y}_{j+2})\end{aligned}$$

Since $\mathbf{d}_1, \ldots, \mathbf{d}_{j+1}$ are $\mathbf{H}$-conjugate, then, by part 1 of Theorem 8.8.3, $\mathbf{d}_{j+1}^t\nabla f(\mathbf{y}_{j+2}) = \mathbf{d}_j^t\nabla f(\mathbf{y}_{j+2}) = 0$. The above equation then implies that

$$\|\nabla f(\mathbf{y}_{j+2})\|^2 = \lambda_{j+1}\mathbf{d}_{j+1}^t\mathbf{H}\nabla f(\mathbf{y}_{j+2}) \tag{8.64}$$

Multiplying $\nabla f(\mathbf{y}_{j+1}) = \nabla f(\mathbf{y}_{j+2}) - \lambda_{j+1}\mathbf{H}\mathbf{d}_{j+1}$ by $\nabla f(\mathbf{y}_{j+1})^t$, and noting that $\mathbf{d}_j^t\mathbf{H}\mathbf{d}_{j+1} = \mathbf{d}_{j+1}^t\nabla f(\mathbf{y}_{j+2}) = \mathbf{d}_j^t\nabla f(\mathbf{y}_{j+2}) = 0$, we get

$$\begin{aligned}\|\nabla f(\mathbf{y}_{j+1})\|^2 &= \nabla f(\mathbf{y}_{j+1})^t[\nabla f(\mathbf{y}_{j+2}) - \lambda_{j+1}\mathbf{H}\mathbf{d}_{j+1}]\\ &= (-\mathbf{d}_{j+1}^t + \alpha_j\mathbf{d}_j^t)[\nabla f(\mathbf{y}_{j+2}) - \lambda_{j+1}\mathbf{H}\mathbf{d}_{j+1}]\\ &= \lambda_{j+1}\mathbf{d}_{j+1}^t\mathbf{H}\mathbf{d}_{j+1}\end{aligned} \tag{8.65}$$

From (8.64) and (8.65), it is obvious that part 3 holds true for $j + 1$.

We have thus shown that if parts 1, 2, and 3 hold for j, then they also hold for $j + 1$. Note that parts 1 and 2 trivially hold for $j = 1$. In addition, using a similar argument to that used in proving that part 3 holds for $j + 1$, it can be easily demonstrated that it holds for $j = 1$. This completes the proof.

The reader should note here that when the function f is quadratic and when exact line searches are performed, the choices of α_j given variously by (8.57), (8.58), and (8.60) all coincide, and thus Theorem 8.8.8 also holds for the Hestenes and Stiefel (HS) and the Polak and Ribiere (PR) choices of α_j. However, for nonquadratic functions,

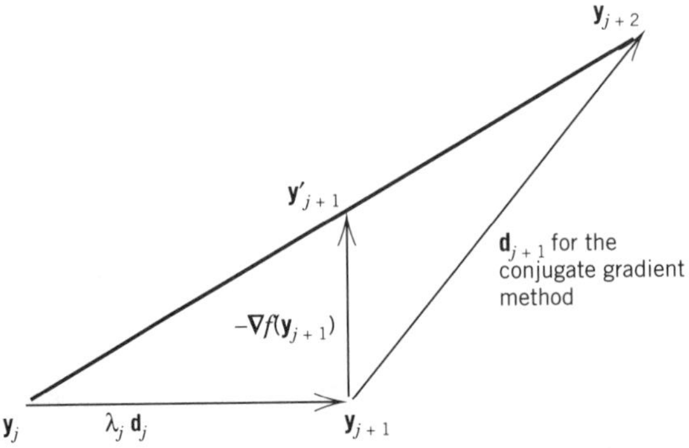

Figure 8.24 Equivalence between the conjugate gradient method and PARTAN.

the choice α_j^{PR} appears to be empirically superior to α_j^{FR}. This is understandable, since the reduction of (8.58) to (8.60) assumes f to be quadratic. In the same vein, when inexact line searches are performed, the choice α_j^{HS} appears to be preferable. Note that even when f is quadratic, if inexact line searches are performed, the conjugacy relationship holds only between consecutive directions. We refer the reader to the Notes and References section for a discussion on some alternate *three-term recurrence relationships* for generating mutually conjugate directions in such a case.

Also, note that we have used $\mathbf{d}_1 = -\mathbf{I}\nabla f(\mathbf{y}_1)$ in the foregoing analysis. In lieu of using the identity matrix here, we could have used some general *preconditioning matrix* $\mathbf{D}$, where $\mathbf{D}$ is symmetric and positive definite. This would have given $\mathbf{d}_1 = -\mathbf{D}\nabla f(\mathbf{y}_1)$, and (8.56b) would have become $\mathbf{d}_{j+1} = -\mathbf{D}\nabla f(\mathbf{y}_{j+1}) + \alpha_j \mathbf{d}_j$ where, for example, in the spirit of (8.57), we have

$$\alpha_j^{HS} = \frac{\mathbf{q}_j \mathbf{D}\nabla f(\mathbf{y}_{j+1})}{\mathbf{q}_j^t \mathbf{d}_j}$$

This corresponds to essentially making a change of variables $\mathbf{y}' = \mathbf{D}^{-1/2}\mathbf{y}$ and using the original conjugate gradient algorithm. Therefore, this motivates the choice of $\mathbf{D}$ from the viewpoint of improving the eigenstructure of the problem, as discussed earlier.

For quadratic functions f, the conjugate gradient step also has an interesting "pattern search" interpretation. Consider Figure 8.24 and suppose that the successive points $\mathbf{y}_j$, $\mathbf{y}_{j+1}$, and $\mathbf{y}_{j+2}$ are generated by the conjugate gradient algorithm. Now suppose that, at the point $\mathbf{y}_{j+1}$ obtained from $\mathbf{y}_j$ by minimizing along $\mathbf{d}_j$, we had instead minimized next along the steepest descent direction $-\nabla f(\mathbf{y}_{j+1})$ at $\mathbf{y}_{j+1}$, leading to the point $\mathbf{y}'_{j+1}$. Then, it can be shown (see Exercise 8.37) that a pattern search step of minimizing the quadratic function f from $\mathbf{y}_j$ along the direction $\mathbf{y}'_{j+1} - \mathbf{y}_j$ would have also led to the same point $\mathbf{y}_{j+2}$. The method, which uses the latter kind of step in general (even for nonquadratic functions), is more popularly known as *PARTAN* (see Exercise 8.49). Note that the global convergence of PARTAN for general functions is tied in to using the negative gradient direction as a spacer step in Theorem 7.3.4 and is independent of any restart conditions, although it is recommended that the method be restarted every n iterations to promote its behavior as a conjugate gradient method.

Memoryless Quasi-Newton Methods

There is an interesting connection between conjugate gradient methods and a simplified variant of the BFGS quasi-Newton method. Suppose that we operate the latter method by updating the inverse Hessian approximation according to $\mathbf{D}_{j+1} = \mathbf{D}_j + \mathbf{C}_j^{\text{BFGS}}$, where the correction matrix $\mathbf{C}_j^{\text{BFGS}}$ is given in (8.48), but assuming that $\mathbf{D}_j \equiv \mathbf{I}$. Hence, we get

$$\mathbf{D}_{j+1} = 1 + \frac{\mathbf{p}_j\mathbf{p}_j^t}{\mathbf{p}_j^t\mathbf{q}_j}\left(1 + \frac{\mathbf{q}_j^t\mathbf{q}_j}{\mathbf{p}_j^t\mathbf{q}_j}\right) - \frac{\mathbf{q}_j\mathbf{p}_j^t + \mathbf{p}_j\mathbf{q}_j^t}{\mathbf{p}_j^t\mathbf{q}_j} \tag{8.66a}$$

We then move along the direction

$$\mathbf{d}_{j+1} = -\mathbf{D}_{j+1}\nabla f(\mathbf{y}_{j+1}) \tag{8.66b}$$

This is akin to "forgetting" the previous approximation $\mathbf{D}_j$ and, instead, updating the identity matrix as might be done at the first iteration of the quasi-Newton method; hence, the name *memoryless quasi-Newton method.* Observe that the storage requirements are similar to that of conjugate gradient methods and that inexact line searches can be performed, so long as $\mathbf{p}_j^t\mathbf{q}_j = \lambda_j\mathbf{d}_j^t[\nabla f(\mathbf{y}_{j+1}) - \nabla f(\mathbf{y}_j)]$ remains positive and $\mathbf{d}_{j+1}$ continues to be a descent direction. Also, note that the loss of positive definiteness of the approximations $\mathbf{D}_j$ in the quasi-Newton method is now no longer of concern. In fact, this scheme has proved to be computationally very effective in conjunction with inexact line searches. We refer the reader to the Notes and References section for a discussion on conjugate gradient methods operated with inexact line searches.

Now, suppose that we do employ exact line searches. Then, we have $\mathbf{p}_j^t\nabla f(\mathbf{y}_{j+1}) = \lambda_j\mathbf{d}_j^t\nabla f(\mathbf{y}_{j+1}) = 0$, and so (8.66) gives

$$\mathbf{d}_{j+1} = -\nabla f(\mathbf{y}_{j+1}) + \frac{\mathbf{q}_j^t\nabla f(\mathbf{y}_{j+1})}{\mathbf{p}_j^t\mathbf{q}_j}\mathbf{p}_j = -\nabla f(\mathbf{y}_{j+1}) + \alpha_j^{\text{HS}}\mathbf{d}_j$$

from (8.57). Hence, the BFGS memoryless update scheme is equivalent to the conjugate gradient method of Hestenes and Stiefel (or Polak and Ribiere) when exact line searches are employed. We mention here that, although this memoryless update can be performed on any other member of the Broyden family as well (see Exercise 8.38), the equivalence with conjugate gradient methods results only for $\phi = 1$ (the BFGS update), and so also does the observed empirical effectiveness of this scheme (see also Exercise 8.39).

Recommendations for Restarting Conjugate Gradient Methods

In several computational experiments using different conjugate gradient techniques, with or without exact line searches, it has time and again been demonstrated that the performance of conjugate gradient methods can be greatly enhanced by employing a proper restart criterion. In particular, a restart procedure suggested by Beale [1972] and augmented by Powell [1977] has proved to be very effective and is invariably implemented, as described below.

Consider the conjugate gradient method formally summarized above in the context of Fletcher and Reeves' choice of α_j. (Naturally, this strategy applies to any other admissible choice of α_j as well.) At some inner loop iteration j of this procedure, having found $\mathbf{y}_{j+1} = \mathbf{y}_j + \lambda_j\mathbf{d}_j$ by searching along $\mathbf{d}_j$ from the point $\mathbf{y}_j$, suppose that we decide to reset. (In the previous description of the algorithm, this decision was made whenever $j = n$.) Let $\tau = j$ denote this restart iteration. For the next iteration, we find the search direction

$$\mathbf{d}_{\tau+1} = -\nabla f(\mathbf{y}_{\tau+1}) + \alpha_\tau\mathbf{d}_\tau \tag{8.67}$$

as usual. Then, at step 3, we replace $\mathbf{y}_1$ by $\mathbf{y}_{\tau+1}$, and we let $\mathbf{x}_{k+1} \equiv \mathbf{y}_{\tau+1}$, $\mathbf{d}_1 = \mathbf{d}_{\tau+1}$, and return to step 1 to continue with the next set of inner loop iterations. However, instead of computing $\mathbf{d}_{j+1} = -\nabla f(\mathbf{y}_{j+1}) + \alpha_j \mathbf{d}_j$ for $j \geq 1$, we now use

$$\mathbf{d}_2 = -\nabla f(\mathbf{y}_2) + \alpha_2 \mathbf{d}_2 \tag{8.68a}$$

and

$$\mathbf{d}_{j+1} = -\nabla f(\mathbf{y}_{j+1}) + \alpha_j \mathbf{d}_j + \gamma_j \mathbf{d}_1 \qquad \text{for } j \geq 2, \text{ where}$$
$$\gamma_j = \frac{\nabla f(\mathbf{y}_{j+1})^t \mathbf{q}_1}{\mathbf{d}_1^t \mathbf{q}_1} \tag{8.68b}$$

and where α_j is computed as before, depending on the method being used. Note that (8.68a) employs the usual conjugate gradient scheme, thereby yielding $\mathbf{d}_1$ and $\mathbf{d}_2$ as $\mathbf{H}$-conjugate when f is quadratic. However, when f is quadratic with a positive definite Hessian $\mathbf{H}$ and $\mathbf{d}_1$ is arbitrarily chosen, then when $j = 2$, for example, the usual choice of α_2 would make $\mathbf{d}_3$ and $\mathbf{d}_2$ $\mathbf{H}$-conjugate, but we would need something additional to make $\mathbf{d}_3$ and $\mathbf{d}_1$ $\mathbf{H}$-conjugate. This is accomplished by the extra term $\gamma_2 \mathbf{d}_1$. Indeed, requiring $\mathbf{d}_3^t \mathbf{H} \mathbf{d}_1 = 0$, where $\mathbf{d}_3$ is given by the expression in (8.68b), and noting $\mathbf{d}_2^t \mathbf{H} \mathbf{d}_1 = 0$, yields $\gamma_2 = \nabla f(\mathbf{y}_3)^t \mathbf{H} \mathbf{d}_1 / \mathbf{d}_1^t \mathbf{H} \mathbf{d}_1 = \nabla f(\mathbf{y}_3)^t \mathbf{q}_1 / \mathbf{d}_1^t \mathbf{q}_1$. Proceeding inductively in this manner, the additional term in (8.68b) ensures the $\mathbf{H}$-conjugacy of all generated directions (see Exercise 8.47).

The foregoing scheme was suggested by Beale with the motivation that whenever a restart is done using $\mathbf{d}_1 = -\nabla f(\mathbf{y}_1)$ instead of $\mathbf{d}_1 = \mathbf{d}_{\tau+1}$ as given by (8.67), we lose important second-order information inherent in $\mathbf{d}_\tau$. Additionally, Powell suggested that after finding $\mathbf{y}_{j+1}$, if any of the following three conditions holds, then the algorithm should be restarted by putting $\tau = j$, computing $\mathbf{d}_{\tau+1}$ via (8.67), and resetting $\mathbf{d}_1 = \mathbf{d}_{\tau+1}$ and $\mathbf{y}_1 = \mathbf{y}_{\tau+1}$:

1. $j = n - 1$
2. $|\nabla f(\mathbf{y}_{j+1})^t \nabla f(\mathbf{y}_j)| \geq 0.2 \|\nabla f(\mathbf{y}_{j+1})\|^2$ for some $j \geq 1$
3. $-1.2 \|\nabla f(\mathbf{y}_{j+1})\|^2 \leq \mathbf{d}_{j+1}^t \nabla f(\mathbf{y}_{j+1}) \leq -0.8 \|\nabla f(\mathbf{y}_{j+1})\|^2$ for some $j \geq 2$

Condition 1 is the usual reset criterion by which, after searching along the direction $\mathbf{d}_{\tau+1} \equiv \mathbf{d}_n$, we will have searched along n conjugate directions for the quadratic case. Condition 2 suggests a reset if a sufficient measure of orthogonality has been lost between $\nabla f(\mathbf{y}_j)$ and $\nabla f(\mathbf{y}_{j+1})$, motivated by the expanding subspace property illustrated in Figure 8.21. (Computationally, instead of using 0.2 here, any constant in the interval [0.1, 0.9] appears to give satisfactory performance.) Condition 3 checks for a sufficient descent along the direction $\mathbf{d}_{j+1}$ at the point $\mathbf{y}_{j+1}$. For similar ideas when employing inexact line searches, we refer the reader to the Notes and References section.

Convergence of Conjugate Direction Methods

As shown in Theorem 8.8.3, if the function under consideration is quadratic, then any conjugate direction algorithm produces an optimal solution in a finite number of steps. We now discuss convergence of these methods if the function is not necessarily quadratic.

In Theorem 7.3.4, we showed that a composite algorithm $\mathbf{A} = \mathbf{CB}$ converges to a point in the solution set Ω if the following properties hold true:

1. $\mathbf{B}$ is closed at ponts not in Ω.
2. If $\mathbf{y} \in \mathbf{B}(\mathbf{x})$, then $f(\mathbf{y}) < f(\mathbf{y})$ for $\mathbf{x} \notin \Omega$.

3. If $\mathbf{z} \in \mathbf{C}(\mathbf{y})$, then $f(\mathbf{z}) \leq f(\mathbf{y})$.
4. The set $\wedge = \{\mathbf{x}: f(\mathbf{x}) \leq f(\mathbf{x}_1)\}$ is compact, where $\mathbf{x}_1$ is the starting point.

For the conjugate direction (quasi-Newton or conjugate gradient) algorithms discussed in this chapter, the map **B** is of the following form. Given **x**, then $\mathbf{y} \in \mathbf{B}(\mathbf{x})$ means that **y** is obtained by minimizing f starting from **x** along the direction $\mathbf{d} = -\mathbf{D}\nabla f(\mathbf{x})$, where **D** is a specified positive definite matrix. In particular, for the conjugate gradient methods, $\mathbf{D} = \mathbf{I}$, and for the quasi-Newton methods, **D** is an arbitrary positive definite matrix. Furthermore, starting from the point obtained by applying the map **B**, the map **C** is defined by minimizing the function f along the directions specified by the particular algorithms. Thus, the map **C** satisfies proeprty 3 above.

Letting $\Omega = \{\mathbf{x}: \nabla f(\mathbf{x}) = \mathbf{0}\}$, we now show that the map **B** satisfies properties 1 and 2 above. Let $\mathbf{x} \in \Omega$ and let $\mathbf{x}_k \to \mathbf{x}$. Furthermore, let $\mathbf{y}_k \in \mathbf{B}(\mathbf{x}_k)$ and let $\mathbf{y}_k \to \mathbf{y}$. We need to show that $\mathbf{y} \in \mathbf{B}(\mathbf{x})$. By definition of $\mathbf{y}_k$, $\mathbf{y}_k = \mathbf{x}_k - \lambda_k \mathbf{D}\nabla f(\mathbf{x}_k)$ for $\lambda_k \geq 0$ such that

$$f(\mathbf{y}_k) \leq f[\mathbf{x}_k - \lambda \mathbf{D}\nabla f(\mathbf{x}_k)] \qquad \text{for all } \lambda \geq 0 \tag{8.69}$$

Since $\nabla f(\mathbf{x}) \neq \mathbf{0}$, then λ_k converges to $\bar{\lambda} = \|\mathbf{y} - \mathbf{x}\|/\|\mathbf{D}\nabla f(\mathbf{x})\| \geq 0$. Therefore, $\mathbf{y} = \mathbf{x} - \bar{\lambda}\mathbf{D}\nabla f(\mathbf{x})$. Taking the limit as $k \to \infty$ in (8.69), then $f(\mathbf{y}) \leq f[\mathbf{x} - \lambda \mathbf{D}\nabla f(\mathbf{x})]$ for all $\lambda \geq 0$, so that **y** is indeed obtained by minimizing f starting from **x** in the direction $-\mathbf{D}\nabla f(\mathbf{x})$. Thus, $\mathbf{y} \in \mathbf{B}(\mathbf{x})$, and **B** is closed. Also part 2 holds by noting that $-\nabla f(\mathbf{x})^t \mathbf{D}\nabla f(\mathbf{x}) < 0$, so that $-\mathbf{D}\nabla f(\mathbf{x})$ is a descent direction. Assuming that the set defined in part 4 is compact, it then follows that the conjugate direction algorithms discussed in this section converge to a point with zero gradient.

The role played by the map **B** described above is akin to that of a *spacer step* as discussed in connection with Theorem 7.3.4. For algorithms that are empirically designed and that may not enjoy theoretical convergence, this can be alleviated by inserting such a spacer step involving a periodic minimization along the negative gradient direction, for example, hence, achieving theoretical convergence.

We now turn our attention to addressing the rate of convergence or local convergence characteristics of the algorithms discussed in this section.

Convergence Rate Characteristics for Conjugate Gradient Methods

Consider the quadratic function $f(\mathbf{x}) = \mathbf{c}^t\mathbf{x} + \frac{1}{2}\mathbf{x}^t\mathbf{H}\mathbf{x}$, where **H** is an $n \times n$ symmetric, positive definite matrix. Suppose that the eigenvalues of **H** are grouped into two sets, of which one set is composed of some m relatively large and perhaps dispersed values, and the other set is a cluster of some $n - m$ relatively smaller eigenvalues. (Such a structure arises, for example, with the use of quadratic penalty functions on linearly constrained quadratic programs, as discussed in Chapter 9.) Let us assume that $(m + 1) < n$, and let α denote the ratio of the largest to the smallest eigenvalue in the latter cluster. Now we know that a standard application of the conjugate gradient method will result in a finite convergence to the optimum in n, or fewer, steps. However, suppose that we operate the conjugate gradient algorithm by restarting with the steepest descent direction every $m + 1$ line searches or steps. Such a procedure is called a *partial conjugate gradient method.*

Starting with a solution $\mathbf{x}_1$, let $\{\mathbf{x}_k\}$ be the sequence thus generated, where for each $k \geq 1$, $\mathbf{x}_{k+1}$ is obtained after applying $m + 1$ conjugate gradient steps upon restarting with $\mathbf{x}_k$ as above. Let us refer to this as an *$(m + 1)$-step process.* As in equation (8.17), let us define an error function $e(\mathbf{x}) = \frac{1}{2}(\mathbf{x} - \mathbf{x}^*)^t\mathbf{H}(\mathbf{x} - \mathbf{x}^*)$, which differs from $f(\mathbf{x})$ by a constant, and which is zero if and only if $\mathbf{x} = \mathbf{x}^*$. Then, it can be shown (see the Notes and References section) that

$$e(\mathbf{x}_{k+1}) \leq \frac{(\alpha - 1)^2}{(\alpha + 1)^2} e(\mathbf{x}_k) \tag{8.70}$$

Hence, this establishes a linear rate of convergence for the above process as in the special case of the steepest descent method for which $m = 0$ [see equation (8.18)]. However, now, the ratio α that governs the convergence rate is independent of the m largest eigenvalues. Thus, the effect of the m largest eigenvalues is eliminated, but at the expense of an $(m + 1)$-step process versus the single-step process of the steepest descent method.

Next, consider the general nonquadratic case to which the usual n-step conjugate gradient process is applied. Intuitively, since the conjugate gradient method accomplishes in n steps what Newton's method does in a single step, by the local quadratic convergence rate of Newton's method, we might similarly expect that the n-step conjugate gradient process also converges quadratically, that is, $\|\mathbf{x}_{k+1} - \mathbf{x}^*\| \leq \beta\|\mathbf{x}_k - \mathbf{x}^*\|^2$. Indeed, it can be shown (see the Notes and References section) that, if the sequence $\{\mathbf{x}_k\} \to \mathbf{x}^*$ and the function under consideration is twice continuously differentiable in some neighborhood of $\mathbf{x}^*$, and the Hessian matrix at $\mathbf{x}^*$ is positive definite, then the n-step process converges superlinearly to $\mathbf{x}^*$. Moreover, if the Hessian matrix satisfies an appropriate Lipschitz condition in some neighborhood of $\mathbf{x}^*$, then the rate of superlinear convergence is n-step quadratic. Again, caution must be exercised in interpreting these results in comparison with, say, the linear convergence rate of steepest descent methods. That is, these are n-step asymptotic results, whereas the steepest descent method is a single-step procedure. Also, given that these methods are usually applied when n is relatively large, it is seldom practical to perform more than $5n$ iterations, or five n-step iterations. Fortunately, empirical results seem to indicate that this does not pose a problem, for reasonable convergence is typically obtained within $2n$ iterations.

Convergence Rate Characteristics for Quasi-Newton Methods

The Broyden class of quasi-Newton methods can also be operated as *partial quasi-Newton methods* by restarting every $m + 1$ iterations with, say, the steepest descent direction. For the quadratic case, the local convergence properties of such a scheme resembles that for conjugate gradient methods as discussed above. Also, for nonquadratic cases, the n-step quasi-Newton algorithm has a local superlinear convergence rate behavior similar to that of the conjugate gradient method. Intuitively, this is because of the identical effect that the n-step process of either method has on quadratic functions. Again, the usual caution of interpreting the value of an n-step superlinear convergence behavior must be adopted. Additionally, we draw the reader's attention to Exercise 8.55 and to the section on scaling quasi-Newton methods, where we discussed the possible ill-conditioning effects resulting from the sequential transformation of the eigenvalues of $\mathbf{D}_{j+1}\mathbf{H}$ to unity for the quadratic case.

Quasi-Newton methods are also sometimes operated as a continuing updating process, without any resets. Although the global convergence of such a scheme requires rather stringent conditions, the local convergence rate behavior is often asymptotically superlinear. For example, for the BFGS update scheme, which has been seen to exhibit a relatively superior empirical performance as mentioned previously, the following result holds (see the Notes and References section). Let $\mathbf{y}^*$ be such that the Hessian $\mathbf{H}(\mathbf{y}^*)$ is positive definite, and that there exists an ε-neighborhood $N_\varepsilon(\mathbf{y}^*)$ of $\mathbf{y}^*$ such that the Lipschitz condition $\|\mathbf{H}(\mathbf{y}) - \mathbf{H}(\mathbf{y}^*)\| \leq L\|\mathbf{y} - \mathbf{y}^*\|$ holds for $\mathbf{y} \in N_\varepsilon(\mathbf{y}^*)$, where L is a positive constant. Then, if a sequence $\{\mathbf{y}_k\}$ generated by a continuing updating quasi-Newton process with a fixed step size of unity converges to such a $\mathbf{y}^*$, the asymptotic rate of convergence is superlinear. Similar superlinear convergence rate results are available for the DFP algorithm, both with exact line searches and with unit step size choices under appropriate conditions. We refer the reader to the Notes and References section for further reading on this subject.

8.9 Subgradient Optimization

Consider Problem P defined as

$$P: \quad \text{Minimize} \quad \{f(\mathbf{x}) : \mathbf{x} \in X\} \tag{8.71}$$

where $f: E_n \to E_1$ is a convex but not necessarily differentiable function, and where X is a nonempty, closed, convex subset of E_n. We assume that an optimal solution exists as would be, for example, if X is bounded, or if $f(\mathbf{x}) \to \infty$ whenever $\|\mathbf{x}\| \to \infty$.

For such a Problem P, we now describe a *subgradient optimization algorithm* that can be viewed as a direct generalization of the steepest descent algorithm in which the negative gradient direction is substituted by a negative subgradient-based direction. However, the latter direction need not necessarily be a descent direction, although, as we shall see, it does result in the new iterate being closer to an optimal solution for a sufficiently small step size. Because of this reason, we do not perform a line search along the negative subgradient direction, but rather prescribe a step size at each iteration that guarantees that the sequence generated will eventually converge to an optimal solution. Also, given an iterate $\mathbf{x}_k \in X$ and adopting a step size λ_k along the direction $\mathbf{d}_k = -\boldsymbol{\xi}_k/\|\boldsymbol{\xi}_k\|$, where $\boldsymbol{\xi}_k$ belongs to the subdifferential $\partial f(\mathbf{x}_k)$ of f at $\mathbf{x}_k$ ($\boldsymbol{\xi}_k \neq \mathbf{0}$, say), the resulting point $\bar{\mathbf{x}}_{k+1} = \mathbf{x}_k + \lambda_k \mathbf{d}_k$ need not belong to X. Consequently, the new iterate $\mathbf{x}_{k+1}$ is obtained by *projecting* $\bar{\mathbf{x}}_{k+1}$ onto X, that is, finding the (unique) closest point in X to $\bar{\mathbf{x}}_{k+1}$. We denote this operation as $\mathbf{x}_{k+1} = P_X(\bar{\mathbf{x}}_{k+1})$, where

$$P_X(\bar{\mathbf{x}}) \equiv \operatorname{argmin}\{\|\mathbf{x} - \bar{\mathbf{x}}\| : \mathbf{x} \in X\} \tag{8.72}$$

The foregoing projection operation should be easy to perform if the method is to be computationally viable. For example, in the context of Lagrangian duality (Chapter 6), wherein subgradient methods and their variants are most frequently used, the set X might simply represent nonnegativity restrictions $\mathbf{x} \geq \mathbf{0}$ on the variables. In this case, we easily obtain $(\mathbf{x}_{k+1})_i = \max\{0, (\bar{\mathbf{x}}_{k+1})_i\}$ for each component $i = 1, \ldots, n$ in (8.72). In other contexts, the set $X = \{\mathbf{x} : l_i \leq x_i \leq u_i,\ i = 1, \ldots, n\}$ might represent simple lower and upper bounds on the variables. In this case, it is again easy to verify that

$$(\mathbf{x}_{k+1})_i = \begin{cases} (\bar{\mathbf{x}}_{k+1})_i & \text{if } l_i \leq (\bar{\mathbf{x}}_{k+1})_i \leq u_i \\ l_i & \text{if } (\bar{\mathbf{x}}_{k+1})_i < l_i \\ u_i & \text{if } (\bar{\mathbf{x}}_{k+1})_i > u_i \end{cases} \quad \text{for } i = 1, \ldots, n \tag{8.73}$$

Also, when an additional knapsack constraint $\boldsymbol{\alpha}^t\mathbf{x} = \beta$ is introduced to define $\mathbf{X} = \{\mathbf{x} : \boldsymbol{\alpha}^t\mathbf{x} = \beta,\ \mathbf{l} \leq \mathbf{x} \leq \mathbf{u}\}$, then, again, $P_X(\bar{\mathbf{x}})$ is relatively easy to obtain (see Exercise 8.56).

Summary of a (Rudimentary) Subgradient Algorithm

Initialization Step Select a starting solution $\mathbf{x}_1 \in X$, let the current upper bound on the optimal objective value be $UB_1 = f(\mathbf{x}_1)$, and let the current incumbent solution be $\mathbf{x}^* = \mathbf{x}_1$. Put $k = 1$, and go to the main step.

Main Step Given $\mathbf{x}_k$, find a subgradient $\boldsymbol{\xi}_k \in \partial f(\mathbf{x}_k)$ of f at $\mathbf{x}_k$. If $\boldsymbol{\xi}_k = \mathbf{0}$, then stop; $\mathbf{x}_k$ (or $\mathbf{x}^*$) solves Problem P. Otherwise, let $\mathbf{d}_k = -\boldsymbol{\xi}_k/\|\boldsymbol{\xi}_k\|$, select a step size $\lambda_k > 0$, and compute $\mathbf{x}_{k+1} = P_X(\bar{\mathbf{x}}_{k+1})$, where $\bar{\mathbf{x}}_{k+1} = \mathbf{x}_k + \lambda_k \mathbf{d}_k$. If $f(\mathbf{x}_{k+1}) < UB_k$, put $UB_{k+1} = f(\mathbf{x}_{k+1})$ and $\mathbf{x}^* = \mathbf{x}_{k+1}$. Otherwise, let $UB_{k+1} \equiv UB_k$. Increment k by 1 and repeat the main step.

Note that the stopping criterion $\boldsymbol{\xi}_k = \mathbf{0}$ may never be realized, even if there exists an interior point optimum and we do find a solution $\mathbf{x}_k$ for which $\mathbf{0} \in \partial f(\mathbf{x}_k)$, for this may not be recognized because of the arbitrary subgradient $\boldsymbol{\xi}_k$ selected by the algorithm. Hence, a practical stopping criterion based on a maximum limit on the number of iterations performed is almost invariably used. Note also that we can terminate the procedure whenever $\mathbf{x}_{k+1} = \mathbf{x}_k$ for any iteration. Alternatively, if the optimal objective value f^* is known as in the problem of finding a feasible solution by minimizing the sum of (absolute) constraint violations, an ε stopping criterion $UB_k \le f^* + \varepsilon$ may be used, for some tolerance $\varepsilon > 0$. (See the Notes and References section for a primal–dual scheme employing a termination criterion based on the duality gap.)

8.9.1 Example

Consider the following Problem P:

$$\text{Minimize } \{f(x, y): -1 \le x \le 1, \ -1 \le y \le 1\}$$
$$\text{where } f(x, y) = \max\{-x, x + y, x - 2y\}$$

By considering $f(x, y) \le c$, where c is a constant, and examining the region bounded by $-x \le c$, $x + y \le c$, and $x - 2y \le c$, we can plot the contours of f as shown in Figure

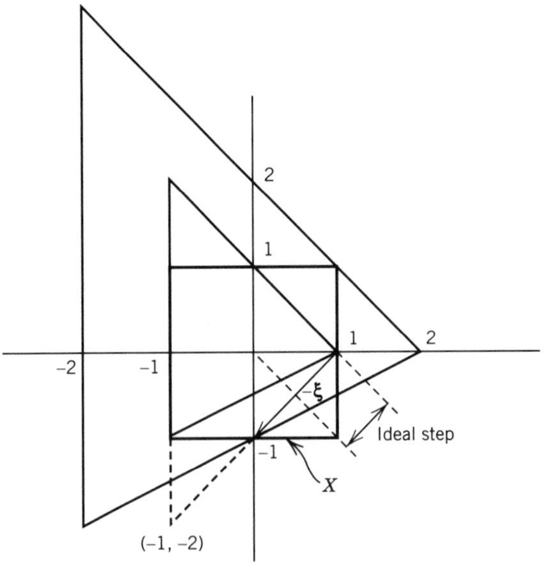

Figure 8.25 Illustration for Example 8.9.1.

8.25. Note that the points of nondifferentiability are of the type $(t, 0)$, $(-t, 2t)$, and $(-t, -t)$, for $t \geq 0$. Also, the optimal solution is $(x, y) = (0, 0)$, at which all the three linear functions defining f tie in value. Hence, although $(0, 0)^t \in \partial f(\mathbf{0})$, we also evidently have $(-1, 0)^t$, $(1, 1)^t$, and $(1, -2)^t$ belonging to $\partial f(\mathbf{0})$.

Now consider the point $(x, y) = (1, 0)$. We have $f(1, 0) = 1$ as determined by the linear functions $x + y$ and $x - 2y$. Hence, $\boldsymbol{\xi} = (1, 1)^t \in \partial f(1, 0)$. Consider the direction $-\boldsymbol{\xi} = (-1, -1)^t$. Note that this is not a descent direction. However, as we begin to move along this direction, we do approach closer to the optimal solution $(0, 0)^t$. Figure 8.25 shows the ideal step that we could take along the direction $\mathbf{d} = -\boldsymbol{\xi}$ to arrive closest to the optimal solution. However, suppose that we take a step length $\lambda = 2$ along $-\boldsymbol{\xi}$. This brings us to the point $(1, 0) - 2(1, 1) = (-1, -2)$. The projection $P_X(-1, -2)$ of $(-1, -2)$ onto X is obtained via (8.73) as $(-1, -1)$. This constitutes one iteration of the foregoing algorithm.

The following result prescribes a step size selection scheme that will guarantee convergence to an optimum.

8.9.2 Theorem

Let Problem P be as defined in (8.71) and assume that an optimum exists. Consider the foregoing subgradient optimization algorithm to solve Problem P, and suppose that the prescribed nonnegative step size sequence $\{\lambda_k\}$ satisfies the conditions $\{\lambda_k\} \to 0^+$ and $\sum_{k=0}^{\infty} \lambda_k = \infty$. Then, either the algorithm terminates finitely with an optimal solution, or else an infinite sequence is generated such that

$$\{\mathrm{UB}_k\} \to f^* \equiv \min \{f(\mathbf{x}) : \mathbf{x} \in X\}$$

Proof

The case of finite termination follows from Theorem 3.4.3. Hence, suppose that an infinite sequence $\{\mathbf{x}_k\}$ is generated along with the accompanying sequence of upper bounds $\{\mathrm{UB}_k\}$. Since $\{\mathrm{UB}_k\}$ is monotone nonincreasing, it has a limit point $\bar{f}$. We show that this limit $\bar{f}$ equals f^* by exhibiting that, for any given value $\alpha > f^*$, the sequence $\{\mathbf{x}_k\}$ enters the level set $S_\alpha = \{\mathbf{x} : f(\mathbf{x}) \leq \alpha\}$. Hence, we cannot have $\bar{f} > f^*$, or else we would obtain a contradiction by taking $\alpha \in (f^*, \bar{f})$, and so we must then have $\bar{f} = f^*$.

Toward this end, consider any $\hat{\mathbf{x}} \in X$ such that $f(\hat{\mathbf{x}}) < \alpha$. (For example, we can take $\hat{\mathbf{x}}$ as an optimal solution to Problem P.) Since $\hat{\mathbf{x}} \in \operatorname{int} S_\alpha$ because f is continuous, there exists a $\rho > 0$ such that $\|\mathbf{x} - \hat{\mathbf{x}}\| \leq \rho$ implies that $\mathbf{x} \in S_\alpha$. In particular, $\mathbf{x}_{Bk} = \hat{\mathbf{x}} + \rho\boldsymbol{\xi}_k/\|\boldsymbol{\xi}_k\|$ lies on the boundary of the ball centered at $\hat{\mathbf{x}}$ with radius ρ and, hence, lies in S_α for all k. But by the convexity of f, we have $f(\mathbf{x}_{Bk}) \geq f(\mathbf{x}_k) + (\mathbf{x}_{Bk} - \mathbf{x}_k)^t\boldsymbol{\xi}_k$ for all k. Hence if, on the contrary, $\{\mathbf{x}_k\}$ never enters S_α, that is, $f(\mathbf{x}_k) > \alpha$ for all k, we shall have $(\mathbf{x}_{Bk} - \mathbf{x}_k)^t\boldsymbol{\xi}_k \leq f(\mathbf{x}_{Bk}) - f(\mathbf{x}_k) < 0$. Substituting for $\mathbf{x}_{Bk}$, this gives $(\hat{\mathbf{x}} - \mathbf{x}_k)^t\boldsymbol{\xi}_k < -\rho\|\boldsymbol{\xi}_k\|$. Hence, using $\mathbf{d}_k = -\boldsymbol{\xi}_k/\|\boldsymbol{\xi}_k\|$, we get

$$(\mathbf{x}_k - \hat{\mathbf{x}})^t\mathbf{d}_k < -\rho \qquad \text{for all } k \tag{8.74}$$

Now, we have

$$\begin{aligned}\|\bar{\mathbf{x}}_{k+1} - \hat{\mathbf{x}}\|^2 &= \|\bar{\mathbf{x}}_{k+1} - \mathbf{x}_{k+1} + \mathbf{x}_{k+1} - \hat{\mathbf{x}}\|^2 \\ &= \|\mathbf{x}_{k+1} - \hat{\mathbf{x}}\|^2 + \|\bar{\mathbf{x}}_{k+1} - \mathbf{x}_{k+1}\|^2 + 2(\bar{\mathbf{x}}_{k+1} - \mathbf{x}_{k+1})^t(\mathbf{x}_{k+1} - \hat{\mathbf{x}})\end{aligned}$$

Hence, by Theorem 2.4.1,

$$\|\mathbf{x}_{k+1} - \hat{\mathbf{x}}\|^2 = \|\bar{\mathbf{x}}_{k+1} - \hat{\mathbf{x}}\|^2 - \|\bar{\mathbf{x}}_{k+1} - \mathbf{x}_{k+1}\|^2 - 2(\bar{\mathbf{x}}_{k+1} - \mathbf{x}_{k+1})^t(\mathbf{x}_{k+1} - \hat{\mathbf{x}})$$
$$\leq \|\bar{\mathbf{x}}_{k+1} - \hat{\mathbf{x}}\|^2$$

Hence, we get

$$\|\mathbf{x}_{k+1} - \hat{\mathbf{x}}\|^2 \leq \|\bar{\mathbf{x}}_{k+1} - \hat{\mathbf{x}}\|^2 = \|\mathbf{x}_k + \lambda_k \mathbf{d}_k - \hat{\mathbf{x}}\|^2$$
$$= \|\mathbf{x}_k - \hat{\mathbf{x}}\|^2 + \lambda_k^2 + 2\lambda_k \mathbf{d}_k^t(\mathbf{x}_k - \hat{\mathbf{x}})$$

Using (8.74), this gives

$$\|\mathbf{x}_{k+1} - \hat{\mathbf{x}}\|^2 \leq \|\mathbf{x}_k - \hat{\mathbf{x}}\|^2 + \lambda_k(\lambda_k - 2\rho)$$

Since $\lambda_k \to 0^+$, there exists a K such that for $k \geq K$, $\lambda_k \leq \rho$; and, hence,

$$\|\mathbf{x}_{k+1} - \hat{\mathbf{x}}\|^2 \leq \|\mathbf{x}_k - \hat{\mathbf{x}}\|^2 - \rho\lambda_k \qquad \text{for all } k \geq K \tag{8.75}$$

Summing the inequalities (8.75) written for $k = K, K + 1, \ldots, K + r$, say, we get

$$\rho \sum_{k=K}^{K+r} \lambda_k \leq \|\mathbf{x}_K - \hat{\mathbf{x}}\|^2 - \|\mathbf{x}_{K+r+1} - \hat{\mathbf{x}}\|^2 \leq \|\mathbf{x}_K - \hat{\mathbf{x}}\|^2 \qquad \text{for all } r \geq 0$$

Since the sum on the left-hand side diverges to infinity as $r \to \infty$, this leads to a contradiction, and the proof is complete.

Note that the proof of the theorem can easily be modified to show that, for each $\alpha > f^*$, the sequence $\{\mathbf{x}_k\}$ enters S_α infinitely often or else, for some K', we would have $f(\mathbf{x}_k) > \alpha$ for all $k \geq K'$, leading to the same contradiction. Hence, whenever $\mathbf{x}_{k+1} \equiv \mathbf{x}_k$ in the foregoing algorithm, $\mathbf{x}_k$ must be an optimal solution.

Furthermore, the above algorithm and proof can be readily extended to solve the problem of minimizing $f(\mathbf{x})$ subject to $\mathbf{x} \in X \cap Q$, where f and X are as above, and where $Q = \{\mathbf{x} : g_i(\mathbf{x}) \leq 0, i = 1, \ldots, m\}$. Here, we assume that each g_i, $i = 1, \ldots, m$, is convex and that $X \cap \text{int}(Q) \neq \emptyset$, so that for each $\alpha > f^*$, defining $S_\alpha = \{\mathbf{x} \in Q : f(\mathbf{x}) \leq \alpha\}$, we have a point $\hat{\mathbf{x}} \in X \cap \text{int}(S_\alpha)$. Now, in the algorithm, if we let $\boldsymbol{\xi}_k$ be a subgradient of f whenever $\mathbf{x}_k \in Q$, and if we let $\boldsymbol{\xi}_k$ be a subgradient of the most violated constraint in Q if $\mathbf{x}_k \notin Q$ (noting that $\mathbf{x}_k$ always lies in X by virtue of the projection operation), we shall again have (8.74) holding true, and the remainder of the convergence proof follows as before.

Choice of Step Sizes

Theorem 8.2.9 guarantees that, so long as the step sizes λ_k satisfy the stated conditions, convergence to an optimal solution will be obtained. Although this is theoretically true, it is unfortunately far from what is realized in practice. For example, choosing $\lambda_k = 1/k$ according to the divergent harmonic series ($\sum_{k=1}^{\infty} (1/k) = \infty$), the algorithm can easily stall and be remote from optimality after thousands of iterations. A careful fine-tuning of the choice of step sizes is usually required to obtain a satisfactory algorithmic performance.

To gain some insight into the choice of step sizes, let $\mathbf{x}_k$ be a nonoptimal iterate with $\boldsymbol{\xi}_k \in \partial f(\mathbf{x}_k)$ and denote by $\mathbf{x}^*$ an optimal solution to Problem (8.71) having objective value $f^* \equiv f(\mathbf{x}^*)$. By the convexity of f, we have $f(\mathbf{x}^*) \geq f(\mathbf{x}_k) + (\mathbf{x}^* - \mathbf{x}_k)^t\boldsymbol{\xi}_k$, or $(\mathbf{x}^* - \mathbf{x}_k)^t(-\boldsymbol{\xi}_k) \geq f(\mathbf{x}_k) - f^* > 0$. Hence, as observed in Example 8.9.1 (see Figure 8.25), although the direction $\mathbf{d}_k = -\boldsymbol{\xi}_k/\|\boldsymbol{\xi}_k\|$ need not be an improving direction, it does lead to points that are closer in Euclidean norm to $\mathbf{x}^*$ than was $\mathbf{x}_k$. In fact, this is the feature that drives the convergence of the algorithm and ensures an eventual improvement in objective function value.

Now, as in Figure 8.25, an ideal step size to adopt might be that which brings us closest to $\mathbf{x}^*$. This step size λ_k^* can be found by requiring that the vector $(\mathbf{x}_k + \lambda_k^* \mathbf{d}_k) - \mathbf{x}^*$ be orthogonal to $\mathbf{d}_k$, or that $\mathbf{d}_k^t[\mathbf{x}_k + \lambda_k^* \mathbf{d}_k - \mathbf{x}^*] = 0$. This gives

$$\lambda_k^* = (\mathbf{x}^* - \mathbf{x}_k)^t \mathbf{d}_k = \frac{(\mathbf{x}_k - \mathbf{x}^*)^t \boldsymbol{\xi}_k}{\|\boldsymbol{\xi}_k\|} \tag{8.76}$$

Of course, the problem with trying to implement the step size λ_k^* is that $\mathbf{x}^*$ is unknown. However, by the convexity of f, we have $f^* = f(\mathbf{x}^*) \geq f(\mathbf{x}_k) + (\mathbf{x}^* - \mathbf{x}_k)^t \boldsymbol{\xi}_k$; and, hence, from (8.76), we have that $\lambda_k^* \geq [f(\mathbf{x}_k) - f^*]/\|\boldsymbol{\xi}_k\|$. Since f^* is also usually unknown, we can recommend using an underestimate $\bar{f}$ in lieu of f^*, noting that the foregoing relationship is a "greater than or equal to" type of inequality. This leads to a choice of step size

$$\lambda_k = \frac{\beta_k [f(\mathbf{x}_k) - \bar{f}]}{\|\boldsymbol{\xi}_k\|} \tag{8.77}$$

where $\beta_k > 0$. In fact, by selecting $\varepsilon_1 < \beta_k \leq 2 - \varepsilon_2$ for all k, for some positive ε_1 and ε_2, and using f^*, itself, instead of $\bar{f}$ in (8.77), it can be shown that the generated sequence $\{\mathbf{x}_k\}$ converges to an optimum $\mathbf{x}^*$. (A linear or geometric convergence rate can be exhibited under some additional assumptions on f.)

A practical way of employing (8.77) that has been empirically found to be computationally attractive is as follows (this is called a *block halving* scheme). First, designate an upper limit N on the number of iterations to be performed. Next, select some $\bar{r} < N$ and divide the potential sequence of iterations $1, \ldots, N$ into $T = \lceil N/\bar{r} \rceil$ blocks, with the first $T - 1$ blocks having $\bar{r}$ iterations, and the final block having the remaining ($\leq \bar{r}$) iterations. Also, for each block t, select a parameter value $\beta(t)$, for $t = 1, \ldots, T$. (Typical values are $N = 200$, $\bar{r} = 75$, $\beta(1) = 0.75$, $\beta(2) = 0.5$, and $\beta(3) = 0.25$, with $T = 3$.) Now, within each block t, compute the first step length using (8.77), with β_k equal to the corresponding $\beta(t)$ value. However, for the remaining iterations within the block, the *step length* is kept the same as for the initial iteration in that block, except that each time the objective function value fails to improve over some $\bar{v}(=10$, say) consecutive iterations, the step length is successively halved. (Alternatively, (8.77) can be used to compute the step length for each iteration, with β_k starting at $\beta(t)$ for block t, and with this β parameter being halved whenever the method experiences $\bar{v}$ consecutive failures as before.) Additionally, at the beginning of a new block, and also whenever the method experiences $\bar{v}$ consecutive failures, the process is reset to the incumbent solution before the modified step length is used. Although some fine-tuning of the foregoing parameter values might be required dependent on the class of problems being solved, the prescribed values work well on reasonably well-scaled problems (see the Notes and References section for empirical evidence using such a scheme).

Subgradient Deflection Methods

It has frequently been observed that the problem with subgradient methods is that, as the iterates progress, the angle between the subgradient-based direction $\mathbf{d}_k$ and the direction $\mathbf{x}^* - \mathbf{x}_k$ toward optimality, although acute, tends to get close to 90°. As a result, the step size needs to shrink considerably before a descent is realized, and this in turn causes the procedure to stall. Hence, it becomes almost imperative to adopt some suitable deflection or rotation scheme to accelerate the convergence behavior.

Toward this end, in the spirit of conjugate gradient methods, one can adopt a direction of search as $\mathbf{d}_1 = -\boldsymbol{\xi}_1$ and $\mathbf{d}_k = -\boldsymbol{\xi}_k + \phi_k \mathbf{d}_{k-1}^a$, where $\mathbf{d}_{k-1}^a \equiv \mathbf{x}_k - \mathbf{x}_{k-1}$ and ϕ_k is an appropriate parameter. (These directions can be normalized and then used in conjunction with the same block-halving step size strategy described above.) Various strategies prompted by theoretical convergence and/or practical efficiency can be designed by choosing ϕ_k appropriately (see the Notes and References section). A simple choice that works reasonably well in practice is the *average direction strategy,* for which $\phi_k = \|\boldsymbol{\xi}_k\|/\|\mathbf{d}_{k-1}^a\|$, so that $\mathbf{d}_k$ bisects the angle between $-\boldsymbol{\xi}_k$ and $\mathbf{d}_{k-1}^a$.

Other viable strategies include the use of $\mathbf{d}_k = -\mathbf{D}_k \boldsymbol{\xi}_k$, where $\mathbf{D}_k$ is a suitable, symmetric, positive definite matrix as in *space dilatation* (or *dilation*) methods that imitate quasi-Newton procedures, or the use of a minimum norm subgradient as in Theorem 6.3.11, but based on an approximation to the subdifferential at $\mathbf{x}_k$, and not the actual subdifferential as in the theorem. For further reading on this subject, refer to the Notes and References section.

Exercises

8.1 For the uniform search method, the dichotomous search method, the golden section method, and the Fibonacci search method, compute the number of functional evaluations required for $\alpha = 0.1, 0.01, 0.001$, and 0.0001, where α is the ratio of the final interval of uncertainty to the length of the initial interval of uncertainty.

8.2 Suppose that θ is differentiable and let $|\theta'| \le a$. Furthermore, suppose that the uniform search method is used to minimize θ. Let $\hat{\lambda}$ be a grid point such that $\theta(\bar{\lambda}) - \theta(\hat{\lambda}) \ge \varepsilon > 0$, for each grid point $\bar{\lambda} \ne \hat{\lambda}$. If the grid length δ is such that $a\delta \le \varepsilon$, show, without assuming strict quasiconvexity, that no point outside the interval $[\hat{\lambda} - \delta, \hat{\lambda} + \delta]$ could provide a functional value less than $\theta(\hat{\lambda})$.

8.3 Show that the golden section method approaches the method of Fibonacci as the number of functional evaluations n approaches ∞.

8.4 Consider the following definitions:

A function $\theta : E_1 \to E_1$ to be *minimized* is said to be *strongly unimodal* over the interval $[a, b]$ if there exists a $\bar{\lambda}$ that minimizes θ over the interval; and for any $\lambda_1, \lambda_2 \in [a, b]$ such that $\lambda_1 < \lambda_2$, we have

$$\lambda_2 \le \bar{\lambda} \quad \text{implies } \theta(\lambda_1) > \theta(\lambda_2)$$

$$\lambda_1 \ge \bar{\lambda} \quad \text{implies } \theta(\lambda_1) < \theta(\lambda_2)$$

A function $\theta : E_1 \to E_1$ to be *minimized* is said to be *strictly unimodal* over the interval $[a, b]$ if there exists a $\bar{\lambda}$ that minimizes θ over the interval; and for $\lambda_1, \lambda_2 \in [a, b]$ such that $\theta(\lambda_1) \ne \theta(\bar{\lambda})$, $\theta(\lambda_2) \ne \theta(\bar{\lambda})$, and $\lambda_1 < \lambda_2$, we have

$$\lambda_2 \le \bar{\lambda} \quad \text{implies } \theta(\lambda_1) > \theta(\lambda_2)$$

$$\lambda_1 \ge \bar{\lambda} \quad \text{implies } \theta(\lambda_1) < \theta(\lambda_2)$$

a. Show that if θ is strongly unimodal over $[a, b]$, then θ is strongly quasiconvex over $[a, b]$. Conversely, show that if θ is strongly quasiconvex over $[a, b]$ and has a minimum in this interval, then it is strongly unimodal over the interval.

b. Show that if θ is strictly unimodal and continuous over $[a, b]$, then θ is strictly quasiconvex over $[a, b]$. Conversely, show that if θ is strictly quasiconvex over $[a, b]$ and has a minimum in this interval, then it is strictly unimodal over this interval.

8.5 Consider the function f defined by $f(\mathbf{x}) = (x_1^3 + x_2)^2 + 2(x_2 - x_1 - 4)^4$. Given a point $\mathbf{x}_1$ and a nonzero direction vector $\mathbf{d}$, let $\theta(\lambda) = f(\mathbf{x}_1 + \lambda\mathbf{d})$.
   a. Obtain an explicit expression for $\theta(\lambda)$.
   b. For $\mathbf{x}_1 = (0, 0)^t$ and $\mathbf{d} = (1, 1)^t$, using the Fibonacci method, find the value of λ that solves the problem to minimize $\theta(\lambda)$ subject to $\lambda \in E_1$.
   c. For $\mathbf{x}_1 = (4, 5)^t$ and $\mathbf{d} = (1, -2)^t$, using the golden section method, find the value of λ that solves the problem to minimize $\theta(\lambda)$ subject to $\lambda \in E_1$.
   d. Repeat parts b and c using the interval bisection method.

8.6 Find the minimum of $e^{-\lambda} + \lambda^2$ by each of the following procedures:
   a. Golden section method.
   b. Dichotomous search method.
   c. Newton's method.
   d. Bisecting search method.

8.7 Consider the problem to minimize $f(\mathbf{x} + \lambda\mathbf{d})$ subject to $\lambda \in E_1$. Show that a necessary condition for a minimum at $\bar{\lambda}$ is that $\mathbf{d}^t\nabla f(\mathbf{y}) = 0$, where $\mathbf{y} = \mathbf{x} + \bar{\lambda}\mathbf{d}$. Under what assumptions is this condition sufficient for optimality?

8.8 Consider the problem to minimize $f(\mathbf{x} + \lambda\mathbf{d})$ subject to $\mathbf{x} + \lambda\mathbf{d} \in S$ and $\lambda \geq 0$, where S is a compact convex set, and f is a convex function. Furthermore, suppose that $\mathbf{d}$ is an improving direction. Show that an optimal solution $\bar{\lambda}$ is given by $\bar{\lambda} = \text{minimum } (\lambda_1, \lambda_2)$, where λ_1 satisfies $\mathbf{d}^t\nabla f(\mathbf{x} + \lambda_1\mathbf{d}) = 0$, and $\lambda_2 = \text{maximum } \{\lambda : \mathbf{x} + \lambda\mathbf{d} \in S\}$.

8.9 Consider the problem to minimize $3\lambda - 2\lambda^2 + \lambda^3 + 2\lambda^4$ subject to $\lambda \geq 0$.
   a. Write a necessary condition for a minimum. Can you make use of this condition to find the global minimum?
   b. Is the function strictly quasiconvex over the region $\{\lambda : \lambda \geq 0\}$? Apply the Fibonacci search method to find the minimum.
   c. Apply both the bisecting search method and Newton's method to the above problem, starting from $\lambda_1 = 6$.

8.10 In Section 8.2, we described Newton's method for finding a point where the derivative of a function vanishes.
   a. Show how the method can be used to find a point where the value of a continuously differentiable function is equal to zero. Illustrate the method for $\theta(\lambda) = \lambda^3 - \lambda$, starting from $\lambda_1 = 5$.
   b. Will the method converge for any starting point? Prove or give a counterexample.

8.11 Show how the line search procedures of Section 8.1 can be used to find a point where a given function assumes the value zero. Illustrate by the function θ defined by $\theta(\lambda) = \lambda^2 - 3\lambda + 2$. (*Hint:* Consider the absolute value function $\hat{\theta} = |\theta|$.)

8.12 In Section 8.2, we discussed the bisecting search method for finding a point where the derivative of a pseudoconvex function vanishes. Show how the method can be used to find a point where a function is equal to zero. Explicitly state the assumptions that the function needs to satisfy. Illustrate by the function θ defined by $\theta(\lambda) = \lambda^3 - \lambda$ defined on the interval $[0.5, 10.0]$.

8.13 It can be verified that, in Example 9.2.4, for a given value of μ, if $\mathbf{x}_\mu = (x_1, x_2)^t$, x_1 satisfies

$$2(x_1 - 2)^3 + \frac{\mu x_1(8x_1^2 - 6x_1 + 1)}{4 + \mu} = 0$$

For $\mu = 0.1, 1.0, 10.0$, and 100.0, find the value of x_1 satisfying the above equation, using a suitable procedure.

8.14 Let $\theta : E_1 \to E_1$ and suppose that we have the three points (λ_1, θ_1), (λ_2, θ_2), and (λ_3, θ_3), where $\theta_j = \theta(\lambda_j)$ for $j = 1, 2, 3$. Show that the quadratic curve q passing through these points is given by

$$q(\lambda) = \frac{\theta_1(\lambda - \lambda_2)(\lambda - \lambda_3)}{(\lambda_1 - \lambda_2)(\lambda_1 - \lambda_3)} + \frac{\theta_2(\lambda - \lambda_1)(\lambda - \lambda_3)}{(\lambda_2 - \lambda_1)(\lambda_2 - \lambda_3)} + \frac{\theta_3(\lambda - \lambda_1)(\lambda - \lambda_2)}{(\lambda_3 - \lambda_1)(\lambda_3 - \lambda_2)}$$

Furthermore, show that the derivative of q vanishes at the point $\bar{\lambda}$ given by

$$\bar{\lambda} = \frac{1}{2}\frac{b_{23}\theta_1 + b_{31}\theta_2 + b_{12}\theta_3}{a_{23}\theta_1 + a_{31}\theta_2 + a_{12}\theta_3}$$

where $a_{ij} = \lambda_i - \lambda_j$ and $b_{ij} = \lambda_i^2 - \lambda_j^2$. Find the quadratic curve passing through the points (1, 3), (2, 1), and (4, 6), and compute $\bar{\lambda}$. Show that if $(\lambda_1, \lambda_2, \lambda_3)$ satisfy the three-point pattern (TPP), then $\lambda_1 < \bar{\lambda} < \lambda_3$. Also,

a. Propose a method for finding $\lambda_1, \lambda_2, \lambda_3$ such that $\lambda_1 < \lambda_2 < \lambda_3$, $\theta_1 \geq \theta_2$, and $\theta_2 \leq \theta_3$.
b. Show that if θ is strictly quasiconvex, then the new interval of uncertainty defined by the revised λ_1 and λ_3 of the quadratic fit line search indeed contains the minimum.
c. Use the procedure described in this exercise to minimize $3\lambda - 2\lambda^2 + \lambda^3 + 2\lambda^4$ over $\lambda \geq 0$.

8.15 Let $\theta : E_1 \rightarrow E_1$ be a continuous strictly quasiconvex function. Let $0 \leq \lambda_1 < \lambda_2 < \lambda_3$ and denote $\theta_j = \theta(\lambda_j)$ for $j = 1, 2, 3$.

a. If $\theta_1 = \theta_2 = \theta_3$, show that this common value coincides with the value of minimum $\{\theta(\lambda) : \lambda \geq 0\}$.
b. Let $\lambda = (\lambda_1, \lambda_2, \lambda_3) \in E_3$ represent a three-point pattern iterate generated by the quadratic fit algorithm described in Section 8.3. Show that the function $\bar{\theta}(\lambda) \equiv \theta(\lambda_1) + \theta(\lambda_2) + \theta(\lambda_3)$ is a continuous function that satisfies the descent property $\bar{\theta}(\lambda_{\text{new}}) < \bar{\theta}(\lambda)$ whenever θ_1, θ_2, and θ_3 are not all equal to each other.

8.16 Let θ be pseudoconvex and continuously twice-differentiable. Consider the algorithm of Section 8.3 with the modification that in case 3, when $\bar{\lambda} = \lambda_2$, we let $\lambda_{\text{new}} = (\lambda_1, \lambda_2, \bar{\lambda})$ if $\theta'(\lambda_2) > 0$, let $\lambda_{\text{new}} = (\lambda_2, \bar{\lambda}, \lambda_3)$ if $\theta'(\lambda_2) < 0$, and stop if $\theta'(\lambda_2) = 0$. Accordingly, if λ_1, λ_2, and λ_3 are not all distinct, let them be said to satisfy the *three-point pattern (TPP)* whenever $\theta'(\lambda_2) < 0$ if $\lambda_1 = \lambda_2 < \lambda_3$, $\theta'(\lambda_2) > 0$ if $\lambda_1 < \lambda_2 = \lambda_3$, and $\theta'(\lambda_2) = 0$, $\theta''(\lambda_2) \geq 0$ if $\lambda_1 = \lambda_2 = \lambda_3$. With this modification, suppose that we use the quadratic interpolation algorithm of Section 8.3 applied to θ given a starting TPP $\lambda = (\lambda_1, \lambda_2, \lambda_3)$, where the quadratic fit matches the two function values and the derivative $\theta'(\lambda_2)$ whenever two of the three points $\lambda_1, \lambda_2, \lambda_3$ are coincident, and where at any iteration, if $\theta'(\lambda_2) = 0$, we put $\lambda^* = (\lambda_2, \lambda_2, \lambda_2)$ and terminate. Define the solution set $\Omega = \{(\lambda, \lambda, \lambda): \theta'(\lambda) = 0\}$.

a. Let $\mathbf{A}$ define the algorithmic map which produces $\lambda_{\text{new}} \in \mathbf{A}(\lambda)$. Show that $\mathbf{A}$ is closed.
b. Show that the function $\bar{\theta}(\lambda) = \theta(\lambda_1) + \theta(\lambda_2) + \theta(\lambda_3)$ is a continuous descent function that satisfies $\bar{\theta}(\lambda_{\text{new}}) < \bar{\theta}(\lambda)$ if $\theta'(\lambda_2) \neq 0$.
c. Hence, show that the defined algorithm either terminates finitely or else generates an infinite sequence whose accumulation points lie in Ω.
d. Comment on the convergence of the algorithm and the nature of the solution obtained if θ is strictly quasiconvex and twice continuously differentiable.

8.17 Define the *percentage test* line search map which determines the step length λ to within $100p\%$, $0 \leq p \leq 1$, of the ideal step λ^* according to $\mathbf{M}(\mathbf{x}, \mathbf{d}) = \{\mathbf{y} : \mathbf{y} = \mathbf{x} + \lambda\mathbf{d}$, where $0 \leq \lambda < \infty$, and $|\lambda - \lambda^*| \leq p\lambda^*\}$, where $\theta(\lambda) \equiv f(\mathbf{x} + \lambda\mathbf{d})$ and $\theta'(\lambda^*) = 0$. Show that if $\mathbf{d} \neq \mathbf{0}$ and θ is continuously differentiable, then $\mathbf{M}$ is closed at $(\mathbf{x}, \mathbf{d})$. Explain how you can use this test in conjunction with the quadratic interpolation method described in Section 8.3.

8.18 Show that as a function of K_k, the expression in (8.14) is maximized when $K_k^2 = \alpha^2$.

8.19 Let $\mathbf{H}$ be an $n \times n$, symmetric, positive definite matrix with condition number α. Then, the *Kantorovich Inequality* asserts that for any $\mathbf{x} \in E_n$, we have

$$\frac{(\mathbf{x}^t\mathbf{x})^2}{(\mathbf{x}^t\mathbf{H}\mathbf{x})(\mathbf{x}^t\mathbf{H}^{-1}\mathbf{x})} \geq \frac{4\alpha}{(1 + \alpha)^2}$$

Justify this inequality, and use it to establish equation (8.18).

8.20 Consider the application of the steepest descent method to minimize $f(\mathbf{x})$ versus the application of this method to minimize $F(\mathbf{x}) = \|\nabla f(\mathbf{x})\|^2$. Assuming that f is quadratic with a positive definite Hessian, compare the rates of convergence of the two schemes and, hence, justify why the equivalent minimization of F is not an attractive strategy.

8.21 Consider the problem to minimize $(x_1^3 - x_2)^2 + 2(x_2 - x_1)^4$. Solve the problem using each of the following methods. Do the methods converge to the same point? If not, explain.

a. The cyclic coordinate method.
b. The method of Hooke and Jeeves.
c. The method of Rosenbrock.
d. The method of Zangwill.
e. The method of steepest descent.
f. The method of Fletcher and Reeves.
g. The method of Davidon–Fletcher–Powell.
h. The method of Broyden–Fletcher–Goldfarb–Shanno (BFGS).

8.22 Consider the problem to minimize $(1 - x_1)^2 + 5(x_2 - x_1^2)^2$. Starting from the point $(2, 0)$, solve the problem by the following procedures:

a. The cyclic coordinate method.
b. The method of Hooke and Jeeves.
c. The method of Rosenbrock.
d. The method of Davidon–Fletcher–Powell.
e. The method of Broyden–Fletcher–Goldfarb–Shanno (BFGS).

8.23 Solve the problem to maximize $x_1 + 2x_2 + 5x_1x_2 - x_1^2 + 3x_2^2$ by the method of Hooke and Jeeves.

8.24 Consider the model $y = \alpha + \beta x + \gamma x^2 + \varepsilon$, where x is the independent variable, y is the observed dependent variable, α, β, and γ are unknown parameters, and ε is a random component representing the experimental error. The following table gives the values of x and the corresponding values of y. Formulate the problem of finding the best estimates of α, β, and γ as an unconstrained optimization problem, by minimizing

a. The sum of squared errors.
b. The sum of the absolute values of the errors.
c. The maximum absolute value of the error.

For each case, find α, β, and γ by a suitable method.

x	0	1	2	3	4	5
y	2	2	−12	−27	−60	−90

8.25 Let $f : E_n \rightarrow E_1$ be differentiable at $\mathbf{x}$ and let the vectors $\mathbf{d}_1, \ldots, \mathbf{d}_n$ in E_n be linearly independent. Suppose that the minimum of $f(\mathbf{x} + \lambda \mathbf{d}_j)$ over $\lambda \in E_1$ occurs at $\lambda = 0$ for $j = 1, \ldots, n$. Show that $\nabla f(\mathbf{x}) = \mathbf{0}$. Does this imply that f has a local minimum at $\mathbf{x}$?

8.26 The following method for generating a set of conjugate directions for minimizing $f : E_n \rightarrow E_1$ is due to Zangwill [1967]:

Initialization Step Choose a termination scalar $\varepsilon > 0$, and choose an initial point $\mathbf{x}_1$. Let $\mathbf{y}_1 = \mathbf{x}_1$, let $\mathbf{d}_1 = -\nabla f(\mathbf{y}_1)$, let $k = j = 1$, and go to the main step.

Main Step

1. Let λ_j be an optimal solution to the problem to minimize $f(\mathbf{y}_j + \lambda \mathbf{d}_j)$ subject to $\lambda \in E_1$, and let $\mathbf{y}_{j+1} = \mathbf{y}_j + \lambda_j \mathbf{d}_j$. If $j = n$, go to step 4; otherwise, go to step 2.
2. Let $\mathbf{d} = -\nabla f(\mathbf{y}_{j+1})$, and let $\hat{\mu}$ be an optimal solution to the problem to minimize $f(\mathbf{y}_{j+1} + \mu \mathbf{d})$ subject to $\mu \geq 0$. Let $\mathbf{z}_1 = \mathbf{y}_{j+1} + \hat{\mu}\mathbf{d}$. Let $i = 1$, and go to step 3.
3. If $\|\nabla f(\mathbf{z}_i)\| < \varepsilon$, stop with $\mathbf{z}_i$. Otherwise, let μ_i be an optimal solution to the problem to minimize $f(\mathbf{z}_i + \mu \mathbf{d}_i)$ subject to $\mu \in E_1$. Let $\mathbf{z}_{i+1} = \mathbf{z}_i + \mu_i \mathbf{d}_i$. If $i < j$, replace i by $i + 1$, and repeat step 3. Otherwise, let $\mathbf{d}_{j+1} = \mathbf{z}_{j+1} - \mathbf{y}_{j+1}$, replace j by $j + 1$, and go to step 1.

4. Let $\mathbf{y}_1 = \mathbf{x}_{k+1} = \mathbf{y}_{n+1}$. Let $\mathbf{d}_1 = -\nabla f(\mathbf{y}_1)$, replace k by $k + 1$, let $j = 1$, and go to step 1.

Note that the steepest descent search in step 2 is used to ensure that $\mathbf{z}_1 - \mathbf{y}_1 \notin L(\mathbf{d}_1, \ldots, \mathbf{d}_j)$ for the quadratic case so that finite convergence is guaranteed.

Illustrate using the problem to minimize $(x_1 - 2)^4 + (x_1 - 2x_2)^2$, starting from the point (0.0, 3.0).

8.27 Consider the following problem:

$$\begin{aligned} \text{Minimize} \quad & x_1 + x_2 \\ \text{subject to} \quad & x_1^2 + x_2^2 = 4 \\ & -2x_1 - x_2 \le 4 \end{aligned}$$

a. Formulate the Lagrangian dual problem by incorporating both constraints into the objective function via the Lagrangian multipliers u_1 and u_2.
b. Using a suitable unconstrained optimization method, compute the gradient of the dual function θ at the point (1, 2).
c. Starting from the point $\bar{\mathbf{u}} = (1, 2)^t$, perform one iteration of the steepest ascent method for the dual problem. In particular, solve the following problem, where $\mathbf{d} = \nabla\theta(\bar{u})$:

$$\begin{aligned} \text{Maximize} \quad & \theta(\bar{\mathbf{u}} + \lambda \mathbf{d}) \\ \text{subject to} \quad & \bar{u}_2 + \lambda d_2 \ge 0 \\ & \lambda \ge 0 \end{aligned}$$

8.28 Suppose that f is continuously twice-differentiable and that the Hessian matrix is invertible everywhere. Given $\mathbf{x}_k$, let $\mathbf{x}_{k+1} = \mathbf{x}_k + \lambda_k \mathbf{d}_k$, where $\mathbf{d}_k = -\mathbf{H}(\mathbf{x}_k)^{-1}\nabla f(\mathbf{x}_k)$ and λ_k is an optimal solution to the problem to minimize $f(\mathbf{x}_k + \lambda \mathbf{d}_k)$ subject to $\lambda \in E_1$. Show that this modification of Newton's method converges to a point in the solution set $\Omega = \{\bar{\mathbf{x}} : \nabla f(\bar{\mathbf{x}})^t \mathbf{H}(\bar{\mathbf{x}})^{-1} \nabla f(\bar{\mathbf{x}}) = 0\}$. Illustrate by minimizing $(x_1 - 2)^4 + (x_1 - 2x_2)^2$ starting from the point $(-2, 3)$.

8.29 Consider the problem in (8.24) and suppose that $\varepsilon_k \ge 0$ is such that $\mathbf{H}(\mathbf{x}_k) + \varepsilon_k \mathbf{I}$ is positive definite. Let $\Delta_k = -[\mathbf{H}(\mathbf{x}_k) + \varepsilon_k \mathbf{I}]^{-1}\nabla f(\mathbf{x}_k)$. Show that $\delta = \mathbf{x}_{k+1} - \mathbf{x}_k$, given by (8.22), and the Lagrange multiplier $\lambda = \varepsilon_k$ satisfy the saddle point optimality conditions for (8.24). Hence, comment on the relationship between the Levenberg–Marquardt method and trust-region methods. Also comment on the case $\varepsilon_k = 0$.

8.30 Let $\mathbf{H}$ be a symmetric $n \times n$ matrix, and let $\mathbf{d}_1, \ldots, \mathbf{d}_n$ be a set of characteristic vectors of $\mathbf{H}$. Show that $\mathbf{d}_1, \ldots, \mathbf{d}_n$ are $\mathbf{H}$-conjugate.

8.31 Let $\mathbf{a}_1, \ldots, \mathbf{a}_n$ be a set of linearly independent vectors in E_n, and let $\mathbf{H}$ be an $n \times n$ symmetric positive definite matrix.

a. Show that the vectors $\mathbf{d}_1, \ldots, \mathbf{d}_n$ defined below are $\mathbf{H}$-conjugate.

$$\mathbf{d}_k = \begin{cases} \mathbf{a}_k & \text{if } k = 1 \\ \mathbf{a}_k - \displaystyle\sum_{i=1}^{k-1} \left[\frac{\mathbf{d}_i^t \mathbf{H} \mathbf{a}_k}{\mathbf{d}_i^t \mathbf{H} \mathbf{d}_i}\right] \mathbf{d}_i & \text{if } k \ge 2 \end{cases}$$

b. Suppose that $\mathbf{a}_1, \ldots, \mathbf{a}_n$ are the unit vectors in E_n, and let $\mathbf{D}$ be the matrix whose columns are the vectors $\mathbf{d}_1, \ldots, \mathbf{d}_n$, defined in part a above. Show that $\mathbf{D}$ is upper triangular with all diagonal elements equal to 1.

c. Illustrate by letting $\mathbf{a}_1 = (1, 0, 0)^t$, $\mathbf{a}_2 = (1, -1, 4)^t$, $\mathbf{a}_3 = (2, -1, 6)^t$, and

$$\mathbf{H} = \begin{bmatrix} 1 & 0 & -1 \\ 0 & 3 & 4 \\ -1 & 4 & 2 \end{bmatrix}$$

d. Illustrate by letting $\mathbf{a}_1$, $\mathbf{a}_2$, and $\mathbf{a}_3$ be the unit vectors in E_3 and $\mathbf{H}$ as given in part c above.

8.32 Consider the quadratic form $f(\mathbf{x}) = \mathbf{c}'\mathbf{x} + \frac{1}{2}\mathbf{x}'\mathbf{H}\mathbf{x}$, where $\mathbf{H}$ is a symmetric $n \times n$ matrix. In many applications, it is desirable to obtain separability in the variables by eliminating the cross-product terms. This could be done by rotating the axes as follows. Let $\mathbf{D}$ be an $n \times n$ matrix whose columns $\mathbf{d}_1, \ldots, \mathbf{d}_n$ are $\mathbf{H}$-conjugate. Letting $\mathbf{x} = \mathbf{D}\mathbf{y}$, verify that the quadratic form is equivalent to $\Sigma_{j=1}^{n} \alpha_j y_j + \frac{1}{2}\Sigma_{j=1}^{n} \beta_j y_j^2$, where $(\alpha_1, \ldots, \alpha_n) = \mathbf{c}'\mathbf{D}$, and $\beta_j = \mathbf{d}_j'\mathbf{H}\mathbf{d}_j$ for $j = 1, \ldots, n$. Furthermore, translating and rotating the axes could be accomplished by the transformation $\mathbf{x} = \mathbf{D}\mathbf{y} + \mathbf{z}$, where $\mathbf{z}$ is any vector satisfying $\mathbf{H}\mathbf{z} + \mathbf{c} = \mathbf{0}$, that is, $\nabla f(\mathbf{z}) = \mathbf{0}$. In this case, show that the quadratic form is equivalent to $(\mathbf{c}'\mathbf{z} + \frac{1}{2}\mathbf{z}'\mathbf{H}\mathbf{z}) + \frac{1}{2}\Sigma_{j=1}^{n} \beta_j y_j^2$. Use the result of this exercise to draw accurate contours of the quadratic form $2x_1 - 4x_2 + x_1^2 + 2x_1x_2 + 3x_2^2$.

8.33 Show that there exists a value of ϕ [as given by (8.50)] for the Broyden correction formula (8.46) that will yield $\mathbf{d}_{j+1} = -\mathbf{D}_{j+1}\nabla f(\mathbf{y}_{j+1}) \equiv \mathbf{0}$. (*Hint:* Use $\mathbf{p}_j = \lambda_j \mathbf{d}_j - \lambda_j \mathbf{D}_j \nabla f(\mathbf{y}_j)$, $\mathbf{q}_j = \nabla f(\mathbf{y}_{j+1}) - \nabla f(\mathbf{y}_j)$, and $\mathbf{d}_j'\nabla f(\mathbf{y}_{j+1}) = \mathbf{p}_j'\nabla f(\mathbf{y}_{j+1}) = \nabla f(\mathbf{y}_j)'\mathbf{D}_j\nabla f(\mathbf{y}_{j+1}) = 0$.)

8.34 Derive the Hessian correction (8.53) for the BFGS update directly, following the scheme used for the update of the Hessian inverse via (8.41) through (8.45).

8.35 Use two sequential applications of the Sherman–Morrison–Woodbury formula given in (8.55) to verify the inverse relationship (8.54) between (8.48) and (8.53).

8.36 Derive a quasi-Newton correction matrix $\mathbf{C}$ for the Hessian approximation $\mathbf{B}_k$ that achieves the minimum Frobenius norm (squared) $\Sigma_i\Sigma_j C_{ij}^2$, where C_{ij} are the elements of $\mathbf{C}$ (to be determined), subject to the quasi-Newton condition $(\mathbf{C} + \mathbf{B}_k)\mathbf{p}_k = \mathbf{q}_k$ amd the symmetry condition $\mathbf{C} = \mathbf{C}'$. (*Hint:* Set up the corresponding optimization problem after enforcing symmetry, and use the KKT conditions. This gives the *Powell–Symmetric Broyden (PSB)* update.)

8.37 Referring to Figure 8.24 and the associated discussion, verify that the minimization of the quadratic function f from $\mathbf{y}_j$ along the pattern direction $\mathbf{d}_P \equiv \mathbf{y}_{j+1}' - \mathbf{y}_j$ will produce the point $\mathbf{y}_{j+2}$. (*Hint:* Let $\mathbf{y}_{j+2}'$ denote the point thus obtained. Using the fact that $\nabla f(\mathbf{y}_{j+1}')'\nabla f(\mathbf{y}_{j+1}) = 0$ and that ∇f is linear, since f is quadratic, show that $\nabla f(\mathbf{y}_{j+2}')$ is orthogonal to both $\nabla f(\mathbf{y}_{j+1})$ and to $\mathbf{d}_P$, so that $\mathbf{y}_{j+2}'$ is a minimizing point in the plane of Figure 8.24. Using part 3 of Theorem 8.8.3, argue now that $\mathbf{y}_{j+2}' \equiv \mathbf{y}_{j+2}$.)

8.38 Show that analogous to (8.66), assuming exact line searches, a memoryless quasi-Newton update performed on a member of the Broyden family (taking $\mathbf{D}_j \equiv \mathbf{I}$) results in a direction $\mathbf{d}_{j+1} = -\mathbf{D}_{j+1}\nabla f(\mathbf{y}_{j+1})$, where

$$\mathbf{D}_{j+1} = \mathbf{I} - (1 - \phi)\frac{\mathbf{q}_j\mathbf{q}_j'}{\mathbf{q}_j'\mathbf{q}_j} - \phi\frac{\mathbf{p}_j\mathbf{q}_j'}{\mathbf{p}_j'\mathbf{q}_j}$$

Observe that the equivalence with conjugate gradient methods results only when $\phi = 1$ (BFGS update).

8.39 Consider the problem to maximize $-x_1^2 - x_2^2 + x_1x_2 - x_1 + 2x_2$. Starting from the origin, solve the problem by the Davidon–Fletcher–Powell method, with $\mathbf{D}_1$ as the identity. Also solve the problem by the Fletcher and Reeves conjugate gradient method. Note that the two procedures generate identical sets of directions. Show that, in general, if $\mathbf{D}_1 = \mathbf{I}$, then the two methods are identical for quadratic functions.

8.40 Consider the problem to minimize $x_1^2 + x_2^2$ subject to $x_1 + x_2 - 2 = 0$.

a. Find the optimal solution to this problem, and verify optimality by the KKT conditions.
b. One approach to solve the problem is to transform it into a problem of the form to minimize $x_1^2 + x_2^2 + \mu(x_1 + x_2 - 2)^2$, where $\mu > 0$ is a large scalar. Solve the unconstrained problem for $\mu = 10$ by a conjugate gradient method, starting from the origin.

8.41 Solve the problem to minimize $x_1 + 2x_2^2 + e^{x_1^2+x_2^2}$, starting with the point $(1, 0)$ and using both the Fletcher and Reeves conjugate gradient method and the BFGS quasi-Newton method.

8.42 Consider the following problem:

$$\text{Minimize } x_1^2 + 2x_1x_2 + 2x_2^2 + x_3^2 - x_2x_3 + x_1 + 3x_2 - x_3$$

Using Exercise 8.31 or any other method, generate three conjugate directions. Starting from the origin, solve the problem by minimizing along these directions.

8.43 Consider the system of simultaneous equations

$$h_i(\mathbf{x}) = 0 \qquad \text{for } i = 1, \ldots, l$$

a. Show how to solve the above system by unconstrained optimization techniques. (*Hint:* Consider the problem to minimize $\sum_{i=1}^{l} |h_i(\mathbf{x})|^p$, where p is a positive integer.)

b. Solve the following system:

$$(x_1 - 2)^4 + (x_1 - 2x_2)^2 - 5 = 0$$

$$x_1^2 - x_2 = 0$$

8.44 Consider the problem to minimize $f(\mathbf{x})$ subject to $h_i(\mathbf{x}) = 0$ for $i = 1, \ldots, l$. A point $\mathbf{x}$ is said to be a KKT point if there exists a vector $\mathbf{v} \in E_l$ such that

$$\nabla f(\mathbf{x}) + \sum_{i=1}^{l} v_i \nabla h_i(\mathbf{x}) = \mathbf{0}$$

$$h_i(\mathbf{x}) = 0 \qquad \text{for } i = 1, \ldots, l$$

a. Show how to solve the above system using unconstrained optimization techniques. (*Hint:* See Exercise 8.43 above.)

b. Find the KKT point for the following problem:

Minimize $(x_1 - 2)^4 + (x_1 - 2x_2)^2$

subject to $x_1^2 - x_2 = 0$

8.45 Consider the problem to minimize $f(\mathbf{x})$ subject to $g_i(\mathbf{x}) \le 0$ for $i = 1, \ldots, m$.

a. Show that the KKT conditions hold at a point $\mathbf{x}$ if there exist u_i and s_i for $i = 1, \ldots, m$ such that

$$\nabla f(\mathbf{x}) + \sum_{i=1}^{m} u_i^2 \nabla g_i(\mathbf{x}) = \mathbf{0}$$

$$g_i(\mathbf{x}) + s_i^2 = 0 \qquad \text{for } i = 1, \ldots, m$$

$$u_i s_i = 0 \qquad \text{for } i = 1, \ldots, m$$

b. Show that unconstrained optimization techniques could be used to find a solution to the above system. (*Hint:* See Exercise 8.43 above.)

c. Use a suitable unconstrained optimization technique to find a KKT point to the following problem:

Minimize $2x_1^2 + x_2^2 - 2x_1x_2 + 2x_1 + 6$

subject to $-2x_1 - x_2 + 3 \le 0$

8.46 Consider the design of a conjugate gradient method in which $\mathbf{d}_{j+1} = -\nabla f(\mathbf{y}_{j+1}) + \alpha_j \mathbf{d}_j$ in the usual notation, and where, for a choice of a scale parameter s_{j+1}, we would like $s_{j+1}\mathbf{d}_{j+1}$ to coincide with the Newton direction $-\mathbf{H}^{-1}\nabla f(\mathbf{y}_{j+1})$, if possible. Equating $s_{j+1}[-\nabla f(\mathbf{y}_{j+1}) + \alpha_j \mathbf{d}_j] = -\mathbf{H}^{-1}\nabla f(\mathbf{y}_{j+1})$, transpose both sides and multiply these by $\mathbf{Hd}_j$, and use the quasi-Newton condition $\lambda_j \mathbf{Hd}_j = \mathbf{q}_j$ to derive

$$\alpha_j = \frac{\nabla f(\mathbf{y}_{j+1})^t \mathbf{q}_j - (1/s_{j+1}) \nabla f(\mathbf{y}_{j+1})^t \mathbf{p}_j}{\mathbf{d}_j^t \mathbf{q}_j}$$

a. Show that with exact line searches, the choice of s_{j+1} is immaterial. Moreover, show that as $s_{j+1} \to \infty$, $\alpha_j \to \alpha_j^{HS}$ in (8.57). Motivate the choice of s_{j+1} by considering the situation in which the Newton direction $-\mathbf{H}^{-1}\nabla f(\mathbf{y}_{j+1})$ is, indeed, contained in the cone spanned by $-\nabla f(\mathbf{y}_{j+1})$ and $\mathbf{d}_j$ but is not coincident with $\mathbf{d}_j$. Hence, suggest a scheme for choosing a value for s_{j+1}.

b. Illustrate, using Example 8.8.2, by assuming that at the previous iteration, $\mathbf{y}_j = (-\frac{1}{2}, 1)^t$, $\mathbf{d}_j = (1, 0)^t$, $\lambda_j = \frac{1}{2}$ (inexact step), so that $\mathbf{y}_{j+1} = (0, 1)^t$, and consider your suggested choice along with the choices (i) $s_{j+1} = \infty$, (ii) $s_{j+1} = 1$, and (iii) $s_{j+1} = \frac{1}{4}$ at the next iteration. Obtain the corresponding directions $\mathbf{d}_{j+1} = -\nabla f(\mathbf{y}_{j+1}) + \alpha_j \mathbf{d}_j$. Which of these can potentially lead to optimality? (Choice (ii) is Perry's [1978] choice. Sherali and Ulular [1990] suggest the scaled version, prescribing a choice for s_{j+1}.)

8.47 Using induction, show that the inclusion of the extra term $\gamma_j \mathbf{d}_1$ in (8.68b), where γ_j is as given therein, ensures the mutual **H**-conjugacy of the directions $\mathbf{d}_1, \ldots, \mathbf{d}_n$ thus generated.

8.48 A problem of the following structure frequently arises in the context of solving a more general nonlinear programming problem:

Minimize $f(\mathbf{x})$

subject to $a_i \leq x_i \leq b_i$ for $i = 1, \ldots, m$

a. Investigate appropriate modifications of the unconstrained optimization methods discussed in this chapter so that the lower and upper bounds on the variables could be handled.

b. Use the results of part a to solve the following problem:

Minimize $(x_1 - 2)^4 + (x_1 - 2x_2)^2$

subject to $3 \leq x_1 \leq 5$

$2 \leq x_2 \leq 6$

8.49 Consider the following method of *parallel tangents* credited to Shah, Buehler, and Kempthorne [1964] for minimizing a differentiable function f of several variables:

Initialization Step Choose a termination scalar $\varepsilon > 0$, and choose a starting point $\mathbf{x}_1$. Let $\mathbf{y}_0 = \mathbf{x}_1$, $k = j = 1$, and go to the main step.

Main Step

1. Let $\mathbf{d} = -\nabla f(\mathbf{x}_k)$ and let $\hat{\lambda}$ be an optimal solution to the problem to minimize $f(\mathbf{x}_k + \lambda\mathbf{d})$ subject to $\lambda \geq 0$. Let $\mathbf{y}_1 = \mathbf{x}_k + \hat{\lambda}\mathbf{d}$. Go to step 2.
2. Let $\mathbf{d} = -\nabla f(\mathbf{y}_j)$, and let λ_j be an optimal solution to the problem to minimize $f(\mathbf{y}_j + \lambda\mathbf{d})$ subject to $\lambda \geq 0$. Let $\mathbf{z}_j = \mathbf{y}_j + \lambda_j\mathbf{d}$, and go to step 3.
3. Let $\mathbf{d} = \mathbf{z}_j - \mathbf{y}_{j-1}$, and let μ_j be an optimal solution to the problem to minimize $f(\mathbf{z}_j + \mu\mathbf{d})$ subject to $\mu \in E_1$. Let $\mathbf{y}_{j+1} = \mathbf{z}_j + \mu_j\mathbf{d}$. If $j < n$, replace j by $j + 1$, and go to step 2. If $j = n$, go to step 4.
4. Let $\mathbf{x}_{k+1} = \mathbf{y}_{n+1}$. If $\|\mathbf{x}_{k+1} - \mathbf{x}_k\| < \varepsilon$, stop. Otherwise, let $\mathbf{y}_0 = \mathbf{x}_{k+1}$, replace k by $k + 1$, let $j = 1$, and go to step 1.

Using Theorem 7.3.4, show that the method converges. Solve the following problems using the method of parallel tangents:

a. Minimize $x_1^2 + x_2^2 + 2x_1x_2 - 2x_1 - 6x_2$.

b. Minimize $x_1^2 + x_2^2 - 2x_1x_2 - 2x_1 - x_2$. (Note that the optimal solution for this problem is unbounded.)

c. Minimize $(x_1 - 2)^2 + (x_1 - 2x_2)^2$.

8.50 Let $f: E_n \rightarrow E_1$ be differentiable. Consider the following procedure for minimizing f:

Initialization Step Choose a termination scalar $\varepsilon > 0$ and an initial step size $\Delta > 0$. Let m be a positive integer denoting the number of allowable failures before reducing step size. Let $\mathbf{x}_1$ be the starting point and let the current upper bound on the optimal objective value be UB $= f(\mathbf{x}_1)$. Let $\nu = 0$, let $k = 1$, and go to the main step.

Main Step

1. Let $\mathbf{d}_k = -\nabla f(\mathbf{x}_k)$, and let $\mathbf{x}_{k+1} = \mathbf{x}_k + \Delta\mathbf{d}_k$. If $f(\mathbf{x}_{k+1}) <$ UB, let $\nu = 0$, $\hat{\mathbf{x}} = \mathbf{x}_{k+1}$, UB $= f(\hat{\mathbf{x}})$, and go to step 2. If, on the other hand, $f(\mathbf{x}_{k+1}) \geq$ UB, then replace ν by $\nu + 1$. If $\nu = m$, go to step 3; and if $\nu < m$, go to step 2.
2. Replace k by $k + 1$, and repeat step 1.

3. Replace k by $k + 1$. If $\Delta < \varepsilon$, stop with $\hat{\mathbf{x}}$ as an estimate of the optimal solution. Otherwise, replace Δ by $\frac{1}{2}\Delta$, let $\nu = 0$, let $\mathbf{x}_k = \hat{\mathbf{x}}$, and go to step 1.

a. Can you prove convergence of the above algorithm for $\varepsilon = 0$?
b. Apply the above algorithm for the three problems in Exercise 8.49.

8.51 Let $\mathbf{H}$ be an $n \times n$ symmetric matrix, and let $f(\mathbf{x}) = \mathbf{c}^t\mathbf{x} + \frac{1}{2}\mathbf{x}^t\mathbf{H}\mathbf{x}$. Consider the following *rank-one correction algorithm* for minimizing f. First, let $\mathbf{D}_1$ be an $n \times n$ positive definite symmetric matrix, and let $\mathbf{x}_1$ be a given vector. For $j = 1, \ldots, n$, let λ_j be an optimal solution to the problem to minimize $f(\mathbf{x}_j + \lambda\mathbf{d}_j)$ subject to $\lambda \in E_1$, and let $\mathbf{x}_{j+1} = \mathbf{x}_j + \lambda_j\mathbf{d}_j$, where $\mathbf{d}_j = -\mathbf{D}_j\nabla f(\mathbf{x}_j)$ and $\mathbf{D}_{j+1}$ is given by

$$\mathbf{D}_{j+1} = \mathbf{D}_j + \frac{(\mathbf{p}_j - \mathbf{D}_j\mathbf{q}_j)(\mathbf{p}_j - \mathbf{D}_j\mathbf{q}_j)^t}{\mathbf{q}_j^t(\mathbf{p}_j - \mathbf{D}_j\mathbf{q}_j)}$$

$$\mathbf{p}_j = \mathbf{x}_{j+1} - \mathbf{x}_j$$

$$\mathbf{q}_j = \mathbf{H}\mathbf{p}_j$$

a. Verify that the matrix added to $\mathbf{D}_j$ to obtain $\mathbf{D}_{j+1}$ is of rank 1.
b. For $j = 1, \ldots, n$, show that $\mathbf{p}_i = \mathbf{D}_{j+1}\mathbf{q}_i$ for $i \le j$.
c. Supposing that $\mathbf{H}$ is invertible, does $\mathbf{D}_{n+1} = \mathbf{H}^{-1}$ hold?
d. Even if $\mathbf{D}_j$ is positive definite, show that $\mathbf{D}_{j+1}$ is not necessarily positive definite. This explains why a line search over the whole real line is used.
e. Are the directions $\mathbf{d}_1, \ldots, \mathbf{d}_n$ necessarily conjugate?
f. Use the above algorithm for minimizing $12x_1 - 6x_2 + x_1^2 + 2x_1x_2 + 2x_2^2$.
g. Suppose that $\mathbf{q}_j$ is replaced by $\nabla f(\mathbf{x}_{j+1}) - \nabla f(\mathbf{x}_j)$. Develop a procedure similar to that of Davidon–Fletcher–Powell for minimizing a nonquadratic function, using the above scheme for updating $\mathbf{D}_j$. Use the procedure to minimize $(x_1 - 2)^4 + (x_1 - 2x_2)^2$.

8.52 Consider the problem to minimize $f(\mathbf{x})$ subject to $\mathbf{x} \in E_n$, and consider the following algorithm credited to Powell [1964]:

Initialization Step Choose a termination scalar $\varepsilon > 0$. Choose an initial point $\mathbf{x}_1$, let $\mathbf{d}_1, \ldots, \mathbf{d}_n$ be the coordinate directions, and let $k = j = i = 1$. Let $\mathbf{z}_1 = \mathbf{y}_1 = \mathbf{x}_1$, and go to the main step.

Main Step

1. Let λ_i be an optimal solution to the problem to minimize $f(\mathbf{z}_i + \lambda\mathbf{d}_i)$ subject to $\lambda \in E_1$, and let $\mathbf{z}_{i+1} = \mathbf{z}_i + \lambda_i\mathbf{d}_i$. If $i < n$, replace i by $i + 1$, and repeat step 1. Otherwise, go to step 2.
2. Let $\mathbf{d} = \mathbf{z}_{n+1} - \mathbf{z}_1$, and let $\hat{\lambda}$ be an optimal solution to the problem to minimize $f(\mathbf{z}_{n+1} + \lambda\mathbf{d})$ subject to $\lambda \in E_1$. Let $\mathbf{y}_{j+1} = \mathbf{z}_{n+1} + \hat{\lambda}\mathbf{d}$. If $j < n$, replace $\mathbf{d}_l$ by $\mathbf{d}_l = \mathbf{d}_{l+1}$ for $l = 1, \ldots, n - 1$, let $\mathbf{d}_n = \mathbf{d}$, let $\mathbf{z}_1 = \mathbf{y}_{j+1}$, let $i = 1$, replace j by $j + 1$, and repeat step 1. Otherwise, $j = n$, and go to step 3.
3. Let $\mathbf{x}_{k+1} = \mathbf{y}_{n+1}$. If $\|\mathbf{x}_{k+1} - \mathbf{x}_k\| < \varepsilon$, stop. Otherwise, let $i = j = 1$, let $\mathbf{z}_1 = \mathbf{y}_1 = \mathbf{x}_{k+1}$, replace k by $k + 1$, and repeat step 1.

a. Suppose that $f(\mathbf{x}) = \mathbf{c}^t\mathbf{x} + \frac{1}{2}\mathbf{x}^t\mathbf{H}\mathbf{x}$, where $\mathbf{H}$ is an $n \times n$ symmetric matrix. After one pass through the main step, show that if $\mathbf{d}_1, \ldots, \mathbf{d}_n$ are linearly independent, then they are also $\mathbf{H}$-conjugate, so that, by Theorem 8.8.3, an optimal solution is produced in one iteration.
b. Consider the following problem credited to Zangwill [1967]:

$$\text{Minimize } (x_1 - x_2 + x_3)^2 + (-x_1 + x_2 + x_3)^2 + (x_1 + x_2 - x_3)^2$$

Apply Powell's method discussed in this exercise, starting from the point $(\frac{1}{2}, 1, \frac{1}{2})$. Note that the procedure generates a set of dependent directions and hence will not yield the optimal point $(0, 0, 0)$.
c. Zangwill [1967] proposed a slight modification of Powell's method to guarantee linear independence of the direction vectors. In particular, in step 2, the point $\mathbf{z}_1$ is obtained from $\mathbf{y}_{j+1}$ by a spacer step application, such as one iteration of the cyclic coordinate

method. Show that this modification indeed guarantees linear independence, and hence by part a, finite convergence for a quadratic function is assured.

d. Apply Zangwill's modified method to solve the problem of part b.

e. If the function is not quadratic, consider the introduction of a spacer step so that in step 3, $\mathbf{z}_1 = \mathbf{y}_1$ is obtained by the application of one iteration of the cyclic coordinate method starting from $\mathbf{x}_{k+1}$. Use Theorem 7.3.4 to prove convergence.

8.53 The method of Rosenbrock can be described by the following map $\mathbf{A}: E_n \times \mathbf{U} \times E_n \rightarrow E_n \times \mathbf{U} \times E_n$. Here, $\mathbf{U} = \{\mathbf{D}:\mathbf{D}$ is an $n \times n$ matrix satisfying $\mathbf{D}^t\mathbf{D} = \mathbf{I}\}$. The algorithmic map $\mathbf{A}$ operates on the triple $(\mathbf{x}, \mathbf{D}, \boldsymbol{\lambda})$, where $\mathbf{x}$ is the current vector, $\mathbf{D}$ is the $n \times n$ matrix whose columns are the directions of the previous iteration, and $\boldsymbol{\lambda}$ is the vector whose components $\lambda_1, \ldots, \lambda_n$ give the distances moved along the directions $\mathbf{d}_1, \ldots, \mathbf{d}_n$. The map $\mathbf{A} = \mathbf{A}_3\mathbf{A}_2\mathbf{A}_1$ is a composite map whose components are discussed in detail below:

1. $\mathbf{A}_1$ is the point-to-point map defined by $\mathbf{A}_1(\mathbf{x}, \mathbf{D}, \boldsymbol{\lambda}) = (\mathbf{x}, \bar{\mathbf{D}})$, where $\bar{\mathbf{D}}$ is the matrix whose columns are the new directions defined by (8.9).
2. The point-to-set map $\mathbf{A}_2$ is defined by $(\mathbf{x}, \mathbf{y}, \bar{\mathbf{D}}) \in \mathbf{A}_2(\mathbf{x}, \bar{\mathbf{D}})$ if minimizing f, starting from $\mathbf{x}$, in the directions $\bar{\mathbf{d}}_1, \ldots, \bar{\mathbf{d}}_n$ leads to $\mathbf{y}$. By Theorem 7.3.5, the map $\mathbf{A}_2$ is closed.
3. $\mathbf{A}_3$ is the point-to-point map defined by $\mathbf{A}_3(\mathbf{x}, \mathbf{y}, \bar{\mathbf{D}}) = (\mathbf{y}, \bar{\mathbf{D}}, \bar{\boldsymbol{\lambda}})$, where $\bar{\boldsymbol{\lambda}} = (\bar{\mathbf{D}})^{-1}(\mathbf{y} - \mathbf{x})$.

a. Show that the map $\mathbf{A}_1$ is closed at $(\mathbf{x}, \mathbf{D}, \boldsymbol{\lambda})$ if $\lambda_j \neq 0$ for $j = 1, \ldots, n$.

b. Is the map $\mathbf{A}_1$ closed if $\lambda_j = 0$ for some j?

$$\left(\textit{Hint: } \text{Consider the sequence } \mathbf{D}_k = \begin{bmatrix} 1 & 0 \\ 0 & 1 \end{bmatrix} \text{ and } \boldsymbol{\lambda}_k = \begin{bmatrix} 1/k \\ 1 \end{bmatrix}.\right)$$

c. Show that $\mathbf{A}_3$ is closed.

d. Verify that the function f could be used as a descent function.

e. Discuss the applicability of Theorem 7.2.3 to prove convergence of Rosenbrock's procedure. (This exercise illustrates that some difficulties could arise in viewing the algorithmic map as a composition of several maps. In Section 8.5, a proof of convergence was provided without decomposing the map $\mathbf{A}$.)

8.54 In this exercise, we describe a modification of the simplex method of Spendley, Hext, and Himsworth [1962] for solving a problem of the form to minimize $f(\mathbf{x})$ subject to $\mathbf{x} \in E_n$. The version of the method described here is credited to Nelder and Mead [1965].

Initialization Step Choose the points $\mathbf{x}_1, \mathbf{x}_2, \ldots, \mathbf{x}_{n+1}$ to form a simplex in E_n. Choose a reflection coefficient $\alpha > 0$, an expansion coefficient $\gamma > 1$, and a positive contraction coefficient $\beta > 1$. Go to the main step.

Main Step

1. Let $\mathbf{x}_r, \mathbf{x}_s \in \{\mathbf{x}_1, \ldots, \mathbf{x}_{n+1}\}$ be such that

$$f(\mathbf{x}_r) = \underset{1 \le j \le n+1}{\text{minimum}}\, f(\mathbf{x}_j) \qquad f(\mathbf{x}_s) = \underset{1 \le j \le n+1}{\text{maximum}}\, f(\mathbf{x}_j)$$

Let $\bar{\mathbf{x}} = \dfrac{1}{n}\sum_{\substack{j=1 \\ j \neq s}}^{n+1} \mathbf{x}_j$, and go to step 2.

2. Let $\hat{\mathbf{x}} = \bar{\mathbf{x}} + \alpha(\bar{\mathbf{x}} - \mathbf{x}_s)$. If $f(\mathbf{x}_r) > f(\hat{\mathbf{x}})$, let $\mathbf{x}_e = \bar{\mathbf{x}} + \gamma(\hat{\mathbf{x}} - \bar{\mathbf{x}})$, and go to step 3. Otherwise, go to step 4.
3. The point $\mathbf{x}_s$ is replaced by $\mathbf{x}_e$ if $f(\hat{\mathbf{x}}) > f(\mathbf{x}_e)$ and by $\hat{\mathbf{x}}$ if $f(\hat{\mathbf{x}}) \le f(\mathbf{x}_e)$ to yield a new set of $n + 1$ points. Go to step 1.
4. If $\max_{1 \le j \le n+1} \{f(\mathbf{x}_j): j \neq s\} \ge f(\hat{\mathbf{x}})$, then $\mathbf{x}_s$ is replaced by $\hat{\mathbf{x}}$ to form a new set of $n + 1$ points, and we go to step 1. Otherwise, go to step 5.
5. Let $\mathbf{x}'$ be defined by $f(\mathbf{x}') = \min\{f(\hat{\mathbf{x}}), f(\mathbf{x}_s)\}$, and let $\mathbf{x}'' = \bar{\mathbf{x}} + \beta(\mathbf{x}' - \bar{\mathbf{x}})$. If $f(\mathbf{x}'') > f(\mathbf{x}')$, replace $\mathbf{x}_j$ by $\mathbf{x}_j + \frac{1}{2}(\mathbf{x}_r - \mathbf{x}_j)$ for $j = 1, \ldots, n + 1$, and go to step 1. If $f(\mathbf{x}'') \le f(\mathbf{x}')$, then $\mathbf{x}''$ replaces $\mathbf{x}_s$ to form a new set of $n + 1$ points. Go to step 1.

a. Let $\mathbf{d}_j$ be an n vector with jth component equal to a and all other components equal to b, where

$$a = \frac{c}{n\sqrt{2}}(\sqrt{n+1} + n - 1) \qquad b = \frac{c}{n\sqrt{2}}(\sqrt{n+1} - 1)$$

and c is a positive scalar. Show that the initial simplex defined by $\mathbf{x}_1, \ldots, \mathbf{x}_{n+1}$ could be chosen by letting $\mathbf{x}_{j+1} = \mathbf{x}_1 + \mathbf{d}_j$, where $\mathbf{x}_1$ is selected arbitrarily. (In particular, show that $\mathbf{x}_{j+1} - \mathbf{x}_1$ for $j = 1, \ldots, n$ are linearly independent. What is the interpretation of c?)

b. Solve the problem to minimize $x_1^2 + 2x_1x_2 + x_3^2 + x_2 - 10x_3$ using the simplex method described in this exercise.

8.55 Consider the quadratic function $f(\mathbf{y}) = \mathbf{c}^t\mathbf{y} + \frac{1}{2}\mathbf{y}^t\mathbf{H}\mathbf{y}$, where $\mathbf{H}$ is an $n \times n$ symmetric, positive definite matrix. Suppose that we use some algorithm for which the iterate $\mathbf{y}_{j+1} = \mathbf{y}_j - \lambda_j\mathbf{D}_j\nabla f(\mathbf{y}_j)$ is generated by an exact line search along the direction $-\mathbf{D}_j\nabla f(\mathbf{y}_j)$ from the previous iterate $\mathbf{y}_j$, where $\mathbf{D}_j$ is some positive definite matrix. Then, if $\mathbf{y}^*$ is the minimizing solution for f, and if $e(\mathbf{y}) = \frac{1}{2}(\mathbf{y} - \mathbf{y}^*)^t\mathbf{H}(\mathbf{y} - \mathbf{y}^*)$ is an error function, show that, at every step j, we have

$$e(\mathbf{y}_{j+1}) \le \frac{(\alpha_j - 1)^2}{(\alpha_j + 1)^2} e(\mathbf{y}_j)$$

where α_j is the ratio of the largest to the smallest eigenvalue of $\mathbf{D}_j\mathbf{H}$.

8.56 Consider the problem of finding the projection $\mathbf{x}^* = P_X(\bar{\mathbf{x}})$ of the point $\bar{\mathbf{x}}$ onto $X = \{\mathbf{x} : \boldsymbol{\alpha}^t\mathbf{x} = \beta, \mathbf{l} \le \mathbf{x} \le \mathbf{u}\}$, where $\mathbf{x}, \bar{\mathbf{x}}, \mathbf{x}^* \in E_n$. The following *variable dimension* algorithm successively projects the current point onto the equality constraint and reduces the problem to an equivalent one in a lower dimensional space, or else stops. Justify the various steps of this algorithm. Illustrate by projecting the point $(-2, 3, 1, 2)^t$ onto $\{\mathbf{x} : x_1 + x_2 + x_3 + x_4 = 1, 0 \le x_i \le 1 \text{ for all } i\}$. (This method is a generalization of the procedures that appear in Bitran and Hax [1976] and in Sherali and Shetty [1980].)

Initialization Set $(\bar{\mathbf{x}}^0, I^0, \mathbf{l}^0, \mathbf{u}^0, \beta^0) = (\bar{\mathbf{x}}, I, \mathbf{l}, \mathbf{u}, \beta)$, where $I = \{i : \alpha_i \ne 0\}$. For $i \notin I$, put $x_i^* = \bar{x}_i$ if $l_i \le \bar{x}_i \le u_i$, $x_i^* = l_i$, if $\bar{x}_i < l_i$ and $x_i^* = u_i$ if $\bar{x}_i > u_i$. Let $k = 0$.

Step 1 Compute the projection $\hat{\mathbf{x}}^k$ of $\bar{\mathbf{x}}^k$ onto the equality constraint in the subspace I^k according to

$$\hat{x}_i^k = \bar{x}_i^k + \left[\frac{\beta^k - \sum_{i \in I^k} \alpha_i \bar{x}_i^k}{\sum_{i \in I^k} \alpha_i^2}\right] \alpha_i \qquad \text{for each } i \in I^k$$

If $l_i^k \le \hat{x}_i^k \le u_i^k$ for all $i \in I^k$, put $x_i^* = \hat{x}_i^k$ for all $i \in I^k$, and stop. Otherwise, proceed to step 2.

Step 2 Define $J_1 = \{i \in I^k : \hat{x}_i^k \le l_i^k\}$, $J_2 = \{i \in I^k : \hat{x}_i^k \ge u_i^k\}$, and compute

$$\gamma = \beta^k + \sum_{i \in J_1} \alpha_i(l_i^k - \hat{x}_i^k) + \sum_{i \in J_2} \alpha_i(u_i^k - \hat{x}_i^k)$$

If $\gamma = \beta^k$, then put $x_i^* = l_i^k$ for $i \in J_1$, $x_i^* = u_i^k$ for $i \in J_2$, and $x_i^* = \hat{x}_i^k$ for $i \in I^k - J_1 \cup J_2$, and stop. Otherwise, define

$$J_3 = \{i \in J_1 : \alpha_i > 0\} \quad \text{and} \quad J_4 = \{i \in J_2 : \alpha_i < 0\} \quad \text{if } \gamma > \beta^k$$

$$J_3 = \{i \in J_1 : \alpha_i < 0\} \quad \text{and} \quad J_4 = \{i \in J_2 : \alpha_i > 0\} \quad \text{if } \gamma < \beta^k$$

Set $x_i^* = l_i^k$ if $i \in J_3$, and $x_i^* = u_i^k$ if $i \in J_4$. (*Note:* $J_3 \cup J_4 \ne \emptyset$.) Update $I^{k+1} = I^k - J_3 \cup J_4$. If $I^{k+1} = \emptyset$, then stop. Otherwise, update $(\bar{x}_i^{k+1} = \hat{x}_i^k$ for $i \in I^{k+1})$, $(l_i^{k+1} = \text{maximum } \{l_i^k, \hat{x}_i^k\}$ if $\alpha_i(\beta^k - \gamma) > 0$ and $l_i^{k+1} = l_i^k$; otherwise, for $i \in I^{k+1})$, $(u_i^{k+1} = \text{minimum } \{u_i^k, \hat{x}_i^k\}$ if $\alpha_i(\beta^k - \gamma) < 0$ and $u_i^{k+1} = u_i^k$, otherwise, for $i \in I^{k+1})$, and $\beta^{k+1} = \beta^k - \sum_{i \in J_3} \alpha_i l_i^k - \sum_{i \in J_4} \alpha_i u_i^k$. Increment k by 1 and return to step 1.

8.57 Solve the example of Exercise 6.25 using the subgradient optimization algorithm starting with the point (0, 4). Re-solve using the conjugate subgradient strategy suggested in Section 8.9.

8.58 Solve the Lagrangian dual problem of Example 6.4.1 using the subgradient algorithm. Resolve using the conjugate subgradient strategy suggested in Section 8.9.

Notes and References

We have discussed several iterative procedures for solving an unconstrained optimization problem. Most of the procedures involve a line search of the type described in Sections 8.1 through 8.3 and, by and large, the efficiency of the line search method greatly affects the overall performance of the solution technique. The Fibonacci search procedure discussed in Section 8.1 is credited to Kiefer [1953]. Several other search procedures, including the golden section method, are discussed in Wilde [1964] and Wilde and Beightler [1967]. These references also show that the Fibonacci search procedure is the best for unimodal functions in that it reduces the maximum interval of uncertainty with the least number of observations.

Another class of procedures uses curve fitting, as discussed in Section 8.3 and illustrated by Exercises 8.14 through 8.16. If a function f of one varaible is to be minimized, the procedures involve finding an approximating quadratic or cube function q. In the quadratic case, the function is selected such that, given three points x_1, x_2 and x_3, the functional values of f and q are equal at these points. In the cubic case, given two points x_1 and x_2, q is selected such that the functional values and derivatives of both functions are the same at these points. In any case, the minimum of q is determined, and this point replaces one of the initial points. Refer to Davidon [1959], Fletcher and Powell [1963], Kowalik and Osborne [1968], Luenberger [1984], Pierre [1969], Powell [1964], and Swann [1964] for a more detailed discussion, particularly on precautions to be taken to ensure convergence. Some limited computational studies on the efficiency of this approach may be found in Himmelblau [1972] and Murtagh and Sargent [1970]. See Armijo [1966] and Luenberger [1984] for further discussions on inexact line searches.

Among the gradient-free methods, the method of Rosenbrock [1960], discussed in Section 8.4, and the method of Zangwill [1967b], discussed in Exercises 8.26 and 8.52, are generally considered quite efficient. As originally proposed, the Rosenbrock method and the procedure of Hooke and Jeeves [1961] do not use line search but employ instead discrete steps along the search directions. Incorporating a line search within Rosenbrock's procedure was suggested by Davies, Swann, and Campey and is discussed by Swann [1964]. An evaluation of this modification is presented by Fletcher [1965] and Box [1966].

There are yet other derivative-free methods for unconstrained minimization. A procedure that is distinctively different, called the sequential simplex search method, is described in Exercise 8.54. The method was proposed by Spendley, Hext, and Himsworth [1962] and modified by Nelder and Mead [1965]. The method essentially looks at the functional values at the extreme points of a simplex. The worst extreme point is rejected and replaced by a new point along the line joining this point and the centroid of the remaining points. The process is repeated until a suitable termination criterion is satisfied. Box [1966], Jacoby, Kowalik, and Pizzo [1972], Kowalik and Osborne [1968], and Parkinson and Hutchinson [1972] compare this method with some of the other methods discussed earlier. Parkinson and Hutchinson [1972] have presented a detailed analysis on the efficiency of the simplex method and its variants. The simplex method seems to be less effective as the dimensionality of the problem increases.

The method of steepest descent proposed by Cauchy in the middle of the nineteenth century continues to be the basis of several gradient-based solution procedures. For example, see Gonzaga [1990] for a polynomial-time, scaled steepest descent algorithm for linear programming. The method of steepest descent uses a first-order approximation of the function being minimized and usually performs poorly as the optimum is reached. On the other hand, Newton's method uses a second-order approximation and usually performs well at points close to the optimum. In general, however, convergence is guaranteed only if the starting point is close to the solution point. For a discussion on Newton–Raphson methods, see Fletcher [1987]. Burns [1991] presents a powerful alternative for solving systems of signomial equations in variables that are restricted to be positive in value. Fletcher [1987] and Dennis and Schnabel [1983] give a good discussion of Levenberg [1944]–Marquardt [1963] type methods used for implementing the modification of Newton's method by replacing $\mathbf{H}(\mathbf{x}_k)$ by $\mathbf{H}(\mathbf{x}_k) + \varepsilon_k\mathbf{I}$ to maintain positive definiteness, and also of the relationship of this to trust-region methods. For a survey and implementation aspects, see Moré [1977]. For a discussion and survey on trust-region methods and related convergence aspects, see Conn et al. [1988]. Ye [1990] presents a method to solve the subproblems appearing in such methods. We also introduced the dog-leg trajectory of Powell [1970a], which compromises between the steepest descent step and Newton's step (see also Fletcher [1987] and Dennis and Schnabel [1983]). For another scheme of combining the steepest descent and Newton's methods, see Luenberger [1984]. Renegar [1988] presents a polynomial-time algorithm for linear programming based on Newton's method. Burns [1989] gives some very interesting graphic plots to illustrate the convergence behavior of some of the aforementioned algorithms.

Among the unconstrained optimization techniques, methods using conjugate directions are considered efficient. For a quadratic function, these methods give the optimum in, at most, n steps. Among the derivative-free methods of this type are the method of Zangwill, discussed in Exercises 8.26 and 8.52, the method of Powell, discussed in Exercise 8.52, and the PARTAN method credited to Shah, Beuhler, and Kempthorne, discussed in Exercise 8.49. Sorenson [1969] has shown that, for quadratic functions, PARTAN is far less efficient than the conjugate gradient methods discussed in Section 8.8.

In yet another class, the direction of movement $\mathbf{d}$ is taken to be $-\mathbf{D}\nabla f(\mathbf{x})$, where $\mathbf{D}$ is a positive definite matrix that approximates the inverse of the Hessian matrix. This class is usually referred to as quasi-Newton methods. (See Davidon et al., 1991 for a note on terminology in this area.) One of the early methods of minimizing a nonlinear function using the approach is that of Davidon [1959], which was simplified and reformulated by Fletcher and Powell [1963] and is referred to as the variable metric method. A useful generalization of the Davidon–Fletcher–Powell method was proposed by Broyden [1967]. Essentially, Broyden introduced a degree of freedom in updating the matrix $\mathbf{D}$. A particular choice of this degree of freedom was then proposed by Broyden [1970], Fletcher [1970], Goldfarb [1970], and Shanno [1970]. This has led to the well-known BFGS update technique. Gill, Murray, and Pitfield [1972], among several others, have shown that this modification performs more efficiently than the original method for most problems. For an extension on updating conjugate directions via the BFGS approach, see Powell [1987].

In 1972, Powell showed that the Davidon–Fletcher–Powell method converges to an optimal solution if the objective function is convex, if the second derivatives are continuous, and if an exact line search is used. Under stronger assumptions, Powell [1971] showed that the method converges superlinearly. In 1973, Broyden, Dennis, and Moré gave local convergence results for the case where the step size is fixed equal to 1 and proved superlinear convergence under certain conditions. Under suitable assumptions, Powell [1976] showed that a version of the variable metric method without exact line search converges to an optimal solution if the objective function is convex. Furthermore, he showed that if the Hessian matrix is positive definite at the solution point, the rate of convergence is superlinear. For further reading on variable metric methods and their convergence characteristics, see Broyden, Dennis, and Moré [1973], Dennis and Moré [1974], Dixon [1972], Fletcher [1987], Gill and Murray [1974], Greenstadt [1970], Huang [1970], and Powell [1971, 1976, 1986, 1987].

The variable metric methods discussed above update the matrix **D** by adding to it two matrices, each having rank 1, and, hence, this class is also referred to as *rank-two correction procedures*. A slightly different strategy for estimating second derivatives is to update the matrix **D** by adding to it a matrix of rank 1. This *rank-one correction* method was briefly introduced in Exercise 8.51. For further details of this procedure, see Broyden [1967], Davidon [1968], Dennis and Schnabel [1983], Fiacco and McCormick [1968], Fletcher [1987], and Powell [1970]. Conn et al. [1991] provide a detailed convergence analysis.

Among the conjugate methods using gradient information, the method of Fletcher and Reeves generates conjugate directions by taking a suitable convex combination of the current gradient and the direction used at the previous iteration. The original idea presented by Hestenes and Stiefel [1952] led to the development of this method, as well as to the conjugate gradient algorithms of Polyak [1969] and Sorenson [1964]. These methods become indispensable when problem size increases. (See Reid [1971] and Fletcher [1987] for some reports on large-scale applications.) Polak and Ribiere [1969] propose another conjugate gradient scheme which Powell [1977] argues to be preferable for nonquadratic functions. Nazareth [1986] also discusses various interesting extensions for conjugate gradient methods. Several authors have investigated the effect of using inexact line searches on the convergence properties of conjugate gradient algorithms. Nazareth [1977] and Dixon et al. [1985] propose alternate three-term recurrence relationships for generating conjugate directions in this case. The reader may also refer to Kawamura and Volz [1973], Klessig and Polak [1972], Lenard [1976], and McCormick and Ritter [1974]. Combination of quasi-Newton concepts with conjugate gradient methods have led to the memoryless quasi-Newton methods (see Nazareth [1979, 1986], Luenberger [1984], and Shanno [1978]). Also, this connection has produced efficient asymptotic ''memoryless'' updates as in Perry [1978] and its scaled version described in Sherali and Ulular [1990]. All these methods greatly benefit by the restart criterion proposed by Beale [1972] and Powell [1977]. For a convergence rate analysis of these methods, see Luenberger [1984], Gill et al. [1981], McCormick and Ritter [1974], and Powell [1986]. Also, for the solution of large-scale problems using *limited memory* (an extension of memoryless) BFGS updates, see Nocedal [1980] and Liu and Nocedal [1989]. For a discussion on using *truncated Newton methods* (in which the Newton direction is solved for inexactly by prematurely truncating a conjugate gradient scheme for solving the associated linear system), see Nash [1985] and Nash and Sofer [1989, 1990, 1991]. For some related computational experiments, see Nocedal [1990].

Several authors have attempted to use unconstrained optimization methods for solving nonlinear problems with constraints. Note that if an unconstrained optimization technique is extended to handle constraints by simply rejecting infeasible points during the search procedure, it would lead to premature termination. A successful and frequently used approach is to define an auxiliary unconstrained problem such that the solution of the unconstrained problem yields the solution of the constrained problem. This is discussed in detail in Chapter 9. A second approach is to use an unconstrained optimization method when we are in the interior of the feasible region and to use one of the suitable constrained optimization methods discussed in Chapter 10 when we are at a point on the boundary of the feasible region. Several authors have also modified the unconstrained optimization methods to handle constraints. Goldfarb [1969] has extended the Davidon–Fletcher–Powell method to handle problems with linear constraints utilizing the concept of gradient projection. The method was generalized by Davies [1970] to handle nonlinear constraints. Coleman and Fenyes [1989] discuss and survey other quasi-Newton methods for solving equality constrained problems. Klingman and Himmelblau [1964] project the search direction in the method of Hooke and Jeeves into the intersection of binding constraints, which leads to a constrained version of the Hooke and Jeeves algorithm. Glass and Cooper [1965] have presented another constrained version of the Hooke and Jeeves algorithm. The method of Rosenbrock has been extended by Davies and Swann [1969] to handle linear constraints. In Exercise 8.54, we discussed the simplex method for solving unconstrained problems. In 1965, Box extended this approach to constrained problems. For other alternative extensions of the simplex method, see Dixon [1973], Friedman and Pinder [1972], Ghani [1972], Guin [1968], Keefer [1973], Mitchell and Kaplan [1968], Paviani and Himmelblau [1969], and Umida and Ichikawa [1971].

For comprehensive surveys of different algorithms for solving unconstrained problems, refer to Dennis and Schnabel [1983], Fletcher [1969, 1987], Gill et al. [1981], Powell [1970], Reklaitis and Phillips [1975], and Zoutendijk [1970]. Furthermore, there are numerous studies reporting computational experience with different algorithms. Most of them study the effectiveness of the methods by solving relatively small-sized test problems of various degrees of complexity. For discussions on the efficiency of the various unconstrained minimization algorithms, see Bard [1970], Cragg and Levy [1969], Fiacco and McCormick [1968], Fletcher [1987], Gill et al. [1981], Himmelblau [1972], Huang and Levy [1970], Murtagh and Sargent [1970], and Sargent and Sebastian [1972]. Computer program listings of some of the algorithms can be found in Brent [1973] and Himmelblau [1972]. The *Computer Journal* and the *Journal of the ACM* also publish computer listings of nonlinear programming algorithms.

Finally, in Section 8.9, we presented the essence of subgradient optimization techniques following Polak [1967, 1969] and Held et al. [1974], with choices of step sizes as in Bazaraa and Sherali [1981], Held et al. [1974], and Sherali and Ulular [1989]. See Allen et al. [1987] for a useful theoretical justification of various practical step-size rules. Bazaraa and Goode [1979] and Sherali and Myers [1988] discuss various aspects of using subgradient optimization in the context of Lagrangian duality. Borwein [1982] addresses the existence of subgradients, and Hiriart-Urruty [1978] discusses optimality conditions under nondifferentiability. Various schemes have been suggested to accelerate the convergence of subgradient methods, but they still require a fine-tuning of parameters to yield an adequate performance, particularly for large-sized problems. Among these, the simplest to implement, and suitable for large problems, are the conjugate subgradient methods discussed in Wolfe [1975], Camerini et al. [1975], and Sherali and Ulular [1989]. Other more efficient methods, but also requiring more effort and storage, and suitable for relatively smaller-sized problems are (1) the space dilatation (or dilation) techniques of Shor [1970, 1975, 1977, 1983], particularly those using dilatation in the direction of the difference between two successive subgradients which imitate variable metric methods (see also Shor [1977] and Minoux [1986] for surveys); (2) the extension of Davidon methods to nondifferentiable problems as in Lemarechal [1975]; and (3) the bundle algorithms described in Lemarechal [1978, 1980], Lemarechal and Mifflin [1978], and Kiwiel [1985, 1989], which attempt to construct descent directions via approximations of subdifferentials. For an insightful connection between the symmetric rank-one quasi-Newton method and space dilatation methods, see Todd [1986].

Chapter 9

Penalty and Barrier Functions

In this chapter, we discuss nonlinear programming problems with equality and inequality constraints. The approach used is to convert the problem into an equivalent unconstrained problem or into a problem with simple constraints, so that the algorithms developed in Chapter 8 can be used. However, in practice, a sequence of problems are solved because of computational considerations, as will be discussed later in the chapter. Basically, there are two alternative approaches. The first is called the penalty or the exterior penalty function method, in which a penalty term is added to the objective function for any violation of the constraints. This method generates a sequence of infeasible points, hence its name, whose limit is an optimal solution to the original problem. The second method is called the barrier or interior penalty function method, in which a barrier term that prevents the points generated from leaving the feasible region is added to the objective function. The method generates a sequence of feasible points whose limit is an optimal solution to the original problem.

The following is an outline of the chapter:

Section 9.1: The Concept of Penalty Functions The concept of penalty functions is introduced. A geometric interpretation of the method is also discussed.

Section 9.2: Exterior Penalty Function Methods The exterior penalty function methods are discussed in detail, and the main convergence theorem is developed. The method is illustrated by means of a numerical example. Computational difficulties associated with this class of algorithms are discussed along with related convergence rate aspects.

Section 9.3: Exact Absolute Value and Augmented Lagrangian Penalty Methods To alleviate the computational difficulties associated with having to take the penalty parameter to infinity in order to recover an optimal solution to the original problem, we introduce the concept of exact penalty functions. Both the absolute value (l_1) and the augmented Lagrangian exact penalty function methods are discussed along with their related computational considerations.

Section 9.4: Barrier Function Methods We discuss the (interior) barrier function methods in detail, and establish their convergence and rate of convergence properties. The method is illustrated by a numerical example.

Section 9.5: A Polynomial-Time Algorithm for Linear Programming Based on a Barrier Function We present a polynomial-time primal–dual path-following algorithm for solving linear programming problems based on the logarithmic barrier function. This algorithm can be extended to solve convex quadratic programming problems in polynomial time as well. Convergence, complexity, and implementation issues are discussed.

9.1 The Concept of Penalty Functions

Methods using penalty functions transform a constrained problem into a single unconstrained problem or into a sequence of unconstrained problems. The constraints are placed into the objective function via a penalty parameter in a way that penalizes any violation of the constraints. To motivate penalty functions, consider the following problem with the single constraint $h(\mathbf{x}) = 0$:

$$\begin{array}{ll} \text{Minimize} & f(\mathbf{x}) \\ \text{subject to} & h(\mathbf{x}) = 0 \end{array}$$

Suppose that this problem is replaced by the following unconstrained problem, where $\mu > 0$ is a large number:

$$\begin{array}{ll} \text{Minimize} & f(\mathbf{x}) + \mu h^2(\mathbf{x}) \\ \text{subject to} & \mathbf{x} \in E_n \end{array}$$

We can intuitively see that an optimal solution to the above problem must have $h^2(\mathbf{x})$ close to zero, because otherwise a large penalty $\mu h^2(\mathbf{x})$ will be incurred.

Now consider the following problem with the single inequality constraint $g(\mathbf{x}) \leq 0$:

$$\begin{array}{ll} \text{Minimize} & f(\mathbf{x}) \\ \text{subject to} & g(\mathbf{x}) \leq 0 \end{array}$$

It is clear that the form $f(\mathbf{x}) + \mu g^2(\mathbf{x})$ is not appropriate, since a penalty will be incurred whether $g(\mathbf{x}) < 0$ or $g(\mathbf{x}) > 0$. Needless to say, a penalty is desired only if the point $\mathbf{x}$ is not feasible, that is, $g(\mathbf{x}) > 0$. A suitable unconstrained problem is therefore given by

$$\begin{array}{ll} \text{Minimize} & f(\mathbf{x}) + \mu \text{ maximum } \{0, g(\mathbf{x})\} \\ \text{subject to} & \mathbf{x} \in E_n \end{array}$$

If $g(\mathbf{x}) \leq 0$, then maximum $\{0, g(\mathbf{x})\} = 0$, and no penalty is incurred. On the other hand, if $g(\mathbf{x}) > 0$, then maximum $\{0, g(\mathbf{x})\} > 0$, and the penalty term $\mu g(\mathbf{x})$ is realized. However, observe that at points $\mathbf{x}$ where $g(\mathbf{x}) = 0$, the foregoing objective function might not be differentiable, even though g is differentiable. If differentiability is desirable in such a case, then one could, for example, consider instead a penalty function term of the type $\mu[\text{maximum } \{0, g(\mathbf{x})\}]^2$.

In general, a suitable penalty function must incur a positive penalty for infeasible points and no penalty for feasible points. If the constraints are of the form $g_i(\mathbf{x}) \leq 0$ for $i = 1, \ldots, m$ and $h_i(\mathbf{x}) = 0$ for $i = 1, \ldots, l$, then a suitable *penalty function* α is defined by

$$\alpha(\mathbf{x}) = \sum_{i=1}^{m} \phi[g_i(\mathbf{x})] + \sum_{i=1}^{l} \psi[h_i(\mathbf{x})] \tag{9.1a}$$

where ϕ and ψ are continuous functions satisfying the following:

$$\begin{aligned} &\phi(y) = 0 \quad \text{if } y \leq 0 \quad \text{and} \quad \phi(y) > 0 \quad \text{if } y > 0 \\ &\psi(y) = 0 \quad \text{if } y = 0 \quad \text{and} \quad \psi(y) > 0 \quad \text{if } y \neq 0 \end{aligned} \tag{9.1b}$$

Typically, ϕ and ψ are of the forms

$$\phi(y) = [\text{maximum } \{0, y\}]^p$$

$$\psi(y) = |\, y|^p$$

where p is a positive integer. Thus, the *penalty function* α is usually of the form

$$\alpha(\mathbf{x}) = \sum_{i=1}^{m} [\text{maximum } \{0, g_i(\mathbf{x})\}]^p + \sum_{i=1}^{l} |h_i(\mathbf{x})|^p$$

We refer to the function $f(\mathbf{x}) + \mu\alpha(\mathbf{x})$ as the *auxiliary function*. Later, we will introduce an *augmented Lagrangian function* in which the Lagrangian function, instead of simply $f(\mathbf{x})$, is augmented with a penalty term.

9.1.1 Example

Consider the following problem:

$$\begin{aligned} &\text{Minimize} && x \\ &\text{subject to} && -x + 2 \leq 0 \end{aligned}$$

Let $\alpha(x) = [\text{maximum } \{0, g(x)\}]^2$. Then,

$$\alpha(x) = \begin{cases} 0 & \text{if } x \geq 2 \\ (-x + 2)^2 & \text{if } x < 2 \end{cases}$$

Figure 9.1 shows the penalty and auxiliary functions α and $f + \mu\alpha$. Note that the minimum of $f + \mu\alpha$ occurs at the point $2 - (1/2\mu)$ and approaches the minimizing point $\bar{x} = 2$ of the original problem as μ approaches ∞.

9.1.2 Example

Consider the following problem:

$$\begin{aligned} &\text{Minimize} && x_1^2 + x_2^2 \\ &\text{subject to} && x_1 + x_2 - 1 = 0 \end{aligned}$$

The optimal solution lies at the point $(\frac{1}{2}, \frac{1}{2})$ and has objective value $\frac{1}{2}$. Now consider the following penalty problem, where $\mu > 0$ is a large number:

$$\begin{aligned} &\text{Minimize} && x_1^2 + x_2^2 + \mu(x_1 + x_2 - 1)^2 \\ &\text{subject to} && (x_1, x_2) \in E_2 \end{aligned}$$

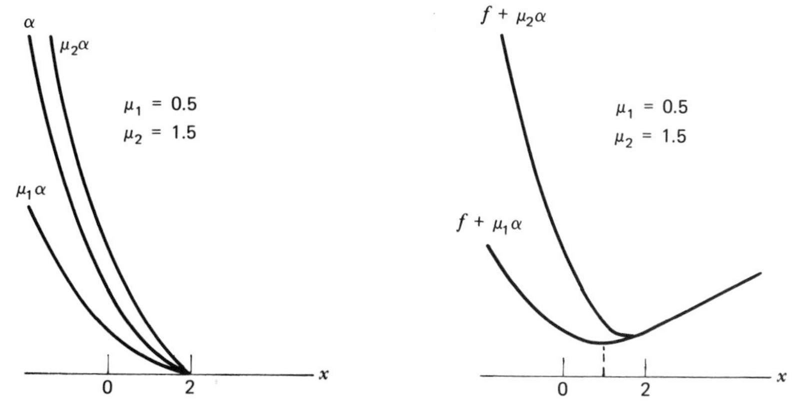

Figure 9.1 Penalty and auxiliary functions.

Note that for any $\mu \geq 0$, the objective function is convex. Thus, a necessary and sufficient condition for optimality is that the gradient of $x_1^2 + x_2^2 + \mu(x_1 + x_2 - 1)^2$ is equal to zero, yielding

$$x_1 + \mu(x_1 + x_2 - 1) = 0$$

$$x_2 + \mu(x_1 + x_2 - 1) = 0$$

Solving these two equations simultaneously, we get $x_1 = x_2 = \mu/(2\mu + 1)$. Thus, the optimal solution of the penalty problem can be made arbitrarily close to the solution of the original problem by choosing μ sufficiently large.

Geometric Interpretation of Penalty Functions

We now use Example 9.1.2 to illustrate the notion of penalty functions geometrically. Suppose that the constraint $h(\mathbf{x}) = 0$ is perturbed so that $h(\mathbf{x}) = \varepsilon$, that is, $x_1 + x_2 - 1 = \varepsilon$. Thus, we get the following problem:

$$\begin{aligned} \nu(\varepsilon) \equiv \text{Minimize} \quad & x_1^2 + x_2^2 \\ \text{subject to} \quad & x_1 + x_2 - 1 = \varepsilon \end{aligned}$$

Substituting $x_2 = 1 + \varepsilon - x_1$ into the objective function, the problem reduces to minimizing $x_1^2 + (1 + \varepsilon - x_1)^2$. The optimum occurs where the derivative vanishes, giving $2x_1 - 2(1 + \varepsilon - x_1) = 0$. Therefore, for any given ε, the optimal solution to the above problem is given by $x_1 = x_2 = (1 + \varepsilon)/2$ and has objective value $\nu(\varepsilon) = (1 + \varepsilon)^2/2$. In addition, for any given ε, the supremum of $x_1^2 + x_2^2$ subject to $x_1 + x_2 - 1 = \varepsilon$ is equal to ∞. Therefore, given any point (x_1, x_2) in E_2 with $x_1 + x_2 - 1 = \varepsilon$, its objective value lies in the interval $[(1 + \varepsilon)^2/2, \infty]$. In other words, the objective values of all points $\mathbf{x}$ in E_2 that satisfy $h(\mathbf{x}) = \varepsilon$ lie between $(1 + \varepsilon)^2/2$ and ∞. In particular, the set $\{[h(\mathbf{x}), f(\mathbf{x})]: \mathbf{x} \in E_2\}$ is shown in Figure 9.2. The lower envelope of this set is given by the parabola $(1 + h)^2/2 = (1 + \varepsilon)^2/2 \equiv \nu(\varepsilon)$. For a fixed $\mu > 0$, the penalty problem is to minimize $f(\mathbf{x}) + \mu h^2(\mathbf{x})$ subject to $\mathbf{x} \in E_2$. The contour $f + \mu h^2 = k$ is illustrated in the (h, f) space of Figure 9.2 by a dotted parabola. The intersection of this parabola on the f axis is equal to k. So if $f + \mu h^2$ is to be minimized, then the parabola must be moved downward as much as possible so that it still has at least one point in common with the shaded set, which describes legitimate combinations of h and f values. This process is continued until the parabola becomes tangential to the

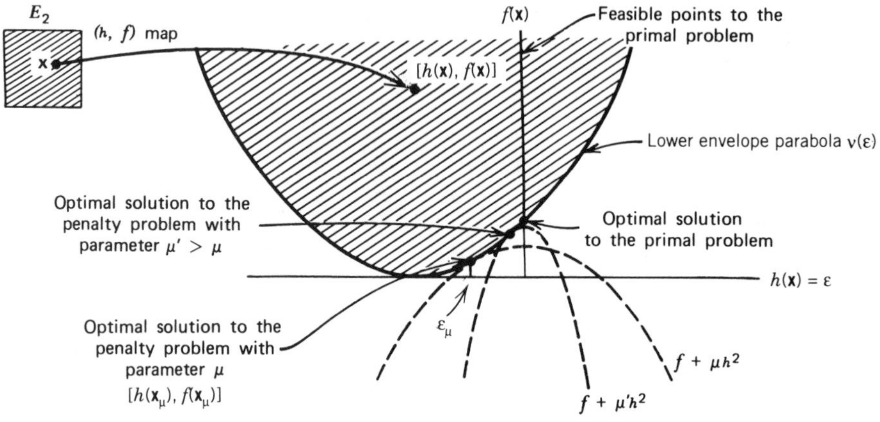

Figure 9.2 Geometry of penalty functions in the (h, f) space.

shaded set, as shown in Figure 9.2. This means that, for this given value of μ, the optimal value of the penalty problem is the intercept of the parabola on the f axis. Note that the optimal solution to the penalty problem is slightly infeasible to the original problem, since $h \neq 0$ at the point of tangency. Furthermore, the optimal objective value of the penalty problem is slightly smaller than the optimal primal objective. Also note that as the value of μ increases, the parabola $f + \mu h^2$ becomes steeper, and the point of tangency approaches the true optimal solution of the original problem.

Nonconvex Problems

In Figure 9.2, we showed that penalty functions can be used to get arbitrarily close to the optimal solution of the convex problem of Example 9.1.2. Figure 9.3 shows a nonconvex case, in which the Lagrangian dual approach would fail to produce an optimal solution of the primal problem because of the presence of a duality gap. Since penalty functions use a nonlinear support, as opposed to the linear support used by the dual function, shown in Figure 9.3, penalty functions can dip through the shaded set and get

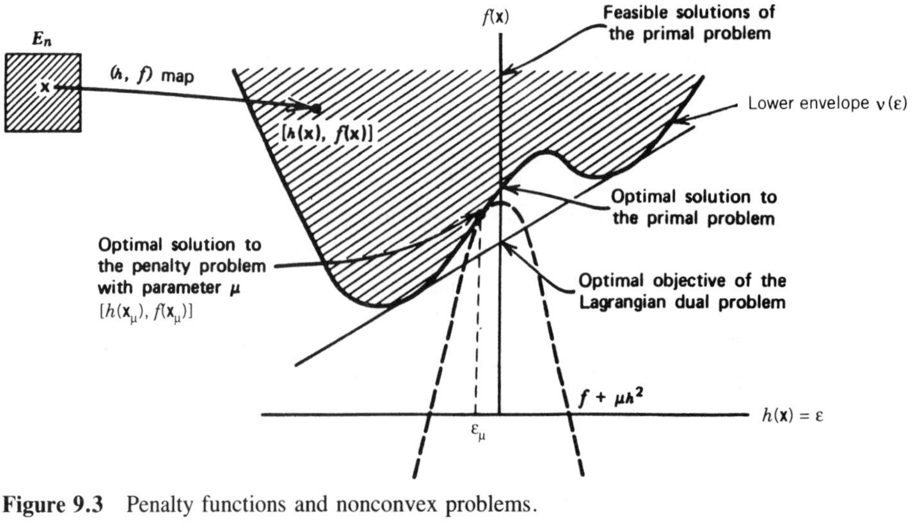

Figure 9.3 Penalty functions and nonconvex problems.

arbitrarily close to the optimal solution of the original problem, provided, of course, that a large penalty parameter μ is used.

Interpretation Via Perturbation Functions

Observe that the function $\nu(\boldsymbol{\epsilon})$ defined above and illustrated in Figs. 9.2 and 9.3 is precisely the perturbation function defined in equation (6.9) of Chapter 6. In fact, for the problem to minimize $f(\mathbf{x})$ subject to $h_i(\mathbf{x}) = 0$ for $i = 1, \ldots, l$, we have, denoting $\boldsymbol{\epsilon} = (\epsilon_1, \ldots, \epsilon_l)^t$ as a vector of perturbations,

$$\begin{aligned}\min_{\mathbf{x}} \{ f(\mathbf{x}) + \mu \sum_{i=1}^{l} h_i^2(\mathbf{x})\} &= \min_{(\mathbf{x},\boldsymbol{\epsilon})} \{ f(\mathbf{x}) + \mu\|\boldsymbol{\epsilon}\|^2 : h_i(\mathbf{x}) = \epsilon_i \text{ for } i = 1, \ldots, l\} \\ &= \min_{\boldsymbol{\epsilon}} [\mu\|\boldsymbol{\epsilon}\|^2 + \min_{\mathbf{x}} \{ f(\mathbf{x}) : h_i(\mathbf{x}) = \epsilon_i \text{ for } i = 1, \ldots, l\}] \\ &\equiv \min_{\boldsymbol{\epsilon}} \{\mu\|\boldsymbol{\epsilon}\|^2 + \nu(\boldsymbol{\epsilon})\} \qquad (9.2)\end{aligned}$$

Thus, we intuitively see that even if ν is nonconvex, as μ increases, the net effect of adding the term $\mu\|\boldsymbol{\epsilon}\|^2$ to $\nu(\boldsymbol{\epsilon})$ is to convexify it; and as $\mu \to \infty$, the minimizing $\boldsymbol{\epsilon}$ in (9.2) approaches zero. This interpretation readily extends to include inequality constraints as well (see Exercise 9.10).

In particular, relating this to Figures 9.2 and 9.3 for the case $l = 1$, if $\mathbf{x}_\mu$ minimizes (9.2) with $h(\mathbf{x}_\mu) = \epsilon_\mu$, assuming that ν is differentiable, we see that for a given $\mu > 0$, $\nu'(\epsilon_\mu) = -2\mu\epsilon_\mu = -2\mu h(\mathbf{x}_\mu)$ at the minimizing solution. Moreover, equating the first and last objective values in (9.2), we obtain $f(\mathbf{x}_\mu) = \nu(\epsilon_\mu)$. Hence, the coordinate $[h(\mathbf{x}_\mu), f(\mathbf{x}_\mu)]$ lies on the graph of $\nu(\epsilon_\mu)$, coinciding with $[\epsilon_\mu, \nu(\epsilon_\mu)]$, and having the slope of ν at ϵ_μ equal to $-2\mu\epsilon_\mu$. Denoting $f(\mathbf{x}_\mu) + \mu h^2(\mathbf{x}_\mu) = k_\mu$, we see that the parabolic function of ϵ given by $f = k_\mu - \mu\epsilon^2$ equals $\nu(\epsilon_\mu)$ in value when $\epsilon = \epsilon_\mu$, and has a slope of $-2\mu\epsilon_\mu$ at this point. Therefore, the solution $[h(\mathbf{x}_\mu), f(\mathbf{x}_\mu)]$ appears as shown in Figures 9.2 and 9.3. Observe also in Figure 9.3 that there does not exist a supporting hyperplane to the epigraph of ν at the point $[0, \nu(0)]$, thereby leading to a duality gap for the associated Lagrangian dual as per Theorem 6.2.7.

9.2 Exterior Penalty Function Methods

In this section, we present and prove an important result that justifies using exterior penalty functions as a means for solving constrained problems. We also discuss some computational difficulties associated with penalty functions and present some approaches geared toward overcoming such problems. Consider the following primal and penalty problems:

Primal Problem

$$\begin{aligned}\text{Minimize} \quad & f(\mathbf{x}) \\ \text{subject to} \quad & \mathbf{g}(\mathbf{x}) \leq \mathbf{0} \\ & \mathbf{h}(\mathbf{x}) = \mathbf{0} \\ & \mathbf{x} \in X\end{aligned}$$

where $\mathbf{g}$ is a vector function with components $g_1, \ldots, g_m$ and $\mathbf{h}$ is a vector function with components $h_1, \ldots, h_l$. Here, $f, g_1, \ldots, g_m, h_1, \ldots, h_l$ are continuous functions on E_n, and X is a nonempty set in E_n. The set X might typically represent simple constraints that could be easily handled explicitly, such as lower and upper bounds on the variables.

Penalty Problem

Let α be a continuous function of the form (9.1a) satisfying the properties stated in (9.1b). The basic penalty function approach attempts to solve the following problem:

$$\begin{array}{ll}\text{Maximize} & \theta(\mu) \\ \text{subject to} & \mu \geq 0\end{array}$$

where $\theta(\mu) = \inf\{f(\mathbf{x}) + \mu\alpha(\mathbf{x}) : \mathbf{x} \in X\}$. The main theorem of this section states that

$$\inf\{f(\mathbf{x}) : \mathbf{x} \in X, \mathbf{g}(\mathbf{x}) \leq \mathbf{0}, \mathbf{h}(\mathbf{x}) = \mathbf{0}\} = \sup_{\mu \geq 0} \theta(\mu) = \lim_{\mu \to \infty} \theta(\mu)$$

From this result, it is clear that we can get arbitrarily close to the optimal objective value of the primal problem by computing $\theta(\mu)$ for a sufficiently large μ. This result is established in Theorem 9.2.2. First, however, the following lemma is needed.

9.2.1 Lemma

Suppose that $f, g_1, \ldots, g_m, h_1, \ldots, h_l$ are continuous functions on E_n, and let X be a nonempty set in E_n. Let α be a continuous function on E_n given by (9.1), and suppose that for each μ, there is an $\mathbf{x}_\mu \in X$ such that $\theta(\mu) = f(\mathbf{x}_\mu) + \mu\alpha(\mathbf{x}_\mu)$. Then, the following statements hold:

1. $\inf\{f(\mathbf{x}) : \mathbf{x} \in X, \mathbf{g}(\mathbf{x}) \leq \mathbf{0}, \mathbf{h}(\mathbf{x}) = \mathbf{0}\} \geq \sup_{\mu \geq 0} \theta(\mu)$, where $\theta(\mu) = \inf\{f(\mathbf{x}) + \mu\alpha(\mathbf{x}) : \mathbf{x} \in X\}$, and where $\mathbf{g}$ is the vector function whose components are $g_1, \ldots, g_m$ and $\mathbf{h}$ is the vector function whose components are $h_1, \ldots, h_l$.
2. $f(\mathbf{x}_\mu)$ is a nondecreasing function of $\mu \geq 0$, $\theta(\mu)$ is a nondecreasing function of μ, and $\alpha(\mathbf{x}_\mu)$ is a nonincreasing function of μ.

Proof

Consider $\mathbf{x} \in X$ with $\mathbf{g}(\mathbf{x}) \leq \mathbf{0}$ and $\mathbf{h}(\mathbf{x}) = \mathbf{0}$, and note that $\alpha(\mathbf{x}) = 0$. Let $\mu \geq 0$. Then,

$$f(\mathbf{x}) = f(\mathbf{x}) + \mu\alpha(\mathbf{x}) \geq \inf\{f(\mathbf{y}) + \mu\alpha(\mathbf{y}) : \mathbf{y} \in X\} = \theta(\mu)$$

Thus, statement 1 follows. To show statement 2, let $\lambda < \mu$. By definition of $\theta(\lambda)$ and $\theta(\mu)$, the following two inequalities hold:

$$f(\mathbf{x}_\mu) + \lambda\alpha(\mathbf{x}_\mu) \geq f(\mathbf{x}_\lambda) + \lambda\alpha(\mathbf{x}_\lambda) \tag{9.3a}$$

$$f(\mathbf{x}_\lambda) + \mu\alpha(\mathbf{x}_\lambda) \geq f(\mathbf{x}_\mu) + \mu\alpha(\mathbf{x}_\mu) \tag{9.3b}$$

Adding these two inequalities and simplifying, we get

$$(\mu - \lambda)[\alpha(\mathbf{x}_\lambda) - \alpha(\mathbf{x}_\mu)] \geq 0$$

Since $\mu > \lambda$, we get $\alpha(\mathbf{x}_\lambda) \geq \alpha(\mathbf{x}_\mu)$. It then follows from (9.3a) that $f(\mathbf{x}_\mu) \geq f(\mathbf{x}_\lambda)$ for $\lambda \geq 0$. By adding and subtracting $\mu\alpha(\mathbf{x}_\mu)$ to the left-hand side of (9.3a), we get:

$$f(\mathbf{x}_\mu) + \mu\alpha(\mathbf{x}_\mu) + (\lambda - \mu)\alpha(\mathbf{x}_\mu) \geq \theta(\lambda)$$

Since $\mu > \lambda$ and $\alpha(\mathbf{x}_\mu) \geq 0$, the above inequality implies that $\theta(\mu) \geq \theta(\lambda)$. This completes the proof.

9.2.2 Theorem

Consider the following problem:

$$\begin{array}{lll} \text{Minimize} & f(\mathbf{x}) & \\ \text{subject to} & g_i(\mathbf{x}) \leq 0 & \text{for } i = 1, \ldots, m \\ & h_i(\mathbf{x}) = 0 & \text{for } i = 1, \ldots, l \\ & \mathbf{x} \in X & \end{array}$$

where $f, g_1, \ldots, g_m, h_1, \ldots, h_l$ are continuous functions on E_n and X is a nonempty set in E_n. Suppose that the problem has a feasible solution, and let α be a continuous function given by (9.1). Furthermore, suppose that for each μ, there exists a solution $\mathbf{x}_\mu \in X$ to the problem to minimize $f(\mathbf{x}) + \mu\alpha(\mathbf{x})$ subject to $\mathbf{x} \in X$, and that $\{\mathbf{x}_\mu\}$ is contained in a compact subset of X. Then,

$$\inf\{f(\mathbf{x}) : \mathbf{g}(\mathbf{x}) \leq \mathbf{0}, \mathbf{h}(\mathbf{x}) = \mathbf{0}, \mathbf{x} \in X\} = \sup_{\mu \geq 0} \theta(\mu) = \lim_{\mu \to \infty} \theta(\mu)$$

where $\theta(\mu) = \inf\{f(\mathbf{x}) + \mu\alpha(\mathbf{x}) : \mathbf{x} \in X\} = f(\mathbf{x}_\mu) + \mu\alpha(\mathbf{x}_\mu)$. Furthermore, the limit $\bar{\mathbf{x}}$ of any convergent subsequence of $\{\mathbf{x}_\mu\}$ is an optimal solution to the original problem, and $\mu\alpha(\mathbf{x}_\mu) \to 0$ as $\mu \to \infty$.

Proof

By part 2 of Lemma 9.2.1, $\theta(\mu)$ is monotone, so that $\sup_{\mu \geq 0} \theta(\mu) = \lim_{\mu \to \infty} \theta(\mu)$. We first show that $\alpha(\mathbf{x}_\mu) \to 0$ as $\mu \to \infty$. Let $\mathbf{y}$ be a feasible point and $\varepsilon > 0$. Let $\mathbf{x}_1$ be an optimal solution to the problem to minimize $f(\mathbf{x}) + \mu\alpha(\mathbf{x})$ subject to $\mathbf{x} \in X$, for $\mu = 1$. If $\mu \geq \frac{1}{\varepsilon}|f(\mathbf{y}) - f(\mathbf{x}_1)| + 2$, then, by part 2 of Lemma 9.2.1, we must have $f(\mathbf{x}_\mu) \geq f(\mathbf{x}_1)$.

We now show that $\alpha(\mathbf{x}_\mu) \leq \varepsilon$. By contradiction, suppose that $\alpha(\mathbf{x}_\mu) > \varepsilon$. Noting part 1 of Lemma 9.2.1, we get

$$\begin{aligned} \inf\{f(\mathbf{x}):\mathbf{g}(\mathbf{x}) \leq \mathbf{0}, \mathbf{h}(\mathbf{x}) = \mathbf{0}, \mathbf{x} \in X\} \geq \theta(\mu) &= f(\mathbf{x}_\mu) + \mu\alpha(\mathbf{x}_\mu) \geq f(\mathbf{x}_1) + \mu\alpha(\mathbf{x}_\mu) \\ &> f(\mathbf{x}_1) + |f(\mathbf{y}) - f(\mathbf{x}_1)| + 2\varepsilon > f(\mathbf{y}) \end{aligned}$$

The above inequality is not possible in view of feasibility of $\mathbf{y}$. Thus, $\alpha(\mathbf{x}_\mu) \leq \varepsilon$ for all $\mu \geq \frac{1}{\varepsilon}|f(\mathbf{y}) - f(\mathbf{x}_1)| + 2$. Since $\varepsilon > 0$ is arbitrary, $\alpha(\mathbf{x}_\mu) \to 0$ as $\mu \to \infty$. Now let $\{\mathbf{x}_{\mu_k}\}$ be any convergent subsequence of $\{\mathbf{x}_\mu\}$, and let $\bar{\mathbf{x}}$ be its limit. Then,

$$\sup_{\mu \geq 0} \theta(\mu) \geq \theta(\mu_k) = f(\mathbf{x}_{\mu_k}) + \mu_k\alpha(\mathbf{x}_{\mu_k}) \geq f(\mathbf{x}_{\mu_k}).$$

Since $\mathbf{x}_{\mu_k} \to \bar{\mathbf{x}}$ and f is continuous, the above inequality implies that

$$\sup_{\mu \geq 0} \theta(\mu) \geq f(\bar{\mathbf{x}}) \tag{9.4}$$

Since $\alpha(\mathbf{x}_\mu) \to 0$ as $\mu \to \infty$, then $\alpha(\bar{\mathbf{x}}) = 0$, that is, $\bar{\mathbf{x}}$ is a feasible solution to the original problem. In view of (9.4) and part 1 of Lemma 9.2.1, it follows that $\bar{\mathbf{x}}$ is an optimal solution to the original problem and that $\sup_{\mu \geq 0} \theta(\mu) = f(\bar{\mathbf{x}})$. Note that $\mu\alpha(\mathbf{x}_\mu) = \theta(\mu) - f(\mathbf{x}_\mu)$. As $\mu \to \infty$, $\theta(\mu)$ and $f(\mathbf{x}_\mu)$ both approach $f(\bar{\mathbf{x}})$, and, hence, $\mu\alpha(\mathbf{x}_\mu)$ approaches zero. This completes the proof.

Corollary

If $\alpha(\mathbf{x}_\mu) = 0$ for some μ, then $\mathbf{x}_\mu$ is an optimal solution to the problem.

Proof

If $\alpha(\mathbf{x}_\mu) = 0$, then $\mathbf{x}_\mu$ is a feasible solution to the problem. Furthermore, since

$$\inf \{f(\mathbf{x}) : \mathbf{g}(\mathbf{x}) \leq \mathbf{0}, \mathbf{h}(\mathbf{x}) = \mathbf{0}, \mathbf{x} \in X\} \geq \theta(\mu) = f(\mathbf{x}_\mu) + \mu\alpha(\mathbf{x}_\mu) = f(\mathbf{x}_\mu)$$

it immediately follows that $\mathbf{x}_\mu$ is an optimal solution.

Note the significance of the assumption that $\{\mathbf{x}_\mu\}$ is contained in a compact subset in X. Obviously, this assumption holds if X is compact. Without this assumption, it is possible that the optimal objective values of the primal and penalty problems are not equal (see Exercise 9.4). This assumption is not restrictive in most practical cases, since the variables usually lie between finite lower and upper bounds.

From the above theorem, it follows that the optimal solution $\mathbf{x}_\mu$ to the problem to minimize $f(\mathbf{x}) + \mu\alpha(\mathbf{x})$ subject to $\mathbf{x} \in X$ can be made arbitrarily close to the feasible region by choosing μ large enough. Furthermore, by choosing μ large enough, $f(\mathbf{x}_\mu) + \mu\alpha(\mathbf{x}_\mu)$ can be made arbitrarily close to the optimal objective value of the original primal problem. As we discuss later in this section, one popular scheme for solving the penalty problem is to solve a sequence of problems of the form

$$\begin{array}{ll} \text{Minimize} & f(\mathbf{x}) + \mu\alpha(\mathbf{x}) \\ \text{subject to} & \mathbf{x} \in X \end{array}$$

for an increasing sequence of penalty parameters. The optimal points $\{\mathbf{x}_\mu\}$ are generally infeasible, but as seen in the proof of Theorem 9.2.2, as the penalty parameter μ is made large, the points generated approach an optimal solution from outside the feasible region. Hence, as mentioned previously, this technique is also referred to as an *exterior penalty function method*.

KKT Lagrange Multipliers at Optimality

Under certain conditions, we can use the solutions to the sequence of penalty problems to recover the KKT Lagrange multipliers associated with the constraints at optimality. Toward this end, suppose that $X = E_n$ for simplicity, and consider the primal problem to minimize $f(\mathbf{x})$ subject to $g_i(\mathbf{x}) \leq 0$, $i = 1, \ldots, m$ and $h_i(\mathbf{x}) = 0$, $i = 1, \ldots, l$. (The following analysis readily generalizes to the case where some inequality and/or equality constraints define X; see Exercise 9.11.) Suppose that the penalty function α is given by (9.1) where, in addition, ϕ and ψ are continuously differentiable with $\phi'(y) \geq 0$ for all y and $\phi'(y) = 0$ for $y \leq 0$. Assuming that the conditions of Theorem 9.2.2 hold true, since $\mathbf{x}_\mu$ solves the problem to minimize $f(\mathbf{x}) + \mu\alpha(\mathbf{x})$, the gradient of the objective function of this penalty problem must vanish at $\mathbf{x}_\mu$. This gives

$$\nabla f(\mathbf{x}_\mu) + \sum_{i=1}^{m} \mu\phi'[g_i(\mathbf{x}_\mu)]\nabla g_i(\mathbf{x}_\mu) + \sum_{i=1}^{l} \mu\psi'[h_i(\mathbf{x}_\mu)]\nabla h_i(\mathbf{x}_\mu) = \mathbf{0} \qquad \text{for all } \mu$$

Now let $\bar{\mathbf{x}}$ be an accumulation point of the generated sequence $\{\mathbf{x}_\mu\}$. Without loss of generality, assume that $\{\mathbf{x}_\mu\}$ itself converges to $\bar{\mathbf{x}}$. Denote $I = \{i : g_i(\bar{\mathbf{x}}) = 0\}$ to be the set of inequality constraints that are binding at $\bar{\mathbf{x}}$. Since $g_i(\bar{\mathbf{x}}) < 0$ for all i L I, by

Theorem 9.2.2, we have $g_i(\mathbf{x}_\mu) < 0$ for μ sufficiently large, yielding $\mu\phi'[g_i(\mathbf{x}_\mu)] = 0$. Hence, we can write the foregoing identity as

$$\nabla f(\mathbf{x}_\mu) + \sum_{i \in I} (\mathbf{u}_\mu)_i \nabla g_i(\mathbf{x}_\mu) + \sum_{i=1}^{l} (\mathbf{v}_\mu)_i \nabla h_i(\mathbf{x}_\mu) = \mathbf{0} \qquad \text{for all } \mu \text{ large enough} \tag{9.5a}$$

where $\mathbf{u}_\mu$ and $\mathbf{v}_\mu$ are vectors having components

$$(\mathbf{u}_\mu)_i \equiv \mu\phi'[g_i(\mathbf{x}_\mu)] \geq 0 \text{ for all } i \in I \quad \text{and } (\mathbf{v}_\mu)_i \equiv \mu\psi'[h_i(\mathbf{x}_\mu)] \quad \text{for all } i = 1, \ldots, l \tag{9.5b}$$

Let us now assume that $\bar{\mathbf{x}}$ is a regular solution as defined in Theorem 4.3.7. Then, we know that there exist *unique* Lagrangian multipliers $\bar{u}_i \geq 0$, $i \in I$ and $\bar{v}_i$, $i = 1, \ldots, l$ such that

$$\nabla f(\bar{\mathbf{x}}) + \sum_{i \in I} \bar{u}_i \nabla g_i(\bar{\mathbf{x}}) + \sum_{i=1}^{l} \bar{v}_i \nabla h_i(\bar{\mathbf{x}}) = \mathbf{0} \tag{9.5c}$$

Since g, h, ϕ, and ψ are all continuously differentiable and since $\{\mathbf{x}_\mu\} \to \bar{\mathbf{x}}$, which is a regular point, we must then have in (9.5) that $(\mathbf{u}_\mu)_i \to \bar{u}_i$, for all $i \in I$ and $(\mathbf{v}_\mu)_i \to \bar{v}_i$, for all $i = 1, \ldots, l$.

Hence, for sufficiently large values of μ, the multipliers given by (9.5b) can be used to estimate the KKT Lagrange multipliers at optimality. For example, if α is the *quadratic penalty function* given by $\alpha(\mathbf{x}) = \sum_{i=1}^{m} [\text{maximum } \{0, g_i(\mathbf{x})\}]^2 + \sum_{i=1}^{l} h_i^2(\mathbf{x})$, then $\phi(y) \equiv [\text{maximum } \{0, y\}]^2$, $\phi'(y) = 2 \text{ maximum } \{0, y\}$, $\psi(y) = y^2$, and $\psi'(y) = 2y$. Hence, from (9.5b), we obtain

$$\begin{aligned} (\mathbf{u}_\mu)_i &= 2\mu \text{ maximum } \{0, g_i(\mathbf{x}_\mu)\} \quad \text{for all } i \in I, \qquad \text{and} \\ (\mathbf{v}_\mu)_i &= 2\mu h_i(\mathbf{x}_\mu) \quad \text{for all } i = 1, \ldots, l \end{aligned} \tag{9.6}$$

In particular, observe that if $\bar{u}_i > 0$ for some $i \in I$, then $(\mathbf{u}_\mu)_i > 0$ for μ large enough, which in turn implies from (9.6) that $g_i(\mathbf{x}_\mu) > 0$. This means that $g_i(\mathbf{x}) \leq 0$ is violated all along the trajectory leading to $\bar{\mathbf{x}}$, and in the limit $g_i(\bar{\mathbf{x}}) = 0$. Hence, if $\bar{u}_i > 0$ for all $i \in I$, $\bar{v}_j \neq 0$ for all j, then all the constraints binding at $\bar{\mathbf{x}}$ are violated along the trajectory $\{\mathbf{x}_\mu\}$ leading to $\bar{\mathbf{x}}$. For instance, in Example 9.1.2, we have $\mathbf{x}_\mu = [\mu/(2\mu + 1), \mu/(2\mu + 1)]$, $h(\mathbf{x}_\mu) = -1/(2\mu + 1)$; and so, $(\mathbf{v}_\mu)_i = -2\mu/(2\mu + 1)$ from (9.6). Note that as $\mu \to \infty$, $(\mathbf{v}_\mu)_i \to -1$, the optimal value of the Lagrange multiplier for this example.

Computational Difficulties Associated with Penalty Functions

The solution to the penalty problem can be made arbitrarily close to the optimal solution of the original problem by choosing μ sufficiently large. However, if we choose a very large μ and attempt to solve the penalty problem, we may get into some computational difficulties of ill-conditioning. With a large μ, more emphasis is placed on feasibility, and most procedures for unconstrained optimization will move quickly toward a feasible point. Even though this point may be far from the optimum, premature termination could occur. To illustrate, suppose that during the course of optimization we reached a feasible point with $\alpha(\mathbf{x}) = 0$. Especially in the presence of nonlinear equality constraints, a movement from $\mathbf{x}$ along any direction $\mathbf{d}$ may result in infeasible points or feasible points with large objective values. In both cases, the value of the auxiliary function $f(\mathbf{x} + \lambda\mathbf{d}) + \mu\alpha(\mathbf{x} + \lambda\mathbf{d})$ is greater than $f(\mathbf{x}) + \mu\alpha(\mathbf{x})$ for noninfinitesimal values of the step size λ. This is obvious in the latter case. In the former case, $\alpha(\mathbf{x} + \lambda\mathbf{d}) > 0$; and since μ is very large, then any reduction in $f(\mathbf{x} + \lambda\mathbf{d})$ over $f(\mathbf{x})$ will be usually offset by the term $\mu\alpha(\mathbf{x} + \lambda\mathbf{d})$. Thus, improvement is only possible if the step size λ is very small such that the term $\mu\alpha(\mathbf{x} + \lambda\mathbf{d})$ would be small, despite the fact that μ is very large. In this case, an improvement in $f(\mathbf{x} + \lambda\mathbf{d})$ over $f(\mathbf{x})$ may offset the fact that $\mu\alpha(\mathbf{x} + \lambda\mathbf{d}) > 0$. The need for using a very small step size may result in slow convergence and premature termination.

The foregoing intuitive discussion also has a formal theoretical basis. To gain insight into this issue, consider the equality constrained problem of minimizing $f(\mathbf{x})$, subject to $h_i(\mathbf{x}) = 0$, for $i = 1, \ldots, l$. Let $F(\mathbf{x}) = f(\mathbf{x}) + \mu\sum_{i=1}^{l} \psi[h_i(\mathbf{x})]$ denote the penalized objective function constructed according to (9.1), where ψ is assumed to be twice-differentiable. Then, denoting by ∇ and ∇^2 the gradient and the Hessian operators, respectively, for the functions F, f, and h_i, $i = 1, \ldots, l$, and denoting the first and second derivatives of ψ by ψ' and ψ'', respectively, we get, assuming twice-differentiability, that

$$\nabla F(\mathbf{x}) = \nabla f(\mathbf{x}) + \mu \sum_{i=1}^{l} \psi'[h_i(\mathbf{x})]\nabla h_i(\mathbf{x})$$

$$\nabla^2 F(\mathbf{x}) = \left[\nabla^2 f(\mathbf{x}) + \sum_{i=1}^{l} \mu\psi'[h_i(\mathbf{x})]\nabla^2 h_i(\mathbf{x})\right] + \mu \sum_{i=1}^{l} \psi''[h_i(\mathbf{x})]\nabla h_i(\mathbf{x})\nabla h_i(\mathbf{x})^t \quad (9.7)$$

Observe that if we also had inequality constraints present in the problem, and we had used the penalty function (9.1a) with $\phi(y) = [\text{maximum } \{0, y\}]^2$, for example, then $\phi'(y) = 2 \text{ maximum } \{0, y\}$, but $\phi''(y)$ would have been undefined at $y = 0$. Hence, $\nabla^2 F(\mathbf{x})$ would be undefined at points having binding inequality constraints. However, if $y > 0$, then $\phi'' = 2$, and so $\nabla^2 F(\mathbf{x})$ would be defined at points violating all the inequality constraints; and in such a case, (9.7) would inherit a similar expression as for the equality constraints.

Now, as we know from Chapter 8, the convergence rate behavior of algorithms used for minimizing F will be governed by the eigenvalue structure of $\nabla^2 F$. To estimate this characteristic, let us examine the eigenstructure of (9.7) as $\mu \to \infty$ and, under the conditions of Theorem 9.2.2, as $\mathbf{x} \equiv \mathbf{x}_\mu \to \bar{\mathbf{x}}$, an optimum to the given problem. Assuming that $\bar{\mathbf{x}}$ is a regular solution, we have, from (9.5), that $\mu\psi'[h_i(\mathbf{x}_\mu)] \to \bar{v}_i$, the optimal Lagrange multiplier associated with the ith constraint, for $i = 1, \ldots, l$. Hence, the term within $[\cdot]$ in (9.7) approaches the Hessian of the Lagrangian function $L(\mathbf{x}) = f(\mathbf{x}) + \sum_{i=1}^{l} \bar{v}_i h_i(\mathbf{x})$. The other term in (9.7), however, is strongly tied in with μ, and is potentially explosive. For example, if $\psi(y) = y^2$ as in the popular quadratic penalty function, then this term equals 2μ times a matrix that approaches $\sum_{i=1}^{l} \nabla h_i(\bar{\mathbf{x}})\nabla h_i(\bar{\mathbf{x}})^t$, a matrix of rank l. It can then be shown (see the Notes and References section) that, as $\mu \to \infty$, we have $\mathbf{x} \equiv \mathbf{x}_\mu \to \bar{\mathbf{x}}$, and $\nabla^2 F$ has l eigenvalues that approach ∞, while $n - l$ eigenvalues approach some finite limits. Consequently, we can expect a severely ill-conditioned Hessian matrix for large values of μ.

Examining the analysis evolving around (8.18), the steepest descent method under such a severe situation would most likely be disastrous. On the other hand, Newton's method or its variants such as conjugate gradient or quasi-Newton methods [operated as, at least, an $(l + 1)$-step process] would be unaffected by the foregoing eigenvalue structure. Superior n-step superlinear (or superlinear) convergence rates might then be achievable, as discussed in Sections 8.6 and 8.8.

9.2.3 Example

Consider the problem of Example 9.1.2. The penalized objective function F is given by $F(\mathbf{x}) = x_1^2 + x_2^2 + \mu(x_1 + x_2 - 1)^2$. Its Hessian, as in (9.7), is given by

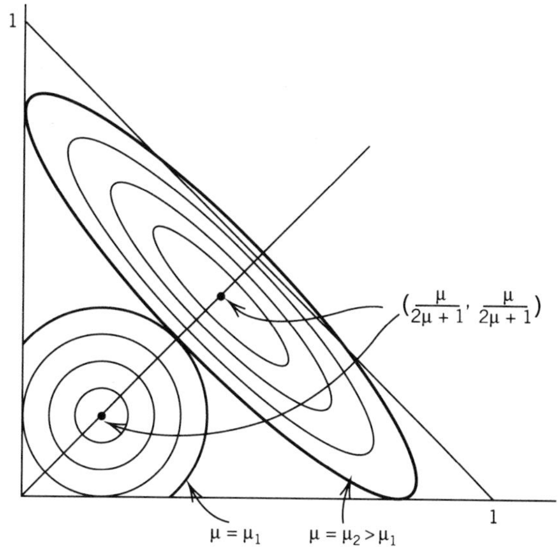

Figure 9.4 Ill-conditioning effect of a large μ value.

$$\nabla^2 F(\mathbf{x}) = \begin{bmatrix} 2(1+\mu) & 2\mu \\ 2\mu & 2(1+\mu) \end{bmatrix}$$

The eigenvalues of this matrix, computed readily via the equation det $|\nabla^2 F(\mathbf{x}) - \lambda I| = 0$, are $\lambda_1 = 2$ and $\lambda_2 = 2(1 + 2\mu)$, with respective eigenvectors $(1, -1)^t$ and $(1, 1)^t$. Note that $\lambda_2 \to \infty$ as $\mu \to \infty$, while $\lambda_1 = 2$ is finite; and, hence, the condition number of $\nabla^2 F$ approaches ∞ as $\mu \to \infty$. Figure 9.4 depicts the contours of F for a particular value of μ. These contours are elliptical with their major and minor axes oriented along the eigenvectors (see Appendix A.1), becoming more and more steep along the direction $(1, 1)^t$ as μ increases. Hence, for a large value of μ, the steepest descent method would severely zigzag to the optimum, unless it is fortunately initialized at a convenient starting solution.

Summary of Penalty Function Methods

As a result of the above difficulties associated with large penalty parameters, most algorithms using penalty functions employ a sequence of increasing penalty parameters. With each new value of the penalty parameter, an optimization technique is employed, starting with the optimal solution corresponding to the previously chosen parameter value. Such an approach is sometimes referred to as a *sequential unconstrained minimization technique (SUMT)*.

We present below a summary of penalty function methods to solve the problem for minimizing $f(\mathbf{x})$ subject to $\mathbf{g}(\mathbf{x}) \leq \mathbf{0}$, $\mathbf{h}(\mathbf{x}) = \mathbf{0}$, and $\mathbf{x} \in X$. The penalty function α used is of the form specified in (9.1). These methods do not impose any restriction on f, $\mathbf{g}$, and $\mathbf{h}$, other than that of continuity. However, they can effectively be used only in those cases where an efficient solution procedure is available to solve the problem specified in step 1 below.

Initialization Step Let $\varepsilon > 0$ be a termination scalar. Choose an initial point $\mathbf{x}_1$, a penalty parameter $\mu_1 > 0$, and a scalar $\beta > 1$. Let $k = 1$, and go to the main step.

Main Step

1. Starting with $\mathbf{x}_k$, solve the following problem:

$$\begin{array}{ll} \text{Minimize} & f(\mathbf{x}) + \mu_k \alpha(\mathbf{x}) \\ \text{subject to} & \mathbf{x} \in X \end{array}$$

Let $\mathbf{x}_{k+1}$ be an optimal solution and go to step 2.

2. If $\mu_k \alpha(\mathbf{x}_{k+1}) < \varepsilon$, stop; otherwise, let $\mu_{k+1} = \beta\mu_k$, replace k by $k + 1$, and go to step 1.

9.2.4 Example

Consider the following problem:

$$\begin{array}{ll} \text{Minimize} & (x_1 - 2)^4 + (x_1 - 2x_2)^2 \\ \text{subject to} & x_1^2 - x_2 = 0 \\ & \mathbf{x} \in X = E_2 \end{array}$$

Note that at iteration k, for a given penalty parameter μ_k, the problem to be solved for obtaining $\mathbf{x}_{\mu_k}$ is, using the quadratic penalty function,

$$\text{minimize } (x_1 - 2)^4 + (x_1 - 2x_2)^2 + \mu_k(x_1^2 - x_2)^2$$

Table 9.1 summarizes the computations using the penalty function method, including the Lagrange multiplier estimate obtained via (9.6). The starting point is taken as $\mathbf{x}_1 = (2.0, 1.0)$, where the objective function value is 0.0. The initial value of the penalty parameter is taken as $\mu_1 = 0.1$, and the scalar β is taken as 10.0. Note that $f(\mathbf{x}_{\mu_k})$ and $\theta(\mu_k)$ are nondecreasing functions and $\alpha(\mathbf{x}_{\mu_k})$ is a nonincreasing function. The procedure could have been stopped after the fourth iteration, where $\alpha(\mathbf{x}_{\mu_k}) = 0.000267$. However, to show more clearly that $\mu_k\alpha(\mathbf{x}_{\mu_k})$ does converge to zero according to Theorem 9.2.2, one more iteration was carried out. At the point $\mathbf{x}' = (0.9461094, 0.8934414)$, the reader can verify that the KKT conditions are satisfied for $v = 3.3632$. Figure 9.5 shows the progess of the algorithm.

9.3 Exact Absolute Value and Augmented Lagrangian Penalty Methods

For the types of penalty functions considered thus far, we have seen that we need to make the penalty parameter infinitely large in a limiting sense to recover an optimal solution. This can cause numerical difficulties and ill-conditioning effects. A natural question to raise, then, is: Can we design a penalty function that is capable of recovering an exact optimal solution for reasonable finite values of the penalty parameter μ without the need for μ to approach infinity? We present below two penalty functions that possess this property and are therefore known as *exact penalty functions*.

The *absolute value or l_1 penalty function*, is an *exact penalty function*, which conforms with the typical form of (9.1) with $p = 1$. Namely, given a penalty parameter $\mu > 0$, the penalized objective function in this case for Problem P to minimize $f(\mathbf{x})$ subject to $g_i(\mathbf{x}) \leq 0$, $i = 1, \ldots, m$, and $h_i(\mathbf{x}) = 0$, $i = 1, \ldots, l$, is given by

Table 9.1 Summary of the Computations for the Penalty Function Method

Iteration k	μ_k	$\mathbf{x}_{k+1} = \mathbf{x}_{\mu_k}$	$f(\mathbf{x}_{k+1})$	$\alpha(\mathbf{x}_{\mu_k}) = h^2(\mathbf{x}_{\mu_k})$	$\theta(\mu_k)$	$\mu_k\alpha(\mathbf{x}_{\mu_k})$	v_{μ_k}
1	0.1	(1.4539, 0.7608)	0.0935	1.8307	0.2766	0.1831	0.270605
2	1.0	(1.1687, 0.7407)	0.5753	0.3908	0.9661	0.3908	1.250319
3	10.0	(0.9906, 0.8425)	1.5203	0.01926	1.7129	0.1926	2.775767
4	100.0	(0.9507, 0.8875)	1.8917	0.000267	1.9184	0.0267	3.266096
5	1000.0	(0.9461094, 0.8934414)	1.9405	0.0000028	1.9433	0.0028	3.363252

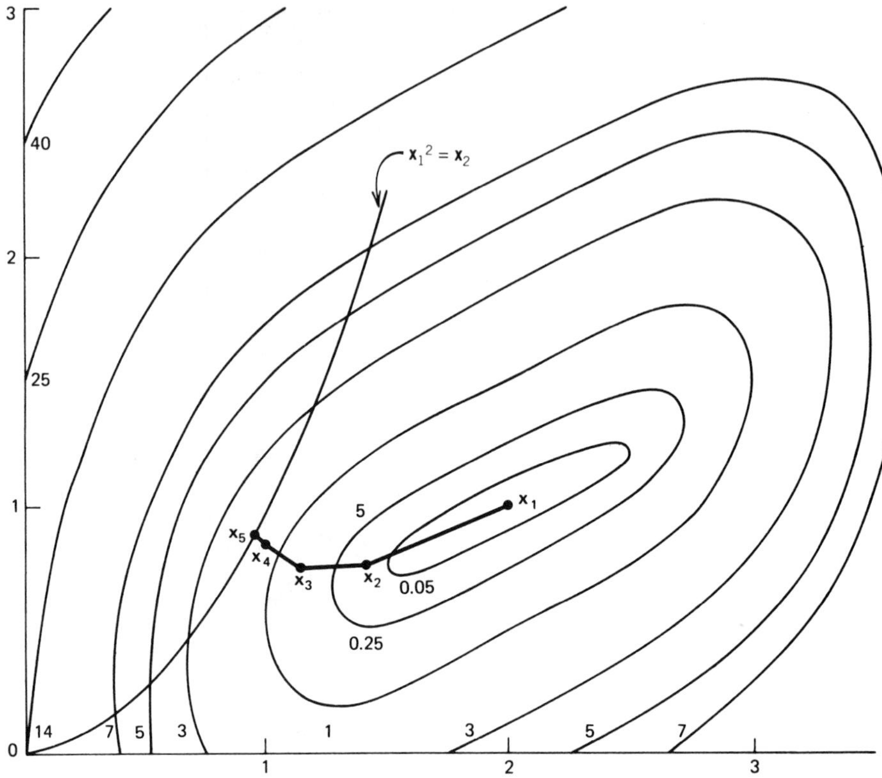

Figure 9.5 Illustration of the penalty function method.

$$F_E(\mathbf{x}) = f(\mathbf{x}) + \mu \left[\sum_{i=1}^{m} \max\{0, g_i(\mathbf{x})\} + \sum_{i=1}^{l} |h_i(\mathbf{x})| \right] \tag{9.8}$$

(For convenience, we suppress the type of constraints $\mathbf{x} \in X$ in our discussion; the analysis readily extends to include such constraints.) The following result shows that, under suitable convexity assumptions (and a constraint qualification), there does exist a finite value of μ that will recover an optimum solution to P via the minimization of F_E. Alternatively, it can be shown that if $\bar{\mathbf{x}}$ satisfies the second-order sufficiency conditions for a local minimum of P as stated in Theorem 4.4.2, then, for μ at least as large as in the theorem below, $\bar{\mathbf{x}}$ will also be a local minimum of F_E (see Exercise 9.12).

9.3.1 Theorem

Consider the following Problem P:

$$\begin{array}{lll} \text{Minimize} & f(\mathbf{x}) & \\ \text{subject to} & g_i(\mathbf{x}) \le 0 & \text{for } i = 1, \ldots, m \\ & h_i(\mathbf{x}) = 0 & \text{for } i = 1, \ldots, l \end{array}$$

Let $\bar{\mathbf{x}}$ be a KKT point with Lagrangian multipliers $\bar{u}_i$, $i \in I$, and $\bar{v}_i$, $i = 1, \ldots, l$ associated with the inequality and the equality constraints, respectively, where $I = \{i \in \{1, \ldots, m\}: g_i(\bar{\mathbf{x}}) = 0\}$ is the index set of binding inequality constraints. Furthermore, suppose that f and g_i, $i \in I$, are convex functions and that h_i, $i = 1, \ldots,$

l are affine functions. Then, for $\mu \geq$ maximum $\{\bar{u}_i, i \in I, |\bar{v}_i|, i = 1, \ldots, l\}$, $\bar{\mathbf{x}}$ also minimizes the exact l_1 penalized objective function F_E defined by (9.8).

Proof

Since $\bar{\mathbf{x}}$ is a KKT point for Problem P, it is feasible to P and satisfies

$$\nabla f(\bar{\mathbf{x}}) + \sum_{i \in I} \bar{u}_i \nabla g_i(\bar{\mathbf{x}}) + \sum_{i=1}^{l} \bar{v}_i \nabla h_i(\bar{\mathbf{x}}) = \mathbf{0}, \qquad \bar{u}_i \geq 0 \text{ for } i \in I \tag{9.9}$$

(Moreover, by Theorem 4.3.8, $\bar{\mathbf{x}}$ solves P.) Now, consider the problem of minimizing $F_E(\mathbf{x})$ over $\mathbf{x} \in E_n$. This can equivalently be restated as follows, for any $\mu \geq 0$:

$$\text{Minimize} \quad f(\mathbf{x}) + \mu \left[\sum_{i=1}^{m} y_i + \sum_{i=1}^{l} z_i \right] \tag{9.10a}$$

$$\text{subject to} \quad y_i \geq g_i(\mathbf{x}) \text{ and } y_i \geq 0 \quad \text{for } i = 1, \ldots, m \tag{9.10b}$$

$$z_i \geq h_i(\mathbf{x}) \text{ and } z_i \geq -h_i(\mathbf{x}) \quad \text{for } i = 1, \ldots, l \tag{9.10c}$$

The equivalence follows easily by observing that for any given $\mathbf{x} \in E_n$, the maximum value of the objective function in (9.10a), subject to (9.10b) and (9.10c), is realized by taking y_i = maximum $\{0, g_i(\bar{\mathbf{x}})\}$ for $i = 1, \ldots, m$ and $z_i = |h_i(\mathbf{x})|$ for $i = 1 \ldots, l$. In particular, given $\bar{\mathbf{x}}$, define $\bar{y}_i$ = maximum $\{0, g_i(\bar{\mathbf{x}})\}$ for $i = 1, \ldots, m$ and $\bar{z}_i = |\mathbf{h}_i(\bar{\mathbf{x}})| \equiv 0$ for $i = 1, \ldots, l$.

Note that, of the inequalities $y_i \geq g_i(\mathbf{x})$, $i = 1, \ldots, m$, only those corresponding to $i \in I$ are binding, while all the other inequalities in (9.10) are binding at $(\bar{\mathbf{x}}, \bar{\mathbf{y}}, \bar{\mathbf{z}})$. Hence, for $(\bar{\mathbf{x}}, \bar{\mathbf{y}}, \bar{\mathbf{z}})$ to be a KKT point for (9.10), we must find Lagrangian multipliers u_i^+, u_i^-, $i = 1, \ldots, m$, and v_i^+, v_i^-, $i = 1, \ldots, l$, associated with the respective pairs of constraints in (9.10b) and (9.10c) such that

$$\nabla f(\bar{\mathbf{x}}) + \sum_{i \in I} u_i^+ \nabla g_i(\bar{\mathbf{x}}) + \sum_{i=1}^{l} (v_i^+ - v_i^-) \nabla h_i(\bar{\mathbf{x}}) = \mathbf{0}$$

$$\mu - u_i^+ - u_i^- = 0 \qquad \text{for } i = 1, \ldots, m$$

$$\mu - v_i^+ - v_i^- = 0 \qquad \text{for } i = 1, \ldots, l$$

$$(u_i^+, u_i^-) \geq 0 \qquad \text{for } i = 1, \ldots, m$$

$$(v_i^+, v_i^-) \geq 0 \qquad \text{for } i = 1, \ldots, l$$

$$u_i^+ = 0 \qquad \text{for } i \notin I$$

Given that $\mu \geq$ maximum $\{\bar{u}_i, i \in I, |\bar{v}_i|, i = 1, \ldots, l\}$, we then have, using (9.9), that $u_i^+ = \bar{u}_i$ for all $i \in I$, $u_i^+ = 0$ for all $i \notin I$, $u_i^- = \mu - u_i^+$ for all $i = 1, \ldots, m$, and $v_i^+ = (\mu + \bar{v}_i)/2$ and $v_i^- = (\mu - \bar{v}_i)/2$ for $i = 1, \ldots, l$ saisfy the foregoing KKT conditions. By Theorem 4.3.8 and the stated convexity assumptions, it follows that $(\bar{\mathbf{x}}, \bar{\mathbf{y}}, \bar{\mathbf{z}})$ solves (9.10), and, so, $\bar{\mathbf{x}}$ minimizes F_E. This completes the proof.

9.3.2 Example

Consider the problem of Example 9.1.2. The Lagrangian multiplier associated with the equality constraints in the KKT conditions at the optimum $\bar{\mathbf{x}} = (\frac{1}{2}, \frac{1}{2})^t$ is $\bar{v} = -2\bar{x}_1 = -2\bar{x}_2 = -1$. The function F_E defined by (9.8) for a given $\mu \geq 0$ is $F_E(\mathbf{x}) = (x_1^2 + x_2^2) + \mu|x_1 + x_2 - 1|$. When $\mu = 0$, this is minimized at (0, 0). For $\mu > 0$, minimizing

$F_E(\mathbf{x})$ is equivalent to minimizing $(x_1^2 + x_2^2 + \mu z)$, subject to $z \geq (x_1 + x_2 - 1)$ and $z \geq (-x_1 - x_2 + 1)$. The KKT conditions for the latter problem require that $2x_1 + (v^+ - v^-) = 0$, $2x_2 + (v^+ - v^-) = 0$, $\mu = v^+ + v^-$, $v^+[z - x_1 - x_2 + 1] = v^-[z + x_1 + x_2 - 1] = 0$ and, moreover, optimality dictates that $z = |x_1 + x_2 - 1|$. Now, if $(x_1 + x_2) < 1$, then we must have $z = -x_1 - x_2 + 1$ and $v^+ = 0$ and, hence, $v^- = \mu$, $x_1 = \mu/2$, and $x_2 = \mu/2$. This is a KKT point, provided that $0 \leq \mu < 1$. On the other hand, if $(x_1 + x_2) = 1$, then $z = 0$, $x_1 = x_2 = \frac{1}{2} = (v^- - v^+)/2$ and, therefore, $v^+ = (\mu - 1)/2$ and $v^- = (\mu + 1)/2$. This is a KKT point, provided that $\mu \geq 1$. However, if $(x_1 + x_2) > 1$, so that $z = x_1 + x_2 - 1$ and $v^- = 0$, we get $x_1 = x_2 = -v^+/2$, while $v^+ = \mu$. Hence, this means that $(x_1 + x_2) = -\mu > 1$, a contradiction to $\mu \geq 0$. Consequently, as μ increases from 0, the minimum of F_E occurs at $(\mu/2, \mu/2)$ until μ reaches the value 1, after which it remains at $(\frac{1}{2}, \frac{1}{2})$, the optimum to the original problem.

Geometric Interpretation for the Absolute Value Penalty Function

The absolute value (l_1) exact penalty function can be given a geometric interpretation similar in spirit to that illustrated in Figure 9.2. Let the perturbation function $v(\epsilon)$ be as illustrated therein. However, in the present case, we are interested in minimizing $f(\mathbf{x}) + \mu|h(\mathbf{x})|$ subject to $\mathbf{x} \in E_2$. For this, we wish to find the smallest value of k so that the contour $f + \mu|h| = k$ maintains contact with the epigraph of v in the (h, f) space. This is illustrated in Figure 9.6. Observe that under the condition of Theorem 9.3.1, $(\bar{\mathbf{x}}, \bar{\mathbf{u}}, \bar{\mathbf{v}})$ is a saddle point by Theorem 6.2.6, and so by Theorem 6.2.7, we have $v(\mathbf{y}) \geq v(\mathbf{0}) - (\bar{\mathbf{u}}^t, \bar{\mathbf{v}}^t)\mathbf{y}$ for all $\mathbf{y} \in E_{m+l}$. In the above example, this translates to the assertion that the hyperplane $f = v(0) - \bar{v}h = v(0) + h$ supports the epigraph of v from below at $[0, v(0)]$. Hence, as seen from Figure 9.6, for $\mu = 1$ (or greater than 1), minimizing F_E recovers the optimum to the original problem.

Although we have overcome the necessity of having to increase the penalty parameter μ to infinity to recover an optimum solution by using the l_1 penalty function, the admissible value of μ as prescribed by Theorem 9.3.1 that accomplishes this is as yet unspecified. As a result, we again need to examine a sequence of increasing μ values until, say, a KKT solution is obtained. Again, if μ is too small, then the penalty problem

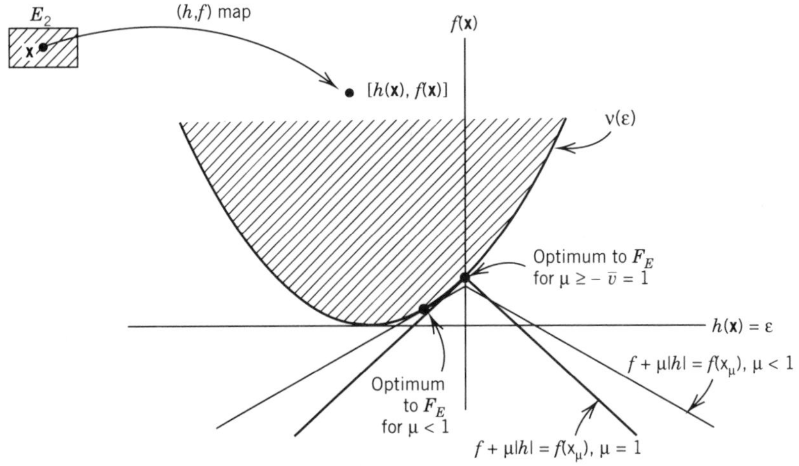

Figure 9.6 Geometric interpretation of the absolute penalty function.

might be unbounded; and if μ is too large, then ill-conditioning occurs. Moreover, a primary difference here is that we need to deal with a nondifferentiable objective function for minimizing F_E, which, as discussed in Section 8.9, does not enjoy solution procedures that are as efficient as for the differentiable case. However, as will be seen in the next chapter, l_1 penalty functions serve a very useful purpose as *merit functions*, which measure sufficient acceptable levels of descent to ensure convergence in other algorithmic approaches (e.g., successive quadratic programming methods) rather than playing a direct role in the direction finding process, itself.

Augmented Lagrangian Penalty Functions

Motivated by our discussion thus far, it is natural to raise the question whether we can design a penalty function that not only recovers an exact optimum for finite penalty parameter values but also enjoys the property of being differentiable. The *augmented Lagrangian penalty function* (ALAG), also known as the *multiplier penalty function*, is one such exact penalty function.

For simplicity, let us begin by discussing the case with only equality constraints, for which augmented Lagrangians were first introduced, and then readily extend the discussion to include inequality constraints as well. Toward this end, consider Problem P of minimizing $f(\mathbf{x})$ subject to $h_i(\mathbf{x}) = 0$ for $i = 1, \ldots, l$. We have seen that if we employ the quadratic penalty function problem to minimize $f(\mathbf{x}) + \mu \sum_{i=1}^{l} h_i^2(\mathbf{x})$, then we typically need to let $\mu \to \infty$ to obtain a constrained optimum for P. We might then be curious whether, if we were to shift the origin of the penalty term to $\boldsymbol{\theta} = (\theta_i, i = 1, \ldots, l)$ and consider the penalized objective function $f(\mathbf{x}) + \mu \sum_{i=1}^{l} [h_i(\mathbf{x}) - \theta_i]^2$ with respect to the problem in which the constraint right-hand sides are perturbed to $\boldsymbol{\theta}$ from $\mathbf{0}$, it could become possible to obtain a constrained minimum to the original problem without letting $\mu \to \infty$. In expanded form, this latter objective function is $f(\mathbf{x}) - \sum_{i=1}^{l} 2\mu\theta_i h_i(\mathbf{x}) + \mu \sum_{i=1}^{l} h_i^2(\mathbf{x}) + \mu \sum_{i=1}^{l} \theta_i^2$. Denoting $v_i = -2\mu\theta_i$ for $i = 1, \ldots, m$ and dropping the final constant term, this can be rewritten as

$$F_{\text{ALAG}}(\mathbf{x}, \mathbf{v}) = f(\mathbf{x}) + \sum_{i=1}^{l} v_i h_i(\mathbf{x}) + \mu \sum_{i=1}^{l} h_i^2(\mathbf{x}) \tag{9.11}$$

Now, observe that if $(\bar{\mathbf{x}}, \bar{\mathbf{v}})$ is a primal-dual KKT solution for P, then indeed at $\mathbf{v} = \bar{\mathbf{v}}$,

$$\nabla_{\mathbf{x}} F_{\text{ALAG}}(\bar{\mathbf{x}}, \bar{\mathbf{v}}) = \left[\nabla f(\bar{\mathbf{x}}) + \sum_{i=1}^{l} \bar{v}_i \nabla h_i(\bar{\mathbf{x}})\right] + 2\mu \sum_{i=1}^{l} h_i(\bar{\mathbf{x}}) \nabla h_i(\bar{\mathbf{x}}) = \mathbf{0} \tag{9.12}$$

for all values of μ; whereas this was not necessarily the case with the quadratic penalty function, unless $\nabla f(\bar{\mathbf{x}})$ was itself zero. Hence, whereas we needed to take $\mu \to \infty$ to recover $\bar{\mathbf{x}}$ in a limiting sense using the quadratic penalty function, it is conceivable that we only need to make μ large enough (under suitable regularity conditions as enunciated below) for the critical point $\bar{\mathbf{x}}$ of $F_{\text{ALAG}}(., \bar{\mathbf{v}})$ to turn out to be its (local) minimizer. In this respect, the last term in (9.11) turns out to be a local convexifier of the overall function.

Observe that the function (9.11) is the ordinary Lagrangian function augmented by the quadratic penalty term, hence the name *augmented Lagrangian penalty function*. Accordingly, (9.11) can be viewed as the usual quadratic penalty function with respect to the following problem that is equivalent to P:

$$\text{Minimize}\left\{f(\mathbf{x}) + \sum_{i=1}^{l} v_i h_i(\mathbf{x}) : h_i(\mathbf{x}) = 0 \text{ for } i = 1, \ldots, l\right\} \tag{9.13}$$

Alternatively, (9.11) can be viewed as a Lagrangian function for the following problem that is also equivalent to P:

$$\text{Minimize}\left\{f(\mathbf{x}) + \mu \sum_{i=1}^{l} h_i^2(\mathbf{x}) : h_i(\mathbf{x}) = 0 \text{ for } i = 1, \ldots, l\right\} \tag{9.14}$$

Since, from this viewpoint, (9.11) corresponds to the inclusion of a "multiplier-based term" in the quadratic penalty objective function, it is also sometimes called a *multiplier penalty function*. As we shall see shortly, these viewpoints lead to a rich theory and algorithmic felicity that is not present in either the pure quadratic penalty function or in the pure Lagrangian duality based approach.

The following result provides the basis by virtue of which the ALAG penalty function can be classified as an *exact penalty function*.

9.3.3 Theorem

Consider Problem P to minimize $f(\mathbf{x})$ subject to $h_i(\mathbf{x}) = 0$ for $i = 1, \ldots, l$, and let the KKT solution $(\bar{\mathbf{x}}, \bar{\mathbf{v}})$ satisfy the second-order sufficiency conditions for a local minimum (see Theorem 4.4.2). Then, there exists a $\bar{\mu}$ such that for $\mu \geq \bar{\mu}$, the ALAG penalty function $F_{\text{ALAG}}(., \bar{\mathbf{v}})$, defined with $\mathbf{v} = \bar{\mathbf{v}}$, also achieves a strict local minimum at $\bar{\mathbf{x}}$. In particular, if f is convex and h_i, $i = 1, \ldots, l$ are affine, then any minimizing solution $\bar{\mathbf{x}}$ for P also minimizes $F_{\text{ALAG}}(., \bar{\mathbf{v}})$ for all $\mu \geq 0$.

Proof

Since $(\bar{\mathbf{x}}, \bar{\mathbf{v}})$ is a KKT solution, we have, from (9.12), that $\nabla_{\mathbf{x}} F_{\text{ALAG}}(\bar{\mathbf{x}}, \bar{\mathbf{v}}) = \mathbf{0}$. Furthermore, letting $G(\bar{\mathbf{x}})$ denote the Hessian of $F_{\text{ALAG}}(., \bar{\mathbf{v}})$ at $\mathbf{x} = \bar{\mathbf{x}}$, we have

$$G(\bar{\mathbf{x}}) = \nabla^2 f(\bar{\mathbf{x}}) + \sum_{i=1}^{l} \bar{v}_i \nabla^2 h_i(\bar{\mathbf{x}}) + 2\mu \sum_{i=1}^{l} [h_i(\bar{\mathbf{x}})\nabla^2 h_i(\bar{\mathbf{x}}) + \nabla h_i(\bar{\mathbf{x}})\nabla h_i(\bar{\mathbf{x}})^t]$$

$$= \nabla^2 L(\bar{\mathbf{x}}) + 2\mu \sum_{i=1}^{l} \nabla h_i(\bar{\mathbf{x}})\nabla h_i(\bar{\mathbf{x}})^t \tag{9.15}$$

where $\nabla^2 L(\bar{\mathbf{x}})$ is the Hessian of the Lagrangian function for P with a multiplier vector $\bar{\mathbf{v}}$ at $\mathbf{x} = \bar{\mathbf{x}}$. From the second-order sufficiency conditions, we know that $\nabla^2 L(\bar{\mathbf{x}})$ is positive definite on the cone $C = \{\mathbf{d} \neq \mathbf{0} : \nabla h_i(\bar{\mathbf{x}})^t \mathbf{d} = 0 \text{ for } i = 1, \ldots, l\}$.

Now, on the contrary, if there does not exist a $\bar{\mu}$ such that $G(\bar{\mathbf{x}})$ is positive definite for $\mu \geq \bar{\mu}$, then it must be the case that, given any $\mu_k = k$, $k = 1, 2, \ldots$, there exists a $\mathbf{d}_k$ with $\|\mathbf{d}_k\| = 1$ such that

$$\mathbf{d}_k^t G(\bar{\mathbf{x}})\mathbf{d}_k = \mathbf{d}_k^t \nabla^2 L(\bar{\mathbf{x}})\mathbf{d}_k + 2k \sum_{i=1}^{l} [\nabla h_i(\bar{\mathbf{x}})^t \mathbf{d}_k]^2 \leq 0 \tag{9.16}$$

Since $\|\mathbf{d}_k\| = 1$ for all k, there exists a convergent subsequence for $\{\mathbf{d}_k\}$ with limit point $\bar{\mathbf{d}}$, where $\|\bar{\mathbf{d}}\| = 1$. Over this subsequence, since the first term in (9.16) approaches $\bar{\mathbf{d}}^t \nabla^2 L(\bar{\mathbf{x}})\bar{\mathbf{d}}$, a constant, we must have $\nabla h_i(\bar{\mathbf{x}})^t \bar{\mathbf{d}} = 0$ for all $i = 1, \ldots, l$ for (9.16) to hold for all k. Hence, $\bar{\mathbf{d}} \in C$. Moreover, since $\mathbf{d}_k^t \nabla^2 L(\bar{\mathbf{x}})\mathbf{d}_k \leq 0$ for all k by (9.16), we have $\bar{\mathbf{d}}^t \nabla^2 L(\bar{\mathbf{x}})\bar{\mathbf{d}} \leq 0$. This contradicts the second-order sufficiency conditions. Consequently, $G(\bar{\mathbf{x}})$ is positive definite for μ exceeding some value $\bar{\mu}$, and so, by Theorem 4.1.4, $\bar{\mathbf{x}}$ is a strict local minimum for $F_{ALAG}(\cdot, \bar{\mathbf{v}})$.

Finally, suppose that f is convex, h_i, $i = 1, \ldots, l$ are affine, and $\bar{\mathbf{x}}$ is optimal to P. By Lemma 5.1.4, there exist a set of Lagrange multipliers $\bar{\mathbf{v}}$ such that $(\bar{\mathbf{x}}, \bar{\mathbf{v}})$ is a KKT solution. As before, we have $\nabla_{\mathbf{x}} F_{ALAG}(\bar{\mathbf{x}}, \bar{\mathbf{v}}) = \mathbf{0}$, and since $F_{ALAG}(\cdot, \bar{\mathbf{v}})$ is convex for any $\mu \geq 0$, this completes the proof.

We remark here that without the second-order sufficiency conditions of Theorem 9.3.3, there might not exist any finite value of μ that will recover an optimum $\bar{\mathbf{x}}$ for Problem P, and it might be that we need to take $\mu \to \infty$ for this to occur. The following example from Fletcher [1987] illustrates this point.

9.3.4 Example

Consider Problem P to minimize $f(\mathbf{x}) = x_1^4 + x_1 x_2$, subject to $x_2 = 0$. Clearly, $\bar{\mathbf{x}} = (0, 0)^t$ is the optimal solution. From the KKT conditions, we also obtain $\bar{v} = 0$ as the unique Lagrangian multiplier. Since

$$\nabla^2 L(\bar{\mathbf{x}}) = \nabla^2 f(\bar{\mathbf{x}}) = \begin{bmatrix} 0 & 1 \\ 1 & 2\mu \end{bmatrix}$$

is indefinite, the second-order sufficiency condition does not hold at $(\bar{\mathbf{x}}, \bar{v})$. Now we have $F_{ALAG}(\mathbf{x}, \bar{v}) = x_1^4 + x_1 x_2 + \mu x_2^2 \equiv F(\mathbf{x})$, say. Note that for any $\mu > 0$

$$\nabla F(\mathbf{x}) = \begin{pmatrix} 4x_1^3 + x_2 \\ x_1 + 2\mu x_2 \end{pmatrix}$$

vanishes at $\bar{\mathbf{x}} = (0, 0)^t$ and at $\hat{\mathbf{x}} = (1/\sqrt{8\mu}, \; -1/(2\mu\sqrt{8\mu}))^t$. Furthermore,

$$\nabla^2 F(\mathbf{x}) = \begin{bmatrix} 12x_1^2 & 1 \\ 1 & 2\mu \end{bmatrix}$$

We see that $\nabla^2 F(\bar{\mathbf{x}})$ is indefinite and, hence, $\bar{\mathbf{x}}$ is not a local minimizer for any $\mu > 0$. However, $\nabla^2 F(\hat{\mathbf{x}})$ is positive definite, and $\hat{\mathbf{x}}$ is in fact the minimizer of F for all $\mu > 0$. Moreover, as $\mu \to \infty$, $\hat{\mathbf{x}}$ approaches the constrained minimum to Problem P.

9.3.5 Example

Consider the problem of Example 9.1.2. We have seen that $\bar{\mathbf{x}} = (\frac{1}{2}, \frac{1}{2})^t$, with $\bar{v} = -1$ is the unique KKT point and optimum for this problem. Furthermore, $\nabla^2 L(\bar{\mathbf{x}}) = \nabla^2 f(\bar{\mathbf{x}})$ is positive definite, and thus the second-order sufficiency condition holds at $(\bar{\mathbf{x}}, \bar{\mathbf{v}})$. Moreover, from (9.11), $F_{ALAG}(\mathbf{x}, \bar{\mathbf{v}}) = (x_1^2 + x_2^2) - (x_1 + x_2 - 1) + \mu(x_1 + x_2 - 1)^2 = (x_1 - \frac{1}{2})^2 + (x_2 - \frac{1}{2})^2 + \mu(x_1 + x_2 - 1)^2 + \frac{1}{2}$, which is clearly uniquely minimized at $\bar{\mathbf{x}} = (\frac{1}{2}, \frac{1}{2})^t$ for all $\mu \geq 0$. Hence, both assertions of Theorem 9.3.3 are verified.

Geometric Interpretation for the Augmented Lagrangian Penalty Function

The ALAG penalty function can be given a geometric interpretation similar to that illustrated in Figures 9.2 and 9.3. Let $v(\boldsymbol{\epsilon}) \equiv$ minimum $\{f(\mathbf{x}) : h(\mathbf{x}) = \boldsymbol{\epsilon}\}$ be the perturbation function as illustrated therein. Assume that $\bar{\mathbf{x}}$ is a regular point and that

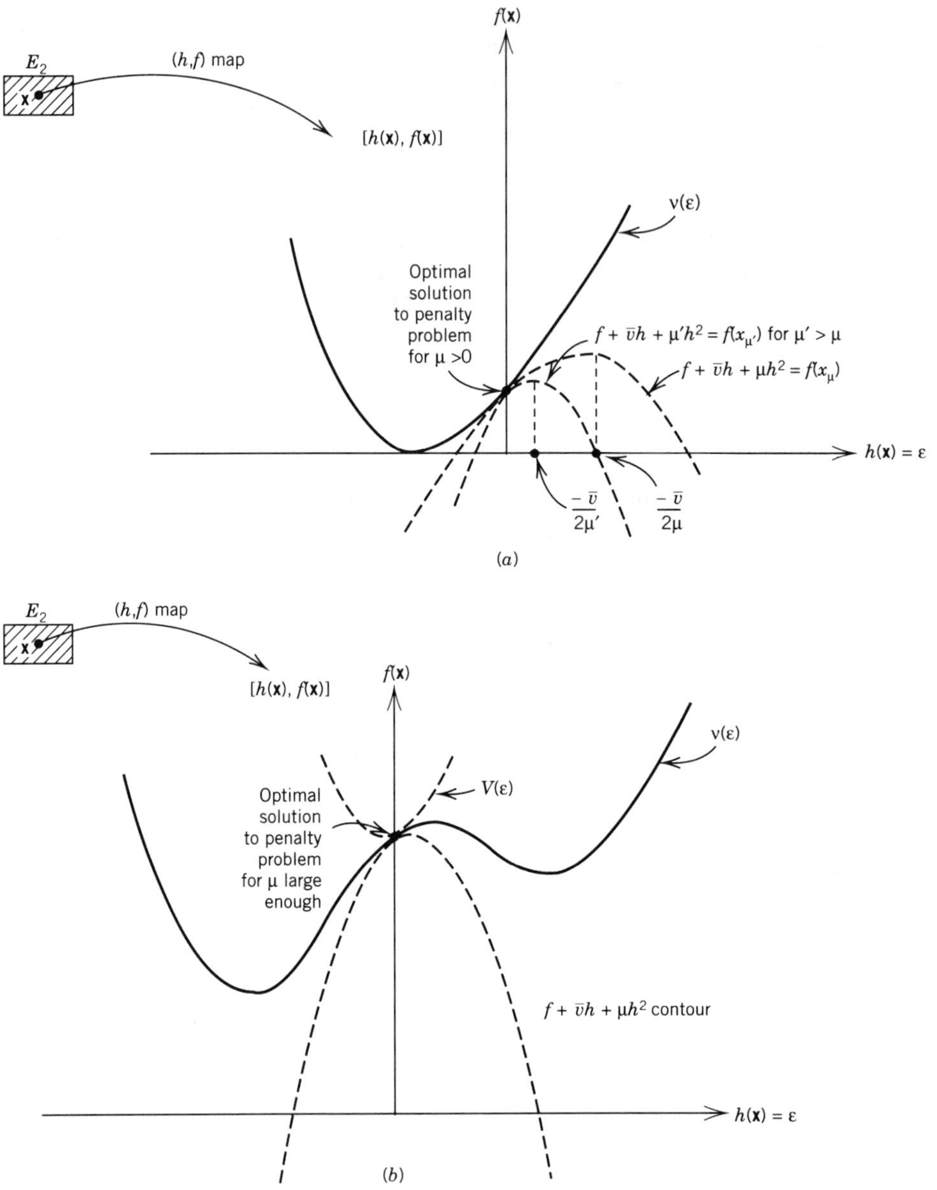

Figure 9.7 Geometric interpretation of the augmented Lagrangian penalty function.

$(\bar{\mathbf{x}}, \bar{\mathbf{v}})$ satisfies the second-order sufficiency condition for a strict local minimum. Then, it readily follows (see Exercise 9.14) in the spirit of Theorem 6.2.7 that $\nabla\nu(\mathbf{0}) = -\bar{\mathbf{v}}$.

Now consider the case of Figure 9.2, which illustrates Examples 9.1.2 and 9.3.5. For a given $\mu > 0$, the augmented Lagrangian penalty problem seeks the minimum of $f(\mathbf{x}) + \bar{v}h(\mathbf{x}) + \mu h^2(\mathbf{x})$ over $\mathbf{x} \in E_2$. This amounts to finding the smallest value of k for which the contour $f + \bar{v}h + \mu h^2 = k$ maintains contact with the epigraph of ν in the (h, f) space. This contour equation can be rewritten as $f = -\mu[h + (\bar{v}/2\mu)]^2 + [k + (\bar{v}^2/4\mu)]$ and represents a parabola whose axis is shifted to $h = -\bar{v}/2\mu$ relative to that at $h = 0$ as in Figure 9.2. Figure 9.7a illustrates this situation. Note that when $h = 0$, we have $f = k$ on this parabola. Moreover, when $k = \nu(0)$, which is the optimal objective value for Problem P, the parabola passes through the point $(0, \nu(0))$ in the (h, f) plane, and the slope of the tangent to the parabola at this point equals $-\bar{v}$. This slope therefore coincides with $\nu'(0) \equiv -\bar{v}$. Hence, as shown in Figure 9.7*a*, for any $\mu > 0$, the minimum of the augmented Lagrangian penalty function coincides with the optimum to Problem P.

For the nonconvex case, under the foregoing assumption that assures $\nabla\nu(\mathbf{0}) = -\bar{\mathbf{v}}$, a similar situation occurs; but in this case, this happens only when μ gets sufficiently large, as shown in Figure 9.7*b*. In contrast, the Lagrangian dual problem leaves a duality gap, and the quadratic penalty function needs to take $\mu \to \infty$ to recover the optimal solution.

To gain further insight, observe that we can write the minimization of the ALAG penalty function for any $\mathbf{v}$ in terms of the perturbation function as

$$\min_{\mathbf{x}} \{ f(\mathbf{x}) + \mathbf{v}^t h(\mathbf{x}) + \mu\|h(\mathbf{x})\|^2\} = \min_{(\mathbf{x},\boldsymbol{\epsilon})} \{ f(\mathbf{x}) + \mathbf{v}^t\boldsymbol{\epsilon} + \mu\|\boldsymbol{\epsilon}\|^2 : h(\mathbf{x}) = \boldsymbol{\epsilon}\}$$

$$= \min_{\boldsymbol{\epsilon}} \{\nu(\boldsymbol{\epsilon}) + \mathbf{v}^t\boldsymbol{\epsilon} + \mu\|\boldsymbol{\epsilon}\|^2\} \tag{9.17}$$

Note that if we take $\mathbf{v} = \bar{\mathbf{v}}$ and define

$$V(\boldsymbol{\epsilon}) = \nu(\boldsymbol{\epsilon}) + \bar{\mathbf{v}}^t\boldsymbol{\epsilon} + \mu\|\boldsymbol{\epsilon}\|^2$$

then, when μ get sufficiently large, V becomes a strictly convex function in the neighborhood of $\boldsymbol{\epsilon} = \mathbf{0}$; and, moreover, $\nabla V(\mathbf{0}) = \nabla\nu(\mathbf{0}) + \bar{\mathbf{v}} = \mathbf{0}$. Hence, we obtain a strict local minimum for V at $\boldsymbol{\epsilon} = \mathbf{0}$.

Schema of an Algorithm Using Augmented Lagrangian Functions: Method of Multipliers

The *method of multipliers* is an approach for solving nonlinear programming problems by using the augmented Lagrangian penalty function in a manner that combines the algorithmic aspects of both Lagrangian duality methods and penalty function methods. However, this is accomplished while gaining from both these concepts without being impaired by their respective shortcomings. The method adopts a dual ascent step similar to the subgradient optimization scheme for optimizing the Lagrangian dual; but, unlike the latter approach, the overall procedure produces both primal and dual solutions. The primal solution is produced via a penalty function minimization; but because of the properties of the ALAG penalty function, this can usually be accomplished without having to make the penalty parameter infinitely large and, hence, having to contend with the accompanying ill-conditioning effects. Moreover, we can employ efficient derivative based methods in minimizing the penalized objective function.

The fundamental schema of this type of algorithm is as follows. Consider the problem of minimizing $f(\mathbf{x})$ subject to the equality constraints $h_i(\mathbf{x}) = 0$ for $i = 1, \ldots, l$. (The extension to incorporating inequality constraints is relatively straightforward and is addressed in the following subsection.) Below, we outline the procedure first, and then provide some interpretations, motivations, and implementation comments. As is typically the case, the augmented Lagrangian function employed is of the form (9.11), except that each constraint is assigned its own specific penalty parameter μ_i, in lieu of a common parameter μ. Hence, constraint violations, and consequent penalizations, can be individually monitored. Accordingly, we replace (9.11) by

$$F_{\text{ALAG}}(\mathbf{x}, \mathbf{v}) = f(\mathbf{x}) + \sum_{i=1}^{l} v_i h_i(\mathbf{x}) + \sum_{i=1}^{l} \mu_i h_i^2(\mathbf{x}) \tag{9.18}$$

Initialization Select some initial Lagrangian multipliers $\mathbf{v} = \bar{\mathbf{v}}$ and positive values $\mu_1, \ldots, \mu_l$ for the penalty parameters. Let $\mathbf{x}_0$ be a null vector, and denote $\text{VIOL}(\mathbf{x}_0) = \infty$, where for any $\mathbf{x} \in E_n$, $\text{VIOL}(\mathbf{x}) = \text{maximum } \{|h_i(\mathbf{x})| : i = 1, \ldots, l\}$ is a measure of constraint violations. Put $k = 1$ and proceed to the "inner loop" of the algorithm.

Inner Loop: Penalty Function Minimization Solve the unconstrained problem to minimize $F_{\text{ALAG}}(\mathbf{x}, \bar{\mathbf{v}})$ subject to $\mathbf{x} \in E_n$, and let $\mathbf{x}_k$ denote the optimal solution obtained. If $\text{VIOL}(\mathbf{x}_k) = 0$, then stop with $\mathbf{x}_k$ as a KKT point. (Practically, one would terminate if $\text{VIOL}(\mathbf{x}_k)$ is lesser than some tolerance $\varepsilon > 0$.) Otherwise, if $\text{VIOL}(\mathbf{x}_k) \le \frac{1}{4}\text{VIOL}(\mathbf{x}_{k-1})$, proceed to the outer loop. On the other hand, if $\text{VIOL}(\mathbf{x}_k) > \frac{1}{4}\text{VIOL}(\mathbf{x}_{k-1})$, then, for each constraint $i = 1, \ldots, l$ for which $|h_i(\mathbf{x}_k)| > \frac{1}{4}\text{VIOL}(\mathbf{x}_{k-1})$, replace the corresponding penalty parameter μ_i by $10\mu_i$ and repeat this inner loop step.

Outer Loop: Lagrange Multiplier Update Replace $\bar{\mathbf{v}}$ by $\bar{\mathbf{v}}_{\text{new}}$, where

$$(\bar{\mathbf{v}}_{\text{new}})_i = \bar{v}_i + 2\mu_i h_i(\mathbf{x}_k) \qquad \text{for } i = 1, \ldots, l \tag{9.19}$$

Increment k by 1, and return to the inner loop.

The inner loop of the foregoing method is concerned with the minimization of the augmented Lagrangian penalty function. For this purpose, we can use $\mathbf{x}_{k-1}$ (for $k \ge 2$) as a starting solution and employ Newton's method (with line searches) in case the Hessian is available, or else use a quasi-Newton method if only gradients are available, or use some conjugate gradient method for relatively large-scale problems. If $\text{VIOL}(\mathbf{x}_k) = 0$, then $\mathbf{x}_k$ is feasible, and, moreover,

$$\nabla_{\mathbf{x}} F_{\text{ALAG}}(\mathbf{x}_k, \bar{\mathbf{v}}) = \nabla f(\mathbf{x}_k) + \sum_{i=1}^{l} \bar{v}_i \nabla h_i(\mathbf{x}_k) + \sum_{i=1}^{l} 2\mu_i h_i(\mathbf{x}_k) \nabla h_i(\mathbf{x}_k) = \mathbf{0} \tag{9.20}$$

implies that $\mathbf{x}_k$ is a KKT point. Whenever the revised iterate $\mathbf{x}_k$ of the inner loop does not improve the measure for constraint violations by the selected factor $\frac{1}{4}$, the penalty parameter is increased by a factor of 10. Hence, the outer loop will be visited after a finite number of iterations when the tolerance $\varepsilon > 0$ is used in the inner loop, since, as in Theorem 9.2.2, we have $h_i(\mathbf{x}_k) \to 0$ as $\mu_i \to \infty$ for $i = 1, \ldots, l$.

Observe that the foregoing argument holds regardless of the dual multiplier update scheme used in the outer loop, and that it is essentially related to using the standard quadratic penalty function approach on the equivalent problem (9.13). In fact, if we adopt this viewpoint, then the Lagrange multiplier estimate associated with the constraints in (9.13) is given by $2\mu_i h_i(\mathbf{x}_k)$ for $i = 1, \ldots, l$, as in (9.6). Since the relationship between the Lagrange multipliers of the original Problem P and its primal equivalent form (9.13) with $\mathbf{v} = \bar{\mathbf{v}}$ is that the Lagrange multiplier vector for P equals $\bar{\mathbf{v}}$ plus the Lagrange multiplier vector for (9.13), equation (9.19) then gives the corresponding estimate for the Lagrange multipliers associated with the constraints of P.

This observation can be reinforced more directly by the following interpretation. Note that having minimized $F_{\text{ALAG}}(\mathbf{x}, \bar{\mathbf{v}})$, we have (9.20) holding true. However, for $\mathbf{x}_k$ and $\bar{\mathbf{v}}$ to be a KKT solution, we want $\nabla_{\mathbf{x}} L(\mathbf{x}_k, \bar{\mathbf{v}}) = \mathbf{0}$, where $L(\mathbf{x}, \mathbf{v}) = f(\mathbf{x}) + \sum_{i=1}^{l} v_i h_i(\mathbf{x})$ is the Lagrangian function for Problem P. Hence, we can choose to revise $\bar{\mathbf{v}}$ to $\bar{\mathbf{v}}_{\text{new}}$ in a manner such that $\nabla f(\mathbf{x}_k) + \sum_{i=1}^{l} (\bar{\mathbf{v}}_{\text{new}})_i \nabla h_i(\mathbf{x}_k) = \mathbf{0}$. Superimposing this identity on (9.20) provides the update scheme (9.19).

Hence, from the viewpoint of the problem (9.13), convergence is obtained above in one of two ways. First, we might finitely determine a KKT point, as is frequently the case. Alternatively, viewing the foregoing algorithm as one of applying the standard quadratic penalty function approach, in spirit, to the equivalent sequence of problems of the type (9.13), each having particular estimates of the Lagrangian multipliers in the objective function, convergence is obtained by letting the penalty parameters approach fininity. In the latter case, the inner loop problems become increasingly ill-conditioned, and second-order methods become imperative.

There is an alternative Lagrangian duality-based interpretation of the update scheme (9.19) when $\mu_i \equiv \mu$ for all $i = 1, \ldots, l$ that leads to an improved procedure with a better overall rate of convergence. Recall that Problem P is also equivalent to the problem (9.14), where this equivalence now holds with respect to both primal and dual solutions. Moreover, the Lagrangian dual function for (9.14) is given by $\theta(\mathbf{v}) \equiv \min_{\mathbf{x}} \{F_{\text{ALAG}}(\mathbf{x}, \mathbf{v})\}$, where $F_{\text{ALAG}}(\mathbf{x}, \mathbf{v})$ is given by (9.11). Hence, at $\mathbf{v} = \bar{\mathbf{v}}$, the inner loop essentially evaluates $\theta(\bar{\mathbf{v}})$, determining an optimal solution $\mathbf{x}_k$. This yields $h(\mathbf{x}_k)$ as a subgradient of θ at $\mathbf{v} = \bar{\mathbf{v}}$. Consequently, the update $\bar{\mathbf{v}}_{\text{new}} = \bar{\mathbf{v}} + 2\mu h(\mathbf{x}_k)$ as characterized by (9.19) is simply a fixed step length subgradient direction-based iteration for the dual function.

This raises the issue that if a quadratically convergent Newton scheme, or a superlinearly convergent quasi-Newton method, is used for the inner loop optimization problems, then the advantage of using a second-order method is lost if we employ a linearly convergent gradient-based update scheme for the dual problem. As can be surmised, the convergence rate for the overall algorithm is intimately connected with that of the dual updating scheme. Assuming that the problem (9.14) has a local minimum at $\mathbf{x}^*$, that $\mathbf{x}^*$ is a regular point with the (unique) Lagrange multiplier vector $\mathbf{v}^*$, and that the Hessian of the Lagrangian with respect to $\mathbf{x}$ is positive definite at $(\mathbf{x}^*, \mathbf{v}^*)$, then it can be shown (see Exercise 9.15) that, in a local neighborhood of $\mathbf{v}^*$, the minimizing solution $\mathbf{x}(\mathbf{v})$ that evaluates $\theta(\mathbf{v})$ when $\mathbf{x}$ is confined to be near $\mathbf{x}^*$ is a continuously differentiable function. Hence, we have $\theta(\mathbf{v}) = F_{\text{ALAG}}[\mathbf{x}(\mathbf{v}), \mathbf{v}]$; and so, since $\nabla_{\mathbf{x}} F_{\text{ALAG}}[\mathbf{x}(\mathbf{v}), \mathbf{v}] = \mathbf{0}$, we have

$$\nabla\theta(\mathbf{v}) = \nabla_{\mathbf{v}} F_{\text{ALAG}}[\mathbf{x}(\mathbf{v}), \mathbf{v}] = h[\mathbf{x}(\mathbf{v})] \tag{9.21}$$

Denoting $\nabla h(\mathbf{x})$ and $\nabla\mathbf{x}(\mathbf{v})$ to be the Jacobians of h and $\mathbf{x}$, respectively, we then have

$$\nabla^2\theta(\mathbf{v}) = \nabla h[\mathbf{x}(\mathbf{v})]\nabla\mathbf{x}(\mathbf{v}) \tag{9.22a}$$

Differentiating the identity $\nabla_{\mathbf{x}} F_{\text{ALAG}}[\mathbf{x}(\mathbf{v}), \mathbf{v}] = \mathbf{0}$ with respect to $\mathbf{v}$, we obtain [see equation (9.12)] $\nabla_{\mathbf{x}}^2 F_{\text{ALAG}}[\mathbf{x}(\mathbf{v}), \mathbf{v}]\nabla\mathbf{x}(\mathbf{v}) + \nabla h[\mathbf{x}(\mathbf{v})]^t = \mathbf{0}$. Solving for $\nabla\mathbf{x}(\mathbf{v})$ (in the neighborhood of $\mathbf{v}^*$) from this equation and substituting into (9.22a) gives

$$\nabla^2\theta(\mathbf{v}) = -\nabla h[\mathbf{x}(\mathbf{v})]\{\nabla_{\mathbf{x}}^2 F_{\text{ALAG}}[\mathbf{x}(\mathbf{v}), \mathbf{v}]\}^{-1}\nabla h[\mathbf{x}(\mathbf{v})]^t \tag{9.22b}$$

At $\mathbf{x}^*$ and $\mathbf{v}^*$, the matrix $\nabla_{\mathbf{x}}^2 F_{\text{ALAG}}[\mathbf{x}^*, \mathbf{v}^*] = \nabla_{\mathbf{x}}^2 L(\mathbf{x}^*) + 2\mu\nabla h(\mathbf{x}^*)^t\nabla h(\mathbf{x}^*)$, and the eigenvalues of this matrix determine the rate of convergence of the stated gradient-based algorithm that uses the update scheme (9.19). Also, a second-order quasi-Newton type of an update scheme can employ an approximation $\mathbf{B}$ for $\{\nabla_{\mathbf{x}}^2 F_{\text{ALAG}}[(\mathbf{x}(\mathbf{v}), \mathbf{v}]\}^{-1}$ and, accordingly, determine $\bar{\mathbf{v}}_{\text{new}}$ as an approximation to $\bar{\mathbf{v}} - [\nabla^2\theta(\mathbf{v})]^{-1}\nabla\theta(\mathbf{v})$. Using (9.21) and (9.22b), this gives

$$\bar{\mathbf{v}}_{\text{new}} = \bar{\mathbf{v}} + [\nabla h(\mathbf{x})\mathbf{B}\nabla h(\mathbf{x})^t]^{-1}h(\mathbf{x}) \tag{9.23}$$

It is also interesting to note that as $\mu \to \infty$, the Hessian in (9.22b) approaches $\frac{1}{2}\mu\mathbf{I}$, and (9.23), then approaches (9.19). Hence, as the penalty parameter increases, the condition number of the dual problem becomes close to unity, implying a very rapid outer loop convergence, while that of the penalty problem in the inner loop becomes increasingly worse.

9.3.6 Example

Consider the problem of Example 9.1.2. Given any v, the inner loop of the method of multipliers evaluates $\theta(v) = \min_{\mathbf{x}} \{F_{ALAG}(\mathbf{x}, v)\}$, where $F_{ALAG}(\mathbf{x}, v) = x_1^2 + x_2^2 + v(x_1 + x_2 - 1) + \mu(x_1 + x_2 - 1)^2$. Solving $\nabla_{\mathbf{x}} F_{ALAG}(\mathbf{x}, v) = \mathbf{0}$ yields $x_1(v) = x_2(v) = (2\mu - v)/2(1 + 2\mu)$. The outer loop then updates the Lagrange multiplier according to (9.19), which gives $v_{new} = v + 2\mu[x_1(v) + x_2(v) - 1] = (v - 2\mu)/(1 + 2\mu)$. Note that as $\mu \to \infty$, $v_{new} \to -1$, the optimal Lagrange multiplier value.

Hence, if we commence this algorithm with $\bar{v} = 0$, and $\mu = 1$, the inner loop will determine $\mathbf{x}(0) = (\frac{1}{3}, \frac{1}{3})^t$, with $\text{VIOL}[\mathbf{x}(0)] = \frac{1}{3}$, and the outer loop will find $v_{new} = -\frac{2}{3}$. Next, at the second iteration, the inner loop solution will be obtained as $\mathbf{x}(-\frac{2}{3}) = (\frac{4}{9}, \frac{4}{9})^t$ with $\text{VIOL}[\mathbf{x}(-\frac{2}{3})] = \frac{1}{9} > \frac{1}{4}\text{VIOL}[\mathbf{x}(0)]$. Hence, we will increase μ to 10, and recompute the revised $\mathbf{x}(-\frac{2}{3}) = (\frac{31}{63}, \frac{31}{63})^t$ with $\text{VIOL}[\mathbf{x}(-\frac{2}{3})] = \frac{1}{63}$. The outer loop will then revise the Lagrange multiplier $\bar{v} = -\frac{2}{3}$ to $\bar{v}_{new} = -\frac{62}{63}$. The iterations will then progress in this fashion, using the foregoing formulas, until the constraint violation at the inner loop solution is acceptably small.

Extension to Include Inequality Constraints in the ALAG Penalty Function

Consider Problem P to minimize $f(\mathbf{x})$ subject to the constraints $g_i(\mathbf{x}) \leq 0$ for $i = 1, \ldots, m$ and $h_i(\mathbf{x}) = 0$ for $i = 1, \ldots, l$. The extension of the foregoing theory of augmented Lagrangians and the method of multipliers to this case, which also includes inequality constraints, is readily accomplished by equivalently writing the inequalities as the equations $g_i(\mathbf{x}) + s_i^2 = 0$ for $i = 1, \ldots, m$. Now suppose that $\bar{\mathbf{x}}$ is a KKT point for Problem P with optimal Lagrange multipliers $\bar{u}_i$, $i = 1, \ldots, m$, and $\bar{v}_i$, $i = 1, \ldots, l$, associated with the inequality and the equality constraints, respectively, and such that the *strict complementary slackness* condition holds, namely, that $\bar{u}_i g_i(\bar{\mathbf{x}}) = 0$ for all $i = 1, \ldots, m$, with $\bar{u}_i > 0$ for each $i \in I(\bar{\mathbf{x}}) = \{i : g_i(\bar{\mathbf{x}}) = 0\}$. Furthermore, suppose that the second-order sufficiency condition of Theorem 4.4.2 holds at $(\bar{\mathbf{x}}, \bar{\mathbf{u}}, \bar{\mathbf{v}})$, namely, that $\nabla^2 L(\bar{\mathbf{x}})$ is positive definite over the cone $C = \{\mathbf{d} \neq \mathbf{0} : \nabla g_i(\bar{\mathbf{x}})^t \mathbf{d} = 0$ for all $i \in I(\bar{\mathbf{x}})$, $\nabla h_i(\bar{\mathbf{x}})^t \mathbf{d} = 0$ for all $i = 1, \ldots, l\}$. (Note that $I^0 = \emptyset$ in Theorem 4.4.2 due to strict complementary slackness.) Then, it can be readily verified (see Exercise 9.16) that the conditions of Theorem 9.3.3 are satisfied for Problem P′ to minimize $f(\mathbf{x})$ subject to the equality constraints $g_i(\mathbf{x}) + s_i^2 = 0$ for $i = 1, \ldots, m$, and $h_i(\mathbf{x}) = 0$ for $i = 1, \ldots, l$, at the solution $(\bar{\mathbf{x}}, \bar{\mathbf{s}}, \bar{\mathbf{u}}, \bar{\mathbf{v}})$, where $\bar{s}_i^2 \equiv -g_i(\bar{\mathbf{x}})$ for all $i = 1, \ldots, m$. Hence, for μ large enough, the solution $(\bar{\mathbf{x}}, \bar{\mathbf{s}})$ will turn out to be a strict local minimizer for the following ALAG penalty function at $(\mathbf{u}, \mathbf{v}) \equiv (\bar{\mathbf{u}}, \bar{\mathbf{v}})$:

$$f(\mathbf{x}) + \sum_{i=1}^{m} u_i [g_i(\mathbf{x}) + s_i^2] + \sum_{i=1}^{l} v_i h_i(\mathbf{x}) + \mu \left[\sum_{i=1}^{m} (g_i(\mathbf{x}) + s_i^2)^2 + \sum_{i=1}^{l} h_i^2(\mathbf{x}) \right] \qquad (9.24)$$

The representation in (9.24) can be simplified into a more familiar form as follows. For a given penalty parameter $\mu > 0$, let $\theta(\mathbf{u}, \mathbf{v})$ represent the minimum of (9.24) over $(\mathbf{x}, \mathbf{s})$ for any given set of Lagrange multipliers $(\mathbf{u}, \mathbf{v})$. Now let us rewrite (9.24) more conveniently as follows:

$$f(\mathbf{x}) + \mu \sum_{i=1}^{m} \left[g_i(\mathbf{x}) + s_i^2 + \frac{u_i}{2\mu} \right]^2 - \sum_{i=1}^{m} \frac{u_i^2}{4\mu} + \sum_{i=1}^{l} v_i h_i(\mathbf{x}) + \mu \sum_{i=1}^{l} h_i^2(\mathbf{x}) \tag{9.25}$$

Hence, in computing $\theta(\mathbf{u}, \mathbf{v})$, we can minimize (9.25) over $(\mathbf{x}, \mathbf{s})$ by first minimizing $[g_i(\mathbf{x}) + s_i^2 + (u_i/2\mu)]$ over s_i in terms of $\mathbf{x}$ for each $i = 1, \ldots, m$ and then minimizing the resulting expression over $\mathbf{x} \in E_n$. The former task is easily accomplished by letting $s_i^2 = -[g_i(\mathbf{x}) + (u_i/2\mu)]$ if this is nonnegative and zero otherwise. Hence, we obtain

$$\begin{aligned} \theta(\mathbf{u}, \mathbf{v}) &= \underset{\mathbf{x}}{\text{minimum}} \left\{ f(\mathbf{x}) + \mu \sum_{i=1}^{m} \max{}^2 \left\{ g_i(\mathbf{x}) + \frac{u_i}{2\mu}, 0 \right\} - \sum_{i=1}^{m} \frac{u_i^2}{4\mu} \right. \\ &\qquad \left. + \sum_{i=1}^{l} v_i h_i(\mathbf{x}) + \mu \sum_{i=1}^{l} h_i^2(\mathbf{x}) \right\} \\ &= \underset{\mathbf{x}}{\text{minimum}} \{F_{\text{ALAG}}(\mathbf{x}, \mathbf{u}, \mathbf{v})\}, \text{ say} \end{aligned} \tag{9.26}$$

Similar to (9.11), the function $F_{\text{ALAG}}(\mathbf{x}, \mathbf{u}, \mathbf{v})$ is sometimes referred to as the *ALAG penalty function* itself in the presence of both inequality and equality constraints. In particular, in the context of the method of multipliers, the inner loop evaluates $\theta(\mathbf{u}, \mathbf{v})$, measures the constraint violations, and revises the penalty parameter(s) in an identical fashion as before. If $\mathbf{x}_k$ minimizes (9.26), then the subgradient component of $\theta(\mathbf{u}, \mathbf{v})$ corresponding to u_i at $(\mathbf{u}, \mathbf{v}) = (\bar{\mathbf{u}}, \bar{\mathbf{v}})$ is given by $2\mu \max \{g_i(\mathbf{x}_k) + (\bar{u}_i/2\mu), 0\}(1/2\mu) - (2\bar{u}_i/4\mu) = (-\bar{u}_i/2\mu) + \max \{g_i(\mathbf{x}_k) + (\bar{u}_i/2\mu), 0\}$. Adopting the fixed step length of 2μ along this subgradient direction as for the equality constrained case revises u_i to $(\bar{u}_{\text{new}})_i = \bar{u}_i + 2\mu [-(\bar{u}_i/2\mu) + \max \{g_i(\mathbf{x}_k) + (\bar{u}_i/2\mu), 0\}]$. Simplifying, this gives

$$(\bar{u}_{\text{new}})_i = \bar{u}_i + \max \{2\mu g_i(\mathbf{x}_k), -\bar{u}_i\} \qquad \text{for } i = 1, \ldots, m \tag{9.27}$$

Alternatively, we can adopt a second-order update scheme as for the case of equality constraints.

9.4 Barrier Function Methods

Similar to penalty functions, barrier functions are also used to transform a constrained problem into an unconstrained problem or into a sequence of unconstrained problems. These functions set a barrier against leaving the feasible region. If the optimal solution occurs at the boundary of the feasible region, the procedure moves from the interior to the boundary. The primal and barrier problems are formulated below.

Primal Problem

$$\begin{array}{ll} \text{Minimize} & f(\mathbf{x}) \\ \text{subject to} & \mathbf{g}(\mathbf{x}) \leq \mathbf{0} \\ & \mathbf{x} \in X \end{array}$$

where $\mathbf{g}$ is a vector function whose components are $g_1, \ldots, g_m$. Here, $f, g_1, \ldots, g_m$ are continuous functions on E_n, and X is a nonempty set in E_n. Note that any equality constraints, if present, are accommodated within the set X. Alternatively, in the case of linear equality constraints, we can possibly eliminate them after solving for some variables in terms of the others, thereby reducing the dimension of the problem. The reason why this treatment is necessary is that barrier function methods require the set $\{\mathbf{x}: g(\mathbf{x}) < \mathbf{0}\}$ to be nonempty, which would obviously not be possible if the equality constraints $h(\mathbf{x}) = \mathbf{0}$ were accommodated within the set of inequalities as $h(\mathbf{x}) \leq \mathbf{0}$ and $h(\mathbf{x}) \geq \mathbf{0}$.

Barrier Problem

$$\begin{array}{ll} \text{Minimize} & \theta(\mu) \\ \text{subject to} & \mu \geq 0 \end{array}$$

where $\theta(\mu) = \inf\{f(\mathbf{x}) + \mu B(\mathbf{x}) : \mathbf{g}(\mathbf{x}) < \mathbf{0}, \mathbf{x} \in X\}$. Here, B is a *barrier function* that is nonnegative and continuous over the region $\{\mathbf{x}: \mathbf{g}(\mathbf{x}) < \mathbf{0}\}$, and approaches ∞ as the boundary of the region $\{\mathbf{x}: \mathbf{g}(\mathbf{x}) \leq \mathbf{0}\}$ is approached from the interior. More specifically, the barrier function B is defined by

$$B(\mathbf{x}) = \sum_{i=1}^{m} \phi[g_i(\mathbf{x})] \tag{9.28a}$$

where ϕ is a function of one variable that is continuous over $\{y: y < 0\}$ and satisfies

$$\phi(y) \geq 0 \text{ if } y < 0 \qquad \text{and} \qquad \lim_{y \to 0^-} \phi(y) = \infty \tag{9.28b}$$

Thus, a typical barrier function is of the form

$$B(\mathbf{x}) = \sum_{i=1}^{m} \frac{-1}{g_i(\mathbf{x})} \qquad \text{or} \qquad B(\mathbf{x}) = -\sum_{i=1}^{m} ln[\min\{1, -g_i(\mathbf{x})\}] \tag{9.29a}$$

Note that the second barrier function in (9.29a) is not differentiable because of the term $\min\{1, -g_i(\mathbf{x})\}$. Actually, since the property (9.28b) for ϕ is essential only in a neighborhood of $y = 0$, it can be shown that the following barrier function, known as *Frisch's logarithmic barrier function*,

$$B(\mathbf{x}) = -\sum_{i=1}^{m} ln\,[-g_i(\mathbf{x})] \tag{9.29b}$$

also admits convergence in the sense of Theorem 9.3.4 given below.

We refer to the function $f(\mathbf{x}) + \mu B(\mathbf{x})$ as the *auxiliary function*. Ideally, we would like the function B to take value zero on the region $\{\mathbf{x}: \mathbf{g}(\mathbf{x}) < \mathbf{0}\}$ and value ∞ on its boundary. This would guarantee that we would not leave the region $\{\mathbf{x}: \mathbf{g}(\mathbf{x}) \leq \mathbf{0}\}$, provided that the minimization problem started at an interior point. However, this discontinuity poses serious difficulties for any computational procedure. Therefore, this ideal construction of B is replaced by the more realistic requirement that B is nonnegative and continuous over the region $\{\mathbf{x}: \mathbf{g}(\mathbf{x}) < \mathbf{0}\}$ and that it approaches infinity as the boundary is approached from the interior. Note that μB approaches the ideal barrier function described above as μ approaches zero. Given $\mu > 0$, evaluating $\theta(\mu) = \inf\{f(\mathbf{x}) + \mu B(\mathbf{x}) : \mathbf{g}(\mathbf{x}) < \mathbf{0}, \mathbf{x} \in X\}$ seems no simpler than solving the original problem because of the presence of the constraint $\mathbf{g}(\mathbf{x}) < \mathbf{0}$. However, as a result of the structure of B, if we start the optimization from a point in the region $S = \{\mathbf{x}: \mathbf{g}(\mathbf{x}) < \mathbf{0}\} \cap X$ and ignore the constraint $\mathbf{g}(\mathbf{x}) < \mathbf{0}$, we will reach an optimal point in S. This results from the fact that as we approach the boundary of $\{\mathbf{x}: \mathbf{g}(\mathbf{x}) \leq \mathbf{0}\}$ from within S, B approaches infinity, which will prevent us from leaving the set S. This is discussed further in the detailed statement of the barrier function method.

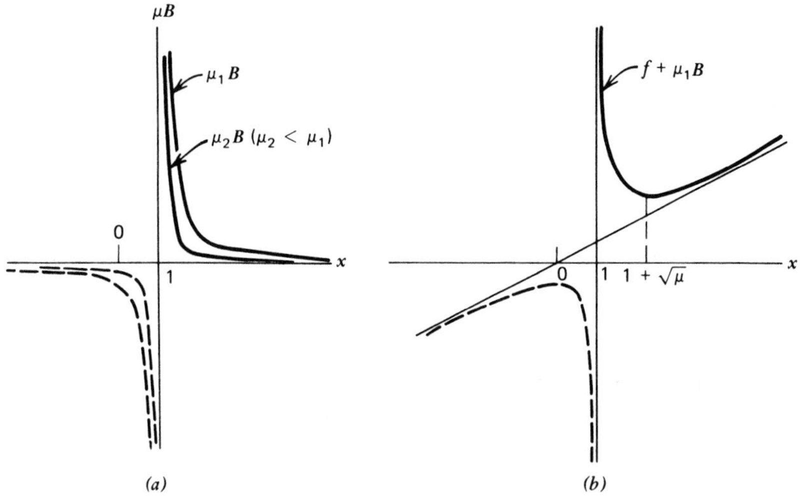

Figure 9.8 Barrier and auxiliary functions.

9.4.1 Example

Consider the following problem:

Minimize x
subject to $-x + 1 \leq 0$

Note that the optimal solution is $\bar{x} = 1$ and that $f(\bar{x}) = 1$. Consider the following barrier function:

$$B(x) = \frac{-1}{-x + 1} \qquad \text{for } x \neq 1$$

Figure 9.8*a* shows μB for various values of $\mu > 0$. Note that, as μ approaches zero, μB approaches a function that has value zero over $x > 1$ and infinity for $x = 1$. Figure 9.8*b* shows the auxiliary function $f(x) + \mu B(x) = x + [\mu/(x - 1)]$. The dotted functions in Figure 9.8 correspond to the set $\{x: g(x) > 0\}$ and do not affect the computational process.

Note that, for any given $\mu > 0$, the barrier problem is to minimize $x + \mu/(x - 1)$ over the region $x > 1$. The function $x + \mu/(x - 1)$ is convex over $x > 1$. Hence, if any of the techniques of Chapter 8 are used to minimize $x + \mu/(x - 1)$, starting with an interior point $x > 1$, we would obtain the optimal point $x_\mu = 1 + \sqrt{\mu}$. Note that $f(x_\mu) + \mu B(x_\mu) = 1 + 2\sqrt{\mu}$. Obviously, as $\mu \to 0$, $x_\mu \to \bar{x}$ and $f(x_\mu) + \mu B(x_\mu) \to f(\bar{x})$.

We now show the validity of using barrier functions for solving constrained problems by converting them into a single unconstrained problem or into a sequence of unconstrained problems. This is done in Theorem 9.4.3, but first the following lemma is needed.

9.4.2 Lemma

Let $f, g_1, \ldots, g_m$ be continuous functions on E_n, and let X be a nonempty closed set in E_n. Suppose that the set $\{\mathbf{x} \in X: \mathbf{g}(\mathbf{x}) < \mathbf{0}\}$ is not empty and that B is a barrier function of the form (9.28) and that it is continuous on $\{\mathbf{x}:\mathbf{g}(\mathbf{x}) < \mathbf{0}\}$. Furthermore, suppose that for any given $\mu > 0$, if $\{\mathbf{x}_k\}$ in X satisfies $\mathbf{g}(\mathbf{x}_k) < \mathbf{0}$ and $f(\mathbf{x}_k) + \mu B(\mathbf{x}_k) \to \theta(\mu)$, then $\{\mathbf{x}_k\}$ has a convergent subsequence.* Then,

1. For each $\mu > 0$, there exists an $\mathbf{x}_\mu \in X$ with $\mathbf{g}(\mathbf{x}_\mu) < \mathbf{0}$ such that
$$\theta(\mu) = f(\mathbf{x}_\mu) + \mu B(\mathbf{x}_\mu) = \inf \{f(\mathbf{x}) + \mu B(\mathbf{x}) : \mathbf{g}(\mathbf{x}) < \mathbf{0},\ \mathbf{x} \in X\}$$
2. $\inf \{f(\mathbf{x}):\mathbf{g}(\mathbf{x}) \leq \mathbf{0},\ \mathbf{x} \in X\} \leq \inf \{\theta(\mu):\mu > 0\}$
3. For $\mu > 0$, $f(\mathbf{x}_\mu)$ and $\theta(\mu)$ are nondecreasing functions of μ, and $B(\mathbf{x}_\mu)$ is a nonincreasing function of μ.

Proof

Fix $\mu > 0$. By definition of θ, there exists a sequence $\{\mathbf{x}_k\}$ with $\mathbf{x}_k \in X$ and $\mathbf{g}(\mathbf{x}_k) < \mathbf{0}$ such that $f(\mathbf{x}_k) + \mu B(\mathbf{x}_k) \to \theta(\mu)$. By assumption, $\{\mathbf{x}_k\}$ has a convergent subsequence $\{\mathbf{x}_k\}_{\mathcal{K}}$ with limit $\mathbf{x}_\mu$ in X. By continuity of $\mathbf{g}$, $\mathbf{g}(\mathbf{x}_\mu) \leq \mathbf{0}$. We show that $\mathbf{g}(\mathbf{x}_\mu) < \mathbf{0}$. If not, then $g_i(\mathbf{x}_\mu) = 0$ for some i; and since the barrier function B satisfies (9.28), for $k \in \mathcal{K}$, $B(\mathbf{x}_k) \to \infty$. Thus, $\theta(\mu) = \infty$, which is impossible, since $\{\mathbf{x}:\mathbf{x} \in X,\ g(\mathbf{x}) < \mathbf{0}\}$ is assumed not empty. Therefore, $\theta(\mu) = f(\mathbf{x}_\mu) + \mu B(\mathbf{x}_\mu)$, where $\mathbf{x}_\mu \in X$ and $\mathbf{g}(\mathbf{x}_\mu) < \mathbf{0}$, so that part 1 holds. Now since $B(\mathbf{x}) \geq 0$ if $\mathbf{g}(\mathbf{x}) < \mathbf{0}$, then, for $\mu \geq 0$, we have

$$\begin{aligned}\theta(\mu) &= \inf \{f(\mathbf{x}) + \mu B(\mathbf{x}) : \mathbf{g}(\mathbf{x}) < \mathbf{0},\ \mathbf{x} \in X\}\\ &\geq \inf \{f(\mathbf{x}) : \mathbf{g}(\mathbf{x}) < \mathbf{0},\ \mathbf{x} \in X\}\\ &\geq \inf \{f(\mathbf{x}) : \mathbf{g}(\mathbf{x}) \leq \mathbf{0},\ \mathbf{x} \in X\}\end{aligned}$$

Since the above inequality holds for each $\mu \geq 0$, part 2 follows. To show part 3, let $\mu > \lambda > 0$. Since $B(\mathbf{x}) \geq 0$ if $\mathbf{g}(\mathbf{x}) < \mathbf{0}$, then $f(\mathbf{x}) + \mu B(\mathbf{x}) \geq f(\mathbf{x}) + \lambda B(\mathbf{x})$ for each $\mathbf{x} \in X$ with $\mathbf{g}(\mathbf{x}) < \mathbf{0}$. Thus, $\theta(\mu) \geq \theta(\lambda)$. Noting part 1, there exist $\mathbf{x}_\mu$ and $\mathbf{x}_\lambda$ such that

$$f(\mathbf{x}_\mu) + \mu B(\mathbf{x}_\mu) \leq f(\mathbf{x}_\lambda) + \mu B(\mathbf{x}_\lambda) \tag{9.30}$$

$$f(\mathbf{x}_\lambda) + \lambda B(\mathbf{x}_\lambda) \leq f(\mathbf{x}_\mu) + \lambda B(\mathbf{x}_\mu) \tag{9.31}$$

Adding (9.30) and (9.31) and rearranging, we get $(\mu - \lambda)[B(x_\mu) - B(\mathbf{x}_\lambda)] \leq 0$. Since $\mu - \lambda > 0$, then $B(\mathbf{x}_\mu) \leq B(\mathbf{x}_\lambda)$. Substituting in (9.31), it follows that $f(\mathbf{x}_\lambda) \leq f(\mathbf{x}_\mu)$. Thus part (3) holds, and the proof is complete.

From the above lemma, θ is a nondecreasing function of μ so that $\inf_{\mu>0} \theta(\mu) = \lim_{\mu\to 0^+} \theta(\mu)$. Theorem 9.4.3 below shows that the optimal solution to the primal problem is indeed equal to $\lim_{\mu\to 0^+} \theta(\mu)$, so that it could be solved by a single problem of the form to minimize $f(\mathbf{x}) + \mu B(\mathbf{x})$ subject to $\mathbf{x} \in X$, where μ is sufficiently small, or it can be solved through a sequence of problems of the above form with decreasing values of μ.

9.4.3 Theorem

Let $f:E_n \to E_1$, and $\mathbf{g}:E_n \to E_m$ be continuous functions, and let X be a nonempty closed set in E_n. Suppose that the set $\{\mathbf{x} \in X:\mathbf{g}(\mathbf{x}) < \mathbf{0}\}$ is not empty. Furthermore, suppose that the primal problem to minimize $f(\mathbf{x})$ subject to $\mathbf{g}(\mathbf{x}) \leq \mathbf{0}$, $\mathbf{x} \in X$ has an optimal

*This assumption holds if $\{\mathbf{x} \in X:\mathbf{g}(\mathbf{x}) \leq \mathbf{0}\}$ is compact.

solution $\bar{\mathbf{x}}$ with the following property. Given any neighborhood N around $\bar{\mathbf{x}}$, there exists an $\mathbf{x} \in X \cap N$ such that $\mathbf{g}(\mathbf{x}) < \mathbf{0}$. Then,

$$\text{minimum}\,\{f(\mathbf{x}) : \mathbf{g}(\mathbf{x}) \leq \mathbf{0}, \mathbf{x} \in X\} = \lim_{\mu \to 0^+} \theta(\mu) = \inf_{\mu > 0} \theta(\mu)$$

Letting $\theta(\mu) = f(\mathbf{x}_\mu) + \mu B(\mathbf{x}_\mu)$, where $\mathbf{x}_\mu \in X$ and $\mathbf{g}(\mathbf{x}_\mu) < \mathbf{0}$,* then the limit of any convergent subsequence of $\{\mathbf{x}_\mu\}$ is an optimal solution to the primal problem and, furthermore, $\mu B(\mathbf{x}_\mu) \to 0$ as $\mu \to 0^+$.

Proof

Let $\bar{\mathbf{x}}$ be an optimal solution to the primal problem satisfying the stated property, and let $\varepsilon > 0$. By continuity of f and by the assumption of the theorem, there is an $\hat{\mathbf{x}} \in X$ with $\mathbf{g}(\hat{\mathbf{x}}) < \mathbf{0}$ such that $f(\bar{\mathbf{x}}) + \varepsilon > f(\hat{\mathbf{x}})$. Then, for $\mu > 0$

$$f(\bar{\mathbf{x}}) + \varepsilon + \mu B(\hat{\mathbf{x}}) > f(\hat{\mathbf{x}}) + \mu B(\hat{\mathbf{x}}) \geq \theta(\mu)$$

Taking the limit as $\mu \to 0^+$, it follows that $f(\bar{\mathbf{x}}) + \varepsilon \geq \lim_{\mu \to 0^+} \theta(\mu)$. Since this inequality holds for each $\varepsilon > 0$, we get $f(\bar{\mathbf{x}}) \geq \lim_{\mu \to 0^+} \theta(\mu)$. In view of part 2 of Lemma 9.4.2, $f(\bar{\mathbf{x}}) = \lim_{\mu \to 0^+} \theta(\mu)$.

For $\mu \to 0^+$, and since $B(\mathbf{x}_\mu) \geq 0$ and $\mathbf{x}_\mu$ is feasible to the original problem, it follows that

$$\theta(\mu) = f(\mathbf{x}_\mu) + \mu B(\mathbf{x}_\mu) \geq f(\mathbf{x}_\mu) \geq f(\bar{\mathbf{x}})$$

Now taking the limit as $\mu \to 0^+$, and noting that $f(\bar{\mathbf{x}}) = \lim_{\mu \to 0^+} \theta(\mu)$, it follows that both $f(\mathbf{x}_\mu)$ and $f(\mathbf{x}_\mu) + \mu B(\mathbf{x}_\mu)$ approach $f(\bar{\mathbf{x}})$. Therefore, $\mu B(\mathbf{x}_\mu) \to 0$ as $\mu \to 0^+$. Furthermore, if $\{\mathbf{x}_\mu\}$ has a convergent subsequence with limit $\mathbf{x}'$, then $f(\mathbf{x}') = f(\bar{\mathbf{x}})$. Since $\mathbf{x}_\mu$ is feasible to the original problem for each μ, it follows that $\mathbf{x}'$ is also feasible and, hence, optimal. This completes the proof.

Note that the points $\{\mathbf{x}_\mu\}$ generated belong to the interior of the set $\{\mathbf{x} : \mathbf{g}(\mathbf{x}) \leq \mathbf{0}\}$ for each μ. It is for this reason that barrier function methods are sometimes also referred to as *interior penalty function methods*.

KKT Lagrange Multipliers at Optimality

Under certain regularity conditions, the barrier interior penalty method also produces a sequence of Lagrange multiplier estimates that converge to an optimal set of Lagrange multipliers. To see this, consider Problem P to minimize $f(\mathbf{x})$ subject $g_i(\mathbf{x}) \leq 0$ for $i = 1, \ldots, m$, and $\mathbf{x} \in X \equiv E_n$. (The case where X might include additional inequality or equality constraints is easily treated in a likewise fashion; see Exercise 9.18.) The barrier function problem is then given by

$$\underset{\mathbf{x}}{\text{minimize}} \left\{ f(\mathbf{x}) + \mu \sum_{i=1}^{m} \phi[g_i(\mathbf{x})] : \mathbf{g}(\mathbf{x}) < \mathbf{0} \right\} \tag{9.32}$$

where ϕ satisfies (9.28). Let us assume that f, $\mathbf{g}$, and ϕ are continuously differentiable, that the conditions of Lemma 9.4.2 and Theorem 9.4.3 hold and, furthermore, that the optimum $\bar{\mathbf{x}}$ to P obtained as an accumulation point of $\{\mathbf{x}_\mu\}$ is a regular point. Without

*Assumptions under which such a point $\mathbf{x}_\mu$ exists are given in Lemma 9.4.2.

loss of generality, assume that $\{\mathbf{x}_\mu\} \to \bar{\mathbf{x}}$ itself. Then, if $I = \{i : g_i(\bar{\mathbf{x}}) = 0\}$ is the index set of active constraints at $\bar{\mathbf{x}}$, we know that there exists a unique set of Lagrange multipliers $\bar{u}_i$, $i = 1, \ldots, m$, such that

$$\nabla f(\bar{\mathbf{x}}) + \sum_{i=1}^{m} \bar{u}_i \nabla g_i(\bar{\mathbf{x}}) = \mathbf{0} \qquad \bar{u}_i \geq 0 \text{ for } i = 1, \ldots, m, \bar{u}_i = 0 \text{ for } i \notin I \tag{9.33}$$

Now, since $\mathbf{x}_\mu$ solves the problem (9.32) with $\mathbf{g}(\mathbf{x}_\mu) < \mathbf{0}$, we have, for all $\mu > 0$,

$$\nabla f(\mathbf{x}_\mu) + \sum_{i=1}^{m} (\mathbf{u}_\mu)_i \nabla g_i(\mathbf{x}_\mu) = \mathbf{0}$$
$$\text{where } (\mathbf{u}_\mu)_i \equiv \mu\phi'[g_i(\mathbf{x}_\mu)], i = 1, \ldots, m \tag{9.34}$$

As $\mu \to 0^+$, we have $\{\mathbf{x}_\mu\} \to \bar{\mathbf{x}}$; and, so, $(\mathbf{u}_\mu)_i \to 0$ for $i \notin I$. Moreover, since $\bar{\mathbf{x}}$ is regular and all the functions f, $\mathbf{g}$, and ϕ are continuously differentiable, we have, from (9.33) and (9.34), that $(\mathbf{u}_\mu)_i \to \bar{u}_i$ for $i \in I$ as well. Hence, $\mathbf{u}_\mu$ provides an estimate for the Lagrange multipliers that approaches the optimal set of Lagrange multipliers $\bar{\mathbf{u}}$ as $\mu \to 0^+$. Therefore, for example, if $\phi(y) = -1/y$, we have $\phi'(y) = 1/y^2$; and hence,

$$(\mathbf{u}_\mu)_i = \frac{\mu}{g_i(\mathbf{x}_\mu)^2} \to \bar{u}_i \qquad \text{for all } i = 1, \ldots, m, \text{ as } \mu \to 0^+ \tag{9.35}$$

Computational Difficulties Associated with Barrier Functions

The use of barrier functions for solving constrained nonlinear programming problems also faces several computational difficulties. First, the search must start with a point $\mathbf{x} \in X$ with $\mathbf{g}(\mathbf{x}) < \mathbf{0}$. For some problems, finding such a point may not be an easy task. In Exercise 9.22, a procedure is described for finding such a starting point. Also, because of the structure of the barrier function B, and for small values of the parameter μ, most search techniques may face serious ill-conditioning and difficulties with round-off errors while solving the problem to minimize $f(\mathbf{x}) + \mu B(\mathbf{x})$ over $\mathbf{x} \in X$, especially as the boundary of the region $\{\mathbf{x} : \mathbf{g}(\mathbf{x}) \leq \mathbf{0}\}$ is approached. In fact, as the boundary is approached, and since search techniques often use discrete steps, a step leading outside the region $\{\mathbf{x} : \mathbf{g}(\mathbf{x}) \leq \mathbf{0}\}$ may indicate a decrease in the value of $f(\mathbf{x}) + \mu B(\mathbf{x})$, a false success. Thus, an explicit check of the value of the constraint function $\mathbf{g}$ is needed to guarantee that we do not leave the feasible region.

To see the potential ill-conditioning effect more formally, we can examine the eigenstructure of the Hessian of the objective function in (9.32) at the optimum $\mathbf{x}_\mu$ as $\mu \to 0^+$. Noting (9.34), and assuming that f, $\mathbf{g}$ and ϕ are twice continuously differentiable, we get that this Hessian is given by

$$\left[\nabla^2 f(\mathbf{x}_\mu) + \sum_{i=1}^{m} (\mathbf{u}_\mu)_i \nabla^2 g_i(\mathbf{x}_\mu)\right] + \mu \sum_{i=1}^{m} \phi''[g_i(\mathbf{x}_\mu)] \nabla g_i(\mathbf{x}_\mu) \nabla g_i(\mathbf{x}_\mu)^t \tag{9.36}$$

As $\mu \to 0^+$, we have $\{\mathbf{x}_\mu\} \to \bar{\mathbf{x}}$ (possibly over a convergent subsequence); and, assuming that $\bar{\mathbf{x}}$ is a regular point, we have $\mathbf{u}_\mu \to \bar{\mathbf{u}}$, the optimal set of Lagrange multipliers. Hence, the term within $[\cdot]$ in (9.36) approaches $\nabla^2 L(\bar{\mathbf{x}})$. The remaining term is potentially problematic. For example, if $\phi(y) = -1/y$, then $\phi''(y) = -2/y^3$; and so, from (9.35), this term becomes

$$-2 \sum_{i \in I} \frac{(\mathbf{u}_\mu)_i}{g_i(\mathbf{x}_\mu)} \nabla g_i(\mathbf{x}_\mu) \nabla g_i(\mathbf{x}_\mu)^t$$

leading to an identical severe ill-conditioning effect as described for the exterior penalty functions. Hence, it is again imperative to use suitable second-order Newton, quasi-Newton, or conjugate gradient methods for solving problem (9.32).

Summary of Barrier Function Methods

We describe below a scheme using barrier functions for optimizing a nonlinear programming problem of the form to minimize $f(\mathbf{x})$ subject to $\mathbf{g}(\mathbf{x}) \leq \mathbf{0}$ and $\mathbf{x} \in X$. The barrier function B used must satisfy (9.28).

The problem stated at step 1 below incorporates the constraint $\mathbf{g}(\mathbf{x}) < \mathbf{0}$. If $\mathbf{g}(\mathbf{x}_k) < \mathbf{0}$, and since the barrier function approaches infinity as the boundary of the region $G = \{\mathbf{x}:\mathbf{g}(\mathbf{x}) < \mathbf{0}\}$ is reached, then the constraint $\mathbf{g}(\mathbf{x}) < \mathbf{0}$ may be ignored, provided that an unconstrained optimization technique is used that will ensure that the resulting optimal point $\mathbf{x}_{k+1} \in G$. However, as most line search methods use discrete steps, if we are close to the boundary, a step could lead to a point outside the feasible region where the value of the barrier function B is a large negative number. Therefore, the problem could be treated as an unconstrained optimization problem only if an explicit check for feasibility is made.

Initialization Step Let $\varepsilon > 0$ be a termination scalar, and choose a point $\mathbf{x}_1 \in X$ with $\mathbf{g}(\mathbf{x}_1) < \mathbf{0}$. Let $\mu_1 > 0$, $\beta \in (0, 1)$, let $k = 1$, and go to the main step.

Main Step

1. Starting with $\mathbf{x}_k$, solve the following problem:

$$\begin{array}{ll} \text{Minimize} & f(\mathbf{x}) + \mu_k B(\mathbf{x}) \\ \text{subject to} & g(\mathbf{x}) < \mathbf{0} \\ & \mathbf{x} \in X \end{array}$$

   Let $\mathbf{x}_{k+1}$ be an optimal solution, and go to step 2.

2. If $\mu_k B(\mathbf{x}_{k+1}) < \varepsilon$, stop. Otherwise, let $\mu_{k+1} = \beta\mu_k$, replace k by $k + 1$, and repeat step 1.

9.4.4 Example

Consider the following problem:

$$\begin{array}{ll} \text{Minimize} & (x_1 - 2)^4 + (x_1 - 2x_2)^2 \\ \text{subject to} & x_1^2 - x_2 \leq 0 \end{array}$$

Here, $X = E_2$. We solve the problem by using the barrier function method with $B(\mathbf{x}) = -1/(x_1^2 - x_2)$. The summary of the computations is shown in Table 9.2, along with the Lagrange multiplier estimates as given by (9.35), and the progress of the algorithm is shown in Figure 9.9. The procedure is started with $\mu_1 = 10.0$, and the unconstrained minimization of the function $\theta(\mu_1)$ was started from a feasible point (0.0, 1.0). The parameter β is taken as 0.10. After six iterations, the point $\mathbf{x}_7^t = (0.94389, 0.89635)$, $u_\mu = 3.385$, is reached, where $\mu_6 B(\mathbf{x}_7) = 0.0184$ and the algorithm is terminated. The reader can verify that this point is very close to the optimum. Noting that μ_k is decreasing, the reader can observe from Table 9.2 that $f(\mathbf{x}_{\mu_k})$ and $\theta(\mu_k)$ are nondecreasing functions of μ_k. Likewise, $B(\mathbf{x}_{\mu_k})$ is a nonincreasing function of μ_k. Furthermore, $\mu_k B(\mathbf{x}_{\mu_k})$ converges to zero as asserted in Theorem 9.4.3.

Table 9.2 Summary of Computations for the Barrier Function Method

Iteration k	μ_k	$\mathbf{x}_{\mu_k} = \mathbf{x}_{k+1}$	$f(\mathbf{x}_{k+1})$	$B(\mathbf{x}_{k+1})$	$\theta(\mu_k)$	$\mu_k B(\mathbf{x}_{\mu_k})$	$u_{\mu k}$
1	10.0	(0.7079, 1.5315)	8.3338	0.9705	18.0388	9.705	9.419051
2	1.0	(0.8282, 1.1098)	3.8214	2.3591	6.1805	2.3591	5.565503
3	0.1	(0.8989, 0.9638)	2.5282	6.4194	3.1701	0.6419	4.120815
4	0.01	(0.9294, 0.9162)	2.1291	19.0783	2.3199	0.1908	3.639818
5	0.001	(0.9403, 0.9011)	2.0039	59.0461	2.0629	0.0590	3.486457
6	0.0001	(0.94389, 0.89635)	1.9645	184.4451	1.9829	0.0184	3.385000

9.5 A Polynomial-Time Algorithm for Linear Programming Based on a Barrier Function

Consider the following pair of primal (P) and dual (D) *linear programming problems* (see Section 2.7):

$$
\begin{array}{llll}
P: \text{ Minimize} & \mathbf{c}'\mathbf{x} & D: \text{ Maximize} & \mathbf{b}'\mathbf{v} \\
\quad\;\; \text{subject to} & \mathbf{A}\mathbf{x} = \mathbf{b} & \quad\;\; \text{subject to} & \mathbf{A}'\mathbf{v} + \mathbf{u} = \mathbf{c} \\
 & \mathbf{x} \geq \mathbf{0} & & \mathbf{u} \geq \mathbf{0}, \mathbf{v} \text{ unrestricted}
\end{array}
$$

where $\mathbf{A}$ is an $m \times n$ matrix and, without loss of generality, has rank $m < n$, and where $\mathbf{v}$ and $\mathbf{u}$ are Lagrange multipliers associated with the equality and the inequality constraints of P, respectively. Let us assume that P has an optimal solution $\mathbf{x}^*$, and let the corresponding optimal Lagrange multipliers be $\mathbf{v}^*$ and $\mathbf{u}^*$. Denoting by $\mathbf{w}$ the triplet $(\mathbf{x}, \mathbf{u}, \mathbf{v})$, we have that $\mathbf{w}^* = (\mathbf{x}^*, \mathbf{u}^*, \mathbf{v}^*)$ satisfies the following KKT conditions for Problem P:

$$\mathbf{A}\mathbf{x} = \mathbf{b} \quad \mathbf{x} \geq \mathbf{0} \tag{9.37a}$$

$$\mathbf{A}'\mathbf{v} + \mathbf{u} = \mathbf{c} \quad \mathbf{u} \geq \mathbf{0}, \mathbf{v} \text{ unrestricted} \tag{9.37b}$$

$$\mathbf{u}'\mathbf{x} = 0 \tag{9.37c}$$

Now let us assume that there exists a $\bar{\mathbf{w}} = (\bar{\mathbf{x}}, \bar{\mathbf{u}}, \bar{\mathbf{v}})$ satisfying (9.37a) and (9.37b) with $\bar{\mathbf{x}} > \mathbf{0}$ and $\bar{\mathbf{u}} > \mathbf{0}$. Consider the following barrier function problem BP based on Frisch's logarithmic barrier function (9.29b), where the equality constraints are used to define the set X:

$$BP: \quad \text{Minimize} \left\{ \mathbf{c}'\mathbf{x} - \mu \sum_{j=1}^{n} \ln(x_j) : \mathbf{A}\mathbf{x} = \mathbf{b}, (\mathbf{x} > \mathbf{0}) \right\} \tag{9.38}$$

The KKT conditions for BP require that we find $\mathbf{x}$ and $\mathbf{v}$ such that $\mathbf{A}\mathbf{x} = \mathbf{b}$, $(\mathbf{x} > \mathbf{0})$, and $\mathbf{A}'\mathbf{v} = \mathbf{c} - \mu[1/x_1, \ldots, 1/x_n]'$. Following (9.34), we can denote $\mathbf{u} = \mu[1/x_1, \ldots, 1/x_n]'$ as our Lagrange multiplier estimate for P, given any $\mu > 0$. Defining the diagonal matrices $\mathbf{X} \equiv \text{diag}\{x_1, \ldots, x_n\}$ and $\mathbf{U} \equiv \text{diag}\{u_1, \ldots, u_n\}$, and denoting $\mathbf{e} = (1, \ldots, 1)'$, we can rewrite the KKT conditions for BP as follows:

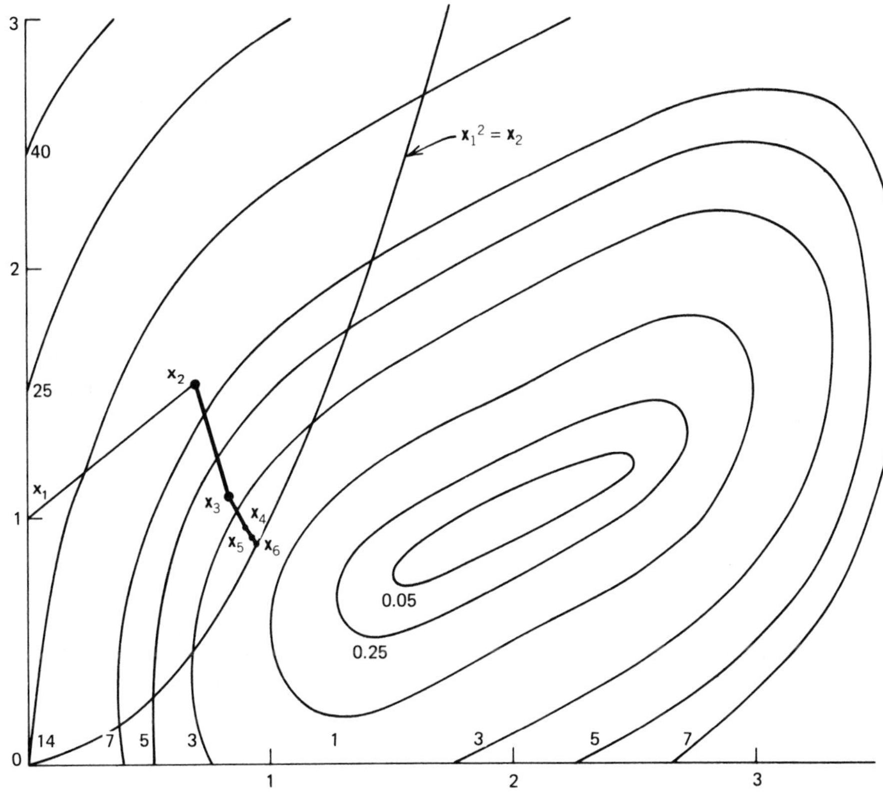

Figure 9.9 Illustration of the barrier function method.

$$\mathbf{Ax} = \mathbf{b} \tag{9.39a}$$

$$\mathbf{A}'\mathbf{v} + \mathbf{u} = \mathbf{c} \tag{9.39b}$$

$$\mathbf{u} = \mu\mathbf{X}^{-1}\mathbf{e} \quad \text{or} \quad \mathbf{XUe} = \mu\mathbf{e} \tag{9.39c}$$

Now, given any $\mu > 0$, by part 1 of Lemma 9.4.2 and by the strict convexity of the objective function of Problem BP over the feasible region, there exists an unique $\mathbf{x}_\mu > \mathbf{0}$ that solves BP. Correspondingly, from (9.39), since $\mathbf{A}'$ has full column rank, we obtain unique accompanying values of $\mathbf{u}_\mu$ and $\mathbf{v}_\mu$. Following Theorem 9.4.3, we can show that the triplet $\mathbf{w}_\mu \equiv (\mathbf{x}_\mu, \mathbf{u}_\mu, \mathbf{v}_\mu)$ approaches an optimal primal–dual solution to P as $\mu \to 0^+$. The trajectory $\mathbf{w}_\mu$, for $\mu > 0$, is known as the *central path* because of the interiority forced by the barrier function. Note that, from (9.39 a, b), the standard linear programming duality gap $\mathbf{c}'\mathbf{x} - \mathbf{b}'\mathbf{v}$ equals $\mathbf{u}'\mathbf{x}$, the violation in the complementary slackness condition. Moreover, from (9.39c), we get $\mathbf{u}'\mathbf{x} = \mu\mathbf{x}'\mathbf{X}^{-1}\mathbf{e} = n\mu$. Hence, we have, from (9.39), that

$$\mathbf{c}'\mathbf{x} - \mathbf{b}'\mathbf{v} = \mathbf{u}'\mathbf{x} = n\mu \tag{9.40}$$

which approaches zero as $\mu \to 0^+$.

Instead of actually finding $\mathbf{w}_\mu$ for each $\mu > 0$ in a sequence approaching zero, we instead start with a $\bar{\mu} > 0$ and a $\bar{\mathbf{w}}$ sufficiently close to $\mathbf{w}_{\bar{\mu}}$ and then revise $\bar{\mu}$ to $\hat{\mu} = \beta\bar{\mu}$ for some $0 < \beta < 1$. Correspondingly, we shall then use a single Newton step to obtain a revised solution $\hat{\mathbf{w}}$ that is also sufficiently close to $\mathbf{w}_{\hat{\mu}}$. Motivated by (9.39) and (9.40), by defining $\mathbf{w} = (\mathbf{x}, \mathbf{u}, \mathbf{v})$ to be "sufficiently close" to $\mathbf{w}_\mu$ whenever

$$\mathbf{A}\mathbf{x} = \mathbf{b} \qquad \mathbf{A}'\mathbf{v} + \mathbf{u} = \mathbf{c} \qquad \|\mathbf{X}\mathbf{U}\mathbf{e} - \mu\mathbf{e}\| \le \theta\mu \quad \text{with } \mathbf{u}'\mathbf{x} = n\mu, \text{ where } 0 \le \theta < 0.5 \tag{9.41}$$

we shall show that such a sequence of iterates $\mathbf{w}$ will then converge to an optimal primal–dual solution.

Toward this end, suppose that we are given a $\bar{\mu} > 0$ and a $\bar{\mathbf{w}} = (\bar{\mathbf{x}}, \bar{\mathbf{u}}, \bar{\mathbf{v}})$ with $\bar{\mathbf{x}} > \mathbf{0}$ and $\bar{\mathbf{u}} > \mathbf{0}$, such that (9.41) holds. (Later, we show how one can obtain such a solution to initialize the algorithm.) Let us now reduce μ to $\hat{\mu} = \beta\bar{\mu}$, where $0 < \beta < 1$, and examine the KKT system (9.39) written for $\mu = \hat{\mu}$. Denote this KKT system as $\mathbf{H}(\mathbf{w}) = \mathbf{0}$. The first-order approximation of this at $\mathbf{w} = \bar{\mathbf{w}}$ is given by $\mathbf{H}(\bar{\mathbf{w}}) + \mathbf{J}(\bar{\mathbf{w}})(\mathbf{w} - \bar{\mathbf{w}}) = \mathbf{0}$, where $\mathbf{J}(\bar{\mathbf{w}})$ is the Jacobian of $\mathbf{H}(\mathbf{w})$ at $\mathbf{w} = \bar{\mathbf{w}}$. Denoting $\mathbf{d}_{\mathbf{w}} = (\mathbf{w} - \bar{\mathbf{w}})$, a Newton step at $\bar{\mathbf{w}}$ will take us to the point $\hat{\mathbf{w}} = \bar{\mathbf{w}} + \mathbf{d}_{\mathbf{w}}$ where $\mathbf{J}(\bar{\mathbf{w}})\mathbf{d}_{\mathbf{w}} = -\mathbf{H}(\bar{\mathbf{w}})$. Writing $\mathbf{d}'_{\mathbf{w}} \equiv (\mathbf{d}'_{\mathbf{x}}, \mathbf{d}'_{\mathbf{u}}, \mathbf{d}'_{\mathbf{v}})$, we have, from (9.39), that the equation $\mathbf{J}(\bar{\mathbf{w}})\mathbf{d}_{\mathbf{w}} = -\mathbf{H}(\bar{\mathbf{w}})$ is given as follows:

$$\mathbf{A}\mathbf{d}_{\mathbf{x}} = \mathbf{0} \tag{9.42a}$$

$$\mathbf{A}'\mathbf{d}_{\mathbf{v}} + \mathbf{d}_{\mathbf{u}} = \mathbf{0} \tag{9.42b}$$

$$\bar{\mathbf{U}}\mathbf{d}_{\mathbf{x}} + \bar{\mathbf{X}}\mathbf{d}_{\mathbf{u}} = \hat{\mu}\mathbf{e} - \bar{\mathbf{X}}\bar{\mathbf{U}}\mathbf{e} \tag{9.42c}$$

The linear system (9.42) can be solved using some stable, factored-form implementation (see Appendix A.2). In explicit form, we get $\mathbf{d}_{\mathbf{u}} = -\mathbf{A}'\mathbf{d}_{\mathbf{v}}$ from (9.42b) and, hence, $\mathbf{d}_{\mathbf{x}} = \bar{\mathbf{U}}^{-1}[\hat{\mu}\mathbf{e} - \bar{\mathbf{X}}\bar{\mathbf{U}}\mathbf{e}] + \bar{\mathbf{U}}^{-1}\bar{\mathbf{X}}\mathbf{A}'\mathbf{d}_{\mathbf{v}}$. Substituting this in (9.42a) gives

$$\mathbf{d}_{\mathbf{v}} = -[\mathbf{A}\bar{\mathbf{U}}^{-1}\bar{\mathbf{X}}\mathbf{A}']^{-1}\mathbf{A}\bar{\mathbf{U}}^{-1}[\hat{\mu}\mathbf{e} - \bar{\mathbf{X}}\bar{\mathbf{U}}\mathbf{e}] \tag{9.43a}$$

$$\mathbf{d}_{\mathbf{u}} = -\mathbf{A}'\mathbf{d}_{\mathbf{v}} \tag{9.43b}$$

$$\mathbf{d}_{\mathbf{x}} = \bar{\mathbf{U}}^{-1}[\hat{\mu}\mathbf{e} - \bar{\mathbf{X}}\bar{\mathbf{U}}\mathbf{e} - \bar{\mathbf{X}}\mathbf{d}_{\mathbf{u}}] \tag{9.43c}$$

where the inverses exist, since $\bar{\mathbf{x}} > \mathbf{0}$, $\bar{\mathbf{u}} > \mathbf{0}$, and rank($\mathbf{A}$) $= m$. The generation of $(\hat{\mu}, \hat{\mathbf{w}})$ from $(\bar{\mu}, \bar{\mathbf{w}})$ in the above fashion describes one step of the algorithm. This procedure can now be repeated until the duality gap $\mathbf{u}'\mathbf{x} = n\mu$ [see equations (9.37), (9.40), and (9.41)] is small enough. The algorithmic steps are summarized below.

Summary of the Primal–Dual Path-Following Algorithm

Initialization Select a starting solution $\bar{\mathbf{w}} = (\bar{\mathbf{x}}, \bar{\mathbf{u}}, \bar{\mathbf{v}})$ with $\bar{\mathbf{x}} > \mathbf{0}$, $\bar{\mathbf{u}} > \mathbf{0}$, and a penalty parameter $\mu = \bar{\mu}$ such that $\bar{\mathbf{w}}$ satisfies (9.41) with $\mu = \bar{\mu}$. (Later, we show how this can be accomplished for a general linear program.) Furthermore, let θ, δ, and β satisfy (9.44) of Theorem 9.5.2 stated below; (for example, let $\theta = \delta = 0.35$ and $\beta = 1 - (\delta/\sqrt{n})$). Put $k = 0$, let $(\mu_0, \mathbf{w}_0) = (\bar{\mu}, \bar{\mathbf{w}})$ and proceed to the main step.

Main Step Let $(\bar{\mu}, \bar{\mathbf{w}}) = (\mu_k, \mathbf{w}_k)$. If $\mathbf{c}'\bar{\mathbf{x}} - \mathbf{b}'\bar{\mathbf{v}} \equiv n\bar{\mu} < \epsilon$ for some tolerance $\epsilon > 0$, then terminate with $\bar{\mathbf{w}}$ as an (ϵ) optimal primal–dual solution. Otherwise, let $\hat{\mu} = \beta\bar{\mu}$, compute $\mathbf{d}'_{\mathbf{w}} = (\mathbf{d}'_{\mathbf{x}}, \mathbf{d}'_{\mathbf{u}}, \mathbf{d}'_{\mathbf{v}})$ by (9.43) [or through (9.42)], and set $\hat{\mathbf{w}} = \bar{\mathbf{w}} + \mathbf{d}_{\mathbf{w}}$. Put $(\mu_{k+1}, \mathbf{w}_{k+1}) = (\hat{\mu}, \hat{\mathbf{w}})$, increment k by 1, and repeat the main step.

9.5.1 Example

Consider the linear program to minimize $\{3x_1 - x_2 : x_1 + 2x_2 = 2, x_1 \ge 0, x_2 \ge 0\}$. The optimal solution is easily seen to be $\mathbf{x}^* = (0, 1)'$, with $\mathbf{v}^* = -0.5$ and $\mathbf{u}^* = (3.5, 0)'$. Suppose that we initialize the algorithm with $\bar{\mathbf{x}} = (\frac{2}{9}, \frac{8}{9})'$, $\bar{\mathbf{u}} = (4, 1)'$, $\bar{v} = -1$, and

$\bar{\mu} = \frac{8}{9}$. This solution can be verified to satisfy (9.39) and, therefore, $\bar{\mathbf{w}} \equiv \mathbf{w}_{\bar{\mu}}$ lies on the central path. Hence, in particular, $(\mu_0, \mathbf{w}_0) \equiv (\bar{\mu}, \bar{\mathbf{w}})$ satisfies (9.41). The present duality gap from (9.40) is given by $\bar{\mathbf{u}}^t\bar{\mathbf{x}} = 2\bar{\mu} = 1.7777777$. Let $\theta = \delta = 0.35$ and $\beta = 1 - (0.35/\sqrt{2}) = 0.7525127$.

Now let us calculate $(\mu_1, \mathbf{w}_1) \equiv (\hat{\mu}, \hat{\mathbf{w}})$ according to $\hat{\mu} = \beta\bar{\mu}$ and $\hat{\mathbf{w}} = \bar{\mathbf{w}} + \mathbf{d}_{\mathbf{w}}$, where $\mathbf{d}_{\mathbf{w}}^t = (\mathbf{d}_{\mathbf{x}}^t, \mathbf{d}_{\mathbf{u}}^t, \mathbf{d}_{\mathbf{v}}^t)$ is the solution to (9.42). Hence, we obtain $\mu_1 = \hat{\mu} = \beta\bar{\mu} = 0.6689001$, and $\mathbf{d}_{\mathbf{w}}$ solves the system

$$\begin{aligned}
d_{x_1} + 2d_{x_2} &= 0 \\
d_v + d_{u_1} &= 0 \\
2d_v + d_{u_2} &= 0 \\
4d_{x_1} + (\tfrac{2}{9})d_{u_1} &= \hat{\mu} - \bar{x}_1\bar{u}_1 = -0.2199887 \\
d_{x_2} + (\tfrac{8}{9})d_{u_2} &= \hat{\mu} - \bar{x}_2\bar{u}_2 = -0.2199887
\end{aligned}$$

Solving, we obtain $d_v = 0.1370698$, $\mathbf{d}_{\mathbf{u}} = (-0.1370698, -0.2741396)^t$, and $\mathbf{d}_{\mathbf{x}} = (-0.0473822, 0.0236909)^t$. This yields $\mathbf{w}_1 = \hat{\mathbf{w}} = (\hat{\mathbf{x}}, \hat{\mathbf{u}}, \hat{\mathbf{v}}) = \bar{\mathbf{w}} + \mathbf{d}_{\mathbf{x}}$, where $\hat{\mathbf{x}} = (0.17484, 0.9125797)^t$, $\hat{\mathbf{u}} = (3.8629302, 0.7258604)^t$, and $\hat{v} = -0.8629302$. Note that the duality gap has reduced to $\hat{\mathbf{u}}^t\hat{\mathbf{x}} = 1.3378 = 2\hat{\mu}$. Also, observe that $\hat{\mathbf{X}}\hat{\mathbf{U}}\mathbf{e} = (\hat{x}_1\hat{u}_1, \hat{x}_2\hat{u}_2)^t = (0.6753947, 0.6624054)^t \neq \hat{\mu}\mathbf{e}$ and, hence, (9.39c) no longer holds and we are not located on the central path. However, $\hat{\mathbf{w}}$ is close enough to $\mathbf{w}_{\hat{\mu}}$ in the sense of (9.41), since $\|\hat{\mathbf{X}}\hat{\mathbf{U}}\mathbf{e} - \hat{\mu}\mathbf{e}\| = 0.009176 \leq \theta\hat{\mu} = 0.234115$.

We ask the reader in Exercise 9.30 to continue the iterations until (near) optimality is attained. Below, we establish the main result.

9.5.2 Theorem

Let $\bar{\mathbf{w}} = (\bar{\mathbf{x}}, \bar{\mathbf{u}}, \bar{\mathbf{v}})$ be such that $\bar{\mathbf{x}} > \mathbf{0}$, $\bar{\mathbf{u}} > \mathbf{0}$, and (9.41) is satisfied with $\mu = \bar{\mu}$. Consider $\hat{\mu} = \beta\bar{\mu}$, where $0 < \beta < 1$ satisfies

$$\beta = 1 - \frac{\delta}{\sqrt{n}} \quad \text{and where} \quad 0 < \delta < \sqrt{n}, \quad \frac{\theta^2 + \delta^2}{2(1 - \theta)} \leq \theta\left(1 - \frac{\delta}{\sqrt{n}}\right), 0 \leq \theta \leq \frac{1}{2} \tag{9.44}$$

(For example, we can take $\theta = \delta = 0.35$.) Then, the solution $\hat{\mathbf{w}} = \bar{\mathbf{w}} + \mathbf{d}_{\mathbf{w}}$ produced by taking a unit step length along the Newton direction $\mathbf{d}_{\mathbf{w}}$ given by (9.43) [or (9.42)] has $\hat{\mathbf{x}} > \mathbf{0}$, $\hat{\mathbf{u}} > \mathbf{0}$, and also satisfies (9.41) with $\mu = \hat{\mu}$. Hence, starting with $(\mu_0, \mathbf{w}_0)$ satisfying (9.41), the algorithm generates a sequence $\{(\mu_k, \mathbf{w}_k)\}$ that satisfies (9.41) at each iteration, and such that any accumulation point of $\{\mathbf{w}_k\}$ solves the original linear program.

Proof

First, note from (9.42a) and (9.42b) that

$$\mathbf{A}\hat{\mathbf{x}} = \mathbf{A}\bar{\mathbf{x}} = \mathbf{b} \qquad \text{and} \qquad \mathbf{A}^t\hat{\mathbf{v}} + \hat{\mathbf{u}} = \mathbf{A}^t\bar{\mathbf{v}} + \bar{\mathbf{u}} = \mathbf{c} \tag{9.45}$$

Now let us show that $\|\hat{\mathbf{X}}\hat{\mathbf{U}}\mathbf{e} - \hat{\mu}\mathbf{e}\| \leq \theta\hat{\mu}$. Denoting $\mathbf{D}_{\mathbf{x}} \equiv \text{diag}\,\{d_{x_1}, \ldots, d_{x_n}\}$, $\mathbf{D}_{\mathbf{u}} \equiv \text{diag}\,\{d_{u_1}, \ldots, d_{u_n}\}$, and $\mathbf{D} = (\bar{\mathbf{X}}^{-1}\bar{\mathbf{U}})^{1/2}$, we have, from (9.42c), that

$$(\hat{\mathbf{X}}\hat{\mathbf{U}}\mathbf{e} - \hat{\mu}\mathbf{e}) \equiv (\bar{\mathbf{X}} + \mathbf{D}_{\mathbf{x}})(\bar{\mathbf{U}} + \mathbf{D}_{\mathbf{u}})\mathbf{e} - \hat{\mu}\mathbf{e} = \mathbf{D}_{\mathbf{x}}\mathbf{D}_{\mathbf{u}}\mathbf{e} \tag{9.46}$$

Moreover, multiplying (9.42c) throughout by $(\bar{\mathbf{X}}\bar{\mathbf{U}})^{-1/2}$, we get

$$[\mathbf{DD_x e} + \mathbf{D}^{-1}\mathbf{D_u e}] = (\bar{\mathbf{X}}\bar{\mathbf{U}})^{-1/2}[\bar{\mu}\mathbf{e} - \bar{\mathbf{X}}\bar{\mathbf{U}}\mathbf{e}] \tag{9.47}$$

Hence,

$$\|\hat{\mathbf{X}}\hat{\mathbf{U}}\mathbf{e} - \hat{\mu}\mathbf{e}\| = \|\mathbf{D_x D_u e}\| = \|(\mathbf{DD_x})(\mathbf{D}^{-1}\mathbf{D_u})\mathbf{e}\| = \left|\sqrt{\sum_j (\pi_j\gamma_j)^2}\right|$$

where $\mathbf{DD_x} \equiv \text{diag}\{\pi_1, \ldots, \pi_n\}$, say, and $\mathbf{D}^{-1}\mathbf{D_u} \equiv \text{diag}\{\gamma_1, \ldots, \gamma_n\}$, say. Using (9.47) and denoting $\overline{xu}_{\min} \equiv \min\{\bar{x}_j\bar{u}_j, j = 1, \ldots, n\}$, we get

$$\begin{aligned}
\|\hat{\mathbf{X}}\hat{\mathbf{U}}\mathbf{e} - \hat{\mu}\mathbf{e}\| &\le \sum_j \pi_j\gamma_j \le \frac{1}{2}\sum_j (\pi_j + \gamma_j)^2 = \frac{1}{2}\|\mathbf{DD_x e} + \mathbf{D}^{-1}\mathbf{D_u e}\|^2 \\
&= \frac{1}{2}\|(\bar{\mathbf{X}}\bar{\mathbf{U}})^{-1/2}[\bar{\mu}\mathbf{e} - \bar{\mathbf{X}}\bar{\mathbf{U}}\mathbf{e}]\|^2 = \frac{1}{2}\sum_j \frac{(\bar{\mu} - \bar{x}_j\bar{u}_j)^2}{\bar{x}_j\bar{u}_j} \le \frac{\Sigma_j(\bar{\mu} - \bar{x}_j\bar{u}_j)^2}{2\overline{xu}_{\min}} \\
&= \frac{\|\bar{\mathbf{X}}\bar{\mathbf{U}}\mathbf{e} - \bar{\mu}\mathbf{e}\|^2}{2\overline{xu}_{\min}}
\end{aligned} \tag{9.48}$$

But by using $\mathbf{e}'[\bar{\mathbf{X}}\bar{\mathbf{U}}\mathbf{e} - \bar{\mu}\mathbf{e}] = \bar{\mathbf{x}}'\bar{\mathbf{u}} - n\bar{\mu} = 0$, we obtain, from (9.41), that

$$\begin{aligned}
\|\bar{\mathbf{X}}\bar{\mathbf{U}}\mathbf{e} - \hat{\mu}\mathbf{e}\|^2 &= \|\bar{\mathbf{X}}\bar{\mathbf{U}}\mathbf{e} - \bar{\mu}\mathbf{e} + (\bar{\mu} - \hat{\mu})\mathbf{e}\|^2 = \|\bar{\mathbf{X}}\bar{\mathbf{U}}\mathbf{e} - \bar{\mu}\mathbf{e}\|^2 + n(\bar{\mu} - \hat{\mu})^2 \\
&\le \theta^2\bar{\mu}^2 + n\bar{\mu}^2(1 - \beta)^2 = \bar{\mu}^2[\theta^2 + n(1 - \beta)^2]
\end{aligned} \tag{9.49}$$

Furthermore, from (9.41), since $\|\bar{\mathbf{X}}\bar{\mathbf{U}}\mathbf{e} - \bar{\mu}\mathbf{e}\| \le \theta\bar{\mu}$ implies that $|\bar{x}_j\bar{u}_j - \bar{\mu}| \le \theta\bar{\mu}$, we get that $\bar{\mu} - \bar{x}_j\bar{u}_j \le \theta\bar{\mu}$ for all $j = 1, \ldots, n$, or that

$$\bar{x}_j\bar{u}_j \ge \bar{\mu}(1 - \theta) \qquad \text{for all } j = 1, \ldots, n, \text{ and so, } \overline{xu}_{\min} \ge \bar{\mu}(1 - \theta) \tag{9.50}$$

Using (9.49) and (9.50) in (9.48), and noting (9.44), we derive

$$\|\hat{\mathbf{X}}\hat{\mathbf{U}}\mathbf{e} - \hat{\mu}\mathbf{e}\| \le \frac{\bar{\mu}^2[\theta^2 + n(1 - \beta)^2]}{2\bar{\mu}(1 - \theta)} = \frac{\bar{\mu}[\theta^2 + \delta^2]}{2(1 - \theta)} \le \beta\bar{\mu}\theta = \hat{\mu}\theta \tag{9.51}$$

Hence, we have $\|\hat{\mathbf{X}}\hat{\mathbf{U}}\mathbf{e} - \hat{\mu}\mathbf{e}\| \le \theta\hat{\mu}$. Let us now show that $\hat{\mathbf{u}}'\hat{\mathbf{x}} = n\hat{\mu}$. Using $\hat{\mathbf{w}} = \bar{\mathbf{w}} + \mathbf{d_w}$, we obtain

$$\begin{aligned}
\hat{\mathbf{u}}'\hat{\mathbf{x}} &= (\bar{\mathbf{u}} + \mathbf{d_u})'(\bar{\mathbf{x}} + \mathbf{d_x}) = \bar{\mathbf{u}}'\bar{\mathbf{x}} + \bar{\mathbf{u}}'\mathbf{d_x} + \mathbf{d_u}'\bar{\mathbf{x}} + \mathbf{d_u}'\mathbf{d_x} \\
&= \mathbf{e}'[\bar{\mathbf{X}}\bar{\mathbf{U}}\mathbf{e} + \bar{\mathbf{U}}\mathbf{d_x} + \bar{\mathbf{X}}\mathbf{d_u}] + \mathbf{d_u}'\mathbf{d_x}
\end{aligned}$$

From (9.42c), the term in $[\cdot]$ equals $\hat{\mu}\mathbf{e}$. Furthermore, from (9.42a, b), we observe that $\mathbf{d_u}'\mathbf{d_x} = -\mathbf{d_v}'\mathbf{A}\mathbf{d_x} = 0$. Hence, this gives $\hat{\mathbf{u}}'\hat{\mathbf{x}} = \mathbf{e}'(\hat{\mu}\mathbf{e}) = n\hat{\mu}$, and so this, along with (9.45) and (9.51), shows that $\hat{\mathbf{w}}$ satisfies (9.41) with $\mu = \hat{\mu}$.

To complete the proof of the first assertion of the theorem, we now need to show that $\hat{\mathbf{x}} > \mathbf{0}$ and $\hat{\mathbf{u}} > \mathbf{0}$. Toward this end, following (9.50) stated for $\hat{\mathbf{w}}$ and $\hat{\mu}$, we have

$$\hat{x}_j\hat{u}_j \ge \hat{\mu}(1 - \theta) > 0 \qquad \text{for all } j = 1, \ldots, n \tag{9.52}$$

Hence, for each $j = 1, \ldots, n$, either $\hat{x}_j > 0$ and $\hat{u}_j > 0$, or else $\hat{x}_j < 0$ and $\hat{u}_j < 0$. Assuming the latter for some j, on the contrary, since $\hat{x}_j = \bar{x}_j + (\mathbf{d_x})_j$ and $\hat{u}_j = \bar{u}_j + (\mathbf{d_u})_j$, where $\bar{x}_j > 0$ and $\bar{u}_j > 0$, we have $(\mathbf{d_x})_j < \hat{x}_j < 0$ and $(\mathbf{d_u})_j < \hat{u}_j < 0$; and, so, from (9.52), we obtain

$$(\mathbf{d_x})_j(\mathbf{d_u})_j > \hat{x}_j\hat{u}_j \ge \hat{\mu}(1 - \theta) \tag{9.53}$$

But from (9.46) and (9.51), we have

$$(\mathbf{d_x})_j(\mathbf{d_u})_j \leq \|\mathbf{D_x D_u e}\| = \|\hat{\mathbf{X}}\hat{\mathbf{U}}\mathbf{e} - \hat{\mu}\mathbf{e}\| \leq \hat{\mu}\theta \tag{9.54}$$

Equations (9.53) and (9.54) imply that $\hat{\mu}(1 - \theta) < \hat{\mu}\theta$, or that $\theta > 0.5$, which contradicts (9.44). Hence, $\hat{\mathbf{x}} > \mathbf{0}$ and $\hat{\mathbf{u}} > \mathbf{0}$.

Finally, observe from (9.41) and the foregoing argument that the algorithm generates a sequence $\mathbf{w}_k = (\mathbf{x}_k, \mathbf{u}_k, \mathbf{v}_k)$ and a sequence μ_k such that

$$\mathbf{Ax}_k = \mathbf{b}, \mathbf{x}_k > \mathbf{0} \qquad \mathbf{A}^t\mathbf{v}_k + \mathbf{u}_k = \mathbf{c}, \mathbf{u}_k > \mathbf{0} \qquad \mathbf{u}_k^t\mathbf{x}_k = n\mu_k = n\mu_0(\beta)^k \tag{9.55}$$

Since $\beta^k \to 0$ as $k \to \infty$, any accumulation point $\mathbf{w}^* = (\mathbf{x}^*, \mathbf{u}^*, \mathbf{v}^*)$ of $\mathbf{w}_k$ satisfies the necessary and sufficient optimality conditions (9.37) for P, and hence yields a primal-dual optimal solution to P. This completes the proof.

Convergence Rate and Complexity Analysis

Observe from (9.40) and (9.55) that the duality gap $\mathbf{c}^t\mathbf{x}_k - \mathbf{b}^t\mathbf{v}_k$ for the pair of primal–dual feasible solutions $\mathbf{x}_k > \mathbf{0}$ and $\mathbf{v}_k$ (with slacks $\mathbf{u}_k > \mathbf{0}$) generated by the algorithm equals $\mathbf{u}_k^t\mathbf{x}_k = n\mu_k = n\mu_0(\beta)^k$, and approaches zero at a geometric (linear) rate of convergence. Moreover, the convergence rate ratio β is given by $\beta = 1 - (\delta/\sqrt{n})$, which, for a fixed value of δ, approaches unity as $n \to \infty$, implying an increasingly slower convergence rate behavior as n increases. Hence, from a practical standpoint, implementations of this algorithm tend to shrink μ faster to zero based on the duality gap (for example, by taking $\mu_{k+1} = (\mathbf{c}^t\mathbf{x}_k - \mathbf{b}^t\mathbf{v}_k)/\phi(n)$, where $\phi(n) = n^2$ for $n \leq 5000$ and $\phi(n) = n\sqrt{n}$ for $n \geq 5000$), and also to conduct line searches along the Newton direction $\mathbf{d_w}$, while maintaining $\mathbf{x}_k > \mathbf{0}$ and $\mathbf{u}_k > \mathbf{0}$ rather than simply taking a unit step size.

The unmodified form of the algorithm possesses the desirable property of having a *polynomial-time complexity*. This means that if the data for P is all integer, and if L denotes the number of binary bits required to represent the data, then while using a number of elementary operations (additions, multiplications, comparisons, etc.) bounded above by a polynomial in the size of the problem as defined by the parameters m, n, and L, the algorithm will *finitely* determine an exact optimum to Problem P. It can be shown (see the Notes and References section) that we can begin with $\|\mathbf{x}_0\| < 2^L$ and $\|\mathbf{u}_0\| < 2^L$, and that once the duality gap $\mathbf{u}_k^t\mathbf{x}_k \leq 2^{-2L}$ for some k, the solution, $\mathbf{w}_k$ can be *purified* (or rounded) in polynomial-time to an exact optimum by the process of seeking a vertex solution with at least as good an objective value as currently obtained. Note that from (9.55) and (9.41), we have,

$$\mathbf{u}_k^t\mathbf{x}_k = n\mu_0(\beta)^k = \mathbf{u}_0^t\mathbf{x}_0(\beta)^k < 2^{2L}(\beta)^k \leq 2^{-2L} \quad \text{when } \beta^k \leq 2^{-4L}, \text{ or } k \geq \frac{[4\,ln\,(2)]L}{-ln\,(\beta)} \tag{9.56}$$

But, by the concavity of $ln\,(\cdot)$ over the positive real line, we have from (9.44) that $ln\,(\beta) = ln\,[1 - (\delta/\sqrt{n})] \leq -(\delta/\sqrt{n})$; and so, $-ln(\beta) \geq (\delta/\sqrt{n})$. Consequently, when $k \geq [4\,ln\,(2)L]/(\delta/\sqrt{n})$, we have (9.56) holding, and we can then purify the available solution finitely in polynomial time to an exact optimum. The number of iterations before this happens is therefore bounded above by a constant times $\sqrt{n}L$; this is denoted as being of *order of complexity* $O(\sqrt{n}L)$. Because each iteration itself requires a number of operations bounded above by a polynomial in m and n, the overall algorithm is of polynomial-time complexity. In the Notes and References section, we refer the reader to a discussion on modifying the algorithm by adaptively varying the parameters from one iteration to the next so that superlinear convergence is realized, without impairing the polynomial-time behavior of the algorithm.

Getting Started

We conclude our discussion of the primal–dual path following algorithm by showing how we can initialize this procedure. Note that the given pair of primal and dual problems P and D need not possess a primal–dual feasible solution $\mathbf{w} = (\mathbf{x}, \mathbf{u}, \mathbf{v})$, with $\mathbf{x} > \mathbf{0}$ and $\mathbf{u} > \mathbf{0}$, and such that (9.41) holds. Hence, we employ artificial variables as in Section 2.7 to accomplish this requirement. Toward this end, let λ and γ be sufficiently large scalars. (Theoretically, we can take $\lambda = 2^{2L}$ and $\gamma = 2^{4L}$ although, practically, these values might be far too large.) Define

$$M_1 = \lambda\gamma \quad \text{and} \quad M_2 = \lambda\gamma(n + 1) - \lambda(\mathbf{e}^t\mathbf{c})$$

Let x_a be a single artificial variable, define an auxiliary variable x_{n+1}, and consider the following (Big-M) artificial primal problem P′ along with its dual D′ (see Section 2.7):

$$\begin{array}{lll} P': & \text{Minimize} & \mathbf{c}^t\mathbf{x} + M_1 x_a \\ & \text{subject to} & \mathbf{A}\mathbf{x} + (\mathbf{b} - \lambda\mathbf{A}\mathbf{e})x_a = \mathbf{b} \\ & & (\mathbf{c} - \gamma\mathbf{e})^t\mathbf{x} - \gamma x_{n+1} = -M_2 \\ & & (\mathbf{x}, x_a, x_{n+1}) \geq \mathbf{0} \end{array}$$

$$\begin{array}{lll} D': & \text{Maximize} & \mathbf{b}^t\mathbf{v} - M_2 v_a \\ & \text{subject to} & \mathbf{A}^t\mathbf{v} + (\mathbf{c} - \gamma\mathbf{e})v_a + \mathbf{u} = \mathbf{c} \\ & & (\mathbf{b} - \lambda\mathbf{A}\mathbf{e})^t\mathbf{v} + u_a = M_1 \\ & & -\gamma v_a + u_{n+1} = 0 \\ & & (\mathbf{u}, u_a, u_{n+1}) \geq \mathbf{0} \\ & & (\mathbf{v}, v_a) \text{ unrestricted} \end{array} \tag{9.57}$$

It is easily verified that the primal–dual pair of solutions

$$(\mathbf{x}^t, x_a, x_{n+1}) = (\lambda\mathbf{e}^t, 1, \lambda) > \mathbf{0}$$

and (9.58)

$$\mathbf{v} = \mathbf{0},\ v_a = 1,\ (\mathbf{u}^t, u_a, u_{n+1}) = (\gamma\mathbf{e}^t, \lambda\gamma, \gamma) > \mathbf{0}$$

are respectively feasible to P' and D', and, moreover, that

$$x_j u_j = \lambda\gamma \text{ for } j = 1, \ldots, n \qquad x_a u_a = \lambda\gamma \qquad x_{n+1}u_{n+1} = \lambda\gamma$$

Consequently, with μ initialized as $\lambda\gamma$, the solution (9.58) lies on the central path, and so (9.41) holds true. Hence, this solution can be used to initialize the algorithm to solve P′ and D′. As with the artificial variable method discussed in Section 2.7, it can be shown that if $x_a = v_a = 0$ at optimality, then the corresponding optimal solutions $\mathbf{x}$ and $\mathbf{v}$ solve P and D, respectively. Otherwise, at termination, if $x_a > 0$ and $v_a = 0$, then P is infeasible; if $x_a = 0$ and $v_a > 0$, then P is unbounded; and if $x_a > 0$ and $v_a > 0$, then P is either infeasible or unbounded. The last case can be resolved by replacing P with its Phase I problem of Section 2.7. Also, since the size of P′ is polynomially related to size of P, the polynomial-time complexity property of the algorithm is preserved.

Exercises

9.1 Consider the following problem:

Minimize $x_1^2 + x_2^2$

subject to $2x_1 + x_2 - 2 \leq 0$

$- x_2 + 1 \leq 0$

a. Find the optimal solution to this problem.

b. Formulate a suitable function, with an initial penalty parameter $\mu = 1$.

c. Starting from the point (2, 6), solve the resulting problem by a suitable unconstrained minimization technique.

d. Replace the penalty parameter μ by 10. Starting from the point you obtained in part c, solve the resulting problem.

9.2 Given the set of inequality constraints $g_i(\mathbf{x}) \leq 0$ for $i = 1, \ldots, m$, any of the following auxiliary functions may be employed:

$$f(\mathbf{x}) + \mu \sum_{i=1}^{m} \text{maximum } \{0, g_i(\mathbf{x})\}$$

$$f(\mathbf{x}) + \mu \sum_{i=1}^{m} [\text{maximum } \{0, g_i(\mathbf{x})\}]^2$$

$$f(\mathbf{x}) + \mu \text{ maximum } \{0, g_i(\mathbf{x}), \ldots, g_m(\mathbf{x})\}$$

$$f(\mathbf{x}) + \mu [\text{maximum } \{0, g_1(\mathbf{x}), \ldots, g_m(\mathbf{x})\}]^2$$

Compare among these forms. What are the advantages and disadvantages of each?

9.3 A new facility is to be placed such that the sum of its squared distance from four existing facilities is minimized. The four facilities are located at the points (1, 2), (−2, 4), (2, 6), and (−6, −3). If the coordinates of the new facility are x_1 and x_2, suppose that x_1 and x_2 must satisfy the restrictions $x_1 + x_2 = 2$, $x_1 \geq 0$, and $x_2 \geq 0$.

a. Formulate the problem.

b. Show that the objective function is convex.

c. Find an optimal solution by making use of the KKT conditions.

d. Solve the problem by a penalty function method using a suitable unconstrained optimization technique.

9.4 Consider the problem to minimize x^3 subject to $x = 1$. Obviously, the optimal solution is $\bar{x} = 1$. Now consider the problem to minimize $x^3 + \mu(x - 1)^2$.

a. For $\mu = 1.0, 10.0, 100.0$, and 1000.0, plot $x^3 + \mu(x - 1)^2$ as a function of x, and for each case, find the point where the derivative of the function vanishes. Also, verify that the optimal solution is unbounded.

b. Show that the optimal solution to the penalty problem is unbounded for any given μ, so that the conclusion of Theorem 9.2.2 does not hold. Discuss.

c. For $\mu = 1.0, 10.0, 100.0$, and 1000.0, find the optimal solution to the penalty problem with the added constraint $|x| \leq 2$.

9.5 Consider the following problem:

Minimize $x_1^3 + x_2^3$

subject to $x_1 + x_2 - 1 = 0$

a. Find an optimal solution to the problem.

b. Consider the following penalty problem:

Minimize $x_1^3 + x_2^3 + \mu(x_1 + x_2 - 1)^2$

For each $\mu > 0$, verify that the optimal solution is unbounded.

c. Note that the optimal solution in parts a and b have different objective values so that the conclusion of Theorem 9.2.2 does not hold. Explain.

d. Add the constraints $|x_1| \leq 1$ and $|x_2| \leq 1$ to the problem, and let $X = \{(x_1, x_2): |x_1| \leq 1, |x_2| \leq 1\}$. The penalty problem becomes

Minimize $x_1^3 + x_2^3 + \mu(x_1 + x_2 - 1)^2$

subject to $|x_1| \leq 1, \quad |x_2| \leq 1$

What is the optimal solution for a given $\mu > 0$? What is the limit of the sequence of optima as $\mu \to \infty$? Note that with the addition of the set X, the conclusion of Theorem 9.2.2 holds true.

9.6 Consider the following problem:

Minimize $e^{x_1} + x_1^2 + 2x_1x_2 + 4x_2^2$
subject to $x_1 + 2x_2 - 6 = 0$

Formulate a suitable exterior penalty function with $\mu = 10$. Starting with the point (1, 1), perform two iterations of the conjugate gradient method.

9.7 The exterior penalty problem can be reformulated as follows: Find $\sup_{\mu \geq 0} \inf_{\mathbf{x} \in X} \{f(\mathbf{x}) + \mu\alpha(\mathbf{x})\}$, where α is a suitable penalty function.

a. Show that the primal problem is equivalent to finding $\inf_{\mathbf{x} \in X} \sup_{\mu \geq 0} \{f(\mathbf{x}) + \mu\alpha(\mathbf{x})\}$. From this, note that the primal and penalty problems can be interpreted as a pair of min–max dual problems.

b. In Theorem 9.2.2, it was shown that

$$\inf_{\mathbf{x} \in X} \sup_{\mu \geq 0} \{ f(\mathbf{x}) + \mu\alpha(\mathbf{x})\} = \sup_{\mu \geq 0} \inf_{\mathbf{x} \in X} \{ f(\mathbf{x}) + \mu\alpha(\mathbf{x})\}$$

without any convexity assumptions regarding f or α. For the Lagrangian dual problem of Chapter 6, however, suitable convexity assumptions had to be made to guarantee equality of the optimal objectives of the primal and dual problems. Comment, relating your discussion to Figure 9.3.

9.8 This exercise describes several strategies for modifying the penalty parameter. Consider the following problem:

Minimize $(x_1 - 6)^2 + (x_2 - 8)^2$
subject to $x_1^2 - x_2 \leq 0$

Using the auxiliary function $(x_1 - 6)^2 + (x_2 - 8)^2 + \mu$ maximum $\{x_1^2 - x_2, 0\}$, and adopting the cyclic coordinate method, solve the above problem starting from the point $\mathbf{x}_1 = (0, -4)^t$ under the following strategies for modifying μ:

a. Starting from $\mathbf{x}_1$, solve the penalty problem for $\mu_1 = 0.1$ resulting in $\mathbf{x}_2$. Then, starting from $\mathbf{x}_2$, solve the problem with $\mu_2 = 100$.

b. Starting from the unconstrained optimal point (6, 8), solve the penalty problem for $\mu_2 = 100$. (This is the limiting case of part a for $\mu_1 = 0$.)

c. Starting from $\mathbf{x}_1$, apply the algorithm described in Section 9.2 by using the successively increasing values of $\mu = 0.1, 1.0, 10.0$, and 100.0.

d. Starting from $\mathbf{x}_1$, solve the penalty problem for $\mu_1 = 100.0$.

Which of the above strategies would you recommend, and why? Also, in each case above, derive an estimate for the Lagrange multiplier associated with the single constraint.

9.9 Consider the following problem:

Minimize $4x_1^2 - 5x_1x_2 + x_2^2$
subject to $x_1^2 - x_2 + 2 \leq 0$
$x_1 + x_2 - 6 \leq 0$
$x_1, x_2 \geq 0$

Solve the above problem by an exterior penalty function method starting from (0, 0) for each of the specifications of X:

a. $X = E_n$
b. $X = \{(x_1, x_2): x_1 \geq 0, x_2 \geq 0\}$
c. $X = \{(x_1, x_2): x_1 + x_2 - 6 \leq 0, x_1 \geq 0, x_2 \geq 0\}$

(Effective methods for handling linear constraints are discussed in Chapter 10.) Compare among the above three alternative approaches. Which one would you recommend?

9.10 Consider the problem to minimize $f(\mathbf{x})$ subject to $g_i(\mathbf{x}) \leq 0$ for $i = 1, \ldots, m$ and $h_i(\mathbf{x}) = 0$ for $i = 1, \ldots, l$. For a given value of $\mu > 0$, provide an interpretation for the problem to $\min_{\mathbf{x}} \{f(\mathbf{x}) + \mu [\sum_{i=1}^m \max^2 \{0, g_i(\mathbf{x})\} + \sum_{i=1}^l h_i^2(\mathbf{x})]\}$ in terms of the problem to $\min_{\mathbf{x}} \{\mu\|\boldsymbol{\epsilon}\|^2 + v(\boldsymbol{\epsilon})\}$, where v is the perturbation function given by equation (6.9).

9.11 Let $X = \{\mathbf{x}: g_i(\mathbf{x}) \leq 0$ for $i = m + 1, \ldots, m + M$, and $h_i(\mathbf{x}) = 0$ for $i = l + 1, \ldots, l + L\}$. Following the derivation of the KKT Lagrange multipliers at optimality when using the exterior penalty function as in Section 9.2, show how and under what conditions one can recover all the Lagrange multipliers at optimality for the problem to minimize $f(\mathbf{x})$, subject to $g_i(\mathbf{x}) \leq 0$ for $i = 1, \ldots, m$, $h_i(\mathbf{x}) = 0$ for $i = 1, \ldots, l$ and $\mathbf{x} \in X$.

9.12 Consider Problem P to minimize $f(\mathbf{x})$ subject $g_i(\mathbf{x}) \leq 0$ for $i = 1, \ldots, m$ and $h_i(\mathbf{x}) = 0$ for $i = 1, \ldots, l$. Let $F_E(\mathbf{x})$ be the exact absolute value penalty function defined by (9.8). Show that if $\bar{\mathbf{x}}$ satisfies the second-order sufficiency conditions for a local minimum of P as stated in Theorem 4.4.2, then, for μ at least as large as given by Theorem 9.3.1, $\bar{\mathbf{x}}$ is also a local minimizer for F_E.

9.13 Solve the following problem using the method of multipliers, starting with the Lagrange multipliers $v_1 = v_2 = 0$ and with the penalty parameters $\mu_1 = \mu_2 = 1.0$:

$$\begin{aligned}\text{Minimize} \quad & x_1 + x_2 - 2x_3 \\ \text{subject to} \quad & x_1^2 + x_2^2 + x_3^2 = 9 \\ & x_1 - 2x_2^2 + x_3 = 2\end{aligned}$$

9.14 Consider the problem to minimize $f(\mathbf{x})$ subject to $h_i(\mathbf{x}) = 0$ for $i = 1, \ldots, l$, where f and h_i, $i = 1, \ldots, l$, are continuously twice-differentiable. Given any $\boldsymbol{\epsilon} = (\epsilon_1, \ldots, \epsilon_l)^t$, define the perturbed problem P $(\boldsymbol{\epsilon})$: minimize $\{f(\mathbf{x}): h_i(\mathbf{x}) = \epsilon_i$ for $i = 1, \ldots, l\}$. Suppose that P has a local minimum $\bar{\mathbf{x}}$ that is a regular point and that $\bar{\mathbf{x}}$ along with the corresponding unique Lagrange multiplier $\bar{\mathbf{v}}$ satisfy the second-order sufficiency conditions for a strict local minimum. Then, show that, for each $\boldsymbol{\epsilon}$ in a neighborhood of $\boldsymbol{\epsilon} = \mathbf{0}$, there exists a solution $\mathbf{x}(\boldsymbol{\epsilon})$ such that (i) $\mathbf{x}(\boldsymbol{\epsilon})$ is a local minimum for P$(\boldsymbol{\epsilon})$; (ii) $\mathbf{x}(\boldsymbol{\epsilon})$ is a continuous function of $\boldsymbol{\epsilon}$, with $\mathbf{x}(\mathbf{0}) = \bar{\mathbf{x}}$; and (iii) $\partial f[\mathbf{x}(\boldsymbol{\epsilon})]/\partial \epsilon_i|_{\boldsymbol{\epsilon}=\mathbf{0}} = \nabla_{\mathbf{x}} f(\bar{\mathbf{x}})^t \nabla_{\epsilon_i} \mathbf{x}(\mathbf{0}) = -\bar{v}_i$ for $i = 1, \ldots, l$. (*Hint:* Use the regularity of $\bar{\mathbf{x}}$ and the second-order sufficiency conditions to show the existence of $\mathbf{x}(\boldsymbol{\epsilon})$. By the chain rule, the required partial derivative then equals $\nabla f(\bar{\mathbf{x}})^t \nabla_{\epsilon_i} \mathbf{x}(\mathbf{0})$, which in turn equals $-\Sigma_j \bar{v}_j \nabla h_j(\bar{\mathbf{x}})^t \nabla_{\epsilon_i} \mathbf{x}(\mathbf{0})$ from the KKT conditions. Now use $\mathbf{h}[\mathbf{x}(\boldsymbol{\epsilon})] = \boldsymbol{\epsilon}$ to complete the derivation.)

9.15 Consider the problem given by (9.14), and suppose that this problem has a local minimum at $\mathbf{x}^*$, where $\mathbf{x}^*$ is a regular point having a unique associated Lagrange multiplier $\mathbf{v}^*$, and that the Hessian of the Lagrangian $f(\mathbf{x}) + \mu \sum_{i=1}^{l} h_i^2(\mathbf{x}) + \sum_{i=1}^{l} v_i^* h_i(\mathbf{x})$ with respect to $\mathbf{x}$ is positive definite at $\mathbf{x} = \mathbf{x}^*$. Define the Lagrangian dual function $\theta(\mathbf{v}) = \text{minimum } \{f(\mathbf{x}) + \mu \sum_{i=1}^{l} h_i^2(\mathbf{x}) + \sum_{i=1}^{l} v_i h_i(\mathbf{x}) : \mathbf{x}$ lies in a sufficiently small neighborhood of $\mathbf{x}^*\}$. Show that for each $\mathbf{v}$ in some neighborhood of $\mathbf{v}^*$, there exists an unique $\mathbf{x}(\mathbf{v})$ which evaluates $\theta(\mathbf{v})$ and, moreover, $\mathbf{x}(\mathbf{v})$ is a continuously differentiable function of $\mathbf{v}$.

9.16 Let P: minimize $\{f(\mathbf{x}): g_i(\mathbf{x}) \leq 0$ for $i = 1, \ldots, m$, and $h_i(\mathbf{x}) = 0$ for $i = 1, \ldots, l\}$, and define P$'$: minimize $\{f(\mathbf{x}): g_i(\mathbf{x}) + s_i^2 = 0$ for $i = 1, \ldots, m$, and $h_i(\mathbf{x}) = 0$ for $i = 1, \ldots, l\}$. Let $\bar{\mathbf{x}}$ be a KKT point for P with Lagrangian multipliers $\bar{\mathbf{u}}$ and $\bar{\mathbf{v}}$ associated with the inequality and the equality constraints, respectively, such that the strict complementary slackness condition holds, namely, that $\bar{u}_i g_i(\bar{\mathbf{x}}) = 0$ for all $i = 1, \ldots, m$ and $\bar{u}_i > 0$ for each $i \in I(\bar{\mathbf{x}}) = \{i: g_i(\bar{\mathbf{x}}) = 0\}$. Furthermore, suppose that the second-order sufficiency conditions of Theorem 4.4.2 hold for Problem P. Write the KKT conditions and the second-order sufficiency conditions for Problem P$'$ and verify that these conditions are satisfied at $(\bar{\mathbf{x}}, \bar{\mathbf{s}}, \bar{\mathbf{u}}, \bar{\mathbf{v}})$, where $\bar{s}_i^2 = -g_i(\bar{\mathbf{x}})$ for all $i = 1, \ldots, m$. Indicate the significance of the strict complementary slackness assumption.

9.17 Consider Problem P to minimize $f(\mathbf{x})$ subject to $g_i(\mathbf{x}) \leq 0$ for $i = 1, \ldots, m$ and $h_i(\mathbf{x}) = 0$ for $i = 1, \ldots, l$. Given $(\bar{\mathbf{u}}, \bar{\mathbf{v}})$, consider the ALAG inner minimization problem (9.26), and suppose that $\mathbf{x}_k$ is a minimizing solution, so that $\nabla_{\mathbf{x}} F_{\text{ALAG}}(\mathbf{x}_k, \bar{\mathbf{u}}, \bar{\mathbf{v}}) = \mathbf{0}$. Show that by requiring $(\mathbf{u}_{\text{new}}, \bar{\mathbf{v}}_{\text{new}})$ to satisfy $\nabla_{\mathbf{x}} L(\mathbf{x}_k, \bar{\mathbf{u}}_{\text{new}}, \bar{\mathbf{v}}_{\text{new}}) = \mathbf{0}$ gives $(\bar{\mathbf{u}}_{\text{new}}, \bar{\mathbf{v}}_{\text{new}})$ as defined by (9.19) and (9.27), where $L(\mathbf{x}, \mathbf{u}, \mathbf{v})$ is the usual Lagrangian function for Problem P.

9.18 Consider the problem to minimize $f(\mathbf{x})$ subject to $g_i(\mathbf{x}) \leq 0$ for $i = 1, \ldots, m$ and $\mathbf{x} \in X$, where $X = \{\mathbf{x}: g_i(\mathbf{x}) \leq 0$ for $i = m + 1, \ldots, m + M$, and $h_i(\mathbf{x}) = 0$ for $i = 1, \ldots, l\}$. Extend the derivation of Lagrange multipliers at optimality for the barrier function approach, as discussed in Section 9.4, to the case where X is as defined above in lieu of $X \equiv E_n$.

9.19 Consider the following problem:

$$\begin{aligned}\text{Minimize} \quad & (x_1 - 5)^2 + (x_2 - 3)^2 \\ \text{subject to} \quad & x_1 + x_2 \le 3 \\ & -x_1 + 2x_2 \le 4\end{aligned}$$

Formulate a suitable barrier problem with the initial parameter equal to 1. Use an unconstrained optimization technique starting with the point (0, 0) to solve the barrier problem. Provide estimates for the Lagrange multipliers.

9.20 This exercise describes several strategies for modifying the barrier parameter μ. Consider the following problem:

$$\begin{aligned}\text{Minimize} \quad & (x_1 - 6)^2 + (x_2 - 8)^2 \\ \text{subject to} \quad & x_1^2 - x_2 \le 0\end{aligned}$$

Using the auxiliary function $(x_1 - 6)^2 + (x_2 - 8)^2 - \mu[1/(x_1^2 - x_2)]$ and adopting the cyclic coordinate method, solve the above problem starting from the point $\mathbf{x}_1 = (0, 12)^t$ under the following strategies for modifying μ:

a. Starting from $\mathbf{x}_1$, solve the barrier problem for $\mu_1 = 10.0$ resulting in $\mathbf{x}_2$. Then starting from $\mathbf{x}_2$, solve the problem with $\mu_2 = 0.01$.

b. Starting from the point (0, 12), solve the barrier problem for $\mu_1 = 0.01$.

c. Apply the algorithm described in Section 9.3 by using the successively decreasing values of $\mu = 10.0, 1.00, 0.10$, and 0.01.

d. Starting from $\mathbf{x}_1$, solve the barrier problem for $\mu_1 = 0.001$.

Which of the above strategies would you recommend, and why? Also, in each case above, derive an estimate for the Lagrange multiplier associated with the single constraint.

9.21 By replacing an equality constraint $h_i(\mathbf{x}) = 0$ by one of the following forms, where $\varepsilon > 0$ is a small scalar,

a. $h_i^2(\mathbf{x}) \le \varepsilon$

b. $|h_i(\mathbf{x})| \le \varepsilon$

c. $h_i(\mathbf{x}) \le \varepsilon, \; -h_i(\mathbf{x}) \le \varepsilon$

barrier functions could be used to handle equality constraints. Discuss in detail the implications of these formulations. Using this approach, solve the following problem with $\varepsilon = 0.05$:

$$\begin{aligned}\text{Minimize} \quad & x_1^2 + x_2^2 \\ \text{subject to} \quad & x_1 + x_2 = 2\end{aligned}$$

9.22 To use a barrier function method, we must find a point $\mathbf{x} \in X$ with $g_i(\mathbf{x}) < 0$ for $i = 1, \ldots, m$. The following procedure is suggested for obtaining such a point.

Initialization Step Select $\mathbf{x}_1 \in X$, let $k = 1$ and go to the main step.

Main Step

1. Let $I = \{i : g_i(\mathbf{x}_k) < 0\}$. If $I = \{1, \ldots, m\}$, stop with $\mathbf{x}_k$ satisfying $g_i(\mathbf{x}_k) < 0$ for all i. Otherwise, select $j \notin I$ and go to step 2.
2. Use a barrier function method to solve the following problem starting with $\mathbf{x}_k$:

$$\begin{aligned}\text{Minimize} \quad & g_j(\mathbf{x}) \\ \text{subject to} \quad & g_i(\mathbf{x}) < 0 \qquad \text{for } i \in I \\ & \mathbf{x} \in X\end{aligned}$$

Let $\mathbf{x}_{k+1}$ be an optimal solution. If $g_j(\mathbf{x}_{k+1}) \ge 0$, stop; the set $\{\mathbf{x} \in X : g_i(\mathbf{x}) < 0 \text{ for } i = 1, \ldots, m\}$ is empty. Otherwise, replace k by $k + 1$ and repeat step 1.

a. Show that the above approach stops in, at most, m iterations with a point $\mathbf{x} \in X$ satisfying $g_i(\mathbf{x}) < 0$ for $i = 1, \ldots, m$ or with the conclusion that no such point exists.

b. Using the above approach, find a point satisfying $x_1 + x_2 < 2$ and $x_1^2 - x_2 < 0$, starting from the point (3, 0).

9.23 Consider the problem to minimize $f(\mathbf{x})$ subject to $\mathbf{x} \in X$, $g_i(\mathbf{x}) \le 0$ for $i = 1, \ldots, m$ and $h_i(\mathbf{x}) = 0$ for $i = 1, \ldots, l$. A *mixed penalty–barrier auxiliary* function is of the form $f(\mathbf{x}) + \mu B(\mathbf{x}) + \frac{1}{\mu}\alpha(\mathbf{x})$, where B is a barrier function which handles the inequality constraints and α is a penalty function which handles the equality constraints. The following result generalizes Theorems 9.2.2 and 9.4.3:

$$\inf\{f(\mathbf{x}) : \mathbf{g}(\mathbf{x}) \le \mathbf{0}, \mathbf{h}(\mathbf{x}) = \mathbf{0}, \mathbf{x} \in X\} = \lim_{\mu \to 0+} \sigma(\mu)$$

$$\mu B(\mathbf{x}_\mu) \to 0, \qquad \tfrac{1}{\mu}\alpha(\mathbf{x}_\mu) \to 0 \qquad \text{as } \mu \to 0^+$$

where

$$\sigma(\mu) = \inf\{f(\mathbf{x}) + \mu B(\mathbf{x}) + \tfrac{1}{\mu}\alpha(\mathbf{x}) : \mathbf{x} \in X, \mathbf{g}(\mathbf{x}) < \mathbf{0}\} = f(\mathbf{x}_\mu) + \mu B(\mathbf{x}_\mu) + \tfrac{1}{\mu}\alpha(\mathbf{x}_\mu)$$

a. Prove the above result after making the appropriate assumptions.

b. State a precise algorithm for solving a nonlinear programming problem using a mixed penalty-barrier function approach, and illustrate by solving the following problem:

Minimize $e^{x_1} - x_1x_2 + x_2^2$
subject to $x_1^2 + x_2^2 = 4$
$2x_1 + x_2 \le 2$

c. Discuss the possibility of using two parameters μ_1 and μ_2 so that the mixed penalty-barrier auxiliary function is of the form $f(\mathbf{x}) + \mu_1 B(\mathbf{x}) + \frac{1}{\mu_2}\alpha(\mathbf{x})$. Using this approach, solve the following problem by a suitable unconstrained optimization technique, starting from the point (0, 0) and initially letting $\mu_1 = 1.0$ and $\mu_2 = 2.0$:

Maximize $-x_1^2 + 2x_1x_2 + x_2^2 - e^{-x_1-x_2}$
subject to $x_1^2 + x_2^2 - 4 = 0$
$x_1 + x_2 \le 1$

(The method described in this exercise is credited to Fiacco and McCormick [1968].)

9.24 Compare the different exterior penalty and the interior barrier function methods in detail. Emphasize the advantages and disadvantages of both methods.

9.25 In the methods discussed in this chapter, the same penalty or barrier parameter was often used for all the constraints. Do you see any advantages for using different parameters for the various constraints? Propose some schemes for updating these parameters. How can you modify Theorems 9.2.2 and 9.4.3 to handle this situation?

9.26 In this exercise, we describe a *parameter-free penalty function* method for solving a problem of the form to minimize $f(\mathbf{x})$ subject to $g_i(\mathbf{x}) \le 0$ for $i = 1, \ldots, m$ and $h_i(\mathbf{x}) = 0$ for $i = 1, \ldots, l$.

Initialization Step Choose a scalar $L_1 < \inf\{f(\mathbf{x}): g_i(\mathbf{x}) \le 0$ for $i = 1, \ldots, m$, and $h_i(\mathbf{x}) = 0$ for $i = 1, \ldots, l\}$. Let $k = 1$, and go to the main step.

Main Step Solve the following problem:

Minimize $\beta(\mathbf{x})$
subject to $\mathbf{x} \in E_n$

where

$$\beta(\mathbf{x}) = [\text{maximum}\{0, f(\mathbf{x}) - L_k\}]^2 + \sum_{i=1}^{m} [\text{maximum}\{0, g_i(\mathbf{x})\}]^2 + \sum_{i=1}^{l} |h_i(\mathbf{x})|^2$$

Let $\mathbf{x}_k$ be an optimal solution, let $L_{k+1} = f(\mathbf{x}_k)$, replace k by $k + 1$, and repeat the main step.

a. Solve the following problem by the above approach starting with $L_1 = 0$, and initiating the optimization process from the point $\mathbf{x} = (0, -1)^t$:

Minimize $(x_1 - 6)^2 + (x_2 - 8)^2$
subject to $x_1^2 - x_2 \le 0$

b. Compare the trajectory of points generated in part a with the points generated in Exercise 9.8.

c. At iteration k, if $\mathbf{x}_k$ is feasible to the original problem, show that it must be optimal.

d. State the assumptions under which the above method converges to an optimal solution, and prove convergence.

9.27 In this exercise, we describe a *parameter-free barrier function method* for solving a problem of the form to minimize $f(\mathbf{x})$ subject to $g_i(\mathbf{x}) \le 0$ for $i = 1, \ldots, m$.

Initialization Step Choose $\mathbf{x}_1$ such that $g_i(\mathbf{x}_1) < 0$ for $i = 1, \ldots, m$. Let $k = 1$, and go to the main step.

Main Step Let $X_k = \{\mathbf{x}: f(\mathbf{x}) - f(\mathbf{x}_k) < 0, g_i(\mathbf{x}) < 0 \text{ for } i = 1, \ldots, m\}$. Let $\mathbf{x}_{k+1}$ be an optimal solution to the following problem:

$$\text{Minimize} \quad \frac{-1}{f(\mathbf{x}) - f(\mathbf{x}_k)} - \sum_{i=1}^{m} \frac{1}{g_i(\mathbf{x})}$$
$$\text{subject to} \quad \mathbf{x} \in X_k$$

Replace k by $k + 1$, and repeat the main step.

(The constraint $\mathbf{x} \in X_k$ could be handled implicitly provided that we started the optimization process from $\mathbf{x} = \mathbf{x}_k$.)

a. Solve the following problem by the above approach starting from $\mathbf{x}_1 = (0, 2)^t$.

$$\text{Minimize} \quad (x_1 - 6)^2 + (x_2 - 8)^2$$
$$\text{subject to} \quad x_1^2 - x_2 \leq 0$$

b. Compare the trajectory of points generated in part a with the points generated in Exercise 9.20.

c. State the assumptions under which the above method converges to an optimal solution, and prove convergence.

9.28 Consider the problem to minimize $f(\mathbf{x})$ subject to $h_i(\mathbf{x}) = 0$ for $i = 1, \ldots, l$, and suppose that it has an optimal solution $\bar{\mathbf{x}}$. The following procedure for solving the problem is credited to Morrison [1968].

Initialization Step Choose $L_1 \leq f(\bar{\mathbf{x}})$, let $k = 1$, and go to the main step.

Main Step

1. Let $\mathbf{x}_k$ be an optimal solution for the problem to minimize $[f(\mathbf{x}) - L_k]^2 + \sum_{i=1}^{l} h_i^2(\mathbf{x})$. If $h_i(\mathbf{x}_k) = 0$ for $i = 1, \ldots, l$, stop with $\mathbf{x}_k$ as an optimal solution to the original problem; otherwise, go to step 2.
2. Let $L_{k+1} = L_k + \nu^{1/2}$, where $\nu = [f(\mathbf{x}_k) - L_k]^2 + \sum_{i=1}^{l} h_i^2(\mathbf{x}_k)$. Replace k by $k + 1$ and repeat step 1.

a. Show that if $h_i(\mathbf{x}_k) = 0$ for $i = 1, \ldots, l$, then $L_{k+1} = f(\mathbf{x}_k) = f(\bar{\mathbf{x}})$, and $\mathbf{x}_k$ is an optimal solution to the original problem.

b. Show that $f(\mathbf{x}_k) \leq f(\bar{\mathbf{x}})$ for each k.

c. Show that $L_k \leq f(\bar{\mathbf{x}})$ for each k and that $L_k \to f(\bar{\mathbf{x}})$.

d. Using the above method, solve the following problem:

$$\text{Minimize} \quad x_1 + x_2 - 2x_3$$
$$\text{subject to} \quad x_1^2 + x_2^2 + x_3^2 = 9$$
$$x_1 - 2x_2^2 + x_3 = 2$$

9.29 Consider the following problem:

$$\text{Minimize} \quad f(\mathbf{x})$$
$$\text{subject to} \quad \mathbf{h}(\mathbf{x}) = \mathbf{0}$$

where $f: E_n \to E_1$ and $\mathbf{h}: E_n \to E_l$ are differentiable. Let $\mu > 0$ be a large penalty parameter, and consider the penalty problem to minimize $q(\mathbf{x})$ subject to $\mathbf{x} \in E_n$, where $q(\mathbf{x}) = f(\mathbf{x}) + \mu \sum_{i=1}^{l} h_i^2(\mathbf{x})$. The following method is suggested to solve the penalty problem (see Luenberger [1984] for a detailed motivation of the method).

Initialization Step Choose a point $\mathbf{x}_1 \in E_n$, let $k = 1$, and go to the main step.

Main Step

1. Let the $l \times n$ matrix $\nabla\mathbf{h}(\mathbf{x}_k)$ denote the Jacobian of $\mathbf{h}$ at $\mathbf{x}_k$. Let $\mathbf{A} = \mathbf{BB}$, where $\mathbf{B} = \nabla\mathbf{h}(\mathbf{x}_k)\nabla\mathbf{h}(\mathbf{x}_k)^t$. Let $\mathbf{d}_k$ be as given below and go to step 2.

$$\mathbf{d}_k = -\frac{1}{2\mu}\nabla\mathbf{h}(\mathbf{x}_k)^t\mathbf{B}^{-1}\nabla\mathbf{h}(\mathbf{x}_k)\nabla q(\mathbf{x}_k)$$

2. Let λ_k be an optimal solution to the problem to minimize $q(\mathbf{x}_k + \lambda\mathbf{d}_k)$ subject to $\lambda \in E_1$, and let $\mathbf{w}_k = \mathbf{x}_k + \lambda_k\mathbf{d}_k$. Go to step 3.
3. Let $\bar{\mathbf{d}} = -\nabla q(\mathbf{w}_k)$, and let α_k be an optimal solution to the problem to minimize $q(\mathbf{w}_k + \alpha\bar{\mathbf{d}}_k)$ subject to $\alpha \in E_1$. Let $\mathbf{x}_{k+1} = \mathbf{w}_k + \alpha_k\bar{\mathbf{d}}_k$, replace k by $k + 1$, and repeat step 1.

a. Apply the above method to solve the following problem by letting $\mu = 100$:

Minimize $x_1^2 + 2x_1x_2 + x_2^2 + 2x_1 + 6x_2$
subject to $x_1 + x_2 = 4$

b. The above algorithm could be easily modified to solve a problem with equality and inequality constraints. In this case, let the penalty problem be to minimize $q(\mathbf{x})$, where $q(\mathbf{x}) = f(\mathbf{x}) + \mu \sum_{i=1}^{l} h_i^2(\mathbf{x}) + \mu \sum_{i=1}^{m} [\text{maximum } \{0, g_i(\mathbf{x})\}]^2$. In the description of the algorithm, $\mathbf{h}(\mathbf{x}_k)$ is replaced by $\mathbf{F}(\mathbf{x}_k)$, where $\mathbf{F}(\mathbf{x}_k)$ consists of the equality constraints and the inequality constraints that are either active or violated at $\mathbf{x}_k$. Use this modified procedure to solve the following problem, by letting $\mu = 100$.

Minimize $x_1^2 + 2x_1x_2 + x_2^2 + 2x_1 + 6x_2$
subject to $x_1 + x_2 \geq 3$
$x_1, \quad x_2 \geq 0$

9.30 Consider the problem of Example 9.5.1. Using (9.43), obtain a simplified closed form expression for $\hat{\mathbf{w}} = (\hat{\mathbf{x}}, \hat{\mathbf{u}}, \hat{\mathbf{v}})$ in terms of $\bar{\mathbf{w}} = (\bar{\mathbf{x}}, \bar{\mathbf{u}}, \bar{\mathbf{v}})$ and $\hat{\mu}$. Hence, obtain the sequence of iterates as it would be generated by the primal–dual path-following algorithm starting with $\bar{\mathbf{x}} = (\frac{2}{9}, \frac{8}{9})^t$, $\bar{\mathbf{u}} = (4, 1)^t$, $\bar{v} = -1$, and $\bar{\mu} = \frac{8}{9}$, using $\theta = \delta = 0.35$.

9.31 Consider the problem of Example 9.5.1. Construct the artificial primal–dual pair of linear Programs P′ and D′ given in (9.57). Using the starting solution (9.58), perform at least two iterations of the primal–dual path-following algorithm.

9.32 Consider the linear programming Problem P to minimize $\mathbf{c}^t\mathbf{x}$ subject to $\mathbf{A}\mathbf{x} = \mathbf{0}$, $\mathbf{e}^t\mathbf{x} = 1$, and $\mathbf{x} \geq \mathbf{0}$, where $\mathbf{A}$ is an $m \times n$ matrix of rank $m < n$, and $\mathbf{e}$ is a vector of n ones. For a given $\mu > 0$, define P_μ to be the problem to minimize $\mathbf{c}^t\mathbf{x} + \mu \sum_{j=1}^{n} x_j \, ln(x_j)$ subject to $\mathbf{A}\mathbf{x} = \mathbf{0}$, and $\mathbf{e}^t\mathbf{x} = 1$, with $\mathbf{x} > \mathbf{0}$ treated implicitly.

a. Show that any linear programming problem of the type to minimize $\mathbf{c}^t\mathbf{x}$ subject to $\mathbf{A}\mathbf{x} = \mathbf{b}$, $\mathbf{x} \geq \mathbf{0}$ that has a bounded feasible region can be transformed into the form of Problem P. (This form of linear program is due to Karmarkar [1984].)

b. What happens to the *negative entropy function* $\mu \sum_{j=1}^{n} x_j \, ln(x_j)$ as some $x_j \to 0^+$? How does this compare with the logarithmic barrier function (9.29b)? Examine the partial derivatives of this negative entropy function with respect to x_j as $x_j \to 0^+$ and, hence, justify why this function might act as a barrier. Hence, show that the optimal solution to P_μ approaches the optimum to P as $\mu \to 0^+$. (This result is due to Fang, [1990].)

c. Consider Problem P to minimize $-x_3$ subject to $x_1 - x_2 = 0$, $x_1 + x_2 + x_3 = 1$, and $\mathbf{x} \geq \mathbf{0}$. For the corresponding Problem P_μ, construct the Lagrangian dual to maximize $\theta(\pi)$, $\pi \in E$, where π is the Lagrange multiplier associated with the constraint $x_1 - x_2 = 0$. Using the KKT conditions, show that this is equivalent to minimizing

$$ln\left[e^{\left(\frac{\pi}{\mu} - 1\right)} + e^{\left(\frac{-\pi}{\mu} - 1\right)} + e^{\left(\frac{1}{\mu} - 1\right)}\right]$$

and, hence, that $\pi = 0$ is optimal. Accordingly, show that P_μ is solved by $x_1 = x_2 = 1/(2 + e^{1/\mu})$ and $x_3 = e^{1/\mu}/(2 + e^{1/\mu})$, and that the limit of this solution as $\mu \to 0^+$ solves P.

9.33 Consider the problem to minimize $\mathbf{c}^t\mathbf{x}$ subject to $\mathbf{A}\mathbf{x} = \mathbf{b}$ and $\mathbf{x} \geq \mathbf{0}$, where $\mathbf{A}$ is an $m \times n$ matrix of rank $m < n$. Suppose that we are given a feasible solution $\bar{\mathbf{x}} > \mathbf{0}$. For a barrier parameter $\mu > 0$, consider Problem BP to minimize $f(\mathbf{x}) \equiv \mathbf{c}^t\mathbf{x} - \mu \sum_{j=1}^{n} ln(x_j)$ subject to $\mathbf{A}\mathbf{x} = \mathbf{b}$, with $\mathbf{x} > \mathbf{0}$ treated implicitly.

a. Find the direction $\mathbf{d}$ that minimizes the second-order Taylor series approximation to $f(\bar{\mathbf{x}} + \mathbf{d})$ subject to $\mathbf{A}\mathbf{d} = \mathbf{0}$. Interpret this as a *projected Newton direction*.

b. Consider the problem $\delta(\bar{\mathbf{x}}, \mu) = \text{minimize}_{(\mathbf{u},\mathbf{v})} \|\frac{1}{\mu} \bar{\mathbf{X}}\mathbf{u} - \mathbf{e}\|$ subject to $\mathbf{A}^t\mathbf{v} + \mathbf{u} = \mathbf{c}$. Find the optimum solution $(\bar{\mathbf{u}}, \bar{\mathbf{v}})$. Interpret this in relation to the dual to P and equation (9.39). Show that the direction of part a satisfies $\mathbf{d} = \bar{\mathbf{x}} - \frac{1}{\mu}\bar{\mathbf{X}}^2\bar{\mathbf{u}}$.

c. Consider the following algorithm due to Roos and Vial [1988]. Start with some $\mu > 0$ and a solution $\bar{\mathbf{x}} > \mathbf{0}$ such that $\mathbf{A}\bar{\mathbf{x}} = \mathbf{b}$ and $\delta(\bar{\mathbf{x}}, \mu) \leq \frac{1}{2}$. Let $\theta = 1/(6\sqrt{n})$.

While the duality gap is not small enough, repeat the main $\mu(1 - \theta)$, computing $\mathbf{d} = \bar{\mathbf{x}} - \frac{1}{\mu}\bar{\mathbf{X}}^2\bar{\mathbf{u}}$ as given by part b, and setting the revised solution $\mathbf{x}$ to $\mathbf{x} + \mathbf{d}$. Illustrate this algorithm for the problem of Example 9.5.1.

Notes and References

The use of penalty functions to solve constrained problems is generally attributed to Courant. Subsequently, Camp [1955] and Pietrgykowski [1962] discussed this approach to solve nonlinear problems. The latter reference also gives a convergence proof. However, significant progress in solving practical problems by the use of penalty methods follows the classic work of Fiacco and McCormick under the title SUMT (sequential unconstrained minimization technique). The interested reader may refer to Fiacco and McCormick [1964, 1966, 1967, 1968] and Zangwill [1967, 1969]. See Himmelblau [1972], Lootsma [1968], and Osborne and Ryan [1970] for the performance of different penalty functions on test problems.

Luenberger [1984] discusses the use of a generalized quadratic penalty function of the form $\mu h(\mathbf{x})^t \Gamma h(\mathbf{x})$, where Γ is a symmetric $l \times l$ positive definite matrix. He also discusses a combination of penalty function and gradient projection methods (see Exercise 9.29). We also refer the reader to Luenberger [1984] for further details on eigenvalue analyses of the Hessians of the various penalty functions and their effect on the convergence characteristics of penalty-based algorithms. Best et al. [1981] discuss how Newton's method can be used to efficiently purify an approximate solution determined by penalty methods to a more accurate optimum. For the use of penalty functions in conducting *sensitivity analyses* in nonlinear programming, see Fiacco [1983].

The barrier function approach was first proposed by Carroll [1961] under the name *created response surface technique*. The approach was used to solve nonlinear inequality constrained problems by Box, Davies, and Swann [1969] and Kowalik [1966]. The barrier function approach has been thoroughly investigated and popularized by Fiacco and McCormick [1964, 1968]. In Exercise 9.23, we introduce the mixed penalty–barrier auxiliary functions studied by Fiacco and McCormick [1968]. Here, equality and inequality constraints are handled, respectively, by a penalty term and a barrier term. Also see Belmore, Greenberg, and Jarvis [1970], Greenberg [1973], and Raghavendra and Rao [1973].

The numerical problem of how to change the parameters of the penalty and barrier functions have been investigated by several authors. See Fiacco and McCormick [1968] and Himmelblau [1972] for a detailed discussion. These same references also give the computational experience for numerous problems. Bazaraa [1975], Lasdon [1972], and Lasdon and Ratner [1973] discuss effective unconstrained optimization algorithms for solving penalty and barrier functions. For eigenvalue analyses relating to the convergence behavior of algorithms applied to the barrier function problem, see Luenberger [1984].

Several extensions to the concepts of penalty and barrier functions have been made. First, in order to avoid the difficulties associated with ill-conditioning as the penalty parameter approaches infinity and as the barrier parameter approaches zero, several parameter-free methods have been proposed. This concept was introduced in Exercises 9.26 and 9.27. For further details of this topic, see Fiacco and McCormick [1968], Huard's method of centers [1967], and Lootsma [1968].

Through the absolute value l_1 penalty function, we introduced the concept of exact penalty functions in which a single unconstrained minimization problem, with a reasonably sized penalty parameter, can yield an optimum solution to the original problem. This was first introduced by Pietrzykowski [1969] and Fletcher [1970], and has been studied by Coleman and Conn [1982a, b], Conn [1985], Han [1977], Evans et al. [1973], Fletcher [1973, 1981, 1984], Gill et al. [1981], and Mayne [1980].

Another popular and useful exact penalty approach uses both a Lagrangian multiplier term and a penalty term in the auxiliary function as in the method of multipliers. This approach was independently proposed by Hestenes [1969] and Powell [1969]. Again, the motivation is to avoid the ill-conditioning difficulties encountered by the classical approach as the penalty parameter approaches infinity. For further details, refer to Bertsekas [1975, 1976a, b], Boggs and Tolle [1980], Fletcher [1975, 1987], Hestenes [1980b], Miele et al. [1971], Pierre and Lowe [1975], Rockafellar [1973a, b, 1974], and Tapia [1977]. Conn et al. [1988] discuss a method for effectively incorporating simple bounds within the constraint set. Fletcher [1984] also suggests that in a mix of nonlinear and linear constraints, it might be worthwhile to incorporate only the nonlinear constraints in the penalty function. Sen and Sherali [1985] and Sherali and Ulular [1989] discuss a primal–dual conjugate subgradient algorithm for solving differentiable or nondifferentiable, decomposable problems using ALAG penalty functions and Lagrangian dual functions in concert with each other. Polyak and Tret'iakov [1972] discuss a finite ALAG approach for linear and quadratic programming problems.

In Section 9.5, we have presented a primal–dual path-following algorithm as developed by Monteiro and Adler [1989a] based on Frisch's [1955] logarithmic barrier function and motivated by Karmarkar's [1984] polynomial-time algorithm for linear programming. This algorithm also readily extends to solve convex quadratic programs with the same order of complexity; see Monteiro and Adler [1989b]. For a discussion on complexity issues in linear programming and purification schemes, we refer the reader to Bazaraa et al. [1990] and Murty [1983]. For an extension of this algorithmic concept to one which employs both first-order and higher-order power series approximations to a weighted barrier path, see Monteiro et al. [1990]. Computational aspects and implementation details of this class of algorithms is discussed in Choi et al. [1990], McShane et al. [1989], Lustig et al. [1989], and Mehrotra [1990]; and many useful ideas such as extrapolation of iterates generated to accelerate convergence can be traced to Fiacco and McCormick [1968]. Tapia and Zhang [1991] present variants that achieve superlinear convergence rates while preserving the polynomial-time behavior. Alternative path-following algorithms are described in Gonzaga [1987], Renegar [1988], Roos and Vial [1988], and Vaidya [1987], and are motivated by Sonnevend's [1985] method of centers and Megiddo's [1986] trajectory to optimality. A nonpolynomial-time algorithm based on using Newton's method with the inverse barrier function is discussed by den Hertog et al. [1991]. Todd [1989] gives an excellent survey of other variants of Karmarkar's algorithm.

Chapter 10

Methods of Feasible Directions

This class of methods solves a nonlinear programming problem by moving from a feasible point to an improved feasible point. The following strategy is typical of feasible directions algorithms. Given a feasible point $\mathbf{x}_k$, a direction $\mathbf{d}_k$ is determined such that for $\lambda > 0$ and sufficiently small, the following two properties are true: (1) $\mathbf{x}_k + \lambda\mathbf{d}_k$ is feasible, and (2) the objective value at $\mathbf{x}_k + \lambda\mathbf{d}_k$ is better than the objective value at $\mathbf{x}_k$. After such a direction is determined, a one-dimensional optimization problem is solved to determine how far to proceed along $\mathbf{d}_k$. This leads to a new point $\mathbf{x}_{k+1}$, and the process is repeated. Since primal feasibility is maintained during this optimization process, these procedures are often referred to as *primal methods*. Methods of this type are frequently shown to converge to KKT solutions or, sometimes, to FJ points. We ask the reader to review Chapter 4 to assess the worth of such solutions.
The following is an outline of the chapter:

Section 10.1: The Method of Zoutendijk This section shows how to generate an improving feasible direction by solving a subproblem that is usually a linear program. Problems with linear constraints and problems with nonlinear constraints are both considered.

Section 10.2: Convergence Analysis of the Method of Zoutendijk We show that the algorithmic map of Section 10.1 is not closed, so that convergence is not guaranteed. A modification of the basic algorithm credited to Topkis and Veinott [1967] that guarantees convergence is presented.

Section 10.3: Successive Linear Programming Approach We describe a penalty-based successive linear programming approach that combines the ideas of sequentially using a linearized feasible direction subproblem along with l_1

penalty function concepts to yield an efficient and convergent algorithm. For vertex optimal solutions, a quadratic convergence rate is possible but, otherwise, convergence can be slow.

Section 10.4: Successive Quadratic Programming or Projected Lagrangian Approach To obtain a quadratic or superlinear convergence behavior for even nonvertex solutions, one can adopt Newton's method or quasi-Newton methods to solve the KKT optimality conditions. This leads to a successive quadratic programming or projected Lagrangian approach in which the feasible direction subproblem is a quadratic problem, with the Hessian of the objective function being that of the Lagrangian function and the constraints representing first-order approximations. We describe both a rudimentary version of this method as well as a globally convergent variant using the l_1 penalty function as either a merit function or, more actively, using it in the subproblem objective function itself. The associated Maratos effect is also discussed.

Section 10.5: The Gradient Projection Method of Rosen This section describes how to generate an improving feasible direction for a problem with linear constraints by projecting the gradient of the objective function on the null space of the gradients of the binding constraints. A convergent variant is also presented.

Section 10.6: The Method of Reduced Gradient of Wolfe and the Generalized Reduced Gradient Method The variables are represented in terms of an independent subset of the variables. For a problem with linear constraints, an improving feasible direction is determined based on the gradient vector in the reduced space. A *generalized reduced gradient* variant for nonlinear constraints is also discussed.

Section 10.7: The Convex–Simplex Method of Zangwill We describe the convex-simplex method for solving a nonlinear program in the presence of linear constraints. The method is identical to the reduced gradient method except that an improving feasible direction is determined by modifying only one nonbasic variable and adjusting the basic variables accordingly. If the objective function is linear, then the convex–simplex method reduces to the simplex method of linear programming.

Section 10.8: Effective First- and Second-Order Variants of the Reduced Gradient Method We unify and extend the reduced gradient and convex simplex methods, introducing the concept of suboptimization through the use of superbasic variables. We also discuss the use of second-order functional approximations for finding a direction of motion in the reduced space of the superbasic variables.

10.1 The Method of Zoutendijk

In this section, we describe the method of feasible directions of Zoutendijk. At each iteration, the method generates an improving feasible direction and then optimizes along that direction. Definition 10.1.1 reiterates the notion of an improving feasible direction from Chapter 4.

10.1.1 Definition

Consider the problem to minimize $f(\mathbf{x})$ subject to $\mathbf{x} \in S$, where $f: E_n \to E_1$ and S is a nonempty set in E_n. A nonzero vector $\mathbf{d}$ is called a *feasible direction* at $\mathbf{x} \in S$ if there exists a $\delta > 0$ such that $\mathbf{x} + \lambda\mathbf{d} \in S$ for all $\lambda \in (0, \delta)$. Furthermore, $\mathbf{d}$ is called an *improving feasible direction* at $\mathbf{x} \in S$ if there exists a $\delta > 0$ such that $f(\mathbf{x} + \lambda\mathbf{d}) < f(\mathbf{x})$ and $\mathbf{x} + \lambda\mathbf{d} \in S$ for all $\lambda \in (0, \delta)$.

The Case of Linear Constraints

We first consider the case where the feasible region S is defined by a system of linear constraints, so that the problem under consideration is of the form

$$\begin{array}{ll} \text{Minimize} & f(\mathbf{x}) \\ \text{subject to} & \mathbf{Ax} \leq \mathbf{b} \\ & \mathbf{Ex} = \mathbf{e} \end{array}$$

Here, $\mathbf{A}$ is an $m \times n$ matrix, $\mathbf{E}$ is an $l \times n$ matrix, $\mathbf{b}$ is an m vector, and $\mathbf{e}$ is an l vector. Lemma 10.1.2 gives a suitable characterization of a feasible direction and a sufficient condition for an improving direction. In particular, $\mathbf{d}$ is an improving feasible direction if $\mathbf{A}_1\mathbf{d} \leq \mathbf{0}$, $\mathbf{Ed} = \mathbf{0}$, and $\nabla f(\mathbf{x})'\mathbf{d} < 0$. The proof of the lemma is straightforward and is left as an exercise to the reader (see Theorem 4.1.2 and Exercise 10.1).

10.1.2 Lemma

Consider the problem to minimize $f(\mathbf{x})$ subject to $\mathbf{Ax} \leq \mathbf{b}$ and $\mathbf{Ex} = \mathbf{e}$. Let $\mathbf{x}$ be a feasible solution, and suppose that $\mathbf{A}_1\mathbf{x} = \mathbf{b}_1$ and $\mathbf{A}_2\mathbf{x} < \mathbf{b}_2$, where $\mathbf{A}'$ is decomposed into $(\mathbf{A}_1', \mathbf{A}_2')$ and $\mathbf{b}'$ is decomposed into $(\mathbf{b}_1', \mathbf{b}_2')$. Then, a nonzero vector $\mathbf{d}$ is a feasible direction at $\mathbf{x}$ if and only if $\mathbf{A}_1\mathbf{d} \leq \mathbf{0}$ and $\mathbf{Ed} = \mathbf{0}$. If $\nabla f(\mathbf{x})'\mathbf{d} < 0$, then $\mathbf{d}$ is an improving direction.

Geometric Interpretation of Improving Feasible Directions

We now illustrate the set of improving feasible directions geometrically by the following example.

10.1.3 Example

Consider the following problem:

$$\begin{array}{ll} \text{Minimize} & (x_1 - 6)^2 + (x_2 - 2)^2 \\ \text{subject to} & -x_1 + 2x_2 \leq 4 \\ & 3x_1 + 2x_2 \leq 12 \\ & -x_1 \leq 0 \\ & -x_2 \leq 0 \end{array}$$

Let $\mathbf{x} = (2, 3)'$ and note that the first two constraints are binding. In particular, the matrix $\mathbf{A}_1$ of Lemma 10.1.2 is given by

$$\mathbf{A}_1 = \begin{bmatrix} -1 & 2 \\ 3 & 2 \end{bmatrix}$$

Therefore, $\mathbf{d}$ is a feasible direction at $\mathbf{x}$ if and only if $\mathbf{A}_1\mathbf{d} \leq \mathbf{0}$, that is, if and only if

$$-d_1 + 2d_2 \leq 0$$

$$3d_1 + 2d_2 \leq 0$$

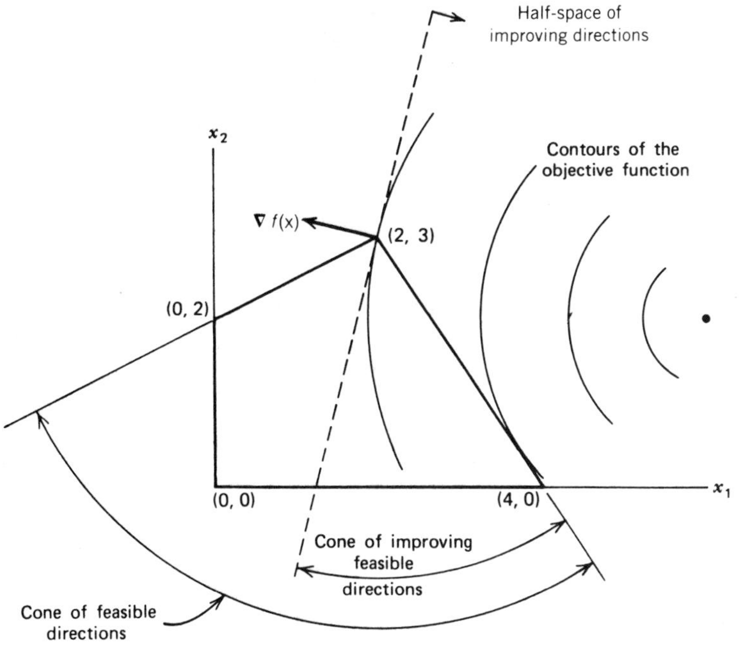

Figure 10.1 Illustration of improving feasible directions.

The collection of these directions, where the origin is translated to a point $\mathbf{x}$ for convenience, forms the cone of feasible directions shown in Figure 10.1. Note that if we move a short distance starting from $\mathbf{x}$ along any vector $\mathbf{d}$ satisfying the above two inequalities, we would remain in the feasible region.

If a vector $\mathbf{d}$ satisfies $0 > \nabla f(\mathbf{x})^t\mathbf{d} = -8d_1 + 2d_2$, then $\mathbf{d}$ is an improving direction. Thus, the collection of improving directions is given by the open half-space $\{(d_1, d_2): -8d_1 + 2d_2 < 0\}$. The intersection of the cone of feasible directions with this half-space gives the set of all improving feasible directions.

Generating Improving Feasible Directions

Given a feasible point $\mathbf{x}$, as shown in Lemma 10.1.2, a nonzero vector $\mathbf{d}$ is an improving feasible direction if $\nabla f(\mathbf{x})^t\mathbf{d} < 0$, $\mathbf{A}_1\mathbf{d} \le \mathbf{0}$, and $\mathbf{Ed} = \mathbf{0}$. A natural method for generating such a direction is to minimize $\nabla f(\mathbf{x})^t\mathbf{d}$ subject to the constraints $\mathbf{A}_1\mathbf{d} \le \mathbf{0}$ and $\mathbf{Ed} = \mathbf{0}$. Note, however, if a vector $\bar{\mathbf{d}}$ such that $\nabla f(\mathbf{x})^t\bar{\mathbf{d}} < 0$, $\mathbf{A}_1\bar{\mathbf{d}} \le \mathbf{0}$, and $\mathbf{E}\bar{\mathbf{d}} = \mathbf{0}$ exists, then the optimal objective value of the foregoing problem is $-\infty$ by considering $\lambda\bar{\mathbf{d}}$, where λ is arbitrarily large. Thus, a constraint that bounds the vector $\mathbf{d}$ or the objective function must be introduced. Such a restriction is usually referred to as a *normalization constraint*. We present below three problems for generating an improving feasible direction. Each problem uses a different normalization constraint.

Problem P1: Minimize $\nabla f(\mathbf{x})^t\mathbf{d}$
subject to

$$\mathbf{A}_1\mathbf{d} \le \mathbf{0}$$
$$\mathbf{Ed} = \mathbf{0}$$
$$-1 \le \mathbf{d}_j \le 1 \qquad \text{for } j = 1, \ldots, n$$

Problem P2: Minimize $\nabla f(\mathbf{x})^t \mathbf{d}$
subject to
$$\mathbf{A}_1\mathbf{d} \le \mathbf{0}$$
$$\mathbf{E}\mathbf{d} = \mathbf{0}$$
$$\mathbf{d}^t\mathbf{d} \le 1$$

Problem P3: Minimize $\nabla f(\mathbf{x})^t \mathbf{d}$
subject to
$$\mathbf{A}_1\mathbf{d} \le \mathbf{0}$$
$$\mathbf{E}\mathbf{d} = \mathbf{0}$$
$$\nabla f(\mathbf{x})^t\mathbf{d} \ge -1$$

Problems P1 and P3 are linear in the variables $d_1, \ldots, d_n$ and can be solved by the simplex method. Problem P2 contains a quadratic constraint but could be considerably simplified (see Exercise 10.31). Since $\mathbf{d} = \mathbf{0}$ is a feasible solution to each of the above problems, and since its objective value is equal to 0, the optimal objective value of Problems P1, P2, and P3 cannot be positive. If the minimal objective function value of Problems P1, P2, and P3 is negative, then, by Lemma 10.1.2, an improving feasible direction is generated. On the other hand, if the minimal objective function value is equal to zero, then $\mathbf{x}$ is a KKT point as shown below.

10.1.4 Lemma

Consider the problem to minimize $f(\mathbf{x})$ subject to $\mathbf{A}\mathbf{x} \le \mathbf{b}$ and $\mathbf{E}\mathbf{x} = \mathbf{e}$. Let $\mathbf{x}$ be a feasible solution such that $\mathbf{A}_1\mathbf{x} = \mathbf{b}_1$ and $\mathbf{A}_2\mathbf{x} < \mathbf{b}_2$, where $\mathbf{A}^t = (\mathbf{A}_1^t, \mathbf{A}_2^t)$ and $\mathbf{b}^t = (\mathbf{b}_1^t, \mathbf{b}_2^t)$. Then, for each $i = 1, 2, 3$, $\mathbf{x}$ is a KKT point if and only if the optimal objective value of Problem Pi is equal to zero.

Proof

The vector $\mathbf{x}$ is a KKT point if and only if there exist a vector $\mathbf{u} \ge \mathbf{0}$ and a vector $\mathbf{v}$ such that $\nabla f(\mathbf{x}) + \mathbf{A}_1^t\mathbf{u} + \mathbf{E}^t\mathbf{v} = \mathbf{0}$. By Corollary 3 to Theorem 2.4.5, this system is solvable if and only if the system $\nabla f(\mathbf{x})^t\mathbf{d} < 0$, $\mathbf{A}_1\mathbf{d} \le \mathbf{0}$, $\mathbf{E}\mathbf{d} = \mathbf{0}$ has no solution, that is, if and only if the optimal objective value of Problems P1, P2, and P3 is equal to zero. This completes the proof.

The Line Search

So far, we have seen how to generate an improving feasible direction or conclude that the current vector is a KKT point. Now let $\mathbf{x}_k$ be the current vector, and let $\mathbf{d}_k$ be an improving feasible direction. The next point $\mathbf{x}_{k+1}$ is given by $\mathbf{x}_k + \lambda_k\mathbf{d}_k$, where the step size λ_k is obtained by solving the following one-dimensional problem:

$$\begin{aligned} \text{Minimize} \quad & f(\mathbf{x}_k + \lambda\mathbf{d}_k) \\ \text{subject to} \quad & \mathbf{A}(\mathbf{x}_k + \lambda\mathbf{d}_k) \le \mathbf{b} \\ & \mathbf{E}(\mathbf{x}_k + \lambda\mathbf{d}_k) = \mathbf{e} \\ & \lambda \ge 0 \end{aligned}$$

Now suppose that $\mathbf{A}^t$ is decomposed into $(\mathbf{A}_1^t, \mathbf{A}_2^t)$ and $\mathbf{b}^t$ is decomposed into $(\mathbf{b}_1^t, \mathbf{b}_2^t)$ such that $\mathbf{A}_1\mathbf{x}_k = \mathbf{b}_1$ and $\mathbf{A}_2\mathbf{x}_k < \mathbf{b}_2$. Then, the above problem could be simplified as follows. First note that $\mathbf{E}\mathbf{x}_k = \mathbf{e}$ and $\mathbf{E}\mathbf{d}_k = \mathbf{0}$, so that the constraint $\mathbf{E}(\mathbf{x}_k + \lambda\mathbf{d}_k) = \mathbf{e}$ is redundant. Since $\mathbf{A}_1\mathbf{x}_k = \mathbf{b}_1$ and $\mathbf{A}_1\mathbf{d}_k \le \mathbf{0}$, then $\mathbf{A}_1(\mathbf{x}_k + \lambda\mathbf{d}_k) \le \mathbf{b}_1$ for all $\lambda \ge 0$. Hence, we only need to restrict λ so that $\lambda\mathbf{A}_2\mathbf{d}_k \le \mathbf{b}_2 - \mathbf{A}_2\mathbf{x}_k$. It thus follows that the above problem reduces to the following line search problem, which could be solved by one of the techniques discussed in Sections 8.1–8.3:

$$\begin{array}{ll} \text{Minimize} & f(\mathbf{x}_k + \lambda \mathbf{d}_k) \\ \text{subject to} & 0 \le \lambda \le \lambda_{\max} \end{array}$$

where

$$\lambda_{\max} = \begin{cases} \text{minimum}\,\{\hat{b}_i/\hat{d}_i : \hat{d}_i > 0\} & \text{if } \hat{\mathbf{d}} \not\leq \mathbf{0} \\ \infty & \text{if } \hat{\mathbf{d}} \le \mathbf{0} \end{cases} \tag{10.1}$$

$$\hat{\mathbf{b}} = \mathbf{b}_2 - \mathbf{A}_2\mathbf{x}_k$$

$$\hat{\mathbf{d}} = \mathbf{A}_2\mathbf{d}_k$$

Summary of the Method of Zoutendijk (Case of Linear Constraints)

We summarize below Zoutendijk's method for minimizing a differentiable function f in the presence of linear constraints of the form $\mathbf{Ax} \le \mathbf{b}$ and $\mathbf{Ex} = \mathbf{e}$.

Initialization Step Find a starting feasible solution $\mathbf{x}_1$ with $\mathbf{Ax}_1 \le \mathbf{b}$ and $\mathbf{Ex}_1 = \mathbf{e}$. Let $k = 1$ and go to the main step.

Main Step

1. Given $\mathbf{x}_k$, suppose that $\mathbf{A}^t$ and $\mathbf{b}^t$ are decomposed into $(\mathbf{A}_1^t, \mathbf{A}_2^t)$ and $(\mathbf{b}_1^t, \mathbf{b}_2^t)$ so that $\mathbf{A}_1\mathbf{x}_k = \mathbf{b}_1$ and $\mathbf{A}_2\mathbf{x}_k < \mathbf{b}_2$. Let $\mathbf{d}_k$ be an optimal solution to the following problem (note that Problem P2 or Problem P3 could be used instead):

$$\begin{array}{lll} \text{Minimize} & \nabla f(\mathbf{x}_k)^t\mathbf{d} & \\ \text{subject to} & \mathbf{A}_1\mathbf{d} \le \mathbf{0} & \\ & \mathbf{Ed} = \mathbf{0} & \\ & -1 \le d_j \le 1 & \text{for } j = 1, \ldots, n \end{array}$$

   If $\nabla f(\mathbf{x}_k)^t\mathbf{d}_k = 0$, stop; $\mathbf{x}_k$ is a KKT point, with the dual variables to the foregoing problem giving the corresponding Lagrange multipliers. Otherwise, go to step 2.

2. Let λ_k be an optimal solution to the following line search problem:

$$\begin{array}{ll} \text{Minimize} & f(\mathbf{x}_k + \lambda \mathbf{d}_k) \\ \text{subject to} & 0 \le \lambda \le \lambda_{\max} \end{array}$$

   where $\lambda_{\max}$ is determined according to (10.1). Let $\mathbf{x}_{k+1} = \mathbf{x}_k + \lambda_k\mathbf{d}_k$, identify the new set of binding constraints at $\mathbf{x}_{k+1}$, and update $\mathbf{A}_1$ and $\mathbf{A}_2$ accordingly. Replace k by $k + 1$ and repeat step 1.

10.1.5 Example

Consider the following problem:

$$\begin{array}{ll} \text{Minimize} & 2x_1^2 + 2x_2^2 - 2x_1x_2 - 4x_1 - 6x_2 \\ \text{subject to} & x_1 + x_2 \le 2 \\ & x_1 + 5x_2 \le 5 \\ & -x_1 \le 0 \\ & -x_2 \le 0 \end{array}$$

Note that $\nabla f(\mathbf{x}) = (4x_1 - 2x_2 - 4, 4x_2 - 2x_1 - 6)^t$. We solve the problem using

Zoutendijk's procedure starting from the initial point $\mathbf{x}_1 = (0, 0)^t$. Each iteration of the algorithm consists of the solution of a subproblem given in step 1 to find the search direction and then a line search along this direction.

Iteration 1

Search Direction At $\mathbf{x}_1 = (0, 0)^t$, we have $\nabla f(\mathbf{x}_1) = (-4, -6)^t$. Furthermore, at the point $\mathbf{x}_1$, only the nonnegativity constraints are binding so that $I = \{3, 4\}$. The direction-finding problem is given by

$$\begin{array}{ll} \text{Minimize} & -4d_1 - 6d_2 \\ \text{subject to} & -d_1 \leq 0 \\ & -d_2 \leq 0 \\ & -1 \leq d_1 \leq 1 \\ & -1 \leq d_2 \leq 1 \end{array}$$

This problem can be solved by the simplex method for linear programming; the optimal solution is $\mathbf{d}_1 = (1, 1)^t$, and the optimal objective value for the direction-finding problem is -10. Figure 10.2 gives the feasible region for the subproblem, and the reader can readily verify geometrically that (1, 1) is indeed the optimal solution.

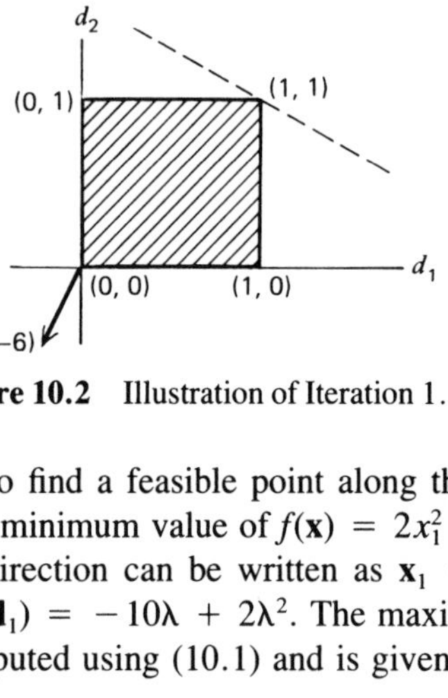

Figure 10.2 Illustration of Iteration 1.

Line Search We now need to find a feasible point along the direction (1, 1) starting from the point (0, 0) with a minimum value of $f(\mathbf{x}) = 2x_1^2 + 2x_2^2 - 2x_1x_2 - 4x_1 - 6x_2$. Any point along this direction can be written as $\mathbf{x}_1 + \lambda\mathbf{d}_1 = (\lambda, \lambda)^t$ and the objective function $f(\mathbf{x}_1 + \lambda\mathbf{d}_1) = -10\lambda + 2\lambda^2$. The maximum value of λ for which $\mathbf{x}_1 + \lambda\mathbf{d}_1$ is feasible is computed using (10.1) and is given by

$$\lambda_{\max} = \text{minimum } \left\{\tfrac{2}{2}, \tfrac{5}{6}\right\} = \tfrac{5}{6}$$

Hence, if $\mathbf{x}_1 + \lambda_1\mathbf{d}_1$ is the new point, the value of λ_1 is obtained by solving the following one-dimensional search problem:

$$\begin{array}{ll} \text{Minimize} & -10\lambda + 2\lambda^2 \\ \text{subject to} & 0 \leq \lambda \leq \tfrac{5}{6} \end{array}$$

Since the objective function is convex and the unconstrained minimum occurs at $\frac{5}{2}$, the solution is $\lambda_1 = \frac{5}{6}$, so that $\mathbf{x}_2 = \mathbf{x}_1 + \lambda_1\mathbf{d}_1 = (\frac{5}{6}, \frac{5}{6})^t$.

Iteration 2

Search Direction At the point $\mathbf{x}_2 = (\frac{5}{6}, \frac{5}{6})^t$, we have $\nabla f(\mathbf{x}_2) = (-\frac{7}{3} - \frac{13}{3})^t$. Furthermore, the set of binding constraints at the point $\mathbf{x}_2$ is given by $I = \{2\}$, so that the direction to move is obtained by solving the following problem:

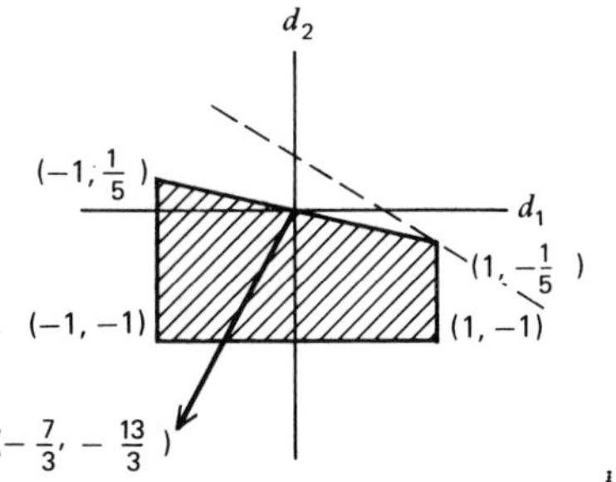

Figure 10.3 Illustration of Iteration 2.

$$\begin{array}{ll} \text{Minimize} & -\frac{7}{3}d_1 - \frac{13}{3}d_2 \\ \text{subject to} & d_1 + 5d_2 \leq 0 \\ & -1 \leq d_1 \leq 1 \\ & -1 \leq d_2 \leq 1 \end{array}$$

The reader can verify from Figure 10.3 that the optimal solution to the above linear program is $\mathbf{d}_2 = (1, -\frac{1}{5})^t$, and the corresponding objective function value is $-\frac{22}{15}$.

Line Search Starting from the point $\mathbf{x}_2$, any point in the direction $\mathbf{d}_2$ can be written as $\mathbf{x}_2 + \lambda\mathbf{d}_2 = (\frac{5}{6} + \lambda, \frac{5}{6} - \frac{1}{5}\lambda)$, and the corresponding objective function value is $f(\mathbf{x}_2 + \lambda\mathbf{d}_2) = -\frac{125}{8} - \frac{22}{15}\lambda + \frac{62}{25}\lambda^2$. The maximum value of λ for which $\mathbf{x}_2 + \lambda\mathbf{d}_2$ is feasible is obtained from (10.1) as

$$\lambda_{\max} = \text{minimum}\left\{\frac{1/3}{4/5}, \frac{5/6}{1/5}\right\} = \frac{5}{12}$$

Therefore, λ_2 is the optimal solution to the following problem:

$$\begin{array}{ll} \text{Minimize} & -\frac{125}{8} - \frac{22}{15}\lambda + \frac{62}{25}\lambda^2 \\ \text{subject to} & 0 \leq \lambda \leq \frac{5}{12} \end{array}$$

The optimal solution is $\lambda_2 = \frac{55}{186}$, the unconstrained minimizer of the objective function, so that $\mathbf{x}_3 = \mathbf{x}_2 + \lambda_2\mathbf{d}_2 = (\frac{35}{31}, \frac{24}{31})^t$.

Iteration 3

Search Direction At $\mathbf{x}_3 = (\frac{35}{31}, \frac{24}{31})^t$, we have $\nabla f(\mathbf{x}_3) = (-\frac{32}{31}, -\frac{160}{31})^t$. Furthermore, the set of binding constraints at the point $\mathbf{x}_3$ is given $I = \{2\}$, so that the direction to move is obtained by solving the following problem:

$$\begin{array}{ll} \text{Minimize} & -\frac{32}{31}d_1 - \frac{160}{31}d_2 \\ \text{subject to} & d_1 + 5d_2 \leq 0 \\ & -1 \leq d_1 \leq 1 \\ & -1 \leq d_2 \leq 1 \end{array}$$

The reader can easily verify from Figure 10.4 that $\mathbf{d}_3 = (1, -\frac{1}{5})^t$ indeed solves the above linear program, with the Lagrange multiplier associated with the first constraint being $\frac{32}{31}$ and zero for the other constraints. The corresponding objective function value is zero, and the procedure is terminated. Furthermore, $\bar{\mathbf{x}} = \mathbf{x}_3 = (\frac{35}{31}, \frac{24}{31})^t$ is a KKT point with the only nonzero Lagrange multiplier being associated with $x_1 + 5x_2 \leq 5$ and equal to $\frac{32}{31}$. (Verify this graphically from Figure 10.5). In this particular problem, f is convex, and by Theorem 4.3.8, $\bar{\mathbf{x}}$ is indeed the optimal solution.

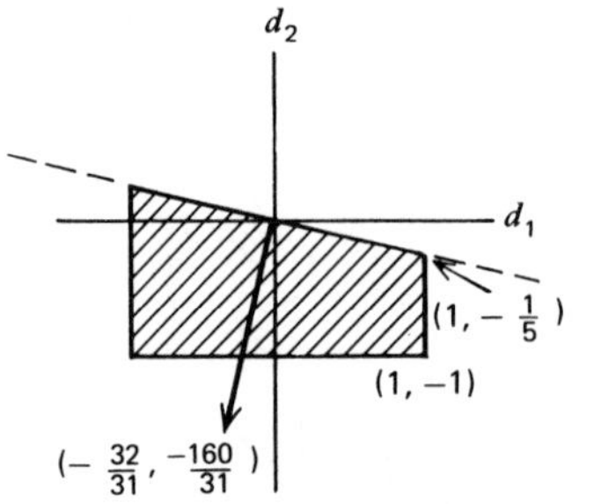

Figure 10.4 Termination at Iteration 3.

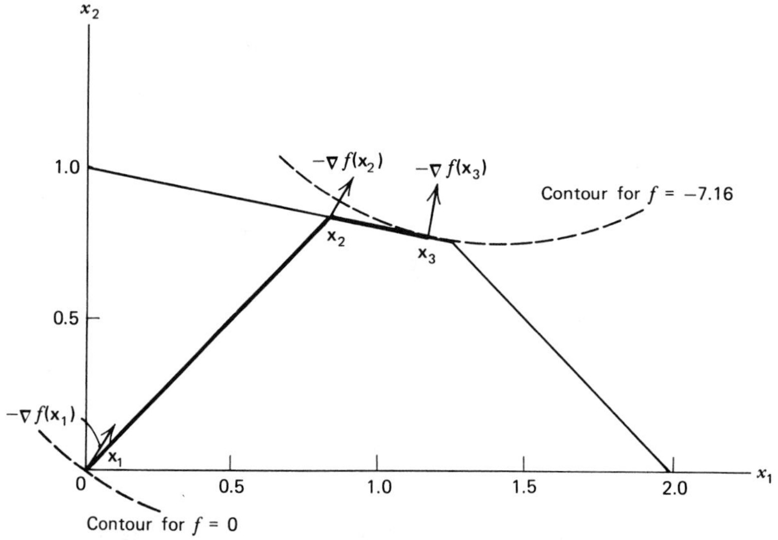

Figure 10.5 Illustration of the method of Zoutendijk for linear constraints.

Table 10.1 below summarizes the computations for solving the problem. The progress of the algorithm is shown in Figure 10.5.

Table 10.1 Summary of Computations for the Method of Zoutendijk with Linear Constraints

Iteration			Search Direction				Line Search		
k	$\mathbf{x}_k$	$f(\mathbf{x}_k)$	$\nabla f(\mathbf{x}_k)$	I	$\mathbf{d}_k$	$\nabla f(\mathbf{x}_k)^t\mathbf{d}_k$	$\lambda_{\max}$	λ_k	$\mathbf{x}_{k+1}$
1	$(0, 0)$	0	$(-4, -6)$	$\{3, 4\}$	$(1, 1)$	-10	$\frac{5}{6}$	$\frac{5}{6}$	$(\frac{5}{6}, \frac{5}{6})$
2	$(\frac{5}{6}, \frac{5}{6})$	-6.94	$(-\frac{7}{3}, -\frac{13}{3})$	$\{2\}$	$(1, -\frac{1}{5})$	$-\frac{22}{15}$	$\frac{5}{12}$	$\frac{55}{186}$	$(\frac{35}{31}, \frac{24}{31})$
3	$(\frac{35}{31}, \frac{24}{31})$	-7.16	$(-\frac{32}{31}, -\frac{160}{31})$	$\{2\}$	$(1, -\frac{1}{5})$	0			

Problems with Nonlinear Inequality Constraints

We now consider the following problem, where the feasible region is defined by a system of inequality constraints that are not necessarily linear:

$$\begin{array}{lll} \text{Minimize} & f(\mathbf{x}) & \\ \text{subject to} & g_i(\mathbf{x}) \leq 0 & \text{for } i = 1, \ldots, m \end{array}$$

Theorem 10.1.6 below gives a sufficient condition for a vector $\mathbf{d}$ to be an improving feasible direction.

10.1.6 Theorem

Consider the problem to minimize $f(\mathbf{x})$ subject to $g_i(\mathbf{x}) \leq 0$ for $i = 1, \ldots, m$. Let $\mathbf{x}$ be a feasible solution, and let I be the set of binding constraints, that is, $I = \{i : g_i(\mathbf{x}) = 0\}$. Furthermore, suppose that f and g_i for $i \in I$ are differentiable at $\mathbf{x}$ and that each g_i for $i \notin I$ is continuous at $\mathbf{x}$. If $\nabla f(\mathbf{x})^t\mathbf{d} < 0$ and $\nabla g_i(\mathbf{x})^t\mathbf{d} < 0$ for $i \in I$, then $\mathbf{d}$ is an improving feasible direction.

Proof

Let $\mathbf{d}$ satisfy $\nabla f(\mathbf{x})^t\mathbf{d} < 0$ and $\nabla g_i(\mathbf{x})^t\mathbf{d} < 0$ for $i \in I$. For $i \notin I$, $g_i(\mathbf{x}) < 0$ and g_i is continuous at $\mathbf{x}$ so that $g_i(\mathbf{x} + \lambda\mathbf{d}) \leq 0$ for $\lambda > 0$ and small enough. By differentiability of g_i for $i \in I$,

$$g_i(\mathbf{x} + \lambda\mathbf{d}) = g_i(\mathbf{x}) + \lambda\nabla g_i(\mathbf{x})^t\mathbf{d} + \lambda\|\mathbf{d}\|\alpha(\mathbf{x};\lambda\mathbf{d})$$

where $\alpha(\mathbf{x};\lambda\mathbf{d}) \to 0$ as $\lambda \to 0$. Since $\nabla g_i(\mathbf{x})^t\mathbf{d} < 0$, then $g_i(\mathbf{x} + \lambda\mathbf{d}) < g_i(\mathbf{x}) = 0$ for $\lambda > 0$ and small enough. Hence, $g_i(\mathbf{x} + \lambda\mathbf{d}) \leq 0$ for $i = 1, \ldots, m$, that is, $\mathbf{x} + \lambda\mathbf{d}$ is feasible for $\lambda > 0$ and small enough. By a similar argument, since $\nabla f(\mathbf{x})^t\mathbf{d} < 0$, we get $f(\mathbf{x} + \lambda\mathbf{d}) < f(\mathbf{x})$ for $\lambda > 0$ and small enough. Hence, $\mathbf{d}$ is an improving and feasible direction. This completes the proof.

Figure 10.6 illustrates the collection of improving feasible directions at $\bar{\mathbf{x}}$. A vector $\mathbf{d}$ satisfying $\nabla g_i(\bar{\mathbf{x}})^t\mathbf{d} = 0$ is tangential to the set $\{\mathbf{x} : g_i(\mathbf{x}) = 0\}$ at $\bar{\mathbf{x}}$. Because of the nonlinearity of g_i, moving along such a vector $\mathbf{d}$ may lead to infeasible points, thus necessitating the strict inequality $\nabla g_i(\bar{\mathbf{x}})^t\mathbf{d} < 0$.

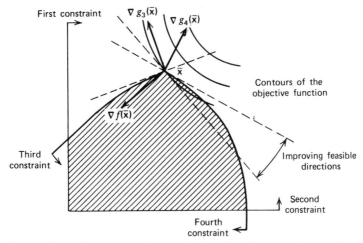

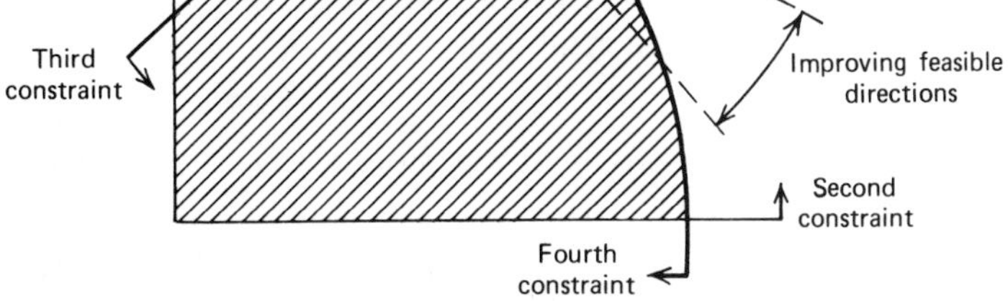

Figure 10.6 Illustration of improving feasible directions for nonlinear constraints.

To find a vector $\mathbf{d}$ satisfying $\nabla f(\mathbf{x})^t\mathbf{d} < 0$ and $\nabla g_i(\mathbf{x})^t\mathbf{d} < 0$ for $i \in I$, it is only natural to minimize the maximum of $\nabla f(\mathbf{x})^t\mathbf{d}$ and $\nabla g_i(\mathbf{x})^t\mathbf{d}$ for $i \in I$. Denoting this maximum by z, and introducing the normalization restrictions $-1 \leq d_j \leq 1$ for each j, we get the following direction-finding problem:

$$\begin{array}{lll} \text{Minimize} & z & \\ \text{subject to} & \nabla f(\mathbf{x})^t\mathbf{d} - z \leq 0 & \\ & \nabla g_i(\mathbf{x})^t\mathbf{d} - z \leq 0 & \text{for } i \in I \\ & -1 \leq d_j \leq 1 & \text{for } j = 1, \ldots, n \end{array}$$

Let $(\bar{z}, \bar{\mathbf{d}})$ be an optimal solution to the above linear program. If $\bar{z} < 0$, then $\bar{\mathbf{d}}$ is obviously an improving feasible direction. If, on the other hand, $\bar{z} = 0$, then the current vector is a Fritz John point, as demonstrated below.

10.1.7 Theorem

Consider the problem to minimize $f(\mathbf{x})$ subject to $g_i(\mathbf{x}) \leq 0$ for $i = 1, \ldots, m$. Let $\mathbf{x}$ be a feasible solution, and let $I = \{i : g_i(\mathbf{x}) = 0\}$. Consider the following direction-finding problem:

$$\begin{array}{lll} \text{Minimize} & z & \\ \text{subject to} & \nabla f(\mathbf{x})^t\mathbf{d} - z \leq 0 & \\ & \nabla g_i(\mathbf{x})^t\mathbf{d} - z \leq 0 & \text{for } i \in I \\ & -1 \leq d_j \leq 1 & \text{for } j = 1, \ldots, n \end{array}$$

Then, $\mathbf{x}$ is a Fritz John point if and only if the optimal objective value to the above problem is equal to zero.

Proof

The optimal objective value to the above problem is equal to zero if and only if the system $\nabla f(\mathbf{x})^t\mathbf{d} < 0$ and $\nabla g_i(\mathbf{x})^t\mathbf{d} < 0$ for $i \in I$ has no solution. By Theorem 2.4.9, this system has no solution if and only if there exist scalars u_0 and u_i for $i \in I$ such that

$$u_0 \nabla f(\mathbf{x}) + \sum_{i \in I} u_i \nabla g_i(\mathbf{x}) = \mathbf{0}$$

$$u_0 \geq 0 \qquad u_i \geq 0 \qquad \text{for } i \in I$$

$$\text{either } u_0 > 0 \text{ or else } u_i > 0 \text{ for some } i \in I$$

These are precisely the Fritz John conditions, and the proof is complete.

Summary of the Method of Zoutendijk (Case of Nonlinear Inequality Constraints)

Initialization Step Choose a starting point $\mathbf{x}_1$ such that $g_i(\mathbf{x}_1) \leq 0$ for $i = 1, \ldots, m$. Let $k = 1$ and go to the main step.

Main Step

1. Let $I = \{i : g_i(\mathbf{x}_k) = 0\}$ and solve the following problem:

$$\begin{array}{lll} \text{Minimize} & z & \\ \text{subject to} & \nabla f(\mathbf{x}_k)^t\mathbf{d} - z \leq 0 & \\ & \nabla g_i(\mathbf{x}_k)^t\mathbf{d} - z \leq 0 & \text{for } i \in I \\ & -1 \leq d_j \leq 1 & \text{for } j = 1, \ldots, n \end{array}$$

Let $(z_k, \mathbf{d}_k)$ be an optimal solution. If $z_k = 0$, stop; $\mathbf{x}_k$ is a Fritz John point. If $z_k < 0$, go to step 2.

2. Let λ_k be an optimal solution to the following line search problem:

$$\begin{array}{ll} \text{Minimize} & f(\mathbf{x}_k + \lambda\mathbf{d}_k) \\ \text{subject to} & 0 \le \lambda \le \lambda_{\max} \end{array}$$

where $\lambda_{\max} = \sup\{\lambda : g_i(\mathbf{x}_k + \lambda\mathbf{d}_k) \le 0 \text{ for } i = 1, \ldots, m\}$. Let $\mathbf{x}_{k+1} = \mathbf{x}_k + \lambda_k\mathbf{d}_k$, replace k by $k + 1$, and return to step 1.

10.1.8 Example

Consider the following problem:

$$\begin{array}{lrl} \text{Minimize} & 2x_1^2 + 2x_2^2 - 2x_1x_2 - 4x_1 - 6x_2 & \\ \text{subject to} & x_1 + 5x_2 & \le 5 \\ & 2x_1^2 - x_2 & \le 0 \\ & -x_1 & \le 0 \\ & -x_2 & \le 0 \end{array}$$

We shall solve the problem using the method of Zoutendijk. The procedure is initiated from a feasible point $\mathbf{x}_1 = (0.00, 0.75)^t$. The reader may note that $\nabla f(\mathbf{x}) = (4x_1 - 2x_2 - 4, 4x_2 - 2x_1 - 6)^t$.

Iteration 1

Search Direction At the point $\mathbf{x}_1 = (0.00, 0.75)^t$, we have $\nabla f(\mathbf{x}_1) = (-5.50, -3.00)^t$, and the binding constraints are defined by $I = \{3\}$. We have $\nabla g_3(\mathbf{x}_1) = (-1, 0)^t$. The direction-finding problem is then as follows:

$$\begin{array}{ll} \text{Minimize} & z \\ \text{subject to} & -5.5d_1 - 3.0d_2 - z \le 0 \\ & -d_1 - z \le 0 \\ & -1 \le d_j \le 1 \qquad \text{for } j = 1, 2 \end{array}$$

Using the simplex method, for example, it can be verified that the optimal solution is $\mathbf{d}_1 = (1.00, -1.00)^t$ and $z_1 = -1.00$.

Line Search Any point starting from $\mathbf{x}_1 = (0.00, 0.75)^t$ along the direction $\mathbf{d}_1 = (1.00, -1.00)^t$ can be written as $\mathbf{x}_1 + \lambda\mathbf{d}_1 = (\lambda, 0.75 - \lambda)^t$, and the corresponding value of the objective function is given by $f(\mathbf{x}_1 + \lambda\mathbf{d}_1) = 6\lambda^2 - 2.5\lambda - 3.375$. The reader can verify that the maximum value of λ for which $\mathbf{x}_1 + \lambda\mathbf{d}_1$ is feasible is given by $\lambda_{\max} = 0.4114$, where the constraint $2x_1^2 - x_2 \le 0$ becomes binding. The value of λ_1 is obtained by solving the following one-dimensional search problem:

$$\begin{array}{ll} \text{Minimize} & 6\lambda^2 - 2.5\lambda - 3.375 \\ \text{subject to} & 0 \le \lambda \le 0.4114 \end{array}$$

The optimal value can readily be found to be $\lambda_1 = 0.2083$. Hence, $\mathbf{x}_2 = (\mathbf{x}_1 + \lambda_1\mathbf{d}_1) = (0.2083, 0.5417)^t$.

Iteration 2

Search Direction At the point $\mathbf{x}_2 = (0.2083, 0.5417)^t$, we have $\nabla f(\mathbf{x}_2) = (-4.2500, -4.2500)^t$. There are no binding constraints, and, hence, the direction-finding problem is

$$\begin{array}{ll} \text{Minimize} & z \\ \text{subject to} & -4.25d_1 - 4.25d_2 - z \le 0 \\ & -1 \le d_j \le 1 \quad \text{for } j = 1, 2 \end{array}$$

The optimal solution is $\mathbf{d}_2 = (1, 1)^t$ and $z_2 = -8.50$.

Line Search The reader can verify that the maximum value of λ for which $\mathbf{x}_2 + \lambda\mathbf{d}_2$ is feasible is $\lambda_{max} = 0.3472$, when the constraint $x_1 + 5x_2 \le 5$ becomes binding. The value of λ_2 is obtained by minimizing $f(\mathbf{x}_2 + \lambda\mathbf{d}_2) = 2\lambda^2 - 8.5\lambda - 3.6354$ subject to $0 \le \lambda \le 0.3472$, and clearly we have $\lambda_2 = 0.3472$ so that $\mathbf{x}_3 = \mathbf{x}_2 + \lambda_2\mathbf{d}_2 = (0.5555, 0.8889)^t$.

Iteration 3

Search Direction At the point $\mathbf{x}_3 = (0.5555, 0.8889)^t$, we have $\nabla f(\mathbf{x}_3) = (-3.5558, -3.5554)^t$, and the binding constraints are defined by $I = \{1\}$. The direction-finding problem is given by

$$\begin{array}{ll} \text{Minimize} & z \\ \text{subject to} & -3.5558d_1 - 3.5554d_2 - z \le 0 \\ & d_1 + 5d_2 - z \le 0 \\ & -1 \le d_j \le 1 \quad \text{for } j = 1, 2 \end{array}$$

The optimal solution is $\mathbf{d}_3 = (1.0000, -0.5325)^t$ and $z_3 = -1.663$.

Line Search The reader can verify that the maximum value of λ for which $\mathbf{x}_3 + \lambda\mathbf{d}_3$ is feasible is $\lambda_{max} = 0.09245$, when the constraint $2x_1^2 - x_2 \le 0$ becomes binding. The value of λ_3 is obtained by minimizing $f(\mathbf{x}_3 + \lambda\mathbf{d}_3) = 1.5021\lambda^2 - 5.4490\lambda - 6.3455$ subject to $0 \le \lambda \le 0.09245$. The optimal solution is $\lambda_3 = 0.09245$, so that $\mathbf{x}_4 = \mathbf{x}_3 + \lambda_3\mathbf{d}_3 = (0.6479, 0.8397)^t$.

Iteration 4

Search Direction At the point $\mathbf{x}_4 = (0.6479, 0.8397)^t$, we have $\nabla f(\mathbf{x}_4) = (-3.0878, -3.9370)^t$, and the binding constraints are defined by $I = \{2\}$. The direction-finding problem is as follows:

$$\begin{array}{ll} \text{Minimize} & z \\ \text{subject to} & -3.0878d_1 - 3.9370d_2 - z \le 0 \\ & 2.5916d_1 - d_2 - z \le 0 \\ & -1 \le d_j \le 1 \quad \text{for } j = 1, 2 \end{array}$$

The optimal solution is $\mathbf{d}_4 = (-0.5171, 1.0000)^t$ and $z_4 = -2.340$.

Line Search The reader can verify that the maximum value of λ for which $\mathbf{x}_4 + \lambda\mathbf{d}_4$ is feasible is $\lambda_{max} = 0.0343$ when the constraint $x_1 + 5x_2 \le 5$ becomes binding. The value of λ_4 is obtained by minimizing $f(\mathbf{x}_4 + \lambda\mathbf{d}_4) = 3.569\lambda^2 - 2.340\lambda - 6.4681$ subject to $0 \le \lambda \le 0.0343$, which gives $\lambda_4 = 0.0343$. Hence, the new point is $\mathbf{x}_5 = \mathbf{x}_4 + \lambda_4\mathbf{d}_4 = (0.6302, 0.8740)^t$. The value of the objective function here is -6.5443 compared with the value of -6.5590 at the optimal point $(0.658872, 0.868226)^t$.

Table 10.2 summarizes the computations for the first four iterations. Figure 10.7 depicts the progress of the algorithm. Note the zigzagging tendency of the algorithm, as might be expected because of the first-order approximations used by this method.

Table 10.2 Summary of Computations for the Method of Zoutendijk with Nonlinear Constraints

Iteration			Search Direction			Line Search		
k	$\mathbf{x}_k$	$f(\mathbf{x}_k)$	$\nabla f(\mathbf{x}_k)$	$\mathbf{d}_k$	$\mathbf{z}_k$	λ_{max}	λ_1	$\mathbf{x}_{k+1}$
1	(0.00, 0.75)	−3.3750	(−5.50, −3.00)	(1.0000, −1.0000)	−1.000	0.4140	0.2083	(0.2083, 0.5417)
2	(0.2083, 0.5477)	−3.6354	(−4.25, −4.25)	(1.0000, 1.0000)	−8.500	0.3472	0.3472	(0.55555, 0.8889)
3	(0.5555, 0.8889)	−6.3455	(−3.5558, −3.5554)	(1.0000, −0.5325)	−1.663	0.09245	0.09245	(0.6479, 0.8397)
4	(0.6479, 0.8397)	−6.4681	(−3.0878, −3.9370)	(−0.5171, 1.0000)	−2.340	0.0343	0.0343	(0.6302, 0.8740)

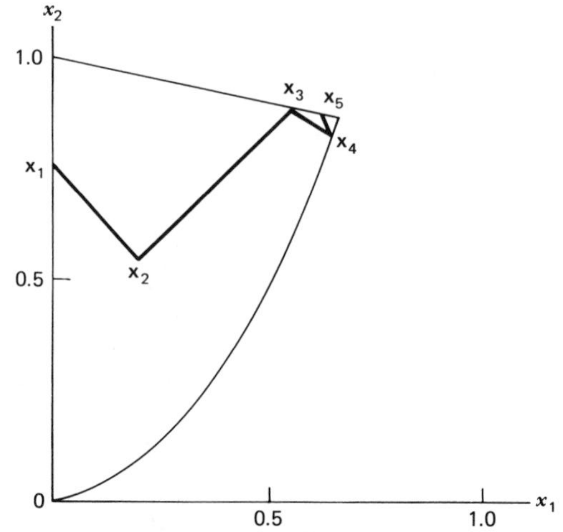

Figure 10.7 Illustration of the method of Zoutendijk for nonlinear inequality constraints.

The Treatment of Nonlinear Equality Constraints

The foregoing method of feasible directions must be modified to handle nonlinear equality constraints. To illustrate, consider Figure 10.8 with a single equality constraint. Given a feasible point $\mathbf{x}_k$, there exists no nonzero direction $\mathbf{d}$ such that $h(\mathbf{x}_k + \lambda\mathbf{d}) = 0$ for $\lambda \in (0, \delta)$, for some positive δ. This difficulty may be overcome by moving along a tangential direction $\mathbf{d}_k$ with $\nabla h(\mathbf{x}_k)^t\mathbf{d}_k = 0$, and then making a corrective move back to the feasible region.

To be more specific, consider the following problem:

$$\begin{array}{lll} \text{Minimize} & f(\mathbf{x}) & \\ \text{subject to} & g_i(\mathbf{x}) \leq 0 & i = 1, \ldots, m \\ & h_i(\mathbf{x}) = 0 & \text{for } i = 1, \ldots, l \end{array}$$

Let $\mathbf{x}_k$ be a feasible point, and let $I = \{i : g_i(\mathbf{x}_k) = 0\}$. Solve the following linear program:

$$\begin{array}{lll} \text{Minimize} & \nabla f(\mathbf{x}_k)^t\mathbf{d} & \\ \text{subject to} & \nabla g_i(\mathbf{x}_k)^t\mathbf{d} \leq 0 & \text{for } i \in I \\ & \nabla h_i(\mathbf{x}_k)^t\mathbf{d} = 0 & \text{for } i = 1, \ldots, l \end{array}$$

The resulting direction $\mathbf{d}_k$ is tangential to the equality constraints and to some of the

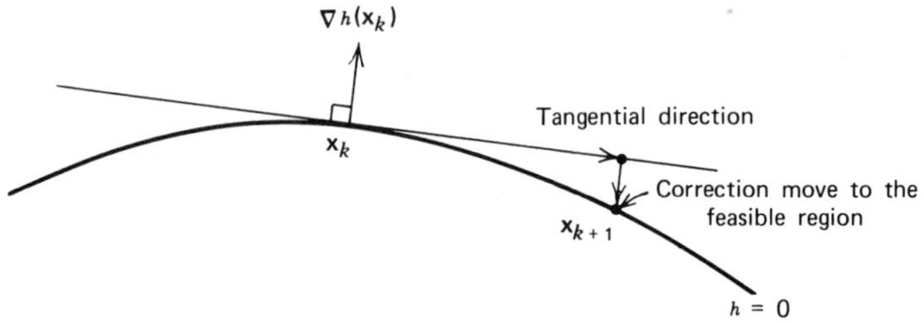

Figure 10.8 Nonlinear equality constraints.

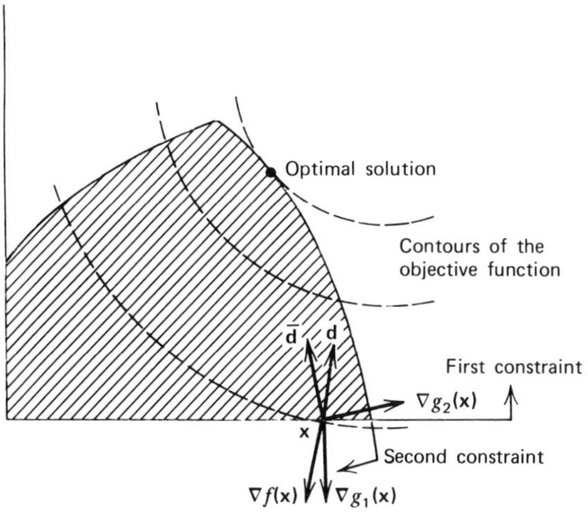

Figure 10.9 Effect of near-binding constraints.

binding nonlinear inequality constraints. A search along $\mathbf{d}_k$ is used, and then a move back to the feasible region leads to $\mathbf{x}_{k+1}$ and the process is repeated.

The Use of Near-Binding Constraints

Recall that the direction-finding problem, both for linear and nonlinear inequality constraints, used only the set of binding constraints. If a given point is close to the boundary of one of the constraints, and if this constraint is not used in the process of finding a direction of movement, it is possible that only a small step could be taken before we hit the boundary of this constraint. In Figure 10.9, the only binding restriction at **x** is the first constraint. However, **x** is close to the boundary of the second constraint. If the set I in the direction-finding problem P is taken as $I = \{1\}$, then the optimal direction will be **d**, and only a small movement can be realized before the boundary of constraint 2 is reached. If, on the other hand, both constraints 1 and 2 are treated as being binding so that $I = \{1, 2\}$, then the direction-finding problem P will produce the direction $\bar{\mathbf{d}}$, thus providing more room to move before reaching the boundary of the feasible region. Therefore, it is suggested to let the set I be the collection of near-binding constraints. More precisely, I is taken as $\{i: g_i(\mathbf{x}) + \varepsilon \geq 0\}$ rather than $\{i: g_i(\mathbf{x}) = 0\}$, where $\varepsilon > 0$ is a suitable small scalar. Of course, some caution is required in such a construct to prevent premature termination. As will be discussed in detail in Section 10.2, the method of feasible directions presented in this section does not necessarily converge to a Fritz John point. This results from the fact that the algorithmic map is not closed. Through a more formal use of the concept of near-binding constraints presented here, closedness of the algorithmic map, and, hence, convergence of the overall algorithm, can be established.

10.2 Convergence Analysis of the Method of Zoutendijk

In this section, we discuss the convergence properties of Zoutendijk's method of feasible directions presented in Section 10.1. As will be learned shortly, the algorithmic map of Zoutendijk's method is not closed, and, hence, convergence is not generally guaranteed.

A modification of the method credited to Topkis and Veinott [1967] assures convergence of the algorithm to a Fritz John point.

Note that the algorithmic map **A** of the method of Zoutendijk is composed of the maps **M** and **D**. The direction-finding map $\mathbf{D}: E_n \to E_n \times E_n$ is defined by $(\mathbf{x}, \mathbf{d}) \in \mathbf{D}(\mathbf{x})$ if **d** is an optimal solution to one of the direction-finding problems P1, P2, or P3 discussed in Section 10.1. The line search map $\mathbf{M}: E_n \times E_n \to E_n$ is defined by $\mathbf{y} \in \mathbf{M}(\mathbf{x}, \mathbf{d})$ if **y** is an optimal solution to the problem to minimize $f(\mathbf{x} + \lambda\mathbf{d})$ subject to $\lambda \geq 0$ and $\mathbf{x} + \lambda\mathbf{d} \in S$, where S is the feasible region. We demonstrate below that the map **D** is not closed in general.

10.2.1 Example (D Is Not Closed)

Consider the following problem:

$$\begin{array}{ll} \text{Minimize} & -2x_1 - x_2 \\ \text{subject to} & x_1 + x_2 \leq 2 \\ & x_1, \quad x_2 \geq 0 \end{array}$$

The problem is illusrated in Figure 10.10. Consider the sequence of vectors $\{\mathbf{x}_k\}$, where $\mathbf{x}_k = (0, 2 - \frac{1}{k})^t$. Noting that at each $\mathbf{x}_k$ the only binding constraint is $x_1 \geq 0$, and the direction-finding problem is given by

$$\begin{array}{ll} \text{Minimize} & -2d_1 - d_2 \\ \text{subject to} & 0 \leq d_1 \leq 1 \\ & -1 \leq d_2 \leq 1 \end{array}$$

The optimal solution $\mathbf{d}_k$ to the above problem is obviously $(1, 1)^t$. At the limit point $\mathbf{x} = (0, 2)^t$, however, both the restrictions $x_1 \geq 0$ and $x_1 + x_2 \leq 2$ are binding, so that the direction-finding problem is given by

$$\begin{array}{ll} \text{Minimize} & -2d_1 - d_2 \\ \text{subject to} & d_1 + d_2 \leq 0 \\ & 0 \leq d_1 \leq 1 \\ & -1 \leq d_2 \leq 1 \end{array}$$

The optimal solution $\bar{\mathbf{d}}$ to the above problem is given by $(1, -1)^t$. Thus,

$$\mathbf{x}_k \to \mathbf{x}$$

$$(\mathbf{x}_k, \mathbf{d}_k) \to (\mathbf{x}, \mathbf{d})$$

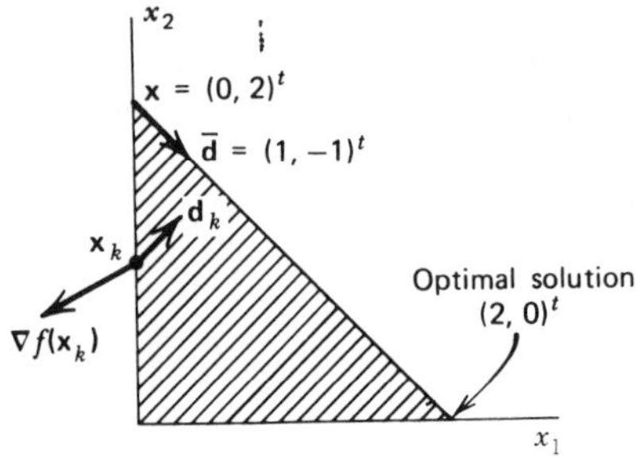

Figure 10.10 The direction map **D** is not closed.

where $\mathbf{d} = (1, 1)^t$. Since $\mathbf{D}(\mathbf{x}) = \{(\mathbf{x}, \bar{\mathbf{d}})\}$, $(\mathbf{x}, \mathbf{d}) \notin \mathbf{D}(\mathbf{x})$. Therefore, the direction-finding map $\mathbf{D}$ is not closed at $\mathbf{x}$.

The following line search map $\mathbf{M}: E_n \times E_n \to E_n$ is used by all feasible direction algorithms for solving a problem of the form to minimize $f(\mathbf{x})$ subject to $\mathbf{x} \in S$. Given a feasible point $\mathbf{x}$ and an improving feasible direction $\mathbf{d}$, $\mathbf{y} \in \mathbf{M}(\mathbf{x}, \mathbf{d})$ means that $\mathbf{y}$ is an optimal solution to the problem to minimize $f(\mathbf{x} + \lambda\mathbf{d})$ subject to $\lambda \geq 0$ and $\mathbf{x} + \lambda\mathbf{d} \in S$. We demonstrate in Example 10.2.2 below that this map is not closed. The difficulty here is that the possible step length that could be taken before leaving the feasible region may approach zero, causing *jamming*.

10.2.2 Example (M Is Not Closed)

Consider the following problem:

$$\begin{array}{ll} \text{Minimize} & 2x_1 - x_2 \\ \text{subject to} & (x_1, x_2) \in S \end{array}$$

where $S = \{(x_1, x_2): x_1^2 + x_2^2 \leq 1\} \cup \{(x_1, x_2): |x_1| \leq 1, 0 \leq x_2 \leq 1\}$. The problem is illustrated in Figure 10.11, and the optimal point $\bar{\mathbf{x}}$ is given by $(-1, 1)^t$. Now consider the sequence $\{(\mathbf{x}_k, \mathbf{d}_k)\}$ formed as follows. Let $\mathbf{x}_1 = (1, 0)^t$ and $\mathbf{d}_1 = (-1/\sqrt{2}, -1/\sqrt{2})^t$. Given $\mathbf{x}_k$, the next iterate $\mathbf{x}_{k+1}$ is given by moving along $\mathbf{d}_k$ until the boundary of S is reached. Given $\mathbf{x}_{k+1}$, the next direction $\mathbf{d}_{k+1}$ is taken as $(\boldsymbol{\xi} - \mathbf{x}_{k+1})/\|\boldsymbol{\xi} - \mathbf{x}_{k+1}\|$, where $\boldsymbol{\xi}$ is the point on the boundary of S that is equidistant from $\mathbf{x}_{k+1}$ and $(-1, 0)^t$.

The sequence $\{(\mathbf{x}_k, \mathbf{d}_k)\}$ is shown in Figure 10.11 and obviously converges to $(\mathbf{x}, \mathbf{d})$, where $\mathbf{x} = (-1, 0)^t$ and $\mathbf{d} = (0, 1)^t$. The line search map $\mathbf{M}$ is defined by $\mathbf{y}_k \in \mathbf{M}(\mathbf{x}_k, \mathbf{d}_k)$ if $\mathbf{y}_k$ is an optimal solution to the problem to minimize $f(\mathbf{x}_k + \lambda\mathbf{d}_k)$ subject to $\lambda \geq 0$ and $\mathbf{x}_k + \lambda\mathbf{d}_k \in S$. Obviously, $\mathbf{y}_k = \mathbf{x}_{k+1}$ and, hence, $\mathbf{y}_k \to \mathbf{x}$. Thus,

$$(\mathbf{x}_k, \mathbf{d}_k) \to (\mathbf{x}, \mathbf{d})$$

$$\mathbf{y}_k \to \mathbf{x} \qquad \text{where } \mathbf{y}_k \in \mathbf{M}(\mathbf{x}_k, \mathbf{d}_k)$$

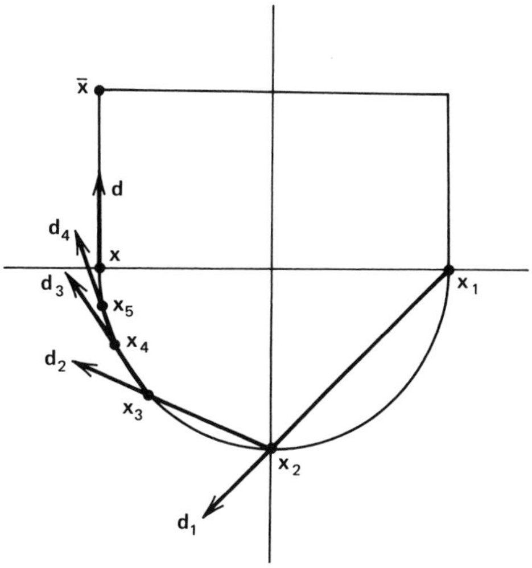

Figure 10.11 The line search map **M** is not closed.

On the other hand, minimizing f starting from $\mathbf{x}$ in the direction $\mathbf{d}$ yields $\bar{\mathbf{x}}$, so that $\mathbf{x} \notin \mathbf{M}(\mathbf{x}, \mathbf{d})$. Thus, $\mathbf{M}$ is not closed at $(\mathbf{x}, \mathbf{d})$.

Wolfe's Counterexample

We demonstrated above that both the direction-finding map and the line search map of Zoutendijk are not closed. Example 10.2.3, credited to Wolfe [1972], shows that Zoutendijk's algorithm may not converge to a KKT point. The difficulty here is that the distance moved along the generated directions tends toward zero, causing *jamming* at a nonoptimal point.

10.2.3 Example (Wolfe [1972])

Consider the following problem:

$$\begin{array}{ll} \text{Minimize} & \frac{4}{3}(x_1^2 - x_1x_2 + x_2^2)^{3/4} - x_3 \\ \text{subject to} & -x_1, -x_2, -x_3 \leq 0 \\ & x_3 \leq 2 \end{array}$$

Note that the objective function is convex and that the optimal solution is achieved at the unique point $\bar{\mathbf{x}} = (0, 0, 2)^t$. We solve this problem using Zoutendijk's procedure, starting from the feasible point $\mathbf{x}_1 = (0, a, 0)^t$, where $a \leq 1/(2\sqrt{2})$. Given a feasible point $\mathbf{x}_k$, the direction of movement $\mathbf{d}_k$ is obtained by solving the following Problem P2:

$$\begin{array}{ll} \text{Minimize} & \nabla f(\mathbf{x}_k)^t\mathbf{d} \\ \text{subject to} & \mathbf{A}_1\mathbf{d} \leq \mathbf{0} \\ & \mathbf{d}^t\mathbf{d} \leq 1 \end{array}$$

where $\mathbf{A}_1$ is the matrix whose rows are the gradients of the binding constraints at $\mathbf{x}_k$. Here, $\mathbf{x}_1 = (0, a, 0)^t$ and $\nabla f(\mathbf{x}_1) = (-\sqrt{a}, 2\sqrt{a}, -1)^t$. Note that

$$\mathbf{A}_1 = \begin{bmatrix} -1 & 0 & 0 \\ 0 & 0 & -1 \end{bmatrix}$$

and that the optimal solution to Problem P2 above is $\mathbf{d}_1 = -\nabla f(\mathbf{x}_1)/\|\nabla f(\mathbf{x}_1)\|$. The optimal solution λ_1 to the line search problem to minimize $f(\mathbf{x}_1 + \lambda\mathbf{d}_1)$ subject to $\lambda \geq 0$ and $\mathbf{x}_1 + \lambda\mathbf{d}_1 \in S$ yields $\mathbf{x}_2 = \mathbf{x}_1 + \lambda_1\mathbf{d}_1 = (\frac{1}{2}a, 0, \frac{1}{2}\sqrt{a})$.
Repeating this process, we obtain the sequence $\{\mathbf{x}_k\}$, where

$$\mathbf{x}_k = \begin{cases} \left[0, (\frac{1}{2})^{k-1}a, \frac{1}{2}\sum_{j=0}^{k-2}\left(\frac{a}{2^j}\right)^{1/2}\right] & \text{if } k \text{ is odd}, k \geq 3 \\ \left[(\frac{1}{2})^{k-1}a, 0, \frac{1}{2}\sum_{j=0}^{k-2}\left(\frac{a}{2^j}\right)^{1/2}\right] & \text{if } k \text{ is even} \end{cases}$$

Note that this sequence converges to the point $\hat{\mathbf{x}} = [0, 0, (1 + \frac{1}{2}\sqrt{2})\sqrt{a}]^t$. Since the optimal solution $\bar{\mathbf{x}}$ is unique, Zoutendijk's method converges to $\hat{\mathbf{x}}$, which is neither optimal nor a KKT point.

Topkis-Veinott's Modification of the Feasible Direction Algorithm

We now describe a modification of Zoutendijk's method of feasible directions. This modification was proposed by Topkis and Veinott [1967] and guarantees convergence to a Fritz John point. The problem under consideration is

$$\begin{array}{lll} \text{Minimize} & f(\mathbf{x}) & \\ \text{subject to} & g_i(\mathbf{x}) \le 0 & \text{for } i = 1, \ldots, m \end{array}$$

Generating a Feasible Direction

Given a feasible point $\mathbf{x}$, a direction is found by solving the following direction-finding linear programming problem DF($\mathbf{x}$):

Problem DF(x):

$$\begin{array}{lll} \text{Minimize} & z & \\ \text{subject to} & \nabla f(\mathbf{x})^t\mathbf{d} - z \le 0 & \\ & \nabla g_i(\mathbf{x})^t\mathbf{d} - z \le -g_i(\mathbf{x}) & \text{for } i = 1, \ldots, m \\ & -1 \le d_j \le 1 & \text{for } j = 1, \ldots, n \end{array}$$

Here, both binding and nonbinding constraints play a role in determining the direction of movement. As opposed to the method of feasible directions of Section 10.1, no sudden change in the direction is encountered when approaching the boundary of a currently nonbinding constraint.

Summary of the Method of Feasible Directions of Topkis and Veinott

A summary of the method of Topkis and Veinott for solving the problem to minimize $f(\mathbf{x})$ subject to $g_i(\mathbf{x}) \le 0$ for $i = 1, \ldots, m$ is given below. As will be shown later, the method converges to a Fritz John point.

Initialization Step Choose a point $\mathbf{x}_1$ such that $g_i(\mathbf{x}_1) \le 0$ for $i = 1, \ldots, m$. Let $k = 1$ and go to the main step.

Main Step

1. Let $(z_k, \mathbf{d}_k)$ be an optimal solution to the following linear programming problem:

$$\begin{array}{lll} \text{Minimize} & z & \\ \text{subject to} & \nabla f(\mathbf{x}_k)^t\mathbf{d} - z \le 0 & \\ & \nabla g_i(\mathbf{x}_k)^t\mathbf{d} - z \le -g_i(\mathbf{x}_k) & \text{for } i = 1, \ldots, m \\ & -1 \le d_j \le 1 & \text{for } j = 1, \ldots, n \end{array}$$

   If $z_k = 0$, stop; $\mathbf{x}_k$ is a Fritz John point. Otherwise, $z_k < 0$ and we go to step 2.

2. Let λ_k be an optimal solution to the following line search problem:

$$\begin{array}{ll} \text{Minimize} & f(\mathbf{x}_k + \lambda\mathbf{d}_k) \\ \text{subject to} & 0 \le \lambda \le \lambda_{\max} \end{array}$$

   where $\lambda_{\max} = \sup\{\lambda : g_i(\mathbf{x}_k + \lambda\mathbf{d}_k) \le 0 \text{ for } i = 1, \ldots, m\}$. Let $\mathbf{x}_{k+1} = \mathbf{x}_k + \lambda_k\mathbf{d}_k$, replace k by $k + 1$, and return to step 1.

10.2.4 Example

Consider the following problem:

$$\begin{array}{ll} \text{Minimize} & 2x_1^2 + 2x_2^2 - 2x_1x_2 - 4x_1 - 6x_2 \\ \text{subject to} & x_1 + 5x_2 \le 5 \\ & 2x_1^2 - x_2 \le 0 \\ & -x_1 \le 0 \\ & -x_2 \le 0 \end{array}$$

We go through five iterations of the algorithm of Topkis and Veinott, starting from the point $\mathbf{x}_1 = (0.00, 0.75)^t$. Note that the gradient of the objective function is given by $\nabla f(\mathbf{x}) = (4x_1 - 2x_2 - 4, 4x_2 - 2x_1 - 6)^t$, and the gradients of the constraint functions are $(1, 5)^t$, $(4x_1, -1)^t$, $(-1, 0)^t$, and $(0, -1)^t$, which are used in defining the direction-finding problem at each iteration.

Iteration 1

Search Direction At $\mathbf{x}_1 = (0.00, 0.75)^t$, we have $\nabla f(\mathbf{x}_1) = (-5.5, -3.0)^t$. Hence, the direction-finding problem is as follows:

$$
\begin{array}{lll}
\text{Minimize} & z & \\
\text{subject to} & -5.5d_1 - 3d_2 - z \le 0 & \\
& d_1 + 5d_2 - z \le 1.25 & \\
& - d_2 - z \le 0.75 & \\
& -d_1 \quad - z \le 0 & \\
& - d_2 - z \le 0.75 & \\
& -1 \le d_j \le 1 & \text{for } j = 1, 2
\end{array}
$$

The right-hand sides of the constraints 2 to 5 are $-g_i(\mathbf{x}_1)$ for $i = 1, 2, 3, 4$. Note that one of the constraints, $-d_2 - z \le 0.75$, is redundant. The optimal solution to the above problem is $\mathbf{d}_1 = (0.7143, -0.03571)^t$ and $z_1 = -0.7143$.

Line Search The reader can readily verify that the maximum value of λ for which $\mathbf{x}_1 + \lambda\mathbf{d}_1$ is feasible is given by $\lambda_{max} = 0.84$ and that $f(\mathbf{x}_1 + \lambda\mathbf{d}_1) = 0.972\lambda^2 - 4.036\lambda - 3.375$. Then, $\lambda_1 = 0.84$ solves the problem to minimize $f(\mathbf{x}_1 + \lambda\mathbf{d}_1)$ subject to $0 \le \lambda \le 0.84$. We then have $\mathbf{x}_2 = \mathbf{x}_1 + \lambda_1\mathbf{d}_1 = (0.60, 0.72)^t$.

Iteration 2

Search Direction At the point $\mathbf{x}_2$, we have $\nabla f(\mathbf{x}_2) = (-3.04, -4.32)^t$. Then, $\mathbf{d}_2$ is the optimal solution to the problem

$$
\begin{array}{lll}
\text{Minimize} & z & \\
\text{subject to} & -3.04d_1 - 4.32d_2 - z \le 0 & \\
& d_1 + 5d_2 - z \le 0.8 & \\
& 2.4d_1 - d_2 - z \le 0 & \\
& -d_1 \quad - z \le 0.6 & \\
& -d_2 - z \le 0.72 & \\
& -1 \le d_j \le 1 & \text{for } j = 1, 2
\end{array}
$$

The optimal solution is $\mathbf{d}_2 = (-0.07123, 0.1167)^t$ and $z_2 = -0.2877$.

Line Search The maximum value of λ such that $\mathbf{x}_2 + \lambda\mathbf{d}_2$ is feasible is given by $\lambda_{max} = 1.561676$. The reader can easily verify that $f(\mathbf{x}_2 + \lambda_2\mathbf{d}_2) = 0.054\lambda^2 - 0.2876\lambda - 5.8272$ attains a minimum over the interval $0 \le \lambda \le 1.561676$ at the point $\lambda_2 = 1.561676$. Hence, $\mathbf{x}_3 = \mathbf{x}_2 + \lambda_2\mathbf{d}_2 = (0.4888, 0.9022)^t$.

The process is then repeated. Table 10.3 summarizes the computations for five iterations. The progress of the algorithm is depicted in Figure 10.12. Note that at the end of five iterations, the point $(0.6548, 0.8575)^t$ is reached with objective function value of -6.5590. Note that the optimal point is $(0.658872, 0.868226)^t$, with objective function value -6.613086. Again, observe the zigzagging of the iterates generated by the algorithm.

Table 10.3 Summary of the Topkis-Veinott Method

Iteration			Search Direction			Line Search		
k	$\mathbf{x}_k$	$f(\mathbf{x}_k)$	$\nabla f(\mathbf{x}_k)$	$\mathbf{d}_k$	z_k	λ_{max}	λ_k	$\mathbf{x}_{k+1}$
1	(0.0000, 0.7500)	−3.3750	(−5.50, −3.00)	(0.7143, −0.03571)	−0.7143	0.84	0.84	(0.6000, 0.7200)
2	(0.6000, 0.7200)	−5.8272	(−3.04, −4.32)	(−0.07123, 0.1167)	−0.2877	1.561676	1.561676	(0.4888, 0.9022)
3	(0.4888, 0.9022)	−6.1446	(−3.8492, −3.3688)	(0.09574, −0.05547)	−0.1816	1.56395	1.56395	(0.6385, 0.8154)
4	(0.6385, 0.8154)	−6.3425	(−5.6308, −4.0154)	(−0.01595, 0.04329)	−0.0840	1.41895	1.41895	(0.6159, 0.8768)
5	(0.6159, 0.8768)	−6.5082	(−3.2900, −3.7246)	(0.02676, −0.01316)	−0.0303	1.45539	1.45539	(0.6548, 0.8575)

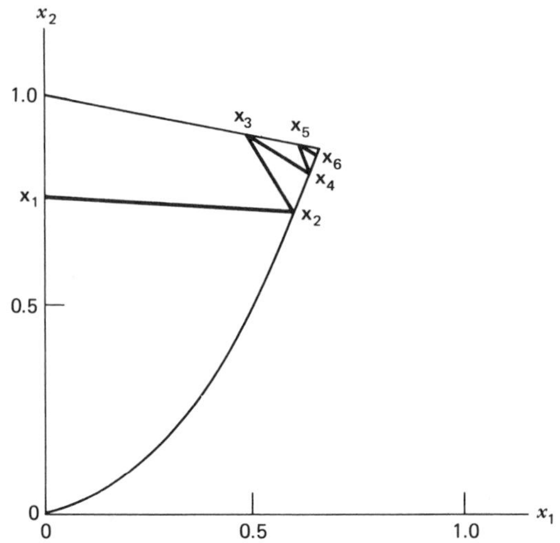

Figure 10.12 Illustration of the method of Topkis and Veinott.

Convergence of the Method of Topkis and Veinott

Theorem 10.2.7 shows convergence of the method of Topkis and Veinott to a Fritz John point. Two intermediate results are needed. Theorem 10.2.5 below provides a necessary and sufficient condition for arriving at a Fritz John point and shows that an optimal solution to the direction-finding problem indeed provides an improving feasible direction.

10.2.5 Theorem

Let $\mathbf{x}$ be a feasible solution to the problem to minimize $f(\mathbf{x})$ subject to $g_i(\mathbf{x}) \le 0$ for $i = 1, \ldots, m$. Let $(\bar{z}, \bar{\mathbf{d}})$ be an optimal solution to problem DF($\mathbf{x}$). If $\bar{z} < 0$, then $\bar{\mathbf{d}}$ is an improving feasible direction. Also, $\bar{z} = 0$, if and only if $\mathbf{x}$ is a Fritz John point.

Proof

Let $I = \{i : g_i(\mathbf{x}) = 0\}$, and suppose that $\bar{z} < 0$. Examining Problem DF($\mathbf{x}$), we note that $\nabla g_i(\mathbf{x})'\bar{\mathbf{d}} < 0$ for $i \in I$. This, together with the fact that $g_i(\mathbf{x}) < 0$ for $i \notin I$, implies that $\mathbf{x} + \lambda\bar{\mathbf{d}}$ is feasible for $\lambda > 0$ and sufficiently small. Thus, $\bar{\mathbf{d}}$ is a feasible direction. Furthermore, $\nabla f(\mathbf{x})'\bar{\mathbf{d}} < 0$, and, hence, $\bar{\mathbf{d}}$ is an improving direction.

Now we prove the second part of the theorem. Noting that $g_i(\mathbf{x}) = 0$ for $i \in I$ and that $g_i(\mathbf{x}) < 0$ for $i \notin I$, it can be easily verified that $\bar{z} = 0$ if and only if the system $\nabla f(\mathbf{x})'\mathbf{d} < 0$ and $\nabla g_i(\mathbf{x})'\mathbf{d} < 0$ for $i \in I$ has no solution. By Theorem 2.4.9, this system has no solution if and only if $\mathbf{x}$ is a Fritz John point, and the proof is complete.

Lemma 10.2.6 will be used to prove Theorem 10.2.7, which asserts convergence of the algorithm of Topkis and Veinott. The lemma essentially states that any feasible direction algorithm cannot generate a sequence of points and directions satisfying properties 1 through 4 stated below.

10.2.6 Lemma

Let S be a nonempty closed set in E_n and let $f: E_n \to E_1$ be continuously differentiable. Consider the problem to minimize $f(\mathbf{x})$ subject to $\mathbf{x} \in S$. Furthermore, consider any feasible direction algorithm whose map $\mathbf{A} = \mathbf{MD}$ is defined as follows. Given $\mathbf{x}$, $(\mathbf{x}, \mathbf{d}) \in \mathbf{D}(\mathbf{x})$ means that $\mathbf{d}$ is an improving feasible direction of f at $\mathbf{x}$. Furthermore, $\mathbf{y} \in \mathbf{M}(\mathbf{x}, \mathbf{d})$ means that $\mathbf{y} = \mathbf{x} + \bar{\lambda}\mathbf{d}$, where $\bar{\lambda}$ solves the line search problem to minimize $f(\mathbf{x} + \lambda\mathbf{d})$ subject to $\lambda \geq 0$ and $\mathbf{x} + \lambda\mathbf{d} \in S$. Let $\{\mathbf{x}_k\}$ be any sequence generated by such an algorithm, and let $\{\mathbf{d}_k\}$ be the corresponding sequence of directions. Then, there cannot exist a subsequence $\{(\mathbf{x}_k, \mathbf{d}_k)\}_{\mathcal{K}}$ satisfying all the following properties:

1. $\mathbf{x}_k \to \mathbf{x}$ for $k \in \mathcal{K}$
2. $\mathbf{d}_k \to \mathbf{d}$ for $k \in \mathcal{K}$
3. $\mathbf{x}_k + \lambda\mathbf{d}_k \in S$ for all $\lambda \in [0, \delta]$ and for each $k \in \mathcal{K}$, for some $\delta > 0$
4. $\nabla f(\mathbf{x})^t\mathbf{d} < 0$

Proof

Suppose, by contradiction, that there exists a subsequence $\{(\mathbf{x}_k, \mathbf{d}_k)\}_{\mathcal{K}}$ satisfying conditions 1 through 4. By condition 4, there exists an $\varepsilon > 0$ such that $\nabla f(\mathbf{x})^t\mathbf{d} = -2\varepsilon$. Since $\mathbf{x}_k \to \mathbf{x}$ and $\mathbf{d}_k \to \mathbf{d}$ for $k \in \mathcal{K}$, and since f is continuously differentiable, there exists a $\delta' > 0$ such that

$$\nabla f(\mathbf{x}_k + \lambda\mathbf{d}_k)^t\mathbf{d}_k < -\varepsilon \qquad \text{for } \lambda \in [0, \delta'] \text{ and for } k \in \mathcal{K} \text{ sufficiently large} \tag{10.2}$$

Now let $\bar{\delta} = \text{minimum } \{\delta', \delta\} > 0$. Consider $k \in \mathcal{K}$ sufficiently large. By condition 3, and by the definition of $\mathbf{x}_{k+1}$, we must have $f(\mathbf{x}_{k+1}) \leq f(\mathbf{x}_k + \bar{\delta}\mathbf{d}_k)$. By the mean value theorem, $f(\mathbf{x}_k + \bar{\delta}\mathbf{d}_k) = f(\mathbf{x}_k) + \bar{\delta}\nabla f(\hat{\mathbf{x}}_k)^t\mathbf{d}_k$, where $\hat{\mathbf{x}}_k = \mathbf{x}_k + \lambda_k\bar{\delta}\mathbf{d}_k$ and $\lambda_k \in [0, 1]$. By (10.2), it then follows that

$$f(\mathbf{x}_{k+1}) < f(\mathbf{x}_k) - \varepsilon\bar{\delta} \qquad \text{for } k \in \mathcal{K} \text{ sufficiently large} \tag{10.3}$$

Since the feasible direction algorithm generates a sequence of points with decreasing objective values, then $\lim_{k\to\infty} f(\mathbf{x}_k) = f(\mathbf{x})$. In particular, both $f(\mathbf{x}_{k+1})$ and $f(\mathbf{x}_k)$ approach $f(\mathbf{x})$ if $k \in \mathcal{K}$ approaches ∞. Thus, from (10.3), we get $f(\mathbf{x}) \leq f(\mathbf{x}) - \varepsilon\bar{\delta}$, which is impossible, since $\varepsilon, \bar{\delta} > 0$. This contradiction shows that no subsequence with properties 1 through 4 could exist.

10.2.7 Theorem

Let $f, g_i: E_n \to E_1$ for $i = 1, \ldots, m$ be continuously differentiable, and consider the problem to minimize $f(\mathbf{x})$ subject to $g_i(\mathbf{x}) \leq 0$ for $i = 1, \ldots, m$. Suppose that the sequence $\{\mathbf{x}_k\}$ is generated by the algorithm of Topkis and Veinott. Then, any accumulation point of $\{\mathbf{x}_k\}$ is a Fritz John point.

Proof

Let $\{\mathbf{x}_k\}_{\mathcal{K}}$ be a convergent subsequence with limit $\mathbf{x}$. We need to show that $\mathbf{x}$ is a Fritz John point. Suppose, by contradiction, that $\mathbf{x}$ is not a Fritz John point, and let $\bar{z}$ be the optimal objective value of Problem DF($\mathbf{x}$). By Theorem 10.2.5, there exists $\varepsilon > 0$ such that $\bar{z} = -2\varepsilon$. For $k \in \mathcal{K}$, consider Problem DF($\mathbf{x}_k$) and let $(z_k, \mathbf{d}_k)$ be an optimal solution. Since $\{\mathbf{d}_k\}_{\mathcal{K}}$ is bounded, there exists a subsequence $\{\mathbf{d}_k\}_{\mathcal{K}'}$ with limit $\mathbf{d}$. Furthermore, since f and g_i for $i = 1, \ldots, m$ are continuously differentiable, and since $\mathbf{x}_k \to \mathbf{x}$ for $k \in \mathcal{K}'$, it follows that $z_k \to \bar{z}$. In particular, for $k \in \mathcal{K}'$ sufficiently large, we must have $z_k < -\varepsilon$. By the definition of Problem DF($\mathbf{x}_k$), we must have

$$\nabla f(\mathbf{x}_k)^t\mathbf{d}_k \le z_k < -\varepsilon \qquad \text{for } k \in \mathcal{K}' \text{ sufficiently large} \tag{10.4}$$

$$g_i(\mathbf{x}_k) + \nabla g_i(\mathbf{x}_k)^t\mathbf{d}_k \le z_k < -\varepsilon \qquad \text{for } k \in \mathcal{K}' \text{ sufficiently large, for } i = 1, \ldots, m \tag{10.5}$$

By continuous differentiability of f, (10.4) implies that $\nabla f(\mathbf{x})^t\mathbf{d} < 0$.

Since g_i is continuously differentiable, from (10.5), there exists a $\delta > 0$ such that the following inequality holds for each $\lambda \in [0, \delta]$:

$$g_i(\mathbf{x}_k) + \nabla g_i(\mathbf{x}_k + \lambda\mathbf{d}_k)^t\mathbf{d}_k < -\frac{\varepsilon}{2} \qquad \text{for } k \in \mathcal{K}' \text{ sufficiently large, for } i = 1, \ldots, m \tag{10.6}$$

Now let $\lambda \in [0, \delta]$. By the mean value theorem, and since $g_i(\mathbf{x}_k) \le 0$ for each k and each i, we get

$$\begin{aligned} g_i(\mathbf{x}_k + \lambda\mathbf{d}_k) &= g_i(\mathbf{x}_k) + \lambda\nabla g_i(\mathbf{x}_k + \alpha_{ik}\lambda\mathbf{d}_k)^t\mathbf{d}_k \\ &= (1 - \lambda)g_i(\mathbf{x}_k) + \lambda[g_i(\mathbf{x}_k) + \nabla g_i(\mathbf{x}_k + \alpha_{ik}\lambda\mathbf{d}_k)^t\mathbf{d}_k] \\ &\le \lambda[g_i(\mathbf{x}_k) + \nabla g_i(\mathbf{x}_k + \alpha_{ik}\lambda\mathbf{d}_k)^t\mathbf{d}_k] \end{aligned} \tag{10.7}$$

where $\alpha_{ik} \in [0, 1]$. Since $\alpha_{ik}\lambda \in [0, \delta]$, from (10.6) and (10.7), it follows that $g_i(\mathbf{x}_k + \lambda\mathbf{d}_k) \le -\lambda\varepsilon/2 \le 0$ for $k \in \mathcal{K}'$ sufficiently large and for $i = 1, \ldots, m$. This shows that $\mathbf{x}_k + \lambda\mathbf{d}_k$ is feasible for each $\lambda \in [0, \delta]$, for all $k \in \mathcal{K}'$ sufficiently large.

To summarize, we have exhibited a sequence $\{(\mathbf{x}_k, \mathbf{d}_k)\}_{\mathcal{K}'}$ that satisfies conditions 1 through 4 of Lemma 10.2.6. By the lemma, however, the existence of such a sequence is not possible. This contradiction shows that $\mathbf{x}$ is a Fritz John point, and the proof is complete.

10.3 Successive Linear Programming Approach

In our foregoing discussion of Zoutendijk's algorithm and its convergent variant as proposed by Topkis and Veinott, we have learned that at each iteration of this method, we solve a direction-finding linear programming problem based on first-order functional approximations in a minimax framework and then conduct a line search along this direction. Conceptually, this is similar to *successive linear programming (SLP)* approaches, also known as *sequential*, or *recursive*, linear programming. Here, at each iteration k, a direction-finding linear program is formulated based on first-order Taylor series approximations to the objective and constraint functons, in addition to appropriate *step bounds* or trust region restrictions on the direction components. If $\mathbf{d}_k = \mathbf{0}$ solves this problem, then the current iterate $\mathbf{x}_k$ is optimal to the first-order approximation, and so, from Theorem 4.2.15, this solution is a KKT point and we terminate the procedure. Otherwise, the procedure either accepts the new iterate $\mathbf{x}_{k+1} = \mathbf{x}_k + \mathbf{d}_k$, or rejects this iterate and reduces the step bounds, and then repeats this process. The decision as to whether to accept or reject the new iterate is typically made based on a *merit function* fashioned around l_1, or the absolute value penalty function [see equation (9.8)].

The philosophy of this approach was introduced by Griffith and Stewart of the Shell Development Company in 1961, and has been widely used since then, particularly in the oil and chemical industries (see Exercise 10.55). The principal advantage of this type of method is its ease and robustness in implementation for large-scale problems, given an efficient and stable linear programming solver. As can be expected, if the optimum is a vertex of the (linearized) feasible region, then a rapid convergence is obtained. Indeed, once the algorithm enters a relatively close neighborhood of such a solution, it essentially behaves like Newton's algorithm applied to the binding constraints (under suitable regularity assumptions), with the Newton iterate being the (unique) linear programming solution, and a quadratic convergence rate obtains. Hence, highly constrained nonlinear programming problems that have nearly as many linearly independent active constraints as variables are very suitable for this class of algorithms. Real-world nonlinear refinery models tend to be of this nature, and problems of up to 1000 rows have been successfully solved. On the negative side, SLP algorithms exhibit slow convergence to nonvertex solutions, and they also have the disadvantage of violating nonlinear constraints en route to optimality.

Below, we describe an SLP algorithm, called the *penalty successive linear programming (PSLP)* algorithm, which employs the l_1 penalty function more actively in the direction-finding problem, itself, rather than as only a merit function, and enjoys good robustness and convergence properties. The problem we consider is of the form:

$$
\begin{aligned}
\textbf{\textit{P:}}\quad &\text{Minimize} && f(\mathbf{x}) && \\
&\text{subject to} && g_i(\mathbf{x}) \leq 0 && \text{for } i = 1, \ldots, m \\
& && h_i(\mathbf{x}) = 0 && \text{for } i = 1, \ldots, l \\
& && \mathbf{x} \in X = \{\mathbf{x} : \mathbf{A}\mathbf{x} \leq \mathbf{b}\} &&
\end{aligned}
$$

where all functions are assumed to be continuously differentiable, where $\mathbf{x} \in E_n$, and where the linear constraints defining the problem have all been accommodated into the set X.

Now let $F_E(\mathbf{x})$ be the l_1, or absolute value, exact penalty function of equation (9.8), restated below for a penalty parameter $\mu > 0$:

$$F_E(\mathbf{x}) = f(\mathbf{x}) + \mu\left[\sum_{i=1}^{m} \max\{0, g_i(\mathbf{x})\} + \sum_{i=1}^{l} |h_i(\mathbf{x})|\right] \tag{10.8}$$

Accordingly, consider the following (linearly constrained) penalty problem PP

$$\textbf{\textit{PP:}}\quad \text{Minimize}\ \{F_E(\mathbf{x}) : \mathbf{x} \in X\} \tag{10.9a}$$

Substituting y_i for $\max\{0, g_i(\mathbf{x})\}$, $i = 1, \ldots, m$, and writing $h_i(\mathbf{x})$ as the difference $z_i^+ - z_i^-$ of two nonnegative variables, where $|h_i(\mathbf{x})| = z_i^+ + z_i^-$, for $i = 1, \ldots, l$, we can equivalently rewrite (10.9a) without the nondifferentiable terms as follows:

$$
\begin{aligned}
\textbf{\textit{PP:}}\quad &\text{Minimize} && f(\mathbf{x}) + \mu\left[\sum_{i=1}^{m} y_i + \sum_{i=1}^{l} (z_i^+ + z_i^-)\right] && \\
&\text{subject to} && y_i \geq g_i(\mathbf{x}) && i = 1, \ldots, m \\
& && z_i^+ - z_i^- = h_i(\mathbf{x}) && i = 1, \ldots, l \\
& && \mathbf{x} \in X, y_i \geq 0 && \text{for } i = 1, \ldots, m \\
& && z_i^+ \text{ and } z_i^- \geq 0 && \text{for } i = 1, \ldots, l
\end{aligned} \tag{10.9b}
$$

Note that, given any $\mathbf{x} \in X$, since $\mu > 0$, the optimal completion $(\mathbf{y}, \mathbf{z}^+, \mathbf{z}^-) = (y_1, \ldots, y_m, z_1^+, \ldots, z_l^+, z_1^- \ldots, z_l^-)$ is determined by letting

$$
\begin{aligned}
& y_i = \max\{0, g_i(\mathbf{x})\} && && i = 1, \ldots, m \\
& z_i^+ = \max\{0, h_i(\mathbf{x})\} && z_i^- = \max\{0, -h_i(\mathbf{x})\} && i = 1, \ldots, l \\
& \text{so that } (z_i^+ + z_i^-) = |h_i(\mathbf{x})| && && i = 1, \ldots, l
\end{aligned} \tag{10.10}
$$

Consequently, (10.9b) is equivalent to (10.9a) and may be essentially viewed as also being a problem in the $\mathbf{x}$-variable space. Moreover, under the condition of Theorem 9.3.1, if μ is sufficiently large and if $\bar{\mathbf{x}}$ is an optimum for P, then $\bar{\mathbf{x}}$ solves PP. Alternatively, as in Exercise 9.12, if μ is sufficiently large and if $\bar{\mathbf{x}}$ satisfies the second-order sufficiency conditions for P, then $\bar{\mathbf{x}}$ is a strict local minimum for PP. In either case, μ must be, at least, as large as the absolute value of any Lagrange multiplier associated with the constraints $\mathbf{g}(\mathbf{x}) \leq \mathbf{0}$ and $\mathbf{h}(\mathbf{x}) = \mathbf{0}$ in P. Note that instead of using a single penalty parameter μ, we can employ a set of parameters $\mu_1, \ldots, \mu_{m+l}$, one associated with each of the penalized constraints. Selecting some reasonably large values for these parameters (assuming a well-scaled problem), we can solve PP; and if an infeasible solution results, then these parameters can be manually increased and the process repeated. We shall assume, however, that we have selected some suitably large, admissible value of a single penalty parameter μ. With this motivation, the algorithm PSLP seeks to solve Problem PP, using a box step or hypercube first-order trust region approach as introduced in Section 8.7.

Specifically, this approach proceeds as follows. Given a current iterate $\mathbf{x}_k \in X$ and a trust region or step bound vector $\mathbf{\Delta}_k \in E_n$, consider the following linearization of PP, given by (10.9a), where we have also imposed a trust region bound based on the l_∞ or sup-norm.

$$\begin{aligned}
\textit{LP}(\mathbf{x}_k,\mathbf{\Delta}_k)\text{:}\quad \text{Minimize}\quad & F_{EL_k}(\mathbf{x}) \equiv f(\mathbf{x}_k) + \nabla f(\mathbf{x}_k)^t(\mathbf{x}-\mathbf{x}_k) \\
& + \mu\left[\sum_{i=1}^{m} \max\{0, g_i(\mathbf{x}_k) + \nabla g_i(\mathbf{x}_k)^t(\mathbf{x}-\mathbf{x}_k)\}\right. \\
& \left. + \sum_{i=1}^{l} |h_i(\mathbf{x}_k) + \nabla h_i(\mathbf{x}_k)^t(\mathbf{x}-\mathbf{x}_k)|\right] \\
\text{subject to}\quad & \mathbf{x} \in X \equiv \{\mathbf{x} : \mathbf{A}\mathbf{x} \leq \mathbf{b}\} \\
& -\mathbf{\Delta}_k \leq \mathbf{x} - \mathbf{x}_k \leq \mathbf{\Delta}_k
\end{aligned} \tag{10.11a}$$

Similar to (10.9) and (10.10), this can be equivalently restated as the following *linear programming problem*, where we have also used the substitution $\mathbf{x} = \mathbf{x}_k + \mathbf{d}$ and have dropped the constant $f(\mathbf{x}_k)$ from the objective function:

$$\begin{aligned}
\textit{LP}(\mathbf{x}_k,\mathbf{\Delta}_k)\text{:}\quad \text{Minimize}\quad & \nabla f(\mathbf{x}_k)^t\mathbf{d} + \mu\left[\sum_{i=1}^{m} y_i + \sum_{i=1}^{l}(z_i^+ + z_i^-)\right] \\
\text{subject to}\quad & y_i \geq g_i(\mathbf{x}_k) + \nabla g_i(\mathbf{x}_k)^t\mathbf{d} \qquad i = 1, \ldots, m \\
& (z_i^+ - z_i^-) = h_i(\mathbf{x}_k) + \nabla h_i(\mathbf{x}_k)^t\mathbf{d} \qquad i = 1, \ldots, l \\
& \mathbf{A}(\mathbf{x}_k + \mathbf{d}) \leq \mathbf{b} \\
& -\Delta_{ki} \leq d_i \leq \Delta_{ki} \qquad i = 1, \ldots, n \\
& \mathbf{y} \geq \mathbf{0}, \mathbf{z}^+ \geq \mathbf{0}, \mathbf{z}^- \geq \mathbf{0}
\end{aligned} \tag{10.11b}$$

The linear program LP($\mathbf{x}_k$, $\mathbf{\Delta}_k$) given by (10.11b) is the *direction-finding subproblem* that yields an optimal solution $\mathbf{d}_k$, say, along with the accompanying values of $\mathbf{y}$, $\mathbf{z}^+$, and $\mathbf{z}^-$, which are given as follows, similar to (10.10):

$$\begin{aligned}
& y_i = \max\{0, g_i(\mathbf{x}_k) + \nabla g_i(\mathbf{x}_k)^t\mathbf{d}_k\} \qquad i = 1, \ldots, m \\
& z_i^+ = \max\{0, h_i(\mathbf{x}_k) + \nabla h_i(\mathbf{x}_k)^t\mathbf{d}_k\} \qquad z_i^- = \max\{0, -[h_i(\mathbf{x}_k) + \nabla h_i(\mathbf{x}_k)^t\mathbf{d}_k]\} \\
& \text{so that } (z_i^+ + z_i^-) \equiv |h_i(\mathbf{x}_k) + \nabla h_i(\mathbf{x}_k)^t\mathbf{d}_k| \qquad i = 1, \ldots, l
\end{aligned} \tag{10.12}$$

As with trust region methods described in Section 8.7, the decision whether to accept or to reject the new iterate $\mathbf{x}_k + \mathbf{d}_k$ and the adjustment of the step bounds $\boldsymbol{\Delta}_k$ is made based on the ratio R_k of the actual decrease ΔF_{E_k} in the l_1 penalty function F_E, and the decrease ΔF_{EL_k} as predicted by its linearized version F_{EL_k}, provided that the latter is nonzero. These quantities are given as follows from (10.8) and (10.11a):

$$\Delta F_{E_k} = F_E(\mathbf{x}_k) - F_E(\mathbf{x}_k + \mathbf{d}_k) \qquad \Delta F_{EL_k} = F_{EL_k}(\mathbf{x}_k) - F_{EL_k}(\mathbf{x}_k + \mathbf{d}_k) \tag{10.13}$$

The principal concepts tying in the development presented thus far are encapsulated by the following result.

10.3.1 Theorem

Consider Problem P and the absolute value (l_1) penalty function (10.8), where μ is assumed to be large enough as in Theorem 9.3.1.

a. If the conditions of Theorem 9.3.1 hold and if $\bar{\mathbf{x}}$ solves Problem P, then $\bar{\mathbf{x}}$ also solves PP of equation (10.9a). Alternatively, if $\bar{\mathbf{x}}$ is a regular point that satisfies the second-order sufficiency conditions for P, then $\bar{\mathbf{x}}$ is a strict local minimum for PP.
b. Consider Problem PP given by (10.9b), where $(\mathbf{y}, \mathbf{z}^+, \mathbf{z}^-)$ are given by (10.10) for any $\mathbf{x} \in X$. If $\bar{\mathbf{x}}$ is a KKT solution for Problem P, then, for μ large enough, as in Theorem 9.3.1, $\bar{\mathbf{x}}$ is a KKT solution for Problem PP. Conversely, if $\bar{\mathbf{x}}$ is a KKT solution for PP and if $\bar{\mathbf{x}}$ is feasible to P, then $\bar{\mathbf{x}}$ is a KKT solution for P.
c. The solution $\mathbf{d}_k = \mathbf{0}$ is optimal for LP($\mathbf{x}_k$, $\boldsymbol{\Delta}_k$) defined by (10.11b) and (10.12) if and only if $\mathbf{x}_k$ is a KKT solution for PP.
d. The predicted decrease ΔF_{EL_k} in the linearized penalty function, as given by (10.13), is nonnegative, and is zero if and only if $\mathbf{d}_k = \mathbf{0}$ solves Problem LP($\mathbf{x}_k$, $\boldsymbol{\Delta}_k$).

Proof

The proof for part a is similar to that of Theorem 9.3.1 and of Exercise 9.12, and is left to the reader in Exercise 10.18. Next, consider part b. The KKT conditions for P require a primal feasible solution $\bar{\mathbf{x}}$, along with Lagrange multipliers $\bar{\mathbf{u}}$, $\bar{\mathbf{v}}$, and $\bar{\mathbf{w}}$ satisfying

$$\begin{gathered}\sum_{i=1}^{m} \bar{u}_i \nabla g_i(\bar{\mathbf{x}}) + \sum_{i=1}^{l} \bar{v}_i \nabla h_i(\bar{\mathbf{x}}) + \mathbf{A}^t\bar{\mathbf{w}} = -\nabla f(\bar{\mathbf{x}}) \\ \bar{\mathbf{u}} \geq \mathbf{0} \qquad \bar{\mathbf{v}} \text{ unrestricted} \qquad \bar{\mathbf{w}} \geq \mathbf{0} \\ \bar{\mathbf{u}}^t\mathbf{g}(\bar{\mathbf{x}}) = 0 \qquad \bar{\mathbf{w}}^t(\mathbf{A}\bar{\mathbf{x}} - \mathbf{b}) = 0\end{gathered} \tag{10.14}$$

Furthermore, $\bar{\mathbf{x}}$ is a KKT point for PP, with $\bar{\mathbf{x}} \in X$ and with $(\mathbf{y}, \mathbf{z}^+, \mathbf{z}^-)$ given accordingly by (10.10), provided Lagrange multipliers $\bar{\mathbf{u}}$, $\bar{\mathbf{v}}$, and $\bar{\mathbf{w}}$ exist satisfying

$$\sum_{i=1}^{m} \bar{u}_i \nabla g_i(\bar{\mathbf{x}}) + \sum_{i=1}^{l} \bar{v}_i \nabla h_i(\bar{\mathbf{x}}) + \mathbf{A}^t\bar{\mathbf{w}} = -\nabla f(\bar{\mathbf{x}}) \tag{10.15a}$$

$$0 \leq \bar{u}_i \leq \mu \quad (\bar{u}_i - \mu)y_i = 0 \quad \bar{u}_i[y_i - g_i(\bar{\mathbf{x}})] = 0 \quad \text{for } i = 1, \ldots, m \tag{10.15b}$$

$$|\bar{v}_i| \leq \mu \quad z_i^+(\mu - \bar{v}_i) = 0 \quad z_i^-(\mu + \bar{v}_i) = 0 \quad \text{for } i = 1, \ldots, l \tag{10.15c}$$

$$\bar{\mathbf{w}}^t(\mathbf{A}\bar{\mathbf{x}} - \mathbf{b}) = 0 \qquad \bar{\mathbf{w}} \geq \mathbf{0} \tag{10.15d}$$

Now let $\bar{\mathbf{x}}$ be a KKT solution for Problem P, with Lagrange multipliers $\bar{\mathbf{u}}$, $\bar{\mathbf{v}}$, and $\bar{\mathbf{w}}$ satisfying (10.14). Defining $(\mathbf{y}, \mathbf{z}^+, \mathbf{z}^-)$ according to (10.10), we get $\mathbf{y} = \mathbf{0}$, $\mathbf{z}^+ = \mathbf{z}^- = \mathbf{0}$; and so, for μ large enough, as in Theorem 9.3.1, $\bar{\mathbf{x}}$ is a KKT solution for PP by (10.15). Conversely, let $\bar{\mathbf{x}}$ be a KKT solution for PP and suppose that $\bar{\mathbf{x}}$ is feasible to Problem P. Then, we again have $\mathbf{y} = \mathbf{0}$, $\mathbf{z}^+ = \mathbf{z}^- = \mathbf{0}$ by (10.10); and so, by (10.15) and (10.14), $\bar{\mathbf{x}}$ is a KKT solution for Problem P. This proves part b.

Part c follows from Theorem 4.2.15, noting that LP$(\mathbf{x}_k, \boldsymbol{\Delta}_k)$ represents a first-order linearization of PP at the point $\mathbf{x}_k$ and that the step bounds $-\boldsymbol{\Delta}_k \leq \mathbf{d} \leq \boldsymbol{\Delta}_k$ are nonbinding at $\mathbf{d}_k = \mathbf{0}$, that is, at $\mathbf{x} = \mathbf{x}_k$.

Finally, consider part d. Since $\mathbf{d}_k$ minimizes LP$(\mathbf{x}_k, \boldsymbol{\Delta}_k)$ in (10.11b), $\mathbf{x} = \mathbf{x}_k + \mathbf{d}_k$ minimizes (10.11a); and so, since $\mathbf{x}_k$ is feasible to (10.11a), we have $F_{EL_k}(\mathbf{x}_k) \geq F_{EL_k}(\mathbf{x}_k + \mathbf{d}_k)$, or that $\Delta F_{EL_k} \geq 0$. By the same token, this difference is zero if and only if $\mathbf{d}_k = \mathbf{0}$ is optimal for LP$(\mathbf{x}_k, \boldsymbol{\Delta}_k)$, and this completes the proof.

Summary of the Penalty Successive Linear Programming (PSLP) Algorithm

Initialization Put the iteration counter $k = 1$, and select a starting solution $\mathbf{x}_k \in X$ feasible to the linear constraints, along with a step bound or trust region vector $\boldsymbol{\Delta}_k > \mathbf{0}$ in E_n. Let $\boldsymbol{\Delta}_{\text{LB}} > \mathbf{0}$ be some small lower bound tolerance on $\boldsymbol{\Delta}_k$. (Sometimes, $\boldsymbol{\Delta}_{\text{LB}} = \mathbf{0}$ is also used.) Also, select a suitable value of the penalty parameter μ (or values for penalty parameters $\mu_1, \ldots, \mu_{m+l}$ as discussed above). Choose values for the scalars $0 < \rho_0 < \rho_1 < \rho_2 < 1$ to be used in the trust region ratio test, and for the step bound adjustment multiplier $\beta \in (0, 1)$. (Typically, $\rho_0 = 10^{-6}$, $\rho_1 = 0.25$, $\rho_2 = 0.75$, and $\beta = 0.5$.)

Step 1 Linear Programming Subproblem Solve the linear program LP$(\mathbf{x}_k, \boldsymbol{\Delta}_k)$ to obtain an optimum $\mathbf{d}_k$. Compute the actual and the predicted decreases ΔF_{E_k} and ΔF_{EL_k}, respectively, in the penalty function as given by (10.13). If $\Delta F_{EL_k} = 0$ (equivalently, by Theorem 10.3.1d, if $\mathbf{d}_k = \mathbf{0}$), then stop. Otherwise, compute the ratio $R_k = \Delta F_{E_k}/\Delta F_{EL_k}$. If $R_k < \rho_0$, then, since $\Delta F_{EL_k} \geq 0$ by Theorem 10.3.1d, the penalty function has either worsened or its improvement is insufficient. Hence, reject the current solution, shrink $\boldsymbol{\Delta}_k$ to $\beta\boldsymbol{\Delta}_k$, and repeat this step. (Zhang, Kim, and Lasdon [1985] show that within a finite number of such reductions, we will have $R_k \geq \rho_0$. Note that, while R_k remains less than ρ_0, components of $\boldsymbol{\Delta}_k$ may shrink below those of $\boldsymbol{\Delta}_{\text{LB}}$.) On the other hand, if $R_k \geq \rho_0$, proceed to step 2.

Step 2 New Iterate and Adjustment of Step Bounds Let $\mathbf{x}_{k+1} = \mathbf{x}_k + \mathbf{d}_k$. If $0 \leq R_k < \rho_1$, then shrink $\boldsymbol{\Delta}_k$ to $\boldsymbol{\Delta}_{k+1} = \beta\boldsymbol{\Delta}_k$, since the penalty function has not improved sufficiently. If $\rho_1 \leq R_k \leq \rho_2$, then retain $\boldsymbol{\Delta}_{k+1} = \boldsymbol{\Delta}_k$. On the other hand, if $R_k > \rho_2$, then amplify the trust region by letting $\boldsymbol{\Delta}_{k+1} = \boldsymbol{\Delta}_k/\beta$. In all cases, replace $\boldsymbol{\Delta}_{k+1}$ by $\max\{\boldsymbol{\Delta}_{k+1}, \boldsymbol{\Delta}_{\text{LB}}\}$, where the max $\{\cdot\}$ is taken componentwise. Increment k by 1 and return to step 1.

A few comments are in order at this point. First, note that the linear program (10.11b) is feasible and bounded ($\mathbf{d} = \mathbf{0}$ is a feasible solution) and that it preserves any sparsity structure of the original problem. Second, if there are any variables that appear linearly in the objective function as well as in the constraints of P, then the corresponding step bounds for such variables can be taken as some arbitrarily large value M and can be retained at that value throughout the procedure. Third, when termination occurs at step 1, then, by Theorem 10.3.1, $\mathbf{x}_k$ is a KKT solution for PP; and if $\mathbf{x}_k$ is feasible to P, then it is also a KKT solution for P. (Otherwise, the penalty parameters may need to be increased as discussed earlier.) Fourth, it can be shown that either the algorithm terminates finitely or, else, an infinite sequence $\{\mathbf{x}_k\}$ is generated such that if the level set $\{\mathbf{x} \in X: F_E(\mathbf{x}) \leq F_E(\mathbf{x}_1)\}$ is bounded, then $\{\mathbf{x}_k\}$ has an accumulation point, and every such accumulation point is a KKT solution for Problem PP. Finally, the stopping criterion of step 1 is usually replaced by several practical termination criteria. For example, if the fractional change in the l_1 penalty function is less than a tolerance ($\varepsilon = 10^{-4}$) for some c (= 3) consecutive iterations, or if the iterate is ε-feasible and either the KKT conditions are satisfied within an ε-tolerance, or if the fractional change in the objective function value for Problem P is less than ε for c consecutive iterations, then the procedure can be terminated. Also, the amplification or reduction in the step bounds are often modified in implementations so as to treat deviations of R_k from unity symmetrically. For example, at step 2, if $|1 - R_k| < 0.25$, then all step bounds are amplified by dividing by $\beta = 0.5$; and if $|1 - R_k| > 0.75$, then all step bounds are reduced by multiplying by β. In addition, if any nonlinear variable remains at the same step bound for c (= 3) consecutive iterations, then its step bound is amplified by dividing by β.

10.3.2 Example

Consider the problem

$$\begin{array}{ll} \text{Minimize} & f(\mathbf{x}) = 2x_1^2 + 2x_2^2 - 2x_1x_2 - 4x_1 - 6x_2 \\ \text{subject to} & g_1(\mathbf{x}) = 2x_1^2 - x_2 \leq 0 \\ & \mathbf{x} \in X = \{\mathbf{x} = (x_1, x_2) : x_1 + 5x_2 \leq 5, \mathbf{x} \geq \mathbf{0}\} \end{array}$$

Figure 10.13*a* provides a sketch for the graphical solution of this problem. Note that this problem has a "vertex" solution, and thus we might expect a rapid convergence behavior. Let us begin with the solution $\mathbf{x}_1 = (0, 1)^t \in X$ and use $\mu = 10$ (which can be verified to be sufficiently large—see Exercise 10.19). Let us also select $\mathbf{\Delta}_1 = (1, 1)^t$, $\mathbf{\Delta}_{LB} = (10^{-6}, 10^{-6})^t$, $\rho_0 = 10^{-6}$, $\rho_1 = 0.25$, $\rho_2 = 0.75$, and $\beta = 0.5$.

The linear program LP($\mathbf{x}_1$, $\mathbf{\Delta}_1$) given by (10.11b) now needs to be solved, as, for example, by the simplex method. To graphically illustrate the process, consider the equivalent problem (10.11a). Noting that $\mathbf{x}_1 = (0, 1)^t$, $\mu = 10$, $f(\mathbf{x}_1) = -4$, $\nabla f(\mathbf{x}_1) = (-6, -2)^t$, $g_1(\mathbf{x}_1) = -1$, and $\nabla g_1(\mathbf{x}_1) = (0, -1)^t$, we have

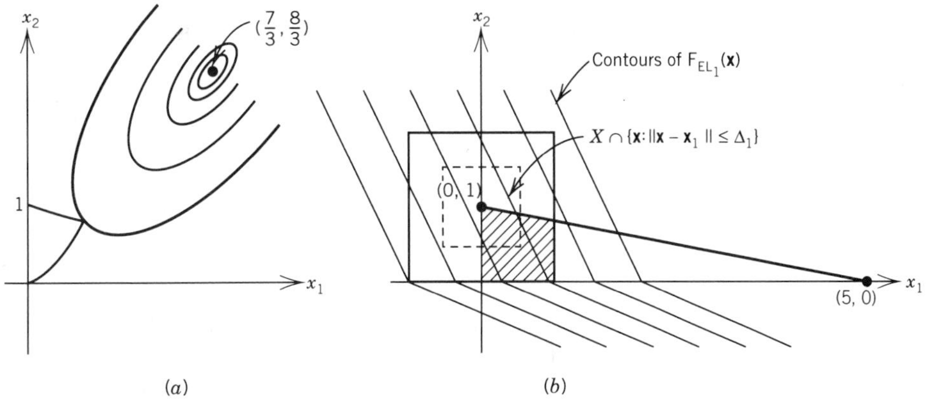

Figure 10.13 Illustration for Example 10.3.2.

$$F_{EL_1}(\mathbf{x}) = -2 - 6x_1 - 2x_2 + 10 \max\{0, -x_2\} \tag{10.16}$$

The solution of LP($\mathbf{x}_1$, $\boldsymbol{\Delta}_1$) via (10.11a) is depicted in Figure 10.13b. The optimum solution is $\mathbf{x} = (1, \frac{4}{5})^t$, so that $\mathbf{d}_1 = (1, \frac{4}{5})^t - (0, 1)^t = (1, -\frac{1}{5})^t$ solves (10.11b). From (10.13), using (10.8) and (10.16) along with $\mathbf{x}_1 = (0, 1)^t$ and $\mathbf{x}_1 + \mathbf{d}_1 = (1, \frac{4}{5})^t$, we get $\Delta F_{E_k} = -4.88$ and $\Delta F_{EL_k} = \frac{28}{5}$. Hence, the penalty function has worsened, and so we reduce the step bounds at step 1 itself and repeat this step with the revised $\boldsymbol{\Delta}_1 = (0.5, 0.5)^t$.

The revised step bound box is shown dotted in Figure 10.13*b*. The corresponding optimal solution is $\mathbf{x} = (0.5, 0.9)^t$, which corresponds to the optimum $\mathbf{d}_1 = (0.5, 0.9)^t - (0, 1)^t = (0.5, -0.1)^t$ for the problem (10.11b). From (10.13), using (10.8) and (10.16) along with $\mathbf{x}_1 = (0, 1)^t$ and $\mathbf{x}_1 + \mathbf{d}_1 = (0.5, 0.9)^t$, we get $\Delta F_{E_k} = 2.18$ and $\Delta F_{EL_k} = 2.8$, which gives $R_k = 2.18/2.8 = 0.7786$. We therefore accept this solution as the new iterate $\mathbf{x}_2 = (0.5, 0.9)^t$ and, since $R_k > \rho_2 = 0.75$, we amplify the trust region by letting $\boldsymbol{\Delta}_2 = \boldsymbol{\Delta}_1/\beta = (1, 1)^t$. We now ask the reader (see Exercise 10.19) to continue this process until a suitable termination criterion is satisfied, as discussed above.

10.4 Successive Quadratic Programming or Projected Lagrangian Approach

We have seen that Zoutendijk's algorithm as well as Topkis and Veinott's modification of this procedure are prone to zigzagging and slow convergence behavior because of the first-order approximations employed. The SLP approach enjoys a quadratic rate of convergence if the optimum occurs at a vertex of the feasible region, because then, the method begins to imitate Newton's method applied to the active constraints. However, for nonvertex solutions, this method again, being essentially a first-order approximation procedure, can succumb to a slow convergence process. To alleviate this behavior, we can employ second-order approximations and derive a *successive quadratic programming approach (SQP)*.

SQP methods, also known as *sequential*, or *recursive, quadratic programming*, employ Newton's method (or quasi-Newton methods) to directly solve the KKT conditions for the original problem. As a result, the accompanying subproblem turns out to be the minimization of a quadratic approximation to the Lagrangian function optimized over a linear approximation to the constraints. Hence, this type of process is also known as a *projected Lagrangian*, or the *Lagrange–Newton*, approach. By its nature, this method produces both primal and dual (Lagrange multiplier) solutions.

To present the concept of this method, consider the equality constrained nonlinear problem, where $\mathbf{x} \in E_n$, and all functions are assumed to be continuously twice-differentiable.

$$\begin{aligned} \textbf{\textit{P}}:\ & \text{Minimize} && f(\mathbf{x}) \\ & \text{subject to} && h_i(\mathbf{x}) = 0 \quad i = 1, \ldots, l \end{aligned} \tag{10.17}$$

The extension for including inequality constraints is motivated by the following analysis for the equality constrained case and is considered subsequently.

The KKT optimality conditions for Problem P require a primal solution $\mathbf{x} \in E_n$ and a Lagrange multiplier vector $\mathbf{v} \in E_l$ such that

$$\begin{aligned} &\nabla f(\mathbf{x}) + \sum_{i=1}^{l} v_i \nabla h_i(\mathbf{x}) = \mathbf{0} \\ &h_i(\mathbf{x}) = 0 \qquad i = 1, \ldots, l \end{aligned} \tag{10.18}$$

Let us write this system of equations more compactly as $\mathbf{W}(\mathbf{x}, \mathbf{v}) = \mathbf{0}$. We now use the Newton–Raphson method to solve (10.18) or, equivalently, use Newton's method to minimize a function for which (10.18) represents the first-order condition that equates the gradient to zero. Hence, given an iterate $(\mathbf{x}_k, \mathbf{v}_k)$, we solve the first-order approximation

$$\mathbf{W}(\mathbf{x}_k, \mathbf{v}_k) + \nabla\mathbf{W}(\mathbf{x}_k, \mathbf{v}_k)\begin{bmatrix} \mathbf{x} - \mathbf{x}_k \\ \mathbf{v} - \mathbf{v}_k \end{bmatrix} = \mathbf{0} \tag{10.19}$$

to the given system to determine the next iterate $(\mathbf{x}, \mathbf{v}) = (\mathbf{x}_{k+1}, \mathbf{v}_{k+1})$, where $\nabla\mathbf{W}$ denotes the Jacobian of $\mathbf{W}$. Defining $\nabla^2 L(\mathbf{x}_k) = \nabla^2 f(\mathbf{x}_k) + \sum_{i=1}^{l} v_{ki} \nabla^2 \mathbf{h}(\mathbf{x}_k)$ to be the usual Hessian of the Lagrangian at $\mathbf{x}_k$ with the Lagrange multiplier vector $\mathbf{v}_k$, and letting $\nabla\mathbf{h}$ denote the Jacobian of $\mathbf{h}$ comprised of rows $\nabla h_i(\mathbf{x})^t$ for $i = 1, \ldots, l$, we have

$$\nabla\mathbf{W}(\mathbf{x}_k, \mathbf{v}_k) = \begin{bmatrix} \nabla^2 L(\mathbf{x}_k) & \nabla\mathbf{h}(\mathbf{x}_k)^t \\ \nabla\mathbf{h}(\mathbf{x}_k) & \mathbf{0} \end{bmatrix} \tag{10.20}$$

Using (10.18) and (10.20), we can rewrite (10.19) as

$$\begin{aligned} &\nabla^2 L(\mathbf{x}_k)(\mathbf{x} - \mathbf{x}_k) + \nabla\mathbf{h}(\mathbf{x}_k)^t(\mathbf{v} - \mathbf{v}_k) = -\nabla f(\mathbf{x}_k) - \nabla\mathbf{h}(\mathbf{x}_k)^t\mathbf{v}_k \\ &\nabla\mathbf{h}(\mathbf{x}_k)(\mathbf{x} - \mathbf{x}_k) = -\mathbf{h}(\mathbf{x}_k) \end{aligned}$$

Substituting $\mathbf{d} = \mathbf{x} - \mathbf{x}_k$, this in turn can be rewritten as

$$\begin{aligned} &\nabla^2 L(\mathbf{x}_k)\mathbf{d} + \nabla\mathbf{h}(\mathbf{x}_k)^t\mathbf{v} = -\nabla f(\mathbf{x}_k) \\ &\nabla\mathbf{h}(\mathbf{x}_k)\mathbf{d} = -\mathbf{h}(\mathbf{x}_k) \end{aligned} \tag{10.21}$$

We can now solve for $(\mathbf{d}, \mathbf{v}) = (\mathbf{d}_k, \mathbf{v}_{k+1})$, say, using this system, if a solution exists. (See the convergence analysis below and Exercise 10.24.) Setting $\mathbf{x}_{k+1} = \mathbf{x}_k + \mathbf{d}_k$, we then increment k by 1 and repeat this process until $\mathbf{d} = \mathbf{0}$ happens to solve (10.21). When this occurs, if at all, noting (10.18), we shall have found a KKT solution for Problem P.

Now, instead of adopting the foregoing process to find *any* KKT solution for P, we can instead employ a quadratic minimization subproblem whose optimality conditions duplicate (10.21), but which might tend to drive the process toward beneficial KKT solutions. Such a quadratic program is stated below, where the constant term $f(\mathbf{x}_k)$ has been inserted into the objective function for insight and convenience.

$$\begin{aligned} \textbf{\textit{QP}}(\mathbf{x}_k, \mathbf{v}_k)\text{:}\quad &\text{Minimize} \quad f(\mathbf{x}_k) + \nabla f(\mathbf{x}_k)^t\mathbf{d} + \tfrac{1}{2}\mathbf{d}^t\nabla^2 L(\mathbf{x}_k)\mathbf{d} \\ &\text{subject to} \quad h_i(\mathbf{x}_k) + \nabla h_i(\mathbf{x}_k)^t\mathbf{d} = 0, \qquad i = 1 \ldots, l \end{aligned} \tag{10.22}$$

Several comments regarding the linearly constrained quadratic subproblem QP($\mathbf{x}_k$, $\mathbf{v}_k$), abbreviated QP wherever unambiguous, are in order at this point. First, note that an optimum to QP, if it exists, is a KKT point for QP and satisfies equations (10.21), where $\mathbf{v}$ is the set of Lagrange multipliers associated with the constraints of QP. However, the minimization process of QP drives the solution toward a desirable KKT point satisfying (10.21) whenever alternatives exist. Second, observe that by the foregoing derivation, the objective function of QP represents not just a quadratic approximation for

$f(\mathbf{x})$ but also incorporates an additional term $\frac{1}{2}\sum_{i=1}^{l} v_{ki}\mathbf{d}^t\nabla^2 \mathbf{h}(\mathbf{x}_k)\mathbf{d}$ to represent the curvature of the constraints. In fact, defining the Lagrangian function $L(\mathbf{x}) = f(\mathbf{x}) + \sum_{i=1}^{l} v_{ki}h_i(\mathbf{x})$, the objective function of QP($\mathbf{x}_k$, $\mathbf{v}_k$) can alternatively be written as follows, noting the constraints:

$$\text{Minimize} \quad L(\mathbf{x}_k) + \nabla_{\mathbf{x}} L(\mathbf{x}_k)^t\mathbf{d} + \tfrac{1}{2}\mathbf{d}^t\nabla^2 L(\mathbf{x}_k)\mathbf{d} \tag{10.23}$$

Observe that (10.23) represents a second-order Taylor series approximation for the Lagrangian function L. In particular, this supports the quadratic convergence rate behavior in the presence of nonlinear constraints (see also Exercise 10.25). Third, note that the constraints of QP represent a first-order linearization at the current iterate $\mathbf{x}_k$. Fourth, observe that QP might be unbounded or infeasible, whereas P is not. Although the first of these unfavorable events can be managed by bounding the variation in $\mathbf{d}$, for instance, the second is more disconcerting. For example, if we have a constraint $x_1^2 + x_2^2 = 1$ and we linearize this at the origin, we obtain an inconsistent restriction requiring $-1 = 0$. We later present a variant of the above scheme which overcomes this difficulty (see also Exercise 10.23). Notwithstanding this problem, and assuming a well-behaved QP subproblem, we are now ready to state a rudimentary SQP algorithm.

Rudimentary SQP Algorithm (RSQP)

Initialization Put the iteration counter $k = 1$ and select a (suitable) starting primal–dual solution ($\mathbf{x}_k$, $\mathbf{v}_k$).

Main Step Solve the quadratic subproblem QP($\mathbf{x}_k$, $\mathbf{v}_k$) to obtain a solution $\mathbf{d}_k$ along with a set of Lagrange multipliers $\mathbf{v}_{k+1}$. If $\mathbf{d}_k = \mathbf{0}$, then, from (10.21), ($\mathbf{x}_k$, $\mathbf{v}_{k+1}$) satisfies the KKT conditions (10.18) for Problem P; stop. Otherwise, put $\mathbf{x}_{k+1} = \mathbf{x}_k + \mathbf{d}_k$, increment k by 1, and repeat the main step.

Convergence Rate Analysis

Under appropriate conditions, we can argue a quadratic convergence behavior for the foregoing algorithm. Specifically, suppose that $\bar{\mathbf{x}}$ is a regular KKT solution for Problem P which, together with a set of Lagrange multipliers $\bar{\mathbf{v}}$, satisfies the second-order sufficiency conditions of Theorem 4.4.2. Then, $\nabla \mathbf{W}(\bar{\mathbf{x}}, \bar{\mathbf{v}}) \equiv \overline{\nabla \mathbf{W}}$, say, defined by (10.20) is nonsingular. To see this, let us show that the system

$$\nabla \mathbf{W}(\bar{\mathbf{x}}, \bar{\mathbf{v}})\begin{bmatrix}\mathbf{d}_1 \\ \mathbf{d}_2\end{bmatrix} = \mathbf{0}$$

has a unique solution given by $(\mathbf{d}_1^t, \mathbf{d}_2^t) = \mathbf{0}$. Consider any solution $(\mathbf{d}_1^t, \mathbf{d}_2^t)$. Since $\bar{\mathbf{x}}$ is a regular solution, $\nabla \mathbf{h}(\bar{\mathbf{x}})^t$ has full column rank; and so, if $\mathbf{d}_1 = \mathbf{0}$, then $\mathbf{d}_2 = \mathbf{0}$ as well. If $\mathbf{d}_1 \neq \mathbf{0}$, then, since $\nabla \mathbf{h}(\bar{\mathbf{x}})\mathbf{d}_1 = \mathbf{0}$, by the second-order sufficiency conditions we have $\mathbf{d}_1^t\nabla^2 L(\bar{\mathbf{x}})\mathbf{d}_1 > 0$. However, since $\nabla^2 L(\bar{\mathbf{x}})\mathbf{d}_1 + \overline{\nabla \mathbf{h}(\bar{\mathbf{x}})\mathbf{d}_2} = \mathbf{0}$, we have $\mathbf{d}_1^t\nabla^2 L(\bar{\mathbf{x}})\mathbf{d}_1 = -\mathbf{d}_2^t\nabla \mathbf{h}(\bar{\mathbf{x}})\mathbf{d}_1 = 0$, a contradiction. Hence, $\overline{\nabla \mathbf{W}}$ is nonsingular; and thus, for ($\mathbf{x}_k$, $\mathbf{v}_k$) sufficienty close to ($\bar{\mathbf{x}}$, $\bar{\mathbf{v}}$), $\nabla \mathbf{W}(\mathbf{x}_k, \mathbf{v}_k)$ is nonsingular. Therefore, the system (10.21), and so Problem QP($\mathbf{x}_k$, $\mathbf{v}_k$), has a well-defined (unique) solution. Consequently, in the spirit of Theorem 8.6.5, when ($\mathbf{x}_k$, $\mathbf{v}_k$) is sufficiently close to ($\bar{\mathbf{x}}$, $\bar{\mathbf{v}}$), a quadratic rate of convergence to ($\bar{\mathbf{x}}$, $\bar{\mathbf{v}}$) is obtained.

Actually, the closeness of $\mathbf{x}_k$ alone to $\bar{\mathbf{x}}$ is sufficient to establish convergence. It can be shown (see the Notes and References section) that, if $\mathbf{x}_1$ is sufficiently close to $\bar{\mathbf{x}}$ and if $\nabla \mathbf{W}(\mathbf{x}_1, \mathbf{v}_1)$ is nonsingular, then the algorithm SQPR converges quadratically to $(\bar{\mathbf{x}}, \bar{\mathbf{v}})$. In this respect, the Lagrange multipliers $\mathbf{v}$, appearing only in the second-order term in QP, do not play as important a role as they do in augmented Lagrangian (ALAG) penalty methods, for example, and inaccuracies in their estimation can be more flexibly tolerated.

Extension to Include Inequality Constraints

We now consider the inclusion of inequality constraints $g_i(\mathbf{x}) \leq 0$, $i = 1, \ldots, m$, in Problem P, where g_i are continuously twice-differentiable for $i = 1, \ldots, m$. This revised problem is restated below:

$$\begin{array}{llll} P\text{:} & \text{Minimize} & f(\mathbf{x}) & \\ & \text{subject to} & g_i(\mathbf{x}) \leq 0 & i = 1, \ldots, m \\ & & h_i(\mathbf{x}) = 0 & i = 1, \ldots, l \end{array} \tag{10.24}$$

For this instance, given an iterate $(\mathbf{x}_k, \mathbf{u}_k, \mathbf{v}_k)$, where $\mathbf{u}_k \geq \mathbf{0}$ and $\mathbf{v}_k$ are, respectively, the Lagrange multiplier estimates for the inequality and the equality constraints, we consider the following quadratic programming subproblem as a direct extension of (10.22):

$$\begin{array}{llll} QP(\mathbf{x}_k, \mathbf{u}_k, \mathbf{v}_k)\text{:} & \text{Minimize} & f(\mathbf{x}_k) + \nabla f(\mathbf{x}_k)^t\mathbf{d} + \frac{1}{2}\mathbf{d}^t\nabla^2 L(\mathbf{x}_k)\mathbf{d} & \\ & \text{subject to} & g_i(\mathbf{x}_k) + \nabla g_i(\mathbf{x}_k)^t\mathbf{d} \leq 0 & i = 1, \ldots, m \\ & & h_i(\mathbf{x}_k) + \nabla h_i(\mathbf{x}_k)^t\mathbf{d} = 0 & i = 1, \ldots, l \end{array} \tag{10.25}$$

where $\nabla^2 L(\mathbf{x}_k) = \nabla^2 f(\mathbf{x}_k) + \sum_{i=1}^m u_{ki}\nabla^2 g_i(\mathbf{x}_k) + \sum_{i=1}^l v_{ki}\nabla^2 h_i(\mathbf{x}_k)$. Note that the KKT conditions for this problem require that, in addition to primal feasibility, we find Lagrange multipliers $\mathbf{u}$ and $\mathbf{v}$ such that

$$\nabla f(\mathbf{x}_k) + \nabla^2 L(\mathbf{x}_k)\mathbf{d} + \sum_{i=1}^m u_i \nabla g_i(\mathbf{x}_k) + \sum_{i=1}^l v_i \nabla h_i(\mathbf{x}_k) = \mathbf{0} \tag{10.26a}$$

$$u_i[g_i(\mathbf{x}_k) + \nabla g_i(\mathbf{x}_k)^t\mathbf{d}] = 0 \qquad i = 1, \ldots, m \tag{10.26b}$$

$$\mathbf{u} \geq \mathbf{0} \qquad \mathbf{v} \text{ unrestricted} \tag{10.26c}$$

Hence, if $\mathbf{d}_k$ solves QP$(\mathbf{x}_k, \mathbf{u}_k, \mathbf{v}_k)$ with Lagrange multipliers $\mathbf{u}_{k+1}$ and $\mathbf{v}_{k+1}$, and if $\mathbf{d}_k = \mathbf{0}$, then $\mathbf{x}_k$ along with $(\mathbf{u}_{k+1}, \mathbf{v}_{k+1})$ yields a KKT solution for the original Problem P. Otherwise, we set $\mathbf{x}_{k+1} = \mathbf{x}_k + \mathbf{d}_k$ as before, increment k by 1, and repeat the process. In a likewise manner, it can be shown that if $\bar{\mathbf{x}}$ is a regular KKT solution which, together with $(\bar{\mathbf{u}}, \bar{\mathbf{v}})$, satisfies the second-order sufficiency conditions, and if $(\mathbf{x}_k, \mathbf{u}_k, \mathbf{v}_k)$ is initialized sufficiently close to $(\bar{\mathbf{x}}, \bar{\mathbf{u}}, \bar{\mathbf{v}})$, then the foregoing iterative process will converge quadratically to $(\bar{\mathbf{x}}, \bar{\mathbf{u}}, \bar{\mathbf{v}})$.

Quasi-Newton Approximations

One disadvantage of the SQP method discussed thus far is that we require second-order derivatives to be calculated and, besides, that $\nabla^2 L(\mathbf{x}_k)$ might not be positive definite. This can be overcome by employing quasi-Newton positive definite approximations for $\nabla^2 L$. For example, given a positive definite approximation $\mathbf{B}_k$ for $\nabla^2 L(\mathbf{x}_k)$ in the algorithm SQPR described above, we can solve the system (10.21) with $\nabla^2 L(\mathbf{x}_k)$ replaced by $\mathbf{B}_k$, to obtain the unique solution $\mathbf{d}_k$ and $\mathbf{v}_{k+1}$, and then set $\mathbf{x}_{k+1} = \mathbf{x}_k + \mathbf{d}_k$. This is equivalent to the iterative step given by

$$\begin{bmatrix} \mathbf{x}_{k+1} \\ \mathbf{v}_{k+1} \end{bmatrix} = \begin{bmatrix} \mathbf{x}_k \\ \mathbf{v}_k \end{bmatrix} - \begin{bmatrix} \mathbf{B}_k & \nabla \mathbf{h}(\mathbf{x}_k)^t \\ \nabla \mathbf{h}(\mathbf{x}_k) & \mathbf{0} \end{bmatrix}^{-1} \begin{bmatrix} \nabla L(\mathbf{x}_k) \\ \mathbf{h}(\mathbf{x}_k) \end{bmatrix}$$

where $\nabla L(\mathbf{x}_k) = \nabla f(\mathbf{x}_k) + \nabla \mathbf{h}(\mathbf{x}_k)^t \mathbf{v}_k$. Then, adopting the popular BFGS update for the Hessian as defined by (8.63), we can compute

$$\mathbf{B}_{k+1} = \mathbf{B}_k + \frac{\mathbf{q}_k \mathbf{q}_k^t}{\mathbf{q}_k^t \mathbf{p}_k} - \frac{\mathbf{B}_k \mathbf{p}_k \mathbf{p}_k^t \mathbf{B}_k}{\mathbf{p}_k^t \mathbf{B}_k \mathbf{p}_k} \tag{10.27a}$$

where

$$\mathbf{p}_k = \mathbf{x}_{k+1} - \mathbf{x}_k \qquad \mathbf{q}_k = \nabla L'(\mathbf{x}_{k+1}) - \nabla L'(\mathbf{x}_k)$$

and where

$$\nabla L'(\mathbf{x}) \equiv \nabla f(\mathbf{x}) + \sum_{i=1}^{l} v_{(k+1)i} \nabla h_i(\mathbf{x}) \tag{10.27b}$$

It can be shown that this modification in the rudimentary process, similar to the quasi-Newton modification of Newton's algorithm, converges superlinearly when initialized sufficiently close to a solution $(\bar{\mathbf{x}}, \bar{\mathbf{v}})$ that satisfies the foregoing regularity and second-order sufficiency conditions. However, this superlinear convergence rate is strongly based on the use of unit step sizes.

A Globally Convergent Variant Using the l_1 Penalty as a Merit Function

A principal disadvantage of the SQP method described thus far is that convergence is guaranteed only when the algorithm is initialized sufficiently close to a desirable solution whereas, in practice, this condition is usually difficult to realize. To remedy this situation and to ensure global convergence, we introduce the idea of a *merit function*. This is a function that, along with the objective function, is simultaneously minimized at the solution of the problem, but one that also serves as a descent function, guiding the iterates and providing a measure of progress. Preferably, it should be easy to evaluate this function, and it should not impair the convergence rate of the algorithm. We describe the use of the popular l_1, or absolute value, penalty function (9.8), restated below, as a merit function for Problem P given in (10.24):

$$F_E(\mathbf{x}) = f(\mathbf{x}) + \mu \left[\sum_{i=1}^{m} \max\{0, g_i(\mathbf{x})\} + \sum_{i=1}^{l} |h_i(\mathbf{x})| \right] \tag{10.28}$$

The following lemma establishes the role of F_E as a merit function. The Notes and References section points out other quadratic and ALAG penalty functions that can be used as merit functions in a similar context.

10.4.1 Lemma

Given an iterate $\mathbf{x}_k$, consider the quadratic subproblem QP given by (10.25), where $\nabla^2 L(\mathbf{x}_k)$ is replaced by any positive definite approximation $\mathbf{B}_k$. Let $\mathbf{d}$ solve this problem with Lagrange multipliers $\mathbf{u}$ and $\mathbf{v}$ associated with the inequality and the equality constraints, respectively. If $\mathbf{d} \neq \mathbf{0}$, and if $\mu \geq \max\{u_1, \ldots, u_m, |v_1|, \ldots, |v_l|\}$, then $\mathbf{d}$ is a descent direction at $\mathbf{x} = \mathbf{x}_k$ for the l_1 penalty function F_E given by (10.28).

Proof

Using the primal feasibility, the dual feasibility, and the complementary slackness conditions (10.25), (10.26a), and (10.26b) for QP, we have

$$\begin{aligned}
\nabla f(\mathbf{x}_k)^t\mathbf{d} &= -\mathbf{d}^t\mathbf{B}_k\mathbf{d} - \sum_{i=1}^{m} u_i\nabla g_i(\mathbf{x}_k)^t\mathbf{d} - \sum_{i=1}^{l} v_i\nabla h_i(\mathbf{x}_k)^t\mathbf{d} \\
&= -\mathbf{d}^t\mathbf{B}_k\mathbf{d} + \sum_{i=1}^{m} u_i g_i(\mathbf{x}_k) + \sum_{i=1}^{l} v_i h_i(\mathbf{x}_k) \\
&\leq -\mathbf{d}^t\mathbf{B}_k\mathbf{d} + \sum_{i=1}^{m} u_i \max\{0, g_i(\mathbf{x}_k)\} + \sum_{i=1}^{l} |v_i|\,|h_i(\mathbf{x}_k)| \\
&\leq -\mathbf{d}^t\mathbf{B}_k\mathbf{d} + \mu\left[\sum_{i=1}^{m} \max\{0, g_i(\mathbf{x}_k)\} + \sum_{i=1}^{l} |h_i(\mathbf{x}_k)|\right]
\end{aligned} \tag{10.29}$$

Now we have, from (10.28), that for a step length $\lambda \geq 0$,

$$\begin{aligned}
F_E(\mathbf{x}_k) - F_E(\mathbf{x}_k + \lambda\mathbf{d}) &= [f(\mathbf{x}_k) - f(\mathbf{x}_k + \lambda\mathbf{d})] \\
&\quad + \mu\left\{\sum_{i=1}^{m} [\max\{0, g_i(\mathbf{x}_k)\} - \max\{0, g_i(\mathbf{x}_k + \lambda\mathbf{d})\}]\right. \\
&\quad \left. + \sum_{i=1}^{l} [|h_i(\mathbf{x}_k)| - |h_i(\mathbf{x}_k + \lambda\mathbf{d})|]\right\}
\end{aligned} \tag{10.30}$$

Letting $O_i(\lambda)$ denote an appropriate function that approaches zero as $\lambda \to 0$, for $i = 0, 1, \ldots, m + l$, we have, for $\lambda > 0$ and sufficiently small,

$$f(\mathbf{x}_k + \lambda\mathbf{d}) = f(\mathbf{x}_k) + \lambda\nabla f(\mathbf{x}_k)^t\mathbf{d} + \lambda O_0(\lambda) \tag{10.31a}$$

Also, $g_i(\mathbf{x}_k + \lambda\mathbf{d}) = g_i(\mathbf{x}_k) + \lambda\nabla g_i(\mathbf{x}_k)^t\mathbf{d} + \lambda O_i(\lambda) \leq g_i(\mathbf{x}_k) - \lambda g_i(\mathbf{x}_k) + \lambda O_i(\lambda)$ from (10.25). Hence,

$$\max\{0, g_i(\mathbf{x}_k + \lambda\mathbf{d})\} \leq (1 - \lambda)\max\{0, g_i(\mathbf{x}_k)\} + \lambda|O_i(\lambda)| \tag{10.31b}$$

Similarly, from (10.25),

$$h_i(\mathbf{x}_k + \lambda\mathbf{d}) = h_i(\mathbf{x}_k) + \lambda\nabla h_i(\mathbf{x}_k) + \lambda O_{m+i}(\lambda) = (1 - \lambda)h_i(\mathbf{x}_k) + \lambda O_{m+i}(\lambda)$$

and hence

$$|h_i(\mathbf{x}_k + \lambda\mathbf{d})| \leq (1 - \lambda)|h_i(\mathbf{x}_k)| + \lambda|O_{m+i}(\lambda)| \tag{10.31c}$$

Using (10.31) in (10.30), we obtain, for $\lambda \geq 0$ and sufficiently small, $F_E(\mathbf{x}_k) - F_E(\mathbf{x}_k + \lambda\mathbf{d}) \geq \lambda[-\nabla f(\mathbf{x}_k)^t\mathbf{d} + \mu\{\sum_{i=1}^{m} \max\{0, g_i(\mathbf{x}_k)\} + \sum_{i=1}^{l} |h_i(\mathbf{x}_k)| + O(\lambda)]$, where $O(\lambda) \to 0$ as $\lambda \to 0$. Hence, by (10.29), this gives $F_E(\mathbf{x}_k) - F_E(\mathbf{x}_k + \lambda\mathbf{d}) \geq \lambda[\mathbf{d}^t\mathbf{B}_k\mathbf{d} + O(\lambda)] > 0$ for all $\lambda \in (0, \delta)$ for some $\delta > 0$ by the positive definiteness of $\mathbf{B}_k$, and this completes the proof.

Lemma 10.4.1 exhibits the flexibility in the choice of $\mathbf{B}_k$ for the resulting direction to be a descent direction for the exact penalty function. This matrix only needs to be positive definite, and may be updated by using any quasi-Newton strategy such as an extension of (10.27), or may even be held constant throughout the algorithm. This descent feature enables us to obtain a globally convergent algorithm under mild assumptions, as shown below.

Summary of the Merit Function SQP Algorithm (MSQP)

Initialization Put the iteration counter at $k = 1$ and select a (suitable) starting solution $\mathbf{x}_k$. Also, select a positive definite approximation $\mathbf{B}_k$ to the Hessian $\nabla^2 L(\mathbf{x}_k)$ defined with respect to some Lagrange multipliers $\mathbf{u}_k \geq \mathbf{0}$ and $\mathbf{v}_k$ associated with the inequality and the equality constraints, respectively, of Problem (10.24). (Note that $\mathbf{B}_k$ might be arbitrary and need not necessarily bear any relationship to $\nabla^2 L(\mathbf{x}_k)$.)

Main Step Solve the quadratic programming subproblem QP given by (10.25) with $\nabla^2 L(\mathbf{x}_k)$ replaced by $\mathbf{B}_k$ and obtain a solution $\mathbf{d}_k$ along with Lagrange multipliers $(\mathbf{u}_{k+1}, \mathbf{v}_{k+1})$. If $\mathbf{d}_k = \mathbf{0}$, then stop with $\mathbf{x}_k$ as a KKT solution for Problem P of (10.24), having Lagrange multipliers $(\mathbf{u}_{k+1}, \mathbf{v}_{k+1})$. Otherwise, find $\mathbf{x}_{k+1} = \mathbf{x}_k + \lambda_k \mathbf{d}_k$, where λ_k minimizes $F_E(\mathbf{x}_k + \lambda \mathbf{d}_k)$ over $\lambda \in E_1$, $\lambda \geq 0$. Update $\mathbf{B}_k$ to a positive definite matrix $\mathbf{B}_{k+1}$ [which might be $\mathbf{B}_k$, itself, or $\nabla^2 L(\mathbf{x}_{k+1})$ defined with respect to $(\mathbf{u}_{k+1}, \mathbf{v}_{k+1})$, or some approximation thereof updated according to a quasi-Newton scheme]. Increment k by 1 and repeat the main step.

The reader may note that the line search above is to be performed with respect to a nondifferentiable function, which obviates the use of the techniques in Sections 8.2 and 8.3, including the popular curve fitting approaches. Below, we sketch the proof of convergence for algorithm MSQP. In Exercise 10.26, we ask the reader to provide a detailed argument.

10.4.2 Theorem

Algorithm MSQP either terminates finitely with a KKT solution to Problem P defined in (10.24), or else an infinite sequence of iterates $\{\mathbf{x}_k\}$ is generated. In the latter case, assume that $\{\mathbf{x}_k\} \subseteq X$, a compact subset of E_n, and that for any point $\mathbf{x} \in X$ and any positive definite matrix $\mathbf{B}$, the quadratic programming subproblem QP (with $\nabla^2 L$ replaced by $\mathbf{B}$) has a unique solution $\mathbf{d}$ (and so this problem is feasible), and has unique Lagrange multipliers $\mathbf{u}$ and $\mathbf{v}$ satisfying $\mu \geq \max \{u_1, \ldots, u_m, |v_1|, \ldots, |v_l|\}$, where μ is the penalty parameter for F_E defined in (10.28). Furthermore, assume that the accompanying sequence $\{\mathbf{B}_k\}$ of positive definite matrices generated lies in a compact subspace, with all accumulation points being positive definite (or with $\{\mathbf{B}_k^{-1}\}$ also being bounded). Then, every accumulation point of $\{\mathbf{x}_k\}$ is a KKT solution for P.

Proof

Let the solution set Ω be composed of all points $\mathbf{x}$ such that the corresponding subproblem QP produces $\mathbf{d} = \mathbf{0}$ at optimality. Note from (10.26) that, given any positive definite matrix $\mathbf{B}$, $\mathbf{x}$ is a KKT solution for P if and only if $\mathbf{d} = \mathbf{0}$ is optimal for QP, that is, $\mathbf{x} \in \Omega$. Now the algorithm MSQP can be viewed as a map **UMD**, where **D** is the direction-finding map that determines the direction $\mathbf{d}_k$ via the subproblem QP defined with respect to $\mathbf{x}_k$ and $\mathbf{B}_k$, **M** is the usual line search map, and **U** is a map that updates $\mathbf{B}_k$ to $\mathbf{B}_{k+1}$. Since the optimality conditions of QP are continuous in the data, the output of QP can readily be seen to be a continuous function of the input. By Theorem 8.4.1, the line search map **M** is also closed, since F_E is continuous. Since the conditions of

Theorem 7.3.2 hold, **MD** is therefore closed. Moreover, by Lemma 10.4.1, if $\mathbf{x}_k \notin \Omega$, then $F_E(\mathbf{x}_{k+1}) < F_E(\mathbf{x}_k)$, thus providing a strict descent function. Since the map **U** does not disturb this descent feature; and since $\{\mathbf{x}_k\}$ and $\{\mathbf{B}_k\}$ are contained within compact sets, with any accumulation point of $\mathbf{B}_k$ being positive definite, the argument of Theorem 7.3.4 holds. This completes the proof.

10.4.3 Example

To illustrate algorithms RSQP and MSQP, consider the following problem:

$$\begin{aligned}
\text{Minimize} \quad & 2x_1^2 + 2x_2^2 - 2x_1x_2 - 4x_1 - 6x_2 \\
\text{subject to} \quad & g_1(\mathbf{x}) = 2x_1^2 - x_2 \le 0 \\
& g_2(\mathbf{x}) = x_1 + 5x_2 - 5 \le 0 \\
& g_3(\mathbf{x}) = -x_1 \le 0 \\
& g_4(\mathbf{x}) = -x_2 \le 0
\end{aligned}$$

A graphical solution of this problem appears in Figure 10.13*a*. Following Example 10.3.2, let us use $\mu = 10$ in the l_1 penalty merit function F_E defined by (10.28). Let us also use $B_k = \nabla^2 L(\mathbf{x}_k)$ itself, and commence with $\mathbf{x}_1 = (0, 1)^t$ and with Lagrange multipliers $\mathbf{u}_1 = (0, 0, 0, 0)^t$. Hence, we have $f(\mathbf{x}_1) = -4 = F_E(\mathbf{x}_1)$, since $\mathbf{x}_1$ happens to be feasible. Also, $g_1(\mathbf{x}_1) = -1$, $g_2(\mathbf{x}_1) = 0$, $g_3(\mathbf{x}_1) = 0$, and $g_4(\mathbf{x}_1) = -1$. The function gradients are $\nabla f(\mathbf{x}_1) = (-6, -2)^t$, $\nabla g_1(\mathbf{x}_1) = (0, -1)^t$, $\nabla g_2(\mathbf{x}_1) = (1, 5)^t$, $\nabla g_3(\mathbf{x}_1) = (-1, 0)^t$, and $\nabla g_4(\mathbf{x}_1) = (0, -1)^t$. The Hessian of the Lagrangian is

$$\nabla^2 L(\mathbf{x}_1) = \nabla^2 f(\mathbf{x}_1) = \begin{bmatrix} 4 & -2 \\ -2 & 4 \end{bmatrix}$$

Accordingly, the quadratic programming subproblem QP defined in (10.25) is as follows:

$$\begin{aligned}
\textbf{\textit{QP:}} \quad \text{Minimize} \quad & -6d_1 - 2d_2 + \tfrac{1}{2}[4d_1^2 + 4d_2^2 - 4d_1d_2] \\
\text{subject to} \quad & -1 - d_2 \le 0, \qquad d_1 + 5d_2 \le 0, \\
& -d_1 \le 0, \qquad -1 - d_2 \le 0
\end{aligned}$$

Figure 10.14 depicts the graphical solution of this problem. At optimality, only the second constraint of QP is binding. Hence, the KKT system gives

$$4d_1 - 2d_2 - 6 + u_2 = 0 \qquad 4d_2 - 2d_1 - 2 + 5u_2 = 0 \qquad d_1 + 5d_2 = 0$$

Solving, we obtain $\mathbf{d}_1 = (\frac{35}{31}, -\frac{7}{31})^t$ and $\mathbf{u}_2 = (0, 1.032258, 0, 0)$ as the primal and dual optimal solutions, respectively, to QP.

Now for algorithm RSQP, we would take a unit step-size to obtain $\mathbf{x}_2 = \mathbf{x}_1 + \mathbf{d}_1 = (1.1290322, 0.7741936)^t$. This completes one iteration. We ask the reader in Exercise 10.27 to continue this process and to examine its convergence behavior.

On the other hand, for algorithm MSQP, we need to perform a line search, minimizing F_E from $\mathbf{x}_1$ along the direction $\mathbf{d}_1$. This line search problem is, from (10.28),

$$\begin{aligned}
\underset{\lambda \ge 0}{\text{Minimize}} \quad & F_E(\mathbf{x}_1 + \lambda \mathbf{d}_1) \\
& = [3.1612897\lambda^2 - 6.3225804\lambda - 4] \\
& + 10[\max\{0, 2.5494274\lambda^2 + 0.2258064\lambda - 1\} \\
& + \max\{0, 0\} + \max\{0, -1.1290322\lambda\} \\
& + \max\{0, -1 + 0.2258064\lambda\}]
\end{aligned}$$

Using the Golden Section Method, for example, we find the step length $\lambda_1 = 0.5835726$.

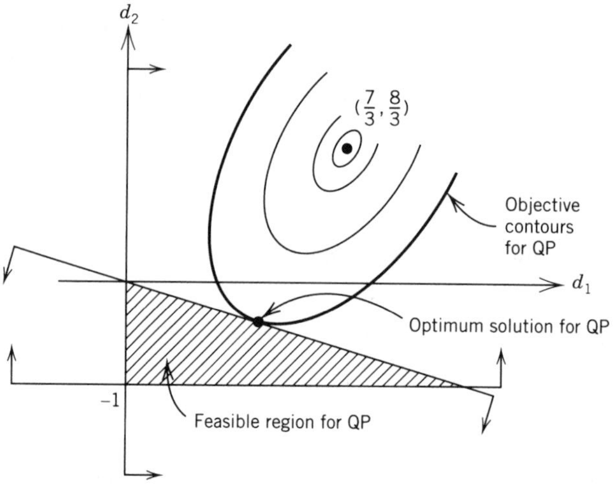

Figure 10.14 Solution of Subproblem QP.

(Note that the unconstrained minimum of $f(\mathbf{x}_1 + \lambda \mathbf{d}_1)$ occurs at $\lambda = 1$; but beyond $\lambda = \lambda_1$, the first max $\{0, \cdot\}$ term starts to become positive and increases the value of F_E, hence, giving λ_1 as the desired step size.) This produces the new iterate $\mathbf{x}_2 = \mathbf{x}_1 + \lambda_1 \mathbf{d}_1 = (0.6588722, 0.8682256)^t$. Observe that, because the generated direction $\mathbf{d}_1$ happened to be leading toward the optimum for P, the minimization of the exact l_1 penalty function (with μ sufficiently large) produced this optimum. We ask the reader in Exercise 10.27 to verify the optimality of $\mathbf{x}_2$ via the corresponding quadratic programming subproblem.

The Maratos Effect

Consider the equality constrained Problem P defined in (10.17). (A similar phenomenon holds for Problem (10.24).) Note that the rudimentary SQP algorithm adopts a unit step size and converges quadratically when $(\mathbf{x}_k, \mathbf{v}_k)$ is initialized close to a regular solution $(\bar{\mathbf{x}}, \bar{\mathbf{v}})$ satisfying the second-order sufficiency conditions. The merit function-based algorithm, however, performs a line search at each iteration to minimize the exact penalty function F_E of (10.28), given that the conditions of Lemma 10.4.1 hold true. Assuming all of the foregoing conditions, one might think that when $(\mathbf{x}_k, \mathbf{v}_k)$ is sufficiently close to $(\bar{\mathbf{x}}, \bar{\mathbf{v}})$, a unit step size would decrease the value of F_E. This statement is incorrect, and its violation is known as the *Maratos effect,* after N. Maratos, who discovered this in relation to Powell's algorithm in 1978.

10.4.4. Example (Maratos Effect)

Consider the following example discussed in Powell (1986).

$$\begin{aligned} \text{Minimize} \quad & f(\mathbf{x}) = -x_1 + 2(x_1^2 + x_2^2 - 1) \\ \text{subject to} \quad & h(\mathbf{x}) = x_1^2 + x_2^2 - 1 = 0 \end{aligned}$$

Clearly, the optimum occurs at $\bar{\mathbf{x}} = (1, 0)^t$. The Lagrange multiplier at this solution is readily obtained from the KKT conditions to be $\bar{v} = -\frac{3}{2}$, and, so, $\nabla^2 L(\bar{\mathbf{x}}) =$

$\nabla^2 f(\bar{\mathbf{x}}) + \bar{v}\nabla^2 h(\bar{\mathbf{x}}) = \mathbf{I}$. Let us take the approximations $\mathbf{B}_k$ to be equal to $\mathbf{I}$ throughout the algorithm.

Now let us select $\mathbf{x}_k$ to be sufficiently close to $\bar{\mathbf{x}}$, but lying on the unit ball defining the constraint. Hence, we can let $\mathbf{x}_k = (\cos\theta, \sin\theta)^t$, where $|\theta|$ is small. The quadratic program (10.22) is given by

$$\text{Minimize} \quad f(\mathbf{x}_k) + (-1 + 4\cos\theta)d_1 + (4\sin\theta)d_2 + \tfrac{1}{2}[d_1^2 + d_2^2]$$

$$\text{subject to} \quad 2\cos\theta d_1 + 2\sin\theta d_2 = 0\}$$

$$\equiv \text{Minimize} \quad \{ f(\mathbf{x}_k) - d_1 + \tfrac{1}{2}(d_1^2 + d_2^2) : \cos\theta d_1 + \sin\theta d_2 = 0\}$$

Writing the KKT conditions for this problem and solving, we readily obtain the optimal solution $\mathbf{d}_k = (\sin^2\theta, -\sin\theta\cos\theta)^t$. Hence, $\mathbf{x}_{k+1} = (\mathbf{x}_k + \mathbf{d}_k) = (\cos\theta + \sin^2\theta, \sin\theta - \sin\theta\cos\theta)^t$. Note that $\|\mathbf{x}_k - \mathbf{x}\|^2 = \sqrt{2(1-\cos\theta)} \simeq \theta$, adopting a second-order Taylor series approximation while, similarly, $\|(\mathbf{x}_k + \mathbf{d}_k) - \bar{\mathbf{x}}\| \simeq \theta^2/2$, thereby attesting to the rapid convergence behavior. However, it is readily verified that $f(\mathbf{x}_k + \mathbf{d}_k) = -\cos\theta + \sin^2\theta$ while $f(\mathbf{x}_k) = -\cos\theta$ and, also, that $h(\mathbf{x}_k + \mathbf{d}_k) = 2\sin^2\theta$ while $h(\mathbf{x}_k) = 0$. Hence, although a unit step makes $\|\mathbf{x}_k + \mathbf{d}_k - \bar{\mathbf{x}}\|$ considerably smaller than $\|\mathbf{x}_k - \bar{\mathbf{x}}\|$, it results in an increase in both f and in the constraint violation, and therefore would increase the value of F_E for any $\mu \geq 0$ or, for that matter, it would increase the value of any merit function.

Several suggestions have been proposed for overcoming the Maratos effect based on tolerating an increase in both f and the constraint violations, or recalculating the step length after correcting for second-order effects, or altering the search direction via modifications in second-order approximations to the objective and the constraint functions. We direct the reader to the Notes and References section for further reading on this subject.

Using the l_1 Penalty in the QP Subproblem—The L_1SQP Approach

In Section 10.3, we presented a superior penalty-based SLP algorithm that adopts trust region concepts and affords a robust and efficient scheme. A similar procedure has been proposed by Fletcher [1981] in the SQP framework, which exhibits a relatively superior computational behavior. Here, given an iterate $\mathbf{x}_k$ and a positive definite approximation $\mathbf{B}_k$ to the Hessian of the Lagrangian function, analogous to (10.11a), this procedure solves the following quadratic subproblem:

$$\begin{aligned} \textit{QP:} \quad \text{Minimize} \quad & [f(\mathbf{x}_k) + \nabla f(\mathbf{x}_k)^t\mathbf{d} + \tfrac{1}{2}\mathbf{d}^t\mathbf{B}_k\mathbf{d}] \\ & + \mu\left[\sum_{i=1}^{m} \max\{0, g_i(\mathbf{x}_k) + \nabla g_i(\mathbf{x}_k)^t\mathbf{d}\}\right. \\ & + \left.\sum_{i=1}^{l} |h_i(\mathbf{x}_k) + \nabla h_i(\mathbf{x}_k)^t\mathbf{d}|\right] \\ \text{subject to} \quad & -\boldsymbol{\Delta}_k \leq \mathbf{d} \leq \boldsymbol{\Delta}_k \end{aligned} \tag{10.32}$$

where $\boldsymbol{\Delta}_k$ is a trust region step bound and, as before, μ is a suitably large penalty parameter. Note that in comparison with Problem (10.25), the constraints have been accommodated into the objective function via an l_1 penalty term and have been replaced by a trust region constraint. Hence, the subproblem QP is always feasible and bounded and has an optimum. To contend with the nondifferentiability of the objective function, the l_1 terms can be retransferred into the constraints as in (10.11b). Similar to the PSLP algorithm, if $\mathbf{d}_k$ solves this problem along with Lagrange multiplier estimates $(\mathbf{u}_{k+1}, \mathbf{v}_{k+1})$, and if $\mathbf{x}_{k+1} = \mathbf{x}_k + \mathbf{d}_k$ is ε-feasible and satisfies the KKT conditions within a tolerance, or if the fractional improvement in the original objective function is not better than a given tolerance over some c consecutive iterations, the algorithm can be terminated. Otherwise, the process is iteratively repeated. This type of procedure enjoys the asymptotic local convergence properties of SQP methods but also achieves global convergence owing to the l_1 penalty function and the trust region features. However, it is also prone to the Maratos effect, and corrective measures are necessary to avoid this phenomenon. The reader should refer to the Notes and References section for further reading on this topic.

10.5 The Gradient Projection Method of Rosen

As we learned in Chapter 8, the direction of steepest descent is that of the negative gradient. In the presence of constraints, however, moving along the steepest descent direction may lead to infeasible points. The gradient projection method of Rosen [1960] projects the negative gradient in such a way that improves the objective function and meanwhile maintains feasibility.

First, consider the following definition of a projection matrix.

10.5.1 Definition

An $n \times n$ matrix $\mathbf{P}$ is called a *projection matrix* if $\mathbf{P} = \mathbf{P}^t$ and $\mathbf{PP} = \mathbf{P}$.

10.5.2 Lemma

Let $\mathbf{P}$ be an $n \times n$ matrix. Then, the following statements are true:

1. If $\mathbf{P}$ is a projection matrix, then $\mathbf{P}$ is positive semidefinite.
2. $\mathbf{P}$ is a projection matrix if and only if $\mathbf{I} - \mathbf{P}$ is a projection matrix.
3. Let $\mathbf{P}$ be a projection matrix and let $\mathbf{Q} = \mathbf{I} - \mathbf{P}$. Then, $L = \{\mathbf{Px}:\mathbf{x} \in E_n\}$ and $\mathbf{L}^\perp = \{\mathbf{Qx}:\mathbf{x} \in E_n\}$ are orthogonal linear subspaces. Furthermore, any point $\mathbf{x} \in E_n$ can be represented uniquely as $\mathbf{p} + \mathbf{q}$, where $\mathbf{p} \in L$ and $\mathbf{q} \in L^\perp$.

Proof

Let $\mathbf{P}$ be a projection matrix, and let $\mathbf{x} \in E_n$ be arbitrary. Then, $\mathbf{x}^t\mathbf{Px} = \mathbf{x}^t\mathbf{PPx} = \mathbf{x}^t\mathbf{P}^t\mathbf{Px} = \|\mathbf{Px}\|^2 \geq 0$, and, hence, $\mathbf{P}$ is positive semidefinite, and part 1 follows.

By Definition 10.5.1, part 2 is obvious. Clearly L and $L^\perp$ are linear subspaces. Note that $\mathbf{P}^t\mathbf{Q} = \mathbf{P}(\mathbf{I} - \mathbf{P}) = \mathbf{P} - \mathbf{PP} = \mathbf{0}$, and hence, L and $L^\perp$ are indeed orthogonal. Now let $\mathbf{x}$ be an arbitrary point in E_n. Then, $\mathbf{x} = \mathbf{Ix} = (\mathbf{P} + \mathbf{Q})\mathbf{x} = \mathbf{Px} + \mathbf{Qx} = \mathbf{p} + \mathbf{q}$, where $\mathbf{p} \in L$ and $\mathbf{q} \in L^\perp$. To show uniqueness, suppose that $\mathbf{x}$ can also be represented as $\mathbf{x} = \mathbf{p}' + \mathbf{q}'$, where $\mathbf{p}' \in L$ and $\mathbf{q}' \in L^\perp$. By subtraction, it follows that $\mathbf{p} - \mathbf{p}' = \mathbf{q}' - \mathbf{q}$. Since $\mathbf{p} - \mathbf{p}' \in L$ and $\mathbf{q}' - \mathbf{q} \in L^\perp$, and since the only point in the intersection of L and $L^\perp$ is the zero vector, it follows that $\mathbf{p} - \mathbf{p}' = \mathbf{q}' - \mathbf{q} = \mathbf{0}$. Thus the representation of $\mathbf{x}$ is unique, and the proof is complete.

Problems with Linear Constraints

Consider the following problem:

$$\begin{array}{ll} \text{Minimize} & f(\mathbf{x}) \\ \text{subject to} & \mathbf{Ax} \le \mathbf{b} \\ & \mathbf{Ex} = \mathbf{e} \end{array}$$

where $\mathbf{A}$ is an $m \times n$ matrix, $\mathbf{E}$ is an $l \times n$ matrix, $\mathbf{b}$ is an m vector, $\mathbf{e}$ is an l vector, and $f: E_n \to E_1$ is a differentiable function. Given a feasible point $\mathbf{x}$, the direction of steepest descent is $-\nabla f(\mathbf{x})$. However, moving along $-\nabla f(\mathbf{x})$ may destroy feasibility. To maintain feasibility, $-\nabla f(\mathbf{x})$ is projected so that we move along $\mathbf{d} = -\mathbf{P}\nabla f(\mathbf{x})$, where $\mathbf{P}$ is a suitable projection matrix. Lemma 10.5.3 gives the form of a suitable projection matrix $\mathbf{P}$ and shows that $-\mathbf{P}\nabla f(\mathbf{x})$ is indeed an improving feasible direction, provided that $-\mathbf{P}\nabla f(\mathbf{x}) \neq \mathbf{0}$.

10.5.3 Lemma

Consider the problem to minimize $f(\mathbf{x})$ subject to $\mathbf{Ax} \le \mathbf{b}$ and $\mathbf{Ex} = \mathbf{e}$. Let $\mathbf{x}$ be a feasible point such that $\mathbf{A}_1\mathbf{x} = \mathbf{b}_1$ and $\mathbf{A}_2\mathbf{x} < \mathbf{b}_2$, where $\mathbf{A}^t = (\mathbf{A}_1^t, \mathbf{A}_2^t)$ and $\mathbf{b}^t = (\mathbf{b}_1^t, \mathbf{b}_2^t)$. Furthermore, suppose that f is differentiable at $\mathbf{x}$. If $\mathbf{P}$ is a projection matrix such that $\mathbf{P}\nabla f(\mathbf{x}) \neq \mathbf{0}$, then $\mathbf{d} = -\mathbf{P}\nabla f(\mathbf{x})$ is an improving direction of f at $\mathbf{x}$. Furthermore, if $\mathbf{M}^t = (\mathbf{A}_1^t, \mathbf{E}^t)$ has full rank, and if $\mathbf{P}$ is of the form $\mathbf{P} = \mathbf{I} - \mathbf{M}^t(\mathbf{MM}^t)^{-1}\mathbf{M}$, then $\mathbf{d}$ is an improving feasible direction.

Proof

Note that

$$\nabla f(\mathbf{x})^t\mathbf{d} = -\nabla f(\mathbf{x})^t\mathbf{P}\nabla f(\mathbf{x}) = -\nabla f(\mathbf{x})^t\mathbf{P}^t\mathbf{P}\nabla f(\mathbf{x}) = -\|\mathbf{P}\nabla f(\mathbf{x})\|^2 < 0$$

By Lemma 10.1.2, $\mathbf{d} = -\mathbf{P}\nabla f(\mathbf{x})$ is an improving direction. Furthermore, if $\mathbf{P} = \mathbf{I} - \mathbf{M}^t(\mathbf{MM}^t)^{-1}\mathbf{M}$, then $\mathbf{Md} = -\mathbf{MP}\nabla f(\mathbf{x}) = \mathbf{0}$, that is, $\mathbf{A}_1\mathbf{d} = \mathbf{0}$ and $\mathbf{Ed} = \mathbf{0}$. By Lemma 10.1.2, $\mathbf{d}$ is a feasible direction, and the proof is complete.

Geometric Interpretation of Projecting the Gradient

Note that the matrix $\mathbf{P}$ of the above lemma is indeed a projection matrix satisfying $\mathbf{P} = \mathbf{P}^t$ and $\mathbf{PP} = \mathbf{P}$. Furthermore, $\mathbf{MP} = \mathbf{0}$, that is, $\mathbf{A}_1\mathbf{P} = \mathbf{0}$ and $\mathbf{EP} = \mathbf{0}$. In other words, the matrix $\mathbf{P}$ projects each row of $\mathbf{A}_1$ and each row of $\mathbf{E}$ into the zero vector. But since the rows of $\mathbf{A}_1$ and $\mathbf{E}$ are the gradients of the binding constraints, $\mathbf{P}$ is the matrix that projects the gradients of the binding constraints into the zero vector. Consequently, in particular, $\mathbf{P}\nabla f(\mathbf{x})$ is the projection of $\nabla f(\mathbf{x})$ onto the nullspace of the binding constraints.

Figure 10.15 illustrates the process of projecting the gradient for a problem with inequality constraints. At the point $\mathbf{x}$, there is only one binding constraint with gradient $\mathbf{A}_1$. Note that the matrix $\mathbf{P}$ projects any vector onto the nullspace of $\mathbf{A}_1$ and that $\mathbf{d} = -\mathbf{P}\nabla f(\mathbf{x})$ is an improving feasible direction.

Resolution of the Case $\mathbf{P}\nabla f(\mathbf{x}) = \mathbf{0}$

We have seen that if $\mathbf{P}\nabla f(\mathbf{x}) \neq \mathbf{0}$, then $\mathbf{d} = -\mathbf{P}\nabla f(\mathbf{x})$ is an improving feasible direction. Now suppose that $\mathbf{P}\nabla f(\mathbf{x}) = \mathbf{0}$. Then,

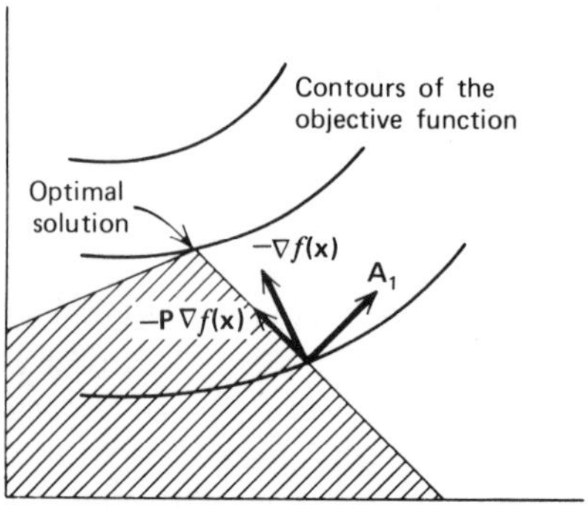

Figure 10.15 Projecting the gradient.

$$\mathbf{0} = \mathbf{P}\nabla f(\mathbf{x}) = [\mathbf{I} - \mathbf{M}^t(\mathbf{M}\mathbf{M}^t)^{-1}\mathbf{M}]\nabla f(\mathbf{x}) = \nabla f(\mathbf{x}) + \mathbf{M}^t\mathbf{w} = \nabla f(\mathbf{x}) + \mathbf{A}_1^t\mathbf{u} + \mathbf{E}^t\mathbf{v}$$

where $\mathbf{w} = -(\mathbf{M}\mathbf{M}^t)^{-1}\mathbf{M}\nabla f(\mathbf{x})$ and $\mathbf{w}^t = (\mathbf{u}^t, \mathbf{v}^t)$. If $\mathbf{u} \geq \mathbf{0}$, then the point $\mathbf{x}$ satisfies the KKT conditions and we may stop. If $\mathbf{u} \ngeq \mathbf{0}$, as Theorem 10.5.4 below shows, a new projection matrix $\hat{\mathbf{P}}$ can be identified such that $\mathbf{d} = -\hat{\mathbf{P}}\nabla f(\mathbf{x})$ is indeed an improving feasible direction.

10.5.4 Theorem

Consider the problem to minimize $f(\mathbf{x})$ subject to $\mathbf{A}\mathbf{x} \leq \mathbf{b}$ and $\mathbf{E}\mathbf{x} = \mathbf{e}$. Let $\mathbf{x}$ be a feasible solution, and suppose that $\mathbf{A}_1\mathbf{x} = \mathbf{b}_1$ and $\mathbf{A}_2\mathbf{x} < \mathbf{b}_2$, where $\mathbf{A}^t = (\mathbf{A}_1^t, \mathbf{A}_2^t)$ and $\mathbf{b}^t = (\mathbf{b}_1^t, \mathbf{b}_2^t)$. Suppose that $\mathbf{M}^t = (\mathbf{A}_1^t, \mathbf{E}^t)$ has full rank, and let $\mathbf{P} = \mathbf{I} - \mathbf{M}^t(\mathbf{M}\mathbf{M}^t)^{-1}\mathbf{M}$. Further, suppose that $\mathbf{P}\nabla f(\mathbf{x}) = \mathbf{0}$, and let $\mathbf{w} = -(\mathbf{M}\mathbf{M}^t)^{-1}\mathbf{M}\nabla f(\mathbf{x})$ and $(\mathbf{u}^t, \mathbf{v}^t) = \mathbf{w}^t$. If $\mathbf{u} \geq \mathbf{0}$, then $\mathbf{x}$ is a KKT point. If $\mathbf{u} \ngeq \mathbf{0}$, let u_j be a negative component of $\mathbf{u}$, and let $\hat{\mathbf{M}}^t = (\hat{\mathbf{A}}_1^t, \mathbf{E}^t)$, where $\hat{\mathbf{A}}_1$ is obtained from $\mathbf{A}_1$ by deleting the row of $\mathbf{A}_1$ corresponding to u_j. Now let $\hat{\mathbf{P}} = \mathbf{I} - \hat{\mathbf{M}}^t(\hat{\mathbf{M}}\hat{\mathbf{M}}^t)^{-1}\hat{\mathbf{M}}$, and let $\mathbf{d} = -\hat{\mathbf{P}}\nabla f(\mathbf{x})$. Then $\mathbf{d}$ is an improving feasible direction.

Proof

By the definition of $\mathbf{P}$, and since $\mathbf{P}\nabla f(\mathbf{x}) = \mathbf{0}$, we get

$$\begin{aligned} \mathbf{0} = \mathbf{P}\nabla f(\mathbf{x}) &= [\mathbf{I} - \mathbf{M}^t(\mathbf{M}\mathbf{M}^t)^{-1}\mathbf{M}]\nabla f(\mathbf{x}) \\ &= \nabla f(\mathbf{x}) + \mathbf{M}^t\mathbf{w} = \nabla f(\mathbf{x}) + \mathbf{A}_1^t\mathbf{u} + \mathbf{E}^t\mathbf{v} \end{aligned} \tag{10.33}$$

In view of (10.33), if $\mathbf{u} \geq \mathbf{0}$, then $\mathbf{x}$ is a KKT point.

Now suppose that $\mathbf{u} \ngeq \mathbf{0}$, and let u_j be a negative component of $\mathbf{u}$. We first show that $\hat{\mathbf{P}}\nabla f(\mathbf{x}) \neq \mathbf{0}$. By contradiction, suppose that $\hat{\mathbf{P}}\nabla f(\mathbf{x}) = \mathbf{0}$. By the definition of $\hat{\mathbf{P}}$ and letting $\hat{\mathbf{w}} = -(\hat{\mathbf{M}}\hat{\mathbf{M}}^t)^{-1}\hat{\mathbf{M}}\nabla f(\mathbf{x})$, we get

$$\mathbf{0} = \hat{\mathbf{P}}\nabla f(\mathbf{x}) = [\mathbf{I} - \hat{\mathbf{M}}^t(\hat{\mathbf{M}}\hat{\mathbf{M}}^t)^{-1}\hat{\mathbf{M}}]\nabla f(\mathbf{x}) = \nabla f(\mathbf{x}) + \hat{\mathbf{M}}^t\hat{\mathbf{w}} \tag{10.34}$$

Note that $\mathbf{A}_1^t\mathbf{u} + \mathbf{E}^t\mathbf{v}$ could be written as $\hat{\mathbf{M}}^t\bar{\mathbf{w}} + u_j\mathbf{r}_j^t$, where $\mathbf{r}_j$ is the jth row of $\mathbf{A}_1$. Thus, from (10.33), we get

$$\mathbf{0} = \nabla f(\mathbf{x}) + \hat{\mathbf{M}}^t\bar{\mathbf{w}} + u_j\mathbf{r}_j^t \tag{10.35}$$

Subtracting (10.35) from (10.34), it follows that $\mathbf{0} = \hat{\mathbf{M}}^t(\hat{\mathbf{w}} - \bar{\mathbf{w}}) - u_j\mathbf{r}_j^t$. This, together with the fact that $u_j \neq 0$, violates the assumption that $\mathbf{M}$ has full rank. Therefore, $\hat{\mathbf{P}}\nabla f(\mathbf{x}) \neq \mathbf{0}$. By Lemma 10.5.3, $\mathbf{d}$ is an improving direction.

Now we show that $\mathbf{d}$ is a feasible direction. Note that $\hat{\mathbf{M}}\hat{\mathbf{P}} = \mathbf{0}$, so that

$$\begin{pmatrix}\hat{\mathbf{A}}_1 \\ \mathbf{E}\end{pmatrix}\mathbf{d} = \hat{\mathbf{M}}\mathbf{d} = -\hat{\mathbf{M}}\hat{\mathbf{P}}\nabla f(\mathbf{x}) = \mathbf{0} \tag{10.36}$$

By Lemma 10.5.3, $\mathbf{d}$ is feasible direction if $\mathbf{A}_1\mathbf{d} \le \mathbf{0}$ and $\mathbf{E}\mathbf{d} = \mathbf{0}$. In view of (10.36), to show that $\mathbf{d}$ is a feasible direction, it suffices to demonstrate that $\mathbf{r}_j\mathbf{d} \le 0$. Premultiplying (10.35) by $\mathbf{r}_j\hat{\mathbf{P}}$, and noting that $\hat{\mathbf{P}}\hat{\mathbf{M}}^t = \mathbf{0}$, it follows that

$$0 = \mathbf{r}_j\hat{\mathbf{P}}\nabla f(\mathbf{x}) + \mathbf{r}_j\hat{\mathbf{P}}(\hat{\mathbf{M}}^t\bar{\mathbf{w}} + u_j\mathbf{r}_j^t) = -\mathbf{r}_j\mathbf{d} + u_j\mathbf{r}_j\hat{\mathbf{P}}\mathbf{r}_j^t$$

By Lemma 10.5.2, $\hat{\mathbf{P}}$ is positive semidefinite, so that $\mathbf{r}_j\hat{\mathbf{P}}\mathbf{r}_j^t \ge 0$. Since $u_j < 0$, then the above equation implies that $\mathbf{r}_j\mathbf{d} \le 0$. This completes the proof.

Summary of the Gradient Projection Method of Rosen (Linear Constraints)

We summarize below Rosen's gradient projection method for solving a problem of the form to minimize $f(\mathbf{x})$ subject to $\mathbf{A}\mathbf{x} \le \mathbf{b}$ and $\mathbf{E}\mathbf{x} = \mathbf{e}$. We assume that for any feasible solution, the set of binding constraints are linearly independent. Otherwise, when the active constraints are dependent, $\mathbf{M}\mathbf{M}^t$ is singular and the main algorithmic step is not defined. Moreover, in such a case, the Lagrange multipliers are nonunique, and an arbitrary choice of dropping a constraint can cause the algorithm to be stuck at a current, non-KKT solution.

Initialization Step Choose a point $\mathbf{x}_1$ with $\mathbf{A}\mathbf{x}_1 \le \mathbf{b}$ and $\mathbf{E}\mathbf{x}_1 = \mathbf{e}$. Suppose that $\mathbf{A}^t$ and $\mathbf{b}^t$ are decomposed into $(\mathbf{A}_1^t, \mathbf{A}_2^t)$ and $(\mathbf{b}_1^t, \mathbf{b}_2^t)$ such that $\mathbf{A}_1\mathbf{x}_1 = \mathbf{b}_1$ and $\mathbf{A}_2\mathbf{x}_1 < \mathbf{b}_2$. Let $k = 1$ and go to the main step.

Main Step

1. Let $\mathbf{M}^t = (\mathbf{A}_1^t, \mathbf{E}^t)$. If $\mathbf{M}$ is vacuous, stop if $\nabla f(\mathbf{x}_k) = \mathbf{0}$, let $\mathbf{d}_k = -\nabla f(\mathbf{x}_k)$, and proceed to step 2. Otherwise, let $\mathbf{P} = \mathbf{I} - \mathbf{M}^t(\mathbf{M}\mathbf{M}^t)^{-1}\mathbf{M}$ and set $\mathbf{d}_k = -\mathbf{P}\nabla f(\mathbf{x}_k)$. If $\mathbf{d}_k \ne \mathbf{0}$, go to step 2. If $\mathbf{d}_k = \mathbf{0}$, compute $\mathbf{w} = -(\mathbf{M}\mathbf{M}^t)^{-1}\mathbf{M}\nabla f(\mathbf{x}_k)$ and let $\mathbf{w}^t = (\mathbf{u}^t, \mathbf{v}^t)$. If $\mathbf{u} \ge \mathbf{0}$, stop; $\mathbf{x}_k$ is a KKT point, with $\mathbf{w}$ yielding the associated Lagrange multipliers. If $\mathbf{u} \not\ge \mathbf{0}$, choose a negative component of $\mathbf{u}$, say, u_j. Update $\mathbf{A}_1$ by deleting the row corresponding to u_j and repeat step 1.
2. Let λ_k be an optimal solution to the following line search problem:

   $$\begin{array}{ll}\text{Minimize} & f(\mathbf{x}_k + \lambda\mathbf{d}_k) \\ \text{subject to} & 0 \le \lambda \le \lambda_{\max}\end{array}$$

   where $\lambda_{\max}$ is given by (10.1). Let $\mathbf{x}_{k+1} = \mathbf{x}_k + \lambda_k\mathbf{d}_k$, and suppose that $\mathbf{A}^t$ and $\mathbf{b}^t$ are decomposed into $(\mathbf{A}_1^t, \mathbf{A}_2^t)$ and $(\mathbf{b}_1^t, \mathbf{b}_2^t)$ such that $\mathbf{A}_1\mathbf{x}_{k+1} = \mathbf{b}_1$ and $\mathbf{A}_2\mathbf{x}_{k+1} < \mathbf{b}_2$. Replace k by $k + 1$ and repeat step 1.

10.5.5 Example

Consider the following problem:

$$\begin{array}{lrl}\text{Minimize} & 2x_1^2 + 2x_2^2 - 2x_1x_2 - 4x_1 - 6x_2 & \\ \text{subject to} & x_1 + x_2 & \le 2 \\ & x_1 + 5x_2 & \le 5 \\ & -x_1 & \le 0 \\ & -x_2 & \le 0\end{array}$$

Note that $\nabla f(\mathbf{x}) = (4x_1 - 2x_2 - 4, 4x_2 - 2x_1 - 6)^t$. We solve this problem using the gradient projection method of Rosen, starting from the point (0, 0). At each iteration, we first find the direction to move by step 1 of the algorithm and then perform a line search along this direction.

Iteration 1

Search Direction At $\mathbf{x}_1 = (0, 0)^t$, we have $\nabla f(\mathbf{x}_1) = (-4, -6)^t$. Furthermore, only the nonnegativity constraints are binding at $\mathbf{x}_1$, so that

$$\mathbf{A}_1 = \begin{bmatrix} -1 & 0 \\ 0 & -1 \end{bmatrix} \qquad \mathbf{A}_2 = \begin{bmatrix} 1 & 1 \\ 1 & 5 \end{bmatrix}$$

We then have

$$\mathbf{P} = \mathbf{I} - \mathbf{A}_1^t(\mathbf{A}_1\mathbf{A}_1^t)^{-1}\mathbf{A}_1 = \begin{bmatrix} 0 & 0 \\ 0 & 0 \end{bmatrix}$$

and $\mathbf{d}_1 = -\mathbf{P}\nabla f(\mathbf{x}_1) = (0, 0)^t$. Noting that we do not have equality constraints, compute

$$\mathbf{w} = \mathbf{u} = (\mathbf{A}_1\mathbf{A}_1^t)^{-1}\mathbf{A}_1\nabla f(\mathbf{x}_1) = (-4, -6)^t$$

Choosing $u_4 = -6$, we delete the corresponding gradient of the fourth constraint from $\mathbf{A}_1$. The matrix $\mathbf{A}_1$ is modified to give $\hat{\mathbf{A}}_1 = (-1, 0)$. The modified projection matrix is then

$$\hat{\mathbf{P}} = \mathbf{I} - \hat{\mathbf{A}}_1^t(\hat{\mathbf{A}}_1\hat{\mathbf{A}}_1^t)^{-1}\hat{\mathbf{A}}_1 = \begin{bmatrix} 0 & 0 \\ 0 & 1 \end{bmatrix}$$

and the direction $\mathbf{d}_1$ to move is given by

$$\mathbf{d}_1 = -\hat{\mathbf{P}}\nabla f(\mathbf{x}_1) = -\begin{bmatrix} 0 & 0 \\ 0 & 1 \end{bmatrix}\begin{pmatrix} -4 \\ -6 \end{pmatrix} = \begin{pmatrix} 0 \\ 6 \end{pmatrix}$$

Line Search Any point $\mathbf{x}_2$ in the direction $\mathbf{d}_1$ starting from the point $\mathbf{x}_1$ can be written as $\mathbf{x}_2 = \mathbf{x}_1 + \lambda\mathbf{d}_1 = (0, 6\lambda)^t$, and the corresponding objective function value is $f(\mathbf{x}_2) = 72\lambda^2 - 36\lambda$. The maximum value of λ for which $\mathbf{x}_1 + \lambda\mathbf{d}_1$ is feasible is obtained from (10.1) as

$$\lambda_{\max} = \text{minimum } \left\{\tfrac{2}{6}, \tfrac{5}{30}\right\} = \tfrac{1}{6}$$

Therefore, λ_1 is the optimal solution to the following problem:

$$\begin{array}{ll} \text{Minimize} & 72\lambda^2 - 36\lambda \\ \text{subject to} & 0 \le \lambda \le \frac{1}{6} \end{array}$$

The optimal solution is $\lambda_1 = \frac{1}{6}$, so that $\mathbf{x}_2 = \mathbf{x}_1 + \lambda_1\mathbf{d}_1 = (0, 1)^t$.

Iteration 2

Search Direction At the point $\mathbf{x}_2 = (0, 1)^t$, we have $\nabla f(\mathbf{x}_2) = (-6, -2)^t$. Furthermore, at this point, constraints 2 and 3 are binding, so that we get

$$\mathbf{A}_1 = \begin{bmatrix} 1 & 5 \\ -1 & 0 \end{bmatrix} \qquad \mathbf{A}_2 = \begin{bmatrix} 1 & 1 \\ 0 & -1 \end{bmatrix}$$

We then have

$$\mathbf{P} = \mathbf{I} - \mathbf{A}_1^t(\mathbf{A}_1\mathbf{A}_1^t)^{-1}\mathbf{A}_1 = \begin{bmatrix} 0 & 0 \\ 0 & 0 \end{bmatrix}$$

and, hence, $-\mathbf{P}\nabla f(\mathbf{x}_2) = (0, 0)^t$. Thus, we compute

$$\mathbf{u} = -(\mathbf{A}_1\mathbf{A}_1^t)^{-1}\mathbf{A}_1\nabla f(\mathbf{x}_2) = (\tfrac{2}{5}, -\tfrac{28}{5})^t$$

Since $u_3 < 0$, the row $(-1, 0)$ is deleted from $\mathbf{A}_1$, which gives the modified matrix $\hat{\mathbf{A}}_1 = [1, 5]$. The projection matrix and the corresponding direction vector are given by

$$\hat{\mathbf{P}} = \mathbf{I} - \hat{\mathbf{A}}_1^t(\hat{\mathbf{A}}_1\hat{\mathbf{A}}_1^t)^{-1}\hat{\mathbf{A}}_1 = \begin{bmatrix} \frac{25}{26} & -\frac{5}{26} \\ -\frac{5}{26} & \frac{1}{26} \end{bmatrix}$$

$$\mathbf{d}_2 = -\hat{\mathbf{P}}\nabla f(\mathbf{x}_2) = \begin{pmatrix} \frac{70}{13} \\ -\frac{14}{13} \end{pmatrix}$$

Since the norm of $\mathbf{d}_2$ is not important, $(\frac{70}{13}, -\frac{14}{13})^t$ is equivalent to $(5, -1)^t$. We therefore let $\mathbf{d}_2 = (5, -1)^t$.

Line Search We are interested in points of the form $\mathbf{x}_2 + \lambda\mathbf{d}_2 = (5\lambda, 1 - \lambda)^t$ and $f(\mathbf{x}_2 + \lambda\mathbf{d}_2) = 62\lambda^2 - 28\lambda - 4$. The maximum value of λ for which $\mathbf{x}_2 + \lambda\mathbf{d}_2$ is feasible is obtained from (10.1) as

$$\lambda_{\max} = \text{minimum } \{\tfrac{1}{4}, \tfrac{1}{1}\} = \tfrac{1}{4}$$

Therefore, λ_2 is the solution to the following problem:

$$\begin{aligned} \text{Minimize} \quad & 62\lambda^2 - 28\lambda - 4 \\ \text{subject to} \quad & 0 \le \lambda \le \tfrac{1}{4} \end{aligned}$$

The optimal solution is $\lambda = \frac{7}{31}$, so that $\mathbf{x}_3 = \mathbf{x}_2 + \lambda_2\mathbf{d}_2 = (\frac{35}{31}, \frac{24}{31})^t$.

Iteration 3

Search Direction At the point $\mathbf{x}_3 = (\frac{35}{31}, \frac{24}{31})^t$, we have $\nabla f(\mathbf{x}_3) = (-\frac{32}{31}, -\frac{160}{31})^t$. Furthermore, the second constraint is binding, so that

$$\mathbf{A}_1 = [1, 5] \qquad \mathbf{A}_2 = \begin{bmatrix} 1 & 1 \\ -1 & 0 \\ 0 & -1 \end{bmatrix}$$

Further, we get

$$\mathbf{P} = \mathbf{I} - \mathbf{A}_1^t(\mathbf{A}_1\mathbf{A}_1^t)^{-1}\mathbf{A}_1 = \frac{1}{26}\begin{bmatrix} 25 & -5 \\ -5 & 1 \end{bmatrix}$$

and the direction $\mathbf{d}_3 = -\mathbf{P}\nabla f(\mathbf{x}_3) = (0, 0)^t$. Thus, we compute

$$\mathbf{u} = -(\mathbf{A}_1\mathbf{A}_1^t)^{-1}\mathbf{A}_1\nabla f(\mathbf{x}_3) = \tfrac{32}{31} \geq 0$$

Hence, the point $\mathbf{x}_3$ is a KKT point. Note that the gradient of the binding constraint points in the direction opposite to $\nabla f(\mathbf{x}_3)$. In particular, $\nabla f(\mathbf{x}_3) + u_2\nabla g_2(\mathbf{x}_3) = \mathbf{0}$ for $u_2 = \frac{32}{31}$, so $\mathbf{x}_3$ is a KKT point. In this particular example, since f is strictly convex, then, by Theorem 4.3.8, the point $\mathbf{x}_3$ is indeed the global optimal solution to the problem.

Table 10.4 summarizes the computations for solving the above problem. The progress of the algorithm is shown in Figure 10.16.

Nonlinear Constraints

So far, we have discussed the gradient projection method with linear constraints. In the linear case, the projection of the gradient of the objective function onto the nullspace of the gradients of the binding constraints, or a subset of the binding constraints, led to an improving feasible direction or to the conclusion that a KKT point was at hand. The same strategy can be used in the presence of nonlinear constraints. The projected gradient will usually not lead to feasible points, since it is only tangential to the feasible region, as illustrated in Figure 10.17. Therefore, a movement along the projected gradient must be coupled with a correction move to the feasible region.

To be more specific, consider the following problem:

$$\begin{array}{lll} \text{Minimize} & f(\mathbf{x}) & \\ \text{subject to} & g_i(\mathbf{x}) \leq 0 & \text{for } i = 1, \ldots, m \\ & h_i(\mathbf{x}) = 0 & \text{for } i = 1, \ldots, l \end{array}$$

Let $\mathbf{x}_k$ be a feasible solution, and let $I = \{i : g_i(\mathbf{x}_k) = 0\}$. Let $\mathbf{M}$ be the matrix whose rows are $\nabla g_i(\mathbf{x}_k)^t$ for $i \in I$ and $\nabla h_i(\mathbf{x}_k)^t$ for $i = 1, \ldots, l$, and let $\mathbf{P} = \mathbf{I} - \mathbf{M}^t(\mathbf{M}\mathbf{M}^t)^{-1}\mathbf{M}$. Note that $\mathbf{P}$ projects any vector onto the nullspace of the gradients of the equality constraints and the binding inequality constraints. Let $\mathbf{d}_k = -\mathbf{P}\nabla f(\mathbf{x}_k)$. If $\mathbf{d}_k \neq \mathbf{0}$, then minimize f starting from $\mathbf{x}_k$ in the direction $\mathbf{d}_k$ and make a correction move to the feasible region. If, on the other hand, $\mathbf{d}_k = \mathbf{0}$, then calculate $(\mathbf{u}^t, \mathbf{v}^t) = -\nabla f(\mathbf{x}_k)^t \cdot \mathbf{M}^t(\mathbf{M}\mathbf{M}^t)^{-1}$. If $\mathbf{u} \geq \mathbf{0}$, then stop with a KKT point $\mathbf{x}_k$. Otherwise, delete the row of $\mathbf{M}$ corresponding to some $u_i < 0$ and repeat the process.

Convergence Analysis of the Gradient Projection Method

Let us first examine the question whether the direction-finding map is closed or not. Note that the direction generated could change abruptly when a new restriction becomes active, or when the projected gradient is the zero vector, necessitating the computation of a new projection matrix. Hence, as shown below, this causes the direction-finding map to be *not* closed.

10.5.6 Example

Consider the following problem:

$$\begin{array}{ll} \text{Minimize} & x_1 - 2x_2 \\ \text{subject to} & x_1 + 2x_2 \leq 6 \\ & x_1, x_2 \geq 0 \end{array}$$

We now illustrate that the direction-finding map of the gradient projection method is not

Table 10.4 Summary of Computations for the Gradient Projection Method of Rosen

Iteration			Search Direction						Line Search		
k	$\mathbf{x}_k$	$f(\mathbf{x}_k)$	$\nabla f(\mathbf{x}_k)$	I	$\mathbf{A}_1$	$\mathbf{P}$	$\mathbf{d}_k$	$\mathbf{u}$	λ_{max}	λ_k	$\mathbf{x}_{k+1}$
1	(0, 0)	0	(−4, −6)	{3, 4}	$\begin{bmatrix} -1 & 0 \\ 0 & -1 \end{bmatrix}$	$\begin{bmatrix} 0 & 0 \\ 0 & 0 \end{bmatrix}$	(0, 0)	(−4, −6)	—	—	—
				{3}	[−1, 0]	$\begin{bmatrix} 0 & 0 \\ 0 & 1 \end{bmatrix}$	(0, 6)	—	$\frac{1}{6}$	$\frac{1}{6}$	(0, 1)
2	(0, 1)	−4.00	(−6, −2)	{2, 3}	$\begin{bmatrix} 1 & 5 \\ -1 & 0 \end{bmatrix}$	$\begin{bmatrix} 0 & 0 \\ 0 & 0 \end{bmatrix}$	(0, 0)	$(\frac{2}{5}, -\frac{28}{5})$	—	—	—
				{2}	[1, 5]	$\begin{bmatrix} \frac{25}{26} & -\frac{5}{26} \\ -\frac{5}{26} & \frac{1}{26} \end{bmatrix}$	$(\frac{70}{13}, -\frac{14}{13})$	—	$\frac{1}{4}$	$\frac{7}{31}$	$(\frac{35}{31}, \frac{24}{31})$
3	$(\frac{35}{31}, \frac{24}{31})$	−7.16	$(-\frac{32}{31}, -\frac{160}{31})$	{2}	[1, 5]	$\begin{bmatrix} \frac{25}{26} & -\frac{5}{26} \\ -\frac{5}{26} & \frac{1}{26} \end{bmatrix}$	(0, 0)	$(\frac{32}{31})$	—	—	—

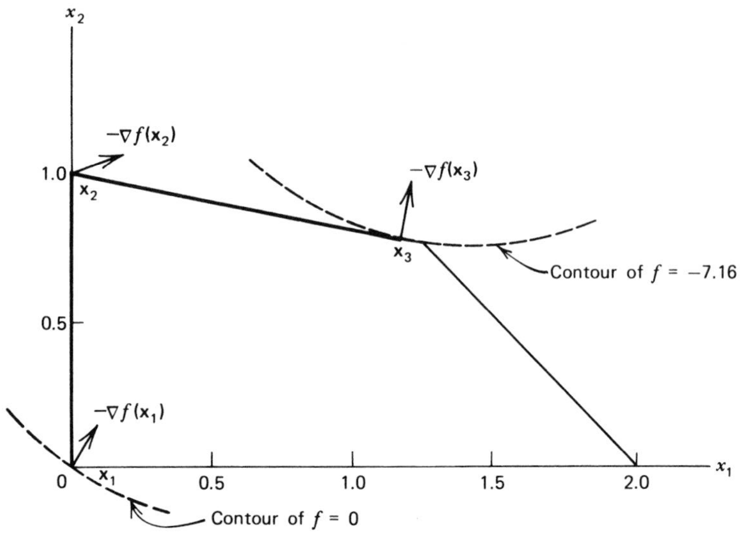

Figure 10.16 Illustration of the gradient projection method of Rosen.

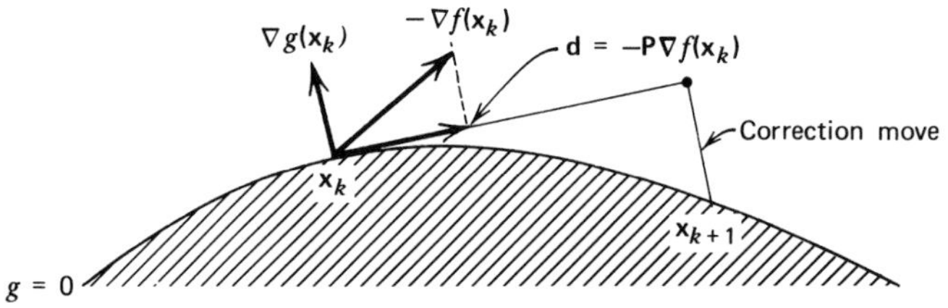

Figure 10.17 Projecting the gradient in the presence of nonlinear constraints.

closed in general. Consider the sequence $\{\mathbf{x}_k\}$, where $\mathbf{x}_k = (2 - \frac{1}{k}, 2)^t$. Note that $\{\mathbf{x}_k\}$ converges to the point $\hat{\mathbf{x}} = (2, 2)^t$. For each k, $\mathbf{x}_k$ is feasible, and the set of binding constraints is empty. Thus, the projection matrix is equal to the identity so that $\mathbf{d}_k = -\nabla f(\mathbf{x}_k) = (-1, 2)^t$. Note, however, that the first constraint is binding at $\hat{\mathbf{x}}$. Here the projection matrix is

$$\mathbf{P} = \begin{bmatrix} \frac{4}{5} & -\frac{2}{5} \\ -\frac{2}{5} & \frac{1}{5} \end{bmatrix}$$

and, hence

$$\mathbf{d} = -\mathbf{P}\nabla f(\hat{\mathbf{x}}) = \begin{pmatrix} -\frac{8}{5} \\ \frac{4}{5} \end{pmatrix}$$

Thus, $\{\mathbf{d}_k\}$ does not converge to $\mathbf{d}$, and the direction-finding map is not closed at $\hat{\mathbf{x}}$. This is illustrated in Figure 10.18. Not only is the direction-finding map not closed but also the line search map that restricts the maximum step length via some feasible set is not closed in general, as seen in Example 10.2.2. Hence, Theorem 7.2.3 cannot be used to prove the convergence of this method. Nonetheless, one can prove that this algorithm converges under the following modification.

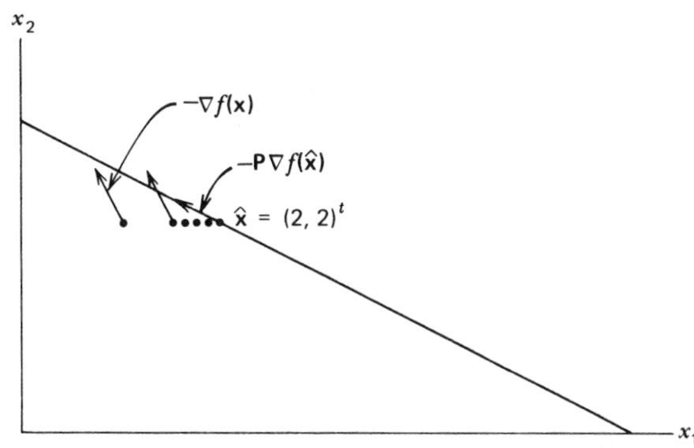

Figure 10.18 The direction-finding map is not closed.

Direction-Finding Routine for a Convergent Variant of the Gradient Projection Method

Consider the following revision of step 1 of the *main step* of the gradient projection method summarized above for linear constraints.

1. Let $\mathbf{M}^t = (\mathbf{A}^t, \mathbf{E}^t)$. If $\mathbf{M}$ is vacuous, then stop if $\nabla f(\mathbf{x}_k) = \mathbf{0}$ or else let $\mathbf{d}_k = -\nabla f(\mathbf{x}_k)$ and proceed to step 2. Otherwise, let $\mathbf{P} = \mathbf{I} - \mathbf{M}^t(\mathbf{M}\mathbf{M}^t)^{-1}\mathbf{M}$ and set $\mathbf{d}_k^{\mathrm{I}} = -\mathbf{P}\nabla f(\mathbf{x}_k)$. Also, compute $\mathbf{w} = -(\mathbf{M}\mathbf{M}^t)^{-1}\mathbf{M}\nabla f(\mathbf{x}_k)$ and let $\mathbf{w}^t = (\mathbf{u}^t, \mathbf{v}^t)$. If $\mathbf{u} \geq \mathbf{0}$, then stop if $\mathbf{d}_k^{\mathrm{I}} = \mathbf{0}$; otherwise, put $\mathbf{d}_k = \mathbf{d}_k^{\mathrm{I}} \neq \mathbf{0}$ and proceed to step 2. On the other hand, if $\mathbf{u} \not\geq \mathbf{0}$, let $u_h = \text{minimum}_j \{u_j\} < 0$, let $\hat{\mathbf{M}}^t = (\hat{\mathbf{A}}_1^t, \mathbf{E}^t)$, where $\hat{\mathbf{A}}_1$ is obtained from $\mathbf{A}_1$ by deleting the row of $\mathbf{A}_1$ corresponding to u_h, construct the projection matrix $\hat{\mathbf{P}} = \hat{\mathbf{I}} - \hat{\mathbf{M}}^t(\hat{\mathbf{M}}\hat{\mathbf{M}}^t)^{-1}\hat{\mathbf{M}}$, and define $\mathbf{d}_k^{\mathrm{II}} = -\hat{\mathbf{P}}\nabla f(\mathbf{x}_k)$. Now, based on some constant scalar $c > 0$, let

$$\mathbf{d}_k = \begin{cases} \mathbf{d}_k^{\mathrm{I}} & \text{if } \|\mathbf{d}_k^{\mathrm{I}}\| > |u_h|c \\ \mathbf{d}_k^{\mathrm{II}} & \text{otherwise} \end{cases} \tag{10.37}$$

and proceed to step 2.

Note that if either $\mathbf{M}$ is vacuous, or if $\mathbf{d}_k^{\mathrm{I}} = \mathbf{0}$ above, the procedural steps are the same as before. Hence, suppose that $\mathbf{M}$ is nonvacuous and that $\mathbf{d}_k^{\mathrm{I}} \neq \mathbf{0}$. While in the previous case, we would have used $\mathbf{d}_k = \mathbf{d}_k^{\mathrm{I}}$ at this point, we now compute $\mathbf{w}$ and switch to using $\mathbf{d}_k^{\mathrm{II}}$ instead, provided that it turns out that $\mathbf{u} \not\geq \mathbf{0}$ *and* that $\|\mathbf{d}_k^{\mathrm{I}}\|$ is "too small" by the measure $\|\mathbf{d}_k^{\mathrm{I}}\| \leq |u_h|c$. In particular, if $c = 0$, then step 1 is identical for the two procedures. The following result establishes that the above step 1 indeed generates an improving feasible direction.

10.5.7 Theorem

Consider the foregoing modification of step 1 of the gradient projection method. Then, either the algorithm terminates with a KKT solution at this step, or else it generates an improving feasible direction.

Proof

By Theorem 10.5.4, if the process stops at this step, it does so with a KKT solution. Also, from the above discussion, the claim follows from Theorem 10.5.4 when $\mathbf{M}$ is vacuous, or if $\mathbf{d}_k^{\mathrm{I}} = \mathbf{0}$, or if $\mathbf{u} \geq \mathbf{0}$, or if $\|\mathbf{d}_k^{\mathrm{I}}\| > |u_h|c$. Hence, suppose that $\mathbf{M}$ is nonvacuous, $\mathbf{u} \not\geq \mathbf{0}$, and that $\mathbf{d}_k^{\mathrm{I}} \neq \mathbf{0}$, but $\|\mathbf{d}_k^{\mathrm{I}}\| \leq |u_h|c$, so that, by (10.37), we use $\mathbf{d}_k = \mathbf{d}_k^{\mathrm{II}}$.

To begin with, note that $\mathbf{d}_k^{\mathrm{II}} = -\hat{\mathbf{P}}\nabla f(\mathbf{x}_k) \neq \mathbf{0}$, or else, by (10.34), we would have $\mathbf{d}_k^{\mathrm{I}} = -\mathbf{P}\nabla f(\mathbf{x}_k) = \mathbf{P}\hat{\mathbf{M}}^t\hat{\mathbf{w}} = \mathbf{0}$, since $\hat{\mathbf{M}}\mathbf{P}^t = \hat{\mathbf{M}}\mathbf{P} = \mathbf{0}$ because $\mathbf{MP} = \mathbf{0}$. This contradicts $\mathbf{d}_k^{\mathrm{I}} \neq \mathbf{0}$. Hence, $\hat{\mathbf{P}}\nabla f(\mathbf{x}_k) \neq \mathbf{0}$; and so, by Lemma 10.5.3, $\mathbf{d}_k^{\mathrm{II}}$ is an improving direction of f at $\mathbf{x}_k$.

Next, let us show that $\mathbf{d}_k^{\mathrm{II}}$ is a feasible direction. As in the proof of Theorem 10.5.4, noting (10.36), it suffices to demonstrate that $\mathbf{r}_h\mathbf{d}_k^{\mathrm{II}} \leq 0$, where $\mathbf{r}_h$ corresponds to the deleted row of $\mathbf{A}_1$. As in (10.33) and (10.35), we have

$$\mathbf{P}\nabla f(\mathbf{x}_k) = \nabla f(\mathbf{x}_k) + \hat{\mathbf{M}}^t\bar{\mathbf{w}} + u_h\mathbf{r}_h^t$$

Premultiplying this by $\mathbf{r}_h\hat{\mathbf{P}}$ gives

$$\mathbf{r}_h\hat{\mathbf{P}}\mathbf{P}\nabla f(\mathbf{x}_k) = -\mathrm{r}_h\mathbf{d}_k^{\mathrm{II}} + \mathbf{r}_h\hat{\mathbf{P}}\hat{\mathbf{M}}^t\bar{\mathbf{w}} + u_h\mathbf{r}_h\hat{\mathbf{P}}\mathbf{r}_h^t \tag{10.38}$$

Since $\hat{\mathbf{M}}\mathbf{P} = \mathbf{0}$, we have $\hat{\mathbf{P}}\mathbf{P} = \mathbf{P}$; and so, $\mathbf{r}_h\hat{\mathbf{P}}\mathbf{P} = \mathbf{r}_h\mathbf{P} = \mathbf{0}$ as $\mathbf{MP} = \mathbf{0}$. Also, since $\hat{\mathbf{P}}\hat{\mathbf{M}}^t = \mathbf{0}$, (10.38) yields $\mathbf{r}_h\mathbf{d}_k^{\mathrm{II}} = u_h\mathbf{r}_h\hat{\mathbf{P}}\mathbf{r}_h^t \leq 0$, since $u_h < 0$, and $\hat{\mathbf{P}}$ is positive semidefinite by Lemma 10.5.2. This completes the proof.

Hence, by Theorem 10.5.7, the various steps of the algorithm are well defined. Although the direction finding and the line search maps are still not closed (Example 10.5.6 continues to apply), Du and Zhang [1989] demonstrate that convergence obtains with the foregoing modification by showing that, if the iterates get too close within a defined ε-neighborhood of a non-KKT solution, then every step changes the active constraint set by 1 until the iterates are forced out of this neighborhood, in a manner which precludes a non-KKT point from becoming a cluster point of the sequence generated. We refer the reader to their paper for further details.

10.6 The Method of Reduced Gradient of Wolfe and the Generalized Reduced Gradient Method

In this section, we describe another procedure for generating improving feasible directions. The method depends upon reducing the dimensionality of the problem by representing all the variables in terms of an independent subset of the variables. The reduced gradient method was first developed by Wolfe [1963] to solve a nonlinear programming problem with linear constraints. The method was later generalized by Abadie and Carpentier [1969] to handle nonlinear constraints. Consider the following problem:

$$\begin{array}{ll} \text{Minimize} & f(\mathbf{x}) \\ \text{subject to} & \mathbf{Ax} = \mathbf{b} \\ & \mathbf{x} \geq \mathbf{0} \end{array}$$

where $\mathbf{A}$ is an $m \times n$ matrix of rank m, $\mathbf{b}$ is an m vector, and f is a continuously differentiable function on E_n. The following *nondegeneracy assumption* is made. Any m columns of $\mathbf{A}$ are linearly independent, and every extreme point of the feasible region has m strictly positive variables. With this assumption, every feasible solution has at least m positive components and, at most, $n - m$ zero components.

Now let $\mathbf{x}$ be a feasible solution. By the nondegeneracy assumption, note that $\mathbf{A}$ can be decomposed into $[\mathbf{B}, \mathbf{N}]$ and $\mathbf{x}^t$ into $[\mathbf{x}_B^t, \mathbf{x}_N^t]$, where $\mathbf{B}$ is an $m \times m$ invertible matrix and $\mathbf{x}_B > \mathbf{0}$. Here, $\mathbf{x}_B$ is called the *basic vector*, and each of its components is strictly positive. The components of the *nonbasic* vector $\mathbf{x}_N$ may either be positive or zero. Let $\nabla f(\mathbf{x})^t = [\nabla_B f(\mathbf{x})^t, \nabla_N f(\mathbf{x})^t]$, where $\nabla_B f(\mathbf{x})$ is the gradient of f with respect to the basic vector $\mathbf{x}_B$, and $\nabla_N f(\mathbf{x})$ is the gradient of f with respect to the nonbasic vector $\mathbf{x}_N$. Recall that a direction $\mathbf{d}$ is an improving feasible direction of f at $\mathbf{x}$ if $\nabla f(\mathbf{x})^t\mathbf{d} < 0$, and if $\mathbf{Ad} = \mathbf{0}$ with $d_j \geq 0$ if $x_j = 0$. We now specify a direction vector $\mathbf{d}$ satisfying these properties. First, $\mathbf{d}^t$ is decomposed into $[\mathbf{d}_B^t, \mathbf{d}_N^t]$. Note that $\mathbf{0} = \mathbf{Ad} = \mathbf{Bd}_B + \mathbf{Nd}_N$ automatically holds if, for any $\mathbf{d}_N$, we let $\mathbf{d}_B = -\mathbf{B}^{-1}\mathbf{Nd}_N$. Let $\mathbf{r}^t = (\mathbf{r}_B^t, \mathbf{r}_N^t) = \nabla f(\mathbf{x})^t - \nabla_B f(\mathbf{x})^t\mathbf{B}^{-1}\mathbf{A} = [\mathbf{0}, \nabla_N f(\mathbf{x})^t - \nabla_B f(\mathbf{x})^t\mathbf{B}^{-1}\mathbf{N}]$ be the *reduced gradient*, and let us examine the term $\nabla f(\mathbf{x})^t\mathbf{d}$:

$$\nabla f(\mathbf{x})^t\mathbf{d} = \nabla_B f(\mathbf{x})^t\mathbf{d}_B + \nabla_N f(\mathbf{x})^t\mathbf{d}_N = [\nabla_N f(\mathbf{x})^t - \nabla_B f(\mathbf{x})^t\mathbf{B}^{-1}\mathbf{N}]\mathbf{d}_N = \mathbf{r}_N^t\mathbf{d}_N$$

We must choose $\mathbf{d}_N$ in such a way that $\mathbf{r}_N^t\mathbf{d}_N < 0$ and that $d_j \geq 0$ if $x_j = 0$.

The following rule is adopted. For each nonbasic component j, let $d_j = -r_j$ if $r_j \leq 0$, and let $d_j = -x_jr_j$ if $r_j > 0$. This ensures that $d_j \geq 0$ if $x_j = 0$, and prevents unduly small step sizes when $x_j > 0$, but small, while $r_j > 0$. This also helps make the direction-finding map closed, thereby enabling convergence. Furthermore, $\nabla f(\mathbf{x})^t\mathbf{d} \leq 0$, where strict inequality holds if $\mathbf{d}_N \neq \mathbf{0}$.

To summarize, we have described a procedure for constructing an improving feasible direction. This fact, as well as the fact that $\mathbf{d} = \mathbf{0}$ if and only if $\mathbf{x}$ is KKT point, is proved in Theorem 10.6.1.

10.6.1 Theorem

Consider the problem to minimize $f(\mathbf{x})$ subject to $\mathbf{Ax} = \mathbf{b}$, $\mathbf{x} \geq \mathbf{0}$, where $\mathbf{A}$ is an $m \times n$ matrix and $\mathbf{b}$ is an m vector. Let $\mathbf{x}$ be a feasible solution such that $\mathbf{x}^t = (\mathbf{x}_B^t, \mathbf{x}_N^t)$ and $\mathbf{x}_B > \mathbf{0}$, where $\mathbf{A}$ is decomposed into $[\mathbf{B}, \mathbf{N}]$ and $\mathbf{B}$ is an $m \times m$ invertible matrix. Suppose that f is differentiable at $\mathbf{x}$, and let $\mathbf{r}^t = \nabla f(\mathbf{x})^t - \nabla_B f(\mathbf{x})^t\mathbf{B}^{-1}\mathbf{A}$. Let $\mathbf{d}^t = (\mathbf{d}_B^t, \mathbf{d}_N^t)$ be the direction formed as follows. For each nonbasic component j, let $d_j = -r_j$ if $r_j \leq 0$ and $-x_jr_j$ if $r_j > 0$, and then let $\mathbf{d}_B = -\mathbf{B}^{-1}\mathbf{Nd}_N$. If $\mathbf{d} \neq \mathbf{0}$, then $\mathbf{d}$ is an improving feasible direction. Furthermore, $\mathbf{d} = \mathbf{0}$ if and only if $\mathbf{x}$ is a KKT point.

Proof

First, note that $\mathbf{d}$ is a feasible direction if and only if $\mathbf{Ad} = \mathbf{0}$, and $d_j \geq 0$ if $x_j = 0$ for $j = 1, \ldots, n$. By definition of $\mathbf{d}_B$, $\mathbf{Ad} = \mathbf{Bd}_B + \mathbf{Nd}_N = \mathbf{B}(-\mathbf{B}^{-1}\mathbf{Nd}_N) + \mathbf{Nd}_N = \mathbf{0}$. If x_j is basic, then $x_j > 0$ by assumption. If x_j is not basic, then d_j could be negative only if $x_j > 0$. Thus, $d_j \geq 0$ if $x_j = 0$, and hence $\mathbf{d}$ is a feasible direction. Furthermore,

$$\nabla f(\mathbf{x})^t\mathbf{d} = \nabla_B f(\mathbf{x})^t\mathbf{d}_B + \nabla_N f(\mathbf{x})^t\mathbf{d}_N = [\nabla_N f(\mathbf{x})^t - \nabla_B f(\mathbf{x})^t\mathbf{B}^{-1}\mathbf{N}]\mathbf{d}_N = \sum_{j \notin I} r_jd_j$$

where I is the index set of basic variables. Noting the definition of d_j, it is obvious that either $\mathbf{d} = \mathbf{0}$ or $\nabla f(\mathbf{x})^t\mathbf{d} < \mathbf{0}$. In the latter case, by Lemma 10.1.2, $\mathbf{d}$ is indeed an improving feasible direction.

Note that $\mathbf{x}$ is a KKT point if and only if there exist vectors $\mathbf{u}^t = (\mathbf{u}_B^t, \mathbf{u}_N^t) \geq (\mathbf{0}, \mathbf{0})$ and $\mathbf{v}$ such that

$$\begin{gathered}[\nabla_B f(\mathbf{x})^t, \nabla_N f(\mathbf{x})^t] + \mathbf{v}^t(\mathbf{B}, \mathbf{N}) - (\mathbf{u}_B^t, \mathbf{u}_N^t) = (\mathbf{0}, \mathbf{0}) \\ \mathbf{u}_B^t\mathbf{x}_B = 0, \ \mathbf{u}_N^t\mathbf{x}_N = 0\end{gathered} \tag{10.39}$$

Since $\mathbf{x}_B > \mathbf{0}$ and $\mathbf{u}_B^t \geq \mathbf{0}$, then $\mathbf{u}_B^t\mathbf{x}_B = 0$ if and only if $\mathbf{u}_B^t = \mathbf{0}$. From the first equation in (10.39), it follows that $\mathbf{v}^t = -\nabla_B f(\mathbf{x})^t\mathbf{B}^{-1}$. Substituting in the second equation in (10.39), it follows that $\mathbf{u}_N^t = \nabla_N f(\mathbf{x})^t - \nabla_B f(\mathbf{x})^t\mathbf{B}^{-1}\mathbf{N}$. In other words, $\mathbf{u}_N = \mathbf{r}_N$. Thus, the KKT conditions reduce to $\mathbf{r}_N \geq \mathbf{0}$ and $\mathbf{r}_N^t\mathbf{x}_N = 0$. By the definition of $\mathbf{d}$, however, note that $\mathbf{d} = \mathbf{0}$ if and only if $\mathbf{r}_N \geq \mathbf{0}$ and $\mathbf{r}_N^t\mathbf{x}_N = 0$. Thus, $\mathbf{x}$ is a KKT point if and only if $\mathbf{d} = \mathbf{0}$, and the proof is complete.

Summary of the Reduced Gradient Algorithm

We summarize below Wolfe's reduced gradient algorithm for solving a problem of the form to minimize $f(\mathbf{x})$ subject to $\mathbf{Ax} = \mathbf{b}$, $\mathbf{x} \geq \mathbf{0}$. It is assumed that all m columns of $\mathbf{A}$ are linearly independent and that every extreme point of the feasible region has m strictly positive components. As we shall shortly show, the algorithm converges to a KKT point, provided that the basic variables are chosen to be the m most positive variables, where a tie is broken arbitrarily.

Initialization Step Choose a point $\mathbf{x}_1$ satisfying $\mathbf{Ax}_1 = \mathbf{b}$, $\mathbf{x}_1 \geq \mathbf{0}$. Let $k = 1$ and go to the main step.

Main Step

1. Let $\mathbf{d}_k^t = (\mathbf{d}_B^t, \mathbf{d}_N^t)$, where $\mathbf{d}_N$ and $\mathbf{d}_B$ are obtained as below from (10.43) and (10.44), respectively. If $\mathbf{d}_k = \mathbf{0}$, stop; $\mathbf{x}_k$ is a KKT point. (The Lagrange multipliers associated with $\mathbf{Ax} = \mathbf{b}$ and $\mathbf{x} \geq \mathbf{0}$ are, respectively, $\nabla_B f(\mathbf{x}_k)^t\mathbf{B}^{-1}$ and $\mathbf{r}$.) Otherwise, go to step 2.

$$I_k = \text{index set of the } m \text{ largest components of } \mathbf{x}_k \tag{10.40}$$

$$\mathbf{B} = \{\mathbf{a}_j : j \in I_k\} \qquad \mathbf{N} = \{\mathbf{a}_j : j \notin I_k\} \tag{10.41}$$

$$\mathbf{r}^t = \nabla f(\mathbf{x}_k)^t - \nabla_B f(\mathbf{x}_k)^t\mathbf{B}^{-1}\mathbf{A} \tag{10.42}$$

$$d_j = \begin{cases} -r_j & \text{if } j \notin I_k \text{ and } r_j \leq 0 \\ -x_j r_j & \text{if } j \notin I_k \text{ and } r_j > 0 \end{cases} \tag{10.43}$$

$$\mathbf{d}_B = -\mathbf{B}^{-1}\mathbf{N}\mathbf{d}_N \tag{10.44}$$

2. Solve the following line search problem:

   Minimize $f(\mathbf{x}_k + \lambda\mathbf{d}_k)$
   subject to $0 \leq \lambda \leq \lambda_{\max}$

   where

$$\lambda_{\max} = \begin{cases} \underset{1 \leq j \leq n}{\text{minimum}} \left\{ \dfrac{-x_{jk}}{d_{jk}} : d_{jk} < 0 \right\} & \text{if } \mathbf{d}_k \ngeq \mathbf{0} \\ \infty & \text{if } \mathbf{d}_k \geq \mathbf{0} \end{cases} \tag{10.45}$$

   and x_{jk}, d_{jk} are the jth components of $\mathbf{x}_k$ and $\mathbf{d}_k$, respectively. Let λ_k be an optimal solution, and let $\mathbf{x}_{k+1} = \mathbf{x}_k + \lambda_k\mathbf{d}_k$. Replace k by $k + 1$ and repeat step 1.

10.6.2 Example

Consider the following problem:

Minimize $2x_1^2 + 2x_2^2 - 2x_1x_2 - 4x_1 - 6x_2$

$$\text{subject to} \quad \begin{aligned} x_1 + \ x_2 + x_3 \quad &= 2 \\ x_1 + 5x_2 \quad + x_4 &= 5 \\ x_1, x_2, x_3, x_4 &\geq 0 \end{aligned}$$

We solve this problem using Wolfe's reduced gradient method starting from the point $\mathbf{x}_1 = (0, 0, 2, 5)^t$. Note that

$$\nabla f(\mathbf{x}) = (4x_1 - 2x_2 - 4, 4x_2 - 2x_1 - 6, 0, 0)^t$$

We shall exhibit the information needed at each iteration in tableau form similar to the simplex tableau of Section 2.7. However, since the gradient vector changes at each iteration, and since the nonbasic variables could be positive, we explicitly give the gradient vector and the complete solution at the top of each tableau. The reduced gradient vector $\mathbf{r}_k$ is shown as the last row of each tableau.

Iteration 1

Search Direction At the point $\mathbf{x}_1 = (0, 0, 2, 5)^t$, we have $\nabla f(\mathbf{x}_1) = (-4, -6, 0, 0)$. By (10.40), we have $I_1 = \{3, 4\}$, so that $\mathbf{B} = [\mathbf{a}_3, \mathbf{a}_4]$ and $\mathbf{N} = [\mathbf{a}_1, \mathbf{a}_2]$. From (10.42), the reduced gradient is given by

$$\mathbf{r}^t = (-4, -6, 0, 0) - (0, 0)\begin{bmatrix} 1 & 1 & 1 & 0 \\ 1 & 5 & 0 & 1 \end{bmatrix} = (-4, -6, 0, 0)$$

Note that the computations for the reduced gradient are similar to the computations for the objective row coefficients in the simplex method of Section 2.7. Also, $r_i = 0$ for $i \in I_1$. The information at this point is summarized in the tableau below.

		x_1	x_2	x_3	x_4
Solution $\mathbf{x}_1$		0	0	2	5
$\nabla f(\mathbf{x}_1)$		-4	-6	0	0
$\nabla_B f(\mathbf{x}_1) = \begin{bmatrix} 0 \\ 0 \end{bmatrix}$	x_3	1	1	1	0
	x_4	1	5	0	1
$\mathbf{r}$		-4	-6	0	0

By (10.16), then, we have $\mathbf{d}_N = (d_1, d_2)^t = (4, 6)^t$. We now compute $\mathbf{d}_B$ using (10.44) to get

$$\mathbf{d}_B = (d_3, d_4)^t = -\mathbf{B}^{-1}\mathbf{N}\mathbf{d}_N = -\begin{bmatrix} 1 & 1 \\ 1 & 5 \end{bmatrix}\begin{pmatrix} 4 \\ 6 \end{pmatrix} = (-10, -34)^t$$

Note that $\mathbf{B}^{-1}\mathbf{N}$ is recorded under the variables corresponding to $\mathbf{N}$, namely x_1 and x_2. The direction vector is, then, $\mathbf{d}_1 = (4, 6, -10, -34)^t$.

Line Search Starting from $\mathbf{x}_1 = (0, 0, 2, 5)^t$, we now wish to minimize the objective function along the direction $\mathbf{d}_1 = (4, 6, -10, -34)^t$. The maximum value of λ such that $\mathbf{x}_1 + \lambda\mathbf{d}_1$ is feasible is computed using (10.45), and we get

$$\lambda_{\max} = \text{minimum } \{\tfrac{2}{10}, \tfrac{5}{34}\} = \tfrac{5}{34}$$

The reader can verify that $f(\mathbf{x}_1 + \lambda\mathbf{d}_1) = 56\lambda^2 - 52\lambda$ so that λ_1 is the solution to the following problem:

$$\begin{array}{ll}\text{Minimize} & 56\lambda^2 - 52\lambda \\ \text{subject to} & 0 \le \lambda \le \frac{5}{34}\end{array}$$

Clearly $\lambda_1 = \frac{5}{34}$, so that $\mathbf{x}_2 = \mathbf{x}_1 + \lambda_1\mathbf{d}_1 = (\frac{10}{17}, \frac{15}{17}, \frac{9}{17}, 0)^t$.

Iteration 2

Search Direction At $\mathbf{x}_2 = (\frac{10}{17}, \frac{15}{17}, \frac{9}{17}, 0)^t$, from (10.40) we have $I_2 = \{1, 2\}$, $\mathbf{B} = [\mathbf{a}_1, \mathbf{a}_2]$ and $\mathbf{N} = [\mathbf{a}_3, \mathbf{a}_4]$. We also have $\nabla f(\mathbf{x}_2) = (-\frac{58}{17}, -\frac{62}{17}, 0, 0)^t$. The information is recorded in the tableau below, where the rows of x_1 and x_2 were obtained by *two pivot operations* on the tableau for iteration 1.

		x_1	x_2	x_3	x_4
Solution $\mathbf{x}_2$		$\frac{10}{17}$	$\frac{15}{17}$	$\frac{9}{17}$	0
$\nabla f(\mathbf{x}_2)$		$-\frac{58}{17}$	$-\frac{62}{17}$	0	0
$\nabla_B f(\mathbf{x}_2) = \begin{bmatrix} -\frac{58}{17} \\ -\frac{62}{17} \end{bmatrix}$	x_1	1	0	$\frac{5}{4}$	$-\frac{1}{4}$
	x_2	0	1	$-\frac{1}{4}$	$\frac{1}{4}$
$\mathbf{r}$		0	0	$\frac{57}{17}$	$\frac{1}{17}$

We have from (10.42),

$$\mathbf{r}^t = (-\tfrac{58}{17}, -\tfrac{62}{17}, 0, 0) - (-\tfrac{58}{17}, -\tfrac{62}{17})\begin{bmatrix} 1 & 0 & \frac{5}{4} & -\frac{1}{4} \\ 0 & 1 & -\frac{1}{4} & \frac{1}{4} \end{bmatrix} = (0, 0, \tfrac{57}{17}, \tfrac{1}{17})$$

From (10.43), then, $d_3 = -(\frac{9}{17})(\frac{57}{17}) = -\frac{513}{289}$ and $d_4 = 0$, so that $\mathbf{d}_N = (-\frac{513}{289}, 0)^t$. From (10.44), we get

$$\mathbf{d}_B = (d_1, d_2)^t = -\begin{bmatrix} \frac{5}{4} & -\frac{1}{4} \\ -\frac{1}{4} & \frac{1}{4} \end{bmatrix}\begin{pmatrix} -\frac{513}{289} \\ 0 \end{pmatrix} = \begin{bmatrix} \frac{2565}{1156} \\ -\frac{513}{1156} \end{bmatrix}$$

The new search direction is therefore given by $\mathbf{d}_2 = (\frac{2565}{1156}, -\frac{513}{1156}, -\frac{513}{289}, 0)^t$.

Line Search Starting from $\mathbf{x}_2 = (\frac{10}{17}, \frac{15}{17}, \frac{9}{17}, 0)^t$, we wish to minimize the objective function along the direction $\mathbf{d}_2 = (\frac{2565}{1156}, -\frac{513}{1156}, -\frac{513}{289}, 0)^t$. The maximum value of λ such that $\mathbf{x}_2 + \lambda\mathbf{d}_2$ is feasible is computed using (10.45), and we get

$$\lambda_{\max} = \text{minimum}\left\{\frac{-\frac{15}{17}}{-\frac{513}{1156}}, \frac{-\frac{9}{17}}{-\frac{513}{289}}\right\} = \frac{17}{57}$$

The reader can verify that $f(\mathbf{x}_2 + \lambda\mathbf{d}_2) = 12.21\lambda^2 - 5.95\lambda - 6.436$, so that λ_2 is obtained by solving the following problem:

$$\begin{array}{ll}\text{Minimize} & 12.21\lambda^2 - 5.95\lambda - 6.436 \\ \text{subject to} & 0 \le \lambda \le \frac{17}{57}\end{array}$$

The reader can verify that $\lambda_2 = \frac{68}{279}$, so that $\mathbf{x}_3 = \mathbf{x}_2 + \lambda_2\mathbf{d}_2 = (\frac{35}{31}, \frac{24}{31}, \frac{3}{31}, 0)^t$.

Iteration 3

Search Direction Now $I_3 = \{1, 2\}$, so that $\mathbf{B} = [\mathbf{a}_1, \mathbf{a}_2]$ and $\mathbf{N} = [\mathbf{a}_3, \mathbf{a}_4]$. Since $I_3 = I_2$, the tableau at iteration 2 can be retained. We have $\nabla f(\mathbf{x}_3) = (-\frac{32}{31}, -\frac{160}{31}, 0, 0)^t$.

			x_1	x_2	x_3	x_4
Solution $\mathbf{x}_3$			$\frac{35}{31}$	$\frac{24}{31}$	$\frac{3}{31}$	0
$\nabla f(\mathbf{x}_3)$			$-\frac{32}{31}$	$-\frac{160}{31}$	0	0
$\nabla_B f(\mathbf{x}_3) =$	$-\frac{32}{31}$	x_1	1	0	$\frac{5}{4}$	$-\frac{1}{4}$
	$-\frac{160}{31}$	x_2	0	1	$-\frac{1}{4}$	$\frac{1}{4}$
$\mathbf{r}$			0	0	0	$\frac{32}{31}$

From (10.42), we get

$$\mathbf{r}^t = (-\tfrac{32}{31}, -\tfrac{160}{31}, 0, 0) - (-\tfrac{32}{31}, -\tfrac{160}{31})\begin{bmatrix} 1 & 0 & \frac{5}{4} & -\frac{1}{4} \\ 0 & 1 & -\frac{1}{4} & \frac{1}{4} \end{bmatrix} = (0, 0, 0, \tfrac{32}{31})$$

From (10.43), then, $\mathbf{d}_N = (d_3, d_4)^t = (0, 0)^t$; and from (10.44), we also get $\mathbf{d}_B = (d_1, d_2)^t = (0, 0)^t$. Hence, $\mathbf{d} = \mathbf{0}$, and the solution $\mathbf{x}_3$ is a KKT solution and, therefore, optimal for this problem. The optimal Lagrange multipliers associated with the equality constraints are $\nabla_B f(\mathbf{x}_3)^t\mathbf{B}^{-1} = (0, -\frac{32}{31})^t$, and those associated with the nonnegativity constraints are $(0, 0, 0, 1)^t$. Table 10.5 gives a summary of the computations, and the progress of the algorithm is shown in Figure 10.19.

Table 10.5 Summary of Computations for the Reduced Gradient Method of Wolfe

Iteration			Search Direction		Line Search	
k	$\mathbf{x}_k$	$f(\mathbf{x}_k)$	$\mathbf{r}_k$	$\mathbf{d}_k$	λ_k	$\mathbf{x}_{k+1}$
1	$(0, 0, 2, 5)$	0.0	$(-4, -6, 0, 0)$	$(4, 6, -10, -34)$	$\frac{5}{34}$	$(\frac{10}{17}, \frac{15}{17}, \frac{9}{17}, 0)$
2	$(\frac{10}{17}, \frac{15}{17}, \frac{9}{17}, 0)$	-6.436	$(0, 0, \frac{57}{17}, \frac{4}{17})$	$(\frac{2565}{1156}, -\frac{513}{1156}, -\frac{513}{289}, 0)$	$\frac{68}{279}$	$(\frac{35}{31}, \frac{24}{31}, \frac{3}{31}, 0)$
3	$(\frac{35}{31}, \frac{24}{31}, \frac{3}{31}, 0)$	-7.16	$(0, 0, 0, \frac{32}{31})$	$(0, 0, 0, 0)$		

Convergence of the Reduced Gradient Method

Theorem 10.6.3 below proves convergence of the reduced gradient method to a KKT point. This is done by a contradiction argument that establishes a sequence satisfying conditions 1 through 4 of Lemma 10.2.6.

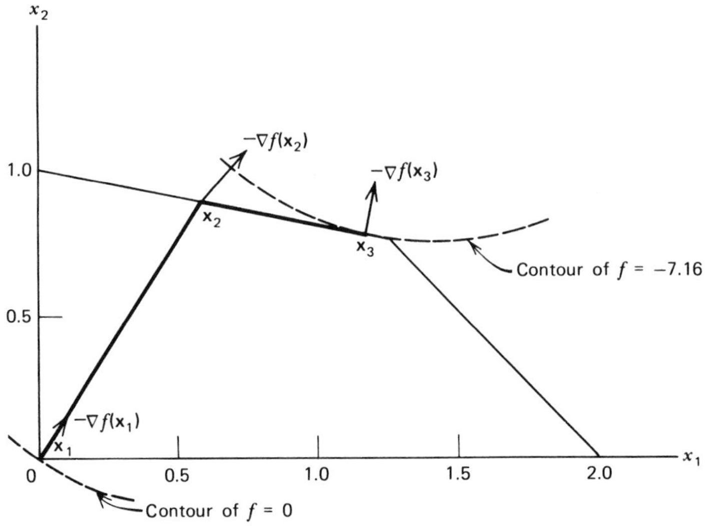

Figure 10.19 Illustration of the reduced gradient method of Wolfe.

10.6.3 Theorem

Let $f: E_n \to E_1$ be continuously differentiable, and consider the problem to minimize $f(\mathbf{x})$ subject to $\mathbf{Ax} = \mathbf{b}$, $\mathbf{x} \geq \mathbf{0}$. Here, $\mathbf{A}$ is an $m \times n$ matrix and $\mathbf{b}$ is an m vector such that all extreme points of the feasible region have m positive components, and any set of m columns of $\mathbf{A}$ are linearly independent. Suppose that the sequence $\{\mathbf{x}_k\}$ is generated by the reduced gradient algorithm. Then, any accumulation point of $\{\mathbf{x}_k\}$ is a KKT point.

Proof

Let $\{\mathbf{x}_k\}_{\mathcal{K}}$ be a convergent subsequence with limit $\hat{\mathbf{x}}$. We need to show that $\hat{\mathbf{x}}$ is a KKT point. Suppose, by contradiction, that $\hat{\mathbf{x}}$ is not a KKT point. We shall exhibit a sequence $\{(\mathbf{x}_k, \mathbf{d}_k)\}_{\mathcal{K}'}$, satisfying conditions 1 through 4 of Lemma 10.2.6, which is impossible.

Let $\{\mathbf{d}_k\}_{\mathcal{K}}$ be the sequence of directions associated with $\{\mathbf{x}_k\}_{\mathcal{K}}$. Note that $\mathbf{d}_k$ is defined by (10.40) through (10.44) at $\mathbf{x}_k$. Letting I_k be the set denoting the indices of the m largest components of $\mathbf{x}_k$ used to compute $\mathbf{d}_k$, then there exists $\mathcal{K}' \subseteq \mathcal{K}$ such that $I_k = \hat{I}$ for each $k \in \mathcal{K}'$, where $\hat{I}$ is the set denoting the indices of the m largest components of $\hat{\mathbf{x}}$. Let $\hat{\mathbf{d}}$ be the direction obtained from (10.40) through (10.44) at $\hat{\mathbf{x}}$, and note, by Theorem 10.6.1, that $\hat{\mathbf{d}} \neq \mathbf{0}$ and $\nabla f(\hat{\mathbf{x}})^t\hat{\mathbf{d}} < 0$. Since f is continuously differentiable, $\mathbf{x}_k \to \hat{\mathbf{x}}$ and $I_k = \hat{I}$ for $k \in \mathcal{K}'$, then, by (10.41) through (10.44), $\mathbf{d}_k \to \hat{\mathbf{d}}$ for $k \in \mathcal{K}'$. To summarize, we have exhibited a sequence $\{(\mathbf{x}_k, \mathbf{d}_k)\}_{\mathcal{K}'}$ satisfying conditions 1, 2, and 4 of Lemma 10.2.6. To complete the proof, we need to show that part 3 also holds true.

From (10.45), recall that $\mathbf{x}_k + \lambda \mathbf{d}_k$ is feasible for each $\lambda \in [0, \delta_k]$, where $\delta_k =$ minimum [minimum $\{-x_{ik}/d_{ik} : d_{ik} < 0\}, \infty] > 0$ for each $k \in \mathcal{K}'$. Suppose that inf $\{\delta_k : k \in \mathcal{K}'\} = 0$. Then, there exists an index set $\mathcal{K}'' \subseteq \mathcal{K}'$ such that $\delta_k = -x_{pk}/d_{pk}$ converges to 0 for $k \in \mathcal{K}''$, where $x_{pk} > 0$, $d_{pk} < 0$, and p is an element of $\{1, \ldots, n\}$. By (10.40) through (10.44), note that $\{d_{pk}\}_{\mathcal{K}''}$ is bounded; and, since $\{\delta_k\}_{\mathcal{K}''}$ converges to 0, then $\{x_{pk}\}_{\mathcal{K}''}$ converges to 0. Thus, $\hat{x}_p = 0$, that is, $p \notin \hat{I}$. But $I_k = \hat{I}$ for $k \in \mathcal{K}''$, and, hence, $p \notin I_k$. Since $d_{pk} < 0$, from (10.43), $d_{pk} = -x_{pk}r_{pk}$. It then follows that $\delta_k = -x_{pk}/d_{pk} = 1/r_{pk}$. This shows that $r_{pk} \to \infty$, which is impossible, since $r_{pk} \to r_p \neq \infty$. Thus, inf $\{\delta_k : k \in \mathcal{K}'\} = \delta > 0$. We have thus shown that there exists a $\delta > 0$ such that $\mathbf{x}_k + \lambda \mathbf{d}_k$ is feasible for each $\lambda \in [0, \delta]$ and for each $k \in \mathcal{K}'$. Therefore, condition 3 of Lemma 10.2.6 holds true, and the proof is complete.

The Generalized Reduced Gradient Method

We can extend the reduced gradient method to handle nonlinear constraints similar to the gradient projection method. This extension is referred to as the *generalized reduced gradient method*, and is briefly sketched below (see also Exercise 10.54 for the originally proposed scheme).

Consider a nonlinear programming problem of the form

$$\text{Minimize } \{f(\mathbf{x}) : \mathbf{h}(\mathbf{x}) = \mathbf{0}, \quad \mathbf{x} \geq \mathbf{0}\}$$

where $\mathbf{h}(\mathbf{x}) = \mathbf{0}$ represents some m equality constraints, $\mathbf{x} \in E_n$, and suitable variable transformations have been used to represent all variables as being nonnegative. Here, any inequality constraint can be assumed to have been written as an equality by introducing a nonnegative slack variable.

Now, given a feasible solution $\mathbf{x}_k$, consider a linearization of $\mathbf{h}(\mathbf{x}) = \mathbf{0}$ given by $\mathbf{h}(\mathbf{x}_k) + \nabla \mathbf{h}(\mathbf{x}_k)(\mathbf{x} - \mathbf{x}_k) = \mathbf{0}$, where $\nabla \mathbf{h}(\mathbf{x}_k)$ is the $m \times n$ Jacobian of $\mathbf{h}$ evaluated at $\mathbf{x}_k$. Noting that $\mathbf{h}(\mathbf{x}_k) = \mathbf{0}$, the set of linear constraints given by $\nabla \mathbf{h}(\mathbf{x}_k)\mathbf{x} = \nabla \mathbf{h}(\mathbf{x}_k)\mathbf{x}_k$ is of the form $\mathbf{Ax} = \mathbf{b}$, where $\mathbf{x}_k \geq \mathbf{0}$ is a feasible solution. Assuming that the Jacobian $\mathbf{A} = \nabla \mathbf{h}(\mathbf{x}_k)$ has full row rank, and partitioning it suitably into $[\mathbf{B}, \mathbf{N}]$ and, accordingly, partitioning $\mathbf{x}^t = (\mathbf{x}_B^t, \mathbf{x}_N^t)$ (where, hopefully, $\mathbf{x}_B > \mathbf{0}$ in $\mathbf{x}_k$), we can compute the reduced gradient $\mathbf{r}$ via (10.42) and, hence, obtain the direction of motion $\mathbf{d}_k$ via (10.43) and (10.44). As before, we obtain $\mathbf{d}_k = \mathbf{0}$ if and only if $\mathbf{x}_k$ is a KKT point, whence the procedure terminates. Otherwise, a line search is performed along $\mathbf{d}_k$.

Earlier versions of this method adopted the following strategy. First, a line search is performed by determining $\lambda_{\max}$ via (10.45) and then finding λ_k as the solution to the line search problem to minimize $f(\mathbf{x}_k + \lambda \mathbf{d}_k)$ subject to $0 \leq \lambda \leq \lambda_{\max}$. This gives $\mathbf{x}' = \mathbf{x}_k + \lambda_k \mathbf{d}_k$. Since $\mathbf{h}(\mathbf{x}') = \mathbf{0}$ is not necessarily satisifed, we need a correction step (see also Exercise 10.10). Toward this end, the Newton–Raphson method is then used to obtain $\mathbf{x}_{k+1}$ satisfying $\mathbf{h}(\mathbf{x}_{k+1}) = \mathbf{0}$, starting with the solution $\mathbf{x}'$ and keeping the components of $\mathbf{x}_N$ fixed at the values $\mathbf{x}'_N$. Hence, $\mathbf{x}_N$ remains at $\mathbf{x}'_N \geq \mathbf{0}$ during this iterative process, but some component(s) of $\mathbf{x}_B$ may tend to become negative. At such a point, a switch is made by replacing a negative basic variable x_r with a nonbasic variable x_q that is preferably positive and that has a significantly nonzero element in the corresponding row r of the column "$\mathbf{B}^{-1}\mathbf{a}_q$." The Newton–Raphson process then continues as above with the revised basis (having now fixed x_r at zero) and the revised linearized system, until a nonnegative solution $\mathbf{x}_{k+1}$ satisfying $\mathbf{h}(\mathbf{x}_{k+1}) = \mathbf{0}$ is finally obtained.

More recent versions of the GRG method adopt a discrete sequence of positive step sizes, and attempt to find a corresponding $\mathbf{x}_{k+1}$ for each such step size sequentially using the foregoing Newton–Raphson scheme. Using the value $f(\mathbf{x}_{k+1})$ at each such point, when a three-point pattern (TPP) of the quadratic interpolation method (see Section 8.3) is obtained, a quadratic fit is used to determine a new step size, for which the corresponding point $\mathbf{x}_{k+1}$ is again determined. The feasible point with the smallest objective value thus found is used as the next iterate. This technique is thought to yield a more reliable algorithm.

As the reader may have surmised, the iterative Newton–Raphson scheme complicates convergence arguments. Indeed, the existing convergence proofs use restrictive and difficult-to-verify assumptions. Nonetheless, this type of algorithm provides quite a robust and efficient scheme for solving nonlinear programming problems.

10.7 The Convex–Simplex Method of Zangwill

The convex–simplex method is identical to the reduced gradient method of Section 10.6, except that only one nonbasic variable is modified while all other nonbasic variables are fixed at their current levels. Of course, the values of the basic variables are modified accordingly to maintain feasibility, so that the method behaves very much like the simplex method for linear programs. The name of the convex–simplex method was coined because the method was originally proposed by Zangwill [1967] for minimizing a convex function in the presence of linear constraints. Below, we state the required modification in the reduced gradient method for the following class of problems:

$$\begin{aligned} \text{Minimize} \quad & f(\mathbf{x}) \\ \text{subject to} \quad & \mathbf{Ax} = \mathbf{b} \\ & \mathbf{x} \geq \mathbf{0} \end{aligned}$$

where $\mathbf{A}$ is an $m \times n$ matrix of rank m, and $\mathbf{b}$ is an m vector.

Summary of the Convex–Simplex Method

We again assume that any m columns of $\mathbf{A}$ are linearly independent and that every extreme point of the feasible region has m strictly positive components. As we shall shortly show, the algorithm converges as before to a KKT point, provided that the basic variables are chosen to be the m most positive variables, where a tie is broken arbitrarily.

Initialization Step Choose a point $\mathbf{x}_1$ such that $\mathbf{Ax}_1 = \mathbf{b}$ and $\mathbf{x}_1 \geq \mathbf{0}$. Let $k = 1$ and go to the main step.

Main Step

1. Given $\mathbf{x}_k$, identify I_k, $\mathbf{B}$, $\mathbf{N}$, and compute $\mathbf{r}$ as follows:

$$I_k = \text{index set of the } m \text{ largest components of } \mathbf{x}_k \tag{10.46}$$

$$\mathbf{B} = \{\mathbf{a}_j : j \in I_k\} \qquad \mathbf{N} = \{\mathbf{a}_j : j \notin I_k\} \tag{10.47}$$

$$\mathbf{r}^t = \nabla f(\mathbf{x}_k)^t - \nabla_B f(\mathbf{x}_k)^t \mathbf{B}^{-1}\mathbf{A} \tag{10.48}$$

Consider (10.49) through (10.55) below. If $\alpha = \beta = 0$, stop; $\mathbf{x}_k$ is a KKT point with Lagrange multipliers $\nabla_B f(\mathbf{x}_k)^t\mathbf{B}^{-1}$ and $\mathbf{r}$, associated with the constraints $\mathbf{Ax} = \mathbf{b}$ and $\mathbf{x} \geq \mathbf{0}$, respectively. If $\alpha > \beta$, compute $\mathbf{d}_N$ from (10.51) and (10.53). If $\alpha < \beta$, compute $\mathbf{d}_N$ from (10.52) and (10.54). If $\alpha = \beta \neq 0$, compute $\mathbf{d}_N$ either from (10.51) and (10.53) or else from (10.52) and (10.54). In all cases, determine $\mathbf{d}_B$ from (10.55) and go to step 2.

$$\alpha = \text{maximum}\{-r_j : r_j \leq 0\} \tag{10.49}$$

$$\beta = \text{maximum}\{x_j r_j : r_j \geq 0\} \tag{10.50}$$

$$\nu = \begin{cases} \text{an index such that } \alpha = -r_\nu & (10.51) \\ \text{an index such that } \beta = x_\nu r_\nu & (10.52) \end{cases}$$

$$d_j = \begin{cases} 0 & \text{if } j \notin I_k, j \neq \nu \\ 1 & \text{if } j \notin I_k, j = \nu \end{cases} \tag{10.53}$$

$$d_j = \begin{cases} 0 & \text{if } j \notin I_k, j \neq \nu \\ -1 & \text{if } j \notin I_k, j = \nu \end{cases} \tag{10.54}$$

$$\mathbf{d}_B = -\mathbf{B}^{-1}\mathbf{N}\mathbf{d}_N \equiv -\mathbf{B}^{-1}\mathbf{a}_\nu d_\nu \tag{10.55}$$

2. Consider the following line search problem:

   Minimize $f(\mathbf{x}_k + \lambda \mathbf{d}_k)$
   subject to $0 \le \lambda \le \lambda_{max}$

   where

$$\lambda_{max} = \begin{cases} \underset{1 \le j \le n}{\text{minimum}} \left\{ \dfrac{-x_{jk}}{d_{jk}} : d_{jk} < 0 \right\} & \text{if } \mathbf{d}_k \not\geq \mathbf{0} \\ \infty & \text{if } \mathbf{d}_k \geq \mathbf{0} \end{cases} \tag{10.56}$$

   and x_{jk}, d_{jk} are the jth components of $\mathbf{x}_k$ and $\mathbf{d}_k$, respectively. Let λ_k be an optimal solution, and let $\mathbf{x}_{k+1} = \mathbf{x}_k + \lambda_k \mathbf{d}_k$. Replace k by $k + 1$ and go to step 1.

Observe that $\alpha = \beta = 0$ if and only if $\mathbf{d}_N = \mathbf{0}$ in the reduced gradient method which, by Theorem 10.6.1, happens if and only if $\mathbf{x}_k$ is a KKT point. Otherwise, $\mathbf{d} \neq \mathbf{0}$ is an improving feasible direction as in the proof of Theorem 10.6.1.

10.7.1 Example

Consider the following problem:

$$\begin{array}{ll} \text{Minimize} & 2x_1^2 + 2x_2^2 - 2x_1x_2 - 4x_1 - 6x_2 \\ \text{subject to} & x_1 + x_2 + x_3 = 2 \\ & x_1 + 5x_2 + x_4 = 5 \\ & x_1, x_2, x_3, x_4 \geq 0 \end{array}$$

We solve the problem using Zangwill's convex simplex method starting from the point $\mathbf{x}_1 = (0, 0, 2, 5)^t$. Note that

$$\nabla f(\mathbf{x}) = (4x_1 - 2x_2 - 4, 4x_2 - 2x_1 - 6, 0, 0)^t$$

As in the reduced gradient method, it is convenient to exhibit the information at each iteration in tableau form, giving the solution vector $\mathbf{x}_k$ and also $\nabla f(\mathbf{x}_k)$.

Iteration 1

Search Direction At the point $\mathbf{x}_1 = (0, 0, 2, 5)^t$, we have $\nabla f(\mathbf{x}_1) = (-4, -6, 0, 0)^t$. From (10.46), we then have $I_1 = \{3, 4\}$, so that $\mathbf{B} = [\mathbf{a}_3, \mathbf{a}_4]$ and $\mathbf{N} = [\mathbf{a}_1, \mathbf{a}_2]$. The reduced gradient is computed using (10.48) as follows:

$$\mathbf{r}^t = (-4, -6, 0, 0) - (0, 0)\begin{bmatrix} 1 & 1 & 1 & 0 \\ 1 & 5 & 0 & 1 \end{bmatrix} = (-4, -6, 0, 0)$$

The tableau at this stage is given below.

			x_1	x_2	x_3	x_4
Solution $\mathbf{x}_1$			0	0	2	5
$\nabla f(\mathbf{x}_1)$			-4	-6	0	0
$\nabla_B f(\mathbf{x}_1) =$	$\begin{bmatrix} 0 \\ 0 \end{bmatrix}$	x_3	1	1	1	0
		x_4	1	5	0	1
$\mathbf{r}$			-4	-6	0	0

Now, from (10.49), α = maximum $\{-r_1, -r_2, -r_3, -r_4\} = -r_2 = 6$. Also from (10.50), β = maximum $\{x_3r_3, x_4r_4\} = 0$; and, hence, from (10.51), $\nu = 2$. Note that $-r_2 = 6$ implies that x_2 can be increased to yield a reduced objective function value. The search direction is given by (10.53) and (10.55). From (10.53), we have $\mathbf{d}_N^t = (d_1, d_2) = (0, 1)$; and from (10.55), we get $\mathbf{d}_B^t = (d_3, d_4) = -(1, 5)$. Note that $\mathbf{d}_B = -\mathbf{B}^{-1}\mathbf{a}_2$ is the negative of the column of x_2 in the above tableau. Hence, $\mathbf{d}_1 = (0, 1, -1, -5)^t$.

Line Search Starting from the point $\mathbf{x}_1 = (0, 0, 2, 5)^t$, we wish to search along the direction $\mathbf{d}_1 = (0, 1, -1, -5)^t$. The maximum value of λ such that $\mathbf{x}_1 + \lambda\mathbf{d}_1$ is feasible is given by (10.56). In this case,

$$\lambda_{max} = \text{minimum } \{\tfrac{2}{1}, \tfrac{5}{5}\} = 1$$

We also have $f(\mathbf{x}_1 + \lambda\mathbf{d}_1) = 2\lambda^2 - 6\lambda$. Hence, we solve the following problem:

$$\begin{array}{ll} \text{Minimize} & 2\lambda^2 - 6\lambda \\ \text{subject to} & 0 \le \lambda \le 1 \end{array}$$

The optimal solution is $\lambda_1 = 1$, so that $\mathbf{x}_2 = \mathbf{x}_1 + \lambda_1\mathbf{d}_1 = (0, 1, 1, 0)^t$.

Iteration 2

Search Direction At the point $\mathbf{x}_2 = (0, 1, 1, 0)^t$, we have, by (10.46), $I_2 = \{2, 3\}$, so that $\mathbf{B} = [\mathbf{a}_2, \mathbf{a}_3]$ and $\mathbf{N} = [\mathbf{a}_1, \mathbf{a}_4]$. The updated tableau obtained by one pivot operation is given below. Note that $\nabla f(\mathbf{x}_2) = (-6, -2, 0, 0)^t$; and, from (10.48), we get

$$\mathbf{r}^t = (-6, -2, 0, 0) - (0, -2)\begin{bmatrix} \frac{4}{5} & 0 & 1 & -\frac{1}{5} \\ \frac{1}{5} & 1 & 0 & \frac{1}{5} \end{bmatrix} = (-\tfrac{28}{5}, 0, 0, \tfrac{2}{5})$$

			x_1	x_2	x_3	x_4
Solution $\mathbf{x}_2$			0	1	1	0
$\nabla f(\mathbf{x}_2)$			-6	-2	0	0
$\nabla_B f(\mathbf{x}_2) =$	$\begin{bmatrix} 0 \\ -2 \end{bmatrix}$	x_3	$\frac{4}{5}$	0	1	$-\frac{1}{5}$
		x_2	$\frac{1}{5}$	1	0	$\frac{1}{5}$
$\mathbf{r}$			$-\frac{28}{5}$	0	0	$\frac{2}{5}$

From (10.49) and (10.50), α = maximum $\{-r_1, -r_2, -r_3\} = -r_1 = \frac{28}{5}$, and β = maximum $\{x_2r_2, x_3r_3, x_4r_4\} = 0$, so that $\nu = 1$. This means that x_1 can be increased. From (10.53) and (10.55), we have $\mathbf{d}_N^t = (d_1, d_4) = (1, 0)$, and $\mathbf{d}_B^t = (d_3, d_2) = (-\frac{4}{5}, -\frac{1}{5})$. Thus, $\mathbf{d}_2 = (1, -\frac{1}{5}, -\frac{4}{5}, 0)^t$.

Line Search Starting from the point $\mathbf{x}_2 = (0, 1, 1, 0)^t$, we wish to search along the direction $\mathbf{d}_2 = (1, -\frac{1}{5}, -\frac{4}{5}, 0)^t$. The maximum value of λ so that $\mathbf{x}_2 + \lambda\mathbf{d}_2$ is feasible is given by (10.56) as follows:

$$\lambda_{\max} = \text{minimum}\left(\frac{1}{\frac{1}{5}}, \frac{1}{\frac{4}{5}}\right) = \frac{5}{4}$$

We also have $f(\mathbf{x}_2 + \lambda\mathbf{d}_2) = 2.48\lambda^2 - 5.6\lambda - 4$. Hence, we solve the problem

$$\begin{array}{ll} \text{Minimize} & 2.48\lambda^2 - 5.6\lambda - 4 \\ \text{subject to} & 0 \le \lambda \le \frac{5}{4} \end{array}$$

The optimal solution is $\lambda_2 = \frac{35}{31}$, so that $\mathbf{x}_3 = \mathbf{x}_2 + \lambda_2\mathbf{d}_2 = (\frac{35}{31}, \frac{24}{31}, \frac{3}{31}, 0)^t$.

Iteration 3

Search Direction At the point $\mathbf{x}_3 = (\frac{35}{31}, \frac{24}{31}, \frac{3}{31}, 0)^t$, from (10.46), we get $I_3 = \{1, 2\}$, so that $\mathbf{B} = [\mathbf{a}_1, \mathbf{a}_2]$ and $\mathbf{N} = [\mathbf{a}_3, \mathbf{a}_4]$. We also have $\nabla f(\mathbf{x}_3) = (-\frac{32}{31}, -\frac{160}{31}, 0, 0)^t$: and from (10.48), we get

$$\mathbf{r}^t = (-\tfrac{32}{31}, -\tfrac{160}{31}, 0, 0) - (-\tfrac{32}{31}, -\tfrac{160}{31})\begin{bmatrix} 1 & 0 & \frac{5}{4} & -\frac{1}{4} \\ 0 & 1 & -\frac{1}{4} & \frac{1}{4} \end{bmatrix} = (0, 0, 0, \tfrac{32}{31})$$

The information is given in the tableau below.

			x_1	x_2	x_3	x_4
Solution $\mathbf{x}_3$			$\frac{35}{31}$	$\frac{24}{31}$	$\frac{3}{31}$	0
$\nabla f(\mathbf{x}_3)$			$-\frac{32}{31}$	$-\frac{160}{31}$	0	0
$\nabla_B f(\mathbf{x}_3) =$	$-\frac{32}{31}$	x_1	1	0	$\frac{5}{4}$	$-\frac{1}{4}$
	$-\frac{160}{31}$	x_2	0	1	$-\frac{1}{4}$	$\frac{1}{4}$
$\mathbf{r}$			0	0	0	$\frac{32}{31}$

In this case, α = maximum $\{-r_1, -r_2, -r_3\} = 0$ and β = maximum $\{x_1r_1, x_2r_2, x_3r_3, x_4r_4\} = 0$. Hence, the point $\mathbf{x}_3 = (\frac{35}{31}, \frac{24}{31}, \frac{3}{31}, 0)^t$ is a KKT solution and, therefore, is optimal for this problem. (The optimal Lagrange multipliers are obtained as in Example 10.6.2.) A summary of the computations is given in Table 10.6. The progress of the algorithm is depicted in Figure 10.20.

Convergence of the Convex–Simplex Method

The convergence of the convex–simplex method to a KKT point can be established similar to Theorem 10.6.3. For the sake of completeness, this argument is sketched below.

Table 10.6 Summary of Computations for the Convex-Simplex Method of Zangwill

Iteration			Search Direction		Line Search	
k	$\mathbf{x}_k$	$f(\mathbf{x}_k)$	$\mathbf{r}$	$\mathbf{d}$	λ_k	$\mathbf{x}_{k+1}$
1	$(0, 0, 2, 5)$	0.0	$(-4, -6, 0, 0)$	$(0, 1, -1, -5)$	1	$(0, 1, 1, 0)$
2	$(0, 1, 1, 0)$	-4.0	$(-\frac{28}{5}, 0, 0, \frac{2}{5})$	$(1, -\frac{1}{5}, -\frac{4}{5}, 0)$	$\frac{35}{31}$	$(\frac{35}{31}, \frac{24}{31}, \frac{3}{31}, 0)$
3	$(\frac{35}{31}, \frac{24}{31}, \frac{3}{31}, 0)$	-7.16	$(0, 0, 0, 1)$			

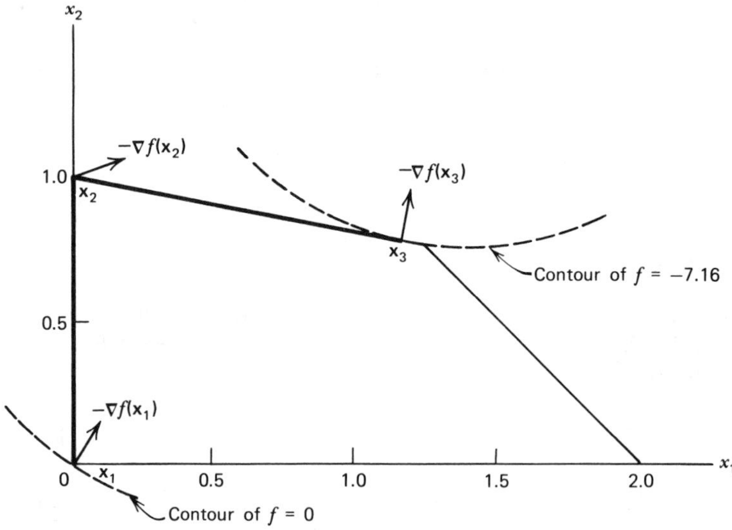

Figure 10.20 Illustration of the convex-simplex method of Zangwill.

10.7.2 Theorem

Let $f: E_n \to E_1$ be continuously differentiable, and consider the problem to minimize $f(\mathbf{x})$ subject to $\mathbf{Ax} = \mathbf{b}$, $\mathbf{x} \geq \mathbf{0}$. Here, $\mathbf{A}$ is an $m \times n$ matrix and $\mathbf{b}$ is an m vector such that all extreme points of the feasible region have m positive components and every m columns of $\mathbf{A}$ are linearly independent. Suppose that the sequence $\{\mathbf{x}_k\}$ is generated by the convex-simplex method. Then, any accumulation point is a KKT point.

Proof

Let $\{\mathbf{x}_k\}_{\mathcal{K}}$ be a convergent subsequence with limit $\hat{\mathbf{x}}$. We need to show that $\hat{\mathbf{x}}$ is a KKT point. Suppose, by contradiction, that $\hat{\mathbf{x}}$ is not a KKT point. We shall exhibit a sequence $\{(\mathbf{x}_k, \mathbf{d}_k)\}_{\mathcal{K}''}$ satisfying conditions 1 through 4 of Lemma 10.2.6, which is impossible.

Let $\{\mathbf{d}_k\}_{\mathcal{K}}$ be the sequence of directions associated with $\{\mathbf{x}_k\}_{\mathcal{K}}$. Note that $\mathbf{d}_k$ is defined by (10.46) through (10.55) at $\mathbf{x}_k$. Letting I_k be the set denoting the indices of the m largest components of $\mathbf{x}_k$ used to compute $\mathbf{d}_k$, then there exists $\mathcal{K}' \subseteq \mathcal{K}$ such that $I_k = \hat{I}$ for each $k \in \mathcal{K}'$, where $\hat{I}$ is the set denoting the indices of the m largest components of $\hat{\mathbf{x}}$. Furthermore, there exists $\mathcal{K}'' \subseteq \mathcal{K}'$ such that $\mathbf{d}_k$ is given either by (10.53) and (10.55) for all $k \in \mathcal{K}''$ or (10.54) and (10.55) for all $k \in \mathcal{K}''$. In the first case, let $\hat{\mathbf{d}}$ be obtained from (10.46), (10.47), (10.48), (10.49), (10.51), (10.53), and (10.55) at $\hat{\mathbf{x}}$, and in the latter case, let $\hat{\mathbf{d}}$ be obtained from (10.46), (10.47), (10.48), (10.50), (10.52),

(10.54), and (10.55) at $\hat{\mathbf{x}}$. In either case, $\mathbf{d}_k = \hat{\mathbf{d}}$ for $k \in \mathcal{K}''$. By continuous differentiability of f, note that $\hat{\mathbf{d}}$ would have been obtained by applying (10.46) through (10.55) at $\hat{\mathbf{x}}$. By assumption, $\hat{\mathbf{x}}$ is not a KKT point, and, hence, $\hat{\mathbf{d}} \neq \mathbf{0}$ and $\nabla f(\hat{\mathbf{x}})^t\mathbf{d} < 0$. To summarize, we have exhibited a sequence $\{(\mathbf{x}_k, \mathbf{d}_k)\}_{\mathcal{K}''}$ satisfying conditions 1, 2, and 4 of Lemma 10.2.6. To complete the proof, we need to show that part 3 also holds true.

Note that $\mathbf{d}_k = \hat{\mathbf{d}}$ for $k \in \mathcal{K}''$. If $\hat{\mathbf{d}} \geq \mathbf{0}$, then $\mathbf{x}_k + \lambda\hat{\mathbf{d}} \geq \mathbf{0}$ for all $\lambda \in [0, \infty)$. If $\hat{\mathbf{d}} \ngeq \mathbf{0}$, and since $\hat{\mathbf{d}}$ is a feasible direction at $\hat{\mathbf{x}}$, then $\hat{\mathbf{x}} + \lambda\hat{\mathbf{d}} \geq \mathbf{0}$ for $\lambda \in [0, 2\delta]$, where $2\delta = \text{minimum}\,\{-\hat{x}_i/\hat{d}_i : \hat{d}_i < 0\}$. Since $x_{ik} \to \hat{x}_i$ and $\mathbf{d}_k = \hat{\mathbf{d}}$, then $\delta_k = \text{minimum}\,\{-x_{ik}/d_{ik} : d_{ik} < 0\} \geq \delta$ for all sufficiently large k in $\mathcal{K}''$. From (10.56), it then follows that $\mathbf{x}_k + \lambda\mathbf{d}_k$ is feasible for all $\lambda \in [0, \delta]$ for large k in $\mathcal{K}''$. Thus, condition 3 of Lemma 10.2.6 holds true, and the proof is complete.

10.8 Effective First- and Second-Order Variants of the Reduced Gradient Method

In both the reduced gradient method of Wolfe and the convex–simplex method of Zangwill, we have seen how, given a feasible solution, we can partition the space into a set of basic variables $\mathbf{x}_B$ and a set of nonbasic variables $\mathbf{x}_N$, and then essentially project the problem onto the space of the nonbasic variables by substituting $\mathbf{x}_B = \mathbf{B}^{-1}\mathbf{b} - \mathbf{B}^{-1}\mathbf{N}\mathbf{x}_N$ (see Exercise 10.50). In this space, the problem under consideration becomes, treating $\mathbf{x}_B \geq \mathbf{0}$ as slack variables in the transformed constraints,

$$\text{Minimize} \quad \{F(\mathbf{x}_N) = f(\mathbf{B}^{-1}\mathbf{b} - \mathbf{B}^{-1}\mathbf{N}\mathbf{x}_N, \mathbf{x}_N) : \mathbf{B}^{-1}\mathbf{N}\mathbf{x}_N \leq \mathbf{B}^{-1}\mathbf{b}, \mathbf{x}_N \geq \mathbf{0}\} \qquad (10.57)$$

Note that

$$[\nabla F(\mathbf{x}_N)]^t = \left[\frac{\partial f}{\partial \mathbf{x}_N} + \frac{\partial f}{\partial \mathbf{x}_B}\frac{\partial \mathbf{x}_B}{\partial \mathbf{x}_N}\right]^t = \nabla_N f(\mathbf{x})^t - \nabla_B f(\mathbf{x})^t\mathbf{B}^{-1}\mathbf{N} = \mathbf{r}_N^t \qquad (10.58)$$

where $\mathbf{r}_N$ is the reduced gradient. Moreover, barring degeneracy, the only binding constraints are those nonnegativity constraints from the set $\mathbf{x}_N \geq \mathbf{0}$ for which $x_j = 0$ currently. Hence, the reduced gradient method constructs the direction-finding subproblem in the nonbasic variable space as follows, where J_N denotes the index set for the nonbasic variables:

$$\underset{\mathbf{d}_N}{\text{Minimize}} \quad \nabla F(\mathbf{x}_N)^t\mathbf{d}_N \equiv \mathbf{r}_N^t\mathbf{d}_N \equiv \sum_{j \in J_N} r_j d_j : \; -x_j|r_j| \leq d_j \leq |r_j| \quad \text{for all } j \in J_N\} \qquad (10.59)$$

Observe that the trivial solution to (10.59) is to let $d_j = |r_j| \equiv -r_j$ if $r_j \leq 0$, and to let $d_j = -x_j|r_j| \equiv -x_jr_j$ if $r_j > 0$. This gives $\mathbf{d}_N$ as in (10.43); and then, using $\mathbf{Ad} = \mathbf{Bd}_B + \mathbf{Nd}_N = \mathbf{0}$, we compute $\mathbf{d}_B = -\mathbf{B}^{-1}\mathbf{N}\mathbf{d}_N$ as in (10.44) to derive the direction of motion $\mathbf{d}^t = (\mathbf{d}_B^t, \mathbf{d}_N^t)$.

The convex–simplex method examines the same direction-finding problem (10.59) but permits only one component of $\mathbf{d}_N$ to be nonzero, namely, the one which has the largest absolute value in the solution of (10.59). This gives the scaled unit vector $\mathbf{d}_N$, and then $\mathbf{d}_B$ is calculated as before according to $\mathbf{d}_B = -\mathbf{B}^{-1}\mathbf{N}\mathbf{d}_N$ to produce $\mathbf{d}$. Hence, whereas the convex–simplex method changes only one component of $\mathbf{d}_N$, moving in a direction parallel to one of the polyhedron edges or axes in the nonbasic variable space, the reduced gradient method permits all components to change as desired according to (10.59). It turns out that while the former strategy is unduly restrictive, the latter strategy also results in a slow progress, as blocking occurs upon taking short steps because of the many components that are changing simultaneously.

Computationally, a compromise between the two foregoing extremes has been found to be beneficial. Toward this end, suppose that we further partition the $n - m$ variables $\mathbf{x}_N$ into $(\mathbf{x}_S, \mathbf{x}_{N'})$ and, accordingly, partition $\mathbf{N}$ into $(\mathbf{S}, \mathbf{N}')$. The variables $\mathbf{x}_S$, indexed by J_S, say, where $0 \leq |J_S| \equiv s \leq n - m$, are called *superbasic variables*, and are usually chosen as a subset of the variables $\mathbf{x}_N$ that are positive (or strictly between lower and upper bounds when both types of bounds are specified in the problem). The remaining variables $\mathbf{x}_{N'}$ are still referred to as *nonbasic variables*. The idea is, therefore, to hold the variables $\mathbf{x}_{N'}$ fixed and to permit the variables $\mathbf{x}_S$ to be the "driving force" in guiding the iterates toward improving feasible points, with the basic variables $\mathbf{x}_B$ following suit as usual. Hence, writing $\mathbf{d}^t = (\mathbf{d}_B^t, \mathbf{d}_S^t, \mathbf{d}_{N'}^t)$, we have $\mathbf{d}_{N'} = \mathbf{0}$; and, from $\mathbf{Ad} = \mathbf{0}$, we get $\mathbf{Bd}_B + \mathbf{Sd}_S = \mathbf{0}$, or $\mathbf{d}_B = -\mathbf{B}^{-1}\mathbf{Sd}_S$. Accordingly, we get

$$\mathbf{d} = \begin{bmatrix} \mathbf{d}_B \\ \mathbf{d}_S \\ \mathbf{d}_{N'} \end{bmatrix} = \begin{bmatrix} -\mathbf{B}^{-1}\mathbf{S} \\ \mathbf{I} \\ \mathbf{0} \end{bmatrix} \mathbf{d}_S \equiv \mathbf{Zd}_S \tag{10.60}$$

where $\mathbf{Z}$ is appropriately defined as an $n \times s$ matrix. Problem (10.59) then reduces to the following direction-finding problem:

$$\begin{aligned} \text{Minimize} \quad & \{\nabla f(\mathbf{x})^t\mathbf{d} \equiv \nabla f(\mathbf{x})^t\mathbf{Zd}_S = [\nabla_S f(\mathbf{x})^t - \nabla_B f(\mathbf{x})^t\mathbf{B}^{-1}\mathbf{S}]\mathbf{d}_S \\ & \quad \equiv \mathbf{r}_S^t\mathbf{d}_S = \sum_{j \in J_S} r_j d_j\} \\ \text{subject to} \quad & -x_j|r_j| \leq d_j \leq |r_j| \qquad \text{for all } j \in J_S \end{aligned} \tag{10.61}$$

Similar to Problem (10.59), the solution to (10.61) yields $d_j = |r_j| = -r_j$ if $r_j \leq 0$, and $d_j = -x_j|r_j| = -x_j r_j$ if $r_j > 0$, for all $j \in J_S$. This gives $\mathbf{d}_S$, and then we obtain $\mathbf{d}$ from (10.60). Note that for the reduced gradient method, we had $\mathbf{S} \equiv \mathbf{N}$, that is, $s = n - m$, whereas for the convex-simplex method, we had $s = 1$.

A recommended practical implementation using the foregoing concept proceeds as follows. (The commercial package MINOS adopts this strategy.) To initialize, based on the magnitude of the components $d_j, j \in J_N$, in the solution to (10.59), some s components $\mathbf{d}_S$ of $\mathbf{d}$ are permitted to change *independently*. (MINOS simply uses the bounds $-|r_j| \leq d_j \leq |r_j|$, $j \in J_N$, in this problem.) This results in some set of s (positive) superbasic variables. The idea now is to execute the reduced gradient method in the space of the $(\mathbf{x}_B, \mathbf{x}_S)$ variables, holding $\mathbf{x}_{N'}$ fixed, and using (10.61) as the direction-finding problem. Accordingly, this technique is sometimes referred to as a *suboptimization strategy*. However, during these iterations, if any component of $\mathbf{x}_B$ or $\mathbf{x}_S$ hits its bound of zero, it is transferred into the nonbasic variable set. Also, a pivot operation is performed only when a basic variable blocks at its bound of zero. (Hence, the method does not necessarily maintain the m most positive components as basic.) Upon pivoting, the basic variable is exchanged for a superbasic variable to give a revised basis, and the leaving basic variable is transferred into the nonbasic set. Noting that we always have $\mathbf{x}_S > \mathbf{0}$, this process continues until either $J_S = \emptyset$ $(s = 0)$, or $\|\mathbf{r}_S\| \leq \varepsilon$, where $\varepsilon > 0$ is some tolerance value. At this point, the procedure enters a *pricing* phase in which the entire vector $\mathbf{r}_N$ is computed. If the KKT conditions are satisfied within an acceptable tolerance, then the procedure stops. Otherwise, an additional variable, or a set of (significantly enterable) additional variables under the option of *multiple pricing*, are transferred from the nonbasic into the superbasic variable set, and the procedure continues. Because of the

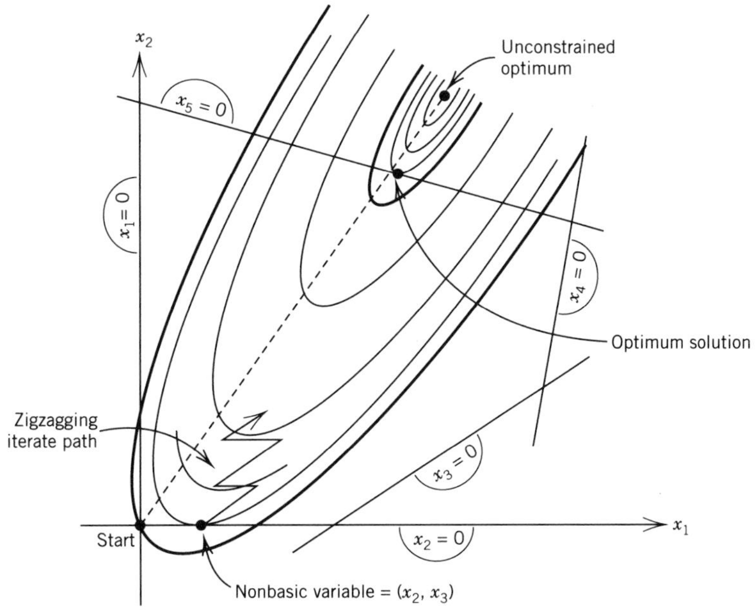

Figure 10.21 Zigzagging of the convex-simplex method.

suboptimization feature, this strategy turns out to be computationally desirable, particularly for large-scale problems that contain several more variables than constraints.

Second-Order Functional Approximations

The direction-finding problem (10.61) adopts a linear approximation to the objective function f. As we know, with steep ridged contours of f, this can be prone to a slow zigzagging convergence behavior. Figure 10.21 illustrates this phenomenon in the context of the convex simplex method ($s = 1$). The reduced gradient method ($s = 2$) would zigzag in a similar fashion, behaving like an unconstrained steepest descent method until some constraint blocks this path (see Exercise 10.52). On the other hand, if we were to adopt a second-order approximation for f in the direction-finding problem, then we can hope to accelerate the convergence behavior.

For example, if the function illustrated in Figure 10.21 was itself quadratic, then such a direction at the origin (using $s = 2$) would point toward its unconstrained minimum as shown by the dashed line in Figure 10.21. This would directly lead us to the point where this dashed line intersects the plane $x_5 = 0$, whence, with $s = 1$ now (x_5 being nonbasic), the next iteration would converge to the optimum (see Exercise 10.52).

The development of such a quadratic direction-finding problem is straightforward. At the current point $\mathbf{x}$, we minimize a second-order approximation to $f(\mathbf{x} + \mathbf{d})$ given by $f(\mathbf{x}) + \nabla f(\mathbf{x})^t\mathbf{d} + \frac{1}{2}\mathbf{d}^t\mathbf{H}(\mathbf{x})\mathbf{d}$ over the linear manifold $\mathbf{Ad} = \mathbf{0}$, where $\mathbf{d}^t = (\mathbf{d}_B^t, \mathbf{d}_S^t, \mathbf{d}_{N'}^t)$, with $\mathbf{d}_{N'} = \mathbf{0}$. This gives $\mathbf{d} = \mathbf{Z}\mathbf{d}_S$ as in (10.60), and therefore the direction-finding problem is given as follows, where we have used (10.61) to write $\nabla f(\mathbf{x})^t\mathbf{d} = \nabla f(\mathbf{x})^t\mathbf{Z}\mathbf{d}_S = \mathbf{r}_S^t\mathbf{d}_S$.

$$\text{Minimize} \quad \{\mathbf{r}_S^t\mathbf{d}_S + \tfrac{1}{2}\mathbf{d}_S^t[\mathbf{Z}^t\mathbf{H}(\mathbf{x})\mathbf{Z}]\mathbf{d}_S : \mathbf{d}_S \in E_s\} \tag{10.62}$$

Note that (10.62) represents an unconstrained minimization of the quadratic approximation to the objective function projected onto the space of the superbasic direction components. Accordingly, the $s \times s$ matrix $\mathbf{Z}^t\mathbf{H}(\mathbf{x})\mathbf{Z}$ is called the *projected Hessian* matrix and can be dense even if $\mathbf{H}$ is sparse. However, hopefully, s is small (see Exercise 10.51). Setting the gradient of the objective function in (10.62) equal to zero, we get

$$[\mathbf{Z}^t\mathbf{H}(\mathbf{x})\mathbf{Z}]\mathbf{d}_S = -\mathbf{r}_S \tag{10.63}$$

Note that $\mathbf{d}_S = \mathbf{0}$ solves (10.63) if and only if $\mathbf{r}_S = \mathbf{0}$. Otherwise, assuming that $\mathbf{Z}^t\mathbf{H}(\mathbf{x})\mathbf{Z}$ is positive definite (which would be the case, for example, if $\mathbf{H}(\mathbf{x})$ is positive definite, since $\mathbf{Z}$ has full column rank), we have $\mathbf{d}_S = -[\mathbf{Z}^t\mathbf{H}(\mathbf{x})\mathbf{Z}]^{-1}\,\mathbf{r}_S \neq \mathbf{0}$; and, moreover, from (10.61) and (10.63), $\nabla f(\mathbf{x})^t\mathbf{d} = \nabla f(\mathbf{x})^t\mathbf{Z}\mathbf{d}_S = \mathbf{r}_S^t\mathbf{d}_S = -\mathbf{d}_S^t[\mathbf{Z}^t\mathbf{H}(\mathbf{x})\mathbf{Z}]\mathbf{d}_S < 0$, so that $\mathbf{d} = \mathbf{Z}\mathbf{d}_S$ is an improving feasible direction. Using this Newton-based direction, we can now perform a line search and proceed using the above suboptimization scheme, with (10.62) replacing (10.61) for the direction-finding step.

Practically, even if the Hessian $\mathbf{H}$ is available and is positive definite, we would most likely not be able to afford to use it in exactly the above fashion. Typically, one maintains a positive definite approximation $\mathbf{B}$ to the projected Hessian $\mathbf{Z}^t\mathbf{H}(\mathbf{x})\mathbf{Z}$ that is updated from one iteration to the next using a quasi-Newton scheme. Note that $\mathbf{H}(\mathbf{x})$ or $\mathbf{Z}^t\mathbf{H}(\mathbf{x})\mathbf{Z}$ are never actually computed, and only a Cholesky factorization $\mathbf{L}\mathbf{L}^t$ of the foregoing quasi-Newton approximation is maintained, while accounting for the variation in the dimension of the superbasic variables. (See Murtagh and Saunders [1978].) Also, $\mathbf{Z}$ is never computed but, rather, an $\mathbf{LU}$ factorization of $\mathbf{B}$ is adopted. This factored form of $\mathbf{B}$ is used in the solution of the system $\boldsymbol{\pi}\mathbf{B} = \nabla_B f(\mathbf{x})^t$, from which $\mathbf{r}_S$ is computed via $\mathbf{r}_S^t = \nabla_S f(\mathbf{x})^t - \boldsymbol{\pi}\mathbf{S}$ as in (10.61), as well as for the solution of $\mathbf{d}_B$ from the system $\mathbf{B}\mathbf{d}_B = -\mathbf{S}\mathbf{d}_S$ as in (10.60), once $\mathbf{d}_S$ is determined.

For problems in which s can get fairly large (≥ 200, say), even a quasi-Newton approach becomes prohibitive. In such a case, a conjugate gradient approach becomes indispensable. Here, the conjugate gradient scheme is applied directly to the projected problem of minimizing $F(\mathbf{d}_S) \equiv f(\mathbf{x} + \mathbf{Z}\mathbf{d}_S)$. Note that in this projected space, $\nabla F(\mathbf{d}_S) = (\partial f/\partial \mathbf{x}_B)(\partial \mathbf{x}_B/\partial \mathbf{d}_S) + (\partial f/\partial \mathbf{x}_S)(\partial \mathbf{x}_S/\partial \mathbf{d}_S) + (\partial f/\partial \mathbf{x}_{N'})(\partial \mathbf{x}_{N'}/\partial \mathbf{d}_S) = [-\nabla_B f(\mathbf{x})^t\mathbf{B}^{-1}\mathbf{S} + \nabla_S f(\mathbf{x})^t\mathbf{I} + \mathbf{0}]^t = \mathbf{r}_S^t$. Hence, the direction $\mathbf{d}_S$ is taken as $-\mathbf{r}_S + \alpha\mathbf{d}_S'$, where $\mathbf{d}_S'$ is the previous direction and α is a multiplier determined by the particular conjugate gradient scheme. Under appropriate conditions, either the quasi-Newton or the conjugate gradient approach leads to a superlinearly convergent process.

Exercises

10.1 For each of the following cases, give a suitable characterization of the set of feasible directions at a point $\mathbf{x} \in S$:

$$S = \{\mathbf{x} : \mathbf{A}\mathbf{x} = \mathbf{b}, \mathbf{x} \geq \mathbf{0}\}$$
$$S = \{\mathbf{x} : \mathbf{A}\mathbf{x} \leq \mathbf{b}, \mathbf{E}\mathbf{x} = \mathbf{e}, \mathbf{x} \geq \mathbf{0}\}$$
$$S = \{\mathbf{x} : \mathbf{A}\mathbf{x} \geq \mathbf{b}, \mathbf{x} \geq \mathbf{0}\}$$

10.2 Consider the problem to minimize $f(\mathbf{x})$ subject to $g_i(\mathbf{x}) \leq 0$ for $i = 1, \ldots, m$. Suppose that $\mathbf{x}$ is a feasible point with $g_i(\mathbf{x}) = 0$ for $i \in I$. Furthermore, suppose that g_i is pseudoconcave at $\mathbf{x}$ for each $i \in I$. Show that the following problem produces an improving feasible direction or concludes that $\mathbf{x}$ is a KKT point:

$$\begin{aligned} \text{Minimize} \quad & \nabla f(\mathbf{x})^t\mathbf{d} \\ \text{subject to} \quad & \nabla g_i(\mathbf{x})^t\mathbf{d} \leq 0 \quad \text{for } i \in I \\ & -1 \leq d_j \leq 1 \quad \text{for } j = 1, \ldots, n \end{aligned}$$

10.3 Consider the problem to minimize $f(\mathbf{x})$ subject to $g_i(\mathbf{x}) \le 0$ for $i = 1, \ldots, m$. Let $\hat{\mathbf{x}}$ be a feasible solution, and let $I = \{i: g_i(\hat{\mathbf{x}}) = 0\}$. Let $(\hat{z}, \hat{\mathbf{d}})$ be an optimal solution to the following problem:

$$\begin{array}{lll} \text{Minimize} & z & \\ \text{subject to} & \nabla f(\hat{\mathbf{x}})\mathbf{d} \le z & \\ & \nabla g_i(\hat{\mathbf{x}})^t\mathbf{d} \le z & \text{for } i \in I \\ & d_j \ge -1 & \text{if } \dfrac{\partial f(\hat{\mathbf{x}})}{\partial x_j} > 0 \\ & d_j \le 1 & \text{if } \dfrac{\partial f(\hat{\mathbf{x}})}{\partial x_j} < 0 \end{array}$$

a. Show that $\hat{z} = 0$ if and only if $\hat{\mathbf{x}}$ is a Fritz John point.
b. Show that, if $\hat{z} < 0$, then $\hat{\mathbf{d}}$ is an improving feasible direction.
c. Show how one of the binding constraints, instead of the objective function, could be used to bound the components of the vector $\mathbf{d}$.

10.4 In Section 10.1, in reference to Zoutendijk's method for linear constraints, we described several normalization constraints such as $\mathbf{d}^t\mathbf{d} \le 1$, $-1 \le d_j \le 1$ for $j = 1, \ldots, n$, and $\nabla f(\mathbf{x}_k)^t\mathbf{d} \ge -1$. Show that each of the following normalization constraints could be used instead:

a. $\sum_{j=1}^{n} |d_j| \le 1$
b. $\underset{1 \le j \le n}{\text{maximum}} |d_j| \le 1$
c. $\mathbf{A}(\mathbf{x}_k + \mathbf{d}) \le \mathbf{b}$, provided the set $\{\mathbf{x}: \mathbf{A}\mathbf{x} \le \mathbf{b}\}$ is bounded
d. $d_j \ge -1$ if $\dfrac{\partial f(\mathbf{x}_k)}{\partial x_j} > 0$ and $d_j \le 1$ if $\dfrac{\partial f(\mathbf{x}_k)}{\partial x_j} < 0$

10.5 Consider the following problem:

$$\begin{array}{ll} \text{Minimize} & (x_1 - 2)^2 + (x_2 - 1)^2 \\ \text{subject to} & x_1^2 - x_2 \le 0 \\ & x_1 - 2x_2 + 1 = 0 \end{array}$$

Starting from $\mathbf{x} = (1, 1)^t$, solve the problem by Zoutendijk's procedure using the following two normalization methods:

a. $|d_j| \le 1$ for $j = 1, 2$
b. $\mathbf{d}^t\mathbf{d} \le 1$

10.6 Solve the following problem by Zoutendijk's method for linear constraints:

$$\begin{array}{ll} \text{Minimize} & x_1^2 + x_1x_2 + 2x_2^2 - 6x_1 - 2x_2 - 12x_3 \\ \text{subject to} & x_1 + x_2 + x_3 = 2 \\ & -x_1 + 2x_2 \le 3 \\ & x_1, x_2, x_3 \ge 0 \end{array}$$

10.7 Solve the following problem using Zoutendijk's method for nonlinear constraints:

$$\begin{array}{ll} \text{Minimize} & x_1^2 + x_1x_2 + 2x_2^2 - 6x_1 - 2x_2 - 12x_3 \\ \text{subject to} & 2x_1^2 + x_2^2 \le 15 \\ & -x_1 + 2x_2 + x_3 \le 3 \\ & x_1, x_2, x_3 \ge 0 \end{array}$$

10.8 Consider the following problem with linear constraints and nonlinear inequality constraints:

$$\begin{array}{lll} \text{Minimize} & f(\mathbf{x}) & \\ \text{subject to} & g_i(\mathbf{x}) \le 0 & \text{for } i = 1, \ldots, m \\ & \mathbf{A}\mathbf{x} \le \mathbf{b} & \\ & \mathbf{E}\mathbf{x} = \mathbf{e} & \end{array}$$

Let $\mathbf{x}$ be a feasible point, and let $I = \{i: g_i(\mathbf{x}) = 0\}$. Furthermore, suppose that $\mathbf{A}_1\mathbf{x} = \mathbf{b}_1$ and $\mathbf{A}_2\mathbf{x} < \mathbf{b}_2$, where $\mathbf{A}^t = [\mathbf{A}_1^t, \mathbf{A}_2^t]$ and $\mathbf{b}^t = (\mathbf{b}_1^t, \mathbf{b}_2^t)$.

a. Show that the following linear program provides an improving feasible direction or concludes that $\mathbf{x}$ is a Fritz John point:

$$\begin{array}{lll} \text{Minimize} & z & \\ \text{subject to} & \nabla f(\mathbf{x})^t\mathbf{d} - z \le 0 & \\ & \nabla g_i(\mathbf{x})^t\mathbf{d} - z \le 0 & \text{for } i \in I \\ & \mathbf{A}_1\mathbf{d} \le \mathbf{0} & \\ & \mathbf{E}\mathbf{d} = \mathbf{0} & \end{array}$$

b. Using this approach, solve the problem in Example 10.1.8, and compare the trajectories generated in both cases.

10.9 In Zoutendijk's procedure, the following problem is solved to generate an improving feasible direction, where $I = \{i : g_i(\mathbf{x}) = 0\}$:

$$\begin{array}{lll} \text{Minimize} & z & \\ \text{subject to} & \nabla f(\mathbf{x})^t\mathbf{d} \le z & \\ & \nabla g_i(\mathbf{x})^t\mathbf{d} \le z & \text{for } i \in I \\ & -1 \le d_j \le 1 & \text{for } j = 1, \ldots, n \end{array}$$

a. Show that the method cannot accommodate nonlinear equality constraints of the form $h_i(\mathbf{x}) = 0$ by replacing each constraint by $h_i(\mathbf{x}) \le 0$ and $-h_i(\mathbf{x}) \le 0$.

b. One way to handle a constraint of the form $h_i(\mathbf{x}) = 0$ is to first replace it by the two constraints $h_i(\mathbf{x}) \le \varepsilon$ and $-h_i(\mathbf{x}) \le \varepsilon$, where $\varepsilon > 0$ is a small scalar, and then to apply the above direction-finding process. Use this method to solve the following problem:

$$\begin{array}{ll} \text{Minimize} & x_1^3 + 2x_2^2x_3 + 2x_3 \\ \text{subject to} & x_1^2 + x_2 + x_3^2 = 4 \\ & x_1^2 - x_2 + 2x_3 \le 2 \end{array}$$

10.10 Consider the following problem, and let $\hat{\mathbf{x}}$ be a feasible point with $g_i(\hat{\mathbf{x}}) = 0$ for $i \in I$:

$$\begin{array}{lll} \text{Minimize} & f(\mathbf{x}) & \\ \text{subject to} & g_i(\mathbf{x}) \le 0 & \text{for } i = 1, \ldots, m \\ & h_i(\mathbf{x}) = 0 & \text{for } i = 1, \ldots, l \end{array}$$

a. Show that $\hat{\mathbf{x}}$ is a KKT point if and only if the optimal objective of the following problem is zero:

$$\begin{array}{lll} \text{Minimize} & \nabla f(\hat{\mathbf{x}})^t\mathbf{d} & \\ \text{subject to} & \nabla g_i(\hat{\mathbf{x}})^t\mathbf{d} \le 0 & \text{for } i \in I \\ & \nabla h_i(\hat{\mathbf{x}})^t\mathbf{d} = 0 & \text{for } j = 1, \ldots, l \\ & -1 \le d_j \le 1 & \text{for } j = 1, \ldots, n \end{array}$$

b. Let $\hat{\mathbf{d}}$ be an optimal solution to the problem in part a. If $\nabla f(\hat{\mathbf{x}})^t\hat{\mathbf{d}} < 0$, then $\hat{\mathbf{d}}$ is an improving direction. Even though $\hat{\mathbf{d}}$ may not be a feasible direction, it is at least tangential to the feasible region at $\hat{\mathbf{x}}$. The following procedure is proposed. Fix $\delta > 0$ and let $\hat{\lambda}$ be an optimal solution to the problem to minimize $f(\hat{\mathbf{x}} + \lambda\hat{\mathbf{d}})$ subject to $0 \le \lambda \le \delta$. Let $\bar{\mathbf{x}} = \hat{\mathbf{x}} + \hat{\lambda}\hat{\mathbf{d}}$. Starting from $\bar{\mathbf{x}}$, a correction move is used to obtain a feasible point. This could be done in several ways.

1. Move along $\mathbf{d} = -\mathbf{A}[\mathbf{A}^t\mathbf{A}]^{-1}\mathbf{F}(\bar{\mathbf{x}})$, where $\mathbf{F}$ is a vector function whose components are h_i for $i = 1, \ldots, l$, and g_i for i such that $g_i(\hat{\mathbf{x}}) \ge 0$, and $\mathbf{A}$ is the matrix whose rows are the transposes of the gradients of the constraints in $\mathbf{F}$.
2. Use a penalty function scheme to minimize total infeasibility starting from $\bar{\mathbf{x}}$.

Using each of the above approaches, solve the problem given in part b of Exercise 10.9.

10.11 Consider the problem to minimize $f(\mathbf{x})$ subject to $\mathbf{Ax} \le \mathbf{b}$, where the region $\{\mathbf{x} : \mathbf{Ax} \le \mathbf{b}\}$ is bounded. Suppose that $\mathbf{x}_k$ is a feasible point, and let $\mathbf{y}_k$ solve the problem to minimize $\nabla f(\mathbf{x}_k)^t\mathbf{y}$ subject to $\mathbf{Ay} \le \mathbf{b}$. Let λ_k be an optimal solution to the problem to minimize $f[\lambda\mathbf{x}_k + (1 - \lambda)\mathbf{y}_k]$ subject to $0 \le \lambda \le 1$, and let $\mathbf{x}_{k+1} = \lambda_k\mathbf{x}_k + (1 - \lambda_k)\mathbf{y}_k$.

a. Show that this procedure can be interpreted as a feasible direction method. Furthermore, show that, in general, the direction $\mathbf{y}_k - \mathbf{x}_k$ cannot be obtained by Problems P1, P2, or P3 discussed in Section 10.1. Discuss any advantages or disadvantages of the above procedure.

b. Solve the problem given in Example 10.1.5 by the above method.
c. Describe the above procedure as the composition of a direction-finding map and a line search map. Using Theorem 7.3.2, show that the composite map is closed. Then, using Theorem 7.2.3, show that the algorithm converges to a KKT point.
d. Compare this method with the successive linear programming algorithm presented in Section 10.3.

(The above procedure is credited to Frank and Wolfe [1956].)

10.12 Consider the following problem with lower and upper bounds on the variables:

Minimize $f(\mathbf{x})$

subject to $a_j \le x_j \le b_j$ for $j = 1, \ldots, n$

Let $\mathbf{x}$ be a feasible point. Let $\nabla_j = \partial f(\mathbf{x})/\partial x_j$, and consider Zoutendijk's procedure for generating an improving feasible direction.

a. Show that an optimal solution to the direction-finding problem, using the normalization constraint $|d_j| \le 1$, is given by

$$d_j = \begin{cases} -1 & \text{if } x_j > a_j \text{ and } \nabla_j \ge 0 \\ 1 & \text{if } x_j < b_j \text{ and } \nabla_j < 0 \\ 0 & \text{otherwise} \end{cases}$$

b. Show that an optimal solution to the direction-finding problem, using the normalization constraint $\mathbf{d}^t\mathbf{d} \le 1$, is given by

$$d_j = \begin{cases} \dfrac{-\nabla_j}{\left(\sum_{i \in I} \nabla_i^2\right)^{1/2}} & \text{if } j \in I \\ 0 & \text{if } j \notin I \end{cases}$$

where $I = \{j : x_j > a_j \text{ and } \nabla_j \ge 0, \text{ or else } x_j < b_j \text{ and } \nabla_j < 0\}$.

c. Using the methods in parts a and b, solve the following problem starting from the point $(-3, -4)$, and compare the trajectories obtained:

Minimize $2x_1^2 - x_1x_2 + 3x_2^2 - 3x_1 - 2x_2$

subject to $-3 \le x_1 \le 0$

$-4 \le x_2 \le 1$

d. Show that both the direction-finding maps in parts a and b are not closed.
e. Prove convergence or give a counterexample showing that the feasible direction algorithms using the direction-finding procedures discussed in parts a and b do not converge to a KKT point.

10.13 Solve the following problem by the Topkis-Veinott method:

Minimize $(1 - x_1)^2 - 10(x_2 - x_1^2)^2 + x_1^2 - 2x_1x_2 + e^{-x_1-x_2}$

subject to $x_1^2 + x_2^2 \le 16$

$(x_2 - x_1)^2 + x_1 \le 6$

$x_1 + x_2 \ge 2$

10.14 Consider the following problem:

Minimize $f(\mathbf{x})$

subject to $g_i(\mathbf{x}) \le 0$ for $i = 1, \ldots, m$

The following is a modification of the direction-finding problem of Topkis and Veinott if g_i is pseudoconcave:

Minimize $\nabla f(\mathbf{x})^t\mathbf{d}$

subject to $g_i(\mathbf{x}) + \nabla g_i(\mathbf{x})^t\mathbf{d} \le 0$ for $i = 1, \ldots, m$

$\mathbf{d}^t\mathbf{d} \le 1$

a. Show that $\mathbf{x}$ is a KKT point if and only if the optimal objective value is equal to zero.
b. Let $\mathbf{d}$ be an optimal solution and suppose that $\nabla f(\mathbf{x})^t\hat{\mathbf{d}} < 0$. Show that $\hat{\mathbf{d}}$ is an improving feasible direction.

c. Can you show convergence to a KKT point of the above modified Topkis and Veinott algorithm?

d. Repeat parts a through c if the normalization constraint is replaced by $-1 \le d_j \le 1$ for $j = 1, \ldots, n$.

e. Using the above approach, solve the problem in Example 10.1.5.

10.15 Consider the following problem with lower and upper bounds on the variables:

Minimize $f(\mathbf{x})$

subject to $a_j \le x_j \le b_j \quad$ for $j = 1, \ldots, n$

Let $\mathbf{x}$ be a feasible solution, let $\nabla_j = \partial f(\mathbf{x})/\partial x_j$, and consider the modified Topkis-Veinott method for generating an improving feasible direction described in Exercise 10.14.

a. Show that an optimal solution to the direction-finding problem using the normalization constraint $|d_j| \le 1$, is given by

$$d_j = \begin{cases} \text{maximum}\,\{a_j - x_j, -1\} & \text{if } \nabla_j \ge 0 \\ \text{minimum}\,\{b_j - x_j, 1\} & \text{if } \nabla_j < 0 \end{cases}$$

b. Show that an optimal solution to the direction-finding problem, using the normalization constraint $\mathbf{d}'\mathbf{d} \le \delta$, is given by

$$d_j = \begin{cases} \text{maximum}\,\{-\nabla_j/\|\nabla f(\mathbf{x})\|, a_j - x_j\} & \text{if } \nabla_j \ge 0 \\ \text{minimum}\,\{-\nabla_j/\|\nabla f(\mathbf{x})\|, b_j - x_j\} & \text{if } \nabla_j < 0 \end{cases}$$

where

$$\delta = \sum_{\nabla_j \ge 0} [\text{maximum}\,\{-\nabla_j/\|\nabla f(\mathbf{x})\|, a_j - x_j\}]^2 + \sum_{\nabla_j < 0} [\text{minimum}\,\{-\nabla_j/\|\nabla f(\mathbf{x})\|, b_j - x_j\}]^2$$

c. Solve the problem in part c of Exercise 10.12 by the methods in parts a and b above, and compare the trajectories obtained.

d. For both the direction-finding maps in parts a and b, show convergence of the method described above to a KKT point.

10.16 Consider the problem to minimize $f(\mathbf{x}) = \mathbf{c}'\mathbf{x} + \frac{1}{2}\mathbf{x}'\mathbf{H}\mathbf{x}$ subject to $\mathbf{A}\mathbf{x} \le \mathbf{b}$. At a point $\mathbf{x}_k$ in the interior of the feasible region, Zoutendijk's procedure of Section 10.1 generates a direction of movement by solving the problem to minimize $\nabla f(\mathbf{x}_k)'\mathbf{d}$ subject to $-1 \le d_j \le 1$ for $j = 1, \ldots, n$. In Chapter 8, we have learned that at interior points where we have an essentially unconstrained problem, conjugate direction methods are effective. The procedure discussed below combines a conjugate direction method with Zoutendijk's method of feasible directions.

Initialization Step Find an initial feasible solution $\mathbf{x}_1$ with $\mathbf{A}\mathbf{x}_1 \le \mathbf{b}$. Let $k = 1$ and go to the main step.

Main Step

1. Starting from $\mathbf{x}_k$, apply Zoutendijk's method yielding $\mathbf{z}$. If $\mathbf{A}\mathbf{z} < \mathbf{b}$, let $\mathbf{y}_1 = \mathbf{x}_k$, $\mathbf{y}_2 = \mathbf{z}$, $\hat{\mathbf{d}}_1 = \mathbf{y}_2 - \mathbf{y}_1$, $\nu = 2$ and go to step 2. Otherwise, let $\mathbf{x}_{k+1} = \mathbf{z}$, replace k by $k + 1$, and repeat step 1.
2. Let $\hat{\mathbf{d}}_\nu$ be an optimal solution to the following program:

   Minimize $\nabla f(\mathbf{y}_\nu)'\mathbf{d}$

   subject to $\hat{\mathbf{d}}_i'\mathbf{H}\mathbf{d} = 0 \quad$ for $i = 1, \ldots, \nu - 1$

   $-1 \le d_j \le 1 \quad$ for $j = 1, \ldots, \nu$

   Let λ_ν be an optimal solution to the following line search problem:

   Minimize $f(\mathbf{y}_\nu + \lambda\hat{\mathbf{d}}_\nu)$

   subject to $0 \le \lambda \le \lambda_{max}$

   where λ_{max} is determined according to (10.1). Let $\mathbf{y}_{\nu+1} = \mathbf{y}_\nu + \lambda_\nu\hat{\mathbf{d}}_\nu$. If $\mathbf{A}\mathbf{y}_{\nu+1} < \mathbf{b}$ and $\nu \le n - 1$, replace ν by $\nu + 1$, and repeat step 2. Otherwise, replace k by $k + 1$, let $\mathbf{x}_k = \mathbf{y}_{\nu+1}$, and go to step 1.

a. Solve the problem in Exercise 10.13 by the procedure discussed above.

b. Using the above procedure, solve the following problem credited to Kunzi, Krelle, and Oettli [1966], starting from the point (0, 0):

$$\begin{aligned} \text{Minimize} \quad & \tfrac{1}{2}x_1^2 + \tfrac{1}{2}x_2^2 - x_1 - 2x_2 \\ \text{subject to} \quad & 2x_1 + 3x_2 \le 6 \\ & x_1 + 4x_2 \le 5 \\ & x_1, x_2 \ge 0 \end{aligned}$$

c. Solve the problem in parts a and b by replacing the Zoutendijk procedure at step 1 of the above algorithm by the modified Topkis–Veinott algorithm discussed in Exercise 10.14.

d. Solve the problem in parts a and b by replacing Zoutendijk's procedure at step 1 of the above algorithm by the gradient projection method.

10.17 Consider Problem P to minimize $f(\mathbf{x})$ subject to $\mathbf{Ax} = \mathbf{b}$ and $\mathbf{x} \ge \mathbf{0}$. Suppose that $\bar{\mathbf{x}}$ is feasible with $\bar{x}_j = 0$ for $j \in J_0$ and $\bar{x}_j > 0$ for $j \in J_+$. Also, assume that the Hessian $\mathbf{H}(\bar{\mathbf{x}})$ is positive definite.

a. Construct a problem for finding a direction $\mathbf{d}$ which minimizes a second-order approximation of $f(\bar{\mathbf{x}} + \mathbf{d})$ over the set of feasible directions at $\bar{x}$, with $\|\mathbf{d}\|_\infty \le 1$.

b. Suppose that $\mathbf{d} = \mathbf{0}$ solves the problem of part a. Show that $\bar{\mathbf{x}}$ is then a KKT point for P.

c. Suppose that $\mathbf{d} = \mathbf{0}$ is *not* an optimum for the problem of part a. Show that the optimum to this problem then yields an improving feasible direction for P.

10.18 Using the proof of Theorem 4.4.2 and Theorem 9.3.1 (see also Exercise 9.12), construct a detailed proof of part a of Theorem 10.3.1

10.19 Consider the problem of Example 10.3.2. Solve the associated KKT conditions to obtain the optimal Lagrange multipliers and, hence, prescribe a suitable value of μ to use in algorithm PSLP. Also, find the eigenvalues and vectors of the Hessian of the objective function, along with the unconstrained minimum and, hence, sketch the contours of the objective function as in Figure 10.13*a*. State a suitable termination criterion for algorithm PSLP and, using the starting iteration solution of Example 10.3.2, continue the algorithm until this termination criterion is satisfied.

10.20 In view of Exercise 9.32, consider the linear programming problem P to minimize $\mathbf{c}^t\mathbf{x}$ subject to $\mathbf{Ax} = \mathbf{0}$, $\mathbf{e}^t\mathbf{x} = 1$, and $\mathbf{x} \ge \mathbf{0}$, where $\mathbf{A}$ is an $m \times n$ matrix of rank m and $\mathbf{e}$ is a vector of n 1's. Defining $\mathbf{Y} = \text{diag}\,\{y_1, \ldots, y_n\}$, this problem can be written as

$$\text{Minimize} \quad \{\mathbf{c}^t\mathbf{Y}^2\mathbf{e} : \mathbf{AY}^2\mathbf{e} = \mathbf{0} \quad \text{and} \quad \mathbf{e}^t\mathbf{Y}^2\mathbf{e} = 1\}, \quad \text{where } \mathbf{x} = \mathbf{Y}^2\mathbf{e}$$

Consider the following algorithm to solve this problem:

Initialization Select a feasible solution $\mathbf{x}_0 > \mathbf{0}$, put $k = 0$, let $\delta \in (0, \sqrt{(n-1)/n})$, and go to the main step.

Main Step Let $y_{kj} = \sqrt{x_{kj}}$ for $j = 1, \ldots, n$ and define $\mathbf{Y}_k = \text{diag}\,\{y_{k1}, \ldots, y_{kn}\}$. Solve the following subproblem (in $\mathbf{y}$):

$$SP\!: \quad \text{Minimize} \quad \{\mathbf{c}^t\mathbf{Y}_k\mathbf{y} : \mathbf{AY}_k\mathbf{y} = \mathbf{0},\ \mathbf{e}^t\mathbf{Y}_k\mathbf{y} = 1, \text{ and } \|\mathbf{Y}_k^{-1}(\mathbf{y} - \mathbf{y}_k)\| \le \delta\}$$

Let $\mathbf{y}_{k+1}$ solve SP. Put $\mathbf{x}'_{k+1} = \mathbf{Y}_k(2\mathbf{y}_{k+1} - \mathbf{y}_k)$, and let $\mathbf{x}_{k+1} = \mathbf{x}'_{k+1}/\mathbf{e}^t\mathbf{x}'_{k+1}$. Increment k by 1 and repeat the main step until a suitable convergence criterion holds.

a. Interpret the derivation of SP in light of an SLP approach to solve P. State a suitable termination criterion. Illustrate by solving the problem to minimize $-2x_1 + x_2 - x_3$ subject to $2x_1 + 2x_2 - 3x_3 = 0$, $x_1 + x_2 + x_3 = 1$, and $\mathbf{x} \ge \mathbf{0}$.

b. Note that, in the above linearization $\boldsymbol{\alpha}^t(\mathbf{y} - \mathbf{y}_k) = 0$ of the constraint $\mathbf{e}^t\mathbf{Y}^2\mathbf{e} = 1$, we have used $\boldsymbol{\alpha}$ equal to the constraint gradient. Suppose, instead, that we use $\boldsymbol{\alpha} = \mathbf{Y}_k^{-1}\mathbf{e}$. Solve the example of part a for the resulting subproblem. (Morshedi and Tapia [1987] show that the procedure of part b is equivalent to *Karmarkar's [1984] algorithm*, and that the procedure of part a is equivalent to the *affine scaling variant* of this algorithm.)

10.21 Derive analogues of the PSLP algorithm described in Section 10.3 by employing (a) a quadratic penalty function and (b) an augmented Lagrangian penalty function in lieu of the exact penalty function. Discuss the applicability, merits, and demerits of your derived procedures.

10.22 Relate the method of Roos and Vial [1988] as described in Exercise 9.33 to an SQP approach.

10.23 Let P: Minimize $\{f(\mathbf{x}): g_i(\mathbf{x}) \leq 0,\ i = 1, \ldots, m\}$ and consider the following quadratic programming direction-finding problem, where $\mathbf{B}_k$ is some positive definite approximation to the Hessian of the Lagrangian at $\mathbf{x} = \mathbf{x}_k$, and where μ is large enough as in Lemma 10.4.1.

$$\begin{aligned} \textit{QP:}\quad & \text{Minimize} \quad \nabla f(\mathbf{x}_k)^t\mathbf{d} + \tfrac{1}{2}\mathbf{d}^t\mathbf{B}_k\mathbf{d} + \mu \sum_{i=1}^{m} z_i \\ & \text{subject to} \quad z_i \geq g_i(\mathbf{x}_k) + \nabla g_i(\mathbf{x}_k)^t\mathbf{d} \qquad \text{for } i = 1, \ldots, m \\ & \qquad\qquad\quad z_1, \ldots, z_m \geq 0 \end{aligned}$$

a. Discuss, in regard to feasibility, the advantage of the above problem QP over (10.25). What is the accompanying disadvantage?

b. Let $\mathbf{d}_k$ solve QP with optimum Lagrange multipliers $\mathbf{u}_k$ associated with the first m constraints. If $\mathbf{d}_k = \mathbf{0}$, then do we have a KKT solution for P? Discuss.

c. Suppose that $\mathbf{d}_k \neq \mathbf{0}$ solves QP. Show that $\mathbf{d}_k$ is then a descent direction for the l_1 penalty function $F_E(\mathbf{x}) = f(\mathbf{x}) + \mu \sum_{i=1}^{m} \max\{0, g_i(\mathbf{x})\}$ at $\mathbf{x} = \mathbf{x}_k$.

d. Extend the above analysis to include equality constraints $h_i(\mathbf{x}) = 0,\ i = 1, \ldots, l$, by considering the following subproblem QP:

$$\begin{aligned} & \text{Minimize} \quad \nabla f(\mathbf{x}_k)^t\mathbf{d} + \tfrac{1}{2}\mathbf{d}^t\mathbf{B}_k\mathbf{d} + \mu\left[\sum_{i=1}^{m} y_i + \sum_{i=1}^{l} (z_i^+ + z_i^-)\right] \\ & \text{subject to} \quad y_i \geq g_i(\mathbf{x}_k) + \nabla g_i(\mathbf{x}_k)^t\mathbf{d} \qquad \text{for } i = 1, \ldots, m \\ & \qquad\qquad\quad z_i^+ - z_i^- = h_i(\mathbf{x}_k) + \nabla h_i(\mathbf{x}_k)^t\mathbf{d} \qquad \text{for } i = 1, \ldots, l \\ & \qquad\qquad\quad \mathbf{y} \geq \mathbf{0},\ \mathbf{z}^+ \geq \mathbf{0},\ \mathbf{z}^- \geq \mathbf{0} \end{aligned}$$

10.24 Consider the system (10.21) and assume that $\nabla^2 L(\mathbf{x}_k)$ is positive definite and that the Jacobian $\nabla\mathbf{h}(\mathbf{x}_k)$ has full row rank. Find an explicit closed-form solution $(\mathbf{d}, \mathbf{v})$ to this system.

10.25 Consider Problem P to minimize $x_1 + x_2$ subject to $x_1^2 + x_2^2 = 2$. Find an optimal primal and dual solution to this problem. Now, consider the quadratic program QP($\mathbf{x}_k$, $\mathbf{v}_k$) defined by (10.22) for any ($\mathbf{x}_k$, $\mathbf{v}_k$), but with the Hessian $\nabla^2 f(\mathbf{x}_k)$ of the objective function *incorrectly* replacing the Hessian $\nabla^2 L(\mathbf{x}_k)$ of the Lagrangian. Comment on the outcome of doing this for the given Problem P. Now, starting at the point $\mathbf{x} = (1, 1)^t$, apply the SQP approach to solve P using a suitable starting Lagrange multiplier $\mathbf{v}$. What happens if $\mathbf{v} = \mathbf{0}$ is chosen as a starting value?

10.26 Provide a detailed proof for Theorem 10.4.2, defining precisely the input and output quantities associated with each algorithmic map, and supporting all the arguments in the proof sketched in Section 10.4.

10.27 Referring to Example 10.4.3, complete its solution using algorithm RSQP. Comment on its convergence behavior. Also, verify the optimality of the iterate $\mathbf{x}_2$ generated by algorithm MSQP using the corresponding quadratic programming subproblem.

10.28 Consider the constraints $\mathbf{Ax} \leq \mathbf{b}$ and let $\mathbf{P} = \mathbf{I} - \mathbf{A}_1^t(\mathbf{A}_1\mathbf{A}_1^t)^{-1}\mathbf{A}_1$, where $\mathbf{A}_1$ represents the gradients of the binding constraints at a given feasible point $\hat{\mathbf{x}}$. What are the implications and geometric interpretation of the following statements?

a. $\mathbf{P}\nabla f(\hat{\mathbf{x}}) = \mathbf{0}$

b. $\mathbf{P}\nabla f(\hat{\mathbf{x}}) = \nabla f(\hat{\mathbf{x}})$

c. $\mathbf{P}\nabla f(\hat{\mathbf{x}}) \neq \mathbf{0}$

10.29 Solve the following problem by the gradient projection method:

$$\begin{aligned} \text{Minimize} \quad & (1 - x_1)^2 - 10(x_2 - x_1)^2 + x_1^2 - 2x_1x_2 + e^{-x_1-x_2} \\ \text{subject to} \quad & 2x_1 + 5x_2 \leq 25 \\ & -x_1 + 2x_2 \leq 8 \\ & x_1, x_2 \geq 0 \end{aligned}$$

10.30 Consider the following problem, where $\mathbf{A}_1$ is a $\nu \times n$ matrix:

Minimize $\quad \|-\nabla f(\mathbf{x}) - \mathbf{d}\|^2$

subject to $\quad \mathbf{A}_1\mathbf{d} = \mathbf{0}$

a. Show that $\bar{\mathbf{d}}$ is an optimal solution to the problem if and only if $\bar{\mathbf{d}}$ is the projection of $-\nabla f(\mathbf{x})$ onto the nullspace of $\mathbf{A}_1$.
(*Hint:* The KKT conditions reduce to $-\nabla f(\mathbf{x}) = \bar{\mathbf{d}} - \mathbf{A}_1^t\mathbf{u}$, $\mathbf{A}_1\bar{\mathbf{d}} = \mathbf{0}$. Note that $\bar{\mathbf{d}} \in L = \{\mathbf{y} : \mathbf{A}_1\mathbf{y} = \mathbf{0}\}$ and that $-\mathbf{A}_1^t\mathbf{u} \in L^\perp = \{\mathbf{A}_1^t\mathbf{v} : \mathbf{v} \in E_\nu\}$.)

b. Suggest a suitable method for solving the KKT system.
(*Hint:* Multiply $-\nabla f(\mathbf{x}) = \bar{\mathbf{d}} - \mathbf{A}_1^t\mathbf{u}$ by $\mathbf{A}_1$. Noting that $\mathbf{A}_1\bar{\mathbf{d}} = \mathbf{0}$, obtain a formula for $\mathbf{u}$, and then substitute to obtain $\bar{\mathbf{d}} = -[\mathbf{I} - \mathbf{A}_1^t(\mathbf{A}_1\mathbf{A}_1^t)^{-1}\mathbf{A}_1]\nabla f(\mathbf{x})$.)

c. Find an optimal solution to the problem, if $\nabla f(\mathbf{x}) = (1, -2, 3)^t$ and $\mathbf{A}_1 = \begin{bmatrix} 1 & 2 & -3 \\ 2 & 1 & 1 \end{bmatrix}$.

10.31 Consider the following problem, where $\mathbf{A}_1$ is a $\nu \times n$ matrix:

Minimize $\quad \nabla f(\mathbf{x})^t\mathbf{d}$

subject to $\quad \mathbf{A}_1\mathbf{d} = \mathbf{0}$

$\quad\quad \mathbf{d}^t\mathbf{d} \le 1$

The KKT conditions are both necessary and sufficient for optimality since a suitable constraint qualification holds at each feasible solution (see Exercise 5.15). In particular, $\bar{\mathbf{d}}$ is an optimal solution if and only if there exist $\mathbf{u}$ and μ such that

$$-\nabla f(\mathbf{x}) = 2\mu\bar{\mathbf{d}} + \mathbf{A}_1^t\mathbf{u}$$

$$A_1\bar{\mathbf{d}} = \mathbf{0}, \qquad \bar{\mathbf{d}}^t\bar{\mathbf{d}} \le 1$$

$$(\bar{\mathbf{d}}^t\bar{\mathbf{d}} - 1)\mu = 0, \qquad \mu \ge 0$$

a. Show that $\mu = 0$ if and only if $-\nabla f(\mathbf{x})$ is in the range-space of $\mathbf{A}_1^t$ or, equivalently, if and only if the projection of $-\nabla f(\mathbf{x})$ onto the nullspace of $\mathbf{A}_1$ is the zero vector. In this case, $\nabla f(\mathbf{x})^t\bar{\mathbf{d}} = 0$.

b. Show that if $\mu > 0$, then an optimal solution $\bar{\mathbf{d}}$ to the above problem points in the direction of the projection of $-\nabla f(\mathbf{x})$ onto the nullspace of $\mathbf{A}_1$.

c. Show that a solution to the KKT system stated above could be immediately obtained as follows. Let $\mathbf{u} = -(\mathbf{A}_1\mathbf{A}_1^t)^{-1}\mathbf{A}_1\nabla f(\mathbf{x})$, and let $\mathbf{d} = -[\mathbf{I} - \mathbf{A}_1^t(\mathbf{A}_1\mathbf{A}_1^t)^{-1}\mathbf{A}_1]\nabla f(\mathbf{x})$. If $\mathbf{d} = \mathbf{0}$, let $\mu = 0$ and $\bar{\mathbf{d}} = \mathbf{0}$. If $\mathbf{d} \ne \mathbf{0}$, let $\mu = \|\mathbf{d}\|/2$ and $\bar{\mathbf{d}} = \mathbf{d}/\|\mathbf{d}\|$.

d. Now consider the problem to minimize $f(\mathbf{x})$ subject to $\mathbf{A}\mathbf{x} \le \mathbf{b}$. Let $\mathbf{x}$ be a feasible solution such that $\mathbf{A}_1\mathbf{x} = \mathbf{b}_1$ and $\mathbf{A}_2\mathbf{x} < \mathbf{b}_2$, where $\mathbf{A}^t = (\mathbf{A}_1^t, \mathbf{A}_2^t)$ and $\mathbf{b}^t = (\mathbf{b}_1^t, \mathbf{b}_2^t)$. Show that $\mathbf{x}$ is a KKT point of the problem to minimize $f(\mathbf{x})$ subject to $\mathbf{A}\mathbf{x} \le \mathbf{b}$ if $\mu = 0$ and $\mathbf{u} \ge \mathbf{0}$.

e. Show that if $\mu = 0$ and $\mathbf{u} \not\ge \mathbf{0}$, then the gradient projection method discussed in Section 10.5 continues by choosing a negative component u_j of $\mathbf{u}$, deleting the associated row from $\mathbf{A}_1$ producing $\mathbf{A}_1'$, and resolving the direction-finding problem to minimize $\nabla f(\mathbf{x})^t\mathbf{d}$ subject to $\mathbf{A}_1'\mathbf{d} = \mathbf{0}$, $\mathbf{d}^t\mathbf{d} \le 1$. Show that the optimal solution to this problem is not equal to zero.

f. Solve the problem of Example 10.5.5 by the gradient projection method, where the projected gradient is found by minimizing $\nabla f(\mathbf{x})^t\mathbf{d}$ subject to $\mathbf{A}_1\mathbf{d} = \mathbf{0}$, $\mathbf{d}^t\mathbf{d} \le 1$.

10.32 Consider the following problem:

Minimize $\quad f(\mathbf{x})$

subject to $\quad \mathbf{A}\mathbf{x} \le \mathbf{b}$

Let $\mathbf{x}$ be a feasible solution and let $\mathbf{A}_1\mathbf{x} = \mathbf{b}_1$, $\mathbf{A}_2\mathbf{x} < \mathbf{b}_2$, where $\mathbf{A}^t = (\mathbf{A}_1^t, \mathbf{A}_2^t)$ and $\mathbf{b}^t = (\mathbf{b}_1^t, \mathbf{b}_2^t)$. Zoutendijk's method of feasible directions finds a direction by solving the problem to minimize $\nabla f(\mathbf{x})^t\mathbf{d}$ subject to $\mathbf{A}_1\mathbf{d} \le \mathbf{0}$ and $\mathbf{d}^t\mathbf{d} \le 1$. In view of Exercise 10.31, the gradient projection method finds a direction by solving the problem to minimize $\nabla f(\mathbf{x})^t\mathbf{d}$ subject to $\mathbf{A}_1\mathbf{d} = \mathbf{0}$, $\mathbf{d}^t\mathbf{d} \le 1$.

a. Compare the methods pointing out their advantages and disadvantages. Also, compare these methods with the successive linear programming algorithm described in Section 10.3.

b. Solve the following problem starting from (0, 0) by the method of Zoutendijk, the gradient projection method, and the algorithm PSLP of Section 10.3, and compare the trajectories:

$$\begin{aligned} \text{Minimize} \quad & x_1^2 + x_1x_2 + 2x_2^2 - 12x_1 - 18x_2 \\ \text{subject to} \quad & -3x_1 + 6x_2 \le 9 \\ & -2x_1 + x_2 \le 1 \\ & x_1, x_2 \ge 0 \end{aligned}$$

10.33 Consider the following problem, where $f: E_n \to E_1$ is differentiable:

$$\begin{aligned} \text{Minimize} \quad & f(\mathbf{x}) \\ \text{subject to} \quad & -\mathbf{x} \le \mathbf{0} \end{aligned}$$

a. Suppose that $\mathbf{x}$ is a feasible solution, and suppose that $\mathbf{x}^t = (\mathbf{x}_1^t, \mathbf{x}_2^t)$, where $\mathbf{x}_1 = \mathbf{0}$ and $\mathbf{x}_2 > \mathbf{0}$. Denote $\nabla f(\mathbf{x})^t$ by (∇_1^t, ∇_2^t). Show that the direction $\mathbf{d}^t$ generated by the gradient projection method is given by $(\mathbf{0}, -\nabla_2^t)$.

b. If $\nabla_2 = \mathbf{0}$, show that the gradient projection method simplifies as follows. If $\nabla_1 \ge \mathbf{0}$, stop; $\mathbf{x}$ is a KKT point. Otherwise, let j be any index such that $x_j = 0$ and $\partial f(\mathbf{x})/\partial x_j < 0$. Then the new direction of movement $\mathbf{d} = (0, \ldots, 0, -\partial f(\mathbf{x})/\partial x_j, 0, \ldots, 0)^t$, where $\partial f(\mathbf{x})/\partial x_j$ appears at position j.

c. Illustrate the method by solving the following problem:

$$\begin{aligned} \text{Minimize} \quad & x_1^2 + 2x_1x_2 + 4x_2^2 + 6x_1 + 2x_2 \\ \text{subject to} \quad & x_1, x_2 \ge 0 \end{aligned}$$

d. Solve the problem in Example 10.2.3 starting from the point (0, 0.1, 0) using the above procedure.

10.34 In the gradient projection method, we often calculate $(\mathbf{A}_1\mathbf{A}_1^t)^{-1}$ in order to compute the projection matrix. Usually, $\mathbf{A}_1$ is updated by deleting or adding a row to $\mathbf{A}_1$. Rather than computing $(\mathbf{A}_1\mathbf{A}_1^t)^{-1}$ from scratch, the old $(\mathbf{A}_1\mathbf{A}_1^t)^{-1}$ could be used to compute the new $(\mathbf{A}_1\mathbf{A}_1^t)^{-1}$.

a. Suppose that $\mathbf{C} = \begin{pmatrix} \mathbf{C}_1 & \mathbf{C}_2 \\ \mathbf{C}_3 & \mathbf{C}_4 \end{pmatrix}$ and $\mathbf{C}^{-1} = \begin{pmatrix} \mathbf{B}_1 & \mathbf{B}_2 \\ \mathbf{B}_3 & \mathbf{B}_4 \end{pmatrix}$. Show that $\mathbf{C}_1^{-1} = \mathbf{B}_1 - \mathbf{B}_2\mathbf{B}_4^{-1}\mathbf{B}_3$. Further, suppose that $\mathbf{C}_1^{-1}$ is known, and show that $\mathbf{C}^{-1}$ could be computed by letting

$$\mathbf{B}_1 = \mathbf{C}_1^{-1} + \mathbf{C}_1^{-1}\mathbf{C}_2\mathbf{C}_0^{-1}\mathbf{C}_3\mathbf{C}_1^{-1}$$

$$\mathbf{B}_2 = -\mathbf{C}_1^{-1}\mathbf{C}_2\mathbf{C}_0^{-1}$$

$$\mathbf{B}_3 = -\mathbf{C}_0^{-1}\mathbf{C}_3\mathbf{C}_1^{-1}$$

$$\mathbf{B}_4 = \mathbf{C}_0^{-1}$$

where $\mathbf{C}_0 = \mathbf{C}_4 - \mathbf{C}_3\mathbf{C}_1^{-1}\mathbf{C}_2$.

b. Simplify the above formulae for the gradient projection method, both when adding and deleting a row. (In the gradient projection method, $\mathbf{C}_1 = \mathbf{A}_1\mathbf{A}_1^t$, $\mathbf{C}_2 = \mathbf{A}_1\mathbf{a}$, $\mathbf{C}_3 = \mathbf{a}^t\mathbf{A}_1^t$, and $\mathbf{C}_4 = \mathbf{a}^t\mathbf{a}$, where $\mathbf{a}^t$ is the row added when $\mathbf{C}_1^{-1}$ is known or deleted when $\mathbf{C}^{-1}$ is known.)

c. Use the gradient projection method, with the scheme described in this exercise for updating $(\mathbf{A}_1\mathbf{A}_1^t)^{-1}$, to solve the following problem:

$$\begin{aligned} \text{Minimize} \quad & x_1^2 + x_1x_2 + 2x_2^2 + 2x_3^2 + 2x_2x_3 + 4x_1 + 6x_2 + 12x_3 \\ \text{subject to} \quad & x_1 + x_2 + x_3 \le 6 \\ & -x_1 - x_2 + 2x_3 \ge 2 \\ & x_1, x_2, x_3 \ge 0 \end{aligned}$$

10.35 Consider the problem to minimize $f(\mathbf{x})$ subject to $\mathbf{Ax} \le \mathbf{b}$. The following modification to Zoutendijk's method and to the gradient projection method of Rosen is proposed. Given a feasible point $\mathbf{x}$, if $-\nabla f(\mathbf{x})$ is feasible, then the direction of movement $\mathbf{d}$ is taken as $-\nabla f(\mathbf{x})$; otherwise, the direction $\mathbf{d}$ is computed according to the respective algorithms.

a. Using the above modification, solve the problem in Example 10.1.5 starting from $\mathbf{x}_1 = (0, 0.75)^t$ by Zoutendijk's method. Compare the trajectory with that obtained in Example 10.1.5.
b. Using the above modification, solve the problem in Example 10.5.5 starting from $\mathbf{x}_1 = (0.0, 0.0)$ by Rosen's gradient projection method. Compare the trajectory with that obtained in Example 10.5.5.

10.36 In the gradient projection method, if $\mathbf{P}\nabla f(\mathbf{x}) = \mathbf{0}$, we throw away from the matrix $\mathbf{A}_1$ a row corresponding to a negative component of the vector $\mathbf{u}$. Suppose that, instead, we throw away all rows corresponding to negative components of the vector $\mathbf{u}$. Show by a numerical example that the resulting projection matrix does not necessarily lead to an improving feasible direction.

10.37 Consider the following problem:

$$\begin{aligned} \text{Minimize} \quad & \mathbf{c}^t\mathbf{x} \\ \text{subject to} \quad & \mathbf{A}\mathbf{x} = \mathbf{b} \\ & \mathbf{x} \geq \mathbf{0} \end{aligned}$$

where $\mathbf{A}$ is an $m \times n$ matrix of rank m. Consider solving the problem by the gradient projection method.

a. Let $\mathbf{x}$ be a basic feasible solution and $\mathbf{d} = -\mathbf{P}\mathbf{c}$, where $\mathbf{P}$ projects any vector onto the nullspace of the gradients of the binding constraints. Show that $\mathbf{d} = \mathbf{0}$.
b. Let $\mathbf{u} = -(\mathbf{M}\mathbf{M}^t)^{-1}\mathbf{M}\mathbf{c}$, where the rows of $\mathbf{M}$ are the transposes of the gradients of the binding constraints. Show that deleting the row corresponding to the most negative u_j associated with the constraint $x_j \geq 0$, forming a new projection matrix $\mathbf{P}'$, and moving along the direction $-\mathbf{P}'\mathbf{c}$ is equivalent to entering the variable x_j in the basis in the simplex method.
c. Using the results of parts a and b, show that the gradient projection method reduces to the simplex method if the objective function is linear.

10.38 Consider the following problem:

$$\begin{aligned} \text{Minimize} \quad & x_1^2 + 2x_2^2 + 3x_3^2 + x_1x_2 - 2x_1x_3 + x_2x_3 - 4x_1 - 6x_2 \\ \text{subject to} \quad & x_1 + 2x_2 + x_3 \leq 4 \\ & x_1, x_2, x_3 \geq 0 \end{aligned}$$

a. Solve the problem by Zoutendijk's method of feasible directions, starting from the origin.
b. Solve the problem by the gradient projection method, starting from the origin.

10.39 In the reduced gradient method, suppose the set I_k defined in (10.40) consists of indices of any m positive variables. Investigate if the direction finding map is closed.

10.40 As originally proposed, given a point $\mathbf{x}$, the reduced gradient method moves along the direction $\mathbf{d}$, where (10.43) is modified as follows:

$$d_j = \begin{cases} -r_j & \text{if } x_j > 0 \text{ or } r_j \leq 0 \\ 0 & \text{otherwise} \end{cases}$$

a. Prove that $\mathbf{d} = \mathbf{0}$ if and only if $\mathbf{x}$ is a KKT point.
b. Show that if $\mathbf{d} \neq \mathbf{0}$, then $\mathbf{d}$ is an improving feasible direction.
c. Using the above direction-finding map, solve the following problem by the reduced gradient method:

$$\begin{aligned} \text{Minimize} \quad & e^{-x_1+x_2} + x_1^2 + 2x_1x_2 + x_2^2 + 2x_1 + 6x_2 \\ \text{subject to} \quad & x_1 + x_2 \leq 4 \\ & -x_1 + x_2 \leq 2 \\ & x_1, x_2 \geq 0 \end{aligned}$$

d. Show that the direction-finding map given above is not closed.

10.41 For both the reduced gradient method and the convex simplex method, it was assumed that each feasible solution has at least m positive components. This exercise gives a necessary and sufficient condition for this to hold. Consider the set $S = \{\mathbf{x}: \mathbf{A}\mathbf{x} = \mathbf{b}, \mathbf{x} \geq \mathbf{0}\}$, where $\mathbf{A}$ is an $m \times n$ matrix of rank m. Show that each $\mathbf{x} \in S$ has at least m

positive components if and only if every extreme point of S has exactly m positive components.

10.42 Consider the following problem:

$$\begin{aligned} \text{Minimize} \quad & x_1^2 + x_1x_2 + 2x_2^2 - 6x_1 - 14x_2 \\ \text{subject to} \quad & x_1 + x_2 + x_3 = 2 \\ & -x_1 + 2x_2 \leq 3 \\ & x_1, x_2, x_3 \geq 0 \end{aligned}$$

a. Solve the problem by the gradient projection method.
b. Solve the problem by the reduced gradient method.
c. Solve the problem by the convex–simplex method.
d. Solve the problem by the PSLP algorithm.
e. Solve the problem by the MSQP algorithm.

10.43 Consider the following problem:

$$\begin{aligned} \text{Minimize} \quad & x_1^3 + 2x_2^3 + x_1 - 2x_2 - x_1^2 \\ \text{subject to} \quad & x_1 + 2x_2 \leq 6 \\ & -x_1 + 2x_2 \leq 3 \\ & x_1, x_2 \geq 0 \end{aligned}$$

Solve the problem by the convex–simplex method. Is the solution obtained a global optimum, a local optimum, or neither?

10.44 Modify the rules of the convex–simplex method such that it handles directly the problem to minimize $f(\mathbf{x})$ subject to $\mathbf{Ax} = \mathbf{c}$, $\mathbf{a} \leq \mathbf{x} \leq \mathbf{b}$. Use the method to solve the following problem:

$$\begin{aligned} \text{Minimize} \quad & e^{-x_1} + x_1^2 - x_1x_2 - 3x_2^2 + 4x_1 - 6x_2 \\ \text{subject to} \quad & 2x_1 + x_2 \leq 8 \\ & -x_1 + x_2 \leq 2 \\ & 1 \leq x_1, x_2 \leq 3 \end{aligned}$$

10.45 Show that the convex–simplex method reduces to the simplex method if the objective function is linear.

10.46 Show that the direction-finding map of the convex–simplex method defined by (10.46) through (10.55) is closed.

10.47 Suppose that in the convex–simplex method, the direction-finding process is modified as follows. The scalar β in (10.50) is computed as

$$\beta = \begin{cases} \text{maximum } \{r_j : x_j > 0 \text{ and } r_j \geq 0\} & \text{if } x_i > 0 \text{ and } r_i \geq 0 \text{ for some } i \\ 0 & \text{otherwise} \end{cases}$$

Furthermore, the index ν is computed as

$$\nu = \begin{cases} \nu \text{ is an index such that } \alpha = -r_\nu & \text{if } \alpha \geq \beta \\ \nu \text{ is an index such that } \beta = r_\nu & \text{if } \alpha < \beta \end{cases}$$

Show that with this modification, the direction-finding map is not necessarily closed.

10.48 Consider the following problem:

$$\begin{aligned} \text{Minimize} \quad & f(\mathbf{x}) \\ \text{subject to} \quad & \mathbf{Ax} = \mathbf{b} \\ & \mathbf{x} \geq \mathbf{0} \end{aligned}$$

Assume that f is a concave function, and that the feasible region is compact, so that, by Theorem 3.4.7, an optimal extreme point exists.

a. Show how the convex–simplex method could be modified so that it only searches among extreme points of the feasible region.
b. At termination, a KKT point is at hand. Is this point necessarily an optimal solution, a local optimal solution, or neither? If the point is not optimal, can you develop a cutting plane method that excludes the current point but not the optimal solution?

c. Illustrate the procedure in parts a and b by solving the following problem, starting from the origin:

$$\begin{aligned} \text{Minimize} \quad & -(x_1 - 2)^2 - (x_2 - 1)^2 \\ \text{subject to} \quad & -x_1 + x_2 \le 4 \\ & 2x_1 + x_2 \le 12 \\ & 3x_1 - x_2 \le 12 \\ & x_1, x_2 \ge 0 \end{aligned}$$

10.49 Consider the problem to minimize $f(\mathbf{x})$ subject to $g_i(\mathbf{x}) \le 0$ for $i = 1, \ldots, m$. The feasible direction methods discussed in this chapter start with a feasible point. This exercise describes a method for obtaining such a point if one is not immediately available. Select an arbitrary point $\hat{\mathbf{x}}$, and suppose that $g_i(\hat{\mathbf{x}}) \le 0$ for $i \in I$ and $g_i(\hat{\mathbf{x}}) > 0$ for $i \notin I$. Now consider the following problem:

$$\begin{aligned} \text{Minimize} \quad & \sum_{i \notin I} y_i \\ \text{subject to} \quad & g_i(\mathbf{x}) \le 0 \quad && \text{for } i \in I \\ & g_i(\mathbf{x}) - y_i \le 0 && \text{for } i \notin I \\ & y_i \ge 0 && i \notin I \end{aligned}$$

a. Show that a feasible solution to the original problem exists if and only if the optimal objective value of the above problem is zero.

b. Let $\mathbf{y}$ be a vector whose components are y_i for $i \notin I$. The above problem could be solved by a feasible direction method sarting from the point $(\hat{\mathbf{x}}, \hat{\mathbf{y}})$, where $\hat{y}_i = g_i(\hat{\mathbf{x}})$ for $i \notin I$. At termination, a feasible solution to the original problem is obtained. Starting from this point, a feasible direction method could be used to solve the original problem. Illustrate this method by solving the following problem starting from the infeasible point (1, 3):

$$\begin{aligned} \text{Minimize} \quad & e^{-x_1 - x_2} + x_1 x_2 + x_2^2 \\ \text{subject to} \quad & e^{-x_1} + x_2^2 \le 4 \\ & 2x_1 + 5x_2 \le 10 \end{aligned}$$

10.50 Consider Problem P to minimize $f(\mathbf{x})$ subject to $\mathbf{Ax} = \mathbf{b}$ and $\mathbf{x} \ge \mathbf{0}$, where $\mathbf{A}$ is an $m \times n$ matrix of rank m. Given a solution $\bar{\mathbf{x}} = (\bar{\mathbf{x}}_B, \bar{\mathbf{x}}_N)$, where $\bar{\mathbf{x}}_B > \mathbf{0}$ and where the corresponding partition $(\mathbf{B}, \mathbf{N})$ of $\mathbf{A}$ has $\mathbf{B}$ nonsingular, write the basic variables $\mathbf{x}_B$ in terms of the nonbasic variables $\mathbf{x}_N$ to derive the following representation of P in the nonbasic variable space:

$$P(\mathbf{x}_N): \quad \text{Minimize} \quad \{F(\mathbf{x}_N) \equiv f(\mathbf{B}^{-1}\mathbf{b} - \mathbf{B}^{-1}\mathbf{N}\mathbf{x}_N, \mathbf{x}_N) : \mathbf{B}^{-1}\mathbf{N}\mathbf{x}_N \le \mathbf{B}^{-1}\mathbf{b}, \mathbf{x}_N \ge \mathbf{0}\}$$

a. Identify the set of binding constraints for $P(\mathbf{x}_N)$ at the current solution $\bar{\mathbf{x}}_N$. Relate $\nabla F(\mathbf{x}_N)$ to the reduced gradient for Problem P.

b. Write a set of necessary and sufficient conditions for $\bar{\mathbf{x}}_N$ to be a KKT point for $P(\mathbf{x}_N)$. Compare this with the result of Theorem 10.6.1.

10.51 Consider the nonlinear programming problem P to minimize $f(\mathbf{x})$ subject to $\mathbf{Ax} = \mathbf{b}$ and $\mathbf{x} \ge \mathbf{0}$, where $\mathbf{A}$ is an $m \times n$ matrix of rank m, and suppose that some $q \le n$ variables appear nonlinearly in the problem. Assuming that P has an optimum, show that there exists an optimum in which the number of superbasic variables s in a reduced gradient approach satisfies $s \le q$. Extend this result to include nonlinear constraints as well.

10.52 a. Consider the problem illustrated in Figure 10.21, and assume that the convex–simplex method of Section 10.7 is used to solve this problem. Use the graph to illustrate a plausible path followed by the algorithm, specifying the set of basic and nonbasic variables, the signs on the components of the reduced gradient $\mathbf{r}_N$, and the result of the line search at each iteration.

b. Repeat to illustrate a conceivable trajectory for the reduced gradient algorithm of Section 10.6.

c. Repeat to illustrate the effect of using the quadratic programming subproblem (10.62), starting at the origin with x_1 and x_2 as superbasic variables, and assuming that the objective function is itself quadratic.

10.53 Consider the following problem:

Minimize $\mathbf{c}^t\mathbf{x} + \frac{1}{2}\mathbf{x}^t\mathbf{H}\mathbf{x}$

subject to $\mathbf{A}\mathbf{x} = \mathbf{b}$

$\mathbf{x} \geq \mathbf{0}$

The constraint $\mathbf{h}(\mathbf{x}) = \mathbf{A}\mathbf{x} - \mathbf{b} = \mathbf{0}$ is handled by a penalty function of the form $\mu\mathbf{h}(\mathbf{x})^t\mathbf{h}(\mathbf{x})$, giving the following problem:

Minimize $\mathbf{c}^t\mathbf{x} + \frac{1}{2}\mathbf{x}^t\mathbf{H}\mathbf{x} + \mu(\mathbf{A}\mathbf{x} - \mathbf{b})^t(\mathbf{A}\mathbf{x} - \mathbf{b})$

subject to $\mathbf{x} \geq \mathbf{0}$

Give the detailed steps of a feasible direction method for solving the above problem. Illustrate the method for the following:

$$\mathbf{H} = \begin{bmatrix} 1 & -1 & 0 \\ -1 & 2 & 0 \\ 0 & 0 & 0 \end{bmatrix} \quad \mathbf{A} = \begin{bmatrix} 1 & 1 & 0 \\ 2 & 1 & 1 \end{bmatrix} \quad \mathbf{c} = \begin{pmatrix} -2 \\ -3 \\ 0 \end{pmatrix} \quad \mathbf{b} = \begin{pmatrix} 3 \\ 5 \end{pmatrix}$$

10.54 This exercise gives a generalization of Wolfe's reduced gradient method for handling nonlinear equality constraints. This generalized procedure was developed by Abadie and Carpentier [1969], and a modified version is given below for brevity. Consider the following problem:

Minimize $f(\mathbf{x})$

subject to $h_i(\mathbf{x}) = 0 \qquad$ for $i = 1, \ldots, l$

$a_j \leq x_j \leq u_j \qquad$ for $j = 1, \ldots, n$

Here, we assume that f and h_i for each i are differentiable. Let $\mathbf{h}$ be the vector function whose components are h_i for $i = 1, \ldots, l$, and, furthermore, let $\mathbf{a}$ and $\mathbf{u}$ be the vectors whose components are a_j and u_j for $j = 1, \ldots, n$. We make the following nondegeneracy assumption. Given any feasible solution $\mathbf{x}^t$, it can be decomposed into $(\mathbf{x}_B^t, \mathbf{x}_N^t)$ with $\mathbf{x}_B \in E_l$ and $\mathbf{x}_N \in E_{n-l}$, where $\mathbf{a}_B < \mathbf{x}_B < \mathbf{u}_B$. Furthermore, the $l \times n$ Jacobian matrix $\nabla\mathbf{h}(\mathbf{x})$ is decomposed accordingly into the $l \times l$ matrix $\nabla_B\mathbf{h}(\mathbf{x})$ and the $l \times (n - l)$ matrix $\nabla_N\mathbf{h}(\mathbf{x})$, such that $\nabla_B\mathbf{h}(\mathbf{x})$ is invertible. The following is an outline of the procedure.

Initialization Step Choose a feasible solution $\mathbf{x}^t$ and decompose it into $(\mathbf{x}_B^t, \mathbf{x}_N^t)$. Go to the main step.

Main Step

1. Let $\mathbf{r}^t = \nabla_N f(\mathbf{x})^t - \nabla_B f(\mathbf{x})^t\nabla_B\mathbf{h}(\mathbf{x})^{-1}\nabla_N\mathbf{h}(\mathbf{x})$. Compute the $n - l$ vector $\mathbf{d}_N$ whose jth component d_j is

$$d_j = \begin{cases} 0 & \text{if } x_j = a_j \text{ and } r_j > 0, \text{ or } x_j = u_j \text{ and } r_j < 0 \\ -r_j & \text{otherwise} \end{cases}$$

If $\mathbf{d}_N = \mathbf{0}$, stop; $\mathbf{x}$ is a KKT point. Otherwise, go to step 2.

2. Find a solution to the nonlinear system $\mathbf{h}(\mathbf{y}, \bar{\mathbf{x}}_N) = \mathbf{0}$ by Newton's method as follows, where $\bar{\mathbf{x}}_N$ is specified below.

Initialization Choose $\varepsilon > 0$ and a positive integer K. Let $\theta > 0$ be such that $\mathbf{a}_N \leq \bar{\mathbf{x}}_N \leq \mathbf{u}_N$, where $\bar{\mathbf{x}}_N = \mathbf{x}_N + \theta\mathbf{d}_N$. Let $\mathbf{y}_1 = \mathbf{x}_B$, let $k = 1$, and go to iteration k below.

Iteration k

(i) Let $\mathbf{y}_{k+1} = \mathbf{y}_k - \nabla_B\mathbf{h}(\mathbf{y}_k, \bar{\mathbf{x}}_N)^{-1}\mathbf{h}(\mathbf{y}_k, \bar{\mathbf{x}}_N)$. If $\mathbf{a}_B \leq \mathbf{y}_{k+1} \leq \mathbf{u}_B$, $f(\mathbf{y}_{k+1}, \bar{\mathbf{x}}_N) < f(\mathbf{x}_B, \mathbf{x}_N)$, and $\|\mathbf{h}(\mathbf{y}_{k+1}, \bar{\mathbf{x}}_N)\| < \varepsilon$, go to (iii); otherwise, go to step (ii).

(ii) If $k = K$, replace θ by $\frac{1}{2}\theta$, let $\bar{\mathbf{x}}_N = \mathbf{x}_N + \theta\mathbf{d}_N$, let $\mathbf{y}_1 = \mathbf{x}_B$, replace k by 1, and repeat step (i). Otherwise, replace k by $k + 1$ and repeat step (i).

(iii) Let $\mathbf{x}^t = (\mathbf{y}_{k+1}^t, \bar{\mathbf{x}}_N^t)$, choose a new basis $\mathbf{B}$, and go to step 1 of the main algorithm.

a. Using the above algorithm, solve the following problem:

Minimize $x_1^2 + 2x_1x_2 + x_2^2 + 12x_1 - 4x_2$

subject to $x_1^2 - x_2 = 0$

$1 \le x_1, x_2 \le 3$

b. Show how the procedure could be modified to handle inequality constraints as well. Illustrate by solving the following problem:

Minimize $x_1^2 + 2x_1x_2 + x_2^2 + 12x_1 - 4x_2$

subject to $x_1^2 + x_2^2 \le 4$

$1 \le x_1, x_2 \le 3$

10.55 This exercise describes the method credited to Griffith and Stewart [1961] for solving a nonlinear programming problem by successively approximating it by a sequence of linear programs. Consider the following problem:

Minimize $f(\mathbf{x})$

subject to $\mathbf{g}(\mathbf{x}) \le \mathbf{0}$

$\mathbf{h}(\mathbf{x}) = \mathbf{0}$

$\mathbf{a} \le \mathbf{x} \le \mathbf{b}$

where $f{:}E_n \to E_1$, $\mathbf{g}{:}E_n \to E_m$, and $\mathbf{h}{:}E_n \to E_l$. Given the point $\mathbf{x}_k$ at iteration k, the algorithm replaces f, $\mathbf{g}$, and $\mathbf{h}$ by their linear approximations at $\mathbf{x}_k$, yielding the following linear programming problem, where $\nabla\mathbf{g}(\mathbf{x}_k)$ is an $m \times n$ matrix denoting the Jacobian of the vector function $\mathbf{g}$, and $\nabla\mathbf{h}(\mathbf{x}_k)$ is an $l \times n$ matrix denoting the Jacobian of the vector function $\mathbf{h}$:

Minimize $f(\mathbf{x}_k) + \nabla f(\mathbf{x}_k)^t(\mathbf{x} - \mathbf{x}_k)$

subject to $\mathbf{g}(\mathbf{x}_k) + \nabla\mathbf{g}(\mathbf{x}_k)(\mathbf{x} - \mathbf{x}_k) \le \mathbf{0}$

$\mathbf{h}(\mathbf{x}_k) + \nabla\mathbf{h}(\mathbf{x}_k)(\mathbf{x} - \mathbf{x}_k) = \mathbf{0}$

$\mathbf{a} \le \mathbf{x} \le \mathbf{b}$

Initialization Step Choose a feasible point $\mathbf{x}_1$, choose a parameter $\delta > 0$ that limits the movement at each iteration, and choose a termination scalar $\varepsilon > 0$. Let $k = 1$, and go to the main step.

Main Step

1. Solve the following linear programming problem:

   Minimize $\nabla f(\mathbf{x}_k)^t(\mathbf{x} - \mathbf{x}_k)$

   subject to $\nabla\mathbf{g}(\mathbf{x}_k)(\mathbf{x} - \mathbf{x}_k) \le -\mathbf{g}(\mathbf{x}_k)$

   $\nabla\mathbf{h}(\mathbf{x}_k)(\mathbf{x} - \mathbf{x}_k) = -\mathbf{h}(\mathbf{x}_k)$

   $\mathbf{a} \le \mathbf{x} \le \mathbf{b}$

   $-\delta \le x_i - x_{ik} \le \delta \qquad \text{for } i = 1, \ldots, n$

   where x_{ik} is the ith component of $\mathbf{x}_k$. Let $\mathbf{x}_{k+1}$ be an optimal solution, and go to step 2.

2. If $\|\mathbf{x}_{k+1} - \mathbf{x}_k\| \le \varepsilon$ and if $\mathbf{x}_{k+1}$ is near-feasible, then stop with $\mathbf{x}_{k+1}$. Otherwise, replace k by $k + 1$ and repeat step 1.

   Even though convergence of the above method is not generally guaranteed, the method is reported to be effective for solving many practical problems.

a. Construct an example that shows that if $\mathbf{x}_k$ is feasible to the original problem, then $\mathbf{x}_{k+1}$ is not necessarily feasible.

b. Now suppose that $\mathbf{h}$ is linear. Show that if $\mathbf{g}$ is concave, then the feasible region to the linear program is contained in the feasible region of the original problem. Furthermore, show that if $\mathbf{g}$ is convex, then the feasible region to the original problem is contained in the feasible region to the linear program.

c. Solve the following problem in Walsh [1975, page 67] both by the method described in this exercise and by Kelley's cutting plane algorithm presented in Exercise 7.22, and compare their trajectories:

Minimize $-2x_1^2 + x_1x_2 - 3x_2^2$

subject to $3x_1 + 4x_2 \le 12$

$x_1^2 - x_2^2 \ge 1$

$0 \le x_1 \le 4$

$0 \le x_2 \le 3$

d. Re-solve the example of part c using the PSLP algorithm of Section 10.3 and compare the trajectories obtained.

10.56 In this exercise, we describe a method credited to Davidon [1959] and developed later by Goldfarb [1969] for minimizing a quadratic function in the presence of linear constraints. The method extends the Davidon-Fletcher-Powell method and retains conjugacy of the search directions in the presence of constraints. Part e of the exercise also suggests an alternative approach. Consider the following problem:

Minimize $\mathbf{c}'\mathbf{x} + \frac{1}{2}\mathbf{x}'\mathbf{H}\mathbf{x}$
subject to $\mathbf{A}\mathbf{x} = \mathbf{b}$

where $\mathbf{H}$ is an $n \times n$ symmetric positive definite matrix and $\mathbf{A}$ is an $m \times n$ matrix of rank m. The following is a summary of the algorithm.

Initialization Step Let $\varepsilon > 0$ be the termination scalar. Choose a feasible point $\mathbf{x}_1$ and an initial symmetric positive definite matrix $\mathbf{D}_1$. Let $k = j = 1$, let $\mathbf{y}_1 = \mathbf{x}_1$, and go to the main step.

Main Step

1. If $\|\nabla f(\mathbf{y}_j)\| < \varepsilon$, stop; otherwise, let $\mathbf{d}_j = -\hat{\mathbf{D}}_j\nabla f(\mathbf{y}_j)$, where

$$\hat{\mathbf{D}}_j = \mathbf{D}_j - \mathbf{D}_j\mathbf{A}'(\mathbf{A}\mathbf{D}_j\mathbf{A}')^{-1}\mathbf{A}\mathbf{D}_j$$

Let λ_j be an optimal solution to the problem to minimize $f(\mathbf{y}_j + \lambda\mathbf{d}_j)$ subject to $\lambda \geq 0$, and let $\mathbf{y}_{j+1} = \mathbf{y}_j + \lambda_j\mathbf{d}_j$. If $j < n$, go to step 2. If $j = n$, let $\mathbf{y}_1 = \mathbf{x}_{k+1} = \mathbf{y}_{n+1}$, replace k by $k + 1$, let $j = 1$, and repeat step 1.

2. Construct $\mathbf{D}_{j+1}$ as follows:

$$\mathbf{D}_{j+1} = \mathbf{D}_j + \frac{\mathbf{p}_j\mathbf{p}_j'}{\mathbf{p}_j'\mathbf{q}_j} - \frac{\mathbf{D}_j\mathbf{q}_j\mathbf{q}_j'\mathbf{D}_j}{\mathbf{q}_j'\mathbf{D}_j\mathbf{q}_j}$$

where $\mathbf{p}_j = \lambda_j\mathbf{d}_j$, and $\mathbf{q}_j = \nabla f(\mathbf{y}_{j+1}) - \nabla f(\mathbf{y}_j)$. Replace j by $j + 1$, and repeat step 1.

a. Show that the points generated by the algorithm are feasible.
b. Show that the search directions are $\mathbf{H}$-conjugate.
c. Show that the algorithm stops in, at most, $n - m$ steps with an optimal solution.
d. Solve the following problem by the method described in this exercise:

Minimize $x_1^2 + x_2^2 + 2x_3^2 + 3x_4^2 + x_1x_2 - x_2x_3 + 2x_3x_4 - 2x_1 - 4x_2 + 6x_3$
subject to $2x_1 + x_2 + x_3 = 6$
$-x_1 + 2x_2 - 2x_3 + x_4 = 8$

e. Consider the following alternative approach. Decompose $\mathbf{x}'$ and $\mathbf{A}$ into $(\mathbf{x}_B', \mathbf{x}_N')$ and $[\mathbf{B}, \mathbf{N}]$, where $\mathbf{B}$ is an invertible $m \times m$ matrix. The system $\mathbf{A}\mathbf{x} = \mathbf{b}$ is equivalent to $\mathbf{x}_B = \mathbf{B}^{-1}\mathbf{b} - \mathbf{B}^{-1}\mathbf{N}\mathbf{x}_N$. Substituting for $\mathbf{x}_B$ in the objective function, a quadratic form involving the $n - m$ vector $\mathbf{x}_N$ is obtained. The resulting function is then minimized using a suitable conjugate direction method such as the Davidon-Fletcher-Powell method. Use this approach to solve the problem given in part d above, and compare the two solution procedures.
f. Modify the above scheme, using the BFGS quasi-Newton update.
g. Extend both the methods discussed in this exercise for handling a general nonlinear objective function.

10.57 Consider the bilinear program to minimize $\phi(\mathbf{x}, \mathbf{y}) = \mathbf{c}'\mathbf{x} + \mathbf{d}'\mathbf{y} + \mathbf{x}'\mathbf{H}\mathbf{y}$ subject to $\mathbf{x} \in X$ and $\mathbf{y} \in Y$, where X and Y are bounded polyhedral sets in E_n and E_m, respectively. Consider the following algorithm.

Initialization Step Select an $\mathbf{x}_1 \in E_n$ and $\mathbf{y}_1 \in E_m$. Let $k = 1$, and go to the main step.

Main Step

1. Solve the linear program to minimize $\mathbf{d}'\mathbf{y} + \mathbf{x}_k'\mathbf{H}\mathbf{y}$ subject to $\mathbf{y} \in Y$. Let $\hat{\mathbf{y}}$ be an optimal solution. Let $\mathbf{y}_{k+1}$ be as specified below and go to step 2.

$$\mathbf{y}_{k+1} = \begin{cases} \mathbf{y}_k & \text{if } \phi(\mathbf{x}_k, \hat{\mathbf{y}}) = \phi(\mathbf{x}_k, \mathbf{y}_k) \\ \hat{\mathbf{y}} & \text{if } \phi(\mathbf{x}_k, \hat{\mathbf{y}}) < \phi(\mathbf{x}_k, \mathbf{y}_k) \end{cases}$$

2. Solve the linear program to minimize $\mathbf{c}^t\mathbf{x} + \mathbf{x}^t\mathbf{H}\mathbf{y}_{k+1}$ subject to $\mathbf{x} \in X$. Let $\hat{\mathbf{x}}$ be an optimal solution. Let $\mathbf{x}_{k+1}$ be as specified below and go to step 3.

$$\mathbf{x}_{k+1} = \begin{cases} \mathbf{x}_k & \text{if } \phi(\hat{\mathbf{x}}, \mathbf{y}_{k+1}) = \phi(\mathbf{x}_k, \mathbf{y}_{k+1}) \\ \hat{\mathbf{x}} & \text{if } \phi(\hat{\mathbf{x}}, \mathbf{y}_{k+1}) < \phi(\mathbf{x}_k, \mathbf{y}_{k+1}) \end{cases}$$

3. If $\mathbf{x}_{k+1} = \mathbf{x}_k$ and $\mathbf{y}_{k+1} = \mathbf{y}_k$, stop, with $(\mathbf{x}_k, \mathbf{y}_k)$ as a KKT point. Otherwise, replace k by $k + 1$ and go to step 1.

a. Using the above algorithm, find a KKT point to the bilinear program to minimize $x_1y_1 + x_2y_2$, subject to $\mathbf{x} \in X$ and $\mathbf{y} \in Y$, where

$$X = \{\mathbf{x}: x_1 + 3x_2 \geq 30, 2x_1 + x_2 \geq 20, 0 \leq x_1 \leq 27, 0 \leq x_2 \leq 16\}$$

$$Y = \{\mathbf{y}: \tfrac{5}{3}y_1 + y_2 \geq 10, y_1 + y_2 \leq 15, 0 \leq y_1 \leq 10, 0 \leq y_2 \leq 10\}.$$

b. Prove that the algorithm converges to a KKT point.

c. Show that if $(\mathbf{x}_k, \mathbf{y}_k)$ is such that $\phi(\mathbf{x}_k, \mathbf{y}_k) \leq \min \{\phi(\mathbf{x}'_k, \mathbf{y}): \mathbf{y} \in Y\}$ for all extreme points $\mathbf{x}'_k$ of X that are adjacent to $\mathbf{x}_k$ (including $\mathbf{x}_k$ itself), and if $\phi(\mathbf{x}_k, \mathbf{y}_k) \leq \min \{\phi(\mathbf{x}, \mathbf{y}'_k): \mathbf{x} \in X\}$ for all extreme points $\mathbf{y}'_k$ of Y that are adjacent to $\mathbf{y}_k$ (including $\mathbf{y}_k$ itself), then $(\mathbf{x}_k, \mathbf{y}_k)$ is a local minimum for the bilinear program. (This result is discussed in Vaish [1974].)

Notes and References

The method of feasible directions is a general concept that is exploited by primal algorithms that proceed from one feasible solution to another. In Section 10.1, we present the methods of Zoutendijk for generating improving feasible directions. It is well known that the algorithmic map used in Zoutendijk's method is not closed, and this is shown in Section 10.2. Furthermore, an example credited to Wolfe [1972] was presented which shows that the procedure does not generally converge to a KKT point. To overcome this difficulty, based on the work of Zoutendijk [1960], Zangwill [1969] presented a convergent algorithm based on the use of the concept of near-binding constraints. In Section 10.2, we describe another approach, credited to Topkis and Veinott [1967]. This method uses all the constraints, both active and inactive, and thereby avoids a sudden change in direction as a new constraint becomes active.

Note that the methods of unconstrained optimization discussed in Chapter 8 could be effectively combined with the method of feasible directions. In this case, an unconstrained optimization method is used at interior points, whereas feasible directions are generated at boundary points by one of the methods discussed in this chapter. An alternative approach is to place additional conditions at interior points which guarantee that the direction generated is conjugate to some of the previously generated directions. This is illustrated in Exercise 10.16. Also refer to Kunzi, Krelli, and Oettli [1965], Zangwill [1967], and Zoutendijk [1960]. Zangwill [1967] developed a procedure for solving quadratic programming problems in a finite number of steps using the convex-simplex method in conjunction with conjugate directions.

In Sections 10.3 and 10.4 we present a very popular and effective class of successive linear (SLP) and quadratic (SQP) programming feasible direction approaches. Griffith and Stewart [1961] introduced the SLP approach at Shell as a method of *approximation programming*. Other similar approaches were developed by Buzby [1974] for a chemical process model at Union Carbide, and by Boddington and Randall [1979] for a blending and refinery problem at the Chevron Oil Company (see also Baker and Ventker [1980]).

Beale [1978] described a combination of SLP ideas with the reduced gradient method, and Palacios-Gomez et al. [1982] present a SLP approach using the l_1 penalty function as a merit function. Although intuitively appealing and popular because of the availability of efficient linear programming-solvers, the foregoing methods are not guaranteed to converge. A first convergent form, which is described in Section 10.3 as the penalty SLP (PSLP) approach, was presented by Zhang et al. [1985] and uses the l_1 penalty function directly in the linear programming subproblem along with trust region ideas, following Fletcher [1981]. Baker and Lasdon [1985] described the use of a simplified version of this algorithm that has been used to solve nonlinear refinery models having up to 1000 rows.

The SQP method concept (or the projected Lagrangian or the Lagrange–Newton method) was first used by Wilson [1963] in his SOLVER procedure as described by Beale [1967]. Han [1976] and Powell [1978a] suggest quasi-Newton approximations to the Hessian of the Lagrangian, and Powell [1978b] provides related superlinear convergence arguments. Han [1975] and Powell [1978a] show how the l_1 penalty function can be used as a merit function to derive globally convergent variants of SQP. This is described as the MSQP method herein. See Crane et al. [1980] for a software description. For the use of other smooth exact penalty (ALAG) functions in this context and related discussion, see Fletcher [1987], Gill et al. [1981], and Schittkowski [1983]. The Maratos [1978] effect is described nicely in Powell [1986], and ways to avoid it have been proposed by permitting increases in both the objective function and constraint violations (Chamberlain et al. [1982], Powell and Yuan [1986], and Schittkowski [1981]) or by altering search directions (Coleman and Conn [1982] and Mayne and Polak [1982]), or through second-order corrections (Fletcher [1982b]) (see also Fukushima [1986]). For ways of handling infeasibility or unbounded quadratic subproblems, see Fletcher [1987] and Burke and Han [1989]. Fletcher [1981, 1987] describes the L_1SQP algorithm mentioned in Section 10.4 which combines the l_1 penalty function with trust region methods to yield a robust and very effective procedure. Tamura and Kobayashi [1991] describe experiences with an actual application of SQP, and Eldersveld [1991] discusses techniques for solving large-scale problems using SQP, along with a comprehensive discussion and computational results. For extensions of SQP methods to nonsmooth problems, see Pshenichnyi [1978] and Fletcher [1987].

In 1960, Rosen developed the gradient projection method for linear constraints and later, in 1961, generalized it for nonlinear constraints. Du and Zhang [1989] provide a comprehensive convergence analysis for a mild modification of this method as stated in Section 10.5. (An earlier analysis appears in Du and Zhang [1986].) In Exercises 10.30 through 10.32, different methods for yielding the projected gradient are presented, and the relationship between Rosen's method and that of Zoutendijk is explored. In 1969, Goldfarb extended the Davidon–Fletcher–Powell method to handle problems with linear constraints utilizing the concept of gradient projection. In Exercise 10.56, we describe how equality constraints could be handled. For the inequality constrained problem, Goldfarb identifies a set of constraints that could be regarded as binding and applies the equality constrained approach. The method was generalized by Davies [1970] to handle nonlinear constraints. Also refer to the work of Sargent and Murtagh [1973] on their variable metric projection method.

The method of reduced gradient was developed by Wolfe [1963] with the direction-finding map discussed in Exercise 10.40. In 1966, Wolfe provided an example to show that the method does not converge to a KKT point. The modified version described in Section 10.6 is credited to McCormick [1969]. The method of generalized reduced gradients was later presented by Abadie and Carpentier [1969] who gave several approaches to handle nonlinear constraints. One such approach is discussed in Section 10.6, and another in Exercise 10.54. Computational experience with the reduced gradient method and its generalization is reported in Abadie and Carpentier [1967], Abadie and Guigon [1970], and Faure and Huard [1965]. Convergence proofs for the GRG method are presented under very restrictive and difficult to verify conditions. (See Smeers, 1974, 1977, and Mokhtar-Kharroubi, 1980.) Improved versions of this method are presented in Abadie [1978], Lasdon et al. [1978], and Lasdon and Warren [1978, 1982]. In Section 10.7, we discuss the convex–simplex method of Zangwill for solving a nonlinear programming problem with linear constraints. This method can be viewed as a reduced gradient method where only one nonbasic variable is changed at a time. A comparison of the convex–simplex method with the reduced gradient method is given by Hans and Zangwill [1972].

In Section 10.8, we present the concept of superbasic variables, which unifies and extends the reduced gradient and convex–simplex methods, and we discuss the use of second-order functional approximations in the superbasic variable space to accelerate the algorithmic convergence. Murtagh and Saunders [1978] present a detailed analysis and algorithmic implementation techniques, along with appropriate factorization methods (see also Gill et al. [1981]). A description of the code MINOS is also presented here, as well as in Murtagh and Saunders [1982, 1983]. Shanno and Marsten [1982] show how the conjugate gradient method can be used to enhance the algorithm and present related restart schemes to maintain second-order information and obtain descent directions.

In this chapter, we discuss search methods for solving a constrained nonlinear programming problem by generating feasible improving directions. Several authors have extended some of the unconstrained optimization techniques to handle simple constraints, such as linear constraints and lower and upper bounds on the variables. One way of handling constraints is to modify an unconstrained optimization technique by simply rejecting infeasible points during the search procedure. However, this approach is not effective, since it leads to premature termination at nonoptimal points. This was demonstrated by results quoted by Friedman and Pinder [1972].

As we discussed earlier in these notes, Goldfarb [1969] and Davies [1970] have extended the Davidon–Fletcher–Powell method to handle linear and nonlinear constraints, respectively. Several gradient-free search methods have also been extended to handle constraints. Glass and Cooper [1965] extended the method of Hooke and Jeeves to deal with constraints. Another attempt to modify the method of Hooke and Jeeves to accommodate constraints is credited to Klingman and Himmelblau [1964]. By projecting the search direction into the intersection of the binding constraints, Davies and Swann [1969] were able to incorporate linear constraints in the method of Rosenbrock with line searches. In Exercise 8.54, we described a variation of the simplex method of Spendley, Hext, and Himmsworth [1962]. Box [1965] developed a constrained version of the simplex method. Several other alternative versions of the method were developed by Ghani [1972], Guin [1968], Friedman and Pinder [1972], Mitchell and Kaplan [1968], and Umida and Ichikawa [1971]. Another method that uses the simplex technique in constrained optimization was proposed by Dixon [1973]. At interior points, the simplex method is used, with occasional quadratic approximations to the function. Whenever constraints are encountered, an attempt is made to move along the boundary. In 1973, Keefer proposed a method in which the basic search uses the Nelder and Mead simplex technique. The lower and upper bound constraints on the variables are dealt with explicitly, while other constraints are handled by a penalty function scheme.

Paviani and Himmelblau [1969] also use the simplex method in conjunction with a penalty function to handle constrained problems. The basic approach is to define a tolerance criterion ϕ_k at iteration k and a penalty function $P(\mathbf{x})$, as discussed in Chapter 9, so that the constraints can be replaced by $P(\mathbf{x}) \leq \phi_k$. In the implementation of the Nelder and Mead method, a point is accepted only if it satisfies this criterion, and ϕ_k is decreased at each iteration. Computational results using this approach are given by Himmelblau [1972].

Several studies for evaluating and testing nonlinear programming algorithms have been made. Stocker [1969] compared five methods for solving 15 constrained and unconstrained optimization test problems of varying degrees of difficulty. In 1970, Colville made a comparative study of many nonlinear programming algorithms. In the study, 34 codes were tested by many participants, with each participant attempting to solve 8 test problems using his own method and code. A summary of the results of this study are reported in Colville [1970]. Computational results are also reported in Himmelblau [1972]. Both studies use a range of nonlinear programming problems with varying degrees of difficulty, including highly nonlinear constraints and objective function, linear constraints, and simple bounds on the variables. Discussions on the comparative performance and evaluation of various algorithms are given in Colville [1970] and Himmelblau [1972]. For further software description and computational comparisons of reduced gradient-type methods, see Gomes and Martinez [1991], Lasdon [1985], Warren et al. [1987], and Wasil et al. [1989].

Chapter 11 Linear Complementary Problem, and Quadratic, Separable, Fractional, and Geometric Programming

In this chapter, we introduce the linear complementarity problem, and then develop some special procedures for solving quadratic, separable, and fractional programming problems. In each case, some variation of the simplex method is used as a solution procedure. For quadratic programming, the KKT system is solved by a complementary pivoting technique that could be used for the more general class of linear complementary problems. Problems that are separable in the variables are approximated by piecewise linearization, and the simplex method is used with a suitable restriction on basis entry. Finally, we discuss geometric programming problems from a Lagrangian duality viewpoint. Such problems find varied applications in engineering design contexts. We also describe two simplex-based methods for solving linear fractional programs.

The following is an outline of the chapter:

Section 11.1: The Linear Complementary Problem We mainly discuss Lemke's algorithm for solving a linear complementary problem (LCP) and show its convergence in a finite number of iterations. Under suitable assumptions, the algorithm either stops with a complementary basic solution or concludes that the original system is inconsistent. Some comments for solving general LCPs are also provided.

Section 11.2: Quadratic Programming We show that the KKT conditions for quadratic programs reduce to a linear complementary problem. The complementary pivoting algorithm is then used to solve the KKT system. Other methods are discussed, and the development of some of these approaches are relegated to the exercises.

Section 11.3: Separable Programming Given a nonlinear programming problem whose objective and constraints are separable in the variables, each function can be approximated by a piecewise linear function using grid points. This is done in such a way that a slight modification of the simplex method can be used to solve the resulting problem. Under suitable convexity assumptions, the optimal objective value to the approximating problem can be made arbitrarily close to that of the original problem. Furthermore, we describe a scheme for generating grid points as needed.

Section 11.4: Linear Fractional Programming Linear fractional programming refers to problems of optimizing the ratio of two linear functions in the presence of linear constraints. We present two procedures for solving the problem. The first method is a simplified version of the convex–simplex method. The second method obtains the optimal solution by solving an equivalent linear program with an additional constraint and an additional variable.

Section 11.5: Geometric Programming This class of problems often arises in engineering applications. We present a technique for solving constrained posynomial geometric programming problems based on the use of Lagrangian duality concepts along with suitable transformations.

11.1 The Linear Complementary Problem

In this section, we briefly introduce the linear complementary problem and present a complementary pivoting algorithm for solving it. Problems of this type arise frequently in engineering applications, game theory, and economics. Also, as will be seen in Section 11.2, the KKT conditions for linear and quadratic programming problems can be written as a linear complementary problem, and, hence, the algorithm presented in this section can be used to solve both linear and quadratic programming problems. Furthermore, the algorithm can be used to solve matrix games.

11.1.1 Definition

Let $\mathbf{M}$ be a given $p \times p$ matrix and let $\mathbf{q}$ be a given p vector. The *linear complementary problem* (*LCP*) is to find vectors $\mathbf{w}$ and $\mathbf{z}$ such that

$$\mathbf{w} - \mathbf{M}\mathbf{z} = \mathbf{q} \tag{11.1}$$

$$w_j \geq 0, z_j \geq 0 \qquad \text{for } j = 1, \ldots, p \tag{11.2}$$

$$w_j z_j = 0 \qquad \text{for } j = 1, \ldots, p \tag{11.3}$$

or to conclude that no such solution exists. Here, (w_j, z_j) is a pair of *complementary variables.* A solution $(\mathbf{w}, \mathbf{z})$ to the above system is called a *complementary feasible solution.* Moreover, such a solution is a *complementary basic feasible solution* if $(\mathbf{w}, \mathbf{z})$ is a basic feasible solution to (11.1) and (11.2) with one variable of the pair (w_j, z_j) basic for each $j = 1, \ldots, p$.

Let $\mathbf{e}_j$ denote a unit vector with a 1 in the jth position, and let $\mathbf{m}_j$ denote the jth column of $\mathbf{M}$ for $j = 1, \ldots, p$. A cone spanned by any p vectors obtained by selecting one vector from each pair $\mathbf{e}_j$, and $-\mathbf{m}_j$ for $j = 1, \ldots, p$, is called a *complementary cone* associated with the matrix $\mathbf{M}$ that defines the system (11.1)–(11.3). Note that there are 2^p such complementary cones, and that the above system has a solution if and only if $\mathbf{q}$ belongs to at least one such cone. Also, observe that if $\mathbf{q}$ belongs to a particular complementary cone and its generators constitute a basis, that is, they are linearly independent, then the corresponding solution is a complementary basic feasible solution, and vice versa. Furthermore, a square matrix $\mathbf{M}$ is called a *Q-matrix* if the corresponding system (11.1)–(11.3) has a solution for each $\mathbf{q} \in E_p$.

Using the concept of complementary cones to characterize a solution to linear complementary problems, we can cast (11.1)–(11.3) as an optimization problem in the following manner. Define a binary variable y_j to take on a value of zero or one accordingly as the variable w_j or z_j is permitted to be positive from the complementary pair (w_j, z_j) for each $j = 1, \ldots, p$, and consider the following *zero–one bilinear programming problem* (BLP):

$$\textbf{\textit{BLP:}} \quad \text{Minimize} \left\{ \sum_{j=1}^{p} y_j w_j + (1 - y_j) z_j : \mathbf{w} - \mathbf{Mz} = \mathbf{q}, \right. \\ \left. \mathbf{w} \geq \mathbf{0}, \mathbf{z} \geq \mathbf{0}, \text{ and } \mathbf{y} \text{ binary} \right\} \tag{11.4}$$

Note that the objective function value for BLP is zero for any feasible solution if and only if $y_j w_j = (1 - y_j) z_j = 0$ for each $j = 1, \ldots, p$, since all the objective terms are nonnegative. Moreover, this happens at optimality if and only if $w_j z_j = 0$ for each $j = 1, \ldots, p$ because of the binariness of $\mathbf{y}$. Hence, a solution $(\mathbf{w}, \mathbf{z})$ is a solution to LCP if and only if it is part of an optimal solution to BLP with a zero objective function value.

Also, observe that we can relax the binary restrictions on the $\mathbf{y}$ variables in (11.4) equivalently to $\mathbf{0} \leq \mathbf{y} \leq \mathbf{1}$, where $\mathbf{1}$ is a vector of p 1's. This follows since, for any partial optimal solution $(\bar{\mathbf{w}}, \bar{\mathbf{z}})$ to BLP, the resulting problem to minimize $\{\sum_{j=1}^{p} [y_j \bar{w}_j + (1 - y_j)\bar{z}_j] : \mathbf{0} \leq \mathbf{y} \leq \mathbf{1}\}$ automatically yields a binary optimal solution for $\mathbf{y}$. Hence, we can also consider (11.4) to be a (continuous) *bilinear programming problem,* which is linear in $\mathbf{y}$ when $(\mathbf{w}, \mathbf{z})$ is fixed in value, and is linear in $(\mathbf{w}, \mathbf{z})$ when $\mathbf{y}$ is fixed in value. Because of the latter property, it follows that, if LCP has a solution, then there exists a solution that is an extreme point of (11.1)–(11.2).

Moreover, using the foregoing characterization of LCP, it can also be cast as one of minimizing a concave objective function $h(\mathbf{y})$ subject to $\mathbf{0} \leq \mathbf{y} \leq \mathbf{1}$, where $h(\mathbf{y}) \equiv \text{minimum} \{\sum_{j=1}^{n} [y_j w_j + (1 - y_j) z_j] : \mathbf{w} - \mathbf{Mz} = \mathbf{q}, \mathbf{w} \geq \mathbf{0}, \text{ and } \mathbf{z} \geq \mathbf{0}\}$ (see Exercise 11.8). Furthermore, assuming that the set defined by (11.1)–(11.2) is bounded, we can linearize BLP into a linear mixed-integer zero-one programming problem (see Exercises 11.8 and 11.9). Hence, we can solve general LCPs using available methods for bilinear programming, concave minimization, or linear integer programming problems. We refer the reader to the Notes and References section for such approaches, including certain specialized techniques such as sequential linear programming methods or interior point approaches when $\mathbf{M}$ possesses particular structural properties. We now proceed to describe a popular simplex type of pivoting method for solving LCPs, which is guaranteed to work under certain nondegeneracy assumptions and when $\mathbf{M}$ satisfies certain properties. However, in practice, it is known to perform well even when these assumptions are violated.

Solving the Linear Complementary Problem

If $\mathbf{q}$ is nonnegative, then we immediately have a solution satisfying (11.1)–(11.3), by letting $\mathbf{w} = \mathbf{q}$ and $\mathbf{z} = \mathbf{0}$. If $\mathbf{q} \ngeq \mathbf{0}$, however, a new column $\mathbf{1}$ and an artificial variable are introduced, leading to the following system, where $\mathbf{1}$ is a vector of p 1's.

$$\mathbf{w} - \mathbf{M}\mathbf{z} - \mathbf{1}z_0 = \mathbf{q} \tag{11.5}$$

$$w_j \geq 0, z_j \geq 0, z_0 \geq 0 \qquad \text{for } j = 1, \ldots, p \tag{11.6}$$

$$w_j z_j = 0 \qquad \text{for } j = 1, \ldots, p \tag{11.7}$$

Letting $z_0 = \text{maximum } \{-q_i : 1 \leq i \leq p\}$, $\mathbf{z} = \mathbf{0}$, and $\mathbf{w} = \mathbf{q} + \mathbf{1}z_0$, we obtain a starting solution to the above system. Through a sequence of pivots, to be specified later, we attempt to drive the artificial variable z_0 to level zero while satisfying (11.5)–(11.7), thus obtaining a solution to the linear complementary problem.

Consider the following definition of an almost complementary basic feasible solution and the definition of an adjacent almost complementary feasible solution. These definitions will be useful in both describing the algorithm and showing its finite convergence.

11.1.2 Definition

Consider the system defined by (11.5)–(11.7). A feasible solution $(\mathbf{w}, \mathbf{z}, z_0)$ to this system is called an *almost complementary basic feasible solution* if

1. $(\mathbf{w}, \mathbf{z}, z_0)$ is a basic feasible solution to (11.5) and (11.6).
2. Neither w_s nor z_s are basic, for some $s \in \{1, \ldots, p\}$.
3. z_0 is basic, and exactly one variable from each complementary pair (w_j, z_j) is basic, for $j = 1, \ldots, p$ and $j \neq s$.

Given an almost complementary basic feasible solution $(\mathbf{w}, \mathbf{z}, z_0)$, where w_s and z_s are both nonbasic, an *adjacent almost complementary basic feasible solution* $(\hat{\mathbf{w}}, \hat{\mathbf{z}}, \hat{z}_0)$ is obtained by introducing either w_s or z_s in the basis if pivoting drives a variable other than z_0 from the basis.

From the above definition, it is clear that each almost complementary basic feasible solution has, at most, two adjacent almost complementary basic feasible solutions. If increasing w_s or z_s drives z_0 out of the basis or produces a ray of the set defined in (11.5) and (11.6), then we have less than two adjacent almost complementary basic feasible solutions.

Summary of Lemke's Complementary Pivoting Algorithm

We summarize below a complementary pivoting algorithm credited to Lemke [1968] for solving the linear complementary problem. A similar scheme due to Cottle and Dantzig

[1968], known as the *principal pivoting method,* is described in Exercise 11.7. Introducing the artificial variable z_0, the former algorithm moves among adjacent almost complementary basic feasible solutions until either a complementary basic feasible solution is obtained or a direction indicating unboundedness of the region defined by (11.5) through (11.7) is found. As will be shown later, under certain assumptions on the matrix $\mathbf{M}$, the algorithm converges in a finite number of steps, with a complementary basic feasible solution.

Initialization Step If $\mathbf{q} \geq \mathbf{0}$, stop; $(\mathbf{w}, \mathbf{z}) = (\mathbf{q}, \mathbf{0})$ is a complementary basic feasible solution. Otherwise, display the system defined by (11.5) and (11.6) in a tableau format. Let $-q_s = \text{maximum}\,\{-q_i : 1 \leq i \leq p\}$, and update the tableau by pivoting at row s and the z_0 column. Thus, the basic variables z_0 and w_j for $j = 1, \ldots, p$ and $j \neq s$ are nonnegative. Let $y_s = z_s$ and go to the main step.

Main Step

1. Let $\mathbf{d}_s$ be the updated column in the current tableau under the variable y_s. If $\mathbf{d}_s \leq \mathbf{0}$, go to step 4. Otherwise, determine the index r by the following minimum ratio test, where $\bar{\mathbf{q}}$ is the updated right-hand side column denoting the values of the basic variables:

$$\frac{\bar{q}_r}{d_{rs}} = \underset{1 \leq i \leq p}{\text{minimum}} \left\{ \frac{\bar{q}_i}{d_{is}} : d_{is} > 0 \right\}$$

   If the basic variable at row r is z_0, go to step 3. Otherwise, go to step 2.

2. The basic variable at row r is either w_l or z_l, for some $l \neq s$. The variable y_s enters the basis and the tableau is updated by pivoting at row r and the y_s column. If the variable that just left the basis is w_l, then let $y_s = z_l$; and if the variable that just left the basis is z_l, then let $y_s = w_l$. Return to step 1.
3. Here y_s enters the basis, and z_0 leaves the basis. Pivot at the y_s column and the z_0 row, producing a complementary basic feasible solution. Stop.
4. Stop with ray termination. A ray $R = \{(\mathbf{w}, \mathbf{z}, z_0) + \lambda \mathbf{d} : \lambda \geq 0\}$ is found such that every point in R satisfies (11.5), (11.6), and (11.7). Here, $(\mathbf{w}, \mathbf{z}, z_0)$ is the almost complementary basic feasible solution associated with the last tableau, and $\mathbf{d}$ is an extreme direction of the set defined by (11.5) and (11.6), having a 1 in the row corresponding to y_s, $-\mathbf{d}_s$ in the rows of the current basic variables and zero everywhere else.

11.1.3 Example (Termination with a Complementary Basic Feasible Solution)

We wish to find a solution to the linear complementary problem defined by

$$\mathbf{M} = \begin{bmatrix} 0 & 0 & -1 & -1 \\ 0 & 0 & 1 & -2 \\ 1 & -1 & 2 & -2 \\ 1 & 2 & -2 & 4 \end{bmatrix} \qquad \mathbf{q} = \begin{bmatrix} 2 \\ 2 \\ -2 \\ -6 \end{bmatrix}$$

Initialization Step Introduce the artificial variable z_0 and form the following tableau:

	w_1	w_2	w_3	w_4	z_1	z_2	z_3	z_4	z_0	RHS
w_1	1	0	0	0	0	0	1	1	-1	2
w_2	0	1	0	0	0	0	-1	2	-1	2
w_3	0	0	1	0	-1	1	-2	2	-1	-2
w_4	0	0	0	1	-1	-2	2	-4	(-1)	-6

Note that the minimum $\{q_i : 1 \leq i \leq 4\} = q_4$, so that we pivot at row 4 and the z_0 column. Go to iteration 1 with $y_s = z_4$.

Iteration 1

	w_1	w_2	w_3	w_4	z_1	z_2	z_3	z_4	z_0	RHS
w_1	1	0	0	-1	1	2	-1	5	0	8
w_2	0	1	0	-1	1	2	-3	6	0	8
w_3	0	0	1	-1	0	3	-4	(6)	0	4
z_0	0	0	0	-1	1	2	-2	4	1	6

Here, $y_s = z_4$ enters the basis. By the minimum ratio test, w_3 leaves the basis; so, for the purpose of the next iteration, $y_s = z_3$. We pivot at the w_3 row and the z_4 column, and we go to iteration 2.

Iteration 2

	w_1	w_2	w_3	w_4	z_1	z_2	z_3	z_4	z_0	RHS
w_1	1	0	$-\frac{5}{6}$	$-\frac{1}{6}$	1	$-\frac{1}{2}$	($\frac{7}{3}$)	0	0	$\frac{14}{3}$
w_2	0	1	-1	0	1	-1	1	0	0	4
z_4	0	0	$\frac{1}{6}$	$-\frac{1}{6}$	0	$\frac{1}{2}$	$-\frac{2}{3}$	1	0	$\frac{2}{3}$
z_0	0	0	$-\frac{2}{3}$	$-\frac{1}{3}$	1	0	$\frac{2}{3}$	0	1	$\frac{10}{3}$

Here, $y_s = z_3$ enters the basis. By the minimum ratio test, w_1 leaves the basis; so, for the purpose of the next iteration, $y_s = z_1$. We pivot at the w_1 row and the z_3 column, and we go to iteration 3.

Iteration 3

	w_1	w_2	w_3	w_4	z_1	z_2	z_3	z_4	z_0	RHS
z_3	$\frac{3}{7}$	0	$-\frac{5}{14}$	$-\frac{1}{14}$	$\frac{3}{7}$	$-\frac{3}{14}$	1	0	0	2
w_2	$-\frac{3}{7}$	1	$-\frac{9}{14}$	$\frac{1}{14}$	$\frac{4}{7}$	$-\frac{11}{14}$	0	0	0	2
z_4	$\frac{2}{7}$	0	$-\frac{1}{14}$	$-\frac{3}{14}$	$\frac{2}{7}$	$\frac{5}{14}$	0	1	0	2
z_0	$-\frac{2}{7}$	0	$-\frac{3}{7}$	$-\frac{2}{7}$	($\frac{5}{7}$)	$\frac{1}{7}$	0	0	1	2

Here, $y_s = z_1$ enters the basis. By the minimum ratio test, z_0 leaves the basis. Pivoting at the z_0 row and the z_1 column gives the complementary basic feasible solution represented by the following tableau:

	w_1	w_2	w_3	w_4	z_1	z_2	z_3	z_4	z_0	RHS
z_3	$\frac{3}{5}$	0	$-\frac{1}{10}$	$\frac{1}{10}$	0	$-\frac{3}{10}$	1	0	$-\frac{3}{5}$	$\frac{4}{5}$
w_2	$-\frac{1}{5}$	1	$-\frac{3}{10}$	$\frac{3}{10}$	0	$-\frac{9}{10}$	0	0	$-\frac{4}{5}$	$\frac{2}{5}$
z_4	$\frac{2}{5}$	0	$\frac{1}{10}$	$-\frac{1}{10}$	0	$\frac{3}{10}$	0	1	$-\frac{2}{5}$	$\frac{6}{5}$
z_1	$-\frac{2}{5}$	0	$-\frac{3}{5}$	$-\frac{2}{5}$	1	$\frac{1}{5}$	0	0	$\frac{7}{5}$	$\frac{14}{5}$

To summarize, the complementary pivoting algorithm produced the point

$$(w_1, w_2, w_3, w_4, z_1, z_2, z_3, z_4) = (0, \tfrac{2}{5}, 0, 0, \tfrac{14}{5}, 0, \tfrac{4}{5}, \tfrac{6}{5})$$

where only one variable from the pair (w_j, z_j) is positive for $j = 1, \ldots, 4$.

11.1.4 Example (Ray Termination)

We wish to find a solution to the linear complementary problem defined by

$$\mathbf{M} = \begin{bmatrix} 0 & 0 & 1 & -1 \\ 0 & 0 & -1 & 2 \\ -1 & 1 & 2 & -2 \\ 1 & -2 & -2 & 2 \end{bmatrix} \qquad \mathbf{q} = \begin{bmatrix} 1 \\ 4 \\ -2 \\ -4 \end{bmatrix}$$

Initialization Step Introduce the artificial variable z_0, leading to the following tableau:

w_1	w_2	w_3	w_4	z_1	z_2	z_3	z_4	z_0	RHS
1	0	0	0	0	0	−1	1	−1	1
0	1	0	0	0	0	1	−2	−1	4
0	0	1	0	1	−1	−2	2	−1	−2
0	0	0	1	−1	2	2	−2	(−1)	−4

Note that minimum $\{q_i: 1 \le i \le 4\} = q_4$, so that we pivot at row 4 and the z_0 column. Go to iteration 1 with $y_s = z_4$.

Iteration 1

	w_1	w_2	w_3	w_4	z_1	z_2	z_3	z_4	z_0	RHS
w_1	1	0	0	−1	1	−2	−3	3	0	5
w_2	0	1	0	−1	1	−2	−1	0	0	8
w_3	0	0	1	−1	2	−3	−4	(4)	0	2
z_0	0	0	0	−1	1	−2	−2	2	1	4

Here, $y_s = z_4$ enters the basis. By the minimum ratio test, w_3 leaves the basis. The tableau is updated by pivoting at the w_3 row and the z_4 column, and we go to iteration 2 with $y_s = z_3$.

Iteration 2

	w_1	w_2	w_3	w_4	z_1	z_2	z_3	z_4	z_0	RHS
w_1	1	0	$-\frac{3}{4}$	$-\frac{1}{4}$	$-\frac{1}{2}$	$\frac{1}{4}$	0	0	0	$\frac{7}{2}$
w_2	0	1	0	-1	1	-2	-1	0	0	8
z_4	0	0	$\frac{1}{4}$	$-\frac{1}{4}$	$\frac{1}{2}$	$-\frac{3}{4}$	-1	1	0	$\frac{1}{2}$
z_0	0	0	$-\frac{1}{2}$	$-\frac{1}{2}$	0	$-\frac{1}{2}$	0	0	1	3

Here, $y_s = z_3$ should enter the basis. However, all the entries under the z_3 column are nonpositive, so we stop with ray termination. We have thus found the ray

$$R = \{(\mathbf{w}, \mathbf{z}, z_0) = (\tfrac{7}{2}, 8, 0, 0, 0, 0, 0, \tfrac{1}{2}, 3) + \lambda(0, 1, 0, 0, 0, 0, 1, 1, 0) : \lambda \geq 0\}$$

where every point on the ray satisfies (11.5)–(11.7).

Finite Convergence of the Complementary Pivoting Algorithm

The following lemma shows that the algorithm must stop in a finite number of iterations, either with a complementary basic feasible solution or with ray termination. Under certain conditions on the matrix **M**, the algorithm stops with a complementary basic feasible solution.

11.1.5 Lemma

Suppose that each almost complementary basic feasible solution of the system (11.5)–(11.7) is nondegenerate; that is, each basic variable is positive. Then, none of the points generated by the complementary pivoting algorithm is repeated, and, furthermore, the algorithm must stop in a finite number of steps.

Proof

Let $(\mathbf{w}, \mathbf{z}, z_0)$ be an almost complementary basic feasible solution, where w_s and z_s are both nonbasic. Then, $(\mathbf{w}, \mathbf{z}, z_0)$ has, at most, two adjacent almost complementary basic feasible solutions, one obtained by introducing w_s in the basis and the other obtained by introducing z_s in the basis.* By the nondegeneracy assumption, each of these solutions is distinct from $(\mathbf{w}, \mathbf{z}, z_0)$.

We now show that none of the almost complementary basic feasible solutions generated by the algorithm is repeated. Let $(\mathbf{w}, \mathbf{z}, z_0)_\nu$ be the point generated at a general iteration ν. By contradiction, suppose that $(\mathbf{w}, \mathbf{z}, z_0)_{k+\alpha} = (\mathbf{w}, \mathbf{z}, z_0)_k$ for some positive integers k and α, where $k + \alpha$ is the smallest index for which a repetition is observed.

*Note that $(\mathbf{w}, \mathbf{z}, z_0)$ may have less than two adjacent almost complementary basic feasible solutions. In this case, the column under w_s or z_s is ≤ 0, or else introducing w_s or z_s in the basis drives z_0 out of the basis, thus producing a complementary basic feasible solution.

By the nondegeneracy assumption, $\alpha > 1$. Furthermore, by the rules of the algorithm, $\alpha > 2$. But since $(\mathbf{w}, \mathbf{z}, z_0)_{k+\alpha-1}$ is adjacent to $(\mathbf{w}, \mathbf{z}, z_0)_{k+\alpha}$, it is adjacent to $(\mathbf{w}, \mathbf{z}, z_0)_k$. If $k = 1$, and since $(\mathbf{w}, \mathbf{z}, z_0)_k$ has exactly one adjacent almost complementary basic feasible solution, $(\mathbf{w}, \mathbf{z}, z_0)_{k+\alpha-1} = (\mathbf{w}, \mathbf{z}, z_0)_{k+1}$, and, hence, a repetition occurs at iteration $k + \alpha - 1$, contradicting our assumption that the first repetition occurs at iteration $k + \alpha$. If $k \geq 2$, then $(\mathbf{w}, \mathbf{z}, z_0)_{k+\alpha-1}$ is adjacent to $(\mathbf{w}, \mathbf{z}, z_0)_k$ and, hence, must be equal to $(\mathbf{w}, \mathbf{z}, z_0)_{k+1}$ or to $(\mathbf{w}, \mathbf{z}, z_0)_{k-1}$. In either case, a repetition occurs at iteration $(\mathbf{w}, \mathbf{z}, z_0)_{k+\alpha-1}$, which contradicts our assumption. Thus, the points generated by the algorithm are distinct.

Since there is only a finite number of almost complementary basic feasible solutions, and since none of them is repeated, the algorithm stops in a finite number of steps with a complementary basic feasible solution or with ray termination. This completes the proof.

To prove the main convergence result specified by Theorem 11.1.8, Lemma 11.1.6 and Definition 11.1.7 are needed. The lemma gives certain implications of ray termination, and the definition introduces the concept of a copositive-plus matrix.

11.1.6 Lemma

Suppose that each almost complementary basic feasible solution of the system defined by (11.5)–(11.7) is nondegenerate. Suppose that the complementary pivoting algorithm is used to solve this system, and, further, suppose that ray termination occurs. In particular, assume that at termination we have the almost complementary basic feasible solution $(\bar{\mathbf{w}}, \bar{\mathbf{z}}, \bar{z}_0)$ and the extreme direction $(\hat{\mathbf{w}}, \hat{\mathbf{z}}, \hat{z}_0)$, giving the ray $R = \{(\bar{\mathbf{w}}, \bar{\mathbf{z}}, \bar{z}_0) + \lambda(\hat{\mathbf{w}}, \hat{\mathbf{z}}, \hat{z}_0) : \lambda \geq 0\}$. Then,

1. $(\hat{\mathbf{w}}, \hat{\mathbf{z}}, \hat{z}_0) \neq (\mathbf{0}, \mathbf{0}, 0)$, $(\hat{\mathbf{w}}, \hat{\mathbf{z}}) \geq \mathbf{0}$, $\hat{z}_0 \geq 0$
2. $\hat{\mathbf{w}} - \mathbf{M}\hat{\mathbf{z}} - \mathbf{1}\hat{z}_0 = \mathbf{0}$
3. $\bar{\mathbf{w}}'\bar{\mathbf{z}} = \bar{\mathbf{w}}'\hat{\mathbf{z}} = \hat{\mathbf{w}}'\bar{\mathbf{z}} = \hat{\mathbf{w}}'\hat{\mathbf{z}} = 0$
4. $\hat{\mathbf{z}} \neq \mathbf{0}$
5. $\hat{\mathbf{z}}'\mathbf{M}\hat{\mathbf{z}} = -\mathbf{1}'\hat{\mathbf{z}}\hat{z}_0 \leq 0$

Proof

Since $(\hat{\mathbf{w}}, \hat{\mathbf{z}}, \hat{z}_0)$ is an extreme direction of the set defined by (11.5) and (11.6), parts 1 and 2 are immediate by Theorem 2.6.6. Recall that every point on the ray R satisfies (11.7), so that $0 = (\bar{\mathbf{w}} + \lambda\hat{\mathbf{w}})'(\bar{\mathbf{z}} + \lambda\hat{\mathbf{z}})$ for each $\lambda \geq 0$. This, together with the nonnegativity of $\bar{\mathbf{w}}$, $\hat{\mathbf{w}}$, $\bar{\mathbf{z}}$, and $\hat{\mathbf{z}}$, imply that

$$\bar{\mathbf{w}}'\bar{\mathbf{z}} = \bar{\mathbf{w}}'\hat{\mathbf{z}} = \hat{\mathbf{w}}'\bar{\mathbf{z}} = \hat{\mathbf{w}}'\hat{\mathbf{z}} = 0 \tag{11.8}$$

Therefore, part 3 holds.

We now show that $\hat{\mathbf{z}} \neq \mathbf{0}$. By contradiction, suppose that $\hat{\mathbf{z}} = \mathbf{0}$. Note that $\hat{z}_0 > 0$, because otherwise $\hat{z}_0 = 0$; and from part 2, we get $\hat{\mathbf{w}} = \mathbf{0}$, contradicting the fact that $(\hat{\mathbf{w}}, \hat{\mathbf{z}}, \hat{z}_0) \neq (\mathbf{0}, \mathbf{0}, 0)$. Thus, $\hat{z}_0 > 0$ and $\hat{\mathbf{w}} = \mathbf{1}\hat{z}_0$.

We have proved that if $\hat{\mathbf{z}} = \mathbf{0}$, then $\hat{z}_0 > 0$ and $\hat{\mathbf{w}} = \mathbf{1}\hat{z}_0$. From (11.8), we get $0 = \hat{\mathbf{w}}'\bar{\mathbf{z}}$. Thus, $\mathbf{1}'\bar{\mathbf{z}} = 0$; and since $\bar{\mathbf{z}} \geq \mathbf{0}$, then $\bar{\mathbf{z}} = \mathbf{0}$. By the nondegeneracy assumption, every component of $\bar{\mathbf{z}}$ is nonbasic. Furthermore, $\bar{z}_0$ is basic, and we must have exactly $p - 1$ basic components of $\bar{\mathbf{w}}$. In particular, since $\bar{\mathbf{w}} - \mathbf{M}\bar{\mathbf{z}} - \mathbf{1}\bar{z}_0 = \mathbf{q}$ and since $\bar{\mathbf{z}} = \mathbf{0}$, then $\bar{z}_0 = \text{maximum } \{-q_i : 1 \leq i \leq p\}$. This shows that the almost complementary basic feasible solution $(\bar{\mathbf{w}}, \bar{\mathbf{z}}, \bar{z}_0)$ is the starting solution, which is impossible by Lemma 11.1.5. Therefore, $\hat{\mathbf{z}} \neq \mathbf{0}$ and part 4 holds. Multiplying $\hat{\mathbf{w}} - \mathbf{M}\hat{\mathbf{z}} - \mathbf{1}\hat{z}_0 = \mathbf{0}$ by $\hat{\mathbf{z}}'$, and noting from (11.8) that $\hat{\mathbf{z}}'\hat{\mathbf{w}} = 0$, then $\hat{\mathbf{z}}'\mathbf{M}\hat{\mathbf{z}} = -\hat{\mathbf{z}}'\mathbf{1}\hat{z}_0 \leq 0$ and part 5 follows. This completes the proof.

11.1.7 Definition

Let $\mathbf{M}$ be a $p \times p$ matrix. Then, $\mathbf{M}$ is said to be *copositive* if $\mathbf{z}'\mathbf{M}\mathbf{z} \geq 0$ for each $\mathbf{z} \geq \mathbf{0}$. Furthermore, $\mathbf{M}$ is said to be *copositive-plus* if it is copositive and if $\mathbf{z} \geq \mathbf{0}$ and $\mathbf{z}'\mathbf{M}\mathbf{z} = 0$ imply that $(\mathbf{M} + \mathbf{M}')\mathbf{z} = \mathbf{0}$.

Theorem 11.1.8 shows that if the system defined by (11.1) and (11.2) is consistent, and if the matrix $\mathbf{M}$ is copositive-plus, then the complementary pivoting algorithm will produce a complementary basic feasible solution in a finite number of steps.

11.1.8 Theorem

Suppose that each almost complementary basic feasible solution to the system defined by (11.5)–(11.7) is nondegenerate, and suppose that $\mathbf{M}$ is copositive-plus. Then, the complementary pivoting algorithm stops in a finite number of steps. In particular, if the system defined by (11.1) and (11.2) is consistent, then the algorithm stops with a complementary basic feasible solution to the system defined by (11.1)–(11.3). On the other hand, if the system defined in (11.1) and (11.2) is inconsistent, then the algorithm stops with ray termination.

Proof

By Lemma 11.1.5, the complementary pivoting algorithm stops in a finite number of steps. Now suppose that the algorithm stops with ray termination. In particular, suppose that $(\bar{\mathbf{w}}, \bar{\mathbf{z}}, \bar{z}_0)$ is the almost complementary basic feasible solution and $(\hat{\mathbf{w}}, \hat{\mathbf{z}}, \hat{z}_0)$ is the extreme direction associated with the final tableau. By Lemma 11.1.6,

$$\hat{\mathbf{z}} \geq \mathbf{0}, \qquad \hat{\mathbf{z}} \neq \mathbf{0}, \qquad \text{and} \qquad \hat{\mathbf{z}}'\mathbf{M}\hat{\mathbf{z}} = -\mathbf{1}'\hat{\mathbf{z}}\hat{z}_0 \leq 0 \tag{11.9}$$

But since $\mathbf{M}$ is copositive-plus, $\hat{\mathbf{z}}'\mathbf{M}\hat{\mathbf{z}} \geq 0$. From (11.9), it follows that $0 = \hat{\mathbf{z}}'\mathbf{M}\hat{\mathbf{z}} = -\mathbf{1}'\hat{\mathbf{z}}\hat{z}_0$. Since $\hat{\mathbf{z}} \neq \mathbf{0}$, $\hat{z}_0 = 0$. But since $(\hat{\mathbf{w}}, \hat{\mathbf{z}}, \hat{z}_0)$ is a direction of the set defined by (11.5) and (11.6), $\hat{\mathbf{w}} - \mathbf{M}\hat{\mathbf{z}} - \mathbf{1}\hat{z}_0 = \mathbf{0}$, and, hence,

$$\hat{\mathbf{w}} = \mathbf{M}\hat{\mathbf{z}} \tag{11.10}$$

Now we show that $\mathbf{q}'\hat{\mathbf{z}} < 0$. Since $\hat{\mathbf{z}}'\mathbf{M}\hat{\mathbf{z}} = 0$ and $\mathbf{M}$ is copositive-plus, $(\mathbf{M} + \mathbf{M}')\hat{\mathbf{z}} = \mathbf{0}$. This, together with part 3 of Lemma 11.1.6 and the fact that $\bar{\mathbf{w}} = \mathbf{q} + \mathbf{M}\bar{\mathbf{z}} + \mathbf{1}\bar{z}_0$, implies

$$0 = \bar{\mathbf{w}}'\hat{\mathbf{z}} = (\mathbf{q} + \mathbf{M}\bar{\mathbf{z}} + \mathbf{1}\bar{z}_0)'\hat{\mathbf{z}} = \mathbf{q}'\hat{\mathbf{z}} - \bar{\mathbf{z}}'\mathbf{M}\hat{\mathbf{z}} + \bar{z}_0\mathbf{1}'\hat{\mathbf{z}} \tag{11.11}$$

From (11.10), $\mathbf{M}\hat{\mathbf{z}} = \hat{\mathbf{w}}$, and, hence, from part 3 of Lemma 11.1.6, it follows that $\bar{\mathbf{z}}'\mathbf{M}\hat{\mathbf{z}} = 0$. Furthermore, $\bar{z}_0 > 0$ and $\mathbf{1}'\hat{\mathbf{z}} > 0$ by (11.9). Substituting in (11.11), it follows that $\mathbf{q}'\hat{\mathbf{z}} < 0$.

To summarize, we have shown that $\mathbf{M}\hat{\mathbf{z}} = \hat{\mathbf{w}} \geq \mathbf{0}$. Since $(\mathbf{M} + \mathbf{M}')\hat{\mathbf{z}} = \mathbf{0}$, then $\mathbf{M}'\hat{\mathbf{z}} = -\mathbf{M}\hat{\mathbf{z}} \leq \mathbf{0}$, $-\mathbf{I}\hat{\mathbf{z}} \leq \mathbf{0}$, and $\mathbf{q}'\hat{\mathbf{z}} < 0$. Thus, the system $\mathbf{M}'\mathbf{y} \leq \mathbf{0}$, $-\mathbf{I}\mathbf{y} \leq \mathbf{0}$, and

$\mathbf{q}'\mathbf{y} < 0$ has a solution, say, $\mathbf{y} = \hat{\mathbf{z}}$. By Theorem 2.4.5, it follows that the system $\mathbf{w} - \mathbf{Mz} = \mathbf{q}$, $\mathbf{w} \geq \mathbf{0}$, $\mathbf{z} \geq \mathbf{0}$ has no solution.

Now if the system defined by (11.1) and (11.2) is consistent, then the algorithm must stop with a complementary basic feasible solution, because otherwise the algorithm would stop with ray termination, which, as we showed above, is only possible if the system (11.1) and (11.2) is inconsistent. If the system defined by (11.1) and (11.2) is inconsistent, then the algorithm obviously could not stop with a complementary basic feasible solution and, hence, must stop with ray termination. This completes the proof.

Corollary

If $\mathbf{M}$ has nonnegative entries, with positive diagonal elements, then the complementary pivoting algorithm stops in a finite number of steps with a complementary basic feasible solution.

Proof

First, note that by the stated assumption on $\mathbf{M}$, the system $\mathbf{w} - \mathbf{Mz} = \mathbf{q}$, $(\mathbf{w}, \mathbf{z}) \geq \mathbf{0}$ has a solution, say, by choosing $\mathbf{z}$ sufficiently large so that $\mathbf{w} = \mathbf{Mz} + \mathbf{q} \geq \mathbf{0}$. The result then follows from the theorem by noting that $\mathbf{M}$ is copositive-plus.

11.2 Quadratic Programming

Quadratic programming represents a special class of nonlinear programming in which the objective function is quadratic and the constraints are linear. In this section, we show that the KKT conditions of a quadratic programming problem reduce to a linear complementary problem. Thus, the complementary pivoting algorithm described in Section 11.1 can be used for solving a quadratic programming problem.

Several other special procedures for solving quadratic programming problems are discussed in the exercises at the end of the chapter. In particular, Exercise 11.20 shows that if the quadratic programming problem is of the form to minimize $\mathbf{c}'\mathbf{x} + \frac{1}{2}\mathbf{x}'\mathbf{H}\mathbf{x}$ subject to only equality constraints $\mathbf{Ax} = \mathbf{b}$, where $\mathbf{A}$ is an $m \times n$ matrix of rank m and $\mathbf{H}$ is positive definite on $\{\mathbf{x}: \mathbf{Ax} = \mathbf{0}\}$, then the unique solution to this problem is obtainable via the solution of a linear system of equations, typically done using an *LU* factorization approach (see Appendix A.2). If inequality constraints are present, we can use the reduced gradient method as described in the previous chapter or, as is most popularly done, we can adopt an *active set* strategy as follows. Here, given a feasible solution, an equality constrained quadratic programming problem is solved to find a correction direction over the nullspace of the active constraints; and either optimality is verified or the set of designated active constraints (to be held as equalities) is modified, at the current solution itself or at a revised solution, and the process is repeated. Exercises 11.21 and 11.22 describe such strategies. For convex quadratic programming problems, we can also extend the primal–dual path-following algorithm described in Chapter 9, using the barrier penalty function algorithm along with Newton's method, to derive a polynomial-time algorithm. The Notes and References section directs the reader to this and to other interior point approaches for convex quadratic programming problems.

When the Hessian of the gradient function is not positive semidefinite, however, finding a global minimum for the underlying quadratic programming problem becomes a difficult task. In fact, quadratic (minimization) problems with even a single negative eigenvalue for the Hessian are known to be NP-Hard. Assuming that an optimum exists, Exercise 11.14 shows how a general quadratic program can be posed as a linear program with complementarity constraints, for which some of the zero–one linearization approaches suggested in the foregoing section and in Exercise 11.8 can be employed.

We now proceed to consider the solution of quadratic programs through linear complementarity problems.

The Karush–Kuhn–Tucker System

Consider the following quadratic programming problem:

$$\begin{array}{ll} \text{Minimize} & \mathbf{c}'\mathbf{x} + \tfrac{1}{2}\mathbf{x}'\mathbf{H}\mathbf{x} \\ \text{subject to} & \mathbf{A}\mathbf{x} \leq \mathbf{b} \\ & \mathbf{x} \geq \mathbf{0} \end{array}$$

where $\mathbf{c}$ is an n vector, $\mathbf{b}$ is an m vector, $\mathbf{A}$ is an $m \times n$ matrix, and $\mathbf{H}$ is an $n \times n$ symmetric matrix. (Note that a more general set of linear constraints can be cast in this format through standard linear transformations. In particular, if the constraints were of the form $\mathbf{A}'\mathbf{x}' = \mathbf{b}'$, $\mathbf{x}' \geq \mathbf{0}$, then the above type of constraints might be an equivalent representation of this region in some nonbasic variable space, using the partitioning scheme described in Section 10.6.) Denoting the Lagrangian multiplier vectors of the constraints $\mathbf{A}\mathbf{x} \geq \mathbf{b}$ and $\mathbf{x} \geq \mathbf{0}$ by $\mathbf{u}$ and $\mathbf{v}$, respectively, and denoting the vector of slack variables by $\mathbf{y}$, the KKT conditions can be written as

$$\begin{aligned} \mathbf{A}\mathbf{x} + \mathbf{y} &= \mathbf{b} \\ -\mathbf{H}\mathbf{x} - \mathbf{A}'\mathbf{u} + \mathbf{v} &= \mathbf{c} \\ \mathbf{x}'\mathbf{v} = 0, \qquad \mathbf{u}'\mathbf{y} &= 0 \\ \mathbf{x}, \mathbf{y}, \mathbf{u}, \mathbf{v} &\geq \mathbf{0} \end{aligned}$$

Now, letting

$$\mathbf{M} = \begin{bmatrix} \mathbf{0} & -\mathbf{A} \\ \mathbf{A}' & \mathbf{H} \end{bmatrix} \qquad \mathbf{q} = \begin{bmatrix} \mathbf{b} \\ \mathbf{c} \end{bmatrix} \qquad \mathbf{w} = \begin{pmatrix} \mathbf{y} \\ \mathbf{v} \end{pmatrix} \qquad \mathbf{z} = \begin{pmatrix} \mathbf{u} \\ \mathbf{x} \end{pmatrix}$$

we can rewrite the KKT conditions as the linear complementary problem $\mathbf{w} - \mathbf{M}\mathbf{z} = \mathbf{q}$, $\mathbf{w}'\mathbf{z} = 0$, $(\mathbf{w}, \mathbf{z}) \geq \mathbf{0}$. Thus the complementary pivoting algorithm discussed in Section 11.1 can be used to find a KKT point of the quadratic programming problem.

11.2.1 Example (Finite Optimal Solution)

Consider the following quadratic programming problem:

$$\begin{array}{ll} \text{Minimize} & -2x_1 - 6x_2 + x_1^2 - 2x_1x_2 + 2x_2^2 \\ \text{subject to} & x_1 + x_2 \leq 2 \\ & -x_1 + 2x_2 \leq 2 \\ & x_1, \quad x_2 \geq 0 \end{array}$$

Note that

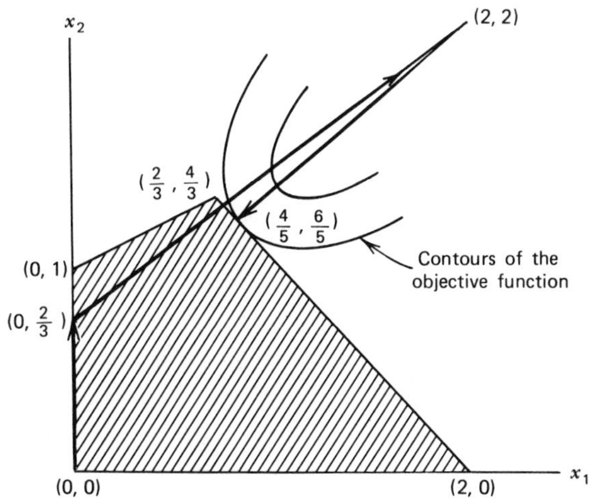

Figure 11.1 Points generated by the complementary pivoting algorithm.

$$\mathbf{A} = \begin{bmatrix} 1 & 1 \\ -1 & 2 \end{bmatrix} \quad \mathbf{H} = \begin{bmatrix} 2 & -2 \\ -2 & 4 \end{bmatrix} \quad \mathbf{b} = \begin{bmatrix} 2 \\ 2 \end{bmatrix} \quad \mathbf{c} = \begin{bmatrix} -2 \\ -6 \end{bmatrix}$$

Denote the vector of slacks by $\mathbf{y}$ and the Lagrangian multiplier vectors for the constraints $\mathbf{Ax} \leq \mathbf{b}$ and $\mathbf{x} \geq \mathbf{0}$ by $\mathbf{u}$ and $\mathbf{v}$. Let

$$\mathbf{M} = \begin{bmatrix} \mathbf{0} & -\mathbf{A} \\ \mathbf{A}^t & \mathbf{H} \end{bmatrix} \quad \mathbf{q} = \begin{bmatrix} \mathbf{b} \\ \mathbf{c} \end{bmatrix} \quad \mathbf{w} = \begin{bmatrix} \mathbf{y} \\ \mathbf{v} \end{bmatrix} \quad \mathbf{z} = \begin{bmatrix} \mathbf{u} \\ \mathbf{x} \end{bmatrix}$$

Then, the KKT conditions reduce to finding a solution to the system $\mathbf{w} - \mathbf{Mz} = \mathbf{q}$, $\mathbf{w}^t\mathbf{z} = 0$ and $(\mathbf{w}, \mathbf{z}) \geq \mathbf{0}$, where

$$\mathbf{M} = \begin{bmatrix} 0 & 0 & -1 & -1 \\ 0 & 0 & 1 & -2 \\ 1 & -1 & 2 & -2 \\ 1 & 2 & -2 & 4 \end{bmatrix} \quad \mathbf{q} = \begin{bmatrix} 2 \\ 2 \\ -2 \\ -6 \end{bmatrix}$$

The problem of finding a complementary basic feasible solution to the above system was solved in Example 11.1.3, producing the KKT point $(x_1, x_2) = (z_3, z_4) = (\frac{4}{5}, \frac{6}{5})$. Reviewing Example 11.1.3, note that the complementary pivoting algorithm started from the point (0, 0), then moved to the point $(0, \frac{2}{3})$, then to the point (2, 2), and then finally to the KKT point $(\frac{4}{5}, \frac{6}{5})$. Since $\mathbf{H}$ is positive definite, the objective function is convex, and so, the KKT point $(\frac{4}{5}, \frac{6}{5})$ is indeed optimal. The path taken by the complementary pivoting algorithm to produce the optimal solution is shown in Figure 11.1.

11.2.2 Example (Unbounded Optimal Solution)

Consider the following quadratic programming problem:

$$\begin{aligned} \text{Minimize} \quad & -2x_1 - 4x_2 + x_1^2 - 2x_1x_2 + x_2^2 \\ \text{subject to} \quad & -x_1 + x_2 \leq 1 \\ & x_1 - 2x_2 \leq 4 \\ & x_1, \quad x_2 \geq 0 \end{aligned}$$

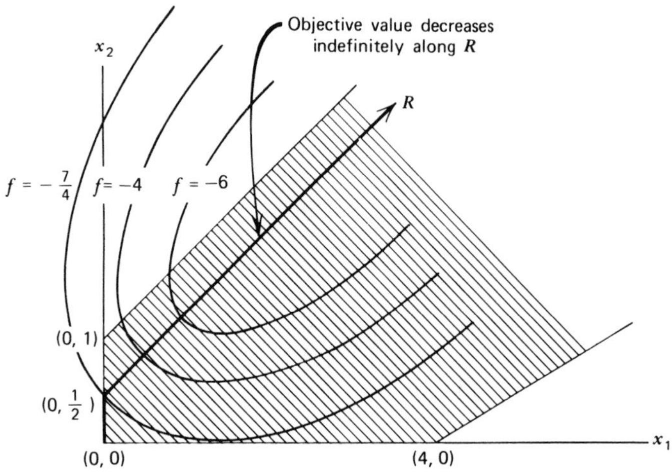

Figure 11.2 Unbounded optimal solution and ray termination.

Note that

$$\mathbf{A} = \begin{bmatrix} -1 & 1 \\ 1 & -2 \end{bmatrix} \quad \mathbf{H} = \begin{bmatrix} 2 & -2 \\ -2 & 2 \end{bmatrix} \quad \mathbf{b} = \begin{bmatrix} 1 \\ 4 \end{bmatrix} \quad \mathbf{c} = \begin{bmatrix} -2 \\ -4 \end{bmatrix}$$

Denote the vector of slacks by $\mathbf{y}$ and the Lagrangian multiplier vectors for the constraints $\mathbf{Ax} \le \mathbf{b}$ and $\mathbf{x} \ge \mathbf{0}$ by $\mathbf{u}$ and $\mathbf{v}$. Let

$$\mathbf{M} = \begin{bmatrix} \mathbf{0} & -\mathbf{A} \\ \mathbf{A}^t & \mathbf{H} \end{bmatrix} \quad \mathbf{q} = \begin{bmatrix} \mathbf{b} \\ \mathbf{c} \end{bmatrix} \quad \mathbf{w} = \begin{pmatrix} \mathbf{y} \\ \mathbf{v} \end{pmatrix} \quad \mathbf{z} = \begin{pmatrix} \mathbf{u} \\ \mathbf{x} \end{pmatrix}$$

Then, solving the KKT conditions reduces to finding a solution to the system $\mathbf{w} - \mathbf{Mz} = \mathbf{q}$, $\mathbf{w}^t\mathbf{z} = 0$ and $(\mathbf{w}, \mathbf{z}) \ge \mathbf{0}$, where

$$\mathbf{M} = \begin{bmatrix} 0 & 0 & 1 & -1 \\ 0 & 0 & -1 & 2 \\ -1 & 1 & 2 & -2 \\ 1 & -2 & -2 & 2 \end{bmatrix} \quad \mathbf{q} = \begin{bmatrix} 1 \\ 4 \\ -2 \\ -4 \end{bmatrix}$$

The problem of finding a complementary basic feasible solution to the above system was solved in Example 11.1.4. As shown in that example, the complementary pivoting algorithm stopped with ray termination and was unable to produce a complementary basic feasible solution. The reason for this is that the optimal solution is unbounded along the ray R produced by the algorithm. Mapped in the (x_1, x_2) space, the ray $R = \{(0, \frac{1}{2}) + \lambda(1, 1) : \lambda \ge 0\}$ leads to an unbounded optimal solution, as shown in Figure 11.2.

Convergence Analysis of the Quadratic Programming Complementary Pivoting Algorithm

In Section 11.1, we showed that, under nondegeneracy, the complementary pivoting algorithm stops in a finite number of steps with either a complementary basic feasible solution or a ray termination. We also showed that if the matrix $\mathbf{M}$ associated with the linear complementary problem is copositive-plus, and the linear constraints are consistent, then the algorithm produces a complementary basic feasible solution. Theorem 11.2.3, gives sufficient conditions for the matrix $\mathbf{M}$ associated with the quadratic problem to be copositive-plus. In addition, Theorem 11.2.4 also gives several conditions under which the complementary pivoting algorithm produces a KKT point and shows that ray termination is only possible if the quadratic programming problem has an unbounded optimal solution.

11.2.3 Theorem

Let $\mathbf{A}$ be an $m \times n$ matrix, and let $\mathbf{H}$ be an $n \times n$ symmetric matrix. If $\mathbf{y}^t\mathbf{H}\mathbf{y} \geq 0$ for each $\mathbf{y} \geq \mathbf{0}$, then the matrix

$$\mathbf{M} = \begin{bmatrix} \mathbf{0} & -\mathbf{A} \\ \mathbf{A}^t & \mathbf{H} \end{bmatrix}$$

is copositive. In addition, if $\mathbf{y} \geq \mathbf{0}$ and $\mathbf{y}^t\mathbf{H}\mathbf{y} = 0$ imply that $\mathbf{H}\mathbf{y} = \mathbf{0}$, then $\mathbf{M}$ is copositive-plus.

Proof

First, we show that $\mathbf{M}$ is copositive. Let $\mathbf{z}^t = (\mathbf{x}^t, \mathbf{y}^t) \geq \mathbf{0}$. Then,

$$\mathbf{z}^t\mathbf{M}\mathbf{z} = (\mathbf{x}^t, \mathbf{y}^t) \begin{bmatrix} \mathbf{0} & -\mathbf{A} \\ \mathbf{A}^t & \mathbf{H} \end{bmatrix} \begin{pmatrix} \mathbf{x} \\ \mathbf{y} \end{pmatrix} = \mathbf{y}^t\mathbf{H}\mathbf{y} \tag{11.12}$$

By assumption, $\mathbf{y}^t\mathbf{H}\mathbf{y} \geq 0$, and, hence, $\mathbf{H}$ is copositive. To show that $\mathbf{M}$ is copositive-plus, suppose that $\mathbf{z} \geq \mathbf{0}$ and $\mathbf{z}^t\mathbf{M}\mathbf{z} = 0$. It suffices to show that $(\mathbf{M} + \mathbf{M}^t)\mathbf{z} = \mathbf{0}$. But

$$\mathbf{M} + \mathbf{M}^t = \begin{bmatrix} \mathbf{0} & \mathbf{0} \\ \mathbf{0} & 2\mathbf{H} \end{bmatrix}$$

and, hence,

$$(\mathbf{M} + \mathbf{M}^t)\mathbf{z} = \begin{bmatrix} \mathbf{0} \\ 2\mathbf{H}\mathbf{y} \end{bmatrix}$$

Since $\mathbf{z}^t\mathbf{M}\mathbf{z} = 0$, then $\mathbf{y}^t\mathbf{H}\mathbf{y} = 0$ by (11.12). By assumption, since $\mathbf{y} \geq \mathbf{0}$ and $\mathbf{y}^t\mathbf{H}\mathbf{y} = 0$, then $\mathbf{H}\mathbf{y} = \mathbf{0}$, and, hence, $(\mathbf{M} + \mathbf{M}^t)\mathbf{z} = \mathbf{0}$, so that $\mathbf{M}$ is copositive-plus. This completes the proof.

Corollary 1

If $\mathbf{H}$ is positive semidefinite, then $\mathbf{y}^t\mathbf{H}\mathbf{y} = 0$ implies that $\mathbf{H}\mathbf{y} = \mathbf{0}$, so $\mathbf{M}$ is copositive-plus.

Proof

It suffices to show that $\mathbf{y}^t\mathbf{H}\mathbf{y} = 0$ implies that $\mathbf{H}\mathbf{y} = \mathbf{0}$. Let $\mathbf{H}\mathbf{y} = \mathbf{d}$, and noting that $\mathbf{H}$ is positive semidefinite, we get

$$0 \leq (\mathbf{y}^t - \lambda\mathbf{d}^t)\mathbf{H}(\mathbf{y} - \lambda\mathbf{d}) = \mathbf{y}^t\mathbf{H}\mathbf{y} + \lambda^2\mathbf{d}^t\mathbf{H}\mathbf{d} - 2\lambda\|\mathbf{d}\|^2$$

Since $\mathbf{y}^t\mathbf{H}\mathbf{y} = 0$, then dividing the above inequality by λ, and letting $\lambda \to 0^+$, it follows that $\mathbf{0} = \mathbf{d} = \mathbf{H}\mathbf{y}$.

Corollary 2

If **H** has nonnegative entries, then **M** is copositive. Further, if **H** has nonnegative elements with positive diagonal elements, then **M** is copositive-plus.

Proof

If $\mathbf{y} \geq \mathbf{0}$ and $\mathbf{y}'\mathbf{H}\mathbf{y} = 0$, then $\mathbf{y} = \mathbf{0}$, and, hence, $\mathbf{H}\mathbf{y} = \mathbf{0}$. By the theorem, **M** is copositive-plus.

11.2.4 Theorem

Consider the problem to minimize $\mathbf{c}'\mathbf{x} + \frac{1}{2}\mathbf{x}'\mathbf{H}\mathbf{x}$ subject to $\mathbf{A}\mathbf{x} \leq \mathbf{b}$, $\mathbf{x} \geq \mathbf{0}$. Suppose that the feasible region is not empty. Further, suppose that the complementary pivoting algorithm described in Section 11.1 is used in an attempt to find a solution to the KKT system $\mathbf{w} - \mathbf{M}\mathbf{z} = \mathbf{q}$, $(\mathbf{w}, \mathbf{z}) \geq \mathbf{0}$, $\mathbf{w}'\mathbf{z} = 0$, where

$$\mathbf{M} = \begin{bmatrix} \mathbf{0} & -\mathbf{A} \\ \mathbf{A}' & \mathbf{H} \end{bmatrix} \qquad \mathbf{q} = \begin{pmatrix} \mathbf{b} \\ \mathbf{c} \end{pmatrix} \qquad \mathbf{w} = \begin{pmatrix} \mathbf{y} \\ \mathbf{v} \end{pmatrix} \qquad \mathbf{z} = \begin{pmatrix} \mathbf{u} \\ \mathbf{x} \end{pmatrix}$$

y is the vector of slack variables, and **u** and **v** are the Lagrangian multiplier vectors associated with the constraints $\mathbf{A}\mathbf{x} \leq \mathbf{b}$ and $\mathbf{x} \geq \mathbf{0}$, respectively. In the absence of degeneracy, under any of the following conditions, the algorithm stops in a finite number of iterations with a KKT point:

1. **H** is positive semidefinite and $\mathbf{c} = \mathbf{0}$.
2. **H** is positive definite.
3. **H** has nonnegative elements with positive diagonal elements.

Furthermore, if **H** is positive semidefinite, then ray termination implies that the optimal solution is unbounded.

Proof

Assume that $\mathbf{H} = \mathbf{H}'$, because otherwise **H** could be replaced by $\frac{1}{2}(\mathbf{H} + \mathbf{H}')$. From Lemma 11.1.5, the complementary pivoting algorithm stops in a finite number of iterations with either a KKT point or a ray termination. If **H** is positive semidefinite or positive definite, or has nonnegative elements with positive diagonal elements, then, by Corollaries 1 and 2 to Theorem 11.2.3, **M** is copositive-plus.

Now suppose that ray termination occurs. By Theorem 11.1.8, since **M** is copositive-plus, ray termination is possible only if the following system has no solution:

$$\begin{aligned} \mathbf{A}\mathbf{x} + \mathbf{y} &= \mathbf{b} \\ -\mathbf{H}\mathbf{x} - \mathbf{A}'\mathbf{u} + \mathbf{v} &= \mathbf{c} \\ \mathbf{x}, \mathbf{y}, \mathbf{u}, \mathbf{v} &\geq \mathbf{0} \end{aligned}$$

By Theorem 2.4.5, the following system must have a solution $(\mathbf{d}, \mathbf{f})$:

$$\mathbf{A}\mathbf{d} \leq \mathbf{0} \tag{11.13}$$

$$\mathbf{A}'\mathbf{f} - \mathbf{H}\mathbf{d} \geq \mathbf{0} \tag{11.14}$$

$$\mathbf{f} \geq \mathbf{0} \tag{11.15}$$

$$\mathbf{d} \geq \mathbf{0} \tag{11.16}$$

$$\mathbf{b}'\mathbf{f} + \mathbf{c}'\mathbf{d} < 0 \tag{11.17}$$

Multiplying (11.14) by $\mathbf{d}' \geq \mathbf{0}$, and noting that $\mathbf{f} \geq \mathbf{0}$ and $\mathbf{Ad} \leq \mathbf{0}$, it follows that

$$0 \leq \mathbf{d}'\mathbf{A}'\mathbf{f} - \mathbf{d}'\mathbf{H}\mathbf{d} \leq 0 - \mathbf{d}'\mathbf{H}\mathbf{d} = -\mathbf{d}'\mathbf{H}\mathbf{d} \tag{11.18}$$

By assumption, there exist $\hat{\mathbf{x}}$ and $\hat{\mathbf{y}}$ such that $\mathbf{A}\hat{\mathbf{x}} + \hat{\mathbf{y}} = \mathbf{b}$, $(\hat{\mathbf{x}}, \hat{\mathbf{y}}) \geq \mathbf{0}$. Substituting for $\mathbf{b}$ in (11.17) and noting (11.14) and that $(\mathbf{f}, \hat{\mathbf{x}}, \hat{\mathbf{y}}) \geq \mathbf{0}$, we get

$$0 > \mathbf{c}'\mathbf{d} + \mathbf{b}'\mathbf{f} = \mathbf{c}'\mathbf{d} + (\hat{\mathbf{y}} + \mathbf{A}\hat{\mathbf{x}})'\mathbf{f} \geq \mathbf{c}'\mathbf{d} + \hat{\mathbf{x}}'\mathbf{A}'\mathbf{f} \geq \mathbf{c}'\mathbf{d} + \hat{\mathbf{x}}'\mathbf{H}\mathbf{d} \tag{11.19}$$

Now suppose that $\mathbf{H}$ is positive semidefinite. By (11.18), it follows that $\mathbf{d}'\mathbf{H}\mathbf{d} = 0$, and by Corollary 1 to Theorem 11.2.3, it follows that $\mathbf{H}\mathbf{d} = \mathbf{0}$. By (11.19), we have $\mathbf{c}'\mathbf{d} < 0$. Since $\mathbf{Ad} \leq \mathbf{0}$ and $\mathbf{d} \geq \mathbf{0}$, $\mathbf{d}$ is a direction of the feasible region, so that $\hat{\mathbf{x}} + \lambda\mathbf{d}$ is feasible for all $\lambda \geq 0$. Now consider $f(\hat{\mathbf{x}} + \lambda\mathbf{d})$, where $f(\mathbf{x}) = \mathbf{c}'\mathbf{x} + \frac{1}{2}\mathbf{x}'\mathbf{H}\mathbf{x}$. Since $\mathbf{H}\mathbf{d} = \mathbf{0}$, we get

$$f(\hat{\mathbf{x}} + \lambda\mathbf{d}) = f(\hat{\mathbf{x}}) + \lambda(\mathbf{c}' + \hat{\mathbf{x}}'\mathbf{H})\mathbf{d} + \tfrac{1}{2}\lambda^2\mathbf{d}'\mathbf{H}\mathbf{d} = f(\hat{\mathbf{x}}) + \lambda\mathbf{c}'\mathbf{d}$$

Since $\mathbf{c}'\mathbf{d} < 0$, $f(\hat{\mathbf{x}} + \lambda\mathbf{d})$ approaches $-\infty$ by choosing λ arbitrarily large; thus we have an unbounded optimal solution.

To complete the proof, we now show that ray termination is not possible under conditions 1, 2, or 3 of the theorem. Now suppose, by contradiction to the conclusion of the theorem, that ray termination occurs under any of these conditions. From (11.18), $\mathbf{d}'\mathbf{H}\mathbf{d} \leq 0$. Under conditions 2 or 3, $\mathbf{d} = \mathbf{0}$, which is impossible in view of (11.19). If condition 1 holds, on the other hand, then $\mathbf{H}\mathbf{d} = \mathbf{0}$. This, together with the assumption that $\mathbf{c} = \mathbf{0}$, contradicts (11.19).

To summarize, we have shown that if $\mathbf{H}$ is positive semidefinite and the algorithm stops with ray termination, then the optimal solution is unbounded. Further, ray termination is impossible under conditions 1, 2, or 3, so the algorithm must produce a KKT point under any of these conditions. This completes the proof.

11.3 Separable Programming

In this section, we discuss the use of the simplex method to obtain solutions to nonlinear programs where the objective function and the constraint functions can be expressed as the sum of functions, each involving only one variable. We denote such a *separable nonlinear program* as *Problem P*, and it can be expressed as follows:

Problem P:

$$\begin{aligned} \text{Minimize} \quad & \sum_{j=1}^{n} f_j(x_j) \\ \text{subject to} \quad & \sum_{j=1}^{n} g_{ij}(x_j) \leq p_i \qquad \text{for } i = 1, \ldots, m \\ & x_j \geq 0 \qquad \text{for } j = 1, \ldots, n \end{aligned} \tag{11.20}$$

Problems of this type arise in numerous applications, including econometric data fitting, electrical network analysis, design and management of water supply systems, logistics, and statistics.

Approximating the Separable Problem

We now discuss how one can define a new problem that approximates the original Problem P. The new problem is obtained by replacing each nonlinear function by an

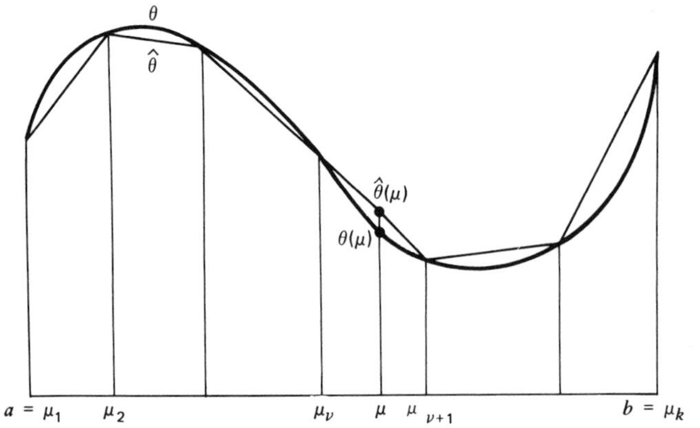

Figure 11.3 Piecewise linear approximation of a function.

approximating piecewise linear function. To see how this can be done, consider a continuous function θ of the variable μ. Suppose that we are interested in values of θ over the interval $[a, b]$. We wish to define a piecewise linear function $\hat{\theta}$ that approximates θ. The interval $[a, b]$ is first partitioned into smaller intervals, via the grid points $a = \mu_1, \mu_2, \ldots, \mu_k = b$, as shown in Figure 11.3. The function θ is approximated in the interval $[\mu_\nu, \mu_{\nu+1}]$ as follows. Let $\mu = \lambda\mu_\nu + (1 - \lambda)\mu_{\nu+1}$ for some $\lambda \in [0, 1]$. Then,

$$\hat{\theta}(\mu) = \lambda\theta(\mu_\nu) + (1 - \lambda)\theta(\mu_{\nu+1}) \tag{11.21}$$

Note that the grid points may or may not be equidistant, and also that the accuracy of the approximation improves as the number of grid points increase. Note, however, that a major difficulty may arise in using the above linear approximation to a function. This is because a given point μ in the interval $[\mu_\nu, \mu_{\nu+1}]$ can be alternatively represented as a convex combination of two or more *nonadjacent* grid points. To illustrate, consider the function θ defined by $\theta(\mu) = \mu^2$. The graph of the function on the interval $[-2, 2]$ is shown in Figure 11.4.

Suppose that we use the grid points -2, -1, 0, 1, and 2. The point $\mu = 1.5$ can be written as $\frac{1}{2}(1) + \frac{1}{2}(2)$ and also as $\frac{1}{4}(0) + \frac{3}{4}(2)$. The value of the function θ at $\mu =$

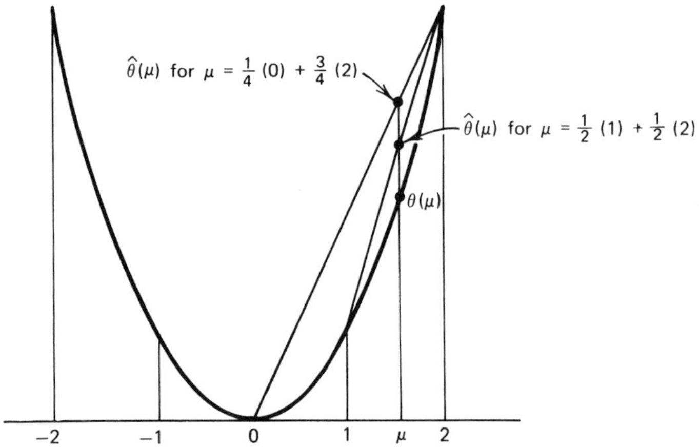

Figure 11.4 Importance of adjacency in approximation.

1.5 is 2.25. The first approximation gives $\hat{\theta}(\mu) = \frac{1}{2}\theta(1) + \frac{1}{2}\theta(2) = 2.5$, whereas the second approximation gives $\hat{\theta}(\mu) = \frac{1}{4}\theta(0) + \frac{3}{4}\theta(2) = 3$. Clearly, the first approximation using adjacent grid points yields a better approximation. In general, therefore, the function θ can be approximated over the interval $[a, b]$ via the grid points $\mu_1, \ldots, \mu_k$ by the piecewise linear function $\hat{\theta}$, defined by

$$\hat{\theta}(\mu) = \sum_{\nu=1}^{k} \lambda_\nu \theta(\mu_\nu), \quad \sum_{\nu=1}^{k} \lambda_\nu = 1, \; \lambda_\nu \geq 0 \qquad \text{for } \nu = 1, \ldots, k \tag{11.22}$$

where at most two λ_ν's are positive, and they must be adjacent. This representation is known as the λ-*form* approximation. An alternative related representation, known as the δ-*form,* is described in Exercises 11.29 and 11.31.

We now present a problem that approximates the separable Problem P defined by (11.19). This is done by considering each variable x_j for which either f_j or g_{ij} is nonlinear for some $i = 1, \ldots, m$ and replacing it with the piecewise linear approximation defined by (11.22). For the sake of clarity, we define a set L as

$$L = \{j : f_j \text{ and } g_{ij} \text{ for } i = 1, \ldots, m \text{ are linear}\}$$

Then, for each $j \notin L$, we consider the interval of interest $[a_j, b_j]$, where $a_j, b_j \geq 0$. We can now define the grid points $x_{\nu j}$ for $\nu = 1, \ldots, k_j$, where $x_{1j} = a_j$ and $x_{k_{jj}} = b_j$. Note that the grid points need not be spaced equally and that different grid lengths could be used for different variables. However, from Theorem 11.3.4, as will be seen later, the maximum grid length used is related to the accuracy of the solution obtained. Using the grid points for each $j \notin L$, from (11.22), the functions f_j and g_{ij} for $i = 1, \ldots, m$ could be replaced by their linear approximations

$$\hat{f}_j(x_j) = \sum_{\nu=1}^{k_j} \lambda_{\nu j} f(x_{\nu j}) \qquad \text{for } j \notin L$$

$$\hat{g}_{ij}(x_j) = \sum_{\nu=1}^{k_j} \lambda_{\nu j} g_{ij}(x_{\nu j}) \qquad \text{for } i = 1, \ldots, m \text{ and } j \notin L$$

$$\sum_{\nu=1}^{k_j} \lambda_{\nu j} = 1 \qquad \text{for } j \notin L$$

$$\lambda_{\nu j} \geq 0 \qquad \text{for } \nu = 1, \ldots, k_j \text{ and } j \notin L$$

By definition, both f_j and g_{ij} for $i = 1, \ldots, m$ are linear for $j \in L$. Hence, no grid points need be defined, and, in this case, the linear approximations are given by

$$\hat{f}_j(x_j) = f_j(x_j), \; \hat{g}_{ij}(x_j) = g_{ij}(x_j) \qquad \text{for } i = 1, \ldots, m \text{ and } j \in L$$

The following Problem AP can then be viewed as the problem that approximates the original Problem P.

Problem AP: Minimize $\displaystyle\sum_{j \in L} f_j(x_j) + \sum_{j \notin L} \hat{f}_j(x_j)$

subject to $\displaystyle\sum_{j \in L} g_{ij}(x_j) + \sum_{j \notin L} \hat{g}_{ij}(x_j) \leq p_i \qquad \text{for } i = 1, \ldots, m$

$x_j \geq 0 \qquad \text{for } j = 1, \ldots, n$

(11.23)

Note that the objective function and constraints in Problem AP are piecewise linear.

However, by using the definition of $\hat{f}_j$ and $\hat{g}_{ij}$ for $j \notin L$, the problem can be restated in an equivalent more manageable form as Problem LAP defined as follows:

$$
\begin{aligned}
&\textbf{\textit{Problem LAP:}}\ \text{Minimize} \sum_{j\in L} f_j(x_j) + \sum_{j\notin L}\sum_{\nu=1}^{k_j} \lambda_{\nu j} f_j(x_{\nu j}) \\
&\qquad \text{subject to} \sum_{j\in L} g_{ij}(x_j) + \sum_{j\notin L}\sum_{\nu=1}^{k_j} \lambda_{\nu j} g_{ij}(x_{\nu j}) \le p_i \quad \text{for } i = 1, \ldots, m \\
&\qquad\qquad \sum_{\nu=1}^{k_j} \lambda_{\nu j} = 1 \quad \text{for } j \notin L \\
&\qquad\qquad \lambda_{\nu j} \ge 0 \quad \text{for } \nu = 1, \ldots, k_j; j \notin L \\
&\qquad\qquad x_j \ge 0 \quad \text{for } j \in L
\end{aligned}
\tag{11.24}
$$

At most, two adjacent $\lambda_{\nu j}$'s are positive for $j \notin L$.

Solving the Approximating Problem

With the exception of the constraint that, at most, two adjacent $\lambda_{\nu j}$'s are positive for $j \notin L$, Problem LAP is a linear program. For solving Problem LAP, one can use the simplex method with the following *restricted basis entry rule*. A nonbasic variable $\lambda_{\nu j}$ is introduced into the basis only if it improves the objective function and if the new basis has no more than two adjacent $\lambda_{\nu j}$'s that are positive for each $j \notin L$. Theorem 11.3.1 below shows that for $j \notin L$, if g_{ij} is convex for $i = 1, \ldots, m$ and if f_j is strictly convex, we can discard the restricted basis entry rule and adopt the simplex method for linear programming as described in Section 2.7.

11.3.1 Theorem

Consider Problem P to minimize $\sum_{j=1}^n f_j(x_j)$ subject to $\sum_{j=1}^n g_{ij}(x_j) \le p_i$ for $i = 1, \ldots, m$, and $x_j \ge 0$ for $j = 1, \ldots, n$. Let $L = \{j: f_j \text{ and } g_{ij} \text{ for } i = 1, \ldots, m \text{ are linear}\}$. Suppose that for $j \notin L$, f_j is strictly convex and that g_{ij} is convex for $i = 1, \ldots, m$. Suppose that for each $j \notin L$, f_j and g_{ij} for $i = 1, \ldots, m$ are replaced by their piecewise linear approximations via the grid points $x_{\nu j}$ for $\nu = 1, \ldots, k_j$, yielding the linear program defined below:

$$
\begin{aligned}
&\text{Minimize} \quad \sum_{j\in L} f_j(x_j) + \sum_{j\notin L}\sum_{\nu=1}^{k_j} \lambda_{\nu j} f_j(x_{\nu j}) \\
&\text{subject to} \quad \sum_{j\in L} g_{ij}(x_j) + \sum_{j\notin L}\sum_{\nu=1}^{k_j} \lambda_{\nu j} g_{ij}(x_{\nu j}) \le p_i \quad \text{for } i = 1, \ldots, m \\
&\qquad\qquad \sum_{\nu=1}^{k_j} \lambda_{\nu j} = 1 \quad \text{for } j \notin L \\
&\qquad\qquad \lambda_{\nu j} \ge 0 \quad \text{for } \nu = 1, \ldots, k_j; j \notin L \\
&\qquad\qquad x_j \ge 0 \quad \text{for } j \in L
\end{aligned}
\tag{11.25}
$$

Let $\hat{x}_j$ for $j \in L$ and $\hat{\lambda}_{\nu j}$ for $\nu = 1, \ldots, k_j$ and $j \notin L$ solve the above problem. Then,

1. For each $j \notin L$, at most two $\hat{\lambda}_{\nu j}$'s are positive, and they must be adjacent.
2. Let $\hat{x}_j = \sum_{\nu=1}^{k_j} \hat{\lambda}_{\nu j} x_{\nu j}$ for $j \notin L$. Then, the vector $\hat{\mathbf{x}}$ whose jth component is $\hat{x}_j$ for $j = 1, \ldots, n$ is feasible to Problem P.

Proof

To prove part 1, it suffices to show that for each $j \notin L$, if $\hat{\lambda}_{lj}$ and $\hat{\lambda}_{pj}$ are positive, the grid points x_{lj} and x_{pj} must be adjacent. By contradiction, suppose that there exist $\hat{\lambda}_{lj}$ and $\hat{\lambda}_{pj} > 0$, where x_{lj} and x_{pj} are not adjacent. Then, there exists a grid point $x_{\gamma j} \in (x_{lj}, x_{pj})$ that can be expressed as $x_{\gamma j} = \alpha_1 x_{lj} + \alpha_2 x_{pj}$, where $\alpha_1, \alpha_2 > 0$ and $\alpha_1 + \alpha_2 = 1$. Now consider the optimal solution to the problem defined by (11.25). Let $u_i \geq 0$ for $i = 1, \ldots, m$ be the optimum Lagrangian multipliers associated with the first m constraints, and for each $j \notin L$, let v_j be the optimal Lagrangian multiplier associated with the constraint $\sum_{\nu=1}^{k_j} \lambda_{\nu j} = 1$. Then, the following subset of the KKT necessary conditions are satisfied:

$$f_j(x_{lj}) + \sum_{i=1}^{m} u_i g_{ij}(x_{lj}) + v_j = 0 \tag{11.26}$$

$$f_j(x_{pj}) + \sum_{i=1}^{m} u_i g_{ij}(x_{pj}) + v_j = 0 \tag{11.27}$$

$$f_j(x_{\nu j}) + \sum_{i=1}^{m} u_i g_{ij}(x_{\nu j}) + v_j \geq 0 \qquad \text{for } \nu = 1, \ldots, k_j \tag{11.28}$$

We show below that the last condition is contradicted for $\nu = \gamma$. By strict convexity of f_j and convexity of g_{ij} and by (11.26) and (11.27), we have

$$\begin{aligned} f_j(x_{\gamma j}) + \sum_{i=1}^{m} u_i g_{ij}(x_{\gamma j}) + v_j &< \alpha_1 f_j(x_{lj}) + \alpha_2 f_j(x_{pj}) \\ &\quad + \sum_{i=1}^{m} u_i[\alpha_1 g_{ij}(x_{lj}) + \alpha_2 g_{ij}(x_{pj})] + v_j = 0 \end{aligned}$$

This contradicts (11.28) for $\nu = \gamma$, and, hence, x_{lj} and x_{pj} must be adjacent, and part 1 of the theorem is proved.

To prove part 2, from convexity of g_{ij} for $j \notin L$ and for each $i = 1, \ldots, m$, and noting that $\hat{x}_j$ for $j \in L$ and $\hat{\lambda}_{\nu j}$ for $\nu = 1, \ldots, k_j$ and $j \notin L$ satisfy the constraints in (11.25), we get

$$\begin{aligned} g_i(\hat{\mathbf{x}}) &= \sum_{j \in L} g_{ij}(\hat{x}_j) + \sum_{j \notin L} g_{ij}(\hat{x}_j) \\ &= \sum_{j \in L} g_{ij}(\hat{x}_j) + \sum_{j \notin L} g_{ij}\left(\sum_{\nu=1}^{k_j} \hat{\lambda}_{\nu j} x_{\nu j}\right) \\ &\leq \sum_{j \in L} g_{ij}(\hat{x}_j) + \sum_{j \notin L} \sum_{\nu=1}^{k_j} \hat{\lambda}_{\nu j} g_{ij}(x_{\nu j}) \\ &\leq p_i \end{aligned}$$

for $i = 1, \ldots, m$. Furthermore, $\hat{x}_j \geq 0$ for $j \in L$, and $\hat{x}_j = \sum_{\nu=1}^{k_j} \hat{\lambda}_{\nu j} x_{\nu j} \geq 0$ for $j \notin L$, since $\hat{\lambda}_{\nu j}, x_{\nu j} \geq 0$ for $\nu = 1, \ldots, k_j$ and $j \notin L$. Hence, $\hat{\mathbf{x}}$ is feasible to Problem P and the proof is complete.

11.3.2 Example

Consider the following separable problem:

$$\begin{array}{lllll} \text{Minimize} & x_1^2 - 6x_1 & + x_2^2 - 8x_2 & - \frac{1}{2}x_3 & \\ \text{subject to} & x_1 + & x_2 & + x_3 & \leq 5 \\ & x_1^2 - & x_2 & & \leq 3 \\ & x_1, & x_2, & x_3 & \geq 0 \end{array}$$

Note that $L = \{3\}$, since there are no nonlinear terms involving x_3, and, hence, we will not take any grid points for x_3. From the constraints, it is clear that both x_1 and x_2 must lie in the interval [0, 5]. Recall that the grid points need not be equally spaced. For the variables x_1 and x_2, we use the grid points 0, 2, 4, and 5, so that $x_{11} = 0$, $x_{21} = 2$, $x_{31} = 4$, and $x_{41} = 5$, and $x_{12} = 0$, $x_{22} = 2$, $x_{32} = 4$, and $x_{42} = 5$. Thus,

$$0\lambda_{11} + 2\lambda_{21} + 4\lambda_{31} + 5\lambda_{41} = x_1$$

$$0\lambda_{12} + 2\lambda_{22} + 4\lambda_{32} + 5\lambda_{42} = x_2$$

$$\lambda_{11} + \lambda_{21} + \lambda_{31} + \lambda_{41} = 1$$

$$\lambda_{12} + \lambda_{22} + \lambda_{32} + \lambda_{42} = 1$$

$$\lambda_{\nu 1}, \lambda_{\nu 2} \geq 0 \qquad \text{for } \nu = 1, 2, 3, 4$$

$$\hat{f}(x) = (-8\lambda_{21} - 8\lambda_{31} + 5\lambda_{41}) + (-12\lambda_{22} - 16\lambda_{32} - 15\lambda_{42}) - \tfrac{1}{2}x_3$$

$$\hat{g}_1(x) = (2\lambda_{21} + 4\lambda_{31} + 5\lambda_{41}) + (2\lambda_{22} + 4\lambda_{32} + 5\lambda_{42}) + x_3 \leq 5$$

$$\hat{g}_2(x) = (4\lambda_{21} + 16\lambda_{31} + 25\lambda_{41}) - (2\lambda_{22} + 4\lambda_{32} + 5\lambda_{42}) \leq 3$$

Introducing the slack variables x_4 and x_5, we get the first tableau given below. We solve the problem using the simplex method with the restricted basis entry rule. The sequence of tableaux obtained are given as follows:

	z	λ_{11}	λ_{21}	λ_{31}	λ_{41}	λ_{12}	λ_{22}	λ_{32}	λ_{42}	x_3	x_4	x_5	RHS
z	1	0	8	8	5	0	12	16	15	$\frac{1}{2}$	0	0	0
x_4	0	0	2	4	5	0	2	4	5	1	1	0	5
x_5	0	0	4	16	25	0	−2	−4	−5	0	0	1	3
λ_{11}	0	1	1	1	1	0	0	0	0	0	0	0	1
λ_{12}	0	0	0	0	0	1	1	(1)	1	0	0	0	1

	z	λ_{11}	λ_{21}	λ_{31}	λ_{41}	λ_{12}	λ_{22}	λ_{32}	λ_{42}	x_3	x_4	x_5	RHS
z	1	0	8	8	5	−16	−4	0	−1	$\frac{1}{2}$	0	0	−16
x_4	0	0	(2)	4	5	−4	−2	0	1	1	1	0	1
x_5	0	0	4	16	25	4	2	0	−1	0	0	1	7
λ_{11}	0	1	1	1	1	0	0	0	0	0	0	0	1
λ_{32}	0	0	0	0	0	1	1	1	1	0	0	0	1

	z	λ_{11}	λ_{21}	λ_{31}	λ_{41}	λ_{12}	λ_{22}	λ_{32}	λ_{42}	x_3	x_4	x_5	RHS
z	1	0	0	-8	-15	0	4	0	-5	$-\frac{7}{2}$	-4	0	-20
λ_{21}	0	0	1	2	$\frac{5}{2}$	-2	-1	0	$\frac{1}{2}$	$\frac{1}{2}$	$\frac{1}{2}$	0	$\frac{1}{2}$
x_5	0	0	0	8	15	12	6	0	-3	-2	-2	1	5
λ_{11}	0	1	0	-1	$-\frac{3}{2}$	2	(1)	0	$-\frac{1}{2}$	$-\frac{1}{2}$	$-\frac{1}{2}$	0	$\frac{1}{2}$
λ_{32}	0	0	0	0	0	1	1	1	1	0	0	0	1

	z	λ_{11}	λ_{21}	λ_{31}	λ_{41}	λ_{12}	λ_{22}	λ_{32}	λ_{42}	x_3	x_4	x_5	RHS
z	1	-4	0	-4	-9	-8	0	0	-3	$-\frac{3}{2}$	-2	0	-22
λ_{21}	0	1	1	1	1	0	0	0	0	0	0	0	1
x_5	0	-6	0	14	24	0	0	0	0	1	1	1	2
λ_{22}	0	1	0	-1	$-\frac{3}{2}$	2	1	0	$-\frac{1}{2}$	$-\frac{1}{2}$	$-\frac{1}{2}$	0	$\frac{1}{2}$
λ_{32}	0	-1	0	1	$\frac{3}{2}$	-1	0	1	$\frac{3}{2}$	$\frac{1}{2}$	$\frac{1}{2}$	0	$\frac{1}{2}$

Note that at the second tableau, λ_{31} could not be introduced into the basis, as it would have violated the restricted basis entry rule. From the final tableau, the optimal solution to the approximating Problem AP is $\hat{\mathbf{x}} = (\hat{x}_1, \hat{x}_2, \hat{x}_3)^t$, where

$$\hat{x}_1 = 2\hat{\lambda}_{21} + 4\hat{\lambda}_{31} + 5\hat{\lambda}_{41} = 2$$

$$\hat{x}_2 = 2\hat{\lambda}_{22} + 4\hat{\lambda}_{32} + 5\hat{\lambda}_{42} = 3$$

$$\hat{x}_3 = 0$$

The corresponding value of the objective function for Problem AP is $\hat{f}(2, 3, 0) = -22$, whereas the value of the objective function for the original Problem P at this point is $f(2, 3, 0) = -23$. Note that the objective function and the constraint functions for this problem satisfy the assumptions of Theorem 11.3.1. Thus, we could have adopted the simplex method without the restricted basis entry rule and yet obtained the above optimal solution.

Relationship Between the Optimal Solutions to the Original and Approximating Problems

As we have seen from Theorem 11.3.1, in the presence of suitable convexity assumptions, an optimal solution to the approximating linear programming problem is a feasible solution to the original problem. We show in Theorem 11.3.4 that, if the grid length is chosen sufficiently small, the optimal objective values to both problems could be made arbitrarily close. To prove this result, the following theorem is needed.

11.3.3 Theorem

Consider Problems P and AP defined in (11.20) and (11.23), respectively. For $j \notin L$, suppose that f_j and g_{ij} for $i = 1, \ldots, m$ are convex; and, furthermore, let $\hat{f}_j$ and $\hat{g}_{ij}$

give their piecewise linear approximations on the interval $[a_j, b_j]$. For $j \notin L$ and for $i = 1, \ldots, m$, let c_{ij} be such that $[g'_{ij}(x_j)| \le c_{ij}$ for $x_j \in [a_j, b_j]$. Further, for $j \notin L$, let c_j be such that $|f'_j(x_j)| \le c_j$ for $x_j \in [a_j, b_j]$. For $j \notin L$, let δ_j be the maximum grid length used for the variable x_j. Then,

$$\hat{f}(\mathbf{x}) \ge f(\mathbf{x}) \ge \hat{f}(\mathbf{x}) - c$$

$$\hat{g}_i(\mathbf{x}) \ge g_i(\mathbf{x}) \ge \hat{g}_i(\mathbf{x}) - c \qquad \text{for } i = 1, \ldots, m$$

where $c = \text{maximum}_{0 \le i \le m} \{\bar{c}_i\}$ and where

$$\bar{c}_0 = \sum_{j \notin L} 2c_j\delta_j \text{ and } \bar{c}_i = \sum_{j \notin L} 2c_{ij}\delta_j \qquad \text{for } i = 1, \ldots, m$$

Proof

We first show that $\hat{f}_j(x_j) \ge f_j(x_j) \ge \hat{f}_j(x_j) - 2c_j\delta_j$ for $j \notin L$. Let $j \notin L$, and let $x_j \in [a_j, b_j]$. Then, there exist grid points μ_k and μ_{k+1} such that $x_j \in [\mu_k, \mu_{k+1}]$. Further, $x_j = \lambda\mu_k + (1 - \lambda)\mu_{k+1}$ for some $\lambda \in [0, 1]$. By the definition of $\hat{f}_j$, and noting the convexity of f_j and that $\lambda \in [0, 1]$, we get

$$\hat{f}_j(x_j) = \lambda f_j(\mu_k) + (1 - \lambda)f_j(\mu_{k+1}) \ge f_j(\lambda\mu_k + (1 - \lambda)\mu_{k+1}) = f_j(x_j)$$

Now we show that $f_j(x_j) \ge \hat{f}_j(x_j) - 2c_j\delta_j$. Note that $\hat{f}_j(x_j)$ can be represented as follows:

$$\hat{f}_j(x_j) = f_j(\mu_k) + (x_j - \mu_k)s \tag{11.29}$$

where $s = [f_j(\mu_{k+1}) - f_j(\mu_k)]/[\mu_{k+1} - \mu_k]$. Further, by Theorem 3.3.3, it follows that

$$f_j(x_j) \ge f_j(\mu_k) + (x_j - \mu_k)f'_j(\mu_k) \tag{11.30}$$

Subtracting (11.30) from (11.29), we get

$$\hat{f}_j(x_j) - f_j(x_j) \le (x_j - \mu_k)[s - f'_j(\mu_k)] \tag{11.31}$$

By the mean value theorem, there exists a $y \in [\mu_k, \mu_{k+1}]$ such that $s = f'_j(y)$. Thus, by assumption, $s - f'_j(\mu_k) \le 2c_j$. Furthermore, $x_j - \mu_k \le \delta_j$, and, hence, from (11.31), we must have $\hat{f}_j(x_j) - f_j(x_j) \le 2c_j\delta_j$. We have thus proved that

$$\hat{f}_j(x_j) \ge f_j(x_j) \ge \hat{f}_j(x_j) - 2c_j\delta_j \qquad \text{for } j \notin L \tag{11.32}$$

for each $x_j \in [a_j, b_j]$. Summing (11.32) over $j \notin L$ and adding $\sum_{j\in L} f_j(x_j)$ to each term, it follows that

$$\hat{f}(\mathbf{x}) \ge f(\mathbf{x}) \ge \hat{f}(\mathbf{x}) - \bar{c}_0 \tag{11.33}$$

In a similar fashion, we get

$$\hat{g}_i(\mathbf{x}) \ge g_i(\mathbf{x}) \ge \hat{g}_i(\mathbf{x}) - \bar{c}_i \qquad \text{for } i = 1, \ldots, m \tag{11.34}$$

By the definition of c, and from (11.33) and (11.34), the result follows.

11.3.4 Theorem

Consider Problem P, defined in (11.20). Let $L = \{j{:}f_j$ and g_{ij} for $i = 1, \ldots, m$ are linear$\}$. For $j \notin L$, let $\hat{f}_j$ and $\hat{g}_{ij}$ be the piecewise linear approximations of f_j and g_{ij}, respectively, for $i = 1, \ldots, m$. Let Problem AP, defined in (11.23), and Problem LAP, defined in (11.24), be the equivalent problems approximating Problem P. For $j \notin L$, suppose that f_j and g_{ij} for $i = 1, \ldots, m$ are convex. Let $\bar{\mathbf{x}}$ be an optimal solution to Problem P. Let $\hat{x}_j$ for $j \in L$ and $\hat{\lambda}_{\nu j}$ for $\nu = 1, \ldots, k_j$ and $j \notin L$ be an optimal solution to Problem LAP such that the vector $\hat{\mathbf{x}}$, whose components are $\hat{x}_j$ for $j \in L$ and $\hat{x}_j = \sum_{\nu=1}^{k_j} \hat{\lambda}_{\nu j} x_{\nu j}$ for $j \notin L$, is an optimal solution to Problem AP. Let $\hat{u}_i \geq 0$ be the optimal Lagrangian multiplier associated with the constraint $\hat{g}_i(\mathbf{x}) \leq 0$ for $i = 1, \ldots, m$. Then,

1. $\hat{\mathbf{x}}$ is a feasible solution to Problem P.
2. $0 \leq f(\hat{\mathbf{x}}) - f(\bar{\mathbf{x}}) \leq c(1 + \sum_{i=1}^{m} \hat{u}_i)$, where c is as defined in Theorem 11.3.3.

Proof

The vector $\hat{\mathbf{x}}$ is feasible to Problem AP, that is, $\hat{g}_i(\hat{\mathbf{x}}) \leq p_i$ for $i = 1, \ldots, m$, and $\hat{\mathbf{x}} \geq \mathbf{0}$. By Theorem 11.3.3, $\hat{g}_i(\hat{\mathbf{x}}) \leq p_i$ implies that $g_i(\hat{\mathbf{x}}) \leq p_i$ for $i = 1, \ldots, m$, and part 1 follows.

The reader can verify that a piecewise linear approximation of a convex function is also convex, so that $\hat{f}_j$ and $\hat{g}_{ij}$ are convex for $i = 1, \ldots, m$ and $j \notin L$. Since the sum of convex functions is also convex, the objective function and constraint functions of Problem AP are convex. Hence, $(\hat{\mathbf{x}}, \hat{\mathbf{u}})$ satisfies the saddle point optimality criteria of Problem AP, given in Theorem 6.2.5, so that

$$\hat{f}(\hat{\mathbf{x}}) \leq \hat{f}(\mathbf{x}) + \hat{\mathbf{u}}^t[\hat{\mathbf{g}}(\mathbf{x}) - \mathbf{p}] \qquad \text{for all } \mathbf{x} \geq \mathbf{0} \tag{11.35}$$

Since $g_i(\bar{\mathbf{x}}) \leq p_i$, by Theorem 11.3.3, $\hat{g}_i(\bar{\mathbf{x}}) - p_i \leq c$ for $i = 1, \ldots, m$. Letting $\mathbf{x} = \bar{\mathbf{x}}$ in (11.35) and noting that $\hat{\mathbf{u}} \geq \mathbf{0}$, it follows that

$$\hat{f}(\hat{\mathbf{x}}) \leq \hat{f}(\bar{\mathbf{x}}) + c \sum_{i=1}^{m} \hat{u}_i \tag{11.36}$$

By part 1 of the theorem, $\hat{\mathbf{x}}$ is feasible to Problem P, and, hence, $f(\hat{\mathbf{x}}) \geq f(\bar{\mathbf{x}})$. From Theorem 11.3.3, $f(\bar{\mathbf{x}}) \geq \hat{f}(\bar{\mathbf{x}}) - c$, and, hence, $f(\hat{\mathbf{x}}) \geq f(\bar{\mathbf{x}}) \geq \hat{f}(\bar{\mathbf{x}}) - c$. From (11.36) and since $\hat{f}(\hat{\mathbf{x}}) \geq f(\hat{\mathbf{x}})$, it follows that

$$f(\hat{\mathbf{x}}) \geq f(\bar{\mathbf{x}}) \geq \hat{f}(\hat{\mathbf{x}}) - c\left(1 + \sum_{i=1}^{m} \hat{u}_i\right) \geq f(\hat{\mathbf{x}}) - c\left(1 + \sum_{i=1}^{m} \hat{u}_i\right)$$

This completes the proof.

In the above theorem, the Lagrangian multipliers $\hat{u}_i$ for $i = 1, \ldots, m$ are immediately available from the optimal simplex tableau for Problem LAP. When the approximating problem is solved, we can use Theorem 11.3.4 to determine the maximum deviation $c(1 + \sum_{i=1}^{m} \hat{u}_i)$ of the true optimal objective value from that at hand. Note that as the grid length is reduced, c will be smaller, and, hence, a better approximation will be obtained. The Notes and References section points the reader to literature on *error estimations* in convex separable programming.

Generation of the Grid Points

It may be noted that the accuracy of the procedure discussed above largely depends upon the number of grid points for each variable. However, as the number of grid points is increased, the number of variables in the approximating linear program LAP also increases. One approach is to use a coarse grid initially and then to use a finer grid around the optimal solution obtained with the coarse grid. An attractive alternative is to generate grid points when necessary. This approach is discussed below. (See Meyer

[1979, 1980] for an alternative that employs a sequence of two segment approximations only.)

Consider Problem LAP, defined in (11.24). Let $x_{\nu j}$ for $\nu = 1, \ldots, k_j$ and $j \notin L$ be the grid points considered so far. Let $\hat{x}_j$ for $j \in L$ and $\hat{\lambda}_{\nu j}$ for $\nu = 1, \ldots, k_j$ and $j \notin L$ solve Problem LAP. Further, let $\hat{u}_i \geq 0$ for $i = 1, \ldots, m$ be the optimal Lagrangian multipliers associated with the first m constraints, and let $\hat{v}_j$ for each $j \notin L$ be the Lagrangian multiplier associated with the constraint $\sum_{\nu=1}^{k_j} \lambda_{\nu j} = 1$. Note that the solution $\hat{x}_j$, $\hat{\lambda}_{\nu j}$, $\hat{u}_i$, and $\hat{v}_j$ satisfy the KKT conditions for Problem LAP. We wish to know whether we need to consider an additional grid point for any of the variables x_j for $j \notin L$ to yield a better piecewise linear approximation in the sense that, if this new grid point were considered in defining Problem LAP, its minimum objective function value would decrease. For some $j \notin L$, suppose we were to consider a grid point $x_{\gamma j}$. The reader may verify that, if

$$f_j(x_{\gamma j}) + \sum_{i=1}^{m} \hat{u}_i g_{ij}(x_{\gamma j}) + \hat{v}_j \geq 0 \tag{11.37}$$

then letting $\hat{\lambda}_{\gamma j} = 0$ will satisfy all the KKT conditions for the revised Problem LAP. However, since we do not know where the new grid point is to be located, we can answer the question whether all x_j satisfying $a_j \leq x_j \leq b_j$ for $j \notin L$ will satisfy (11.37) by solving the following subproblem PS for each $j \notin L$.

Problem PS: Minimize $\quad f_j(x_j) + \sum_{i=1}^{m} \hat{u}_i g_{ij}(x_j) + \hat{v}_j$

subject to $\quad a_j \leq x_j \leq b_j$

If the minimum objective function value is nonnegative for all $j \notin L$, then we cannot find a new grid point contradicting (11.37). Theorem 11.3.5 asserts that, if this is the case, the current solution is optimal to the original Problem P and that if the minimum objective value is negative for some $j \notin L$, we can get a better approximation to Problem P. Furthermore, the theorem provides bounds on the optimum objective function value for Problem P at each iteration.

11.3.5 Theorem

Consider Problem P defined in (11.20). Let $L = \{j: f_j$ and g_{ij} for $i = 1, \ldots, m$ are linear$\}$. Suppose, without loss of generality, that $f_j(x_j)$ is of the form $c_j x_j$ and $g_{ij}(x_j)$ is of the form $a_{ij} x_j$ for $i = 1, \ldots, m$ and for $j \in L$. Using the grid points $x_{\nu j}$, $\nu = 1, \ldots, k_j$ for $j \notin L$, let Problem LAP be defined as in (11.24). For $j \notin L$, suppose that f_j and g_{ij} are convex for $i = 1, \ldots, m$. Let $\hat{x}_j$ for $j \in L$ and $\hat{\lambda}_{\nu j}$ for $\nu = 1, \ldots, k_j$ and $j \notin L$ be optimal to Problem LAP with a corresponding objective function value $\hat{z}$. Let $\hat{u}_i \geq 0$ for $i = 1, \ldots, m$, be the Lagrangian multipliers corresponding to the first m constraints, and let $\hat{v}_j$ for $j \notin L$ be the Lagrangian multipliers associated with the constraints $\sum_{\nu=1}^{k_j} \lambda_{\nu j} = 1$ in Problem LAP. Now, for each $j \notin L$, consider the following problem:

Minimize $\quad f_j(x_j) + \sum_{i=1}^{m} \hat{u}_i g_{ij}(x_j)$

subject to $\quad a_j \leq x_j \leq b_j$

where $[a_j, b_j]$ with $a_j, b_j \geq 0$ is the interval of interest for x_j. Let $\bar{z}_j$ be the optimal objective function value to the above problem. Then, the following holds:

1. $\sum_{j\notin L} \bar{z}_j - \sum_{i=1}^{m} \hat{u}_i p_i \leq \sum_{j=1}^{n} f_j(\bar{x}_j) \leq \sum_{j=1}^{n} f_j(\hat{x}_j) \leq \hat{z}$

   where $\hat{x}_j = \sum_{\nu=1}^{k_j} \hat{\lambda}_{\nu j} x_{\nu j}$ for $j \notin L$, and $\bar{\mathbf{x}} = (\bar{x}_1, \ldots, \bar{x}_n)^t$ is an optimal solution to Problem P.
2. If $\bar{z}_j + \hat{v}_j \geq 0$ for $j \notin L$, then $\hat{\mathbf{x}} = (\hat{x}_1, \ldots, \hat{x}_n)$ is an optimal solution to Problem P. Furthermore, $\sum_{j=1}^{n} f_j(\hat{x}_j) = \hat{z}$.
3. If $\bar{z}_j + \hat{v}_j < 0$ for some $j \notin L$, let $x_{\gamma j}$ be the optimal solution that yielded $\bar{z}_j < -\hat{v}_j$. Then, adding the grid point $x_{\gamma j}$ in defining Problem LAP will give a new approximating Problem LAP with a minimum objective function value not higher than $\hat{z}$.

Proof

Since $\hat{u}_i$ and $\hat{v}_j$ are the optimal Lagrangian multipliers associated with Problem LAP, the reader can verify that the following subset of the KKT conditions hold true:

$$c_j + \sum_{i=1}^{m} \hat{u}_i a_{ij} \geq 0 \qquad \text{for } j \in L$$

Multiplying by $x_j \geq 0$, and noting that $f_j(x_j) = c_j x_j$ and $g_{ij}(x_j) = a_{ij} x_j$, we get

$$f_j(x_j) + \sum_{i=1}^{m} \hat{u}_i g_{ij}(x_j) \geq 0 \qquad \text{for } j \in L \text{ and for all } x_j \geq 0 \tag{11.38}$$

Further, from the definition of $\bar{z}_j$, we have

$$f_j(x_j) + \sum_{i=1}^{m} \hat{u}_i g_{ij}(x_j) \geq \bar{z}_j \qquad \text{for } j \notin L \text{ and for all } a_j \leq x_j \leq b_j \tag{11.39}$$

Summing (11.38) over $j \in L$ and (11.39) over $j \notin L$, and subtracting $\sum_{i=1}^{m} \hat{u}_i p_i$ from the resulting sum, we get

$$\sum_{j=1}^{n} f_j(x_j) + \sum_{i=1}^{m} \hat{u}_i \left[\sum_{j=1}^{n} g_{ij}(x_j) - p_i \right] \geq \sum_{j\notin L} \bar{z}_j - \sum_{i=1}^{m} \hat{u}_i p_i \qquad \text{for all } a_j \leq x_j \leq b_j \tag{11.40}$$

Noting that $a_j \leq \bar{x}_j \leq b_j$, $\sum_{j=1}^{n} g_{ij}(\bar{x}_j) \leq p_i$, and that $\hat{u}_i \geq 0$, then (11.40) implies that $\sum_{j=1}^{n} f_j(\bar{x}_j) \geq \sum_{j\notin L} \bar{z}_j - \sum_{i=1}^{m} \hat{u}_i p_i$, which is the first inequality in part 1 of the theorem. Now, by Theorem 11.3.4, $\hat{\mathbf{x}} = (\hat{x}_1, \ldots, \hat{x}_n)^t$ is feasible to Problem P, so that $\sum_{j=1}^{n} f_j(\bar{x}_j) \leq \sum_{j=1}^{n} f_j(\hat{x}_j)$. Finally, by the convexity of f_j for $j \notin L$, we have

$$\begin{aligned}
\sum_{j=1}^{m} f_j(\hat{x}_j) &= \sum_{j\in L} f_j(\hat{x}_j) + \sum_{j\notin L} f_j(\hat{x}_j) \\
&= \sum_{j\in L} f_j(\hat{x}_j) + \sum_{j\notin L} f_j\left[\sum_{\nu=1}^{k_j} \hat{\lambda}_{\nu j} x_{\nu j} \right] \\
&\leq \sum_{j\in L} f_j(\hat{x}_j) + \sum_{j\notin L} \sum_{\nu=1}^{k_j} \hat{\lambda}_{\nu j} f_j(x_{\nu j}) \\
&= \hat{z}
\end{aligned}$$

Hence, part 1 of the theorem holds.

To prove part 2, consider Problem LAP defined in (11.24). The reader can verify that the complementary slackness conditions of the KKT optimality conditions provide

$$f_j(\hat{x}_j) + \sum_{i=1}^{m} \hat{u}_i g_{ij}(\hat{x}_j) = 0 \qquad \text{for } j \in L \tag{11.41}$$

$$\hat{\lambda}_{\nu j}\left[f_j(x_{\nu j}) + \sum_{i=1}^{m} \hat{u}_i g_{ij}(x_{\nu j}) + \hat{v}_j\right] = 0 \qquad \text{for } \nu = 1, \ldots, k_j;\ j \notin L \tag{11.42}$$

$$\hat{u}_i\left[\sum_{j \in L} g_{ij}(\hat{x}_j) + \sum_{j \notin L}\sum_{\nu=1}^{k_j} \hat{\lambda}_{\nu j} g_{ij}(x_{\nu j}) - p_i\right] = 0 \qquad \text{for } i = 1, \ldots, m \tag{11.43}$$

Summing (11.41) over $j \in L$ and (11.42) over $\nu = 1, \ldots, k_j$, and $j \notin L$, we get

$$\left[\sum_{j\in L} f_j(\hat{x}_j) + \sum_{j\notin L}\sum_{\nu=1}^{k_j} \hat{\lambda}_{\nu j} f_j(x_{\nu j})\right] + \sum_{i=1}^{m} \hat{u}_i\left[\sum_{j\in L} g_{ij}(\hat{x}_j) + \sum_{j\notin L}\sum_{\nu=1}^{k_j} \hat{\lambda}_{\nu j} g_{ij}(x_{\nu j})\right] + \sum_{j\notin L}\sum_{\nu=1}^{k_j} \hat{\lambda}_{\nu j}\hat{v}_j = 0 \tag{11.44}$$

But the first term in (11.44) is precisely $\hat{z}$ by definition, and the second term is equal to $\sum_{i=1}^{m} \hat{u}_i p_i$ by (11.43). Further, $\sum_{\nu=1}^{k_j} \hat{\lambda}_{\nu j} = 1$ for $j \notin L$, since $\hat{\lambda}_{\nu j}$ is feasible to Problem LAP defined in (11.24). Hence,

$$\hat{z} + \sum_{i=1}^{m} \hat{u}_i p_i + \sum_{j\notin L} \hat{v}_j = 0 \tag{11.45}$$

Also from part 1 of the theorem,

$$\sum_{j\notin L} \bar{z}_j - \sum_{i=1}^{m} \hat{u}_i p_i \leq \sum_{j=1}^{n} f_j(\bar{x}_j) \tag{11.46}$$

Adding (11.45) to (11.46), we get $\sum_{j\notin L} (\bar{z}_j + \hat{v}_j) + \hat{z} \leq \sum_{j=1}^{n} f_j(\bar{x}_j)$. But by assumption in part 2, $\bar{z}_j + \hat{v}_j \geq 0$ for $j \notin L$. Hence, $\hat{z} \leq \sum_{j=1}^{n} f_j(\bar{x}_j)$; and, using part 1 of the theorem, we get, $\hat{z} \leq \sum_{j=1}^{n} f_j(\bar{x}_j) \leq \sum_{j=1}^{n} f_j(\hat{x}_j) \leq \hat{z}$. This implies that $\sum_{j=1}^{n} f_j(\bar{x}_j) = \sum_{j=1}^{n} f_j(\hat{x}_j)$. Since $\hat{\mathbf{x}} = (\hat{x}_1, \ldots, \hat{x}_n)^t$ is feasible to Problem P, then part 2 follows.

To prove part 3, suppose that $x_{\gamma j}$ is the optimal solution that yielded $\bar{z}_j < -v_j$. We then have $f_j(x_{\gamma j}) + \sum_{i=1}^{m} \hat{u}_i g_{ij}(x_{\gamma j}) + \hat{v}_j < 0$. But if the grid point $x_{\gamma j}$ were included in defining the approximating Problem LAP, then one of the KKT conditions, namely $f_j(x_{\gamma j}) + \sum_{i=1}^{m} \hat{u}_i g_{ij}(x_{\gamma j}) + \hat{v}_j \geq 0$ would be violated. The reader can easily verify that introducing $x_{\gamma j}$ in the basis will yield an objective value in Problem LAP not higher than $\hat{z}$, and the proof is complete.

Summary of the Grid Point Generation Procedure

The procedure described below can be used to solve a problem of the form to minimize $\sum_{j=1}^{n} f_j(x_j)$ subject to $\sum_{j=1}^{n} g_{ij}(x_j) \leq 0$ for $i = 1, \ldots, m$ and $x_j \geq 0$ for $j = 1, \ldots, n$. Let $L = \{j: f_j$ and g_{ij} for $i = 1, \ldots, m$ are linear$\}$. The procedure will yield an optimal solution using the simplex method without the restricted basis entry if g_{ij} is convex for $i = 1, \ldots, m$ and $j \notin L$, and f_j is strictly convex for $j \notin L$.

Initialization Step Define $a_j, b_j \geq 0$ such that all feasible points satisfy $x_j \in [a_j, b_j]$ for $j \notin L$. For each $j \notin L$, select a set of grid points. Set k_j equal to the number of grid points for $j \notin L$ and go to the main step.

Main Step

1. Solve Problem LAP defined in (11.24). Let the optimal solution be $\hat{x}_j$ for $j \in L$ and $\hat{\lambda}_{\nu j}$ for $\nu = 1, \ldots, k_j$ and $j \notin L$. Let $\hat{u}_i$ be the Lagrangian multipliers

associated with the first m constraints, and let $\hat{v}_j$ for $j \notin L$ be the Lagrangian multipliers associated with $\sum_{\nu=1}^{k_j} \lambda_{\nu j} = 1$. Go to step 2.

2. For each $j \notin L$, solve the problem to minimize $f_j(x_j) + \sum_{i=1}^{m} \hat{u}_i g_{ij}(x_j)$, subject to $a_j \le x_j \le b_j$. Let the optimum objective function value be $\bar{z}_j$ for $j \notin L$. If $\bar{z}_j + \hat{v}_j \ge 0$ for all $j \notin L$, stop; the optimal solution to the original problem is $\hat{\mathbf{x}}$, whose components are given by $\hat{x}_j$ for $j \in L$ and $\hat{x}_j = \sum_{\nu=1}^{k_j} \hat{\lambda}_{\nu j} x_{\nu j}$. Otherwise, go to step 3.
3. Let $\bar{z}_p + \hat{v}_p = \text{minimum}_{j \notin L} (\bar{z}_j + \hat{v}_j) < 0$. Let $x_{\nu p}$ be the optimum solution yielding $\bar{z}_p < -\hat{v}_p$. Let $\nu = k_p + 1$, replace k_p by $k_p + 1$, and go to step 1.

11.3.6 Example

Consider the following separable problem:

$$
\begin{array}{lllll}
\text{Minimize} & x_1^2 - 6x_1 & + x_2^2 - 8x_2 & - \frac{1}{2}x_3 & \\
\text{subject to} & x_1 & + x_2 & + x_3 & \le 5 \\
 & x_1^2 & - x_2 & & \le 3 \\
 & x_1, & x_2, & x_3 & \ge 0
\end{array}
$$

Iteration 1

Since the objective and constraint functions associated with x_3 are linear, we let $L = \{3\}$. We start the grid generation procedure with the initial grid points $x_{11} = x_{12} = 0$. The corresponding columns are $(0, 0, 1, 0)^t$ and $(0, 0, 0, 1)^t$, and the corresponding objective values are both equal to zero. Letting x_4 and x_5 be the slack variables, we get the first tableau given below. At this stage, x_3 enters the basis and x_4 leaves the basis, giving the second tableau.

	z	λ_{11}	λ_{12}	x_3	x_4	x_5	RHS
z	1	0	0	0.5	0	0	0
x_4	0	0	0	(1)	1	0	5
x_5	0	0	0	0	0	1	3
λ_{11}	0	1	0	0	0	0	1
λ_{12}	0	0	1	0	0	0	1

	z	λ_{11}	λ_{12}	x_3	x_4	x_5	RHS
z	1	0	0	0	−0.5	0	−2.5
x_3	0	0	0	1	1	0	5
x_5	0	0	0	0	0	1	3
λ_{11}	0	1	0	0	0	0	1
λ_{12}	0	0	1	0	0	0	1

Note that $\hat{x}_j = \sum_\nu \hat{\lambda}_{\nu j} x_{\nu j}$ for $j = 1, 2$. From the second tableau, $\hat{\lambda}_{11} = \hat{\lambda}_{12} = 1$, so that

$\hat{x}_1 = \hat{x}_2 = 0$. Therefore, the current solution $\hat{\mathbf{x}} = (0, 0, 5)^t$ and $f(\hat{\mathbf{x}}) = -2.5$. Note that the Lagrangian multipliers $\hat{u}_1$ and $\hat{u}_2$ associated with the constraints $x_1 + x_2 + x_3 \le 5$ and $x_1^2 - x_2 \le 3$ are the negatives of the entries in row 0 and under x_4 and x_5 so that $\hat{u}_1 = 0.5$ and $\hat{u}_2 = 0$. The Lagrangian multipliers $\hat{v}_1$ and $\hat{v}_2$ associated with the constraints $\Sigma_\nu \lambda_{\nu 1} = 1$ and $\Sigma_\nu \lambda_{\nu 2} = 1$ are the negatives of the entries in row 0 under λ_{11} and λ_{12} so that $\hat{v}_1 = \hat{v}_2 = 0$. To find whether a new grid point is needed, we solve the following two problems:

$$\text{Minimize} \quad f_1(x_1) + \sum_{i=1}^{2} \hat{u}_i g_{i1}(x_1) = x_1^2 - 5.5x_1 \quad \text{subject to} \quad 0 \le x_1 \le 5$$

$$\text{Minimize} \quad f_2(x_2) + \sum_{i=1}^{2} \hat{u}_i g_{i2}(x_2) = x_2^2 - 7.5x_2 \quad \text{subject to} \quad 0 \le x_2 \le 5$$

For the first problem, the optimal solution is $\bar{x}_1 = 2.75$ with optimal objective value $\bar{z}_1 = -7.56$. Thus, $\bar{z}_1 + \hat{v}_1 = -7.56 < 0$, and the grid point $\bar{x}_1 = 2.75$ would improve the objective value if introduced. For the second problem, the optimal solution is $\bar{x}_2 = 3.75$ with optimal objective value $\bar{z}_2 = -14.06$. Thus, $\bar{z}_2 + \hat{v}_2 = -14.06 < 0$, and the grid point $\bar{x}_2 = 3.75$ would also improve the objective function if introduced. Since minimum $\{\bar{z}_1 + \hat{v}_1, \bar{z}_2 + \hat{v}_2\} = \bar{z}_2 + \hat{v}_2 = -14.06$, we introduce the grid point $x_{22} = \bar{x}_2 = 3.75$. The variable associated with the grid point x_{22} is λ_{22}. (Computationally, $\bar{x}_1$ may be temporarily stored and entered sequentially if it remains enterable following the chosen pivot operation.)

Iteration 2

Note that $g_{12}(x_{22}) = 3.75$ and $g_{22}(x_{22}) = -3.75$, so that the column associated with x_{22} is $(3.75, -3.75, 0, 1)^t$. This column needs to be updated by premultiplying it by the basis inverse $\mathbf{B}^{-1}$. From the last tableau, $\mathbf{B}^{-1} = \mathbf{I}$, and, hence, the updated column for λ_{22} is $(3.75, -3.75, 0, 1)^t$. The updated coefficient in row 0 is given by $-(\bar{z}_2 + \hat{v}_2) = 14.06$. The associated tableau is given below, and λ_{22} enters the basis giving the second tableau.

	z	λ_{11}	λ_{12}	λ_{22}	x_3	x_4	x_5	RHS
z	1	0	0	14.06	0	-0.5	0	-2.5
x_3	0	0	0	3.75	1	1	0	5
x_5	0	0	0	-3.75	0	0	1	3
λ_{11}	0	1	0	0	0	0	0	1
λ_{12}	0	0	1	(1)	0	0	0	1

	z	λ_{11}	λ_{12}	λ_{22}	x_3	x_4	x_5	RHS
z	1	0	-14.06	0	0	-0.5	0	-16.56
x_3	0	0	-3.75	0	1	1	0	1.25
x_5	0	0	3.75	0	0	0	1	6.75
λ_{11}	0	1	0	0	0	0	0	1
λ_{22}	0	0	1	1	0	0	0	1

From the last tableau, $\hat{\lambda}_{11} = \hat{\lambda}_{22} = 1$ and $\hat{\lambda}_{12} = 0$. Noting that $\hat{x}_j = \sum_\nu \hat{\lambda}_{\nu j} x_{\nu j}$ for $j = 1, 2$ it follows that $\hat{x}_1 = 0$ and $\hat{x}_2 = 3.75$. Since $\hat{x}_3 = 1.25$, the current solution is $\hat{\mathbf{x}} = (0, 3.75, 1.25)^t$ and $f(\hat{\mathbf{x}}) = -17.19$. From the above tableau, $\hat{u}_1 = 0.5$, $\hat{u}_2 = 0$, $\hat{v}_1 = 0$, and $\hat{v}_2 = 14.06$. Since the values of $\hat{u}_1$ and $\hat{u}_2$ do not change from those at iteration 1, $\bar{x}_1 = 2.75$ and $\bar{x}_2 = 3.75$ remain optimal. Note that $\bar{z}_1 = -7.56$ and $\bar{z}_2 = -14.06$, so that minimum $\{\bar{z}_1 + \hat{v}_1, \bar{z}_2 + \hat{v}_2\} = \bar{z}_1 + \hat{v}_1 = -7.56$. Thus, we introduce the grid point $x_{21} = \bar{x}_1 = 2.75$. The variable corresponding to x_{21} is λ_{21}.

Iteration 3

Note that $g_{11}(x_{21}) = 2.75$ and $g_{21}(x_{21}) = 7.56$, so that the column associated with x_{21} is $(2.75, 7.56, 1, 0)^t$. From the last tableau, the basis inverse $\mathbf{B}^{-1}$ is given by

$$\mathbf{B}^{-1} = \begin{bmatrix} 1 & 0 & 0 & -3.75 \\ 0 & 1 & 0 & 3.75 \\ 0 & 0 & 1 & 0 \\ 0 & 0 & 0 & 1 \end{bmatrix}$$

Hence, the updated column for λ_{21} is $\mathbf{B}^{-1}(2.75, 7.56, 1, 0)^t = (2.75, 7.56, 1, 0)^t$. The entry in row 0 under λ_{21} is given by $-(\bar{z}_1 + \hat{v}_1) = 7.56$. The associated tableau is given below, and λ_{21} enters the basis giving the second tableau.

	z	λ_{11}	λ_{21}	λ_{12}	λ_{22}	x_3	x_4	x_5	RHS
z	1	0	7.56	−14.06	0	0	−0.5	0	−16.56
x_3	0	0	(2.75)	−3.75	0	1	1	0	1.25
x_5	0	0	7.56	3.75	0	0	0	1	6.75
λ_{11}	0	1	1	0	0	0	0	0	1
λ_{22}	0	0	0	1	1	0	0	0	1

	z	λ_{11}	λ_{21}	λ_{12}	λ_{22}	x_3	x_4	x_5	RHS
z	1	0	0	−3.78	0	−2.72	−3.22	0	−19.96
λ_{21}	0	0	1	−1.36	0	0.36	0.36	0	0.45
x_5	0	0	0	14.03	0	−2.72	−2.72	1	3.35
λ_{11}	0	1	0	1.36	0	−0.36	−0.36	0	0.55
λ_{22}	0	0	0	1	1	0	0	0	1

From the above tableau, $\hat{\lambda}_{11} = 0.55$, $\hat{\lambda}_{21} = 0.45$, $\hat{\lambda}_{12} = 0$, and $\hat{\lambda}_{22} = 1$. Therefore, $\hat{x}_1 = 1.25$ and $\hat{x}_2 = 3.75$. The current solution is thus $\hat{\mathbf{x}} = (1.25, 3.75, 0)^t$ and $f(\hat{\mathbf{x}}) = -21.88$. From the last tableau, $\hat{u}_1 = 3.22$, $\hat{u}_2 = 0$, $\hat{v}_1 = 0$, and $\hat{v}_2 = 3.78$. To find whether a new point is needed, we solve the following two problems:

$$\text{Minimize} \quad f_1(x_1) + \sum_{i=1}^{2} \hat{u}_i g_{i1}(x_1) = x_1^2 - 2.78x_1 \quad \text{subject to} \quad 0 \le x_1 \le 5$$

$$\text{Minimize} \quad f_2(x_2) + \sum_{i=1}^{2} \hat{u}_i g_{i2}(x_2) = x_2^2 - 4.78x_2 \quad \text{subject to} \quad 0 \le x_2 \le 5$$

The optimal solution to the first problem is $\bar{x}_1 = 1.39$ and the optimal objective value is $\bar{z}_1 = -1.93$. The optimal solution to the second problem is $\bar{x}_2 = 2.39$ and the optimal objective value is $\bar{z}_2 = -5.71$. Thus, minimum $\{\bar{z}_1 + \hat{v}_1, \bar{z}_2 + \hat{v}_2\} = \bar{z}_1 + \hat{v}_1 = \bar{z}_2 + \hat{v}_2 = -1.93$. Therefore, we can introduce either the grid point $\bar{x}_1 = 1.39$ or the grid point $\bar{x}_2 = 2.39$. Note that

$$\sum_{j=1}^{2} \bar{z}_j - \sum_{i=1}^{2} \hat{u}_i p_i = -23.74 \qquad \text{and} \qquad f(\hat{\mathbf{x}}) = -21.88.$$

By part 1 of Theorem 11.3.5, the optimal objective value of the original problem lies between -23.74 and -21.88. Thus, if we stop the algorithm at this stage, we would have a feasible solution $\hat{\mathbf{x}} = (1.25, 3.75, 0)'$ whose objective value is -21.88. and we would also know that a lower bound on the optimal objective value to the original problem is -23.74. If more accuracy is desired, the process would continue by introducing the new grid point $x_{31} = 1.39$ or the new grid point $x_{32} = 2.39$.

11.4 Linear Fractional Programming

In this section, we consider a problem in which the objective function is the ratio of two linear functions and the constraints are linear. Such problems are called *linear fractional programming problems* and can be stated precisely as follows:

$$\begin{aligned} \text{Minimize} \quad & \frac{\mathbf{p}'\mathbf{x} + \alpha}{\mathbf{q}'\mathbf{x} + \beta} \\ \text{subject to} \quad & \mathbf{A}\mathbf{x} = \mathbf{b} \\ & \mathbf{x} \geq \mathbf{0} \end{aligned}$$

where $\mathbf{p}$ and $\mathbf{q}$ are n vectors, $\mathbf{b}$ is an m vector, $\mathbf{A}$ is an $m \times n$ matrix, and α and β are scalars. As we shall soon observe, if an optimal solution for a linear fractional program exists, then an extreme point optimum exists. Furthermore, every local minimum is a global minimum. Hence, a procedure that moves from one extreme point to an adjacent one seems very attractive for solving such a problem. Lemma 11.4.1 below gives some important properties of the objective function.

11.4.1 Lemma

Let $f(\mathbf{x}) = (\mathbf{p}'\mathbf{x} + \alpha)/(\mathbf{q}'\mathbf{x} + \beta)$, and let S be a convex set such that $\mathbf{q}'\mathbf{x} + \beta \neq 0$ over S. Then, f is both pseudoconvex and pseudoconcave over S.

Proof

First, note that either $\mathbf{q}'\mathbf{x} + \beta > 0$ for all $\mathbf{x} \in S$ or $\mathbf{q}'\mathbf{x} + \beta < 0$ for all $\mathbf{x} \in S$. Otherwise, there exist an $\mathbf{x}_1$ and $\mathbf{x}_2$ in S such that $\mathbf{q}'\mathbf{x}_1 + \beta > 0$ and $\mathbf{q}'\mathbf{x}_2 + \beta < 0$; and, hence, for some convex combination $\mathbf{x}$ of $\mathbf{x}_1$ and $\mathbf{x}_2$, $\mathbf{q}'\mathbf{x} + \beta = 0$, contradicting our assumption. We first show that f is pseudoconvex. Suppose that $\mathbf{x}_1, \mathbf{x}_2 \in S$ with $(\mathbf{x}_2 - \mathbf{x}_1)'\nabla f(\mathbf{x}_1) \geq 0$. We need to show that $f(\mathbf{x}_2) \geq f(\mathbf{x}_1)$. Note that

$$\nabla f(\mathbf{x}_1) = \frac{(\mathbf{q}'\mathbf{x}_1 + \beta)\mathbf{p} - (\mathbf{p}'\mathbf{x}_1 + \alpha)\mathbf{q}}{(\mathbf{q}'\mathbf{x}_1 + \beta)^2}$$

Since $(\mathbf{x}_2 - \mathbf{x}_1)'\nabla f(\mathbf{x}_1) \geq 0$ and since $(\mathbf{q}'\mathbf{x}_1 + \beta)^2 > 0$, it follows that

$$0 \le (\mathbf{x}_2 - \mathbf{x}_1)^t[(\mathbf{q}^t\mathbf{x}_1 + \beta)\mathbf{p} - (\mathbf{p}^t\mathbf{x}_1 + \alpha)\mathbf{q}]$$
$$= (\mathbf{p}^t\mathbf{x}_2 + \alpha)(\mathbf{q}^t\mathbf{x}_1 + \beta) - (\mathbf{q}^t\mathbf{x}_2 + \beta)(\mathbf{p}^t\mathbf{x}_1 + \alpha)$$

Therefore, $(\mathbf{p}^t\mathbf{x}_2 + \alpha)(\mathbf{q}^t\mathbf{x}_1 + \beta) \ge (\mathbf{q}^t\mathbf{x}_2 + \beta)(\mathbf{p}^t\mathbf{x}_1 + \alpha)$. But since $\mathbf{q}^t\mathbf{x}_1 + \beta$ and $\mathbf{q}^t\mathbf{x}_2 + \beta$ are both either positive or negative, dividing by $(\mathbf{q}^t\mathbf{x}_1 + \beta)(\mathbf{q}^t\mathbf{x}_2 + \beta) > 0$, we get

$$\frac{\mathbf{p}^t\mathbf{x}_2 + \alpha}{\mathbf{q}^t\mathbf{x}_2 + \beta} \ge \frac{\mathbf{p}^t\mathbf{x}_1 + \alpha}{\mathbf{q}^t\mathbf{x}_1 + \beta}, \quad \text{that is, } f(\mathbf{x}_2) \ge f(\mathbf{x}_1)$$

Therefore, f is pseudoconvex. Similarly, it can be shown that $(\mathbf{x}_2 - \mathbf{x}_1)^t\nabla f(\mathbf{x}_1) \le 0$ implies that $f(\mathbf{x}_2) \le f(\mathbf{x}_1)$, and, hence, f is pseudoconcave, and the proof is complete.

Several implications of the above lemma for a linear fractional programming problem may be noted.

1. Since the objective function is both pseudoconvex and pseudoconcave over S, then by Theorem 3.5.11, it is also quasiconvex, quasiconcave, strictly quasiconvex, and strictly quasiconcave.
2. Since the objective function is both pseudoconvex and pseudoconcave, then, by Theorem 4.3.8, a point satisfying the KKT conditions for a minimization problem is also a global minimum over the feasible region. Likewise, a point satisfying the KKT conditions for a maximization problem is also a global maximum over the feasible region.
3. Since the objective function is strictly quasiconvex and strictly quasiconcave, then, by Theorem 3.5.6, a local minimum is also a global minimum over the feasible region. Likewise, a local maximum is also a global maximum over the feasible region.
4. Since the objective function is quasiconcave and quasiconvex, if the feasible region is bounded, then, by Theorem 3.5.3, the objective function has a minimum at an extreme point of the feasible region and also has a maximum at an extreme point of the feasible region.

The above facts about the objective function f give very useful results that can be used to develop suitable computational procedures for solving the fractional programming problem. In particular, we may search among the extreme points of the polyhedral set $\{\mathbf{x}: \mathbf{A}\mathbf{x} = \mathbf{b}, \mathbf{x} \ge \mathbf{0}\}$ until a KKT point is reached. We now show that the convex–simplex method gives a convenient solution procedure.

Minimization by the Convex–Simplex Method

Because of the special structure of the objective function f, the convex–simplex method simplifies into a minor modification of the simplex method of linear programming. Suppose that we are given an extreme point of the feasible region with basis $\mathbf{B}$ such that $\mathbf{x}_B = \mathbf{B}^{-1}\mathbf{b} > \mathbf{0}$ and $\mathbf{x}_N = \mathbf{0}$. Recall from Section 10.7 that the convex–simplex method increases or decreases one of the nonbasic variables and then modifies the basic variables accordingly. Since the current point is an extreme point with $\mathbf{x}_N = \mathbf{0}$, decreasing a nonbasic variable is not permitted, as it would violate the nonnegativity restriction. Thus, the direction-finding process simplifies as follows. Let $\mathbf{r}_N$ denote the nonbasic components of the reduced gradient vector $\mathbf{r}^t = \nabla f(\mathbf{x})^t - \nabla_B f(\mathbf{x})^t\mathbf{B}^{-1}\mathbf{A}$, so that

$$\mathbf{r}_N^t = \nabla_N f(\mathbf{x})^t + \nabla_B f(\mathbf{x})^t \mathbf{B}^{-1}\mathbf{N}$$

By Theorem 10.5.1, if $\mathbf{r}_N \geq \mathbf{0}$, then the current point is a KKT point, and we must stop. Otherwise, let $-r_j$ = maximum $\{-r_i : r_i \leq 0\}$, where r_i is the ith component of $\mathbf{r}_N$. The nonbasic variable x_j is increased, and the basic variables are modified to maintain feasibility. This is equivalent to moving along the direction $\mathbf{d}$ whose nonbasic and basic components $\mathbf{d}_N$ and $\mathbf{d}_B$ are given as follows. The direction $\mathbf{d}_N$ is a vector of zeros, except for a 1 at the jth position, and $\mathbf{d}_B = -\mathrm{B}^{-1}\mathbf{a}_j$, where $\mathbf{a}_j$ is the jth column of $\mathbf{A}$. By Theorem 10.6.1, $\mathbf{d}$ is an improving feasible direction. As we shall see by Lemma 11.4.2, no line search along the direction $\mathbf{d}$ is needed. Indeed, due to the special structure of the objective function, if $\nabla f(\mathbf{x})^t\mathbf{d} < 0$, then the function f continues to decrease by moving along $\mathbf{d}$. Thus, we must move along $\mathbf{d}$ as far as possible. Since moving along $\mathbf{d}$ is equivalent to increasing a nonbasic variable and adjusting the basic variables, we move along $\mathbf{d}$ until a basic variable drops to zero and leaves the basis, producing an adjacent extreme point. The whole process is then repeated.

11.4.2 Lemma

Let $f(\mathbf{x}) = (\mathbf{p}^t\mathbf{x} + \alpha)/(\mathbf{q}^t\mathbf{x} + \beta)$, and let S be a convex set. Furthermore, suppose that $\mathbf{q}^t\mathbf{x} + \beta \neq 0$ on S. Given $\mathbf{x} \in S$, let $\mathbf{d}$ be such that $\nabla f(\mathbf{x})^t\mathbf{d} < 0$. Then, $f(\mathbf{x} + \lambda\mathbf{d})$ is a decreasing function of λ.

Proof

Note that

$$\nabla f(\mathbf{y}) = \frac{(\mathbf{q}^t\mathbf{y} + \beta)\mathbf{p} - (\mathbf{p}^t\mathbf{y} + \alpha)\mathbf{q}}{(\mathbf{q}^t\mathbf{y} + \beta)^2} \tag{11.47}$$

Letting $\mathbf{y} = \mathbf{x} + \lambda\mathbf{d}$, $s = [\mathbf{q}^t(\mathbf{x} + \lambda\mathbf{d}) + \beta]^2 > 0$, and $s' = (\mathbf{q}^t\mathbf{x} + \beta)^2 > 0$, we get

$$\nabla f(\mathbf{x} + \lambda\mathbf{d}) = \frac{[\mathbf{q}^t(\mathbf{x} + \lambda\mathbf{d}) + \beta]\mathbf{p} - [\mathbf{p}^t(\mathbf{x} + \lambda\mathbf{d}) + \alpha]\mathbf{q}}{s}$$

$$= \frac{s'}{s}\nabla f(\mathbf{x}) + \frac{\lambda}{s}[(\mathbf{q}^t\mathbf{d})\mathbf{p} - (\mathbf{p}^t\mathbf{d})\mathbf{q}]$$

Therefore,

$$\begin{aligned}\nabla f(\mathbf{x} + \lambda\mathbf{d})^t\mathbf{d} &= \frac{s'}{s}\nabla f(\mathbf{x})^t\mathbf{d} + \frac{\lambda}{s}[(\mathbf{q}^t\mathbf{d})(\mathbf{p}^t\mathbf{d}) - (\mathbf{p}^t\mathbf{d})(\mathbf{q}^t\mathbf{d})] \\ &= \frac{s'}{s}\nabla f(\mathbf{x})^t\mathbf{d}\end{aligned} \tag{11.48}$$

Now let $\theta(\lambda) = f(\mathbf{x} + \lambda\mathbf{d})$. Then, by (11.48), $\theta'(\lambda) = \nabla f(\mathbf{x} + \lambda\mathbf{d})^t\mathbf{d} < 0$ for all λ, and the result follows.

To summarize, given the extreme point $\mathbf{x}$ and the direction $\mathbf{d}$ with $\nabla f(\mathbf{x})^t\mathbf{d} < 0$, no minimization of f along $\mathbf{d}$ is necessary, since $f(\mathbf{x} + \lambda\mathbf{d})$ is a decreasing function of λ. Therefore, we move along $\mathbf{d}$ as much as possible, that is, until an adjacent extreme point is reached, and we then repeat the process. A precise summary of the algorithm utilizing a tableau format for updating the extreme points generated is presented below.

Summary of the Fractional Programming Algorithm of Gilmore and Gomory

We present below a method credited to Gilmore and Gomory [1963] for solving a linear fractional program of the form to minimize $(\mathbf{p}^t\mathbf{x} + \alpha)/(\mathbf{q}^t\mathbf{x} + \beta)$ subject to $\mathbf{x} \in S = \{\mathbf{x}:\mathbf{A}\mathbf{x} = \mathbf{b}, \mathbf{x} \geq \mathbf{0}\}$. We will assume that the set S is bounded and that $\mathbf{q}^t\mathbf{x} + \beta \neq 0$ for all $\mathbf{x} \in S$.

Initialization Step Find a starting basic feasible solution $\mathbf{x}_1$ to the system $\mathbf{A}\mathbf{x} = \mathbf{b}$, $\mathbf{x} \geq \mathbf{0}$. Form the corresponding tableau represented by $\mathbf{x}_B + \mathbf{B}^{-1}\mathbf{N}\mathbf{x}_N = \mathbf{B}^{-1}\mathbf{b}$. Let $k = 1$ and go to the main step.

Main Step

1. Compute the vector $\mathbf{r}_N^t = \nabla_N f(\mathbf{x}_k)^t - \nabla_B f(\mathbf{x}_k)^t\mathbf{B}^{-1}\mathbf{N}$. If $\mathbf{r}_N \geq \mathbf{0}$, stop; the current point $\mathbf{x}_k$ is an optimal solution. Otherwise, go to step 2.
2. Let $-r_j = \text{maximum } \{-r_i : r_i \leq 0\}$, where r_i is the ith component of $\mathbf{r}_N$. Determine the basic variable $\mathbf{x}_{B_r}$ to leave the basis by the following minimum ratio test:

$$\frac{\bar{b}_r}{y_{rj}} = \underset{1\leq i\leq m}{\text{minimum}} \left\{ \frac{\bar{b}_i}{y_{ij}} : y_{ij} > 0 \right\}$$

   where $\bar{\mathbf{b}} = \mathbf{B}^{-1}\mathbf{b}$, $\mathbf{y}_j = \mathbf{B}^{-1}\mathbf{a}_j$, and $\mathbf{a}_j$ is the jth column of $\mathbf{A}$. Go to step 3.
3. Replace the variable x_{B_r} by the variable x_j. Update the tableau correspondingly by pivoting at y_{rj}. Let the current solution be $\mathbf{x}_{k+1}$. Replace k by $k + 1$ and go to step 1.

Exercise 11.38 shows that the reduced gradient $\mathbf{r}_N$ could be readily computed if two additional rows, one corresponding to $\mathbf{p}^t\mathbf{x} + \alpha$ and the other corresponding to $\mathbf{q}^t\mathbf{x} + \beta$ are introduced and carried forward at each iteration.

Finite Convergence

We now assume that $\mathbf{x}_B > \mathbf{0}$ for each extreme point. The algorithm moves from one extreme point to another. By Lemma 11.4.2 and the nondegeneracy assumption, the objective function strictly decreases at each iteration so that the extreme points generated are distinct. There is only a finite number of these points, and, hence, the algorithm stops in a finite number of steps. At termination, the reduced gradient is nonnegative resulting in a KKT point; and, by Lemma 11.4.1, this point is indeed an optimal point.

11.4.3 Example

Consider the following problem:

$$\begin{aligned}
\text{Minimize} \quad & \frac{-2x_1 + x_2 + 2}{x_1 + 3x_2 + 4} \\
\text{subject to} \quad & -x_1 + x_2 \leq 4 \\
& x_2 \leq 6 \\
& 2x_1 + x_2 \leq 14 \\
& x_1, x_2 \geq 0
\end{aligned}$$

Figure 11.5 shows the feasible region with the extreme points (0, 0), (0, 4), (2, 6), (4, 6), and (7, 0). The objective at these points is 0.5, 0.375, 0.167, 0.0, and −1.09, respectively, and, hence, the optimal point is (7, 0).

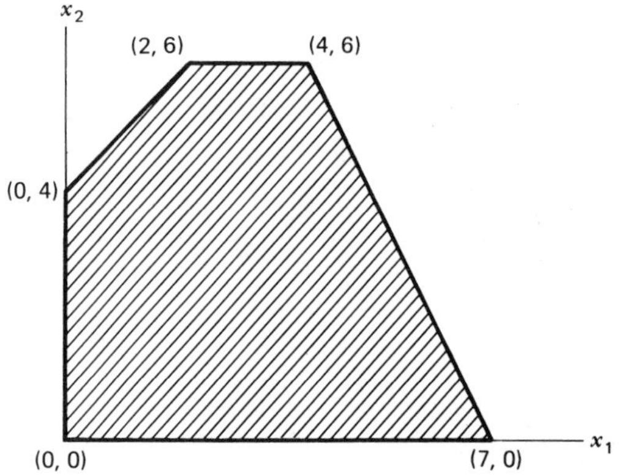

Figure 11.5 The feasible region for Example 11.4.3.

Introducing the slack variables x_3, x_4, and x_5, we get the initial extreme point $\mathbf{x}_1 = (0, 0, 4, 6, 14)^t$.

Iteration 1

The tableau below summarizes the computations for this iteration.

	x_1	x_2	x_3	x_4	x_5	RHS
$\nabla f(\mathbf{x}_1)$	$-\frac{10}{16}$	$-\frac{2}{16}$	0	0	0	—
x_3	-1	1	1	0	0	4
x_4	0	1	0	1	0	6
x_5	2	1	0	0	1	14
$\mathbf{r}$	$-\frac{10}{16}$	$-\frac{2}{16}$	0	0	0	—

We have $\mathbf{q}^t\mathbf{x}_1 + \beta = 4$ and $\mathbf{p}^t\mathbf{x}_1 + \alpha = 2$. Hence, from (11.47), we get $\nabla f(\mathbf{x})^t = (-\frac{10}{16}, -\frac{2}{16}, 0, 0, 0)$. $\nabla_N f(\mathbf{x})^t = (-\frac{10}{16}, -\frac{2}{16})$ and $\nabla_B f(\mathbf{x})^t = (0, 0, 0)$. The columns of x_1 and x_2 give $\mathbf{B}^{-1}\mathbf{N}$, and we get

$$\mathbf{r}_N^t = (r_1, r_2) = \nabla_N f(\mathbf{x}_1)^t - \nabla_B f(\mathbf{x}_1)^t\mathbf{B}^{-1}\mathbf{N}$$

$$= \left(-\frac{10}{16}, -\frac{2}{16}\right) - (0, 0, 0)\begin{bmatrix} -1 & 1 \\ 0 & 1 \\ 2 & 1 \end{bmatrix} = \left(-\frac{10}{16}, -\frac{2}{16}\right)$$

Note that $\mathbf{r}_N^t = (r_3, r_4, r_5) = (0, 0, 0)$. Since maximum $\{-r_1, -r_2, -r_3, -r_4, -r_5\} = \frac{10}{16}$, x_1 enters the basis. By the minimum ratio test, x_5 leaves the basis.

Iteration 2

The computations for this iteration are summarized below.

	x_1	x_2	x_3	x_4	x_5	RHS
$\nabla f(\mathbf{x}_2)$	$-\frac{10}{121}$	$\frac{47}{121}$	0	0	0	—
x_3	0	$\frac{3}{2}$	1	0	$\frac{1}{2}$	11
x_4	0	1	0	1	0	6
x_1	1	$\frac{1}{2}$	0	0	$\frac{1}{2}$	7
$\mathbf{r}$	0	$\frac{52}{121}$	0	0	$\frac{5}{121}$	—

When x_1 replaces x_5 in the basis, we get the point $\mathbf{x}_2^t = (7, 0, 11, 6, 0)$. Now $\mathbf{q}^t\mathbf{x}_2 + \beta = 11$ and $\mathbf{p}^t\mathbf{x}_2 + \alpha = -12$, so that, from (11.47), we get $\nabla f(\mathbf{x}_2)^t = (-\frac{10}{121}, \frac{47}{121}, 0, 0, 0)$. $\mathbf{B}^{-1}\mathbf{N}$ is given by the columns of x_2 and x_5 in the tableau, and we then get

$$\mathbf{r}_N^t = (r_2, r_5) = \nabla_N f(\mathbf{x}_2)^t - \nabla_B f(\mathbf{x}_2)^t\mathbf{B}^{-1}\mathbf{N}$$

$$= (\tfrac{47}{121}, 0) - (0, 0, -\tfrac{10}{121})\begin{bmatrix} \frac{3}{2} & \frac{1}{2} \\ 1 & 0 \\ \frac{1}{2} & \frac{1}{2} \end{bmatrix}$$

$$= (\tfrac{52}{121}, \tfrac{5}{121})$$

Since $\mathbf{r}_N \geq \mathbf{0}$, we stop with the optimal solution $x_1 = 7$ and $x_2 = 0$. The corresponding objective function value is -1.09.

The Method of Charnes and Cooper [1962]

We now describe another procedure using the simplex method for solving a linear fractional programming problem. Consider the following problem:

$$\begin{aligned} \text{Minimize} \quad & \frac{\mathbf{p}^t\mathbf{x} + \alpha}{\mathbf{q}^t\mathbf{x} + \beta} \\ \text{subject to} \quad & \mathbf{A}\mathbf{x} \leq \mathbf{b} \\ & \mathbf{x} \geq \mathbf{0} \end{aligned}$$

Suppose that the set $S = \{\mathbf{x}: \mathbf{A}\mathbf{x} \leq \mathbf{b} \text{ and } \mathbf{x} \geq \mathbf{0}\}$ is compact, and suppose that $\mathbf{q}^t\mathbf{x} + \beta > 0$ for each $\mathbf{x} \in S$. Letting $z = 1/(\mathbf{q}^t\mathbf{x} + \beta)$ and $\mathbf{y} = z\mathbf{x}$, the above problem leads to the following linear program:

$$\begin{aligned} \text{Minimize} \quad & \mathbf{p}^t\mathbf{y} + \alpha z \\ \text{subject to} \quad & \mathbf{A}\mathbf{y} - \mathbf{b}z \leq \mathbf{0} \\ & \mathbf{q}^t\mathbf{y} + \beta z = 1 \\ & \mathbf{y} \geq \mathbf{0} \\ & z \geq 0 \end{aligned}$$

First, note that, if $(\mathbf{y}, z)$ is a feasible solution to the above problem, then $z > 0$. This follows since, if $z = 0$, then $\mathbf{y} \neq \mathbf{0}$ must be such that $\mathbf{A}\mathbf{y} \leq \mathbf{0}$ and $\mathbf{y} \geq \mathbf{0}$, which means that $\mathbf{y}$ is a direction of S, violating the compactness assumption. We now demonstrate that, if $(\bar{\mathbf{y}}, \bar{z})$ is an optimal solution to the above linear program, then $\bar{\mathbf{x}} = \bar{\mathbf{y}}/\bar{z}$ is an optimal solution to the fractional program.

Note that $\mathbf{A}\bar{\mathbf{x}} \leq \mathbf{b}$ and $\bar{\mathbf{x}} \geq \mathbf{0}$, so that $\bar{\mathbf{x}}$ is a feasible solution to the fractional program. To show optimality of $\bar{\mathbf{x}}$, let $\mathbf{x}$ be such that $\mathbf{A}\mathbf{x} \leq \mathbf{b}$ and $\mathbf{x} \geq \mathbf{0}$. Note that $\mathbf{q}'\mathbf{x} + \beta > 0$ by assumption, and that the vector $(\mathbf{y}, z)$ is a feasible solution to the linear program, where $\mathbf{y} = \mathbf{x}/(\mathbf{q}'\mathbf{x} + \beta)$ and $z = 1/(\mathbf{q}'\mathbf{x} + \beta)$. Since $(\bar{\mathbf{y}}, \bar{z})$ is an optimal solution to the linear program, $\mathbf{p}'\bar{\mathbf{y}} + \alpha\bar{z} \leq \mathbf{p}'\mathbf{y} + \alpha z$. Substituting for $\bar{\mathbf{y}}$, $\mathbf{y}$, and z, this inequality gives $\bar{z}(\mathbf{p}'\bar{\mathbf{x}} + \alpha) \leq (\mathbf{p}'\mathbf{x} + \alpha)/(\mathbf{q}'\mathbf{x} + \beta)$. The result immediately follows by dividing the left-hand side by $1 = \mathbf{q}'\bar{\mathbf{y}} + \beta\bar{z}$.

Now, if $\mathbf{q}'\mathbf{x} + \beta < 0$ for all $\mathbf{x} \in S$, then letting $-z = 1/(\mathbf{q}'\mathbf{x} + \beta)$ and $\mathbf{y} = z\mathbf{x}$ gives the following linear program:

$$
\begin{array}{ll}
\text{Minimize} & -\mathbf{p}'\mathbf{y} - \alpha z \\
\text{subject to} & \mathbf{A}\mathbf{y} - \mathbf{b}z \leq \mathbf{0} \\
& -\mathbf{q}'\mathbf{y} - \beta z = 1 \\
& \mathbf{y} \geq \mathbf{0} \\
& z \geq 0
\end{array}
$$

In a fashion similar to that above, if $(\bar{\mathbf{y}}, \bar{z})$ solves the above linear program, then $\bar{\mathbf{x}} = \bar{\mathbf{y}}/\bar{z}$ solves the fractional programming problem.

To summarize, we have shown that a fractional linear program could be solved by a linear programming problem with one additional variable and one additional constraint. The form of the linear program used depends on whether $\mathbf{q}'\mathbf{x} + \beta > 0$ for all $\mathbf{x} \in S$ or $\mathbf{q}'\mathbf{x} + \beta < 0$ for all $\mathbf{x} \in S$. If there exist $\mathbf{x}_1, \mathbf{x}_2 \in S$ such that $\mathbf{q}'\mathbf{x}_1 + \beta > 0$ and $\mathbf{q}'\mathbf{x}_2 + \beta < 0$, then the optimal solution to the fractional program is unbounded.

11.4.4 Example

Consider the following problem:

$$
\begin{array}{ll}
\text{Minimize} & \dfrac{-2x_1 + x_2 + 2}{x_1 + 3x_2 + 4} \\
\text{subject to} & -x_1 + x_2 \leq 4 \\
& 2x_1 + x_2 \leq 14 \\
& x_2 \leq 6 \\
& x_1, x_2 \geq 0
\end{array}
$$

The feasible region for this problem is shown in Figure 11.5. We solve this problem using the method of Charnes and Cooper. Note that the point (0, 0) is feasible and that at this point $-x_1 + 3x_2 + 4 > 0$. Hence, the denominator is positive over the entire feasible region. The equivalent linear program is given by:

$$
\begin{array}{ll}
\text{Minimize} & -2y_1 + y_2 + 2z \\
\text{subject to} & -y_1 + y_2 - 4z \leq 0 \\
& 2y_1 + y_2 - 14z \leq 0 \\
& y_2 - 6z \leq 0 \\
& y_1 + 3y_2 + 4z = 1 \\
& y_1, y_2, z \geq 0
\end{array}
$$

The reader can verify that $y_1 = \frac{7}{11}$, $y_2 = 0$, and $z = \frac{1}{11}$ is an optimal solution to the above linear program. Hence, the optimal solution to the original problem is $x_1 = y_1/z_1 = 7$ and $x_2 = y_2/z_2 = 0$.

11.5 Geometric Programming

In this section we consider problems of the type

$$\begin{aligned} GP: \quad & \text{Minimize} && f(\mathbf{x}) \\ & \text{subject to} && g_i(\mathbf{x}) \le 1 \qquad \text{for } i = 1, \ldots, m \\ &&& \mathbf{x} > \mathbf{0} \end{aligned}$$

where each of the functions f and g_i is a posynomial of $\mathbf{x} \in E_n$ and where the variables $\mathbf{x}$ assume strictly positive values by the nature of the problem itself. A *posynomial* is a function composed of terms of the type

$$T_k = \alpha_k \prod_{j=1}^{n} x_j^{a_{kj}} \tag{11.49}$$

where $\alpha_k > 0$, and where the exponents a_{kj}, $j = 1, \ldots, n$, are rational numbers which can be of either sign. In particular, for $\mathbf{x} > \mathbf{0}$, we have $T_k > 0$ as well. Hence, the objective and constraint functions can be written as

$$f(\mathbf{x}) = \sum_{k \in J_0} T_k \qquad g_i(\mathbf{x}) = \sum_{k \in J_i} T_k \qquad \text{for } i = 1, \ldots, m \tag{11.50a}$$

where the collection of index sets $J_0, J_1, \ldots, J_m$ are mutually disjoint, and where

$$J_0 \cup J_1 \cup \ldots \cup J_m \equiv \{1, 2, \ldots, M\} \tag{11.50b}$$

represents a total of M terms, each of the type (11.49). Problems GP of this type are called *posynomial programming problems*. When the coefficients α_k are permitted to be negative, then the functions (11.50a) are called *signomials,* and Problem GP is then known as a *signomial programming problem*. In either case, the problem is called a *geometric programming problem,* the name arising from the geometric–arithmetic mean inequality (see Exercise 4.25) used in the original analysis presented by Duffin, Peterson, and Zener [1967] to transform the problem into a simpler equivalent form.

Geometric programming problems arise frequently in engineering applications where the decision variables $\mathbf{x}$ are design variables that are required to take on positive values to be meaningful, and where the objective and constraint functions model fundamental physical or economical relationships that, by their nature, turn out to be posynomials or may be transformed into such functions (see Exercises 11.41, 11.42, and 11.43). Ominous as these problems might appear, there exists a transformation that considerably simplifies this problem, often rendering it solvable as a linear system of equations, or as a manageable, linearly constrained problem. This transformation involves two steps of using a change of variables, interposed by an application of Lagrangian duality concepts from Chapter 6.

To introduce the first change of variables, let us substitute

$$y_j = ln\,(x_j) \qquad \text{for } j = 1, \ldots, n \tag{11.51}$$

so that the term T_k in (11.49) becomes the following function τ_k of $\mathbf{y}$:

$$\tau_k = \alpha_k \prod_{j=1}^{n} (e^{y_j})^{a_{kj}} = \alpha_k e^{\mathbf{a}_k^t \mathbf{y}} \qquad \text{for } k = 1, \ldots, M \tag{11.52}$$

where $\mathbf{a}_k = (a_{k1}, \ldots, a_{kn})^t$ for $k = 1, \ldots, M$. Furthermore, in addition to using the substitution (11.51) in Problem GP, let us also equivalently write the objective function as one of minimizing $ln\ [f(\mathbf{x})]$ and the constraints as $ln\ [g_i(\mathbf{x})] \leq 0$ for $i = 1, \ldots, m$, noting the monotonicity of the logarithmic function and the positivity of the objective and constraint functions. Hence, we can write GP equivalently as

$$\begin{aligned} &\text{Minimize} && ln\ [F(\mathbf{y})] && && (11.53a)\\ &\text{subject to} && ln\ [G_i(\mathbf{y})] \leq 0 && \text{for } i = 1, \ldots, m && (11.53b)\\ & && \mathbf{y} \text{ unrestricted in sign} \end{aligned}$$

where, from (11.50)–(11.52), we have

$$F(\mathbf{y}) \equiv \sum_{k \in J_0} \tau_k \qquad G_i(\mathbf{y}) \equiv \sum_{k \in J_i} \tau_k \qquad \text{for } i = 1, \ldots, m \tag{11.53c}$$

The following result establishes an extremely useful characterization of Problem (11.53).

11.5.1 Lemma

Given the posynomial geometric programming problem GP, consider the equivalent problem (11.53) obtained under the transformation (11.51). Then, the objective and constraint functions of this problem are all convex and, hence, (11.53) is a convex programming problem.

Proof

First, consider any term τ_k, which is a function of $\mathbf{y}$, as defined by (11.52). Denoting $\mathbf{a}_k = (a_{k1}, \ldots, a_{kn})^t$, we have $\nabla\tau_k = \tau_k\mathbf{a}_k$, and $\nabla^2\tau_k = \tau_k\mathbf{a}_k\mathbf{a}_k^t$, where ∇ and ∇^2 denote the gradient and Hessian operators, respectively. Now, denote $h(\mathbf{y}) = ln\ [F(\mathbf{y})]$. We have $\nabla h(\mathbf{y}) = \nabla F(\mathbf{y})/F(\mathbf{y})$ and

$$\nabla^2 h(\mathbf{y}) = \frac{[F(\mathbf{y})\nabla^2 F(\mathbf{y}) - \nabla F(\mathbf{y})\nabla F(\mathbf{y})^t]}{[F(\mathbf{y})]^2}$$

$$= \frac{\left[\sum_{k \in J_0} \tau_k\right]\left[\sum_{k \in J_0} \tau_k\mathbf{a}_k\mathbf{a}_k^t\right] - \left[\sum_{k \in J_0} \tau_k\mathbf{a}_k\right]\left[\sum_{k \in J_0} \tau_k\mathbf{a}_k^t\right]}{[F(\mathbf{y})]^2}$$

Using (11.53c) and the foregoing expressions for $\nabla\tau_k$ and $\nabla^2\tau_k$, the numerator of $\nabla^2 h(\mathbf{y})$ equals

$$\sum_{k \in J_0}\sum_{l \in J_0} \tau_k\tau_l\mathbf{a}_k\mathbf{a}_k^t - \sum_{k \in J_0}\sum_{l \in J_0} \tau_k\tau_l\mathbf{a}_k\mathbf{a}_l^t = \sum\sum_{k<l \text{ in } J_0} \tau_k\tau_l[\mathbf{a}_k\mathbf{a}_k^t + \mathbf{a}_l\mathbf{a}_l^t]$$

$$- \sum\sum_{k<l \text{ in } J_0} \tau_k\tau_l[\mathbf{a}_k\mathbf{a}_l^t + \mathbf{a}_l\mathbf{a}_k^t] = \sum\sum_{k<l \text{ in } J_0} \tau_k\tau_l(\mathbf{a}_k - \mathbf{a}_l)(\mathbf{a}_k - \mathbf{a}_l)^t$$

Consequently, as $\tau_k > 0$ and $\tau_l > 0$, and $(\mathbf{a}_k - \mathbf{a}_l)(\mathbf{a}_k - \mathbf{a}_l)^t$ is positive semidefinite, we have that $ln\ [F(\mathbf{y})]$ is a convex function. Similarly, $ln\ [G_i(\mathbf{y})]$ is a convex function for each $i = 1, \ldots, m$, and this completes the proof.

Now, assuming that a suitable constraint qualification holds (such as the interiority constraint qualification of Theorem 6.2.4), we can invoke Theorem 6.2.4 to assert that there is no duality gap between (11.53) and its Lagrangian dual stated below:

$$\textit{LD:}\quad \text{Maximize} \quad \{\theta(\mathbf{u}) : \mathbf{u} \geq \mathbf{0}\} \tag{11.54a}$$

where

$$\theta(\mathbf{u}) = \underset{\mathbf{y}}{\text{minimum}} \{L(\mathbf{y}, \mathbf{u})\} \tag{11.54b}$$

and where $L(\mathbf{y}, \mathbf{u})$ is the Lagrangian function given by

$$L(\mathbf{y}, \mathbf{u}) = ln\,[F(\mathbf{y})] + \sum_{i=1}^{m} u_i \, ln\,[G_i(\mathbf{y})] \tag{11.54c}$$

Since for any $\mathbf{u} \geq \mathbf{0}$, $\theta(\mathbf{u})$ equals that value of $L(\mathbf{y}, \mathbf{u})$ evaluated at the point $\mathbf{y}$ for which $\nabla_{\mathbf{y}} L(\mathbf{y}, \mathbf{u})$ equals zero by Lemma 11.5.1, we can equivalently write the Lagrangian dual (11.54) as

$$\text{Maximize } \{L(\mathbf{y}, \mathbf{u}) : \nabla_{\mathbf{y}} L(\mathbf{y}, \mathbf{u}) = \mathbf{0}, \mathbf{u} \geq \mathbf{0}, \mathbf{y} \text{ unrestricted}\} \tag{11.55}$$

To further simplify (11.55), note that, as in the proof of lemma 11.5.1, we have

$$\begin{aligned} \nabla_{\mathbf{y}} L(\mathbf{y}, \mathbf{u}) &= \frac{\nabla F(\mathbf{y})}{F(\mathbf{y})} + \sum_{i=1}^{m} u_i \frac{\nabla G_i(\mathbf{y})}{G_i(\mathbf{y})} \\ &= \frac{1}{F(\mathbf{y})} \sum_{k \in J_0} \tau_k \mathbf{a}_k + \sum_{i=1}^{m} \frac{u_i}{G_i(\mathbf{y})} \left[\sum_{k \in J_i} \tau_k \mathbf{a}_k \right] \end{aligned} \tag{11.56}$$

We now employ a second transformation. Define $\delta_1, \ldots, \delta_M$ according to

$$\delta_k = \frac{\tau_k}{F} \text{ for all } k \in J_0 \qquad \text{and} \qquad \delta_k = \frac{u_i \tau_k}{G_i} \qquad \text{for all } k \in J_i, \ i = 1, \ldots, m \tag{11.57}$$

Note that we have dropped the argument ($\mathbf{y}$) for notational convenience, recognizing the dependence of F, G_i, τ_k, and δ_k for all $i = 1, \ldots, m$, $k = 1, \ldots, M$ on $\mathbf{y}$. However, we would now like to treat $\boldsymbol{\delta} = (\delta_1, \ldots, \delta_M)^t$ as a set of *variables*, and write (11.55) in terms of ($\boldsymbol{\delta}$, $\mathbf{u}$) by eliminating $\mathbf{y}$ from the problem.

Note that, from (11.53c) and (11.57), we must have

$$\sum_{k \in J_0} \delta_k = 1 \qquad \text{and} \qquad \sum_{k \in J_i} \delta_k = u_i \qquad \text{for } i = 1, \ldots, m \tag{11.58}$$

Constraints (11.58) are called *normalization* constraints and, together with $\boldsymbol{\delta} \geq \mathbf{0}$, restrict the values that $\boldsymbol{\delta}$ can assume in (11.57). Furthermore, using (11.56), the equality constraint in (11.55) can be written as follows under the transformation (11.57):

$$\sum_{k=1}^{M} \delta_k \mathbf{a}_k = \mathbf{0} \tag{11.59}$$

Constraint (11.59) is known as the *orthogonality* constraint, since it asserts that $\boldsymbol{\delta}$ is orthogonal to each of the n rows of the $n \times M$ matrix $\mathbf{A}$ with columns $\mathbf{a}_1, \ldots, \mathbf{a}_M$.

In the transformed problem, we impose the relationships (11.58) and (11.59) on ($\boldsymbol{\delta}$, $\mathbf{u}$), along with nonnegativity restrictions. However, given any ($\boldsymbol{\delta}$, $\mathbf{u}$) feasible to these conditions, since there does not necessarily exist a $\mathbf{y}$ that satisfies (11.57), we need some further analysis to justify the utility of the transformed problem derived below.

Toward this end, let us also simplify the objective function of (11.55) under (11.57) and (11.58). Consider the term $u_i \, ln\,[G_i]$ for any $i \in \{1, \ldots, m\}$. Assuming that $u_i > 0$, and writing this term as $u_i \, ln\,(u_i) + u_i \, ln\,[G_i/u_i]$, we get, upon using (11.58), (11.57), and (11.52), in turn, that

$$u_i \; ln \; [G_i] = u_i \; ln \; (u_i) + u_i \; ln \left[\frac{G_i}{u_i}\right] = u_i \; ln \; (u_i) + \sum_{k \in J_i} \delta_k \; ln \left[\frac{G_i}{u_i}\right]$$

$$= u_i \; ln \; (u_i) + \sum_{k \in J_i} \delta_k \; ln \left[\frac{\tau_k}{\delta_k}\right] = u_i \; ln \; (u_i) + \sum_{k \in J_i} \delta_k \; ln \left[\frac{\alpha_k}{\delta_k} e^{\mathbf{a}_k^t \mathbf{y}}\right] \tag{11.60a}$$

$$= u_i \; ln \; (u_i) + \sum_{k \in J_i} \delta_k \; ln \left[\frac{\alpha_k}{\delta_k}\right] + \sum_{k \in J_i} \delta_k \mathbf{a}_k^t \mathbf{y}$$

Similarly, we have

$$ln \; [F] = \sum_{k \in J_0} \delta_k \; ln \left[\frac{\alpha_k}{\delta_k}\right] + \sum_{k \in J_0} \delta_k \mathbf{a}_k^t \mathbf{y} \tag{11.60b}$$

Hence, noting (11.50b) and that $\sum_{k=1}^{M} \delta_k \mathbf{a}_k^t \mathbf{y} = 0$ by (11.59), we observe, from (11.54c) and (11.60), that the objective function in (11.55) is given by

$$\sum_{k=1}^{M} \delta_k \; ln \left[\frac{\alpha_k}{\delta_k}\right] + \sum_{i=1}^{m} u_i \; ln \; (u_i) \tag{11.61}$$

Noting that $u_i \; ln \; (u_i) \to 0$ as $u_i \to 0^+$, and that $\delta_k \; ln \; [\alpha_k/\delta_k] \to 0$ as $\delta_k \to 0^+$, we finally use (11.58), (11.59), and (11.61) to replace (11.55) with the following *dual geometric program* (*DGP*) in the variables ($\boldsymbol{\delta}$, $\mathbf{u}$), where the separable objective function terms are defined as zeros for $\delta_k = 0$ or $u_i = 0$:

DGP: Maximize
$$\sum_{k=1}^{M} \delta_k \; ln \left[\frac{\alpha_k}{\delta_k}\right] + \sum_{i=1}^{m} u_i \; ln \; (u_i) \tag{11.62a}$$

subject to
$$\sum_{k=1}^{M} \delta_k \mathbf{a}_k \equiv \mathbf{A}\boldsymbol{\delta} = \mathbf{0} \tag{11.62b}$$

$$\sum_{k \in J_0} \delta_k = 1 \tag{11.62c}$$

$$\sum_{k \in J_i} \delta_k = u_i \qquad \text{for } i = 1, \ldots, m \tag{11.62d}$$

$$\begin{aligned} \delta_k &\geq 0 \qquad \text{for } k = 1, \ldots, M \\ u_i &\geq 0 \qquad \text{for } i = 1, \ldots, m \end{aligned} \tag{11.62e}$$

Note that Problem DGP is a linearly constrained problem with a separable, concave objective function (see Exercise 11.39) and is therefore a convex programming problem that is readily solvable by using the methods of Chapter 10 and Section 11.3. Note from (11.62d) that we can write the variables u_i, $i = 1, \ldots, m$, in terms of the $\boldsymbol{\delta}$-variables. Furthermore, assuming that the $(n + 1)$ constraints (11.62b) and (11.62c) are linearly independent, we can solve for some $(n + 1)$ variables δ_k in terms of the remaining $(M - n - 1)$ $\boldsymbol{\delta}$-variables. The resulting degrees of freedom in the problem due to (11.62b)–(11.62d) is called the

$$\textit{Degree of difficulty } (\text{DD}) \equiv \text{number of terms } (M) - \text{number of variables } (n) - 1 \tag{11.63}$$

In general, the degree of difficulty DD equals M minus the number of linearly independent constraints in (11.62b, c). Note that if DD $= 0$, as is sometimes the case, then the solution to DGP, if it exists, is uniquely determined by (11.62b)–(11.62d), itself. Otherwise, a linearly constrained problem embedded essentially in dimension DD needs to be solved. The following result prescribes the recovery of the optimum to GP from an optimum to DGP.

11.5.2 Theorem

Consider the dual geometric program DGP, and suppose that $(\boldsymbol{\delta}^*, \mathbf{u}^*) > \mathbf{0}$ solves this problem with an optimal objective function value ν(DGP). Furthermore, let $\mathbf{v}^* = (v_1^*, \ldots, v_n^*)^t$, and let $\mathbf{w}^* = (w_0^*, w_1^*, \ldots, w_m^*)^t$ be the corresponding optimal Lagrange multiplier values associated with the constraints (11.62b) and (11.62c, d), respectively. Then, an optimum $\mathbf{y}^*$ to Problem (11.53) is given by

$$y_j^* = v_j^* \qquad \text{for } j = 1, \ldots, n \tag{11.64a}$$

with the optimal objective value of this problem being ν(DGP) and with $\mathbf{u}^*$ being the set of optimal Lagrange multipliers associated with (11.53b). Moreover,

$$x_j^* = e^{y_j^*} \qquad j = 1, \ldots, n \tag{11.64b}$$

solves Problem GP.

Proof

Since $(\boldsymbol{\delta}^*, \mathbf{u}^*) > \mathbf{0}$ at optimality, by the differentiability of the objective function at this point and the linearity of the constraints, by Lemma 5.1.4 we must have a solution to the KKT system given by (11.62b)–(11.62e), along with the complementary slack dual feasibility conditions

$$\left[ln\left(\frac{\alpha_k}{\delta_k^*}\right) - 1\right] + \mathbf{a}_k^t\mathbf{v}^* + w_i^* = 0 \qquad \text{for all } k \in J_i,\ i = 0, 1, \ldots, m \tag{11.65}$$

$$[ln\,(u_i^*) + 1] - w_i^* = 0 \qquad \text{for all } i = 1, \ldots, m \tag{11.66}$$

where $\mathbf{v}^*$ and $\mathbf{w}^*$ are as defined in the theorem. Substituting for w_i from (11.66) into (11.65) for $i = 1, \ldots, m$, we get

$$\mathbf{a}_k^t\mathbf{v}^* = ln\left[\frac{\delta_k^*}{\alpha_k u_i^*}\right] \qquad \text{for all } k \in J_i,\ i = 1, \ldots, m \tag{11.67}$$

Now consider $\mathbf{y}^*$ as given by (11.64). We then have from (11.52), (11.53c), (11.62d), and (11.67) that

$$G_i(\mathbf{y}^*) = \sum_{k \in J_i} \tau_k = \sum_{k \in J_i} \alpha_k e^{\mathbf{a}_k^t\mathbf{y}^*}$$

$$= \sum_{k \in J_i} \alpha_k \left[\frac{\delta_k^*}{\alpha_k u_i^*}\right] = 1 \qquad \text{for } i = 1, \ldots, m \tag{11.68}$$

Moreover, from (11.52) and (11.65)–(11.67), we have

$$\tau_k \equiv \alpha_k e^{\mathbf{a}_k^t\mathbf{y}^*} = \frac{\delta_k^*}{u_i^*} \qquad \text{for all } k \in J_i,\ i = 1, \ldots, m \tag{11.69a}$$

Furthermore, we have

$$\tau_k \equiv \alpha_k e^{\mathbf{a}_k^t \mathbf{y}^*} = \alpha_k e^{[1 - w_0^* + \ln[\delta_k^*/\alpha_k]]} = \delta_k^* e^{(1 - w_0^*)} \qquad \text{for all } k \in J_0$$

But $F(\mathbf{y}^*) \equiv \sum_{k \in J_0} \tau_k = e^{(1-w_0^*)} \sum_{k \in J_0} \delta_k^* = e^{1-w_0^*}$ from (11.62c). Hence, we have

$$\tau_k = F(\mathbf{y}^*)\delta_k^* \qquad \text{for all } k \in J_0 \tag{11.69b}$$

Substituting (11.69) into (11.56) and using (11.62b) and (11.68), we get

$$\nabla_{\mathbf{y}} L(\mathbf{y}^*, \mathbf{u}^*) = \sum_{k \in J_0} \delta_k^* \mathbf{a}_k + \sum_{i=1}^{m} \sum_{k \in J_i} \delta_k^* \mathbf{a}_k = \mathbf{0} \tag{11.70}$$

Consequently, from (11.62e), (11.68), and (11.70) the primal–dual solution $(\mathbf{y}^*, \mathbf{u}^*)$ satisfies the KKT conditions for Problem (11.53) and, hence, using Lemma 11.5.1, solves this problem. Moreover, noting (11.54c), (11.60), (11.61), (11.68), and (11.69), we have

$$\nu(\text{DGP}) = L(\mathbf{y}^*, \mathbf{u}^*) = \ln\ [F(\mathbf{y}^*)]$$

Finally, by the equivalence of GP and Problem (11.53) under the transformation (11.51), we also have that $x_j^* = e^{y_j^*}$, $j = 1, \ldots, n$, solves GP, and this completes the proof.

Observe that, given a positive optimal solution to Problem DGP, we are able to claim that GP has an optimum and, moreover, recover an optimum to this problem via (11.64). On the other hand, if GP has an optimum, and if the interior point constraint qualification of Theorem 6.2.4 holds, then it can be shown that DGP also has an optimum $(\boldsymbol{\delta}^*, \mathbf{u}^*)$ with the same objective function value ν^*, and that an optimum to Problem (11.53) can be recovered by solving the system

$$\mathbf{a}_k^t \mathbf{y} = \ln\ (\delta_k^* e^{\nu^*}/\alpha_k) \qquad \text{for } k \in J_0 \tag{11.71a}$$

$$\mathbf{a}_k^t \mathbf{y} = \ln\ [\delta_k^*/(u_i^* \alpha_k)] \qquad \text{for } k \in J_i \text{ and } i \in \{1, \ldots, m\} \text{ such that } u_i^* > 0 \tag{11.71b}$$

This system arises from (11.52) and (11.57), noting that $\ln\ [F(\mathbf{y}^*)] = \nu^*$ and that $G_i(\mathbf{y}^*) = 1$ for the active constraints that have $u_i^* > 0$ (see Duffin, Peterson, and Zener [1967]). Note, from (11.52) and (11.69), that the proof of Theorem 11.52 verifies this system to yield $\mathbf{y}^*$ in terms of the (primal) solution to DGP under the conditions of Theorem 11.5.2. Hence, (11.71) provides an alternative to (11.64a) for recovering a primal optimal solution to Problem GP via (11.64b).

11.5.3 Example (Zero Degrees of Difficulty)

Suppose that we wish to construct a right circular cylinder of radius r and height h that is closed at both ends, has a volume of at least V, and uses the least amount of material. Hence, the problem we wish to solve is to minimize the total surface area $2\pi r^2 + 2\pi rh$, so that the volume $\pi r^2 h$ is, at least, V. Rewriting the constraint in standard form, this gives

$$\textit{GP:}\quad \text{Minimize}\left\{2\pi r^2 + 2\pi rh : \frac{V}{\pi} r^{-2} h^{-1} \le 1, r > 0, h > 0\right\}$$

Note that the number of terms in the problem is $M = 3$, and that the number of variables is $n = 2$, namely, r and h. Hence, the degree of difficulty equals zero from (11.63).

The α coefficients for the three terms are given by $\alpha_1 = 2\pi$, $\alpha_2 = 2\pi$, and $\alpha_3 = V/\pi$. The respective exponent vectors are

$$\mathbf{a}_1^t = (2, 0) \qquad \mathbf{a}_2^t = (1, 1) \qquad \mathbf{a}_3^t = (-2, -1)$$

The corresponding orthogonality and normalization constraints (11.62b, c, d) are as follows, noting that $J_0 = \{1, 2\}$, and $J_1 = \{3\}$:

$$\begin{aligned} 2\delta_1 + \delta_2 - 2\delta_3 &= 0 \\ \delta_2 - \delta_3 &= 0 \\ \delta_1 + \delta_2 \qquad &= 1 \\ \delta_3 &= u_1 \end{aligned}$$

Solving, we obtain $\delta_1^* = \frac{1}{3}$, $\delta_2^* = \delta_3^* = u_1^* = \frac{2}{3}$. Note that $(\boldsymbol{\delta}^*, \mathbf{u}^*) > \mathbf{0}$, so that the condition of Theorem 11.5.2 holds true. The optimal objective function value of Problem DGP is $v^* = \frac{1}{3} \ln [6\pi] + \frac{2}{3} \ln [3\pi] + \frac{2}{3} \ln [3V/2\pi] + \frac{2}{3} \ln [\frac{2}{3}] = \ln [(54\pi V^2)^{1/3}]$. Hence, $e^{v^*} = (54\pi V^2)^{1/3}$. Consequently, from (11.71), we get

$$2y_1 = \ln \left[\frac{1}{3} \frac{(54\pi V^2)^{1/3}}{2\pi} \right] = \ln \left[\frac{V}{2\pi} \right]^{2/3}$$

$$y_1 + y_2 = \ln \left[\frac{2}{3} \frac{(54\pi V^2)^{1/3}}{2\pi} \right] = \ln \left[\frac{2V^2}{\pi^2} \right]^{1/3}$$

$$-2y_1 - y_2 = \ln \left[\frac{\pi}{V} \right]$$

(Note that the third equation above is redundant.) Solving, we get $y_1 = \ln (V/2\pi)^{1/3}$ and $y_2 = \ln (4V/\pi)^{1/3}$. Hence, from (11.64b), we get $r^* = (V/2\pi)^{1/3}$, and $h^* = (4V/\pi)^{1/3} \equiv 2r^*$ as the optimum for Problem GP.

11.5.4 Example (Degree of Difficulty = 1)

Consider Example 11.5.3, and suppose now that we also need to connect a wire of length h joining the centers of the base and the top of the cylinder. The ratio of the cost per unit length (cm) of this wire to the cost per unit surface area (cm^2) of the cylinder is 2π. Also, the volume is required to be, at least, $V \equiv (1024\pi/135)\ \text{cm}^3$.

The Problem GP now has the form

$$\text{Minimize} \quad \left\{ 2\pi r^2 + 2\pi rh + 2\pi h : \left(\frac{V}{\pi}\right) r^{-2} h^{-1} \le 1, r > 0, h > 0 \right\}$$

Here, we now have $m = 1$, $n = 2$, $M = 4$, $\text{DD} = M - n - 1 = 1$, $\alpha_1 = 2\pi$, $\alpha_2 = 2\pi$, $\alpha_3 = 2\pi$, $\alpha_4 = (V/\pi)$, $\mathbf{a}_1^t = (2, 0)$, $\mathbf{a}_2^t = (1, 1)$, $\mathbf{a}_3^t = (0, 1)$, $\mathbf{a}_4^t = (-2, -1)$, with $J_0 = \{1, 2, 3\}$ and $J_1 = \{4\}$. The orthogonality and normalization constraints (11.62b)–(11.62d) give

$$2\delta_1 + \delta_2 - 2\delta_4 = 0 \qquad \delta_2 + \delta_3 - \delta_4 = 0 \qquad \delta_1 + \delta_2 + \delta_3 = 1 \qquad \delta_4 = u_1$$

Solving for all variables in terms of δ_4, which represents the single degree of freedom, we obtain

$$\delta_1 = (1 - \delta_4) \qquad \delta_2 = (4\delta_4 - 2) \qquad \delta_3 = (2 - 3\delta_4) \qquad u_1 = \delta_4 \qquad (11.72)$$

The nonnegativity constraints (11.62e) then imply that $\frac{1}{2} \le \delta_4 \le \frac{2}{3}$. Hence, Problem DGP projected onto the space of the variable δ_4 is given as follows, where we have used $\delta_4 \ln(\alpha_4/\delta_4) + u_1 \ln(u_1) \equiv \delta_4 \ln(\alpha_4)$, since $u_1 = \delta_4$ in (11.72):

$$\text{Minimize} \quad (1 - \delta_4) \ln\left[\frac{2\pi}{1 - \delta_4}\right] + (4\delta_4 - 2) \ln\left[\frac{2\pi}{4\delta_4 - 2}\right] + (2 - 3\delta_4) \ln\left[\frac{2\pi}{2 - 3\delta_4}\right] + \delta_4 \ln\left(\frac{1024}{135}\right)$$

$$\text{subject to} \quad \tfrac{1}{2} \le \delta_4 \le \tfrac{2}{3}$$

(Note that we could have opted to solve (11.62) directly, without projecting it first into a one-dimensional problem.) Now differentiating the objective function of DGP and setting it equal to zero gives $\delta_4 = \frac{7}{12}$. Since the objective function is concave and this value is feasible, it solves Problem DGP. Using (11.72), we obtain

$$\delta_1^* = \tfrac{5}{12} \qquad \delta_2^* = \tfrac{1}{3} \qquad \delta_3^* = \tfrac{1}{4} \qquad \delta_4^* = \tfrac{7}{12} \qquad u_1^* = \tfrac{7}{12}$$

which satisfy the condition of Theorem 11.5.2. The optimal objective function value is $\nu(\text{DGP}) \equiv \nu^* = \ln[18553.59\pi]$. Hence, $e^{\nu^*} = 18553.59\pi$. Consequently, from (11.71) we obtain $2y_1^* = \ln[(\frac{5}{12})(18553.59\pi)(\frac{1}{2}\pi)]$ using $k = 1$, and we get $y_2^* = \ln[(\frac{1}{4})(18553.59\pi)(\frac{1}{2}\pi)]$ using $k = 3$. (The other equations in (11.71) are redundant.) Using (11.64b), this finally yields

$$r^* = e^{y_1^*} = 62.171794 \text{ cm} \qquad h^* = e^{y_2^*} = 2319.199 \text{ cm}$$

In Exercise 11.40, we ask the reader to study the sensitivity of the solution to the specified cost ratio factor in the objective function of Problem GP.

Exercises

11.1 Consider the linear complementary problem to find $(\mathbf{w}, \mathbf{z})$ such that $\mathbf{w} - \mathbf{Mz} = \mathbf{q}$, $\mathbf{w}'\mathbf{z} = 0$, and $\mathbf{w}, \mathbf{z} \ge \mathbf{0}$, where

$$\mathbf{M} = \begin{bmatrix} 1 & -1 & 0 & 0 \\ 2 & 2 & 0 & 0 \\ 0 & 1 & -1 & -2 \\ 1 & 0 & 1 & -2 \end{bmatrix} \qquad \mathbf{q} = \begin{bmatrix} -1 \\ 3 \\ -2 \\ -4 \end{bmatrix}$$

a. Is the matrix $\mathbf{M}$ copositive-plus?

b. Apply Lemke's algorithm discussed in Section 11.1 to the above problem.

11.2 Find a complementary basic feasible solution to the system $\mathbf{w} - \mathbf{Mz} = \mathbf{q}$, $\mathbf{w}'\mathbf{z} = 0$, and $\mathbf{w}, \mathbf{z} \ge \mathbf{0}$ by using Lemke's algorithm. Here,

$$\mathbf{M} = \begin{bmatrix} 1 & 1 & 3 & 4 \\ 5 & 3 & 1 & 1 \\ 2 & 1 & 2 & 2 \\ 1 & 4 & 1 & 1 \end{bmatrix} \qquad \mathbf{q} = \begin{bmatrix} -1 \\ 2 \\ 1 \\ -3 \end{bmatrix}$$

11.3 Consider the following linear programming problem:

$$\begin{aligned} \text{Minimize} \quad & \mathbf{c}'\mathbf{x} \\ \text{subject to} \quad & \mathbf{Ax} = \mathbf{b} \\ & \mathbf{x} \ge \mathbf{0} \end{aligned}$$

a. Write the KKT system for this problem.

b. Use the complementary pivoting algorithm to solve the KKT system for the following problem:

$$\begin{aligned} \text{Minimize} \quad & -x_1 - 3x_2 \\ \text{subject to} \quad & x_1 + x_2 \leq 6 \\ & -x_1 + x_2 \leq 4 \\ & x_1, x_2 \geq 0 \end{aligned}$$

c. Repeat part b if the first constraint is replaced by $x_2 \leq 6$.

11.4 In Section 11.1, we showed constructively in Theorem 11.1.8 that if the system $\mathbf{w} - \mathbf{Mz} = \mathbf{q}$, $(\mathbf{w}, \mathbf{z}) \geq \mathbf{0}$ is consistent and if $\mathbf{M}$ is copositive-plus, then the system defined by (11.1), (11.2), and (11.3) is solvable. Prove this fact directly.

11.5 In a *bimatrix game,* there are two players, I and II. For player I, there exist m possible strategies, and for player II, there exist n possible strategies. If player I chooses strategy i, and player II chooses strategy j, then player I loses a_{ij} and player II loses b_{ij}. Let the loss matrices of players I and II be $\mathbf{A}$ and $\mathbf{B}$, where a_{ij} and b_{ij} are the ijth entries of $\mathbf{A}$ and $\mathbf{B}$, respectively. If player I chooses to play strategy i with probability x_i, and player II chooses to play strategy j with probability y_j, then the expected loss of the two players is $\mathbf{x}^t\mathbf{Ay}$ and $\mathbf{x}^t\mathbf{By}$, respectively. The strategy pair $(\bar{\mathbf{x}}, \bar{\mathbf{y}})$ is said to be an equilibrium point if

$$\bar{\mathbf{x}}^t\mathbf{A}\bar{\mathbf{y}} \leq \mathbf{x}^t\mathbf{A}\bar{\mathbf{y}} \qquad \text{for all } \mathbf{x} \geq \mathbf{0} \text{ such that } \sum_{i=1}^{m} x_i = 1$$

$$\bar{\mathbf{x}}^t\mathbf{B}\bar{\mathbf{y}} \leq \bar{\mathbf{x}}^t\mathbf{B}\mathbf{y} \qquad \text{for all } \mathbf{y} \geq \mathbf{0} \text{ such that } \sum_{j=1}^{n} y_j = 1$$

a. Show how an equilibrium pair $(\bar{\mathbf{x}}, \bar{\mathbf{y}})$ is obtained by formulating a suitable linear complementary problem of the form $\mathbf{w} - \mathbf{Mz} = \mathbf{q}$, $\mathbf{w}^t\mathbf{z} = 0$, and $\mathbf{w}, \mathbf{z} \geq \mathbf{0}$.

b. Investigate the properties of the matrix $\mathbf{M}$. Verify whether the complementary problem has a solution.

c. Find an equilibrium pair for the following loss matrices:

$$\mathbf{A} = \begin{bmatrix} 2 & 4 & 3 \\ 3 & 3 & 2 \end{bmatrix} \qquad \mathbf{B} = \begin{bmatrix} 1 & 3 & 5 \\ 4 & 2 & 1 \end{bmatrix}$$

11.6 The following problem is usually referred to as the *nonlinear complementary problem.* Find a point $\mathbf{x} \in E_n$ such that $\mathbf{x} \geq \mathbf{0}$, $\mathbf{g}(\mathbf{x}) \geq \mathbf{0}$, and $\mathbf{x}^t\mathbf{g}(\mathbf{x}) = 0$, where $\mathbf{g}: E_n \to E_n$ is a continuous vector function.

a. Show that the linear complementary problem is a special case of the above nonlinear problem.

b. Show that the KKT conditions for optimality for a nonlinear programming problem could be written as a nonlinear complementary problem.

c. Show that if $\mathbf{g}$ satisfies the following strong monotonicity property, then there exists a unique solution to the nonlinear complementary problem. (Detailed proof is given in Karamardian [1969].)

$\mathbf{g}$ is said to be *strongly monotone* if there exists an $\varepsilon > 0$ such that

$(\mathbf{y} - \mathbf{x})^t [\mathbf{g}(\mathbf{y}) - \mathbf{g}(\mathbf{x})] \geq \varepsilon \|\mathbf{y} - \mathbf{x}\|^2$

d. Can you devise a computational scheme for solving the nonlinear complementary problem?

11.7 In this exercise, we describe the *principal pivoting method* credited to Cottle and Dantzig [1968] for solving the following linear complementary problem:

$$\begin{aligned} \mathbf{w} - \mathbf{Mz} &= \mathbf{q} \\ \mathbf{w}, \mathbf{z} &\geq \mathbf{0} \\ \mathbf{w}^t\mathbf{z} &= 0 \end{aligned}$$

If the system has a solution, if $\mathbf{M}$ is positive definite, and if every basic solution to the above system is nondegenerate, then the algorithm stops in a finite number of steps with a complementary basic feasible solution.

Initialization Step Consider the basic solution $\mathbf{w} = \mathbf{q}$, $\mathbf{z} = \mathbf{0}$, and form the associated tableau. Go to the main step.

Main Step

1. Let $(\mathbf{w}, \mathbf{z})$ be a complementary basic solution with $\mathbf{z} \geq \mathbf{0}$. If $\mathbf{w} \geq \mathbf{0}$, stop; $(\mathbf{w}, \mathbf{z})$ is a complementary basic feasible solution. Otherwise, let $w_k < 0$. Let v be the variable complementary to w_k and go to step 2.
2. Increase v until either w_k reaches value zero or some positive basic variable decreases to zero. In the former case, go to step 1 after pivoting to update the tableau. In the latter case, pivot to update the tableau, and let v be the variable complementary to that just removed from the basis. Repeat step 2.

a. Show that at each iteration of step 2, w_k increases until it reaches the value zero.
b. Prove finite convergence of the algorithm to a complementary basic feasible solution.
c. Can the method be used to solve a quadratic program where the objective function is strictly convex?

11.8 Consider the LCP problem of finding a solution, if one exists, to the system $\mathbf{w} - \mathbf{M}\mathbf{z} = \mathbf{q}$, $\mathbf{w} \geq \mathbf{0}$, $\mathbf{z} \geq \mathbf{0}$, and $\mathbf{w}'\mathbf{z} = 0$, where $\mathbf{M}$ is a $p \times p$ matrix. Define

$$h(\mathbf{y}) = \text{minimum}\left\{ \sum_{j=1}^{n} y_j w_j + (1 - y_j) z_j : \mathbf{w} - \mathbf{M}\mathbf{z} = \mathbf{q}, \mathbf{w} \geq \mathbf{0}, \text{and } \mathbf{z} \geq \mathbf{0} \right\} \quad (11.73)$$

a. Show that h is a concave function of $\mathbf{y}$ over $\mathbf{0} \leq \mathbf{y} \leq \mathbf{1}$, where $\mathbf{1}$ is a vector of p ones.
b. Show that LCP is equivalent to minimizing h over either $\mathbf{0} \leq \mathbf{y} \leq \mathbf{1}$ or $\mathbf{y}$ restricted to take on binary values.
c. Assume that the set $\mathbf{Z} = \{\mathbf{z}: -\mathbf{M}\mathbf{z} \leq \mathbf{q}, \mathbf{z} \geq \mathbf{0}\}$ is nonempty and bounded, with $0 \leq z^+_k \equiv \{z_k : \mathbf{z} \in \mathbf{Z}\} < \infty$ for all $k = 1, \ldots, p$. Let $\mathbf{M}_j$ denote the jth row of $\mathbf{M}$, $j = 1, \ldots, p$. Construct the set Z_p by multiplying each of inequalities in Z with each y_j, and $(1 - y_j)$, for $j = 1, \ldots, p$. Hence, construct the problem

$$\textit{LCP}': \quad \text{Minimize} \quad \left\{ \sum_{j=1}^{n} y_j(q_j + \mathbf{M}_j\mathbf{z}) + \sum_{j=1}^{n} (1 - y_j) z_j : (\mathbf{y}, \mathbf{z}) \in Z_p, \mathbf{y} \text{ binary} \right\}$$

Now, linearize LCP′ by substituting x_{ij} in place of the product $y_i z_j$ for all $i, j = 1, \ldots, p$, and, hence, obtain a resulting linear mixed-integer zero-one programming problem MIP in continuous variables $\mathbf{z}$ and $\mathbf{x}$, and in binary variables $\mathbf{y}$. Show that the constraints of MIP imply that

$$0 \leq x_{ij} \leq z^+_j y_i, \quad \text{and} \quad z_j - z^+_j(1 - y_i) \leq x_{ij} \leq z_j \quad \text{for all } i, j = 1, \ldots, p$$

Hence, show that solving MIP is equivalent to solving LCP.

d. Discuss how you might use parts b and c to derive a solution method for solving LCP. (Sherali et al. [1991a, b] discuss this transformation and related algorithms.)

11.9 Consider the LCP problem of finding a solution, if one exists, to the system $(\mathbf{M}\mathbf{z} + \mathbf{q}) \geq \mathbf{0}$, $\mathbf{z} \geq \mathbf{0}$, and $(\mathbf{M}\mathbf{z} + \mathbf{q})'\mathbf{z} = 0$, where $\mathbf{M}$ is a $p \times p$ matrix. Consider the following linear mixed integer programming problem:

$$\begin{aligned} \textit{MIP}: \quad & \text{Minimize} \quad \alpha \\ & \text{subject to} \quad \mathbf{0} \leq \mathbf{M}\mathbf{x} + \alpha\mathbf{q} \leq \mathbf{1} - \mathbf{y} \\ & \qquad\qquad\quad \mathbf{0} \leq \mathbf{x} \leq \mathbf{y} \\ & \qquad\qquad\quad \mathbf{y} \text{ binary}, 0 \leq \alpha \leq 1 \end{aligned}$$

Show that if MIP has an optimum solution $(\alpha^*, \mathbf{x}^*, \mathbf{y}^*)$ with objective value $\alpha^* > 0$, then $\mathbf{z} = \mathbf{x}/\alpha^*$ solves LCP. On the other hand, if $\alpha^* = 0$ at optimality, then show that LCP has no solution. (This formulation is due to Pardalos and Rosen [1988].)

11.10 Consider the LCP problem of finding a solution, if one exists, to the system $\mathbf{w} - \mathbf{M}\mathbf{z} = \mathbf{q}$, $\mathbf{w} \geq \mathbf{0}$, $\mathbf{z} \geq \mathbf{0}$, and $\mathbf{w}'\mathbf{z} = 0$, where $\mathbf{M}$ is a $p \times p$ matrix. Define $Z = \{\mathbf{z} \geq \mathbf{0} : \mathbf{M}\mathbf{z} + \mathbf{q} \geq \mathbf{0}\}$, and $W = \{\mathbf{w} : \mathbf{0} \leq \mathbf{w} \leq K\mathbf{1}\}$, where K is a large number and $\mathbf{1}$ is a vector of p ones. Consider the problem

$$\textit{LCP}': \quad \text{Minimize} \quad \left\{ \sum_{j=1}^{n} [\min\{0, w_j\} + z_j] : \mathbf{z} \in Z, \mathbf{w} \in W, \mathbf{w} + \mathbf{z} = \mathbf{q} + \mathbf{M}\mathbf{z} \right\}$$

Discuss the structure of LCP′ and its equivalence with respect to solving LCP. (This formulation is proposed by Bard and Falk [1982].)

11.11 Solve the KKT system for the following problem by the complementary pivoting algorithm:

$$\begin{array}{ll} \text{Minimize} & 5x_1 + 6x_2 - 12x_3 + 2x_1^2 + 4x_2^2 + 6x_3^2 - 2x_1x_2 - 6x_1x_3 + 8x_1x_3 \\ \text{subject to} & x_1 + 2x_2 + x_3 \geq 6 \\ & x_1 + x_2 + x_3 \leq 16 \\ & -x_1 + 2x_2 \leq 4 \\ & x_1, x_2, x_3 \geq 0 \end{array}$$

11.12 Use the complementary pivoting algorithm to solve the following quadratic programming problem:

$$\begin{array}{ll} \text{Maximize} & 2x_1 - 3x_2 - 2x_1^2 - 3x_1x_2 - 2x_2^2 \\ \text{subject to} & x_1 + 2x_2 \leq 6 \\ & -x_1 + 2x_2 \leq 4 \\ & x_1, \quad x_2 \geq 0 \end{array}$$

11.13 Consider the following problem:

$$\begin{array}{ll} \text{Minimize} & \mathbf{c}'\mathbf{x} + \frac{1}{2}\mathbf{x}'\mathbf{H}\mathbf{x} \\ \text{subject to} & \mathbf{A}\mathbf{x} = \mathbf{b} \\ & \mathbf{x} \geq \mathbf{0} \end{array}$$

a. Write the KKT conditions.
b. Suppose that a point $\hat{\mathbf{x}}$ satisfies the KKT conditions. Is it necessarily true that $\hat{\mathbf{x}}$ is a global or local minimum?
c. Show that, if $\mathbf{H}$ is positive semidefinite on the cone of feasible directions at $\hat{\mathbf{x}}$, then $\hat{\mathbf{x}}$ is a global optimal solution.

11.14 Consider the problem to minimize $\mathbf{c}'\mathbf{x} + \frac{1}{2}\mathbf{x}'\mathbf{H}\mathbf{x}$ subject to $\mathbf{A}\mathbf{x} = \mathbf{b}$, $\mathbf{x} \geq \mathbf{0}$. Now consider the following problem:

$$\begin{array}{ll} \text{Minimize} & \mathbf{c}'\mathbf{x} - \mathbf{b}'\mathbf{u} \\ \text{subject to} & \mathbf{A}\mathbf{x} = \mathbf{b} \\ & \mathbf{H}\mathbf{x} + \mathbf{A}'\mathbf{u} - \mathbf{v} = -\mathbf{c} \\ & \mathbf{v}'\mathbf{x} = \mathbf{0} \\ & \mathbf{x}, \mathbf{v} \geq \mathbf{0} \qquad \mathbf{u} \text{ unrestricted} \end{array}$$

a. Show that an optimal solution to the problem stated above gives a point with minimal objective value among all the KKT points. Does this imply that the optimal solution of the problem is a global minimum?
b. Give an interpretation of the objective function of the above problem.
c. Suggest a procedure for solving the above problem using the technique of Exercise 11.8, for example, and illustrate by solving the following problem:

$$\begin{array}{lrcl} \text{Minimize} & -(x_1 - 1)^2 - (x_2 - 1)^2 & & \\ \text{subject to} & -x_1 + 2x_2 + x_3 & = & 4 \\ & x_1 + x_2 + x_4 & = & 4 \\ & 3x_1 - 2x_2 + x_5 & = & 8 \\ & x_1, x_2, x_3, x_4, x_5 & \geq & 0 \end{array}$$

11.15 In this exercise, we describe a procedure that is a modified version of a similar procedure credited to Wolfe [1959] for solving a quadratic programming problem of the form: minimize $\mathbf{c}'\mathbf{x} + \frac{1}{2}\mathbf{x}'\mathbf{H}\mathbf{x}$ subject to $\mathbf{A}\mathbf{x} = \mathbf{b}$, $\mathbf{x} \geq \mathbf{0}$, where $\mathbf{A}$ is an $m \times n$ matrix of rank m. The KKT conditions for this problem can be written as follows:

$$\begin{aligned} \mathbf{A}\mathbf{x} \qquad\qquad &= \mathbf{b} \\ \mathbf{H}\mathbf{x} + \mathbf{A}'\mathbf{u} - \mathbf{v} &= -\mathbf{c} \\ \mathbf{x}, \mathbf{u} &\geq \mathbf{0} \\ \mathbf{v}'\mathbf{x} &= 0 \end{aligned}$$

The method first finds a starting basic feasible solution to the system $\mathbf{A}\mathbf{x} = \mathbf{b}$, $\mathbf{x} \geq \mathbf{0}$. Using this solution, and denoting $\mathbf{A}$ by $[\mathbf{B}, \mathbf{N}]$ and $\mathbf{H}$ by $[\mathbf{H}_1, \mathbf{H}_2]$, where $\mathbf{B}$ is the basis, the above system can be rewritten as follows:

$$\mathbf{x}_B + \mathbf{B}^{-1}\mathbf{N}\mathbf{x}_N = \mathbf{B}^{-1}\mathbf{b}$$
$$[\mathbf{H}_2 - \mathbf{H}_1\mathbf{B}^{-1}\mathbf{N}]\mathbf{x}_N + \mathbf{A}'\mathbf{u} - \mathbf{v} = -\mathbf{H}_1\mathbf{B}^{-1}\mathbf{b} - \mathbf{c}$$
$$\mathbf{v}'\mathbf{x} = 0$$
$$\mathbf{x}_B, \mathbf{x}_N, \mathbf{v} \geq \mathbf{0} \qquad \mathbf{u} \text{ unrestricted}$$

Here, to start, we introduce n artificial variables in the last n constraints, with coefficient $+1$ if $(\mathbf{H}_1\mathbf{B}^{-1}\mathbf{b} + \mathbf{c})_i \leq 0$ and with coefficient -1 if $(\mathbf{H}_1\mathbf{B}^{-1}\mathbf{b} + \mathbf{c})_i > 0$. We then have a basic feasible solution to the above system with the initial basis consisting of $\mathbf{x}_B$ and the artificial variables. The simplex method is then used to find a KKT point by minimizing the sum of the artificial variables. In order to maintain complementary slackness, the following restricted basis entry rule is adopted. If x_j is basic, then v_j cannot enter the basis, unless the minimum ratio test drives x_j out of the basis; conversely, if v_j is in the basis, then x_j cannot enter the basis unless the minimum ratio test drives v_j out of the basis.

a. What modifications are required if the constraint $\mathbf{Ax} = \mathbf{b}$ is replaced by $\mathbf{Ax} \leq \mathbf{b}$?

b. Use the above method to find a KKT point to the following quadratic program:

$$\begin{aligned} \text{Minimize} \quad & x_1^2 + x_1x_2 + 6x_2^2 - 2x_1 + 8x_2 \\ \text{subject to} \quad & x_1 + 2x_2 \leq 4 \\ & 2x_1 + x_2 \leq 5 \\ & x_1, x_2 \geq 0 \end{aligned}$$

c. Show that, in the absence of degeneracy and under any of the following conditions, the above method produces a KKT point in a finite number of steps:
   (i) $\mathbf{H}$ is positive semidefinite and $\mathbf{c} = \mathbf{0}$.
   (ii) $\mathbf{H}$ is positive definite.
   (iii) $\mathbf{H}$ has nonnegative elements with strictly positive diagonal elements.

d. Show that if $\mathbf{H}$ is positive semidefinite and if, at termination, the sum of the artificial variables is not equal to zero, then the quadratic program has an unbounded optimal solution.

e. Solve the following quadratic program by Wolfe's method:

$$\begin{aligned} \text{Minimize} \quad & -4x_1 - 6x_2 + x_1^2 - 2x_1x_2 + x_2^2 \\ \text{subject to} \quad & 2x_1 + x_2 \leq 2 \\ & -x_1 + x_2 \leq 4 \\ & x_1, x_2 \geq 0 \end{aligned}$$

11.16 Consider the quadratic programming problem to minimize $\mathbf{c}'\mathbf{x} + \frac{1}{2}\mathbf{x}'\mathbf{Hx}$ subject to $\mathbf{Ax} \leq \mathbf{b}$, $\mathbf{x} \geq \mathbf{0}$. For simplicity, suppose that $\mathbf{b} \geq \mathbf{0}$. The KKT conditions can be written as follows:

$$\mathbf{Ax} + \mathbf{y} = \mathbf{b}$$
$$-\mathbf{Hx} - \mathbf{A}'\mathbf{u} + \mathbf{v} = \mathbf{c}$$
$$\mathbf{v}'\mathbf{x} = 0, \ \mathbf{u}'\mathbf{y} = 0$$
$$\mathbf{x}, \mathbf{y}, \mathbf{u}, \mathbf{v} \geq \mathbf{0}$$

Now introduce the artificial variable z and consider the following problem:

$$\begin{aligned} \text{Minimize} \quad & z \\ \text{subject to} \quad & \mathbf{Ax} + \mathbf{y} = \mathbf{b} \\ & -\mathbf{Hx} - \mathbf{A}'\mathbf{u} + \mathbf{v} + \mathbf{e}z = \mathbf{c} \\ & \mathbf{x}, \mathbf{y}, \mathbf{u}, \mathbf{v} \geq \mathbf{0} \end{aligned}$$

where the ith component e_i of $\mathbf{e}$ is given by

$$e_i = \begin{cases} -1 & \text{if } c_i < 0 \\ 0 & \text{otherwise} \end{cases}$$

We summarize below a modification of Wolfe's method described in Exercise 11.15 above for solving the KKT system.

Step 1 Start with $\mathbf{y}$ and $\mathbf{v}$ as basic variables and note that some components of $\mathbf{v}$ may be negative. Let v_i be the most negative component of $\mathbf{v}$. Pivot at the z column and the v_r row, so that v_r is removed from the basis. We now have a basic solution with $z > 0$ and all variables nonnegative. Note that $x_jv_j = 0$ and that $u_iy_i = 0$.

Step 2 Minimize z by the simplex method using a restricted basic entry so that $v_j x_j = 0$ for $j = 1, \ldots, n$ and $u_i y_i = 0$ for $i = 1, \ldots, m$.

a. Solve the problem defined in Example 11.2.1 by the above procedure.

b. Suppose that $\mathbf{H}$ is positive semidefinite. Show that the above algorithm gives an optimal solution to the original problem or indicates that the problem is unbounded.

c. Show that, if we delete the objective function row, the complementary pivoting algorithm discussed in Section 11.1 could be used to solve the KKT system. In this case, a variable enters the basis automatically if its complementary variable drops from the basis in the preceding iteration. Here, x_j and v_j are complementary variables, and u_i and y_i are complementary variables.

11.17 This exercise describes a method credited to Dantzig [1963] for solving a quadratic programming problem of the form: minimize $\frac{1}{2}\mathbf{x}^t\mathbf{H}\mathbf{x}$ subject to $\mathbf{A}\mathbf{x} = \mathbf{b}$, $\mathbf{x} \geq \mathbf{0}$, where $\mathbf{H}$ is positive semidefinite. The KKT conditions for the above problem are

$$\begin{aligned} \mathbf{A}\mathbf{x} &= \mathbf{b} \\ \mathbf{H}\mathbf{x} + \mathbf{A}^t\mathbf{u} - \mathbf{v} &= \mathbf{0} \\ v_j x_j &= 0 \qquad \text{for } j = 1, \ldots, n \\ \mathbf{x}, \mathbf{v} &\geq \mathbf{0} \end{aligned}$$

The procedure always satisfies the first two conditions in addition to nonnegativity of $\mathbf{x}$. The restriction $\mathbf{v} \geq \mathbf{0}$ is satisfied only at optimality. Further, at each iteration, $v_j x_j = 0$ for all j except for at most one index.

Initialization Step Let $(\mathbf{x}_B^t, \mathbf{x}_N^t)$ be a basic feasible solution to $\mathbf{A}\mathbf{x} = \mathbf{b}$, $\mathbf{x} \geq \mathbf{0}$, and let $\mathbf{v}^t = (\mathbf{v}_B^t, \mathbf{v}_N^t)$. Consider the basic solution to the system with the basic vectors $\mathbf{x}_B$, $\mathbf{u}$, $\mathbf{v}_N$. Note that the solution satisfies all the constraints except possibly $\mathbf{v} \geq \mathbf{0}$. Since $\mathbf{u}$ is unrestricted and since the algorithm relaxes $\mathbf{v} \geq \mathbf{0}$, as a variable enters the basis, only x_j variables are eligible to leave the basis.

Main Step

1. If $\mathbf{v} \geq \mathbf{0}$, stop. The current solution is optimal. Otherwise, let v_j = minimum $\{v_i : v_i < 0\}$. Go to step 2.
2. Introduce x_j into the basis. If v_j drops, repeat step 1. Otherwise, x_r drops for some r. Go to step 3.
3. Introduce v_r into the basis. If v_j drops, go to step 1. If another variable x_k drops, repeat step 3 with v_r replace by v_k.

a. Solve the following problem using the above method:

$$\begin{aligned} \text{Minimize} \quad & x_1^2 + 2x_2^2 - x_1x_2 \\ \text{subject to} \quad & -x_1 + x_2 \leq 0 \\ & x_1 + x_2 \geq 4 \\ & 2x_1 + x_2 \leq 10 \\ & x_1, \quad x_2 \geq 0 \end{aligned}$$

b. Prove that the above method converges to an optimal solution in a finite number of steps.

c. Consider the following problem credited to Finkbeiner and Kall [1973]:

$$\begin{aligned} \text{Minimize} \quad & \tfrac{1}{2}x_1^2 + \tfrac{1}{2}x_2^2 + 3x_1 + 7x_3 + x_4 \\ \text{subject to} \quad & x_1 + 2x_2 + x_3 = 8 \\ & x_1 + 2x_2 + x_4 = 5 \\ & x_1, x_2, x_3, x_4 \geq 0 \end{aligned}$$

Starting with the basic variables $x_1 = 2$, $x_2 = 3$, $u_1 = 2$, $u_2 = 7$, $v_3 = 9$, and $v_4 = -6$, apply the above algorithm. Note that after one iteration, the variable v_1 should enter the basis but no appropriate variable could leave the basis, so that the method fails in the presence of linear terms in the objective function.

d. Consider the following modification in step 3 of the above procedure suggested by Finkbeiner and Kall [1973]:

If no variable drops from the basis when v_r is introduced, increase v_r if v_j does not decrease, or decrease v_r if v_j decreases without violating nonnegativity of the $\mathbf{x}$ vector.

Solve the problem in part c by this method, and show that the procedure works in general.

11.18 In Section 11.2, we described a complementary pivoting procedure for solving a quadratic programming problem of the form to minimize $\mathbf{c}'\mathbf{x} + \frac{1}{2}\mathbf{x}'\mathbf{H}\mathbf{x}$ subject to $\mathbf{A}\mathbf{x} = \mathbf{b}$, $\mathbf{x} \geq \mathbf{0}$. We showed that the method produces an optimal solution if $\mathbf{H}$ is positive definite, or if $\mathbf{H}$ is positive semidefinite and $\mathbf{c} = \mathbf{0}$. The following modification of the procedure to handle the case where $\mathbf{H}$ is positive semidefinite is similar to that given by Wolfe [1959].

Step 1 Apply the complementary pivoting algorithm, where $\mathbf{c}$ is replaced by the zero vector. By Theorem 11.2.4, we obtain a complementary basic feasible solution to the following system:

$$\begin{aligned} \mathbf{A}\mathbf{x} &= \mathbf{b} \\ \mathbf{H}\mathbf{x} + \mathbf{A}'\mathbf{u} - \mathbf{v} &= \mathbf{0} \\ v_j x_j &= 0 \qquad \text{for } j = 1, \ldots, n \\ \mathbf{x}, \mathbf{v} &\geq \mathbf{0} \qquad \mathbf{u} \text{ unrestricted} \end{aligned}$$

Step 2 Starting with the solution obtained in step 1 above, solve the following problem using the simplex method with the restricted basis entry so that v_j and x_j are never in the basis simultaneously:

$$\begin{aligned} \text{Maximize} \quad & z \\ \text{subject to} \quad & \mathbf{A}\mathbf{x} = \mathbf{b} \\ & \mathbf{H}\mathbf{x} + \mathbf{A}'\mathbf{u} - \mathbf{v} + z\mathbf{c} = \mathbf{0} \\ & \mathbf{x} \geq \mathbf{0},\ \mathbf{v} \geq \mathbf{0},\ z \geq 0 \qquad \mathbf{u} \text{ unrestricted} \end{aligned}$$

At optimality, either $\bar{z} = 0$ or $\bar{z} = \infty$ along an extreme direction. In the former case, the optimal solution to the quadratic program is unbounded. In the latter case, an optimal solution to the quadratic program is determined by letting $z = 1$ along the ray giving rise to the unbounded solution.

a. Show that, if the optimal objective to the problem in step 2 is finite, then it must be zero. Show in this case that the optimal objective value of the original problem is unbounded.

b. Show that, if the optimal objective value $\bar{z} = \infty$, then the solution along the optimal ray with $z = 1$ still maintains complementary slackness and, hence, gives an optimal solution to the original problem.

c. Solve the problem of Example 11.2.1 by the above procedure.

11.19 In this exercise, we describe the method of Frank and Wolfe [1956] for solving a quadratic programming problem. This method generalizes a similar procedure by Barankin and Dorfman [1955]. Consider the problem to minimize $\mathbf{c}'\mathbf{x} + \frac{1}{2}\mathbf{x}'\mathbf{H}\mathbf{x}$ subject to $\mathbf{A}\mathbf{x} \leq \mathbf{b}$, $\mathbf{x} \geq \mathbf{0}$, where $\mathbf{H}$ is positive semidefinite.

a. Show that the KKT conditions can be stated as follows:

$$\begin{aligned} \mathbf{A}\mathbf{x} + \mathbf{x}_s &= \mathbf{b} \\ \mathbf{H}\mathbf{x} - \mathbf{u} + \mathbf{A}'\mathbf{v} &= -\mathbf{c} \\ \mathbf{u}'\mathbf{x} + \mathbf{v}'\mathbf{x}_s &= 0 \\ \mathbf{x}, \mathbf{x}_s, \mathbf{u}, \mathbf{v} &\geq \mathbf{0} \end{aligned}$$

The system can be rewritten as $\mathbf{E}\mathbf{y} = \mathbf{d}$, $\mathbf{y} \geq \mathbf{0}$, $\mathbf{y}'\tilde{\mathbf{y}} = 0$, where

$$\mathbf{E} = \begin{bmatrix} \mathbf{A} & \mathbf{I} & \mathbf{0} & \mathbf{0} \\ \mathbf{H} & \mathbf{0} & -\mathbf{I} & \mathbf{A}' \end{bmatrix} \qquad \mathbf{d} = \begin{pmatrix} \mathbf{b} \\ -\mathbf{c} \end{pmatrix}$$

$$\begin{aligned} \mathbf{y}' &= (\mathbf{x}', \mathbf{x}_s', \mathbf{u}', \mathbf{v}') \\ \tilde{\mathbf{y}}' &= (\mathbf{u}', \mathbf{v}', \mathbf{x}', \mathbf{x}_s') \end{aligned}$$

b. Consider the following problem:

$$\begin{aligned} \text{Minimize} \quad & \mathbf{y}'\tilde{\mathbf{y}} \\ \text{subject to} \quad & \mathbf{E}\mathbf{y} = \mathbf{d} \\ & \mathbf{y} \geq \mathbf{0} \end{aligned}$$

Show that a feasible point $\mathbf{y}$ satisfying $\mathbf{y}'\tilde{\mathbf{y}} = 0$ yields a KKT point to the original problem.

c. Use the Frank-Wolfe method discussed in Exercise 10.11 to solve the problem stated in part b, and show that the algorithm simplifies as follows. Suppose that at iteration k, we have a basic feasible solution $\mathbf{y}_k$ for the above constraints and a feasible solution $\mathbf{w}_k$ to the same system, which is not necessarily basic. Starting with $\mathbf{y}_k$, solve the following linear program:

$$\begin{aligned} \text{Minimize} \quad & \tilde{\mathbf{w}}_k'\mathbf{y} \\ \text{subject to} \quad & \mathbf{E}\mathbf{y} = \mathbf{d} \\ & \mathbf{y} \geq \mathbf{0} \end{aligned}$$

A sequence of solutions is obtained ending with a point $\mathbf{y} = \mathbf{g}$, where either $\mathbf{g}'\tilde{\mathbf{g}} = 0$ or $\mathbf{g}'\tilde{\mathbf{w}}_k \leq \frac{1}{2}\mathbf{w}_k'\tilde{\mathbf{w}}_k$. In the former case, we stop with $\mathbf{g}$ as an optimal solution. In the latter case, set $\mathbf{y}_{k+1} = \mathbf{g}$ and let $\mathbf{w}_{k+1}$ be the convex combination of $\mathbf{w}_k$ and $\mathbf{y}_{k+1}$, which minimizes the objective function $\mathbf{y}'\tilde{\mathbf{y}}$. Replace k by $k + 1$ and repeat the process. Show that this procedure converges to an optimal solution, and illustrate it by solving the following problem:

$$\begin{aligned} \text{Minimize} \quad & -2x_1 - 6x_2 + x_1^2 + x_2^2 \\ \text{subject to} \quad & x_1 + 2x_2 \leq 5 \\ & x_1 + x_2 \leq 3 \\ & x_1, x_2 \geq 0 \end{aligned}$$

d. Use the procedure of Frank and Wolfe described in Exercise 10.11 to solve the quadratic programming problem directly without first formulating the KKT conditions. Illustrate by solving the numerical problem in part c, and compare the trajectories.

11.20 Consider the quadratic program

$$QP: \quad \text{Minimize} \quad \{\mathbf{c}'\mathbf{x} + \tfrac{1}{2}\mathbf{x}'\mathbf{H}\mathbf{x} : \mathbf{A}\mathbf{x} = \mathbf{b}\}$$

where $\mathbf{A}$ is an $m \times n$ matrix of rank m and $\mathbf{H}$ is positive definite on $\{\mathbf{x} : \mathbf{A}\mathbf{x} = \mathbf{0}\}$, that is, $\mathbf{x}'\mathbf{H}\mathbf{x} > 0$ for all $\mathbf{x} \neq \mathbf{0}$ satisfying $\mathbf{A}\mathbf{x} = \mathbf{0}$.

a. Show that the matrix $\begin{bmatrix} \mathbf{H} & \mathbf{A}' \\ \mathbf{A} & \mathbf{0} \end{bmatrix}$ is nonsingular.

b. Hence, show that the linear equations of the KKT system for QP yield a unique solution.

c. Assuming that $\mathbf{H}$ is positive definite and hence nonsingular, derive an explicit closed-form expression for the optimal solution to QP.

11.21 Consider the quadratic programming problem to minimize $\mathbf{c}'\mathbf{x} + \frac{1}{2}\mathbf{x}'\mathbf{H}\mathbf{x}$ subject to $\mathbf{A}\mathbf{x} \leq \mathbf{b}$, where $\mathbf{H}$ is an $n \times n$ positive definite matrix and $\mathbf{A}$ is an $m \times n$ matrix. For any subset $S \subseteq \{1, \ldots, m\}$ of the constraint indices, let $\mathbf{x}_S$ be a minimum of $\mathbf{c}'\mathbf{x} + \frac{1}{2}\mathbf{x}'\mathbf{H}\mathbf{x}$ subject to the constraints in S as binding, and let $V(\mathbf{x}_S)$ be the set of constraints violated by $\mathbf{x}_S$.

a. Show that if $V(\mathbf{x}_S) \neq \emptyset$, then S could be a subset of the set of binding constraints $\hat{S}$ at optimality only if there exists some $h \in \hat{S} \cap V(\mathbf{x}_S)$.

b. Show that if $V(\mathbf{x}_S) = \emptyset$, then $\mathbf{x}_S$ is an optimal solution to the original problem if and only if $h \in V(\mathbf{x}_{S-h})$ for each $h \in S$.

c. From parts a and b, show that the following *active set* strategy credited to Theil and Van de Panne [1961] solves the quadratic problem. First, solve the unconstrained problem so that $S = \emptyset$. If $V(\mathbf{x}_\emptyset) = \emptyset$, then $\mathbf{x}_\emptyset$ is an optimal solution. Otherwise, form sets of the type S_1, where $S_1 = \{h\}$ and $h \in V(\mathbf{x}_\emptyset)$. Find $\mathbf{x}_{S_1}$ for each such S_1. If $V(\mathbf{x}_{S_1}) = \emptyset$ for some S_1, check by part b whether $\mathbf{x}_{S_1}$ is optimal. If no candidate problems of the form S_1 could produce an optimal solution, form sets of the type S_2 with two binding constraints, where $S_2 = S_1 \cup \{h\}$, S_1 is a set with one binding constraint such that $V(\mathbf{x}_{S_1}) \neq \emptyset$ and $h \in V(\mathbf{x}_{S_1})$. The process is repeated by finding $\mathbf{x}_{S_2}$ that solves the original problem, or by forming sets of the type S_3 containing three binding constraints.

d. Illustrate the method of Theil and Van de Panne by solving the problem stated in Example 11.2.1.

e. Can you generalize the method to the following convex programming problem where f is strictly convex and g_i is convex for $i = 1, \ldots, m$?

Minimize $f(\mathbf{x})$

subject to $g_i(\mathbf{x}) \leq 0$ for $i = 1, \ldots, m$

11.22 Consider the quadratic programming problem

$$QP: \text{ Minimize } \{\mathbf{c}'\mathbf{x} + \tfrac{1}{2}\mathbf{x}'\mathbf{H}\mathbf{x} : \mathbf{A}_i'\mathbf{x} = b_i \text{ for } i \in E, \mathbf{A}_i'\mathbf{x} \leq b_i \text{ for } i \in I\}$$

where $\mathbf{H}$ is positive definite, and where the index sets E and I record the equality and inequality constraints in the problem, respectively. (Nonnegativities, if present, are included in the set indexed by I.) Consider the following *active set method* for solving QP. Given a feasible solution $\mathbf{x}_k$, define the working index set $W_k = E \cup I_k$, where $I_k \equiv \{i \in I : \mathbf{A}_i'\mathbf{x}_k = b_i\}$ represents the binding inequality constraints, and consider the following direction-finding problem:

$$QP(x_k): \text{ Minimize } \{(\mathbf{c} + \mathbf{H}\mathbf{x}_k)'\mathbf{d} + \tfrac{1}{2}\mathbf{d}'\mathbf{H}\mathbf{d} : \mathbf{A}_i'\mathbf{d} = \mathbf{0} \quad \text{for all } i \in W_k\}$$

Let $\mathbf{d}_k$ be the optimum obtained (see Exercise 11.20).

a. Show that $(\mathbf{x}_k + \mathbf{d}_k)$ solves the problem

$$\text{Minimize } \{\mathbf{c}'\mathbf{x} + \tfrac{1}{2}\mathbf{x}'\mathbf{H}\mathbf{x} : \mathbf{A}_i'\mathbf{x} = b_i \quad \text{for all } i \in W_k\} \tag{11.74}$$

b. If $\mathbf{d}_k = \mathbf{0}$, let v_i^*, $i \in W_k$, denote the optimum Lagrange multipliers for QP($\mathbf{x}_k$). If $v_i^* \geq 0$ for all $i \in I_k$, then show that $\mathbf{x}_k$ is optimal to QP. On the other hand, if the value minimum $\{v_i^* : i \in I_k\} \equiv v_q^* < 0$, then let $I_{k+1} = I_k - \{q\}$, $W_{k+1} = E \cup I_{k+1}$, and $\mathbf{x}_{k+1} = \mathbf{x}_k$.

c. If $\mathbf{d}_k \neq \mathbf{0}$ and if $(\mathbf{x}_k + \mathbf{d}_k)$ is feasible to QP, put $\mathbf{x}_{k+1} = (\mathbf{x}_k + \mathbf{d}_k)$ and $W_{k+1} = W_k$. On the other hand, if $(\mathbf{x}_k + \mathbf{d}_k)$ is not feasible to QP, let $\alpha_k < 1$ be the maximum step length along $\mathbf{d}_k$ that maintains feasibility as given by,

$$\alpha_k = \underset{i \in I_k : \mathbf{A}_i'\mathbf{d}_k > 0}{\text{minimum}} \left\{ \frac{b_i - \mathbf{A}_i'\mathbf{x}_k}{\mathbf{A}_i'\mathbf{d}_k} \right\} = \frac{b_q - \mathbf{A}_q'\mathbf{x}_k}{\mathbf{A}_q'\mathbf{d}_k}$$

Put $\mathbf{x}_{k+1} = \mathbf{x}_k + \alpha_k\mathbf{d}_k$, $I_{k+1} = I_k \cup \{q\}$, and $W_{k+1} = E \cup I_{k+1}$. From parts b and c, having determined $\mathbf{x}_{k+1}$ and W_{k+1}, increment k by 1 and reiterate. Let $\{\mathbf{x}_k\}$ be the sequence thus generated. Provide a finite convergence argument for this procedure by showing that, while the solution to (11.74) is infeasible to QP, the method continues to add active constraints until the solution to (11.74) becomes feasible to QP, and then either verifies optimality or provides a strict descent in the objective function value. Hence, using the fact that the number of possible working sets is finite, establish finite convergence of the procedure.

d. Illustrate by solving the problem of Example 11.2.1, starting at the origin.

11.23 Let $f(\mathbf{x}) = \mathbf{c}'\mathbf{x} + \frac{1}{2}\mathbf{x}'\mathbf{H}\mathbf{x}$, where $\mathbf{H}$ is a symmetric positive semidefinite matrix. Show that $f : E_n \to E_1$ is *unbounded* from below if and only if $\mathbf{c} + \mathbf{H}\mathbf{x} = \mathbf{0}$ has no solution.

11.24 Consider the following problem:

$$\begin{aligned} \text{Minimize} \quad & e^{x_1} + e^{x_2} + x_1 + 2x_2^2 + x_1^2 \\ \text{subject to} \quad & x_1 + 2x_2 \leq 6 \\ & x_1 - x_2 \leq 0 \\ & x_1, x_2 \geq 0 \end{aligned}$$

a. Show that the objective function is strictly convex and that the constraints are convex. Hence, the restricted basis entry can be dropped if the problem is to be solved by the separable programming method discussed in Section 11.3.

b. Use suitable grid points and solve the problem.

11.25 Consider the following problem:

$$\begin{aligned} \text{Minimize} \quad & 2x_1 - 6x_2 + x_1^2 - x_1x_2 + 2x_2^2 \\ \text{subject to} \quad & x_1 + 2x_2 \leq 12 \\ & x_1^2 + x_2^2 = 8 \\ & x_1, \quad x_2 \geq 0 \end{aligned}$$

Make a suitable change of variables such that the problem becomes separable. Choose suitable grids for partitioning, set up the initial simplex tableau, and then solve the approximating problem. If you were to solve the problem over again, how can you make use of your answer to obtain a better partitioning?

11.26 Does Theorem 11.3.1 hold if $f_1, \ldots, f_n$ are convex rather than strictly convex? If not, modify the statement of the theorem so that it handles the convex case.

11.27 Does the simplex method with the restricted basis entry provide an optimal solution to the approximating Problem LAP in the nonconvex case? Prove or give a counterexample.

11.28 Solve the following problem using the method discussed in Section 11.3:

$$\begin{aligned} \text{Minimize} \quad & \frac{1}{x_1 + 1} + x_2^3 \\ \text{subject to} \quad & x_1^2 - x_2^3 \le 5 \\ & x_1, x_2 \ge 10 \end{aligned}$$

11.29 Consider the following alternative method for approximating a function θ in the interval $[a, b]$. The interval $[a, b]$ is divided into smaller subintervals via the grid points $a = \mu_1, \ldots, \mu_k = b$. Let $\Delta_i = \mu_{i+1} - \mu_i$, and let $\Delta\theta_i = \theta(\mu_{i+1}) - \theta(\mu_i)$ for $i = 1, \ldots, k - 1$. Now consider x in the interval $[\mu_\nu, \mu_{\nu+1}]$. Then, x can be represented as $x = \mu_1 + \sum_{i=1}^{\nu} \delta_i \Delta_i$, and $\theta(x)$ can be approximated by $\hat{\theta}(x) = \theta_1 + \sum_{i=1}^{\nu} \delta_i \Delta\theta_i$, where $\delta_\nu \in [0, 1]$, $\delta_i = 1$ for $i = 1, \ldots, \nu - 1$, and $\theta_1 = \theta(x_1)$.

a. Interpret this approximation of θ geometrically.

b. Show how this approximation can be used to solve the following separable problem by the simplex method, with a suitable restricted basis entry:

$$\begin{aligned} \text{Minimize} \quad & \sum_{j=1}^{n} f_j(x_j) \\ \text{subject to} \quad & \sum_{j=1}^{n} g_{ij}(x_j) \le 0 \qquad \text{for } i = 1, \ldots, m \\ & a_j \le x_j \le b_j \qquad \text{for } j = 1, \ldots, n \end{aligned}$$

Hint: Let $x_{\nu j}$ for $\nu = 1, \ldots, k_j + 1$ be the grid points used for x_j, and consider the following problem:

$$\begin{aligned} \text{Minimize} \quad & \sum_{j=1}^{n} \sum_{\nu=1}^{k_j} (\Delta f_{\nu j})\delta_{\nu j} + \sum_{j=1}^{n} f_j(a_j) \\ \text{subject to} \quad & \sum_{j=1}^{n} \sum_{\nu=1}^{k_j} (\Delta g_{ij\nu})\delta_{\nu j} + \sum_{j=1}^{n} g_{ij}(a_j) \le 0 \qquad \text{for } i = 1, \ldots, m \\ & 0 \le \delta_{\nu j} \le 1 \qquad \nu = 1, \ldots, k_j;\ j = 1, \ldots, n \\ & \delta_{\nu j} > 0 \Rightarrow \delta_{lj} = 1 \qquad \text{for } l < \nu;\ j = 1, \ldots, n \end{aligned}$$

where $\Delta f_{\nu j} = f(x_{\nu+1,j}) - f(x_{\nu j})$

$\Delta g_{ij\nu} = g_j(x_{\nu+1,j}) - g_{ij}(x_{\nu j})$

c. Use the procedure developed in part b to solve the following problem:

$$\begin{aligned} \text{Maximize} \quad & 2x_1 + 5x_2 - 2x_1^2 - x_2^2 \\ \text{subject to} \quad & 2x_1 + 6x_2 \le 9 \\ & -x_1 + x_2 \ge -3 \\ & x_1, \quad x_2 \ge 0 \end{aligned}$$

11.30 Consider the following problem:

$$\begin{aligned} \text{Minimize} \quad & e^{x_1} + x_1^2 + 4x_1 + 2x_2^2 - 6x_2 + 2x_3 \\ \text{subject to} \quad & x_1^2 + e^{x_2} + 6x_3 \le 15 \\ & x_1^4 - x_2 + 5x_3 \le 25 \\ & 0 \le x_1 \le 4 \\ & 0 \le x_2 \le 2 \\ & 0 \le x_3 \end{aligned}$$

a. Using the grid points 0, 2, and 4 for x_1, and 0, 1, and 2 for x_2, solve the above problem by the separable programming algorithm.

b. Starting from the optimal solution obtained in part a, use the grid point generation scheme to generate three more grid points to obtain a better solution.

c. Using the optimal point obtained in part b, give lower and upper bounds on the optimal objective value to the original problem.

11.31 In Section 11.3, we approximated a separable programming problem using the λ-form. An alternative, called the δ-*form,* was considered in Exercise 11.29. Consider a variable x in the interval $[a, b]$ and grid points $\mu_1 = a, \mu_2, \ldots, \mu_k = b$. Then, using the λ- and δ-forms, x can be represented as

1. $x = \sum_{j=1}^{k} \lambda_j \mu_j \qquad \sum_{j=1}^{k} \lambda_j = 1 \qquad \lambda_j \geq 0 \qquad \text{for } j = 1, \ldots, k$
   where $\lambda_p \lambda_q = 0$ if μ_p and μ_q are not adjacent.
2. $x = \mu_1 + \sum_{j=1}^{k-1} \Delta_j \delta_j \qquad 0 \leq \delta_j \leq 1 \qquad \text{for } j = 1, \ldots, k-1$
   $\delta_i > 0 \Rightarrow \delta_j = 1 \qquad \text{for } j < i$

   Show that the two forms are related by the relationship

$$\lambda_j = \begin{cases} \delta_{j-1} - \delta_j & \text{if } j = 1, \ldots, k-1 \\ \delta_{j-1} & \text{if } j = k \end{cases}$$

where $\delta_0 = 1$. In particular, show that this relationship could be written in vector form as $\boldsymbol{\lambda} = \mathbf{T}\boldsymbol{\delta}$, where $\mathbf{T}$ is an upper triangular matrix.

11.32 Solve the following problem by the two linear fractional programming algorithms discussed in Section 11.4:

$$\begin{array}{ll} \text{Minimize} & \dfrac{-2x_1 + 3x_2 + 5x_3 + 2}{x_1 + 2x_2 + x_3 + 2} \\ \text{subject to} & 2x_1 + 3x_2 + x_3 \leq 12 \\ & x_1 \quad 2x_2 \qquad \geq 2 \\ & x_1 \qquad + x_3 \leq 8 \\ & x_1, \quad x_2, \quad x_3 \geq 0 \end{array}$$

11.33 Consider the following problem:

$$\begin{array}{ll} \text{Maximize} & \dfrac{8x_1 + 6x_2 - 5}{-4x_1 + 2x_2 - 40} \\ \text{subject to} & x_1 + x_2 \leq 10 \\ & 3x_1 - 5x_2 \leq 6 \\ & x_1, x_2 \geq 0 \end{array}$$

a. Solve the problem by the method of Gilmore and Gomory.
b. Solve the problem by the method of Charnes and Cooper.

11.34 Suppose that the region $\{\mathbf{x} : \mathbf{A}\mathbf{x} = \mathbf{b}, \mathbf{x} \geq \mathbf{0}\}$ is unbounded. Further, suppose that an improving feasible direction $\mathbf{d}$ is found while minimizing a linear fractional function over the above region. In particular, suppose that $\mathbf{d}_N$ consists of a vector of zeros, except for a 1 at position j and $\mathbf{d}_B = -\mathbf{B}^{-1}\mathbf{a}_j \geq \mathbf{0}$. Is it necessarily true that the optimal objective is unbounded by moving from the current extreme point in the direction $\mathbf{d}$? If not, what are the possible cases that can be encountered?

11.35 Consider the function f defined by

$$f(\mathbf{x}) = \frac{x_1 + 2x_2 - 6}{3x_1 - x_2 + 2}$$

a. Sketch the following sets in the (x_1, x_2) plane, and determine whether they are convex:
   $S = \{(x_1, x_2) : f(\mathbf{x}) \leq 2\}$
   $S_1 = S \cap \{(x_1, x_2) : 3x_1 - x_2 + 2 > 0\}$
   $S_2 = S \cap \{(x_1, x_2) : 3x_1 - x_2 + 2 < 0\}$
b. Is your conclusion in part a inconsistent with the fact that f is quasiconvex on the region $\{(x_1, x_2) : 3x_1 - x_2 + 2 \neq 0\}$? Discuss.

11.36 Let

$$f(\mathbf{x}) = \frac{\mathbf{p}'\mathbf{x} + \alpha}{\mathbf{q}'\mathbf{x} + \beta}$$

and let $S = \{\mathbf{x} : \mathbf{q}'\mathbf{x} + \beta > 0\}$. Show directly that f is quasiconvex, quasiconcave, strictly quasiconvex, and strictly quasiconcave on S.

11.37 Let $f : E_n \to E_1$ be quasiconcave, and let $\theta(\lambda) = f(\mathbf{x} + \lambda\mathbf{d})$, where $\mathbf{x}$ is a given vector and $\mathbf{d}$ is a given direction.

a. Show that θ is a quasiconcave function in λ.

b. Consider the problem to minimize $\theta(\lambda)$ subject to $\lambda \in [a, b]$. Show that, if $\nabla f(\mathbf{x})^t\mathbf{d} < 0$, then $\lambda = b$ is an optimal solution to the above problem.

c. Letting $f(\mathbf{x}) = (\mathbf{p}^t\mathbf{x} + \alpha)/(\mathbf{q}^t\mathbf{x} + \beta)$, use the result in part b to show that no line search is needed for solving linear fractional programs by the convex–simplex method.

11.38 In solving a linear fractional programming problem, suppose that we add the following two rows to the initial tableau:

$$z_1 - \mathbf{p}^t\mathbf{x} = \alpha$$
$$z_2 - \mathbf{q}^t\mathbf{x} = \beta$$

As the problem is solved by the convex–simplex method, the coefficients of the basic vector $\mathbf{x}_B$ in these rows are equal to zero so that the updated rows are given by

$$z_1 - (\mathbf{p}_N^t - \mathbf{p}_B^t\mathbf{B}^{-1}\mathbf{N})\mathbf{x}_N = \alpha + \mathbf{p}_B^t\mathbf{B}^{-1}\mathbf{b}$$
$$z_2 - (\mathbf{q}_N^t - \mathbf{q}_B^t\mathbf{B}^{-1}\mathbf{N})\mathbf{x}_N = \beta + \mathbf{q}_B^t\mathbf{B}^{-1}\mathbf{b}$$

Show that the reduced gradient vector $\mathbf{r}_N$ is given by

$$\mathbf{r}_N^t = \frac{(\mathbf{p}_N^t - \mathbf{p}_B^t\mathbf{B}^{-1}\mathbf{N})\bar{z}_2 - (\mathbf{q}_N^t - \mathbf{q}_B^t\mathbf{B}^{-1}\mathbf{N})\bar{z}_1}{\bar{z}_2^2}$$

where $\bar{z}_1 = \alpha + \mathbf{p}_B^t\mathbf{B}^{-1}\mathbf{b}$ and $\bar{z}_2 = \beta + \mathbf{q}_B^t\mathbf{B}^{-1}\mathbf{b}$. Note that each term in the expression for $\mathbf{r}_N$ is immediately available from the updated tableau. Solve the problem in Example 11.4.3 using the above procedure for computing $\mathbf{r}_N$.

11.39 Verify that the separable objective function (11.62a) of the dual geometric program (DGP) is concave.

11.40 Consider the geometric programming problem of Example 11.5.4, and let C denote the ratio of the cost per unit length (cm) of the wire to the cost per unit surface area (cm^2) of the cylinder. Analyze this problem to study the sensitivity of the optimal dimensions of the cylinder to this cost factor C.

11.41 Consider the problem to minimize $f_1(\mathbf{x}) + [f_2(\mathbf{x})]^a f_3(\mathbf{x})$ where f_i, $i = 1, 2, 3$ are posynomials and where $a > 0$. Show that this is equivalent to the standard posynomial geometric program GP to minimize $f_1(\mathbf{x}) + x_0^a f_3(\mathbf{x})$ subject to $x_0^{-1} f_2(\mathbf{x}) \le 1$, where x_0 is an additional variable. Illustrate by solving the problem to minimize $x_1^{-1/2}x_2^{1/8} + [\frac{4}{5}x_1^{1/2}x_2^{2/3} + \frac{2}{5}x_1^{1/3}x_2]^{1/2}x_1^{1/4}x_2^{-1/2}$.

11.42 Consider the problem to minimize

$$f_1(\mathbf{x}) + \frac{f_2(\mathbf{x})}{[f_3(\mathbf{x}) - f_4(\mathbf{x})]^a}$$

where f_i, $i = 1, \ldots, 4$, are posynomials but f_3 has only one term, and where $a > 0$. Show that this is equivalent to the standard posynomial geometric program to minimize $f_1(\mathbf{x}) + f_2(\mathbf{x})x_0^{-a}$ subject to $x_0/f_3(\mathbf{x}) + f_4(\mathbf{x})/f_3(\mathbf{x}) \le 1$.

11.43 Consider the problem to minimize $f_1(\mathbf{x}) - f_2(\mathbf{x})$, where f_1 and f_2 are posynomials, f_2 has only one term, and the optimal value is known to be negative. Show that this can be equivalently solved as the standard posynomial geometric program to minimize x_0^{-1} subject to $[x_0/f_2(\mathbf{x})] + [f_1(\mathbf{x})/f_2(\mathbf{x})] \le 1$.

11.44 Suppose that a metal wire frame has to be constructed for a rectangular box having a skeleton and dimensions (in cm) as shown below.

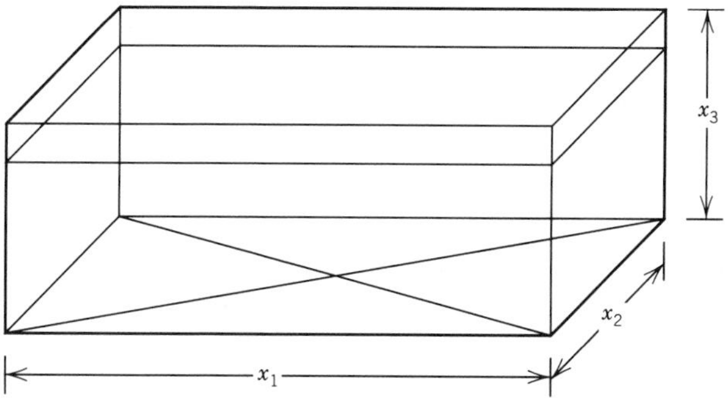

Formulate the problem of minimizing the total length of wire used subject to the volume being, at least, 10 cm^3 as a standard posynomial geometric programming problem. What is the degree of difficulty? Solve this problem.

11.45 Consider the geometric programming problem to

Minimize $40x_1x_2 + 20x_2x_3$

subject to $\frac{1}{5}x_1^{-1}x_2^{-1/2} + \frac{3}{5}x_2^{-1}x_3^{-2/3} \leq 1$

$\mathbf{x} > \mathbf{0}$

What is its degree of difficulty? Solve this problem.

11.46 Consider the geometric programming problem to

Minimize $40x_1^{-1}x_2^{-1/2}x_3^{-1} + 20x_1x_3 + 40x_1x_2x_3$

subject to $\frac{1}{3}x_1^{-2}x_2^{-2} + \frac{4}{3}x_2^{1/2}x_3^{-1} \leq 1$

$\mathbf{x} > \mathbf{0}$

What is its degree of difficulty? Solve this problem.

11.47 Re-solve the geometric programming problem of Example 11.5.3, assuming that the cylinder is open at one end.

Notes and References

In Section 11.1, we introduced the linear complementary problem. The KKT optimality conditions for linear and quadratic programs can be expressed as linear complementary problems. The problem also arises in several contexts such as bimatrix games and engineering optimization. The interested reader may refer to Cottle and Dantzig [1968], Dennis [1959], Du Val [1940], Kilmister and Reeve [1966], Lemke [1965, 1968], Lemke and Howson [1964], and Murty [1976, 1988]. LCP's also arise in finite difference schemes for fractional value problems (Cryer [1971]) and in electronic circuit stimulation problems (Van Bokhoven [1980]). The general LCP has been shown to be NP-complete in the strong sense, and so we may not expect even a pseudo-polynomial time algorithm unless P = NP (see Garey and Johnson [1979]). In 1968, Lemke proposed a complementary pivoting algorithm, discussed in Section 11.1, for solving the linear complementary problem. Lemke proved that the method converges in a finite number of steps to a complementary basic feasible solution to the system if the matrix **M** is copositive-plus. Eaves [1971] extended the result to a more general class of matrices. In 1974, van de Panne developed a varient of Lemke's method for solving the linear complementary problem. Mangasarian [1976] and Solow and Sengupta [1985] give other optimization-based approaches, Kostreva [1978] provides an algebraic approach, and Cottle and Pang [1978] give a topological scheme. Cottle and Dantzig [1968] have given a *principal pivoting algorithm,* which was introduced in Exercise 11.7. In particular, if **M** is positive semidefinite, or if **M** is a *P* matrix (having all principal minors positive), then Murty [1972] has shown that the principal pivoting methods of Dantzig and Cottle [1967] or Lemke's algorithm [1968] can solve this problem. Also, see Murty [1988], and Rohn [1990] for further discussion on principal pivoting methods. Todd [1974] presents a general pivoting system that provides a natural setting for the study of complementary pivoting algorithms.

When **M** is positive semidefinite, Chung and Murty [1981] and Kozlov et al. [1979] have shown that Khachian's [1979] algorithm can be modified to polynomially solve LCP. Cottle and Veinott [1969], Mangasarian [1977], Kojima et al. [1988], and Ye [1988] also discuss polynomially solvable cases, where **M** is a *Z* matrix, or is positive semidefinite, or is skew-symmetric, or belongs to a restricted class of *P* matrices. Ye and Pardalos [1989] employ an interior point potential reduction algorithm to develop another class of LCP's that are polynomially solvable, suggesting that positive semidefiniteness of **M** may not be the fundamental demarcation between P and NP in the context of LCP's.

Several attempts have also been made to solve LCP's as a single linear program (Mangasarian [1976, 1978, 1979]) or as a sequence of linear programs (see Roy and Solow [1985] and Shiau [1983]).

Al-Khayyal [1987] develops a branch-and-bound algorithm for solving general LCP's. Pardalos and Rosen [1988] show how LCP can be formulated as a mixed-integer linear programming problem, and develop an efficient heuristic. Al-Khayyal [1990] also shows how LCP's can be solved as bilinear programming problems (see Al-Khayyal and Falk [1983], Konno and Yajima [1989], Sherali and Shetty [1980], Sherali and Alameddine [1991], and Vaish and Shetty [1976, 1977] for competitive approaches). Concave minimization approaches (Bard and Falk [1972], Sherali et al. [1992a]) and other linear mixed-integer zero–one approaches (see Sherali et al. [1992b]) have also been suggested. (See Parker and Rardin [1988] and Nemhauser and Wolsey [1989] for a discussion on solving mixed-integer problems.) Kostreva and Wiecek [1989] draw interesting interrelationships between LCP's and multiple objective programming problems.

The linear complementary problem has been extended to the nonlinear case and was briefly introduced in Exercise 11.6. The KKT conditions for a general nonlinear programming problem can be expressed as a nonlinear complementary problem. There has been a considerable amount of research on the existence of solutions to such a problem, but little has been done in the area of development of computational schemes for finding such solutions. See Cottle [1966], Eaves [1971], Habetler and Price [1971, 1973], Karamardian [1969, 1971, 1972], and Murty [1988].

There are several approaches for solving a quadratic programming problem. The methods of feasible directions discussed in Chapter 10 could be used to solve the problem. One such implementation is Beale's method [1955, 1959], which is essentially a specialization of the convex–simplex method. Another popular procedure that has been used is a combinatorial approach to iteratively determine the set of binding constraints at optimality, and is known as an *active set strategy*. This is done by solving a sequence of equality-constrained problems. See Boot [1961, 1964], Goldfarb and Idnani [1981], Powell [1985], Theil and van de Panne [1960], and van de Panne [1974]. (Also, see Luenberger [1983] and Fletcher [1989] and Exercises 11.21 and 11.22.) Yet another approach adopted by Houthaker [1960] is to solve a restricted problem by adding a constraint of the form $\Sigma x_j \leq \beta$ and successively increasing β.

One of the popular schemes for solving a quadratic program is to solve the KKT system as proposed by Barankin and Dorfman [1955] and by Markowitz [1956]. There are several methods for solving the KKT system. Wolfe [1959] developed a slight modification of the simplex method to solve the KKT system where dual feasibility is relaxed. This method was briefly discussed in Exercise 11.15. As we discussed earlier in these notes, complementary pivoting methods for solving a linear complementary problem can also be used to solve the KKT system. In Sections 11.1 and 11.2, we discussed Lemke's method for solving a quadratic program, where primal and dual feasibilities are relaxed. In Exercises 11.7, 11.16, 11.7, and 11.19, we present several alternative methods for solving the KKT system. For more details, see Cottle and Dantzig [1968], Dantzig [1963], Frank and Wolfe [1956], and Shetty [1963]. Poljak and Tret'iakov [1972] present a finite algorithm based on the ALAG penalty function approach presented in Chapter 9. Polynomial-time algorithms have also been developed for convex quadratic programs (see Ben Daya and Shetty, 1988, and Ye, 1989 for a survey). Primal-Dual path-following algorithms in the spirit of the algorithm presented in Chapter 9 have also been proposed (see Anstreicher, 1990, and Monteiro et al., 1990). A polynomial-time algorithm, specialized to box-constrained quadratic programs is presented by Han et al. [1989]. See Ye [1990] for quadratic minimization over a sphere.

The methods discussed above deal with convex quadratic programs. Extensions to the nonconvex case have been studied by several researchers. In Exercise 11.14, the problem of finding an optimal solution is posed as the minimization of a linear objective function subject to the constraints representing a linear complementary problem. One approach for solving such problems is the use of cutting plane methods, as discussed in Balas [1972], Balas and Burdet [1973], Burdet [1977], Ritter [1966], and Tui [1964]. Alternative approaches may be found in Cabot and Francis [1970], Mueller [1970], Mylander [1971], Taha [1973], and Zwart [1974]. Horst and Tuy [1990] and Pardalos and Rosen [1987] survey other recent, competitive methods. Sherali [1989] discusses nonconvex quadratic programming duality. Pardalos and Vavasis [1991] show that such problems are NP-Hard, even if the Hessian has a single negative eigenvalue (for minimization problems). An algorithm for a problem that significantly generalizes quadratic programming to the case when the objective and constraint functions are general polynomials has been developed by Sherali and Tuncbilek [1992].

In Section 11.3, we discuss the simplex method with restricted basis entry for solving separable programming problems. Applications include economic data fitting (Bachem and Korte [1977]), electrical networks (Rockafellar [1976]), water supply system design (Collins et al., [1978] and Meyer [1980], in which problems having more than 600 constraints and 900 variables have been solved), logistics (Saaty, [1970]), and statistics (Teng [1978]). This approach is found in the works of Charnes and Cooper [1957], Dantzig, Johnson, and White [1958], and Markowitz and Manne [1957]. For further discussion on this approach, see Miller [1963] and Wolfe [1963]. Myer [1980] discusses a novel two-segment approximation approach, and Meyer [1980] and Thakur [1978] discuss bounds on error upon early termination. In the nonconvex case, even though optimality cannot be claimed with the restricted basis entry rule, good solutions are produced. In the convex case, we showed that by choosing a small grid, we can obtain a solution sufficiently close to the global optimal solution. In Section 11.3, we also discussed the grid generation scheme of Wolfe [1963]. Here, grid points are not fixed beforehand but are generated as needed.

In Section 11.4, we discussed the methods of Charnes and Cooper [1962] and of Gilmore and Gomory [1963] for solving a linear fractional programming problem. The first approach makes a transformation of variables and solves an equivalent linear program. The second approach is an adaptation of the convex–simplex method. Algorithms in this category are closely related to the original work of Isbell and Marlow [1956]. Dorn [1962] presents a procedure for solving the problem that can be viewed as a generalization of the dual simplex method. For other algorithms in this general class, see Abadie and Williams [1968], Avriel et al. [1988], Bitran and Novaes [1973], Konno and Kuno [1989], Martos [1964, 1975], and Schaible [1989].

The linear fractional programming problem has been extended to the case where the objective function is the ratio of two nonlinear functions. Properties of such fractional functions are discussed in Exercises 3.57 and 3.58. Several algorithms for solving nonlinear fractional programs are developed. The interested reader may refer to Almogy and Levin [1971], Bector [1968], Dinkelbach [1967, Mangasarian [1969] and Swarup [1965].

Geometric programming problems, discussed in Section 11.5, arise frequently in engineering applications (see Duffin et al. [1967], Bradley and Clyne [1976], and Dembo and Avriel [1978], for example). An excellent pioneering exposition appears in Duffin, Peterson, and Zener [1967]. Exercises 11.41 through 11.46 are presented in this work, along with many other examples. We principally discuss posynomial geometric programs, following a Lagrangian duality approach (see Fletcher [1989]; see also Duffin et al. [1967] for a generalization of Theorem 11.5.2). Duffin and Peterson [1972, 1973] provide further discussions, Peterson [1976] gives a survey of approaches to a wider class of geometric programs, and Dembo [1978] and Ecker [1980] give an excellent discussion on implementation and computational aspects of solving geometric programming problems. Dembo [1979] also presents details of an efficient second-order Newton-type method for solving DGP. Geometric programming problems that involve the optimization of general polynomial objective functions subject to polynomial constraints are discussed by Floudas and Visweswaran [1991], Shor [1990], and Sherali and Tuncbilek [1990]. Some test problems appear in Dembo [1976].

Appendix A Mathematical Review

This appendix reviews notation, basic definitions, and results related to vectors and matrices and real analysis that are used throughout the text. For more details, see Bartle [1976], Berge [1963], Berge and Ghouila-Houri [1965], Buck [1965], Cullen [1972], Flet [1966], and Rudin [1964].

A.1 Vectors and Matrices

Vectors

An n vector $\mathbf{x}$ is an array of n scalars $x_1, x_2, \ldots, x_n$. Here, x_j is called the jth *component*, or *element*, of the vector $\mathbf{x}$. The notation $\mathbf{x}$ represents a *column vector*, whereas the notation $\mathbf{x}^t$ represents a *row vector*. Vectors are denoted by lowercase boldface letters, such as $\mathbf{a}$,$\mathbf{b}$, $\mathbf{c}$, $\mathbf{x}$, and $\mathbf{y}$. The collection of all n vectors forms the *n-dimensional Euclidean space*, which is denoted by E_n.

Special Vectors

The *zero vector*, denoted by $\mathbf{0}$, is a vector consisting entirely of zeros. The *sum vector* is denoted by $\mathbf{1}$ or $\mathbf{e}$ and has each component equal to 1. The ith *coordinate vector*, also referred to as the ith *unit vector*, is denoted by $\mathbf{e}_i$ and consists of zeros except for a 1 at the ith position.

Vector Addition and Multiplication by a Scalar

Let $\mathbf{x}$ and $\mathbf{y}$ be two n vectors. The *sum* of $\mathbf{x}$ and $\mathbf{y}$ is written as the vector $\mathbf{x} + \mathbf{y}$. The jth component of the vector $\mathbf{x} + \mathbf{y}$ is $x_j + y_j$. The *product* of a vector $\mathbf{x}$ and a scalar α is denoted by $\alpha\mathbf{x}$ and is obtained by multiplying each element of $\mathbf{x}$ by α.

Linear, and Affine Independence

A collection of vectors $\mathbf{x}_1, \ldots, \mathbf{x}_k$ in E_n is considered *linearly independent* if $\sum_{j=1}^{k} \lambda_j \mathbf{x}_j = \mathbf{0}$ implies that $\lambda_j = 0$ for $j = 1, \ldots, k$. A collection of vectors $\mathbf{x}_0, \mathbf{x}_1, \ldots, \mathbf{x}_k$ in E_n is considered to be *affinely independent* if $(\mathbf{x}_1 - \mathbf{x}_0), \ldots, (\mathbf{x}_k - \mathbf{x}_0)$ are linearly independent.

Linear, Affine, and Convex Combinations and Hulls

A vector $\mathbf{y}$ in E_n is said to be a *linear combination* of the vectors $\mathbf{x}_1, \ldots, \mathbf{x}_k$ in E_n if $\mathbf{y}$ can be written as $\mathbf{y} = \sum_{j=1}^{k} \lambda_j \mathbf{x}_j$ for some scalars $\lambda_1, \ldots, \lambda_k$. If, in addition, $\lambda_1, \ldots, \lambda_k$ are restricted to satisfy $\sum_{j=1}^{k} \lambda_j = 1$, then $\mathbf{y}$ is said to be an *affine combination* of $\mathbf{x}_1, \ldots, \mathbf{x}_k$. Furthermore, if we also restrict $\lambda_1, \ldots, \lambda_k$ to be nonnegative, then this is known as a *convex combination* of $\mathbf{x}_1, \ldots, \mathbf{x}_k$. The *linear, affine,* or *convex hull* of a set $S \subseteq E_n$ is respectively the set of all linear, affine, or convex combinations of points within S.

Spanning Vectors

A collection of vectors $\mathbf{x}_1, \ldots, \mathbf{x}_k$ in E_n is said to *span* E_n if any vector in E_n can be represented as a linear combination of $\mathbf{x}_1, \ldots, \mathbf{x}_k$. The *cone spanned* by a collection of vectors $\mathbf{x}_1, \ldots, \mathbf{x}_k$ is the set of nonnegative linear combinations of these vectors.

Basis

A collection of vectors $\mathbf{x}_1, \ldots, \mathbf{x}_k$ in E_n is called a *basis* of E_n if it spans E_n and if the deletion of any of the vectors prevents the remaining vectors from spanning E_n. It can be shown that $\mathbf{x}_1, \ldots, \mathbf{x}_k$ form a basis of E_n if and only if $\mathbf{x}_1, \ldots, \mathbf{x}_k$ are linearly independent and if, in addition, $k = n$.

Inner Product

The *inner product* of two vectors $\mathbf{x}$ and $\mathbf{y}$ in E_n is defined by $\mathbf{x}'\mathbf{y} = \sum_{j=1}^{n} x_j y_j$. If the inner product of two vectors is equal to zero, then the two vectors are said to be *orthogonal*.

The Norm of a Vector

The *norm* of a vector $\mathbf{x}$ in E_n is denoted by $\|\mathbf{x}\|$ and defined by $\|\mathbf{x}\| = (\mathbf{x}'\mathbf{x})^{1/2} = (\sum_{j=1}^{n} x_j^2)^{1/2}$.

Schwartz Inequality

Let $\mathbf{x}$ and $\mathbf{y}$ be two vectors in E_n and let $|\mathbf{x}'\mathbf{y}|$ denote the absolute value of $\mathbf{x}'\mathbf{y}$. Then, the following inequality, referred to as the *Schwartz inequality*, holds:

$$|\mathbf{x}'\mathbf{y}| \leq \|\mathbf{x}\| \, \|\mathbf{y}\|$$

Matrices

A *matrix* is a rectangular array of numbers. If the matrix has m rows and n columns, it is called an $m \times n$ matrix. Matrices are denoted by boldface capital letters, such as $\mathbf{A}$,

B, and **C**. The entry in row i and column j of a matrix **A** is denoted by a_{ij}, its ith row is denoted by $\mathbf{A}_i$, and its jth column is denoted by $\mathbf{a}_j$.

Special Matrices

An $m \times n$ matrix whose elements are all equal to zero is called a *zero matrix* and is denoted by **0**. A square $n \times n$ matrix is called the *identity* matrix if $a_{ij} = 0$ for $i \neq j$ and $a_{ii} = 1$ for $i = 1, \ldots, n$. The $n \times n$ identity matrix is denoted by **I** and sometimes by $\mathbf{I}_n$ to highlight its dimension. An $n \times n$ *permutation* matrix **P** is one which has the same rows of $\mathbf{I}_n$ but which are permuted in some order. An orthogonal $m \times n$ matrix **Q** is one that satisfies $\mathbf{Q}^t\mathbf{Q} = \mathbf{I}_n$ or $\mathbf{Q}\mathbf{Q}^t = \mathbf{I}_m$. In particular, if **Q** is square, then $\mathbf{Q}^{-1} = \mathbf{Q}^t$. Note that a permutation matrix **P** is an orthogonal square matrix.

Addition of Matrices and Scalar Multiplication of a Matrix

Let **A** and **B** be two $m \times n$ matrices. The sum of **A** and **B**, denoted by $\mathbf{A} + \mathbf{B}$, is the matrix whose ij entry is $a_{ij} + b_{ij}$. The *product* of a matrix **A** by a scalar α is the matrix whose ij entry is αa_{ij}.

Matrix Multiplication

Let **A** be an $m \times n$ matrix and **B** be an $n \times p$ matrix. Then the *product* **AB** is defined to be the $m \times p$ matrix **C** whose ij entry c_{ij} is given by

$$c_{ij} = \sum_{k=1}^{n} a_{ik}b_{kj} \qquad \text{for } i = 1, \ldots, m \qquad j = 1, \ldots, p$$

Transposition

Let **A** be an $m \times n$ matrix. The *transpose* of **A**, denoted by $\mathbf{A}^t$, is the $n \times m$ matrix whose ij entry is equal to a_{ji}. A square matrix **A** is said to be *symmetric* if $\mathbf{A} = \mathbf{A}^t$. It is said to be *skew symmetric* if $\mathbf{A}^t = -\mathbf{A}$.

Partitioned Matrices

A matrix can be partitioned into submatrices. For example, the $m \times n$ matrix **A** could be partitioned as follows:

$$\mathbf{A} = \left[\begin{array}{c|c} \mathbf{A}_{11} & \mathbf{A}_{12} \\ \hline \mathbf{A}_{21} & \mathbf{A}_{22} \end{array}\right]$$

where $\mathbf{A}_{11}$ is $m_1 \times n_1$, $\mathbf{A}_{12}$ is $m_1 \times n_2$, $\mathbf{A}_{21}$ is $m_2 \times n_1$, $\mathbf{A}_{22}$ is $m_2 \times n_2$, $m = m_1 + m_2$, and $n = n_1 + n_2$.

Determinant of a Matrix

Let **A** be an $n \times n$ matrix. The *determinant* of **A**, denoted by det [**A**], is defined iteratively as follows:

$$\det[\mathbf{A}] = \sum_{i=1}^{n} a_{i1}\mathbf{A}_{i1}$$

Here, $\mathbf{A}_{i1}$ is the *cofactor* of a_{i1} defined as $(-1)^{i+1}$ times the determinant of the submatrix of **A** formed by deleting the ith row and the first column, and where the determinant of any scalar is the scalar itself. Similar to the use of the first column here, the determinant can be expressed in terms of any row or column.

Inverse of a Matrix

A square matrix **A** is called *nonsingular* if there is a matrix $\mathbf{A}^{-1}$ called the *inverse* matrix such that $\mathbf{A}\mathbf{A}^{-1} = \mathbf{A}^{-1}\mathbf{A} = \mathbf{I}$. The inverse of a square matrix, if it exists, is unique. Furthermore, a square matrix has an inverse if and only if its determinant is not equal to zero.

The Rank of a Matrix

Let **A** be an $m \times n$ matrix. The *rank* of **A** is the maximum number of linearly independent rows or, equivalently, the maximum number of linearly independent columns of the matrix **A**. If the rank of **A** is equal to minimum $\{m, n\}$, then **A** is said to have *full rank*.

Norm of a Matrix

Let **A** be an $n \times n$ matrix. Most commonly, the *norm* of **A**, denoted by $\|\mathbf{A}\|$, is defined by

$$\|\mathbf{A}\| = \underset{\|\mathbf{x}\|=1}{\text{maximum}} \frac{\|\mathbf{A}\mathbf{x}\|}{\|\mathbf{x}\|}$$

where $\|\mathbf{A}\mathbf{x}\|$ and $\|\mathbf{x}\|$ are the usual Euclidean (l_2) norms of the corresponding vectors. Hence, for any vector $\mathbf{z}$, $\|\mathbf{A}\mathbf{z}\| \le \|\mathbf{A}\|\,\|\mathbf{z}\|$. A similar use of an l_p norm $\|\cdot\|_p$ induces a corresponding matrix norm $\|\mathbf{A}\|_p$. In particular, the above matrix norm, sometimes denoted $\|\mathbf{A}\|_2$, is equal to the [maximum eigenvalue of $\mathbf{A}^t\mathbf{A}]^{1/2}$. Also, the *Frobenius* norm of **A** is given by

$$\|\mathbf{A}\|_F = \left[\sum_{i=1}^{n}\sum_{j=1}^{n} |a_{ij}|^2\right]^{1/2}$$

and is simply the l_2 norm of the vector whose elements are all the elements of **A**.

Eigenvalues and Eigenvectors

Let **A** be an $n \times n$ matrix. A scalar λ and a nonzero vector $\mathbf{x}$ satisfying the equation $\mathbf{A}\mathbf{x} = \lambda\mathbf{x}$ are called, respectively, an *eigenvalue* and an *eigenvector* of **A**. To compute the eigenvalues of **A**, we solve the equation $\det[\mathbf{A} - \lambda\mathbf{I}] = 0$. This yields a polynomial equation in λ that can be solved for the eigenvalues of **A**. If **A** is symmetric, then it has n (possibly nondistinct) eigenvalues. The eigenvectors associated with distinct eigenvalues are necessarily orthogonal, and for any collection of some p coincident eigenvalues, there exists a collection of p orthogonal eigenvectors. Hence, given a symmetric matrix **A**, we can construct an orthogonal basis **B** for E_n, that is, a basis of orthogonal column vectors, each representing an eigenvector of **A**. Furthermore, let us assume that each column of **B** has been normalized to have a unit norm. Hence, $\mathbf{B}^t\mathbf{B} = \mathbf{I}$, so that $\mathbf{B}^{-1} \equiv \mathbf{B}^t$. Such a matrix is said to be an *orthogonal* matrix.

Now consider the *quadratic form* $\mathbf{x}^t\mathbf{A}\mathbf{x}$, where **A** is an $n \times n$ symmetric matrix. Let $\lambda_1, \ldots, \lambda_n$ be the eigenvalues of **A**, let $\mathbf{\Lambda} = \operatorname{diag}\{\lambda_1, \ldots, \lambda_n\}$ be a *diagonal matrix* comprised of diagonal elements $\lambda_1, \ldots, \lambda_n$ and zeros elsewhere, and let **B** be the orthogonal eigenvector matrix comprised of the orthogonal, normalized eigenvectors $\mathbf{b}_1, \ldots, \mathbf{b}_n$ as its columns. Define the linear transformation $\mathbf{x} = \mathbf{B}\mathbf{y}$ that writes any vector $\mathbf{x}$ in terms of the eigenvectors of $\mathbf{A}$. Under this transformation, the given quadratic form becomes

$$\mathbf{x}^t\mathbf{A}\mathbf{x} = \mathbf{y}^t\mathbf{B}^t\mathbf{A}\mathbf{B}\mathbf{y} = \mathbf{y}^t\mathbf{B}^t\mathbf{\Lambda}\mathbf{B}\mathbf{y} = \mathbf{y}^t\mathbf{\Lambda}\mathbf{y} \equiv \sum_{i=1}^{n} \lambda_i y_i^2$$

This is called a *diagonalization process.*

Observe also that we have $\mathbf{AB} = \mathbf{B\Lambda}$, so that, because $\mathbf{B}$ is orthogonal, we get $\mathbf{A} = \mathbf{B\Lambda B}^t = \sum_{i=1}^{n} \lambda_i \mathbf{b}_i\mathbf{b}_i^t$. This representation is called the *spectral decomposition* of $\mathbf{A}$. For an $m \times n$ matrix $\mathbf{A}$, a related factorization $\mathbf{A} = \mathbf{U\Sigma V}^t$, where $\mathbf{U}$ is an $m \times m$ orthogonal matrix, $\mathbf{V}$ is an $n \times n$ orthogonal matrix, and $\mathbf{\Sigma}$ is an $m \times n$ matrix with elements $\Sigma_{ij} = 0$ for $i \neq j$, $\Sigma_{ij} \geq 0$ for $i = j$, is known as a *singular-value decomposition (SVD)* of $\mathbf{A}$. Here, the columns of $\mathbf{U}$ and $\mathbf{V}$ are normalized eigenvectors of $\mathbf{AA}^t$ and $\mathbf{A}^t\mathbf{A}$, respectively. The Σ_{ij} values are the (absolute) square roots of the eigenvalues of $\mathbf{AA}^t$ if $m \leq n$ or of $\mathbf{A}^t\mathbf{A}$ if $m \geq n$. The number of nonzero Σ_{ii} values equals the rank of $\mathbf{A}$.

Definite and Semidefinite Matrices

Let $\mathbf{A}$ be an $n \times n$ symmetric matrix. Here, $\mathbf{A}$ is said to be *positive definite* if $\mathbf{x}^t\mathbf{A}\mathbf{x} > 0$ for all nonzero $\mathbf{x}$ in E_n and is said to be *positive semidefinite* if $\mathbf{x}^t\mathbf{A}\mathbf{x} \geq 0$ for all $\mathbf{x}$ in E_n. Similarly, if $\mathbf{x}^t\mathbf{A}\mathbf{x} < 0$ for all nonzero $\mathbf{x}$ in E_n, then $\mathbf{A}$ is called *negative definite*; and if $\mathbf{x}^t\mathbf{A}\mathbf{x} \leq 0$ for all $\mathbf{x}$ in E_n, then $\mathbf{A}$ is called *negative semidefinite*. By the foregoing diagonalization process, the matrix $\mathbf{A}$ is positive definite, positive semidefinite, negative definite, and negative semidefinite if and only if its eigenvalues are positive, nonnegative, negative, and nonpositive, respectively. (Note that the superdiagonalization algorithm discussed in Chapter 3 is a more efficient method for ascertaining definiteness properties.) Also, by the definition of $\mathbf{\Lambda}$ and $\mathbf{B}$ above, if $\mathbf{A}$ is positive definite, then its *square root* $\mathbf{A}^{1/2}$ is the matrix satisfying $\mathbf{A}^{1/2}\mathbf{A}^{1/2} = \mathbf{A}$ and is given by $\mathbf{A}^{1/2} = \mathbf{B\Lambda}^{1/2}\mathbf{B}^t$.

A.2 Matrix Factorizations

Let $\mathbf{B}$ be a nonsingular $n \times n$ matrix, and consider the system of equations $\mathbf{Bx} = \mathbf{b}$. The solution given by $\mathbf{x} = \mathbf{B}^{-1}\mathbf{b}$ is seldom computed by directly finding the inverse $\mathbf{B}^{-1}$. Instead, a factorization or decomposition of $\mathbf{B}$ into multiplicative components is usually employed, whereby $\mathbf{Bx} = \mathbf{b}$ is solved in a numerically stable fashion, often through the solution of triangular systems via back-substitution. This becomes very relevant particularly in ill-conditioned situations when $\mathbf{B}$ is nearly singular, or when we wish to verify positive definiteness of $\mathbf{B}$ as in quasi-Newton or Levenberg–Marquardt methods. Several useful factorizations are discussed below. For more details, including schemes for updating such factors in the context of iterative methods, we refer the reader to Bartels et al. [1970], Bazaraa et al. [1990], Dennis and Schnabel [1983], Dongarra et al. [1979], Gill et al. [1974, 1976], Golub and Van Loan [1983, 1989], Murty [1983], and Stewart [1973], along with the many accompanying references cited therein. Standard software such as LINPACK, MATLAB, and the Harwell Library routines are also available to perform these factorizations efficiently.

LU and *PLU* Factorization for a Basis B

In the *LU* factorization, we reduce **B** to an upper triangular form **U** through a series of permutations and Gaussian pivot operations. At the *i*th stage of this process, having reduced **B** to $\mathbf{B}^{(i-1)}$, say, which is upper triangular in columns $1, \ldots, i-1$ (where $\mathbf{B}^0 \equiv \mathbf{B}$), we first premultiply $\mathbf{B}^{(i-1)}$ by a *permutation matrix* $\mathbf{P}_i$ to exchange row *i* with that row in $\{i, i+1, \ldots, n\}$ of $\mathbf{B}^{(i-1)}$ which has the largest absolute valued element in column *i*. This is done to ensure that the (i, i) element of $\mathbf{P}_i\mathbf{B}^{(i-1)}$ is significantly nonzero. Using this as a pivot element, we perform row operations to zero out the elements in rows $i+1, \ldots, n$ of column *i*. This *triangularization* can be represented as a premultiplication with a suitable *Gaussian pivot matrix* $\mathbf{G}_i$, which is a *unit lower triangular* matrix, having 1's on the diagonal and suitable possibly nonzero elements in rows $i+1, \ldots, n$ of column *i*. This gives $\mathbf{B}^{(i)} = (\mathbf{G}_i\mathbf{P}_i)\,\mathbf{B}^{(i-1)}$. Hence, we get, after some $r \leq (n-1)$ such operations,

$$(\mathbf{G}_r\mathbf{P}_r)\ldots(\mathbf{G}_2\mathbf{P}_2)(\mathbf{G}_1\mathbf{P}_1)\mathbf{B} = \mathbf{U} \tag{A.1}$$

The system $\mathbf{Bx} = \mathbf{b}$ can now be solved by computing $\bar{\mathbf{b}} = (\mathbf{G}_r\mathbf{P}_r)\ldots(\mathbf{G}_1\mathbf{P}_1)\mathbf{b}$ and then solving the triangular system $\mathbf{Ux} = \bar{\mathbf{b}}$ by back-substitution. If no permutations are performed, then $\mathbf{G}_r\ldots\mathbf{G}_1$ is lower triangular and, denoting its (lower triangular) inverse as **L**, we have the factored form $\mathbf{B} = \mathbf{LU}$ for **B**, hence, its name. Also, if $\mathbf{P}^t$ is a permutation matrix that is used to *a priori* rearrange the rows of **B**, and then we apply the Gaussian triangularization operation to derive $\mathbf{L}^{-1}\mathbf{P}^t\mathbf{B} = \mathbf{U}$, we can write $\mathbf{B} = (\mathbf{P}^t)^{-1}\mathbf{LU} = \mathbf{PLU}$, noting that $\mathbf{P}^t = \mathbf{P}^{-1}$. Hence, this factorization is sometimes called a *PLU* decomposition. If **B** is sparse, $\mathbf{P}^t$ can be used to make $\mathbf{P}^t\mathbf{B}$ nearly upper triangular (assuming that the columns of **B** have been appropriately permuted) and then only a few and sparse Gaussian pivot operations will be required to obtain **U**. This method is therefore very well suited for sparse matrices.

QR and *QRP* Factorization for a Basis B

This factorization is most suitable and frequently used for solving dense equation systems. Here, the matrix **B** is reduced to an upper triangular form **R** by premultiplying it with a sequence of square, symmetric orthogonal matrices $\mathbf{Q}_i$. Given $\mathbf{B}^{(i-1)} \equiv \mathbf{Q}_{i-1}\ldots\mathbf{Q}_1\mathbf{B}$ that is upper triangular in columns $1, \ldots, i-1$ (where $\mathbf{B}^{(0)} \equiv \mathbf{B}$), we use $\mathbf{Q}_i$ so that $\mathbf{Q}_i\mathbf{B}^{(i-1)} \equiv \mathbf{B}^{(i)}$ is upper triangular in column *i* as well, while columns $1, \ldots, i-1$ remain unaffected. The matrix $\mathbf{Q}_i$ is a square, symmetric orthogonal matrix of the form $\mathbf{Q}_i \equiv \mathbf{I} - \gamma_i\mathbf{q}_i\mathbf{q}_i^t$, where $\mathbf{q}_i = (0, \ldots, 0, q_{ii}, \ldots, q_{ni})^t$, and $\gamma_i \in E_1$ are suitably chosen to perform the foregoing operation. Such a matrix $\mathbf{Q}_i$ is called a *Householder transformation* matrix. If the elements in rows $i, \ldots, n$ of column *i* of $\mathbf{B}^{(i-1)}$ are denoted by $(\alpha_i, \ldots, \alpha_n)^t$, then we have $q_{ii} = \alpha_i + \theta_i$, $q_{ji} = \alpha_j$ for $j = i+1, \ldots, n$, $\gamma_i = 1/\theta_i q_{ii}$, where $\theta_i = sign\,(\alpha_i)[\alpha_i^2 + \cdots + \alpha_n^2]^{1/2}$, and where $sign\,(\alpha_i) = 1$ if $\alpha_i > 0$ and -1 otherwise. Defining $\mathbf{Q} = \mathbf{Q}_{n-1}\ldots\mathbf{Q}_1$, we see that **Q** is also a symmetric orthogonal matrix and that $\mathbf{QB} = \mathbf{R}$, or that $\mathbf{B} = \mathbf{QR}$, since $\mathbf{Q} = \mathbf{Q}^t = \mathbf{Q}^{-1}$, that is, **Q** is an *involutory* matrix.

Now, to solve $\mathbf{Bx} = \mathbf{b}$, we equivalently solve $\mathbf{QRx} = \mathbf{b}$ or $\mathbf{Rx} = \mathbf{Qb}$ by finding $\bar{\mathbf{b}} = \mathbf{Qb}$ first and then solving the upper triangular system $\mathbf{Rx} = \bar{\mathbf{b}}$ via back-substitution. Note that since $\|\mathbf{Qv}\| = \|\mathbf{v}\|$ for any vector **v**, we have $\|\mathbf{R}\| = \|\mathbf{QR}\| = \|\mathbf{B}\|$, so that **R** preserves the relative magnitudes of the elements in **B**, maintaining stability. This is its principal advantage.

Also, a permutation matrix **P** is sometimes used to postmultiply $\mathbf{B}^{(i-1)}$ before applying $\mathbf{Q}_i$ to it, so as to move that column which has the largest value of the sum of squares below row $(i - 1)$ into the ith column position (see the computation of θ_i above). Since the product of permutation matrices is also a permutation matrix, and since a permutation matrix is orthogonal, this leads to the decomposition $\mathbf{B} = \mathbf{QRP}$ via the operation sequence $\mathbf{Q}_{n-1} \cdots \mathbf{Q}_1 \mathbf{B} \mathbf{P}_1 \mathbf{P}_2 \cdots \mathbf{P}_{n-1} = \mathbf{R}$.

Cholesky Factorization LL^t and LDL^t for Symmetric, Positive Definite Matrices B

The Cholesky factorization of a symmetric, positive definite matrix **B** represents this matrix as $\mathbf{B} = \mathbf{LL}^t$, where **L** is a lower triangular matrix of the form

$$\mathbf{L} = \begin{bmatrix} \ell_{11} & & & & \\ \ell_{21} & \ell_{22} & & \mathbf{0} & \\ \ell_{31} & \ell_{32} & \ell_{33} & & \\ \cdot & \cdot & & \cdot & \\ \cdot & \cdot & & & \cdot \\ \cdot & \cdot & & & \cdot \\ \ell_{n1} & \ell_{n2} & & & \ell_{nn} \end{bmatrix}$$

so that

$$\mathbf{LL}^t = \begin{bmatrix} \ell_{11}^2 & & & & \\ \ell_{21}\ell_{11} & (\ell_{21}^2 + \ell_{22}^2) & & \text{(SYMMETRIC)} & \\ \ell_{31}\ell_{11} & (\ell_{21}\ell_{31} + \ell_{22}\ell_{32}) & (\ell_{31}^2 + \ell_{32}^2 + \ell_{33}^2) & & \\ \cdot & \cdot & \cdot & \cdot & \\ \cdot & \cdot & \cdot & \cdot & \\ \cdot & \cdot & \cdot & \cdot & \\ \ell_{n1}\ell_{11} & (\ell_{21}\ell_{n1} + \ell_{22}\ell_{n2}) & (\ell_{31}\ell_{n1} + \ell_{32}\ell_{n2} + \ell_{33}\ell_{n3}) & \cdots & (\ell_{n1}^2 + \cdots + \ell_{nn}^2) \end{bmatrix}$$

By directly equating the elements of **B** to those in $\mathbf{LL}^t$, we obtain the system of equations

$$\begin{array}{lllll} \ell_{11}^2 = b_{11}, & \ell_{21}\ell_{11} = b_{21}, & \ell_{31}\ell_{11} = b_{31}, & \ldots, & \ell_{n1}\ell_{11} = b_{n1} \\ & \ell_{21}^2 + \ell_{22}^2 = b_{22}, & \ell_{21}\ell_{31} + \ell_{22}\ell_{32} = b_{32}, & \ldots, & \ell_{21}\ell_{n1} + \ell_{22}\ell_{n2} = b_{n2} \\ & & \ell_{31}^2 + \ell_{32}^2 + \ell_{33}^2 = b_{33}, & \ldots, & \ell_{31}\ell_{n1} + \ell_{32}\ell_{n2} + \ell_{33}\ell_{n3} = b_{n3} \\ & & & \cdot & \cdot \\ & & & \cdot & \cdot \\ & & & \cdot & \cdot \\ & & & & \ell_{n1}^2 + \cdots + \ell_{nn}^2 = b_{nn} \end{array}$$

These equations can be used sequentially to compute the unknowns l_{ij} in the order l_{11}, $l_{21}, \ldots, l_{n1}, l_{22}, l_{32}, \ldots, l_{n2}, l_{33}, \ldots, l_{n3}, \ldots, l_{nn}$, by using the equation for b_{ij} to compute l_{ij} for $j = 1, \ldots, n$, $i = j, \ldots, n$. Note that these equations are well defined for a symmetric, positive definite matrix **B**, and that $\mathbf{LL}^t$ is positive definite if and only if $l_{ii} > 0$ for all $i = 1, \ldots, n$.

The equation system $\mathbf{Bx} = \mathbf{b}$ can now be solved via $\mathbf{L}(\mathbf{L}^t\mathbf{x}) = \mathbf{b}$ through the solution of two triangular systems of equations. We first find **y** to satisfy $\mathbf{Ly} = \mathbf{b}$, and then compute **x** via the system $\mathbf{L}^t\mathbf{x} = \mathbf{y}$.

Sometimes, the Cholesky factorization is also represented as $\mathbf{B} = \mathbf{LDL}^t$, where **L** is a lower triangular matrix and **D** is a diagonal matrix, both having positive diagonal entries. Writing $\mathbf{B} = \mathbf{LDL}' = (\mathbf{LD}^{1/2})(\mathbf{LD}^{1/2})' = \mathbf{L}'\mathbf{L}''$, we see that the two representations are equivalently related. The advantage of the representation $\mathbf{LDL}'$ is that, $\mathbf{D}$ can be used to avoid the square root operation associated with the diagonal system of equations and, hence, this improves the accuracy of computations. (For example, the diagonal components of L can be made unity.)

Also, if $\mathbf{B}$ is a general basis matrix then, since $\mathbf{BB}'$ is symmetric and positive definite, it has a Cholesky factorization $\mathbf{BB}' = \mathbf{LL}'$. In such a case, $\mathbf{L}$ is referred to as the *Cholesky factor associated with* $\mathbf{B}$. Note that we can determine $\mathbf{L}$ in this case by finding the QR decomposition for $\mathbf{B}'$, so that $\mathbf{BB}' \equiv \mathbf{R}'\mathbf{Q}'\mathbf{QR} = \mathbf{R}'\mathbf{R}$ and, therefore, $\mathbf{L} \equiv \mathbf{R}'$. Whenever this is done, note that the matrix $\mathbf{Q}$ or its components $\mathbf{Q}_i$ need not be stored, since we are only interested in the resulting upper triangular matrix $\mathbf{R}$.

A.3 Sets and Sequences

A *set* is a collection of elements or objects. A set may be specified by listing its elements, or by specifying the properties that the elements must satisfy. For example, the set $S = \{1, 2, 3, 4\}$ can be represented alternatively as $S = \{x : 1 \leq x \leq 4, x \text{ integer}\}$. If $\mathbf{x}$ is a member of S, we write $\mathbf{x} \in S$, and if $\mathbf{x}$ is not a member of S, we write $\mathbf{x} \notin S$. Sets are denoted by capital letters, such as S, X, and Λ. The *empty set*, denoted by $\emptyset$, has no elements.

Unions, Intersection, and Subsets

Given two sets, S_1 and S_2, the set consisting of elements that belong to either S_1 or S_2 is called the *union* of S_1 and S_2 and is denoted by $S_1 \cup S_2$. The elements belonging to both S_1 and S_2 form the *intersection* of S_1 and S_2 and is denoted by $S_1 \cap S_2$. If S_1 is a *subset* of S_2, that is, if each element of S_1 is also an element of S_2, we write $S_1 \subseteq S_2$ or $S_2 \supseteq S_1$. Thus, we write $S \subseteq E_n$ to denote that all elements in S are points in E_n. A *strict containment* $S_1 \subseteq S_2$, $S_1 \neq S_2$, is denoted by $S_1 \subset S_2$.

Closed and Open Intervals

Let a and b be two real numbers. The closed interval $[a, b]$ denotes all real numbers satisfying $a \leq x \leq b$. Real numbers satisfying $a \leq x < b$ are represented by $[a, b)$, while those satisfying $a < x \leq b$ are denoted by $(a, b]$. Finally, the set of points x with $a < x < b$ is represented by the open interval (a, b).

Greatest Lower Bound and Least Upper Bound

Let S be a set of real numbers. Then, the *greatest lower bound*, or the *infimum*, of S is the largest possible scalar α satisfying $\alpha \leq x$ for each $x \in S$. The infimum is denoted by $\inf\{x : x \in S\}$. The *least upper bound*, or the *supremum*, of S is the smallest possible scalar α satisfying $\alpha \geq x$ for each $x \in S$. The supremum is denoted by $\sup\{x : x \in S\}$.

Neighborhoods

Given a point $\mathbf{x} \in E_n$ and an $\varepsilon > 0$, the *ball* $N_\varepsilon(\mathbf{x}) = \{\mathbf{y} : \|\mathbf{y} - \mathbf{x}\| \leq \varepsilon\}$ is called an *ε-neighborhood* of $\mathbf{x}$. The inequality in the definition of $N_\varepsilon(\mathbf{x})$ is sometimes replaced by a strict inequality.

Interior Points and Open Sets

Let S be a subset of E_n, and let $\mathbf{x} \in S$. Then, $\mathbf{x}$ is called an *interior* point of S if there is an ε-neighborhood of $\mathbf{x}$ that is contained in S, that is, if there exists an $\varepsilon > 0$ such that $\|\mathbf{y} - \mathbf{x}\| \leq \varepsilon$ implies that $\mathbf{y} \in S$. The set of all such points is called the interior of S and is denoted by int S. Furthermore, S is called *open* if $S = \text{int } S$.

Relative Interior

Let $S \subset E_n$, and let aff (S) denote the *affine hull* of S. Although int $(S) = \varnothing$, the interior of S as viewed in the space of its affine hull may be nonempty. This is called the *relative interior* of S, and is denoted by relint (S). Specifically, relint $(S) = \{\mathbf{x} \in S : N_\varepsilon(\mathbf{x}) \cap \text{aff } (S) \subset S \text{ for some } \varepsilon > 0\}$. Note that if $S_1 \subseteq S_2$, then relint (S_1) is not necessarily contained within relint (S_2), although int $(S_1) \subseteq$ int (S_2). For example, if $S_1 = \{\mathbf{x} : \boldsymbol{\alpha}'\mathbf{x} = \beta\}$, $\boldsymbol{\alpha} \neq \mathbf{0}$, and $S_2 = \{\mathbf{x} : \boldsymbol{\alpha}'\mathbf{x} \leq \beta\}$, then $S_1 \subseteq S_2$, int $(S_1) = \varnothing \subseteq$ int $(S_2) = \{\mathbf{x} : \boldsymbol{\alpha}'\mathbf{x} < \beta\}$, but relint $(S_1) \equiv S_1$ N relint $(S_2) \equiv$ int (S_2).

Bounded Sets

A set $S \subset E_n$ is said to be *bounded* if it can be contained within a ball of finite radius.

Closure Points and Closed Sets

Let S be a subset of E_n. The *closure* of S, denoted by cl S, is the set of all points that are arbitrarily close to S. In particular, $\mathbf{x} \in$ cl S if for each $\varepsilon > 0$, $S \cap N_\varepsilon(\mathbf{x}) \neq \varnothing$, where $N_\varepsilon(\mathbf{x}) = \{\mathbf{y} : \|\mathbf{y} - \mathbf{x}\| \leq \varepsilon\}$. The set S is said to be *closed* if $S = \text{cl } S$.

Boundary Points

Let S be a subset of E_n. Then, $\mathbf{x}$ is called a *boundary* point of S if for each $\varepsilon > 0$, $N_\varepsilon(\mathbf{x})$ contains a point in S and a point not in S, where $N_\varepsilon(\mathbf{x}) = \{\mathbf{y} : \|\mathbf{y} - \mathbf{x}\| \leq \varepsilon\}$. The set of all boundary points is called the boundary of S and is denoted by ∂S.

Sequences and Subsequences

A *sequence* of vectors $\mathbf{x}_1, \mathbf{x}_2, \mathbf{x}_3, \ldots,$ is said to *converge* to the *limit point* $\bar{\mathbf{x}}$ if $\|\mathbf{x}_k - \bar{\mathbf{x}}\| \to 0$ as $k \to \infty$, that is, if for any given $\varepsilon > 0$, there is a positive integer N such that $\|\mathbf{x}_k - \bar{\mathbf{x}}\| < \varepsilon$ for all $k \geq N$. The sequence is usually denoted by $\{\mathbf{x}_k\}$, and the limit point $\bar{\mathbf{x}}$ is represented by either $\mathbf{x}_k \to \bar{\mathbf{x}}$ as $k \to \infty$ or by $\lim_{k\to\infty} \mathbf{x}_k = \bar{\mathbf{x}}$. Any converging sequence has a unique limit point.

By deleting certain elements of a sequence $\{\mathbf{x}_k\}$, we obtain a *subsequence*. A subsequence is usually denoted as $\{\mathbf{x}_k\}_{\mathcal{K}}$ where $\mathcal{K}$ is a subset of all positive integers. To illustrate, let $\mathcal{K}$ be the set of all even positive integers. Then, $\{\mathbf{x}_k\}_{\mathcal{K}}$ denotes the subsequence $\{\mathbf{x}_2, \mathbf{x}_4, \mathbf{x}_6, \ldots\}$.

Given a subsequence $\{\mathbf{x}_k\}_{\mathcal{K}}$, the notation $\{\mathbf{x}_{k+1}\}_{\mathcal{K}}$ denotes the subsequence obtained by adding 1 to the indices of all elements in the subsequence $\{\mathbf{x}_k\}_{\mathcal{K}}$. To illustrate, if $\mathcal{K} = \{3, 5, 10, 15, \ldots,\}$, then $\{\mathbf{x}_{k+1}\}_{\mathcal{K}}$ denotes the subsequence $\{\mathbf{x}_4, \mathbf{x}_6, \mathbf{x}_{11}, \mathbf{x}_{16}, \ldots\}$.

A sequence $\{\mathbf{x}_k\}$ is called a *Cauchy sequence* if, for any given $\varepsilon > 0$, there is a positive integer N such that $\|\mathbf{x}_k - \mathbf{x}_m\| < \varepsilon$ for all $k, m \geq N$. A sequence in E_n has a limit if and only if it is Cauchy.

Let $\{x_n\}$ be a bounded sequence in E_1. The *limit superior* of $\{x_n\}$, denoted limsup (x_n) or $\overline{\lim}(x_n)$, equals the infimum of all numbers $q \in E_1$ for which at most a finite number of the elements of $\{x_n\}$ (strictly) exceed q. Similarly, the *limit inferior* of $\{x_n\}$ is given by liminf $(x_n) \equiv \underline{\lim}\ (x_n) \equiv \sup\ \{q\text{: at most a finite number of elements of } \{x_n\} \text{ are (strictly) less than } q\}$. A bounded sequence always has a unique $\overline{\lim}$ and $\underline{\lim}$.

Compact Sets

A set S in E_n is said to be *compact* if it is closed and bounded. For every sequence $\{\mathbf{x}_k\}$ in a compact set S, there is a convergent subsequence with a limit in S.

A.4 Functions

A *real-valued function* f defined on a subset S of E_n associates with each point $\mathbf{x}$ in S a real number $f(\mathbf{x})$. The notation $f: S \to E_1$ denotes that the domain of f is S and that the range is a subset of the real numbers. If f is defined everywhere on E_n or if the domain is not important, the notation $f: E_n \to E_1$ is used. A collection of real-valued functions $f_1, \ldots, f_m$ can be viewed as a single *vector function* $\mathbf{f}$ whose jth component is f_j.

Continuous Functions

A function $f: S \to E_1$ is said to be *continuous at* $\bar{\mathbf{x}} \in S$ if, for any given $\varepsilon > 0$, there is a $\delta > 0$ such that $\mathbf{x} \in S$ and $\|\mathbf{x} - \bar{\mathbf{x}}\| < \delta$ imply that $|f(\mathbf{x}) - f(\bar{\mathbf{x}})| < \varepsilon$. Equivalently, f is continuous at $\bar{\mathbf{x}} \in S$, if, for any sequence $\{\mathbf{x}_n\} \to \bar{\mathbf{x}}$ such that $\{f(\mathbf{x}_n)\} \to \bar{f}$, we have $f(\bar{\mathbf{x}}) = \bar{f}$ as well. A vector-valued function is said to be continuous at $\bar{\mathbf{x}}$ if each of its components is continuous at $\bar{\mathbf{x}}$.

Upper and Lower Semicontinuity

Let S be a nonempty set in E_n. A function $f: S \to E_1$ is said to be *upper semicontinuous* at $\bar{\mathbf{x}} \in S$, if for each $\varepsilon > 0$ there exists a $\delta > 0$ such that $\mathbf{x} \in S$ and $\|\mathbf{x} - \bar{\mathbf{x}}\| < \delta$ imply that $f(\mathbf{x}) - f(\bar{\mathbf{x}}) < \varepsilon$. Similarly, a function $f: E_n \to E_1$ is called *lower semicontinuous* at $\bar{\mathbf{x}} \in S$, if for each $\varepsilon > 0$ there exists a $\delta > 0$ such that $\mathbf{x} \in S$ and $\|\mathbf{x} - \bar{\mathbf{x}}\| < \delta$ imply that $f(\mathbf{x}) - f(\bar{\mathbf{x}}) > -\varepsilon$. Equivalently, f is *upper semicontinuous* at $\bar{\mathbf{x}} \in S$, if, for any sequence $\{\mathbf{x}_n\} \to \bar{\mathbf{x}}$ such that $\{f(\mathbf{x}_n)\} \to \bar{f}$, we have $f(\bar{\mathbf{x}}) \geq \bar{f}$. Similarly, if $f(\bar{\mathbf{x}}) \leq \bar{f}$ for any such sequence, then f is said to be *lower semicontinuous* at $\bar{\mathbf{x}}$. Hence, a function is *continuous* at $\bar{\mathbf{x}}$ if and only if it is both upper and lower semicontinuous at $\bar{\mathbf{x}}$. A vector-valued function is called upper or lower semicontinuous if each of its components is upper or lower semicontinuous, respectively.

Minima and Maxima of Semicontinuous Functions

Let S be a nonempty compact set in E_n and suppose that $f: E_n \to E_1$. If f is lower semicontinuous, then it assumes a minimum over S; that is, there exists an $\bar{\mathbf{x}} \in S$ such that $f(\mathbf{x}) \geq f(\bar{\mathbf{x}})$ for each $\mathbf{x} \in S$. Similarly, if f is upper semicontinuous, then it assumes a maximum over S. Since a continuous function is both lower and upper semicontinuous, it achieves both a minimum and a maximum over any nonempty compact set.

Differentiable Functions

Let S be a nonempty set in E_n, $\bar{\mathbf{x}} \in \text{int } S$ and let $f : S \to E_1$. Then, f is said to be *differentiable* at $\bar{\mathbf{x}}$ if there is a vector $\nabla f(\bar{\mathbf{x}})$ in E_n called the *gradient* of f at $\bar{\mathbf{x}}$ and a function β satisfying $\beta(\bar{\mathbf{x}}; \mathbf{x}) \to 0$ as $\mathbf{x} \to \bar{\mathbf{x}}$ such that

$$f(\mathbf{x}) = f(\bar{\mathbf{x}}) + \nabla f(\bar{\mathbf{x}})^t(\mathbf{x} - \bar{\mathbf{x}}) + \|\mathbf{x} - \bar{\mathbf{x}}\| \, \beta(\bar{\mathbf{x}}; \mathbf{x}) \qquad \text{for each } \mathbf{x} \in S$$

The gradient vector consists of the partial derivatives, that is,

$$\nabla f(\bar{\mathbf{x}})^t = \left(\frac{\partial f(\bar{\mathbf{x}})}{\partial x_1}, \frac{\partial f(\bar{\mathbf{x}})}{\partial x_2}, \ldots, \frac{\partial f(\bar{\mathbf{x}})}{\partial x_n} \right)$$

Furthermore, f is called *twice-differentiable* at $\bar{\mathbf{x}}$ if, in addition to the gradient vector, there exist an $n \times n$ symmetric matrix $\mathbf{H}(\bar{\mathbf{x}})$, called the *Hessian matrix* of f at $\bar{\mathbf{x}}$, and a function β satisfying $\beta(\bar{\mathbf{x}}; \mathbf{x}) \to 0$ as $\mathbf{x} \to \bar{\mathbf{x}}$ such that

$$f(\mathbf{x}) = f(\bar{\mathbf{x}}) + \nabla f(\bar{\mathbf{x}})^t(\mathbf{x} - \bar{\mathbf{x}}) + \tfrac{1}{2}(\mathbf{x} - \bar{\mathbf{x}})^t\mathbf{H}(\bar{\mathbf{x}})(\mathbf{x} - \bar{\mathbf{x}}) + \|\mathbf{x} - \bar{\mathbf{x}}\|^2 \, \beta(\bar{\mathbf{x}}; \mathbf{x}) \qquad \text{for each } \mathbf{x} \in S$$

The element in row i and column j of the Hessian matrix is the second partial $\partial^2 f(\bar{\mathbf{x}})/\partial x_i \partial x_j$.

A vector-valued function is differentiable if each of its components is differentiable and is twice-differentiable if each of its components is twice-differentiable.

In particular, for a differentiable vector function $\mathbf{h} : E_n \to E_l$ where $\mathbf{h}(\mathbf{x}) = (h_1(\mathbf{x}), \ldots, h_l(\mathbf{x}))^t$, the *Jacobian* of $\mathbf{h}$, denoted by the gradient notation $\nabla \mathbf{h}(\mathbf{x})$, is given by the $l \times n$ matrix

$$\nabla \mathbf{h}(\mathbf{x}) = \begin{bmatrix} \nabla h_1(\mathbf{x})^t \\ \vdots \\ \nabla h_l(\mathbf{x})^t \end{bmatrix}_{l \times n}$$

whose rows correspond to the transpose of the gradients of $h_1, \ldots, h_l$, respectively.

Mean Value Theorem

Let S be a nonempty open convex set in E_n, and let $f : S \to E_1$ be differentiable. The mean value theorem can be stated as follows. For every $\mathbf{x}_1$ and $\mathbf{x}_2$ in S, we must have

$$f(\mathbf{x}_2) = f(\mathbf{x}_1) + \nabla f(\mathbf{x})^t(\mathbf{x}_2 - \mathbf{x}_1)$$

where $\mathbf{x} = \lambda \mathbf{x}_1 + (1 - \lambda)\mathbf{x}_2$ for some $\lambda \in (0, 1)$.

Taylor's Theorem

Let S be a nonempty open convex set in E_n, and let $f : S \to E_1$ be twice-differentiable. The second-order form of *Taylor's theorem* can be stated as follows. For every $\mathbf{x}_1$ and $\mathbf{x}_2$ in S, we must have

$$f(\mathbf{x}_2) = f(\mathbf{x}_1) + \nabla f(\mathbf{x}_1)^t(\mathbf{x}_2 - \mathbf{x}_1) + \tfrac{1}{2}(\mathbf{x}_2 - \mathbf{x}_1)^t\mathbf{H}(\mathbf{x})(\mathbf{x}_2 - \mathbf{x}_1)$$

where $\mathbf{H}(\mathbf{x})$ is the Hessian of f at $\mathbf{x}$ and $\mathbf{x} = \lambda \mathbf{x}_1 + (1 - \lambda)\mathbf{x}_2$ for some $\lambda \in (0, 1)$.

Appendix B Summary of Convexity, Optimality Conditions, and Duality

This appendix gives a summary of the relevant results from Chapters 2 through 6 on convexity, optimality conditions, and duality. *It is intended to provide the minimal background needed for an adequate coverage of Chapters 8 through 11, excluding convergence analysis.*

B.1 Convex Sets

A set S in E_n is said to be *convex* if, for each $\mathbf{x}_1, \mathbf{x}_2 \in S$, the *line segment* $\lambda\mathbf{x}_1 + (1 - \lambda)\mathbf{x}_2$ for $\lambda \in [0, 1]$ belongs to S. Points of the form $\mathbf{x} = \lambda\mathbf{x}_1 + (1 - \lambda)\mathbf{x}_2$ for $\lambda \in [0, 1]$ are called *convex combinations* of $\mathbf{x}_1$ and $\mathbf{x}_2$.

Figure B.1 illustrates an example of a convex set and an example of a nonconvex set.

We present below some examples of convex sets frequently encountered in mathematical programming.

1. ***Hyperplane.*** $S = \{\mathbf{x} : \mathbf{p}^t\mathbf{x} = \alpha\}$, where $\mathbf{p}$ is a nonzero vector in E_n, called the *normal* to the hyperplane, and α is a scalar.
2. ***Half-space.*** $S = \{\mathbf{x} : \mathbf{p}^t\mathbf{x} \leq \alpha\}$, where $\mathbf{p}$ is a nonzero vector in E_n and α is a scalar.
3. ***Open Half-space.*** $S = \{\mathbf{x} : \mathbf{p}^t\mathbf{x} < \alpha\}$, where $\mathbf{p}$ is a nonzero vector in E_n and α is a scalar.

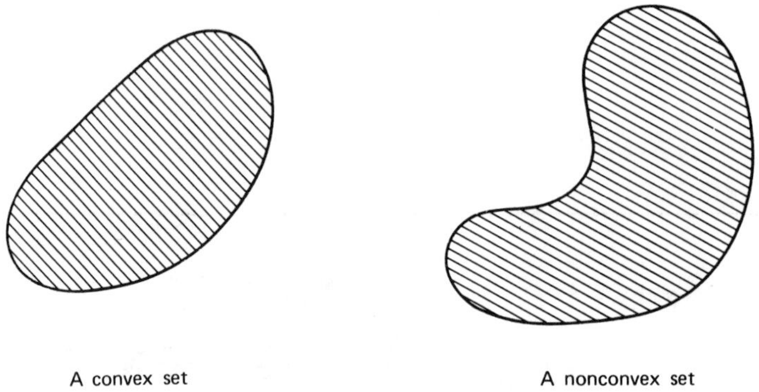

Figure B.1 Illustration of convexity.

4. ***Polyhedral set.*** $S = \{\mathbf{x} : \mathbf{Ax} \leq \mathbf{b}\}$, where $\mathbf{A}$ is an $m \times n$ matrix and $\mathbf{b}$ is an m vector.
5. ***Polyhedral cone.*** $S = \{\mathbf{x} : \mathbf{Ax} \leq \mathbf{0}\}$, where $\mathbf{A}$ is an $m \times n$ matrix.
6. ***Cone spanned by a finite number of vectors.*** $S = \{\mathbf{x} : \mathbf{x} = \sum_{j=1}^{m} \lambda_j \mathbf{a}_j,\ \lambda_j \geq 0$ for $j = 1, \ldots, m\}$, where $\mathbf{a}_1, \ldots, \mathbf{a}_m$ are given vectors in E_n.
7. ***Neighborhood.*** $S = \{\mathbf{x} : \|\mathbf{x} - \bar{\mathbf{x}}\| \leq \varepsilon\}$, where $\bar{\mathbf{x}}$ is a fixed vector in E_n and $\varepsilon > 0$.

Given two nonempty sets S_1 and S_2 in E_n such that $S_1 \cap S_2 = \varnothing$, there exists a hyperplane $H = \{\mathbf{x} : \mathbf{p}'\mathbf{x} = \alpha\}$ that separates them, that is,

$$\mathbf{p}'\mathbf{x} \leq \alpha \text{ for all } \mathbf{x} \in S_1 \qquad \text{and} \qquad \mathbf{p}'\mathbf{x} \geq \alpha \text{ for all } \mathbf{x} \in S_2.$$

Here, H is called a *separating hyperplane* whose normal is the nonzero vector $\mathbf{p}$.

Closely related to the above concept is the notion of a *supporting hyperplane*. Let S be a nonempty convex set in E_n, and let $\bar{\mathbf{x}}$ be a boundary point. Then, there exists a hyperplane $H = \{\mathbf{x} : \mathbf{p}'\mathbf{x} = \alpha\}$ that supports S at $\bar{\mathbf{x}}$, that is,

$$\mathbf{p}'\bar{\mathbf{x}} = \alpha \qquad \text{and} \qquad \mathbf{p}'\mathbf{x} \leq \alpha \text{ for all } \mathbf{x} \in S$$

In Figure B.2, the separating and supporting hyperplanes are illustrated.

The following two theorems are used in proving optimality conditions and duality relationships and in developing termination criteria for algorithms.

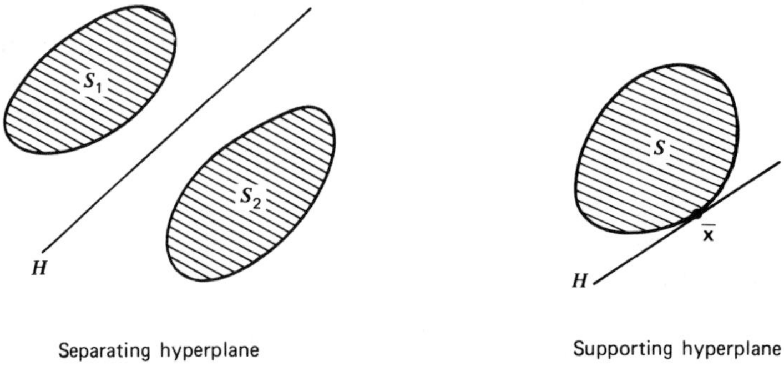

Figure B.2 Illustration of separating and supporting hyperplanes.

Farkas' Theorem

Let $\mathbf{A}$ be an $m \times n$ matrix and $\mathbf{c}$ be an n vector. Then, exactly one of the following two systems has a solution:

System 1 $\mathbf{Ax} \leq \mathbf{0}$, $\mathbf{c}^t\mathbf{x} > \mathbf{0}$ for some $\mathbf{x} \in E_n$

System 2 $\mathbf{A}^t\mathbf{y} = \mathbf{c}$, $\mathbf{y} \geq \mathbf{0}$ for some $\mathbf{y} \in E_m$

Gordan's Theorem

Let $\mathbf{A}$ be an $m \times n$ matrix. Then, exactly one of the following systems has a solution.

System 1 $\mathbf{Ax} < \mathbf{0}$ for some $\mathbf{x} \in E_n$

System 2 $\mathbf{A}^t\mathbf{y} = \mathbf{0}$, $\mathbf{y} \geq \mathbf{0}$ for some $\mathbf{y} \in E_m$

An important concept in convexity is that of an extreme point. Let S be a nonempty convex set in E_n. A vector $\mathbf{x} \in S$ is called an *extreme point* of S if $\mathbf{x} = \lambda\mathbf{x}_1 + (1 - \lambda)\mathbf{x}_2$ with $\mathbf{x}_1, \mathbf{x}_2 \in S$ and $\lambda \in (0, 1)$ implies that $\mathbf{x} = \mathbf{x}_1 = \mathbf{x}_2$. In other words, $\mathbf{x}$ is an extreme point if it cannot be represented as a strict convex combination of two distinct points in S. In particular, if the set $S = \{\mathbf{x} : \mathbf{Ax} = \mathbf{b}, \mathbf{x} \geq \mathbf{0}\}$, where $\mathbf{A}$ is an $m \times n$ matrix of rank m and $\mathbf{b}$ is an m vector, then $\mathbf{x}$ is an extreme point of S if and only if the following conditions hold. The matrix $\mathbf{A}$ can be decomposed into $[\mathbf{B}, \mathbf{N}]$, where $\mathbf{B}$ is an $m \times m$ invertible matrix, and $\mathbf{x}^t = (\mathbf{x}_B^t, \mathbf{x}_N^t)$, where $\mathbf{x}_B = \mathbf{B}^{-1}\mathbf{b} \geq \mathbf{0}$ and $\mathbf{x}_N = \mathbf{0}$.

Another concept that is used in the case of an unbounded convex set is that of a direction of the set. Specifically, if S is an unbounded closed convex set, then a vector $\mathbf{d}$ is a *direction* of S if $\mathbf{x} + \lambda\mathbf{d} \in S$ for each $\lambda \geq 0$ and for each $\mathbf{x} \in S$.

B.2 Convex Functions and Extensions

Let S be a nonempty convex set in E_n. The function $f : S \to E_1$ is said to be *convex* on S if

$$f[\lambda\mathbf{x}_1 + (1 - \lambda)\mathbf{x}_2] \leq \lambda f(\mathbf{x}_1) + (1 - \lambda)f(\mathbf{x}_2)$$

for each $\mathbf{x}_1, \mathbf{x}_2 \in S$ and for each $\lambda \in [0, 1]$. The function f is said to be *strictly convex* on S if the above inequality holds as a strict inequality for each distinct $\mathbf{x}_1, \mathbf{x}_2 \in S$ and for each $\lambda \in (0, 1)$. The function f is said to be *concave (strictly concave)* if $-f$ is convex (strictly convex).

Figure B.3 shows some examples of convex and concave functions.

The following are some examples of convex functions. By taking the negatives of these functions, we get some examples of concave functions.

1. $f(x) = 3x + 4$
2. $f(x) = |x|$
3. $f(x) = x^2 - 2x$
4. $f(x) = -x^{1/2}$ if $x \geq 0$
5. $f(x_1, x_2) = 2x_1^2 + x_2^2 - 2x_1x_2$
6. $f(x_1, x_2, x_3) = x_1^4 + 2x_2^2 + 3x_3^2 - 4x_1 - 4x_2x_3$

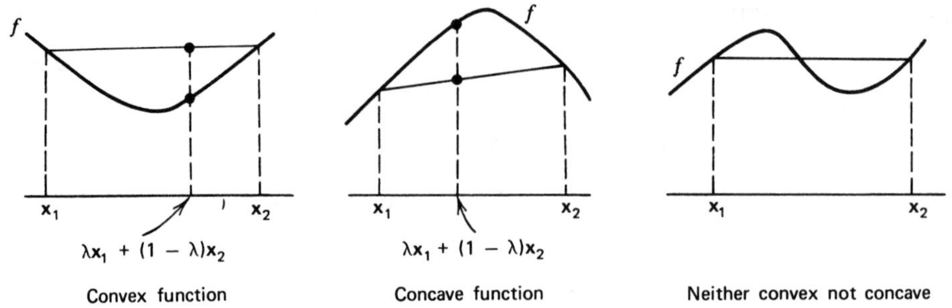

Figure B.3 Examples of convex and concave functions.

In many cases, the assumption of convexity of a function can be relaxed to the weaker notions of quasiconvex and pseudoconvex functions.

Let S be a nonempty convex set in E_n. The function $f: S \to E_1$ is said to be *quasiconvex* on S if, for each $\mathbf{x}_1, \mathbf{x}_2 \in S$, the following inequality holds:

$$f[\lambda \mathbf{x}_1 + (1-\lambda)\mathbf{x}_2] \leq \text{maximum } \{f(\mathbf{x}_1), f(\mathbf{x}_2)\} \qquad \text{for each } \lambda \in (0, 1)$$

The function f is said to be *strictly quasiconvex* on S if the above inequality holds as a strict inequality, provided that $f(\mathbf{x}_1) \neq f(\mathbf{x}_2)$. The function f is said to be *strongly quasiconvex* on S if the above inequality holds as a strict inequality for $\mathbf{x}_1 \neq \mathbf{x}_2$.

Let S be a nonempty open convex set in E_n. The function $f: S \to E_1$ is said to be *pseudoconvex* if, for each $\mathbf{x}_1, \mathbf{x}_2 \in S$ with $\nabla f(\mathbf{x}_1)^t(\mathbf{x}_2 - \mathbf{x}_1) \geq 0$, we must have $f(\mathbf{x}_2) \geq f(\mathbf{x}_1)$. The function f is said to be *strictly pseudoconvex* on S if, whenever $\mathbf{x}_1$ and $\mathbf{x}_2$ are distinct points in S with $\nabla f(\mathbf{x}_1)^t(\mathbf{x}_2 - \mathbf{x}_1) \geq 0$, we have $f(\mathbf{x}_2) > f(\mathbf{x}_1)$.

The above generalizations of convexity extend to the concave case by replacing f above by $-f$. Figure B.4 illustrates the above concepts. Figure B.5 summarizes the relationships among different types of convexity.

We now give a summary of important properties for various types of convex functions. Here, $f: S \to E_1$, where S is a nonempty convex set in E_n.

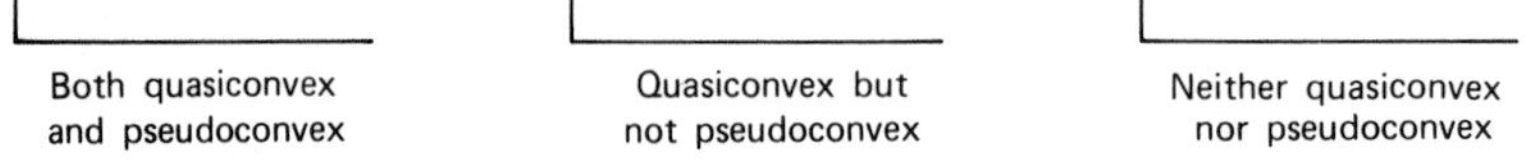

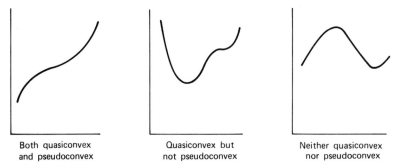

Figure B.4 Illustration of quasiconvexity and pseudoconvexity.

Strictly Convex Functions

1. The function f is continuous on the interior of S.
2. The set $\{(\mathbf{x}, y) : \mathbf{x} \in S, y \geq f(\mathbf{x})\}$ is convex.
3. The set $\{\mathbf{x} \in S : f(\mathbf{x}) \leq \alpha\}$ is convex for each real α.
4. A diferentiable function f is strictly convex on S if and only if $f(\mathbf{x}) > f(\bar{\mathbf{x}}) + \nabla f(\bar{\mathbf{x}})^t(\mathbf{x} - \bar{\mathbf{x}})$ for each distinct $\mathbf{x}, \bar{\mathbf{x}} \in S$.

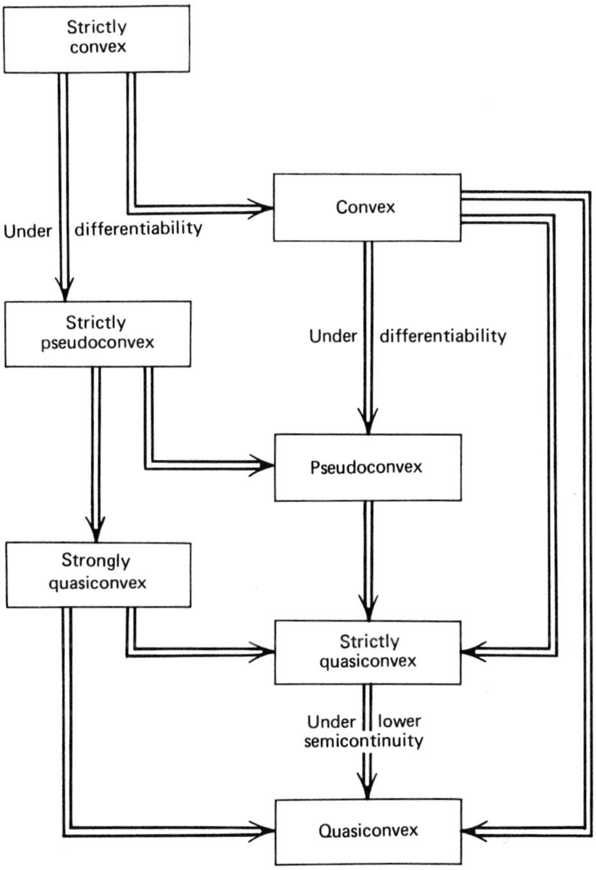

Figure B.5 Relationship among various types of convexity.

5. Let f be twice-differentiable. Then, if the Hessian $\mathbf{H}(\mathbf{x})$ is positive definite for each $\mathbf{x} \in S$, f is strictly convex on S. Further, if f is strictly convex on S, then the Hessian $\mathbf{H}(\mathbf{x})$ is positive semidefinite for each $\mathbf{x} \in S$.
6. Every local minimum of f over a convex set $X \subseteq S$ is the unique global minimum.
7. If $\nabla f(\bar{\mathbf{x}}) = \mathbf{0}$, then $\bar{\mathbf{x}}$ is the unique global minimum of f over S.
8. The maximum of f over a nonempty compact polyhedral set $X \subseteq S$ is achieved at an extreme point of X.

Convex Functions

1. The function f is continuous on the interior of S.
2. The function f is convex on S if and only if the set $\{(\mathbf{x}, y) : \mathbf{x} \in S, y \geq f(\mathbf{x})\}$ is convex.
3. The set $\{\mathbf{x} \in S : f(\mathbf{x}) \leq \alpha\}$ is convex for each real α.
4. A differentiable function f is convex on S if and only if $f(\mathbf{x}) \geq f(\bar{\mathbf{x}}) + \nabla f(\bar{\mathbf{x}})^t(\mathbf{x} - \bar{\mathbf{x}})$ for each $\mathbf{x}, \bar{\mathbf{x}} \in S$.
5. A twice-differentiable function f is convex on S if and only if the Hessian $\mathbf{H}(\mathbf{x})$ is positive semidefinite for each $\mathbf{x} \in S$.

6. Every local minimum of f over a convex set $X \subseteq S$ is a global minimum.
7. If $\nabla f(\bar{\mathbf{x}}) = \mathbf{0}$, then $\bar{\mathbf{x}}$ is a global minimum of f over S.
8. A maximum of f over a nonempty compact polyhedral set $X \subseteq S$ is achieved at an extreme point of X.

Pseudoconvex Functions

1. The set $\{\mathbf{x} \in S : f(\mathbf{x}) \leq \alpha\}$ is convex for each real α.
2. Every local minimum of f over a convex set $X \subseteq S$ is a global minimum.
3. If $\nabla f(\bar{\mathbf{x}}) = \mathbf{0}$, then $\bar{\mathbf{x}}$ is a global minimum of f over S.
4. A maximum of f over a nonempty compact polyhedral set $X \subseteq S$ is achieved at an extreme point of X.
5. This characterization and the next relate to twice-differentiable functions f defined on an open convex set $S \subseteq E_n$, with Hessian $\mathbf{H}(\mathbf{x})$. f is pseudoconvex on S if $\mathbf{H}(\mathbf{x}) + r(\mathbf{x})\nabla f(\mathbf{x})\nabla f(\mathbf{x})^t$ is positive semidefinite for all $\mathbf{x} \in S$, where $r(\mathbf{x}) = \frac{1}{2}[\delta - f(\mathbf{x})]$ for some $\delta > f(\mathbf{x})$. Moreover, this condition is both necessary and sufficient if f is quadratic.
6. Define the $(n + 1) \times (n + 1)$ *bordered Hessian* $\mathbf{B}(\mathbf{x})$ of f as follows, where $\mathbf{H}(\mathbf{x})$ is "bordered" by an additional row and column:

$$\mathbf{B}(\mathbf{x}) = \begin{bmatrix} \mathbf{H}(\mathbf{x}) & \nabla f(\mathbf{x}) \\ \nabla f(\mathbf{x})^t & 0 \end{bmatrix}$$

Given any $k \in \{1, \ldots, n\}$, and $\gamma = \{i_1, \ldots, i_k\}$ composed of some k distinct indices $1 \leq i_1 < i_2 < \ldots < i_k \leq n$, the *principal submatrix* $\mathbf{B}_{\gamma,k}(\mathbf{x})$ is a $(k + 1) \times (k + 1)$ submatrix of $\mathbf{B}(\mathbf{x})$ formed by picking the elements of $\mathbf{B}(\mathbf{x})$ that intersect in the rows $i_1, \ldots, i_k, (n + 1)$ and the columns $i_1, \ldots, i_k, (n + 1)$ of $\mathbf{B}(\mathbf{x})$. The *leading principal submatrix* of $\mathbf{B}(\mathbf{x})$ is denoted by $\mathbf{B}_k(\mathbf{x})$ and equals $\mathbf{B}_{\gamma,k}$ for $\gamma \equiv \{1, \ldots, k\}$. Similarly, let $\mathbf{H}_{\gamma,k}(\mathbf{x})$ and $\mathbf{H}_k(\mathbf{x})$ be the $k \times k$ principal submatrix and the leading principal submatrix, respectively, of $\mathbf{H}(\mathbf{x})$. Then, f is pseudoconvex on S if, for each $\mathbf{x} \in S$, we have (i) det $\mathbf{B}_{\gamma,k}(\mathbf{x}) \leq 0$ for all γ, $k = 1, \ldots, n$ *and* (ii) if det $\mathbf{B}_{\gamma,k}(\mathbf{x}) = 0$ for any γ, k, then det $\mathbf{H}_{\gamma,k} \geq 0$. Moreover, if f is quadratic, then these conditions are both necessary and sufficient. Also, the condition det $\mathbf{B}_k(\mathbf{x}) < 0$ for all $k = 1, \ldots, n$, $\mathbf{x} \in S$ is sufficient for f to be pseudoconvex on S.
7. Let $f : S \subseteq E_n \rightarrow E_1$ be quadratic, where S is a convex subset of E_n. Then, [f is pseudoconvex on S] $\Leftrightarrow$ [the bordered Hessian $\mathbf{B}(\mathbf{x})$ has exactly one simple negative eigenvalue for all $\mathbf{x} \in S$] $\Leftrightarrow$ [for each $\mathbf{y} \in E_n$ such that $\nabla f(\mathbf{x})^t\mathbf{y} = 0$, we have $\mathbf{y}^t\mathbf{H}(\mathbf{x})\mathbf{y} \geq 0$ for all $\mathbf{x} \in S$]. Moreover, [f is strictly pseudoconvex on S] $\Leftrightarrow$ [for all $\mathbf{x} \in S$, and for all $k = 1, \ldots, n$, we have (i) det $\mathbf{B}_k(\mathbf{x}) \leq 0$ *and* (ii) if det $\mathbf{B}_k(\mathbf{x}) = 0$, then det $\mathbf{H}_k > 0$].

Quasiconvex Functions

1. The function f is quasiconvex over S if and only if $\{\mathbf{x} \in S : f(\mathbf{x}) \leq \alpha\}$ is convex for each real α.
2. A maximum of f over a nonempty compact polyhedral set $X \subseteq S$ is achieved at an extreme point of X.
3. A differentiable function f on S is quasiconvex over S if and only if $\mathbf{x}_1, \mathbf{x}_2 \in S$ with $f(\mathbf{x}_1) \leq f(\mathbf{x}_2)$ implies that $\nabla f(\mathbf{x}_2)^t(\mathbf{x}_1 - \mathbf{x}_2) \leq 0$.

4. Let $f : S \subseteq E_n \rightarrow E_1$, where f is twice-differentiable and S is a *solid* (i.e., having a nonempty interior) convex subset of E_n. Define the bordered Hessian of f and its submatrices, as in property 6 of pseudoconvex functions. Then, a sufficient condition for f to be quasiconvex on S is that for each $\mathbf{x} \in S$, det $\mathbf{B}_k(\mathbf{x}) < 0$ for all $k = 1, \ldots, n$. (Note that this condition actually implies that f is pseudoconvex.) On the other hand, a necessary condition for f to be quasiconvex on S is that, for each $\mathbf{x} \in S$, det $\mathbf{B}_k(\mathbf{x}) \leq 0$ for all $k = 1, \ldots, n$.
5. Let $f : S \subseteq E_n \rightarrow E_1$ be a quadratic function where $S \subseteq E_n$ is a solid (nonempty interior) convex subset of E_n. Then, f is quasiconvex on S if and only if f is pseudoconvex on int (S).

A local minimum of a strictly quasiconvex function over a convex set $X \subseteq S$ is also a global minimum. Furthermore, if the function is strongly quasiconvex then the minimum is unique. If a function f is both strictly quasiconvex and lower semicontinuous, then it is quasiconvex, so that the above properties hold.

B.3 Optimality Conditions

Consider the following problem:

Problem P: Minimize $f(\mathbf{x})$

subject to
$$g_i(\mathbf{x}) \leq 0 \quad \text{for } i = 1, \ldots, m$$
$$h_i(\mathbf{x}) = 0 \quad \text{for } i = 1, \ldots, l$$
$$\mathbf{x} \in X$$

where f, g_i, $h_i : E_n \rightarrow E_1$ and X is a nonempty open set in E_n. We give below the *Fritz John necessary optimality conditions*. If a point $\bar{\mathbf{x}}$ is a local optimal solution to the above problem, then there must exist a nonzero vector $(u_0, \mathbf{u}, \mathbf{v})$ such that

$$u_0\nabla f(\bar{\mathbf{x}}) + \sum_{i=1}^{m} u_i \nabla g_i(\bar{\mathbf{x}}) + \sum_{i=1}^{l} v_i \nabla h_i(\bar{\mathbf{x}}) = \mathbf{0}$$

$$u_i g_i(\bar{\mathbf{x}}) = 0 \qquad \text{for } i = 1, \ldots, m$$

$$u_0 \geq 0, u_i \geq 0 \qquad \text{for } i = 1, \ldots, m$$

where $\mathbf{u}$ and $\mathbf{v}$ are m and l vectors whose ith components are u_i and v_i, respectively. Here, u_0, u_i, and v_i are referred to as the *Lagrange* or *Lagrangian multipliers* associated respectively with the objective function, the ith inequality constraint $g_i(\mathbf{x}) \leq 0$, and the ith equality constraint $h_i(\mathbf{x}) = 0$. The condition $u_i g_i(\bar{\mathbf{x}}) = 0$ is called the *complementary slackness condition* and stipulates that either $u_i = 0$ or $g_i(\bar{\mathbf{x}}) = 0$. Thus, if $g_i(\bar{\mathbf{x}}) < 0$, then $u_i = 0$. By letting I be the set of binding inequality constraints at $\bar{\mathbf{x}}$, that is, $I = \{i : g_i(\bar{\mathbf{x}}) = 0\}$, the Fritz John conditions could be written in the following equivalent form. If $\bar{\mathbf{x}}$ is a local optimal solution to Problem P above, then there exists a nonzero vector $(u_0, \mathbf{u}_I, \mathbf{v})$ satisfying the following, where $\mathbf{u}_I$ is the vector of Lagrangian multipliers associated with $g_i(\mathbf{x}) \leq 0$ for $i \in I$:

$$u_0\nabla f(\bar{\mathbf{x}}) + \sum_{i \in I} u_i \nabla g_i(\bar{\mathbf{x}}) + \sum_{i=1}^{l} v_i \nabla h_i(\bar{\mathbf{x}}) = \mathbf{0}$$

$$u_0 \geq 0, u_i \geq 0 \qquad \text{for } i \in I$$

If $u_0 = 0$, then the Fritz John conditions become virtually useless, since essentially they merely state that the gradients of the binding inequality constraints and the gradients of the equality constraints are linearly dependent. Under suitable assumptions, referred to as *constraint qualifications*, u_0 is guaranteed to be positive, and the Fritz John conditions reduce to the KKT conditions. A typical constraint qualification is that the gradients of the inequality constraints for $i \in I$ and the gradients of the equality constraints at $\bar{\mathbf{x}}$ are linearly independent.

The KKT necessary optimality conditions can be stated as follows. If $\bar{\mathbf{x}}$ is a local optimal solution to Problem P, and under a suitable constraint qualification, there exists a vector $(\mathbf{u}, \mathbf{v})$ such that

$$\nabla f(\bar{\mathbf{x}}) + \sum_{i=1}^{m} u_i \nabla g_i(\bar{\mathbf{x}}) + \sum_{i=1}^{l} v_i \nabla h_i(\bar{\mathbf{x}}) = \mathbf{0}$$

$$u_i g_i(\bar{\mathbf{x}}) = 0 \qquad \text{for } i = 1, \ldots, m$$

$$u_i \geq 0 \qquad \text{for } i = 1, \ldots, m$$

Again, u_i and v_i are the *Lagrangian multipliers* associated with the constraints $g_i(\mathbf{x}) \leq 0$ and $h_i(\mathbf{x}) = 0$, respectively. Further, $u_i g_i(\bar{\mathbf{x}}) = 0$ is referred to as a *complementary slackness condition*. If we let $I = \{i : g_i(\bar{\mathbf{x}}) = 0\}$, then the above conditions could be rewritten as

$$\nabla f(\bar{\mathbf{x}}) + \sum_{i \in I} u_i \nabla g_i(\bar{\mathbf{x}}) + \sum_{i=1}^{l} v_i \nabla h_i(\bar{\mathbf{x}}) = \mathbf{0}$$

$$u_i \geq 0 \qquad \text{for } i \in I$$

Under suitable convexity assumptions, the KKT conditions are also *sufficient* for optimality. In particular, suppose that $\bar{\mathbf{x}}$ is a feasible solution to Problem P and that the KKT conditions stated below hold:

$$\nabla f(\bar{\mathbf{x}}) + \sum_{i \in I} u_i \nabla g_i(\bar{\mathbf{x}}) + \sum_{i=1}^{l} v_i \nabla h_i(\bar{\mathbf{x}}) = \mathbf{0}$$

$$u_i \geq 0 \qquad \text{for } i \in I$$

where $I = \{i : g_i(\bar{\mathbf{x}}) = 0\}$. If f is pseudoconvex, g_i is quasiconvex for $i \in I$; and if h_i is quasiconvex if $v_i > 0$ and quasiconcave if $v_i < 0$, then $\bar{\mathbf{x}}$ is an optimal solution to Problem P.

To illustrate the KKT conditions, consider the following problem:

$$\begin{array}{ll} \text{Minimize} & (x_1 - 3)^2 + (x_2 - 2)^2 \\ \text{subject to} & x_1^2 + x_2^2 \leq 5 \\ & x_1 + 2x_2 \leq 4 \\ & -x_1 \leq 0 \\ & -x_2 \leq 0 \end{array}$$

The problem is illustrated in Figure B.6. Note that the optimal point is $\bar{\mathbf{x}} = (2, 1)^t$. We first verify that the KKT conditions hold at $\bar{\mathbf{x}}$. Here, the set of binding inequality constraints is $I = \{1, 2\}$, so that $u_3 = u_4 = 0$ to satisfy the complementary slackness conditions. Note that

$$\nabla f(\bar{\mathbf{x}}) = (-2, -2)^t \qquad \nabla g_1(\bar{\mathbf{x}}) = (4, 2)^t \qquad \nabla g_2(\bar{\mathbf{x}}) = (1, 2)^t$$

Thus, $\nabla f(\bar{\mathbf{x}}) + u_1 \nabla g_1(\bar{\mathbf{x}}) + u_2 \nabla g_2(\bar{\mathbf{x}}) = \mathbf{0}$ holds by letting $u_1 = \frac{1}{3}$ and $u_2 = \frac{2}{3}$, so that

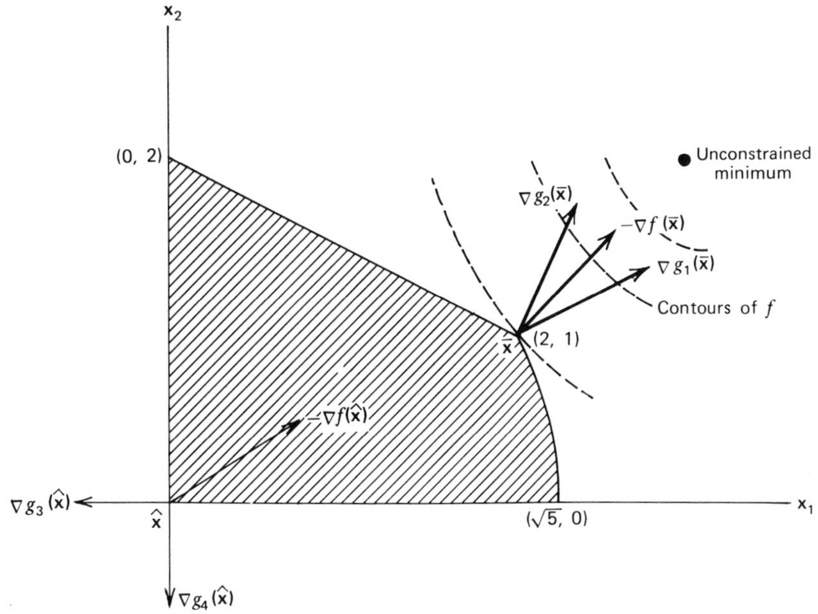

Figure B.6 Illustration of the KKT conditions.

the KKT conditions are satisfied at $\bar{\mathbf{x}}$. Noting that f, g_1, and g_2 are convex, then $\bar{\mathbf{x}}$ is indeed an optimal solution by sufficiency of the KKT conditions.

Now let us check whether the KKT conditions hold at the point $\hat{\mathbf{x}} = (0, 0)^t$. Here, $I = \{3, 4\}$, so that $u_1 = u_2 = 0$ to satisfy complementary slackness. Note that

$$\nabla f(\hat{\mathbf{x}}) = (-6, -4)^t \qquad \nabla g_3(\hat{\mathbf{x}}) = (-1, 0)^t \qquad \nabla g_4(\hat{\mathbf{x}}) = (0, -1)^t$$

Thus, $\nabla f(\hat{\mathbf{x}}) + u_3\nabla g_3(\hat{\mathbf{x}}) + u_4\nabla g_4(\hat{\mathbf{x}}) = \mathbf{0}$ holds true by letting $u_3 = -6$ and $u_4 = -4$, violating the nonnegativity of the Lagrangian multipliers. This shows that $\hat{\mathbf{x}}$ is not a KKT point and, hence, could not be a candidate for an optimal solution.

In Figure B.6, the gradients of the objective function and the binding constraints are illustrated for both $\bar{\mathbf{x}}$ and $\hat{\mathbf{x}}$. Note that $-\nabla f(\bar{\mathbf{x}})$ lies in the cone spanned by the gradients of the binding constraints at $\bar{\mathbf{x}}$, whereas $-\nabla f(\hat{\mathbf{x}})$ does not lie in the corresponding cone. Indeed, the KKT conditions for a problem with inequality constraints could be interpreted geometrically as follows. A vector $\bar{\mathbf{x}}$ is a KKT point if and only if $-\nabla f(\bar{\mathbf{x}})$ lies in the cone spanned by the gradients of the binding constraints at $\bar{\mathbf{x}}$.

Let Problem P be as defined above, where all objective and constraint functions are continuously twice-differentiable, and let $\bar{\mathbf{x}}$ be a KKT solution with associated Lagrange multipliers $(\bar{\mathbf{u}}, \bar{\mathbf{v}})$. Define the (restricted) Lagrangian function $L(\mathbf{x}) = f(\mathbf{x}) + \bar{\mathbf{u}}^t\mathbf{g}(\mathbf{x}) + \bar{\mathbf{v}}^t\mathbf{h}(\mathbf{x})$ and let $\nabla^2 L(\bar{\mathbf{x}})$ denote its Hessian at $\bar{\mathbf{x}}$. Let C denote the cone $\{\mathbf{d} : \nabla g_i(\bar{\mathbf{x}})^t\mathbf{d} = 0$ for all $i \in I^+$, $\nabla g_i(\bar{\mathbf{x}})^t\mathbf{d} \leq 0$ for all $i \in I^0$, and $\nabla h_i(\bar{\mathbf{x}})^t\mathbf{d} = 0$ for all $i = 1, \ldots, l\}$, where $I^+ = \{i \in \{1, \ldots, m\} : \bar{u}_i > 0\}$, and $I^0 = \{1, \ldots, m\} - I^+$. Then, we have the following *second-order sufficient conditions* holding: If $\nabla^2 L(\bar{\mathbf{x}})$ is positive definite on C, that is, $\mathbf{d}^t\nabla^2 L(\bar{\mathbf{x}})\,\mathbf{d} > 0$ for all $\mathbf{d} \in C$, $\mathbf{d} \neq \mathbf{0}$, then $\bar{\mathbf{x}}$ is a strict local minimum for Problem P.

We also remark that if $\nabla^2 L(\mathbf{x})$ is positive semidefinite for all feasible $\mathbf{x}$ (respectively for all feasible $\mathbf{x}$ in $N_\varepsilon(\bar{\mathbf{x}})$ for some $\varepsilon > 0$), then $\bar{\mathbf{x}}$ is a global (local) minimum for P. Conversely, suppose that $\bar{\mathbf{x}}$ is a local minimum for P, and let the gradients $\nabla g_i(\bar{\mathbf{x}})$,

$i \in I$, $\nabla h_i(\bar{\mathbf{x}})$, $i = 1, \ldots, l$ be linearly independent, where $I = \{i \in \{1, \ldots, m\} : g_i(\bar{\mathbf{x}}) = 0\}$. Define $C' = \{\mathbf{d} \neq \mathbf{0} : \nabla g_i(\bar{\mathbf{x}})^t\mathbf{d} \leq 0$ for all $i \in I$, $\nabla h_i(\bar{\mathbf{x}})^t\mathbf{d} = 0$ for all $i = 1, \ldots, l\}$. Then, $\bar{\mathbf{x}}$ is a KKT point with associated Lagrange multipliers $(\bar{\mathbf{u}}, \bar{\mathbf{v}})$. Moreover, defining the (restricted) Lagrangian function $L(\mathbf{x}) = f(\mathbf{x}) + \bar{\mathbf{u}}^t\mathbf{g}(\mathbf{x}) + \bar{\mathbf{v}}^t\mathbf{h}(\mathbf{x})$, the *second-order necessary condition* is that $\nabla^2 L(\bar{\mathbf{x}})$ is positive semidefinite on C'.

B.4 Lagrangian Duality

Given a nonlinear programming problem, called the *primal problem*, there exists a problem that is closely associated with it, called the *Lagrangian dual problem*. These two problems are given below.

Primal Problem P:

$$\begin{array}{lll} \text{Minimize} & f(\mathbf{x}) & \\ \text{subject to} & g_i(\mathbf{x}) \leq 0 & \text{for } i = 1, \ldots, m \\ & h_i(\mathbf{x}) = 0 & \text{for } i = 1, \ldots, l \\ & \mathbf{x} \in X & \end{array}$$

where $f, g_i, h_i : E_n \to E_1$, and X is a nonempty set in E_n. Let $\mathbf{g}$ and $\mathbf{h}$ be the m and l vector functions whose ith components are, respectively, g_i and h_i.

Lagrangian Dual Problem D:

$$\begin{array}{ll} \text{Maximize} & \theta(\mathbf{u}, \mathbf{v}) \\ \text{subject to} & \mathbf{u} \geq \mathbf{0} \end{array}$$

where $\theta(\mathbf{u}, \mathbf{v}) = \inf \{f(\mathbf{x}) + \sum_{i=1}^m u_i g_i(\mathbf{x}) + \sum_{i=1}^l v_i h_i(\mathbf{x}) : \mathbf{x} \in X\}$. Here, the vectors $\mathbf{u}$ and $\mathbf{v}$ belong to E_m and E_l, respectively. The ith component u_i of $\mathbf{u}$ is referred to as the dual variable or Lagrangian multiplier associated with the constraint $g_i(\mathbf{x}) \leq 0$, and the ith component v_i of $\mathbf{v}$ is referred to as the dual variable or Lagrangian multiplier associated with the constraint $h_i(\mathbf{x}) = 0$. It may be noted that θ is a *concave* function, even in the absence of any convexity or concavity assumptions on f, g_i or h_i, or convexity of the set X.

We summarize below some important relationships between the primal and dual problems:

1. If $\mathbf{x}$ is feasible to Problem P and if $(\mathbf{u}, \mathbf{v})$ is feasible to Problem D, then $f(\mathbf{x}) \geq \theta(\mathbf{u}, \mathbf{v})$. Thus,

$$\inf \{f(\mathbf{x}) : \mathbf{g}(\mathbf{x}) \leq \mathbf{0}, \mathbf{h}(\mathbf{x}) = \mathbf{0}, \mathbf{x} \in X\} \geq \sup \{\theta(\mathbf{u}, \mathbf{v}) : \mathbf{u} \geq \mathbf{0}\}$$

   This result is called the *weak duality theorem*.
2. If $\sup \{\theta(\mathbf{u}, \mathbf{v}) : \mathbf{u} \geq \mathbf{0}\} = \infty$, then there exists no point $\mathbf{x} \in X$ such that $\mathbf{g}(\mathbf{x}) \leq \mathbf{0}$ and $\mathbf{h}(\mathbf{x}) = \mathbf{0}$, so that the primal problem is infeasible.
3. If $\inf \{f(\mathbf{x}) : \mathbf{g}(\mathbf{x}) \leq \mathbf{0}, \mathbf{h}(\mathbf{x}) = \mathbf{0}, \mathbf{x} \in X\} = -\infty$, then $\theta(\mathbf{u}, \mathbf{v}) = -\infty$ for each $(\mathbf{u}, \mathbf{v})$ with $\mathbf{u} \geq \mathbf{0}$.
4. If there exists a feasible $\mathbf{x}$ to the primal problem and a feasible $(\mathbf{u}, \mathbf{v})$ to the dual problem such that $f(\mathbf{x}) = \theta(\mathbf{u}, \mathbf{v})$, then $\mathbf{x}$ is an optimal solution to Problem P, and $(\mathbf{u}, \mathbf{v})$ is an optimal solution to Problem D. Furthermore, the complementary slackness condition $u_i g_i(\mathbf{x}) = 0$ for $i = 1, \ldots, m$ holds.
5. Suppose that X is convex, that $f, g_i : E_n \to E_1$ for $i = 1, \ldots, m$ are convex, and that $\mathbf{h}$ is of the form $\mathbf{h}(\mathbf{x}) = \mathbf{A}\mathbf{x} - \mathbf{b}$, where $\mathbf{A}$ is an $m \times n$ matrix and $\mathbf{b}$ is an m vector. Under a suitable constraint qualification, the optimal objective

values of both Problems P and D are equal, that is,

$$\inf\ \{f(\mathbf{x}) : \mathbf{x} \in X,\ \mathbf{g}(\mathbf{x}) \leq \mathbf{0},\ \mathbf{h}(\mathbf{x}) = \mathbf{0}\} = \sup\ \{\theta(\mathbf{u}, \mathbf{v}) : \mathbf{u} \geq \mathbf{0}\}$$

Further, if the inf is finite, then the sup is achieved at $(\bar{\mathbf{u}}, \bar{\mathbf{v}})$ with $\bar{\mathbf{u}} \geq \mathbf{0}$. Also, if the inf is achieved at $\bar{\mathbf{x}}$, then $\bar{u}_i g_i(\bar{x}) = 0$ for $i = 1, \ldots, m$. This result is referred to as the *strong duality theorem*.

Bibliography

Abadie, J. (Ed.), *Nonlinear Programming*, North Holland Publishing Company, Amsterdam, 1967a.

Abadie, J., "On the Kuhn–Tucker Theorem," in *Nonlinear Programming*, J. Abadie (Ed.), 1967b.

Abadie, J. (Ed.), *Integer and Nonlinear Programming*, North Holland Publishing Company, Amsterdam, 1970a.

Abadie, J., "Application of the GRG Algorithm to Optimal Control," in *Integer and Nonlinear Programming*, J. Abadie (Ed.), 1970b.

Abadie, J., "The GRG Method for Nonlinear Programming," in *Design and Implementation of Optimization Software*, H. J. Greenberg (Ed.), Sijthoff and Noordhoff, Netherlands, pp. 335–362, 1978.

Abadie, J., "Un Nouvel Algorithme pour la Programmation Non-Linéarire," *R.A.I.R.O. Recherche Opérationnelle*, 12(2), pp. 233–238, 1978.

Abadie, J., and J. Carpentier, "Some Numerical Experiments with the GRG Method for Nonlinear Programming," Paper HR 7422, Electricité de France, 1967.

Abadie, J., and J. Carpentier, "Generalization of the Wolfe Reduced Gradient Method to the Case of Nonlinear Constraints," in *Optimization*, R. Fletcher (Ed.), 1969.

Abadie, J., and J. Guigou, "Numerical Experiments with the GRG Method," in *Integer and Nonlinear Programming*, J. Abadie (Ed.), 1970.

Abadie, J., and A. C. Williams, "Dual and Parametric Methods in Decomposition," in *Recent Advances in Mathematical Programming*, R. L. Graves and P. Wolfe (Eds.), 1968.

Abou-Taleb, N., I. Megahed, A. Moussa, and A. Zaky, "A New Approach to the Solution of Economic Dispatch Problems," *Winter Power Meeting*, New York, 1974.

Adachi, N., "On Variable Metric Algorithms," *J. Optimization Theory and Applications*, 7, pp. 391–410, 1971.

Adams, N., F. Beglari, M. A. Laughton, and G. Mitra, "Math Programming Systems in Electrical Power Generation, Transmission and Distribution Planning," in *Proc. 4th Power Systems Computation Conference*, 1972.

Adhigama, S. T., E. Polak, and R. Klessig, "A Comparative Study of Several General Convergence Conditions for Algorithms Modeled by Point-to-set-Maps," in *Point-to-set Maps and Mathematical Programming, Mathematical Programming Study*, 10, P. Huard (Ed.), North Holland, pp. 172–190, 1970.

Adler, I., and Monteiro, R., "An Interior Point Algorithm Applied to a Class of Convex Separable Programming Problems," *TIMS/ORSA National Meeting*, Nashville, Tenn., May 12–15, 1991.

Afriat, S. N., "The Progressive Support Method for Convex Programming," *SIAM J. Numerical Analysis*, 7, pp. 447–457, 1970.
Afriat, S. N., "Theory of Maxima and the Method of Lagrange," *SIAM J. Applied Mathematics*, 20, pp. 343–357, 1971.
Agunwamba, C. C., "Optimality Condition: Constraint Regularization," *Mathematical Programming*, 13, pp. 38–48, 1977.
Akgul, M., "An Algorithmic Proof of the Polyhedral Decomposition Theorem," *Naval Research Logistics Quarterly*, 35, pp. 463–472, 1988.
Al-Baali, M., "Descent Property and Global Convergence of the Fletcher–Reeves Method with Inexact Line Search," *IMA J. Numerical Analysis*, 5, pp. 121–124, 1985.
Al-Baali, M., and R. Fletcher, "An Efficient Line Search for Nonlinear Least Squares," *J. Optimization Theory Applications*, 48, pp. 359–378, 1986.
Al-Khayyal, F., "On Solving Linear Complementarity Problems as Bilinear Programs," *Arabian J. for Science and Engineering*, forthcoming (manuscript 1990).
Al-Khayyal, F. A., "An Implicit Enumeration Procedure for the General Linear Complementarity Problem," *Mathematical Programming Study*, 31, pp. 1–20, 1987.
Al-Khayyal, F. A., "Linear, Quadratic, and Bilinear Programming Approaches to the Linear Complementarity Problem," *European J. of Operational Research*, 24, pp. 216–227, 1986.
Al-Khayyal, F. A., and J. E. Falk, "Jointly Constrained Biconvex Programming," *Mathematics of Operations Research*, 8, pp. 273–286, 1983.
Ali, H. M., A. S. J. Batchelor, E. M. L. Beale, and J. F. Beasley, *Mathematical Models to Help Manage the Oil Resources of Kuwait*, Internal Report, Scientific Control Systems Ltd., 1978.
Allen, E., R. Helgason, J. Kennington, and B. Shetty, "A Generalization of Polyak's Convergence Result from Subgradient Optimization," *Mathematical Programming*, 37, pp. 309–317, 1987.
Almogy, Y., and O. Levin, "A Class of Fractional Programming Problems," *Operations Research*, 19, pp. 57–67, 1971.
Altman, M., "A General Separation Theorem for Mappings, Saddle-Points, Duality, and Conjugate Functions," *Studia Mathematica*, 36, pp. 131–166, 1970.
Anderson, D., "Models for Determining Least-Cost Investments in Electricity Supply," *Bell*, 3, pp. 267–299, 1972.
Anderssen, R. S., L. Jennings, and D. Ryan (Eds.), *Optimization*, University of Queensland Press, St. Lucia, Queensland, 1972.
Anstreicher, K. M., "On Long Step Path Following and SUMT for Linear and Quadratic Programming," Department of Operations Research, Yale University, New Haven, Conn., 1990.
Aoki, M., *Introduction to Optimization Techniques*, Macmillan, New York, 1971.
Argaman Y., D. Shamir, and E. Spivak, "Design of Optimal Sewage Systems," *J. Environmental Engineering Division American Society of Civil Engineers*, 99, pp. 703–716, 1973.
Armijo, L., "Minimization of Functions Having Lipschitz Continuous First-Partial Derivatives," *Pacific J. Mathematics*, 16, 1, pp. 1–3, 1966.
Arrow, K. J., and A. C. Enthoven, "Quasi-Concave Programming," *Econometrica*, 29, pp. 779–800, 1961.
Arrow, K. J., F. J. Gould, and S. M. Howe, "A General Saddle Point Result for Constrained Optimization," *Mathematical Programming*, 5, pp. 225–234, 1973.
Arrow, K. J., L. Hurwicz, and H. Uzawa (Eds.), *Studies in Linear and Nonlinear Programming*, Stanford University Press, Stanford, 1958.
Arrow, K. J., L. Hurwicz, and H. Uzawa, "Constraint Qualifications in Maximization Problems," *Naval Research Logistics Quarterly*, 8, pp. 175–191, 1961.
Arrow, K. J., and H. Uzawa, "Constraint Qualifications in Maximization Problems, II," *Tech. Report No.* 84, Institute of Mathematical Studies in Social Sciences, Stanford, 1960.
Asaadi, J., "A Computational Comparison of Some Nonlinear Programs," *Mathematical Programming*, 4, pp. 144–156, 1973.
Asimov, M., *Introduction to Design*, Prentice Hall, Englewood Cliffs, N.J., 1962.
Aspvall, B., and R. E. Stone, "Khachiyan's Linear Programming Algorithm," *J. of Algorithms*, 1, pp. 1–13, 1980.

Avila, J. H., and P. Concus, "Update Methods for Highly Structured Systems for Nonlinear Equations," *SIAM J. Numerical Analysis*, 16, pp. 260–269, 1979.

Avis, D., and V. Chvatal, "Notes on Bland's Pivoting Rule," *Mathematical Programming*, 8, pp. 24–34, 1978.

Avriel, M., "Fundamentals of Geometric Programming," in *Applications of Mathematical Programming Techniques*, E. M. L. Beale (Ed.), 1970.

Avriel, M., "r-Convex Functions," *Mathematical Programming*, 2, pp. 309–323, 1972.

Avriel, M., "Solution of Certain Nonlinear Programs Involving r-Convex Functions," *J. Optimization Theory and Applications*, 11, pp. 159–174, 1973.

Avriel, M., *Nonlinear Programming: Analysis and Methods*, Prentice Hall, Englewood Cliffs, N.J., 1976.

Avriel, M., and R. S. Dembo (Eds.), "Engineering Optimization," *Mathematical Programming Study*, 11, 1979.

Avriel, M., W. E. Diewert, S. Schaible, and I. Zang, *"Generalized Concavity,"* Plenum Press, New York, 1988.

Avriel, M., M. J. Rijkaert, and D. J. Wilde (Eds.), *Optimization and Design*, Prentice Hall, Englewood Cliffs, N.J., 1973.

Avriel, M., and A. C. Williams, "Complementary Geometric Programming," *SIAM J. Applied Mathematics*, 19, pp. 125–141, 1970a.

Avriel, M., and A. C. Williams, "On the Primal and Dual Constraint Sets in Geometric Programming," *J. Mathematical Analysis and Applications*, 32, pp. 684–688, 1970b.

Avriel, M., and I. Zang, "Generalized Convex Functions with Applications to Nonlinear Programming," in *Mathematical Programs for Activity Analysis*, P. Van Moeseki (Ed.), 1974.

Bachem, A., and B. Korte, "Quadratic Programming Over Transportation Polytopes," Report No. 7767-OR, Institute für Okonometrie und Operations Research, Bonn, 1977.

Baker, T. E., and L. S. Lasdon, "Successive Linear Programming at Exxon," *Management Science*, 31, pp. 264–274, 1985.

Baker, T. E., and R. Ventker, "Successive Linear Programming in Refinery Logistic Models," presented at ORSA/TIMS Joint National Meeting, Colorado Springs, Colorado, 1980.

Balakrishnan, A. V. (Ed.), *Techniques of Optimization*, Academic Press, New York, 1972.

Balas, E., "Disjunctive Programming: Properties of the Convex Hull of Feasible Points," Management Science Research Report 348, GSIA, Carnegie Mellon University, 1974.

Balas, E., "Disjunctive Programming and a Hierachy of Relaxations for Discrete Optimization Problems," *SIAM J. on Algebraic and Discrete Methods*, 6(3), pp. 466–486, 1985.

Balas, E., and C. A. Burdet, "Maximizing a Convex Quadratic Function Subject to Linear Constraints," Management Science Research Report #299, 1973.

Balas, E., "Nonconvex Quadratic Programming via Generalized Polars," *SIAM J. Applied Mathematics*, 28, pp. 335–349, 1975.

Balinski, M. L. (Ed.), *Pivoting and Extensions*, Mathematical Programming Study, No. 1, American Elsevier, New York, 1974.

Balinski, M. L., and W. J. Baumol, "The Dual in Nonlinear Programming and Its Economic Interpretation," *Review Economic Studies*, 35, pp. 237–256, 1968.

Balinski, M. L., and E. Helleman (Eds.), *Computational Practice in Mathematical Programming*, Mathematical Programming Study, No. 4, American Elsevier, New York, 1975.

Balinski, M. L., and C. Lemarechal (Eds.), "Mathematical Programming in Use," *Mathematical Programming Study*, 9, 1978.

Balinski, M. L., and P. Wolfe (Eds.), *Nondifferentiable Optimization*, Mathematical Programming Study, No. 2, American Elsevier, New York, 1975.

Bandler, J. W., and C. Charalambous, "Nonlinear Programming Using Minimax Techniques," *J. Optimization Theory and Applications*, 13, pp. 607–619, 1974.

Barankin, E. W., and R. Dorfman, "On Quadratic Programming," University of California Publications in Statistics, 2, pp. 285–318, 1958.

Bard, Y., "On Numerical Instability of Davidon-like Methods," *Mathematics of Computation*, 22, pp. 665–666, 1968.

Bard, Y., "Comparison of Gradient Methods for the Solution of Nonlinear Parameter Estimation Problems," *SIAM J. Numerical Analysis*, 7, pp. 157–186, 1970.

Bard, J. F., and J. E. Falk, "A Separable Programming Approach to the Linear Complementarity Problem," *Computers and Operations Research*, 9, pp. 153–159, 1982.

Bartels, R. H., "A Penalty Linear Programming Method Using Reduced-Gradient Basis-Exchange Techniques," *Linear Algebra and its Applications*, 29, pp. 17–32, 1980.

Bartels, R. H., and A. R. Conn, "Linearly Constrained Discrete l_1 Problems," *ACM Trans. Mathematics, Software*, 6, pp. 594–608, 1980.

Bartels, R. H., G. Golub, and M. A. Saunders, "Numerical Techniques in Mathematical Programming," in *Nonlinear Programming*, J. B. Rosen O. L. Mangasarian, and K. Ritter (Eds.), Academic Press, New York, pp. 123–176, 1970.

Bartholomew-Biggs, M. C., "Recursive Quadratic Programming Methods for Nonlinear Constraints," in *Nonlinear Optimization*, M. J. D. Powell (Ed.), Academic Press, London, pp. 213–221, 1981.

Bartle, R. G., *The Elements of Real Analysis* (2nd Edition), John Wiley & Sons, New York, 1976.

Batt, J. R., and R. A. Gellatly, "A Discretized Program for the Optimal Design of Complex Structures," *AGARD Lecture Series M70*, NATO, 1974.

Bauer, F. L., "Optimally Scaled Matrices," *Numerical Mathematics*, 5, pp. 73–87, 1963.

Bazaraa, M. S., "A Theorem of the Alternative with Application to Convex Programming: Optimality, Duality, and Stability," *J. Mathematical Analysis and Applications*, 41, pp. 701–715, 1973a.

Bazaraa, M.S., "Geometry and Resolution of Duality Gaps," *Naval Research Logistics Quarterly*, 20, pp. 357–365, 1973b.

Bazaraa, M. S., "An Efficient Cyclic Coordinate Method for Constrained Optimization," *Naval Research Logistics Quarterly*, 22, pp. 399–404, 1975.

Bazaraa, M. S., and J. J. Goode, "Necessary Optimality Criteria in Mathematical Programming in the Presence of Differentiability," *J. Mathematical Analysis and Applications*, 40, pp. 509–621, 1972.

Bazaraa, M. S., and J. J. Goode, "On Symmetric Duality in Nonlinear Programming," *Operations Research*, 21, pp. 1–9, 1973a.

Bazaraa, M. S., and J. J. Goode, "Necessary Optimality Criteria in Mathematical Programming in Normed Linear Spaces," *J. Optimization Theory and Applications*, 11, pp. 235–244, 1973b.

Bazaraa, M. S., and J. J. Goode, "Extension of Optimality Conditions via Supporting Functions," *Mathematical Programming*, 5, pp. 267–285, 1973c.

Bazaraa, M. S., and J. J. Goode, "The Travelling Salesman Problem: A Duality Approach," *Mathematical Programming*, 13, pp. 221–237, 1977.

Bazaraa, M. S., and J. J. Goode, "A Survey of Various Tactics for Generating Lagrangian Multipliers in the Context of Lagrangian Duality," *European J. of Operational Research*, 3, pp. 322–338, 1979.

Bazaraa, M. S., J. J. Goode, and C. M. Shetty, "Optimality Criteria Without Differentiability," *Operations Research*, 19, pp. 77–86, 1971a.

Bazaraa, M. S., J. J. Goode, and C. M. Shetty, "A Unified Nonlinear Duality Formulation," *Operations Research*, 19, pp. 1097–1100, 1971b.

Bazaraa, M. S., J. J. Goode, and C. M. Shetty, "Constraint Qualifications Revisited," *Management Science*, 18, pp. 567–573, 1972.

Bazaraa, M. S., J. J. Jarvis, and H. D. Sherali, *Linear Programming and Network Flows* (2nd Edition), John Wiley and Sons, New York, 1990.

Bazaraa, M. S., and H. D. Sherali, "On the Choice of Step Sizes in Subgradient Optimization," *European J. of Operational Research*, 17(2), pp. 380–388, 1981.

Bazaraa, M. S., and H. D. Sherali, "On the Use of Exact and Heuristic Cutting Plane Methods for the Quadratic Assignment Problem," *J. of the Operational Research Society*, 33(11), pp. 999–1003, 1982.

Bazaraa, M. S., and C. M. Shetty, *Foundations of Optimization*, Lecture Notes in Economics and Mathematical Systems, No. 122, Springer-Verlag, New York, 1976.

Beale, E. M. L., "On Minimizing a Convex Function Subject to Linear Inequalities," *J. Royal Statistical Society*, Ser. B 17, pp. 173–184, 1955.

Beale, E. M. L., "On Quadratic Programming," *Naval Research Logistics Quarterly*, 6, pp. 227–244, 1959.

Beale, E. M. L., "Numerical Methods," in *Nonlinear Programming*, J. Abadie (Ed.), 1967.

Beale, E. M. L., "Nonlinear Optimization by Simplex-Like Methods," in *Optimization*, R. Fletcher (Ed.), 1969.

Beale, E. M. L., "Computational Methods for Least Squares," in *Integer and Nonlinear Programming*, J. Abadie (Ed.), 1970a.

Beale, E. M. L. (Ed.), *Applications of Mathematical Programming Techniques*, English Universities Press, London, 1970b.

Beale, E. M. L., "Advanced Algorithmic Features for General Mathematical Programming Systems," in *Integer and Nonlinear Programming*, J. Abadie (Ed.), 1970c.

Beale, E. M. L., "Nonlinear Programming Using a General Mathematical Programming System," in *Design and Implementation of Optimization Software*, H. J. Greenberg (Ed.), Sijthoff and Noordhoff, Netherlands, pp. 259–279, 1978.

Beckenbach, E. F., and R. Bellman, *Inequalities*, Springer-Verlag, Berlin, 1961.

Beckman, F. S., "The Solution of Linear Equations by the Conjugate Gradient Method," in *Mathematical Methods for Digital Computers*, A. Ralston and H. Wilf (Eds.), John Wiley & Sons, New York, 1960.

Beckmann, M. J., and K. Kapur, "Conjugate Duality: Some Applications to Economic Theory," *J. Economic Theory*, 5, pp. 292–302, 1972.

Bector, C. R., "Programming Problems with Convex Fractional Functions," *Operations Research*, 16, pp. 383–391, 1968.

Bector, C. R., "Some Aspects of Quasi-Convex Programming," *Zeitschrift für Angewandte Mathematik und Mechanik*, 50, pp. 495–497, 1970.

Bector, C. R., "Duality in Nonlinear Fractional Programming," *Zeitschrift für Operations Research*, 17, pp. 183–193, 1973a.

Bector, C. R., "On Convexity, Pseudo-Convexity and Quasi-Convexity of Composite Functions," *Cahiers Centre Études Recherche Operationnelle*, 15, pp. 411–428, 1973b.

Beglari, F., and M. A. Laughton, "The Combined Costs Method for Optimal Economic Planning of an Electrical Power System," *IEEE Transactions Power Apparatus and Systems*, PAS-94, pp. 1935–1942, 1975.

Bellman, R. (Ed.), *Mathematical Optimization Techniques*, University of California Press, Berkeley, 1963.

Bellmore, M., H. J. Greenberg, and J. J. Jarvis, "Generalized Penalty Function Concepts in Mathematical Optimization," *Operations Research*, 18, pp. 229–252, 1970.

Beltrami, E. J., "A Computational Approach to Necessary Conditions in Mathematical Programming," *Bull. International Journal of Computer Mathematics*, 6, pp. 265–273, 1967.

Beltrami, E. J., "A Comparison of Some Recent Iterative Methods for the Numerical Solution of Nonlinear Programs," in *Computing Methods in Optimization Problems*, Lecture Notes in Operations Research and Mathematical Economics, No. 14, Springer-Verlag, New York, 1969.

Beltrami, E. J., *An Algorithmic Approach to Nonlinear Analysis and Optimization*, Academic Press, New York, 1970.

Benveniste, R., "A Quadratic Programming Algorithm Using Conjugate Search Directions," *Mathematical Programming*, 16, pp. 63–80, 1979.

Ben-Daya, M., and C. M. Shetty, "Polynomial Barrier Function Algorithms for Convex Quadratic Programming," Report Series J88-5, School of Industrial and Systems Engineering, Georgia Institute of Technology, Atlanta, GA, 1988.

Ben-Tal, A., "Second-Order and Related Extremality Conditions in Nonlinear Programming," *J. of Optimization Theory and Applications*, 31, pp. 143–165, 1980.

Ben-Tal, A., and J. Zowe, "A Unified Theory of First and Second-Order Conditions for Extremum Problems in Topological Vector Spaces," *Mathematical Programming Study*, 19, pp. 39–76, 1982.

Bereanu, B., ''A Property of Convex, Piecewise Linear Functions with Applications to Mathematical Programming,'' *Unternehmensforschung*, 9, pp. 112–119, 1965.

Bereanu, B., ''On the Composition of Convex Functions,'' *Revue Roumaine Mathématiques Pures et Appliquées*, 14, pp. 1077–1084, 1969.

Bereanu, B., ''Quasi-Convexity, Strict Quasi-Convexity and Pseudo-Convexity of Composite Objective Functions,'' *Revue Francaise Automatique, Informatique Recherche Operationnelle*, 6, R-1, pp. 15–26, 1972.

Berge, C., *Topological Spaces*, Macmillan, New York, 1963.

Berge, C., and A. Ghoulia-Houri, *Programming, Games, and Transportation Networks*, John Wiley and Sons, New York, 1965.

Berman, A., *Cones, Metrics and Mathematical Programming*, Lecture Notes in Economics and Mathematical Systems, No. 79, Springer-Verlag, New York, 1973.

Berna, R. J., M. H. Locke, and A. W. Westerberg, ''A New Approach to Optimization of Chemical Processes,'' *AICHE J.*, 26(2), 37, 1980.

Bertsekas, D. P., ''On Penalty and Multiplier Methods for Constrained Minimization,'' in *Nonlinear Programming-2.*, O. L. Mangasarian, R. Meyer, and S. M. Robinson (Eds.), Academic Press, New York, 1975.

Bertsekas, D. P., *Nondifferentiable Optimization*, North Holland Publishing Company, Amsterdam, 1975.

Bertsekas, D. P., ''Necessary and Sufficient Conditions for a Penalty Function to be Exact,'' *Mathematical Programming*, 9, pp. 87–99, 1975a.

Bertsekas, D. P., ''Combined Primal–Dual and Penalty Methods for Constrained Minimization,'' *SIAM J. Control and Optimization*, 13, pp. 521–544, 1975b.

Bertsekas, D. P., ''Multiplier Methods: A Survey,'' *Automatica*, 12, pp. 133–145, 1976a.

Bertsekas, D. P., ''On Penalty and Mutiplier Methods for Constrained Minimization,'' *SIAM J. Control and Optimization*, 14, pp. 216–235, 1976b.

Bertsekas, D. P., ''*Constrained Optimization and Lagrange Multiplier Methods*,'' Academic Press, New York, 1982.

Bertsekas, D. P., and S. K. Mitter, ''A Descent Numerical Method for Optimization Problems with Nondifferentiable Cost Functionals,'' *SIAM J. Control*, 11, pp. 637–652, 1973.

Best, M. J., ''A Method to Accelerate the Rate of Convergence of a Class of Optimization Algorithms,'' *Mathematical Programming*, 9, pp. 139–160, 1975.

Best, M. J., ''A Quasi-Newton Method can be Obtained from a Method of Conjugate Directions,'' *Mathematical Programming*, 15, pp. 189–199, 1978.

Best, M. J., J. Brauninger, K. Ritter, and S. M. Robinson, ''A Globally and Quadratically Convergent Algorithm for General Nonlinear Programming Problems,'' *Computing*, 26, pp. 141–153, 1981.

Beveridge, G., and R. Schechter, *Optimization: Theory and Practice*, McGraw-Hill, New York, 1970.

Bhatia, D., ''A Note on Duality Theorem for a Nonlinear Programming Problem,'' *Management Science*, 16, pp. 604–606, 1970.

Bhatt, S. K., and S. K. Misra, ''Sufficient Optimality Criteria in Nonlinear Programming in the Presence of Convex Equality and Inequality Constraints,'' *Zeitschrift für Operations Research*, 19, pp. 101–105, 1975.

Biggs, M. C., ''Constrained Minimization Using Recursive Equality Quadratic Programming,'' in *Numerical Methods for Non-Linear Optimization*, F. A. Lootsma (Ed.), Academic Press, London and New York, pp. 411–428, 1972.

Biggs, M.C., ''Constrained Minimization Using Recursive Quadratic Programming: Some Alternative Subproblem Formulations,'' in *Towards Global Optimization*, L. C. W. Dixon and G. P. Szego (Eds.), pp. 341–349, North Holland, Amsterdam, 1975.

Biggs, M. C., ''On the Convergence of Some Constrained Minimization Algorithms Based on Recursive Quadratic Programming,'' *J. Inst. Mathematical Applications*, 21, pp. 67–81, 1978.

Bitran, G., and A. Hax, ''On the Solution of Convex Knapsack Problems with Bounded Variables,'' *Proc. of IXth International Symposium on Mathematical Programming*, Budapest, pp. 357–367, 1976.

Bitran, G. R., and T. L. Magnanti, "Duality and Sensitivity Analysis for Fractional Programs," *Operations Research*, 24, pp. 657–699, 1976.

Bitran, G. R., and A. G. Novaes, "Linear Programming with a Fractional Objective Function," *Operations Research*, 21, pp. 22–29, 1973.

Bjorck, A., "Stability Analysis of the Method of Semi-Normal-Equations For Linear Least Squares Problems," Linkoping Univ. Report LiTH-MATH-R-1985-08, 1985.

Bland, R. C., "New Finite Pivoting Rules for the Simplex Method," *Mathematics of Operations Research*, 2, pp. 103–107, 1977a.

Bloom, J. A., "Solving an Electric Generating Capacity Expansion Planning Problem by Generalized Benders, Decomposition," *Operations Research*, 31, pp. 84–100, 1983.

Bloom, J. A., M. C. Caramanis, and L. Charny, "Long Range Generation Planning Using Generalized Benders Decomposition: Implementation and Experience," *Operations Research*, 32, pp. 314–342, 1984.

Blum, E., and W. Oettli, "Direct Proof of the Existence Theorem for Quadratic Programming," *Operations Research*, 20, pp. 165–167, 1972.

Blum, E., and W. Oettli, *Mathematische Optimierung-Grundlager und Verfahren*, Econometrics and Operations Research, No. 20, Springer-Verlag, New York, 1975.

Boddington, C. E., and W. C. Randall, "Nonlinear Programming for Product Blending," Joint National TIMS/ORSA Meeting, New Orleans, May 1979.

Boggs, P. T., and J. W. Tolle, "Augmented Lagrangians which are Quadratic in the Multiplier," *J. Optimization Theory Applications*, 31, pp. 17–26, 1980.

Boggs, P. T., and J. W. Tolle, "A Family of Descent Functions for Constrained Optimization," *SIAM J. Numerical Analysis*, 21, pp. 1146–1161, 1984.

Boggs, P. T., J. W. Tolle, and P. Wang, "On the Local Convergence of Quasi-Newton Methods for Constrained Optimization," *SIAM J. Control and Optimization*, 20, pp. 161–171, 1982.

Boot, J. C. G., "Notes on Quadratic Programming: The Kuhn–Tucker and Theil–van de Panne Conditions, Degeneracy and Equality Constraints," *Management Science*, 8, pp. 85–98, 1961.

Boot, J. C. G., "On Trivial and Binding Constraints in Programming Problems," *Management Science*, 8, pp. 419–441, 1962.

Boot, J. C. G., "Binding Constraint Procedures of Quadratic Programming," *Econometrica*, 31, pp. 464–498, 1963a.

Boot, J. C. G., "On Sensitivity Analysis in Convex Quadratic Programming Problems," *Operations Research*, 11, pp. 771–786, 1963b.

Boot, J. C. G., *Quadratic Programming*, North Holland Publishing Company, Amsterdam, 1964.

Borwein, J. M., "A Note on the Existence of Subgradients," *Mathematical Programming*, 24, pp. 225–228, 1982.

Box, M. J., "A Comparison of Several Current Optimization Methods, and the Use of Transformations in Constrained Problems," *Computer Journal*, 9, pp. 67–77, 1966.

Box, M. J., "A New Method of Constrained Optimization and a Comparison with Other Methods," *Computer Journal*, 8, pp. 42–52, 1965.

Box, M. J., D. Davies, and W. H. Swann, *Nonlinear Optimization Techniques*, I.C.I. Monograph, Oliver and Boyd, Edinburgh, 1969.

Bracken, J., and G. P. McCormick, *Selected Applications of Nonlinear Programming*, John Wiley and Sons, New York, 1968.

Bram, J., "The Lagrange Multiplier Theorem for Max-Min with Several Constraints," *SIAM J. Applied Mathematics*, 14, pp. 665–667, 1966.

Bradley, J., and H. M. Clyne, "Applications of Geometric Programming to Building Design Problem," in *Optimization in Action*, L. C. W. Dixon (Ed.), Academic Press, London, 1976.

Braswell, R. N., and J. A. Marban, "Necessary and Sufficient Conditions for the Inequality Constrained Optimization Problem Using Directional Derivatives," *International J. Systems Science*, 3, pp. 263–275, 1972.

Brayton, R. K., and J. Cullum, "An Algorithm for Minimizing a Differentiable Function Subject to Box Constraints and Errors," *J. Optimization Theory and Applications*, 29, pp. 521–558, 1979.

Brent, R. P., *Algorithms for Minimization without Derivatives*, Prentice Hall, Englewood Cliffs, N.J., 1973.

Brodlie, K. W., "An Assessment of Two Approaches to Variable Metric Methods," *Mathematical Programming*, 12, pp. 344–355, 1977.

Brøndsted, A., and R. T. Rockafeller, "On the Subdifferential of Convex Functions," *Proceedings of American Mathematical Society*, 16, pp. 605–611, 1965.

Brooke, A., D. Kendrick, and A. Mieerans, "*GAMS-A Users Guide*," Scientific Press, Redwood City, CA, 1988.

Brooks, R., and A. Geoffrion, "Finding Everett's Lagrange Multipliers by Linear Programming," *Operations Research*, 16, pp. 1149–1152, 1966.

Brooks, S. H., "A Discussion of Random Methods for Seeking Maxima," *Operations Research*, 6, pp. 244–251, 1958.

Brooks, S. H., "A Comparison of Maximum Seeking Methods," *Operations Research*, 7, pp. 430–457, 1959.

Brown, K. M., and J. E. Dennis, "A New Algorithm for Nonlinear Least Squares Curve Fitting," in *Mathematical Software*, J. R. Rice (Ed.), Academic Press, New York, 1971.

Broyden, C. G., "A Class of Methods for Solving Nonlinear Simultaneous Equations," *Mathematics of Computation*, 19, pp. 577–593, 1965.

Broyden, C. G., "Quasi-Newton Methods and Their Application to Function Minimization," *Mathematics of Computation*, 21, pp. 368–381, 1967.

Broyden, C. G., "The Convergence of a Class of Double Rank Minimization Algorithms 2. The New Algorithm," *J. Institute of Mathematics and Its Applications*, 6, pp. 222–231, 1970.

Broyden, C. G., J. E. Dennis, and J. J. More, "On the Local and Superlinear Convergence of Quasi-Newton Methods," *J. Inst. Mathematics Applications*, 12, pp. 223–246, 1973.

Buck, R. C., *Mathematical Analysis*, McGraw-Hill, New York, 1965.

Buckley, A. G., "A Combined Conjugate-Gradient Quasi-Newton Minimization Algorithm," *Mathematical Programming*, 15, pp. 200–210, 1978.

Bullard, S., H. D. Sherali, and D. Klemperer, "Estimating Optimal Thinning and Rotation for Mixed Species Timber Stands," *Forest Science*, 13(2), pp. 303–315, 1985.

Bunch, J. R., and L. C. Kaufman, "A Computational Method for the Indefinite Quadratic Programming Problems," *Linear Algebra and Its Applications*, 34, pp. 341–370, 1980.

Buras, N., *Scientific Allocation of Water Resources*, American Elsevier, New York, 1972.

Burdet, C. A., "Elements of a Theory in Nonconvex Programming," *Naval Research Logistics Quarterly*, 24, pp. 47–66, 1977.

Burke, J. V., and S.-P. Han, "A Robust Sequential Quadratic Programming Method," *Mathematical Programming*, 43, pp. 277–303, 1989.

Burke, J. V., J. J. More, and G. Toraldo, "Convergence Properties of Trust Region Methods for Linear and Convex Constraints," *Mathematical Programming*, 47, pp. 305–336, 1990.

Burley, D. M., *Studies in Optimization*, John Wiley and Sons, New York, 1974.

Burns, S. A., "Graphical Representation of Design Optimization Processes," *Computer-Aided Design*, 21(1), pp. 21–25, 1989.

Buys, J. D., and R. Gonin, "The Use of Augmented Lagrangian Functions for Sensitivity Analysis in Nonlinear Programming," *Mathematical Programming*, 12, pp. 281–284, 1977.

Buzby, B. R., "Techniques and Experience Solving Really Big Nonlinear Programs," *Optimization Methods for Resource Allocation*, R. Cottle and J. Krarup (Eds.), English University Press, pp. 227–237, 1974.

Cabot, V. A., and R. L. Francis, "Solving Certain Nonconvex Quadratic Minimization Problems by Ranking Extreme Points," *Operations Research*, 18, pp. 82–86, 1970.

Camerini, P. M., L. Fratta, and F. Maffioli, "On Improving Relaxation Methods by Modified Gradient Techniques," in *Nondifferentiable Optimization*, M. L. Balinski and P. Wolfe (Eds.), 1975. (See *Math. Prog. Study*, 3, pp. 26–34, 1975.)

Camp, G. D., "Inequality-Constrained Stationary-Value Problems," *Operations Research*, 3, pp. 548–550, 1955.

Candler, W., and R. J. Townsley, "The Maximization of a Quadratic Function of Variables Subject to Linear Inequalities," *Management Science*, 10, pp. 515–523, 1964.

Canon, M. D., and C. D. Cullum, "A Tight Upper Bound on the Rate of Convergence of the Frank–Wolfe Algorithm," *SIAM J. Control*, 6, pp. 509–516, 1968.

Canon, M. D., C. D. Cullum, and E. Polak, "Constrained Minimization Problems in Finite Dimensional Spaces," *SIAM J. Control*, 4, pp. 528–547, 1966.

Canon, M. D., C. Cullum, and E. Polak, *Theory of Optimal Control and Mathematical Programming*, McGraw-Hill, New York, 1970.

Canon, M. D., and J. H. Eaton, "A New Algorithm for a Class of Quadratic Programming Problems, with Application to Control," *SIAM J. Control*, 4, pp. 34–44, 1966.

Cantrell, J. W., "Relation Between the Memory Gradient Method and the Fletcher-Reeves Method," *J. Optimization Theory and Applications*, 4, pp. 67–71, 1969.

Carnillo, M. J., "A Relaxation Algorithm for the Minimization of a Quasiconcave Function on a Convex Polyhedron," *Mathematical Programming*, 13, pp. 69–80, 1977.

Carroll, C. W., "The Created Response Surface Technique for Optimizing Nonlinear Restrained Systems," *Operations Research*, 9, pp. 169–184, 1961.

Cass, D., "Duality: A Symmetric Approach from the Economist's Vantage Point," *J. Economic Theory*, 7, pp. 272–295, 1974.

Chamberlain, R. M., "Some Examples of Cycling in Variable Metric Methods for Constrained Minimization," *Mathematical Programming*, 16, pp. 378–383, 1979.

Chamberlain, R. M., C. Lemarechal, H. C. Pedersen, and M. J. D. Powell, "The Watchdog Technique for Forcing Convergence in Algorithms for Constrained Optimization," in *Algorithms for Constrained Minimization of Smooth Nonlinear Functions*, A. G. Buckley and J. L. Goffin (Eds.), *Mathematical Programming Study*, 16, North Holland, Amsterdam, 1982.

Charnes, A., and W. W. Cooper, "Nonlinear Power of Adjacent Extreme Point Methods of Linear Programming,"*Econometrica*, 25, pp. 132–153, 1957.

Charnes, A., and W. W. Cooper, "Chance Constrained Programming," *Management Science*, 6, pp. 73–79, 1959.

Charnes A., and W. W. Cooper, *Management Models and Industrial Applications of Linear Programming*, 2 volumes, John Wiley & Sons, New York, 1961.

Charnes, A., and W. W. Cooper, "Programming with Linear Fractionals," *Naval Research Logistics Quarterly*, 9, pp. 181–186, 1962.

Charnes, A., and W. W. Cooper, "Deterministic Equivalents for Optimizing and Satisficing Under Chance Constraints," *Operations Research*, 11, p. 18–39, 1963.

Charnes, A., W. W. Cooper, and K. O. Kortanek, "A Duality Theory for Convex Programs with Convex Constraints," *Bull. American Mathematical Society*, 68, pp. 605–608, 1962.

Charnes, A., M. J. L. Kirby, and W. M. Raike, "Solution Theorems in Probablistic Programming: A Linear Programming Approach," *J. Mathematical Analysis and Applications*, 20. pp. 565–582, 1967.

Choi, I. C., C. L. Monma, and D. F. Shanno, "Further Development of a Primal–Dual Interior Point Method," *ORSA J. on Computing*, 2, pp. 304–311, 1990.

Chung, S. J., "NP-Completeness of the Linear Complementarity Problem," *J. of Optimization Theory and Applications*, 60, pp. 393–399, 1989.

Chung, S. J., and K. G. Murty, "Polynomially Bounded Ellipsoid Algorithm for Convex Quadratic Programming," in *Nonlinear Programming 4*, O. L. Mangasarian, R. R. Meyer, and S. M. Robinson (Eds.), Academic Press, New York, pp. 439–485, 1981.

Chvatal, V., *Linear Programming*, W. H. Freeman and Company, New York/San Francisco, 1980.

Citron, S. J., *Elements of Optimal Control*, Holt, Rinehart, and Winston, New York, 1969.

Cohen, A., "Rate of Convergence of Several Çonjugate Gradient Algorithms," *SIAM J. Numerical Analysis*, 9, pp. 248–259, 1972.

Cohen, G., and D. L. Zhu, "Decomposition-Coordination Methods in Large Scale Optimization Problems: The Nondifferentiable Case and the Use of Augmented Lagrangians," in *Advances in Large Scale Systems, Theory and Applications*, Vol. 1, J. B. Cruz, Jr. (Ed.), JAI Press Inc., Greenwich, Conn., 1983.

Cohn, M. Z. (Ed.), *An Introduction to Structural Optimization*, University of Waterloo Press, 1969.

Coleman, T. F., and A. R. Conn, "Nonlinear Programming via an Exact Penalty Function: Asymptotic Analysis," *Mathematical Programming*, 24, pp. 123–136, 1982a.

Coleman, T. F., and A. R. Conn, "Nonlinear Programming via an Exact Penalty Function: Global Analysis," *Mathematical Programming*, 24, pp. 137–161, 1982b.

Coleman, T., and P. A. Fenyes, "Partitioned Quasi-Newton Methods for Nonlinear Equality Constrained Optimization," Report No. 88-14, Cornell Computational Optimization Project, 311 Upson Hall, Cornell University, Ithaca, New York, 1989.

Collins, M., L. Cooper, R. Helgason, J. Kennington, and L. Le Blanc, "Solving the Pipe Network Analysis Problem Using Optimization Techniques," *Management Science*, 24, pp. 747–760, 1978.

Colville, A. R., "A Comparative Study of Nonlinear Programming Codes," in *Proceedings of the Princeton Symposium on Mathematical Programming*, H. Kuhn (Ed.), 1970.

Conn, A. R., "Constrained Optimization Using a Nondifferential Penalty Function," *SIAM J. Numerical Analysis*, 10, pp. 760–784, 1973.

Conn, A. R., "Linear Programming via a Non-differentiable Penalty Function," *SIAM J. Numerical Analysis*, 13, pp. 145–154, 1976.

Conn, A. R., "Nonlinear Programming, Exact Penalty Functions and Projection Techniques for Non-smooth Functions," in *Numerical Optimization 1984*, P. T. Boggs (Ed.), SIAM, Philadelphia, 1985.

Conn, A. R., "Penalty Function Methods," in *Nonlinear Optimization 1981*, Academic Press, 1982.

Conn, A. R., N. I. M. Gould, and Ph. L. Toint, "A Globally Convergent Augmented Lagrangian Algorithm for Optimization with General Constraints and Simple Bounds," Report 88/23, Department of Computer Sciences, University of Waterloo, Waterloo, Canada, 1988–

Conn, A. R., N. I. M. Gould, and Ph. L. Toint, "Global Convergence of a Class of Trust Region Algorithms for Optimization with Simple Bounds," *SIAM J. on Numerical Analysis*, 25(2), pp. 433–460, 1988. (See also errata in *SIAM J. on Numerical Analysis*, 26(3), pp. 764–767, 1989.)

Conn, A. R., N. I. M. Gould, and Ph. L. Toint, "Convergence of Quasi-Newton Matrices Generated by the Symmetric Rank-One Update," *Mathematical Programming*, 50(2), pp. 177–196, 1991.

Conn, A. R., and T. Pietrzykowski, "A Penalty Function Method Converging Directly to a Constrained Optimum," *SIAM J. Numerical Analysis*, 14, pp. 348–375, 1977.

Conte, S. D., and C. de Boor, *Elementary Numerical Analysis: An Algorithmic Approach* (3rd Edition), McGraw-Hill, New York, 1980.

Conti, R., and A. Ruberti (Eds.), *5th Conference on Optimization Techniques*, Part 1, Lecture Notes in Computer Science, No. 3, Springer-Verlag, New York, 1973.

Coope, I. D., and R. Fletcher, "Some Numerical Experience with a Globally Convergent Algorithm for Nonlinearly Constrained Optimization," *J. Optimization Theory Applications*, 32, pp. 1–16, 1980.

Cottle, R. W., "A Theorem of Fritz John in Mathematical Programming," *RAND Corporation Memo*, RM-3858-PR, 1963a.

Cottle, R. W., "Symmetric Dual Quadratic Programs," *Quart. Applied Mathematics*, 21, pp. 237–243, 1963b.

Cottle, R. W., "Note on a Fundamental Theorem in Quadratic Programming," *SIAM J. Applied Mathematics*, 12, pp. 663–665, 1964.

Cottle, R. W., "Nonlinear Programs with Positively Bounded Jacobians," *SIAM J. Applied Mathematics*, 14, pp. 147–158, 1966.

Cottle, R. W., "On the Convexity of Quadratic Forms over Convex Sets," *Operations Research*, 15, pp. 170–172, 1967.

Cottle, R. W., "The Principal Pivoting Method of Quadratic Programming," in *Mathematics of the Decision Sciences*, G. B. Dantzig and A. F. Veinott (Eds.), 1968.

Cottle, R. W., and G. B. Dantzig, "Complementary Pivot Theory of Mathematical Programming," *Linear Algebra and Applications*, 1, pp. 103–125, 1968.

Cottle, R. W., and G. B. Dantzig, "A Generalization of the Linear Complementarity Problem," *J. Combinatorial Theory*, 8, pp. 79–90, 1970.

Cottle, R. W., and J. A. Ferland, "Matrix-Theoretic Criteria for the Quasi-Convexity and Pseudo-Convexity of Quadratic Functions," *J. Linear Algebra and Applications*, 5, pp. 123–136, 1972.

Cottle, R. W., and C. E. Lemke (Eds.), *Nonlinear Programming*, American Mathematical Society, Providence, R.I., 1976.

Cottle, R. W., and J. S. Pang, "On Solving Linear Complementarity Problems as Linear Programs," *Mathematical Programming Study, 7*, pp. 88–107, 1978.

Cottle, R. W., and A. F. Veinott, Jr., "Polyhedral Sets Having a Least Element," *Mathematical Programming*, 3, pp. 238–249, 1969.

Crabill, T. B., J. P. Evans, and F. J. Gould, "An Example of an Ill-Conditioned NLP Problem," *Mathematical Programming*, 1, pp. 113–116, 1971.

Cragg, E. E., and A. V. Levy, "Study on a Supermemory Gradient Method for the Minimization of Functions," *J. Optimization Theory and Applications*, 4, pp. 191–205, 1969.

Crane, R. L., K. E., Hillstrom, and M. Minkoff, "Solution of the General Nonlinear Programming Problem with Subroutine VMCON," Argonne National Laboratory, 9700 South Cass Avenue, Argonne, Ill., Mathematics and Computers (UC-32), July 1980.

Craven, B. D., "A Generalization of Lagrange Multipliers," *Bull. Australian Mathematical Society*, 3, pp. 353–362, 1970.

Crowder, H., and P. Wolfe, "Linear Convergence of the Conjugate Gradient Method," *IBM J. Research and Development*, 16, pp. 407–411, 1972.

Cryer, C. W., "The Solution of a Quadratic Programming Problem Using Systematic Overrelaxation," *SIAM J. on Control*, 9, pp. 385–392, 1971.

Cullen, C. G., *Matrices and Linear Transformations* (2nd Edition), Addison-Wesley, Reading, Mass., 1972.

Cullum, J., "An Explicit Procedure for Discretizing Continuous Optimal Control Problems," *J. Optimization Theory and Applications*, 8, pp. 15–34, 1971.

Cunningham, K., and L. Schrage, "*The LINGO Modeling Language*," Lindo Systems, Inc., Chicago, IL, 1989.

Curry, H. B., "The Method of Steepest Descent for Nonlinear Minimization Problems," *Quart. Applied Mathematics*, 2, pp. 258–263, 1944.

Curtis A. R., and J. K. Reid, "On the Automatic Scaling of Matrices for Gaussian Elimination," *J. Inst. Mathematical Applications*, 10, 118–124, 1972.

Cutler, C. R., and R. T. Perry, "Real Time Optimization With Multivariable Control is Required to Maximize Profits," *Computers and Chemical Engineering*, 7, pp. 663–667, 1983.

Dajani, J. S., R. S. Gemmel, and E. K. Morlok, "Optimal Design of Urban Waste Water Collection Networks," *J. Sanitary Engineering Division, Am Soc. Civil Engineering*, 98-SAG, pp. 853–867, 1972.

Daniel, J., "Global Convergence for Newton Methods in Mathematical Programming," *J. Optimization Theory and Applications*, 12, pp. 233–241, 1973.

Danskin, J. W., *The Theory of Max-Min and Its Applications to Weapons Allocation Problems*, Springer-Verlag, New York, 1967.

Dantzig, G. B., "Maximization of a Linear Function of Variables Subject to Linear Inequalities," in *Activity Analysis of Production and Allocation*, Koopman (Ed.), Cowles Commission Monograph, 13, John Wiley & Sons, New York, 1951.

Dantzig, G. B., "Linear Programming Under Uncertainty," *Management Science*, 1, pp. 197–206, 1955.

Dantzig, G. B., "General Convex Objective Forms," in *Mathematical Methods in the Social Sciences*, K. Arrow, S. Karlin, and P. Suppes (Eds.), Stanford University Press, Stanford, 1960.

Dantzig, G. B., *Linear Programming and Extensions*, Princeton University Press, Princeton, N.J., 1963.

Dantzig, G. B., "Linear Control Processes and Mathematical Programming," *SIAM J. Control*, 4, pp. 56–60, 1966.

Dantzig, G. B., E. Eisenberg, and R. W. Cottle, "Symmetric Dual Nonlinear Programs," *Pacific J. Mathematics*, 15, pp. 809–812, 1965.

Dantzig, G. B., S. M. Johnson, and W. B. White, "A Linear Programming Approach to the Chemical Equilibrium Problem," *Management Science*, 5, pp. 38–43, 1958.

Dantzig, G. B., and A. Orden, "Duality Theorems," RAND Report RM-1265, *The RAND Corporation*, Santa Monica, Calif., 1953.

Dantzig, G. B., and A. F. Veinott (Eds.), *Mathematics of the Decision Sciences*, Parts 1, 2, Lectures in Applied Mathematics, No. 11, 12, American Mathematical Society, Providence, R.I., 1968.

Davidon, W. C., "Variable Metric Method for Minimization," *AEC Research Development Report*, ANL-5990, 1959.

Davidon, W. C., "Variance Algorithms for Minimization," in *Optimization*, R. Fletcher (Ed.), 1969.

Davidon, W. C., R. B. Mifflin, and J. L. Nazareth, "Some Comments on Notation for Quasi-Newton Methods," *OPTIMA*, 32, pp. 3–4, 1991.

Davies, D., "Some Practical Methods of Optimization," in *Integer and Nonlinear Programming*, J. Abadie (Ed.), 1970.

Davies, D., and W. H. Swann, "Review of Constrained Optimization," in *Optimization*, R. Fletcher (Ed.), 1969.

Deb, A. K., and A. K. Sarkar, "Optimization in Design of Hydraulic Networks," *J. Sanitary Engineering Division, American Society Civil Engineers* 97-SA2, pp. 141–159, 1971.

Dembo, R. S., "A Set of Geometric Programming Test Problems and Their Solutions," *Mathematical Programming*, 10, p. 192, 1976.

Dembo, R. S., "Current State of the Art of Algorithms and Computer Software for Geometric Programming," *J. Optimization Theory and Applications*, 26, pp. 149–184, 1978.

Dembo, R. S., and M. Avriel, "Optimal Design of a Membrane Separation Process Using Geometric Programming," *Mathematical Programming*, 15, pp. 12–25, 1978.

Dembo, R. S., "Second-Order Algorithms for the Polynomial Geometric Programming Dual, Part I: Analysis," *Mathematical Programming*, 17, pp. 156–175, 1979.

Dembo, R. S., S. C. Eisenstat, and T. Steinhaug, "Inexact Newton Methods," *SIAM J. Numerical Analysis*, 19(2), pp. 400–408, 4/82.

Dembo, R. S., and J. G. Klincewicz, "A Scaled Reduced Gradient Algorithm for Network Flow Problems with Convex Separable Costs," in *Network Models and Applications*, D. Klingman and J. M. Mulvey (Eds.), *Mathematical Programming Study 15*, North Holland, Amsterdam, 1981.

Demyanov, V. F., "Algorithms for some Minimax Problems," *J. Computer and System Sciences*, 2, pp. 342–380, 1968.

Demyanov, V. F., "On the Maximization of a Certain Nondifferentiable Function," *J. Optimization Theory and Applications*, 7, pp. 75–89, 1971.

Demyanov, F. F., and D. Pallaschke, "Nondifferentiable Optimization Motivation and Applications," *Proc. of the IIASA Workshop on Nondifferentiable Optimization*, Sopron, Hungary, September 1984, *Lecture Notes in Economic and Math Systems*, 255, 1985.

Demyanov, V. F., and A. M. Rubinov, The Minimization of a Smooth Convex Functional on a Convex Set," *J. SIAM Control*, 5, pp. 280–294, 1967.

Demyanov, V. F., and L. V. Vasiler, *Nondifferentiable Optimization*, 1985.

den Hertog, D., C. Roos, and T. Terlaky, "Inverse Barrier Methods for Linear Programming," Delft University of Technology, ISSN 0922-5641, Reports of the Faculty of Technical Mathematics and Informatics No. 91-27, Delft, The Netherlands, 1991.

Dennis, J. B., *Mathematical Programming and Electrical Networks*, M.I.T. Press and John Wiley, New York, 1959.

Dennis, J. E., ''A Brief Survey of Convergence Results for Quasi-Newton Methods,'' in *Nonlinear Programming, SIAM-AMS Proceedings*, Vol. 9, R. W. Cottle and C. E. Lemki (Eds.), pp. 185–199, 1976.

Dennis, J. E., Jr., ''A Brief Introduction of Quasi-Newton Methods,'' in *Numerical Analysis*, G. H. Golub and J. Oliger (Eds.), AMS, Providence, R. I., pp. 19–52, 1978.

Dennis, J. E., D. M. Gay, and R. E. Welsch, ''An Adaptive Nonlinear Least-Squares Algorithm,'' *ACM Trans. Mathematical Software*, 7, pp. 348–368, 1981.

Dennis, J. E., Jr., and E. S. Marwil, ''Direct Secant Updates of Matrix Factorizations,'' *Mathematical Computations*, 38, pp. 459–474, 1982.

Dennis, J. E., Jr., and H. H. W. Mei, ''Two New Unconstrained Optimization Algorithms Which Use Function and Gradient Values,'' *J. Optimization Theory and Applications*, 28, pp. 453–482, 1979.

Dennis, J. E. and J. J. More, ''A Characterization of Superlinear Convergence and its Application to Quasi-Newton Methods,'' *Mathematics of Computation*, 28(126), pp. 549–560, 1974

Dennis, J. E., Jr., and J. J. More, ''Quasi-Newton Methods: Motivation and Theory,'' *SIAM Review*, 19, pp. 46–89, 1977.

Dennis, J. E., Jr., and R. E. Schnabel, ''Least Change Secant Updates for Quasi-Newton Methods,'' *SIAM Review*, 21, pp. 443–469, 1979.

Dennis, J. E., Jr., and R. E. Schnabel, ''A New Derivation of Symmetric Positive Definite Secant Updates,'' in *Nonlinear Programming, 4*, O. L. Mangasarian, R. R. Meyer, and S. M. Robinson (Eds.), Academic Press, London and New York, pp. 167–199, 1981.

Dennis, J. E., Jr., and R. B. Schnabel, *''Numerical Methods for Unconstrained Optimization and Nonlinear Equations*,'' Prentice Hall, Englewood Cliffs, N.J., 1983.

Dinkel, J. J., ''An Implementation of Surrogate Constraint Duality,'' *Operations Research*, 26(2), pp. 358–364, 1978.

Di Pillo, G., and L. Grippo, ''A New Class of Augmented Lagrangians in Nonlinear Programming,'' *SIAM J. Control Optimization*, 17, pp. 618–628, 1979.

Dinkelbach, W., ''On Nonlinear Fractional Programming,'' *Management Science*, 13, pp. 492–498, 1967.

Dixon, L. C. W., ''Quasi-Newton Algorithms Generate Identical Points,'' *Mathematical Programming*, 2, pp. 383–387, 1972a.

Dixon, L. C. W., ''Quasi-Newton Techniques Generate Identical Points II: The Proofs of Four New Theorems,'' *Mathematical Programming*, 3, pp. 345–358, 1972b.

Dixon, L. C. W., ''The Choice of Step Length, A Crucial Factor in the Performance of Variable Metric Algorithms,'' in *Numerical Methods for Nonlinear Optimization*, F. A. Lootsma (Ed.), 1972c.

Dixon, L. C. W., *Nonlinear Optimization*, The English Universities Press, London, 1972d.

Dixon, L. C. W., ''Variable Metric Algorithms: Necessary and Sufficient Conditions for Identical Behavior of Nonquadratic Functions,'' *J. Optimization Theory and Applications*, 10, pp. 34–40, 1972e.

Dixon, L. C. W., ''ACSIM—An Accelerated Constrained Simplex Techniques,'' *Computer Aided Design*, 5, pp. 23–32, 1973a.

Dixon, L. C. W. (Ed.), *Optimization in Action*, Academic Press, New York, 1976.

Dixon, L. C. W., ''Conjugate Directions Without Line Searches,'' *J. Institute Mathematics Applications*, 11, pp. 317–328, 1973b.

Dongarra, J. J., J. R. Bunch, C. B. Moler, and G. W. Stewart, *LINPACK Users Guide*, SIAM Publications, Philadelphia, 1979.

Dorfman, R., P. A. Samuelson, and R. M. Solow, *Linear Programming and Economic Analysis*, McGraw-Hill, New York, 1958.

Dorn, W. S., ''Duality in Quadratic Programming,'' *Quart. Applied Mathematics*, 18, pp. 155–162, 1960a.

Dorn, W. S., ''A Symmetric Dual Theorem for Quadratic Programs,'' *J. Operations Research Society of Japan*, 2m, pp. 93–97, 1960b.

Dorn, W. S., "Self-Dual Quadratic Programs," *J. of the Society for Industrial and Applied Mathematics*, 9, pp. 51–54, 1961a.

Dorn, W. S., "On Lagrange Multipliers and Inequalities," *Operations Research 9*, pp. 95–104, 1961b.

Dorn, W. S., "Linear Fractional Programming," *IBM Research Report*, RC-830, 1962.

Dorn, W. S., "Nonlinear Programming—A Survey," *Management Science*, 9, pp. 171–208, 1963.

Drud, A., *CONOPT–A GRGT Code for Large Sparse Dynamic Nonlinear Optimization Problems*, Technical Note No. 21, Development Research Center, The World Bank, 1818 H Street N.W., Washington, D.C. 20433, March 1984.

Drud, D., "CONOPT A GRG—Code for Large Sparse Dynamic Nonlinear Optimization Problems," *Mathematical Programming*, 31, pp. 153–191, 1985.

Du, D.-Z., and X.-S. Zhang, "A Convergence Theorem for Rosen's Gradient Projection Method," *Mathematical Programming*, 36, pp. 135–144, 1986.

Du, D.-Z, and X.-S. Zhang, "Global Convergence of Rosen's Gradient Project Method," *Mathematical Programming*, 44, pp. 357–366, 1989.

Du Val, P., "The Unloading Problem for Plane Curves," *American J. Mathematics*, 62, pp. 307–311, 1940.

Dubois, J., "Theorems of Convergence for Improved Nonlinear Programming Algorithms," *Operations Research*, 21, pp. 328–332, 1973.

Dubovitskii, M. D., and A. A. Milyutin, "Extremum Problems in the Presence of Restriction," *USSR Computational Mathematics and Mathematical Physics*, 5, pp. 1–80, 1965.

Duffin, R. J. "Convex Analysis Treated by Linear Programming," *Mathematical Programming*, 4, pp. 125–143, 1973.

Duffin, R. J., and E. L. Peterson, "The Proximity of (Algebraic) Geometric Programming to Linear Programming," *Mathematical Programming*, 3, pp. 250–253, 1972.

Duffin, R. J., and E. L. Peterson, "Geometric Programming with Signomials," *J. Optimization Theory and Application*, 11, pp. 3–35, 1973.

Duffin, R. J., E. L. Peterson, and C. Zener, *Geometric Programming*, John Wiley & Sons, New York, 1967.

Eaves, B. C., "On the Basic Theorem of Complementarity," *Mathematical Programming*, 1, pp. 68–75, 1971a.

Eaves, B. C., "The Linear Complementarity Problem," *Management Science*, 17, pp. 612–634, 1971b.

Eaves, B. C., "On Quadratic Programming," *Management Science*, 17, pp. 698–711, 1971c.

Eaves, B. C., "Computing Kakutani Fixed Points," *SIAM J. Applied Mathematics*, 21, pp. 236–244, 1971d.

Eaves, B. C., and W. I. Zangwill, "Generalized Cutting Plane Algorithms," *SIAM J. Control*, 9, pp. 529–542, 1971.

Ecker, J. G., "Geometric Programming: Methods, Computations and Applications," *SIAM Review*, 22, pp. 338–362, 1980.

Eckhardt, U., "Pseudo-complementarity Algorithms for Mathematical Programming," in *Numerical Methods for Nonlinear Optimization*, F. A. Lootsma (Ed.), 1972.

Eggleston, H. G., *Convexity*, Cambridge University Press, Cambridge, 1958.

Eisenberg, E., "Supports of a Convex Foundation," *Bull. American Mathematical Society*, 68, pp. 192–195, 1962.

Eisenberg, E., "On Cone Functions," in *Recent Advances in Mathematical Programming*, R. L. Graves and P. Wolfe (Eds.), 1963.

Eisenberg, E., "A Gradient Inequality for a Class of Nondifferentiable Functions,", *Operations Research*, 14, pp. 157–163, 1966.

El-Attar, R. A., M. Vidyasagar, and S. R. K. Dutta, "An Algorithm for l_1-Norm Minimization With Application to Nonlinear l_1 Approximation," *SIAM J. Numerical Analysis*, 16, pp. 70–86, 1979.

Eldersveld, S. K., *Large-Scale Sequential Quadratic Programming*, SOL91, Department of Operations Research, Stanford University, Stanford, Calif. 94305-4022, 1991.

Elmaghraby, S. E., "Allocation Under Uncertainty When the Demand Has Continuous d.f.," *Management Science*, 6, pp. 270–294, 1960.

Elzinga, J., and T. G. Moore, "A Central Cutting Plane Algorithm for the Convex Programming Problem," *Mathematical Programming*, 8, pp. 134–145, 1975.

Evans, J. P., "On Constraint Qualifications in Nonlinear Programming, " *Naval Research Logistics Quarterly*, 17, pp. 281–286, 1970.

Evans, J. P., and F. J. Gould, "Stability in Nonlinear Programming," *Operations Research*, 18, pp. 107–118, 1970.

Evans, J. P., and F. J. Gould, "On Using Equality-Constraint Algorithms for Inequality Constrained Problems," *Mathematical Programming*, 2, pp. 324–329, 1972a.

Evans, J. P., and F. J. Gould, "A Nonlinear Duality Theorem Without Convexity," *Econometrica*, 40, pp. 487–496, 1972b.

Evans, J. P., and F. J. Gould, "A Generalized Lagrange Multiplier Algorithm for Optimum or Near Optimum Production Scheduling," *Management Science*, 18, pp. 299–311, 1972c.

Evans, J. P., and F. J. Gould, "An Existence Theorem for Penalty Function Theory," *SIAM J. Control*, 12, pp. 509–516, 1974.

Evans, J. P., F. J. Gould, and S. M. Howe, "A Note on Extended GLM," *Operations Research*, 19, pp. 1079–1080, 1971.

Evans, J. P., Gould, F. J., and Tolle, J. W., "Exact Penalty Functions in Nonlinear Programming," *Mathematical Programming*, 4, pp. 72–97, 1973.

Everett, H., "Generalized LaGrange Multiplier Method for Solving Problems of Optimum Allocation of Research," *Operations Research*, 11, pp. 399–417, 1963.

Evers, W. H., "A New Model for Stochastic Linear Programming," *Management Science*, 13, pp. 680–693, 1967.

Fadeev, D. K., and V. N. Fadeva, *Computational Methods of Linear Algebra*, W. H. Freeman, San Francisco, 1963.

Falk, J. E. "Lagrange Multipliers and Nonlinear Programming," *J. Mathematical Analysis and Applications,* 19, pp. 141–159, 1967.

Falk, J. E., "Lagrange Multipliers and Nonconvex Programs," *SIAM J. Control*, 7, pp. 534–545, 1969.

Falk, J. E., "Conditions for Global Optimality in Nonlinear Programming," *Operations Research*, 21, pp. 337–340, 1973.

Falk, J. E., and K. L. Hoffman, "A Successive Underestimating Method for Concave Minimization Problems," *Math of Operations Research*, 1, pp. 251–259, 1976.

Fang, S. C., *A New Unconstrained Convex Programming Approach to Linear Programming*, OR Research Report, North Carolina State University, February 1990.

Farkas, J., "Über die Theorie der einfachen Ungleichungen," *J. für die Reine und Angewandte Mathematick*, 124, pp. 1–27, 1902.

Faure, P., and P. Huard, "Résolution des Programmes Mathématiques à Fonction Nonlinéarire par la Méthode der Gradient Reduit," *Revue Francaise de Recherche Operationelle*, 9, pp. 167–205, 1965.

Fenchel, W., "On Conjugate Convex Functions," *Canadian J. Mathematics*, 1, pp. 73–77, 1949.

Fenchel, W., "Convex Cones, Sets, and Functions," Lecture Notes (mimeographed), Princeton University, 1953.

Ferland, J. A., "Mathematical Programming Problems with Quasi-Convex Objective Functions," *Mathematical Programming*, 3, pp. 296–301, 1972.

Fiacco, A. V., "A General Regularized Sequential Unconstrained Minimization Technique," *SIAM J. Applied Mathematics*, 17, pp. 1239–1245, 1969.

Fiacco, A. V., "Penalty Methods for Mathematical Programming in E^n with General Constraint Sets," *J. Optimization Theory and Applications*, 6, pp. 252–268, 1970.

Fiacco, A. V., "Convergence Properties of Local Solutions of Sequences of Mathematical Programming Problems in General Spaces," *J. Optimization Theory and Applications*, 13, pp. 1–12, 1974.

Fiacco, A. V., "Sensitivity Analysis for Nonlinear Programming Using Penalty Methods," *Mathematical Programming*, 10, pp. 287–311, 1976.

Fiacco, A. V., "Introduction to Sensitivity and Stability Analysis in Nonlinear Programming," *Mathematics in Science and Engineering*, 165, R. Bellman (Ed.), Academic Press, 1983.

Fiacco, A. V., and G. P. McCormick, "The Sequential Unconstrained Minimization Technique for Nonlinear Programming, A Primal-Dual Method," *Management Science*, 10, pp. 360–366, 1964a.

Fiacco, A. V., and G. P. McCormick, "Computational Algorithm for the Sequential Unconstrained Minimization Technique for Nonlinear Programming," *Management Science*, 10, pp. 601–617, 1964b.

Fiacco, A. V., and G. P. McCormick, "Extensions of SUMT for Nonlinear Programming: Equality Constraints and Extrapolation," *Management Science*, 12, pp. 816–828, 1966.

Fiacco, A. V., and G. P. McCormick, "The Slacked Unconstrained Minimization Technique for Convex Programming," *SIAM J. Applied Mathematics*, 15, pp. 505–515, 1967a.

Fiacco, A. V., and G. P. McCormick, "The Sequential Unconstrained Minimization Technique (SUMT), without Parameters," *Operations Research*, 15, pp. 820–827, 1967b.

Fiacco, A. V., and G. P. McCormick, *Nonlinear Programming: Sequential Unconstrained Minimization Techniques*, John Wiley & Sons, New York, 1968.

Finetti, B. De, "Sulla stratificazoni convesse," *Annali di Matematica Pura ed Applicata*, 30, 141, pp. 173–183, 1949.

Finkbeiner, B., and P. Kall, "Direct Algorithms in Quadratic Programming," *Zeitschrift für Operations Research*, 17, pp. 45–54, 1973.

Fisher, M. L., and F. J. Gould, "A Simplicial Algorithm for the Nonlinear Complementarity Problem," *Mathematical Programming*, 6, pp. 281–300, 1974.

Fisher, M. L., W. D. Northup, and J. F. Shapiro, "Using Duality to Solve Discrete Optimization Problems: Theory and Computational Experience," in *Nondifferentiable Optimization*, M. L. Balinski and P. Wolfe (Eds.), 1975.

Flet, T., *Mathematical Analysis*, McGraw-Hill, New York, 1966.

Fletcher, R., "Function Minimization without Evaluating Derivatives—A Review," *Computer Journal*, 8, pp. 33–41, 1965.

Fletcher, R. (Ed.), *Optimization*, Academic Press, London, 1969a.

Fletcher, R., "A Review of Methods for Unconstrained Optimization," in *Optimization*, R. Fletcher (Ed.), pp. 1–12, 1969b.

Fletcher, R., "A New Approach to Variable Metric Algorithms," *Computer Journal*, 13, pp. 317–322, 1970a.

Fletcher, R., "A Class of Methods for Nonlinear Programming with Termination and Convergence Properties," in *Integer and Nonlinear Programming*, J. Abadie (Ed.), 1970b.

Fletcher, R., "A General Quadratic Programming Algorithm," *J. Institute of Mathematics and Its Applications*, 7, pp. 76–91, 1971.

Fletcher, R., "A Class of Methods for Nonlinear Programming III: Rates of Convergence," in *Numerical Methods for Nonlinear Optimization*, F.A. Lootsma (Ed.), 1972a.

Fletcher, R., "Minimizing General Functions Subject to Linear Constraints," in *Numerical Methods for Nonlinear Optimization*, F. A. Lootsma (Ed.), 1972b.

Fletcher, R., "An Algorithm for Solving Linearly Constrained Optimization Problems," *Mathematical Programming*, 2, pp. 133–161, 1972c.

Fletcher, R., "An Exact Penalty Function for Nonlinear Programming with Inequalities," *Mathematical Programming*, 5, pp. 129–150, 1973.

Fletcher, R., "An Ideal Penalty Function for Constrained Optimization," *J. of the Institute of Mathematics and Its Application*, 15, pp. 319–342, 1975.

Fletcher, R., "On Newton's Method For Minimization," in *Proc. IX International Symposium on Mathematical Programming*, A. Prekopa (Ed.), Academiai Kiado, Budapest, 1978.

Fletcher, R., *Practical Methods of Optimization* (2nd Edition), John Wiley & Sons, New York, 1987.

Fletcher, R., *Practical Methods of Optimization 1: Unconstrained Optimization*, John Wiley & Sons, Chichester, 1980.

Fletcher, R., *Practical Methods of Optimization 2: Constrained Optimization*, John Wiley & Sons, Chichester, 1981.

Fletcher, R., "Numerical Experiments with an L_1 Exact Penalty Function Method," in *Nonlinear Programming, 4*, O. L. Mangasarian, R. R. Meyer, and S. M. Robinson (Eds.), Academic Press, New York, 1981.

Fletcher, R., "A Model Algorithm for Composite Nondifferentiable Optimization Problems," in *Nondifferential and Variational Techniques in Optimization*, D. C. Sorensen and R. J.-B. Wets (Eds.), *Mathematical Programming Study, 17*, North Holland, Amsterdam, 1982a.

Fletcher, R., "Second Order Corrections for Nondifferentiable Optimization," in *Numerical Analysis*, "Dundee 1981," G. A. Watson (Ed.), *Lecture Notes in Mathematics*, 912, Springer-Verlag, Berlin, 1982b.

Fletcher, R., "Penalty Functions," in *Mathematical Programming: The State of The Art*, A. Bachem, M. Grotschel, and B. Korte (Eds.), Springer-Verlag, Berlin, Heidelberg, New York, and Tokyo, pp. 87–114, 1983.

Fletcher, R., "An l_1 Penalty Method for Nonlinear Constraints," in *Numerical Optimization*, P. T. Boggs, R. H. Byrd, and R. B. Schnabel (Eds.), *SIAM Publications*, Philadelphia, 1985.

Fletcher, R., and T. L. Freeman, "A Modified Newton Method for Minimization," *J. Optimization Theory Applications*, 23, pp. 357–372, 1977.

Fletcher, R., and S. Lill, "A Class of Methods for Nonlinear Programming II: Computational Experience," in *Nonlinear Programming*, J. B. Rosen, O. L. Mangasarian, and K. Ritter (Eds.), 1971.

Fletcher, R., and A. McCann, "Acceleration Techniques for Nonlinear Programming," in *Optimization*, R. Fletcher (Ed.), 1969.

Fletcher, R., and M. Powell, "A Rapidly Convergent Descent Method for Minimization," *Computer Journal*, 6, pp. 163–168, 1963.

Fletcher, R., and C. Reeves, "Function Minimization by Conjugate Gradients," *Computer Journal*, 7, pp. 149–154, 1964.

Fletcher, R., and E. Sainz de la Maza, "Nonlinear Programming and Nonsmooth Optimization by Successive Linear Programming," University of Dundee, Department of Mathematical Science, Report NA/100, 1987.

Fletcher, R., and G. A. Watson, "First- and Second-Order Conditions for a Class of Nondifferentiable Optimization Problems," *Math Programming*, 18, pp. 291–307; abridged from a University of Dundee Department of Mathematics Report NA/28 (1978), 1980.

Floudas, C. A., and V. Visweswaren, "A Primal-Relaxed Dual Global Optimization Approach," Department of Chemical Engineering, Princeton University, Princeton, N.J., 1991.

Forsythe, G., and T. Motzkin, "Acceleration of the Optimum Gradient Method," *Bull. American Mathematical Society*, 57, pp. 304–305, 1951.

Fourer, R., D. Gay, and B. Kernighan, "A Modeling Language for Mathematical Programming," *Management Science*, 36(5), pp. 519–554, 1990.

Fox, R. L., "Mathematical Methods in Optimization," in *An Introduction to Structural Optimization*, M. Z. Cohn (Ed.), University of Waterloo, 1969.

Fox, R. L., *Optimization Methods for Engineering Design*, Addison-Wesley, Reading, Mass., 1971.

Frank, M., and P. Wolfe, "An Algorithm for Quadratic Programming," *Naval Research Logistics Quarterly*, 3, pp. 95–110, 1956.

Fruend, R. J., "The Introduction of Risk with a Programming Model," *Econometrica*, 24, pp. 253–263, 1956.
Friedman, P., and K. L. Pinder, "Optimization of Simulation Model of a Chemical Plant," *Industrial and Engineering Chemistry Product Research and Development*, 11, pp. 512–520, 1972.
Frisch, K. R., "The Logarithmic Potential Method of Convex Programming," University Institute of Economics, Oslo, Norway, 1955 (unpublished manuscript).
Fujiwara, O., B. Jenmchaimahakoon, and N. C. P. Edirisinghe, "A Modified Linear Programming Gradient Method for Optimal Design of Looped Water Distribution Networks," *Water Resources Research*, 23(6), pp. 977–982, June 1987.
Fukushima, M., "A Successive Quadratic Programming Algorithm with Global and Superlinear Convergence Properties," *Mathematical Programming*, 35, pp. 253–264, 1986.
Gabay, D., "Reduced Quasi-Newton Methods With Feasibility Improvement for Nonlinearly Constrained Optimization," *Mathematical Programming Study*, 16, pp. 18–44, 1982.
Gacs, P., and L. Lovasz, "Khachiyan's Algorithm for Linear Programming," *Mathematical Programming Study*, 14, pp. 61–68, 1981.
Garcia, C. B., and W. I. Zangwill, *Pathways to Solutions, Fixed Points, and Equilibria*, Prentice Hall, Englewood Cliffs, N. J., 1981.
Garey, M. R., and D. S. Johnson, "*Computers and Intractability: A Guide to the Theory of NP-Completeness*," W. H. Freeman & Co., New York, 1979.
Garstka, S. J., "Regularity Conditions for a Class of Convex Programs," *Management Science*, 20, pp. 373–377, 1973.
Gay, D. M., "Some Convergence Properties of Broyden's Method," *SIAM J. Numerical Analysis*, 16, pp. 623–630, 1979.
Gehner, K. R., "Necessary and Sufficient Optimality Conditions for the Fritz John Problem with Linear Equality Constraints," *SIAM J. Control*, 12, pp. 140–149, 1974.
Geoffrion, A. M., "Strictly Concave Parametric Programming, I, II," *Management Science*, 13, pp. 244–253, 1966; and 13, pp. 359–370, 1967a.
Geoffrion, A. M., "Reducing Concave Programs with Some Linear Constraints," *SIAM J. Applied Mathematics*, 15, pp. 653–664, 1967b.
Geoffrion, A. M., "Stochastic Programming with Aspiration or Fractile Criteria," *Management Science*, 13, pp. 672–679, 1967c.
Geoffrion, A. M., "Proper Efficiency and the Theory of Vector Maximization," *J. Mathematical Analysis and Applications*, 22, 618–630, 1968.
Geoffrion, A. M., "A Markovian Procedure for Strictly Concave Programming with Some Linear Constraints," in *Proceedings of the Fourth International Conference on Operational Research*, Wiley-Interscience, New York, 1969.
Geoffrion, A. M., "Primal Resource-Directive Approaches for Optimizing Nonlinear Decomposable Systems," *Operations Research*, 18, pp. 375–403, 1970a.
Geoffrion, A. M., "Elements of Large-Scale Mathematical Programming, I, II," *Management Science*, 16, pp. 652–675, 676–691, 1970b.
Geoffrion, A. M., "Large-Scale Linear and Nonlinear Programming," in *Optimization Methods for Large-Scale Systems*, D. A. Wismer (Ed.), 1971a.
Geoffrion, A. M., "Duality in Nonlinear Programming: A Simplified Applications-Oriented Development," *SIAM Review*, 13, pp. 1–37, 1971b.
Geoffrion, A. M., "Generalized Benders Decomposition," *J. Optimization Theory and Applications*, 10, pp. 237–260, 1972a.
Geoffrion, A. M. (Ed.), *Perspectives on Optimization*, Addison-Wesley, Reading, Mass., 1972b.
Geoffrion, A. M., "Objective Function Approximations in Mathematical Programming," *Mathematical Programming*, 13, pp. 23–27, 1977.
Gerencser, L., "On a Close Relation Between Quasi-Convex and Convex Functions and Related Investigations," *Mathematische Operationsforschung und Statist*, 4, pp. 201–211, 1973.

Ghani, S. N., "An Improved Complex Method of Function Minimization," *Computer Aided Design*, 4, pp. 71–78, 1972.

Gilbert, E. G., "An Iterative Procedure for Computing the Minimum of a Quadratic Form on a Convex Set," *SIAM J. Control*, 4, pp. 61–80, 1966.

Gill, P. E., G. H. Golub, W. Murray, and M. A. Saunders, "Methods for Modifying Matrix Factorizations," *Mathematics of Comp.*, 28, pp. 505–535, 1974.

Gill, P. E., N. I. M. Gould, W. Murray, M. A. Saunders, and M. H. Wright, "Weighted Gram-Schmidt Method for Convex Quadratic Programming," *Mathematical Programming*, 30, 176–195.

Gill, P. E., and W. Murray, "Quasi-Newton Methods for Unconstrained Optimization," *J. Institute of Mathematics and Its Applications*, 9, pp. 91–108, 1972.

Gill, P. E., and W. Murray, "Newton-Type Methods for Unconstrained and Linearly Constrained Optimization," *Mathematical Programming*, 7, pp. 311–350, 1974a.

Gill, P. E., and W. Murray, *Numerical Methods for Constrained Optimization*, Academic Press, New York, 1974b.

Gill, P. E., and W. Murray, "Numerically Stable Methods For Quadratic Programming," *Mathematical Programming*, 14, pp. 349–372, 1978.

Gill, P. E., and W. Murray, "The Computation of Lagrange Multiplier Estimates for Constrained Minimization," *Mathematical Programming*, 17, pp. 32–60, 1979b.

Gill, P. E., W. Murray, W. M. Pickens, and M. H. Wright, "The Design and Structure of a Fortran Program Library for Optimization," *ACM Trans. Mathematical Software*, 5, pp. 259–283, 1979.

Gill, P. E., W. Murray, and P. A. Pitfield, "The Implementation of Two Revised Quasi-Newton Algorithms for Unconstrained Optimization," Report NAC-11, *National Physical Lab.*, 1972.

Gill, P. E., W. Murray, and M. A. Saunders, "Methods for Computing and Modifying the LDV Factors of a Matrix," *Math. Comp.*, 29, pp. 1051–1077, 1975.

Gill, P. E., W. Murray, M. A. Saunders, J. A. Tomlin, and M. H. Wright, "On Projected Newton Barrier Methods for Linear Programming and an Equivalence to Karmarkar's Method," *Mathematical Programming*, 36, 183–209, 1986.

Gill, P. E., W. Murray, M. A. Saunders, and M. H. Wright, "QP-Based Methods for Large-Scale Nonlinearly Constrained Optimization," *Nonlinear Programming, 4*, O. L. Mangasarian, R. R. Meyer, and S. M. Robinson (Eds.), Academic Press, London and New York, 1981b.

Gill, P. E., W. Murray, M. A. Saunders, and M. H. Wright, "User's Guide for QPSOL (Version 3.2): A Fortran Package for Quadratic Programming," Report SOL 84-6, Department of Operations Research, Stanford University, Calif., 1984a.

Gill, P. E., W. Murray, M. A. Saunders, and M. H. Wright, "User's Guide for NPSOL (Version 2.1): A Fortran Package for Nonlinear Programming," Report SOL 84-7, Department of Operations Research, Stanford University, Calif., 1984b.

Gill, P. E., W. Murray, M. A. Saunders, and M. Wright, "Sparse Matrix Methods in Optimization," *SIAM J. Science Stat. Comp.*, 5, pp. 562–589, 1984c.

Gill, P. E., W. Murray, M. A. Saunders, and M. H. Wright, "Procedures For Optimization Problems with a Mixture of Bounds and General Linear Constraints," *ACM Transactions on Mathematical Software*, 10, pp. 282–298, 1984d.

Gill, P. E., W. Murray, M. A. Saunders, and M. H. Wright, "Software and Its Relationship to Methods," Report SOL 84-10, Department of Operations Research, Stanford University, Calif., 1984e.

Gill, P. E., W. Murray, M. A. Saunders, and M. H. Wright, "Some Theoretical Properties of an Augmented Lagrangian Merit Function," Systems Optimization Lab. Report SOL 86-6, Stanford University, Calif., 1986.

Gill, P. E., W. Murray, M. A. Saunders, and M. H. Wright, "Model Building and Practical Aspects of Nonlinear Programming," *NATO ASI Series*, F15, *Computational Mathematical Programming*, K. Schittkowski (Ed.), Springer-Verlag, Berlin and Heidelberg, pp. 209–247, 1985.

Gill, P. E., W. Murray, and M. H. Wright, *Practical Optimization*, Academic Press, London and New York, 1981.

Gillmore, P. C., and R. E. Gomory, "A Linear Programming Approach to the Cutting Stock Problem, Part II," *Operations Research*, pp. 863–888, 1963.

Girsanov, I. V., *Lectures on Mathematical Theory of Extremum Problems*, Lecture Notes in Economics and Mathematical Systems, No. 67, Springer-Verlag, New York, 1972.

Gittleman, A., "A General Multiplier Rule," *J. Optimization Theory and Applications*, 7, pp. 29–38, 1970.

Glad, S. T., "Properties of Updating Methods for the Multipliers in Augmented Lagrangians," *J. Optimization Theory and Applications*, 28, pp. 135–156, 1979.

Glad, S. T., and Polak, E., "A Multiplier Method with Automatic Limitation of Penalty Growth," *Mathematical Programming*, 17, pp. 140–155, 1979.

Glass, H., and L. Cooper, "Sequential Search: A Method for Solving Constrained Optimization Problems," *J. Association Computing Machinery*, 12, pp. 71–82, 1965.

Goffin, J. L., "On Convergence Rates of Subgradient Optimization Methods," *Mathematical Programming*, 13, pp. 329–347, 1977.

Goffin, J. L., "Convergence Results for a Class of Variable Metric Subgradient Methods," in *Nonlinear Programming, 4*, O. L. Mangasarian, R. R. Meyer, and S. M. Robinson (Eds.), Academic Press, London and New York, 1980.

Goffin, J. L., "The Relaxation Method for Solving Systems of Linear Inequalities," *Mathematics of Operations Research*, 5(3), pp. 388–414, 1980.

Goldfarb, D., "Extension of Davidon's Variable Metric Method to Maximization Under Linear Inequality and Equality Constraints," *SIAM J. Applied Mathematics*, 17, pp. 739–764, 1969a.

Goldfarb, D., "Sufficient Conditions for the Convergence of a Variable Metric Algorithm," in *Optimization*, R. Fletcher (Ed.), 1969b.

Goldfarb, D., "A Family of Variable Metric Methods Derived by Variational Means," *Mathematics of Computation*, 24, pp. 23–26, 1970.

Goldfarb, D., "Extensions of Newton's Method and Simplex Methods for Solving Quadratic Programs," in *Numerical Methods for Nonlinear Optimization*, F. A. Lootsma (Ed.), 1972.

Goldfarb, D., "Curvilinear Path Steplength Algorithms for Minimization Which Use Directions of Negative Curvature," *Mathematical Programming*, 18, pp. 31–40, 1980.

Goldfarb, D., and A. Idnani, "A Numerically Stable Dual Method for Solving Strictly Convex Quadratic Programs," *Mathematical Programming*, 27, pp. 1–33, 1983.

Goldfarb, D., and L. Lapidus, "A Conjugate Gradient Method for Nonlinear Programming," *Industrial and Engineering Chemistry Fundamentals*, 7, pp. 142–151, 1968.

Goldfarb, D., and S. Liu, "An $O(n^3L)$ Primal Interior Point Algorithm for Convex Quadratic Programming," *Mathematical Programming*, 49(3), pp. 325–340, 1990/1991.

Goldfarb, D., and J. K. Reid, "A Practicable Steepest-edge Algorithm," *Mathematical Programming*, 12, pp. 361–371, 1977.

Goldfeld, S. M., R. E. Quandt, and M. F. Trotter, "Maximization by Improved Quadratic Hill Climbing and Other Methods," *Econ. Res. Memo 95*, Princeton University Research Program, 1968.

Goldstein, A. A., "Cauchy's Method of Minimization," *Numerische Mathematik*, 4, pp. 146–150, 1962.

Goldstein, A. A., "Convex Programming and Optimal Control," *SIAM J. Control*, 3, pp. 142–146, 1965a.

Goldstein, A. A., "On Steepest Descent," *SIAM J. Control*, 3, pp. 147–151, 1965b.

Goldstein, A. A., "On Newton's Method," *Numerische Mathematik*, 7, pp. 391–393, 1965c.

Goldstein, A. A., and J. F. Price, "An Effective Algorithm for Minimization," *Numerische Mathematik*, 10, pp. 184–189, 1967.

Golub, G. H., and M. A. Saunders, "Linear Least Squares and Quadratic Programming," in *Nonlinear Programming*, J. Abadie (Ed.), 1970.

Golub, G. H., and C. Van Loan, *Matrix Computations*, The Johns Hopkins University Press, 1983 (2nd Edition, 1989).

Gomes, H. S., and J. M. Martinez, "A Numerically Stable Reduced-Gradient Type Algorithm for Solving Large-Scale Linearly Constrained Minimization Problems," *Computers and Operations Research*, 18(1), pp. 17–31, 1991.

Gomory, R., "Large and Nonconvex Problems in Linear Programming," *Proc. Symposium Applied Mathematics*, 15, pp. 125–139, American Mathematical Society, Providence, R.I. 1963.

Gonzaga, C. C., "An Algorithm for Solving Linear Programming in $O(n^3L)$ Operations," in *Progress in Mathematical Programming-Interior-Point and Related Methods*, Nimrod Megiddo (Ed.), Springer-Verlag, pp. 1–28, 1989 (Manuscript 1987).

Gonzaga, C. C., "Polynomial Affine Algorithms for Linear Programming," *Mathematical Programming*, 49, 7–21, 1990.

Gottfred, B. S., and J. Weisman, *Introduction to Optimization Theory*, Prentice Hall, Englewood Cliffs, N.J., 1973.

Gould, F. J., "Extensions of Lagrange Multipliers in Nonlinear Programming," *SIAM J. Applied Mathematics*, 17, pp. 1280–1297, 1969.

Gould, F. J., "A Class of Inside-Out Algorithms for General Programs," *Management Science*, 16, pp. 350–356, 1970.

Gould, F. J., "Nonlinear Pricing: Applications to Concave Programming," *Operations Research*, 19, pp. 1026–1035, 1971.

Gould, F. J., and J. W. Tolle, "A Necessary and Sufficient Qualification for Constrained Optimization," *SIAM J. Applied Mathematics*, 20, pp. 164–172, 1971.

Gould, F. J., and J. W. Tolle, "Geometry of Optimality Conditions and Constraint Qualifications," *Mathematical Programming*, 2, pp. 1–18, 1972.

Graves, R. L., "A Principal Pivoting Simplex Algorithm for Linear and Quadratic Programming," *Operations Research*, 15, pp. 482–494, 1967.

Graves, R. L., and P. Wolfe, *Recent Advances in Mathematical Programming*, McGraw-Hill, New York, 1963.

Greenberg, H. J., "A Lagrangian Property for Homogeneous Programs," *J. Optimization Theory and Applications*, 12, pp. 99–10, 1973a.

Greenberg, H. J. "The Generalized Penalty-Function/Surrogate Model," *Operations Research*, 21, pp. 162–178, 1973b.

Greenberg, H. J., "Bounding Nonconvex Programs by Conjugates," *Operations Research*, 21, pp. 346–348, 1973c.

Greenberg, H. J., and W. P. Pierskalla, "Symmetric Mathematical Programs," *Management Science*, 16, pp. 309–312, 1970.

Greenberg, H. J., and W. P. Pierskalla, "Surrogate Mathematical Programming," *Operations Research*, 18, pp. 924–939, 1970a.

Greenberg, H. J., and W. P. Pierskalla, "A Review of Quasi-Convex Functions," *Operations Research*, 29, pp. 1553–1570, 1971b.

Greenberg, H. J., and W. P. Pierskalla, "Extensions of the Evans-Gould Stability Theorems for Mathematical Programs," *Operations Research*, 20, pp. 143–153, 1972.

Greenstadt, J., "On the Relative Efficiencies of Gradient Methods," *Mathematics of Computation*, 21, pp. 360–367, 1967.

Greenstadt, J., "Variations on Variable Metric Methods," *Mathematics of Computation*, 24, pp. 1–22, 1970.

Greenstadt, J., "A Quasi-Newton Method with No Derivatives," *Mathematics of Computation*, 26, pp. 145–166, 1972.

Griewank, A. O., and Ph. L. Toint, "Partitioned Variable Metric Updates for Large Sparse Optimization Problems," *Numerical Mathematics*, 39, 119–37, 1982a.

Griewank, A. O., and Ph. L. Toint, "Local Convergence Analysis for Partitioned Quasi-Newton Updates in the Broyden Class," *Numerical Mathematics*, 39, 1982b.

Griffith, R. E., and R. A. Stewart, "A Nonlinear Programming Technique for the Optimization of Continuous Processing Systems," *Management Science*, 7, pp. 379–392, 1961.
Grinold, R. C., "Lagrangian Subgradients," *Management Science*, 17, pp. 185–188, 1970.
Grinold, R. C., "Mathematical Programming Methods of Pattern Classification," *Management Science*, 19, pp. 272–289, 1972a.
Grinold, R. C., "Steepest Ascent for Large-Scale Linear Programs," *SIAM Review*, 14, pp. 447–464, 1972b.
Grotzinger, S. J., "Supports and Convex Envelopes," *Mathematical Programming*, 31, pp. 339–347, 1985.
Grünbaum, B., *Convex Polytopes*, John Wiley & Sons, New York, 1967.
Gue, R. L., and M. E. Thomas, *Mathematical Methods in Operations Research*, The Macmillan Company, Collier-Macmillan Ltd., London, 1968.
Guignard, M., "Generalized Kuhn-Tucker Conditions for Mathematical Programming Problems in a Banach Space," *SIAM J. Control*, 7, pp. 232–241, 1969.
Guin, J. A., "Modification of the Complex Method of Constrained Optima," *Computer Journal*, 10, pp. 416–417, 1968.
Haarhoff, P. C., and J. D. Buys, "A New Method for the Optimization of a Nonlinear Function Subject to Nonlinear Constraints," *The Computer Journal*, 13, pp. 178–184, 1970.
Habetler, G. J., and A. L. Price, "Existence Theory for Generalized Nonlinear Complementarity Problems," *J. Optimization Theory and Applications*, 7, pp. 223–239, 1971.
Habetler, G. J., and A. L. Price, "An Iterative Method for Generalized Nonlinear Complementarity Problems," *J. Optimization Theory and Applications*, 11, pp. 36–48, 1973.
Hadamard, J., "Étude sur les propriétés des fonctions entières et en particulier d'une fonction considérée par Riemann," *J. Mathematiques Pures et Appliquées*, 58, pp. 171–215, 1893.
Hadley, G., *Linear Programming*, Addison-Wesley, Reading, Mass., 1962.
Hadley, G., *Nonlinear and Dynamic Programming*, Addison-Wesley, Reading, Mass., 1964.
Hadley, G., and T. M. Whitin, *Analyses of Inventory Systems*, Prentice Hall, Englewood Cliffs, N.J., 1963.
Haimes, Y. Y., "Decomposition and Multi-level Approach in Modeling and Management of Water Resources Systems," in *Decomposition of Large-Scale Problems*, D. M. Himmelblau (Ed.), 1973.
Haimes, Y. Y., *Hierarchical Analyses of Water Resources Systems: Modeling and Optimization of Large-Scale Systems*, McGraw-Hill, New York, 1977.
Haimes, Y. Y., and W. S. Nainis, "Coordination of Regional Water Resource Supply and Demand Planning Models," *Water Resources Research*, 10, pp. 1051–1059, 1974.
Hald, J., and K. Madsen, "Combined LP and Quasi-Newton Methods for Minimax Optimization," *Mathematical Programming*, 20, pp. 49–62, 1981.
Hald, J., and K. Madsen, "Combined LP and Quasi-Newton Methods for Nonlinear l_1 Optimization," *SIAM J. Numerical Analysis*, 22, pp. 68–80, 1985.
Halkin, H., and L. W. Neustadt, "General Necessary Conditions or Optimization Problems," *Proc. National Academy of Sciences, USA*, 56, pp. 1066–1071, 1966.
Hammer, P. L., and G. Zoutendijk (Eds.), *Mathematical Programming in Theory and Practice*, Proceedings of the Nato Advanced Study Institute, Portugal, North Holland Publishing Company, New York, 1974.
Han, S.-P., "A Globally Convergent Method for Nonlinear Programming," Report TR 75-257, Computer Science, Cornell University, Ithaca, N.Y., 1975.
Han, S. P., "Penalty Lagrangian Methods in a Quasi-Newton Approach," Report TR 75-252, Computer Science, Cornell University, Ithaca, N.Y., 1975.
Han, S. P., "Superlinearly Convergent Variable Metric Algorithms For General Nonlinear Programming Problems," *Mathematical Programming*, 11, pp. 263–282, 1976.
Han, S. P., and O. L. Magasarian, "Exact Penalty Functions in Nonlinear Programming," *Mathematical Programming*, 17, pp. 251–269, 1979.

Han, C.-G., P. M. Pardalos, and Y. Ye, "Computational Aspects of an Interior Point Algorithm for Quadratic Programming Problems with Box Constraints," Computer Science Department, Pennsylvania State University, University Park, Penn., 1989.

Hancock, H., "Theory of Maxima and Minima," Dover Publications, New York (original publication 1917), 1960.

Handler, G. Y., and P. B. Mirchandani, *Location on Networks: Theory and Algorithms,* MIT Press, Cambridge, Mass., 1979.

Hans Tjian, T. Y., and W. I. Zangwill, "Analysis and Comparison of the Reduced Gradient and the Convex Simplex Method for Convex Programming," Paper presented at ORSA 41st Nat'l Meeting, New Orleans, April 1972.

Hanson, M. A., "A Duality Theorem in Nonlinear Programming with Nonlinear Constraints," *Australian J. Statistics*, 3, pp. 64–72, 1961.

Hanson, M. A., "An Algorithm for Convex Programming," *Australian J. Statistics*, 5, pp. 14–19, 1963.

Hanson, M. A., "Duality and Self-Duality in Mathematical Programming," *SIAM J. Applied Mathematics*, 12, pp. 446–449, 1964.

Hanson, M. A., "On Sufficiency of the Kuhn–Tucker Conditions," *J. of Mathematical Analysis & Applications*, 80, pp. 545–550, 1981.

Hanson, M. A., and B. Mond, "Further Generalization of Convexity in Mathematical Programming," *J. of Information Theory and Optimization Sciences*, pp. 25–32, 1982.

Hanson, M. A., and B. Mond, "Necessary and Sufficient Conditions in Constrained Optimization," *Mathematical Programming*, 37, pp. 51–58, 1987.

Hardy, G. H., J. E. Littlewood, and G. Polya, *Inequalities*, Cambridge University Press, Cambridge, England, 1934.

Hartley, H. O., "Nonlinear Programming by the Simplex Method," *Econometrica*, 29, pp. 223–237, 1961.

Hartley, H. O., and R. R. Hocking, "Convex Programming by Tangential Approximation," *Management Science*, 9, pp. 600–612, 1963.

Hartley, H. O., and R. C. Pfaffenberger, "Statistical Control of Optimization," in *Optimizing Methods in Statistics*, J. S. Rustagi (Ed.), Academic Press, New York, 1971.

Hausdorff, F., *Set Theory*, Chelsea, New York, 1962.

Hearn, D. W., and S. Lawphongpanich, "Lagrangian Dual Ascent by Generalized Linear Programming," *Operations Research Letters*, 8, pp. 189–196, 1989.

Hearn, D. W., and S. Lawphongpanich, "A Dual Ascent Algorithm for Traffic Assignment Problems," *Transportation Research-B*, 248(6), pp. 423–430, 1990.

Held, M., and R. M. Karp, "The Travelling Salesman Problem and Minimum Spanning Trees," *Operation Research*, 18, pp. 1138–1162, 1970.

Held, M., P. Wolfe, and H. Crowder, "Validation of Subgradient Optimization," *Mathematical Programming*, 6, pp. 62–88, 1974.

Hensgen, C., "Process Optimization by Non-Linear Programming," *Institut Belge de Regulation et d'Automatisme. Revue A.*, 8, pp. 99–104, 1966.

Hestenes, M. R., *Calculus of Variations and Optimal Control Theory*, John Wiley & Sons, New York, 1966.

Hestenes, M. R., "Multiplier and Gradient Methods," *J. Optimization Theory and Applications*, 4, pp. 303–320, 1969.

Hestenes, M. R., *Conjugate-Direction Methods in Optimization*, Springer-Verlag, Berlin, 1980.

Hestenes, M. R., "Augmentability in Optimization Theory," *J. Optimization Theory Applications*, 32, pp. 427–440, 1980b.

Hestenes, M. R., and E. Stiefel, "Methods of Conjugate Gradients for Solving Linear Systems," *J. Research National Bureau of Standards*, 49, pp. 409–436, 1952.

Heyman, D. P., "Another Way to Solve Complex Chemical Equilibria," *Operations Research*, 38(2), pp. 355–358, 1980.

Hildreth, C., "A Quadratic Programming Procedure," *Naval Research Logistics Quarterly*, 4, pp. 79–85, 1957.
Himmelblau, D. M., *Applied Nonlinear Programming*, McGraw-Hill, New York, 1972a.
Himmelblau, D. M., "A Uniform Evaluation of Unconstrained Optimization Techniques," in *Numerical Methods for Nonlinear Optimization*, F. A. Lootsma (Ed.), pp. 69–97, 1972b.
Himmelblau, D. M. (Ed.), *Decomposition of Large-Scale Problems*, North Holland, Amsterdam, 1973.
Hiriart-Urruty, J. B., "On Optimality Conditions in Nondifferentiable Programming," *Mathematical Programming*, 14, pp. 73–86, 1978.
Hock, W., and K. Schittkowski, "Test Examples For Nonlinear Programming," in *Lecture Notes in Economics and Mathematical Systems*, Vol. 187, Springer-Verlag, Berlin, Heidelberg, and New York, 1981.
Hogan, W. W., "Directional Derivatives for Extremal-Value Functions with Applications to the Completely Convex Case," *Operations Research*, 21, pp. 188–209, 1973a.
Hogan, W. W., "The Continuity of the Perturbation Function of a Convex Program," *Operations Research*, 21, pp. 351–352, 1973b.
Hogan, W. W., "Applications of a General Convergence Theory for Outer Approximation Algorithms," *Mathematical Programming*, 5, pp. 151–168, 1973c.
Hogan, W. W., "Point-to-Set Maps in Mathematical Programming," *SIAM Review*, 15, pp. 591–603, 1973d.
Hohenbalken, B. von, "Simplicial Decomposition in Nonlinear Programming Algorithms," *Mathematical Programming*, 13, pp. 49–68, 1977.
Hölder, O., Ūber einen Mittelwertsatz, *Nachrichten von der Gesellschaft der Wisenschaften zu Göttingen*, pp. 38–47, 1889.
Holloway, C. A., "A Generalized Approach to Dantzig-Wolfe Decomposition for Concave Programs," *Operations Research*, 21, pp. 210–220, 1973.
Holloway, C. A., "An Extension of the Frank and Wolfe Method of Feasible Directions," *Mathematical Programming*, 6, pp. 14–27, 1974.
Holt, C. C., F. Modigliani, J. F. Muth, and H. A. Simon, *Planning Production, Inventories, and Work Force*, Prentice Hall, Englewood Cliffs, N.J., 1960.
Hooke, R., and T. A. Jeeves, "Direct Search Solution of Numerical and Statistical Problems," *J. Association Computer Machinery*, 8, pp. 212–229, 1961.
Horst, R., and H. Tuy, "*Global Optimization: Deterministic Approaches*," Springer-Verlag, Berlin and Heidelberg, 1990.
Houthaker, H. S., "The Capacity Method of Quadratic Programming," *Econometrica*, 28, pp. 62–87, 1960.
Howe, S., "New Conditions for Exactness of a Simple Penalty Function," *SIAM J. Control*, 11, pp. 378–381, 1973.
Howe, S., "A Penalty Function Procedure for Sensitivity Analysis of Concave Programs," *Management Science*, 21, pp. 341–347, 1976.
Huang, H. Y., "Unified Approach to Quadratically Convergent Algorithms for Function Minimization," *J. Optimization Theory and Applications*, 5, pp. 405–423, 1970.
Huang, H. Y., and J. P. Chamblis, "Quadratically Convergent Algorithms and One-Dimensional Search Schemes," *J. Optimization Theory and Applications*, 11, pp. 175–188, 1973.
Huang, H. Y., and A. V. Levy, "Numerical Experiments on Quadratically Convergent Algorithms for Function Minimization," *J. Optimization Theory and Applications*, 6, pp. 269–282, 1970.
Huard, P., "Resolution of Mathematical Programming with Nonlinear Constraints by the Method of Centres," in *Nonlinear Programming*, J. Abadie (Ed.), 1967.
Huard, P., "Optimization Algorithms and Point-to-Set Maps," *Mathematical Programming*, 8, pp. 308–331, 1975.
Huard, P., "Extensions of Zangwill's Theorem," in *Point-to-set Maps, Mathematical Programming Study, Vol. 10*, P. Huard (Ed.), North Holland, pp. 98–103, 1979.

Hwa, C. S., "Mathematical Formulation and Optimization of Heat Exchanger Networks Using Separable Programming," *Proceedings of the Joint American Institute of Chemical Engineers/Institution of Chemical Engineers, London Symposium*, June, 4, pp. 101–106, 1965.

Ignizio, J. P., *Goal Programming and Extensions*, Lexington Books, D. C. Health & Co., 1976.

Intriligator, M. D., *Mathematical Optimization and Economic Theory*, Prentice Hall, Englewood Cliffs, N.J., 1971.

Iri, M., and K. Tanabe (Eds.), *Mathematical Programming: Recent Developments and Applications*, Kluwer Academic Publishers, Tokyo, Japan, 1989.

Isbell, J. R., and W. H. Marlow, "Attrition Games," *Naval Research Logistics Quarterly*, 3, pp. 71–94, 1956.

Jacoby, S. L. S., "Design of Optimal Hydraulic Networks," *J. Hydraulics Division American Society of Civil Engineers*, 94-HY3, pp. 641–661, 1968.

Jacoby, S. L. S., J. S. Kowalik, and J. T. Pizzo, *Iterative Methods for Nonlinear Optimization Problems*, Prentice Hall, Englewood Cliffs, N.J., 1972.

Jacques, G., "A Necessary and Sufficient Condition to Have Bounded Multipliers in Nonconvex Programming," *Mathematical Programming*, 12, pp. 136–138, 1977.

Jagannathan, R., "A Simplex-Type Algorithm for Linear and Quadratic Programming—A Parametric Procedure," *Econometrica*, 34, pp. 460–471, 1966a.

Jagannathan, R., "On Some Properties of Programming Problems in Parametric Form Pertaining to Fractional Programming," *Management Science*, 12, pp. 609–615, 1966b.

Jagannathan, R., "Duality for Nonlinear Fractional Programs," *Zeitschrift für Operations Research A*, 17, pp. 1–3, 1973.

Jagannathan, R., "A Sequential Algorithm for a Class of Programming Problems with Nonlinear Constraints," *Management Science*, 21, pp. 13–21, 1974.

Jefferson, T. R., and C. H. Scott, "The Analysis of Entropy Models with Equality and Inequality Constraints," *Transportation Research*, 13B, pp. 123–132, 1979.

Jensen, J. L. W. V., "Om Konvexe Funktioner og Uligheder mellem Middelvaerdier," *Nyt Tidsskrift for Matematik*, 16B, pp. 49–69, 1905.

Jensen, J. L. W. V., "Sur les fonctions convexes et les inégalités entre les valeurs moyennes," *Acta Mathematica*, 30, pp. 175–193, 1906.

John, F., "Extremum Problems with Inequalities as Side Conditions," in *Studies and Essays, Courant Anniversary Volume*, K. O. Friedrichs, O. E. Neugebauer, and J. J. Stoker (Eds.), Wiley-Interscience, New York, 1948.

Johnson, R. C., *Optimum Design of Mechanical Systems*, John Wiley & Sons, New York, 1961.

Johnson, R. C., *Mechanical Design Synthesis with Optimization Examples*, Van Nostrand Reinhold, New York, 1971.

Kall, P., *Stochastic Linear Programming*, Lecture Notes in Economics and Mathematical Systems, No. 21, Springer-Verlag, New York, 1976.

Kapur, K. C., "On Max-Min Problems," *Naval Research Logistics Quarterly*, 20, pp. 639–644, 1973.

Karamardian, S., "Strictly Quasi-Convex (Concave) Functions and Duality in Mathematical Programming," *J. Mathematical Analysis and Applications*, 20, pp. 344–358, 1967.

Karamardian, S., "The Nonlinear Complementarity Problem with Applications," I, II, *J. Optimization Theory and Applications*, 4, pp. 87–98, pp. 167–181, 1969.

Karamardian, S., "Generalized Complementarity Problem," *J. Optimization Theory and Applications*, 8, pp. 161–168, 1971.

Karamardian, S., "The Complementarity Problem," *Mathematical Programming*, 2, pp. 107–129, 1972.

Karamardian, S., and S. Schaible, "Seven Kinds of Monotone Maps," Working Paper 90-3, Graduate School of Management, University of California at Riverside, Calif., 1990.

Karlin, S., *Mathematical Methods and Theory in Games, Programming, and Economics*, Vol. II, Addison-Wesley, Reading, Mass., 1959.

Karmarkar, N., "A New Polynomial-Time Algorithm for Linear Programming," *Combinatorics*, 4, pp. 373–395, 1984.

Karush, W., "Minima of Functions of Several Variables with Inequalities as Side Conditions," M.S. Thesis, Department of Mathematics, University of Chicago, 1939.

Karwan, M. H., and R. L. Rardin, "Some Relationships Between Langrangian and Surrogate Duality in Integer Programming," *Mathematical Programming*, 17, pp. 320–334, 1979.

Karwan, M. H., and R. L. Rardin, "Searchability of the Composite and Multiple Surrogate Dual Functions," *Operations Research*, 28, pp. 1251–1257, 1980.

Kawamura, K., and R. A. Volz, "On the Rate of Convergence of the Conjugate Gradient Reset Methods with Inaccurate Linear Minimizations," *IEEE Transactions on Auutomatic Control*, 18, pp. 360–366, 1973.

Keefer, D. L., "SIMPAT: Self-bounding Direct Search Method for Optimization," *J. Industrial and Engineering Chemistry Products Research and Development*, 12, No. 1, 1973.

Keller, E. L., "The General Quadratic Optimization Problem," *Mathematical Programming*, 5, pp. 311–337, 1973.

Kelley, J. E., "The Cutting Plane Method for Solving Convex Programs," *SIAM J. Industrial and Applied Mathmematics*, 8, pp. 703–712, 1960.

Khachiyan, L. G., "A Polynomial Algorithm in Linear Programming," *Soviet Math. Dokl.*, 20(1), pp. 191–194, 1979.

Khachiyan, L. G., "A Polynomial Algorithm in Linear Programming," *Doklady Akad. Nauk. USSR*, 244, pp. 1093–1096, 1979.

Kiefer, J., "Sequential Minimax Search for a Maximum," *Proceedings of the American Mathematical Society*, 4, pp. 502–506, 1953.

Kilmister, C. W., and J. E. Reeve, *Rational Mechanics*, American Elsevier, New York, 1966.

Kim, S., and H. Ahn, "Convergence of a Generalized Subgradient Method for Nondifferentiable Convex Optimization," *Mathematical Programming*, 50, pp. 75–80, 1991.

Kirchmayer, L. K., *Economic Operation of Power Systems*, John Wiley & Sons, New York, 1958.

Kiwiel, K., *Methods of Descent for Nondifferentiable Optimization,* Springer-Verlag, 1985.

Kiwiel, K. C., "A Survey of Bundle Methods for Nondifferentiable Optimization," Systems Research Institute, Polish Academy of Sciences, Newelska 6, 01-447 Warsaw, Poland, 1989.

Kiwiel, K. C., "A Tilted Cutting Plane Proximal Bundle Method for Convex Nondifferentiable Optimization," *Operations Research Letters*, 10, pp. 75–81, 1991.

Klee, V., "Separation and Support Properties of Convex Sets—A Survey," in *Calculus of Variations and Optimal Control*, A. V. Balakrishnan (Ed.), pp. 235–303, 1969.

Klessig, R., "A General Theory of Convergence for Constrained Optimization Algorithms that Use Antizigzagging Provisions," *SIAM J. Control*, 12, pp. 598–608, 1974.

Klessig, R., and E. Polak, "Efficient Implementation of the Polak-Ribiere Conjugate Gradient Algorithm," *SIAM J. Control*, 10, pp. 524–549, 1972.

Klingman, W. R., and D. M. Himmelblau, "Nonlinear Programming with the Aid of Multiplier Gradient Summation Technique," *J. Association for Computing Machinery*, 11, pp. 400–415, 1964.

Kojima, M., "A Unification of the Existence Theorem of the Nonlinear Complementarity Problem," *Mathematical Programming*, 9, pp. 257–277, 1975.

Kojima, M., N. Megiddo, and Y. Ye, "An Interior Point Potential Reduction Algorithm for the Linear Complementarity Problem," Research Report RJ 6486, IBM Almaden Research Center, San Jose, Calif., 1988.

Kojima, M., S. Mizuno, and A. Yoshise, "An $O(\sqrt{\mathbf{n}}\ L)$ Iteration Potential Reduction Algorithm for Linear Complementarity Problems," Research Report on Information Sciences B-217, Tokyo Institute of Technology, Tokyo, Japan, 1988.

Konno, H., and Y. Yajima, "Solving Rank Two Bilinear Programs by Parametric Simplex Algorithms," Manuscript, Tokyo University, Tokyo, Japan, 1989.

Konno, H., and T. Kuno, "Generalized Linear Multiplicative and Fractional Programming," Manuscript, Tokyo University, Tokyo, Japan, 1989.

Kostreva, M. M., "Block Pivot Methods for Solving the Complementarity Problem," *Linear Algebra and its Application*, 21, pp. 207–215, 1978.

Kostreva, M. M., and M. Wiecek, "Linear Complementarity Problems and Multiple Objective Programming," Department of Math Sciences, Technical Report #578, Clemson University, Clemson, South Carol., 1989.

Kowalik, J., "Nonlinear Programming Procedures and Design Optimization," *Acta Polytechica Scandinavica*, 13, pp. 1–47, 1966.

Kowalik, J., and M. R. Osborne, *Methods for Unconstrained Optimization Problems*, American Elsevier, New York, 1968.

Kozlov, M. K., S. P. Tarasov, and L. G. Khachiyan, "Polynomial Solvability of Convex Quadratic Programming," *Doklady Akademiia Nauk SSSR*, 5, pp. 1051–1053, 1979 (Translated in: *Soviet Mathematics Doklady*, 20, pp. 108–111, 1979.)

Kuester, J. L., and J. H. Mize, *Optimization Techniques With Fortran*, McGraw-Hill, New York, 1973.

Kuhn, H. W., "Duality, in Mathematical Programming," *Mathematical Systems Theory and Economics I* (Lecture Notes in Operations Research and Mathematical Economics, No. 11), pp. 67–91, Springer-Verlag, New York, 1969.

Kuhn, H. W. (Ed.), *Proceedings of the Princeton Symposium on Mathematical Programming*, Princeton University Press, Princeton, N.J., 1970.

Kuhn, H. W., "Nonlinear Programming: A Historical View," in *Nonlinear Programming*, R. W. Cottle and C. E. Lemke (Eds.), 1976.

Kuhn, H. W., and A. W. Tucker, "Nonlinear Programming," in *Proc. 2nd Berkeley Symposium on Mathematical Statistics and Probability*, J. Neyman (Ed.), University of California Press, Berkeley, Calif., 1951.

Kuhn, H. W., and A. W. Tucker (Eds.), "Linear Inequalities and Related Systems," *Ann. Math. Study*, 38, Princeton University Press, Princeton, N.J., 1956.

Kunzi, H. P., W. Krelle, and W. Oettli, *Nonlinear Programming*, Blaisdell, Amsterdam, 1966.

Kuo, M. T., and D. I. Rubin, "Optimization Study of Chemical Processes," *Canadian J. Chemical Engineering*, 40, pp. 152–156, 1962.

Kyparisis, J., "On Uniqueness of Kuhn–Tucker Multipliers in Nonlinear Programming," *Mathematical Programming*, 32, pp. 242–246, 1985.

Lasdon, L. S., "Duality and Decomposition in Mathematical Programming," *IEEE Transactions, Systems Science and Cybernetics*, 4, pp. 86–100, 1968.

Lasdon, L. S., *Optimization Theory for Large Systems*, Macmillan, New York, 1970.

Lasdon, L. S., "An Efficient Algorithm for Minimizing Barrier and Penalty Functions," *Mathematical Programming*, 2, pp. 65–106, 1972.

Lasdon, L. S., "Nonlinear Programming Algorithm-Applications, Software and Comparison," in *Numerical Optimization 1984*, P. T. Boggs, R. H. Byrd, and R. B. Schnabel (Eds.), SIAM Publications, Philadelphia, 1985.

Lasdon, L. S., and P. O. Beck, "Scaling Nonlinear Programs," *Operations Research Letters*, 1(1), pp. 6–9, 1981.

Lasdon, L. S., and M. W. Ratner, "An Efficient One-Dimensional Search Procedure for Barrier Functions," *Mathematical Programming*, 4, pp. 279–296, 1973.

Lasdon, L. S., and A. D. Waren, "Generalized Reduced Gradient Software for Linearly and Nonlinearly Constrained Problems," *Design and Implementation of Optimization Software*, H. J. Greenberg (Ed.), Sijthoff and Noordhoff, Holland, pp. 335–362, 1978.

Lasdon, L. S., and A. D. Waren, "A Survey of Nonlinear Programming Applications," *Operations Research*, 28(5), pp. 34–50, 1980.

Lasdon, L. S., and A. D. Waren, "GRG2 User's Guide," Dept. of General Business, School of Business Administration, University of Texas, Austin, May 1982.

Lasdon, L. S., A. D. Waren, A. Jain, and M. Ratner, "Design and Testing of a GRG Code for Nonlinear Optimization," *ACM Trans. Mathematical Software*, 4, pp. 34–50, 1978.

Lavi, A., and T. P. Vogl (Eds.), *Recent Advances in Optimization Techniques*, John Wiley & Sons, New York, 1966.

Lawson, C. L., and R. J. Hanson, *Solving Least-Squares Problems*, Prentice Hall, Englewood Cliffs, N.J., 1974.

Leitmann, G. (Ed.), *Optimization Techniques*, Academic Press, New York, 1962.

Lemarechal, C., "Note on an Extension of Davidon Methods to Nondifferentiable Functions," *Mathematical Programming*, 7, pp. 384–387, 1974.

Lemarechal, C., "An Extension of Davidon Methods to Non-Differentiable Problems," *Mathematical Programming Study*, 3, pp. 95–109, 1975.

Lemarechal, C., "Bundle Methods in Nonsmooth Optimization," in *Nonsmooth Optimization*, C. Lemarechal and R. Mifflin (Eds.), *IIASA Proceedings 3*, Pergamon, Oxford, 1978.

Lemarechal, C., "Nondifferential Optimization," in *Nonlinear Optimization, Theory and Algorithms*, Dixon, Spedicato, and Szego (Eds.), Birkhauser, Boston, 1980.

Lemarechal, C., and R. Mifflin (Eds.), "Nonsmooth Optimization," in *IIASA Proceedings 3*, Pergamon, Oxford, 1978.

Lemke, C. E., "A Method of Solution for Quadratic Programs," *Management Science*, 8, pp. 442–455, 1962.

Lemke, C. E., "Bimatrix Equilibrium Points and Mathematical Programming," *Management Science*, 11, pp. 681–689, 1965.

Lemke, C. E., "On Complementary Pivot Theory," in *Mathematics of the Decision Sciences*, G. B. Dantzig and A. F. Veinott (Eds.), 1968.

Lemke, C. E., "Recent Results on Complementarity Problems," in *Nonlinear Programming*, J. B. Rosen, O. L. Mangasarian, and K. Ritter (Eds.), 1970.

Lemke, C. E., and J. T. Howson, "Equilibrium Points of Bi-matrix Games," *SIAM J. Applied Mathematics*, 12, pp. 412–423, 1964.

Lenard, M. L., "Practical Convergence Conditions for Unconstrained Optimization," *Mathematical Programming*, 4, pp. 309–323, 1973.

Lenard, M. L., "Practical Convergence Condition for the Davidon–Fletcher–Powell Method," *Mathematical Programming*, 9, pp. 69–86, 1975.

Lenard, M. L., "Convergence Conditions for Restarted Conjugate Gradient Methods with Inaccurate Line Searches," *Mathematical Programming*, 10, pp. 32–51, 1976.

Lenard, M. L., "A Computational Study of Active Set Strategies in Nonlinear Programming with Linear Constraints," *Mathematical Programming*, 16, pp. 81–97, 1979.

Lenstra, J. K., A. H. G. Rinnooy Kan, and A. Schrijver, *History of Mathematical Programming: A Collection of Personal Reminiscences*, CWI-North Holland, Amsterdam, 1991.

Leon, A., "A Comparison Among Eight Known Optimizing Procedures," in *Recent Advances in Optimization Techniques*, A. Lavi and T. P. Vogl (Eds.), 1966.

Levenberg, K., "A Method for the Solution of Certain Problems in Least Squares," *Quarterly of Applied Mathematics*, 2, 164–168, 1944.

Liebman, J., L. Lasdon, L. Schrage, and A. Waren, *Modeling and Optimization With GINO*, Scientific Press, Palo Alto, Calif., 1986.

Lill, S. A., "A Modified Davidon Method for Finding the Minimum of a Function Using Difference Approximations for Derivatives," *Computer Journal*, 13, pp. 111–113, 1970.

Lill, S. A., "Generalization of an Exact Method for Solving Equality Constrained Problems to Deal with Inequality Constraints," in *Numerical Methods for Nonlinear Optimization*, F. A. Lootsma (Ed.), 1972.

Liu, D. C., and J. Nocedal, "On the Limited Memory BFGS Method for Large-Scale Optimization," *Mathematical Programming*, 45, pp. 503–528, 1989.

Loganathan, G. V., H. D. Sherali, and M. P. Shah, "A Two-Phase Network Design Heuristic for Minimum Cost Water Distribution Systems Under a Reliability Constraint," *Engineering Optimization*, Vol. 15, pp. 311–336, 1990.

Lootsma, F. A., "Constrained Optimization Via Parameter-Free Penalty Functions," *Philips Research Reports*, 23, pp. 424–437, 1968a.

Lootsma, F. A., "Constrained Optimization Via Penalty Functions," *Philips Research Reports*, 23, pp. 408–423, 1968b.

Lootsma, F. A. (Ed.), *Numerical Methods for Nonlinear Optimization*, Academic Press, New York, 1972a.

Lootsma, F. A., "A Survey of Methods for Solving Constrained Minimization Problems via Unconstrained Minimization," in *Numerical Methods for Nonlinear Optimization*, F. A. Lootsma (Ed.), 1972b.

Love, R. F., J. G. Morris, and G. O. Wesolowsky, *Facility Location: Models and Methods*, North Holland Publishing Company, Amsterdam, 1988.

Luenberger, D. G., "Quasi-Convex Programming," *SIAM J. Applied Mathematics*, 16, pp. 1090–1095, 1968.

Luenberger, D. G., *Optimization by Vector Space Methods*, John Wiley & Sons, New York, 1969.

Luenberger, D. G., "The Conjugate Residual Method for Constrained Minimization Problems," *SIAM J. Numerical Analysis*, 7, pp. 390–398, 1970.

Luenberger, D. G., "Convergence Rate of a Penalty-Function Scheme," *J. Optimization Theory and Applications*, 7, pp. 39–51, 1971.

Luenberger, D. G., "Mathematical Programming and Control Theory: Trends of Interplay," in *Perspectives of Optimization*, A. M. Geoffrion (Ed.), pp. 102–133, 1972.

Luenberger, D. G., *Introduction to Linear and Nonlinear Programming*, Addison-Wesley, Reading, Mass., 1973a.

Luenberger, D. G., "An Approach to Nonlinear Programming," *J. Optimization Theory and Applications*, 11, pp. 219–227, 1973b.

Luenberger, D. G., "A Combined Penalty Function and Gradient Projection Method for Nonlinear Programming," *J. Optimization Theory and Applications*, 14, pp. 477–495, 1974.

Lustig, I. J., R. E. Marsten, and D. F. Shanno, "The Primal-Dual Interior Point Method on the Cray Supercomputer," in *Large-Scale Numerical Optimization*, T. F. Coleman and Y. Li (Eds.), SIAM, Philadelphia, pp. 70–80, 1990.

Lyness, J. N., "Has Numerical Differentiation a Future?" *Proc. of the 7th Manitoba Conference on Numerical Mathematics*, pp. 107–129, 1977.

Maass, A., M. M. Hufschmidt, R. Dorfman, H. A. Thomas Jr., S. A. Marglin, and G. M. Fair, *Design of Water-Resource Systems*, Harvard University Press, Cambridge, Mass., 1967.

Madansky, A., "Some Results and Problems in Stochastic Linear Programming," *The RAND Corporation Paper P-1596*, 1959.

Madansky, A., "Methods of Solution of Linear Programs Under Uncertainty," *Operations Research*, 10, pp. 463–471, 1962.

Magnanti, T. L., "Fenchel and Lagrange Duality are Equivalent," *Mathematical Programming*, 7, pp. 253–258,1974.

Mahajan, D. G., and M. N. Vartak, "Generalization of Some Duality Theorems in Nonlinear Programming," *Mathematical Programming*, 12, pp. 293–317, 1977.

Mahidhara, D., and L. S. Lasdon, "An SQP Agorithm for Large Sparse Nonlinear Programs," Working Paper, MSIS Dept., School of Business, The University of Texas, Austin, TX, 1990.

Majid, K. I., *Optimum Design of Structures*, John Wiley & Sons, New York, 1974.

Majthay, A., "Optimality Conditions for Quadratic Programming," *Mathematical Programming*, 1, pp. 359–365, 1971.

Mangasarian, O. L., "Duality in Nonlinear Programming," *Quarterly of Applied Mathematics*, 20, pp. 300–302, 1962.

Mangasarian, O. L., "Nonlinear Programming Problems with Stochastic Objective Functions," *Management Science*, 10, pp. 353–359, 1964.

Mangasarian, O. L., "Pseudo-Convex Functions," *SIAM J. Control*, 3, pp. 281–290, 1965.

Mangasarian, O. L., *Nonlinear Programming*, McGraw-Hill, New York, 1969a.
Mangasarian, O. L., "Nonlinear Fractional Programming," *J. Operations Research Society of Japan*, 12, pp. 1–10, 1969b.
Mangasarian, O. L., "Optimality and Duality in Nonlinear Programming," in *Proceedings of Princeton Symposium on Mathematical Programming*, H. W. Kuhn (Ed.), pp. 429–443, 1970a.
Mangasarian, O. L., "Convexity, Pseudo-Convexity and Quasi-Convexity of Composite Functions," *Cahiers Centre Études Recherche Opperationelle*, 12, pp. 114–122, 1970b.
Mangasarian, O. L., "Linear Complementarity Problems Solvable by a Single Linear Program," *Mathematical Programming*, 10, pp. 263–270, 1976.
Mangasarian, O. L., "Characterization of Linear Complementarity Problems as Linear Programs," *Mathematical Programming Study,* 7, pp. 74–87, 1978.
Mangasarian, O. L., "Simplified Characterization of Linear Complementarity Problems Solvable as Linear Programs," *Mathematics of Operations Research*, 4(3), pp. 268–273, 1979.
Mangasarian, O. L., "A Simple Characterization of Solution Sets of Convex Program," *OR Letters*, 7(1), pp. 21–26, 1988.
Mangasarian, O. L., and S. Fromovitz, "The Fritz John Necessary Optimality Conditions in the Presence of Equality and Inequality Constraints," *J. Mathematical Analysis and Applications*, 17, pp. 37–47, 1967.
Mangasarian, O. L., R. R. Meyer, and S. M. Johnson (Eds.), *Nonlinear Programming*, Academic Press, New York, 1975.
Mangasarian, O. L., and J. Ponstein, "Minimax and Duality in Nonlinear Programming," *J. Mathematical Analysis and Applications*, 11, pp. 504–518, 1965.
Maratos, N., "Exact Penalty Function Algorithms for Finite Dimensional and Control Optimization Problems," Ph.D. Thesis, Imperial College Science Technology, University of London, 1978.
Markowitz, H. M., "Portfolio Selection," *Journal of Finance*, 7, pp. 77–91, 1952.
Markowtiz, H. M., "The Optimization of a Quadratic Function Subject to Linear Constraints," *Naval Research Logistics Quarterly*, 3, pp. 111–133, 1956.
Markowitz, H. M., and A. S. Manne, "On the Solution of Discrete Programming Problems," *Econometrica*, 25, pp. 84–110, 1957.
Marquardt, D. W., "An Algorithm for Least Squares Estimation of Nonlinear Parameters," *SIAM J. of Industrial & Applied Mathematics*, 11, pp. 431–441, 1963.
Marsten, R. E., "The Use of the Boxstep Method in Discrete Optimization," in *Nondifferentiable Optimization*, M. L. Balinski and P. Wolfe (Eds.), *Mathematical Programming Study*, 3, North Holland, Amsterdam, 1975.
Martensson, K., "A New Approach to Constrained Function Optimization," *J. Optimization Theory and Applications*, 12, pp. 531–554, 1973.
Martos, B., "Hyperbolic Programming," *Naval Research Logistics Quarterly*, 11, pp. 135–155, 1964.
Martos, B., "The Direct Power of Adjacent Vertex Programming Methods," *Management Science*, 12, pp. 241–252, 1965; errata, *ibid.*, 14, pp. 255–256, 1967a.
Martos, B., "Quasi-Convexity and Quasi-Monotonicity in Nonlinear Programming," *Studia Scientiarum Mathematicarum Hungarica*, 2, pp. 265–273, 1967b.
Martos, B., "Subdefinite Matrices and Quadratic Forms," *SIAM J. Applied Mathematics*, 17, pp. 1215–1233, 1969.
Martos, B., "Quadratic Programming with a Quasiconvex Objective Function," *Operations Research*, 19, pp. 87–97, 1971.
Martos, B., *Nonlinear Programming: Theory and Methods*, American Elsevier, New York, 1975.
Massam, H., and S. Zlobec, "Various Definitions of the Derivative in Mathematical Programming," *Mathematical Programming*, 7, pp. 144–161, 1974.
Matthews, A., and D. Davies, "A Comparison of Modified Newton Methods for Unconstrained Optimization," *Computer Journal*, 14, pp. 293–294, 1971.

Mayne, D. Q., "On the Use of Exact Penalty Functions to Determine Step Length in Optimization Algorithms," in *Numerical Analysis*, "Dundee 1979," G. A. Watson (Ed.), *Lecture Notes in Mathematics*, 773, Springer-Verlag, Berlin, 1980.

Mayne, D. Q., and N. Maratos, "A First-Order, Exact Penalty Function Algorithm for Equality Constrained Optimization Problems," *Mathematical Programming*, 16, pp. 303–324, 1979.

Mayne, D. Q., and E. Polak, "A Superlinearly Convergent Algorithm for Constrained Optimization Problems," *Mathematical Programming Study,* 16, pp. 45–61, 1982.

McCormick, G. P., "Second Order Conditions for Constrained Minima," *SIAM J. Applied Mathematics*, 15, pp. 641–652, 1967.

McCormick, G. P., "Anti-Zig-Zagging by Bending," *Management Science*, 15, pp. 315–320, 1969a.

McCormick, G. P., "The Rate of Convergence of the Reset Davidon Variable Metric Method," MRC Technical Report #1012, Mathematics Research Center, University of Wisconsin, 1969b.

McCormick, G. P., "The Variable Reduction Method for Nonlinear Programming," *Management Science Theory*, 17, pp. 146–160, 1970a.

McCormick, G. P., "A Second Order Method for the Linearly Constrained Nonlinear Programming Problems," in *Nonlinear Programming*, J. B. Rosen, O. L. Mangasarian, and K. Ritter (Eds.), 1970b.

McCormick, G. P., "Penalty Function Versus Non-Penalty Function Methods for Constrained Nonlinear Programming Problems," *Mathematical Programming*, 1, pp. 217–238, 1971.

McCormick G. P., "Attempts to Calculate Global Solutions of Problems that may have Local Minima," in *Numerical Methods for Nonlinear Optimization*, F. A. Lootsma (Ed.), 1972.

McCormick, G. P., "A Modification of Armijo's Step-Size Rule for Negative Curvature," *Mathematical Programming*, 13, pp. 111–115, 1977.

McCormick, G. P., *Nonlinear Programming*, Wiley-Interscience, New York, 1983.

McCormick, G. P., and J. D. Pearson, "Variable Metric Method and Unconstrained Optimization," in *Optimization*, R. Fletcher (Ed.), 1969.

McCormick, G. P., and K. Ritter, "Methods of Conjugate Direction versus Quasi-Newton Methods," *Mathematical Programming*, 3, pp. 101–116, 1972.

McCormick, G. P., and K. Ritter, "Alternative Proofs of the Convergence Properties of the Conjugate Gradient Methods," *J. Optimization Theory and Applications*, 13, pp. 497–515, 1974.

McLean, R. A., and G. A. Watson, "Numerical Methods of Nonlinear Discrete L_1 Approximation Problems," in *Numerical Methods of Approximation Theory*, L. Collatz, G. Meinardus, and H. Warner (Eds.), ISNM 52, Birkhauser-Verlag, Basle, 1980.

McMillan, C. Jr., *Mathematical Programming*, John Wiley & Sons, New York, 1970.

McShane, K. A., C. L. Monma, and D. F. Shanno, "An Implementation of a Primal-Dual Interior Point Method for Linear Programming," *ORSA J. on Computing*, 1, pp. 70–83, 1989.

Megiddo, N., "Pathways to the Optimal Set in Linear Programming," in *Proc. of the Mathematical Programming Symposium of Japan*, Nagoya, Japan, pp. 1–36, 1986. (See also *Progress in Mathematical Programming-Interior-Point and Related Methods*, Nimrod Megiddo (Ed.), Springer-Verlag, pp. 131–158, 1989.)

Mehndiratta, S. L., "General Symmetric Dual Programs," *Operations Research*, 14, pp. 164–172, 1966.

Mehndiratta, S. L., "Symmetry and Self-Duality in Nonlinear Programming," *Numerische Mathematik*, 10, pp. 103–109, 1967a.

Mehndiratta, S. L., "Self-Duality in Mathematical Programming," *SIAM J. Applied Mathematics*, 15, pp. 1156–1157, 1967b.

Mehndiratta, S. L., "A Generalization of a Theorem of Sinha on Supports of a Convex Function," *Australian J. Statistics*, 11, pp. 1–6, 1969.

Mehrotra, S., "On the Implementation of a (Primal-Dual) Interior-Point Method," Technical Report 90-03, Department of Industrial Engineering and Management Sciences, Northwestern University, Evanston, Ill., 1990.

Mereau, P., and J. G. Paquet, "A Sufficient Condition for Global Constrained Extrema," *International J. Control*, 17, pp. 1065–1071, 1973a.

Mereau, P., and J. G. Paquet, "The Use of Pseudo-Convexity and Quasi-Convexity in Sufficient Conditions for Global Constrained Extrema," *International J. Control*, 18, pp. 831–838, 1973b.

Mereau, P., and J. G. Paquet, "Second Order Conditions for Pseudo-Convex Functions," *SIAM J. Applied Mathematics*, 27, pp. 131–137, 1974.

Messerli, E. J., and E. Polak, "On Second Order Necessary Conditions of Optimality," *SIAM J. Control*, 7, pp. 272–291, 1969.

Meyer, G. G. L., "A Derivable Method of Feasible Directions," *SIAM J. Control*, 11, pp. 113–118, 1973.

Meyer, G. G. L., "Nonwastefulness of Interior Iterative Procedures," *J. Mathematical Analysis and Applications*, 45, pp. 485–496, 1974.

Meyer, G. G. L., "Accelerated Frank-Wolfe Algorithms," *SIAM J. Control*, 12, pp. 655–663, 1974.

Meyer, R. R., "The Validity of a Family of Optimization Methods," *SIAM J. Control*, 8, pp. 41–54, 1970.

Meyer, R. R., "Sufficient Conditions for the Convergence of Monotonic Mathematical Programming Algorithms," *J. Computer and System Sciences*, 12, pp. 108–121, 1976.

Meyer, R. R., "Two-Segment Separable Programming," *Management Science*, 25(4), pp. 385–395, 1979.

Meyer, R. R., "Computational Aspects of Two-Segment Separable Programming," Computer Sciences Technical Report #382, University of Wisconsin–Madison, March, 1980.

Miele, A., and J. W. Cantrell, "Study on a Memory Gradient Method for the Minimization of Functions," *J. Optimization Theory and Applications*, 3, pp. 459–470, 1969.

Miele, A., E. E. Cragg, R. R. Iyer, and A. V. Levy, "Use of the Augmented Penalty Function in Mathematical Programming Problems: Part I," *J. Optimization Theory and Applications*, 8, pp. 115–130, 1971.

Miele, A., E. E. Cragg, and A. V. Levy, "Use of the Augmented Penalty Function in Mathematical Programming: Part 2," *J. Optimization Theory and Applications*, 8, pp. 131–153, 1971.

Miele, A., P. Moseley, A. V. Levy, and G. H. Coggins, "On the Method of Multipliers for Mathematical Programming Problems," *J. Optimization Theory and Applications*, 10, pp. 1–33, 1972.

Mifflin, R., "A Superlinearly Convergent Algorithm for Minimization Without Evaluating Derivatives," *Mathematical Programming*, 9, pp. 100–117, 1975.

Miller, C. E., "The Simplex Method for Local Separable Programming," in *Recent Advances in Mathematical Programming*, R. L. Graves and P. Wolfe (Eds.), 1963.

Minch, R. A., "Applications of Symmetric Derivatives in Mathematical Programming," *Mathematical Programming*, 1, pp. 307–320, 1974.

Minhas, B. S., K. S. Parikh, and T. N. Srinivasan, "Toward the Structure of a Production Function for Wheat Yields with Dated Inputs of Irrigation Water," *Water Resources Research*, 10, pp. 383–393, 1974.

Minkowski, H., *Gesammelte Abhandlungen*, Teubner, Berlin, 1911.

Minoux, M., "Subgradient Optimization and Benders Decomposition in Large Scale Programming," *Mathematical Programming*, R. W. Cottle, M. L. Kelmanson, and B. Korte (Eds.), North Holland, pp. 271–288, 1984.

Minoux, M., *Mathematical Programming: Theory and Algorithms*, John Wiley & Sons, New York, 1986.

Minoux, M., and J. Y. Serreault, "Subgradient Optimization and Large Scale Programming: Application to Multicommodity Network Synthesis with Security Constraints," *RAIRO*, 15(2), pp. 185–203, 1980.

Mitchell, R. A., and J. L. Kaplan, "Nonlinear Constrained Optimization by a Nonrandom Complex Method," *J. Research of the National Bureau of Standards, Section C, Engineering and Instrumentation*, 72-C, pp. 249–258, 1968.

Mobasheri, F., "Economic Evaluation of a Water Resources Development Project in a Developing Economy," Contribution 126 *Water Resources Center*, University of California, Berkeley, 1968.

Moeseke, van P. (Ed.), *Mathematical Programs for Activity Analysis*, North Holland, Amsterdam, 1974.

Mokhtar-Kharroubi, H., "Sur la Convergence Théorique de la Méthode du Gradient Réduit Généralisé," *Numerische Mathematik*, 34, pp. 73–85, 1980.

Mond, B., "A Symmetric Dual Theorem for Nonlinear Programs," *Quarterly of Applied Mathematics*, 23, pp. 265–269, 1965.

Mond, B., "On a Duality Theorem for a Nonlinear Programming Problem," *Operations Research*, 21, pp. 369—370, 1973.

Mond, B., "A Class of Nondifferentiable Mathematical Programming Problems," *J. Mathematical Analysis and Applications*, 46, pp. 169–174, 1974.

Mond, B., and R. W. Cottle, "Self-Duality in Mathematical Programming," *SIAM J. Applied Mathematics*, 14, pp. 420–423, 1966.

Monteiro, R. D. C., and I. Adler, "Interior Path Following Primal-Dual Algorithms—Part I: Linear Programming," *Mathematical Programming*, 44, pp. 27–42, 1989.

Monteiro, R. D. C., and I. Adler, "Interior Path Following Primal-Dual Algorithms—Part II: Convex Quadratic Programming," *Mathematical Programming*, 44, pp. 43–66, 1989.

Monteiro, R. D. C., I. Adler, and M. G. C. Resende, "A Polynomial-Time Primal-Dual Affine Scaling Algorithm for Linear and Convex Quadratic Programming and its Power Series Extension," *Mathematics of Operations Research*, 15(2), pp. 191–214, 1990.

Moré, J. J., "Class of Functions and Feasibility Conditions in Nonlinear Complementarity Problems," *Mathematical Programing*, 6, pp. 327–338, 1974.

Moré, J. J., "The Levenberg–Marquardt Algorithm: Implementation and Theory," in *Numerical Analysis*, G. A. Watson (Ed.), *Lecture Notes in Mathematics*, 630, Springer-Verlag, Berlin, pp. 105–116, 1977.

Moré, J. J., "Implementation and Testing of Optimization Software," in *Performance Evaluation of Numerical Software*, L. D. Fosdick (Ed.), North Holland, Amsterdam, pp. 253–266, 1979b.

Moré, J. J., and D. C. Sorensen, "On The Use of Directions of Negative Curvature in a Modified Newton Method," *Mathematical Programming*, 15, pp. 1–20, 1979.

Moreau, J. J., "Convexity and Duality," in *Functional Analysis and Optimization*, E .R. Caianiello (Ed.), Academic Press, New York, 1966.

Morgan, D. R., and I. C. Goulter, "Optimal Urban Water Distribution Design," *Water Research*, 21(5), pp. 642–652, May 1985.

Morshedi, A. M., and R. A. Tapia, "Karmarkar as a Classical Method," Rice University Technical Report 87-7, Houston, Tex., March 1987.

Motzkin, T. S., *Beiträge zur Theorie der Linearen Ungleichungen*, Dissertation, University of Basel, Jerusalem, 1936.

Mueller, R. K., "A Method for Solving the Indefinite Quadratic Programming Problem," *Management Science*, 16, pp. 333–339, 1970.

Murphy, F. H., "Column Dropping Procedures for the Generalized Programming Algorithm," *Management Science*, 19, pp. 1310–1321, 1973a.

Murphy, F. H., "A Column Generating Algorithm for Nonlinear Programming," *Mathematical Programming*, 5, pp. 286–298, 1973b.

Murphy, F. H., "A Class of Exponential Penalty Functions," *SIAM J. Control*, 12, pp. 679–687, 1974.

Murphy, F. H., H. D. Sherali, and A. L. Soyster, "A Mathematical Programming Approach for Determining Oligopolistic Market Equilibria," *Mathematical Programming*, 25(1), pp. 92–106, 1982.

Murray, W. (Ed.), *Numerical Methods for Unconstrained Optimization*, Academic Press, London, 1972a.

Murray, W., "Failure, the Causes and Cures," in *Numerical Methods for Unconstrained Optimization*, W. Murray (Ed.), 1972b.

Murray, W., and M. L. Overton, "A Projected Lagrangian Algorithm for Nonlinear Minimax Optimization," *SIAM J. Science Stat. Comput.*, 1, pp. 345–370, 1980a.

Murray, W., and M. L. Overton, "A Projected Lagrangian Algorithm for Nonlinear l_1 Optimization," Report SOL 80-4, Department of Operations Research, Stanford University, Calif., 1980b.

Murray, W., and M. H. Wright, "Computations of the Search Direction in Constrained Optimization Algorithms," *Mathematical Programming Study*, 16, pp. 63–83, 1980.

Murtagh, B. A., "*Advanced Linear Programming: Computation and Practice*," McGraw-Hill, New York, 1981.

Murtagh, B. A., and R. W. H. Sargent, "A Constrained Minimization Method with Quadratic Convergence," in *Optimization*, R. Fletcher (Ed.), 1969.

Murtagh, B. A., and R. W. H. Sargent, "Computational Experience with Quadratically Convergent Minimization Methods," *Computer Journal*, 13, pp. 185–194, 1970.

Murtagh, B. A., and M. A. Saunders, "A Projected Lagrangian Algorithm and its Implementation for Sparse Nonlinear Constraints," *Mathematical Programming Study* 16, pp. 84–117, 1982.

Murtagh, B. A., and M. A. Saunders, "Large-Scale Linearly Constrained Optimization," *Mathematical Programming*, 14, pp. 41–72, 1978.

Murtagh, B. A., and M. A. Saunders, "Minos 5.0 Users Guide," Technical Report SOL 83.20, Systems Optimization Laboratory, Stanford University, Stanford, Calif., 1983.

Murty, K., *Linear and Combinatorial Programming*, John Wiley & Sons, New York, 1976.

Murty, K. G., "On the Number of Solutions to the Complementarity Problem and Spanning Properties of Complementarity Cones," *Linear Algebra and its Application*, 5, pp. 65–108, 1972.

Murty, K. G., *Linear Programming*, John Wiley & Sons, New York, 1983.

Murty, K., *Linear Complementarity, Linear and Nonlinear Programming*, Heldermann Verlag, Berlin, 1988.

Murty, K., "On Checking Unboundedness of Functions," Department of Industrial Engineering, University of Michigan, Ann Arbor, Mich., March 1989.

Myers, G., "Properties of the Conjugate Gradient and Davidon Methods," *J. Optimization Theory and Applications*, 2, pp. 209–219, 1968.

Myers, R. H., *Response Surface Methodology*, Virginia Polytechnic Institute and State University Press, Blacksburg, Virginia, 1976.

Mylander, W. C., "Nonconvex Quadratic Programming by a Modification of Lemké's Methods," Report No. RAC-TP-414, Research Analysis Corporation, 1971.

Mylander, W. C., "Finite Algorithms for Solving Quasiconvex Quadratic Programs," *Operations Research*, 20, pp. 167–173, 1972.

Nakayama, H., H. Sayama, and Y. Sawaragi, "A Generalized Lagrangian Function and Multiplier Method," *J. Optimization Theory and Applications*, 17(3/4), pp. 211–227, 1975.

Nash, S. G., "Preconditioning of Truncated-Newton Methods," *SIAM J. of Science and Statistical Computations*, 6, pp. 599–616, 1985.

Nash, S. G., and A. Sofer, "Block Truncated-Newton Methods for Parallel Optimization," *Mathematical Programming*, 45, pp. 529–546, 1989.

Nash, S. G., and A. Sofer, "Truncated-Newton Method for Constrained Optimization," TIMS/ORSA National Meeting, Nashville, Tenn., May 12–15, 1991.

Nash, S. G., and A. Sofer, "A General-Purpose Parallel Algorithm for Unconstrained Optimization," Technical Report No. 63, Center for Computational Statistics, George Mason University, Fairfax, Virginia, June 1990.

Nashed, M. Z., "Supportably and Weakly Convex Functionals with Applications to Approximation Theory and Nonlinear Programming," *J. Mathematical Analysis and Applications*, 18, pp. 504–521, 1967.

Nazareth, J. L., "A Conjugate Direction Algorithm Without Line Searches," *J. Optimization Theory and Applications*, 23(3), pp. 373–387, 1977.

Nazareth, J. L., "A Relationship Between the BFGS and Conjugate-Gradient Algorithms and Its Implications for New Algorithms," *SIAM J. Numerical Analysis*, 26, pp. 794–800, 1979.

Nazareth, J. L., "Conjugate Gradient Methods Less Dependency on Conjugacy," *SIAM Review*, 28(4), pp. 501–511, 1986.

Nelder, J. A., and R. Mead, "A Simplex Method for Function Minimization," *Computer Journal*, 7, pp. 308–313, 1964.

Nelder, J. A., and R. Mead, "A Simplex Method for Function Minimization—Errata," *Computer Journal*, 8, p. 27, 1965.

Nemhauser, G. L., and W. B. Widhelm, "A Modified Linear Program for Columnar Methods in Mathematical Programming," *Operations Research*, 19, pp. 1051–1060, 1971.

Nemhauser, G. L., and L. A. Wolsey, "*Integer and Combinatorial Optimization*," John Wiley and Sons, New York, 1988.

Neustadt, L. W., "A General Theory of Extremals," *J. Computer and System Sciences*, 3, pp. 57–92, 1969.

Neustadt, L. W., *Optimization*, Princeton University Press, Princeton, N.J., 1974.

Nguyen, V. H., J. J. Strodiot, and R. Mifflin, "On Conditions to Have Bounded Multipliers in Locally Lipschitz Programming," *Mathematical Programming*, 18, pp. 100–106, 1980.

Nikaido, H., *Convex Structures and Economic Theory*, Academic Press, New York, 1968.

Nocedal, J., "The Performance of Several Algorithms for Large-Scale Unconstrained Optimization," in *Large-Scale Numerical Optimization*, T. F. Coleman and Y. Li (Eds.), SIAM, Philadelphia, pp. 138–151, 1990.

Nocedal, J., and M. L. Overton, "Projected Hessian Updating Algorithms for Nonlinearly Constrained Optimization," *SIAM J. Numerical Analysis*, 22, pp. 821–850, 1985.

O'Laoghaine, D. T., and D. M. Himmelblau, *Optimal Expansion of a Water Resources System*, Academic Press, New York, 1974.

O'Learly, D. P., "Estimating Matrix Condition Numbers," *SIAM J. Science Stat. Comput.*, 1, pp. 205–209, 1980.

Oliver, J., "An Algorithm for Numerical Differentiation of a Function of One Real Variable," *J. Comput. Applied Mathematics*, 6, pp. 145–160, 1980.

Oliver, J. and A. Ruffhead, "The Selection of Interpolation Points in Numerical Differentiation," *Nordisk Tidskr. Informationbehandling (BIT)*, 15, pp. 283–295, 1975.

Orchard-Hays, W., "History of Mathematical Programming Systems," in *Design and Implementation of Optimization Software*, H. J. Greenberg (Ed.), Sijthoff and Noordhoff, Netherlands, pp. 1–26, 1978a.

Orchard-Hays, W., "Scope of Mathematical Programming Software," in *Design and Implementation of Optimization Software*, H. J. Greenberg (Ed.), Sijthoff and Noordhoff, Netherlands, pp. 27–40, 1978b.

Orchard-Hays, W., "Anatomy of a Mathematical Programming System," in *Design and Implementation of Optimization Software*, H. J. Greenberg (Ed.), Sijthoff and Noordhoff, Netherlands, pp. 41–102, 1978c.

Orden, A., "Stationary Points of Quadratic Functions Under Linear Constraints,"*Computer Journal,* 7, pp. 238–242, 1964.

Oren, S. S., "On the Selection of Parameters in Self-Scaling Variable Metric Algorithms," *Mathematical Programming,* 7, pp. 351–367, 1974a.

Oren, S. S., "Self-Scaling Variable Metric (SSVM) Algorithms II: Implementation and Experiments," *Management Science,* 20, pp. 863–874, 1974b.

Oren, S. S., and E. Spedicato, "Optimal Conditioning of Self-Scaling and Variable Metric Algorithms," *Mathematical Programming,* 10, pp. 70–90, 1976.

Ortega, J. M., and W. C. Rheinboldt, *Interactive Solution of Nonlinear Equations in Several Variables,* Academic Press, New York, 1970.

Ortega, J. M., and W. C. Rheinboldt, "A General Convergence Result for Unconstrained Minimization Methods," *SIAM J. Numerical Analysis*, 9, pp. 40–43, 1972.

Osborne, M. R., and D. M. Ryan, "On Penalty Function Methods for Nonlinear Programming Problems," *J. Mathematical Analysis and Applications*, 31, pp. 559–578, 1970.

Osborne, M. R., and D. M. Ryan, "A Hybrid Algorithm for Nonlinear Programming," in *Numerical Methods for Nonlinear Optimization*, F. A. Lootsma (Ed.), 1972.

Palacios-Gomez, R., L. Lasdon, and M. Engquist, "Nonlinear Optimization by Successive Linear Programming," *Management Science*, 28(10), pp. 1106–1120, 1982.
Panne, C. van de, *Methods for Linear and Quadratic Programming*, North Holland, Amsterdam, 1974.
Panne, C. van de, "A Complementary Variant of Lemke's Method for the Linear Complementary Problem," *Mathematical Programming*, 7, pp. 283–310, 1976.
Panne, C. van de, and A. Whinston, "Simplicial Methods for Quadratic Programming," *Naval Research Logistics Quarterly*, 11, pp. 273–302, 1964a.
Panne, C. van de, and A. Whinston, "The Simplex and the Dual Method for Quadratic Programming," *Operational Research Quarterly*, 15, pp. 355–388, 1964b.
Panne, C. van de, and A. Whinston, "A Parametric Simplicial Formulation of Houthakker's Capacity Method," *Econometrica*, 34, pp. 354–380, 1966a.
Panne, C. van de, and A. Whinston, "A Comparison of Two Methods for Quadratic Programming," *Operations Research*, 14, pp. 422–441, 1966b.
Panne, C. van de, and A. Whinston, "The Symmetric Formulation of the Simplex Method for Quadratic Programming," *Econometrica*, 37, pp. 507–527, 1969.
Pardalos, P. M., and J. B. Rosen, "Constrained Global Optimization: Algorithms and Applications," *Lecture Notes in Computer Science*, 268, G. Goos and J. Hartmann's (Eds.), Springer-Verlag, 1987.
Pardalos, P. M., and J. B. Rosen, "Global Optimization Approach to the Linear Complementarity Problem," *SIAM J. on Scientific and Statistical Computing*, 9(2), pp. 341–353, 1988.
Pardalos, P. M., and S. A. Vavasis, "Quadratic Programs with One Negative Eigen-Value is NP-Hard," *J. of Global Optimization*, 1(1), pp. 15–22, 1991.
Parikh, S. C., "Equivalent Stochastic Linear Programs," *SIAM J. Applied Mathematics*, 18, pp. 1–5, 1970.
Parker, R. G., and R. L. Rardin, "*Discrete Optimization*," Academic Press, San Diego, Calif., 1988.
Parkinson, J. M., and D. Hutchinson, "An Investigation into the Efficiency of Variants on the Simplex Method," in *Numerical Methods for Nonlinear Optimization*, F. A. Lootsma (Ed.), pp. 115–136, 1972a.
Parkinson, J. M., and D. Hutchinson, "A Consideration of Nongradient Algorithms for the Unconstrained Optimization of Function of High Dimensionality," in *Numerical Methods for Nonlinear Optimization*, F. A. Lootsma (Ed.), 1972b.
Parsons, T. D., and A. W. Tucker, "Hybrid Programs: Linear and Least-Distance," *Mathematical Programming*, 1, pp. 153–167, 1971.
Paviani, D. A., and D. M. Himmelblau, "Constrained Nonlinear Optimization By Heuristic Programming," *Operations Research*, 17, pp. 872–882, 1969.
Pearson, J. D., "Variable Metric Methods of Minimization," *Computer Journal*, 12, pp. 171–178, 1969.
Perry, A., "A Modified Conjugate Gradient Algorithm," *Operations Research*, 26(6), pp. 1073–1078, 1978.
Peterson, D. W., "A Review of Constraint Qualifications in Finite-Dimensional Spaces," *SIAM Review*, 15, pp. 639–654, 1973.
Peterson, E. L., "An Economic Interpretation of Duality in Linear Programming," *J. Mathematical Analysis and Applications*, 30, pp. 172–196, 1970.
Peterson, E. L., "An Introduction to Mathematical Programming," in *Optimization and Design*, M. Avriel, M. J. Rijckaert, D. J. Wilde (Eds.), 1973a.
Peterson, E. L., "Geometric Programming and Some of Its Extensions," in *Optimization and Design*, M. Avriel, M. J. Rijckaert, and D. J. Wilde (Eds.), 1973b.
Peterson, E. L., "Geometric Programming," *SIAM Review*, 18, pp. 1–15, 1976.
Phelan, R. M., *Fundamentals of Mechanical Design*, McGraw-Hill, New York, 1957.
Pierre, D. A., *Optimization Theory with Applications*, John Wiley & Sons, New York, 1969.
Pierre, D. A., and M. J. Lowe, *Mathematical Programming via Augmented Lagrangians: An Introduction with Computer Programs*, Addison-Wesley, Reading, Mass., 1975.

Pierskalla, W. P., "Mathematical Programming with Increasing Constraint Functions," *Management Science*, 15, pp. 416–425, 1969.

Pietrgykowski, T., "Application of the Steepest Descent Method to Concave Programming," *Proc. of International Federation of Information Processing Societies Congress* (Munich), North Holland, Amsterdam, pp. 185–189, 1962.

Pironneau, O., and E. Polak, "Rate of Convergence of a Class of Methods of Feasible Directions," *SIAM J. Numerical Analysis*, 10, 161–174, 1973.

Polak, E., "On the Implementation of Conceptual Algorithms," in *Nonlinear Programming*, J. B. Rosen, O. L. Mangasarian, and K. Ritter (Eds.), 1970.

Polak, E., *Computational Methods in Optimization*, Academic Press, New York, 1971.

Polak, E., "A Survey of Feasible Directions for the Solution of Optimal Control Problems," *IEEE Transactions Automatic Control*, AC-17, pp. 591–596, 1972.

Polak, E., "An Historical Survey of Computational Methods in Optimal Control," *SIAM Review*, 15, pp. 553–584, 1973.

Polak, E., "A Modified Secant Method for Unconstrained Minimization," *Mathematical Programming*, 6, pp. 264–280, 1974.

Polak, E., and M. Deparis, "An Algorithm for Minimum Energy Control," *IEEE Transactions Automatic Control*, AC-14, pp. 367–377, 1969.

Polak, E., and G. Ribiere, "Note sur la Convergence de Methods de Directions Conjugres," *Revue Francaise Informat, Recherche Operationelle*, 16, pp. 35–43, 1969.

Polyak, B. T., "A General Method for Solving Extremum Problems," *Soviet Mathematics*, 8, pp. 593–597, 1967.

Polyak, B. T., "The Method of Conjugate Gradient in Extremum Problems," *USSR Computational Mathematics and Mathematical Physics* (English Translation), 9, pp. 94–112, 1969.

Polyak, B. T., "Subgradient Methods: A Survey of Soviet Research," in *Nonsmooth Optimization*, C. Lemarechal and R. Mifflin (Eds.), Pergamon Press, pp. 5–30, 1978.

Polyak, B. T., and N. W. Tret'iakov, "An Iterative Method for Linear Programming and Its Economic Interpretation," *Ekonomika i Matematicheskie Metody, Matekon*, 5, pp. 81–100, 1972.

Ponstein, J., "An Extension of the Min-Max Theorem," *SIAM Review*, 7, pp. 181–188, 1965.

Ponstein, J., "Seven Kinds of Convexity," *SIAM Review*, 9, pp. 115–119, 1967.

Powell, M. J. D., "An Efficient Method for Finding the Minimum of a Function of Several Variables Without Calculating Derivatives," *Computer Journal*, 7, pp. 155–162, 1964.

Powell, M. J. D., "A Method for Nonlinear Constraints in Minimization Problems," in *Optimization*, R. Fletcher (Ed.), 1969.

Powell, M. J. D., "Rank One Methods for Unconstrained Optimization," in *Integer and Nonlinear Programming*, J. Abadie (Ed.), 1970a.

Powell, M. J. D., "A Survey of Numerical Methods for Unconstrained Optimization," *SIAM Review*, 12, pp. 79–97, 1970b.

Powell, M. J. D., "A Hybrid Method for Nonlinear Equations," in *Numerical Methods for Nonlinear Algebraic Equations*, P. Rabinowitz (Ed.), Gordon and Breach, London, pp. 87–114, 1970c.

Powell, M. J. D., "Recent Advances in Unconstrained Optimization," *Mathematical Programming*, 1, pp. 26–57, 1971a.

Powell, M. J. D., "On the Convergence of the Variable Metric Algorithm," *J. Institute Mathematics and Its Applications*, 7, pp. 21–36, 1971b.

Powell, M. J. D., "Quadratic Termination Properties of Minimization Algorithms I, II," *J. Institute of Mathematics and Its Applications*, 10, pp. 333–342, 343–357, 1972.

Powell, M. J. D., "On Search Directions for Minimization Algorithms," *Mathematical Programming*, 4, pp. 193–201, 1973.

Powell, M. J. D., "Introduction to Constrained Optimization," in *Numerical Methods for Constrained Optimization*, P. E. Gill and W. Murray (Eds.), Academic Press, London and New York, pp. 1–28, 1974.

Powell, M. J. D., "Some Global Convergence Properties of a Variable Metric Algorithm for Minimization Without Exact Line Searches," in *Nonlinear Programming, SIAM, AMS Proceedings, Vol. IX*, R. W. Cottle and C. E. Lemke (Eds.), New York, pp. 53–72, 1976.

Powell, M. J. D., "Quadratic Termination Properties of Davidon's New Variable Metric Algorithm," *Mathematical Programming*, 12, pp. 141–147, 1977a.

Powell, M. J. D., "Restart Procedures for the Conjugate Gradient Method," *Mathematical Programming*, 12, pp. 241–254, 1977b.

Powell, M. J. D., "Algorithms for Nonlinear Constraints That Use Lagrangian Functions," *Mathematical Programming*, 14, 224–248, 1978.

Powell, M. J. D., "A Fast Algorithm for Nonlinearly Constrained Optimization Calculations," in *Numerical Analysis*, "Dundee 1977," G. A. Watson (Ed.), *Lecture Notes in Mathematics*, 630, Springer-Verlag, Berlin, 1978a.

Powell, M. J. D., "The Convergence of Variable Metric Methods of Nonlinearly Constrained Optimization Calculations," in *Nonlinear Programming, 3*, O. L. Mangasarian, R. R. Meyer, and S. M. Robinson (Eds.), Academic Press, New York, 1978b.

Powell, M. J. D., "A Note on Quasi-Newton Formulae for Sparse Second Derivative Matrices," *Mathematical Programming*, 20, pp. 144–151, 1981.

Powell, M. J. D., "Variable Metric Methods for Constrained Optimization," in *Mathematical Programming: The State of the Art*, A. Bachem, M. Grotschel, and B. Korte (Eds.), Springer-Verlag, Berlin, Heidelberg, New York and Tokyo, pp. 288–311, 1983a.

Powell, M. J. D., "On the Quadratic Programming Algorithm of Goldfarb and Idnani," in *Mathematical Programming Essays in Honor of George B. Dantzig, Part II*, R. W. Cottle (Ed.), *Mathematical Programming Study*, 25, North Holland, Amsterdam, 1985a.

Powell, M. J. D., "The Performance of Two Subroutines for Constrained Optimization on Some Difficult Test Problems," in *Numerical Optimization 1984*, P. T. Boggs, R. H. Byrd, and R. B. Schnabel (Eds.), *SIAM Publications*, Philadelphia, 1985b.

Powell, M. J. D., "How Bad Are the BFGS and DFP Methods When the Objective Function is Quadratic?", University of Cambridge DAMTP Report 85/NA4, 1985c.

Powell, M. J. D., "Convergence Properties of Algorithms for Nonlinear Optimization," *SIAM Review*, 28(4), pp. 487–500, 1986.

Powell, M. J. D., "Updating Conjugate Directions by the BFGS Formula," *Mathematical Programming*, 38, pp. 29–46, 1987.

Powell, M. J. D., and P. L. Toint, "On the Estimation of Sparse Hessian Matrices," *SIAM J. Numerical Analysis*, 16, pp. 1060–1074, 1979.

Powell, M. J. D., and Y. Yuan, "A Recursive Quadratic Programming Algorithm that Uses Differentiable Penalty Functions," *Mathematical Programming*, 35, pp. 265–278, 1986.

Prager, W., "Mathematical Programming and Theory of Structures," *SIAM J. Applied Mathematics*, 13, pp. 312–332, 1965.

Prince, L., B. Purrington, J. Ramsey, and J. Pope, "Gasoline Blending at Texaco Using Nonlinear Programming," Paper presented at TIMS/ORSA Joint National Meeting, Chicago, Ill., April 25–27, 1983.

Pshenichnyi, B. N., "Nonsmooth Optimization and Nonlinear Programming in Nonsmooth Optimization," C. Lemarechala and R. Mifflin (Eds.), *IIASA Proceedings 3*, Pergamon, Oxford, 1978.

Pugh, G. E., "Lagrange Multipliers and the Optimal Allocation of Defense Resources," *Operations Research*, 12, pp. 543–567, 1964.

Pugh, R. E., "A Language for Nonlinear Programming Problems," *Mathematical Programming*, 2, pp. 176–206, 1972.

Raghavendra, V., and K. S. P. Rao, "A Note on Optimization Using the Augmented Penalty Function," *J. Optimization Theory and Applications*, 12, pp. 320–324, 1973.

Rani, O., and R. N. Kaul, "Duality Theorems for a Class of Nonconvex Programming Problems," *J. Optimization Theory and Applications*, 11, pp. 305–308, 1973.

Ratschek, H., and J. Rokne, ''*New Computer Methods for Global Optimization*,'' Ellis Horwood, Chichester, 1988.

Rauch, S. W., ''A Convergence Theory for a Class of Nonlinear Programming Problems,'' *SIAM J. Numerical Analysis*, 10, pp. 207–228, 1973.

Reddy, P. J., H. J. Zimmermann, and A. Husain, ''Numerical Experiments on DFP-Method, A Powerful Function Minimization Technique,'' *J. of Computational and Applied Mathematics*, 4, pp. 255–265, 1975.

Reid, J. K., ''On the Method of Conjugate Gradients for the Solution of Large Sparse Systems of Equations,'' in *Large Sparse Sets of Linear Equations*, J. K. Reid (Ed.), Academic Press, London, 1971.

Reklaitis, G. V., and D. T. Phillips, ''A Survey of Nonlinear Programming,'' *AIIE Transactions*, 7, pp. 235–256, 1975.

Reklaitis, G. V., and D. J. Wilde, ''Necessary Conditions for a Local Optimum Without Prior Constraint Qualifications,'' in *Optimizing Methods in Statistics*, J. S. Rustagi (Ed.), Academic Press, New York, 1971.

Renegar, J., ''A Polynomial-Time Algorithm, Based on Newton's Method for Linear Programming,'' *Mathematical Programming*, 40, pp. 59–93, 1988.

Rissanen, J., ''On Duality Without Convexity,'' *J. Mathematical Analysis and Applications*, 18, pp. 269–275, 1967.

Ritter, K., ''A Method for Solving Maximum Problems with a Nonconcave Quadratic Objective Function,'' *Z. Wahrscheinlichkeitstheorie und Verwandte Gebiete*, 4, pp. 340–351, 1966.

Ritter, K., ''A Method of Conjugate Directions for Unconstrained Minimization,'' *Operations Research Verfahren*, 13, pp. 293–320, 1972.

Ritter, K., ''A Superlinearly Convergent Method for Minimization Problems with Linear Inequality Constraints,'' *Mathematical Programming*, 4, pp. 44–71, 1973.

Roberts, A. W., and D. E. Varberg, *Convex Functions*, Academic Press, New York, 1973.

Robinson, S. M., ''A Quadratically-Convergent Algorithm for General Nonlinear Programming Problems,'' *Mathematical Programming*, 3, pp. 145–156, 1972.

Robinson, S. M., ''Computable Error Bounds for Nonlinear Programming,'' *Mathematical Programming*, 5, pp. 235–242, 1973.

Robinson, S. M., ''Perturbed Kuhn–Tucker Points and Rates of Convergence for a Class of Nonlinear Programming Algorithms,'' *Mathematical Programming*, 7, pp. 1–16, 1974.

Robinson, S. M., ''Generalized Equations and Their Solutions, Part II: Applications to Nonlinear Programming,'' in *Optimality and Stability in Mathematical Programming*, M. Guignard (Ed.), *Mathematical Programming Study*, 19, North Holland, Amsterdam, 1982.

Robinson, S. M., and R. H. Day, ''A Sufficient Condition for Continuity of Optimal Sets in Mathematical Programming,'' *J. Mathematical Analysis and Applications*, 45, pp. 506–511, 1974.

Robinson, S. M., and R. R. Meyer, ''Lower Semicontinuity of Multivalued Linearization Mappings,'' *SIAM J. Control*, 11, pp. 525–533, 1973.

Rockafeller, R. T., ''Minimax Theorems and Conjugate Saddle Functions,'' *Mathematica Scandinavica*, 14, pp. 151–173, 1964.

Rockafeller, R. T., ''Extension of Fenchel's Duality Theorem for Convex Functions,'' *Duke Mathematical Journal*, 33, pp. 81–90, 1966.

Rockafeller, R. T., ''A General Correspondence Between Dual Minimax Problems and Convex Programs,'' *Pacific J. Mathematics*, 25, pp. 597–612, 1968.

Rockafeller, R. T., ''Duality in Nonlinear Programming,'' in *Mathematics of the Decison Sciences*, G. B. Dantzig and A. Veinott (Eds.), American Mathematical Society, Providence, R.I., 1969.

Rockafeller, R. T., *Convex Analysis*, Princeton University Press, Princeton, N.J., 1970.

Rockafeller, R. T., ''A Dual Approach to Solving Nonlinear Programming Problems by Unconstrained Optimization,'' *Mathematical Programming*, 5, pp. 354–373, 1973a.

Rockafeller, R. T., ''The Multiplier Method of Hestenes and Powell Applied to Convex Programming,'' *J. Optimization Theory and Applications*, 12, pp. 555–562, 1973b.

Rockafeller, R. T., "Augmented Lagrange Multiplier Functions and Duality in Nonconvex Programming," *SIAM J. Control*, 12, pp. 268–285, 1974.

Rockafellar, R. T., "Lagrange Multipliers in Optimization," in *Nonlinear Programming, SIAM, AMS Proceedings, Vol. IX*, R. W. Cottle and C. E. Lemke (Eds), New York, March 23–24, 1975.

Rockafellar, R. T., "Optimization in Networks," Lecture Notes, University of Washington, 1976.

Rockafellar, R. T., *The Theory of Subgradients and its Applications to Problems of Optimization, Convex and Nonconvex Functions*, Heldermann Verlag, Berlin, 1981.

Rohn, J., "A Short Proof of Finiteness of Murty's Principal Pivoting Algorithm," *Mathematical Programming*, 46, pp. 255–256, 1990.

Roode, J. D., "Generalized Lagrangian Functions in Mathematical Programming," Ph.D. Thesis, The University of Leiden, The Netherlands, 1968.

Roode, J. D., "Generalized Lagrangian Functions and Mathematical Programming," in *Optimization*, R. Fletcher (Ed.), 1969.

Roos, C., and J.-P. Vial, "A Polynomial Method of Approximate Centers for Linear Programming," Report, Delft University of Technology, 1988 (to appear in *Mathematical Programming*).

Rosen, J. B., "The Gradient Projection Method for Nonlinear Programming, Part I, Linear Constraints," *SIAM J. Applied Mathematics*, 8, pp. 181–217, 1960.

Rosen, J. B., "The Gradient Projection Method for Nonlinear Programming Part II: Nonlinear Constraints," *SIAM J. Applied Mathematics*, 9, pp. 514–553, 1961.

Rosen J. B., and J. Kreuser, "A Gradient Projection Algorithm for Nonlinear Constraints," in *Numerical Methods for Nonlinear Optimization*, F. A. Lootsma (Ed.), 1972.

Rosen, J. B., O. L. Mangasarian, and K. Ritter (Eds.), *Nonlinear Programming*, Academic Press, New York, 1970.

Rosen, J. B., and S. Suzuki, "Construction of Nonlinear Programming Test Problems," *Communications of Association for Computing Machinery*, 8, p. 113, 1965.

Rosenbrock, H. H., "An Automatic Method for Finding the Greatest or Least Value of a Function," *Computer Journal*, 3, pp. 175–184, 1960.

Rothfarb, B., H. Frank, D. M. Rosenbaum, K. Steiglitz, and D. J. Kleitman, "Optimal Design of Offshore Natural-Gas Pipeline Systems," *Operations Research*, 18, pp. 992–1020, 1970.

Roy, S., and D. Solow, "A Sequential Linear Programming Approach for Solving the Linear Complementarity Problem," Department of Operations Research, Case Western Reserve University, Cleveland, Ohio, 1985.

Rozvany, G. I. N., *Optimal Design of Flexural Systems: Beams, Grillages, Slabs, Plates and Shells*, Pergamon Press, New York, 1976.

Rubio, J. E., "Solution of Nonlinear Optimal Control Problems in Hilbert Space by Means of Linear Programming Techniques," *J. Optimization Theory and Applications*, 30(4), pp. 643–661, 1980.

Rudin, W., *Principles of Mathematical Analysis* (2nd Edition), McGraw-Hill, New York, 1964.

Rupp, R. D., "On the Combination of the Multiplier Method of Hestenes and Powell with Newton's Method," *J. Optimization Theory and Applications*, 15, pp. 167–187, 1975.

Russell, D. L., *Optimization Theory*, W. A. Benajmin, New York, 1970.

Rvačev, V. L., "On the Analytical Description of Certain Geometric Objects," *Soviet Mathematics*, 4, pp. 1750–1753, 1963.

Sargent, R. W. H., "Minimization Without Constraints," in *Optimization and Design*, M. Avriel, M. J. Rijckaert, and D. V. Wilde (Eds.), 1973.

Sargent, R. W. H., and B. A. Murtagh, "Projection Methods for Nonlinear Programming," *Mathematical Programming*, 4, pp. 245–268, 1973.

Sargent, R. W. H., and D. J. Sebastian, "Numerical Experience with Algorithms for Unconstrained Minimizations," in *Numerical Methods for Nonlinear Optimization*, F. A. Lootsma (Ed.), 1972.

Sargent, R. W. H., and D. J. Sebastian, "On the Convergence of Sequential Minimization Algorithms," *J. Optimization Theory and Applications*, 12, pp. 567–575, 1973.

Sarma, P. V. L. N., and G. V. Reklaitis, "Optimization of a Complex Chemical Process Using an Equation Oriented Model," presented at Tenth International Symposium on Mathematical Programming, Montreal, Aug. 27–31, 1979.

Sasai, H., "An Interior Penalty Method for Minimax Problems with Constraints," *SIAM J. Control*, 12, pp. 643–649, 1974.

Sasson, A. M., "Nonlinear Programming Solutions for Load Flow, Minimum-Loss and Economic Dispatching Problems," *IEEE Transactions on Power Apparatus Systems*, PAS-88, pp. 399–409, 1969a.

Sasson, A. M., "Combined Use of the Powell and Fletcher–Powell Nonlinear Programming Methods for Optimal Load Flows," *IEEE Transactions on Power Apparatus and Systems*, PAS-88, p. 1530–1537, 1969b.

Sasson, A. M., F. Aboytes, R. Carenas, F. Gome, and F. Viloria, "A Comparison of Power Systems Static Optimization Techniques," in *Proc. of 7th Power Industry Computer Applications Conference*, Boston, pp. 329–337, 1971.

Sasson, A. M., and H. M. Merrill, "Some Applications of Optimization Techniques to Power Systems Problems," *Proc. of the IEEE*, 62, pp. 959–972, 1974.

Savage, S. L., "Some Theoretical Implications of Local Optimization," *Mathematical Programming*, 10, pp. 354–366, 1976.

Schaible, S., "Quasi-Convex Optimization in General Real Linear Spaces," *Zeitschrift für Operations Research A*, 16, pp. 205–213, 1972.

Schaible, S., "Quasi-Concave, Strictly Quasi-Concave and Pseudo-Concave Functions," *Operations Research Verfahren*, 17, pp. 308–316, 1973a.

Schaible, S., "Quasi-Concavity and Pseudo-Concavity of Cubic Functions," *Mathematical Programming*, 5, pp. 243–247, 1973b.

Schaible, S., "Parameter-Free Convex Equivalent and Dual Programs of Franctional Programming Problems," *Zeitschrift für Operations Research A*, 18, pp. 187–196, 1974a.

Schaible, S., "Maximization of Quasi-Concave Quotients and Products of Finitely Many Functionals," *Cahiers Centre Études Recherche Opperationelle*, 16, pp. 45–53, 1974b.

Schaible, S., "Duality in Fractional Programming: A Unified Approach," *Operations Research*, 24, pp. 452–461, 1976.

Schaible, S., "Generalized Convexity of Quadratic Functions," in *Generalized Concavity in Optimization and Economics*, S. Schaible and W. T. Ziemba (Eds.), Academic Press, New York, pp. 183–197, 1981a.

Schaible, S., "Quasiconvex, Pseudoconvex, and Strictly Pseudoconvex Quadratic Functions," *J. of Optimization Theory & Applications*, 35, pp. 303–338, 1981b.

Schaible, S., "Multi-Ratio Fractional Programming Analysis and Applications," Working Paper 90-4, Graduate School of Management, University of California at Riverside, 1989.

Schaible, S., and W. T. Ziemba, *Generalized Concavity in Optimization and Economics*, Academic Press, San Francisco, 1981.

Schechter, S., "Minimization of a Convex Function by Relaxation," in *Integer and Nonlinear Programming*, J. Abadie (Ed.), 1970.

Schittkowski, K., "Nonlinear Programming Codes—Information, Tests, Performance," in *Lecture Notes in Economics and Mathematical Systems*, Vol. 183, Springer-Verlag, Berlin, Heidelberg, New York (1980).

Schittkowski, K., "The Nonlinear Programming Method of Wilson, Han, and Powell with an Augmented Lagrangian Type Line Search Function—Part I: Convergence Analysis," *Numerical Mathematics*, 38, pp. 83–114, 1981.

Schittkowski, K., "On the Convergence of a Sequential Quadratic Programming Method With an Augmented Lagrangian Line Search Function," *Math. Operationsforsch. u. Statist., Ser. Optimization*, 14, pp. 197–216, 1983.

Schrage, L., *Linear, Integer, and Quadratic Programming With LINDO*, The Scientific Press, Palo Alto, Calif., 1984.

Scott, C. H., and T. R. Jefferson, "Duality for Minmax Programs," *J. of Mathematical Analysis and Applications*, 100(2), pp. 385–392, 1984.

Scott, C. H., and T. R. Jefferson, "Conjugate Duality in Generalized Fractional Programming," *J. of Optimization Theory and Applications*, 60(3), pp. 475–487, 1989.

Sen, S., and H. D. Sherali, "On the Convergence of Cutting Plane Algorithms for a Class of Nonconvex Mathematical Problems," *Mathematical Programming*, 31(1), pp. 42–56, 1985.

Sen, S., and H. D. Sherali, "A Branch and Bound Algorithm for Extreme Point Mathematical Programming Problems," *Discrete Applied Mathematics*, 11, pp. 265–280, 1985.

Sen, S., and H. D. Sherali, "Facet Inequalities From Simple Disjunctions in Cutting Plane Theory," *Mathematical Programming*, 34(1), pp. 72–83, 1986.

Sen, S., and H. D. Sherali, "A Class of Convergent Primal-Dual Subgradient Algorithms for Decomposable Convex Programs," *Mathematical Programming*, 35(3), pp. 279–297, 1986.

Sengupta, J. K., *Stochastic Programming-Methods and Applications*, American Elseiver, New York, 1972.

Sengupta, J. K., and J. H. Portillo–Campbell, "A Fractile Approach to Linear Programming Under Risk," *Management Science*, 16, pp. 298–308, 1970.

Sengupta, J. K., G. Tintner, and C. Millham, "On Some Theorems in Stochastic Linear Programming with Applications," *Management Science*, 10, pp. 143–159, 1963.

Shah, B. V., R. J. Beuhler, and O. Kempthorne, "Some Algorithms for Minimizing a Function of Several Variables," *SIAM J. Applied Mathematics*, 12, pp. 74–92, 1964.

Shamir, D., "Optimal Design and Operation of Water Distribution Systems," *Water Resources Research*, 10, pp. 27–36, 1974.

Shanno, D. F., "Conditioning of Quasi-Newton Methods for Function Minimizations," *Mathematics of Computation*, 24, pp. 641–656, 1970.

Shanno, D. F., "Conjugate Gradient Methods with Inexact Line Searches," *Mathematics of Operations Research*, 3, pp. 244–256, 1978.

Shanno, D. F., "On Variable Metric Methods for Sparse Hessians," *Mathematics of Computation*, 34, pp. 499–514, 1980.

Shanno, D. F., and R. E. Marsten, "Conjugate Gradient Methods for Linearly Constrained Nonlinear Programming," *Mathematical Programming Study*, 16, pp. 149–161, 1982.

Shanno, D. F., and K.-H. Phua, "Matrix Conditioning and Nonlinear Optimization," *Mathematical Programming*, 14, pp. 149–160, 1978.

Shanno, D. F., and K. H. Phua, "Numerical Comparison of Several Variable Metric Algorithms," *J. Optimization Theory and Applications*, 25, pp. 507–518, 1978b.

Shapiro, J. F., *Mathematical Programming; Structures and Algorithms*, John Wiley and Sons, New York, 1979.

Sharma, I. C., and K. Swarup, "On Duality in Linear Fractional Functionals Programming," *Zeitschrift für Operations Research A*, 16, pp. 91–100, 1972.

Sherali, H. D., "Algorithmic Insights and a Convergence Analysis for a Karmarkar-type of Algorithm for Linear Programs," *Naval Research Logistics Quarterly*, 34, pp. 399–416, 1987.

Sherali, H. D., "A Multiple Leader Stackelberg Model and Analysis," *Operations Research*, 32(2), pp. 390–404, 1984.

Sherali, H. D., "A Restriction and Steepest Descent Feasible Directions Approach to a Capacity Expansion Problem," *European J. of Operational Research*, 19(3), pp. 345–361, 1985.

Sherali, H. D., "A Constructive Proof of the Representation Theorem for Polyhedral Sets Based on Fundamental Definitions," *American J. of Mathematical and Management Sciences*, 7(3/4), pp. 253–270, 1987.

Sherali, H. D., "Dorn's Duality for Quadratic Programs Revisited: The Nonconvex Case," *European J. of Operational Research*, to appear, 1991.

Sherali, H. D., and W. P. Adams, "A Decomposition Algorithm for a Discrete Location–Allocation Problem," *Operations Research*, 32(4), pp. 878–900, 1984.

Sherali, H. D., and A. Alameddine, "An Explicit Characterization of the Convex Envelope of a Bivariate Bilinear Function Over Special Polytopes," *Annals for Operations Research*, Computational Methods in Global Optimization, P. Pardalos and J. B. Rosen (Eds.), Vol. 25, pp. 197–210, 1990.

Sherali, H. D., and A. Alameddine, "A New Reformulation-Linearization Algorithm for Solving Bilinear Programming Problems," *J. of Global Optimization*, to appear (1991).
Sherali, H. D., and S. E. Dickey, "An Extreme Point Ranking Algorithm for the Extreme Point Mathematical Programming Problem," *Computers and Operations Research*, 13(4), pp. 465–475, 1986.
Sherali, H. D., R. Krishnan, and F. A. Al-Khayyal, "A Parametric Concave Programming Approach for Linear Complementarity Problems," Department of Industrial and Systems Engineering, VPI&SU, Blacksburg, Virginia, 1992a.
Sherali, H. D., R. Krishnan, and F. A. Al-Khayyal, "A Reformulation-Linearization-Enumeration Approach for Linear Complementarity Problems," Department of Industrial and Systems Engineering, VPI&SU, Blacksburg, Virginia, 1992b.
Sherali, H. D., and D. C. Myers, "The Design of Branch and Bound Algorithms for a Class of Nonlinear Integer Programs," *Annals of Operations Research*, Special issue on *Algorithms and Software for Optimization*, C. Monma (Ed.), 5, pp. 463–484, 1986.
Sherali, H. D., and D. C. Myers, "Dual Formulations and Subgradient Optimization Strategies for Linear Programming Relaxations of Mixed-Integer Programs," *Discrete Applied Mathematics*, 20, pp. 51–68, 1988.
Sherali, H. D., and S. Sen, "A Disjunctive Cutting Plane Algorithm for the Extreme Point Mathematical Programming Problem," *Opsearch* (*Theory*), 22(2), pp. 83–94, 1985.
Sherali, H. D., and S. Sen, "On Generating Cutting Planes form Combinatorial Disjunctions," *Operations Research*, 33(4), pp. 928–933, 1985.
Sherali, H. D., and C. M. Shetty, "A Finitely Convergent Algorithm for Bilinear Programming Problems Using Polar Cuts and Disjunctive Face Cuts," *Mathematical Programming*, 19, pp. 14–31, 1980a.
Sherali, H. D., and C. M. Shetty, "On the Generation of Deep Disjunctive Cutting Planes," *Naval Research Logistics Quarterly*, 27(3), pp. 453–475, 1980b.
Sherali, H. D., and C. M. Shetty, *Optimization with Disjunctive Constraints* (Series in Economics and Mathematical System, Publication Number 181), Springer-Verlag, Berlin–Heidelberg–New York, 1980c.
Sherali, H. D., and C. M. Shetty, "A Finitely Convergent Procedure for Facial Disjunctive Programs," *Discrete Applied Mathematics*, 4, pp. 135–148, 1982.
Sherali, H. D., and C. M. Shetty, "Nondominated Cuts for Disjunctive Programs and Polyhedral Annexation Methods," *Opsearch* (*Theory*), 20(3), pp. 129–144, 1983.
Sherali, H. D., B. O. Skarpness, and B. Kim, "An Assumption-Free Convergence Analysis for a Perturbation of the Scaling Algorithm for Linear Programming, With Application to the L_1 Estimation Problem," *Naval Research Logistics Quarterly*, 35, pp. 473–492, 1988.
Sherali, H. D., and A. L. Soyster, "Analysis of Network Structured Models for Electric Utility Capacity Planning and Marginal Cost Pricing Problems," in *Energy Models and Studies: Studies in Management Science and Systems Series*, B. Lev (Ed.), North Holland Publishing Company, pp. 113–134, 1983.
Sherali, H. D., A. L. Soyster, and F. H. Murphy, "Stackelberg–Nash–Cournot–Equilibria: Characterization and Computations," *Operations Research*, 31(2), pp. 253–276, 1983.
Sherali, H. D., and K. Staschus, "A Nonlinear Hierarchical Approach for Incorporating Solar Generation Units in Electric Utility Capacity Expansion Plans," *Computers and Operations Research*, 12(2), pp. 181–199, 1985.
Sherali, H. D., and G. H. Tuncbilek, "A Global Optimization Algorithm for Polynomial Programming Problems Using A Reformulation-Linearization Technique," *Journal of Global Optimization*, Vol. 2, pp. 101–112, 1992.
Sherali, H. D., and O. Ulular, "A Primal-Dual Conjugate Subgradient Algorithm for Specially Structured Linear and Convex Programming Problems," *Applied Mathematics and Optimization*, 20, pp. 193–221, 1989.
Sherali, H. D., and O. Ulular, "Conjugate Gradient Methods Using Quasi-Newton Updates with Inexact Line Searches," *J. of Mathematical Analysis and Applications*, 150(2), pp. 359–377, 1990.

Sherman, A. H., "On Newton Iterative Methods for the Solution of Systems of Nonlinear Equations," *SIAM J. Numerical Analysis*, 15, pp. 755–771, 1978.

Shetty, C. M., "A Simplified Procedure for Quadratic Programming," *Operations Research*, 11, pp. 248–260, 1963.

Shetty, C. M., and H. D. Sherali, "Rectilinear Distance Location-Allocation Problem: A Simplex Based Algorithm," *Proc. of the International Symposium on Extremal Methods and Systems Analyses*, Vol. 174, pp. 442–464, Springer-Verlag, Berlin/Heidelberg, 1980.

Shiau, T.-H., "Iterative Linear Programming for Linear Complementarity and Related Problems," Computer Sciences Technical Report #507, University of Wisconsin-Madison, August 1983.

Shor, N. Z., "On the Rate of Convergence of the Generalized Gradient Method," *Kibernetika*, 4(3), pp. 98–99, 1968.

Shor, N. Z., "Convergence Rate of the Gradient Descent Method With Dilatation of the Space," *Cybernetics*, 6(2), pp. 102–108, 1970.

Shor, N. Z., "Convergence of Gradient Method With Space Dilatation in the Direction of the Difference Between Two Successive Gradients," *Kibernetika*, 11(4), pp. 48–53, 1975.

Shor, N. Z., "New Development Trends in Nondifferentiable Optimization," Translated from *Kibernetika*, 6, pp. 87–91, 1977.

Shor, N. Z., "The Cut-Off Method with Space Dilation for Solving Convex Programming Problems," *Kibernetika*, 13, pp. 94–95, 1977.

Shor, N. Z., *Minimization Methods for Non-Differentiable Functions* (Translated from Russian), Springer-Verlag, Berlin, 1985.

Shor, N. Z., "Dual Quadratic Estimates in Polynomial and Boolean Programming," *Annals of Operations Research*, 25(1–4), pp. 163–168, 1990.

Siddal, J. N., *Analytical Decision-Making in Engineering Design,* Prentice Hall, Englewood Cliffs, N.J., 1972.

Simonnard, M., *Linear Programming* (translated by W. S. Jewell), Prentice Hall, Englewood Cliffs, N.J., 1966.

Sinha, S. M., "An Extension of a Theorem on Supports of a Convex Function," *Management Science*, 12, pp. 380–384, 1966.

Sinha, S. M., "A Duality Theorem for Nonlinear Programming," *Management Science*, 12, pp. 385–390, 1966.

Sinha, S. M., and K. Swarup, "Mathematical Programming: A Survey," *J. Mathematical Sciences*, 2, pp. 125–146, 1967.

Sion, M., "On General Minmax Theorems," *Pacific J. Mathematics*, 8, pp. 171–176, 1958.

Slater, M., "Lagrange Multipliers Revisited: A Contribution to Nonlinear Programming," *Cowles Commission Discussion Paper*, Mathematics, 403, 1950.

Smeers, Y., "A Convergence Proof of a Special Version of the Generalized Reduced Gradient Method," *(GRGS), R.A.I.R.O.*, 5(3), 1974.

Smeers, Y., "Generalized Reduced Gradient Method as an Extension of Feasible Directions Methods," *J. of Optimization Theory and Applications*, 22(2), pp. 209–226, 1977.

Smith, S., and L. Lasdon, "Solving Large Sparse Nonlinear Programs Using GRG," *ORSA J. on Computing*, 4(1), pp. 2–15, 1992.

Soland, R. M., "An Algorithm for Separable Nonconvex Programming Problems, II," *Management Science*, 17, pp. 759–773, 1971.

Soland, R. M., "An Algorithm for Separable Piecewise Convex Programming Problems," *Naval Research Logistics Quarterly*, 20, pp. 325–340, 1973.

Solow, D., and P. Sengupta, "A Finite Descent Theory for Linear Programming, Piecewise Linear Minimization and the Linear Complementarity Problem," *Naval Research Logistics Quarterly*, 32, pp. 417–431, 1985.

Sonnevand, Gy., "An Analytic Centre for Polyhedrons and New Classes of Global Algorithms for Linear (Smooth, Convex) Programming," Preprint, Department of Numerical Analysis, Institute of Mathematics, Eotvos University, Budapest, Hungary, 1985.

Sorensen, D. C., *Trust Region Methods for Unconstrained Optimization, Nonlinear Optimization*, Academic Press, New York, 1982.

Sorensen, D. C., "Newton's Method With a Model Trust Region Modification," *SIAM J. Numerical Analysis*, 19, pp. 409–426, 1982.

Sorenson, H. W., "Comparison of Some Conjugate Directions Procedures for Function Minimization," *J. Franklin Institute*, 288, pp. 421–441, 1969.

Spendley, W., "Nonlinear Least Squares Fitting Using a Modified Simplex Minimization Method," in *Optimization*, R. Fletcher (Ed.), 1969.

Spendley, W., G. R. Hext, and F. R. Himsworth, "Sequential Application of Simplex Designs of Optimization and Evolutionary Operations," *Technometrics*, 4, pp. 441–461, 1962.

Steur, R. E., *Multiple Criteria Optimization: Theory, Computation, and Application*, John Wiley & Sons, New York, 1986.

Stewart, G. W., III, "A Modification of Davidon's Minimization Method to Accept Difference Approximations of Derivatives," *J. Association for Computing Machinery*, 14, pp. 72–83, 1967.

Stewart, G. W., *Introduction to Matrix Computations*, Academic Press, London and New York, 1973.

Stocker, D. C., A Comparative Study of Nonlinear Programming Codes, M.S. Thesis, The University of Texas, Austin, Texas, 1969.

Stoer, J., "Duality in Nonlinear Programming and the Minimax Theorem," *Numerische Mathematik*, 5, pp. 371–379, 1963.

Stoer, J., "Foundations of Recursive Quadratic Programming Methods for Solving Nonlinear Programs," *Nato ASI Series, F125, Computational Mathematical Programming*, K. Schittkowski (Ed.), Springer-Verlag, Berlin/Heidelberg, pp., 1985.

Stoer, J., and C. Witzgall, *Convexity and Optimization in Finite Dimensions I*, Springer-Verlag, New York, 1970.

Straeter, T. A., and J. E. Hogge, "A Comparison of Gradient Dependent Techniques for the Minimization of an Unconstrained Function of Several Variables," *J. American Institute of Aeronautics and Astronautics*, 8, pp. 2226–2229, 1970.

Strodiot, J. J., and V. H. Nguyen, "Kuhn–Tucker Multipliers and NonSmooth Programs," in *Mathematical Programming Study, 19, Optimality, Duality, and Stability*, M. Guignard (Ed.), pp. 222–240, 1982.

Swann, W. H., "Report on the Development of a New Direct Search Method of Optimization," *Imperial Chemical Industries Ltd. Central Instr. Res. Lab. Research Note* 64/3, London, 1964.

Swarup, K., "Linear Fractional Functionals Programming," *Operations Research*, 13, pp. 1029–1035, 1965.

Swarup, K., "Programming With Quadratic Fractional Functions," *Opsearch*, 2, pp. 23–30, 1966.

Szego, G. P. (Ed.), *Minimization Algorithms—Mathematical Theories and Computer Results*, Academic Press, New York, 1972.

Tabak, D., "Comparative Study of Various Minimization Techniques Used in Mathematical Programming," *IEEE Transactions on Automatic Control*, AC-14, p. 572, 1969.

Tabak, D., and B. C. Kuo, *Optimal Control by Mathematical Programming*, Prentice Hall, Englewood Cliffs, N.J., 1971.

Taha, H. A., "Concave Minimization Over a Convex Polyhedron," *Naval Research Logistics Quarterly*, 20, pp. 533–548, 1973.

Takahashi, I., "Variable Separation Principle for Mathematical Programming," *J. Operations Research Society of Japan*, 6, pp. 82–105, 1964.

Tamir, A., "Line Search Techniques Based on Interpolating Polynomials Using Function Values Only," *Management Science*, 22(5), pp. 576–586, 1976.

Tamura, M., and Y. Kobayashi, "Application of Sequential Quadratic Programming Software Program to an Actual Problem," *Mathematical Programming*, 52(1), pp. 19–28, 1991.

Tanabe, K., "An Algorithm for the Constrained Maximization in Nonlinear Programming," *J. Operations Research Society of Japan*, 17, pp. 184–201, 1974

Tapia, R. A., "Newton's Method for Optimization Problems with Equality Constraints," *SIAM J. Numerical Analysis*, 11, pp. 874–886, 1974a.

Tapia, R. A., "A Stable Approach to Newton's Method for General Mathematical Programming Problems in R^n," *J. Optimization Theory and Applications*, 14, pp. 453–476, 1974b.

Tapia, R. A., "Diagonalized Multiplier Methods and Quasi-Newton Methods for Constrained Optimization," *J. Optimization Theory and Applications*, 22, pp. 135–194, 1977.

Tapia, R. A., "Quasi-Newton Methods for Equality Constrained Optimization: Equivalents of Existing Methods and New Implementation," *Symposium on Nonlinear Programming III*, O. Mangasarian, R. Meyer, and S. Robinson (Eds.), Academic Press, New York, pp. 125–164, 1978.

Tapia, R. A., Y. Zhang, "A Polynomial and Superlinearly Convergent Primal-Dual Interior Algorithm for Linear Programming,"*Joint National TIMS/ORSA Meeting,* Nashville, Tenn., May 12–15, 1991.

Taylor, A. E., and W. R. Mann, *Advanced Calculus* (3rd Edition), John Wiley & Sons, New York, 1983.

Teng, J. Z., "Exact Distribution of the Kruksal–Wallis *H* Test and the Asymptotic Efficiency of the Wilcoxon Test with Ties," Ph.D. Thesis, University of Wisconsin—Madison, 1978.

Thakur, L. S., "Error Analysis for Convex Separable Programs: The Piecewise Linear Approximation and the Bounds on the Optimal Objective Value,"*SIAM J. on Applied Mathematics,* 34, pp. 704–714, 1978.

Theil, H., and C. van de Panne, "Quadratic Programming as an Extension of Conventional Quadratic Maximization," *Management Science*, 7, pp. 1–20, 1961.

Thompson, W. A., and D. W. Parke, "Some Properties of Generalized Concave Functions," *Operations Research*, 21, pp. 305–313, 1973.

Todd, M. J., "A Generalized Complementary Pivoting Algorithm," *Mathematical Programming*, 6, pp. 243–263, 1974.

Todd, M. J., "The Symmetric Rank-One Quasi-Newton Method Is a Space-Dilation Subgradient Algorithm," *Operations Research Letters*, 5(5), pp. 217–220, 1986.

Todd, M. J., "Recent Developments and New Directions in Linear Programming," in *Mathematical Programming*, M. Iri and K. Tanabe (Eds.), KTK Scientific Publishers, Tokyo, Japan, pp. 109–157, 1989.

Toint, P.L., "On the Superlinear Convergence of an Algorithm for Solving a Sparse Minimization Problem," *SIAM J. Numerical Analysis*, 16, pp. 1036–1045, 1979.

Tomlin, J. A., "On Scaling Linear Programming Problems," in *Computational Practice in Mathematical Programming*, M. L. Balinski and E. Hellerman (Eds.), 1975 (manuscript, 1973).

Tone, K., "Revisions of Constraint Approximations in the Successive QP Method for Nonlinear Programming," *Mathematical Programming*, 26, pp. 144–152, 1983.

Topkis, D. M., and A. F. Veinott, "On the Convergence of Some Feasible Direction Algorithms for Nonlinear Programming," *SIAM J. Control*, 5, pp. 268–279, 1967.

Torsti, J. J., and A. M. Aurela, "A Fast Quadratic Programming Method for Solving Ill-Conditioned Systems of Equations," *J. Mathematical Analysis and Applications*, 38, pp. 193–204, 1972.

Tripathi, S. S., and K. S. Narendra, "Constrained Optimization Problems Using Multiplier Methods," *J. Optimization Theory and Applications*, 9, pp. 59–70, 1972.

Tucker, A. W., "Linear and Nonlinear Programming," *Operations Research*, 5, pp. 244–257, 1957.

Tucker, A. W., A Least-Distance Approach to Quadratic Programming," in *Mathematics of the Decision Sciences*, G. B. Dantzig and A. F. Veinott (Eds.), 1968.

Tucker, A. W., "A Least Distance Programming," in *Proc. of Princeton Conference on Mathematical Programming*, H. W. Kuhn (Ed.), 1970.

Tui, H., "Concave Programming under Linear Constraints" (Russian), English transation in *Soviet Mathematics*, 5, pp. 1437–1440, 1964.

Umida, T., and A. Ichikawa, "A Modified Complex Method for Optimization," *J. Industrial and Engineering Chemistry Products Research and Development*, 10, pp. 236–243, 1971.

Uzawa, H., "The Kuhn–Tucker Theorem in Concave Programming," in *Studies in Linear and Nonlinear Programming*, K. J. Arrow, L. Hurwicz, and H. Uzawa (Eds.), 1958a.

Uzawa, H., "Gradient Method for Concave Programming, II," in *Studies in Linear and Nonlinear Programming*, K. J. Arrow, L. Hurwicz, and H. Uzawa (Eds.), 1958b.

Uzawa, H., "Iterative Methods for Concave Programming," in *Studies in Linear and Nonlinear Programming*, K. J. Arrow, L. Hurwicz, and H. Uzawa (Eds.), 1958c.

Uzawa, H., "Market Mechanisms and Mathematical Programming," *Econometrica*, 28, pp. 872–880, 1960.

Uzawa, H., "Duality Principles in the Theory of Cost and Production," *International Economic Review*, 5, pp. 216–220, 1964.

Vaidya, P.M., "An Algorithm for Linear Programming Which Requires 0 $(((m + n)n^2 + (m + n)^{1.5}n)L)$ Arithmetic Operations," Preprint, AT&T Bell Laboratories, Murray Hill, N.J., 1987 (to appear in *Mathematical Programming*).

Vaish, H., "Nonconvex Programming with Applications to Production and Location Problems," Ph.D. Dissertation, Georgia Institute of Technology, 1974.

Vaish, H., and C. M. Shetty, "The Bilinear Programming Problem," *Naval Research Logistics Quarterly*, 23, pp. 303–309, 1976.

Vaish, H., and C. M. Shetty, "A Cutting Plane Algorithm for the Bilinear Programming Problem, *Naval Research Logistics Quarterly*, 24, pp. 83–94, 1977.

Vajda, S., *Mathematical Programming*, Addison-Wesley, Reading, Mass., 1961.

Vajda, S., "Nonlinear Programming and Duality," in *Nonlinear Programming*, J. Abadie (Ed.), 1967.

Vajda, S., "Stochastic Programming," in *Integer and Nonlinear Programming*, J. Abadie (Ed.), 1970.

Vajda, S., *Probabilistic Programming*, Academic Press, New York, 1972.

Vajda, S., *Theory of Linear and Non-Linear Programming*, Longman, London, 1974a.

Vajda, S., "Tests of Optimality in Constrained Optimization," *J. Institute of Mathematics and Its Applications*, 13, pp. 187–200, 1974b.

Valentine, F. A., *Convex Sets*, McGraw-Hill, New York, 1964.

Van Bokhoven, W. M. G., "Macromodelling and Simulation of Mixed Analog-Digital Networks by a Piecewise Linear System Approach," *IEEE 1980 Circuits and Computers*, pp. 361–365, 1980.

Vandergraft, J. S., *Introduction to Numerical Computations*, Academic Press, Orlando, Fla., 1983.

Varaiya, O., "Nonlinear Programming in Banach Spaces," *SIAM J. Applied Mathematics*, 15, pp. 284–293, 1967.

Varaiya, P. P., *Notes on Optimization*, Van Nostrand Reinhold, New York, 1972.

Veinott, A. F., "The Supporting Hyperplane Method for Unimodal Programming," *Operations Research*, 15, pp. 147–152, 1967.

Von Neumann, J., "Zur Theorie der Gesellschaftsspiele," *Mathematische Annalen*, 100, pp. 295–320, 1928.

Von Neumann, J., and O. Morgenstern, *Theory of Games and Economic Behavior*, Princeton University Press, Princeton, N.J., 1947.

Wall, T. W., D. Greening, and R. E. D. Woolsey, "Solving Complex Chemical Equilibria Using a Geometric-Programming Based Technique," *Operations Research*, 34, pp. 345–355, 1986.

Walsh, G. R., *Methods of Optimization*, John Wiley & Sons, New York, 1975.

Walsh, S., and L. C. Brown, "Least Cost Method for Sewer Design," *J. Environmental Engineering Division, American Society of Civil Engineers*, 99-EE3, pp. 333–345, 1973.

Waren, A. D., M. S. Hung, and L. S. Lasdon, "The Status of Nonlinear Programming Software: An Update," *Operations Research*, 35(4), pp. 489–503, 1987.

Wasil, E., B. Golden, and L. Liu, "State-of-the-Art in Nonlinear Optimization Software for the Microcomputer," *Computers and Operations Research*, 16(6), pp. 497–512, 1989.

Watanabe, N., Y. Nishimura, and M. Matsubara, "Decomposition in Large System Optimization Using the Method of Multipliers," *J. of Optimization Theory and Applications*, 25(2), pp. 181–193, 1978.

Watson, G. A., "A Class of Programming Problems Whose Objective Function Contains A Norm," *J. Approximation Theory*, 23, pp. 401–411, 1978.

Watson, G. A., "The Minimax Solution of an Overdetermined System of Nonlinear Equations," *J. Inst. Mathematics Applications*, 23, pp. 167–180, 1979.
Weatherwax, R., "General Lagrange Multiplier Theorems," *J. Optimization Theory and Applications*, 14, pp. 51–72, 1974.
Wets, R. J. B., "Programming Under Uncertainty: The Equivalent Convex Program," *SIAM J. Applied Mathematics*, 14, pp. 89–105, 1966a.
Wets, R. J. B., "Programming Under Uncertainty: The Complete Problem," *Z. Wahrscheinlichkeits-Theorie und Verwandte Gebiete*, 4, pp. 316–339, 1966b.
Wets, R. J. B., "Necessary and Sufficient Conditions for Optimality: A Geometric Approach," *Operations Research Verfahren*, 8, pp. 305–311, 1970.
Wets, R. J. B., "Characterization Theorems for Stochastic Programs," *Mathematical Programming*, 2, pp. 165–175, 1972.
Whinston, A., "A Dual Decomposition Algorithm for Quadratic Programming," *Cahiers Centre Études Recherche Opperationelle*, 6, pp. 188–201, 1964.
Whinston, A., "The Bounded Variable Problem—An Application of the Dual Method for Quadratic Programming," *Naval Research Logistics Quarterly*, 12, pp. 315–322, 1965.
Whinston, A., "Some Applications of the Conjugate Function Theory to Duality," in *Nonlinear Programming*, J. Abadie (Ed.), 1967.
Whittle, P., *Optimization Under Constraints*, Wiley-Interscience, London, 1971.
Wilde, D. J., *Optimum Seeking Methods*, Prentice Hall, Englewood Cliffs, N.J., 1964.
Wilde, D. J., and C. S. Beightler, *Foundations of Optimization*, Prentice Hall, Englewood Cliffs, N.J., 1967.
Williams, A. C., "On Stochastic Linear Programming," *SIAM J. Applied Mathematics*, 13, pp. 927–940, 1965.
Williams, A. C., "Approximation Formulas for Stochastic Linear Programming," *SIAM J. Applied Mathematics*, 14, pp. 666–677, 1966.
Williams, A. C., "Nonlinear Activity Analysis," *Management Science*, 17, pp. 127–139, 1970.
Wilson, R. B., "A Simplicial Algorithm for Convex Programming," Ph.D. Dissertation, Harvard University, Graduate School of Business Administration, 1963.
Wismer, D. A. (Ed.), *Optimization Methods for Large Scale Systems*, McGraw-Hill, New York, 1971.
Wolfe, P., "The Simplex Method for Quadratic Programming," *Econometrica*, 27, pp. 382–398, 1959.
Wolfe, P., "A Duality Theorem for Nonlinear Programming," *Quarterly of Applied Mathematics*, 19, pp. 239–244, 1961.
Wolfe, P., "Some Simplex-Like Nonlinear Programming Procedures," *Operations Research*, 10, pp. 438–447, 1962.
Wolfe, P., "Methods of Nonlinear Programming," in *Recent Advances in Mathematical Programming*, R. L. Graves and P. Wolfe (Eds.), 1963.
Wolfe, P., "Methods of Nonlinear Programming," in *Nonlinear Programming*, J. Abadie (Ed.), 1967.
Wolfe, P., "Convergence, Theory in Nonlinear Programming," in *Integer and Nonlinear Programming*, J. Abadie (Ed.), 1970.
Wolfe, P., "On the Convergence of Gradient Methods Under Constraint," *IBM J. Research and Development*, 16, pp. 407–411, 1972.
Wolfe, P., "Note on a Method of Conjugate Subgradients for Minimizing Nondifferentiable Functions," *Mathematical Programming*, 7, pp. 380–383, 1974.
Wolfe, P., "A Method of Conjugate Subgradients for Minimizing Nondifferentiable Functions," in *Nondifferentiable Optimization*, M. L. Balinski and P. Wolfe (Eds.), 1976. (See also *Math. Prog. Study*, 3, pp. 145–173, 1975.)
Wolfe, P., "The Ellipsoid Algorithm," *Optima (Mathematic Programming Soc. Newsletter)*, Number 1, 1980a.
Wolfe, P., "A Bibliography for the Ellipsoid Algorithm," Report RC 8237, IBM Yorktown Heights Research Center, 1980b.

Womersley, R. S., "Optimality Conditions for Piecewise Smooth Functions,"in *Nondifferential and Variational Techniques in Optimization,* D. C. Sorensen and R. J.-B. Wets (Eds.), *Mathematical Programming Study, 17,* North Holland, Amsterdam, 1982.

Womersley, R. S., and R. Fletcher, "An Algorithm for Composite Nonsmooth Optimization Problems," *J. Optimization Theory and Applications*, 48, pp. 493–523, 1986.

Wood, D. J., and C. O. Charles, "Minimum Cost Design of Water Distribution Systems,"OWRR, B-017-DY(3), Report No. 62, Kentucky University Water Resources Research Institute, Lexington, 1973.

Yefimov, N. V., *Quadratic Forms and Matrices: An Introductory Approach* (Trans. by A. Shenitzer), Academic Press, New York, 1964.

Ye, Y., "A Further Result on the Potential Reduction Algorithm for the P-Matrix Linear Complementarity Problem," Department of Management Sciences, Univerisity of Iowa, Iowa City, Iowa, 1988.

Ye, Y., "Interior Point Algorithms for Quadratic Programming," Paper Series No. 89-29, Department of Management Sciences, University of Iowa, Iowa City, Iowa, 1989.

Ye, Y., "A New Complexity Result on Minimization of a Quadratic Function Over a Sphere Constraint," Working Paper Series No. 90-23, Department of Management Sciences, The University of Iowa, Iowa City, Iowa, 1990.

Ye, Y., and P. Pardalos, "A Class of Linear Complementarity Problems Solvable in Polynomial Time," Department of Management Sciences, University of Iowa, Iowa City, Iowa, 1989.

Yu, W., and Y. Y. Haimes, "Multi-level Optimization for Conjunctive Use of Ground Water and Surface Water," *Water Resources Research*, 10, pp. 625–636, 1974.

Yuan, Y., "An Example of Only Linear Convergence of Trust Region Algorithms for Nonsmooth Optimization," *IMA J. Numerical Analysis*, 4, pp. 327–335, 1984.

Yuan, Y., "Conditions for Convergence of Trust Region Algorithms for Nonsmooth Optimization," *Mathematical Programming*, 31, pp. 220–228, 1985a.

Yuan, Y., "On the Superlinear Convergence of a Trust Region Algorithm for Nonsmooth Optimization," *Mathematical Programming*, 31, pp. 269–285, 1985b.

Yudin, D. E., and A. S. Nemirovsky, "Computational Complexity and Efficiency of Methods of Solving Convex Extremal Problems," *Ekonomika i Matem. Metody*, XII, n°2, pp. 357–369 (in Russian), 1976.

Zadeh, L. A., L. W. Neustadt, and A. V. Balakrishnan (Eds.), *Computing Methods in Optimization Problems 2*, Academic Press, New York, 1969.

Zangwill, W. I., "The Convex Simplex Method," *Management Science*, 14, pp. 221–283, 1967a.

Zangwill, W. I., "Minimizing a Function Without Calculating Derivatives," *Computer Journal*, 10, pp. 293–296, 1967b.

Zangwill, W. I., "Nonlinear Programming Via Penalty Functions," *Management Science*, 13, pp. 344–358, 1967c.

Zangwill, W. I., "The Piecewise Concave Function," *Management Science*, 13, pp. 900–912, 1967d.

Zangwill, W. I., *Nonlinear Programming: A Unified Approach*, Prentice Hall, Englewood Cliffs, N.J., 1969.

Zeleny, M., *Linear Multi-Objective Programming*, Lecture Notes in Economics and Mathematical Systems, No. 95, Springer-Verlag, New York, 1974.

Zeleny, M, and J. L. Cochrane (Ed.)., *Multiple Criteria Decision Making*, University of South Carolina, Columbia, 1973.

Zhang, J., N. H. Kim, and L. S. Lasdon, "An Improved Successive Linear Programming Algorithm," *Management Science*, 31(10), pp. 1312–1331, 1985.

Ziemba, W. T., "Computational Algorithms for Convex Stochastic Programs with Simple Recourse," *Operations Research*, 18, pp. 414–431, 1970.

Ziemba, W. T., "Transforming Stochastic Dynamic Programming Problems into Nonlinear Programs," *Management Science*, 17, pp. 450–462, 1971.

Ziemba, W. T., "Stochastic Programs with Simple Recourse," in *Mathematical Programming in Theory and Practice*, P. L. Hammer and G. Zoutendijk (Eds.), 1974.
Ziemba, W. T., and R. G. Vickson (Eds.), *Stochastic Optimization Models in Finance*, Academic Press, New York, 1975.
Zionts, S., "Programming with Linear Fractional Functions," *Naval Research Logistics Quarterly*, 15, pp. 449–452, 1968.
Zoutendijk, G., *Methods of Feasible Directions*, Elsevier, Amsterdam, and D. Van Nostrand, Princeton, N.J., 1960.
Zoutendijk, G., "Nonlinear Programming: A Numerical Survey," *SIAM J. Control*, 4, pp. 194–210, 1966.
Zoutendijk, G., "Computational Methods in Nonlinear Programming," *Studies in Optimization 1*, Society for Industrial and Applied Mathematics, Philadelphia, 1970a.
Zoutendijk, G., "Nonlinear Programming, Computational Methods," in *Integer and Nonlinear Programming*, J. Abadie (Ed.), 1970b.
Zoutendijk, G., "Some Algorithms Based on the Principle of Feasible Directions," in *Nonlinear Programming*, J. B. Rosen, O. L. Mangasarian, and K. Ritter (Eds.), 1970c.
Zoutendijk, G., "Some Recent Developments in Nonlinear Programming," in *The 5th Conference on Optimization Techniques*, R. Conti and A. Ruberti (Eds.), 1973.
Zoutendijk, G., *Mathematical Programming Methods*, North Holland, Amsterdam, 1976.
Zwart, P. B., "Nonlinear Programming: Global Use of the Lagrangian," *J. Optimization Theory and Applications*, 6, pp. 150–160, 1970a.
Zwart, P. B., "Nonlinear Programming: A Quadratic Analysis of Ridge Paralysis," *J. Optimization Theory and Applications*, 6, pp. 331–339, 1970b.
Zwart, P. B., "Nonlinear Programming—The Choice of Direction by Gradient Projection," *Naval Research Logistics Quarterly*, 17, pp. 431–438, 1970c.
Zwart, P. B., "Global Maximization of a Convex Function with Linear Inequality Constraints," *Operations Research*, 22, pp. 602–609, 1974.

Index